Steel Designers' Manual

STEEL DESIGNERS' MANUAL

FIFTH EDITION

The Steel Construction Institute

Edited by

Graham W. Owens
Director, The Steel Construction Institute
and the late
Peter R. Knowles
Consultant Engineer, Peter Knowles & Associates

with the assistance of an Advisory Committee, chaired by
Patrick J. Dowling
British Steel Professor of Steel Structures and Head of Civil Engineering Department, Imperial College, London

Oxford
BLACKWELL SCIENTIFIC PUBLICATIONS
London Edinburgh Boston
Melbourne Paris Berlin Vienna

Copyright © The Steel Construction Institute, 1992

Blackwell Scientific Publications
Editorial Offices:
Osney Mead, Oxford OX2 0EL
25 John Street, London WC1N 2BL
23 Ainslie Place, Edinburgh EH3 6AJ
238 Main Street, Cambridge,
   Massachusetts 02142, USA
54 University Street, Carlton
   Victoria 3053, Australia

All rights reserved. No part of this
publication may be reproduced, stored
in a retrieval system, or transmitted,
in any form or by any means,
electronic, mechanical, photocopying,
recording or otherwise without the
prior permission of the publisher.

First Edition published by
Crosby Lockwood & Son Ltd, 1955
Reprinted 1956 (twice), 1957, 1959
Second Edition 1960
Reprinted with further revision 1962
Third Edition 1966
Reprinted with minor corrections 1967
Fourth Edition published by Granada
Publishing Limited in
Crosby Lockwood Staples 1972
Reprinted 1974, 1975, 1977, 1978 (twice),
1979, 1981
Fourth Edition (Revised) published by
Granada Publishing Limited 1983
Reprinted 1984
Reprinted by Collins Professional and
Technical Books 1985, 1987
Reprinted by BSP Professional Books,
1989, 1990
Fifth Edition published by Blackwell Scientific
Publications, 1992
Printed and bound in Great Britain
at the University Press, Cambridge

DISTRIBUTORS

Marston Book Services Ltd
PO Box 87
Oxford OX2 0DT
(*Orders*: Tel: 0865 791155
          Fax: 0865 791927
          Telex: 837515)

USA
American Society of Civil Engineers
345 East 4th Street
New York 10017
(*Orders*: Tel: 212 705−7420
          Fax: 212 705−7712)

Canada
Oxford University Press
70 Wynford Drive
Don Mills
Ontario M3C 1J9
(*Orders*: Tel: 416 441−2941)

Australia
Blackwell Scientific Publications
(Australia) Pty Ltd
54 University Street
Carlton, Victoria 3053
(*Orders*: Tel: 03 347−0300)

British Library
Cataloguing in Publication Data
A catalogue record for this book is
available from the British Library

ISBN 0−632−02488−7

Library of Congress Cataloging in Publication
Data available

Although care has been taken to ensure, to the best of our knowledge, that all data and information contained herein are accurate to the extent that they relate to either matters of fact or accepted practice or matters of opinion at the time of publication, the Steel Construction Institute assumes no responsibility for any errors in or misinterpretations of such data and/or information or any loss or damage arising from or related to their use.

Extracts from British Standards are reproduced with the permission of BSI. Complete copies of the standards quoted can be obtained by post from BSI Sales, Linford Wood, Milton Keynes, MK14 6LE.

Contents

Foreword to fifth edition	xi
Introduction to fifth edition	xiii
List of contributors	xvii
Notation	xxv

SECTION 1: DESIGN SYNTHESIS

1 **Single-storey buildings** — 1
   Range of building types; Anatomy of structure; Loading; Design of common structural forms; Worked example

2 **Multi-storey buildings** — 36
   Introduction; Factors influencing choice of form; Anatomy of structure; Worked example

3 **Industrial steelwork** — 81
   Range of structures and scale of construction; Anatomy of structure; Loading; Structure in its wider context

4 **Bridges** — 111
   Introduction; Selection of span; Selection of type; Codes of practice; Traffic loading; Other actions; Steel grades; Overall stability and articulation; Initial design; Worked examples

5 **Other structural applications of steel** — 157
   Towers and masts; Space frames; Cable structures; Steel in housing; Atria

SECTION 2: STEEL TECHNOLOGY

6 **Applied metallurgy of steel** — 199
   Introduction; Chemical composition; Heat treatment; Manufacture and effect on properties; Engineering properties and mechanical tests; Fabrication effects and service performance; Summary

7 **Fatigue** — 226
   Introduction; Loadings for fatigue; The nature of fatigue; Fracture mechanics analysis; Improvement techniques; Fatigue-resistant design

8 **Brittle fracture** — 242
   Introduction; Ductile and brittle behaviour; Linear elastic fracture mechanics; General yielding fracture mechanics; Material testing; Fracture-safe design

SECTION 3: DESIGN THEORY

9 Introduction to manual and computer analysis — 255
Introduction; Element analysis; Line elements; Plates; Analysis of skeletal structures; Finite element method

10 Beam analysis — 294
Simply-supported beams; Propped cantilevers; Fixed, built-in or encastré beams; Continuous beams; Plastic failure of single members; Plastic failure of propped cantilevers

11 Plane frame analysis — 311
Formulae for rigid frames; Portal frame analysis

12 Applicable dynamics — 323
Introduction; Fundamentals of dynamic behaviour; Distributed parameter systems; Damping; Finite element analysis

SECTION 4: ELEMENT DESIGN

13 Local buckling and cross-section classification — 341
Introduction; Cross-sectional dimensions and moment–rotation behaviour; Effect of moment–rotation behaviour on approach to design and analysis; Classification table; Economic factors

14 Tension members — 352
Introduction; Types of tension member; Design for axial tension; Combined bending and tension; Eccentricity of end connections; Other considerations; Cables; Worked examples

15 Columns and struts — 374
Introduction; Common types of member; Design considerations; Cross-sectional considerations; Compressive resistance; Torsional and flexural-torsional buckling; Effective lengths; Special types of strut; Economic points; Worked examples

16 Beams — 403
Common types of beam; Cross-section classification and moment capacity: M_c; Basic design; Lateral bracing; Bracing action in bridges – U-frame design; Design for restricted depth; Cold-formed sections as beams; Beams with web openings; Worked examples

17 Plate girders — 444
Introduction; Advantages and disadvantages; Initial choice of cross-section in buildings; Design of plate girders used in buildings to BS 5950: Part 1; Initial choice of cross-section for plate girders used in bridges; Design of steel bridges to BS 5400: Part 3; Worked example

18	**Members with compression and moments**	491
	Occurrence of combined loading; Types of response – interaction; Effect of moment gradient loading; Selection of type of cross-section; Basic design procedure; Cross-section classification under compression and bending; Special design methods for members in portal frames; Worked examples	
19	**Trusses**	528
	Common types of trusses; Guidance on overall concept; Effects of load reversal; Selection of elements and connections; Guidance on methods of analysis; Detailed design considerations for elements; Factors dictating the economy of trusses; Other applications of trusses; Rigid-jointed Vierendeel girders; Worked example	
20	**Composite deck slabs**	568
	Introduction; Deck types; Normal and lightweight concretes; Selection of floor system; Basic design; Fire resistance; Diaphragm action; Other constructional features; Worked example	
21	**Composite beams**	592
	Applications of composite beams; Economy; Guidance on span-to-depth ratios; Types of shear connection; Span conditions; Analysis of composite section; Basic design; Worked examples	
22	**Composite columns**	642
	Introduction; Design of encased composite columns; Design of concrete-filled tubes; Worked example	

SECTION 5: CONNECTION DESIGN

23	**Bolts**	664
	Types of bolt; Methods of tightening and their application; Geometric considerations; Methods of analysis of bolt groups; Design strengths; Tables of strengths	
24	**Welds**	677
	Full penetration and partial penetration welds; Geometric considerations; Methods of analysis of weld groups; Design strengths	
25	**Plate and stiffener elements in connections**	689
	Dispersion of load through plates and flanges; Stiffeners; Plates loaded perpendicular to their plane; Plates loaded in-plane	
26	**Design of connections**	707
	Conceptual design of connections; Types of connection; Worked examples	
27	**Foundations and holding-down systems**	779
	Foundations; Connection of the steelwork; Analysis; Holding-down systems; Worked examples	

SECTION 6: OTHER ELEMENTS

28 Bearings and joints 807
Introduction; Bearings; Joints; Bearings and joints – other considerations

29 Steel piles 833
Bearing piles; Sheet piles; Worked example

30 Floors and orthotropic decks 870
Steel plate floors; Open-grid flooring; Orthotropic decks

SECTION 7: CONSTRUCTION

31 Tolerances 881
Introduction; Standards; Implications of tolerances; Fabrication tolerances; Erection tolerances

32 Fabrication 911
Introduction; Economy of fabrication; Welding; Bolting; Cutting and machining; Handling and routing of steel; Quality management

33 Erection 941
Introduction; The method statement; Programme; Cranes and craneage; Use of sub-assemblies; Safety; Site practice; Special structures

34 Fire protection and fire engineering 985
Introduction; Standards and Building Regulations; Structural performance in fire; Methods of protection; Fire testing; Fire engineering

35 Corrosion resistance 998
The corrosion process; Effect of the environment; Design and corrosion; Surface preparation; Metallic coatings; Paint coatings; Application of paints; Weathering steels; The protective treatment specification

Appendix
Steel technology
Elastic properties of steel 1023
Extracts from
  BS 4360 and BS EN 10 025: 1990 Weldable structural steel plate 1024
  BS 970: Part 1 Austenitic stainless steels 1025

Design theory
Bending moment, shear and deflection tables for
  cantilevers 1026
  simply-supported beams 1028
  built-in beams 1036
  propped cantilevers 1043

Bending moment and reaction tables for continuous beams	1051
Influence lines for continuous beams	1054
Second moments of area of	
two flanges	1066
rectangular plates	1068
a pair of unit areas	1072
Geometrical properties of plane sections	1074
Plastic modulus of	
two flanges	1077
rectangles	1078
Formulae for rigid frames	1080

Element design

Explanatory notes	
1 General	1098
2 Dimension of sections	1099
3 Section properties	1101
Tables of dimensions and properties	
Universal beams	1106
Universal columns	1112
Joists	1115
Universal bearing piles	1116
Circular hollow sections	1118
Square hollow sections	1120
Rectangular hollow sections	1122
Channels	1124
Compound struts – two channels laced	1128
Compound struts – two channels back to back	1129
Equal angles	1130
Unequal angles	1132
Compound equal angles: legs back to back	1133
Compound unequal angles:	
short legs back to back	1134
long legs back to back	1135
Castellated universal beams	1136
Castellated universal columns	1140
Castellated joists	1142
Structural tees cut from universal beams	1144
Structural tees cut from universal columns	1148
Extracts from BS 5950: Part 1: 1990	
Deflection limits (Section two: Table 5)	1150
Design strengths for steel (Section three: Table 6)	1150
Section classification (Section three: Table 7)	1151
Bending strengths (Section four: Tables 11 and 12)	1152
Strut table selection (Section four: Table 25)	1155
Compressive strength (Section four: Table 27)	1156

Connection design

Hole sizes	1164
Bolt strengths	1164
Bolt capacities in tension	1165
Spacing, end and edge distances	1165
Maximum centres of fasteners	1165
Maximum edge distances	1166
Back marks in channel flanges	1168
Back marks in angles	1168
Cross centres through flanges	1169
Grade 4.6 bolts in shear and bearing	1170
Grade 8.8 bolts in shear and bearing	1172
Bolts to BS 4395, slip resistance	1178
Bolts to BS 4395, maximum external tension capacity	1178
Bolts to BS 4395, bearing strength	1179
Bolts to BS 4395, long joints	1179
Bolt group moduli — fasteners in the plane of the force	1180
Bolt group moduli — fasteners not in the plane of the force	1184
Weld group moduli — welds in the plane of the force	1186
Capacities of fillet welds	1190
Weld group moduli — welds not in the plane of the force	1191

Other elements

Piling information

Frodingham steel sheet piling	1194
Larssen steel sheet piling	1202
Floor plate design tables	1204

Construction

Fire information sheets	1206
Limiting temperatures	1222
Section factors for	
universal beams	1224
universal columns	1225
circular hollow sections	1226
rectangular hollow sections	1227
rectangular hollow sections (square)	1228
Minimum thickness of spray protection	1229
Basic data on corrosion	1230

Miscellaneous

Conversion tables	1234
British Standards covering the design and construction of steelwork	1248

Index   1253

Foreword to fifth edition

At the instigation of the Iron and Steel Federation, the late Bernard Godfrey began work in 1952 on the first edition of the *Steel Designers' Manual*. As principal author he worked on the manuscript almost continually for a period of two years. On many Friday evenings he would meet with his co-authors, Charles Gray, Lewis Kent and W.E. Mitchell to review progress and resolve outstanding technical problems. A remarkable book emerged. Within approximately 900 pages it was possible for the steel designer to find everything necessary to carry out the detailed design of most conventional steelwork. Although not intended as an analytical treatise, the book contained the best summary of methods of analysis then available. The standard solutions, influence lines and formulae for frames could be used by the ingenious designer to disentangle the analysis of the most complex structure. Information on element design was intermingled with guidance on the design of both overall structures and connections. It was a book to dip into rather than read from cover to cover. However well one thought one knew its contents, it was amazing how often a further reading would give some useful insight into current problems. Readers forgave its idiosyncrasies, especially in the order of presentation. How could anyone justify slipping a detailed treatment of angle struts between a very general discussion of space frames and an overall presentation on engineering workshop design?

The book was very popular. It ran to four editions with numerous reprints in both hard and soft covers. Special versions were also produced for overseas markets. Each edition was updated by the introduction of new material from a variety of sources. However, the book gradually lost the coherence of its original authorship and it became clear in the 1980s that a more radical revision was required.

After thirty-six very successful years it was decided to rewrite and re-order the book, while retaining its special character. This decision coincided with the formation of the Steel Construction Institute and it was given the task of co-ordinating this activity. An Advisory Committee was established comprising:

Patrick J. Dowling	Imperial College of Science, Technology & Medicine (Chairman)
David Dibb-Fuller	Gifford & Partners, previously Conder Southern
Alan Hart	Ove Arup & Partners
Alan Hayward	Cass Hayward & Partners

Peter Knowles   (Editor)
Jim Mathys   H.L. Watermans
Roger Pope   Blight & White Ltd
Graham Owens   Steel Construction Institute (Editor).

The committee endorsed a complete restructuring of the book, with more material on overall design and a new section on construction. The analytical material was condensed because it is now widely available elsewhere, but all the design data were retained in order to maintain the practical usefulness of the book as a day-to-day design manual. Earlier editions had been based on allowable stress design concepts (in accordance with the relevant codes BS 449 and BS 153). This edition is the first to be entirely based on limit state design concepts. It gives relevant background to BS 5950 for buildings and BS 5400 for bridges. Design examples are to the more appropriate of these two codes for each particular application. Thirty authors were commissioned to produce individual chapters and Peter Knowles was commissioned to carry out detailed editorial work.

As principal editor, I would like to pay tribute to all those who have made this fifth edition possible; the Advisory Committee for the direction that they provided at the outset, the authors who gave so freely of their time and were patient with the editorial process as the book was welded into a coherent whole, and the staff at the Institute, particularly Ip-Luew Lee, Derek Mullett and David Waters, who have worked so hard on the design data and the worked examples. Finally, I would like to pay tribute to the invaluable contribution from Peter Knowles. Sadly, Peter died in late 1991, having completed a meticulous piece of editing, but without seeing the final results of his and our labours.

By kind permission of the British Standards Institution, references are made to British Standards throughout the manual. The tables of fabrication and erection tolerances in Chapter 31 are taken from the *National Structural Steelwork Specification*, second edition. Much of the text and illustrations for Chapter 33 are taken from *Steelwork Erection* by Harry Arch. Both these sources are used by kind permission of the British Constructional Steelwork Association, the publishers. These permissions are gratefully acknowledged.

Graham Owens

Further information on the SCI is given in a prospectus available free on request from: The Membership Secretary, The Steel Construction Institute, Silwood Park, Ascot, Berkshire SL5 7QN. Telephone: (0990) 23345, Fax: (0990) 22944, Telex: 846843.

Introduction to fifth edition

It is just over one hundred years since steel was first used in the construction of a major bridge, Britain's Forth Railway Bridge, and the first all steel building, the Second Rand McNally Building in Chicago. Since then it has demonstrated its versatility in almost the full range of structural applications.

The recent renaissance in steel construction which has been centred on the United Kingdom has benefited greatly from the many improvements made in the various aspects of steel technology, the advances in steel fabrication techniques, the improved understanding of structural behaviour, and the upgrading in the standard of structural steel design, as well as the wide dissemination of excellent material on which the training of the users of structural steel can be based.

The manufacture of the raw material has improved in a number of significant ways:

- Basic oxygen steel-making and continuous casting produce a fine-grained weldable material with consistent strength and toughness.
- Modern rolling mills manufacture a wide range of shapes of close tolerance, good surface finish and consistent, homogeneous composition.
- Heat treatments during production permit greater control and enhancement of mechanical properties.

The range of commercially available structural steels has been widened. The higher yield strength steels are attractive because they can show economies in design. In appropriate atmospheric conditions it is possible to leave weather-resistant steels unpainted throughout the life of a structure. Stainless steels are produced in a wide range of compositions. In the thinner gauge products, coatings can be applied to produce attractive, corrosion-resistant cladding and roofing sheeting. And thin gauge strip material produced by cold rolling can be shaped into a variety of cold-formed sections.

The fabrication of structural steelwork has also shown significant increases in quality in recent years. The production of fabricated steelwork is now very much a manufacturing operation with the emphasis on accurate cutting, shaping, drilling and welding of components on a production line which makes considerable use of numerically-controlled machine tools. Substantial use has been made of automation in a range of pre-engineered building systems now available. Protective anti-corrosion treatment has been much improved by the introduction of automatic shot blasting and painting plants.

Design, aided by research, has become more refined with the result that designers are able, with confidence, to propose structures which make more economical use of steel. Many design improvements have been incorporated in Codes of Practice, such as the new Eurocodes which draw together the best information available not only throughout the European Community, but also worldwide.

Steel structures have always had the advantages of lightness, stiffness and strength and lend themselves to rapid construction compared to other construction materials. Developments which have made steel even more competitive in the building field include the use of metal deck as permanent shuttering to concrete slabs, composite steel and concrete construction and lightweight fire protection. Many of these design improvements have had the effect of further facilitating speedy construction. Steel also comes into its own in refurbishment of existing buildings because of its flexibility. The application of steel to long-span bridges is, of course, well established, but its advantages for short- to medium-span bridges are becoming more and more obvious to bridge engineers as is the application of steel to motorway bridge widening and upgrading schemes.

Developments in structural steel are still possible, a century after the first use of the material. Some of the possible areas of improvement include:

- Developments in steel production technology which will lead to higher strength steels becoming available at a competitive price.
- An increasingly large range of both hot- and cold-rolled products to permit greater flexibility in design.
- Greater use of the tensile strength of steel in cable structures.
- Improvements in fire engineering and corrosion-resistant steel, giving the material greater opportunity for expansion.
- An increasing use of computers leading to greater refinement of design to reduce fabrication and construction costs.
- Greater emphasis on prefabrication to reduce the amount of site work.
- Greater use of standard connections.
- A more sophisticated use of composite construction over a wide range of applications.

The authors of each chapter have been chosen on the basis of their known expertise in the area and present the best relevant information likely to be useful to the modern structural steelwork designer. By putting this knowledge at the disposal of the reader they hope to share with them some of the excitement and satisfaction they have had in designing much of Britain's new evolving infrastructure using that most versatile of building materials, steel.

Introduction xv

Scope

The *Steel Designers' Manual* is aimed at designers of structural steelwork of all kinds. It is divided into seven main sections:
 Design synthesis
 Steel technology
 Design theory
 Element design
 Connection design
 Other elements
 Construction.

Design synthesis: Chapters 1–5

A description of the nature of the process by which design solutions are arrived at for a wide range of steel structures including:
- Single- and multi-storey buildings (Chapters 1 and 2)
- Heavy industrial frames (Chapter 3)
- Bridges (Chapter 4)
- Other diverse structures such as space frames, cable structures, towers and masts, atria and steel in housing (Chapter 5).

Steel technology: Chapters 6–8

Background material sufficient to inform designers of the important problems inherent in the production and use of steel and methods of overcoming them in practical design.
- Applied metallurgy (Chapter 6)
- Fatigue (Chapter 7)
- Brittle fracture (Chapter 8).

Design theory: Chapters 9–12

A résumé of analytical methods for determining the forces and moments in structures subject to static or dynamic loads, both manual and computer-based. Comprehensive tables for a wide variety of beams and frames are given in the Appendix.
- Manual and computer analysis (Chapter 9)
- Beam analysis (Chapter 10)
- Frame analysis (Chapter 11)
- Applicable dynamics (Chapter 12).

Element design: Chapters 13–22

A comprehensive treatment of the design of steel elements, singly, in combination or acting compositely with concrete.
- Local buckling and cross-section classification (Chapter 13)

- Tension members (Chapter 14)
- Columns and struts (Chapter 15)
- Beams (Chapter 16)
- Plate girders (Chapter 17)
- Members with compression and moments (Chapter 18)
- Trusses (Chapter 19)
- Composite floors (Chapter 20)
- Composite beams (Chapter 21)
- Composite columns (Chapter 22).

Connection design: Chapters 23–27

The general basis of design of connections is surveyed and amplified by consideration of specific connection methods.
- Bolts (Chapter 23)
- Welds (Chapter 24)
- Plate and stiffener elements (Chapter 25)
- Design of connections (Chapter 26)
- Foundations and holding-down systems (Chapter 27).

Other elements: Chapters 28–30

- Bearings and joints (Chapter 28)
- Piles (Chapter 29)
- Floors and orthotropic decks (Chapter 30).

Construction: Chapters 31–35

Important aspects of steel construction about which a designer must be informed if he is to produce structures which can be economically fabricated and erected and which will have a long and safe life.
- Tolerances (Chapter 31)
- Fabrication (Chapter 32)
- Erection (Chapter 33)
- Fire protection and fire engineering (Chapter 34)
- Corrosion resistance (Chapter 35).

Appendix

A comprehensive collection of data of direct use to the practising designer.

Patrick Dowling

Contributors

Harry Arch

Harry Arch graduated from Manchester Faculty of Technology. For many years he worked for Sir William Arrol, where he became a director, responsible for all outside construction activities including major bridges, power stations and steelworks construction. In 1970 he joined Redpath Dorman Long International, working on offshore developments. Now retired, he is still widely in demand as a lecturer on fabrication and erection.

Hubert Barber

Hubert Barber joined Redpath Brown in 1948 and for five years gained a wide experience in steel construction. The remainder of his working life was spent in local government, first at Manchester and then in Yorkshire where he became Chief Structural Engineer of West Yorkshire. He also lectured part-time for fourteen years at the University of Sheffield.

Michael Burdekin

Michael Burdekin graduated from Cambridge University in 1959. After fifteen years of industrial research and design experience he went to UMIST, where he is now Professor of Civil and Structural Engineering. His specific expertise is the field of welded steel structures, particularly in the application of fracture mechanics to fracture and fatigue failure.

Brian Cheal

Brian Cheal graduated from Brighton Technical College in 1951 with an External Degree of the University of London. He was employed with W.S. Atkins and

Partners from 1951 to 1986, becoming a Technical Director in 1979, and specialized in the analysis and design of steel-framed structures, including heavy structural framing for power stations and steelworks. He has written design guides and given lectures on various aspects of connection design and is co-author of *Structural Steelwork Connections*.

David Dibb-Fuller

David Dibb-Fuller started his career with the Cleveland Bridge and Engineering Company in London. His early bridge related work gave a strong emphasis to heavy fabrication; in later years he moved on to building structures. As Technical Director for Conder Southern in Winchester his strategy was to develop close links between design for strength and design for production. Currently he is a partner with Gifford and Partners in Southampton where he continues to exercise his skills in the design of steel structures.

Patrick Dowling

Patrick Dowling graduated from University College Dublin in 1960. After post-graduate training and four years at BCSA, he joined the Civil Engineering Department of Imperial College in 1968 where he is now British Steel Professor of Steel Structures and Head of Department. The author of 175 papers and editor of three books, his numerous outside activities include the chairmanship of the Eurocode Committee on Design of Steel Structures.

Ian Duncan

Ian Duncan joined the London office of Ove Arup and Partners in 1966 after graduating from Surrey University. From 1975 he taught for four years at University College Cardiff before joining Buro Happold. He now runs his own practice in Bristol.

Michael Green

Michael Green graduated from Liverpool University in 1971. After an early career in general civil engineering, he joined Buro Happold, where he is now an executive partner. He has worked on a wide variety of building projects, developing a specialist expertise in atria and large-span structures.

Alan Hart

Alan Hart graduated from the University of Newcastle-upon-Tyne in 1968 and joined Ove Arup and Partners. During his career he has been involved in the design of a number of major award-winning buildings, including Carlsberg Brewery, Northampton; Cummins Engine Plant, Shotts, Lanarkshire; the Hongkong and Shanghai Bank, Hong Kong. He is a project director of Ove Arup and Partners.

Alan Hayward

Alan Hayward is a partner of Cass Hayward and Partners, specialist structural steel consultants, who design and devise the erection methodology for all kinds of steel bridges, many built on a design:construct basis. Projects include London Docklands Light Railway viaducts, the M25/M4 and M25/A1(M) interchanges, the Festival Way Flyover (Stoke-on-Trent) and a number of roll-on/roll-off berthing structures. He is a former Chief Examiner for the Institution of Structural Engineers and chairman of the Wales branch. He became a visiting professor at the Polytechnic of Central London in 1991.

Eric Hindhaugh

Eric Hindhaugh trained as a structural engineer in design and constructional steel work, timber and lightweight roll-formed sections. He then branched into promotional and marketing activities. Most recently he has been a Market Development Manager in Construction for British Steel Strip Products, where he has been involved in Colorcoat and the widening use of lightweight steel sections for structural steel products.

Ken Johnson

Ken Johnson is head of corrosion and coatings at British Steel's Swinden Laboratories. His early experience was in the paint industry but he has since worked in steel for over twenty-five years, dealing with the corrosion and protection aspects of the whole range of British Steel's products, including plates, section, piling, strip products, tubes, stainless steels, etc. He represents the steel industry on several BSI and European Committees and is a Council member of the Paint Research Association.

Mark Lawson

A graduate of Imperial College, and the University of Salford, where he worked in the field of cold-formed steel, Mark Lawson spent his early career at Ove Arup and Partners and the Construction Industry Research and Information Association. In 1987 he joined the newly formed Steel Construction Institute as Research Manager for steel in buildings, with particular reference to composite construction, fire engineering and cold-formed steel. He is a member of the Eurocode 4 project team on fire-resistant design.

Ian Liddell

After leaving Cambridge, Ian Liddell joined Ove Arup and Partners to work on the roof of the Sydney Opera House and on the South Bank Art Centre. His early career encompassed a wide range of projects, with particular emphasis on shell structures and lightweight tension and fabric structures. Since 1976 he has been a partner of Buro Happold and has been responsible for a wide range of projects, many with special structural engineering features, including mosques, auditoriums, mobile and temporary structures, stadiums and retail atria.

Stephen Matthews

Stephen Matthews graduated from the University of Nottingham in 1974 and completed postgraduate studies at Imperial College in 1976–7. His early professional experience was gained with Rendel Palmer and Tritton in their structural steelwork design office, and on the Thames Barrier site. During subsequent employment with Fairfield Mabey and Cass Hayward and Partners, he worked on the design of several large composite bridges including the Tweed Bridge, Berwick; Simon De Montfort Bridge, Evesham; and viaducts on the Island Gardens Branch of the Docklands Light Railway. He is currently associate director of the Structures Division of Frank Graham Consulting Engineers.

Rangachari Narayanan

Rangachari Narayanan graduated in civil engineering from Annamalai University (India) in 1951. In a varied professional career spanning over forty years, he has held senior academic positions at the Universities of Delhi, Manchester and Cardiff. He is the recipient of several awards including the Benjamin Baker Gold Medal and George Stephenson Gold Medal, both from the Institution of Civil

Engineers. Currently he heads the Education and Publication Divisions at the Steel Construction Institute.

David Nethercot

Since graduating from the University of Wales, Cardiff, David Nethercot has completed twenty-five years of teaching, research and specialist advisory work in the area of structural steelwork. The author of over 100 technical papers, he has lectured frequently on post-experience courses, he is vice-chairman of the BSI Committee responsible for BS 5950 and is a frequent contributor to technical initiatives associated with the structural steelwork industry. Since 1989 he has been professor of civil engineering at the University of Nottingham.

Gerard Parke

Gerry Parke is a lecturer in structural engineering at the University of Surrey specializing in the analysis and design of steel structures. His particular interests lie in assessing the collapse behaviour of both steel industrial buildings and large-span steel space structures.

Alan Pottage

Alan Pottage graduated from the University of Newcastle-upon-Tyne in 1976 and gained a Masters degree in structural steel design from Imperial College in 1984. He has gained experience in all forms of steel construction, particularly portal frame and multi-rise structures, and has contributed to various code committees, and SCI guides on composite design and connections. He is presently employed as Technical Director at J.N. Rowen Ltd.

Graham Raven

Graham Raven graduated from King's College, London in 1963 and joined Ove Arup and Partners. After thirteen years with consulting engineers he joined a software house pioneering work in structural steelwork design and detailing systems. This experience took him to Ward Building Systems Ltd where, in addition to design responsibilities, he was closely associated with the development of cold-formed cladding and rail systems and the increased use of welded sections for portals and composite beams in commercial buildings. He has recently joined the Steel Construction Institute's staff as Manager of Construction.

John Righiniotis

John Righiniotis graduated from the University of Thessalonika in 1987 and obtained an MSc in structural steel design from Imperial College in 1988. He worked at the Steel Construction Institute on a wide range of projects until June 1990 when he was required to return to Greece to carry out his military service.

John Roberts

John Roberts graduated from the University of Sheffield in 1969 and was awarded a PhD there in 1972 for research on the impact loading of steel structures. His professional career has included working on site for Alfred McAlpine. In 1981 he joined Allott and Lomax, consulting engineers, becoming director in charge of structural engineering in 1985. He has been involved with many major industrial projects, including power stations. He serves as a member of council of the Institution of Structural Engineers and represents the Association of Consulting Engineers on the Constructional Steelwork QA scheme.

Jef Robinson

Jef Robinson graduated in metallurgy from Durham University in 1962 where he undertook a research project into high temperature material behaviour for the NASA space programme. His early career in the steel industry included designing high-strength notch-tough steels for super-tankers, North Sea drilling platforms and large bridges. Currently the Market Development Manager for the structural sections division of British Steel he chaired the BSI Committee which formulated BS 5950: Part 8 *Fire-resistant design* for structural steelwork.

Robert Simpson

Robert Simpson graduated from the University of Surrey in 1974. In 1977 he obtained an MSc in offshore engineering from Cranfield Institute of Technology and undertook research into weldability of offshore steels. He spent four years with the Steel Construction Institute and more recently moved to the Health and Safety Executive as research manager.

Dick Stainsby

Dick Stainsby was for many years chief designer with Redpath Dorman Long, Middlesbrough, and previously Dorman Long and Co, dealing with 'design & build' contracts for which that company was famous. He is now Managing Director of Cleveland Technical Services, engaged in the design and detailing of steel structures particularly using computer techniques. He is author of the *National Steelwork Specification for Building Structures*, and is currently compiling *Joints in Simple Structures – Practical Applications*.

Paul Tasou

Paul Tasou graduated from Queen Mary College, London in 1978 and subsequently obtained an MSc in structural steel design from Imperial College, London. He spent eleven years at Rendel Palmer and Tritton working on a wide range of bridge, building and civil engineering projects. He is now principal partner in Tasou Associates.

Colin Taylor

Colin Taylor graduated from Cambridge in 1959. He started his professional career in steel fabrication, initially in the West Midlands and subsequently in South India. After eleven years he moved into consultancy where, besides practical design, he became involved with graduate training, the use of computers for design and drafting, company technical standards and drafting work for British Standards and for Eurocodes as editorial secretary for Eurocode 3. Moving to the Steel Construction Institute on its formation as manager of the Codes and Advisory Division, he also became involved with the European standard for steel fabrication and erection *Execution of Steel Structures*.

John Tyrrell

John Tyrrell graduated from Aston University in 1965 and immediately joined Ove Arup and Partners. He has worked for them on a variety of projects in the UK, Australia and West Africa; he is now a Project Director. He has been responsible for the design of a wide range of towers and guyed masts. He currently leads the Industrial Structures Group covering diverse fields of engineering from telecommunications and broadcasting to the power industry.

Dennis Waite

Dennis Waite has spent thirty-nine years in the steel industry, involved initially in the design of structural steelwork and later in the design of steel piled structures. In 1985, he became an independent consulting engineer specializing in foundations and earth retaining structures. He is a member of BSI code drafting committee CSB/2, *Code of Practice for Earth Retaining Structures* and CSB/17, *Code of Practice for Maritime Structures*. He is chairman of the Institution of Civil Engineers' Piling Panel.

Peter Wickens

Having graduated from Nottingham University in 1971, Peter Wickens spent much of his early career in civil engineering, designing bridges and Metro stations. In 1980, he changed to the building structures field and was Project Engineer for the Billingsgate Development, one of the first of the new generation of steel composite buildings. He is currently manager of the Structural Division and Head of Discipline for Building Structures at Mott MacDonald.

Michael Willford

Michael Willford graduated from Cambridge in 1975. He has worked with Ove Arup and Partners since graduation, becoming a Technical Director in 1986. Between 1980 and 1983 he played a major role in the design of the Hongkong and Shanghai Bank Building, Hong Kong. Recently he has been associated with many major building and industrial structures, specializing in design for dynamic and environmental loads.

Notation

Several different notations are adopted in steel design; different specializations frequently give different meanings to the same symbol. These differences have been maintained in this book. To do otherwise would be to separate this text both from other literature on a particular subject and from common practice. The principal definitions for symbols are given below. For conciseness, only the most commonly adopted subscripts are given; others are defined adjacent to their usage.

A		Area
	or	End area of pile
	or	Constant in fatigue equations
A_e		Effective area
A_g		Gross area
A_s		Shear area of a bolt
A_t		Tensile stress area of a bolt
A_v		Shear area of a section
a		Spacing of transverse stiffeners
	or	Effective throat size of weld
	or	Crack depth
	or	Distance from central line of bolt to edge of plate
	or	Shaft area of pile
B		Breadth
B		Transformation matrix
b		Outstand
	or	Width of panel
	or	Distance from centreline of bolt to toe of fillet weld or to half of root radius as appropriate
b_e		Effective breadth or effective width
b_1		Stiff bearing length
C		Crack growth constant
C		Transformation matrix
C_v		Charpy impact value
C_y		Damping coefficient
c		Bolt cross-centres
	or	Cohesion of clay soils

D		Depth of section
		Diameter of section or hole
$\mathbf{D}$		Elasticity matrix
D_r		Profile height for metal deck
D_s		Slab depth
d		Depth of web
	or	Nominal diameter of fastener
	or	Depth
d_e		Effective depth of slab
E		Modulus of elasticity of steel (Young's modulus)
e		End distance
e_y		Material yield strain
F_c		Compressive force due to axial load
F_s		Shear force (bolts)
F_t		Tensile force
F_v		Shear force (sections)
f		Flexibility coefficient
f_a		Longitudinal stress in flange
f_c		Compressive stress due to axial load
f_{cu}		Cube strength of concrete
f_m		Force per unit length on weld group from moment
f_r		Resultant force per unit length on weld group from applied concentric load
f_v		Force per unit length on weld group from shear
		or Shear stress
G		Shear modulus of steel
g		Gravitational acceleration
H		Warping constant of section
	or	Horizontal reaction
h		Height
	or	Stud height
	or	Depth of overburden
I_o		Polar second moment of area of bolt group
I_{oo}		Polar second moment of area of weld group of unit throat about polar axis
I_x		Second moment of area about major axis
I_{xx}		Polar second moment of area of weld group of unit throat about xx axis
I_y		Second moment of area about minor axis
I_{yy}		Polar second moment of area of weld group of unit throat about yy axis
K		Degree of shear connection
	or	Stiffness
$\mathbf{K}$		Stiffness matrix

K_s		Curvature of composite section from shrinkage
	or	Constant in determining slip resistance of HSFG bolts
K_1, K_2, K_3		Empirical constants defining the strength of composite columns
k_a		Coefficient of active pressure
k_d		Empirical constant in composite slab design
k_p		Coefficient of passive resistance
L		Length of span or cable
L_y		Shear span length of composite slab
M		Moment
	or	Larger end moment
M_{ax}, M_{ay}		Maximum buckling moment about major or minor axis in presence of axial load
M_b		Buckling resistance moment (lateral–torsional)
M_E		Elastic critical moment
M_o		Mid-length moment on a simply-supported span equal to unrestrained length
M_{pc}		Plastic moment capacity of composite section
M_{rx}, M_{ry}		Reduced moment capacity of section about major or minor axis in the presence of axial load
M_x, M_y		Applied moment about major or minor axis
$\overline{M}_x, \overline{M}_y$		Equivalent uniform moment about major or minor axis
M_1, M_2		End moments for a span of a continuous composite beam
m		Equivalent uniform moment factor
	or	Empirical constant in fatigue equation
	or	Number of vertical rows of bolts
m_d		Empirical constant in composite slab design
N		Number of cycles to failure
N_c, N_q, N_γ		Constants in Terzaghi's equation for the bearing resistance of clay soils
n		Slenderness correction factor
	or	Crack growth constant
	or	Number of shear studs per trough in metal deck
	or	Number of horizontal rows of bolts
	or	Distance from bolt centreline to plate edge
P		Force in structural analysis
	or	Load per unit surface area on cable net
	or	Crushing resistance of web
P_{bb}		Bearing capacity of a bolt
P_{bg}		Bearing capacity of parts connected by friction-grip fasteners
P_{bs}		Bearing capacity of parts connected by ordinary bolts
P_c		Compression resistance
P_{cx}, P_{cy}		Compression resistance considering buckling about major or minor axis only
P_o		Minimum shank tension for preloaded bolt
P_s		Shear capacity of a bolt

xxviii *Notation*

P_{sL}		Slip resistance provided by a friction-grip fastener
P_t		Tension capacity of a member or fastener
P_u		Compressive strength of stocky composite column
P_v		Shear capacity of a section
p		Ratio of cross-sectional area of profile to that of concrete in a composite slab
p_b		Bending strength
p_{bb}		Bearing strength of a bolt
p_{bg}		Bearing strength of parts connected by friction-grip fasteners
p_{bs}		Bearing strength of parts connected by ordinary bolts
p_c		Compressive strength
p_E		Euler strength
p_o		Minimum proof stress of a bolt
p_s		Shear strength of a bolt
p_t		Tension strength of a bolt
p_w		Design strength of a fillet weld
p_y		Design strength of steel
Q		Prying force
q		Ultimate bearing capacity
q_b		Basic shear strength of a web panel
q_{cr}		Critical shear strength of a web panel
q_e		Elastic critical shear strength of a web panel
q_f		Flange-dependent shear strength factor
R		Reaction
	or	Load applied to bolt group
	or	Radius of curvature
R_c		Compressive capacity of concrete section in composite construction
R_q		Capacity of shear connectors between point of contraflexure and point of maximum negative moment in composite construction
R_r		Tensile capacity in reinforcement in composite construction
R_s		Tensile capacity in steel section in composite construction
R_w		Compression in web section in composite construction
r		Root radius in rolled section
r_r		Reduction factor in composite construction
r_x, r_y		Radius of gyration of a member about its major or minor axis
S		Span of cable
S_R		Applied stress range
S_x, S_y		Plastic modulus about major or minor axis
s		Spacing
	or	Leg length of a fillet weld
T		Thickness of a flange or leg
	or	Tension in cable
t		Thickness of web
U		Elastic energy
U_s		Specified minimum ultimate tensile strength of steel
u		Buckling parameter of the section

V		Shear force
	or	Shear resistance per unit length of beam in composite construction
V_b		Shear buckling resistance of stiffened web utilizing tension field action
V_{cr}		Shear buckling resistance of stiffened or unstiffened web without utilizing tension field action
v		Slenderness factor for beam
W		Point load
	or	Foundation mass
	or	Load per unit length on a cable
	or	Energy required for crack growth
w		Lateral displacement
	or	Effective width of flange per bolt
	or	Uniformly distributed load on plate
X_e		Elastic neutral axis depth in composite section
x		Torsion index of section
x_p		Plastic neutral axis depth in composite section
Y		Correction factor in fracture mechanics
Y_s		Specified minimum yield stress of steel
Z_c		Elastic section modulus for compression
Z_{oo}		Elastic modulus for weld group of unit throat subject to torsional load
Z_x, Z_y		Elastic modulus about major or minor axis
z		Depth of foundation
α		Coefficient of linear thermal expansion
α_e		Modular ratio
β		Ratio of smaller to larger end moment
	or	Coefficient in determination of prying force
γ		Ratio M/M_o, i.e. ratio of larger end moment to mid-length moment on simply-supported span equal to unrestrained length
	or	Bulk density of soil
	or	Coefficient in determination of prying force
γ_f		Overall load factor
γ_m		Material strength factor
Δ		Displacements in vector
δ		Deflection
	or	Elongation
δ_c		Deflection of composite beam at serviceability limit state
δ_{ic}		Deflection of composite beam at serviceability limit state in presence of partial shear connection
δ_o		Deflection of steel beam at serviceability limit state
δ_{oo}		Deflection in continuous composite beam at serviceability limit state
ε		Constant $(275/p_y)^{1/2}$
	or	Strain

η		Load ratio for composite columns
λ		Slenderness, i.e. effective length divided by radius of gyration
λ_{cr}		Elastic critical load factor
λ_{LO}		Limiting equivalent slenderness
λ_{LT}		Equivalent slenderness
λ_o		Limiting slenderness
μ		Slip factor
μ_x, μ_y		Reduction factors on moment resistance considering axial load
σ		Stress
σ_ε		Tensile stress
ϕ		Diameter of composite column
	or	Angle of friction in granular soil

Chapter 1
Single-storey buildings

by GRAHAM RAVEN and ALAN POTTAGE

1.1 Range of building types

It is estimated that around 50% of the hot-rolled constructional steel used in the UK is fabricated into single-storey buildings being some 40% of the total steel used in them. The remainder is light-gauge steel cold-formed into purlins, rails, cladding and accessories. Over 90% of single-storey non-domestic buildings have steel frames demonstrating the dominance of steel construction for this class of building. These relatively light, long-span, durable structures are simply and quickly erected and developments in steel cladding have enabled architects to design economical buildings of attractive appearance to suit a wide range of applications and budgets.

The traditional image was a dingy industrial shed with a few exceptions such as aircraft hangars and exhibition halls. Changes in retailing and the replacement of traditional heavy industry with electronics-based products have led to a demand for increased architectural interest and enhancement.

Clients expect their buildings to have the potential for easy change of layout several times during the building's life. This is true for both institutional investors and owner users. The primary feature is therefore flexibility of planning which, in general terms, means as few columns as possible consistent with economy. The ability to provide spans up to 60 m, but most commonly around 30 m, gives an extremely popular structural form for the supermarkets, do-it-yourself stores and the like, which are now surrounding towns in the UK. The development of steel cladding in a wide variety of colours and shapes has enabled distinctive and attractive forms and house styles to be created.

Improved reliability of steel-intensive roofing systems has contributed to their acceptability in buildings used by the public and perhaps more importantly in 'high-tech' buildings requiring controlled environments. The structural form will vary according to span, aesthetics, integration with services, cost and suitability for the proposed activity. A cement manufacturing building will clearly have different requirements from a warehouse, food processing plant or computer factory.

The growth of the leisure industry has provided a challenge to designers, and buildings vary from the straightforward requirement of cover for bowls, tennis, etc., to an exciting environment which encourages people to spend days of their holidays indoors at water centres and similar controlled environments suitable for year round recreation.

2 Single-storey buildings

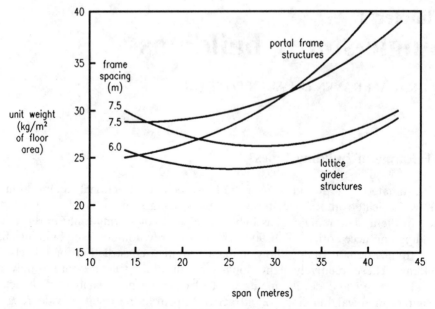

Fig. 1.1 Comparison of bare frame weights for portal and lattice structures

In all instances the requirement is to provide a covering to allow a particular activity to take place; the column spacing is selected to give as much freedom of use of the space consistent with economy. The normal span range will be from 12 m to 50 m, but larger spans are feasible for hangars and enclosed sports stadia.

Figure 1.1 shows how steel weight varies with structural form and span.[1]

1.2 Anatomy of structure

A typical single-storey building consisting of cladding, secondary steel and a frame structure is shown in Fig. 1.2.

1.2.1 Cladding

Cladding is required to be weathertight, to provide insulation, to have penetrations for daylight and access, to be aesthetically pleasing and to last the maximum time with a minimum of maintenance consistent with the budget.

The requirements for the cladding to roofs and walls are somewhat different. The weathertightness of the roof is clearly paramount, particularly as the demand

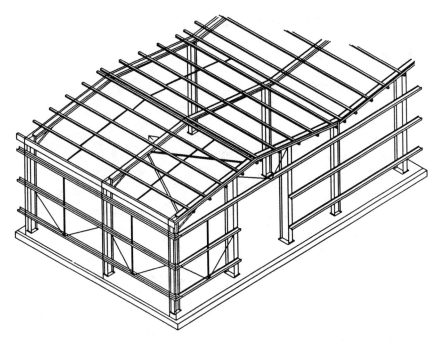

Fig. 1.2 Structural form for portal-frame building (some rafter bracing omitted for clarity)

for lower pitches increases, whereas aesthetics dominate in the choice of walling. Over the last 20 years metal cladding has emerged as the most popular choice for both areas. The two major substrates are steel and aluminium. Steel is more economical and, with a much lower coefficient of expansion, has practical advantages but it does depend on the integrity of the coatings to maintain corrosion-resistance and this has led to the selection of aluminium in some sensitive cases. However current British Steel formulations (HP200 plastisol, and PVF2) have Agrément Certificates with life expectancy to first maintenance of 10–25 years depending on environment.

The sheeting (Fig. 1.3) from which the cladding is formed consists of a steel substrate with layers of galvanizing, primer and colour coating. The inner surface is normally finished in white polyester with the outer coating in PVF2, PVC (plastisol) or polyester. Plastisol is the most popular, providing a robust 200 micron thick coating but with restrictions on use in marine environments or in latitudes less than 49°. Light colours are recommended for roofs to minimize heat absorption.

The construction of the cladding skin can take several forms. Those most commonly seen in the UK are:

(1) single-skin trapezoidal
(2) double trapezoidal shell

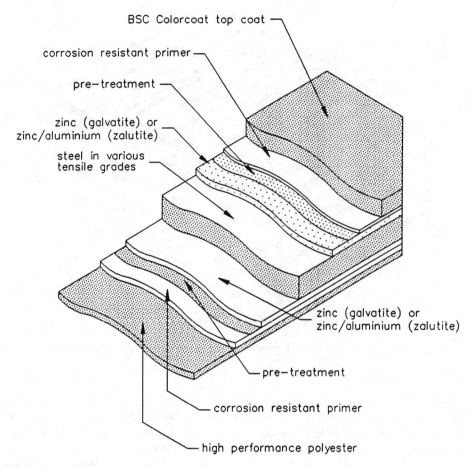

Fig. 1.3 Cladding sheet strata

(3) standing seam and concealed fix
(4) composite.

Further information on cladding is given in section 1.4.7.

1.2.2 Secondary elements

In the normal single-storey building the cladding is supported on secondary members which transmit the loads back to main structural steel frames. The spacing of the frames, determined by the overall economy of the building, is normally in the range 5 m−8 m, with 6 m and 7.5 m as the most common spacings.

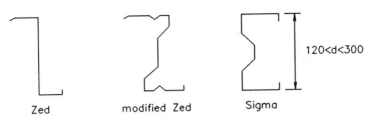

Fig. 1.4 Popular purlin and frame sections

A combination of cladding performance, erectability and the restraint requirements for economically-designed main frames dictates that the purlin and rail spacing should be 1.5–2 m.

For this range the most economic solution has proved to be cold-formed light-gauge sections of proprietary shape and volume produced to order on computer numerically controlled (CNC) rolling machines. These have proved to be extremely efficient since the components are delivered to site pre-engineered to the exact requirements, which minimizes fabrication and erection times and eliminates material wastage. Because of the high volumes, manufacturers have been encouraged to develop and test all material-efficient sections. These fall into three main categories: Zed, modified Zed and Sigma sections. Figure 1.4 illustrates the range.

The Zed section was the first shape to be introduced. It is material-efficient but the major disadvantage is that the principal axes are inclined to the web. If subject to unrestrained bending in the plane of the web, out-of-plane displacements occur; if these are restrained, out-of-plane forces are generated.

More complicated shapes have to be rolled rather than press braked. This is a feature of the UK where the market is supplied by relatively few manufacturers and the volumes produced by each allow the advanced manufacturing techniques to be employed giving competitive products and service.

As roof pitches become lower, modified Zed sections have been developed with the inclination of the principal axis considerably reduced, so enhancing overall performance. Stiffening has been introduced, improving material efficiency.

The Sigma shape in which the shear centre is approximately coincident with the load application line has advantages. One manufacturer now produces, using rolling, a second-generation product of this configuration, which is economical.

1.2.3 Primary frames

The frame supports the cladding but, with increasing architectural and service demands, other factors are important. The basic structural form has developed against the background of achieving the lowest cost envelope by enclosing the minimum volume. Plastic design of portal frames brings limitations on the spacing

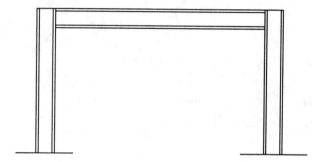

Fig. 1.5 Simplest single-storey structure

of restraints of around 1.8−2 m. The cladding profiles are economic in this range; they can support local loads and satisfy drainage requirements. The regime is therefore for the loads to be transferred from the sheeting on to the purlins and rails which in turn must be supported on a primary structure. Figure 1.5 shows the simplest possible type of structure with vertical columns and a horizontal spanning beam. There is a need for a fall in the roof finish to provide drainage, but for small spans the beam can be effectively horizontal with the fall being created in the finishes or by a nominal slope in the beam. The minimum slope is also a function of weatherproofing requirements of the roof material.

The simple form shown would be a mechanism unless restraint to horizontal forces is provided. This is achieved either by the addition of bracing in both plan and vertical planes or by the provision of redundancies in the form of moment-resisting joints. The important point is that all loads must be transmitted to the foundations in a coherent fashion even in the simplest of buildings, whatever their size.

The range of frame forms is discussed in more detail in later sections but Fig. 1.6 shows the structural solutions commonly used. The most common is the portal shape with pinned bases, although this gives a slightly heavier frame than the fixed-base option. The overall economy, including foundations, is favourable. The portal form is both functional and economic with overall stability being derived from the provision of moment-resisting connections at eaves and apex.

The falls required to the roof are provided naturally with the cladding being carried on purlins which in turn are supported by the main frame members. Architectural pressures have led to the use of flatter slopes compatible with weathertightness; the most common is around 6°, but slopes as low as 1° are used, which means deflection control is increasingly important.

Traditionally, portal frames have been fabricated from compact rolled sections and designed plastically. More recently the adoption of automated welding techniques has led to the introduction of welded tapered frames which have been

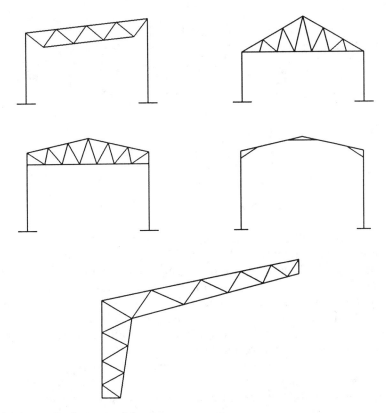

Fig. 1.6 A range of structural forms

extensively used for many years in the USA. For economy these frames have deep slender sections and are designed elastically. In addition to material economies, the benefit is in the additional stiffness and reduced deflections.

Although the portal form is inherently pleasing to the eye, given a well-proportioned and detailed design, the industrial connotation, together with increased service requirements, has encouraged the use of lattice trusses for the roof structure. They are used both in the simple forms with fixed column bases and as portal frames with moment-resisting connections between the tops of the columns for long-span structures such as aircraft hangars, exhibition halls and enclosed sports facilities.

1.2.4 Resistance to sway forces

Most of the common forms provide resistance to sidesway forces in the plane of the frame. It is essential also to provide resistance to out-of-plane forces; these

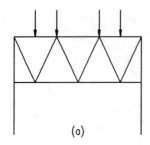 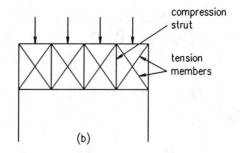

Fig. 1.7 Roof bracing

are usually transmitted to the foundations with a combination of horizontal and vertical girders. The horizontal girder in the plane of the roof can be of two forms as shown in Fig. 1.7. Type (a) is formed from members, often tubes, capable of carrying tension or compression. One of the benefits is in the erection stage as the braced bay can be erected first and lined and levelled to provide a square and safe springboard for the erection of the remainder.

Type (b) uses less material but more members are required. The diagonals are tension-only members (wire rope has been used) and the compression is taken in the orthogonal strut which has the shortest possible effective length. It may be possible to use the purlins, strengthened where necessary, for this purpose.

Similar arrangements must be used in the wall to carry the forces down to foundation level. If the horizontal and vertical girders are not in the same bay, care must be taken to provide suitable connecting members.

1.3 Loading

1.3.1 External gravity loads

The dominant gravity load is from snow. The general case is the application of a basic uniform load of 0.6 kN/m^2, but with sloping roofs having multiple spans and parapets, the action of drifting snow has to be considered. Design information is given in BS 6399[2] and Building Research Establishment (BRE) Digest 332.[3] The basic loading is variable according to location and is derived from the snow map which has been incorporated into BS 5502.[4] The main frame design for portals can be carried out using the uniform load case but the variable loads caused by drifting are to be applied to cladding and purlins. The effects of drifting are idealized into triangular loadings with formulae given for the various effects of valleys, parapets, upstands, etc. Early tests carried out at the BRE established that equivalent uniformly distributed loads can be used for the purlin design. In the areas of high local load, consideration has to be given as to whether to reduce purlin spacing or to increase the gauge. Where practicable the reduction of spacing is preferable as it prevents the dangers and disruption involved with identification and production of different thicknesses of purlin supplied to one job.

The load factor for snow loading is 1.6 for both secondary and main frame elements unless in combination with crane or similar loading. The factors to be used for various combinations are given in BS 5950: Part 1.[5]

For portal frames the frame strength will usually be determined by the snow load case, unless the eaves height is large in relation to span.

1.3.2 Wind loads

With lightweight cladding and purlins and rails, wind loads are important. Cladding and its fasteners are designed for the local pressure coefficient as given in BS 6399: Part 3.[2] Purlins and main frames are designed using the relevant statistical factors, but not additional local coefficients. Care must be taken to include the total effect of both internal and external pressure coefficients.

When designing to BS 5950 the load factor for the wind force is 1.4 acting against dead loads having a factor 1.0.

1.3.3 Internal gravity loads

Service loads for lighting, etc., are reasonably included in the global 0.6 kN/m^2. As service requirements have increased, it has become necessary to consider carefully the provision to be made.

Most purlin manufacturers can provide proprietary clips for hanging limited point loads to give flexibility of layout. Where services and sprinklers are required, it is normal to design the purlins for a global service load of 0.1–0.2 kN/m^2 with a reduced value for the main frames to take account of likely spread. Particular items of plant must be treated individually. The specifying engineer should make a realistic assessment of the need as the elements are sensitive and while the loads may seem low, they represent a significant percentage of the total and affect design economy accordingly.

1.3.4 Cranes

The most common form of craneage is the overhead type running on beams supported by the columns. The beams are carried on cantilever brackets or in heavier cases by providing dual columns.

In addition to the weight of the cranes and their load, the effects of acceleration and deceleration have to be catered for. This is simplified by a quasi-static approach with enhanced load factors being used. The allowances to be made are given in BS 6399: Part 1. For simple forms of crane gantry these are:

(1) for loads acting vertically, the maximum static wheel loads increased by 25% for an electric overhead crane or 10% for a hand-operated crane;
(2) for the horizontal force acting transverse to the rails, the following percentage of the combined weight of the crab and the load lifted:
    (a) 10% for an electric overhead crane, or
    (b) 5% for a hand-operated crane;
(3) for the horizontal forces acting along the rails, 5% of the static wheel loads which can occur on the rails for overhead cranes which are either electric or hand-operated.

For heavy, high-speed or multiple cranes the allowances should be specially calculated with reference to the manufacturers.

The combination load factors for design are given in BS 5950: Part 1. The constant movement of a crane gives rise to a fatigue condition. This is, however, restricted to the local areas of support, i.e. the crane beam itself, the support bracket and its connection to the main column. It is not normal to design the whole frame for fatigue as the stress levels due to the crane travel are relatively low.

1.4 Design of common structural forms

1.4.1 Beam and column

The cross section shown in Fig. 1.5 is undoubtedly the simplest framing solution which can be used to provide structural integrity to single-storey buildings. Used predominantly in spans of up to 10 m, where flat roof construction is acceptable, the frame comprises standard hot-rolled sections having simple or moment-resisting joints.

Flat roofs are notoriously difficult to weatherproof, since deflections of the horizontal cross-beam induce ponding of rainwater on the roof which tends to penetrate the laps of traditional cladding profiles and, indeed, any weakness of the exterior roofing fabric. To counteract this, either the cross-member is cambered to provide the required fall across the roof, or the cladding itself is laid to a predetermined fall, again facilitating drainage of surface water off the roof.

Due to the need to control excessive deflections, the sections tend to be somewhat heavier than those required for strength purposes alone, particularly if the cross-beam is designed as simply-supported. In its simplest form, the cross-beam is designed as spanning between columns, which, for gravity loadings, are in

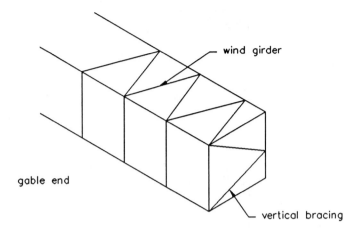

Fig. 1.8 Simple wind bracing system

direct compression apart from a small bending moment at the top of the column due to the eccentricity of the beam connection. The cross-beam acts in bending due to the applied gravity loads, the compression flange being restrained either by purlins, which support the roof sheet, or by a proprietary roof deck which may span between the main frames and which must be adequately fastened. The columns are treated as vertical cantilevers for in-plane wind loads.

Resistance to lateral loads is achieved by the use of a longitudinal wind girder, usually situated within the depth of the cross-beam. This transmits load from the top of the columns to bracing in the vertical plane, and thence to the foundation. The bracing is generally designed as a pin-jointed frame, in keeping with the simple joints used in the main frame. Details are shown in Fig. 1.8.

Buildings which employ the use of beam-and-column construction often have brickwork cladding in the vertical plane. With careful detailing, the brickwork can be designed to provide the vertical sway bracing, acting in a similar manner to the shear walls of a multi-storey building.

Resistance to lateral loading can also be achieved either by the use of rigid connections at the column/beam joint or by designing the columns as fixed-base cantilevers. The latter point is covered in more detail in the following sub-section relating to the truss and stanchion framing system.

1.4.2 Truss and stanchion

The truss and stanchion system is essentially an extension of the beam-and-column solution, providing an economic means of increasing the useful span.

Single-storey buildings

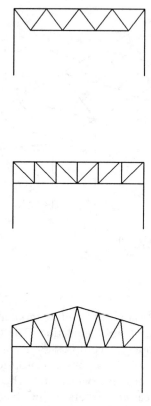

Fig. 1.9 Truss configurations

Typical truss shapes are shown in Fig. 1.9.

Members of lightly-loaded trusses are generally hot-rolled angles as the web elements, and either angles or structural tees as the boom and rafter members, the latter facilitating ease of connection without the use of gusset plates. More heavily loaded trusses comprise universal beam and column sections and hot-rolled channels, with connections invariably employing the use of heavy gusset plates.

In some instances there may be a requirement for alternate columns to be omitted for planning requirements. In this instance load transmission to the foundations is effected by the use of long-span eaves beams carrying the gravity loads of the intermediate truss to the columns; lateral loading from the intermediate truss is transmitted to points of vertical bracing, or indeed vertical cantilevers by means of longitudinal bracing as detailed in Fig. 1.10. The adjacent frames must be designed for the additional loads.

Considering the truss and stanchion frame shown in Fig. 1.11, the initial assumption is that all joints are pinned, i.e. they have no capacity to resist

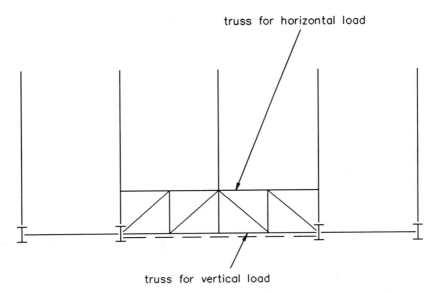

Fig. 1.10 Additional framing where edge column is omitted

bending moment. The frame is modelled in a structural analysis package or by hand calculation, and for the load cases considered, applied loads are assumed to act at the node points. It is clear from Fig. 1.11 that the purlin positions and nodes are not coincident; consequently, due account must be taken of the bending moment induced in the rafter section. The rafter section is analyzed as a continuous member from eaves to apex, the node points being assumed as the supports, and the purlin positions as the points of load application (Fig. 1.12).

The rafter is sized by accounting for bending moment and axial loads, the web members and bottom chord of the truss being initially sized on the basis of axial load alone.

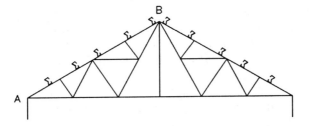

Fig. 1.11 Truss with purlins offset from nodes

14 Single-storey buildings

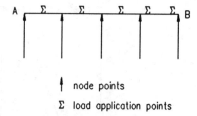

Fig. 1.12 Rafter analyzed for secondary bending

Use of structural analysis packages allows the engineer to rapidly analyze any number of load combinations. Typically, dead load, live load and wind load cases are analyzed separately, and their factored combinations are then investigated to determine the worst loading case for each individual member. Most software packages provide an envelope of forces on the truss for all load combinations, giving maximum tensile and compressive forces in each individual member, thus facilitating rapid member design.

Under gravity loading the bottom chord of the truss will be in tension and the rafter chords in compression. In order to reduce the slenderness of the compression members, lateral restraint must be provided along their length, which in the present case is provided by the purlins which support the roof cladding. In the case of load reversal, the bottom chord is subject to compression and must be restrained. A typical example of restraint to the bottom chord is the use of ties, which run the length of the building at a spacing governed by the slenderness limits of the compression member; they are restrained by a suitable end bracing system. Another solution is to provide a compression strut from the chord member to the roof purlin, in a similar manner to that used to restrain compression flanges of rolled sections used in portal frames. The sizing of all restraints is directly related to the compressive force in the primary member, usually expressed as a percentage of the compressive force in the chord. Care must be taken in this instance to ensure that, should the strut be attached to a thin-gauge purlin, bearing problems in thin-gauge material are accounted for. Examples of restraints are shown in Fig. 1.13.

Connections are initially assumed as pins, thereby inferring that the centroidal axes of all members intersecting at a node point are coincident. Practical considerations invariably dictate otherwise, and it is quite common for member axes to be eccentric to the assumed node for reasons of fit-up and the physical constraints that are inherent in the truss structure. Such eccentricities induce secondary bending stresses of the node points which must be accounted for not only by local bending and axial load checks at the ends of all constituent members, but also in connection design. Typical truss joints are detailed in Fig. 1.14.

It is customary to calculate the net bending moment at each node point due to any eccentricities, and proportion this moment to each member connected to the node in relation to member stiffness.

In heavily-loaded members secondary effects may be of such magnitude as to

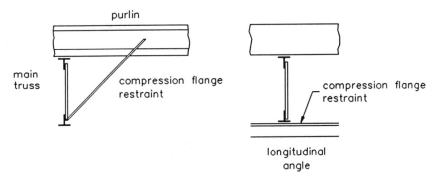

Fig. 1.13 Restraints to bottom chord members

require member sizes to be increased quite markedly above those required when considering axial load effects alone. In such instances, consideration should be given to the use of gusset plates, which can be used to ensure that member centroids are coincident at node points, as shown in Fig. 1.15. Types of truss connections are very much dependent on member size and loadings. For lightly-loaded members, welding is most commonly used with bolted connections in the chords if the truss is to be transported to site in pieces and then erected. In heavily-loaded members, using gusset plates, either bolting, welding or a combination of the two may be used. However, the type of connection is generally based on the fabricator's own reference.

Where the roof truss has a small depth at the eaves, lateral loading is resisted either by longitudinal wind girders in the plane of the bottom boom and/or rafter, or by designing the columns as vertical fixed-base cantilevers, as shown in Fig. 1.16. Where the truss has a finite depth at the eaves, benefit can be obtained by developing a moment connection at this position with the booms designed for appropriate additional axial loads. This latter detail may allow the column base to be designed as a pin, rather than fixed, depending on the magnitude of the applied loading, and the serviceability requirements for deflection.

Longitudinal stability is provided by a wind girder in the plane of the truss boom and/or rafter at the gable wall; the load from the gable being transmitted to the foundations by vertical bracing as shown in Fig. 1.17.

1.4.3 Portal frames

By far the most common form of structure for single-storey structures is the portal frame, the principal types being shown in Fig. 1.18.

Spans of up to 60 m can be achieved by this form of construction, the frame generally comprising hot-rolled universal beam sections. However, with the

16 *Single-storey buildings*

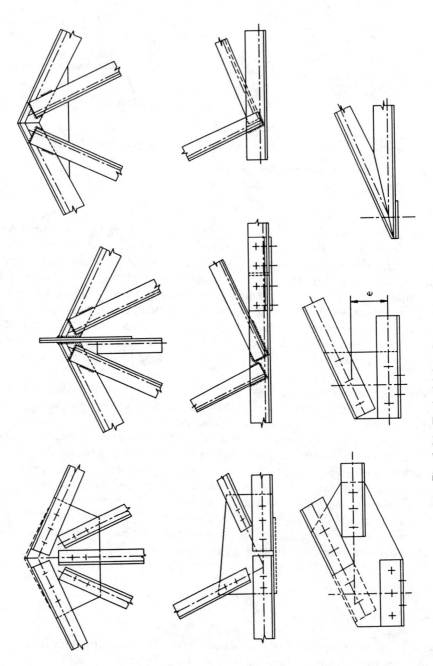

Fig. 1.14 Typical joints in trusses

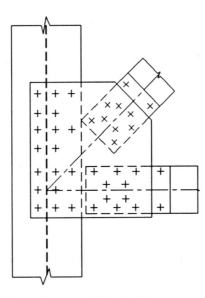

Fig. 1.15 Ideal joint with all member centroids coincident

increase in understanding of how slender plate elements react under combined bending moment, axial load and shear force, several fabricators now offer a structural frame fabricated from plate elements. These frames use tapered stanchions and rafters to provide an economic structural solution for single-storey buildings, the frame being 'custom designed' for each particular loading criterion.

Roof slopes for portal frames are generally of the order of 6° but slopes as low as 1° are becoming increasingly popular with the advent of new cladding systems such as standing seam roofs. It should be noted, however, that frame deflections at low slopes must be carefully controlled, and due recognition must be taken of the large horizontal thrusts that arise at the base.

Frame centres are commonly of the order of 6–7.5 m, with eaves heights ranging from 6–15 m in the case of aircraft hangars or similar structures.

Resistance to lateral loading is provided by moment-resisting connections at the eaves, stanchion bases being either pinned or fixed. Frames which are designed on the basis of having pinned bases are heavier than those having fixity at the bases,

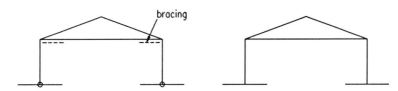

Fig. 1.16 Sway resistance for truss roofs

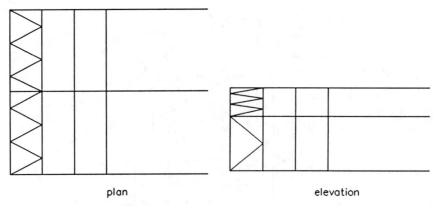

Fig. 1.17 Gable-end bracing systems

although the increase in frame cost is offset by the reduced foundation size for the pinned-base frames.

Parallel-flange universal sections, subject to meeting certain physical constraints regarding breadth-to-thickness ratios of both flanges and webs, lend themselves to rapid investigation by the plastic methods of structural analysis. The basis of the plastic method is the need to determine the load applied to the frame which will induce a number of 'plastic hinges' within the frame, thereby causing failure of the frame as a mechanism.

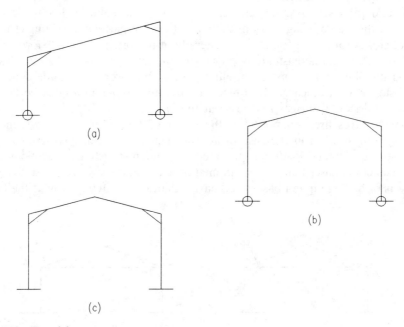

Fig. 1.18 Portal-frame structures

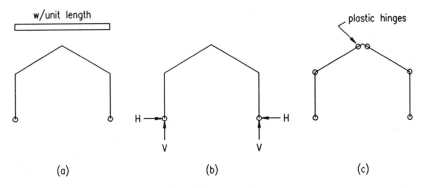

Fig. 1.19 Structural behaviour of pinned-base portal

This requirement is best illustrated by the following simple example.

Considering the pinned-base frame shown in Fig. 1.19(a), subject to a uniform vertical load, w, per unit length, the reactions at the foundations are shown in Fig. 1.19(b). The frame has one degree of indeterminacy. In order that the frame fails as a mechanism, at least two plastic hinges must form (i.e. the (degree of indeterminacy + 1)) as shown in Fig. 1.19(c). (It should be noted that although four hinges are shown in Fig. 1.19(c), due to 'theoretical' symmetry only one pair either side of the apex will in fact form, due to the obvious imperfections in both loading and erection conditions.)

In many structures, other than the most simple, it is not clear where the plastic hinges will form. There are several methods available to the design engineer which greatly assist the location of these hinge positions, not least the abundance of proprietary software packages specifically relating to this form of analysis. Prior to the use of these packages, however, it is imperative that the engineer fully understands the fundamentals of plastic analysis by taking time to calculate, by hand, several design examples. The example which follows uses a graphical construction as a means of illustrating the applications of the method to a simple portal frame. Further information on plastic analysis is given in Chapter 11.

The frame shown in Fig. 1.20(a) has one degree of indeterminacy. It is made statically determinate by assuming a roller at the right-hand base as shown in Fig. 1.20(b), and the free-bending moment diagram drawn as shown in Fig. 1.20(c). The reactant line for the horizontal force 'removed' to achieve a statically determinate structure must now be drawn as follows:

(1) Considering member DC, the bending moment due to the reaction R at point D, $M_D = Rh_1$. Similarly, at point C, $M_C = R(h_1 + h_2)$.

(2) The gradient of the reactant line along member DC,

$$m = \frac{R(h_1 + h_2) - R(h_1)}{L_1} = \frac{Rh_2}{L_1}$$

20   Single-storey buildings

(3) The reactant line must pass through a point O as shown in Fig. 1.20(c). By similar triangles,

$$\frac{Rh_1}{x} = \frac{Rh_2}{L_1} \quad \text{giving} \quad x = \frac{L_1 h_1}{h_2}$$

Therefore, by rotating the reactant line through point O, the positions of the bending moment of equal magnitude in the positive and negative regions can be found and the member sized accordingly based on this value of bending moment.

Having found the positions of the reactant line which gives the number of hinges required for a mechanism to form, in this case two, the reactant line for the stanchions can be drawn through point O, a distance h_1 from the end of the free-bending moment diagram. The unknown reaction, H, is then calculated as $H = M_D/h_1$.

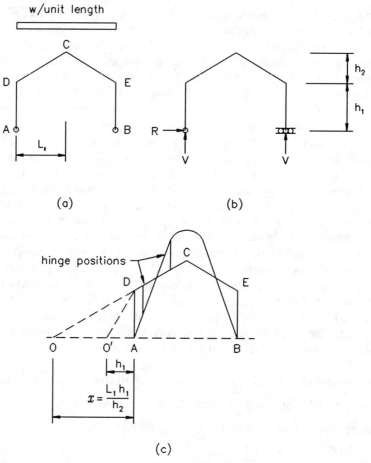

Fig. 1.20 Application of graphical method

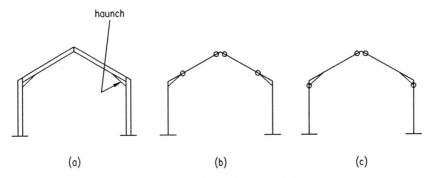

Fig. 1.21 Alternative hinge locations for haunched portal frames

In a majority of instances, portal frames are constructed with a haunch at the eaves, as shown in Fig. 1.21(a).

Depending on the length/depth of the haunch, the plastic hinges required for a mechanism to form are shown in Figs 1.21(b) and (c). The dimensional details for the haunch can be readily investigated by the graphical method by superimposing the dimensions of the haunch on the free-bending moment diagram. The reactant line can then be rotated accordingly until the required mechanism is achieved and members sized accordingly.

Haunches are generally fabricated from parallel beam sections (Fig. 1.22). In all cases, the haunch must remain in the elastic region. A detailed check is required along its length, from a stability point of view, in accordance with the requirements laid down in the relevant code of practice. In some instances, bracing of the bottom flange must be provided from a purlin position within its length, as shown in Fig. 1.23.

Portal frames analyzed by the plastic methods of structural analysis tend to be more economical in weight than their elastically designed counterparts. However, engineers should be aware that minimum weight sections, and by inference minimum depth sections, have to be connected together to withstand the moments and forces induced by the applied loading. Particular attention should be made at an *early stage* in the design process as to the economics of the connection. Cost penalties may be induced by having to provide, for example, gusset plates between pairs of bolts should the section be so shallow as to necessitate a small number of highly-stressed bolts. In addition, end plates should not be much thicker than the flanges of the sections to which they are attached.

Providing the engineer is prepared to consider the implications of his calculated member sizes on the connections inherent in the structure at an early stage, an economic solution will undoubtedly result. Leaving connection design solely to the fabricator, without any consideration as to the physical constraints of providing a number of bolts in an extremely shallow depth, will undoubtedly result in a connection which is both difficult to design and fabricate, and is costly.

22 *Single-storey buildings*

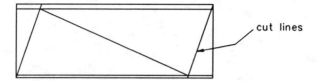

Fig. 1.22 Haunch fabrication

Having sized the members based on the previous procedure, it is imperative that an analysis at Serviceability Limit State is carried out (i.e. unfactored loads) to check deflections at both eaves and apex. This check is required, not only to ascertain whether deflections are excessive, but also as a check to ensure that the deflections and accompanying frame movement can be accommodated by the

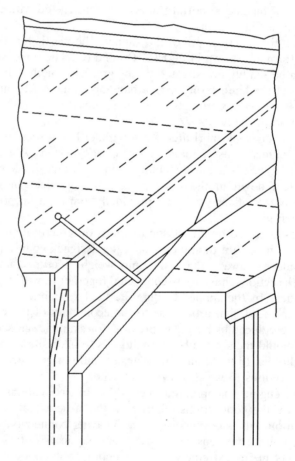

Fig. 1.23 Bottom flange restraint

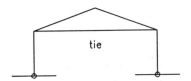

Fig. 1.24 Tied portal frame

building envelope without undue cracking of any brickwork or tearing of metal cladding sheets at fixing positions. Excessive lateral deflections can be reduced by increasing the rafter size and/or by fixing the frame bases. It should be noted that the haunch has a significant effect on frame stiffness due to its large section properties in regions of high bending moment.

1.4.4 Tied portal

The tied portal, a variation of the portal frame, is illustrated in Fig. 1.24.

Economy of material can be achieved, albeit at the expense of reducing the allowable headroom of the building, by provision of a tie at eaves level. Under vertical loading, the eaves spread is reduced due to the induced tensile force in the tie. However, although the rafter size is reduced, because under vertical loading the only possible mode of failure of the rafter is that of a fixed-end beam, deflections are more critical. Lateral loading due to wind further complicates the problem since in some cases the tie may not act in tension and becomes redundant. It is common, therefore, for tied portals to have fixed bases, which provide greater stability and resistance to horizontal loading.

1.4.5 Stressed-skin design

In addition to providing the weathertight membrane, the steel sheeting can also be utilized as a structural element itself. Correct detailing of connections between individual cladding sheets (and their connections to the support steelwork) will induce a stressed-skin effect which can offer both resistance to transverse wind loads, and restraint to the compression flanges of the main frame elements.

It is clear that when vertical load is applied to a pitched roof portal frame, the apex tends to deflect downwards, and the eaves spread horizontally. This displacement cannot occur without some deformation of the cladding. Since the cladding is fixed to the purlins, the behaviour of the cladding and purlins together is analogous to that of a deep plate girder, i.e. the purlins resist the bending moment, in the form of axial forces, and the steel sheeting resists the applied shear, as shown in Fig. 1.25.

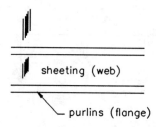

Fig. 1.25 Stressed-skin action

As a consequence of this action, the load applied over the 'length' of the 'plate girder' has to be resisted at the ends of the span. In the case of stressed-skin action being used to resist transverse wind loads on the gable end of the structure, adequate connection must be present over the end bay to transmit the load from the assumed 'plate girder' into the braced bay, and into the foundations, as shown in Fig. 1.26.

Some degree of stressed-skin action is present in all portal frame structures where cladding is fixed to the supporting members by mechanical fasteners. Claddings which are either brittle (i.c. fibre cement) or are attached to the supporting structure by clips (i.e. standing seam systems) are not suitable for stressed-skin applications.

Correct detailing and correct fixing is essential to the integrity of stressed-skin applications:

(1) suitable bracing members must be used to allow the applied loads from diaphragm action to be transmitted to the foundations.
(2) at the position of laps in the sheeting, suitable mechanical fasteners must be present to allow continuity of load between sheets. These fasteners can be

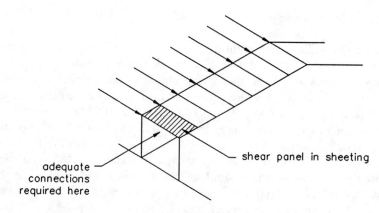

Fig. 1.26 Stressed-skin action for gable-end bracing

screws or rivets, which must be capable of withstanding both shear and pull out due to the stressed-skin effect and applied loads respectively.
(3) the ends of the sheet must similarly be connected to the supporting members (i.e. purlins).
(4) all stressed-skin panels must be connected to adequately designed edge members, which must be capable of transmitting the axial forces induced by bending, in addition to the forces induced by imposed and wind load effects.

1.4.6 Purlins and siderails

Due to the increasing awareness of the load-carrying capabilities of sections formed from thin-gauge material, proprietary systems for both purlins and siderails have been developed by several manufacturers in the United Kingdom. Consequently, unless a purlin or siderail is to be used in a long-span or high-load application (when a hot-rolled angle or channel may be used) a cold-formed section is the most frequently used cladding support member for single-storey structures.

Cold-formed sections manufactured from thin-gauge material are particularly prone to twisting and buckling due to several factors which are directly related to the section's shape. The torsional constant of all thin-gauge sections is low (it is a function of the cube of the thickness); in the case of lipped channels the shear centre is eccentric to the point of application of load thus inducing a twist on the section; in the case of Zeds the principal axes are inclined to the plane of the web thus inducing bi-axial bending effects. These effects affect the load-carrying capacity of the section.

When used in service the support system is subject to downward loading, due to dead and live loads such as cladding weight, snow, services, etc., and uplift if the design wind pressure is greater than the dead load of the system. Therefore, for a typical double-span system as shown in Fig. 1.27, the compression flange is restrained against rotation by the cladding for downward loading, but it is not so restrained in the case of load reversal.

In supporting the external fabric of the building, the purlins and siderails gain some degree of restraint against twisting and rotation from the type of cladding used and the method of its fastening to the supporting member. In addition, the

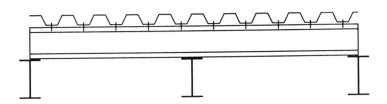

Fig. 1.27 Typical double-span purlin with cladding restraint

connection of the support member to the main frame also has a significant effect on the load-carrying capacity of the section. Economical design, therefore, must take account of the above effects.

There are four possible approaches to the design of a purlin system.

(1) Design by calculation based on an elastic analysis as detailed in the relevant code of practice BS 5950: Part 5.[6] This approach neglects any beneficial effect of cladding restraint for the wind uplift case.
(2) Empirical design based on approximate procedures for Zeds as given in the codes of practice. This approach leads to somewhat uneconomic design.
(3) Design by calculation based on a rational analysis which accounts for the stabilizing influence of the cladding, plasticity in the purlin as the ultimate load is approached and the behaviour of the cleat at the internal support. The effects, however, are difficult to quantity.
(4) Design on the basis of full-scale testing.

Manufacturers differ in the methods; the example used here is from one manufacturer who has published the method used.[7]

For volume production, design by testing is the approach which is used. Although this approach is expensive, maximum economy of material can be achieved and the cost of the testing can be spread over several years of production.

Design by testing involves the 'fine-tuning' of theoretical expressions for the collapse load of the system. The method is based on the mechanism shown in Fig. 1.28 for a two-span system.

From the above mechanism it can be shown that:

$$\text{the collapse load, } W_c = f(M_1, M_2, x, L)$$
$$x = f(M_1, M_2, L)$$
$$\theta_1 = f(W_c, M_1, L)$$

The performance of, for example, a two-span system is considerably enhanced if some redistribution of bending moment from the internal support is taken into account. The moment–rotation characteristic at the support is very much dependent on the cleat detail and the section shape. The characteristics of the central support can be found by testing a simply-supported beam subject to a central

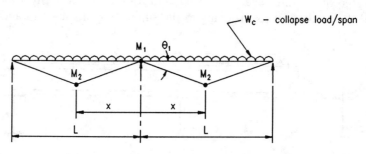

Fig. 1.28 Collapse mechanism for a two-span purlin system

point load, so as to simulate the behaviour of the central support of a double-span system.

From this test, the load–deflection characteristics can be plotted well beyond the deflection at which first yield occurs. A lower bound empirical expression can then be found for the support moment, M_1, based on an upper limit rotational capacity. A similar expression can be found for M_2, the internal span moment, again on the basis of a test on a simply-supported beam subject to a uniformly distributed load, applied by the use of a vacuum rig, or perhaps sand bags.

The design expressions can then be confirmed by the execution of numerous full-scale tests on double-span systems employing pairs of purlins supporting proprietary cladding.

As described earlier, load reversal under wind loading invariably occurs thereby inducing compressive forces to the flange in the internal span which is not restrained by the cladding, as is the case in the downward load case. Anti-sag rods or the like are placed within the internal span, thereby reducing the overall buckling length of the member. The system is again tested in load reversal conditions and, as before, the design expressions can be further refined.

In some instances, sheeting other than conventional trapezoidal cladding (which is invariably through fixed to the purlin by self-drilling fastenings in alternate troughs) will not afford the full restraint to the compression flange; examples are: standing seam roofs, brittle cladding, etc. The amount of restraint afforded by these latter types of cladding cannot easily be quantified. For the reasons outlined above, further full-scale testing is carried out and similar procedures of verification of design expressions are carried out.

The results of the full-scale tests are then condensed into easy to use load–span tables which are given in the purlin manufacturers' design and detail literature. The use of these tables is outlined in Table 1.1.

The tabular format is typical of that contained in all purlin manufacturers' technical literature. The table is generally prefaced by explanatory notes regarding fixing condition and lateral restraint requirements, the latter being particularly

Table 1.1 Typical load table

| Span (m) | Section | UDL (kN)[a] | Purlin centres (mm) | | | | |
			1000	1200	1400	1600	1800
6.0	CF170160	9.50	1.58	1.32	1.13	0.99	0.88
	CF170170	10.75	1.79	1.49	1.28	1.12	1.00
	CF170180	11.75	1.96	1.63	1.40	1.22	1.09
	CF200160	12.25	2.04	1.70	1.46	1.28	1.13
	CF200180	13.50	2.25	1.88	1.61	1.41	1.25

[a] UDL is the *total* uniformly distributed load on a single purlin of a 6 m double-span system which would produce a failure of the section.

Loads given beneath the columns under purlin centres give the allowable uniformly distributed load on the purlin at a given spacing. The figure is arrived at by dividing the UDL by the product of span and purlin centres, i.e. for CF170180, 6 m span, 1800 mm centre,

allowable load = $11.75/(6 \times 1.8) = 1.09$ kN/m^2

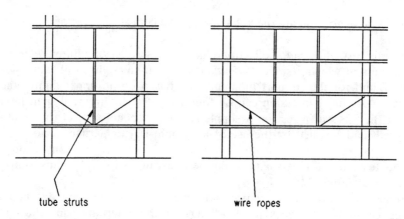

Fig. 1.29 Anti-sag systems for side wall rails

relevant to the load-reversal case. Conditions which arise in practice, and which are not covered in the technical literature, are best dealt with by the manufacturers' Technical Services Department which should be consulted for all non-standard cases.

Siderail design is essentially identical to that for purlins; load capacities are again arrived at after test procedures.

Self-weight deflections of siderails due to bending about the weak axis of the section are overcome by a tensioned wire system incorporating tube struts, typically at mid-span for spans of 6–7 m, and third-points of the span for spans of 7–8 m and above (Fig. 1.29).

1.4.7 Cladding

In addition to the visual requirements the cladding system has to provide the weatherproofing and insulation requirements suited to the building. It has to be strong enough to withstand the applied loads, normally snow, wind and traffic, during erection and maintenance. It must also provide the necessary lateral stability to the supporting purlins and rails and occasionally will form part of the structure itself in a stressed-skin design.

Cladding profiles are roll-formed and produced in volume so giving excellent economy and this means that, apart from an extremely rare occasion, the chosen profile will be from a manufacturer's published range. The loads that the profile can withstand are determined by the manufacturer and published in his brochure. Until recently they would have been determined by test, but calculation methods have been developed which are satisfactory for trapezoidal shapes. These give slightly conservative answers but in general, profiles that are walkable will have capacities in excess of those required. Sheeting can be erected satisfactorily without the need to walk indiscriminately over the whole area. The crowns at mid-span are the most susceptible to foot traffic but these positions can easily be avoided, even if crawl boards are not used.

In double-skin construction it is normal to assume all loads are taken on the outer sheet and the inner liner merely has to be erectable and stiff enough to prevent noticeable sag once installed.

The specifier, therefore, has to select the profile appropriate to the design requirement and check the strength and deflection criteria against the manufacturer's published load tables. Care must also be taken to ensure the fasteners are adequate.

A variety of systems is available to suit environmental and financial constraints. The most common are listed below.

Single skin trapezoidal roofing

This was widely used in the past with plasterboard or similar material as the lining material, and fibreglass insulation in the sandwich. The construction is susceptible to the plasterboard becoming damp due to condensation. An alternative is the use of rigid insulation boards, which are impervious to damp, supported on tee bars between the purlins. Unless the joints are sealed, which is difficult to achieve, condensation is likely to form. Although inexpensive, this type is therefore limited in its applicability.

The minimum slope is governed by the need to provide watertight joints and fasteners. If manufacturers' instructions on the use of sealants and stitching to laps are rigorously followed, this type can be used down to slopes of approximately 4°.

Double shell roof construction

In this form of construction the plasterboard has been replaced by a steel liner sheet of 0.4 mm thickness with some stiffening corrugations. The lining is first installed and fastened to the purlins, followed by the spacing Zeds, insulation and outer sheet. The liner tray is not designed to take full wind and erection loads and, therefore, large areas should not be erected in advance of the outer skin. The liner tray is normally supplied in white polyester finish providing a pleasing internal finish. The weatherproofing criteria are the same as for single-skin systems and generally the minimum slope is 4°. Differing thicknesses of insulation are accommodated by varying spacer depths. The norm is 80 mm of fibreglass giving a nominal U value of 0.44 W/m^2 °C.

Standing seam systems

The traditional forms of construction described above suffer from the inherent disadvantage of having to be fixed by screw-type fasteners penetrating the sheet.

Traditional fixing methods also limit the length of sheet that can be handled even if, in theory, long lengths can be rolled; thus laps are required.

The need for weathertightness at the lap constrains the minimum slope. A 5000 m² traditional roof has 20 000 through fasteners and has to resist around 1 million gallons of water a year. The difficulty in ensuring that this large number of fasteners is watertight demonstrates the desirability of minimizing the number of penetrations. This has led to the development of systems having concealed fastenings and the ability to roll and fix long lengths. In order to cater for the thermal expansion in sheets, which may be 30 m long, the fastenings are in the form of clips which, while holding down the sheeting, allow it to move longitudinally. As discussed elsewhere, this may reduce the restraint available to the purlins and affect their design. When used in double-skin configuration the liner panel is normally conventionally fastened and provides sufficient restraint. The available permutations are too numerous to give general rules but purlin manufacturers will give advice.

It is necessary to fasten the sheets to the structure at one point to resist downslope forces and progressive movement during expansion and contraction. With the through fasteners reduced to the minimum and laps eliminated or specially detailed roof slopes as low as 1°, (after deflection) can be utilized. The roofs must be properly maintained since accumulation of debris is more likely and ponding leads to a reduced coating life.

Standing seam systems are used to replace the traditional trapezoidal outer sheets in single- and double-skin arrangements as described earlier.

Composite panels

This most recent development in cladding systems provides solutions for many of the potential problems with metal roofing. The insulating foam is integral with the sheets and so totally fills the cavity, and with good detailing at the joints condensation can be eliminated in most environments.

The strength of the panel is dependent on the composite action of the two metal skins in conjunction with the foam. Theoretical calculations are possible although there are no codified design procedures. Since both steel and foam properties can vary and these are predetermined by the manufacturer, it is a question of selecting the panels from load tables provided rather than individual design. In addition to having to resist external loads, the effects of temperature differential must be taken into account. The critical combinations are wind suction with summer temperatures and snow acting with winter temperatures. The range of temperature considered is dependent on the colour and hence heat absorption of the outer skin; darker colours for roofs should only be considered in conjunction with the manufacturer, if at all.

Both standing seam and traditional trapezoidal forms are available with the same slope restrictions as non-composite forms.

A particular advantage is the erectability of the panels, which is a one-pass operation and, therefore, a rapid process. This is combined with inherent robustness and walkability.

Since the integrity of the panel is important, and it is difficult to inspect the foam and its adhesion once manufactured, quality control of the materials and manufacturing environment in terms of temperature and dust control is vital. Reputable manufacturers should, therefore, be specified and their manufacturing methods ascertained.

External firewall

Where buildings are close to the site boundary the Building Regulations require that the construction is such that reasonable steps are taken to prevent fire spreading to adjacent property. It has been demonstrated by tests that walls of double-skin steel construction with fibreglass or mineral wool insulation can achieve a four hour fire rating. The siderails and fixings require special details which were included in the test arrangements of the particular manufacturer and it is important that these are followed closely. They include such things as providing slotted holes to allow expansion of the rails rather than induce buckling, which may allow gaps to open in the sheeting at joints.

References to Chapter 1

1. Horridge J.F. (1985) *Design of Industrial Buildings*. Civil Engineering Steel Supplement, November.
2. British Standards Institution (1988) Part 3: *Code of practice for imposed roof loads*. BS 6399, BSI, London.
3. Building Research Establishment (1988) *Loads on roofs from snow drifting against vertical obstructions and in valleys*. BRE Digest 332, May. BRE, Watford.
4. British Standards Institution (1987) Part 22: *Code of practice for design, construction and loading of buildings and structures for agriculture*. BS 5502, BSI, London.
5. British Standards Institution (1990) Part 1: *Code of practice for design in simple and continuous construction: hot rolled sections*. BS 5950, BSI, London.
6. British Standards Institution (1987) Part 5: *Code of practice for design of cold formed sections*. BS 5950, BSI, London.
7. Davies J.M. & Raven G.K. (1986) Design of cold formed purlins. *IABSE Colloquium, Thin Walled Metal Structures in Buildings*, Stockholm.

A worked example follows which is relevant to Chapter 1.

The Steel Construction Institute Silwood Park, Ascot, Berks SL5 7QN	Subject *SINGLE STOREY DESIGN*	Chapter ref. *1*
	Design code *BS5950: Part 1*	Made by *AVP* Checked by *GWO*	Sheet no. *1*

Problem

Design the column and rafter of a portal frame using the graphical method to determine the position of the plastic hinges.

30 m span portal frame, 6° pitch
6 m frame centres
6 m to underside of haunch
Foundations 450 mm below finished floor level, pinned bases

Loading

Dead (including self weight)	$0.20 \ kN/m^2$
Services	$0.10 \ kN/m^2$
Imposed	$0.60 \ kN/mm^2$

For (Dead & Imposed) case at ultimate limit state,

w_u = $1.40 \ (0.20 + 0.10) + 1.60 \times 0.60 \ kN/m^2$
 = $1.38 \ kN/m^2$

Load to frame, W = 6×1.38 = $8.28 \ kN/m$

Free bending moment = $\dfrac{8.28 \times 30^2}{8}$ = $931.5 \ kNm$

Draw free bending moment diagram and reactant lines

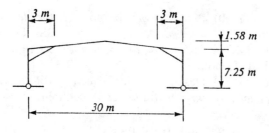

i.e. $L_1 = 15.00 \ m$ $h_1 = 7.25 \ m$ $h_2 = 1.58 \ m$

∴ $L_1 h_1 / h_2$ = $68.83 \ m$

Worked example

The Steel Construction Institute Silwood Park, Ascot, Berks SL5 7QN	Subject **SINGLE STOREY DESIGN**	Chapter ref. **1**
Design code BS5950: Part 1	Made by **AVP** Checked by **GWO**	Sheet no. **2**

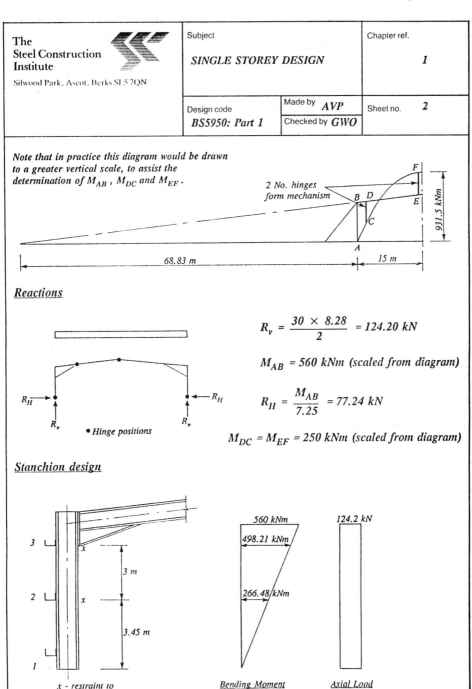

Note that in practice this diagram would be drawn to a greater vertical scale, to assist the determination of M_{AB}, M_{DC} and M_{EF}.

Reactions

$$R_v = \frac{30 \times 8.28}{2} = 124.20 \text{ kN}$$

$M_{AB} = 560$ kNm (scaled from diagram)

$$R_H = \frac{M_{AB}}{7.25} = 77.24 \text{ kN}$$

$M_{DC} = M_{EF} = 250$ kNm (scaled from diagram)

Stanchion design

x - restraint to compression flange

Try 533 × 210 × 82 UB (Grade 43) stanchion

BS 5950:
Part 1:
1990

The Steel Construction Institute Silwood Park, Ascot, Berks SL5 7QN	Subject **SINGLE STOREY DESIGN**	Chapter ref. *1*
	Design code *BS5950: Part 1* Made by *AVP* Checked by *GWO*	Sheet no. *3*

Section capacity check

F_v = 77.24 kN

P_v = $0.6\, p_y\, A_v$ = $0.6 \times 275 \times 528.3 \times 9.6 \times 10^{-3}$ 4.2.3
= 836.83 kN

$F_v < 0.6\, P_y$, no reduction in bending capacity due to shear   4.2.5

M_{cx} = $S_x\, p_y$ = $2060 \times 275/10^3$ = 566.50 kN

P = $A_g\, p_y$ = $104 \times 275/10$ = 2860 kN

$$\frac{F}{P} + \frac{M_x}{M_{cx}} = \frac{124.20}{2860} + \frac{498.21}{566.50}$$ 4.8.3.2

= 0.923 < 1.0 *satisfactory*

Overall buckling check - simplified approach   4.8.3.3

$$\frac{F}{A_g\, p_c} + \frac{m\, M_x}{M_b} \le 1$$

Length 1 - 2

m = 0.57, $m\, M_x$ = 0.57×266.48 Table 18
= 151.90 kNm

For L_e = 3.45 m, M_b = 364 kNm   *Steelwork Design Guide Vol. 1*

$\dfrac{m\, M_x}{M_b}$ = 0.417

For L_e = 3.45 m $A_g\, p_c$ = 1320 kN

$\dfrac{F}{A_g\, p_c}$ = 0.094   *Steelwork Design Guide Vol. 1*

The Steel Construction Institute Silwood Park, Ascot, Berks SL5 7QN	Subject **SINGLE STOREY DESIGN**		Chapter ref. 1
	Design code BS5950: Part 1	Made by AVP Checked by GWO	Sheet no. 4

$$\therefore \quad \frac{F}{A_g \, p_c} + \frac{m \, M_x}{M_b} = 0.511 < 1.00 \quad \underline{satisfactory}$$

A similar exercise for length 2 - 3 gives

$$\frac{F}{A_g \, p_c} + \frac{m \, M_x}{M_b} = \frac{124.20}{1420} + \frac{386.17}{453.29} = 0.939 < 1.00 \quad \underline{satisfactory}$$

Rafter design

$M_p = 250 \text{ kNm}$

$\text{Minimum, } S_x = \dfrac{250 \times 10^3}{275} = 909.09 \text{ cm}^3$

Use $406 \times 178 \times 54$ UB, $\quad S_x = 1050 \text{ cm}^3$

Clearly, S_x provided exceeds that required, and economy would be achieved by reducing the haunch length.

Local capacity checks, and overall bucking checks are carried out as shown above, with due cognizance being paid to hinge positions, which should be restrained.

Checks for members containing plastic hinges are covered in BS 5950 Clause 5.3.5 and/or Appendix G.

It is common practice to use proprietary computer programs in the design of portal frames. The above example was analysed with the following results:

Stanchion moment (6.45 m from base) = 485 kNm
Rafter moment (3.0 m position) = 256 kNm

Most programs have the great benefit of performing all the required stability, buckling and deflection checks automatically, together with providing the foundation loads.

Chapter 2
Multi-storey buildings

by ALAN HART

2.1 Introduction

The term multi-storey building encompasses a wide range of building forms. This chapter reviews some of the factors that should be considered when designing the type of multi-storey buildings commonly found in Europe, namely those less than 15 storeys in height. Advice on designing taller buildings may be found in the references to this chapter.[1-5]

2.1.1 The advantages of steel

In recent years the development of steel-framed buildings with composite metal deck floors has transformed the construction of multi-storey buildings in the UK. During this time, with the growth of increasingly sophisticated requirements for building services, the very efficiency of the design has led to the steady decline of the cost of the structure as a proportion of the overall cost of the building, yet the choice of the structural system remains a key factor in the design of successful buildings.

The principal reasons for the appeal of steel for multi-storey buildings are noted below.

- Steel frames are fast to erect.
- The construction is lightweight, particularly in comparison with traditional concrete construction.
- The elements of the framework are prefabricated and manufactured under controlled, factory conditions to established quality procedures.
- The accuracy implicit in the manufacturing process by which the elements are produced enables the designer to take a confident view of the geometric properties of the erected framework.
- The dryness of the form of construction results in less on-site activities, plant, materials and labour.
- The framework is not susceptible to drying-out movement or delays due to slow strength gain.
- Steel frames have potential for adaptability inherent in their construction.

Later modification to a building can be achieved relatively easily by unbolting a connection: with traditional concrete construction such modifications would be expensive, and more extensive and disruptive.
- The use of steel makes possible the creation of large, column-free internal spaces which can be divided by partitions and, by eliminating the external wall as a load-bearing element, allows the development of large window areas incorporated in prefabricated cladding systems.

2.1.2 Design aims

For the full potential of the advantages of steel-frame construction to be realized, the design of multi-storey buildings requires a considered and disciplined approach by the architects, engineers and contractors involved in the project. They must be aware of the constraints imposed on the design programme by the lead time between placing a contract for the supply of the steel frame and the erection of the first pieces on site. The programme should include such critical dates on information release as are necessary to ensure that material order and fabrication can progress smoothly.

The designer must recognize that the framework is the skeleton around which every other element of the building will be constructed. The design encompasses not only the structure but also the building envelope, services and internal finishes. All these elements must be co-ordinated by a firm dimensional discipline, which recognizes the modular nature of the components, to ensure maximum repetition and standardization. Consequently it is impossible to consider the design of the framework in isolation. It is vital to see the frame as part of an integrated building design from the outset: the most efficient solution for the structure may not be effective in achieving a satisfactory solution for the total building.

In principle, the design aims can be considered under three headings:

- Technical
- Architectural
- Financial.

Technical aims

The designer must ensure that the framework, its elements and connections are strong enough to withstand the applied loads to which the framework will be subjected throughout its design life. The system chosen on this basis must be sufficiently robust to prevent the progressive collapse of the building or a significant part of it under accidental loading. This is the primary technical aim. However, as issues related to strength have become better understood and techniques for the strength design of frameworks have been formalized, designers have progressively used lighter and stronger materials. This has generated a greater need to consider serviceability, including dynamic floor response, as part of the development of the structural concept.

Other important considerations are to ensure adequate resistance to fire and corrosion. The design should aim to minimize the cost, requirements and intrusion of the protection systems on the efficiency of the overall building.

Architectural aims

For the vast majority of buildings the most effective structural steel frame is the one which is least obtrusive. In this way it imposes least constraint on internal planning, and produces maximum usable floor area, particularly for open-plan offices. It also provides minimal obstruction to the routing of building services. This is an important consideration, particularly since building services are becoming more extensive and demanding on space and hence on the building framework.

Occasionally the structure is an essential feature of the architectural expression of the building. Under these circumstances the frame must achieve, among other aims, a balance between internal planning efficiency and an expressed structural form. However, these buildings are special, not appropriate to this manual, and will not be considered in more detail, except to give a number of references.

Financial aims

The design of a steel frame should aim to achieve minimum overall cost. This is a balance between the capital cost of the frame and the improved revenue from early occupation of the building through fast erection of the steel frame: a more expensive framework may be quicker to build and for certain uses would be more economic to a client in overall terms. Commercial office developments are a good example of this balance. Figure 2.1 shows a breakdown of construction costs for a typical development.

2.1.3 Influences on overall design concept

Client brief

Clients specify their requirements through a brief. It is essential for effective design to understand exactly the intentions of the client: the brief is the way in which the client expresses and communicates these intentions. As far as the frame designer is concerned, the factors which are most important are intended use, budget cost limits, time to completion and quality. Once these are understood a

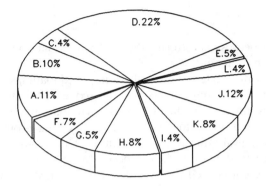

```
Fabric
A  Demolition. Piling foundations and concrete work.......11%
B  Steel Frame.  Deck and Fire Protection..................10%
C  Brickwork and Drywalling........................................ 4%
D  External and Internal Cladding and Sunscreens..........22%
E  Roofing and Rooflights............................................ 5%
                                                                              52%

Finishes                          Services
F  Ceilings and Floors.. 7%      I  Plumbing and Sprinklers.. 4%
G  Stone ................. 5%     J  H.V.A.C......................12%
H  Others ................ 8%     K  Electrics..................... 8%
                        20%      L  Lifts ......................... 4%
                                                                28%
```

Fig. 2.1 Typical cost breakdown

realistic basis for producing the design will be established. The designer should recognize however that in practice the brief is likely to evolve as the design develops.

Statutory constraints

The design of all buildings is subject to some form of statutory constraint. Multi-storey buildings, particularly those in an urban environment, are subject to a high level of constraint which will generally be included in the conditions attached to the granting of Outline Planning Permission. The form and degree that this may take could have significant impact on the frame design. For example, street patterns and lighting restrictions may result in a non-rectilinear plan with a 'stepped-back' structure. If an appropriate layout is to be provided it is vital to understand these constraints from the outset.

Physical factors

The building must be designed to suit the parameters determined by its intended use and its local environment.

Its intended use will dictate the intensity of imposed loadings, the fire protection and corrosion resistance requirements and the scope of building services.

The local environment will dictate the lateral load requirements but more importantly it will determine the nature of the existing ground conditions. To achieve overall structural efficiency it is essential that the structural layout of the frame is responsive to the constraints imposed by these ground conditions.

These factors are considered in more detail in the next section.

2.2 Factors influencing choice of form

Environmental

There are a number of factors which influence the choice of structural form that are particular to the site location. These can have a dominant effect on the framing arrangement for the structure.

The most obvious site-dependent factors are related to the ground conditions. A steel-framed building is likely to be about 60% of the weight of a comparable reinforced concrete building. This difference will result in smaller foundations with a consequent reduction on costs. In some cases this difference in weight enables simple pad foundations to be used for the steel frame where the equivalent reinforced concrete building would require a more complex and expensive solution. For non-uniformly loaded structures it will also reduce the magnitude of differential settlements and for heavily loaded structures may make possible the use of a simple raft foundation in preference to a large capacity piled solution (Fig. 2.2).

Difficult ground conditions may dictate the column grid. Long spans may be required to bridge obstructions in the ground. Such obstructions could include, for example, buried services, underground railways or archaeological remains. Generally, a widely spaced column grid is desirable since it reduces the number of foundations and increases the simplicity of construction in the ground.

Other site-dependent constraints are more subtle. In urban areas they relate to the physical constraints offered by the surrounding street plan, and the rights of light of adjoining owners. They also relate to the planning and architectural objectives for specific sites. The rights of light issues or planning considerations may dictate that upper floors are set back from the perimeter resulting in stepped construction of the upper levels. Invariably the resulting framing plan is not rectilinear and may have skew grids, cantilevers and re-entrant corners.

These constraints need to be identified early in the design in order that they are accommodated efficiently into the framing. For example, wherever possible,

Factors influencing choice of form   41

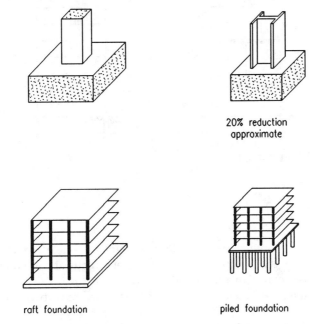

20% reduction
approximate

raft foundation   piled foundation

Fig. 2.2   Foundation savings

stepped-back façades should be arranged so that steps take place on the column grid and hence avoid the need for heavy bridging structures. In other situations the designer should always investigate ways in which the impact of lack of uniformity in building form can be contained within a simple structural framing system which generates a minimum of element variations and produces simple detailing.

Building use

The building use will dictate the planning module of the building which will in turn determine the span and column grids. Typical grids may be based on a planning module of 600/1200 mm or 500/1500 mm. However, the use has much wider impact, particularly on floor loadings and building services. The structural arrangement, and depth selected, must satisfy and accommodate these requirements.

For example, financial-dealing floors require clear open spaces located on the lower floors which would dictate a different structural solution to the rest of the building. This may necessitate the use of a transfer structure to carry the upper floors on an economical column grid (Fig. 2.3).

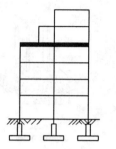 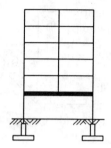

Fig. 2.3 Typical load transfer systems

Floor loadings

Because steel-framed buildings are relatively light in weight, excessive imposed loadings will have a greater effect on the sizing of structural components, particularly floor beams, than with reinforced concrete structures.

The floor loadings to be supported by the structure have two components:

- The permanent or dead loading comprising the self-weight of the flooring and the supporting structure together with the weight of finishes, raised flooring, ceiling, air-conditioning ducts and equipment.
- The imposed loading which is the load that the floor is likely to sustain during its life and will depend on the building use. Imposed floor loads for various types of building are governed by BS 6399 but the standard loading for office buildings is usually 4 kN/m^2 with an additional allowance of 1 kN/m^2 for movable partitioning.

For normal office loadings, dead and imposed loadings are roughly equal in proportion but higher imposed load allowances will be necessary in plantrooms or to accommodate special requirements such as storage or heavy equipment.

Floor beams will be designed to limit deflection under the imposed loadings. British Standard BS 5950 governing the design of structural steelwork sets a limit for deflection under imposed loading of (span/200) generally and (span/360) where there are brittle finishes. Edge beams supporting cladding will be subject to restriction on deflection of 10–15 mm. Deflections may be noticeable in the ceiling layout and should be taken into account when determining the available clearance for service routes. The designer should therefore check the cumulative effect of deflections in the individual members of a floor system although the actual maximum displacement is in practice almost always less than that predicted. In some instances, vibrations of floor components may cause discomfort or affect sensitive equipment and the designer should check the fundamental frequency of the floor system. The threshold of perceptible vibrations in building is difficult to define and present limits are rather arbitrary. There is some evidence that modern long span, lightweight floors can be sensitive to dynamic loads which may have an effect on delicate equipment.

Building services and finishes

In buildings requiring anything other than minimal electrical services distribution, the inter-relationship of the structure, the mechanical and electrical services and the building finishes will need to be considered together from the outset.

It is essential to co-ordinate the details of the building services, cladding and structure at an early stage of the project in order to produce a building which is simple to fabricate and quick to erect. Apparently minor variations to the steelwork, brought about by services and finishes requirements, defined after a steel fabrication contract has been let, can have a disproportionate effect on the progress of fabrication and erection. Steel buildings impose a strict discipline on the designer in terms of the early production of final design information. If the designer fails to recognize this, the advantages of steel-framed building cannot be realized.

The integration of the building services with the structure is an important factor in the choice of an economic structural floor system. The overall depth of the floor construction will depend on the type and distribution of services in the ceiling void. The designer may choose to separate the structural and services zones or accommodate the services by integrating them with the structure allowing for the structural system to occupy the full depth of the floor construction. (See Fig. 2.4).

Separation of zones usually requires confining the ducts, pipes and cables to a horizontal plane below the structure resulting in either a relatively deep overall floor construction or close column spacings. Integration of services with structure requires either deep perforated structural components or vertical zoning of the services and structure.

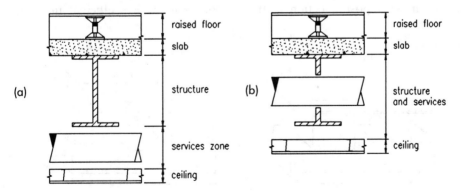

Fig. 2.4 Building services and floor structure: (a) separation of services and structure; (b) integration of services and structure

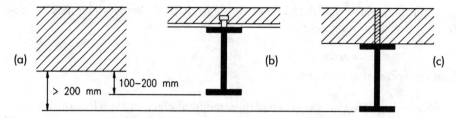

Fig. 2.5 Overall floor depths: (a) R.C. flat slab; (b) composite; (c) non-composite

For the range of structural grids used in conventional building, traditional steel floor construction is generally deeper than the equivalent reinforced concrete flat slab: the difference is generally 100–200 mm for floor structures which utilize composite action and greater for non-composite floors (Fig. 2.5). The increased depth is only at the beam position: elsewhere, between beams, the depth is much less and the space between them may accommodate services, particularly if the beams may be penetrated (Fig. 2.6). The greater depth of steel construction does not therefore necessarily result in an increase in building height if the services are integrated within the zone occupied by the structure. A number of possible solutions exist for integrated systems, particularly in long-span structures utilizing castellated or stub-girder beams. (See Fig. 2.7.)

Recently a number of solutions have been developed which allow long spans to co-exist with separation of the building services by profiling the steel beam to provide space for services, either at the support or in the span. The concept of this approach is not new; the innovation has come from the use of automated plate cutting and welding techniques to produce economical profiled plate girders. (See Fig. 2.8.)

Overall depth may be reduced by utilizing rigid or semi-rigid rather than simple connections at the ends of the beams. This reduces the maximum bending moment and deflection. However, such solutions are not as efficient as would first appear since the non-composite section at the support is much less efficient than the

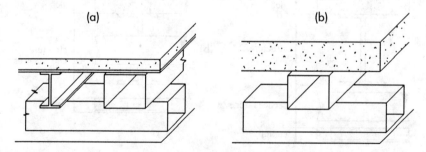

Fig. 2.6 Ceiling voids: (a) steel frame: variable void height; (b) concrete slab: constant void height

Factors influencing choice of form 45

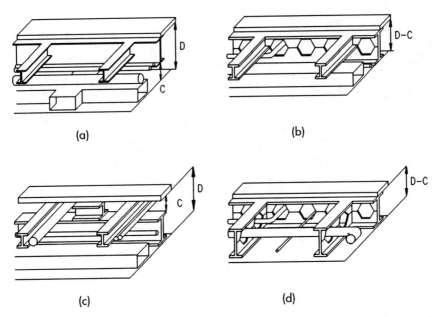

Fig. 2.7 Integration of services: (a) separated (traditional); (b) integrated (short-medium span 'secondary' beams; (c) integrated (long span 'primary' beam – stub girder; (d) integrated (long span 'secondary' beams)

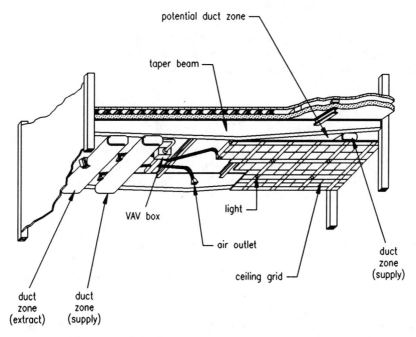

Fig. 2.8 Tapered beams and services

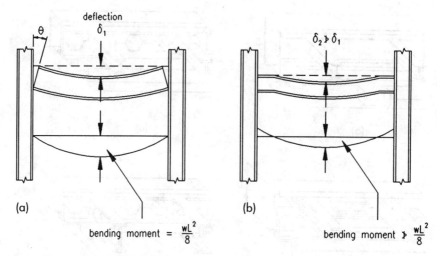

Fig. 2.9 Floor depth, (a) simple and (b) rigid

composite section at mid-span. Indeed, if the support bending moments are large in comparison to the span bending moments the depth may be greater than the simply-supported composite beam. This is an expensive fabrication in comparison to straight rolled beam sections. (See Fig. 2.9.) In addition, the use of rigid joints can increase column sizes considerably.

The overall depth may also be reduced by using higher-strength steel, but this is only of advantage where the element design is controlled by strength. The stiffness characteristics of both steels are the same; hence, where deflection or vibration govern, no advantage is gained by using the stronger steel.

External wall construction

The external skin of a multi-storey building is supported off the structural frame. In most high quality commercial buildings the cost of external cladding systems greatly exceeds the cost of the structure. This influences the design and construction of the structural system in a number of ways:

- The perimeter structure must provide a satisfactory platform to support the cladding system and be sufficiently rigid to limit deflections of the external wall.
- A reduction to the floor zone may be more cost-effective than an overall increase in the area of cladding.
- Fixings to the structure should facilitate rapid erection of cladding panels.
- A reduction in the weight of cladding at the expense of cladding cost will not necessarily lead to a lower overall construction cost.

Lateral stiffness

Steel buildings must have sufficient lateral stiffness and strength to resist wind and other lateral loads. In tall buildings the means of providing sufficient lateral stiffness forms the dominant design consideration. This is not the case for low- to medium-rise buildings.

Most multi-storey buildings are designed on the basis that wind forces acting on the external cladding are transmitted to the floors which form horizontal diaphragms transferring the lateral load to rigid elements and then to the ground. These rigid elements are usually either braced-bay frames, rigid-jointed frames or reinforced concrete shear walls.

British Standard BS 5950 sets a limit on lateral deflection of columns as height/300 but height/600 may be a more reasonable figure for buildings where the external envelope consists of sensitive or brittle materials such as stone facings.

Accidental loading

A series of incidents in the 1960s culminating in the partial collapse of a system-built tower block at Ronan Point in 1968 led to a fundamental reappraisal of the approach to structural stability in building.

Traditional load-bearing masonry buildings have many in-built elements providing inherent stability which are lacking in modern steel-framed buildings. Modern structures can be refined to a degree where they can resist the horizontal and vertical design loadings with the required factor of safety but may lack the ability to cope with the unexpected.

It is this concern with the safety of the occupants and the need to limit the extent of any damage in the event of unforeseen or accidental loadings that has led to the concept of robustness in building design. Any element in the structure that supports a major part of the building either must be designed for blast loading or must be capable of being supported by an alternative load path. In addition, suitable ties should be incorporated in the horizontal direction in the floors and in the vertical direction through the columns. The designer should be aware of the consequences of the sudden removal of key elements of the structure and ensure that such an event does not lead to the progressive collapse of the building or a substantial part of it. In practice, most modern steel structures can be shown to be adequate without any modification.

Cost considerations

The time taken to realize a steel building from concept to completion is generally less than that for a reinforced concrete alternative. This reduces time-related building costs, enables the building to be used earlier and produces an earlier return on the capital invested.

To gain full benefit from the 'factory' process and particularly the advantages of

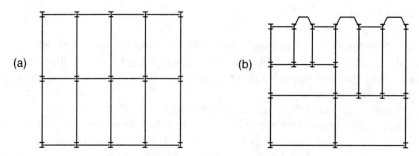

Fig. 2.10 Structural costs: (a) economic and (b) uneconomic layouts

speed of construction, prefabrication, accuracy and lightness, the cladding and finishes of the building must have similar attributes. The use of heavy, slow and in situ finishing materials is not compatible with the lightweight, prefabricated and fast construction of a steel framework.

The cost of steel frameworks is governed to a great extent by the degree of simplicity and repetition embodied in the frame components and connections. This also applies to the other elements which complete the building.

The criterion for the choice of an economic structural system will not necessarily be to use the minimum weight of structural steel. Material costs represent only 30–40% of the total cost of structural steelwork. The remaining 60–70% is accounted for in the design, detailing, fabrication, erection and protection. Hence a choice which needs a larger steel section to avoid, say, plate stiffeners around holes or allows greater standardization will reduce fabrication costs and may result in the most economic overall system.

Because a steel framework is made up of prefabricated components produced in a factory, repetition of dimensions, shapes and details will streamline the manufacturing process and are major factors in economic design (Fig. 2.10).

Fabrication

The choice of structural form and method of connection detailing have a significant impact on the cost and speed of fabrication and erection. Simple braced frameworks with bolted connections are considered the most economic and the fastest to build for low- to medium-rise buildings.

Economy is generally linked to the use of standard rolled sections but, with the advent of automated cutting and welding equipment, special fabricated sections are becoming economic if there is sufficient repetition.

The development of efficient, automated, cold-sawing techniques and punching and drilling machines has led to the fabrication of building frameworks with bolted assemblies. Welded connections involve a greater amount of handling in the fabrication shop, with consequent increases in labour and cost.

Site-welded connections require special access, weather protection, inspection and temporary erection supports. By comparison, on-site bolted connections enable the components to be erected rapidly and simply into the frame and require no further handling.

The total weight of steel used in rigid frames is less than in simple frames, but the connections for rigid frames are more complex and costly to fabricate and erect. On balance, the cost of a rigid frame structure is greater, but there may be other considerations which offset this cost differential. For example, in general the overall structural depth of rigid frames is less. This may reduce the height of the building or improve the distribution of building services, both of which could reduce the overall cost of the building.

Corrosion protection to internal building elements is an expensive and time-consuming activity. Experience has shown that it is unnecessary for most internal locations and consequently only steelwork in risk areas should require any protection.

Construction

A period of around 10–14 weeks is usual between placing a steel order and the arrival of the first steel components on site. Site preparation and foundation construction generally take a similar or longer period (see Fig. 2.11). Hence, by progressing fabrication in parallel with site preparation, significant on-site construction time may be saved, as commencement of shop fabrication is equivalent to start-on-site for an in situ concrete-framed building. By manufacturing the frame in a factory, the risks of delay caused by bad weather or insufficient or inadequate construction resources in the locality of the site are significantly reduced.

Structural steel frameworks should generally be capable of being erected without temporary propping or scaffolding, although temporary bracing will be required, especially for welded frames. This applies particularly to the construction of the concrete slab which should be self-supporting at all stages of erection. Permanent metal or precast concrete shutters should be used to support the in situ concrete.

In order to allow a rapid start to construction, the structural steelwork frame should commence at foundation level and preference should be given to single foundations for each column rather than raft or shared foundations (Fig. 2.12).

Speed of erection is directly linked to the number of crane hours available. To reduce the number of lifts required on site, the number of elements forming the framework should be minimized within the lifting capacity of the craneage provided on site for other building components. For similar sized buildings, the one with the longer spans and fewer elements will be the fastest to erect. However, as has been mentioned earlier, longer spans require deeper, heavier elements which will increase the cost of raw materials and pose a greater obstruction to the distribution of building services thereby requiring the element to be perforated or shaped and hence increasing the cost of fabrication.

50   Multi-storey buildings

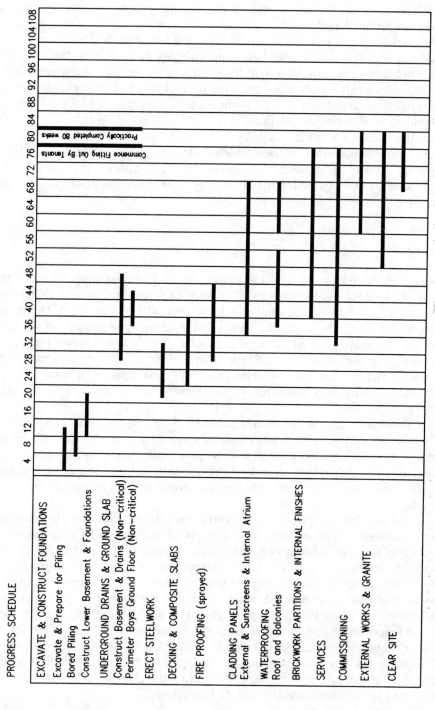

Fig. 2.11 Typical progress schedule (in weeks)

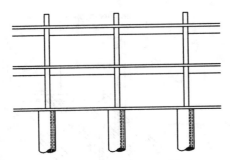

Fig. 2.12 Columns on large diameter bored piles

Columns are generally erected in multi-storey lengths; two is common and three is not unusual. The limitation on longer lengths is related more to erection than restrictions on transportation, although for some urban locations, length is a major consideration for accessibility.

To provide rapid access to the framework the staircases should follow the erection of the frame. This is generally achieved by using prefabricated stairs which are detailed as part of the steel frame.

The speed of installation of the following building elements is hastened if their connection and fixing details are considered at the same time as the structural steel frame design. In this way the details can either be incorporated in the framework or separated from it, whichever is the most effective overall: it is generally more efficient to separate the fixings and utilize the high inherent accuracy of the frame to use simple post-fixed details, provided these do not require staging or scaffolding to give access.

Finally, on-site painting extends the construction period and provides potential compatibility problems with following applied fire protection systems. Painting should therefore only be specified when absolutely necessary.

2.3 Anatomy of structure

In simple terms, the vertical load-carrying structure of a multi-storey building comprises a system of vertical column elements interconnected by horizontal beam elements which support floor-element assemblies. The resistance to lateral loads is provided by diagonal bracing elements, or wall elements, introduced into the vertical rectangular panels bounded by the columns and beams to form vertical trusses, or walls, called 'no-sway' frames. Alternatively, lateral resistance may be provided by developing a rigid-frame action between the beams and columns, called 'sway' frames. The floor-element assemblies provide the resistance to lateral loads in the horizontal plane.

In summary, the components of a building structure are columns, beams, floors and bracing systems (Fig. 2.13).

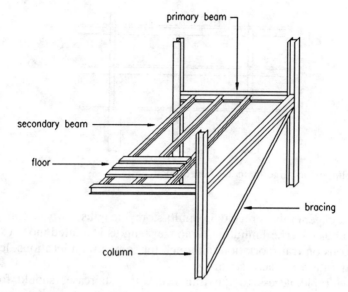

Fig. 2.13 Steel frame components

2.3.1 Columns

These are generally standard, universal column, hot-rolled sections. They provide a compact, efficient section for normal building storey heights. Also, because of the section shape, they give unobstructed access for beam connections to either the flange or web. For a given overall width and depth of section, there is a range of weights which enable the overall dimensions of structural components to be maintained for a range of loading intensities.

Where the loading requirements exceed the capacity of standard sections, additional plates may be welded to the section to form plated columns, or fabricated columns may be formed by welding plates together to form a plate-column (Fig. 2.14).

The use of circular or rectangular tubular elements marginally improves the load-carrying efficiency of components as a result of their higher stiffness-to-weight ratio. However, connections to beams become more complicated.

2.3.2 Beams

Structural steel floor systems consist of prefabricated standard components, and columns should be laid out on a repetitive grid which establishes a standard structural bay. For most multi-storey buildings, functional requirements will deter-

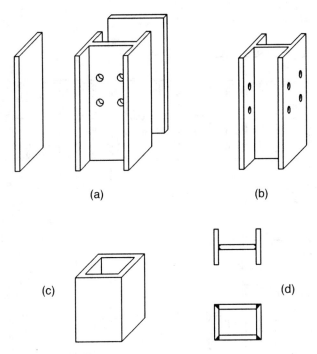

Fig. 2.14 Types of column: (a) plated (by addition of plates to U.C. section); (b) universal; (c) tabular; (d) fabricated plate

mine the column grid which will dictate spans where the limiting criterion will be stiffness rather than strength (Fig. 2.15).

Steel components are uni-directional and consequently orthogonal structural column and beam grids have been found to be the most efficient. The most efficient floor plan is rectangular, not square, in which main, or 'primary', beams span the shorter distance between columns and closely-spaced 'secondary' floor beams span the longer distance between main beams. The spacing of the floor beams is controlled by the spanning capability of the concrete floor construction (Fig. 2.16).

Having decided on the structural grid, the designer must choose an economic structural system to satisfy all the design constraints. The choice of system and its depth depends on the span of the floor (Fig. 2.17). The minimum depth is fixed by practical considerations such as fitting practical connections. As the span increases, the depth will be determined by the bending strength of the member and, for longer spans, by the stiffness necessary to prevent excessive deflection under imposed load or excessive sensitivity to induced vibrations (Fig. 2.18). For spans up to 10–12 m, simple universal beams with precast floors or composite metal deck floors are likely to be most economic. A range of section capacities for each depth enables a constant depth of construction to be maintained for a range

Multi-storey buildings

Steelwork Plan Level 3

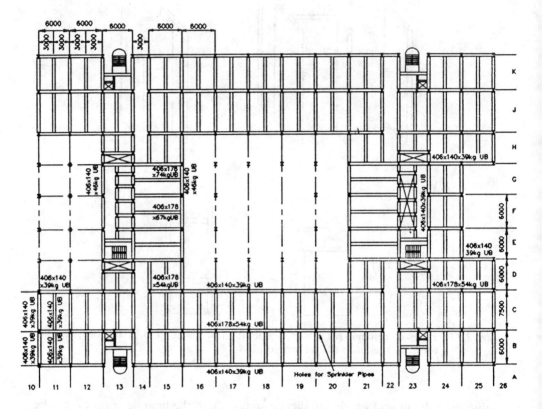

Fig. 2.15 Typical floor layout

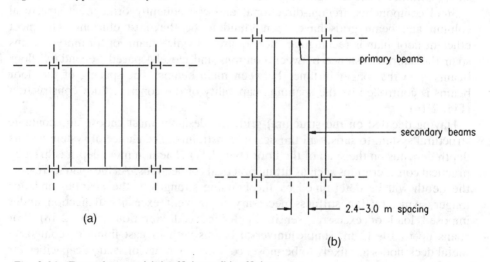

Fig. 2.16 Beam layout: (a) inefficient; (b) efficient

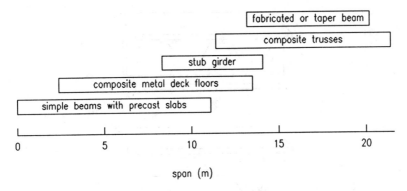

Fig. 2.17 Choice of floor system

of spans and loading. As with column components, plated-beams and fabricated girders may be used for spans above 10–12 m. They are particularly appropriate where heavier loading is required and overall depth is limited. For medium- to lightly-loaded floors and long spans, beams may also take the form of castellated beams fabricated from standard sections. Above 15 m, composite steel trusses may be economic. As the span increases the depth and weight of the structural floor increase and above 15 m spans, depth predominates because of the need to achieve adequate stiffness.

Castellated beam sections

Castellated beams (Fig. 2.19) have been used for many years to increase the bending capacity of the beam section and to provide limited openings for services. These openings are rarely of sufficient size for ducts to penetrate without significant

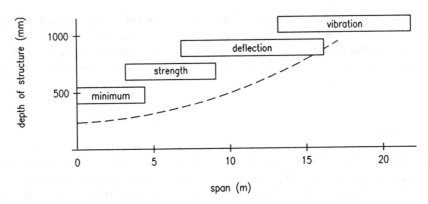

Fig. 2.18 Structural criteria governing choice

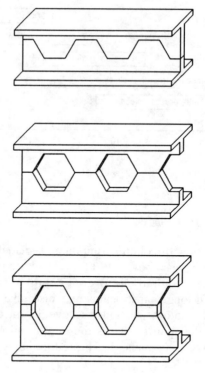

Fig. 2.19 Castellated beams

modification and fabrication costs are high. The cellform concept is a development of castellated beams that provides circular spans and greater shear capacity.

Fabricated plate girders

Conventional universal beams span a maximum of about 15 m. Recent advances in automatic and semi-automatic fabrication techniques have allowed the economic production of plate girders for longer span floors. Particularly if a non-symmetric plate girder is used, it is possible to achieve economic construction well in excess of 15 m. (Fig. 2.20.)

Taper beams

Taper beams (Fig. 2.20) are similar to fabricated plate girders except that their depth varies from a maximum in mid-span to a minimum at supports thus achieving a highly efficient structural configuration. For simply-supported composite taper beams in buildings the integration of the services can be accommodated by

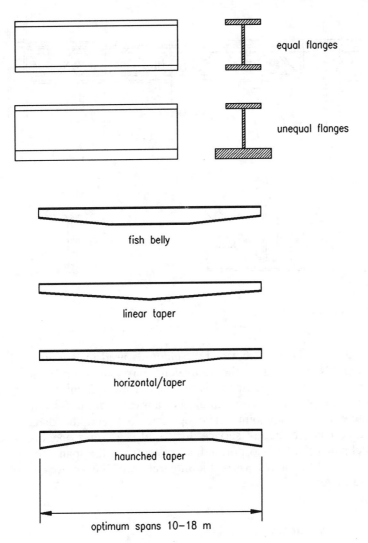

Fig. 2.20 Fabricated plate girders and taper beams

locating the main ducts close to the columns. Alternative taper beam configurations can be used to optimize the integration of the building services.

Composite steel floor trusses

Use of composite steel floor trusses as primary beams in the structural floor system permits much longer spans than would be possible with conventional universal beams. The use of steel trusses for flooring systems is common for multi-

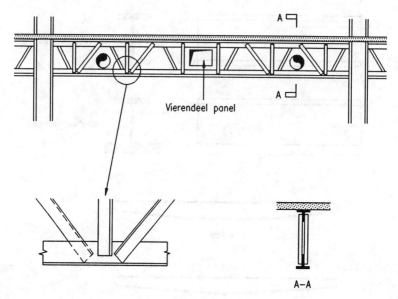

Fig. 2.21 Composite truss

storey buildings in North America but seldom is used in Britain. Although they are considerably lighter than the equivalent universal beam section the cost of fabrication is very much greater as is the cost of fireproofing the truss members. For maximum economy, trusses should be fabricated from T-sections and angles using simple welded lap joints. The openings between the diagonal members should be designed to accept service ducts and if a larger opening is required a Vierendeel panel can be incorporated at the centre of the span. Because a greater depth is required for floor trusses, the integration of the services is always within the structural zone. (Fig. 2.21.)

Stub girder construction

Stub girders were developed in North America in the 1970s as an alternative form of construction for intermediate range spans of between 10–14 m. They have not been used significantly in the UK. Figure 2.22 shows a typical stub girder with a bottom chord consisting of a compact universal column section which supports the secondary beams at approximately 3-metre centres. Between the secondary beams a steel stub is welded on to the bottom chord to provide additional continuity and to support the floor slab. The whole system acts as a composite Vierendeel truss. A disadvantage of stub girders is that the construction needs to be propped while the concrete is poured and develops strength. Arguably, a deep universal beam with large openings provides a more cost-effective alternative to the stub girder because of the latter's high fabrication content.

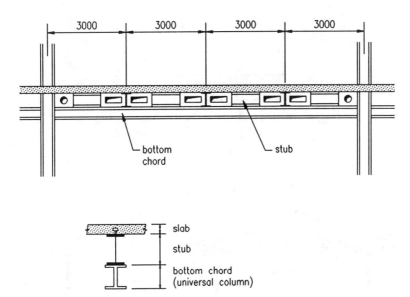

Fig. 2.22 Stub girder

2.3.3 Floors

These take the form of concrete slabs of various forms of construction spanning between steel floor beams (Fig. 2.23). The types generally found are:

- in situ concrete slab cast on to permanent profiled metal decking, acting compositely with the steel floor beams;
- precast concrete slabs acting non-compositely with the floor beams;
- in situ concrete slab, with conventional removable shuttering, acting compositely with the floor beams;
- in situ concrete slab cast on thin precast concrete slabs to form a composite slab, which in turn acts compositely with the floor beams.

The most widely used construction internationally is profiled metal decking. Composite action with the steel beam is provided by shear connectors welded through the metal decking on to the beam flange. Precast concrete systems are, however, still used extensively in the UK.

Metal deck floor construction

Composite action enables the floor slab to work with the beam enhancing its strength and reducing deflection (Fig. 2.24). Because composite action works by

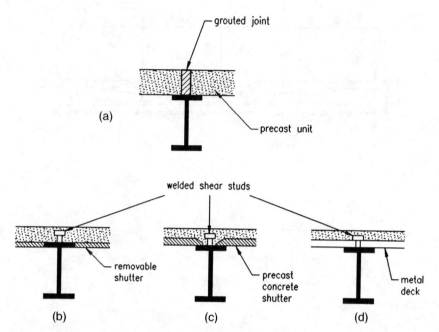

Fig. 2.23 Floor construction: (a) precast (non-composite); (b) in situ (composite); (c) in situ/precast (composite); (d) in situ/metal decking (composite)

allowing the slab to act as the compression flange of the combined steel and concrete system, the advantage is greatest when the beam is sagging. Consequently composite floor systems are usually designed as simply supported.

Experience has shown that the most efficient floor arrangements are those using metal decking spanning about 3 m between floor beams. For these spans the metal decking does not normally require propping during concreting and the concrete thicknesses are near the practical minimum for consideration of strength and fire separation.

Steel studs are welded through the decking on the flange of the beams below to form a connection between steel beam and concrete slab. Concrete, which may be either lightweight or normal weight, is then poured on to the decking usually by pumping to make up the composite system. Metal decking acts both as permanent formwork for the concrete and as tensile reinforcement for the slab. There are

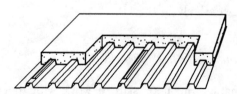

Fig. 2.24 Metal deck floor slabs

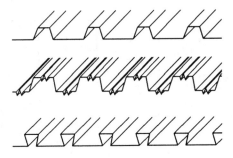

Fig. 2.25 Metal deck profiles

many types of steel decking available (Fig. 2.25) but perhaps the most commonly used is the re-entrant profile type which provides a flat soffit and facilitates fixings for building services and ceilings.

Some of the advantages of composite metal deck floor construction are:

- Steel decking acts as permanent shuttering which can eliminate the need for slab reinforcement and, due to its high stiffness and strength, propping of the construction while the wet concrete develops strength.
- Composite action reduces the overall depth of structure.
- It provides up to 2 hours fire resistance without additional fire protection and 4 hours with added thickness or extra surface protection.
- It is a light, adaptable system that can be easily manhandled on site, cut to awkward shapes and drilled or cut out for additional service requirements.
- Lightweight construction reduces frame loadings and foundation costs.
- It allows simple, rapid construction techniques.

Figure 2.26 illustrates alternative arrangements of primary and secondary beams for an optimum deck span of 3 m.

Precast floor systems

Universal beams supporting precast prestressed floor units have some advantages over other forms of construction. Although of heavier construction than comparable composite metal deck floors, this system offers the following advantages.

- Fewer floor beams since precast floor units can span up to 6–8 m without difficulty.
- No propping is required.
- Shallow floor construction can be obtained by supporting precast floor units on shelf angles or on wide plates attached to the bottom flanges of universal columns acting as beams (Slim floor).

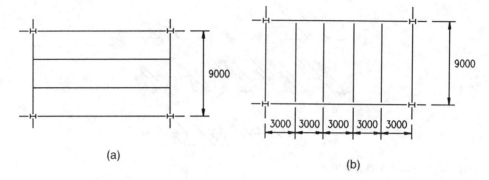

Fig. 2.26 Alternative framing systems for floors: (a) long span secondary beams; (b) long span primary beams

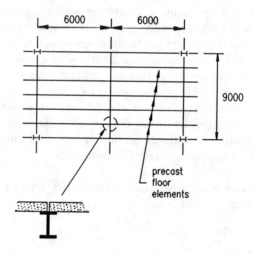

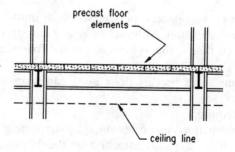

Fig. 2.27 Precast concrete floors

- Fast construction because no time is needed for curing and the development of concrete strength.

On the other hand the disadvantages are:

- Composite action is not readily achieved without a structural floor screed.
- Heavy floor units are difficult to erect in many locations and require the use of a tower crane which may have implications for the construction programme.

2.3.4 Bracings

Three structural systems are used to resist lateral loads: rigidly-jointed frames, reinforced concrete walls and braced-bay frames (Fig. 2.28). Combinations of these systems may also be used.

Rigid frames

Rigidly-jointed frames, or 'sway' frames, are those with full moment resisting connections between beam and columns. It is not necessary that all connections in a building are detailed in this way; only sufficient frames to satisfy the performance requirements of the building.

The advantage of the rigid frame is:

- Provides total internal adaptability with no bracings between columns or walls to obstruct circulation.

However, the disadvantages are:

- Increased fabrication for complex framing connections
- Increased site connection work, particularly if connections are welded
- Columns are larger to resist bending moments
- Generally, less stiff than other bracing systems.

Shear walls

Reinforced concrete walls constructed to enclose lift, stair and service cores generally possess sufficient strength and stiffness to resist the lateral loading.

Cores should be located to avoid eccentricity between the line of action of the lateral load and the centre of stiffness of the core arrangement. However, the core locations are not always ideal because they may be irregularly shaped, located at one end of the building or are too small. In these circumstances,

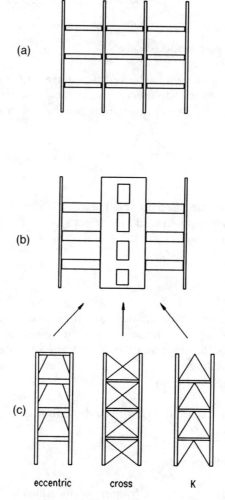

Fig. 2.28 Bracing structures: (a) rigid frame; (b) reinforced concrete wall; (c) braced bay frames

additional braced bays or rigidly-jointed frames should be provided at other locations. (Fig. 2.29.)

Although shear walls are normally constructed in in situ reinforced concrete they may also be constructed of either precast concrete or brickwork.

The advantages of shear walls are:

- The beam-to-column connections throughout the frame are simple, easily fabricated and rapidly erected.
- Concrete walls tend to be thinner than other bracing systems and hence save space in congested areas such as service and lift cores.

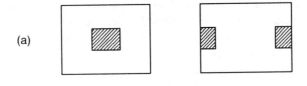

(a)

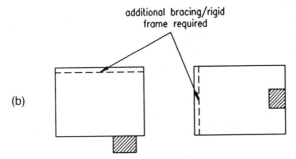

(b)

Fig. 2.29 Core locations: (a) efficient; (b) inefficient

- They are very rigid and highly effective.
- They act as fire compartment walls.

The disadvantages are:

- The construction of walls, particularly in low- and medium-rise buildings, is slow and less accurate than steelwork.
- The walls are difficult to modify if alterations to the building are required in the future.
- They are a separate form of construction which is likely to delay the contract programme.
- It is difficult to provide connections between steel and concrete to transfer the large forces generated.

Braced-bay frames

Braced-bays are positioned in similar locations to reinforced concrete walls, so they have minimal impact upon the planning of the building. They act as vertical trusses which resist the wind loads by cantilever action.

The bracing members can be arranged in a variety of forms designed to carry solely tension or alternatively tension and compression. When designed to take tension only, the bracing is made up of crossed diagonals. Depending on the wind direction, one diagonal will take all the tension while the other remains inactive.

Tensile bracing is smaller than the equivalent strut and is usually made up of flat-plate, channel or angle sections. When designed to resist compression, the bracings become struts and the most common arrangement is the 'K' brace.

The advantages of braced-bay frames are:

- All beam-to-column connections are simple
- The braced bays are concentrated in location on plan
- The bracing configurations may be adjusted to suit planning requirements (eccentric bracing)
- The system is adjustable if building modifications are required in the future
- Bracing can be arranged to accommodate doors and openings for services
- Bracing members can be concealed in partition walls
- They provide an efficient bracing system.

A disadvantage is:

- Diagonal members with fire proofing can take up considerable space.

2.3.5 Connections

The most important aspect of structural steelwork for buildings is the design of the connections between individual frame components.

The selection of a component should be governed not only by its capability to support the applied load, but also by its ease of connection to other components.

Basically there are three types of connection, each defined by its structural behaviour; simple, rigid and semi-rigid (Fig. 2.30):

(1) Simple connections transmit negligible bending moment across the joint; the connection is detailed to allow the beam end to rotate. The beam behaves as a simply supported beam.
(2) Rigid connections are designed to transmit shear force and bending moment across the joint. The connection is detailed to ensure a monolithic joint. Beam end moments are transmitted into the column itself and any beam framing into the column on the opposite side.
(3) Semi-rigid connections are designed to transmit the shear force and a proportion of the bending moment across the joint. The principle of these connections is to provide a partial restraint to beam end-rotation without introducing complicated fabrication to the joint. However, the design of such joints is complex and their wider application will be dependent to a large extent on the development of simple design procedures based upon experimental evidence. In the absence of these procedures, many details which may exhibit semi-rigid behaviour are used as simple connections.

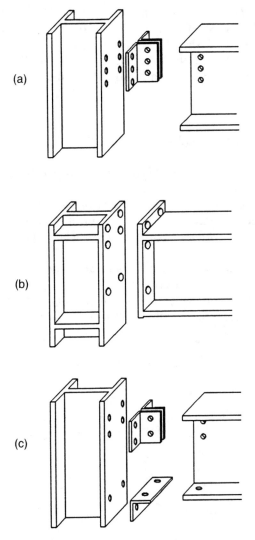

Fig. 2.30 Connections: (a) simple; (b) rigid; (c) semi-rigid

References to Chapter 2

1. Hart F., Henn W., Sontag H. & Godfrey G.B. (Ed.) (1985) *Multi-Storey Buildings in Steel*, 2nd edn. Collins, London.
2. National Economic Development Office and Economic Development Committee for Constructional Steelwork (1985) *Efficiency in the Construction of Steel Framed Multi-Storey Buildings*. NEDO, Sept.

3. Owens G. (1987) *Trends and Developments in the Use of Structural Steel for Multi-Storey Buildings*. Steel Construction Institute, Ascot, Berks.
4. McGuire W. (1968) *Steel Structures*. Prentice-Hall.
5. Zunz G.J. & Glover M.J. (1986) *Advances in Tall Buildings*. Council on Tall Buildings and Urban Habitat. Van Nostrand Reinhold.

Further reading for Chapter 2

Brett P. & Rushton J. (1990) *Parallel beam approach – a design guide*. The Steel Construction Institute, Ascot, Berks.
Lawson R.M. & Rackham J.W. (1989) *Design of haunched composite beams in buildings*. The Steel Construction Institute, Ascot, Berks.
Mullett D.L. (1991) *Slim floor design and construction*. The Steel Construction Institute, Ascot, Berks.
Owens G.W. (1989) *Design of fabricated composite beams in buildings*. The Steel Construction Institute, Ascot, Berks.
Ward J.K. (1990) *Design of composite and non-composite cellular beams*. The Steel Construction Institute, Ascot, Berks.

A worked example follows which is relevant to Chapter 2.

Worked example 69

The Steel Construction Institute Silwood Park, Ascot, Berks SL5 7QN	Subject **MULTI-STOREY DESIGN EXAMPLE**		Chapter ref. **2**
	Design code **BS5950: Part 1**	Made by **AJK** Checked by **GWO**	Sheet no. **1**

Building geometry

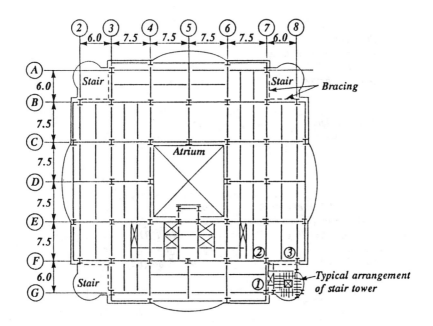

Typical floor plan

Building use: Office building with basement car parking and high level plant room. Imposed loading for office floors exceeds minimum statutory loading at client's request. Design example illustrates design of elements in braced towers provided in four corners of building to achieve lateral stability. Floor plate is generally 130 mm lightweight aggregate concrete on metal decking which acts compositely with decking and floor beams. Fire protection is achieved with sprayed vermiculite cement.

The Steel Construction Institute Silwood Park, Ascot, Berks SL5 7QN	Subject **MULTI-STOREY DESIGN EXAMPLE**	Chapter ref. **2**
	Design code **BS5950: Part 1**	Made by **AJK** — Sheet no. **2** Checked by **GWO**

Loading

kN/m^2

Roof:

Steelwork + metal deck	0.27
Concrete (130 mm lw)	1.8
Finishes	2.0
Services below	0.3
	4.37

BS 6399   Imposed load   1.5

Plant room/B1:

Steelwork + metal deck	0.49
150 mm lw concrete slab	2.15
Suspended ceiling	0.2 *(not B1)*
Services	0.3 *(not B1)*
	3.14 2.64 (B1)

Imposed load   7.5   2.50 (B1)

Office floors:

Steelwork + metal deck	0.27
130 mm lw concrete slab	1.80
Suspended ceiling	0.2 *(not ground)*
Raised floor	0.3
Services	0.3 *(not ground)*
	2.87 2.37 (ground)

Imposed load   5.0
Partitions   1.0

The Steel Construction Institute Silwood Park, Ascot, Berks SL5 7QN	Subject **MULTI-STOREY DESIGN EXAMPLE**		Chapter ref. **2**
	Design code BS5950: Part 1	Made by **AJK** Checked by **GWO**	Sheet no. **3**

Perimeter loads

              kN/m

Roof   1.5
Plant room   18.8
General   9.6

Length supported by column 1 = 1.65 m (G7)
                                             2 = 0   (F7)
                                             3 = 1.65 m (F8)

Imposed stair loading   = $4.0 \; kN/m^2$
Total stair area   = $158.0 \; m^2$

	* Area supported (m^2)		
	Col.1	Col.2	Col.3
Floor	24.8	43.8	20.9
Stairs	0.8	1.6	0

* *The above areas apply to all levels.*

  Total floor area = $1688 \; m^2$

72   Worked example

The Steel Construction Institute Silwood Park, Ascot, Berks SL5 7QN	Subject **MULTI-STOREY DESIGN EXAMPLE**		Chapter ref. 2	
	Design code BS5950: Part 1	Made by AJK Checked by GWO	Sheet no. 4	

Column Imposed loads

Level	Imposed load kN/m^2	% Imposed load reduction	Column load (kN)			* Cumulative loads in columns (kN)		
			Col.1	Col.2	Col.3	Col.1	Col.2	Col.3
R	1.5	0	40.4	72.1	31.4	40.4	72.1	31.4
P	7.5	10.0	189.2 (1)	334.9 (2)	156.8	206.6	366.3	169.4
9.0	6.0	20.0	152	269.2	125.4	305.3	541.0 (3)	250.9
8.0	6.0	30.0	152	269.2	125.4	373.5	661.8	307.3
7.0	6.0	40.0	152	269.2	125.4	411.4	728.8	338.6
6.0	6.0	40.0	152	269.2	125.4	502.6	890.3	413.9
5.0	6.0	40.0	152	269.2	125.4	593.8	1051.8	489.1
4.0	6.0	40.0	152	269.2	125.4	685.0	1213.3	564.4
3.0	6.0	40.0	152	269.2	125.4	776.2	1374.8	639.6
2.0	6.0	40.0	152	269.2	125.4	867.4	1536.4	714.8
1.0	6.0	40.0	152	269.2	125.4	958.6	1697.9	790.1
G	6.0	50.0	152	269.2	125.4	874.8	1549.5	721.1
B1	6.0	50.0	152	269.2	125.4	950.8	1684.1	783.8
B2	2.5	50.0	65.2	115.9	52.3	983.4	1742.1	810.0

Note: The above loads are based on the areas for floors and stairs shown in the Table on Sheet no.3.

* values include % Imposed load reduction.

Examples:

(1) $(7.5 \times 24.8) + (4.0 \times 0.8)$ = 189.2 kN
(2) $(7.5 \times 43.8) + (4.0 \times 1.6)$ = 334.9 kN
(3) $(72.1 + 334.9 + 269.2) \, 0.8$ = 541.0 kN

		Subject		Chapter ref.
The Steel Construction Institute Silwood Park, Ascot, Berks SL5 7QN		**MULTI-STOREY DESIGN EXAMPLE**		2
		Design code **BS5950: Part 1**	Made by AJK Checked by GWO	Sheet no. 5

Column Dead loads

		Unit loads		Col. Dead loads			Cumulative Dead loads		
		Floor kN/m²	Perim.R kN/m	1 kN	2 kN	3 kN	1 kN	2 kN	3 kN
R	8	4.37	1.5	114.4	198.4 (2)	93.8	114.4	198.4	93.8
P	4	3.14	18.8	(1) 111.4	142.6	96.6	225.8	(3) 341.0	190.4
10	4	2.87	9.6	89.3	130.3	75.8	315.1	471.3	266.2
9	4	2.87	9.6	89.3	130.3	75.8	404.4	601.6	342.0
8	4	2.87	9.6	89.3	130.3	75.8	493.7	731.9	417.8
7	4	2.87	9.6	89.3	130.3	75.8	583.0	862.2	493.6
6	4	2.87	9.6	89.3	130.3	75.8	672.3	992.5	569.4
5	4	2.87	9.6	89.3	130.3	75.8	761.6	1122.8	645.2
4	4	2.87	9.6	89.3	130.3	75.8	850.9	1253.1	721.0
3	4	2.87	9.6	89.3	130.3	75.8	940.2	1383.4	796.8
2	4	2.87	9.6	89.3	130.3	75.8	1029.5	1513.7	872.6
1	6	2.87	9.6	89.3	130.3	75.8	1118.8	1644.0	948.4
G	4	2.37	-	60.7	107.6	49.5	1179.5	1751.6	997.9
B1	4	2.64	-	67.6	119.9	55.2	1247.1	1871.5	1053.1
B2									

Examples:

(1) $(3.14 \times 24.8) + (18.8 \times 1.65) + (3.14 \times 0.8)$ = 111.4 kN
(2) $(4.37 \times 43.8) + (4.37 \times 1.6)$ = 198.4 kN
(3) $198.4 + 142.6$ = 341.0 kN

				Subject		Chapter ref.
The Steel Construction Institute				MULTI-STOREY DESIGN EXAMPLE		2
Silwood Park, Ascot, Berks SL5 7QN						
				Design code BS5950: Part 1	Made by AJK Checked by GWO	Sheet no. 6

Wind load

CP3 Ch. V: Part 2: 1972
Basic wind speed V_B = 38 m/s (London)
$S_1 = S_3 = 1.0$ (no topography/statistical considerations)

$C_f = 0.95$ Building width = 48.0 m

$V_s = S_1 S_2 S_3 V_B$ $q = 0.613 V_s^2$ (N/m²)

h	S_2	q (kN/m²)	A (m²)	Force P (kN)	Force F (kN)
54	1.04	0.957	384		175
				349	
46	1.02	0.921	192		258
				168	
42	1.01	0.903	192		167
				165	
38	1.00	0.885	192		163
				161	
34	0.98	0.850	192		158
				155	
30	0.97	0.833	192		154
				152	
26	0.94	0.782	192		148
				143	
22	0.91	0.733	192		139
				134	
18	0.88	0.685	192		130
				125	
14	0.81	0.581	192		116
				106	
10	0.74	0.485	192		94.8
				83.5	
6	0.67	0.397	288		96.3
				109	
					54.5

F is the total force applied at that level.

Loads are divided by 4 in the analysis to represent the force applied to one braced tower.

The Steel Construction Institute Silwood Park, Ascot, Berks SL5 7QN	Subject **MULTI-STOREY DESIGN EXAMPLE**		Chapter ref. **2**
	Design code **BS5950: Part 1**	Made by **AJK** Checked by **GWO**	Sheet no. **7**

<u>*Notional Horizontal Forces (NHF):*</u>

To account for practical imperfections such as lack of verticality, notional horizontal forces are considered.

At each level,

F_{NHF} = 0.5% *factored dead and imposed load at that level*

(the imposed loads are reduced by the overall reduction factor of 0.5)

e.g. *for level 9:*

$$\text{stair area} = 158 \ m^2$$

$$\text{D.L.} = 2.87 \underset{floor}{(1688 + 158)} + \underset{perimeter}{(9.6 \times 164.6)} = 6878 \ kN$$

Level	NHF (kN)
R	17.74
P	28.84
10	23.10
9	23.10
8	23.10
7	23.10
6	23.10
5	23.10
4	23.10
3	23.10
2	23.10
1	23.10
G	18.42
B1	13.38
B2	-

reduced imposed load

$= [(1688 \times 6.0) + (158 \times 6.0)] \ 0.5 = 5538 \ kN$

Note: *For this case, it is convenient to keep the floor and stair imposed loads equal to 6.0 kN/m².*

$$\therefore \ NHF \ per \ tower \ = \ \frac{0.5}{100 \times 4} \ [(1.4 \times 6878) + (1.6 \times 5538)]$$

$$= 23.1 \ kN$$

The NHF load case is considered as acting simultaneously with dead and applied loads when no wind forces are acting.

Worked example

The Steel Construction Institute Silwood Park, Ascot, Berks SL5 7QN	Subject **MULTI-STOREY DESIGN EXAMPLE**	Chapter ref. **2**
	Design code **BS5950: Part 1**	Made by **AJK** — Sheet no. **8** Checked by **GWO**

Because of symmetry one of the four braced towers only was modelled as a space frame and analysed using beam finite elements on a personal computer, subject to a quarter of total horizontal loading.

Basic load cases

1. Wind loading (y direction)
2. Notional Horizonatal Forces (NHF)
3. Dead loads
4. Reduced Imposed loads

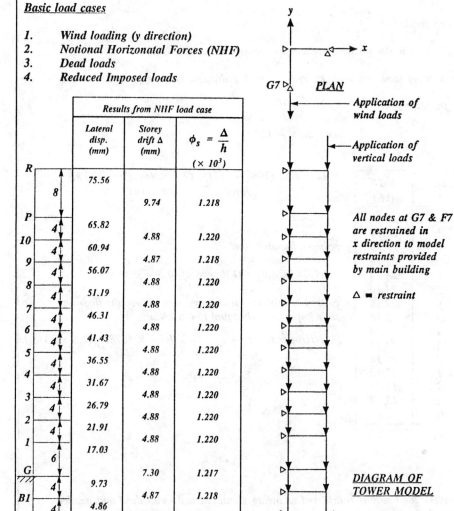

All nodes at G7 & F7 are restrained in x direction to model restraints provided by main building

$\triangle$ = restraint

DIAGRAM OF TOWER MODEL

	Results from NHF load case		
	Lateral disp. (mm)	Storey drift Δ (mm)	$\phi_s = \dfrac{\Delta}{h}$ ($\times 10^3$)
R	75.56		
		9.74	1.218
P	65.82		
10		4.88	1.220
	60.94		
9		4.87	1.218
	56.07		
8		4.88	1.220
	51.19		
7		4.88	1.220
	46.31		
6		4.88	1.220
	41.43		
5		4.88	1.220
	36.55		
4		4.88	1.220
	31.67		
3		4.88	1.220
	26.79		
2		4.88	1.220
	21.91		
1		4.88	1.220
	17.03		
G		7.30	1.217
	9.73		
B1		4.87	1.218
	4.86		
B2		4.86	1.215

The Steel Construction Institute Silwood Park, Ascot, Berks SL5 7QN	Subject **MULTI-STOREY DESIGN EXAMPLE**	Chapter ref. **2**
	Design code **BS5950: Part 1**	Made by **AJK**
	Checked by **GWO**	Sheet no. **9**

P - Δ effects

The amplified sway method was used to account for P - Δ (secondary) effects.

F.2.3
5.6.3 (b)

Elastic critical factor is estimated from the max. sway index under notional horizontal forces.

$$\lambda_{cr} = \frac{1}{200\ \phi_{s\ max}} = \frac{1}{200 \times 1.22 \times 10^{-3}} = 4.1$$

Amplified sway factor

$$m = \frac{\lambda_{cr}}{\lambda_{cr} - 1} = 1.32$$

P - Δ effects occur when real horizontal loads are applied (i.e. wind loads but not NHF).

Load case combinations

1.4 Dead + 1.6 Imposed
1.2 (Dead + Imposed ± m × wind)
1.4 Dead ± 1.4 m × wind
1.0 Dead ± 1.4 m × wind
1.4 Dead + 1.6 Imposed ± 1.0 NHF

2.4.2.3

The Steel Construction Institute Silwood Park, Ascot, Berks SL5 7QN		Subject **MULTI-STOREY DESIGN EXAMPLE**	Chapter ref. **2**
Design code **BS5950: Part 1**		Made by **AJK** Checked by **GWO**	Sheet no. **10**

Proposed section sizes

Steel Grade 50

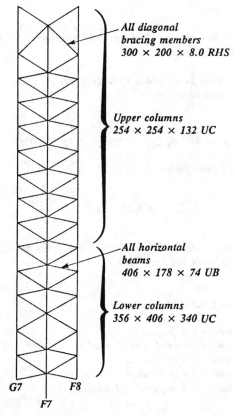

Lateral deflection

Under wind load case:

Level	Deflection (mm)	h/Δ
R	156	364
P	134	364
10	123	333
9	111	333
8	99	333
7	87	462
6	74	364
5	63	333
4	51	400
3	41	444
2	32	444
1	23	500
G	11	571
B1	4	100
B2	0	-

Maximum differential storey

$$\text{deflection} = \frac{h}{333} < \frac{h}{300}$$

∴ *deflection is acceptable.*

N.B. *The magnification factor may be re-evaluated for each combination considered, based on the factored gravity load acting in that combination.*

Table 5

	Subject	Chapter ref.
The Steel Construction Institute Silwood Park, Ascot, Berks SL5 7QN	**MULTI-STOREY DESIGN EXAMPLE**	**2**
	Design code **BS5950: Part 1**	Made by **AJK**	Sheet no. **11**
		Checked by **GWO**	

Confirmation of element sizes

Critical member forces determined from an examination of all load cases are:

	Compression load case		Tension load case		Stress ratio max.
	F_c	M_x	F_T	M_x	
A. Upper column	3426	4	1222	10	0.86
B. Lower column	6666	40	3682	40	0.58
C. Lower column ($L_e = 6.0$)	5112	31	2424	31	0.70
D. Diagonal	906	-	897	-	0.73
E. Horizontal	633	-	631	-	0.85

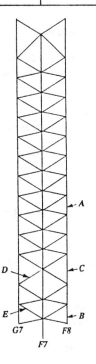

By inspection compression is critical in each case. All sections are adequate.

Local capacity check: 4.8.3.2

$$\frac{F_c}{A_g \, p_y} + \frac{M_x}{M_{cx}} \leq 1$$

overall buckling check: 4.8.3.3

$$\frac{F_c}{A_g \, p_c} + \frac{m \, M_x}{M_b} \leq 1$$

Section summary

UPPER COLUMNS	: $254 \times 254 \times 132$ UC
LOWER COLUMNS (LEVEL B2 - 3)	: $356 \times 406 \times 340$ UC
DIAGONAL BRACING	: $300 \times 200 \times 8.0$ RHS
HORIZONTAL BRACING	: $406 \times 178 \times 74$ UB

The Steel Construction Institute Silwood Park, Ascot, Berks SL5 7QN	Subject **MULTI-STOREY DESIGN EXAMPLE**	Chapter ref. **2**
	Design code BS5950: Part 1	Made by AJK Checked by GWO	Sheet no. **12**

Robustness of frame

To allow for practical imperfections such as lack of verticality in columns, column ties should be capable of transmitting the greater of:

2.4.5.3

(1) Generally, the tying force for internal floor ties is given by $0.5 \, w_f \, s_t \, L_a$ but not less than 75 kN, where:

w_f = factored dead and imposed unit load on floor
$s_t \, L_a$ = floor area per tie

tying force = $0.5 \times 7.5 \times 7.5 \times [(2.87 \times 1.4) + (6.0 \times 1.6)]$
= 383 kN > 75 kN

or

(2) At the periphery, 1% of the factored vertical load in the column or $0.25 \, w_f \, s_t \, L_a$ or 75 kN whichever is the greater.

Max. column force = 7309 kN ∴ tie force = 731 kN (max. by inspection)

457 × 191 × 82 UB: F/P_{cy} = 731/837 = 0.87

∴ horizontal tie members are adequate.

(NB these forces should not be considered as additive to other loads.)

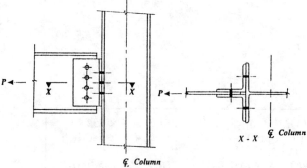

Design connection for P = ± 731 kN

Chapter 3
Industrial steelwork

by JOHN ROBERTS

3.1 Range of structures and scale of construction

3.1.1 Introduction

Structural steelwork for industrial use is characterized by its function, which is primarily concerned with the support, protection and operation of plant and equipment. In scale it ranges from simple support frameworks for single tanks, motors or similar equipment, to some of the largest integrated steel structures, for example, complete electric power-generating facilities.

Whereas conventional single- and multi-storey structures provide environmental protection to space enclosed by walls and roof and, for multi-storey buildings, support of suspended floor areas, these features are never dominant in industrial steelwork. Naturally, in many industrial structures, the steel framework also provides support for wall and roof construction to give weather protection, but where this does occur the wall and roof profiles are designed to fit around and suit the industrial plant and equipment, frequently providing lower or different standards of protection in comparison with conventional structures. Many plant installations are provided only with rain shielding; high levels of insulation are unusual and some plant and equipment are able to function and operate effectively without any weather protection at all. In such circumstances the requirements for operational and maintenance personnel dictate the provision of cladding, sheeting or decking.

Similarly, most industrial steelwork structures have some areas of conventional floor construction, but this is not a primary requirement and the flooring is incidental to the plant and equipment installation.

Floors are provided to allow access to and around the installation, being arranged to suit particular operational features. They are therefore unlikely to be constructed at constant vertical spacing or to be laid out on plan in any regular repetitive pattern. Steelwork designers must be particularly careful not to neglect the importance of two factors. First, floors cannot automatically be assumed to provide a horizontal wind girder or diaphragm to distribute lateral loadings to vertical-braced or framed bays; openings, missing sections or changes in levels can each destroy this action. Secondly, column design is similarly hampered by the lack of frequent and closely-spaced two-directional lateral support commonly available in normal multi-storey structures.

Floors require further consideration regarding the choice of construction (see section 3.2.3) and loading requirements (see section 3.3.1).

3.1.2 Power station structures

Industrial steelwork for electrical generating plants varies considerably depending on the size of station and the fuel being used. These variations are most marked in boiler house structures; whereas coal-fired and oil-fired boilers are similar, nuclear power station boilers (reactors) are generally constructed in concrete for biological shielding purposes, steel being used normally in a secondary building envelope role. Turbine halls are, in principle, largely independent of fuel type and many of the other plant structures (mechanical annexes, electrical switchgear buildings, coal hoppers, conveyors, pump-houses) are common in style to other industrial uses and so brief descriptions of the salient design features are of general interest.

Boiler houses (coal- or oil-fired) (Fig. 3.1) have to solve one overriding design criterion and as a result can be considered exercises in pure structural design.

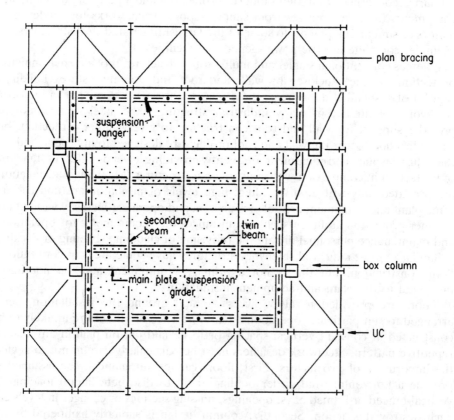

boiler shown shaded

vertical bracing not shown

Fig. 3.1 Plan of boiler house framing

Modern boilers are huge single pieces of plant with typical dimensions of 20 m × 20 m × 60 m high for a single 500 MW coal boiler. Where poor-quality coal is burnt or higher capacities are required, the dimensions can be even larger, up to about 25 m × 25 m × 80 m high for 900 MW size sets. As may be anticipated, the weights are equally massive, typically in the range of 7000–10 000 t for the plant sizes noted above. Boilers are always *top-suspended* from the supporting structures and not built directly from foundation level upwards, nor carried by a combination of top and bottom support. This is because the thermal expansion of the boiler prevents dual support systems; unsurmountable buckling and stability problems on the thin-walled-tube structure of the casing would arise if the boiler was bottom-supported and hence in compression, rather than top-suspended and hence in tension. For obvious reasons no penetrations of the boiler can be acceptable and therefore no internal columns can be provided. The usual structural system is to provide an extremely deep and stiff system of suspension girders (plate or box) spanning across the boiler with an extensive framework of primary, secondary and tertiary beams terminating in individual suspension rods or hangers which support the perimeter walls and roof of the boiler itself.

Columns are massively loaded from the highest level and so are usually constructed from welded box-sections since their loading will be considerably above the capacity of any rolled section. It is usual practice for a perimeter strip some 5–10 m wide to be built around the boiler itself allowing a structural grid to be provided with an adequate bay width for bracing to be installed for lateral stability. It also provides support for ancillary plant and equipment adjacent to specific zones of the boiler, for pipework and valves, for personnel access walkways or floor zones.

The pipework support requirements are often onerous and in particular the pipework designers may require restrictive deflection limitations that are sometimes set as low as 50 mm maximum deflection under wind loadings at the top of 90 m high structures.

Turbine halls (Fig. 3.2) support and house turbo-generating machines that operate on steam produced by the boiler, converting heat energy into mechanical energy of rotation and then into electrical energy by electromagnetic induction. Turbo-generators are linear in layout, built around a single rotating shaft, typically some 25 m long for 500 MW units. The function of a steel-framed turbine hall is to protect and allow access to the generator, to support steam supply and condensed water return pipework and numerous other items of ancillary plant and equipment. Heavy crane capacity is usually provided since generators are working machines that require routine servicing as well as major overhaul and repairs. They are probably the largest rotating machines in common use and dynamic analysis of turbine generator support steelwork is imperative. Fortunately they operate at a sensibly constant speed and so design of the support steelwork is amenable to an analytical examination of dynamic frequencies of motion of the whole support structure with plant loading in each possible mode, with similar consideration of any local frequency effect, such as vibration of individual elements of the structure or its framework.

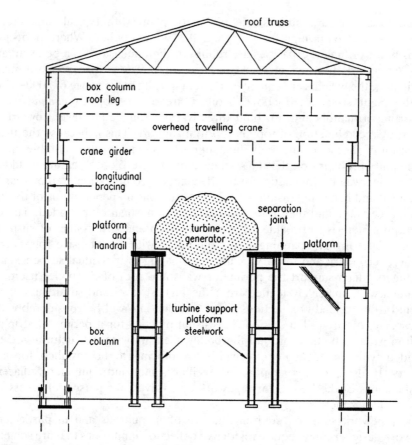

Fig. 3.2 Cross section through turbine hall

These frequencies of response are compared to the forcing frequency of motion of the machine, the design being adjusted so that no response frequencies exist in a band either side of the forcing frequency, to avoid resonance effects.

A complete physical separation is provided between the generator support steelwork and any adjoining main frame, secondary floor or support steel to isolate vibration effects. Zones on adjoining suspended floors are set aside for strip-down and servicing and specific 'laydown' loading rates are allowed in the design of these areas to cater for heavy point and distributed loads.

3.1.3 Process plant steelwork

Steelwork for process and manufacturing plants varies across a wide spectrum of different industrial uses. Here it is considered to be steelwork that is intimately connected with the support and operation of plant and equipment, rather than a steel-framed building envelope constructed over a process plant.

Although the processes and plant vary widely, the essential features of this type of industrial steelwork are common to many applications and are conveniently examined by reference to some typical specific examples. There are many similarities with power station boiler house and turbine house steelwork described in section 3.1.2.

Cement manufacturing plant. Typical cement plants are an assembly of functional structures arranged in a manufacturing flow sequence, with many short-term storage and material transfer facilities incorporated into the processes. The physical height and location of the main drums are likely to dictate the remaining plant orientation.

Vehicle assembly plants. Substantial overhead services to the various assembly lines characterize vehicle assembly plants. It is normal to incorporate a heavy-duty and closely-spaced grid roof structure which also will support the roof covering. Reasonably large spans are needed to allow flexibility in arranging assembly line layouts without being constrained by column locations. Automation of the assembly process brings with it stiffness requirements to allow use of robots for precise operations such as welding and bonding. Open trusses in two directions are likely to satisfy most of these requirements, providing structural depth for deflection control and a zone above bottom boom level that can be used for service runs. Building plans are normally regular with rectangular type plan forms and uniformly regular roof profiles.

Nuclear fuel process and treatment plants. Steelwork for nuclear fuel process and treatment plants is highly dependent on the actual process involved, and often has to incorporate massive concrete sections for biological radiation shielding purposes. Particular points to note are the importance of the paint or other finishes, both from the point of view of restricted access in certain locations, leading to maintenance problems, and also from the necessity for finishes in some areas to be capable of being decontaminated. Specialized advice is needed for the selection of suitable finishes or to give guidance on whether, for example, structural stainless steels would be appropriate. In addition, certain nuclear facilities must be designed for extreme events, the most relevant of which is seismic loading set by the statutory regulatory body. Frequently designs will have to be undertaken to comply with well-established Codes of Practice for seismically active zones in, for example, America. Seismic design requires the establishment of ductile structures to allow high levels of energy absorption prior to collapse, and local stability and ductile connection behaviour become of critical importance. It follows that joint and connection design has to be fully integrated with the structural steel design generally; the normal responsibilities of joint design assumed by the steelwork fabricator may have to be altered on these projects.

Petrochemical plants. Petrochemical plants tend to be open structures with little or no weather protection. The steelwork required is dedicated to providing support to plant and pipework, and support for access walkways, gangways, stairs and ladders. Plant layouts are relatively static over long periods of use and the steelwork is relatively economic in relation to the equipment costs. It is normal therefore to design the steelwork in a layout exactly suited to the plant and equipment without regard for a uniform structural grid. Benefits can still be gained from standardization of sizes or members and from maintaining an orthogonal grid to avoid connection problems. The access floors and interconnecting walkways and stairways must be carefully designed for all-weather access; use of grid flooring is almost universal. Protection systems for the steelwork should acknowledge both the threat from the potentially corrosive local environment due to liquid or gaseous emissions and the requirements to prevent closing down the facility for routine maintenance of the selected protection system.

Careful account needs to be taken of wind loading in terms of the loads that occur on an open structure and the lack of well-defined horizontal diaphragms from conventional floors. These two points are discussed further in sections 3.3.4 and 3.2.2.

3.1.4 Conveyors, handling and stacking plants

Many industrial processes need bulk or continuous handling of materials with a typical sequence as follows:

(1) loading from bulk delivery or direct from mining or quarrying work
(2) transportation from bulk loading area to short-term storage (stacking or holding areas)
(3) reclaim from short-term storage and transport to process plant.

Structural steelwork for the industrial plant which is utilized in these operations is effectively part of a piece of working machinery (Fig. 3.3). Design and construction standards must recognize the dynamic nature of the loadings and particularly must cater for out-of-balance running, overload conditions and plant fault or machinery failure conditions, any of which can cause stresses and deflections significantly higher than those resulting from normal operation. Most designers would adopt slightly lower factors of safety on loading for these conditions, but decisions need to be based on engineering judgement as to the relative frequency and duration of these types of loadings. The plant designer may well be unaware that such design decisions can be made for the steelwork and frequently will provide single maximum loading parameters that could incorporate a combination of all such possible events rather than a separate tabulation; the structural steelwork designer who takes the trouble to understand the operation of the plant can therefore ask for the appropriate information and use it to the best advantage.

A further feature of this type of steelwork is the requirement for frequent relocation of the loading area and hence the transportation equipment. Practical experience suggests that precise advance planning for specific future relocation is rarely feasible, so attention should be directed to both design and detailing so that future moves can cause the least disruption.

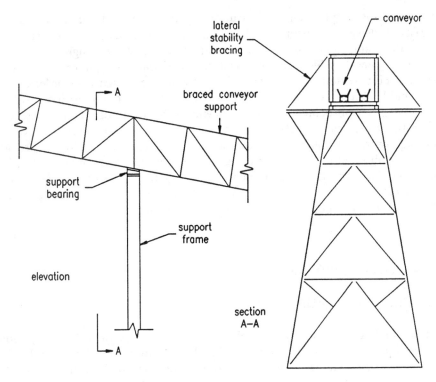

Fig. 3.3 Typical details of conveyor support

Foundation levels can be set at constant heights; or, if variations have to occur, then modular steps above or below a standard height should be adopted. The route should utilize standard plan angles between straight sections, and uniform vertical sloping sections between horizontal runs. Common base plate details and foundation bolt details can be utilized, even where this may be uneconomic for the initial layout installation. Some consideration should be given to allowing the supporting structure to be broken down into conveniently handled sections rather than into individual elements; the break-down joints can be permanently identified, for example, by painting in different colours.

3.1.5 General design requirements

Fatigue loadings must be considered, but even where fatigue turns out not to be a design criterion, it is vital to protect against vibration and consequent loosening of bolted assemblies. In this respect HSFG bolts are commonly used as they dispense with the necessity to provide a lock-nut; when correctly specified and installed they also display good fatigue-resistance.

However, just as plant engineers can display a lack of understanding about the necessity to design supporting steelwork afresh for each different structure even though the 'same' plant is being built, structural steelwork engineers must be aware that seemingly identical or repeat pieces of plant or equipment can in fact vary in significant details. Most plant installations are designed to order and the layout and loadings are often provided initially to the structural engineer in terms of estimated or approximate values. It is vital to be aware of the accuracy of the information being used for design at any stage, and to avoid carrying out designs at an inappropriately high level. A considerable margin should be allowed, provided that it is established that the plant designers have not allowed a margin already in estimating the plant loadings. Experience shows that loadings are often over-estimated at the preliminary design stage, but that this is compensated for by new loadings at new locations that were not originally envisaged.

3.2 Anatomy of structure

3.2.1 Gravity load paths

Vertical loadings on industrial steelwork can be extraordinarily heavy; some individual pieces of plant have a mass of 10 000 t or more. Furthermore, by their very nature these loadings generally act as discrete point loads or line loads rather than uniformly distributed loadings. Load values are often ill-defined (see section 3.3.1) at the steelwork design stage and frequently additional vertical loadings are introduced at new locations late in the design process.

For these reasons, the gravity load paths must be established at an early stage to provide a simple, logical and well-defined system. The facility should exist to cater for a new load location within the general area of the equipment without the need to alter all existing main structural element locations. Typically this means that it is best to provide a layered system of beams or trusses with known primary span directions and spacings, and then with secondary (or sometimes in complex layouts, tertiary as well) beams, which actually provide vertical support to the plant.

While simplicity and a uniform layout of structure are always attractive to a structural designer, the non-uniform loadings and layout of plant mean that supporting columns may have to be positioned in other than a completely regular grid to provide the most direct and effective load path to the foundations. It is certainly preferable to compromise on a layout that gives short spans and a direct and simple route of gravity load to columns than to proceed with designing on a regular grid of columns only to lead to a large range of member sizes and even types of beams, girders or trusses (Fig. 3.4).

Similarly, although it is clearly preferable for columns to run consistently down to foundation level, interference at low levels by further plant or equipment is quite common, which may make it preferable to transfer vertical loading to an

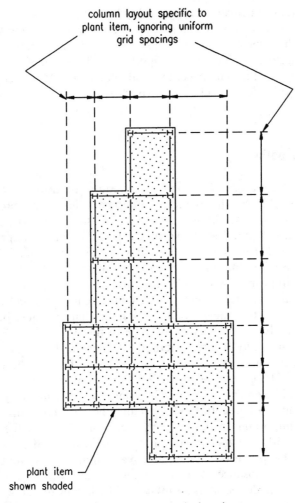

Fig. 3.4 Support steelwork for an unsymmetrical plant item

offset column rather than use much larger spans at all higher levels in the structure. A conscious effort to be familiar with all aspects of the industrial process, and close liaison with the plant designers so that they are aware of the importance of an early and inviolate scheme for column locations, are both necessary to overcome this problem.

It is worth remembering that different plant designers may be handling the equipment layout at differing levels in the structure and ensuring that a briefing from all parties is received before the crucial decision on column spacings is made.

If a widely variable layout of potential loading attachment points exists, then consideration must be given to hanging or top-supporting certain types of plant. A typical example of this is a vehicle assembly line structure, where the overhead assembly line system of conveyors can sensibly be supported on a deep truss roof with a regular column system. If this is contrasted with the multitude of columns and foundations that would be needed to support such plant from below and the prospect of having to reposition these supports during the design stage as more accurate information on the plant becomes available, then it can be an appropriate, if unorthodox, method of establishing a gravity load path.

3.2.2 Sway load paths

For two fundamental reasons sway load paths require particular consideration in industrial steelwork. First, the plant or equipment itself can produce lateral loading on the structure in addition to wind or seismic load where applicable. Therefore, higher total lateral loadings can exist whose points of application may differ from the usual cladding and floor intersection locations. The type and magnitudes of such additional loads are covered in section 3.3.3.

Second, many industrial steelwork structures lack a regular and complete (in plan) floor construction that provides a convenient and effective horizontal diaphragm. Therefore, the lateral load transfer mechanism must be considered deliberately and carefully at a very early stage in the design (Fig. 3.5).

Naturally, there are many different methods of achieving lateral stability. Where floor construction is reasonably complete and regular through the height of the structure, then the design can be based on horizontal diaphragms transferring load to vertical stiff elements at intervals along the structure length or width. Where large openings or penetrations exist in otherwise conventional concrete floors, then it is important to design both the floor itself and the connections between the floor and steelwork for the forces acting, rather than simply relying automatically on the provision of an effective diaphragm as would often and justifiably occur in the absence of such openings.

It can be worthwhile deliberately to influence the layout to allow for at least a reasonable width of floor along each external face of building, say of the order of $\frac{1}{10}$ of the horizontal spacing between braced bays or other vertical stiff elements.

When concrete floors as described above are wholly or sensibly absent, then other types of horizontal girders or diaphragms can be developed. If solid plate or open mesh steel flooring is used, then it is possible, in theory at least, to design such flooring as a horizontal diaphragm, usually by incorporating steel beams as boom members of an idealized girder where the floor steel acts as the web. However, in practice, this is usually inadvisable as the flooring plate fixings are rarely found to be adequate for load transfer and indeed it may be a necessary criterion that some or all of the plates can be removable for operational or similar purposes. A further factor is that any line of beams used as an assumed truss boom must be checked for additional direct compression or tension loadings; the end connections also have to be designed to transfer these axial forces.

It is therefore normal to provide plan bracing in steelwork in the absence of

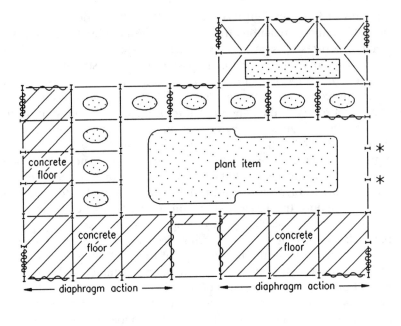

Fig. 3.5 Establishment of sway load paths

concrete floor construction. The influence that this will have on plant penetrations, pipework routing and many other factors must be considered at an early stage in the liaison with the plant engineers. Naturally, the design considerations of steel beams serving a dual function as truss boom members are exactly as set out above and must not be neglected. Indeed, it may be preferable to separate totally the lateral load restraint steelwork provided on plan from other steelwork in the horizontal plane. This will avoid such clashes of purposes and clearly signal to the plant designers the function of the steelwork, as a result preventing its misuse or abuse at a later stage. Many examples exist where plan bracing members have been removed due to subsequent plant modifications. A practical suggestion to further minimize the possibility of this happening is to paint such steelwork a completely different colour as well as to separate it completely from any duality of function with respect to plant support or restraint.

Where it proves impossible to provide any type of horizontal girder, then each and every frame can be vertically braced or rigid-framed down to foundations. It is best not to mix these two systems if possible, since they have markedly differing stiffnesses and will thus deflect differently under loading.

If diaphragms or horizontal girders are used, then the vertical braced bays that receive lateral loading as reactions from them are usually braced in steelwork. In conventional structures tension-only 'X' bracing is frequently used and whereas this may be satisfactory for some straightforward industrial structures such as tank support frames and conveyor support legs, it often proves necessary to design combined tension/compression bracing in 'N', 'K', 'M' or similar layouts depending on the relative geometry of the height to width of the bay and on what obstruction the bracing members cause to plant penetrations. Indeed, it is sound advice initially to provide significantly more bracing than may be considered necessary, for example, by bracing in two, three or more bays on one line. When later developments in the plant and equipment layout mean that perhaps one or more panels must be altered or even removed, it is then still possible to provide lateral stability by redesigning the bracing using linear elastic plane frame computing techniques, without a major change in bracing location.

One particular aspect of vertical bracing design that requires care is in the evaluation of uplift forces in the tension legs of braced frames. Where plant and equipment provide a significant proportion of the total dead load, then it is important that *minimum* dead weights of the plant items are used in calculations. For virtually every other design requirement it is likely that rounding-up or contingency additions to loadings will have been made, especially at the early stages of the design. It is also important to establish whether part of the plant loading is a variable contents weight and, if so, to deduct this when examining stability of braced frames. In some cases, for example, in hoppers, silos, tanks, this is obvious, but boilers or turbines which normally operate on steam may have weights expressed in a hydraulic test condition when flooded with water. The structural engineer has to be aware of these plant design features in order to seek out the correct data for design (see section 3.3.2).

3.2.3 Floors

As has already been discussed in section 3.2.2, full conventional concrete floors are not usual in industrial steelwork structures. The following floor constructions are in common use:

(1) In situ concrete (cast on to removable formwork)
(2) In situ concrete (cast on to metal deck formwork)
(3) Fully precast flooring (no topping)
(4) Precast concrete units with in situ concrete topping
(5) Raised pattern 'Durbar' solid plate
(6) Flat solid steel plate
(7) Open grid steel flooring.

Some typical sections of floor construction are shown in Fig. 3.6.

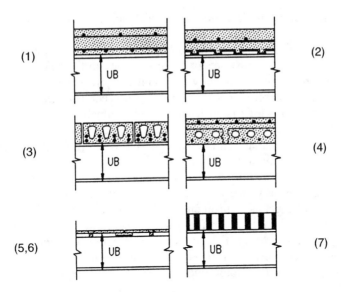

Fig. 3.6 Typical sections of floor construction

The selection of floor type depends on the functional requirements and anticipated usage of the floor areas. Solid steel plates are used where transit or infrequent access is required or where the floor must be removable for future access to plant or equipment. They are normally used only internally (at least in the UK) to avoid problems with wet or waterlogged surfaces.

Open grid flooring is used inside for similar functions as solid steel plate, but with additional functional requirements where the flooring is subjected to spillage of liquids or where air flow through the floor is important. It is also used for stair treads and landings. Consideration should be given to making, say, landings and access strips from solid plate at intervals to assist in promoting a feeling of security among users. It is also in common use externally due to its excellent performance in wet weather.

Concrete floors are in use where heavy-duty non-removable floor areas are necessary. It is not normally advisable to use beam-and-pot-type floorings in any heavy industrial environment due to the damage that can be suffered by lightweight thin walled blocks. For similar reasons, where precast concrete floors require a topping for finishes or to act as a structural diaphragm, it is advisable to use a fully-bonded small aggregate concrete topping with continuous mesh reinforcement, with a typical minimum thickness of 75–100 mm.

A particular advantage of both precast flooring and metal deck permanent formwork is that in many industrial structures the floor zones are irregular in plan and elevation and therefore cheap repetitive formwork is often not practicable.

Even where formwork can sensibly be used, the early installation of major plant items as steel frame erection proceeds can complicate in situ concrete formwork.

Floors must be able to accept holes, openings and plant penetrations on a random layout, and often must accept them very late in the design stage or as an alteration after construction. This provides significant problems for certain flooring, particularly precast concrete. In situ concrete can accept in a convenient manner most types of openings prior to construction, but it may be prudent deliberately to allow for randomly positioned holes up to a certain size by oversizing reinforcement in both directions to act as trimming around holes within the specified units without extra reinforcement.

Metal deck formwork is not so adaptable as regards large openings and penetrations, since it is usually one-way spanning, especially where the formwork is of a type that can also act as reinforcement. If there is sufficient depth of concrete above the top of the metal deck profile, then conventional reinforcing bars can be used to trim openings. Many designers use bar reinforcement as a matter of course with metal deck formwork to overcome this problem and also to overcome fire protection problems that sometimes occur with unprotected metal decks used as reinforcement.

Steel-plate flooring should be designed to span two ways where possible, adding to its flexibility in coping with openings since it can be altered to span in one direction locally if required. Open-grid flooring is less adaptable in this respect as it only spans one way and therefore openings will usually need special trimming and support steelwork. Early agreement with plant and services engineers is vital to establish the likely maximum random opening required, and any structure-controlled restraint on location.

Other policy matters that should be agreed at an early stage are the treatment of edges of holes, edges of floor areas, transition treatments between different floor constructions and plant plinth or foundation requirements.

This last item is extremely important as many plant items have fixings or bearings directly on to steel and the exact interface details and limit of supply of structural steelwork must be agreed. Tolerances of erected structural steelwork are sometimes much larger than anticipated by plant and equipment designers and some method of local adjustment in both position and level must normally be provided. Where plant sits on to areas of concrete flooring then plinths are usually provided to raise equipment above floor level for access and pipe or cable connections. It is convenient to cast plinths later than the main floor but adequate connection, for example, by means of dowelled vertical bars and a scabbled or hacked surface, should always be provided. It is usually more practical to drill and fix subsequently all dowel bars, anchor bolts for small-scale steelwork and for holding-down plant items, and similar fixings, than to attempt to cast them into the concrete floor.

Except in particularly aggressive environments, floor areas are usually left unfinished in industrial structures. For concrete floors hard trowelled finishes, floated finishes and ground surfaces are all used; selection depends on the use and wear that will occur. Steel plate (solid or open-grid) is normally supplied with either paint or hot-dip galvanized finishes depending on the corrosiveness of the environment; further guidance is given in BS 4592: Part 1[1] and by floor plate manufacturers.

Steel-plate floors are fixed to supporting steelwork by countersunk set screws, by countersunk bolts where access to nuts on the underside is practicable for removal of plates or by welding where plates are permanent features and unlikely to require replacement following damage.

Alternatively, proprietary clip fixings can be used for open-grid flooring plates to secure the plates to the underside of support beam or joist flanges.

3.2.4 Main and secondary beams

The plan arrangement of main and secondary beams in industrial steelwork structures normally follows both from the layout of the main items of plant and from the column locations. Thus a sequence of design decisions often occurs in which main beam locations dictate the column locations and not vice versa. If major plant occurs at more than one level then some compromise on column position and hence beam layout may be needed.

Since large plant items normally impose a line or point loading there are clear advantages in placing main or secondary beams directly below plant support positions. Brackets, plinths or bearings may be fitted directly to steelwork and for major items of plant this is preferable to allowing the plant to sit on a concrete or steel floor. Where plant or machinery requires a local floor zone around its perimeter for access or servicing, it is common practice to leave out the flooring below the plant for access or because the plant protrudes below the support level.

Deflection requirements between support points should be ascertained. They may well control the beam design since stringent limits, for example, relative deflections of 1 in 1000 of support spans, may apply. In addition when piped services are connected to the plant then total deflections of the support structure relative to the beam-to-column intersections may also need to be limited. Relative deflections can best be controlled by the use of deep beams in lower-grade steel, grade 43 rather than 50; total deflections by placing columns as close as possible to the support positions.

It is preferable to avoid the necessity for load-bearing stiffeners at support points unless the plant dimensions are fixed before steelwork design and detailing take place. Where this is not possible then stiffened zones to prevent secondary bending of top (or bottom if supports are hung) flanges should be provided, even if the design requirements do not require load-bearing stiffeners. This then allows a measure of tolerance for aligning the support positions without causing local overstressing problems.

The stiffness of major plant items should be considered, at least qualitatively, in the steelwork design. Deep-walled tanks, bunkers or silos for example may well be an order of magnitude stiffer than the steel supporting structure.

96   *Industrial steelwork*

The loading distribution given by the plant design engineers will automatically assume fully stiff (zero deflection) supports. When the stiffness of the supporting structure is not uniform in relation to the support point locations and loadings, then significant redistribution of loads can take place as the structure deflects. Where the plant support positions and loads and the structural steel layout are symmetrical then engineering judgement can be applied without quantitative evaluation. In extreme cases, however, a plant−structure interaction analysis may be required to establish the loadings accurately.

When hanger supports are required then pairs of beams or channels are a convenient solution which allows for random hanger positioning in the longitudinal direction (Fig. 3.7). Many hanger supports have springs bearings to minimize variations in support conditions due to plant temperature changes or to avoid the plant stiffness interactions described above.

The method of installation or removal of major items of plant frequently requires that beams above or, less commonly, below the plant must be designed to cater for hoisting or jacking-up the installed plant sections. Structural designers should query the exact method and route of plant installation to ensure that the temporary hoisting, jacking, rolling or set-down loads are catered for by the steel beam framework. Experience suggests that these data are not provided as a matter of course, and plant designers commonly believe that the steel structure can support these loads anywhere. Where, to ease the problems, plant installation occurs during steel erection then the method of removing plant during the building life-span may be the most significant temporary loading for the beam framework.

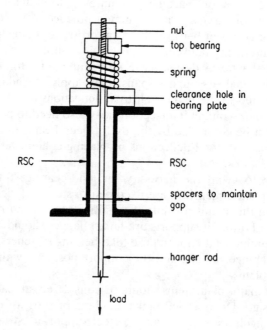

Fig. 3.7 Detail at hanger support

When main or secondary beams are specifically designed for infrequent but heavy lifting operations, it is good practice to fit a lifting connection to the beam to give positive location to the lifting position and to allow it to be marked with a safe working load. For design, smaller load safety factors are appropriate in these circumstances.

3.2.5 Columns

Column location in industrial buildings must be decided on practical considerations. Although regular grid layouts are desirable in normal structures, it is sometimes impossible to avoid an irregular layout which reflects the major plant and equipment location, the irregular floor or walkway layouts and an envelope with irregular wall and roof profiles to suit. Obviously some degree of regularity is of considerable benefit in standardizing as many secondary members as possible; a common method of achieving this is to lay out the columns on a line-grid basis with a uniform spacing between lines. This compromise will allow standard lengths for beam or similar components in one direction while giving the facility to vary spans and provide direct plant support at least in the other direction. The line-grids should be set out perpendicular to the longer direction of the structure if possible.

When vertical loadings are high and the capacity of rolled sections is exceeded, several types of built-up columns are available. Where bending capacity is also of importance, large plate I-sections are appropriate, for example, in frameworks where rigid frame action is required in one direction. However, if high vertical loads dominate then fabricated box columns are often employed (Fig. 3.8). Design of box columns is principally constrained by practical fabrication and erection considerations. Internal access during fabrication is usually necessary for fitting of internal stiffeners and similarly internal access may be needed during erection for making splice connections between column lengths, or for beam-to-column connections. Preferred minimum dimensions are of the order of 1 m with absolute minimum dimensions of about 900 mm. Whenever possible, column plates should be sized to avoid the necessity of longitudinal and transverse stiffeners to control plate buckling. The simpler fabrication that results from the use of thick plates without stiffeners should lead to overall economies and the increased weight of the member is not a serious penalty to pay in columns; the same argument does not apply to long-span box girders where increases in self-weight may well be of overriding importance. Under most conditions of internal exposure no paint protection to the box interior is necessary. If erection access is needed then simple fixed ladders should be provided. Transverse diaphragms are necessary at intervals (say 3–4 times the minimum column dimension) to assist in maintaining a straight, untwisted profile, and also at splices and at major beam-to-column intersections even if, as is usually the case, rigid connections are not being used. Diaphragms should be welded to all four box sides and be provided with manhole cut-outs if internal access is needed.

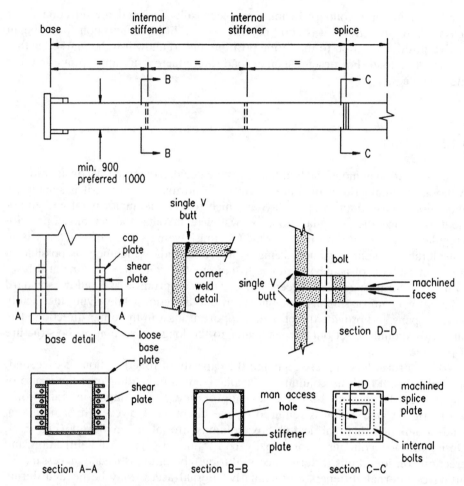

Fig. 3.8 Typical details of box columns

3.2.6 Connections

Connections between structural elements are similar to those in general structural practice. Specific requirements relating to industrial steel structures are considered here.

On occasions industrial plant and equipment may impose significant load variations on the structure and so consideration must be given to possible fatigue effects. Since basic steel members themselves are not susceptible to fatigue failure in normal conditions, attention must be focused on fatigue-susceptible details, particularly those relating to welded and other connections. Specific guidance for certain types of structure is available; and where this is not relevant, general fatigue design guidance can be used.

Of general significance is the question of vibration and the possible damage to bolted connections that this can cause. It should be common practice for steelwork in close contact with any moving machinery to have vibration-resistant fixings. For main steelwork connections there is a choice between using HSFG bolts which are inherently vibration-resistant, or using normal bolts with lock-nuts or lock-washer systems. A wide variety of locking systems is available which can be selected after consultation with the various manufacturers.

Connection design for normally-sized members should not vary from established practice, but for the large box and plate I-section members that are used in major industrial steel structures, connections must be designed to suit both the member type and the design assumptions about the joints. For particularly deep beam members, where plate girders are several times deeper than the column dimensions, assumed pin or simple connections must be carefully detailed to prevent inadvertent moment capacity. If this care is not taken, significant moments can be introduced into column members even by notional simple connections due to the relative scale of the beam depth.

In certain cases it will be necessary to load a column centrally to restrict bending on it; a typical example is where deep suspension girders on power station boilers apply very high vertical loadings to their supporting columns. Here, a rocker cap plate detail is often used to assure centroidal load transfer into the column (Fig. 3.9). Conventional connections on smaller scale members would

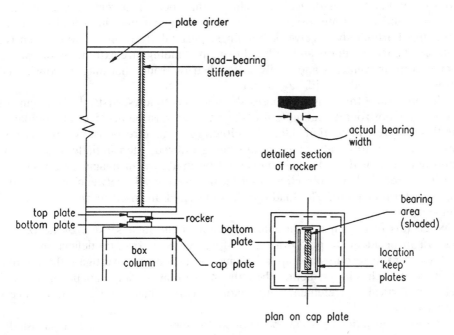

Fig. 3.9 Rocker bearing – plate girder to box column

not usually require such a precise connection as load eccentricities would be allowed for in the design.

3.2.7 Bracing, stiff walls or cores

Section 3.2.2 discusses the particular features of industrial steel structures in relation to achieving a horizontal or sway load path and describes the various methods by which horizontal loads can be satisfactorily transferred to braced bays or other vertical stiff elements. General design requirements and some practical suggestions are also given in section 3.2.2 for braced steel bay design.

The layout on plan of vertically stiff elements is frequently difficult even in conventional and regularly framed structures. General guiding principles are that the centre of resistance of the bracing system in any direction should be coincident with the centre of action of the horizontal forces in that direction. In practice this means that the actions and resistances should be evaluated, initially qualitatively, in the two directions perpendicular to the structural frame layout.

Another desirable feature which is also common to many structures is that the braced bays or stiff cores should be located centrally on the plan rather than at the extremities. This is to allow for expansion and contraction of the structure without undue restraint from the stiff bays, and applies equally to a single structure or to an independent part of a structure separated by movement joints from other parts. It is difficult to achieve this ideal in a regular and uniform structure, and almost impossible in a typically highly irregular industrial steelwork structure. Fortunately, steelwork buildings, particularly those without extensive reinforced concrete floors and with lightweight cladding, are extremely tolerant of temperature movements and rarely suffer distress from what may be considered to be a less than ideal stiff bay layout.

The procedure for design purposes should be as follows; first a basic means of transferring horizontal loading to foundation level must be decided, and guidance on this is given in section 3.2.2. In either or both directions, where discrete braced bays or stiff walls or cores are being utilized, then initially a geometric apportionment of the total loading should be made by an imaginary division of the structure on plan into sections that terminate centrally between the vertically stiff structural element. The loadings thus obtained are used to design each stiff element.

When this process has been completed and if the means exist, by horizontal diaphragm or adequate plan bracing, to force equal horizontal deflections on to each element, a second-stage appraisal may be needed to investigate the relative stiffness of each stiff element. Then the horizontal loading can be distributed between vertical stiff elements on a more accurate basis and the step process repeated.

Considerable judgement can be applied to this procedure, since it is usually only of significance when fundamentally differing stiff elements are used together

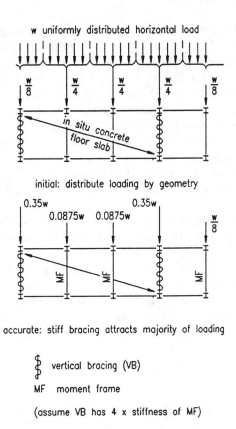

Fig. 3.10 Apportioning horizontal loads to vertical stiff elements

in one direction on the same structure. For example, where a horizontal diaphragm or plan bracing exists and where some of the stiff elements are braced steelwork and some are rigid frames, it will normally be found that the braced frames are relatively stiffer and will therefore carry proportionately more load than the rigid frames. Similarly, where a combination of concrete shear walls (or cores) and braced frames is used, then a relative stiffness distribution will be needed if an effective horizontal diaphragm or plan bracing ensures sensibly constant horizontal deflection or sway (Fig. 3.10).

3.3 Loading

3.3.1 General

Two difficulties exist in defining the appropriate loadings on industrial buildings. First, the actual weight, and particularly the details of the position and method of load application, of items of plant or equipment must be established. Even for routine or replicated plant this information is difficult to obtain in a form that suits the structural engineer. When the plant or equipment is being custom-built then the problem becomes one of timing; information from the plant designer may not be available early enough for the structural design.

The second difficulty that frequently occurs is in the choice of a general, uniformly distributed imposed loading for any remaining floor areas not occupied by items of plant or equipment. Guidance from Codes of Practice must be used carefully when it is applied only to circulation spaces between all the fixed items of plant which are known and whose loading has been evaluated and is considered separately as dead load. Further advice on both these problems is given in section 3.3.2.

The structural steelwork designer must take an open-minded approach to the loading information. It must be appreciated that early information from the plant designers will represent only estimated loadings, and an uncertainty allowance may well already be included in the values supplied, so that further allowances may be unnecessary. However, it frequently turns out that many secondary plant items are added at a later stage in the information process. The loadings from these should be satisfactorily absorbed into an initial uniformly distributed loading allowance provided that a logical scheme for dealing with loading is established at an early stage.

3.3.2 Process plant and equipment

The most important advice to the structural designer of steelwork for an industrial purpose is to ensure reasonable familiarity with the entire process or operation involved. Existing facilities can be visited and the plant designers and operators will normally be more than willing to give a briefing which provides the opportunity to describe the form of structure envisaged to the plant designer at an early stage so avoiding later misunderstandings.

Some attempt should also be made to understand the jargon of the industry in question in order to gain the confidence of the plant designers and to allow effective intercommunication.

Particular examples of structural requirements which are not easily recognized as important by plant designers, and which should therefore be fully explained, are as follows:

(1) The physical space requirements of bracing members and the fact that they cannot be moved locally to give clearances.
(2) The actual size of finished steelwork taking account of splice plates, bolt heads and fittings projecting from the section sizes noted on drawings.
(3) The fact that a steel structure is not 100% stiff and that all loads cause deflections.

(4) The fact that a steel structure may interact with a dynamic loading and that dynamic overload multipliers calculated or allowed on the assumption of a fully rigid or infinite mass support are not always appropriate.

The basis of loading information for plant and equipment must be critically examined. Frequently loadings are given as a single all-up value, the components of which may not all act together or have only a very small probability of so doing. Alternatively, the maximum loadings may represent a peak testing condition or a fault, overload condition, whereas normal operating loadings may be considerably less in value. Worthwhile and justifiable savings in steelwork can be made if a statistically based examination of the expected frequency and duration of such unusual conditions is made, leading to the adoption of reduced load factors without reducing the overall factor of safety. Comparisons with, for example, wind loading, can be used to establish on a reasonably logical basis the appropriate load factors.

By definition, items of fixed plant can be treated as dead load in accordance with BS 6399: Part 1^2 when specific location and loads are known. At the early stages of design it is usual to adopt a relatively large imposed loading which will have to cater for fixed items of plant or equipment, the existence of which may not even be known at this stage and certainly not the location and loading data. The choice of what imposed loading to use at this stage can be assisted by the following guidance:

very light industrial processes	7.5 kN/m^2
medium/average industrial processes	10–15 kN/m^2
very heavy industrial processes	20–30 kN/m^2

One factor which will influence the choice within these values is the timing of the release of final plant and equipment design data in relation to the steelwork design and fabrication detailing process. If a second-stage design is possible then a lower imposed load can be allowed since local variations needed to account for specific items of fixed plant which exceed the imposed plan loading allowance can be accommodated.

Under these conditions it can also be worthwhile, particularly for designers with previous experience of the industrial process, to design columns and foundations for a lower imposed loading than beams. In certain layouts with long span main beams or girders at wide spacings this preliminary reduction can also be used for these members. It should be stressed that these proposals are not intended to contradict or override the particular reduced loading clauses in BS 6399,[2] but are a practical suggestion for the preliminary design stage.

When detailed plant layouts with location and loading data become available, fixed plant and equipment can be considered as dead load and subject therefore to the appropriate load factor, 1.4 instead of 1.6. Remaining zones of floor space without major items of plant should be allocated an imposed load which should reflect only the access and potential use of the floor and may therefore very well be reduced from the preliminary imposed loading, often in the range of 5–10 kN/m².

Specific laydown areas for removal, replacement or maintenance of heavy plant are the only likely exceptions to this range of loading.

An alternative scheme for dealing with the second stage of loading information is to institute a checking procedure where the equivalent loading intensity of plant and equipment is calculated for each item as follows:

$$\frac{1.4}{1.6} \times \frac{\text{weight (kN)}}{\text{plan area (m}^2\text{)}}$$

the factor 1.4/1.6 being introduced to cater for the reclassification from imposed load to dead load.

Only in bays where this equivalent loading intensity exceeds the preliminary imposed loading will further evaluation be required, unless a scheme of reduced imposed loading for long span beams and columns has been adopted in which case a rigorous check on actual loading intensity must be carried out.

Many items of plant contain moving parts which are always liable to exert dynamic or vibratory loadings on to the structure. Assessing the structural response to these loadings is never easy, particularly when as is usually the case, the plant and the structure are being designed out of sequence and by different designers.

Many rule-of-thumb approximations exist, in which dynamic effects are allowed for by percentage increases to dead load. Some instances of this are codified, for example, crane vertical loadings are factored by 1.25 (see BS 6399: Part 1[2]). Similar multipliers to static loading can be provided by plant suppliers but these must be used with great care and the limitations of such an approach should be appreciated. The dynamic load from one-off impact-type actions or from successive load applications at irregular time intervals depends both on the characteristics of the plant itself *and* the mass and stiffness of the structure to which they are applied. Any dynamic load calculated independently from a knowledge of the structure will be based on assumed structure properties and often on limit values such as zero mass or infinite stiffness. Providing that the structure remains elastic and that resonance or similar frequency-related amplification does not occur the approximate dynamic factors normally represent upper limits. If more accurate evaluations are warranted for major items of plant, a plant–structure interaction dynamic analysis must be undertaken.

Vibrating plant or equipment will transmit vibrations to the structure, the effects of which can be dramatic. The frequency of vibration of the plant can induce resonance in the structure if the natural frequency of vibration of the whole or any part of the structure subjected to the vibrating force is equal or very close to the forcing frequency. Considerable judgement is needed to identify all possible modes of natural vibration that could be excited by the vibrating equipment or machinery. To safeguard against a resonant response there should be no natural frequency of vibration within the range of 0.5–1.5 times the forcing frequency. Useful guidance on calculating natural frequencies of structures is given by Bolton.[3]

A useful initial step is to calculate the lowest natural frequency of vertical motion of the floor construction and to ensure that it is higher than 3–4 Hz (cycles per second). This should avoid problems of response to human-induced vibration, the so-called 'springiness' of floors, caused by resonant amplification of footfalls, which lie in the range of 1–2 Hz. Then only plant-induced vibrations of lower than 1 Hz or higher than 2 Hz need to be investigated. Anti-vibration mountings for plant can be specified but these are *not* an automatic success as they only filter the induced forces and some form of variable loading will still be transmitted to the structure. Various types and grades of anti-vibration mountings are available; information on their suitability can be obtained from the manufacturers, based on details of both the item of plant and the structure.

Many of these devices are intended to reduce vibration of the plant or equipment itself. Care is required to ensure that the altered forcing vibrations which are imposed on to the structure do not have a secondary adverse effect of inducing resonant vibrations into it.

3.3.3 Lateral loadings from plant

Lateral loadings imposed by plant on the structure derive from three sources. These are considered separately although there is a common theme throughout that the operating process undergoes a change in regime which is the cause of loading. Many of the actions which give rise to horizontal loads also cause vertical loads or at least vertical loading components which must be incorporated into the design. However, whereas vertical loadings are usually readily understood and allowed for in loadings provided by plant designers, one feature of the horizontal components of such loadings that causes confusion is the fundamental concept of equilibrium. Notwithstanding exotic situations where masses (projectiles) leave or impinge on a structure, or where motion energy is dissipated as heat of friction, equilibrium considerations dictate that lateral components of forces are in balance and consequentially they are often ignored. This is not satisfactory as balancing components may act a considerable distance apart (the load path must be examined in detail) or indeed the balancing components may act at different levels, leading to a more conventionally understood requirement to transfer lateral loading.

The three causes are as follows:

(1) temperature-induced restraints
(2) restraint against rotational or (more rarely) linear motion
(3) restraint against hydraulic or gaseous pressures.

Where plant undergoes a significant change in temperature, plant designers will typically assume that the structure is fully rigid and so can absorb the forces generated by application of restraints at the structure connection points. They will then design the plant itself for the additional stresses that are caused by preventing free thermal expansion or contraction. This is a safe upper bound procedure since the forces generated in both structure and plant represent maxima, with any deflection at the support reducing forces in both elements. Naturally the structural steelwork designer must be made aware both of the assumption of zero deformation so that the support can be made as stiff as possible in the required direction, and of the forces that are thus imposed.

In spite of the apparent complications of this approach, it is frequently adopted by plant designers for convenience on small items, and to avoid complexities in interconnection between plant items and with piped and ducted services on larger items.

The alternative approach, common on major plant subjected to significant thermal variation such as boilers and ovens, is to assume completely free supports with zero restraint against expansion or contraction. This is a lower bound solution which needs some rational assessment of possible forces that could result from bearing or guide misalignment, malfunction or simple inefficiency. Where large plant items are involved the forces even at such guides or bearings can be significant if the balancing reaction is, for example, at ground level or even outside the structure itself.

The lateral forces from constant speed, rotating machinery, which are usually relatively low, are generally balanced by an essentially equal and opposite set of forces coming from the source of motive power. Nevertheless, their point of application must be considered and a load path established which either transfers them back to balance each other or alternatively down to foundation level in the conventional way. The structural steel designer must have a completely clear and unambiguous understanding of the source and effect of all the moving plant forces, to ensure that all of them are accounted for in the crucial interface between plant and equipment.

The start-up forces when inertia of the plant mass is being overcome, and the fault or 'jamming' loads which can apply when rotating or linear machinery is brought to a rapid halt, need consideration. They are obviously of short duration and so can justifiably be treated as special cases with a lower factor of safety. At the same time the load combinations that can actually co-exist should be established to avoid any loss of economy in design.

Pressure pipe loadings, significant in many plant installations, occur wherever there are changes in direction of pipework and associated pipework supports or restraints. Thermal change must also be considered. Pipework designers may well combine the effects to provide a schedule of the total forces acting at each support position. Since pipework forces are normally *not* reversible in direction, then on major installations the pipework designer may wish to make the installation by 'forcing' the pipework configuration, deliberately making sections too short or too long and then prestressing the pipework so that when installation is complete the operating conditions take the internal pipework stresses through a neutral stress zone and then reverse them.

Forcing is a complex procedure sensitive to lack of fit at the pipework supports and restraints and often to other factors such as the ambient temperature during installation and the exact sequence of connections. Where major pipe installations are planned and where the pipework designers are adopting these techniques, the structural steelwork designer should acknowledge that the loading values quoted may not be achieved in practice and make a further allowance.

3.3.4 Wind loadings

Wind loadings on fully enclosed industrial structures do not differ from wind loadings on conventional structures. The only special consideration that must be given applies to the assessment of pressure or force coefficients on irregular or unusual shaped buildings. A number of sources give guidance on this topic and specific advice can be sought from the Building Research Establishment (BRE) Advisory Service.

However on partly or wholly open structures with exposed plant or equipment great care must be exercised in dealing with wind loading.

It is frequently the case that the total wind loads are higher than on a fully clad building of the same size, due to two causes. First, small structural elements attract a higher force than equivalent exposed areas which form part of a large façade. Secondly, repetitive structural elements of plant items which are nominally shielded by any particular wind direction are not actually shielded and each element is subjected individually to a wind load. The procedure for carrying out this assessment is set out in Section 8 of CP3: Chapter V: Part 2.[4]

Loading on particularly large individual pieces of plant or equipment exposed to wind can be calculated by considering them to be small buildings and deriving overall force coefficients that relate to their size.

For smaller or more complex shapes, such as ductwork, conveyors and individual smaller plant items, it is sensible to take a conservative and easy to apply rule-of-thumb and use an overall force coefficient $C_f = 2.0$ applied to the projected exposed area. The point of application of wind loadings from plant items on to the structure may be different from the vertical loading transfer points if sliding bearings or guides are being used.

3.3.5 Blast loadings

This section deals only with blast loadings from industrial processes and not with any generalized design requirements to survive blast loadings from unspecified sources or of unspecified values.

Varying requirements exist for blast loadings. Typical examples are:

(1) transformers, where the requirement is usually to deflect any blast away from other vulnerable pieces of plant but where frequently one or more walls and

the roof are open, serving to dissipate much of the energy discharged
(2) dust or fine particle enclosures, which are often wholly inside enclosed buildings.

The decisions to be made are as follows.

(1) Can the potential source of the blast be relocated outside the building altogether, in a separate enclosure?
(2) If not, can it be placed against the external wall with arrangements to have a major permanent vented area or a specially designed blow-off panel, both of which will limit loadings on the remaining structure?
(3) Where the location cannot be controlled, it must be established which direction or directions require full protection against damage and which can tolerate certain degrees of damage.

Loading data given by plant designers are usually stated in terms of peak pressures to be applied to projected areas in line with the potential source of the blast. The validity of the data must be treated as being highly suspect since blast loadings are classic examples of true dynamic loading where the time-dependent response of the structure actually determines the loading that is imposed. Quoted blast pressures, which are probably derived from theoretical considerations of high-rigidity high-mass targets fully enclosing the source, may be invalid for steel structures which have tremendous capacity to deform rapidly, absorbing energy and thereby reducing and smoothing out peak blast pressures. Whatever results are obtained from analysis or calculation, it is good practice to use Grade 43 steel which has such good ductility, a lower yield stress from which to commence ductile behaviour and a long and reliable extensibility prior to fracture. While this will ensure reasonable material behaviour, overall ductility of the structure also depends on stability against premature buckling and the ductile behaviour of connections.

Steelwork designers should acknowledge the very imprecise nature of most blast loading data, even when the potential source of the blast is precisely located and specified. They should thus direct their attention to ensuring that collapse does not occur until major deflections and rotations have occurred, following normal guidelines for achieving plastic behaviour.

3.3.6 Thermal effects

This section deals with thermal effects from environmental factors; specific consideration of plant induced thermal effects is given in section 3.3.3. Conventional guidance on the provision of structural expansion or contraction joints is often inappropriate and impractical to implement, and indeed joints frequently fail to perform as intended.

The key to avoiding damage or problems from thermal movements is to consider carefully the detailing of vulnerable finishes (for example, brickwork, blockwork, concrete floors, large glazing areas and similar rigid or brittle materials). Provided that conventional guidance is followed in the movement provisions for these materials, then structural joints in steel frames can usually be avoided unless there is particularly severe restraint between foundations and low-level steelwork.

Many industrial structures with horizontal dimensions of 100–200 m or more have been constructed without thermal movement joints, usually with lightweight, non-brittle cladding and roofing, and a lack of continuous suspended concrete floors.

When high restraint close to foundations or vulnerable plant or finishes are present, a thermal analysis can be carried out on the steel framework to examine the induced stresses and deflections and to evaluate options such as the introduction of joints or altering the structural restraints. Where steelwork is externally exposed in the UK, the conditions vary locally but a minimum of −5°C to +35°C suffices for an initial sensitivity study. For many of the likely erection conditions a median temperature of 15°C can be assumed, and a range of ±20°C can thus be examined. Where the effects of initial investigations based on these values highlight a potential problem, more specific consideration can be given to the actual characteristic minimum and maximum temperatures (advice in the UK is available from the Meteorological Office), and steps may need to be taken to control erection, particularly foundation fixing, to take place at specified median temperatures.

3.4 Structure in its wider context

Industrial structural steelwork is inherently inflexible, being purpose-designed for a particular function or process, and indeed often being detailed to suit quite specific items of major plant and equipment. Nevertheless, it is important to try to cater for at least local flexibility to allow minor alterations in layout, upgrading or replacement of plant items. The most appropriate way to ensure this is to repeat the advice that has been given on numerous occasions already in this chapter. The designer must understand the industrial process involved and be aware of both structural and layout solutions that have been adopted elsewhere for similar processes. Previous structural solutions may not be right, but it is preferable to be aware of them and positively to reject them for a logical reason, than to reinvent the wheel at regular intervals.

General robustness in industrial buildings may be difficult to achieve by the normal route of adopting simple, logical shapes and structural forms, with well-defined load-paths and frequent effective bracing or other stability provisions. Instead of these provisions, then, it is sensible to ensure that a reasonable margin exists on element and connection design. Typically, planning for a 60–80% capacity utilization at the initial design stages will be appropriate, so that even when these allowances are reduced, as so frequently occurs, during the final design and checking stages, adequate spare capacity still exists to ensure that no individual element or joint can weaken disproportionately the overall structural strength of the building.

References to Chapter 3

1. British Standards Institution (1989) *Industrial type metal flooring, walkways and stair treads*. Part 1: *Specification for open bar gratings*. BS 4592, BSI, London.
2. British Standards Institution (1984) *Loading for buildings*. Part 1: *Code of practice for dead and imposed loads*. BS 6399, BSI, London.
3. Bolton A. (1987) Natural frequencies of structures for designers. *The Structural Engineer*, **56A**, No. 9, Sept.
4. British Standards Institution (1984) *Code of basic data for the design of buildings*. Chapter V: Part 2. CP3, BSI, London.

Further reading for Chapter 3

Booth E.D., Pappin J.W. & Evans J.J.B. (1988) Computer aided analysis methods for the design of earthquake resistant structures − a review. *Proc. Instn Civ. Engrs*, **84**, Part 1, Aug., 671−91.
Fisher J.M. & Buckner D.R. (1979) *Light and Heavy Industrial Buildings*. American Institute of Steel Construction, Chicago, USA, Sept.
Forzey E.J. & Prescott N.J. (1989) Crane supporting girders in BS 15 − a general review. *The Structural Engineer*, **67**, No. 11, 6th June.
Jordan G.W. & Mann A.P. (1990) THORP receipt and storage − design and construction. *The Structural Engineer*, **68**, No. 1, 9th Jan.
Kuwamura H. & Hanzawa M. (1987) Inspection and repair of fatigue cracks in crane runway girders. *J. Struct. Engng, ASCE*, **113**, No. 11, Nov.
Mann A.P. & Brotton D.M. (1989) The design and construction of large steel framed buildings. *Proc. Second East Asia Pacific Conference on Structural Engineering and Construction*, Chaing Mai, Thailand, **2**, Jan., 1342−7.
Morris L.J. (Ed.) (1983) Instability and plastic collapse of steel structures. *Proc. M.R. Horne Conference*. Granada, St Albans.
Taggart R. (1986) Structural steelwork fabrication. *The Structural Engineer*, **64A**, No. 8, Aug.

Chapter 4
Bridges

by ALAN HAYWARD

4.1 Introduction

Use of structural steel in bridges exploits its advantageous properties of economically carrying heavy loads over long spans with the minimum dead weight. Steel is however suitable for all span ranges, categorized in Table 4.1.

Table 4.1 Span ranges of bridges

Short	Up to 30 m
Medium	30 m to 150 m
Long	150 m

For long spans steel has been the natural solution since 1890 when the Firth of Forth cantilever railway bridge, the world's first major steel bridge was completed. For short and medium spans concrete bridges held a monopoly from 1950 to 1980 because of the introduction of prestressing and precasting. Developments in steel during this period such as higher tensile strength and improved welding techniques were applied mainly to long spans. However, improvements in construction methods from 1980 have enabled steel to restore its market share within Europe and other continents including North America.[1] Contributing factors to this trend are shown in Table 4.2.

Table 4.2 Factors contributing to the improved market share for steel in short and medium spans from 1980

Factor	Reasons
(1) Stability in price of rolled steel products	Stability worldwide
(2) Automation of fabrication processes	Mechanized equipment for preparation, welding and girder assembly
(3) Faster erection with larger components	Availability of high-capacity cranes
(4) Improved design codes	Easier methods of design based on research
(5) Evidence of durability problems with concrete	Life expiry of concrete bridges under 30 years old
(6) Better education in steel design	Reversal of earlier trend which favoured concrete

Where traffic disruption during construction must be minimized then steel is always suitable. Most bridgework is likely to be carried out under these conditions in the future so that structural steel will continue as a primary choice. Steel has an advantage where speed of construction is vital; it is no coincidence that this usually results in cost economies. If rapidly erected steelwork is used as a skeleton from which the slab and finishes can be carried out without need for falsework then the advantages (summarized in Table 4.3) are fully realized.

For short and medium span highway bridges composite deck construction is economic because the slab contributes to the capacity of the primary members.[2] For continuous spans it uses the attributes of steel and concrete to best advantage. In cases where construction depth is restricted, for example, in developed areas, then half-through or through construction is convenient; this is common for railway and pedestrian bridges.

For long spans, including suspension or cable-stayed bridges, orthotropic plate decks are used. Although the intrinsic costs of a steel orthotropic plate are higher (about four times more) than an equivalent concrete slab, the advantage in dead weight reduction (approximately 1:3 ratio) may more than offset this when the overall economy is considered. Steel decks are also employed when erection must be completed in limited occupations, such as for railway bridges on existing

Table 4.3 Advantages of Steel Bridges

Feature	... leading to ...	Advantages
(1) Low weight of deck	Smaller foundations. Typical 30%–50% reduction of weight compared with concrete decks.	Economy
(2) Light units for erection	Erection by mobile cranes	Rapid construction
(3) Bolted site connections	Minimal site inspection	Flexible site planning
(4) Prefabrication in factory	Effective quality control	More reliable product
(5) Modern methods of protective treatment	Application of treatment before erection. Long life to first maintenance.	Predictable whole life cost of structure
(6) Use of weathering steel		Minimum whole life cost
(7) Shallow construction depth	Minimum length of approach grades	Overall economy of highway. Slender appearance.
(8) Self-supporting steel	Elimination of falsework	Easier construction especially for high structures
(9) Continuous spans	Fewer bearings and joints	Slender appearance. Reduced maintenance cost.

routes. Moving bridges of swing, lift, or rolling-lift ('bascule') type usually employ steel decks to rationalize the amount of counterweighting and the mechanical equipment. For long spans of suspension or cabled-stayed form then special considerations affect the design including aerodynamic behaviour, the feasibility of deep foundations in estuarial conditions, cables with anchorages, non-linear structural behaviour and the absolute necessity to include the effects of the erection procedure in the design process.

This chapter on bridges gives emphasis to initial design, an important stage of the process, because the basic decisions as to member proportions, spacings and splice positions vitally affect economy of the structure. It is essential that the detailed analysis is based upon optimized sizes which are as accurate as possible. If this is not achieved then the detailed design will be inefficient because time consuming repetitive work will have been expended, adversely affecting the economy of the design and the construction costs. Guidance is given on the initial design of highway bridges using composite deck construction, which is a significant proportion of the number of steel bridges built.

4.2 Selection of span

The majority of bridges fall within the category of short span because for many crossings of rivers, railways or secondary highways a single span of less than 30 m is sufficient. For multiple-span viaducts a decision on span length must be made, which depends on factors shown in Table 4.4.

For long viaducts it is necessary to carry out comparative estimates for different spans to determine the optimum choice, as shown in Fig. 4.1.

Table 4.4 Factors which decide choice of span for viaducts

Factor	Reasons
Location of obstacles	Pier positions are often dictated by rivers, railway tracks and buried services
Construction depth	Span length may be limited by the maximum available construction depth
Relative superstructure and substructure costs	Poor ground conditions require expensive foundations; economy favours longer spans
Feasibility of constructing intermediate piers in river crossings	(a) Tidal or fast-flowing rivers may preclude intermediate piers (b) For navigable waterways, accidental ship impact may preclude mid-river piers
Height of deck above ground	Where the height exceeds about 15 m, costs of piers are significant, encouraging longer spans
Loading	Heavier loadings such as railways encourage shorter spans

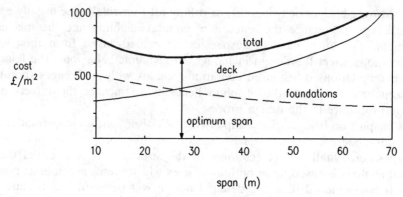

Fig. 4.1 Choice of optimum span for viaducts

Typical optimum spans are shown in Table 4.5.

Table 4.5 Typical optimum span ranges (m) for viaducts

Conditions	Highway	Railway
Simple foundations (spread footing or short piles)	25–45	20–30
Difficult foundations (piles 20 m long)	35–55	25–40
Piers 15 m high	45–65	30–45

Bridges of medium span exceeding 100 m and those of long span are usually only adopted when a larger number of shorter spans are precluded by restriction of the site. This is because the material content and cost per unit length of long spans is much greater. Therefore the bridging of a wide river or estuary should generally use a number of medium spans. Only if there is a high risk of shipping collision, or if the depth and speed of water flow is such as to make foundation construction very difficult, should a long-span bridge be chosen.

4.3 Selection of type

Suspension or cable-stayed bridges are suitable for the longest spans, but are less suitable to support heavy loading across short or medium spans. At the same time medium-span footbridges can appropriately be suspension or cable-stayed types because concentrated loading is absent. Some of the considerations given to long spans such as aerodynamics need to be applied to footbridges. Other bridge types such as arches or portals may be suitable in special locations. For example, an arch is a logical solution for a medium span across a steep-sided ravine.

Through trusses are suitable where the available construction depth is limited. The vast majority of short and medium span highway bridges are formed with composite construction because the highway profile can be arranged to suit the depth available.

For short and medium spans the most important factor which influences the type is the available construction depth. Where depth is limited then types such as arch, portal or truss offer an alternative solution. Types of steel bridge are shown in Fig. 4.2 with their normal economic span range and the world's longest. Each is briefly described below.

4.3.1 Suspension bridges

Suspension bridges (Fig. 4.3) are used for the longest spans across river estuaries where intermediate piers are not feasible. The cables form catenaries supporting both sides of the deck and are tied to the ground usually by gravity foundations sometimes combined with rock anchors. Thus ground conditions with rock at or close to the surface of the ground are essential. Towers are usually of twin steel box members which are braced together above the roadway level. They are designed so as to be freestanding under wind loading during construction until the cables are installed. Cables are either a compacted bundle of parallel high tensile steel strands (commonly 5 mm diameter) installed progressively by 'spinning' or may be formed from a group of wire ropes. Deck hangers are wire ropes (or round steel rods for light loading as for a footbridge) clamped to the cable and connected to the deck at a spacing equal to the length of each deck unit erected, typically 18 m. The construction process for suspension bridges is more time consuming than for other types because the deck cannot be installed until the towers, anchorages, cable and hangers are completed.

Depending upon ground conditions, the cables can be catenaries supporting the side spans. Cables may alternatively be straight from tower top to the ground anchorages and merely support a main span, side spans being non-existent or formed as short-span viaducts. Decks are usually trusses with a steel orthotropic plate floor spanning between. Footways are often cantilevered outside the two sets of cables.

Aerodynamic behaviour must be considered in design because of the tendency for the deck and cables to oscillate in flexure and torsion under 'vortex shedding' and other wind effects. This is due to the flexible nature and light weight of suspension bridges illustrated by the collapse of the USA Tacoma Narrows Bridge in 1940 which had a very flexible narrow deck consisting of twin plate girders forming a torsionally weak deck of 'bluff' shape prone to wind vortex shedding. Aerodynamic considerations often justify wind tunnel testing of models. In the Severn Bridge of 1966 the deck is a steel box girder with integral orthotropic floor having an aerofoil cross section which discourages vortex shedding and reduces the drag coefficient. This and the Bosporus Bridge of 1970 use inclined hangers

116 *Bridges*

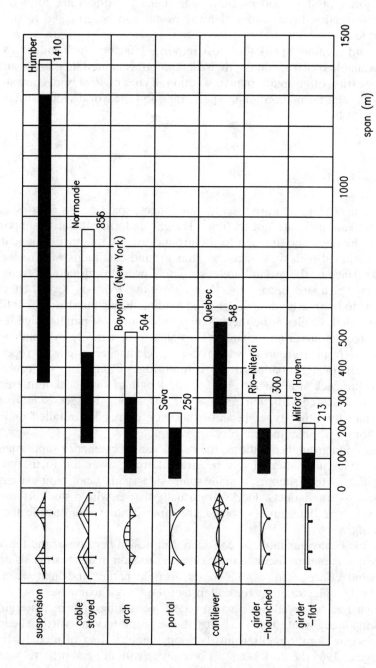

Fig. 4.2 Normal span range of bridge types

Selection of type 117

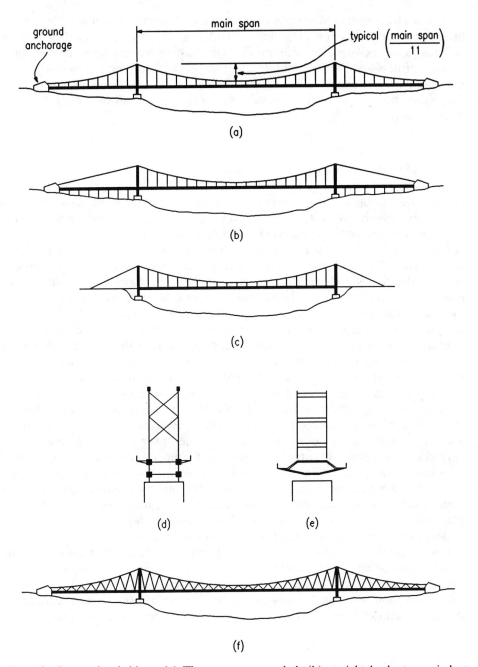

Fig. 4.3 Suspension bridges. (a) Three-span suspended; (b) straight back stays, viaduct side spans; (c) straight back stays, no side spans; (d) truss deck; (e) aerofoil box deck; (f) inclined hangers

that form a truss system which helps to reduce any tendency to oscillate. Suspension bridges behave as non-linear structures under asymmetric deck loading so that deflections may be large. Behaviour under such loading depends upon the combined gravity stiffness and the flexural rigidity of the deck or stiffening girder. The type is less suitable for heavy loading such as railway traffic, especially for short spans. Suspension bridges are sometimes suitable for medium spans carrying pedestrian or light traffic.

4.3.2 Cable-stayed bridges

Cable-stayed bridges (Fig. 4.4) are of a suspension form using straight cables which are directly connected to the deck. The structure is self-anchoring and therefore less dependent upon good foundation conditions, but the deck must be designed for the significant axial stress from the horizontal component of the cable forces. The construction process is quicker than for a suspension bridge because the cables and deck are erected at the same time and the amount of temporary works is reduced. Either twin sets of cables are used or alternatively for dual carriageways a single plane of cables and tower can be located in the central reserve space. Two basic forms of cable configurations are used, either 'fan' or 'harp'. A fan layout minimizes bending effects in the structure due to its better triangulation but anchorages can be less easy to incorporate into the towers. The harp form is often preferred where there are more than say, four cables. The number of cables depends on the span and cable size, which is often selected such that each fabricated length of deck (say, 20 m) contains an anchorage at one end to suit a cantilever erection method. Bridges either have two towers and are symmetrical in elevation or have a single tower as suited to the site.

Decks are generally an orthotropic steel plate but composite slabs can be used for spans up to about 250 m. A box girder is essential for bridges having a single plane of cables to achieve torsional stability, but otherwise either box girders or plate girders are suitable. Aerodynamic oscillation is a much less serious problem than with suspension bridges but must be considered. Some bridges with plate girders incorporate non-structural aerodynamic edge fairings.

It is essential to use cables of maximum strength and modulus at a high working stress so that sag due to self-weight, which produces non-linear effects, is negligible. Cables are normally of parallel wires or prestretched locked coil wire rope. During erection the cable lengths are adjusted or prestressed so as to counteract the dead load deflections of the deck arising from extension of the cables.

4.3.3 Arch bridges

Arch bridges (Fig. 4.5) are suitable in particular site conditions. An example is a medium single span over a ravine where an arch with spandrel columns will

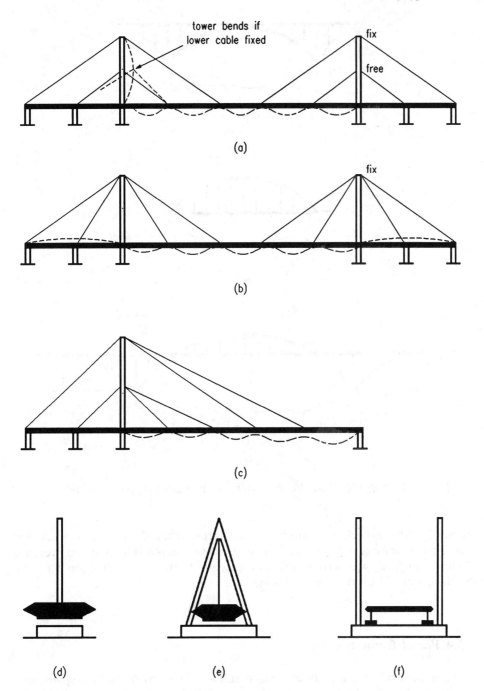

Fig. 4.4 Cable-stayed bridges. (a) Harp; (b) fan; (c) single tower; (d) single plane; (e) single plane; (f) twin plane

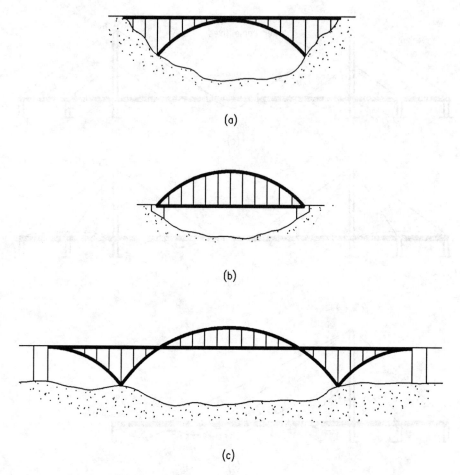

Fig. 4.5 Arch bridges. (a) Spandrel post arch; (b) tied arch; (c) part tied arch

efficiently carry a deck with the horizontal thrust taken directly to rock. A tied arch is suitable for a single span where construction depth is limited and presence of curved highway geometry or other obstruction to the approaches conflicts with the back stays of a cable-stayed bridge.

4.3.4 Portal frame bridges

Portal frame bridges (Fig. 4.6) are mainly suitable for short or medium spans. In a three-span form with sloping legs they can provide an economic solution by offering a reduction in span and have an attractive appearance. The risk of shipping collision with sloping legs must be considered for bridges over navigable

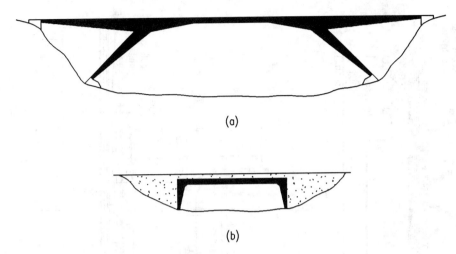

Fig. 4.6 Portal frame bridges. (a) Three-span inclined legs; (b) single span

rivers. Portal bridges tend to be less economic than beam bridges because the foundations must be designed to resist horizontal thrust, complex details can be required at the joints and erection is more complicated.

4.3.5 Truss bridges

Through trusses are used for medium spans where a limited construction depth precludes use of a composite deck bridge. They are suitable in flat terrain to reduce the height and length of approach embankments and for railway bridges where existing gradients cannot be modified. A truss may be unacceptable visually; a bowstring truss is an alternative solution. For short spans and medium spans up to 50 m, trusses are generally less economic than girders because of higher fabrication cost. They are therefore adopted only where the available construction depth is not sufficient for composite beams. (See Fig. 4.7.)

4.3.6 Girder type

Girder bridges (Fig. 4.8) predominate over the previously described types for short and medium spans, and generally provide the most economic solution. For highway bridges composite deck construction is generally used unless the depth is very critical in which case half-through girders or through trusses may be necessary. Railway bridges frequently require to be of half-through or through form because of depth limitations and are more suitable, being generally narrower. Pedestrian

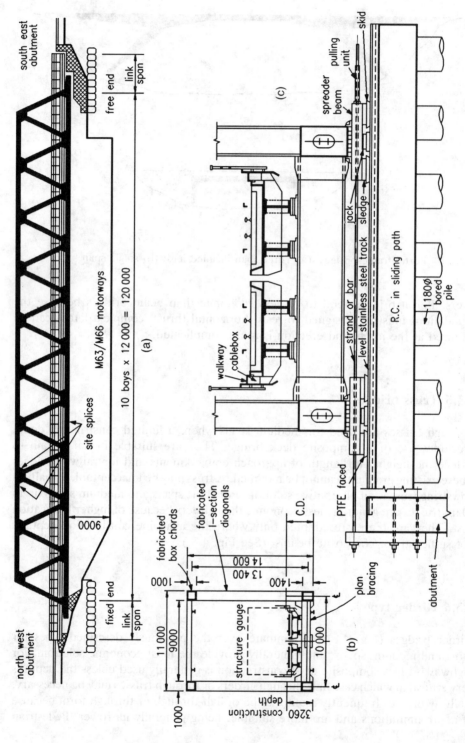

Fig. 4.7 Truss bridge: (a) elevation (b) cross section (c) arrangements for sliding into position

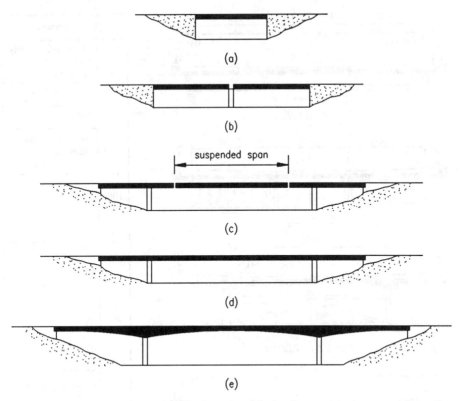

Fig. 4.8 Girder type bridges. (a) Single span; (b) simply-supported spans; (c) cantilever and suspended span; (d) continuous; (e) continuous − curved soffit

bridges are similarly suited to a half-through form. Highway, railway and pedestrian bridges of girder type for short and medium spans are further described below.

4.3.6.1 Highway bridges − composite deck construction
(see also Chapter 17)

Composite deck construction (Fig. 4.9) should be used wherever the construction depth will permit. If possible multiple spans should be made continuous over the intermediate supports so reducing the number of bearings and expansion joints. Continuity gives economies throughout the structure and reduces traffic disruption arising from the maintenance needs of these vulnerable elements. A number of options are available for maintaining continuity over intermediate supports. If ground conditions are poor such that the predicted differential settlement of the supports is significant (say, exceeding (span/1000)) then to avoid overstress the structure should be made statically determinate by use of simply-supported spans.

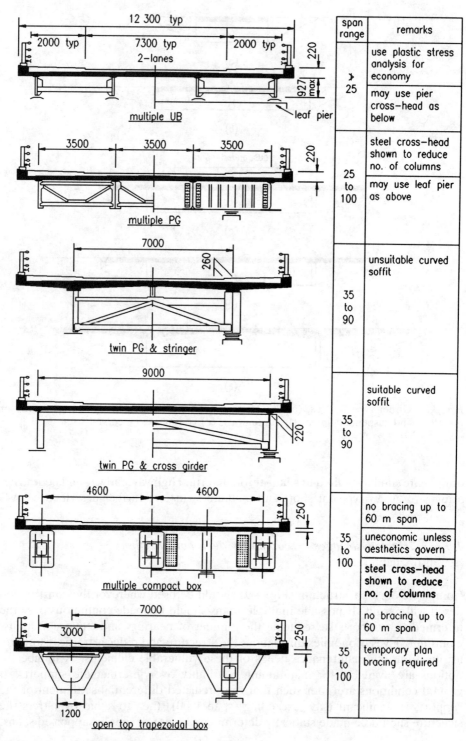

Fig. 4.9 Highway bridges – composite deck construction

Cantilever and suspended spans are alternative options which retain some of the advantages of continuity. A girder depth of (span/20) (girder depth excludes floor slab) is generally economic although shallow girders can be used down to a depth of (span/30) or less.

Rolled sections, available up to 914 mm × 419 mm serial size, are appropriate for short spans up to 25 m span if simply supported and about 33 m when continuous. For continuous spans exceeding about 22 m, fabricated plate girders will show economy because lighter flanges and webs can be inserted in the mid-span regions. Automated manufacture of plate girders means that they are highly economic when compared with box girders. Normally a girder spacing of 2.5–3.5 m is optimum with a floor slab of about 200–240 mm thick. Edge cantilevers should not exceed 50% of the beam spacing and to simplify falsework should where possible be less than 1.5 m. An even number of girders (i.e. 2, 4, 6, 8, etc.) achieves better optimization for material ordering and permits girders to be braced in pairs for erection.

For medium spans exceeding 40 m where adequate construction depth is available it may be economic to use twin girders only. A number of variants are available as shown in Fig. 4.9, using a thickened haunched slab, longitudinal stringers and cross girders to support the slab intermediately. For narrow bridges the complete precasting of composite floors may offer advantages in speed of construction.[3] Use of girders with a curved soffit becomes economical for medium spans exceeding 45 m and efficiently achieves maximum headroom if required over the central portions of a span (see Fig. 4.9). Plate girder flanges should be proportioned so as to be as wide as possible consistent with outstand limit to reduce the number of intermediate bracings. For practical reasons a desirable minimum flange width is about 400 mm to accommodate shear connections and to permit the possible use of permanent formwork. A maximum flange thickness of 65 mm is recommended as a guide to avoid heavy butt welds, but thicker flanges can be used where necessary, up to say 100 mm maximum.

For medium spans exceeding 100 m, box girders (see Fig. 4.10) are more suitable than plate girders for which flange sizes would be excessive. Other reasons for using box girders at these spans may include a need to improve aerodynamic stability, the presence of severe plan curvature, a requirement for single column supports or very limited construction depth. However for spans less than 100 m, box girders are generally less economic because, although a reduction in flange sizes may be possible due to superior load distribution properties, this is more than offset by the amount of internal diaphragms and stiffening and the extra costs in manufacture. Fabrication costs are high because the assembly and welding processes are less amenable to automation than with plate girders. Also access must be permitted inside box girders for welding, protective treatment processes, and permanent inspection. However erection of box girders is often easier because they require minimal external bracing to maintain overall stability. Multiple compact section box girders have proved to be economic for spans of up to 50 m in particular situations where longitudinal stiffeners can be eliminated. Open-top trapezoidal box girders (known as 'bath tubs') are widely used in North

126 *Bridges*

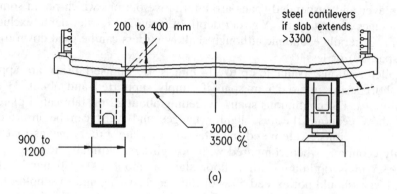

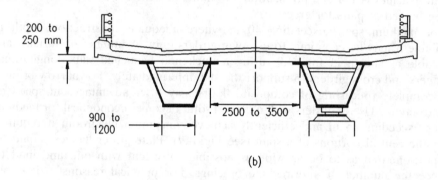

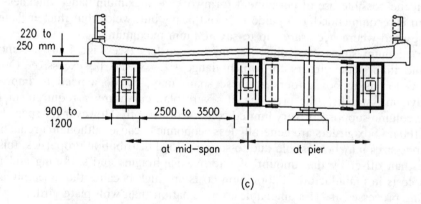

Fig. 4.10 Highway bridges – box girders. (a) Twin box and cross girders (spans 40–150 m); (b) open top box (spans 40–100 m); (c) multiple box (spans 30–60 m)

America and possess some of the advantages of plate girders. However temporary bracing is required during construction to maintain shape and relative twist of the sections until the concrete slab is placed and the full torsional rigidity achieved. Following a number of problems this subject has been researched in Canada.[4]

4.3.6.2 Railway bridges – girder type (Fig. 4.11)

Many existing small-span railway bridges weather cast or wrought iron girders to which rails are fixed directly without use of ballast or half-through girders with trough floors. As these bridges have reached the end of their lives they have been replaced in steelwork in modern form with ballasted track. The legacy of the original decks with their very shallow construction depth has influenced modern under-line bridge practice in replacements and new bridges. The majority of railway underline bridges therefore tend to be of half-through type. The direct fastening of track without ballast gives the cheapest possible form of railway bridge and is appropriate for carrying sidings where accuracy and adjustment of the track are less critical.

Girder depths are generally greater than for highway bridges due to the heavier loading and because limits for deformation are necessary on high-speed routes. A span-to-girder depth ratio of 12 to 15 is typical. For short spans half-through plate girders are used with composite steel cross girders forming rigid U-frames and supporting a concrete floor with ballasted single or double track. Through construction is appropriate for medium spans exceeding 50 m using trusses. For spans up to 39 m, British Rail use a standard box girder design with a steel ribbed floor of minimal depth which is achieved by spanning between the inner webs of trapezoidal box girders proportioned so as to fit closely within the station platform space. The type has advantages in using components entirely of steel which are bolted together at site and commissioned during temporary possession of existing tracks. Where sufficient depth is available then deck construction is preferable and more economic, with either twin plate girders or a box girder beneath each rail track.

Simply-supported spans are widely used for railway bridges because:

(a) individual spans can be erected or replaced quickly during temporary track possession,
(b) uplift is more likely to occur if spans are unequal under heavy railway loading,
(c) fatigue is less critical.

4.3.6.3 Pedestrian bridges (Fig. 4.12)

A minimum clear deck width of 1.8 m is usual, increased to 3.0 m or 4.0 m in busy areas or if a cycleway is also present. Steel provides an efficient solution

128 *Bridges*

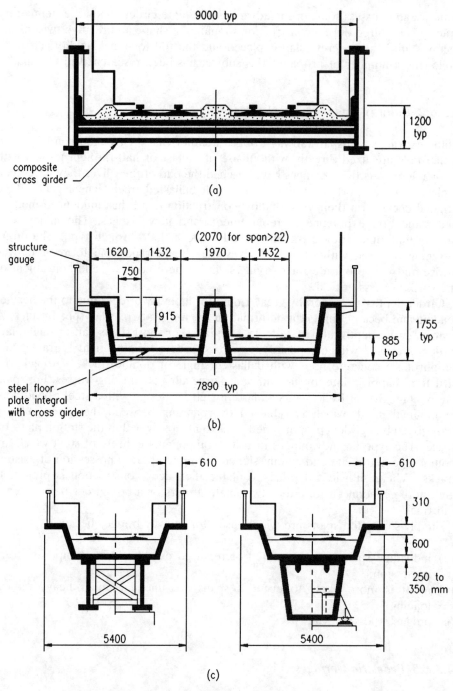

Fig. 4.11 Railway bridges. (a) Half-through plate girders; (b) half-through box girders; (c) composite deck type (spans > 30 m)

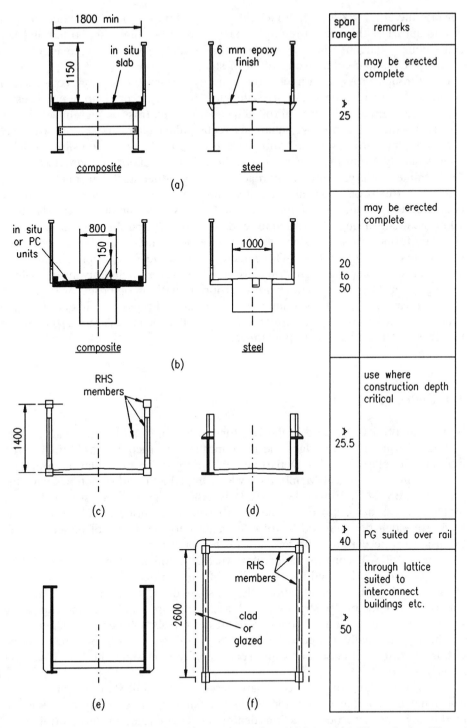

Fig. 4.12 Pedestrian bridges. (a) Twin universal beam; (b) box; (c) Vierendeel or Warren girder; (d) half-through universal beam; (e) half-through plate girder; (f) through lattice

because the entire cross section including parapets may be fabricated and erected in one unit. For this reason multiple spans tend to be simply supported and not continuous. Staircase and sloping or stepped ramp spans are also erected as complete units and columns are often of steel.

Half-through cross sections are often used because the shallow construction depth is able to provide the shortest lengths of staircase or ramp approaches in urban areas. Either half-through rolled beams, Warren truss or Vierendeel girders are used with rolled hollow (square or rectangular) members. Floors are steel stiffened plate often surfaced with a factory-applied epoxy non-slip surfacing approximately 5 mm thick. Rigid connections between floor and girders provide for U-frame stability. Staircase approaches may be either steel stringers supporting steel plate treads or a central spine box with cantilever treads of steel or precast concrete. Ramps may be similarly formed, but where the span exceeds about 10 m then half-through construction tends to be used. Spiral approach ramps are popular, formed in steel using twin rolled sections curved in plan supporting a composite or steel plate floor. Where adequate construction depth is available, for example, when a pedestrian bridge is required across a motorway or railway cutting, then deck cross sections are appropriate with a composite or steel plate floor. Precast concrete floor units are also suitable. Primary members may be a single box girder, twin plate girders or twin rolled beams. A single box girder provides a structure of neat appearance.

4.4 Codes of Practice

At present BS 5400[5] is used in the UK for the design of all bridges. Part 3 deals with the design of steel bridges and if composite construction is used then Part 5 must also be referred to. The Department of Transport has particular requirements in the form of Technical Memoranda which implement and sometimes vary the clauses in BS 5400. Eurocode 3 (EC3) for the design of steel structures is in preparation. Other Eurocodes relevant to composite bridges will be EC2 and EC4. Although the Eurocodes will refer mainly to the design of buildings their extension to the design of bridges is a further objective.

Eurocode 3 will contain basic rules which, although intended primarily for buildings, can also be used for the design of steel bridges. Though the rules for bridges need further amplification, the existing draft design rules are already applicable to some bridge structures such as simply-supported steel and composite spans, simple trusses and for connections. Rules for plate girders exceeding 1.5 m depth, longitudinally stiffened girders, box girders, orthotropic steel floors, U-frames, restraints at supports and serviceability limit state checks are not yet available.

Both BS 5400 and the Eurocodes are based on the limit state design concept, which means that the verification is carried out for serviceability and ultimate limit states. The Eurocode on actions deals with general rules for the determination

Table 4.6 Eurocode 1: Actions on Structures

List of contents	Structure of the chapters
Chapter 1 *General rules* 2 *Density of materials* 3 *Self-weight* 4 Soil and water pressure 5 Imposed deformations 6 Imposed loads on floors and roofs 7 *Snow* and ice 8 *Wind loads* 9 Water and wave loads 10 Thermal actions 11 Silo and tank loads 12 Road bridges 13 Railway bridges 14 Crane actions 15 Actions from machinery 16 Construction loads 17 Impact 18 Explosion 19 Seismic action	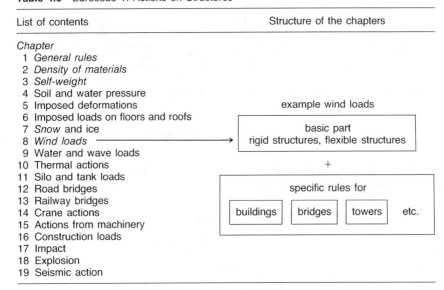

of the design loads and the load combinations and gives information on the characteristic values of the loads (see Table 4.6).

The first steps for the preparation of the rules for the traffic loads for bridges, particularly for road bridges, have been undertaken. For the traffic loads on railway bridges the harmonized uniform load model UIC 71[6] has been previously adopted by the different national railway authorities.

4.5 Traffic loading

4.5.1 Highway bridges

Highway bridges in the UK are currently designed for HA loading (a uniformly distributed loading plus knife edge load) together with HB (abnormal vehicle) loading for structures carrying main highways. For design to Eurocodes it is intended to adopt a unified loading model.

Although great differences in the national regulations for loads, resistances and safety factors exist, these differences appear to compensate one another and the resulting designs in all countries are generally similar.

4.5.2 Railway bridges

The uniform load model UIC 71[6] used by the European national railway authorities is shown in Fig. 4.13; it is also used in BS 5400: Part 2. Where specified this

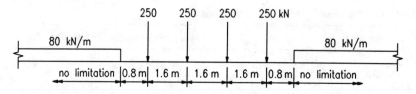

Fig. 4.13 UIC and BS 5400 load model (excluding impact)

loading is multiplied by a factor for bridges on lines carrying heavier or lighter traffic.

Dynamic factors are applied as shown in Table 4.7. Dimension L is the length (m) of the influence line for deflection of the element under consideration.

Table 4.7 Dynamic factors for railway loading

Dimension L	Dynamic factor	
(m)	Bending moment	Shear
$\not> 3.6$	2.00	1.67
>3.6 to 67	$0.73 + \sqrt{\left(\dfrac{2.16}{L - 0.2}\right)}$	$0.82 + \dfrac{1.44}{\sqrt{(L - 0.2)}}$
>67	1.00	1.00

Other loads arising from railway traffic are:

- centrifugal forces on curved track
- nosing – 100 kN force acting transversely at rail level
- traction and braking
- derailed vehicles (for overturning consider 80 kN/m over 20 m lengths acting on the edge of the structure).

Fatigue is important for the design of railway bridges because a higher proportion of regular traffic attains the maximum loading compared with highway bridges and the vehicles are constrained to run in the same lateral paths. Typically it is necessary to reduce the working stresses of grade 50 steel for main girders of spans less than about 24 m so that grade 43 or even lower stresses are used. For floor members including cross girders, trimmers, rail bearers and floor plating it is nearly always necessary to use reduced working stresses in a similar manner.

4.6 Other actions

Actions other than traffic loading which may need to be considered in one or more combinations are shown in Table 4.8.

Table 4.8 Summary of actions other than those due to traffic loading

Action	Comments
Dead loads[a]	Weight of the structure
Superimposed dead loads[a]	Finishes and surfacings Services
Wind	Transverse, longitudinal and vertical Consider in presence of live load or otherwise
Temperature	Restraint and movement (e.g. flexure of columns) Frictional restraint of bearings Effect of temperature difference
Differential settlement	Foundation movements
Earth pressure	Vertical and horizontal pressures from retained material
Erection effects[a]	Strength during construction, e.g. stability of steelwork before composite floor slab cast
Snow load	May be relevant to moving bridges
Seismic	As may be specified by the national authority
Water flow	Flow against bridge supports

[a] These actions must be considered during initial design. For most bridge decks actions other than those from vertical traffic loading, dead and superimposed dead loads are unlikely to have a fundamental effect.

4.7 Steel grades

Steels to BS EN 10025 grade 50 are usual for bridges as they offer a lower cost-to-strength ratio than grade 43. All parts subject to tensile stress are required to achieve a specified notch toughness, depending upon design minimum temperature, stress level and material thickness.

Special consideration may be needed if tensile stresses occur only during erection, for example, arising from lateral bending in girders due to wind loading, or during lifting of components. Judgement is necessary but it is often accepted that modest values of tensile stress can be permitted in such cases without the need for specific notch toughness provided that the work is not carried out during very low temperatures. In all other cases appropriate grades having specified notch toughness are necessary.

Weather-resistant steel

To eliminate the need for painting, weather-resistant (WR) grades in BS 4360 should be considered. Although it can be shown that the commuted costs of repainting are less than 1% of the initial bridge cost, weather-resistant steel is particularly useful in eliminating maintenance where access is difficult – over a railway for example. Weather-resistant steel is not suitable at or near the coast (i.e. within about one mile of the sea). The Department of Transport requires sacrificial thickness to be added to all exposed surfaces for possible long-term corrosion of 2 mm per face in a severe industrial environment, 1 mm otherwise.

4.8 Overall stability and articulation

It is important to consider overall stability of the bridge and its articulation under temperature effects (see Fig. 4.14). For simply-supported bridges each span must transfer longitudinal and transverse loads to the foundations while being able to accommodate movement with suitable bearings and expansion joints. For continuous spans the deck must be pinned at one support with free bearings elsewhere. Normally only one bearing within the deck width should be pinned longitudinally so that each girder is free to articulate under traffic loading independently, unless the pier consists of a separate column beneath each bearing which gives flexibility. A number of choices are open to the designer but the system used will affect the design of bearings, bearing stiffeners, expansion joints and the foundations.

For temperature movements:
Movement range: $(12 \times 10^{-6}$ per °C$) \times$ temperature range $\times$ length from pinned bearing.
For typical conditions in Europe, for steel or composite decks, ultimate movement $= \pm 4.5$ mm per 10 m of length from mean temperature.

4.9 Initial design

4.9.1 Suspension bridges

For initial design Fig. 4.15 gives approximate formulae[7] for making first estimates of cable size and bending of the stiffening girders. The most severe condition for the stiffening girder is with approximately one half of the main span loaded asymmetrically.

4.9.2 Cable-stayed bridges

Cable-stayed bridges virtually behave as continuous beams with elastic supports at the cable anchorage points. Cable lengths are adjusted during construction so that the effect of cable extension under dead load conditions is cancelled out. Provided that high tensile material is used for the cables then non-linear effects

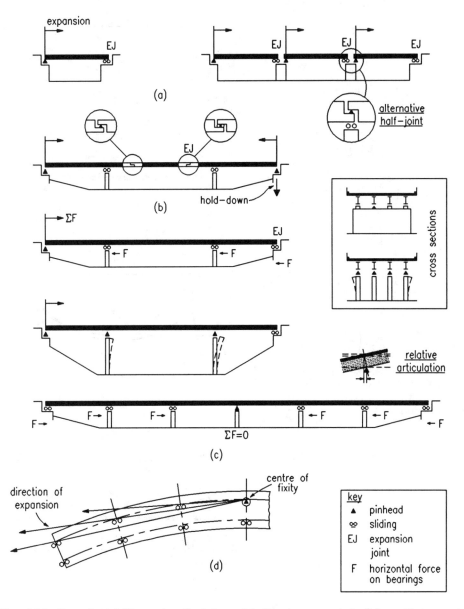

Fig. 4.14 Overall stability and articulation. (a) Simply supported, (b) cantilever and suspended span, (c) continuous, (d) curved viaducts

due to self-weight can be shown to be negligible.[8] For initial design the deck may be proportioned as a continuous beam, as shown in Fig. 4.16, to which are added:

(a) deck deflections at the cable positions due to extension,
(b) axial forces due to horizontal component of the cable forces.

$\ell = L\left(1 + \dfrac{8}{3}\left(\dfrac{d}{L}\right)^2\right)$

$H = \dfrac{wL^2}{8d}$

$T = H\left(1 + \dfrac{64\,d^2\,x^2}{L^4}\right)^{1/2}$

$\Delta\ell = \dfrac{H\ell}{AE}\left(1 + \dfrac{16}{3}\dfrac{d^2}{L^2}\right)$

$\Delta d = \dfrac{\Delta\ell}{\dfrac{16}{15}\dfrac{d}{L}\left(5 - \dfrac{24d^2}{L^2}\right)}$

$\Delta d = \dfrac{\Delta L\left(15 - 40\dfrac{d^2}{L^2} + 288\dfrac{d^4}{L^4}\right)}{16\dfrac{d}{L}\left(5 - 24\dfrac{d^2}{L^2}\right)}$

$M_{max} = 0.161\,pL^2\left(\dfrac{4}{\alpha}\dfrac{EId}{wL^4}\right)^{1/2}$

where α is the equivalent modulus of foundation, approximately 1100 kN/m²

M_{max} occurs when $\dfrac{2a}{L} = \dfrac{\pi}{4}\left(\dfrac{4}{\alpha}\dfrac{EId}{wL^4}\right)^{1/2}$

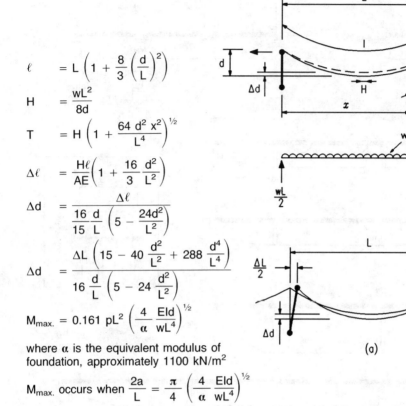

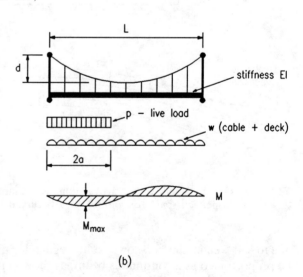

Fig. 4.15 Suspension bridges, approximate formulae (Pugsley). (a) Neglecting stiffness of deck; (b) including deck stiffness

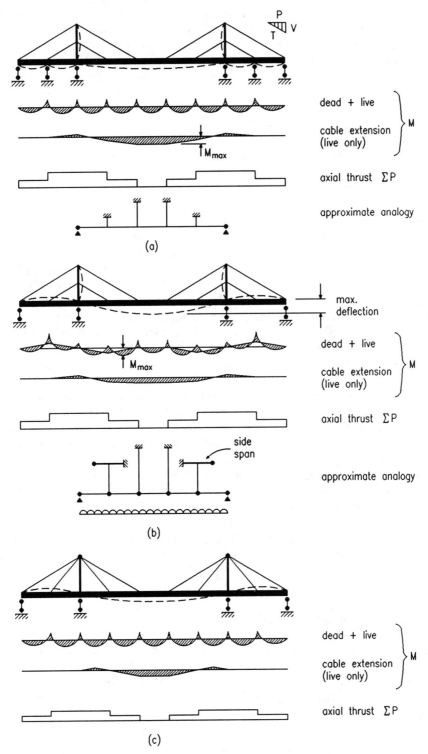

Fig. 4.16 Cable-stayed bridges, initial analysis. (a) Anchored side spans (harp shown); (b) harp; (c) fan

Where a single plane of cables is used then an initial design check must be made to ensure that sufficient torsional rigidity is provided (using single or twin box girders) to control transverse tilt of the deck at mid-span when traffic loading occupies one carriageway.

4.9.3 Highway bridges − composite deck construction

4.9.3.1 General

Bridges with composite deck construction (Fig. 4.17) represent a high proportion of the total number of steel bridges built in Europe since 1960. Generally a floor slab 220 mm to 250 mm thick is used composite with rolled sections or plate girders at spacings up to 3.5 m and depth between (span/20) and (span/30). For spans exceeding 40 m where adequate construction depth is available then twin girders can offer advantages with typical depth (span/18) to (span/25), and the floor slab is either haunched over the girders or supported by subsidiary stringers or cross girders. The slab thickness is determined by its requirement to resist local bending and punching shear effects from heavy wheel loads and needs to be reinforced in both directions; elastic design charts by Pucher are suitable.[9] In the hogging regions of continuous spans the slab will crack and be ineffective in overall flexure unless it is prestressed. The longitudinal reinforcement may however be assumed to act as part of the composite section. Most composite bridges are designed as 'unpropped' i.e. the erected steelwork supports its own weight and the concrete slab (including formwork allowance) until hardened with composite action only being assumed for superimposed dead and live loads. Box girders tend to be used for the longer spans especially those exceeding 100 m. Popular forms include twin box girders, multiple compact boxes or open top trapezoidal boxes.

Plate girder flanges should be made as wide as possible, consistent with outstand limitations, to give the best achievable stability during erection and to reduce the number of intermediate bracings. For practical reasons a desirable minimum width is about 400 mm. A maximum flange thickness of 65 mm is recommended to avoid heavy welds.

4.9.3.2 Intermediate supports

Intermediate supports often take the form of reinforced concrete walls, columns or portals. Steel supports may alternatively be used and tubular columns are efficient, especially if filled with concrete and designed compositely. Where fewer columns are required for multiple girders then integral steel crossheads at the supports are sometimes used.

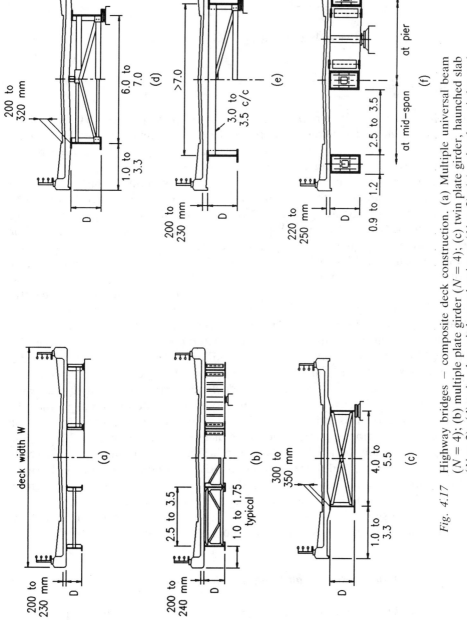

Fig. 4:17 Highway bridges – composite deck construction. (a) Multiple universal beam ($N = 4$); (b) multiple plate girder ($N = 4$); (c) twin plate girder, haunched slab ($N = 2$); (d) twin plate girder and stringer ($N = 2$); (e) twin plate girder and cross girder ($N = 2$); (f) multiple box ($N = 6$)

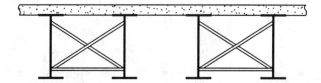

Fig. 4.18 Lateral bracing of bridge girders

4.9.3.3 Bracings

For rolled beam or plate girder bridges, lateral bracings are necessary for stability during erection and concreting of the slab. The bracings are required at all supports and within the hogging regions of continuous spans. If required by the designer they may be assumed to contribute to the transverse rigidity of the deck when carrying out an analysis of the transverse distribution of concentrated live loads. At the abutments the bracing can be a rolled section trimmer composite with the slab and supporting its free end. Over intermediate supports a channel section can be used between each pair of girders up to about 1.2 m depth. For deeper girders triangulated bracings are necessary.

Intermediate bracings in hogging regions are typically spaced at about 12 × (bottom flange width). Where bracing is provided across the full width of the bridge, i.e. between all girders, it increases transverse stiffness significantly. Because of this stiffness, such bracing will attract high stresses under loading that varies across the width of the bridge. The effect of this behaviour on the fatigue life of the bracing needs to be considered. Alternatively bracing should only be provided between neighbouring pairs of girders. This reduces the transverse stiffness considerably and alleviates the problem of fatigue. Such a structure is also likely to be easier to erect (Fig. 4.18). If the bridge is curved in plan with girders fabricated in straight chords they should be located adjacent to the site splices where torsion is induced. Bracings may be of a triangulated form or of single channel sections between each pair of girders where up to 1.2 m depth. Bracings are usually bolted to vertical web stiffeners in the main girders which may need to be increased in size to accommodate the bolted connections. Angle sections are commonly used with lapped single shear connections which permit tolerance in accommodating camber difference between adjacent girders.

Plan bracing systems may be required for spans exceeding 55 m for temporary stability; they may be removed after the floor slab is cast.

4.9.3.4 Locations of splices and change of section

Where rolled beams up to 1.0 m deep are used then for the maximum span range of about 33 m it is convenient to use a constant section with one splice within each span located at about 0.1 to 0.2 × span from the internal supports. The beam size

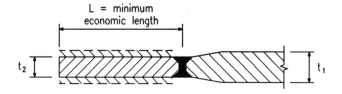

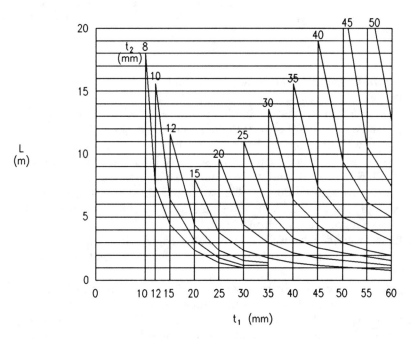

Fig. 4.19 Economy of flange and web thickness changes. The figure gives an indication of the minimum length (L) for which a selected thickness change will be economic for flanges and webs of girders. Below this length it will be more economic to continue the thicker plate (t_1)

will be determined by the maximum bending and shear effects at the supports where the slab is cracked.

For plate or box girders the component lengths for shop fabrication should be the maximum possible consistent with delivery and site restrictions to reduce the amount of site assembly. For spans up to about 55 m two splices per span are generally suitable, located at 0.15 to 0.25 × span from the internal supports, at which changes of flanges and web thickness should be made. A minimum number of other flange or web workshop joints should be made, consistent with plate length availability. The decision whether to introduce thickness changes within a fabricated length should take account of the cost of butt welds compared with the potential for material saving. Figure 4.19 indicates a basis for considering this optimization.

4.9.3.5 Curved bridge decks

Bridges which are curved in plan may be formed using straight fabricated girders, with direction changes introduced at each site splice. Alternatively the steel girders can also be truly curved in plan in which case the secondary stresses which arise from torsion must be evaluated. It is likely that additional transverse bracings will be needed to reduce these stresses.

4.9.3.6 Initial sizes – composite plate girders

For economic design an analysis should be carried out of the transverse distribution of concentrated live loads between the main girders of the cross section. The floor slab together with any continuous transverse bracing will significantly redistribute the maximum load applied to the most severely loaded girders. Prior to this the designer will wish to select initial sizes and make an estimate of the total weight of structural steel. Figures 4.20–4.24[10] provide initial estimates of flange area, web thickness and overall unit weight of steelwork (kg/m^2) for typical composite bridge cross sections as shown in Fig. 4.17.

The figures were derived from approximate designs using simplifying assumptions for loads and transverse distribution and to achieve correlation with actual UK bridges.

The sizes indicated do not represent final design, which must be checked by a form of distribution analysis using the actual specified loads. A two-dimensional grid analysis is usually employed with any out-of-plane effects being ignored. Comparison with three-dimensional analyses has shown that out-of-plane effects are generally negligible for most composite bridge decks. The following assumptions apply to use of Figs 4.20–4.24.

Deck slab 230 mm average thickness (5.75 kN/m^2).
Superimposed dead loads equivalent to 100 mm of finishes (2.40 kN/m^2).
Formwork weight 0.50 kN/m^2 of slab soffit area.
Steel grade 50 (yield strength 355 N/mm^2).
Span-to-girder-depth ratios of 20 and 30.
Webs have vertical stiffeners at approximately 2.0 m centres where such stiffening is required.

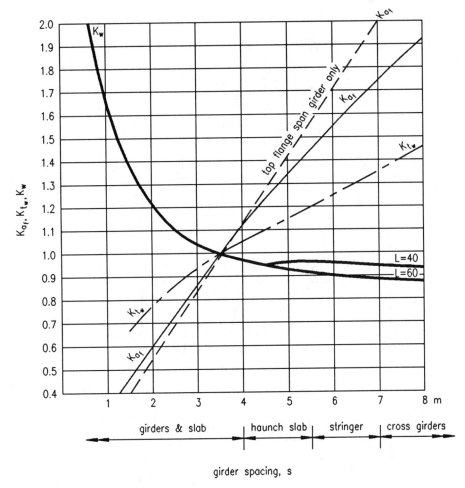

Fig. 4.20 Girder spacing factors

Elastic stress analysis is used for plate girders. If however the plastic modulus is used for compact cross sections, then economies are possible.

Steelwork is unpropped during casting of the floor slab.

Sufficient transverse bracings are used such that stresses are not significantly reduced due to buckling criteria.

Top flanges in sagging regions are dictated by a maximum stress during concreting allowing for formwork and concreting effects.

Live loading is approximately equivalent to United Kingdom 45HB or HA as indicated.

Continuous spans are approximately equal in length.

144 *Bridges*

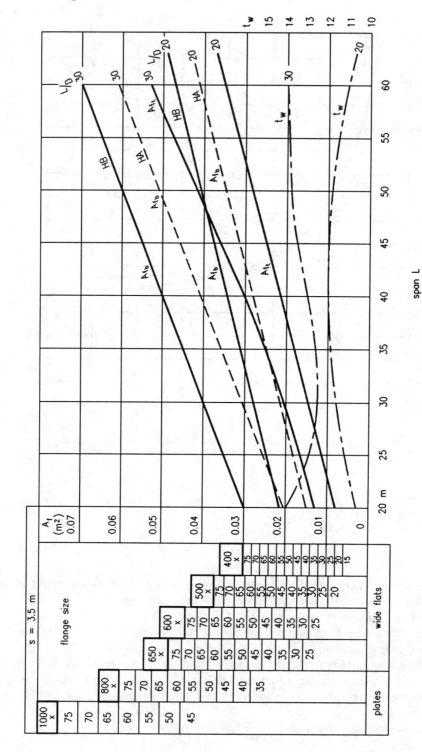

Fig. 4.21 Flange and web sizes – simply-supported bridges

Initial design   145

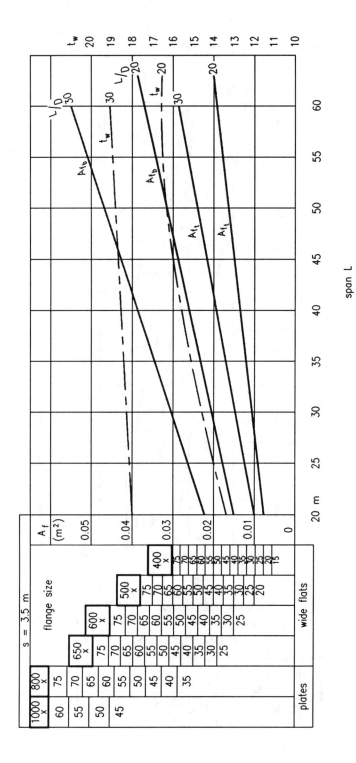

Fig. 4.22 Flange and web sizes – continuous bridges, pier girders

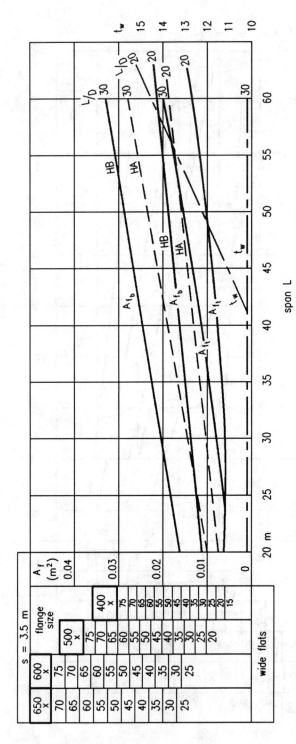

Fig. 4.23 Flange and web sizes – continuous bridges span girders

Initial design 147

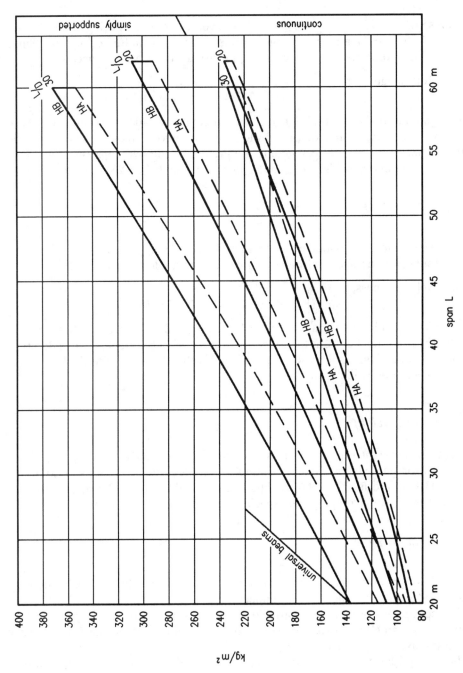

Fig. 4.24 Overall unit weights – plate girder bridges

Flange sizes

Figures 4.21–4.23 are applicable to an average girder spacing s of 3.5 m. Figure 4.20 gives a girder spacing factor K_{a_f} which is multiplied by the flange areas, obtained above, to give values appropriate to the actual spacing,

e.g. top flange area, $A_{f_t} = A_{f_t}$ $\qquad \times K_{a_f}$
$\qquad\qquad\qquad\quad$ (Figs 4.21–4.23) $\quad$ (Fig. 4.20)

The figures also show actual flange sizes, ranging from 400 × 15 mm to 1000 × 75 mm. The flange area of pier girders of continuous unequal spans may be approximately estimated by assuming the greater of the two adjacent spans. End spans of continuous bridges may be estimated using $L = 1.25 \times$ actual span.

Web thickness

Web thicknesses are obtained using Figs 4.21–4.23 applicable to $s = 3.5$ m. Adjustment for the actual girder spacing s is obtainable from Fig. 4.20,

i.e. web thickness, $t_w = t_w$ $\qquad \times K_{t_w}$
$\qquad\qquad\qquad\qquad$ (Figs 4.21–4.23) $\quad$ (Fig. 4.20)

The thickness obtained may be regarded as typical. However, designers may prefer to opt for thicker webs to reduce the number of web stiffeners. Consideration should also be given to the use of unstiffened webs of appropriate thickness, i.e. $d/t \not< 60\text{–}100$ depending on shear.

Overall unit weight

Overall unit weight (kg/m² of gross deck area) is read against the span L from Fig. 4.24 for simply-supported or continuous bridges with L/D ratios of 20 or 30, under HB or alternatively HA loading and applicable to $s = 3.5$ m.

Adjustment for average girder spacing s other than 3.5 m is obtainable from Fig. 4.20,

i.e. kg/m² = kg/m² $\qquad \times K_w$
$\qquad\qquad$ (Fig. 4.24) $\quad$ (Fig. 4.20)

The unit weight provides an approximate first estimate of steelwork weight allowing for all stiffeners, bracings, shear connectors, etc.

For continuous bridges with variable depth, Figs 4.21 and 4.23 may be used to provide a rough guide, assuming a span-to-depth ratio (L/D) for each span based upon the average girder depth.

For box-girder bridges a rough estimate may be obtained by replacing each box girder with an equivalent pair of plate girders.

The mean span for use in Fig. 4.24 should be determined as follows:

$$\text{mean span, } L = \sqrt[+]{\left(\frac{L_1^4 + L_2^4 + \ldots + L_n^4}{n}\right)}$$

where n is the number of spans.

4.9.3.7 Initial sizes – rolled section beams

An estimate of size for simply-supported spans only may be obtained from Figs 4.25 and 4.26[10] for elastic or plastic stress analysis respectively, using universal beams up to 914 × 419 × 388 kg/m size. For an estimate of the total weight of structural steel a factor of 1.1 applied to the main beams provides a reasonable allowance for bracings, bearing stiffeners and shear connectors. Figures 4.25 and 4.26 are based upon a concrete strength of 37.5 N/mm^2, and show the required mass per metre of universal beam. Table 4.9 gives reference to the relevant serial size.

Table 4.9 Universal beam sizes

Reference Figs 4.25 & 4.26	Universal beam size		Actual depth (mm)
	Serial size	Mass per metre (kg)	
388	914 × 419	388	920.5
343		343	911.4
289	914 × 305	289	926.6
253		253	918.5
224		224	910.3
201		201	903.0
226	838 × 292	226	850.9
194		194	840.7
176		176	834.9
197	762 × 267	197	769.6
173		173	762.0
147		147	753.9
170	686 × 254	170	692.9
152		152	687.6
140		140	683.5
125		125	677.9
238	610 × 305	238	633.0
179		179	617.5
149		149	609.6
140	610 × 229	140	617.0
125		125	611.9
113		113	607.3
101		101	602.2

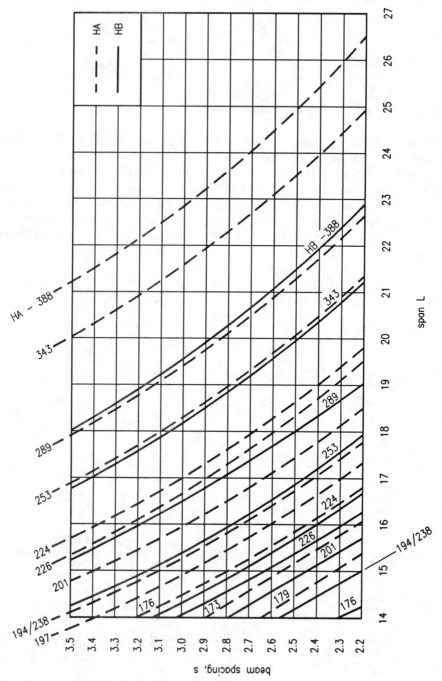

Fig. 4.25 Universal beam sizes — simply-supported bridges — elastic design

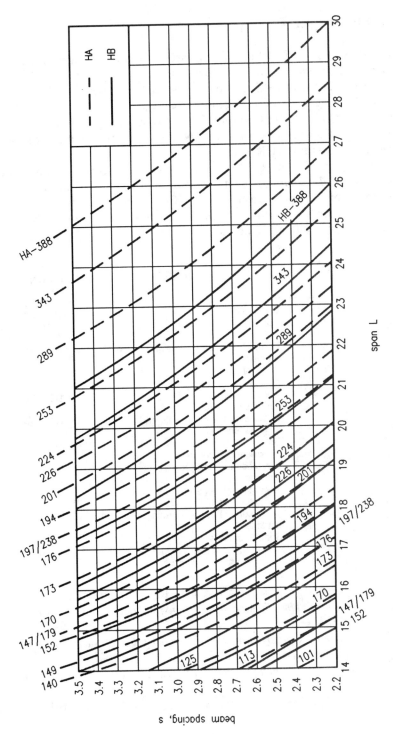

Fig. 4.26 Universal beam sizes – simply-supported bridges – plastic design

References to Chapter 4

1. Hayward A.C.G. (1988) State of the art – highway bridges. *ECCS/BCSA International Symposium on Steel Bridges.*
2. Johnson R.P. & Buckby R.J. (1986) *Composite Structures in Steel and Concrete, Volume 2: Bridges*, 2nd edn. Collins, London.
3. Hayward A.C.G. (1987) Composite pedestrian and cycle bridge at Welham Green. *Steel Construction Today*, **1**, No. 1, Feb. (The journal of the Steel Construction Institute, UK.)
4. Branco F.A. & Green R. (1982) Construction bracing for composite box girder bridges. *International Conference on Short and Medium Span Bridges*, Toronto, Canada. Canadian Society for Civil Engineering
5. British Standards Institution. *Steel, concrete and composite bridges: Parts 1–10*. BS 5400, BSI, London.

BS 5400: Part	Date	Title	UK Department of Transport Standard
1	1988	General statement	BD15/82
2	1978	Specification for loads	BD14/82 & IRLS (Sept. 1983)
3	1982	Code of practice for design of steel bridges	BD13/82
4	1990	Code of practice for design of concrete bridges	BD24/84
5	1979	Code of practice for design of composite bridges	BD16/82
6	1980	Specification for materials and workmanship, steel	BD11/82
7	1978	Specification for materials and workmanship, concrete, reinforcement and prestressing tendons	–
8	1978	Recommendations for materials and workmanship, concrete, reinforcement and prestressing tendons	–
9	1983	Bridge bearings	BD10/83
10	1980	Code of practice for fatigue	BD9/81

6. UIC (1971) Leaflet 702. UIC, 14 rue Jean-Ray F., 75015 Paris, France.
7. Pugsley A. (1968) *The Theory of Suspension Bridges*, 2nd edn. Edward Arnold, London.
8. Podolny W. (1976) *Construction and Design of Cable-Stayed Bridges*. John Wiley & Sons, New York.
9. Pucher A. (1973) *Influence Surfaces of Elastic Plates*, 4th edn. Springer-Verlag, New York.
10. Hayward A.C.G. (1988) *Composite Steel Highway Bridges*. British Steel plc.

A worked example follows which is relevant to Chapter 4.

Worked example

The Steel Construction Institute Silwood Park, Ascot, Berks SL5 7QN	Subject **INITIAL DESIGN OF HIGHWAY BRIDGE**	Chapter ref. **4**
	Design code BS5400	Made by ACGH	Sheet no. 1
		Checked by GWO	

Problem

A composite highway bridge has 3 continuous spans of 24, 40 and 32 m as shown. Overall deck width is 12 m and it carries 45 units of HB loading. There are 4 plate girders in the cross-section of 1.75 m depth. Estimate the main girder sizes and the weight of structural steel.

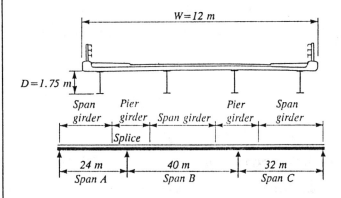

Average girder spacing 's' = $\dfrac{12}{4}$ = 3.0 m

Flange and web sizes

From Figure 4.20 k_{af} =   0.85 (top flange span girders)
       k_{af} =   0.87 (generally)
       k_{tw} =   0.95

SPAN A:

This is an end span so take

L   =   1.25 × 24 m   =   30 m

therefore L/D   = $\dfrac{30 \text{ m}}{1.75 \text{ m}}$ =   17

so assume L/D   =   20

The Steel Construction Institute Silwood Park, Ascot, Berks SL5 7QN		Subject INITIAL DESIGN OF HIGHWAY BRIDGE		Chapter ref. 4
		Design code BS5400	Made by ACGH Checked by GWO	Sheet no. 2

Top Flange, A_{ft} = A_f (from Figure 4.23) × K_{af}

                     = 0.006×0.85 = $0.0051 \, m^2$ <u>400 × 15</u>

Bottom Flange, A   = A_f (from Figure 4.23) × K_{af}

                     = 0.014×0.87 = $0.012 \, m^2$ <u>500 × 25</u>

Web, t_w = t_w (from Figure 4.23) × K_{tw}

                     = 10×0.95 = $9.5 \, mm$ <u>10 mm web</u>

<u>SPAN B:</u>

Span girder

L/D   = $\dfrac{40 \, m}{1.75 \, m}$ =   22.9

Top Flange, A_{ft} = A_f (from Figure 4.23) × K_{af}

                     = 0.009×0.85 = $0.0077 \, m^2$ <u>400 × 20</u>

Bottom Flange, A_{fb} = A_f (from Figure 4.23) × K_{af}

                     = 0.020×0.87 = $0.0174 \, m^2$ <u>500 × 35</u>

Web, t_w = t_w (see Figure 4.23) × K_{tw}

                     = 10×0.95 = $9.5 \, mm$ <u>10 mm web</u>

<u>SPAN C:</u>

This is an end span so take

L   = $1.25 \times 32 \, m$ = $40 \, m$

Therefore sizes as Span B.

The Steel Construction Institute Silwood Park, Ascot, Berks SL5 7QN	Subject **INITIAL DESIGN OF HIGHWAY BRIDGE**	Chapter ref. **4**
	Design code **BS5400**	Made by **ACGH** Sheet no. **3**
		Checked by **GWO**	

Pier Girders

Take L as the greater of the two adjacent spans
i.e. assume L = 40 m at both supports.

Therefore L/D	=	40/1.75	=	22.9	
Top Flange, A_{ft}	=	A_{ft} (see Figure 4.22) × K_{af}			
	=	0.015 × 0.87	=	0.0131 m²	<u>400 × 35</u>
Bottom Flange, A_{fb}	=	A_{fb} (see Figure 4.22) × K_{af}			
	=	0.030 × 0.87	=	0.026 m²	<u>500 × 55</u>
Web, t_w	=	t_w (see Figure 4.22) × K_{tw}			
	=	16.5 × 0.95	=	15.7 mm	<u>18 mm web</u>

Steel Weight

Span A:	L	=	1.25 × 24 m	=	30 m
Span B:	L	=	40 m		
Span C:	L	=	1.25 × 32 m	=	40 m

$$\text{Mean Span} = \left[\frac{L^4 + L_2^4 \ldots L_n^4}{n} \right]^{\frac{1}{4}} = \left[\frac{30^4 + 40^4 + 40^4}{3} \right]^{\frac{1}{4}}$$

= <u>37.5</u>

L/D = 37.5/1.75 = 21

kg/m² = kg/m² (from Figure 4.24) × K_w (from Figure 4.20)

= 142 kg/m² × 1.03 = <u>146 kg/m²</u>

Therefore steel weight

= $\frac{146 \text{ kg/m}^2}{1000}$ × (24 m + 40 m + 32 m) × 12 m wide = <u>168 tonnes</u>

156   *Worked example*

The Steel Construction Institute Silwood Park, Ascot, Berks SL5 7QN	Subject **INITIAL DESIGN OF HIGHWAY BRIDGE**	Chapter ref. **4**
	Design code **BS5400** — Made by **ACGH** / Checked by **GWO**	Sheet no. **4**

<u>Problem</u>

A composite bridge has a simply supported span of 24 m. Overall deck width is 9.6 m and it carries HA loading only. Estimate the beam sizes and total weight of structural steel assuming that there are 4 beams in the cross-section.

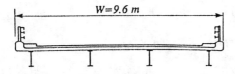

a)   Elastic Design

For 4 beams,   s   =   9.6/4   =   2.4 m

From Figure 4.25 use section no. 343
(914 × 419 × 343 kg/m universal beam)

Total steel weight,
343 × 4 × 24 × 1.08/1000   =   36 tonnes   =   154 kg/m²

(1.08 factor allows for bracing and stiffeners)

b)   Plastic Design

From Figure 4.26 use section no. 253
(914 × 305 × 253 kg/m universal beam)

Total steel weight,
253 × 4 × 24 × 1.08/1000   =   26 tonnes   =   114 kg/m²

Plastic design gives a significant reduction in the steel weight, subject to a satisfactory check of the serviceability limit state.

Chapter 5
Other structural applications of steel

Edited by IAN DUNCAN with contributions from MICHAEL GREEN, ERIC HINDHAUGH, IAN LIDDELL, GERARD PARKE and JOHN TYRRELL

5.1 Towers and masts

5.1.1 Introduction

Self-supporting and guyed towers have a wide variety of uses, from broadcasting of television and radio, telecommunications for telephone and data transmission to overhead power lines, industrial structures, such as chimneys and flares, and miscellaneous support towers for water supply, observation or lighting. These structures range from minor lighting structures, where collapse might have almost no further consequences, to major telecommunications links passing thousands of telephone calls or flare structures on which the safety of major chemical plant can depend. The term 'mast' describes a tower which depends for its stability on cable guys.

5.1.2 Structural types

Steel towers can be constructed in a number of ways but the most efficient use of material is achieved by using an open steel lattice. Typical arrangements for microwave radio and transmission towers are shown in Fig. 5.1. The use of an open lattice avoids presenting the full width of structure to the wind but enables the construction of extremely lightweight and stiff structures. Most power transmission, telecommunication and broadcasting structures fall into this class.

Lattice towers are typically square or triangular and have low redundancy. The legs are braced by the main bracings; both of these are often propped by additional secondary bracing to reduce the effective buckling lengths. The most common forms of main bracing are shown in Fig. 5.2.

Lattice towers for most purposes are made of bolted angles. Tubular legs and bracings can be economic, especially when the stresses are low enough to allow relatively simple connections. Towers with tubular members may be less than half the weight of angle towers because of the reduced wind load on circular sections. However the extra cost of the tube and the more complicated connection details can exceed the saving of steel weight and foundations.

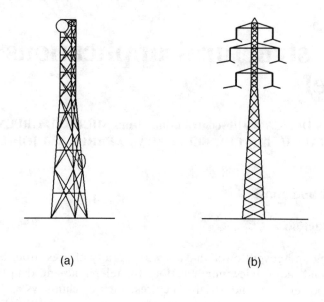
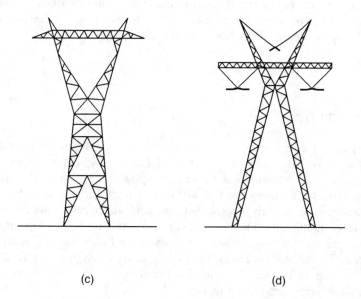

Fig. 5.1 Lattice towers: (a) microwave tower, (b), (c) and (d) transmission towers

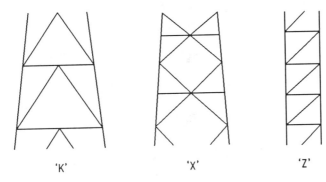

Fig. 5.2 Main bracing arrangements

Connections are usually arranged to allow site bolting and erection of relatively small components. Angles can be cut to length and bolt holes punched by machines as part of the same operation. Where heavy-lift cranes are available much larger segments of a tower can be erected but often even these are site bolted together.

Guyed towers provide height at a much lower material cost than self-supporting towers due to the efficient use of high-strength steel in the guys. Guyed towers are normally guyed in three directions over an anchor radius of typically $\frac{2}{3}$ of the tower height and have a triangular lattice section for the central mast. Tubular masts are also used, especially where icing is very heavy and lattice sections would ice up fully. A typical example of a guyed tower is shown in Fig. 5.3.

The range of structural forms is wide and varied. Other examples are illustrated in Figs 5.4 and 5.5. Figure 5.4 is a modular tower arrangement capable of extension for an increased number of antennas. The arrangement shown in Fig. 5.5 is adopted for supporting flare risers where maintenance of the flare tip is carried out at ground level.

A significant influence on the economics of tower construction is the method of erection which should be carefully considered at the design stage.

5.1.3 Environmental loading

The primary environmental loads on tower structures are usually due to wind and ice, sometimes in combination. Earthquakes can be important in some parts of the world for structures of high mass, such as water towers. Loading from climatic temperature variations is not normally significant but solar radiation may induce local stresses or cause significant deflections and temperatures can influence the choice of ancillary materials.

Most wind codes use a simple quasi-static method of assessing the wind loads which has some limitations for calculating the along-wind responses but is adequate

160   *Other structural applications of steel*

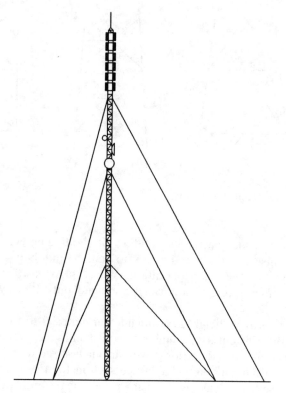

Fig. 5.3   Guyed tower

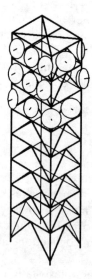

Fig. 5.4   Modular tower

Towers and masts 161

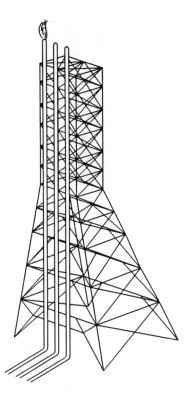

Fig. 5.5 Flare tower

for the majority of structures. Tower structures with aerodynamically solid sections and some individual members can be subject to aeroelastic wind forces caused by vortex shedding, galloping, flutter and a variety of other mechanisms which are either poorly covered or ignored by current codes of practice. Such factors have been responsible for more tower collapses and serviceability failures worldwide than any shortfall in resistance to along-wind loads.

CP3[1] specifies wind loads in the UK in terms of design gust wind speeds that will recur on average once in a 50-year period (i.e. with an annual probability of 2%). Guidance is given on wind shape factors for typical sections and lattices. There is no consideration of dynamic response to the wind. It is assumed that the loads will be used with working stresses in the members.

BS 8100[2] is a more recent code in a limit state format specifically written for towers. Wind loads are specified in terms of a '50-year return' mean hourly wind pressure together with gust factors which convert the forces to an equivalent static gust. The overall wind forces calculated using BS 8100 are substantially similar to those that would be obtained using CP3 but forces near the tops of towers are relatively higher due to an allowance for dynamic response. The code also gives guidance on means of allowing for the importance of particular structures by adjusting the partial factor on the design wind speed.

Guidance is limited for structures that have a significant dynamic response at their natural frequencies and gust factors for guyed towers are specifically excluded from the scope of the code. Part 4 of BS 8100 is intended to address these aspects.

The influence of height and topography on wind speed can be significant; this is covered in some detail in both codes. Ice loads and types of ice are also covered but neither mentions the very significant influence of topography on the formation of ice. This has not yet been subject to systematic study but some hill sites are known to be subject to icing well in excess of the code requirements. The combination of wind and ice loads is even less well understood although some guidance is given.

5.1.4 Analysis

In the analysis of towers the largest uncertainty is accurate knowledge of the wind loads. Highly sophisticated methods of analysis cannot improve this. A static linear three-dimensional structural analysis is sufficient for almost all lattice tower structures.

For transmission towers, line break conditions can also be critical. Line breakage will in general induce dynamic loads in addition to any residual static loads. Detailed consideration of transmission tower loading is outside the scope of this section.

For lattice towers with large complicated panel bracing, the secondary bracing forces can be significantly altered by non-linear effects caused by curvature of the panels under the influence of the design loads. Generally the rules in the codes are sufficient, but where structures are of particular importance or where there is much repetition of a design, a non-linear analysis may be necessary.

Dynamic analyses of self-supporting lattice towers are rarely necessary unless there are special circumstances such as high masses at the top, use as a viewing platform or circular or almost solid sections of mast which could be responsive to vortex shedding or galloping. Knowledge of the dynamic response is also necessary for assessment of fatigue of joints if this is significant.

For guyed towers the non-linear behaviour of the guys is a primary influence and cannot be ignored. The choice of initial tension, for example, can have a very great effect on the deflections (and dynamic behaviour). The effects of the axial loads in the mast on column stiffness can be significant. Methods of static analysis are given in the main international codes for the design of guyed towers. Guyed towers can also be particularly sensitive to dynamic wind effects especially those with cylindrical or solid sections.[2]

General guidance on the dynamic responses and aerodynamic instabilities of towers can be obtained from References 3, 4, 5 and 6.

5.1.5 Serviceability

Serviceability requirements vary greatly depending on the purpose of the structure and its location.

Steel towers and connections are normally galvanized and are also painted with a durable paint system if the environment is likely to be polluted or otherwise corrosive. It is important that regular maintenance is carried out; climbing access is normally provided for inspections.

Deflections of towers are generally significant only if they would result in a loss of serviceability. This can be critical for the design of telecommunication structures using dish antennas. In the past signal losses due to deflection have often been assessed on the misunderstanding that the deflections under the design wind storm would occur sufficiently often to affect the signals. Studies have demonstrated that short periods of total loss of signal during storms smaller than the design wind storm have a negligible effect on the reliability of microwave links compared to losses due to regular atmospheric conditions.

5.1.6 Masts and towers in building structures

Consideration of a masted solution arises from the need to provide a greater flexibility in the plan or layout of the building coupled with its aesthetic value to the project as a whole. At the same time it offers the opportunity to utilize structural materials in their most economic and effective tensile condition. The towers or masts can also provide high-level access for maintenance and plant support for services. The plan form resulting from a mast structure eliminates the need for either internal support or a deeper structure to accommodate the clear span. By providing span assistance via suspension systems the overall structural depth is minimized giving a reduction in the clad area of the building perimeter. The concentration of structural loads to the mast or towers can also benefit substructure particularly in poor ground conditions where it is cost effective to limit the extent of substructures (Figs 5.6 and 5.7). However, differential settle-

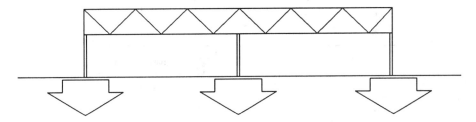

Fig. 5.6 Traditional long-span structure

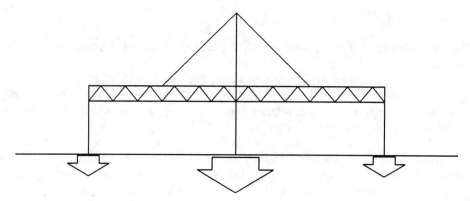

Fig. 5.7 Suspension structure

ment can have a significant effect on the structure by relaxing ties on suspension systems. The consequent load redistribution must be considered.

Most tension structure building forms consist of either central support, perimeter support or a mixture of the two. Any other solutions are invariably a variation on a theme. Plan form tends to be either linear or a series of repetitive squares.

The forces and loads experienced by towers and masts are illustrated in Figs 5.8 and 5.9. In all cases it is advantageous but not essential that forces are balanced

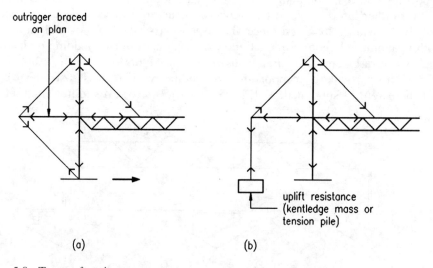

Fig. 5.8 Types of perimeter support

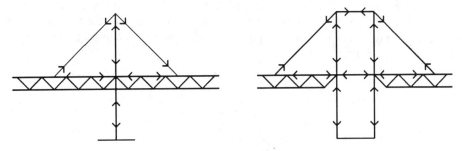

Fig. 5.9 Central support

about the mast. Out-of-balance loads will obviously generate variable horizontal and vertical forces which require resolution in the assessment of suitable structural sections.

Suspension ties must be designed not only to resist tension but also the effects of vibration, ice build-up and catenary sag. Ties induce additional compressive forces in the members they assist. These forces require careful consideration, often necessitating additional restraints in the roof plane in either the open sections or top chord of any truss.

Longitudinal stability is created by either twinning the masts and creating a vertical truss or by cross bracing preferably to ground (Fig. 5.10).

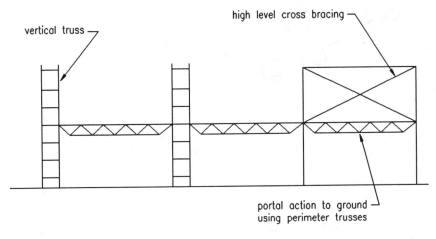

Fig. 5.10 Longitudinal stability

166   *Other structural applications of steel*

If outriggers are used as is the case with the majority of masted structures then the lateral stability of the outrigger can be resolved in a similar form to the masts by a stiff truss or Vierendeel section in plan or by plan or diagonal bracing at the extremity. (Fig. 5.11.)

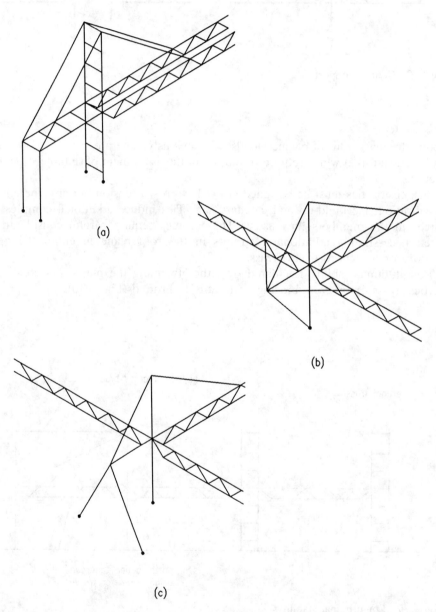

Fig. 5.11 Outrigger stability: (a) Vierendeel, (b) plan bracing, (c) diagonal bracing

5.2 Space frames

5.2.1 Introduction

Steel skeletal space frames are three-dimensional structures capable of very large column-free spans. These structures constructed from either individual elements or prefabricated modules possess a high strength-to-weight ratio and stiffness. Steel space frames may be used efficiently to form roofs, walls and floors for projects such as shopping arcades but their real supremacy is in providing roof cover for sports stadia, exhibition halls, aircraft hangars and similar major structures.

5.2.2 Structural types

Space frames are classified as single-, double- or multi-layered structures which may be flat, resulting in grid structures, or may be curved in one or two directions, forming barrel vaults and dome structures. Grid structures can be further categorized into lattice and space grids in which the members may run in two, three or four principal directions. In double-layer lattice grids the top and bottom grids are identical, with the top layer positioned directly over the bottom layer. Double-layer space grids are usually formed from pyramidal units with triangular or square bases resulting in either identical parallel top and bottom grids offset horizontally to each other, or parallel top and bottom grids each with a different configuration interconnected at the node points by inclined web members to form a regular stable structure.

Single-layer grids are primarily subject to flexural moments, whereas the members in double- and triple-layer grids are almost entirely subject to axial tensile or compressive forces. These characteristics of single-, double- and triple-layer grids determine to a very large extent their structural performance. Single-layer grids, developing high flexural stresses are suitable for clear spans up to 15 m while double-layer grids have proved to be economical for clear spans in excess of 100 m. The main types of double-layer grids in common use are shown in Fig. 5.12.

Skeletal space frames curved in one direction forming single- or double-layer barrel vaults also provide elegant structures capable of covering large clear spans. Single-layer vaults are suitable for column-free spans of up to 40 m which may be substantially increased by incorporating selected areas of double-layer structure forming stiffening rings. Double-layer barrel vaults are normally capable of clear spans in excess of 120 m. Figure 5.13 shows the main types of bracing used for single-layer barrel vaults.

Dome structures present a particularly efficient and graceful way of providing cover to large areas. Single-layer steel domes have been constructed from tubular members with spans in excess of 50 m while double-layer dome structures have been constructed with clear spans slightly greater than 200 m. Skeletal dome

168 *Other structural applications of steel*

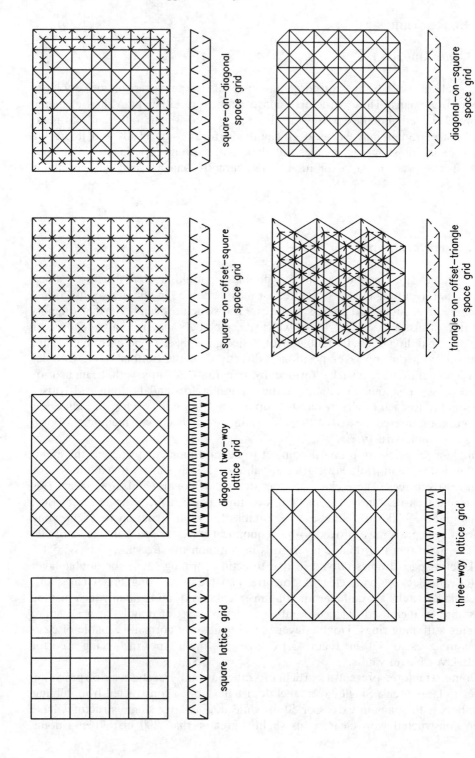

Fig. 5.12 Lattice and space grids

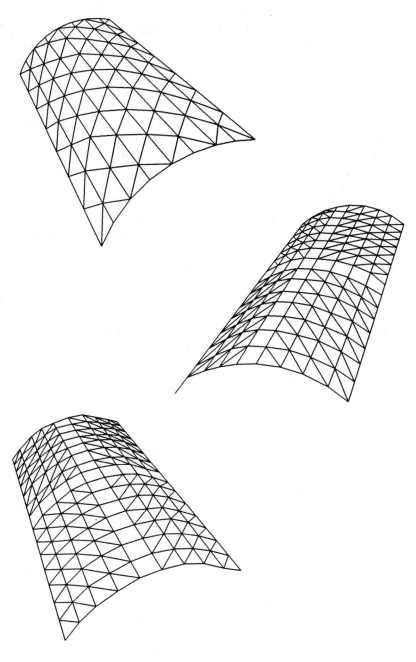

Fig. 5.13 Bracing of single-layer barrel vaults

structures can be classified into several categories depending on the orientation and position of the principal members. The four most popular types usually constructed in steel are ribbed domes, Schwedler domes, three-way grid domes and parallel lamella domes.

Ribbed domes, as the name suggests, are formed from a number of identical rib members which follow the meridian line of the dome and span from the foundations up to the top of the structure. The individual rib members may be of tubular lattice construction and are usually interconnected at the crown of the dome using a small diameter ring beam.

The Schwedler dome is also formed from a series of meridional ribs but unlike the ribbed dome, these members are interconnected along their length by a series of horizontal rings. In order to resist unsymmetric loads the structure is braced by diagonal members positioned on the surface of the dome bisecting each trapezium formed by the meridional ribs and horizontal rings.

Three-way grid domes are formed from three principal sets of members arranged to form a triangular space lattice. This member topology is ideally suited to both single-layer and double-layer domes and numerous beautiful large-span steel three-way domes have been constructed throughout the world.

The steel lamella dome is formed from a number of 'lozenge'-shaped lamella units which are interconnected together to form a diamond or rhombus arrangement. The spectacular Houston Astrodome is an excellent example of this type of construction. This impressive steel double-layer dome was constructed from lamella units 1.52 m deep and has an outside diameter of 217 m with an overall height of 63.4 m. Figure 5.14 shows the four main dome configurations now in prominent use worldwide.

5.2.3 Special features

The inherent characteristics of steel skeletal space frames facilitate their ease of fabrication, transportation and erection on site. There are two main groups into which the majority of space frames may be classified for assembly purposes; the particular structure may be assembled from a number of individual members connected together by purpose-made nodes or alternatively may be constructed by joining together modular units which have been accurately fabricated in a factory before transportation to site.

There are numerous examples of 'chord and joint' space frame systems available for immediate construction. These systems offer full flexibility of member lengths and intersecting angles required in the construction of skeletal dome structures. Many jointing systems are available; Figs 5.15 and 5.16 show a typical spherical node used in the MERO system and a cast steel node used in the NODUS system.

The 'modular' systems are usually based on pyramidal units which are prefabricated from channel, angle, circular hollow section or solid bars. The individual

Space frames 171

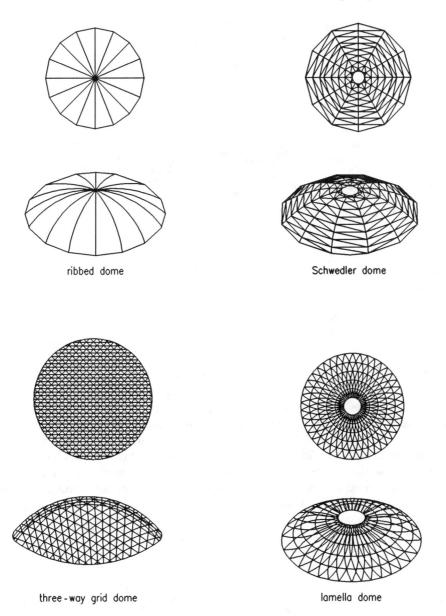

Fig. 5.14 Dome configurations

units are designed to nest together to facilitate storage and transportation by road or sea. Most manufacturers of modular systems hold standard units in stock which greatly enhances the speed of erection. Figure 5.17 shows typical details of the prefabricated steel modular inverted pyramidal units used in the Space Deck System.

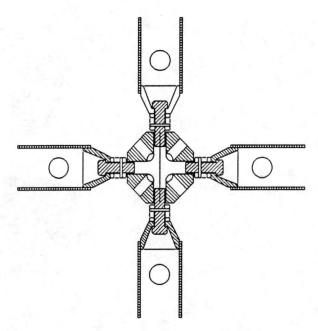

Fig. 5.15 MERO node connector

Steel space frames are generally erected rapidly without the use of falsework. Double-layer grids of substantial span can be constructed entirely at ground level including services and cladding and subsequently lifted or jacked up into the final position. Dome structures can be assembled from the top downwards using a central climbing column or tower. A novel approach adopted for the erection of a dome with a major axis of 110 m and a minor axis of 70 m involved fabrication of the dome on the ground in five sections which were temporarily pinned to each other. The central section was then lifted and the remaining segments of the dome locked into position as shown in Fig. 5.18.

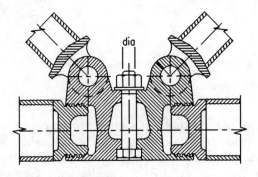

Fig. 5.16 NODUS node connector

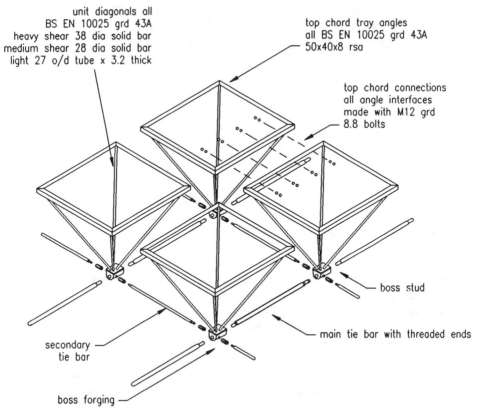

Fig. 5.17 Exploded view of 2.4 m square section of Space Deck

5.2.4 Analysis

The analysis of space frames results in the production of large sets of linear simultaneous equations which must inevitably require the use of a computer for their solution. For these equations to be formulated it is necessary to input into the computer significant amounts of information relating to the topology of the structure and properties of the individual members. This operation can be very time-consuming unless modern pre-processing techniques which allow rapid data generation are adopted.[7]

The members forming double-layer space frames are principally stressed in either axial tension or axial compression. Consequently it is usual in the analysis of double-layer space trusses to assume that the members in the structure are pin-jointed, irrespective of the actual joint rigidity. This assumption may lead to the overestimation of node deflections but because only three degrees of freedom are

174 *Other structural applications of steel*

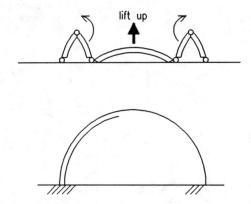

Fig. 5.18 Novel method of dome erection

permitted at each node this approach minimizes computer storage requirements and processing time.

Single-layer grid and barrel vault structures carry the imposed load by flexure of the members so it is important to include in the analysis the flexural, torsional and shear rigidity of the members. Linear analysis only is required for the majority of skeletal space structures but to ensure stability of shallow domes it is essential to undertake a non-linear analysis of these structures. Large-span space structures may benefit from a full collapse analysis where the ultimate load-carrying capacity and collapse behaviour of the structure are determined.[8]

The behaviour of both single-layer and double-layer space trusses is influenced to a great extent by the support positions of the structure. The effects of joint and overall frame rigidity also have a commanding influence and affect the buckling behaviour of the compression members within the structure. Great care must be taken in assessing the effective lengths of compression members and it is unlikely that internal and edge compression members will exhibit similar critical buckling loads.

5.3 Cable structures

5.3.1 Range of applications

5.3.1.1 Introduction

Most structural elements are able to carry bending forces as well as tension and compression, and are hence able to withstand reversals in the direction of loading. Tension elements are unique in that they can only carry tension. In compressive or bending elements, the loading capacity is often reduced by buckling effects while tension elements can work up to the full tensile stress of the material. Consequently full advantage can be taken of high-strength materials to create light, efficient and cost-effective long-span structures.

To create useful spanning or space-enclosing structures the tension elements have to work in conjunction with compression elements. From an architectural point of view, the separation of the tension, compression and bending elements leads to a visual expression of the way the structure carries the loads, or at least one set of loads. Tension structures come in a wide range of forms which can be broadly categorized as follows:

(1) two-dimensional   − suspension bridges
                                    − draped cables
                                    − cable-stayed beams
                                    − cable trusses
(2) three-dimensional − cable truss systems
(3) surface-stressed   − pneumatically-stressed
                                    − prestressed.

5.3.1.2 Structural forms

Suspension bridges (draped cable)

A suspension bridge (Fig. 5.19) is essentially a catenary cable prestressed by dead weight only. Early suspension bridges with flexible decks suffered from large deflections and sometimes from unstable oscillation under wind. A system of inclined hangers proposed originally to reduce deflection under live load has been employed recently also to counteract wind effects. The suspension cable is taken over support towers to ground anchors. The stiffened deck is supported primarily by the vertical or inclined hangers. The system is ideally suited to resisting uniform downward loads. The principle has been used for buildings, mostly as draped cable structures.

Cable-stayed beams

Cables assist the deck beam by supporting its self-weight. Compression is taken in the deck beam so that ground anchors are not required (Fig. 5.20). The cable-stayed principle has recently been developed for single-storey buildings (see section 5.1 of this chapter). The cable system is designed for and primarily resists gravity loads; in buildings with lightweight roof construction the uplift forces, which are of similar magnitude, are resisted by bending of the stiffening girders. The system is suitable for spans of 30−90 m and has recently been widely used for industrial and sports buildings.

176   Other structural applications of steel

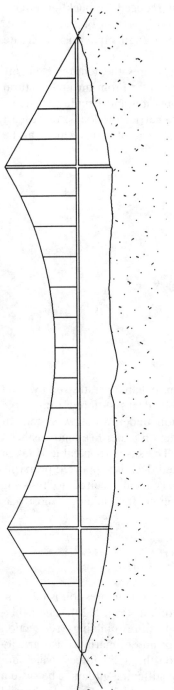

Fig. 5.19 Suspension bridge

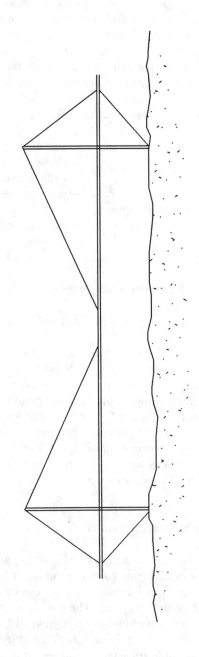

Fig. 5.20 Cable-stayed beam

Cable truss

The hanging cable resists downloads and the hogging cable resists upload (Fig. 5.21). If diagonal bracing is used, non-uniform load can be resisted without large deflections but with larger fluctuation of force in the cables.

Three-dimensional cable truss

The classic form of this structure is the bicycle wheel roof, in which a circular ring beam is braced against buckling by a radial cable system. These radial cables are divided into an upper and lower set, providing support to the central hub (Fig. 5.22). The system is suitable for spans of 20–60 m diameter. This system has recently been developed (by David Geiger) into a cable dome, having two or three rings of masts (Fig. 5.23). The radial forces at the bases of the masts are resisted by circumferential cables. The masts are also cross-cabled circumferentially to maintain their stability. These structures can span up to 200 m.

Surface-stressed structures

A cable network can be arranged to have a doubly-curved surface either by giving it a boundary geometry which is out-of-plane or by inflating it with air pressure. The cable net must be prestressed either by tensioning the cables to the boundary points or by the inflation pressure. The effect of the double curvature and the prestress is to stiffen the structure to prevent undue deflection and oscillation under loads. Cable net structures can create dramatic wide-span roofs very economically. They can be clad with fabric, transparent foil, metal decking or timber boarding, insulation and tiles.

Low-profile air-supported roofs can provide the most economical structure for covering very large areas (10–50 acres). In designing these structures the aerodynamic profile must be taken into consideration, as must snow loading and the methods of installation and maintenance.

5.3.2 Special features

5.3.2.1 Elementary cable mathematics

Load/extension relationship

$$\text{Extension} \quad e = \frac{TL}{AE}$$

where T = load in cable
L = length of cable
A = cross-sectional area
E = Young's modulus.

178 *Other structural applications of steel*

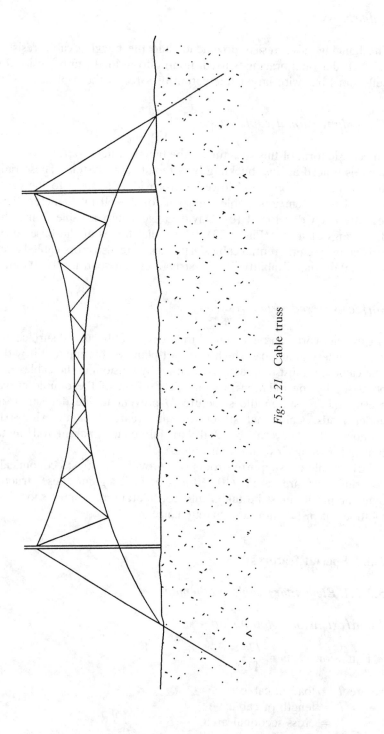

Fig. 5.21 Cable truss

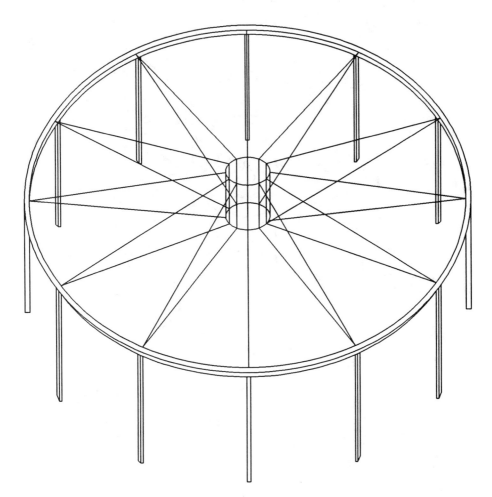

Fig. 5.22 Bicycle wheel roof

Typical values for materials are given in Table 5.1.

Table 5.1 Material properties

Material	E (kN/mm^2)	Ultimate tensile strength (N/mm^2)
Solid steel	210.0	400–2000
Strand	150.0	2000
Wire rope	112.0	2000
Polyester fibres	7.5	910
Aramid fibres	112.0	2800

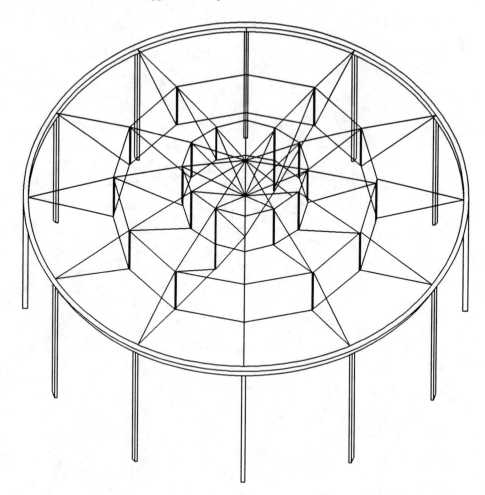

Fig. 5.23 Cable dome

The *E* value for wire rope applies after the construction stretch has been pulled out of wire rope by load cycling to 50% of the ultimate tensile strength. In wire rope the construction stretch can be as much as 0.5%. This is of the same order of magnitude as the elastic stretch in the cable at maximum working load.

Circular arc loaded radially (Fig. 5.24)

Tension $T = PR$

where P = load/unit length radial to cable
      R = radius of cable.

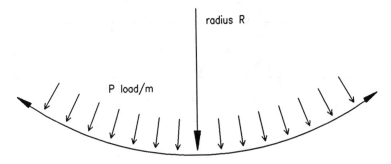

Fig. 5.24 Circular arc loaded radially

Radius of circular arc, R (Fig. 5.25)

Radius $\quad R = \dfrac{S^2}{8d} + \dfrac{d}{2}$

where S = span
$\quad\quad d$ = dip.

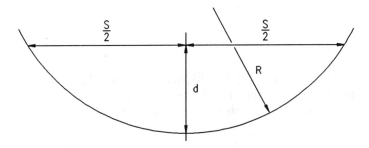

Fig. 5.25 Radius of circular arc

Catenary loaded vertically (Fig. 5.26)

Horizontal force $\quad H = \dfrac{WS}{8d}$

Vertical force $\quad V = \dfrac{WS}{2}$

Maximum tension $\quad T = (H^2 + V^2)^{\frac{1}{2}}$

where S = span
$\quad\quad d$ = dip
$\quad\quad W$ = vertical load/unit length.

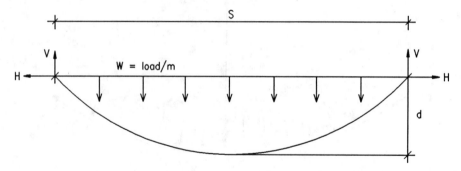

Fig. 5.26 Catenary loaded vertically

These formulae permit initial estimates of forces in cables to be made. For full and accurate analysis it is necessary to use a non-linear computer analysis which takes into account the change of curvature caused by stretch. For well curved cables the hand analysis is accurate enough and gives a useful guide to the forces involved and hence the sizes of cables and fittings.

Prestressed cable

The straight cable (or flat fabric) is a special problem. To be straight, the cable must have an initial or prestress tension and theoretically zero weight. In order to carry load the cable must stretch and sag to a radius R.
For

span	$= S$
load/unit length	$= W$
stiffness	$= EA$
pre-tension	$= T_0$
tension under load	$= T$

Equilibrium equation $\quad T = RW$ (5.1)

New length $\quad L = 2R \sin^{-1}(S/2R)$ (5.2)

Strain $\quad = (L - S)/S$ (5.3)

Tension $\quad T = T_0 + EA(L - S)/S$ (5.4)

Eliminating R between Equations (5.1) and (5.2), substituting for L in Equation (5.4) and rearranging gives:

$$\frac{T - T_0}{EA} = \frac{2T}{SW} \sin^{-1} \frac{SW}{2T} - 1 \quad\quad (5.5)$$

This equation can be solved iteratively for T. The deflection can be found from the earlier formulae for T and R for a circular arc. It should be noted that:

(1) the extension of a tie with low initial tension is considerably more than TL/EA.
(2) straight cables can be used for load carrying, but either the tension will be very high or there will be large deflections.

Two-way cable net (Fig. 5.27)

Approximate calculations can be carried out by hand in a similar way to the single cable calculations above. The basic equilibrium equation for two opposing cables is:

$$\frac{t_1}{R_1} + \frac{t_2}{R_2} = P$$

where P = load/surface area and t_1 and t_2 are the tensions/unit width, i.e.

$$t_1 = \frac{T_1}{a_1} \qquad t_2 = \frac{T_2}{a_2}$$

where a_1 and a_2 are the cable spacings and T_1 and T_2 are the cable tensions.

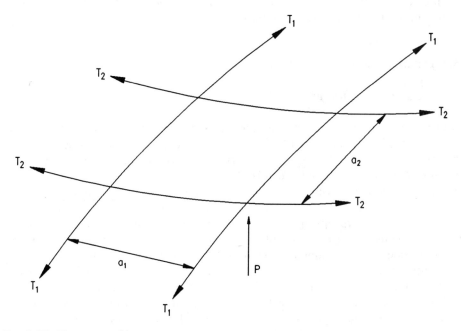

Fig. 5.27 Two-way cable net

The loads on ridge cables and boundaries can be estimated from resolution of the cable forces if the geometry is known.

The full analysis of cable net structures is a specialized and complicated process which requires specially written computer programs. The procedure involves the following stages:

(1) Formfinding: In this stage the cables are treated as constant tension elements and the geometry is allowed to move into its equilibrium position. On completion of the formfinding stage the cable net geometry under the prestress forces will be defined.
(2) Load analysis: The prestress model is converted into an elastic model. To do this the slack lengths, l, of all elements must be set so that

$$l_0 = l - \frac{T}{EA}$$

The model is then analyzed for the defined dead, wind and snow loads.
(3) Cutting pattern definitions: Cable net structures are fully prefabricated with exact cable lengths so that the prestressed form can be realized. In this stage the form model will be refined so that the mesh lengths are exactly equal, etc. The offsets for the boundary clamps must be allowed for. The stressed cable lengths can then be defined.

During prefabrication of the cables they must be prestretched to eliminate construction stretch; they must then be marked at the specified tensions. Tolerances in fabrication are of the order of 0.02%.

5.3.3 Detailing and construction

5.3.3.1 Cables and fittings

Wire rope cables are spun from high tensile wire. For structural work the cables are multi-strand, typically 6 × 19 or 6 × 37 with independent wire rope core and galvanized to Class A. For increased corrosion-resistance, the largest diameter wire should be used and cables can be filled with zinc powder in a slow-setting polyurethane varnish during the spinning process. For even greater corrosion-resistance, filled strand or locked-coil strand can be used to which a shrunk-on polyurethane or polypropylene sleeve can be fitted. Stainless steel, although apparently highly corrosion-resistant, is affected by some aggressive atmospheres if air is excluded; the resulting corrosion can be more severe than with mild steel.

The simplest and cheapest type of termination is a swaged Talurit Eye made round a thimble (Fig. 5.28(a)) and connected into a clevis type connection or on to the pin of a shackle. The neatest and most streamlined fitting is a swaged eye or jaw end termination (Fig. 5.28(b)). Hot-poured zinc terminations have to be used for very heavy cables of greater diameter than 50 mm (Fig. 5.28(c)). Epoxy

Cable structures 185

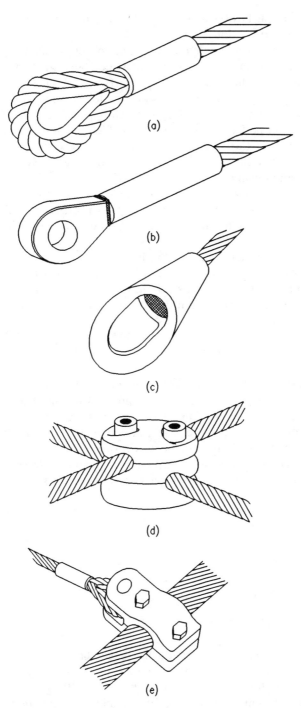

Fig. 5.28 Cable fittings: (a) swaged eye, (b) swaged terminator, (c) hot-poured white metal eye, (d) cross clamp, (e) net and boundary connection

resin with steel balls can be used as a filler in place of zinc, offering an improvement in fatigue life at the termination.

On-site connections can be made with bulldog clips but they are ugly and damage the rope. For cable net construction the standard detail is a three-part forged steel clamp, of which the two outer parts are identical (see Fig. 5.28(d)). Forging is expensive for small numbers and so for smaller structures machined aluminium components may be preferred. Double cables can have a swaged aluminium extrusion prefixed to each pair of cables which can then be connected with a single bolt. For the attachment of net cables to edge cables, forged steel clamps are generally used (Fig. 5.28(e)). Lower cost alternatives are bent plate or machined aluminium clamps.

Cable life is reduced by corrosion and fatigue. Galvanized cables under cover suffer very little corrosion; external cables properly protected should have a life of 50 years. Plastic sheathing has the great disadvantage of making inspection of the cable impossible.

Fatigue investigations have shown that it is wise to limit the maximum tension in a cable to 40% of its ultimate strength for long-life structures. For structures with a design life of up to ten years a limit of 50% is acceptable. Flexing of the cables at clamps or end termination will cause rapid fatigue damage.

5.3.3.2 Rods as tension members

Steel rods are often used as tension members in external situations since they are stiffer and can be given better protection against corrosion. The rods are usually threaded at the ends and screwed into special end fittings. In the case of long rods which can be vibrated by the wind, the end connection must be free to move in two directions, otherwise fatigue damage will occur.

Since tension members are usually critical components of the structural system, consideration should be given to using rods or cables in pairs to provide additional safety.

5.4 Steel in housing

5.4.1 Range of application

Although lightweight steel construction has never had a large share of the UK housing market it has twice in this century expanded considerably to meet demand arising from housing and material shortages at the ends of the two World Wars. Steel houses built just after 1919 are still in occupation and post-1945 prefabs exist in some numbers, although originally intended to be only for short-term use. Nevertheless, for a variety of reasons, interest in steel-framed housing has waned after the initial enthusiasm of the fruitful periods.

Steel frames have many advantages over other kinds of construction, some of which are:

(1) They lead to rapid enclosure of the building. Frames can be erected in one day and it is feasible to get the roof waterproofed the following day. Trades can therefore begin interior work within a day or two of the start of building. In contrast the roof of a masonry house cannot be put up until the bricks and blocks have been laid layer by layer up to eaves level.
(2) Because of (1) the total house-build time can be reduced by several weeks which means that the builder gets a quicker return on his capital.
(3) They are dimensionally stable and are manufactured to tight geometric tolerances.

The obvious competitor in the framed house market is the timber-framed house which shares the same advantage of rapid construction. However timber is very much affected by dimensional changes caused by moisture, which have to be considered in detailing. In contrast steel does not shrink or creep; it is therefore possible to design the fabric more simply as allowance does not have to be made for dimensional change, other than for completely predictable deflections within the elastic limit. The lengthy maintenance period during which defects caused by movement and shrinkage are put right, inevitable in traditional housing, is not necessary for a steel-framed house.

5.4.2 Current systems

Most houses built today are for sale; the purchase is generally funded by mortgage. With few exceptions the mortgage making agencies (building societies, banks) will not fund new construction unless it has been approved by the National House Building Council (NHBC). The NHBC has not published design and construction codes for steel-framed houses. Any steel system has therefore to be assessed and approved by the British Board of Agrément (BBA).

The cost of approval is quite large, which acts as a deterrent to poorly thought out systems, but does to some extent restrict the number of entrants to the market. The quality control and restricted number are beneficial as they guarantee that future steel houses will not exhibit the weaknesses of some earlier systems. The major faults were corrosion of the frame and internal condensation, the latter often the cause of the former. Proper protective coating of the steel and adequate thermal insulation external to the frame are now seen to be essential.

Figure 5.29 shows a modern steel-framed house under construction and Fig. 5.30 shows the attractive overall appearance that can be achieved.

188 *Other structural applications of steel*

Fig. 5.29 Frame during construction

Fig. 5.30 Completed house

5.4.3 Other uses of steel in housing

Development programmes are in hand for the following:

(1) soffits, fascias and bargeboards
(2) suspended ground floors
(3) door frames and doors
(4) window sills and surrounds
(5) window frames
(6) infill panels
(7) insulation systems
(8) roofing systems.

The steel-framed housing systems of the past and present are, in many ways, substitutes for timber frames and so do not exploit the properties of steel to the full. Studies are in hand to examine the possibilities of a panel-based building system.

5.5 Atria

5.5.1 Introduction

5.5.1.1 General

Structural engineers, architects, environmental and fire safety engineers all have a significant contribution to make when planning the concept and subsequent detail of intermediate spaces such as atria, arcades, malls and covered courtyards. It is important that all disciplines take a wide view and understand the total problem to ensure a balanced design consistent with all of the technical criteria, without compromising the fundamental requirements of providing an agreeable, comfortable and safe environment.

A number of criteria affect the design:

(1) The enclosure to be readily maintained
(2) Environmental design to minimize or eliminate major mechanical plant and hence servicing
(3) The safety of potentially large numbers of the public
(4) Fire safety
(5) Erection of structure and cladding
(6) Internal and external cleaning.

It may not be essential to design for minimum cost since the result must reflect the visual requirements of the enclosure elegantly, economically and in a form which

190 *Other structural applications of steel*

is readily constructed. Nevertheless it is important to understand the philosophy behind the funding of the enclosure in developing designs which can be radically different if the improvement of the surrounding accommodation is the aim rather than the provision of additional accommodation. If for example an atrium deck is simply a space with no direct revenue-earning capacity, its cost must be offset by the enhancement of the value of the surrounding accommodation.

5.5.1.2 Structural aspects

The structures in Figs 5.31 to 5.35 are very simple in concept, but effectively illustrate a series of solutions in which there is a generally decreasing reliance on the perimeter structure.

As a group the arch, the beam and the dome are potentially the most economic solutions, but rely totally on the support provided by the perimeter structure. This is simple for a new development, which can be designed relatively easily to support the resultant vertical and horizontal loads, but for existing buildings it may not be so straightforward, since strengthening works or evidence that increased loads can be supported will be required. In the case of an existing building, unless considerable party wall structures are available for support, it may be easier to set the roof at a lower level where the percentage increase in vertical load will be

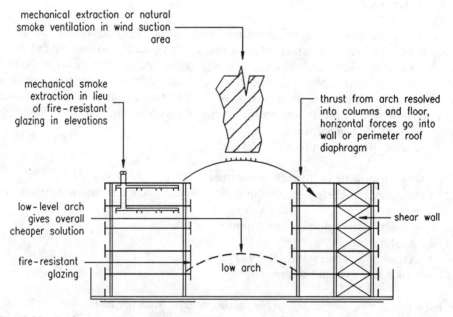

Fig. 5.31 Arch

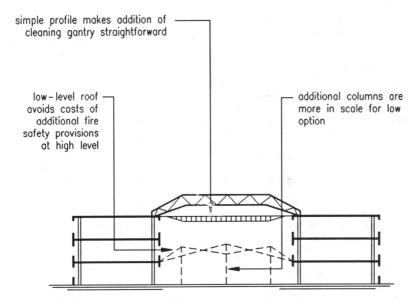

Fig. 5.32 Beam or truss

much smaller or where horizontal thrusts from an arch will have a smaller resultant overturning effect.

As our town centres and streets are redeveloped to compete with out-of-town shopping the need for enclosure is a common solution often considered. In these cases, it is often not practical to gain support from a wide variety of constructions

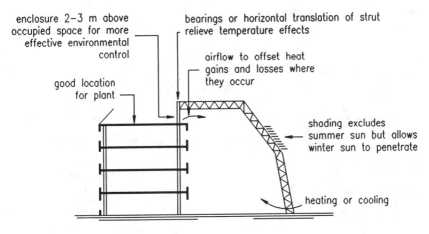

Fig. 5.33 Half portal

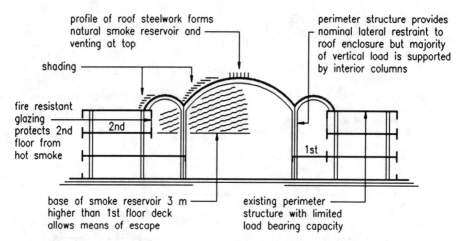

Fig. 5.34 Beam and column

of various heights and condition and therefore the solutions defined in Fig. 5.34 are more appropriate. The perimeter structure is required to support a small proportion of the vertical load and relatively small horizontal loads from wind and stabilizing forces for the new columns, which does not generally pose a problem

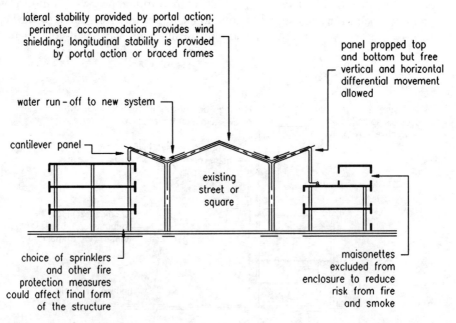

Fig. 5.35 Portal frame

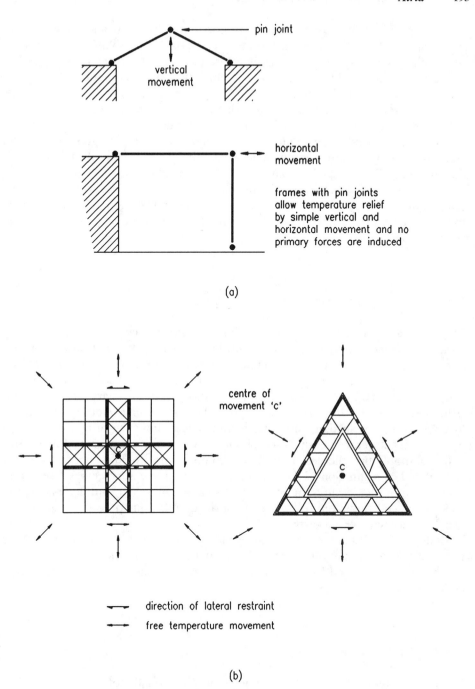

Fig. 5.36 Movements caused by temperature changes

since there is usually some redundant horizontal capacity due to the wind-shielding effect of the new enclosure. The portal style solution Fig. 5.35 is substantially self-supporting, and is therefore likely to be expensive compared to the arch or beam solution but may be the only satisfactory option when an existing street is to be covered.

The provision of restraint for stability or to resist wind loads needs careful study since this restraint can be incompatible with the need to allow movement of the steelwork caused by large temperature differentials. Temperatures in the steel could be 30°C higher in black painted steel compared to white if unshaded and directly exposed to the sun's radiation and also depend on the extent of ventilation provisions. One approach is to allow the movement to take place without inducing temperature stress (Fig. 5.36(a)). Alternatively, lateral restraints can be arranged to allow the roof to expand freely about centre of movement yet provide adequate support (Fig. 5.36(b)).

5.5.1.3 Form of members and connections

The shape and connections of the individual structural members which form the basic framework can be a significant contributory feature to an enclosure and are also an important aspect of the architectural treatment. This framework will be required to perform a range of functions over and above the need to span and carry the primary loads. Ideally, the modulation of the primary and secondary structure should be organized to support glazing/cleaning gantries, walkways, lighting, as well as shading and other equipment, without the provision of layers of additional structure which would obscure a well-designed primary framework.

It is desirable that the design of these 'clip-on' items should be under the direct control and co-ordination of the overall fabrication drawing process is essential to some extent if the design of individual items passes on to sub-contractors. Careful control and co-ordination to the overall fabrication drawing process is essential to avoid unnecessary visual distractions.

Methods of jointing steelwork to meet the structural, architectural and aesthetic requirements are numerous. Their development requires a degree of creativity, which can be improved by a study of the actual projects, the technical journals and the architectural press.

5.5.1.4 Paint systems

The paint system is an important factor in the visual quality of the structure. The choice of a paint system is always difficult due to a number of conflicting requirements. Paint technology provides an ever-increasing number of options. However, the choice can be guided by consideration of the following criteria:

(1) the steelwork fabricator who delivers and erects the steel will inevitably damage any paint applied at the works no matter how carefully the protection measures are prepared
(2) a contractor who has a very tight programme may like to see primer, undercoat and finish coat on the steelwork applied at the works since it may be very difficult to apply the finish on high-level steelwork without extensive temporary works
(3) very hard paint finishes are more difficult to damage although when the inevitable damage occurs they are usually more difficult to repair to a good standard
(4) if the finish-coat is applied on site, some parts of the steelwork may be difficult to reach and the standard of finish will not be as good as that achieved in the paint shop
(5) unless touching up to damaged areas is very minor, the result is unlikely to be satisfactory and therefore a full site applied decorative finish may prove necessary
(6) the choice of system will also depend on the fabricator's paint shop and its relationship with the paint supplier.

5.5.2 Special features

5.5.2.1 Fire engineering

The addition of a roof to enclose a number of levels linked by a central void through which fire and smoke can spread is alien to traditional fire safety regulations and therefore each case requires careful and individual attention if a safe solution is to be achieved. The Building Regulations are not directly applicable, there are no current British Standards and therefore reliance is placed on research papers and the practical experience of the designers and controlling officers to determine a safe solution.

The considerations for fire safety can influence the structure in two ways:

(1) the effect of fire on the structure itself must be considered by reviewing the risk and extent of failure and the resultant consequences of that failure
(2) the fire engineering requirements on the surrounding accommodation can influence the level and form of an enclosure or alternatively the shape can be affected if there is a requirement for smoke control.

It takes considerable build-up of heat or a weak local condition to cause failure of an unprotected atrium roof structure during a fire and therefore the risk of failure is low. However, even though the Building Regulations do not require a roof to have fire-resistance, it is sensible to review each case to limit the risks and consequences of failure. No definitive standard has been set for the required performance of such a roof although a supporting strategy for the provision of an unprotected roof structure can be prepared from the following set of recommendations.

(1) For small enclosed areas everyone within the space is likely to be fully aware of a fire and be able to escape accordingly. The complexity of the planning and the means of escape influence this decision. However, 2000 m^2 is suggested as a preliminary maximum for the application of this principle, although variations could be expected depending on the complexity of the space.
(2) Smoke control schemes and the failure of the glazing can prevent a general build-up of heat and inhibit global failure of the structure.
(3) Direct flame impingement on the structure from a fire on the atrium deck or high-level perimeter accommodation can be limited by the provision of sprinklers which reduce the risk of spread of fire. Alternatively fire-resistant construction or glazing can be used to isolate the steel from the fire.
(4) Provide a structure for which local failure will not cause collapse of the whole.
(5) If it is felt necessary, due to exceptionally large fire loads, and very large numbers of people at risk, it is possible to assess the fire-resistance of unprotected steelwork by analytical techniques.

The fire engineering design can also influence the height and shape of the roof enclosure when the interaction of fire protection measures for the enclosure and surrounding accommodation are considered. For every level attached to the enclosure there is a direct cost which can be directly attributed to fire safety provisions such as ductwork, fans, louvres, fire-resistant glazing, detection and control systems. It would be cheaper and safer to exclude residential accommodation from the enclosure whereas it may be desirable to enclose multi-level shopping facilities thus providing a high profile to form a smoke reservoir for venting to allow safe means of escape.

5.5.2.2 The influence of environmental engineering

The provision of intermediate space as an extra attraction with earning power requires that the building physics be considered and an appropriate environment achieved. At the simplest level the enclosure may only provide protection from the rain or at the other extreme may be fully conditioned to performance criteria similar to those of the surrounding accommodation. However, it is in the intermediate environment, somewhere between indoors and outdoors, commonly called a buffer zone, that the total engineering technology starts to interface with the fire and structural design of the enclosure.

The height and form of the roof, and the arrangement of shading, will influence the solar heat gains and the cooling effect of the natural ventilation and consequently the risk of condensation affecting the atrium floor. Internal shading or walkway shading systems are easier to maintain but are not as effective as the more expensive external shading systems.

The natural light and the acoustic performance also contribute to a stimulating and fresh environment. The complexity of the permutations of how for example natural ventilation can be influenced by the smoke ventilation, which in turn affects the structural form, ensures that every design is different. Similarly, the thermal physics design may indicate that structural fabrics which generally allow a larger structural modulation (e.g. PVC-coated polyester, PTFE-coated fibreglass, ET foil) meet the performance specification most suitably, although care must be taken to ensure that the fire performance of these fabrics is satisfactory.

References to Chapter 5

1. British Standards Institution (1972) *Loading*. Chapter V: Part 2: *Wind loads*. CP3, BSI, London.
2. British Standards Institution (1986) *Lattice towers and masts*. Part 1: *Code of practice for loading*. BS 8100, BSI, London.
3. Construction Industry Research and Information Association (CIRIA) (1980) *Wind Engineering in the Eighties*. CIRIA Conference Report 12/13 Nov.
4. Engineering Sciences Data Unit. Wind engineering sub-series (4 volumes). ESDU International, London.
5. Vickery B.J. & Basu R.I. (1983) Across wind vibrations of structures of circular cross section. *Journal of Wind Engineering and Industrial Aerodynamics*, **12**.
6. International Association for Shell and Spatial Structures (IASS) (1981) *IASS Recommendations for Guyed Masts*. IASS, Madrid.
7. Nooshin H. & Disney P. (1989) Elements of Formian. *Proceedings of 4th Intl Conf. Civ. and Struct. Engng Computing* (Ed. by H. Nooshin), pp. 528–32. University of Surrey, Guildford, UK.
8. Parke G.A.R. (1990) Collapse analysis and design of double-layer grids. In *Studies in Space Structures* (Ed. by H. Nooshin). Multi-Science Publishers.

Further reading for Chapter 5

Section 5.2

Bell A.J. & Ho T.Y. (1984) NODUS spaceframe roof construction in Hong Kong. *Proceedings 3rd International Conference on Space Structures* (Ed. by H. Nooshin), pp. 1010–15. University of Surrey, Guildford, UK.

Bunni U.K., Disney P. & Makowski Z.S. (1980) *Multi-Layer Space Frames*. Constrado, London.

Makowski Z.S. (1984) *Analysis, Design and Construction of Braced Domes*. Granada, St Albans.
Makowski Z.S. (1985) *Analysis, Design and Construction of Braced Barrel Vaults*. Elsevier Applied Science Publishers, Barking, Essex.
Parke G.A.R. & Walker H.B. (1984) A limit state design of double-layer grids. *Proceedings 3rd International Conference on Space Structures* (Ed. by H. Nooshin), pp. 528–32. University of Surrey, Guildford, UK.
Supple W.J. & Collins I. (1981) Limit state analysis of double-layer grids. *Analysis, Design and Construction of Double-Layer Grids* (Ed. by Z.S. Makowski). pp. 93–117. Applied Science Publishers, Barking Essex.

Section 5.3

Liddell W.I. (1988) Structural fabric and foils. *Kerensky Memorial Conference – Tension Structures*, June, Institution of Structural Engineers.
Troitsky M.S. (1977) *Cable Stayed Bridges Theory and Design*. Crosby Lockwood Staples, London.

For surface stressed structures refer to publications of the Institut für Leicht Flächentragwerke, Universität Stuttgart, Pfaffenwaldring 14 7000, Stuttgart 80
 IL 5   Convertible roofs
 IL 8   Nets in nature and technics
 IL 15   Air hall handbook

Section 5.5

Baker N. (1983) Atria and conservatories, Parts 2 and 3. *Architects' Journal*, 11/18/25 May.
Bednar M.J. (1986) *The New Atrium*. McGraw-Hill.
DeCicco P.R. (1983) *Life Safety Considerations in Atrium Buildings*. Fire Prevention 164.
Dickson M.G.T. & Green M.G. (1986) *Providing Intermediate Space*. National Structural Steel Conference. BCSA.
Hawkes D. (1983) Atria and conservatories, Part 1. *Architects' Journal*, 11/18/25 May.
Land District Surveyors Association (1989) *Fire Safety in Atrium Buildings*.
Law M. & O'Brien T. (1989) *Fire Safety of Bare External Structural Steel*. Steel Construction Institute, Ascot, Berks.
Lloyds Chambers (1983) *Framed in Steel*, No. 11, Nov., BSC.

Chapter 6
Applied metallurgy of steel

by MICHAEL BURDEKIN

6.1 Introduction

The versatility of steel for structural applications rests on the fact that it can be readily supplied at a relatively cheap price in a wide range of different product forms, and with a useful range of material properties. The key to understanding the versatility of steel lies in its basic metallurgical behaviour. Steel is an efficient material for structural purposes because of its good strength-to-weight ratio. A diagram of strength-to-weight ratio against cost per unit weight for various structural materials is shown in Fig. 6.1. Steel can be supplied with strength levels from about 250 N/mm^2 up to about 2000 N/mm^2 for common structural applications,

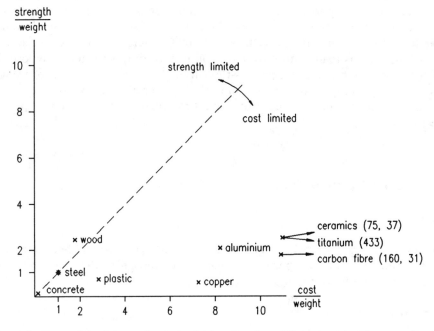

Fig. 6.1 Strength/weight and cost/weight ratios for different materials normalized to steel (1,1)

although the strength requirements may limit the product form. The material is normally ductile with good fracture toughness for most practical applications. Product forms range from thin sheet material, through optimized structural sections and plates, to heavy forgings and castings of intricate shape. Although steel can be made to a wide range of strengths it generally behaves as an elastic material with a high (and relatively constant) value of the elastic modulus up to the yield or proof strength. It also usually has a high capacity for accepting plastic deformation beyond the yield strength, which is valuable for drawing and forming of different products, as well as for general ductility in structural applications.

Steel derives its mechanical properties from a combination of *chemical composition*, *heat treatment* and *manufacturing processes*. While the major constituent of steel is always iron the addition of very small quantities of other elements can have a marked effect upon the type and properties of steel. These elements also produce a different response when the material is subjected to heat treatments involving cooling at a prescribed rate from a particular peak temperature. The manufacturing process may involve combinations of heat treatment and mechanical working which are of critical importance in understanding the subsequent performance of steels and what can be done satisfactorily with the material after the basic manufacturing process.

Although steel is such an attractive material for many different applications, two particular problems which must be given careful attention are those of corrosion behaviour and fire resistance, which are dealt with in detail in Chapters 35 and 34 respectively. Corrosion performance can be significantly changed by choice of a steel of appropriate chemical composition and heat treatment, as well as by corrosion protection measures. Although normal structural steels retain their strength at temperatures up to about 300°C, there is a progressive loss of strength above this temperature so that in an intense fire, bare steel may lose the major part of its structural strength. Although the hot strength and creep strength of steels at high temperature can be improved by special chemical formulation, it is usually cheaper to provide fire protection for normal structural steels by protective cladding.

6.2 Chemical composition

6.2.1 General

The key to understanding the effects of chemical composition and heat treatment on the metallurgy and properties of steels is to recognize that the properties depend upon the following factors:

(1) microstructure
(2) grain size

(3) non-metallic inclusions
(4) precipitates within grains or at grain boundaries
(5) the presence of absorbed or dissolved gases.

Steel is basically iron with the addition of small amounts of carbon up to a maximum of 1.67% by weight, and other elements added to provide particular mechanical properties. Above 1.67% carbon the material generally takes the form of cast iron. As the carbon level is increased, the effect is to raise the strength level, but reduce the ductility and make the material more sensitive to heat treatment. The cheapest and simplest form is therefore a plain carbon steel commonly supplied for the steel reinforcement in reinforced concrete structures, for wire ropes, for some general engineering applications in the form of bars or rods, and for some sheet/strip applications. However, plain carbon steels at medium to high carbon levels give rise to problems where subsequent fabrication/manufacturing takes place, particularly where welding is involved, and more versatility can be obtained by keeping carbon to a relatively low level and adding other elements in small amounts. When combined with appropriate heat treatments, addition of these other elements produces higher strength while retaining good ductility, fracture toughness, and weldability, or the development of improved hot strength, or improved corrosion-resistance. The retention of good fracture toughness with increased strength is particularly important for thick sections, and for service applications at low temperatures where brittle fracture may be a problem. Hot strength is important for service applications at high temperatures such as pressure vessels and piping in the power generation and chemical process plant industries. Corrosion-resistance is important for any structures exposed to the environment, particularly for structures immersed in sea water. Weathering grades of steel are designed to develop a tight adherent oxide layer which slows down and stifles continuing corrosion under normal atmospheric exposure of alternate wet and dry conditions. Stainless steels are designed to have a protective oxide surface layer which re-forms if any damage takes place to the surface, and these steels are therefore designed not to corrode under oxidizing conditions. Stainless steels find particular application in the chemical industry.

6.2.2 Added elements

The addition of small amounts of carbon to iron increases the strength and the sensitivity to heat treatment (or hardenability, see later). Other elements which also affect strength and hardenability, although to a much lesser extent than carbon, are *manganese, chromium, molybdenum, nickel* and *copper*. Their effect is principally on the microstructure of the steel, enabling the required strength to be obtained for given heat treatment/manufacturing conditions, while keeping the carbon level very low. Refinement of the grain structure of steels leads to an

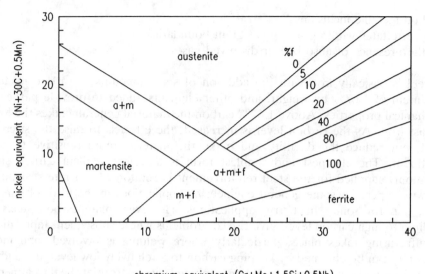

Fig. 6.2 Constitution (Schaeffler) diagram for stainless steels

increase in yield strength and improved fracture toughness and ductility at the same time, and this is therefore an important route for obtaining enhanced properties in steels. Although heat treatment and in particular cooling rate are key factors in obtaining grain refinement, the presence of one or more elements which promote grain refinement by aiding the nucleation of new grains during cooling is also extremely beneficial. Elements which promote grain refinement, and which may be added in small quantities up to about 0.050%, are *niobium*, *vanadium* and *aluminium*.

The major elements which may be added for hot strength and also for corrosion-resistance are chromium, nickel and molybdenum. Chromium is particularly beneficial in promoting corrosion-resistance as it forms a chromium oxide surface layer on the steel which is the basis of stainless steel corrosion protection in oxidizing environments. When chromium and nickel are added in substantial quantities with chromium levels in the range 12% to 25%, and nickel content up to 20%, different types of stainless steel can be made. As with the basic effect of carbon in the iron matrix, certain other elements can have a similar effect to chromium or nickel but on a lesser scale. From the point of view of the effects of chemical composition, the type of stainless steel formed by different combinations of chromium and nickel can be shown on the Schaeffler diagram in Fig. 6.2. The three basic alternative types of stainless steel are ferritic, austenitic and martensitic stainless steels, which have different inherent lattice crystal structures and micro-structures, and hence may show significantly different performance characteristics.

6.2.3 Non-metallic inclusions

The presence of non-metallic inclusions has to be carefully controlled for particular applications. Such inclusions arise as a residue from the ore in the steelmaking process, and special steps have to be taken to reduce them to the required level. The commonest impurities are *sulphur* and *phosphorus*, high levels of which lead to reduced resistance to ductile fracture and the possibility of cracking problems in welded joints. For weldable steels the sulphur and phosphorus levels must be kept less than 0.050%, and with modern steel-making practice should now preferably be less than 0.010%. They are not always harmful however and in cases where welding or fracture toughness are not important, deliberate additions of sulphur may be made up to about 0.15% to promote free machining qualities of steel, and small additions of phosphorus may be added to non-weldable weathering grade steels. Other elements which may occur as impurities and may sometimes have serious detrimental effects in steels are *tin*, *antimony* and *arsenic* which in certain steels may promote a problem known as temper embrittlement, in which the elements migrate to grain boundaries if the steel is held in a temperature range between about 500°C and 600°C for any length of time. At normal temperature steels in this condition can have very poor fracture toughness, with failure occurring by inter-granular fracture. It is particularly important to ensure that this group of *tramp elements* is eliminated from low alloy steels.

Steels with a high level of dissolved gases, particularly oxygen and nitrogen, can behave in a brittle manner. The level of dissolved gases can be controlled by addition of small amounts of elements with a particular affinity for them so that the element combines with the gas and either floats out in the liquid steel at high temperature or remains as a distribution of solid non-metallic inclusions. A steel with no such additions to control oxygen level is known as a *rimming steel*, but for most structural applications the elements *silicon* and/or *aluminium* are added as deoxidants. Aluminium also helps in controlling the free nitrogen level which it is important to keep to low levels in cases where the phenomenon of strain ageing embrittlement may be important.

6.3 Heat treatment

6.3.1 Effect on microstructure and grain size

During the manufacture of steel the required chemical composition is achieved while it is in the liquid state at high temperature. As the steel cools, it solidifies at the melting temperature at about 1350°C, but substantial changes in structure take place during subsequent cooling and may also be affected by further heat treatments. If the steel is cooled slowly, it is able to take up the equilibrium type of lattice crystal structure and microstructure appropriate to the temperature and chemical composition.

These conditions can be summarized on a phase or equilibrium diagram for the particular composition; the equilibrium diagram for the iron−iron carbide system is shown in Fig. 6.3. Essentially this is a diagram of temperature against percentage of carbon by weight in the iron matrix. At 6.67% carbon, an inter-metallic compound called *cementite* is formed which is an extremely hard and brittle material. At the left-hand end of the diagram, with very low carbon contents, the equilibrium structure at room temperature is *ferrite*. At carbon contents between these limits the equilibrium structure is a mixture of ferrite and cementite in proportion depending on the carbon level. On cooling from the melting temperature, at low carbon levels a phase known as *delta ferrite* is formed first, which then transforms to a different phase called *austenite*. At higher carbon levels, the melting temperature drops with increasing carbon level and the initial transformation may be direct to austenite. The austenite phase has a face-centred cubic lattice crystal structure which is maintained down to the lines AE and BE on Fig. 6.3. As cooling proceeds slowly the austenite then starts to transform to the mixture of ferrite and cementite which results at room temperature. However point E on the diagram represents a eutectoid at a composition of 0.83% carbon at which ferrite

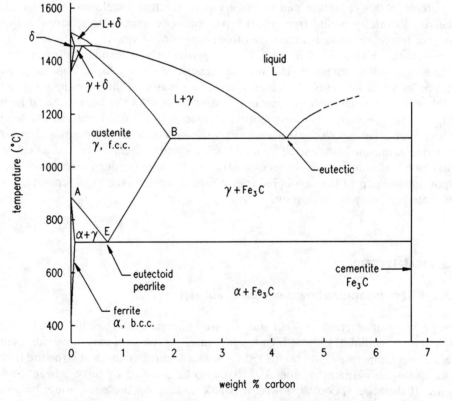

Fig. 6.3 Equilibrium phase diagram for iron − iron carbide system (f.c.c. face-centred cubic; b.c.c. body-centred cubic)

and cementite precipitate alternately in thin laths to form a structure known as *pearlite*. At compositions less than 0.83% carbon, the type of microstructure formed on slow cooling transformation from austenite is a mixture of ferrite and pearlite. Each type of phase present at its appropriate temperature has its own grain size, and the ferrite/pearlite grains tend to precipitate in a network within and based on the previous austenite grain boundary structure. The lattice crystal structure of the ferrite material which forms the basic matrix is essentially a body-centred cubic structure. Thus in cooling from the liquid condition, complex changes in both lattice crystal structure and microstructure take place dependent on the chemical composition. For the equilibrium diagram conditions to be observed, cooling must be sufficiently slow to allow time for the transformations in crystal structure and for the diffusion/migration of carbon to take place to form the appropriate microstructures.

If a steel is cooled from a high temperature and held at a lower constant temperature for sufficient time, different conditions may result; these are represented on a diagram known as the *isothermal transformation diagram*. The form of the diagram depends on the chemical analysis and in particular on the carbon or related element content. In plain carbon steels the isothermal transformation diagram typically has the shape of two letters C each with a horizontal bottom line as shown in Fig. 6.4. The left-hand/upper curve on the diagram of temperature against time represents the start of transformation, and the right-hand/lower curve represents the completion of transformation with time. For steels with a carbon content below the eutectoid composition of 0.83% carbon, holding at a temperature to produce isothermal transformation through the top half of the letter C, leads to the formation of a ferrite/pearlite microstructure. If the transformation temperature is lowered to pass through the lower part of the C curves, but above the bottom horizontal lines, a new type of microstructure is obtained, which is called *bainite*, which is somewhat harder and stronger than pearlite, but also tends to have poorer fracture toughness. If the transformation temperature is dropped further to lie below the two horizontal lines, transformation takes place to a very hard and brittle substance called *martensite*. In this case the face-centred cubic lattice crystal structure of the austenite is not able to transform to the body-centred cubic crystal structure of the ferrite, and the crystal structure becomes locked into a distorted form known as a body-centred tetragonal lattice. Bainite and martensite do not form on equilibrium cooling but result from quenching to give insufficient time for the equilibrium transformations to take place.

The position and shape of the C curves on the time axis depend on the chemical composition of the steel. Higher carbon contents move the C curve to the right on the time axis, making the formation of martensite possible at slower cooling rates. Alloying elements change the shape of the C curves and an example for a low-alloy steel is shown in Fig. 6.5. Additional effects on microstructure, grain size and resultant properties can be obtained by combinations of mechanical work at appropriate temperatures during manufacture of the basic steel.

In addition to the effect of cooling rate on microstructure, the grain size is significantly affected by time at high temperatures and subsequent cooling rate.

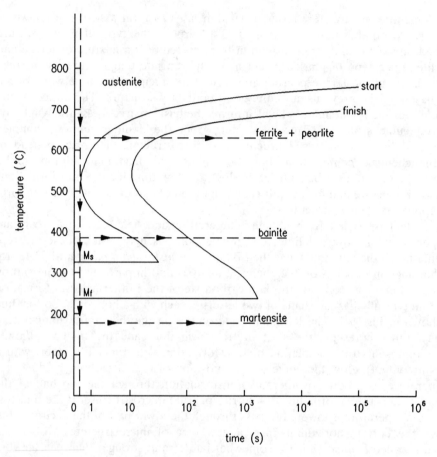

Fig. 6.4 Isothermal transformation diagram for 0.2% C, 0.9% Mn steel

Long periods of time at higher temperatures within a particular phase lead to the merging of the grain boundaries and growth of larger grains. For ferritic crystal structures, grain growth starts at temperatures above about 600°C and hence long periods in the temperature range 600°C to 850°C with slow cooling will tend to promote coarse grain size ferrite/pearlite microstructures. Faster cooling through the upper part of the C curves will give a finer grain structure but still of ferrite/pearlite microstructure.

The type of microstructure present in a steel can be shown and examined by the preparation of carefully polished and etched samples viewed through a microscope. Etching with particular types of reagent attacks different parts of the microstructure preferentially and the etched parts are characteristic of the type of microstructure. Examples of some of the more common types of microstructure mentioned above are shown in Fig. 6.6. The basic microstructure of the steel is usually shown by examination in the microscope to magnifications of from 100 to about 500 times.

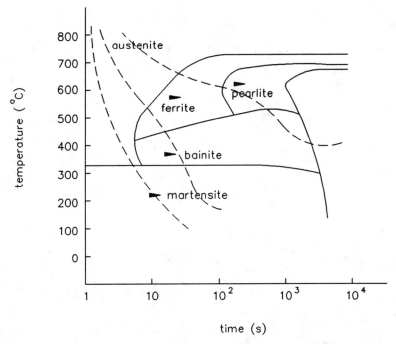

Fig. 6.5 Continuous cooling transformation diagram for 0.4% C, 0.8% Mn, 1% Cr, 0.2% Mo steel

Where it is necessary to examine the effects of very fine precipitates or grain boundary effects, it may be necessary to go to higher magnifications. With the electron microscope it is possible to reach magnifications of many thousands and, with specialized techniques, to reach the stage of seeing dislocations and imperfections in the crystal lattice itself.

6.3.2 Heat treatment in practice

In practical steelmaking or fabrication procedures cooling occurs continuously from high temperatures to lower temperatures. The response of the steel to this form of cooling can be shown on the continuous cooling transformation diagram (CCT diagram) of Fig. 6.7. This resembles the isothermal transformation diagram, but the effect of cooling rate can be shown by lines of different slopes on the diagram. For example, slow cooling, following line (a) on Fig. 6.7 passes through the top part of the C curve and leads to the formation of a ferrite/pearlite mixture. Cooling at an intermediate rate, following line (b), passes through pearlite/ferrite transformation at higher temperatures, but changes to bainite

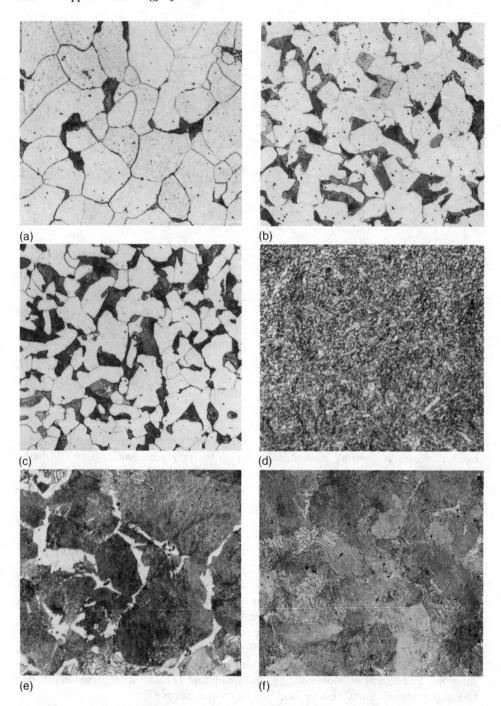

Fig. 6.6 Examples of common types of microstructure in steel (magnification × 500) (courtesy of Manchester Materials Science Centre, UMIST)

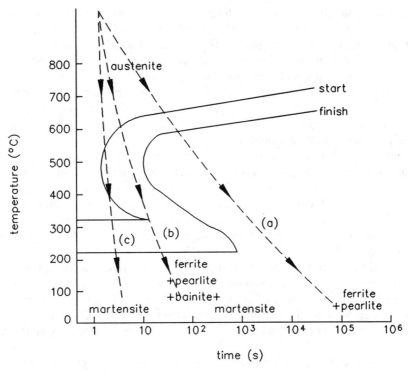

Fig. 6.7 Continuous cooling transformation diagram for 0.2% C, 0.9% Mn steel

transformation at lower temperatures so that a mixture of pearlite and bainite results. Rapid cooling following line (c) misses the C curves completely and passes through the two horizontal lines to show transformation to martensite. Thus in practice for any given composition of steel different microstructures and resultant properties can be produced by varying the cooling rate.

The microstructure and properties of a steel can be changed by carefully chosen heat treatments after the original manufacture of the basic product form. A major group of heat treatments is effected by heating the steel to a temperature such that it transforms back to austenite, this temperature being normally in the range 850°C to 950°C. It is important to ensure that the temperature is sufficient for full transformation to austenite, otherwise a very coarse-grained ferritic structure may result. It is also important that the austenitizing temperature is not too high, and that the time at this temperature is not too long otherwise a coarse-grained austenite structure will form, making subsequent transformation to fine grains more difficult. A heat treatment in which cooling is slow and essentially carried out in a furnace is known as *annealing*. This tends to lead to a relatively coarse-grained final structure, as predicted by the basic equilibrium phase diagram and is used to put materials into their softest condition. If the steel is allowed to cool freely in air from the austenitizing temperature, the heat treatment is known as *normalizing* which gives a finer grain size and hence tends to higher yield strength and better toughness for a given composition of steel. Normalizing may be combined with rolling of a particular product form over a relatively narrow band of temperatures, followed by natural cooling in air, in which case it is known as *controlled rolling*. When the steel product form is cooled more rapidly by immersing it directly into oil or water, the heat treatment is known as *quenching*. Quenching into a water bath is generally more severe than quenching into an oil bath.

A second stage heat treatment to temperatures below the austenitizing range is frequently applied, known as *tempering*. This has the effect of giving more time for the transformation processes which were previously curtailed to develop further, and can permit changes in the precipitation of carbides allowing them to merge together and develop into larger or spheroidal forms. These thermally activated events are highly dependent on temperature and time for particular compositions. The net effect of tempering is to soften previously hardened structures and make them tougher and more ductile.

Both plain carbon and low alloy steels can be supplied in the quenched and tempered condition for plates and engineering sections to particular specifications. The term 'hardenability' is used to describe the ability of steel to form martensite to greater depths from the surface, or greater section sizes. There are, therefore, practical limits of section thickness or size at which particular properties can be obtained.

In BS 970 a range of compositions of engineering steels is given, together with the choice of heat treatments and limiting section sizes for which different properties can be supplied. The heat treatment condition is represented by a letter in the range P to Z. The more commonly supplied conditions are in the range P to T. It should be noted that the term 'hardenability' does not refer to the absolute hardness level which can be achieved, but to the ability to develop uniform hardening throughout the cross section. Cooling rates vary at different positions in the cross section as heat is conducted away in a quenching operation from the surface. Examples are given in the Appendix *Physical chemical and mechanical properties of steel* of some of the more common compositions and heat treatment conditions of general engineering steels in BS 970.

It is sometimes necessary to apply heat treatment to components or structures after fabrication, particularly when they have been welded. The aim is mainly to relieve residual stresses but heat treatment may also be required to produce controlled metallurgical changes in the regions where undesirable effects of welding have occurred. Applications at high temperatures may also lead to metallurgical changes taking place in service. It is vitally important that where any form of heat treatment is applied the possible metallurgical effects on the particular type of steel are taken into account.

Heat treatments are sometimes applied to produce controlled changes in shape or correction of distortion and again temperatures and times involved in these heat treatments must be carefully chosen and controlled for the particular type of steel being used.

6.4 Manufacture and effect on properties

6.4.1 Steelmaking

Manufacture of steel takes place mainly in massive integrated steelworks. The first stage starts with iron ore and coke which are mixed and heated to produce a sinter. This mixture then has limestone added to form the burden or raw material fed into a blast furnace. Reactions which take place at high temperature in the blast furnace lead to the formation of iron; the molten iron is tapped continuously from the bottom of the blast furnace. The molten metal at this stage is approximately 90% to 95% iron, the remainder being impurities which have to be removed or reduced to acceptable levels at the next stage, that of steelmaking. This material is fed together with recovered scrap iron or steel into the steelmaking furnace, the common types of which are known as either a basic oxygen furnace or an electric arc furnace. In the basic oxygen furnace oxygen is blown on to the molten metal by a water-cooled lance. In the electric arc furnace heat is produced by an arc between electrodes over the metal surface and the molten metal itself conducting electricity. Chemical reactions take place following additions of selected materials to the molten metal, which lead to the reduction of the impurities and to the achievement of the required controlled chemical composition of the steel. The impurities are reduced by addition of elements which combine and float out to the surface of the molten metal in the slag or dross waste material on the surface. Deoxidation or killing of the steel takes place in the final stages before the furnace is tapped. Older steel manufacturing practice was to tap the steel from the furnace into ladles and then pour the molten steel into large moulds to produce ingots. These ingots would normally be allowed to solidify and cool before reprocessing at a later stage by rolling into the required product form. Modern steelmaking practice has now moved much more to a process known as *continuous casting*, in which molten steel is poured at a steady rate into a mould to form a continuous solid strand from which lengths of semi-finished product are cut for subsequent processing. Semi-finished products take the form of *slabs*, *billets* or *blooms*. Continuous casting has the advantage of eliminating the reheating and first stage rolling required in the ingot production route, and is generally more efficient, but ingot production is still required for some product forms.

6.4.2 Casting and forging

If the final product form is a casting the liquid steel is poured direct into a mould of the required geometry and shape. Steel castings provide a versatile way of achieving the required finished product, particularly where either many items of the same type are required and/or complex geometries are involved. Special skills are required in the design and manufacture of the moulds in order to ensure that good quality castings are obtained with the required mechanical properties and freedom from significant imperfections or defects. High-integrity castings for structural applications have been successfully supplied for critical components in bridges, such as the major cable saddles for suspension bridges, cast node and tubular sections for offshore structures, and the pump bowl casings for pressurized water reactor systems. The size of component which can be made in cast form is limited to a maximum of some 30 t – 50 t however, and only a small proportion of total steel production is completed as castings for direct application.

Another specialist route to the finished steel product is by forging, in which a bloom is heated to the austenitizing temperature range and formed by repeated mechanical pressing in different directions to achieve the required shape. The combination of temperature and mechanical work enables high-quality products with good mechanical properties to be obtained. An example of high-integrity forgings is the production of steel rings to form the shell/barrel of the reactor pressure vessel in a pressurized water reactor system. Again the proportion of steel production as forgings is a relatively small and specialized part of overall steel production.

6.4.3 Rolling

By far the largest amount of finished steel production is achieved by rolling. The semi-finished products cast from the steelmaking furnace are reheated to the austenitizing range and passed through a series of mills with rolls of the required profile to force the hot steel into the finished shape. An example of the distribution of steel products produced by British Steel (BS) in 1987/88 is shown in the pie chart of Fig. 6.8. It can be seen that a major part of the BS's production goes into strip or sheet material which is produced by continuous rolling from slabs down to sheet of the required width and thickness which is first collected as a coil at the end of the rolling process, and subsequently cut into required lengths. This sheet material is typically used in the motor car industry, for containers and packaging including the food industry, and for domestic household equipment. The sheet material can be supplied either in bare steel form or with different types of coating. For example, it is possible to obtain steel sheet with a continuous galvanizing (zinc) coating for corrosion protection and it is also possible to obtain it with integral plastic coatings of different colours and patterns for decorative finish as well as corrosion protection.

For the structural industry steel slabs can be rolled into plates of the required thickness, or into structural sections such as universal columns, universal beams, angle sections, rail sections, etc. Round blooms or ingots can be processed by a seamless tube rolling mill into seamless tubes of different diameters and thicknesses or solid bar subsequently drawn out into wire. Tubes can be used either for carrying fluids in small-diameter pipelines or as structural hollow sections of circular or rectangular shape. The shape of engineering structural sections is determined by the required properties of the cross section such as cross-sectional

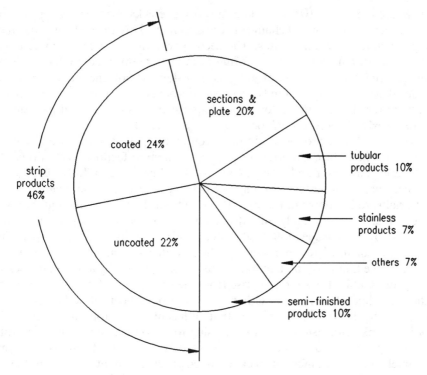

Fig. 6.8 British Steel distribution of turnover by product group 1987/88

area and moments of inertia about different axes to give an effective distribution of the weight of the material for structural purposes. Rolled structural sections are supplied in a standard range of shapes detailed in BS 4: Part 1 and a selection of typical shapes and section properties is given in the Appendix *Geometrical properties of plane sections*.

6.4.4 Defects

In any bulk manufacturing process, such as the manufacture of steel, it is inevitable that a small proportion of the production will have imperfections which may or may not be harmful from the point of view of intended service performance of the product. In general the appropriate applications Standards have clauses which limit any such imperfections to acceptable and harmless levels. In castings a particular family of imperfections can occur which are dependent on the material and the geometry being manufactured. The most serious types of imperfection are cracks caused by shrinkage stresses during cooling, particularly at sharp changes in cross section. A network of fine shrinkage cracks or tears can sometimes develop, again particularly at changes in cross section where the metal is subjected to a range of different cooling rates. The second type of imperfection in castings is solid inclusions, particularly in the form of sand where this medium has been used to form the moulds. Porosity, or gaseous inclusions, is not uncommon in castings to some degree and again tends to occur at changes in cross section. There is usually appreciable tolerance for minor imperfections such as sand inclusions or porosity provided these do not occur to extreme levels.

In rolled or drawn products, the most common types of defect are either cold laps or rolled-in surface imperfections. A lap is an imperfection which forms when the material has been rolled back on to itself but has not fully fused at the interface. Surface imperfections may occur from the same cause where a tongue of material is rolled down but does not fuse fully to the underlying material. Both of these faults are normally superficial and in any serious cases ought to be eliminated by final inspection at the steel mills. A third form of imperfection which can occur in plates, particularly when produced from the ingot route, is a *lamination*: the failure of the material to fuse together, usually at the mid-thickness of the plate. Laminations tend to arise from the rolling-out of pipes, or separation on the centreline of an ingot at either top or bottom which formed at the time of casting the ingot. Normal practice is that sufficient of the top and bottom of an ingot is cut off before subsequent processing to prevent laminations being rolled into subsequent products, but nevertheless they do occur from time to time. Fortunately the development of cracks in rolled products is relatively rare although it may occasionally occur in drawn products or as a result of quenching treatments in heat-treated products.

Since much of the manufacture of steels involves processing at, and subsequent cooling from, high temperatures it will be appreciated that high thermal stresses can develop during differential cooling and these can lead to residual stresses in the finished product. In many cases these residual stresses are of no significance to the subsequent performance of the product but there are situations where their effect must be taken into account. The two in which residual stresses from the steel manufacture are most likely to be of importance are where close tolerance machining is required, or where compression loading is being applied to slender structural sections. For the machining case it may be necessary to apply a stress relief treatment, or alternatively to carry out the machining in a series of very fine cuts. The effect of the inherent manufacturing residual stresses on structural sections is taken account of in the design codes such as BS 5950 for steel buildings, by giving a varying factor on the limiting permissible stresses depending on the product shape. These factors have been determined by a series of research programmes on the buckling behaviour of different shaped sections coupled to measurements of the inherent residual stresses present.

6.5 Engineering properties and mechanical tests

As part of the normal quality control procedures of the steel manufacturer, and as laid down in the different Specifications for manufacture of steel products, tests are carried out on samples representing each batch of steel and the results recorded on a test certificate. At the stage when the chemical analysis of the steel is being adjusted in the steelmaking furnace, samples are taken from the liquid steel melt at different stages to check the analysis results. Samples are also taken from the melt just before the furnace is tapped, and the analysis of these test results is taken to represent the chemical composition of the complete cast. The results of this analysis are given on Test Certificates for all products which are subsequently made from the same initial cast. The Test Certificate will normally give analysis results for C, Mn, Si, S and P for all steels, and where the Specification requires particular elements to be present in a specific range the results for these elements will also be given. Even when additional elements are not specified, the steel manufacturer will often provide analysis results for residual elements which may have been derived from scrap used or which could affect subsequent fabrication of performance during fabrication particularly welding. Thus steel supplied to the *Specification for weldable structural steels*, BS 4360, will often have Test Certificates giving Cr, Ni, Cu, V, Mb and Al, as well as the main basic five elements.

In some Specifications, the requirement is given for additional chemical analysis testing on each item of the final product form and this is presented on the Test Certificates as product analysis in addition to the cast analysis. This does however incur additional costs. The *Specification for weldable structural steels* gives the opportunity for requiring the steelmaker to supply information on the carbon equivalent to assist the fabricator on deciding about precautions during welding (see later). In low-alloy and stainless steels, the Test Certificates will of course give the percentage of the alloying elements such as chromium, nickel, etc.

The Test Certificates should also give the results of mechanical tests on samples selected to represent each product range in accordance with the appropriate Specification. The mechanical test results provided will normally include tensile tests giving the yield strength, ultimate strength and elongation to failure. In structural steels, where the fracture toughness is important, Specifications include requirements for Charpy V-notch impact tests to BS 131: Part 2. The Charpy test is a standard notched bar impact test of 10 mm square cross section with a 2 mm deep V-notch in one face. A series of specimens is tested under impact loading either at one Specification temperature or over a range of temperatures, and the energy required to break the sample is recorded. In the Eurocode and BS 4360 *Specification for weldable structural steels*, these notch ductility requirements are specified by letter grades A to F. Essentially these requirements are that the steel should show a minimum of 27 J energy absorption at a specified testing temperature corresponding to the letter grade. Grade A is no requirement, grade B requires testing at room temperature (+20°C), C at 0°C, D at −20°C, E at −50°C, and F at −60°C.

Charpy test requirements are also included in some of the general engineering steel Specifications (BS 970) for pressure vessel steels and other important structural applications, particularly where welded structures are used.

The Specifications normally require the steelmaker to extract specimens with their length parallel to the main rolling direction. In fact it is unlikely that the steel will be wholly isotropic, and significant differences in material properties may occur under different testing directions, which would not be evident from the normal Test Certificates unless special tests were carried out. It is possible in some Specifications to have material tests carried out both transverse to the main rolling direction, and in the through-thickness direction of rolled products. Testing in the through-thickness direction is particularly important where the material may in fact be loaded in this direction in service by welded attachments. Since such tests are additional to the normal routine practice of the steel manufacturer, and cost extra both for the tests themselves and for the disruption to main production, it is not unexpected that steels required to be tested to demonstrate properties in other directions are more expensive than the basic quality of steel tested in one standard direction only. The quality-control system at the steel manufacturers normally puts markings in the form of stamped numbers or letters on each length or batch of products so that it can be traced back to its particular cast and manufacturing route. In critical structural applications it is important that this numbering system is transferred on through fabrication to the finished structure so that each piece can be identified and confirmed as being of the correct grade and quality. The Test Certificate for each batch of steel is therefore a most important document to the steel manufacturer, to the fabricator, and to the subsequent purchaser of the finished component or structure. In addition to the chemical composition and mechanical properties, the Test Certificate should also record details of the steelmaking route and any heat treatments applied to the material by the steel manufacturer.

It is not uncommon for some semi-finished products to be sold by the steel manufacturer to other product finishers, or to stockholders. Unless these parties retain careful records of the supply of the material it may be difficult to trace specific details of the properties of steel bought from them subsequently, although some stockholders do maintain such records.

Where products are manufactured from semi-finished steel and subsequently given heat treatment for sale to the end user, the intermediate manufacturer should produce his own Test Certificates detailing both the chemical analysis of the steel and the mechanical properties of the finished product. For example, bolts used for structural connections are manufactured from bar material and are normally stamped with markings indicating the grade and type of bolt. Samples of bolts are taken from manufactured batches after heat treatment and subjected to mechanical tests to give reassurance that the correct strength of steel and heat treatment have been used.

6.6 Fabrication effects and service performance

Basic steel products supplied from the steel manufacturer are rarely used directly without some subsequent fabrication. The various processes involved in fabrication may influence the suitability for service of the steel and over the years established procedures of good practice have been developed which are acceptable for particular industries and applications.

6.6.1 Cutting, drilling, forming and drawing

Basic requirements in the fabrication of any steel component are likely to be cutting and drilling. In thin sections, such as sheet material, steel can be cut satisfactorily by guillotine shearing and although this may form a hardened edge it is usually of little or no consequence. Thicker material in structural sections up to about 15 mm thickness can also be cut by heavy-duty shears, useful for small part pieces such as gussets, brackets, etc. Heavier section thicknesses will usually have to be cut by cold saw or abrasive wheel or by flame cutting. Cold saw and abrasive wheel cutting produce virtually no detrimental effects and give good clean cuts to accurate dimensional tolerances. Flame cutting is carried out using an oxyacetylene torch to burn the steel away in a narrow slit and this is widely used for cutting of thicker sections in machine-controlled cutting equipment. The intense heating in flame cutting does subject the edge of the metal to rapid heating and cooling cycles and so produces the possibility of a hardened edge in some steels. This can be controlled by either preheating just ahead of the cutting torch, or using slower cutting speeds, or alternatively, if necessary, any hardened edge can be removed by subsequent machining. In recent years laser cutting has become a valuable additional cutting method for thin material, in that intricate shapes and patterns can be cut out rapidly by steering a laser beam around the required shape.

Drilling of holes presents little problem and there are now available numerical/computer controlled systems which will drill multiple groups of holes to the required size and spacing. For thinner material hole punching is commonly used and although this, like shearing, can produce a hardened edge, provided the punch is sharp no serious detrimental effects occur in thinner material.

It is sometimes necessary to bend, form or draw steel into different shapes. Reinforcing steel for reinforced concrete structures commonly has to be bent into the form of hooks and stirrups. The curved sections of tubular members of offshore structures or cylindrical parts of pressure vessels are often rolled from flat plate to the required curvature. In these cases yielding and plastic strain take place as the material is deformed beyond its elastic limit. This straining moves the material condition along its basic stress/strain curve, and it is therefore important to limit the amount of plastic strain used up in the fabrication process so that that available for subsequent service is not diminished to an unacceptable extent. The important variable in limiting the amount of plastic strain which occurs during cold forming is usually the ratio of the radius of any bend to the thickness or diameter of the material. Provided this ratio is kept high the amount of strain will be limited. Where the amount of cold work which has been introduced during fabrication is excessive, it may be necessary to carry out a reheat treatment in order to restore the condition of the material to give its required properties.

In the manufacture of wire, the steel is drawn through a series of dies gradually reducing its diameter and increasing the length from the initial rod sample. This cold drawing is equivalent to plastic straining and has the effect of both increasing the strength of the material and reducing its remaining ductility as the material moves along its stress/strain curve. In certain types of wire manufacture, intermediate heat treatments are necessary in order to remove damaging effects of cold work and enhance and improve the final mechanical properties.

6.6.2 Welding

One of the most important fabrication processes for use with steel is welding. There are many different types of welding and this subject is itself a fascinating multi-disciplinary world involving combined studies in physics, chemistry, electronics, metallurgy, and mechanical, electrical and structural engineering. Most welding processes involve fusion of the material being joined, by raising the temperature to the melting point of the material, either with or without the addition of separate filler metal. Although there is a huge variety of different welding processes, probably the most common and most important ones for general applications are the group of arc welding processes and the group of resistance welding processes. Among the newer processes are the high energy density beam processes such as electron beam and laser welding.

Arc welding processes involve the supply of an intense heat source from an electric arc which melts the parent material locally, and may provide additional filler metal by the melting of a consumable electrode. These processes are extensively used in the construction industry, and for any welding of material thicknesses above the range of sheets. The resistance group of welding processes involve the generation of heat at the interface between two pieces of material by the passage of very heavy current directly between opposing electrodes on each face. The resistance processes do not involve additional filler metal, and can be used to produce local joints as spot welds, or a series of such welds to form a continuous seam. This group of processes is particularly suitable for sheet material and is widely used in the automotive and domestic equipment markets.

It will be appreciated that fusion welding processes involve rapid heating and cooling locally at the position where a joint is to be made. The temperature gradients associated with welding are intense, and high thermal stresses and subsequent residual stresses on cooling are produced. The residual stresses associated with welding are generally much more severe than those which result during the basic steel manufacturing process itself as the temperature gradients are more localized and intense. Examples of the residual stress distribution resulting from the manufacture of a butt weld between two plates, and a T-butt weld with one member welded on to the surface of a second, are shown in Fig. 6.9. As will be seen from other chapters, residual stresses can be important in the performance of steel structures because of their possible effects on brittle fracture, fatigue and distortion. If a steel material has low fracture toughness, and is operating below its transition temperature, residual stresses may be very important in contributing to failure by brittle fracture at low applied stresses. If on the other hand the material is tough and yields extensively before failure, residual stresses will be of little importance in the overall structural strength. These effects are summarized in Fig. 6.10.

In fatigue-loaded structures residual stresses from welding are important in altering the mean stress and stress ratio. Although these are secondary factors compared to the stress range in fatigue, the residual stress effect is sufficiently important that it is now commonly assumed that the actual stress range experienced at a weld operates with an upper limit of the yield strength due to locked-in residual stresses at this level. Thus although laboratory experiments demonstrate different fatigue performance for the same stress range at different applied mean stress in plain unwelded material, the trend in welded joints is for the applied stress ratio effect to be overridden by locked-in residual stresses.

The effect of welding residual stresses on distortion can be significant both at the time of fabrication, and in any subsequent machining which may be required. The forces associated with shrinkage of welds are enormous, and will produce overall shrinkage of components and bending/buckling deformations out of a flat plane. These effects have to be allowed for either by pre-setting in the opposite direction to compensate for any out-of-plane deformations, or by making allowances with components initially over-length to allow for shrinkage.

Just as the basic steel manufacturing process can lead to the presence of imperfections, welding also can lead to imperfections which may be significant. The types of imperfection can be grouped into three main areas: planar discontinuities, non-planar (volumetric) discontinuities, and profile imperfections. By far the most serious of these are planar discontinuities, as these are sharp and can be of a significant size. There are four main types of weld cracking which can occur in steels as planar discontinuities. These are *solidification (hot) cracking*, *hydrogen-induced (cold) cracking*, *lamellar tearing*, and *reheat cracking*. Examples of these are shown in Fig. 6.11. *Hot cracking* occurs during the solidification of a weld due to the rejection of excessive impurities to the centreline. Impurities responsible are usually sulphur and phosphorus, and the problem is controlled by keeping them to a low level and avoiding deep narrow weld beads. *Cold cracking* is due to the combination of a susceptible hardened microstructure and the effects of hydrogen in the steel lattice. The problem is avoided by control of the steel chemistry, arc energy heat input, preheat level, quenching effect of the thickness of joints being welded, and by careful attention to electrode coatings to keep hydrogen potential to very low levels. Guidance on avoiding this type of cracking in the heat affected zones of weldable structural steels is given in BS 5135.

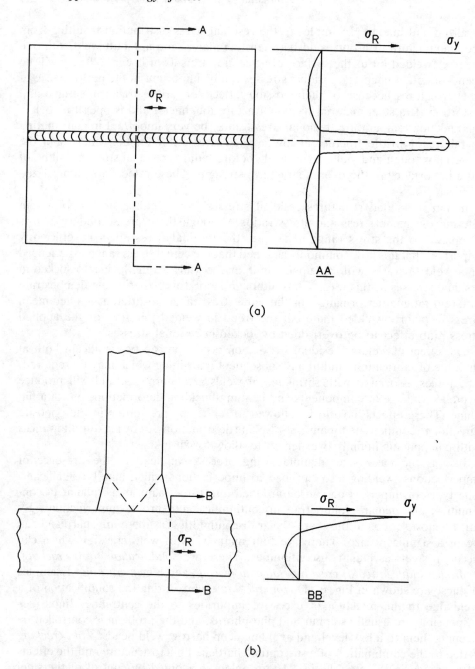

Fig. 6.9 Typical weld residual stress distributions: (a) butt weld, (b) T-butt weld (σ_R residual stress; σ_y yield stress)

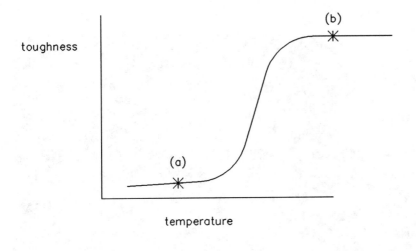

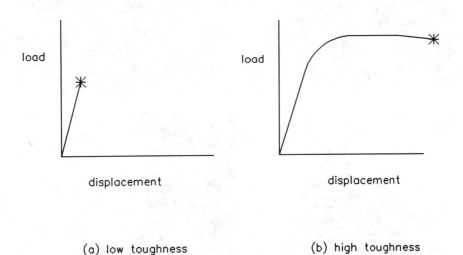

(a) low toughness   (b) high toughness

Fig. 6.10 Effects of toughness and residual stresses on strength in tension

Lamellar tearing is principally due to the presence of excessive non-metallic inclusions in rolled steel products resulting in the splitting open of these inclusions under the shrinkage forces of welds made on the surface. The non-metallic inclusions usually responsible are either sulphides or silicates; manganese sulphides are probably the most common. The problem is avoided by keeping the impurity content low, particularly the sulphur level to below about 0.010%, and by specifying tensile tests in the through-thickness direction to show a minimum ductility by reduction of area dependent on the amount of weld shrinkage anticipated (i.e. size of welded attachment, values of R of A of 10% to 20% are usually adequate).

222　*Applied metallurgy of steel*

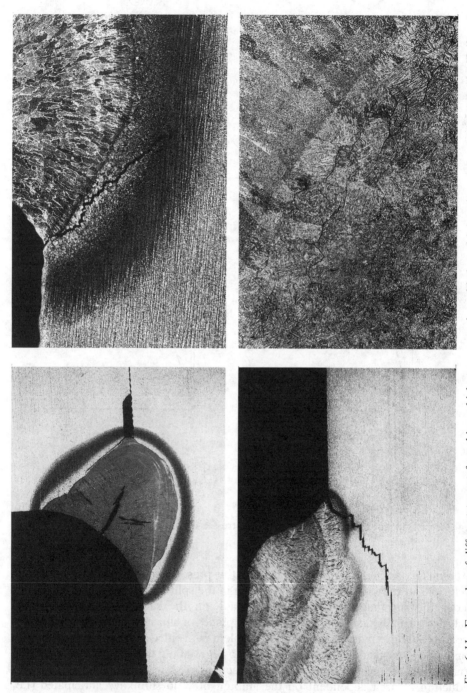

Fig. 6.11 Examples of different types of cracking which may occur in welded steel joints (courtesy of The Welding Institute)

Reheat cracking is a form of cracking which can develop during stress relief heat treatment or during high temperature service in particular types of steel (usually molybdenum or vanadium bearing) where secondary precipitation of carbides develops before relaxation of residual stresses has taken place.

Other forms of planar defect in welds are the operator or procedure defects of *lack of penetration* and *lack of fusion*. The volumetric/non-planar imperfections divide into the groups of *solid inclusions* and *gaseous inclusions*. The solid inclusions are usually slag from the electrode/flux coating and the gaseous inclusions result from porosity trapped during the solidification of the weld. In general the non-planar defects are much less critical than planar defects of the same size and are usually limited in their effect because their size is inherently limited by their nature.

6.7 Summary

6.7.1 Criteria influencing choice of steel

The basic requirement in the choice of a particular steel is that it must be fit for the product application and design conditions required. It must be available in the product form and shape required and it should be at the minimum cost for the required application. Clearly before the generic type of material is chosen as steel, it must be shown to be advantageous to use steel over other contending materials, and therefore the strength-to-weight ratio and cost ratios must be satisfactory.

The steel must have the required strength, ductility and long-term service life in the required environmental service conditions. For structural applications the steel must also have adequate fracture toughness, this requirement being implemented by standard Charpy test quality control levels.

Where the steel is to be fabricated into components or structures, its ability to retain its required properties in the fabricated condition must be clearly established. One of the most important factors in this respect for a number of industries is the weldability of steel, and in this respect the chemical composition of the steel must be controlled within tight limits, and the welding processes and procedures adopted must be compatible with the material chosen.

The corrosion-resistance and potential fire-resistance/high temperature performance of the steel may be important factors in some applications. A clear decision has to be taken at the design stage as to whether resistance to these effects is to be achieved by external or additional protection measures, or inherently by the chemical composition of the steel itself. Stainless steels with high quantities of chromium and nickel are significantly more expensive than ferritic carbon or carbon manganese steels. Particular application standards generally specify the range of material types which are considered suitable for their particular application.

Increased strength of steels can be obtained by various routes, including increased alloying content, heat treatment, or cold working. In general as the strength increases so does the cost and there may be little advantage in using high-strength steels in situations where either fatigue or buckling are likely to be ruling modes of failure. It should not be overlooked that although there is some increase in cost of the basic raw material with increasing strength, there is likely to be a significant increase in fabrication costs with additional precautions necessary for the more sophisticated types of higher-strength material.

Certain product forms are only available in certain grades of steel. It may not be possible to achieve high strength in some product shapes and retain dimensional requirements through the stage of heat treatment because of distortion problems.

Wherever possible, guidance should be sought on the basis of similar previous experience or prototype trials to ensure that the particular material chosen will be suitable for its required application.

6.7.2 Steel specifications and choice of grade

Structural steelwork, comprising rolled products of plate, sections and hollow sections, is normally of a weldable carbon or carbon-manganese structural steel to BS 4360 or to the new European based standard, BS EN 10025. Two strength grades are most commonly used, grade 43 (or Fe 430) and grade 50 (or Fe 510), having yield strengths typically of 275 N/mm^2 or 355 N/mm^2 respectively. In buildings grade 43 is more common, though columns and composite beams in multi-storey buildings will often be grade 50. In bridges and other major structures grade 50 is more common. The design codes for buildings and bridges require that structural steelwork subject to tensile stresses have adequate notch toughness at the *minimum service temperature* of the structure. This temperature is specified as −5°C for internal building steelwork, −15°C for external building steelwork and typically between −15°C and −25°C for bridge steelwork, depending on location and exposure. The subgrade of steel needed to achieve the required toughness is then usually C or D (i.e. grades 43C, 43D, or Fe 430C, Fe 510D), or possibly higher subgrades for thicker material.

Weather resistant steels in the form of rolled sections and plate are also covered by BS 4360.

Cold reduced hot-rolled steel is produced in accordance with BS 1449: Part 1, but for most structural applications it is used in the form of galvanized cold formed steel supplied in accordance with BS 2989 and BS EN 10142. Two strength grades are most commonly specified, grades Z28 and Z35, with yield strengths of 280 N/mm^2 and 350 N/mm^2 respectively. Cold formed steel is extensively used as profiled cladding sheets and its support members.

Stainless steel is specified by BS 1449: Part 2 for flat products and by BS 970: Part 1 for bars. Simple sections can be formed by cold rolling; hot rolled sections are not commonly available in stainless steel. For structural use, austenitic steels are required, usually of grades 304 or 316. The former is more common; the latter has superior corrosion resistance and is used in more aggressive environments. Stainless steel generally has good notch toughness, even at very low temperatures.

Castings are manufactured in accordance with BS 3100. This specification covers all type of casting for engineering purposes, including carbon and stainless steels. Castings can be welded, subject to appropriate post-welding heat treatment.

Wrought or forged steels are required where particularly thick sections or higher strength components are needed. BS 970 covers a wide range of alloys, though only a small number of them are appropriate for normal structural use. Components are normally machined from the solid bar, and since the high strengths depend on special heat treatment, little or no welding is possible on such wrought steel parts.

Steel wire ropes are normally made with cold drawn galvanised steel wires, which are manufactured according to BS 2763. The ultimate strengths of individual wires can be as high as 1700 N/mm^2. The design of wire ropes depends on their usage and is covered by a range of design standards.

Further reading for Chapter 6

Burgan B.A., Dier A.F. & Baddon N. (1992) *Concise Guide to the Structural Design of Stainless Steel*. The Steel Construction Institute, Ascot, Berks.
Dieter G.E. (1986) *Mechanical Metallurgy*, 3rd edn. McGraw-Hill.
Gaskell D. (1981) *Introduction to Metallurgical Thermodynamics*, 2nd edn. McGraw-Hill, New York.
Honeycombe R.W.K. (1981) *Steels: Microstructure and Properties*. Edward Arnold.
Lancaster J.F. (1987) *Metallurgy of Welding*, 4th edn. Allen and Unwin.
Porter D.A. & Easterling K.E. (1981) *Phase Transformations in Metals and Alloys*. Van Nostrand, London.
Smallman R.E. (1970) *Modern Physical Metallurgy*, 3rd edn. Butterworths.
Szekely J. & Themelis N.J. (1971) *Rate Phenomena in Process Metallurgy*. Wiley.

Chapter 7
Fatigue

by ROBERT SIMPSON

7.1 Introduction

A component or structure which survives a single application of load may fracture if the application is repeated a large number of times. This would be classed as fatigue failure. Fatigue failure can be defined as the number of cycles and hence the time taken to reach a pre-defined failure criterion. Fatigue failure is by no means a rigorous science and the idealizations and approximations inherent in it prevent the calculation of an absolute fatigue life for even the simplest joint.

In the analysis of a structure for fatigue there are three main areas of difficulty in prediction:

(1) The operational environment of a structure and the relationship between the environment and the actual forces on it
(2) The internal stresses at a critical point in the structure induced by external forces acting on the structure
(3) The time to failure due to the accumulated stress history at the critical point.

There are two basic approaches for the assessment of fatigue life of structural components. The first method, which is currently in general use, relies on empirically derived relationships between applied stress ranges and fatigue life commonly called the $S-N$ approach. The second method, based on fracture mechanics, considers the growth rate of an existing defect at each stage in its propagation.

7.2 Loadings for fatigue

Examples of structures and the loads which can cause fatigue are:

Bridges: commercial vehicles, goods trains
Cranes: lifting, rolling and inertial loads
Offshore structures: waves
Slender towers: wind gusting.

The designer's objective is to anticipate the sequence of service loading throughout the structure's life. The magnitude of the peak load, which is vital for static design

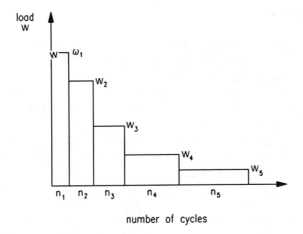

Fig. 7.1 Typical load spectrum for design

purposes, is generally of little concern as it only represents one cycle in millions. For example, highway bridge girders may experience 100 million significant cycles in their lifetime. The sequence is important because it affects the stress range, particularly if the structure is loaded by more than one independent load system.

For convenience, loadings are usually simplified into a load spectrum, which defines a series of bands of constant load levels and the number of times that each band is experienced, as shown in Fig. 7.1.

Slender structures, with natural frequencies low enough to respond to the loading frequency, may suffer dynamic magnification of stress. This can shorten the life considerably.

References 1–4 are useful sources of information on fatigue loading.

7.3 The nature of fatigue

7.3.1 Introduction

Materials subject to a cyclically variable stress of a sufficient magnitude change their mechanical properties. In practice a very high percentage of all engineering failures are due to fatigue. Most of these failures can be attributed to poor design or subsequent manufacture. The fatigue process is not a single mechanism but the result of several mechanisms operating in sequence during the life of a structure: initiation of a microscopic defect, slow incremental crack propagation and final unstable fracture. In most steel structures which are welded the initiation phase is not necessary for crack growth as existing defects are already likely to be present.

7.3.2 S–N curves

The most common form of presentation of fatigue data is the $S-N$ curve, where the total cyclic stress range (S) is plotted against the number of cycles to failure (N). A typical curve is shown in Fig. 7.2.

These data are obtained experimentally. For a particular welded joint configuration, a series of specimens is subject to cycles of constant amplitude to failure. Sufficient specimens are tested for statistical analysis to be carried out to determine both mean fatigue strength and its standard deviation. Depending on the design philosophy adopted, design strength is taken as mean minus an appropriate number of standard deviations.

Logarithmic scales are conventionally used for both axes, and the $S-N$ curves are based on test data. To carry out fatigue life predictions a linear fatigue damage mode is used in conjunction with the relevant $S-N$ curve. One such fatigue damage model is that postulated by Miner. In essence, the model assumes that irreversible damage accumulates linearly to a fixed level, at which failure

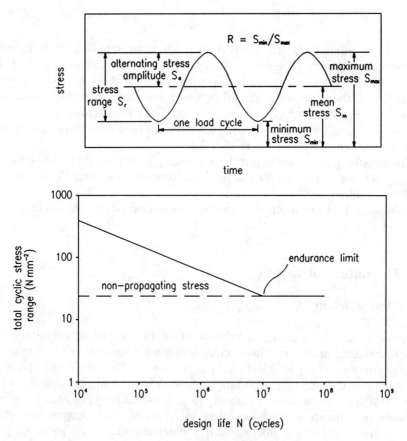

Fig. 7.2 Typical $S-N$ curve and nomenclature used in fatigue

occurs. For welded joints the relationship between fatigue life and applied stress range is linear over a wide range of stress and takes the form

$$NS_r^m = A \tag{7.1}$$

where N = number of cycles to failure, S_r = applied stress range and m, A = constants depending on joint type. Figure 7.3 presents a series of $S-N$ curves which have been defined by taking different values of m and A, applicable to most common welded joint configurations. The curves have been established empirically from a large number of small-scale laboratory tests. For the purposes of fatigue design constructional details are grouped in eight classes designated B, C, D, E, F, F2, G and W. Details of these groupings are given in BS 5400: Part 10.

Examples of the classification of some joints are given in Table 7.1. In any welded joint there are at least five potential locations at which fatigue cracks may develop. These are at the weld toe in each of the two parts joined, at the two ends, and in the weld itself. Each must be classified separately.

Table 7.1 Fatigue: joint classification

Type 1   Material free from welding

Notes on potential modes of failure

In plain steel, fatigue cracks initiate at the surface, usually either at surface irregularities or at corners of the cross section. In welded construction, fatigue failure will rarely occur in a region of plain material since the fatigue strength of the welded joints will usually be much lower. In steel with rivet or bolt holes or other stress concentrations arising from the shape of the member, failure will usually initiate at the stress concentration.

Type and description	Class	Explanatory comments	Examples, including failure modes
1.1 Plain steel			
(a) In the as-rolled condition, or with cleaned surfaces but with no flame-cut edges of re-entrant corners.	B	Beware of using Class B for a member which may acquire stress concentrations during its life, e.g. as a result of rust pitting. In such an event Class C would be more appropriate.	
(b) As (a) but with any flame-cut edges subsequently ground or machined to remove all visible sign of the drag lines.	B	Any re-entrant corners in flame-cut edges should have a radius greater than the plate thickness.	
(c) As (a) but with the edges machine flame-cut by a controlled procedure to ensure that the cut surface is free from cracks.	C	Note, however, that the presence of a re-entrant corner implies the existence of a stress concentration so that the design stress should be taken as the net stress multiplied by the relevant stress concentration factor.	

Fatigue

Table 7.1 Continued

Type 2 Continuous welds essentially parallel to the direction of applied stress

Notes on potential modes of failure

With the excess weld metal dressed flush, fatigue cracks would be expected to initiate at weld defect locations. In the as-welded condition, cracks might initiate at stop–start positions or, if these are not present, at weld surface ripples.

General comments

(1) Backing strips
If backing strips are used in making these joints: (a) they must be continuous, and (b) if they are attached by welding those welds must also comply with the relevant Class requirements (note particularly that tack welds, unless subsequently ground out or covered by a continuous weld, would reduce the joint to Class F, (see joint Type 6.5 (not shown here)).

(2) Edge distance
An edge distance criterion exists to limit the possibility of local stress concentrations occurring at unwelded edges as a result, for example, of undercut, weld spatter, or accidental overweave in manual fillet welding (see also notes on joint Type 4). Although an edge distance can be specified only for the 'width' direction of an element, it is equally important to ensure that no accidental undercutting occurs on the unwelded corners of, for example, cover plates or box girder flanges. If it does occur it should subsequently be ground smooth.

Type and description	Class	Explanatory comments	Examples, including failure modes
2.1 Full or partial penetration butt welds, or fillet welds: parent or weld metal in members, without attachments, built up of plates or sections, and joined by continuous welds.			applied stress
(a) Full penetration butt welds with the weld overfill dressed flush with the surface and finish-machined in the direction of stress, and with the weld proved free from significant defects by non-destructive examination.	B	The significance of defects should be determined with the aid of specialist advice and/or by the use of fracture mechanics analysis. The NDT technique must be selected with a view to ensuring the detection of such significant defects.	
(b) Butt or fillet welds with the welds made by an automatic submerged or open arc process and with no stop–start positions within the length.	C	If an accidental stop–start occurs in a region where Class C is required, remedial action should be taken so that the finished weld has a similar surface and root profile to that intended.	
(c) As (b) but with the weld containing stop–start positions within the length.	D	For situation at the *ends* of flange cover plates see joint Type 6.4 (not shown here).	edge distance from weld toe to edge of flange > 10 mm

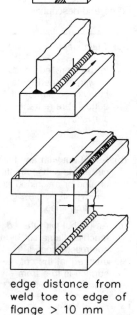

Table 7.1 Continued

Type 3   Transverse butt welds in plates (i.e. essentially perpendicular to the direction of applied stress)

Notes on potential modes of failure

With the weld ends machined flush with the plate edges, fatigue cracks in the as-welded condition normally initiate at the weld toe, so that the fatigue strength depends largely upon the shape of the weld overfill. If this is dressed flush the stress concentration caused by it is removed and failure is then associated with weld defects. In welds made on a permanent backing strip, fatigue cracks initiate at the weld metal/strip junction, and in partial penetration welds (which should not be used under fatigue conditions) at the weld root.

Welds made entirely from one side, without a permanent backing, require care to be taken in the making of the root bead in order to ensure a satisfactory profile.

Design stresses

In the design of butt welds of Types 3.1 or 3.2 which are not aligned, the stresses must include the effect of any eccentricity. An approximate method of allowing for eccentricity in the thickness direction is to multiply the normal stress by $(1 + 3e/t)$, where

e is the distance between centres of thickness of the two abutting members; if one of the members is tapered, the centre of the untapered thickness must be used; and

t is the thickness of the thinner member.

With connections which are supported laterally, e.g. flanges of a beam which are supported by the web, eccentricity may be neglected.

Type and description	Class	Explanatory comments	Examples, including failure modes
3.1 Parent metal adjacent to, or weld metal in, full penetration butt joints welded from both sides between plates of equal width and thickness or where differences in width and thickness are machined to a smooth transition not steeper than 1 in 4.		Note that this includes butt welds which do not completely traverse the member, such as circular welds used for inserting infilling plates into temporary holes.	
(a) With the weld overfill dressed flush with the surface and with the weld proved free from significant defects by non-destructive examination.	C	The significance of defects should be determined with the aid of specialist advice and/or by the use of fracture mechanics analysis. The NDT technique must be selected with a view to ensuring the detection of such significant defects.	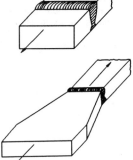
(b) With the welds made, either manually or by an automatic process other than submerged arc, provided all runs are made in the downhand position.	D	In general welds made by the submerged arc process, or in positions other than downhand, tend to have a poor reinforcement shape, from the point of view of fatigue strength. Hence such welds are downgraded from D to E.	

Table 7.1 Continued

Type and description	Class	Explanatory comments	Examples, including failure modes
Type 3.1 cont. (c) Welds made other than in (a) or (b).	E	In both (b) and (c) the corners of the cross section of the stressed element at the weld toes should be dressed to a smooth profile. Note that step changes in thickness are in general not permitted under fatigue conditions, but that where the thickness of the thicker member is not greater than $1.15 \times$ the thickness of the thinner member, the change can be accommodated in the weld profile without any machining. Step changes in width lead to large reductions in strength (see joint Type 3.3).	
3.2 Parent metal adjacent to, or weld metal in, full penetration butt joints made on a permanent backing strip between plates of equal width and thickness or with differences in width and thickness machined to a smooth transition not steeper than 1 in 4.	F	Note that if the backing strip is fillet welded or tack welded to the member the joint could be reduced to Class G (joint Type 4.2).	no tack welds
3.3 Parent metal adjacent to, or weld metal in, full penetration butt welded joints made from both sides between plates of unequal width, with the weld ends ground to a radius not less than 1.25 times the thickness t.	F2	Step changes in width can often be avoided by the use of shaped transition plates, arranged so as to enable butt welds to be made between plates of equal width. Note that for this detail the stress concentration has been taken into account in the joint classification.	$r \geqslant 1.25t$

Table 7.1 Continued

Type 4   Welded attachments on the surface or edge of a stressed member

Notes on potential modes of failure

When the weld is parallel to the direction of the applied stress fatigue cracks normally initiate at the weld ends, but when it is transverse to the direction of stressing they usually initiate at the weld toe; for attachments involving a single, as opposed to a double, weld cracks may also initiate at the weld root. The cracks then propagate into the stressed member. When the welds are on or adjacent to the edge of the stressed member the stress concentration is increased and the fatigue strength is reduced; this is the reason for specifying an 'edge distance' in some of these joints (see also note on edge distance in joint Type 2).

Type and description	Class	Explanatory comments	Examples, including failure modes
4.1 Parent metal (of the stressed member) adjacent to toes or ends of bevel-butt or fillet welded attachments, regardless of the orientation of the weld to the direction of applied stress, and whether or not the welds are continuous round the attachment.		Butt welded joints should be made with an additional reinforcing fillet so as to provide a similar toe profile to that which would exist in a fillet welded joint.	
(a) With attachment length (parallel to the direction of the applied stress) ≤150 mm and with edge distance ≥10 mm.	F	The decrease in fatigue strength with increasing attachment length is because more load is transferred into the longer gusset giving an increase in stress concentration.	edge distance
(b) With attachment length (parallel to the direction of the applied stress) >150 mm and with edge distance ≤10 mm.	F2		
4.2 Parent metal (of the stressed member) at the toes or the ends of butt or fillet welded attachments on or within 10 mm of the edges or corners of a stressed member, and regardless of the shape of the attachment.	G	Note that the classification applies to all sizes of attachment. It would therefore include, for example, the junction of two flanges at right angles. In such situations a low fatigue classification can often be avoided by the use of a transition plate (see also joint Type 3.3).	edge distance

Table 7.1 Continued

Type and description	Class	Explanatory comments	Examples, including failure modes
4.3 Parent metal (of the stressed member) at the toe of a butt weld connecting the stressed member to another member slotted through it.		Note that this classification does not apply to fillet welded joints (see joint Type 5.1b (not shown here)). However it does apply to loading in either direction (L or T in the sketch).	
(a) With the length of the slotted-through member, parallel to the direction of the applied stress, ≤150 mm and with edge distance ≥10 mm.	F		
(b) With the length of the slotted-through member, parallel to the direction of the applied stress, >150 mm and with edge distance ≥10 mm.	F2		
(c) With edge distance <10 mm.	G		

Type 5 Load-carrying fillet and t butt welds

Notes on potential modes of failure

Failure in cruciform or T joints with full penetration welds will normally initiate at the weld toe, but in joints made with load-carrying fillet or partial penetration butt welds cracking may initiate either at the weld toe and propagate into the plate or at the weld root and propagate through the weld. In welds parallel to the direction of the applied stress, however, weld failure is uncommon; cracks normally initiate at the weld end and propagate into the plate perpendicular to the direction of applied stress. The stress concentration is increased, and the fatigue is therefore reduced, if the weld end is located on or adjacent to the edge of a stressed member rather than on its surface.

7.3.3 Variable-amplitude loading

For constant-amplitude loading, the permissible stress range can be obtained directly from Fig. 7.3 by considering the required design life. In practice it is more common for structures to be subjected to a loading spectrum of varying amplitudes or random vibrations. In such cases use is made of Miner's rule. For a joint subjected to a number of repetitions, n_i, of each of several stress ranges S_{ri}, the value of n_i corresponding to each S_{ri} should be determined from stress spectra measured on similar equipment or by making reasonable assumptions as to the expected service history. The permissible number of cycles, N_i, at each stress range, S_{ri}, should then be determined from Fig. 7.3 for the relevant joint class, and the stress range adjusted so that the cumulative damage summation

The nature of fatigue 235

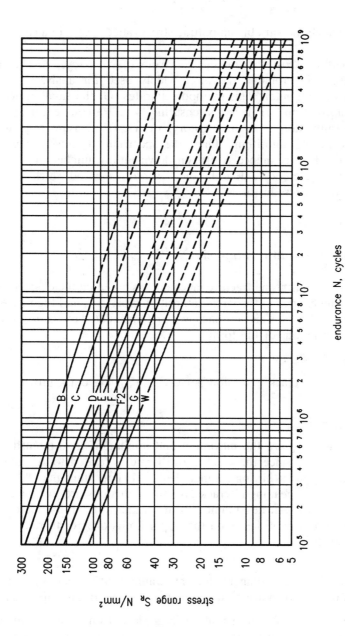

Fig. 7.3 Design S–N curves used for steel

$$\frac{n_1}{N_1} + \frac{n_2}{N_2} + \frac{n_3}{N_3} + \frac{n_4}{N_4} + \ldots = \Sigma \frac{n}{N} < 1.0 \tag{7.2}$$

The order in which the variable amplitude stress ranges occur in a structure is not a relevant factor.

Various methods exist to sum the spectrum of stress cycles. The 'rainflow' counting method is probably the most widely used for analyzing long stress histories using a computer. This method separates out the small cycles that are often superimposed on larger cycles, ensuring both are counted. The procedure involves the simulation of a time history, or use of measure sequences, with appropriate counting algorithms. Once the spectrum of stress cycles has been determined, the load sequence is broken down into a number of constant load range segments. Reference 2 describe the 'reservoir' method which is easy to use by hand for short stress histories.

7.4 Fracture mechanics analysis

Fatigue life assessment using fracture mechanics is based on the observed relationship between the change in the stress intensity factor, ΔK, and the rate of growth of fatigue cracks, da/dN. This approach was first formulated by Paris who proposed the power law relations of the form:

$$da/dN = C(\Delta K)^n \tag{7.3}$$

where da/dN is the crack extension per cycle, C, n are crack growth constants, and

$$\Delta K = K_{max} - K_{min}$$

where K_{max} and K_{min} are the maximum and minimum stress intensities respectively in each cycle. Since the crack growth rate is related to ΔK raised to an exponent, the exponent having a range 3–4, it is important that ΔK should be known accurately if meaningful crack growth predictions are to be made.

Values of C and n can be obtained from specific tests on the materials under consideration. From published data the upper limits on growth rates for ferritic steels are given by

$n = 3$
$C = 3 \times 10^{-13}$ for non-aggressive environments at temperatures up to 100°C
$C = 3 \times 10^{-12}$ for marine environments at temperatures up to 20°C

If da/dN versus ΔK for an actual crack is plotted on a logarithmic scale, an approximate sigmoidal curve results as shown in Fig. 7.4. Below a threshold stress intensity factor range, ΔK_{th}, no growth occurs. For intermediate values of ΔK, growth rate is idealized by a straight line.

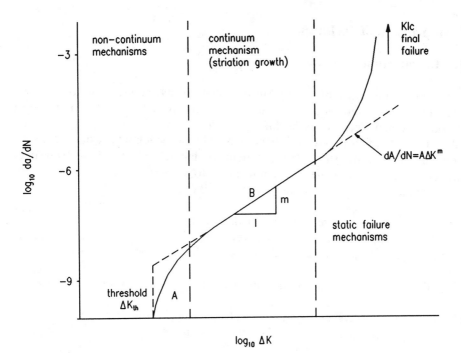

Fig. 7.4 Schematic presentation of crack growth

For a crack at the toe of a welded joint

$$\Delta K = M_K Y \Delta S_r \sqrt{(\pi a)} \qquad (7.4)$$

where S_r = applied stress range
a = crack depth
Y = a correction factor dependent on crack size, shape and loading
M_K = a function which allows for the stress concentration effect of the welded joint and depends on crack site, plate thickness, joint geometry and loading.

Substituting Equation (7.4) in Equation (7.1), rearranging and integrating gives:

$$\int_{a_i}^{a_f} \frac{da}{[M_K Y \sqrt{(\pi a)}]^a} = CAS_r^n N \qquad (7.5)$$

where a_i is the initial crack depth and a_f is the final crack depth corresponding to failure. Thus, if a welded joint contains a crack or crack-like flaw, equations can be used to predict its fatigue strength, assuming that the life consists of crack growth from a pre-existing crack and the initial crack size is known.

7.5 Improvement techniques

7.5.1 Introduction

The fatigue performance of a joint can be enhanced by the use of weld improvement techniques. There is a large amount of data available on the influence of weld improvement techniques on fatigue life but as yet little progress has been made into developing practical design rules. Modern steelmaking has led to the production of structural steels with excellent weldability. The low fatigue strength of a welded connection is generally attributed to a short crack initiation period.

An extended crack initiation life can be achieved by:

(1) reducing the stress concentration of the weld,
(2) removing crack-like defects at the weld toe,
(3) reducing tensile welding residual stresses or introducing compressive stresses.

The methods employed fall broadly into two categories as illustrated below:

weld geometry improvement: grinding; weld dressing; profile control
residual stress reduction: peening; thermal stress relief

Most of the current information relating to weld improvement has been obtained from small-scale specimens. When considering actual structures one important factor is size. In a large structure long range residual stresses due to the assembly of the members are present and will influence the fatigue life. In contrast to small joints where peak stress is limited to the weld toe, the peak stress region in a large multi-pass joint may include several weld beads and cracks may initiate anywhere in this highly stressed area.

However, it is good practice not to seek benefit from improvement techniques in the design office.

7.5.1.1 Grinding

The improvement of the weld toe profile and the removal of slag inclusions can be achieved by grinding either with a rotary burr or with a disc. To obtain the maximum benefit from this type of treatment it is important to extend the grinding to a sufficient depth to remove all small undercuts and inclusions. The degree of improvement achieved increases with the amount of machining carried out and the care taken by the operator to produce a smooth transition.

The performance of toe-ground cruciform specimens is fully investigated in Reference 1. Under freely corroding conditions the benefit from grinding is minimal. However, in air, the results appear to fall on the safe side of the mean curve; endurance is altered by a factor of 2.2. It is therefore recommended that an increase in fatigue life by a factor of 2.2 can be taken if controlled local machining or grinding is carried out.

7.5.1.2 Weld toe remelting

Weld toe remelting by TIG and plasma arc dressing are performed by remelting the toe region with a torch held at an angle of 50° or 90° to the plate (without the addition of filler material). The difference between TIG and plasma dressing is that the latter requires a higher heat input.

Weld toe remelting can result in large increases in fatigue strength, due to the effect of providing low contact angle in the transition area between the weld and the plate, and by the removal of slag inclusions and undercuts at the toe.

7.5.1.3 Hammer peening

Improved fatigue properties of peened welds are obtained by extensive cold working of the toe region. These improved fatigue properties are due to:

(1) introduction of high compressive residual stresses,
(2) a flattening of crack-like defects at the toe,
(3) an improved toe profile.

It can be shown that weld improvement techniques greatly improve the fatigue life of weldments. For weldments subject to bending and axial loading, peening appears to offer the greatest improvement in fatigue life, followed by grinding and TIG dressing.

7.6 Fatigue-resistant design

The nature of fatigue is well understood and the analytical tools are available to calculate the fatigue life of complex structures. The accuracy of any fatigue life calculation is highly dependent on a good understanding of the expected loading sequence during the whole life of a structure. Once a global pattern has been developed then a more detailed inspection of particular areas of a structure, where the effects of loading may be more important, due to the geometries of joints for example, should be carried out.

Data have been gathered for many years on the performance of bridges, towers, cranes and offshore structures where fatigue is a major design consideration. Codes of Practice such as BS 5400 for bridges and the various offshore codes give details for the estimation of fatigue lives. Where a structure is subjected to fatigue it is important that welded joints are considered carefully. Fatigue and brittle fractures can be initiated at discontinuities of shape, at notches and cracks which give rise to high local stress. Avoidance of local structural and notch peak stresses by good design is the most effective means of increasing fatigue life. It is important that during the design process consideration is given to the manner in which the structure is to be fabricated. Acute-angled welds of less than 30° are difficult to fabricate, particularly in tubular structures. This could lead to defects in the weld. Furthermore it is also difficult to carry out non-destructive testing on such welds. Despite these problems, some recently designed structures have incorporated such features.

Repairs carried out to structures in service are expensive, and in the worst case may require that a facility is closed down temporarily. Care needs to be taken when specifying secondary attachments to main primary steelwork. These are often not considered in detail during the design process as they themselves are not complex. However, there have been a number of failures in offshore structures due to fatigue crack growth resulting from a welded attachment. The following general suggestions can assist in the development of an appropriate design of a welded structure with respect to fatigue strength:

(1) Adopt butt or single and double bevel butt welds in preference to fillet welds
(2) Use double-sided in preference to single-sided fillet welds
(3) Aim to place weld, particularly toe, root and weld end in area of low stress
(4) Ensure good welding procedures are adopted and adequate non-destructive testing (NDT) undertaken
(5) Consider the effects of localized stress concentration factors
(6) Consider potential effects of residual stresses.

References to Chapter 7

1. American Society for Testing and Materials (1982) *Design of Fatigue and Fracture Resistant Structures*. ASTM STP 761.
2. British Standards Institution (1980) *Steel, concrete and composite bridges*. Part 10: *Code of practice for fatigue*. BS 5400, BSI, London.
3. Department of Energy (1990) *Offshore Installations: Guidance Design Construction and Certification*, 4th edn. HMSO.
4. British Standards Institution (1983 & 1980) *Rules for the design of cranes*. Part 1: *Specification for classification, stress calculations and design criteria for structures*. Part 2: *Specification for classification, stress calculations and design of mechanisms*. BS 2573, BSI, London.

Further reading for Chapter 7

Gray T.F.G., Spence J. & North T.H. (1975) *Rational Welding Design*, 1st edn. Newnes-Butterworth (2nd edn 1982).
Gurney T.R. (1979) *Fatigue of Welded Structures*, 2nd edn. Cambridge University Press.
Pellini W.S. (1983) *Guidelines for Fracture-Safe and Fatigue-Reliable Design of Steel Structures*. The Welding Institute.
Radaj D. (1990) *Design and Analysis of Fatigue Resistant Welded Structures*. Abington Publishing.

Chapter 8
Brittle fracture

by ROBERT SIMPSON

8.1 Introduction

The term brittle fracture is used to describe fast unstable fractures, in contrast to stable fractures of the fatigue type or slow unstable fractures such as plastic yielding. Some metals, such as copper and aluminium, have a crystalline structure that enables them to resist fast fracture under all loading conditions and at all temperatures. This is not the case for many ferrous alloys, particularly structural steels. The consequence of a brittle fracture may be a catastrophic failure; an understanding of the fundamentals of this subject is therefore important for all structural engineers.

The introduction and development of fracture mechanics technology allows the engineer to examine the susceptibility of steel structures, especially their welded joints, to failure assuming a given defect size and operating conditions.

8.2 Ductile and brittle behaviour

Ductile fracture is normally preceded by extensive plastic deformation. Ductile fracture is slow, and generally results from the formation and coalescence of voids. These voids are often formed at inclusions due to the large tensile stresses set up at the inclusion/metal interface, as seen in Fig. 8.1(a). Ductile fracture usually goes through the grains but, if the density of inclusions or of pre-existing holes is higher on grain boundaries than it is within the grains, then the fracture path may follow the boundaries, giving a fibrous or ductile intergranular fracture. In cases where inclusions are absent, it has been found that voids are formed in severely deformed regions via the presence of localized slip bands and macroscopic instabilities, resulting in either necking or the formation of zones of concentrated shear, as depicted in Fig. 8.1(b).

The fracture path of a ductile crack is often irregular and the presence of a large number of small voids gives the fracture surface a dull fibrous appearance.

The capacity of most metals of engineering interest for plastic deformation and work hardening is extremely valuable as a safeguard against design oversight, accidental overloads or failure by cracking due to fatigue, corrosion or creep.

Brittle fracture is often thought to refer to rapid propagation of cracks without

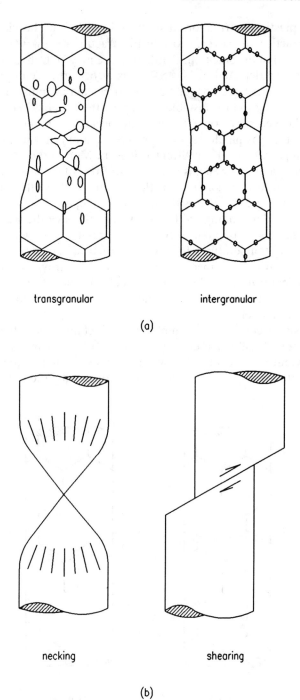

Fig. 8.1 Plastic deformation (a) by voids growth and coalescence, (b) by necking or shearing

244 Brittle fracture

any excessive plastic deformation at a stress level below the yield stress of the material. In practice, however, most brittle fractures show very limited plastic deformation ahead of the crack tip. Brittle fracture may be transgranular (cleavage) or intergranular, as depicted in Fig. 8.2. It is also worth mentioning that metals which often show ductile behaviour can, under certain circumstances, behave in a brittle fashion leading to fast unstable crack growth. This has been clearly demonstrated over the years by some unfortunate accidents involving ships, bridges, offshore structures, gas pipelines, pressure vessels and other major constructions. The Liberty ships and the King Street bridge in Melbourne, Australia, the Sea Gem drilling rig for North Sea gas and the more recent collapse of the Alexander Kielland oil rig are a few examples of the casualties of brittle fracture.

An important feature of steels is the transition temperature between ductile and brittle fracture. Understanding the factors which influence the transition temperature allows designers to be able to select a material which will be ductile at the required operating temperatures for a given structure. This transition is shown in Fig. 8.3: the variation on impact values for small Charpy specimens[1] is plotted against test temperature for a low carbon steel used in offshore applications. BS EN 10025[2] gives temperatures for minimum toughness values to be obtained for a range of structural steel grades.

Impact transition curves are a simple way of defining the effect that variables e.g. heat treatment, alloying elements, effects of welding, etc., have on the fracture behaviour of a steel. Charpy values are useful for quality control but more sophisticated tests[3,4] are required to categorize fully the failure behaviour of

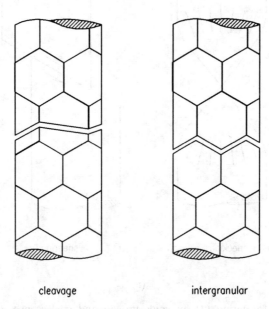

Fig. 8.2 Transgranular and intergranular brittle fracture

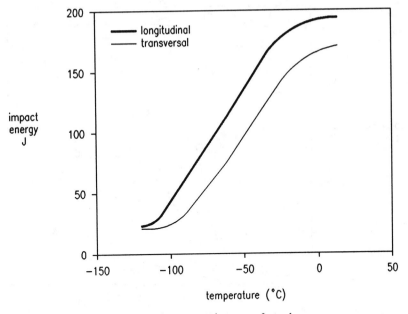

Fig. 8.3 Effects of temperature on impact resistance of steel

a material. Figure 8.4 demonstrates the effect of carbon content on the impact values.

An understanding of the fracture behaviour of steel is particularly important when considering welded structures as welding can considerably reduce the toughness of plate in regions close to the fusion line. This coupled with the likelihood of defects in the weld area can, if not properly controlled, lead to cracking and eventual failure of a joint.

8.3 Linear elastic fracture mechanics

A crack in a solid can be stressed in three different modes as shown in Fig. 8.5. Normal stresses give rise to the opening mode denoted as Mode 1. The displacements of the crack surfaces are perpendicular to the plane of the crack. Mode 1 crack opening is the most important in the study of crack growth. If factor K, the crack tip stress intensity factor, can be derived, it gives the amplitude of the characteristic stress pattern K_1 for Mode 1 crack opening which varies according to applied load, crack size and other geometrical features and can be represented by:

$$K_1 = \sigma \sqrt{(\pi a)} \tag{8.1}$$

246   *Brittle fracture*

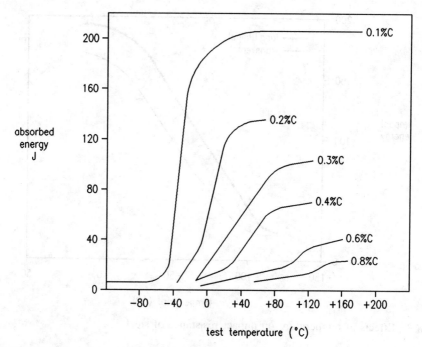

Fig. 8.4   The change in mechanical properties of carbon steels with carbon content

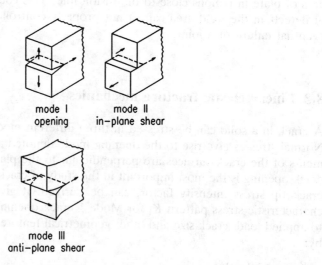

Fig. 8.5   Crack opening modes

where σ is the tensile stress normal to the crack forces and a is the crack length.

Crack extension will occur when the stresses and strains at the crack tip reach a critical value. This means that fracture will be expected to occur when K_1 reaches a critical value K_{1c}.

The fundamental work on fracture mechanics was undertaken by Griffith who published his work in 1924.[5] The basis of his work was to explain why the observed strength of glass was many times lower than the theoretical ultimate strength based on atomic bonding. He postulated that this reduction in strength was due to the presence of small crack-like defects invisible to the naked eye.

Consider an infinite cracked plate of unit thickness with a central transverse crack of length $2a$. The plate is stressed to σ and fixed at its ends as in Fig. 8.6. The elastic energy contained in the plate is represented by the area OAB. If the crack extends over a length da the stiffness of the plate will drop (line OC) which

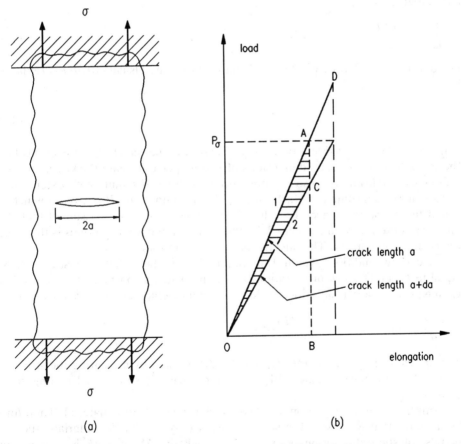

Fig. 8.6 The Griffith criterion for fixed grips. (a) Cracked plate with fixed ends, (b) elastic energy of plate in (a)

means that some load will be relaxed since the ends of the plate are fixed. Consequently, the elastic energy content will drop to a magnitude represented by area OCB. Crack propagation from a to $a + da$ will result in an elastic energy release equal in magnitude to area OAC.

If the plate is stressed at a higher stress there is a larger energy release if the crack grows an amount da. Griffith stated that crack propagation occurs if the energy released upon crack growth is sufficient to provide all the energy that is required for crack growth. If the latter is not the case the stress has to be raised. The triangle ODE represents the amount of energy available if the crack grows.

The condition for crack growth is:

$$\frac{dU}{da} = \frac{dW}{da} \tag{8.2}$$

where U is the elastic energy and W is the energy required for crack growth. Based upon stress field calculations for an elliptical flaw by Inglis, Griffiths calculated dU/da as

$$\frac{dU}{da} = \frac{2\pi\sigma^2 a}{E} \tag{8.3}$$

per unit plate thickness, where E is Young's modulus. Usually dU/da is replaced by

$$G = \frac{\pi\sigma^2 a}{E} \tag{8.4}$$

where G is the so-called 'elastic energy release rate' per crack tip. It is also called the crack driving force: its dimensions of energy per unit plate thickness and per unit crack extension are also the dimensions of force per unit crack extension.

The energy consumed in crack propagation is denoted by $R = dW/da$ which is called the crack resistance. To a first approximation it can be assumed that the energy required to produce a crack by the decohesion of atomic bonds is the same for each increment da. This means that R is a constant.

The energy condition given in Equation (8.2) now states that G must be at least equal to R before crack propagation can occur. If R is a constant, this means that G must exceed a certain critical value G_{1c}. Hence crack growth occurs when

$$\frac{\pi\sigma_c^2 a}{E} = G_{1c} \quad \text{or} \quad \sigma_c = \sqrt{\left(\frac{EG_{1c}}{\pi a}\right)} \tag{8.5}$$

The critical energy release rate G_{1c} can be determined by measuring the stress σ_c required to fracture a plate with a crack of size $2a$, and by calculating from Equation (8.5).

Griffith derived his equation for glass which is a very brittle material. Therefore he assumed that R consisted of surface energy only. In ductile materials, such as metals, plastic deformation occurs at the crack tip. Much work is required in producing a new plastic zone at the tip of the advancing crack. Since this plastic zone has to be produced upon crack growth the energy for its formation can be considered as energy required for crack propagation. This means that for metals, R is mainly plastic energy; the surface energy is so small that it can be neglected. The energy criterion is necessary for crack extension. It need not be a sufficient criterion. Even if sufficient energy for crack propagation can be provided, the crack will not propagate unless the material at the crack tip is ready to fail: the material should be at the end of its capacity to take load and to undergo further straining. However, the latter criterion is equivalent to the energy criterion and therefore

$$G = \frac{K^2}{E} \tag{8.6}$$

Apparently the stress criterion and the energy criterion are fulfilled simultaneously. Hence Equations (8.1) and (8.5) are equivalent. It can be shown that Equation (8.6) is valid for plane stress and that a term $(1 - v^2)$ has to be added in the case of plane strain, leading to

$$G_1 = \frac{K_1^2}{(1 - v^2)E} \quad \text{and} \quad G_{1c} = \frac{K_{1c}^2}{(1 - v^2)E} \tag{8.7}$$

It is found experimentally that the critical value K_{1c} is dependent on geometric effects of thickness, reaching a minimum value for thick materials under plane strain conditions. The least thickness to give this minimum value is found to be, for material yield stress σ_y,

$$B \geq 2.5 \left(\frac{K_{1c}}{\sigma_y}\right)^2$$

8.4 General yielding fracture mechanics

The need to consider fracture resistance of materials outside the limits of validity of plane strain linear elastic fracture mechanics (LEFM) where the plastic zone size becomes large compared to the critical crack size is important for most engineering designs. To obtain valid K_{1c} results for relatively tough materials it would be necessary to use a test piece of dimensions so large that they would not be representative of the sections actually in use.

Significant yielding at a crack tip leads to the physical separation of the surfaces of a crack, and the magnitude of this separation is termed crack tip opening displacement (CTOD), and has been given the symbol δ. The CTOD approach enables critical toughness test measurements to be made in terms of δ_c, and then applied to determine allowable defect sizes for structural components. CTOD design curves have been developed using

$$\frac{\delta_c}{2\pi e_y a} = \left(\frac{e}{e_y}\right)^2 \quad \text{for } \frac{e}{e_y} \leq 0.5$$

$$= \frac{e}{e_y} - 0.25 \quad \text{for } \frac{e}{e_y} > 0.5$$

where δ_c = critical crack tip opening displacement
 a = allowable crack size
 e_y = material yield strain.
Reference 6 provides procedures for implementing this method.

8.5 Material testing

8.5.1 Charpy test

Reference 1 specifies the procedure for the Charpy V-notch impact test. The test consists of measuring the energy absorbed in breaking a notched bar specimen by one blow from a pendulum as shown in Fig. 8.7. The test can be carried out at a range of temperatures to determine the transition between ductile and brittle behaviour for the material.

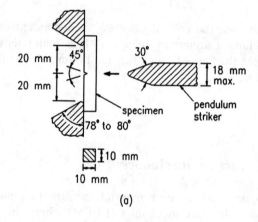

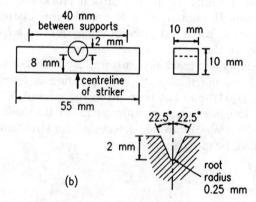

Fig. 8.7 Charpy test. (a) Test arrangement; (b) specimen

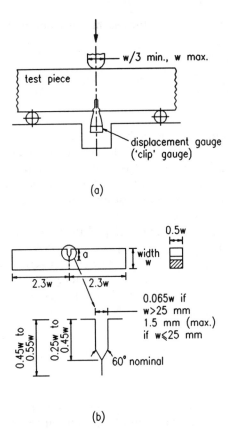

Fig. 8.8 CTOD test. (a) Test arrangement (preferred); (b) typical specimen dimensions

8.5.2 Crack tip opening displacement

Reference 4 specifies methods for CTOD tests, including the requirements for specimens, equipment, data analysis and recording of results. Figure 8.8 gives the test specimen dimensions.

The object of the test is to determine crack opening displacement values at the tip of a defect under various loads. The CTOD value is obtained indirectly by a 'clip' gauge suitably calibrated and linked to a data recording facility. The test is carried out in three-point bending over a loading span of $4w$.

CTOD tests are conducted on pre-cracked specimens. The procedure for obtaining the crack is outlined in Section 7 of BS 5762: 1979. Once the fatigue crack has been formed it is inspected to ensure compliance with the Code. The test piece is then loaded, typically at a rate of increase of stress intensity factor in the range 15–80 N/mm$^{-3/2}$ during the initial elastic stage. The test yields a plot of load versus clip gauge reading, from which the CTOD values can be calculated.

On completion of the test the fracture surface is examined to determine the type of fracture.

8.5.3 Other tests

There are other tests that can be carried out such as the wide plate test used for testing welded plate joints which was developed by Wells at the Welding Institute.[7,8] A large full-thickness plate, typically one metre square, is butt welded using the process and treatments to be used in the production weld. This test has the advantage of representing failure of an actual welded joint without the need for machining prior to testing.

While CTOD testing was being developed in the UK an alternative approach, the J integral method, was developed in the USA,[3] mainly for the nuclear industry where operating temperatures are well above the transition temperatures of the materials concerned. A particular difference between the two methods is in the specimen size. The small-scale sample for J integral tests may not adequately predict brittle fracture in real structures.

8.5.4 Test specimens

As has already been described each test procedure requires a sample of a certain size. Besides this feature the position of a sample in relation to a weld or, in the case of a thick plate, its position through the thickness is important. In modern structural steels toughness of the parent plate is rarely a problem. However, once a weld is deposited the toughness of the plate surrounding the weld, particularly in the heat-affected zone (HAZ), will be reduced. Although lower than for the parent plate the C_v or CTOD values that can be obtained should still provide adequate toughness at all standard operating temperatures. In the case of thicker joints then appropriate post-weld heat treatments should be carried out.

8.6 Fracture-safe design

The design requirements for steel structures in which brittle fracture is a consideration are given in most structural codes. There are several key factors which need to be considered when determining the risk of brittle fracture in a structure. These are:

(1) minimum operating temperature
(2) loading, in particular, rate of loading

(3) metallurgical features, i.e. parent plate, weld metal, HAZ
(4) thickness of material to be used.

Each of these factors influences the likelihood of brittle fracture occurring.

From experimental data and parametric studies examining the maximum tolerable defect sizes in steels under various operating conditions, Codes such as BS 5950, BS 5400 and EC3 provide tables giving the maximum permissible thickness of steel at given operating temperatures. Where a design does not fit into these broad categories attempts have been made to use fracture mechanics to provide a criterion for material selection in terms of Charpy test energy absorption. These relationships have been developed for structural steels, as at normal operating temperatures and slow rates of loading valid K_{Ic} values are not normally obtained. Critical CTOD tests have been widely used in the offshore and nuclear industries. However, all forms of fracture mechanics testing are expensive compared to routine quality control tests such as Charpy testing.

In general, the fracture toughness of structural steel increases with increasing temperature and decreasing loading rates. The effect of temperature on fracture toughness is well known. The effect of loading rate may be equally important not only in designing new structures but in understanding the behaviour of existing ones which may have been built from material with low toughness at their service temperature. The shift in the ductile/brittle transition temperature for structural steels can be considerable when comparing loading rates used in slow bend tests to those in Charpy tests. Results from experimental work have shown that the transition temperature for a BS 4360 grade 50 type steel can change from around 0°C to −60°C with decreasing loading rate.

Materials Standards set limits for the transition temperatures of various steel grades based on Charpy tests. When selecting an appropriate steel for a given structure it must be remembered that the Charpy values noted in the Standards apply to parent plate. Material toughness varies in the weld and heat affected zones of welded joints and where necessary these should be checked for adequate toughness. Furthermore, larger defects may be present in the weld area; appropriate procedures should be adopted to ensure that a welded structure will perform as designed. These could involve non-destructive testing including visual examination. In situations where defects are found, procedures such as those in Reference 6 can be used to assess their significance.

References to Chapter 8

1. British Standards Institution (1990) *Charpy impact test on metallic materials*. BS EN 10045−1, BSI, London.
2. British Standards Institution (1990) *Hot rolled products of non-alloy structural steels*. BS EN 10025, BSI, London.
3. American Society for Testing and Materials (1981) *The Standard Test for J_{Ic}, a*

Measure of Fracture Toughness. ASTM E813–81.
4. British Standards Institution (1979) *Methods for crack opening displacement (COD) testing.* BS 5762, BSI, London.
5. Griffith A.A. (1924) The theory of rupture. *Proceedings of International Congress on Applied Mechanics.* Delft.
6. British Standards Institution (1991) *Guidance on some methods for the derivation of acceptance levels for defects in fusion welded joints.* BS PD 6493, BSI, London.
7. Knott J.F. (1976) *Fundamentals of Fracture Mechanics.* Butterworths.
8. Meguid S.A. (1989) *Engineering Fracture Mechanics.* Elsevier Applied Science, Barking, Essex.

Chapter 9
Introduction to manual and computer analysis

by RANGACHARI NARAYANAN

9.1 Introduction

The analysis of structures consists essentially of mathematical modelling of the response of a structure to the applied loading. Such models are based on idealizations of the structural behaviour of the material and of the components. They are, therefore, imperfect to a larger or smaller degree, depending upon the extent of inaccuracy built into the assumptions in modelling. This is not to imply that the calculations are meaningless, rather to emphasize the fact that the assessment of structural responses is the best estimate that can be obtained in the light of the assumptions implicit in the modelling of the system. Some of these assumptions are necessary in the light of inadequate data; others are introduced to simplify the calculation procedure to economic levels.

There are several idealizations introduced in the modelling process.

- Firstly, the physical dimensions of the structural components are idealized. For example, skeletal structures are represented by a series of line elements and joints are assumed to be of negligible size. The imperfections in the member straightness are ignored or at best idealized.
- Material behaviour is simplified. For example, the stress—strain characteristic is assumed to be linearly elastic, and then perfectly plastic. No account is taken of the variation of yield stress along or across the member. The influence of residual stresses due to thermal processes (such as hot rolling and flame cutting), as well as that due to cold working and roller straightening, is ignored.
- The implications of actions which are included in the analytical process itself are frequently ignored. For example, the development of local plasticity at connections or possible effects of change of geometry causing local instability are rarely, if ever, accounted for in the analysis.

However it must be recognized that the design loads employed in assessing structural response are themselves approximate. The analysis chosen should therefore be adequate for the purpose and should be capable of providing the solutions at an economical cost.

The fundamental concepts employed in mathematical modelling are discussed next.

9.1.1 Equations of static equilibrium

From Newton's law of motion, the conditions under which a body remains in static equilibrium can be expressed as follows:

- The sum of the components of all forces acting on the body, resolved along any arbitrary direction is equal to zero. This condition is completely satisfied if the components of all forces resolved along the x, y, z directions individually add up to zero. (This can be represented by $\Sigma P_x = 0$, $\Sigma P_y = 0$, $\Sigma P_z = 0$, where P_x P_y and P_z represent forces resolved in the x, y, z directions.) These three equations represent the condition of zero translation.
- The sum of the moments of all forces resolved in any arbitrarily chosen plane about any point in that plane is zero. This condition is completely satisfied when all the moments resolved into xy, yz and zx planes all individually add up to zero. ($\Sigma M_{xy} = 0$, $\Sigma M_{yz} = 0$ and $\Sigma M_{zx} = 0$.) These three equations provide for zero rotation about the three axes.

If a structure is planar and is subjected to a system of coplanar forces, the conditions of equilibrium can be simplified to *three* equations as detailed below:

- The components of all forces resolved along the x and y directions will individually add up to zero ($\Sigma P_x = 0$ and $\Sigma P_y = 0$).
- The sum of the moments of all the forces about any arbitrarily chosen point in the plane is zero (i.e. $\Sigma M = 0$).

9.1.2 The principle of superposition

This principle is only applicable when the displacements are linear functions of applied loads. For structures subjected to multiple loading, the total effect of several loads can be computed as the sum of the individual effects calculated by applying the loads separately. This principle is a very useful tool in computing the combined effects of many load effects (e.g. moment, deflection, etc.). These can be calculated separately for each load and then summed.

9.2 Element analysis

Any complex structure can be looked upon as being built up of simpler units or components termed 'members' or 'elements'. Broadly speaking, these can be classified into *three* categories:

- *Skeletal structures* consisting of members whose one dimension (say, length) is much larger than the other two (viz. breadth and height). Such a *line element* is variously termed as a bar, beam, column or tie. A variety of structures are obtained by connecting such members together using rigid or hinged joints. Should all the axes of the members be situated in one plane, the structures so produced are termed *plane structures*. Where all members are not in one plane, the structures are termed *space structures*.

- Structures consisting of members whose two dimensions (viz. length and breadth) are of the same order but much greater than the thickness, fall into the second category. Such structural elements are called *plated structures*. Such structural elements are further classified as plates and shells depending upon whether they are plane or curved. In practice these units are used in combination with beams or bars. Slabs supported on beams, cellular structures, cylindrical or spherical shells are all examples of plated structures.

- The third category consists of structures composed of members having all the three dimensions (viz. length, breadth and depth) of the same order. The analysis of such structures is extremely complex, even when several simplifying assumptions are made. Dams, massive raft foundations, thick hollow spheres, caissons are all examples of three-dimensional structures.

For the most part the structural engineer is concerned with skeletal structures. Increasing sophistication in available techniques of analysis has enabled the economic design of plated structures in recent years. Three-dimensional analysis of structures is only rarely carried out. Under incremental loading, the initial deformation or displacement response of a steel member is elastic. Once the stresses caused by the application of load exceed the yield point, the cross section gradually yields. The gradual spread of plasticity results initially in an elasto-plastic response and then in plastic response, before ultimate collapse occurs.

9.3 Line elements

The deformation response of a line element is dependent on a number of cross-sectional properties such as area, A, second moment of area ($I_{xx} = \int y^2 dA$; $I_{yy} = \int x^2 dA$) and the product moment of area ($I_{xy} = \int xy\, dA$). The two axes xx and yy are orthogonal. For doubly symmetric sections, the axes of symmetry are those for which $\int xy\, dA = 0$. These are known as *principal axes*. For a plane area, the principal axes may be defined as a pair of rectangular axes in its plane and passing through its centroid, such that the product moment of area $\int xy\, dA = 0$, the co-ordinates referring to the principal axes. If the plane area has an axis of symmetry, it is obviously a principal axis (by symmetry $\int xy\, dA = 0$). The other axis is at right angles to it, through the centroid of the area.

Tables of properties of the section (including the centroid and shear centre of the section) are available as published data (e.g. SCI Steelwork Design Guide, Vol. 1).[1]

If the section has no axis of symmetry (e.g. an angle section) the principal axes

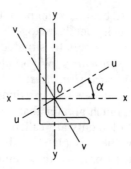

Fig. 9.1 Angle section (no axis of symmetry)

will have to be determined. Referring to Fig. 9.1, if *uou* and *vov* are the principal axes, the angle α between the *uu* and *xx* axes is given by

$$\tan 2\alpha = \frac{-2 I_{xy}}{I_{xx} - I_{yy}}$$

$$I_{uu} = \frac{I_{xx} + I_{yy}}{2} + \frac{I_{xx} - I_{yy}}{2} \cos 2\alpha - I_{xy} \sin 2\alpha \tag{9.1}$$

$$I_{vv} = \frac{I_{xx} + I_{yy}}{2} - \frac{I_{xx} - I_{yy}}{2} \cos 2\alpha + I_{xy} \sin 2\alpha \tag{9.2}$$

The values of α, I_{uu} and I_{vv} are available in published Steel Design Guides (e.g. Reference 1).

9.3.1 Elastic analysis of line elements under axial loading

When a cross section is subjected to a compressive or tensile axial load, P, the resulting stress is given by the load/area of the section, i.e. P/A. Axial load is defined as one acting at the centroid of the section. When loads are introduced into a section in a uniform manner (e.g. through a heavy end-plate), this represents the state of stress throughout the section. On the other hand, when a tensile load is introduced via a bolted connection, there will be regions of the member where stress concentrations occur and plastic behaviour may be evident locally, even though the mean stress across the section is well below yield.

If the force P is not applied at the centroid, the longitudinal direct stress distribution will no longer be uniform. If the force is offset by eccentricities of e_x and e_y measured from the centroidal axes in the y and x directions, the equivalent set of actions are (1) an axial force P, (2) a bending moment $M_x = P\,e_x$ in the yz plane and (3) a bending moment $M_y = P\,e_y$ in the zx plane (see Fig. 9.2). The method of evaluating the stress distribution due to an applied moment is given in

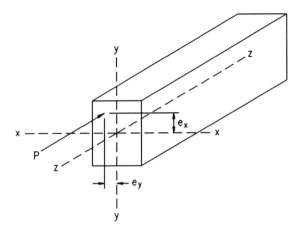

Fig. 9.2 Compressive force applied eccentrically with reference to the centroidal axis

a later section. The total stress at any section can be obtained as the algebraic sum of the stresses due to P, M_x and M_y.

9.3.2 Elastic analysis of line elements in pure bending

For a section having at least one axis of symmetry and acted upon by a bending moment in the plane of symmetry, the Bernoulli equation of bending may be used as the basis to determine both stresses and deflections within the elastic range. The assumptions which form the basis of the theory are:

- The beam is subjected to a pure moment (i.e. shear is absent). (Generally the deflections due to shear are small compared with those due to flexure; this is not true of deep beams.)
- Plane sections before bending remain plane after bending.
- The material has a constant value of modulus of elasticity (E) and is linearly elastic.

The following equation results (see Fig. 9.3).

$$\frac{M}{I} = \frac{f}{y} = \frac{E}{R} \tag{9.3}$$

where M is the applied moment; I is the second moment of area about the neutral axis; f is the longitudinal direct stress at any point within the cross section; y is the distance of the point from the neutral axis; E is the modulus of elasticity; R is the radius of curvature of the beam at the neutral axis.

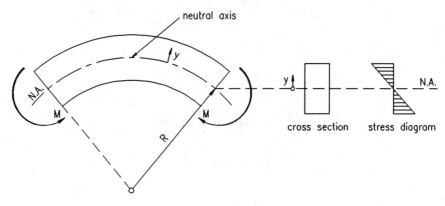

Fig. 9.3 Pure bending

From the above, the stress at any section can be obtained as

$$f = \frac{M\,y}{I}$$

For a given section (having a known value of I) the stress varies linearly from zero at the neutral axis to a maximum at extreme fibres on either side of the neutral axis.

$$f_{max} = \frac{M\,y_{max}}{I} = \frac{M}{Z} \tag{9.4}$$

where $Z = \dfrac{I}{y_{max}}$.

The term Z is known as the 'elastic section modulus' and is tabulated in section tables.[1] The elastic moment capacity of a given section may be found directly as the product of the elastic section modulus, Z, and the maximum allowable stress.

If the section is doubly symmetric, then the neutral axis is mid-way between the two extreme fibres. Hence, the maximum tensile and compressive stresses will be equal. For an unsymmetric section, this will not be the case as the value of y for the two extreme fibres will be different.

For a monosymmetric section, such as the T-section shown in Fig. 9.4, subjected to a moment acting in the plane of symmetry, the elastic neutral axis will be the centroidal axis. The above equations are still valid. The values of y_{max} for the two extreme fibres (one in compression and the other in tension) are different. For an applied sagging (positive) moment shown in Fig. 9.4, the extreme fibre stress in the flange will be compressive and that in the stalk will be tensile. The numerical values of the maximum tensile and compressive stresses will differ. In the case sketched in Fig. 9.4, the magnitude of the tensile stress will be greater, as y_{max} in tension is greater than that in compression.

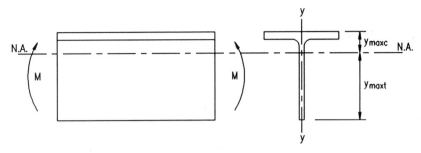

Fig. 9.4 Monosymmetric section subjected to bending

Caution has to be exercised in extending the pure bending theory to asymmetric sections. There are two special cases where no twisting occurs:

- Bending about a principal axis in which no displacement perpendicular to the plane of the applied moment results.
- The plane of the applied moment passes through the shear centre of the cross section.

When a cross section is subjected to an axial load and a moment such that no twisting occurs, the stresses may be determined by resolving the moment into components M_{uu} and M_{vv} about the principal axes uu and vv and combining the resulting longitudinal stresses with those resulting from axial loading:

$$f_{u,v} = \pm \frac{P}{A} \pm \frac{M_{uu} \, v}{I_{uu}} \pm \frac{M_{vv} \, u}{I_{vv}} \qquad (9.5)$$

For a section having two axes of symmetry (see Fig. 9.2) this simplifies to

$$f_{x,y} = \pm \frac{P}{A} \pm \frac{M_{xx} \, y}{I_{xx}} \pm \frac{M_{yy} \, x}{I_{yy}}$$

Pure bending does *not* cause the section to twist. When the shear force is applied eccentrically in relation to the shear centre of the cross section, the section twists and initially plane sections no longer remain plane. The response is complex and consists of a twist and a deflection with components in and perpendicular to the plane of the applied moment. This is not discussed in this chapter. A simplified method of calculating the elastic response of cross sections subjected to twisting moments is given in an SCI publication.[2]

9.3.3 Elastic analysis of line elements subject to shear

Pure bending discussed in the preceding section implies that the shear force applied on the section is zero. Application of transverse loads on a line element

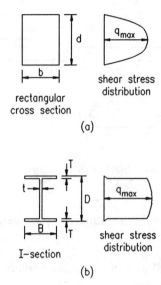

Fig. 9.5 Shear stress distribution: (a) in a rectangular cross section and (b) in an I-section

will, in general, cause a bending moment which varies along its length, and hence a shear force which also varies along the length is generated.

If the member remains elastic and is subjected to bending in a plane of symmetry (such as the vertical plane in a doubly symmetric or monosymmetric beam), then the shear stresses caused vary with the distance from the neutral axis.

For a narrow rectangular cross section of breadth b and depth d, subjected to a shear force V and bent in its strong direction (see Fig. 9.5(a)), the shear stress varies parabolically from zero at the lower and upper surfaces to a maximum value, q_{max}, at the neutral axis given by

$$q_{max} = \frac{3V}{2bd}$$

i.e. 50% higher than the average value.

For an I-section (Fig. 9.5(b)), the shear distribution can be evaluated from

$$q = \frac{V}{IB} \int_{y=h}^{y=h_{max}} by \, dy \qquad (9.6)$$

where B is the breadth of the section at which shear stress is evaluated. The integration is performed over that part of the section remote from the neutral axis, i.e. from $y = h$ to $y = h_{max}$ with a general variable width of b.

Clearly, for the I-(or T-) section, at the web/flange interface, the value of the integral will remain constant. As the section just inside the web becomes the section just inside the flange, the value of the vertical shear abruptly changes as B changes from web thickness to flange width.

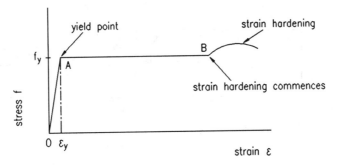

Fig. 9.6 Idealized stress–strain relationship for mild steel

9.3.4 Elements stressed beyond the elastic limit

The most important characteristic of structural steels (possessed by no other material to the same degree), is their capacity to withstand considerable deformation without fracture. A large part of this deformation occurs during the process of 'yielding', when the steel extends at a constant and uniform stress known as the 'yield stress'.

Figure 9.6 shows, in its idealized form, the stress–strain curve for structural steels subjected to direct tension. The line 0A represents the elastic straining of the material in accordance with Hooke's law. From A to B, the material yields while the stress remains constant and is equal to the yield stress, f_y. The strain occurring in the material during yielding remains after the load has been removed and is called 'plastic strain'. It is important to note that this plastic strain AB is at least ten times as large as the elastic strain, ε_y, at yield point.

When subjected to compression, various grades of structural steel behave in a similar manner and display the same property of yield. This characteristic is known as *ductility of steel*.

9.3.5 Bending of beams beyond the elastic limit

For simplicity, the case of a beam symmetrical about both axes is considered first. The fibres of the beam subjected to bending are stressed in tension or compression according to their position relative to the neutral axis and are strained as shown in Fig. 9.7.

While the beam remains entirely elastic, the stress in every fibre is proportional to its strain and to its distance from the neutral axis. The stress, f, in the extreme fibres cannot exceed the yield stress, f_y.

When the beam is subjected to a moment slightly greater than that which first

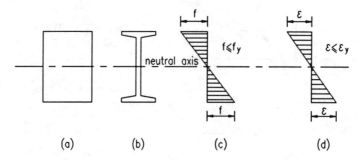

Fig. 9.7 Elastic distribution of stress and strain in a symmetric beam. (a) Rectangular section, (b) I-section, (c) stress distribution for (a) or (b), (d) strain distribution for (a) or (b)

produces yield in the extreme fibres, it does not fail. Instead, the outer fibres yield at constant stress, f_y, while the fibres nearer to the neutral axis sustain increased elastic stresses. Figure 9.8 shows the stress distribution for beams subjected to such moments. Such beams are said to be 'partially plastic' and those portions of their cross sections which have reached the yield stress are described as 'plastic zones'.

The depths of the plastic zones depend upon the magnitude of the applied moment. As the moment is increased, the plastic zones increase in depth, and, it is assumed that plastic yielding will continue to occur at yield stress, f_y, resulting in two stress blocks, one zone yielding in tension and one in compression. Figure 9.9 represents the stress distribution in beams stressed to this stage. The plastic zones occupy the whole area of the sections, which are then described as

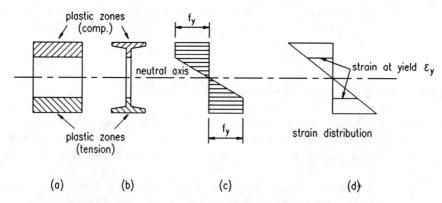

Fig. 9.8 Distribution of stress and strain beyond the elastic limit for a symmetric beam. (a) Rectangular section, (b) I-section, (c) stress distribution for (a) or (b), (d) strain distribution for (a) or (b)

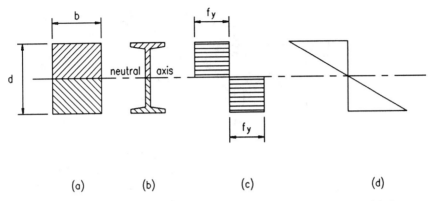

Fig. 9.9 Distribution of stress and strain in a fully plastic cross section. (a) Rectangular section, (b) I-section, (c) stress distribution for (a) or (b), (d) strain distribution for (a) or (b)

being 'fully plastic'. When the cross section of a member is fully plastic under a bending moment, any attempt to increase this moment will cause the member to act as if hinged at that point. This point is then described as a 'plastic hinge'.

The bending moment producing a plastic hinge is called the 'fully plastic moment' and is denoted by M_p. As the total compressive force and the total tensile force on the cross section must be equal, it follows that the plastic neutral axis is also the equal area axis, i.e. half the area of section is plastic in tension and the other half is plastic in compression. This is true for monosymmetric or unsymmetrical sections as well.

Shape factor

As described previously there will be two stress blocks, one in tension, the other in compression, each at yield stress. For equilibrium of the cross section, the areas in compression and tension must be equal. For a rectangular section the plastic moment can be calculated as

$$M_p = 2b \frac{d}{2} \frac{d}{4} f_y = \frac{bd^2}{4} f_y$$

which is 1.5 times the elastic moment capacity.

It will be noted that in developing this increased moment, there is large straining in the external fibres of the section together with large rotations and deflections. The behaviour may be plotted as a moment–rotation curve. Curves for various sections are shown in Fig. 9.10.

The ratio of the plastic modulus, S, to the elastic modulus, Z, is known as the shape factor, v, and it will govern the point in the moment–rotation curve when

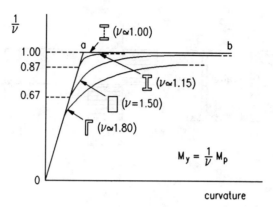

Fig. 9.10 Moment–rotation curves

non-linearity starts. For the ideal section in bending, i.e. two flange plates, this will have a value of unity. The value increases for more material at the centre of the section. For a universal beam, the value is about 1.15 increasing to 1.5 for a rectangle.

Plastic hinges and rigid plastic analysis

In deciding the manner in which a beam may fail it is desirable to understand the concept of how plastic hinges form when the beam becomes fully plastic. The number of hinges necessary for failure does not vary for a particular structure subject to a given loading condition, although a part of a structure may fail independently by the formation of a smaller number of hinges. The member or structure behaves in the manner of a hinged mechanism and in doing so, adjacent hinges rotate in opposite directions.

As the plastic deformations at collapse are considerably larger than elastic ones, it is assumed that the line element remains rigid between supports and hinge positions i.e. all plastic rotation occurs at the plastic hinges.

Considering a simply-supported beam subjected to a point load at mid-span (Fig. 9.11), the maximum strain will take place at the centre of the span where a plastic hinge will be formed at yield of full section. The remainder of the beam will remain straight: thus the entire energy will be absorbed by the rotation of the plastic hinge.

Work done at the plastic hinge $= M_p (2\theta)$

Work done by the displacement of the load $= W \left(\dfrac{L}{2} \theta\right)$

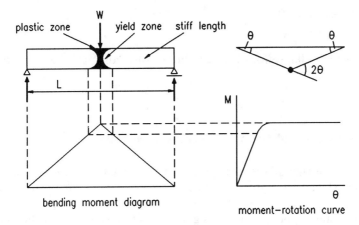

Fig. 9.11 Centrally-loaded simply-supported beam

At collapse, these two must be equal

$$2M_p \theta = \frac{WL}{2} \theta$$

$$W = 4M_p/L \quad \text{or} \quad M_p = WL/4$$

The moment at collapse of an encastré beam with a uniformly distributed load ($w = W/L$), is worked out in a manner similar to the above from Fig. 9.12.

Work done at the three plastic hinges $= M_p (\theta + 2\theta + \theta) = 4M_p\theta$

Work done by the displacement of the load (W) $= \dfrac{W}{L} \dfrac{L}{2} \dfrac{L}{2} \theta = \dfrac{WL}{4} \theta$

Equating the two,

$$\frac{WL}{4} \theta = 4M_p \theta$$

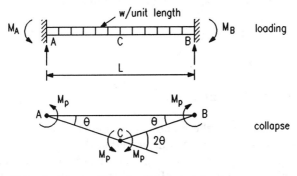

Fig. 9.12 Encastré beam with a uniformly distributed load

$$W = 16 M_p/L$$
or $$M_p = WL/16$$

The moments at collapse for other conditions of loading can be worked out by a similar procedure.

9.3.6 Load factor and theorems of plastic collapse

The load factor at rigid plastic collapse, λ_p, is defined as the lowest multiple of the design loads which will cause the whole structure, or any part of it, to become a mechanism.

In the limit-state approach, the designer seeks to ensure that at the appropriate factored loads the structure will *not* fail. Thus the rigid plastic load factor, λ_p, must not be less than unity, under factored loads.

The number of independent mechanisms, n, is related to the number of possible plastic hinge locations, h, and the degree of redundancy, r, of the skeletal structure, by the equation

$$n = h - r$$

The three theorems of plastic collapse are given below for reference

(1) *Lower bound or static theorem*
    A load factor, λ_s, computed on the basis of an *arbitrarily assumed* bending moment diagram which is *in equilibrium with the applied loads and where the fully plastic moment of resistance is nowhere exceeded*, will always be less than, or at best equal to, the load factor at rigid plastic collapse, λ_p.
    λ_p is the highest value of λ_s which can be found.
(2) *Upper bound or kinematic theorem*
    A load factor, λ_k, computed on the basis of an *arbitrarily assumed* mechanism will always be greater than, or at best equal to, the load factor at rigid plastic collapse, λ_p.
    λ_p is the lowest value of λ_k which can be found.
(3) *Uniqueness theorem*
    If *both* the above criteria ((1) and (2)) are satisfied, then $\lambda = \lambda_p$.

9.3.7 Effect of axial load and shear

If a member is subjected to the combined action of bending moment and axial force, the plastic moment capacity will be reduced.

The presence of an axial load implies that the sum of the tension and compression forces in the section is not zero (see Fig. 9.13). This means that the neutral axis

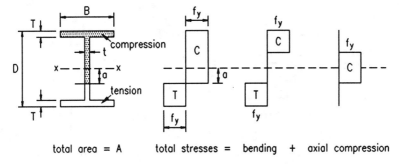

Fig. 9.13 The effect of combined bending and compression

moves away from the equal area axis providing an additional area in tension or compression depending on the type of axial load. The presence of shear forces will also reduce the moment capacity. For the beam sketched in Fig. 9.13,

axial load resisted $= 2a\, t\, f_y$

Defining $\quad n = \dfrac{\text{axial force resisted}}{\text{axial capacity of section}} = \dfrac{2at}{A}$,

$a = \dfrac{n A}{2t}$

For a given cross section, the plastic moment capacity, M_p, can be evaluated as explained previously. The reduced moment capacity, M'_p, in the presence of the axial load can be calculated as follows:

$$M'_p = M_p - t\, a^2\, f_y$$
$$= M_p - t\, \dfrac{n^2 A^2}{4t^2} f_y = \left(S - \dfrac{n^2 A^2}{4t}\right) f_y \tag{9.7}$$

where S is the plastic modulus of the section.

Section tables provide the moment capacity for available steel sections using the approach given above.[1] Similar expressions will be obtained for minor axis bending.

9.3.8 Plastic analysis of beams subjected to shear

Once the material in a beam has started to yield in a longitudinal direction, it is unable to sustain applied shear. When a shear, V, and an applied moment, M, are applied simultaneously to an I-section, a simplifying assumption is employed to reduce the complexity of calculations; shear resistance is assumed to be provided by the web, hence the shear stress in the web is obtained as a constant value of V divided by the web area (see Fig. 9.14). The longitudinal direct stress to cause

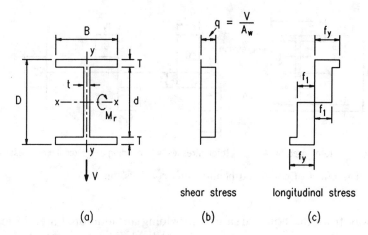

Fig. 9.14 Combined bending and shear

yield, f_1, in the presence of this shear stress, q, is obtained by using the von Mises yield criterion.[3]

$$f_y^2 = f_1^2 + 3q^2$$

The reduced plastic moment capacity is given by

$$M_r = M_p - \left(\frac{f_y - f_1}{f_y}\right) M_{pw} \qquad (9.8)$$

where M_{pw} is the fully plastic moment of resistance of the web.

The addition of an axial load to the above condition can be dealt with by shifting the neutral axis as was done in Fig. 9.13. The web area required to carry the axial load is now given by P/f_1 and the depth of the web, d_a, corresponding to this is given by

$$d_a = \frac{P}{f_1 t_w}$$

A further reduction in moment due to the introduction of the axial load is given by

$$\frac{t_w d_a^2}{4} f_1$$

Hence the reduced moment capacity of the section is given by

$$M_1 = M_p - \left(\frac{f_y - f_1}{f_y}\right) M_{pw} - \frac{t_w d_a^2}{4} f_1 \qquad (9.9)$$

where $f_1 = \sqrt{(f_y^2 - 3q^2)}$.

9.3.9 Plastic analysis for more than one condition of loading

When more than one condition of loading is to be applied to a line element, it may not always be obvious which is critical. It is necessary then to perform separate calculations, one for each loading condition, the section being determined by the solution requiring the largest plastic moment.

Unlike the elastic method of design in which moments produced by different loading systems can be added together, the opposite is true for the plastic theory. Plastic moments obtained by different loading systems *cannot* be combined, i.e. the plastic moment calculated for a given set of loads is only valid for that loading condition. This is because the 'principle of superposition' becomes *invalid* when parts of the structure have yielded.

9.4 Plates

Most steel structures consist of members which can be idealized as line elements. However, structural components having significant dimensions in two directions (viz. plates) are also encountered frequently. In steel structures, plates occur as components of I-, H-, T- or channel sections as well as in structural hollow sections. Sheets used to enclose lift shafts or walls or cladding in framed structures are also examples of plates.

With plane sheets, the stiffness and strength in all directions is identical and the plate is termed 'isotropic'. This is no longer true when stiffeners or corrugations are introduced in one direction. The stiffnesses of the plate in the x and y directions are substantially different. Such a plate is termed 'orthotropic'.

The x and y axes for the analysis of the plate are usually taken in the plane of the plate, as shown in Fig. 9.15, while the z axis is perpendicular to that plane. An element of the plate will be subjected to six stress components, viz. *three* direct stresses (σ_x, σ_y and σ_z) and *three* shear stresses (τ_{xy}, τ_{yz} and τ_{zx}). There are

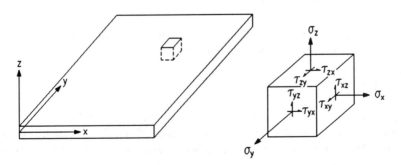

Fig. 9.15 Stress components on an element

six corresponding strains, viz. three direct strains (ϵ_x, ϵ_y and ϵ_z) and three shear strains (γ_{xy}, γ_{yz} and γ_{zx}). These stresses and strains are related in the elastic region by the material properties Young's modulus (E) and Poisson's ratio (v).

When considering the response of the plate, the approach customarily employed is termed 'plane stress idealization'. As the thickness, t, of the plate is small compared with its other two dimensions in the x and y directions, the *stresses* having components in the z direction are negligible (i.e. σ_z, τ_{yz} and τ_{xz} are all zero). This implies that the out-of-plane displacement is *not* zero and this condition is referred to as plane stress idealization.

For an isotropic plate, the general equation relating the displacement, w, perpendicular to the plane of the plate element is given by

$$\frac{\partial^4 w}{\partial x^4} + 2\frac{\partial^4 w}{\partial x^2 \partial y^2} + \frac{\partial^4 w}{\partial y^4} = \frac{q}{D} \tag{9.10}$$

where q is the normal applied load per unit area in the z direction which will, in general, vary with x and y. The term D is the flexural rigidity of the plate, given by

$$D = \frac{E t^3}{12 (1 - v^2)} \tag{9.11}$$

The main difficulty in using this approach lies in the choice of a suitable displacement function, w, which satisfies the boundary conditions. For loading conditions other than the simplest, an exact solution of this differential equation is virtually impossible. Hence approximate methods (e.g. multiple Fourier series) are utilized. Once a satisfactory displacement function, w, is obtained, the moments per unit width of the plate may be derived from

$$M_x = -D \left(\frac{\partial^2 w}{\partial x^2} + v \frac{\partial^2 w}{\partial y^2} \right)$$

$$M_y = -D \left(\frac{\partial^2 w}{\partial y^2} + v \frac{\partial^2 w}{\partial x^2} \right) \tag{9.12}$$

$$M_{xy} = -M_{yx} = D (1 - v) \frac{\partial^2 w}{\partial x \partial y}$$

For orthotropic plates, the stiffness in x and y directions is different and the equations are suitably modified as given below

$$D_x \frac{\partial^4 w}{\partial x^4} + 2D_{xy} \frac{\partial^4 w}{\partial x^2 \partial y^2} + D_y \frac{\partial^4 w}{\partial y^4} = q \tag{9.13}$$

where D_x and D_y are the flexural rigidities in the two directions.

In view of the difficulty of using classical methods for the solution of plate problems, finite element methods have been developed in recent years to provide satisfactory answers.

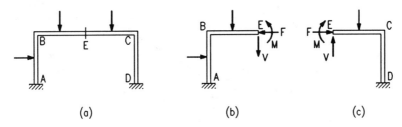

Fig. 9.16 Free body diagram

9.5 Analysis of skeletal structures

The evaluation of the stress resultants in members of skeletal frames involves the solution of a number of simultaneous equations. When a structure is in equilibrium, every element or constituent part of it is also in equilibrium. This property is made use of in developing the concept of the *free body diagram* for elements of a structure.

The portal frame sketched in Fig. 9.16 will now be considered for illustrating the concept. Assuming that there is an imaginary cut at E on the beam BC, the part ABE continues to be in equilibrium if the two forces and moment which existed at section E of the uncut frame are applied externally. The internal forces which existed at E are given by (1) an axial force F, (2) a shear force V and (3) a bending moment M. These are known as 'stress resultants'. The external forces on ABE, together with the forces F, V and M, keep the part ABE in equilibrium; Fig. 9.16(b) is called the free body diagram. On a rigid jointed plane frame there are three stress resultants at each imaginary cut. The part ECD must also remain in equilibrium. This consideration leads to a similar set of forces F, V and M shown in Fig. 9.16(c). It will be noted that the forces acting on the cut face E are equal and opposite. If the two free body diagrams are moved towards each other, it is obvious the internal forces F, V and M cancel out and the structure is restored to its original state of equilibrium. As previously stated, equilibrium implies $\Sigma P_x = 0$; $\Sigma P_y = 0$; $\Sigma M = 0$ for a planar structure. These equations can be validly applied by considering the structure as a whole, or by considering the free body diagram of a part of a structure.

In a similar manner, it can be seen that a three-dimensional rigid-jointed frame has six stress resultants across each section. These are, the axial force, two shears in two mutually perpendicular directions and three moments, as shown in Fig. 9.17.

With pin-jointed frames, be they two- or three-dimensional, there is only one stress resultant per member, viz. its axial load. When forces act on an elastic structure, it undergoes deformations, causing displacements at every point within the structure.

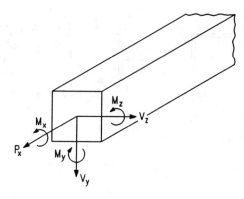

Fig. 9.17 Force and moments in x, y and z directions

The solution of forces in the frames is accomplished by relating the stress resultants to the displacements. The number of equations needed is governed by the 'degrees of freedom', i.e. the number of possible component displacements. At one end of the member of a pin-jointed plane frame, the member displacement has translational components in the x and y directions only, and no rotational displacement. The number of degrees of freedom is two. By similar reasoning it will be apparent that the number of degrees of freedom for a rigid-jointed plane frame member is three. For a member of a three-dimensional pin-jointed frame, it is also three, and for a similar rigid-jointed frame, it is six.

9.5.1 Stiffness and flexibility

Forces and displacements have a vital and interrelated role in the analysis of structures. Forces cause displacements and the occurrence of displacements implies the existence of forces. The relationship between forces and displacements is defined in one of two ways, viz. *flexibility* and *stiffness*.

Flexibility gives a measure of displacements associated with a given set of forces acting on the structure. This concept will be illustrated by considering the example of a spring loaded at one end by a static load P (see Fig. 9.18).

As the spring is linearly elastic, the extension, Δ, produced is directly proportional to the applied load, P. The deflection produced by a unit load (defined as the *flexibility* of the spring) is obviously Δ/P. Figure 9.18(b) illustrates the deflection response of a beam to an applied load P. Once again the flexibility of the beam is Δ/P.

In the simple cases considered above, flexibility simply gives the load–displacement response at a point. A more generalized definition applicable to the displacement response at a number of locations will now be obtained by considering a beam sketched in Fig. 9.19.

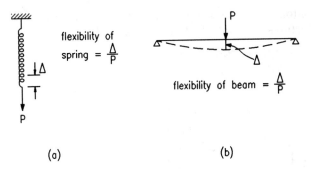

Fig. 9.18 Flexibility

Considering a unit load acting at point 1 (Fig. 9.19(b)), the corresponding deflections at points 1, 2 and 3 are denoted as f_{11}, f_{21} and f_{31} (the first subscript denotes the point at which the deflection is measured; the second subscript refers to the point at which the unit load is applied). The terms f_{11}, f_{21}, f_{31} are called flexibility coefficients. Figure 9.19(c) and (d) give the corresponding flexibility

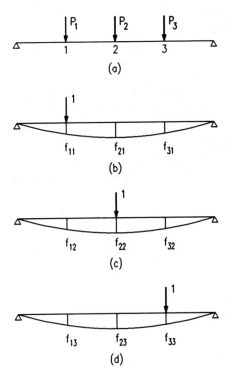

Fig. 9.19 Flexibility coefficients for a loaded beam

coefficients for load positions 2 and 3 respectively. By the principle of superposition, the total deflections at points 1, 2 and 3 due to P_1, P_2 and P_3 can be written as

$$\Delta_1 = P_1 f_{11} + P_2 f_{12} + P_3 f_{13}$$
$$\Delta_2 = P_1 f_{21} + P_2 f_{22} + P_3 f_{23}$$
$$\Delta_3 = P_1 f_{31} + P_2 f_{32} + P_3 f_{33}$$

Written in matrix form, this becomes

$$\begin{Bmatrix} \Delta_1 \\ \Delta_2 \\ \Delta_3 \end{Bmatrix} = \begin{bmatrix} f_{11} & f_{12} & f_{13} \\ f_{21} & f_{22} & f_{23} \\ f_{31} & f_{32} & f_{33} \end{bmatrix} \begin{Bmatrix} P_1 \\ P_2 \\ P_3 \end{Bmatrix} \quad (9.14)$$

or $\quad \{\Delta\} = [F]\{P\}$

where $\{\Delta\}$ = displacement matrix
$\quad\quad\quad [F]$ = flexibility matrix relating displacements to forces
$\quad\quad\quad \{P\}$ = force matrix

Hence $\{P\} = [F]^{-1}\{\Delta\}$

Stiffness is the inverse of flexibility and gives a measure of the forces corresponding to a given set of displacements. Considering the spring illustrated in Fig. 9.18(a), it is noted that the deflection response is directly proportional to the applied load, P. The force corresponding to unit displacement is obviously P/Δ. Likewise in Fig. 9.18(b), the load to be applied on the beam to cause a unit displacement at a point below the load is P/Δ. In its simplest form, stiffness coefficient refers to the load corresponding to a unit displacement at a given point and can be seen to be the reciprocal of flexibility. The concept is explained further using Fig. 9.20.

First the locations 2 and 3 are restrained from movement and a unit displacement is given at 1. This implies a downward force k_{11} at 1, an upward force k_{21} at 2 and a downward force k_{31} at 3. The forces at points 2 and 3 are necessary as otherwise there will be displacements at the locations 2 and 3.

The forces k_{11}, k_{21} and k_{31} are designated as stiffness coefficients. In a similar manner, the stiffness coefficients corresponding to unit displacements at points 2 and 3 are obtained.

The stiffness coefficients and the corresponding forces are linked by the following equations

$$P_1 = k_{11}\Delta_1 + k_{12}\Delta_2 + k_{13}\Delta_3$$
$$P_2 = k_{21}\Delta_1 + k_{22}\Delta_2 + k_{23}\Delta_3$$
$$P_3 = k_{31}\Delta_1 + k_{32}\Delta_2 + k_{33}\Delta_3$$

or $\quad \begin{Bmatrix} P_1 \\ P_2 \\ P_3 \end{Bmatrix} = \begin{bmatrix} k_{11} & k_{12} & k_{13} \\ k_{21} & k_{22} & k_{23} \\ k_{31} & k_{32} & k_{33} \end{bmatrix} \begin{Bmatrix} \Delta_1 \\ \Delta_2 \\ \Delta_3 \end{Bmatrix} \quad (9.15)$

or $\quad \{P\} = [K]\{\Delta\}$

where $[K]$ is the stiffness matrix relating forces and displacements.

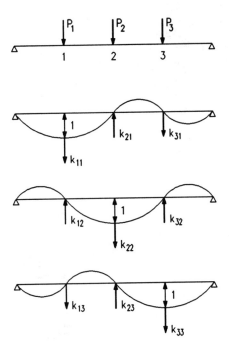

Fig. 9.20 Stiffness coefficients

9.5.2 Introduction to statically indeterminate skeletal structures

A structure for which the external reactions and internal forces and moments can be computed by using only the three equations of statics ($\Sigma P_x = 0$, $\Sigma P_y = 0$ and $\Sigma M = 0$) is known as 'statically determinate'. A structure for which the forces and moments cannot be computed from the principles of statics alone is 'statically indeterminate'. Examples of statically determinate skeletal structures are shown in Fig. 9.21.

In structures shown in Fig. 9.21(a), (b) and (c), the supporting forces and moments are just sufficient in number to withstand the external loading. For example, if one of the supports of (b) were to fail or if one of the members of (c) were to be removed, the structure would collapse.

However, when the beam or frame is provided with additional supports (see Fig. 9.21(d), (e)) or if the pin-jointed truss has more members than are required to make it 'perfect' (Fig. 9.21b), the structure becomes statically indeterminate.

The degree of indeterminacy (also termed 'the degree of redundancy') is obtained by the number of member forces or reaction components (viz. moments or forces) which should be 'released' to convert a statically indeterminate structure to a determinate one. If n forces or moments are required to be so released, the

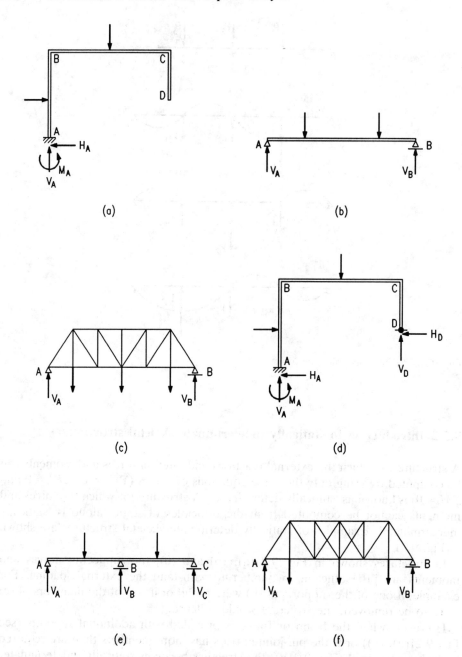

Fig. 9.21 Statically determinate and indeterminate skeletal structures. (a), (b) and (c) are determinate; (d), (e) and (f) are indeterminate

degree of indeterminacy is n. We need n independent equations (in addition to three equations of statics for a planar structure) to solve for forces and moments at all locations in the structure. The additional equations are usually written by considering the deformations or displacements of the structure. This means that the section properties (viz. area, second moment of area, etc.) have an important effect in evaluating the forces and moments of an indeterminate structure. Also, the settlement of a support or a slight lack of fit in a pin-jointed structure contributes materially to the internal forces and moments of an indeterminate structure.

9.5.3 The area moment method

The simplest technique of analyzing a beam which is indeterminate to a low degree is by the area moment method. The method is based on two theorems (see Fig. 9.22):

- *Area Moment Theorem 1*: The change in slope (in radians) between two points of the deflection curve in a loaded beam is numerically equal to the area under the *M/EI* diagram between these two points.
- *Area Moment Theorem 2*: The vertical intercept on any chosen line between the tangents drawn to the ends of any portion of a loaded beam, which was originally straight and horizontal, is numerically equal to the first moment of the area under the *M/EI* diagram between the two ends taken about that vertical line.

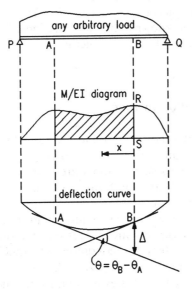

Fig. 9.22 Area moment theorems

$$\theta_B - \theta_A = \text{area of } \frac{M}{EI} \text{ diagram between A and B}$$

$$= \int_A^B \frac{M}{EI} dx \tag{9.16}$$

Δ = moment of the $\frac{M}{EI}$ diagram between A and B taken about the vertical line RS

$$= \int_A^B \frac{Mx}{EI} dx \tag{9.17}$$

(*Caution*: The vertical intercept is *not* the deflection of the beam from its original position.)

The area moment method can be used for solving problems like encastré beams, propped cantilevers, etc. The procedure is as follows:

(1) The redundant supports are removed thereby releasing the redundant forces and moments. The statically determinate M/EI diagram for externally applied loads can then be drawn.
(2) The externally applied loads are removed and the redundant forces and moments are introduced one at a time and the M/EI diagrams corresponding to each of these forces and moments are drawn.
(3) The slopes at supports and intercepts on a vertical axis passing through the supports are then calculated.
(4) A number of expressions are obtained. These are then equated to known values of slopes or displacements at supports. The equations so obtained can then be solved for the unknown redundant reactions. This enables the evaluation of the forces and moments in the structure.

9.5.4 The slope–deflection method

The slope–deflection method can be used to analyze all types of statically indeterminate beams and rigid frames. In this method all joints are considered rigid and the angles between members at the joints are considered not to change as the loads are applied. When beams or frames are deformed, the rigid joints are considered to rotate as a whole.

In the slope–deflection method, the rotations and translations of the joints are the unknowns. All end moments are expressed in terms of end rotations and translations. In order to satisfy the conditions of equilibrium, the sum of end moments acting on a rigid joint must total zero. Using this equation of equilibrium, the unknown rotation of each joint is evaluated, from which the end moments are computed.

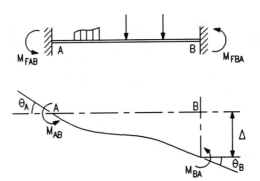

Fig. 9.23 Slope–deflection method

For the span AB shown in Fig. 9.23, the object is to express the end moments M_{AB} and M_{BA} in terms of end rotations θ_A and θ_B and translation Δ.

With the applied loading on the member, fixed end moments M_{FAB} and M_{FBA} are required to hold the tangents at the ends fixed in direction. (Counter clockwise end moments and rotations are taken as positive.) The slope–deflection equations for the case sketched in Fig. 9.23 are

$$M_{AB} = M_{FAB} + \frac{2EI}{L}\left(-2\theta_A - \theta_B + \frac{3\Delta}{L}\right)$$

$$M_{BA} = M_{FBA} + \frac{2EI}{L}\left(-2\theta_B - \theta_A + \frac{3\Delta}{L}\right)$$

(9.18)

where M_{FAB}, M_{FBA} are the fixed end moments at A and B due to loading and settlement of supports (counter clockwise positive); θ_A and θ_B are end rotations; and Δ is the downward settlement of support B relative to support A.

9.5.5 The moment-distribution method

The moment-distribution method can be employed to analyze continuous beams or rigid frames. Essentially it consists of solving the simultaneous equations in the slope–deflection method by successive approximations. Since the solution is by successive iteration, it is not even necessary to determine the degree of redundancy.

Two facets of the method must be appreciated (see Fig. 9.24):

- When a stiff joint in a structural system absorbs an applied moment with rotational movement only (i.e. no translation), Fig. 9.24(a), the moment resisted by the various members meeting at the joint is in proportion to their respective stiffnesses (Fig. 9.24(b)).
- When a member is fixed at one end and a moment, M, is applied at the other freely supported end, the moment induced at the fixed end is half the applied

282   *Introduction to manual and computer analysis*

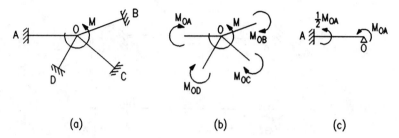

(a)   (b)   (c)

Fig. 9.24   Moment-distribution method

moment and acts in the same direction as *M*. (This is frequently referred to as the carry over.) (Fig. 9.24(c).)

Figure 9.25 shows the sign conventions employed in the moment-distribution method and Fig. 9.26 illustrates the moment-distribution procedure.

The moment-distribution method consists of locking all joints first and then releasing them one at a time. To begin with, all joints are locked, which implies that the fixed end moments due to applied loading will be applied at each joint. By releasing one joint at a time, the unbalanced moment at each joint is distributed to the various members meeting at the joint. Half of these applied moments are then carried over to the other end of each member. This creates a further imbalance at each joint and the unbalanced moments are once again distributed to all members meeting at each joint in proportion to their respective stiffnesses.

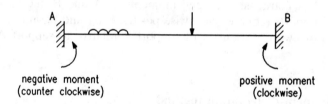

negative moment
(counter clockwise)

positive moment
(clockwise)

convention: clockwise moments are positive

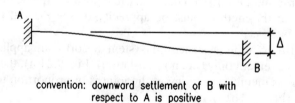

convention: downward settlement of B with
respect to A is positive

Fig. 9.25   Sign conventions used in moment-distribution method

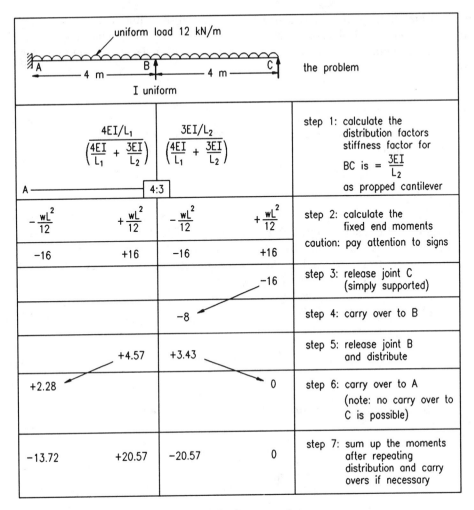

Fig. 9.26 An example of moment-distribution procedure

This procedure is repeated until the totals of all moments at each joint are sufficiently close to zero. At this stage the moment-distribution process is stopped and the final moments are obtained by summing up all the numbers in the respective columns.

9.5.6 Unit load method

Energy methods provide powerful tools for the analysis of structures. The unit load method can be directly derived from the complementary Energy Theorem which states that for any elastic structure in equilibrium under loads $P_1, P_2, \ldots$, the corresponding displacements $x_1, x_2, \ldots$ are given by the partial derivatives of the complementary energy, C, with respect to the loads P_1, P_2, etc. In other words,

$$\frac{\partial C}{\partial P_1} = x_1 \qquad \frac{\partial C}{\partial P_2} = x_2 \qquad (9.19)$$

For a linearly elastic system, the complementary energy is equal to the strain energy, U, hence

$$x_1 = \frac{\partial U}{\partial P_1} \qquad x_2 = \frac{\partial U}{\partial P_2} \qquad (9.20)$$

The total strain energy of an elastic system is given by the sum of the strain energies stored in each member due to bending, shear, torsion and axial loading.

The use of this will be illustrated by considering a simply-supported beam of length l subject to an external loading (see Fig. 9.27). Strain energy stored in the beam is predominantly flexural and is given by

$$\int_0^l \frac{M^2\,dx}{2EI}$$

$$x_1 = \text{deflection under } P_1 = \frac{\partial}{\partial P_1}\left(\int_0^l \frac{M^2 dx}{2EI}\right) = \int_0^l \frac{M}{EI}\frac{\partial M}{\partial P_1}\,dx \qquad (9.21)$$

$\dfrac{\partial M}{\partial P_1}$ is the bending moment due to a unit load and is denoted by m

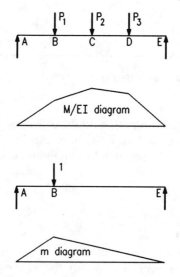

Fig. 9.27 Unit load method

Hence the procedure of the unit load method can be outlined (see Fig. 9.27):

(1) $\dfrac{M}{EI}$ diagram due to the external loading is obtained.
(2) The external loads are now removed and the moment diagram (m) due to a unit load applied at the point of required deflection is drawn.
(3) These two diagrams should now be integrated; in other words, the ordinates of the two diagrams are multiplied to obtain the deflection, given by

$$x = \int_0^l \frac{M}{EI} m \, dx$$

The same principle can be employed to determine the displacement due to other causes, viz. axial load or shear or torsion.

9.6 Finite element method

The advent of high-speed electronic digital computers has given tremendous impetus to numerical methods for solving engineering problems. Finite element methods form one of the most versatile classes of such methods which rely strongly on the matrix formulation of structural analysis. The application of finite elements dates back to the mid-1950s with the pioneering work by Argyris,[4] Clough and others.

The finite element method was first applied to the solution of plane stress problems and subsequently extended to the analysis of axisymmetric solids, plate bending problems and shell problems. A useful listing of elements developed in the past is documented in text books on finite element analysis.[5]

Stiffness matrices of finite elements are generally obtained from an assumed displacement pattern. Alternative formulations are equilibrium elements and hybrid elements. A more recent development is the so-called strain based elements. The formulation is based on the selection of simple independent functions for the linear strains or change of curvature; the strain–displacement equations are integrated to obtain expressions for the displacements.

The basic assumption in the finite element method of analysis is that the response of a continuous body to a given set of applied forces is equivalent to that of a system of discrete elements into which the body may be imagined to be subdivided. From the energy point of view, the equivalence between the body and its finite element model is therefore exact if the strain energy of the deformed body is equal to that of its discrete model.

The energy due to straining of the element, U, written in two-dimensional form is

$$U = \frac{1}{2} \iint (\varepsilon_x \sigma_x + \varepsilon_y \sigma_y + \gamma_{xy} \tau_{xy}) \, d(\text{vol})$$

or in matrix form

$$U = \frac{1}{2} \iint \{\varepsilon\}^T \{\sigma\} \, d(\text{vol}) \qquad (9.22)$$

in which $\{\varepsilon\}^T = \{\varepsilon_x, \varepsilon_y, \gamma_{xy}\}$
$$\{\sigma\} = \{\sigma_x, \sigma_y, \tau_{xy}\}^T \qquad (9.23)$$
and $\{\varepsilon\} = [f] \{\delta^e\}$ $\qquad\qquad\qquad\qquad\qquad\qquad(9.24)$
$\{\delta^e\} = \{u_1, u_2, \ldots, u_n\}^T$

where $\{\delta\}$ is the nodal displacement vector, $[f]$ is a function defining the strain distribution and e refers to a typical finite element. When the strain distribution within the model is exactly the same as that prevailing within the body, then the energy equation will be exactly satisfied.

The exact determination of the strain distribution function $[f]$ in Equation (9.24) presents considerable difficulties, since this can only be done by a rigorous solution of the equations of linear elasticity. It may not always be possible to obtain an *exact* shape function for the solution; however, a suitable function which is *adequate* to model strains can usually be selected. The derivation of simple membrane elements for plate problems is presented in the following pages.

9.6.1 Finite element procedure

As mentioned above, the basic concept of the finite element method is the idealization of the continuum as an assemblage of discrete structural elements. The stiffness properties of each element are then evaluated and the stiffness properties of the complete structure are obtained by superposition of the individual element stiffnesses. This gives a system of linear equations in terms of nodal point loads and displacements whose solution yields the unknown nodal point displacements.

The idealization governs the type of element which must be used in the solution. In many cases only one type of element is used for a given problem, but sometimes it is more convenient to adopt a 'mixed' subdivision in which more than one type of element is used.

The elements are assumed to be interconnected at a discrete number of nodal points or nodes. The nodal degrees of freedom normally refer to the displacement functions and their first partial derivatives at a node but very often may include other terms such as stresses, strains and second or even higher partial derivatives. For example, the triangular and rectangular membrane elements of Fig. 9.28(a) and (b) have 6 and 8 degrees of freedom respectively, representing the translations u and v at the corner nodes in the x and y directions.

A displacement function in terms of the co-ordinate variables x, y and the nodal displacement parameters (e.g. u_i, v_i, or δ_i) are chosen to represent the displacement variations within each element. By using the principle of virtual work or the

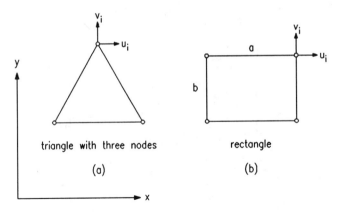

Fig. 9.28 Triangular and rectangular membrane elements

principle of minimum total potential energy, a stiffness matrix relating the nodal forces to the nodal displacements can be derived. Hence the choice of suitable displacement functions is the most important part of the whole procedure. A good displacement function leads to an element of high accuracy with converging characteristics; conversely, a wrongly chosen displacement function yields poor or non-converging results.

A displacement function may conveniently be established from simple polynomials or interpolation functions. The displacement field in each element must be expressed as a function of nodal point displacements only, and this must be done in such a way as to maintain inter-element compatibility, since this condition is necessary to establish a bound on the strain energy. Therefore, the displacement pattern and the nodal point degrees of freedom must be selected properly for each problem considered.

In general, it is not *always* necessary that the compatibility must be satisfied in order to achieve convergence to the true solution. If complete compatibility is not achieved, there exists an uncertainty as to the bound on the strain energy of the system. Therefore, in order to justify the performance and the ability of these elements to converge to the true solution, a critical test which indicates the performance in the limit must be carried out; that is, a convergence test with decreasing mesh size. If the performance is adequate, then convergence to the true solution is achieved within a reasonable computational effort, and the requirement for the complete compatibility can be relaxed.

Besides satisfying the compatibility requirement, the assumed displacement functions should include the following properties:

(1) Rigid body modes
(2) Constant strain and curvature states
(3) Invariance of the element stiffnesses.

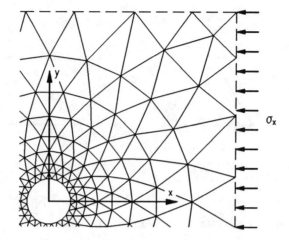

Fig. 9.29 Circular hole in uniform stress field

9.6.2 Idealization of the structure

The finite element idealization should represent the real structure as closely as possible with regard to geometrical shape, loading and boundary conditions. The geometrical form of the structure is the major factor to be considered when deciding the shape of elements to be used. In two-dimensional analyses the most frequently used elements are triangular or rectangular shapes. The triangular element has the advantage of simplicity in use and the ability to fit into irregular boundaries. Figure 9.29 shows an example using triangular elements and Fig. 9.30 shows a combination of triangular and rectangular elements. These figures also demonstrate the need to use relatively small elements in areas where high stress gradients occur. In many such cases the geometrical shape of the structure is such that a fine mesh of elements is required in order to match this shape.

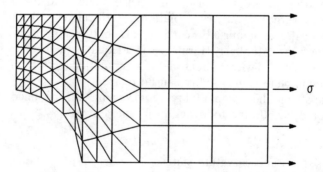

Fig. 9.30 Perforated tension strip (plane stress) – a quarter of a plate is analyzed

9.6.3 Procedure for evaluating membrane element stiffness

The formulation of the stiffness matrix $[K^e]$ of the membrane element is briefly discussed below.

The energy due to straining of the element is given by Equation (9.22). The stress components are shown in Fig. 9.31 and ε_x, ε_y, γ_{xy} are the corresponding strain components. For the state of plane stress, the stress components are related to the strain components as given below

$$\begin{Bmatrix} \sigma_x \\ \sigma_y \\ \tau_{xy} \end{Bmatrix} = \frac{E}{1-v^2} \begin{bmatrix} 1 & v & 0 \\ v & 1 & 0 \\ 0 & 0 & \frac{1-v}{2} \end{bmatrix} \begin{Bmatrix} \varepsilon_x \\ \varepsilon_y \\ \gamma_{xy} \end{Bmatrix} \tag{9.25}$$

where E and v are Young's modulus and Poisson's ratio respectively.

$$\therefore \{\sigma\} = [D]\{\varepsilon\}$$

The matrix $[D]$ is referred to as the elasticity or property matrix. The strain energy of the element is given by Equation (9.22) as

$$U = \frac{1}{2} \iint \{\varepsilon\}^T [D] \{\varepsilon\} \, d(\text{vol}) \tag{9.26}$$

Denoting the generalized nodal displacement by the vector $\{\delta^e\}$,

$$\{\delta^e\} = [C]\{A\} \tag{9.27}$$

where $\{A\} = [a_1, a_2, \ldots, a_n]^T$, which is a vector of polynomial constants, and $[C]$ is a transformation matrix.

The strains are obtained through making appropriate differentiations of the displacement function with respect to the relevant co-ordinate variable x or y. Thus,

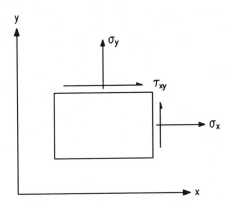

Fig. 9.31 Stress components on a plane element

$$\{\varepsilon\} = \begin{Bmatrix} \varepsilon_x \\ \varepsilon_y \\ \gamma_{xy} \end{Bmatrix} = \begin{Bmatrix} \partial u/\partial x \\ \partial v/\partial y \\ \dfrac{\partial u}{\partial y} + \dfrac{\partial v}{\partial x} \end{Bmatrix} \quad (9.28)$$

$$\therefore \quad \{\varepsilon\} = [B]\{A\} = [B][C]^{-1}\{\delta^e\} \quad (9.29)$$

where $[B]$ is the transformation matrix.

Using Equations (9.26) and (9.29), the following expression for U in terms of the nodal point displacement vector $\{\delta^e\}$ is obtained:

$$U = \frac{1}{2}\{\delta^e\}^T[C]^{-1,T}\left[\iint [B]^T[D][B]\mathrm{d}(\mathrm{vol})\right][C]^{-1}\{\delta^e\} \quad (9.30)$$

Differentiation of U with respect to the nodal displacements yields the stiffness matrix $[K^e]$.

$$[K^e] = t\,[C]^{-1,T}\left[\int_A [B]^T[D][B]\mathrm{d}A\right][C]^{-1} \quad (9.31)$$

where t is the thickness of the element (assumed constant).

The calculation of the stiffness matrices is generally carried out in two stages. The first stage is to calculate the terms inside the square brackets of Equation (9.31) i.e. the integration part. The second stage is to multiply the resulting integrations by the inverse of the transformation matrix $[C]$ and its transpose.

Equation (9.31) can now be written as

$$[K^e] = t\,[C]^{-1,T}[Q][C]^{-1} \quad (9.32)$$

where $[Q] = \int_A [B]^T[D][B]\,\mathrm{d}A$

The simplest elements for plane stress analysis have nodal points at the corners only and have two degrees of kinematic freedom at each nodal point, i.e. u and v. This type of element proves simple to derive and has been widely used. The simplest elements of this type are rectangular and triangular in shape.

A triangular element with nodal points at the corners is shown in Fig. 9.28(a). The displacement function of this element has two degrees of freedom at each nodal point and the displacements are assumed to vary linearly between nodal points. This results in constant values of the three strain components over the entire element; the displacement functions are

$$u = a_1 + a_2 x + a_3 y \\ v = a_4 + a_5 x + a_6 y \quad (9.33)$$

The rectangular element with sides a and b, shown in Fig. 9.28(b), is used with the following displacement functions

$$u = a_1 + a_2 x + a_3 y + a_4 xy$$
$$v = a_5 + a_6 x + a_7 y + a_8 xy \tag{9.34}$$

9.6.4 Procedure for evaluating plate bending element stiffness

The energy due to straining (bending) of the element is

$$U_b = \frac{1}{2} \int_A (\chi_x M_x + \chi_y M_y + 2 \chi_{xy} M_{xy}) \, dA \tag{9.35}$$

or in matrix form

$$U_b = \frac{1}{2} \int_A \{\chi_x, \chi_y, 2\chi_{xy}\} \begin{Bmatrix} M_x \\ M_y \\ M_{xy} \end{Bmatrix} dA$$

$$= \frac{1}{2} \int_A [\chi]^T \{M\} \, dA$$

where the moments M_x, M_y, M_{xy} are given in Fig. 9.32 and χ_x, χ_y and χ_{xy} are the corresponding curvatures, i.e.

$$\chi_x = \frac{\partial^2 w}{\partial x^2} \qquad \chi_y = \frac{\partial^2 w}{\partial y^2} \qquad \chi_{xy} = \frac{\partial^2 w}{\partial x \partial y} \tag{9.36}$$

where w is the transverse displacement of the plate element.

The conventional relationship between curvatures and moment is

$$\begin{Bmatrix} M_x \\ M_y \\ M_{xy} \end{Bmatrix} = [\overline{D}] \begin{Bmatrix} \chi_x \\ \chi_y \\ 2\chi_{xy} \end{Bmatrix} \tag{9.37}$$

or

$$\{M\} = [\overline{D}]\{\chi\}$$

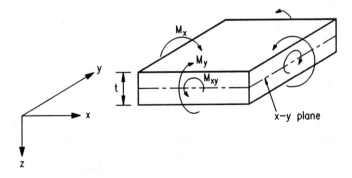

Fig. 9.32 Moments acting on a plane element

For an isotropic plate, the rigidity matrix, $[\bar{D}]$, may be written as

$$[\bar{D}] = \begin{bmatrix} D_1 & vD_1 & 0 \\ vD_1 & D_1 & 0 \\ 0 & 0 & D_{xy} \end{bmatrix}$$

and

$$D_x = D_y = D_1 = \frac{Et^3}{12(1-v^2)}$$

$$D_{xy} = \frac{1}{2}(1-v)D_1$$

Finally the strain energy can be written as

$$U_b = \frac{1}{2} \int_A [\chi]^T [\bar{D}][\chi] \, dA \tag{9.38}$$

Denoting the generalized nodal displacement by the vector $\{\bar{\delta}^e\}$,

$$\{\bar{\delta}^e\} = [\bar{C}]\{\bar{A}\} \tag{9.39}$$

where $\{\bar{A}\}$ is a vector of polynomial constants.
From Equations (9.36) and (9.39),

$$\{\chi\} = [\bar{B}]\{\bar{A}\} = [\bar{B}][\bar{C}]^{-1}\{\bar{\delta}^e\} \tag{9.40}$$

Using Equations (9.38) and (9.40), the following equation for U_b in terms of nodal point displacement parameters $\{\bar{\delta}^e\}$ is obtained

$$U_b = \frac{1}{2} \{\bar{\delta}^e\}^T [\bar{C}]^{-1,T} \left[\int_A [\bar{B}]^T [\bar{D}][\bar{B}] dA \right] [\bar{C}]^{-1} \{\bar{\delta}^e\} \tag{9.41}$$

Differentiation of U with respect to the nodal displacements yields the stiffness matrix $[\bar{K}^e]$

$$[\bar{K}^e] = [\bar{C}]^{-1,T} \left[\int_A [\bar{B}]^T [\bar{D}][\bar{B}] dA \right] [\bar{C}]^{-1} \tag{9.42}$$

Equation (9.42) can now be written as

$$[\bar{K}^e] = [\bar{C}]^{-1,T} [\bar{Q}][\bar{C}]^{-1} \tag{9.43}$$

where

$$[\bar{Q}] = \int_A [\bar{B}]^T [\bar{D}][\bar{B}] dA \tag{9.44}$$

As mentioned before, the accuracy is dependent on choosing a large number of elements. Many refined elements giving greater accuracy are described in standard books on finite element methods, which also provide details of assembling the elements and analyzing the structure.[5-7] These methods are used for solving a wide range of problems.

References to Chapter 9

1. The Steel Construction Institute (SCI) (1992) *Steelwork Design Guide to BS 5950: Part 1: 1990, Vol. 1: Section Properties Member Capacities* (3rd edn) SCI, Ascot, Berks.
2. Nethercot D.A., Salter P.R. & Malik A.S. (1989) *Design of Members Subject to Combined Bending and Torsion*. The Steel Construction Institute, Ascot, Berks.
3. Timoshenko S. (1965) *Strength of Materials – Part 2*. Van Nostrand & Co., New Jersey, USA.
4. Argyris J.H. (1960) *Energy Theorems and Structural Analysis*. Butterworths, London.
5. Zienkiewicz O.C. & Cheung Y.K. (1967) *The Finite Element Method in Structural and Continuum Mechanics*. McGraw-Hill, London.
6. Coates R.C., Coutie M.G. & Kong F.K. (1977) *Structural Analysis*. Thomas Nelson & Sons, Walton-on-Thames, UK.
7. Nath B. (1974) *Fundamentals of Finite Elements for Engineers*. Athlow Press, London.

Chapter 10
Beam analysis

by JOHN RIGHINIOTIS

10.1 Simply-supported beams

The calculations required to obtain the shear forces (SF) and bending moments (BM) in simply-supported beams form the basis of many other calculations required for the analysis of built-in beams, continuous beams and other indeterminate structures.

Appropriate formulae for simple beams and cantilevers under various types of loads are presented in the Appendix *Bending moment, shear and deflection tables for cantilevers* and *simply-supported beams*.

In the case of simple beams it is necessary to calculate the support reactions before the bending moments can be evaluated; the procedure is reversed for built-in or continuous beams. The following rules relate to the SF and BM diagrams for beams:

(1) the shear force at any section is the algebraic sum of normal forces acting to one side of the section
(2) shear is considered positive when the shear force calculated as above is upwards to the left of the section
(3) the BM at any section is the algebraic sum of the moments about that section of all forces to one side of the section
(4) moments are considered positive when the middle of a beam sags with respect to its ends or when tension occurs in the lower fibres of the beam
(5) for point loads only the SF diagram will consist of a series of horizontal and vertical lines, while the BM diagram will consist of sloping straight lines, changes of slope occurring only at the loads
(6) for uniformly distributed loads (UDL) the SF diagram will consist of sloping straight lines, while the BM diagram will consist of second-degree parabolas
(7) the maximum BM occurs at the point of zero shear, where such exists, or at the point where the shear force curve crosses the base line.

10.2 Propped cantilevers

Beams which are built-in at one end and simply-supported at the other are known as propped cantilevers. Normally, the ends of the beams are on the same level, in

Propped cantilevers 295

which case bending moments and reactions may be derived in two ways: by employing the Theorem of Three Moments or by deflection formulae. Appropriate formulae for propped cantilevers under various types of loading are presented in the Appendix *Bending moment, shear and deflection tables for propped cantilevers*.

10.2.1 Solution by the Theorem of Three Moments

Consider the propped cantilever AB in Fig. 10.1.

The bending moment at B may be found by using the Theorem of Three Moments, and assuming that AB is one span of a two-span continuous beam ABC which is symmetrical in every way about B.

Then the loads on AB and BC will produce free BM diagrams whose areas are A_1 and A_2 respectively, the centres of gravity (CG) of the areas being distances x_1 and x_2 from A and C respectively.

Now $M_A L_1 + 2M_B(L_1 + L_2) + M_C L_2 = 6 \left(\dfrac{A_1 x_1}{L_1} + \dfrac{A_2 x_2}{L_2} \right)$

where M_A, M_B and M_C are the numerical values of the hogging moments at the supports A, B and C.

But $L_1 = L_2 = L$
$M_A = M_C = 0$
$A_1 = A_2 = A$
and $x_1 = x_2 = x$

Hence $2M_B(2L) = 6 \times 2 \left(\dfrac{Ax}{L} \right)$

and $M_B = \dfrac{3Ax}{L^2}$

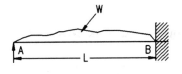

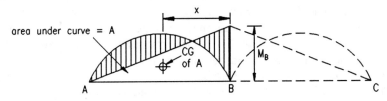

Fig. 10.1 Bending moment diagram for propped cantilever

Therefore the moment at the fixed end of a propped cantilever

$= 3Ax/L^2$

where A = the area of the free BM diagram, AB being considered as a simply-supported beam

x = the distance from the prop to the CG of the free BM diagram

and L = the span.

The reactions at each support may be found by employing a modified form of the formula used for beams built-in at both ends

$$SF_A = \text{the simple support reaction at A} = -\frac{M_B}{L}$$

$$SF_B = \text{the simple support reaction at B} = +\frac{M_B}{L}$$

where A is the propped end and B is built-in.

10.2.2 Sinking of supports

When the supports for a loaded propped cantilever do not maintain the same relative levels as in the unloaded condition, the BM and SF may be obtained by using the deflection method (Fig. 10.2). When the prop, B, sinks the load which it takes is reduced, while the fixing moment at the other end is increased. Two special cases arise: the first when the prop sinks so much that no load is taken by the prop, and the second when the built-in end sinks so much that the fixing moment is reduced to zero, i.e. the cantilever resembles a simple support beam. The two special cases are shown in Fig. 10.2.

10.3 Fixed, built-in or encastré beams

When the ends of a beam are firmly held so that they cannot rotate under the action of the superimposed loads, the beam is known as a fixed, built-in or encastré beam. The BM diagram for such a beam is in two parts: the free or positive BM diagram, which would have resulted had the ends been simply-supported, i.e. free to rotate, and the fixing or negative BM diagram which results from the restraints imposed upon the ends of the beam.

Normally, the supports for built-in beams are at the same level and the ends of the beams are horizontal. This type will be considered first.

Fixed, built-in or encastré beams 297

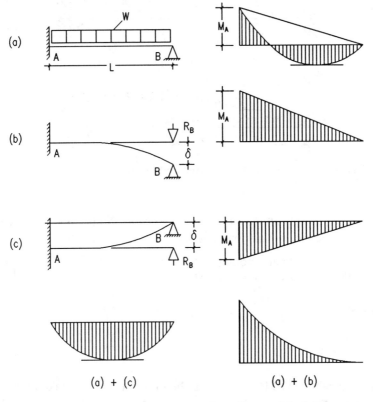

Fig. 10.2 Bending moment diagram for propped cantilever with sinking support

10.3.1 Beams with supports at the same level

The two conditions for solution, derived by Mohr, are:

(1) the area of the fixing or negative BM diagram is equal to that of the free or positive BM diagram
(2) the centres of gravity of the two diagrams lie in the same vertical line, i.e. are equidistant from a given end of the beam.

Figure 10.3 shows a typical BM diagram for a built-in beam.
ACDB is the diagram of the free moment M_s and the trapezium AEFB is the diagram of the fixing moment M_i, the portions shaded representing the final diagram.

Let A_s = the area of the free BM diagram
and A_i = the area of the fixing moment diagram.

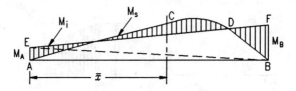

Fig. 10.3 Bending moment diagram for fixed-end beam

Then from condition (1) above, $A_s = A_i$, while from condition (2) their centres of gravity lie in the same vertical line, say, distance $\bar{x}$ from the left-hand support A.

Now   AE = the fixing moment M_A
and   BF = the fixing moment M_B.

Therefore
$$\frac{M_A + M_B}{2} \times L = A_i$$

and
$$M_A + M_B = \frac{2A_i}{L} \tag{10.1}$$

Divide the trapezium AEFB by drawing the diagonal EB and take area moments about the support A.

Then
$$A_i \bar{x} = \left(\frac{M_A \times L}{2} \times \frac{L}{3}\right) + \left(\frac{M_B \times L}{2} \times \frac{2L}{3}\right)$$

$$= \frac{L^2}{6}(M_A + 2M_B)$$

$$M_A + 2M_B = \frac{6A_i \bar{x}}{L^2} \tag{10.2}$$

But
$$M_A + M_B = \frac{2A_i}{L} \tag{10.3}$$

Also $A_s = A_i$

Subtracting Equation (10.1) from Equation (10.2) and substituting A_s for A_i gives

$$M_B = \frac{6A_s \bar{x}}{L^2} - \frac{2A_s}{L}$$

Similarly
$$M_A = \frac{4A_s}{L} - \frac{6A_s \bar{x}}{L^2}$$

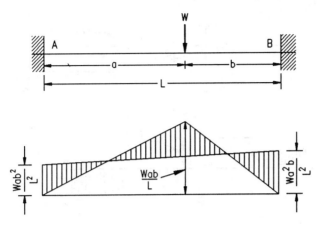

Fig. 10.4 Bending moment diagram for point load on fixed-end beam

It will be seen, therefore, that the fixing moments for any built-in beam on level supports can be calculated provided that the area of the free BM diagram and the position of its centre of gravity are known.

For point loads, however, the principle of reciprocal moments provides the simplest solution.

With reference to Fig. 10.4,

$$M_A = \frac{Wab}{L} \times \frac{b}{L} = \frac{Wab^2}{L^2}$$

$$M_B = \frac{Wab}{L} \times \frac{a}{L} = \frac{Wa^2b}{L^2}$$

i.e. the fixing moments are in reciprocal proportion to the distances of the ends of the beam from the point load.

In the case of several isolated loads, this principle is applied to each load in turn and the results summed.

It should be noted that appropriate formulae for built-in beams are given in the Appendix *Bending moment, shear and deflection tables for built-in beams*.

10.3.2 Beams with supports at different levels

The ends are assumed, as before, to be horizontal.

The bent form of the unloaded beam as shown in Fig. 10.5 is similar to the bent form of two simple cantilevers which can be achieved by cutting the beam at the centre C, and placing downward and upward loads at the free ends of the cantilevers such that the deflection at the end of each cantilever is $d/2$.

Beam analysis

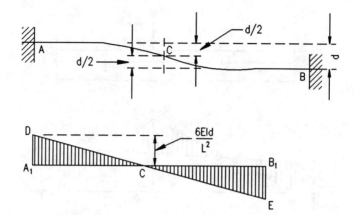

Fig. 10.5 Bending moment diagram for fixed-end beam with supports at different levels

Therefore $\dfrac{d}{2} = \dfrac{P(L/2)^3}{3EI}$ (being the standard deflection formula)

or $P = \dfrac{12EId}{L^3}$

This load would cause a BM at A or B equal to

$$P \times \dfrac{L}{2} = \dfrac{12EId}{L^3} \times \dfrac{L}{2} = \dfrac{6EId}{L^2}$$

The solution in any given case consists of adding to the ordinary diagram of BM, the BM diagram A_1DCEB_1.

Shear forces in fixed beams

In the case of fixed beams it is necessary to evaluate the BM before the SF can be determined. This is the converse of the procedure for the case of simply-supported beams.

The SF at the ends of a beam is found in the following manner:

$$SF_A = \text{the simple support reaction at A} = +\dfrac{M_A - M_B}{L}$$

$$SF_B = \text{the simple support reaction at B} = +\dfrac{M_B - M_A}{L}$$

where M_A and M_B are the numerical values of the moments at the ends of the beam.

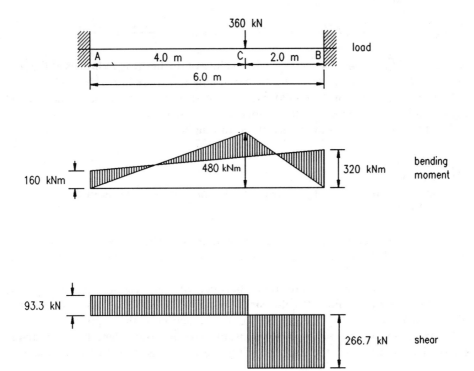

Fig. 10.6 Bending moment and shear force diagrams for fixed-end beam

These formulae must be followed exactly with respect to the signs shown since if M_A is smaller than M_B the signs will adjust themselves.

It will be seen that for symmetrical loads, where $M_A = M_B$, the reactions will be the same as for simply-supported beams.

An example of bending moment and shear force diagrams for a built-in beam carrying a point load is given in Fig. 10.6.

10.4 Continuous beams

The solution of this type of beam consists, in the first instance, of the evaluation of the fixing or negative moments at the supports.

The most general method is the use of Clapeyron's Theorem of Three Moments.

The theorem applies only to any two adjacent spans in a continuous beam and in its simplest form deals with a beam which has all the supports at the same level, and has a constant section throughout its length.

The proof of the theorem results in the following expression:

$$M_A \times L_1 + 2M_B(L_1 + L_2) + M_C \times L_2 = 6\left(\frac{A_1 \times x_1}{L_1} + \frac{A_2 \times x_2}{L_2}\right)$$

where M_A, M_B and M_C are the numerical values of the hogging moments at the supports A, B and C respectively, and the remaining terms are illustrated in Fig. 10.7.

In a continuous beam the conditions at the end supports are usually known, and these conditions provide starting points for the solution.

The types of end conditions are three in number:

(1) simply supported
(2) partially fixed, e.g. a cantilever
(3) completely fixed, i.e. the end of the beam is horizontal as in the case of a fixed beam.

The SF at the end of any span is calculated after the support moments have been evaluated, in the same manner as for a fixed beam, each span being treated separately.

It is essential to note the difference between SF and reaction at any support, e.g. with reference to Fig. 10.7 the SF at support B due to span AB added to the SF at B due to span BC is equal to the total reaction at the support.

If the section of the beam is not constant over its whole length, but remains constant for each span, the expression for the moments is rewritten as follows:

$$M_A \times \frac{L_1}{I_1} + 2M_B\left(\frac{L_1}{I_1} + \frac{L_2}{I_2}\right) + M_C \times \frac{L_2}{I_2} = 6\left(\frac{A_1 \times x_1}{L_1 \times I_1} + \frac{A_2 \times x_2}{L_2 \times I_2}\right)$$

in which I_1 is the second moment of area for span L_1 and I_2 is the second moment of area for span L_2.

Example

A two-span continuous beam ABC, of constant cross section, is simply supported at A and C and loaded as shown in Fig. 10.8.

Applying Clapeyron's theorem,

$L_1 = 2.0$ m

$L_2 = 3.0$ m

$A_1 = \dfrac{10 \times 2}{8} \times \dfrac{2 \times 2}{3} = \dfrac{10}{3}$ kN m^2

$A_2 = \dfrac{200 \times 1 \times 2}{3} \times \dfrac{3}{2} = 200$ kN m^2

$x_1 = 1.0$ m

$x_2 = \dfrac{4.0}{3}$ m

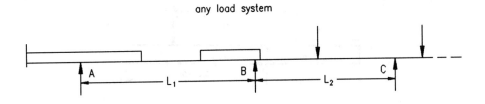

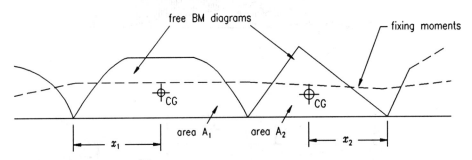

Fig. 10.7 Clapeyron's Theorem of Three Moments

Therefore $M_A \times 2 + 2M_B(2 + 3) + M_C \times 3 = 6\left(\dfrac{10 \times 1}{3 \times 2} + \dfrac{200 \times 4}{3 \times 3}\right)$

Since A and C are simple supports

$M_A = M_C = 0$

Therefore $M_B = \dfrac{6}{10}\left(\dfrac{10}{6} + \dfrac{800}{9}\right) = \dfrac{6}{10}(90.56) = 54.33 \text{ kN m}$

$SF_A = 5 + \dfrac{0 - 54.33}{2} = 5 - 27.17 = -22.17 \text{ kN}$

SF_B for span AB $= 5 + 27.17 = 32.17 \text{ kN}$

$SF_C = \dfrac{200 \times 2}{3} + \dfrac{0 - 54.33}{3} = 133.33 - 18.11 = 115.22 \text{ kN}$

SF_B for span BC $= \dfrac{200}{3} + 18.11 = 66.67 + 18.11 = 84.78 \text{ kN}$

Note that the negative reaction at A means that the end A will tend to lift off its support and will have to be held down.

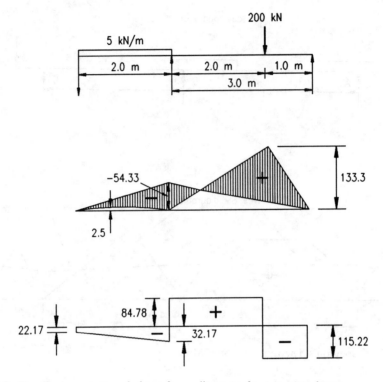

Fig. 10.8 Bending moment and shear force diagrams for two-span beam

10.5 Plastic failure of single members

The concept of the plastic hinge, capable of undergoing large rotation once the applied moment has reached the limiting value M_p, constitutes the basis of plastic design. This concept may be illustrated by examining the development of the collapse mode of a fixed-end beam subjected to a uniformly distributed load of increasing intensity w (Fig. 10.9(a)). Such a member is statically indeterminate, having three redundancies which however reduce to two unknowns if the axial thrust in the member is assumed to be zero. It will be assumed that the two unknown quantities are the fixing moments M_A and M_B. As the load increases, the beam initially behaves in an elastic manner and the value of the redundant moments can be derived by applying the three general conditions used in elastic structural analysis, namely those of

(1) equilibrium (application of statics)
(2) moment–curvature ($EI\ d^2y/dx^2 = M$)
(3) compatibility condition (continuity, including geometric conditions at the supports).

Plastic failure of single members

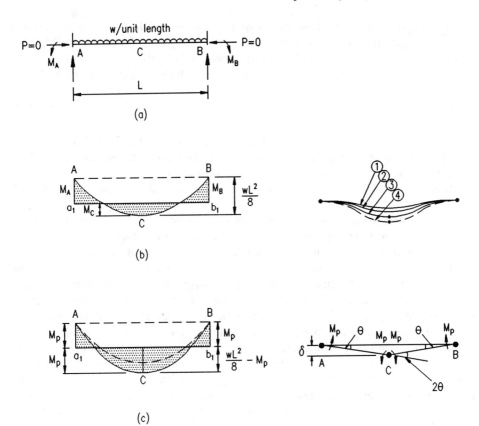

Fig. 10.9 Plastic failure of a fixed-end beam

For the beam in Fig. 10.9(a), the first condition (equilibrium) is satisfied by drawing the bending moment diagram (shaded in Fig. 10.9(b)) as a superposition of the simply-supported sagging parabolic moment diagram ACB (peak value of $wL^2/8$) on the uniform moment diagram Aa_1b_1B due to the end hogging moments M_A and M_B. From the moment–curvature relation applied over the full length of the beam, it is readily deduced, by any one of a number of standard methods, that the rotations of the end sections at A (θ_A clockwise) and at B (θ_B anti-clockwise) and the central deflection Δ_C are given by

$$\theta_A = \frac{wL^3}{24EI} - \frac{M_A L}{3EI} - \frac{M_B L}{6EI} \tag{10.4}$$

$$\theta_B = \frac{wL^3}{24EI} - \frac{M_A L}{6EI} - \frac{M_B L}{3EI} \tag{10.5}$$

$$\Delta_C = \frac{5}{384}\frac{wL^4}{EI} - \frac{M_A L^2}{16EI} - \frac{M_B L^2}{16EI} \tag{10.6}$$

The compatibility conditions that now have to be satisfied are the directional restraint of the end sections of the beam, i.e. $\theta_A = \theta_B = 0$, giving the well-known result $M_A = M_B = wL^2/12$. Equilibrium considerations now lead to the derivation of the central sagging moment, M_C. It follows from Fig. 10.9(b) that

$$M_C = \frac{wL^2}{8} - \left(\frac{M_A}{2} + \frac{M_B}{2}\right) \qquad (10.7)$$

whence $M_C = wL^2/24$, i.e. the end moments are twice the central moment. Finally, application of Equation (10.3) gives the central deflection as $wL^4/384EI$.

Suppose the load intensity w is increased until the fibres yield at some point in the beam. It is assumed that the cross section has an idealized moment–curvature relationship, i.e. the shape factor is unity (see Section 9.3.5). Up to this stage the ratio of the end to central moments remains at 2, as represented by the dashed line in Fig. 10.9(c). Due to the symmetry of the structure and loading, plastic hinges will develop simultaneously at the fixed ends, i.e. at the points of maximum moment. At the moment when the hinges form at the ends, the fixing moments have become equal to the M_p of the beam and the loading intensity w has reached a value of $12M_p/L^2$. A slight load increase then causes the plastic hinges to rotate while sustaining this constant moment M_p. This means that thereafter the beam behaves as a simply-supported beam with constant end moments of M_p. The structure is now statically determinate, and therefore the two degrees of redundancy in the original problem no longer exist. In other words, two plastic hinges have formed, eliminating a corresponding number of redundancies. At the same time two compatibility requirements have been eliminated; the condition that end slopes are zero is no longer correct because the ends of the member are now rotating as plastic hinges.

The central deflection at this stage, derived from Equation (10.6) has the value

$$\Delta_C = \frac{5}{384}\left(\frac{12M_p}{L^2}\right)\frac{L^4}{EI} - \frac{M_pL^2}{8EI} = \frac{M_pL^2}{32EI}$$

and, from Equation (10.7), the central sagging moment M_C becomes $[(wL^2/8) - M_p]$. Further increases in the load intensity cause a third plastic hinge to form at mid-span, see the full line in Fig. 10.9(c). This means that the moment M_C attains a value of M_p and therefore at C

$$M_p = \frac{wL^2}{8} - M_p$$

hence

$$M_p = wL^2/16 \quad \text{or} \quad w = 16M_p/L^2$$

The ratio of the end and central moments is now unity; as a result of the formation of the plastic hinges there has been a redistribution of moments.

Substituting $w = 16M_p/L^2$ and $M_A = M_B = M_p$ in Equation (10.6) gives a central deflection of $\Delta_C = M_pL^2/12EI$. When the final hinge has formed the

Plastic failure of single members   307

deflection increases rapidly without any further increase in the load. The beam is said to have failed as a hinged mechanism.

Now consider the same fixed-end beam with an initial settlement of $wL^4/144EI$ at end A (see Fig. 10.10(a)). The results derived from moment–curvature considerations [Equations (10.1)–(10.3)] can still be applied provided θ_A, θ_B and Δ_C are taken with reference to the chord AB between the ends of the member, the chord having rotated through an angle $wL^3/144EI$ relative to its initial position A_0B. Thus the compatibility conditions at the ends of the member are now

$$\theta_A = -\theta_B = wL^3/144EI$$

whence, by substitution in Equations (10.4), (10.5) and (10.7),

$$M_A = wL^2/24 \qquad M_B = wL^2/8 \qquad M_C = wL^3/24$$

The largest elastic moment occurs at B and therefore the first hinge starts there at a load intensity given by $M_p = wL^2/8$, i.e. $w = 8M_p/L^2$. Equation (10.6) shows that the central deflection is $M_pL^2/144EI$.

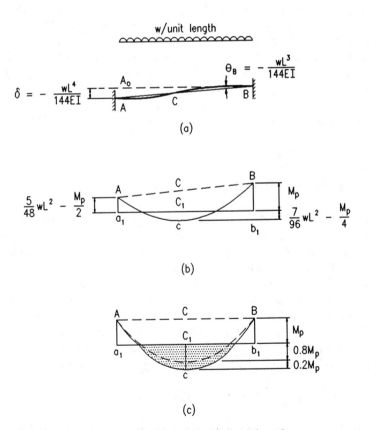

Fig. 10.10 Plastic failure of a fixed-end beam with initial settlement

The number of redundancies has now been reduced by one (since $M_B = M_p$) while the compatibility condition represented by Equation (10.7) no longer applies. Substitution of $\theta_A = wL^3/144EI$ and $M_B = M_p$ in Equations (10.4) and (10.7) gives

$$M_A = \frac{5wL^2}{48} - \frac{M_p}{2}$$

$$M_C = \frac{7wL^2}{96} - \frac{M_p}{4}$$

Using these new values for M_A and M_C the bending-moment diagram Fig. 10.10(b) is obtained and inspection of the diagram shows that the second plastic hinge can be expected to occur at end A. Putting $M_A = M_p$ gives

$$w = \frac{3M_p}{2} \times \frac{48}{5L^2} = \frac{14.4M_p}{L^2} \qquad M_c = 0.8M_p$$

while Equation (10.3) gives $\Delta_C = M_p L^2/16EI$. The beam is now statically determinate with $M_A = M_B = M_p$, and eventually the third hinge forms at C when w reaches $16M_p/L^2$ (see Fig. 10.10(c)) and $\Delta_C = M_p L^2/12EI$. The important point to note is that, despite the difference in initial conditions, the failure pattern, the failure load and the deflection at the point of failure (relative to the ends) are the same for the two cases. The uniqueness of the plastic limit load, i.e. its independence of initial conditions of internal stress or settlement of supports, is a general feature of plastic analysis. Deflections at the point of collapse can however be affected, as indeed they are in the second case just described, when considered relative to the original support position A_0B (Fig. 10.10(a)).

10.6 Plastic failure of propped cantilevers

The case of the propped cantilever under a uniformly distributed load, Fig. 10.11(a), cannot be solved quite so simply and both upper and lower bound methods described in Chapter 9 have to be used. A possible equilibrium condition is shown in Fig. 10.11(b) where the reactant line a_1B has been arranged so that the co-ordinate at the left-hand support is equal to the co-ordinate of the resultant moment diagram at mid-span. At the right-hand support the condition of zero resultant moment has to be satisfied. If the equal moments at A and C are regarded as plastic hinge (M_p) values, the mechanism condition is satisfied. Hence M_p is an upper bound (unsafe) value and should be denoted by M_u. Considering the geometry of the moment diagram at mid-span,

$$\frac{\lambda w L^2}{8} = Cc_1 + c_1c = \frac{M_u}{2} + M_u \qquad (M_u = M_p)$$

where λ is the load factor at rigid plastic collapse (see Section 9.3.6)

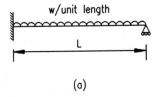

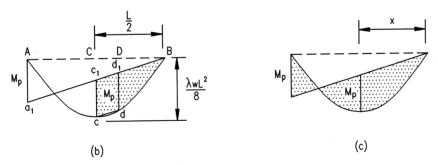

Fig. 10.11 Plastic failure of a propped cantilever

whence

$$M_c = \frac{\lambda w L^2}{12} = 0.0833 \lambda w L^2$$

By a closer inspection of the resultant moment diagram, it is readily shown that the maximum resultant moment does not occur at the mid-span, but between C and D, at $5L/12$ from the propped support. The value of this moment is $\lambda w L^2/11.52$ ($= 0.0868\lambda w L^2$), which is in excess of M_u. Note that this value is based on the moment at the fixed end being $\lambda w L^2/12$. By making the plastic moment of the beam equal to this higher value, a bending moment diagram satisfying the equilibrium and plastic moment condition is derived, i.e. a static solution $M_1 = 0.0868\lambda w L^2$. The required M_p lies between these limits, M_1 and M_u, i.e.

$$0.0833\lambda w L^2 \le 0.0868\lambda w L^2$$

To obtain the exact value of M_p, the sagging hinge is positioned at an unknown distance x from the right-hand support (Fig. 10.11(c)). Considering the total ordinate of the free-moment diagram at this hinge position, then

$$\frac{x}{L} M_p + M_p = \frac{\lambda w L}{2} x - \frac{\lambda w}{2} x^2$$

whence

$$M_p = \frac{\lambda w L}{2} x \left(\frac{L - x}{L + x} \right) \tag{10.8}$$

As the mechanism condition is satisfied, then any value of x inserted into Equation (10.8) will give an upper bound solution, but the safest design will be achieved when M_p is a maximum. By differentiating with respect to x, the solution $x = (\sqrt{2} - 1)L = 0.414L$ is obtained; substituting for x in Equation (10.5) gives $M_p = \lambda w L^2/11.66 \; (= 0.0858\lambda w L^2)$. Note that this exact value of M_p lies within the range indicated by the upper and lower bounds previously calculated. This particular result is useful and can be applied directly to continuous beam problems where an end span carries a uniformly distributed load.

Further reading for Chapter 10

Horne M.R. & Morris L.J. (1981) *Plastic Design of Low-Rise Frames*. Constrado Monograph, Collins, London.
Kleinlogel A. (1931) *Mehrstielige Rahmen*. Ungar, New York.
Neal B.G. (1956) *Plastic Methods of Structural Analysis*. Chapman & Hall.

Chapter 11
Plane frame analysis

by JOHN RIGHINIOTIS

11.1 Formulae for rigid frames

11.1.1 General

The formulae given in this section are based on Professor Kleinlogel's *Rahmenformeln* and *Mehrstielige Rahmen*.[1] The formulae are applicable to frames which are symmetrical about a central vertical axis, and in which each member has constant second moment of area.

Formulae are given for the following types of frame:

Frame I   Hingeless rectangular portal frame.
Frame II   Two-hinged rectangular portal frame.
Frame III   Hingeless gable frame with vertical legs.
Frame IV   Two-hinged gable frame with vertical legs.

The loadings are so arranged that dead, snow and wind loads may be reproduced on all the frames. For example, wind suction acting normal to the sloping rafters of a building may be divided into horizontal and vertical components, for which appropriate formulae are given, although all the signs must be reversed because the loadings shown in the tables act inwards, not outwards as in the case of suction.

It should be noted that, with few exceptions, the loads between node or panel points are uniformly distributed over the whole member. It is appreciated that it is normal practice to impose loads on frames through purlins, siderails or beams. By using the coefficients in Fig. 11.1, however, allowance can be made for many other symmetrically placed loads on the cross-beams of frames I and II shown, where the difference in effect is sufficient to warrant the corrections being made. The indeterminate BMs in the whole frame are calculated as though the loads were uniformly distributed over the beam being considered, and then all are adjusted by multiplying by the appropriate coefficient in Fig. 11.1. It may be of interest to state why these adjustments are made. In any statically indeterminate structure the indeterminate moments vary directly with the value of the following quantity:

$$\frac{\text{area of the free BM diagram}}{EI}$$

Where the loaded member is of constant cross section, EI may be ignored.

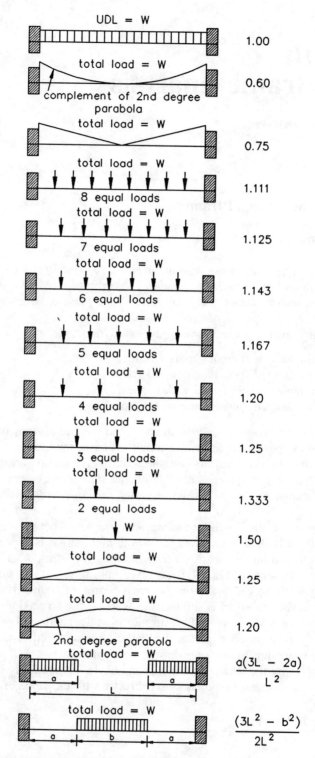

Fig. 11.1 Conversion coefficients for symmetrical loads

Consider, as an example, the case of an encastré beam of constant cross section and of length L carrying a UDL of W. Then the area of the free BM diagram is

$$\frac{WL}{8} \times \frac{2L}{3} = \frac{WL^2}{12}$$

If, however, W is a central point load, the area of the free BM diagram is

$$\frac{WL}{4} \times \frac{L}{2} = \frac{WL^2}{8}$$

The fixed end moments (FEM) due to the two types of loadings are $WL/12$ and $WL/8$ respectively, thus demonstrating that the indeterminate moments vary with the area of the free BM diagram and proving that the indeterminate moments are in the proportion of $1:1.5$.

No rules can be laid down for the effect on the reactions of a change in the mode of application of the load, although sometimes they will vary with the indeterminate moments. Consider a simple rectangular portal with hinged feet. If a UDL placed over the whole of the beam is replaced by a central point load of the same magnitude, then the knee moments will increase by 50% with a corresponding increase in the horizontal thrusts H, while the vertical reactions V will remain the same.

Although the foregoing remarks relating to the indeterminate moments resulting from symmetrical loads apply to all rectangular portals, the rule applies for asymmetrical loads imposed upon the cross-beam of a rectangular portal frame with hinged feet. If a vertical UDL on the cross-beam is replaced by any vertical load of the same magnitude, then the indeterminate moments vary with the areas of the respective free BM diagram.

No doubt readers who use the tables frequently will learn short cuts, but it is not inappropriate to mention some. For example, if a UDL of W over the whole of a single-bay symmetrical frame is replaced by a UDL of the same magnitude of W over either the left-hand or right-hand half of the frame, the horizontal thrust at the feet is unaltered. If the frame has a pitched roof then the ridge moment will also be unaltered.

The charts in the Appendix have been prepared to assist in the design of rectangular frames or frames with a roof pitch of 1 in 5.

11.1.2 Arrangement of formulae

Each set of formulae is treated as a separate section. The data required for each frame, together with the constants to be used in the various formulae, are given on the first page. This general information is followed by the detailed formulae for the various loading conditions, each of which is illustrated by two diagrams placed side-by-side; the left-hand diagram giving a loading condition and the right-hand one giving the appropriate BM and reaction diagram. It should be noted, however, that some BMs change their signs as the frames change their proportions. This will be appreciated by examining the charts.

For simple frames, i.e. for single-storey frames, the formulae for reactions immediately follow the formulae for BMs for each load.

Considering the simple frames only, the type of formula depends on the degree of indeterminacy and the shape of the frame. Auxiliary coefficients X are introduced whenever the direct expressions become complicated or for other reasons of expediency.

No hard and fast rules can be laid down for the notation and it must be noted that each set of symbols and constants applies only to the particular frame under consideration, although, of course, an attempt has been made to produce similarity in the types of symbols.

11.1.3 Sign conventions

All computations must be carried out algebraically, hence every quantity must be given its correct sign. The results will then be automatically correct in sign and magnitude.

The direction of the load or applied moment shown in the left-hand diagram for each load condition is considered to be positive. If the direction of the load or moment is reversed, the signs of all the results obtained from the formulae as printed must be reversed.

For simple frames, the moments causing tension on the inside faces of the frame are considered to be positive. Upward vertical reactions and inward horizontal reactions are also positive.

For multi-storey or multi-bay frames the same general rules apply to moments and vertical reactions.

11.1.4 Checking calculations for indeterminate frames

Calculations for indeterminate frames may be checked by using some other method of analysis, but it is also possible to check any frame or portion of a frame, such as that above the line AB in Fig. 11.2, by ensuring that the following rules are obeyed:

(1) the three fundamental statical equations, i.e. $\Sigma H = 0$, $\Sigma V = 0$ and $\Sigma M = 0$, have been satisfied, and, in addition, either that
(2) the sum of the areas of the M/EI diagram above any line, such as AB, is zero if A and B are fully fixed; or

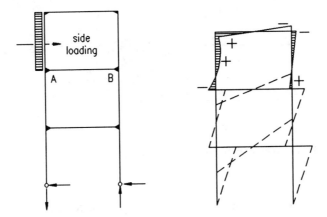

Fig. 11.2 Checking an indeterminate frame

(3) the sum of the moments, with respect to the base AB, of the areas of the M/EI diagram above the line AB is zero if A and B are partially restrained (as shown in Fig. 11.2) or are hinged.

The underlying principles in rules (2) and (3) above are those used in the application of the 'column analogy' method of analysis.

As an example of rule (2), consider the frame in Fig. 11.3 where EI is constant.

Then the sum of the areas of the M/EI diagram, considering the legs first is,

$$\frac{2}{EI}\left[\frac{(+0.0736 - 0.0826) \times 6.0}{2}\right] + \frac{2}{EI}\left[\frac{(-0.0826 + 0.0893) \times 8.078}{2}\right]$$

$$= \frac{-0.054 + 0.054}{EI} = 0$$

Thus demonstrating that the moments calculated are correct.

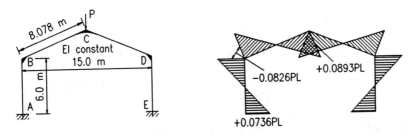

Fig. 11.3 Checking a single-bay portal – rule (2)

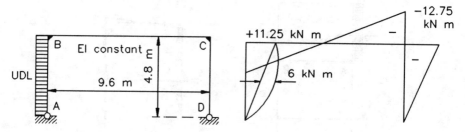

Fig. 11.4 Checking a rectangular portal – rule (3)

Now consider the frame in Fig. 11.4 as an example for rule (3).
Then the sum of the moments of the areas of the M/EI diagram, working from A round to D, is

$$\frac{1}{EI}\left\{\left[\frac{11.25 \times 4.8}{2} \times \frac{2 \times 4.8}{3}\right] + \left[\frac{6 \times 4.8 \times 2}{3} \times \frac{4.8}{2}\right]\right.$$
$$+ \left[\left(\frac{11.25 - 12.75}{2}\right) 9.6 \times 4.8\right] + \left[\frac{-12.75 \times 4.8}{2} \times \frac{2 \times 4.8}{3}\right]\right\} = 0$$

$$\frac{1}{EI}\left[\frac{4.8^2 \times 2}{2 \times 3}(11.25 + 6 - 4.5 - 12.75)\right] = 0$$

Demonstrating again that the calculations are correct.

11.2 Portal frame analysis

The design process explained here assumes that the reader is familiar with the basic concepts of limit state steel design, and the basic analysis and design of continuous beams. These notes take the form of a worked example. A method of determining preliminary member sizes has been included.

11.2.1 Methods of analysis

BS 5950: Part 1^2 allows two main methods of analysis of a structure.

(1) *Linear elastic.* The frame is analyzed by either hand or computer assuming linear elastic behaviour. Once the forces, moments and shears have been derived by elastic analysis the ultimate capacity of each section is checked using the rules given in Section 4 of the Code.
(2) *Simple plastic theory.* The frame is analyzed using the basic principles of

simple plastic theory. Once the forces, moments and shears have been derived by analysis the member capacities are checked. Those containing plastic hinges are checked in accordance with Section 4.

Elastic analysis

Using elastic analysis it is to be expected that the structure will be heavier than that designed by plastic methods, but less stability bracing will be needed. It may well be that the final details will also be more simple.

It will remain the engineer's responsibility to ensure that stability is provided both locally and in the overall condition. BS 5950: Part 1^2 provides no specific rules regarding the stability of the frame as a whole; this means that the engineer must ensure that the stability is checked using the general rules for all frames. He must also check that the movement of the frame under all loading cases is not sufficient to cause damage to adjacent construction, i.e. brick walls or cladding, the serviceability limit state of deflection.

Plastic analysis

The method of calculating the ultimate load of a portal frame is described in many publications. The main essence of the method is to assume that plastic 'hinges' occur at points in the frame where the value of M/M_p is at its highest value; the load being considered as increasing proportionally until the failure or ultimate state is reached. Because of the straining at the hinge points it is essential that the local buckling and lateral distortion do not occur before failure. Failure is deemed to have taken place when sufficient hinges have formed to create a mechanism.

The member capacities are calculated using the rules given in Section 4 of the Code but with the additional restrictions applied to hinge positions. In addition, positive requirements are put on checking frame stability for both single-bay and multi-bay frames. Plastic designed frames are lighter than elastic designed frames, providing deflection is not a governing point; however, additional bracing may well be required.

11.2.2 Stability

With the use of lighter frames, various aspects of stability take a more prominent part in the design procedures. As far as portal frames are concerned the following areas are important:

(1) overall frame stability, in that the strength of the frame should not be affected by changes in geometry during loading ($P\Delta$ effect)

(2) snap-through stability, in multi-bay frames (three or more), where the effects of continuity can result in slender rafters
(3) plastic hinge stability, where the member must be prevented from moving out of plane or rotating at plastic hinges
(4) rafter stability, ensuring that the rafter is stable in bending as an unrestrained beam
(5) leg stability, where the leg below the plastic hinge must be stable
(6) haunch stability, where the tapered member is checked to ensure that the inner (compression flange) is stable.

11.2.3 Selecting suitable members for a trial design

The design of a portal frame structure is in reality a process of selecting suitable members and then proving their ability to perform in a satisfactory manner. Inexperienced engineers can be given some guidance to estimate initial member sizes. In order to speed the initial selection of members, three graphs have been produced to enable simple pin-based frames to be sized quickly.

These graphs have been prepared making the following assumptions:

(1) plastic hinges are formed at the bottom of the haunch in the leg and near the apex in the rafter, the exact position being determined by the frame geometry
(2) the depth of the rafter is approximately span/55 and the depth of the haunch below the eaves intersection is 1.5 times rafter depth
(3) the haunch length is 10% of the span of the frame, a limit generally regarded as providing a balance between economy and stability
(4) the moment in the rafter at the top of the haunch is $0.87M_p$, i.e. it is assumed that the haunch area remains elastic
(5) the calculations assume that the calculated values of M_p are provided exactly by the sections and that there are no stability problems. Clearly these conditions will not be met and it is the engineer's responsibility to ensure that the chosen sections are fully checked for all aspects of behaviour.

The graphs cover the range of span/eaves height between 2 and 5 and rise/span of 0 to 0.2 (where 0 is a flat roof). Interpolation is permissible but extrapolation is not. The three graphs give:

Figure 11.5: the horizontal force at the feet of the frame as a proportion of the total factored load wL, where w is the load/unit length of rafter and L is the span of the frame
Figure 11.6: the value of the moment capacity required in the rafters as a proportion of the load times span wL^2
Figure 11.7: the value of the moment capacity required in the legs as a proportion of the load times span wL^2.

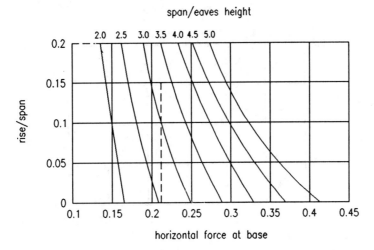

Fig. 11.5 Rise/span against horizontal force at base for various span/eaves heights

The graphs are non-dimensional and may be used with any consistent set of units. In the worked example kilonewtons and metres are used.

Method of use of the graphs

(1) Determine the ratio span/height to eaves.
(2) Determine the ratio rise/span.

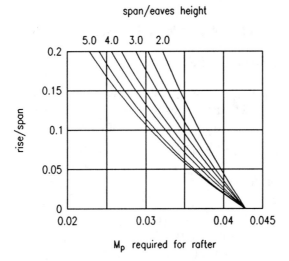

Fig. 11.6 Rise/span against required M_p of rafter for various span/eaves heights

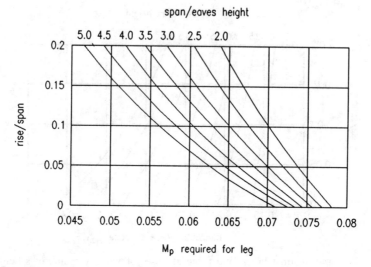

Fig. 11.7 Rise/span against required M_p of leg for various span/eaves heights

(3) Calculate wL (total load) and wL^2.
(4) Look up the values from the graphs.
(5) Horizontal force at root of frame = value from Fig. 11.5 × wL.
(6) M_p required in rafter = value from Fig. 11.6 × wL^2.
(7) M_p required in leg = value from Fig. 11.7 × wL^2.

11.2.4 Worked example of plastic design (see Fig. 11.8)

Determination of member sizes

Although the engineer may use his experience or other methods to determine preliminary member sizes, the graphs in Figs 11.5, 11.6 and 11.7 will be used for this example. In order to use these graphs, four parameters are required:

span/height to eaves	= 25/7.6	= 3.29
rise/span	= 3.75/25	= 0.15
wL (total load on frame)	= 9.48 × 25	= 237 kN
wL^2	= 9.48 × 25^2	= 5925 kN m

Horizontal thrust at feet of frame (Fig. 11.5)
= 0.21 × 237 = 49.8 kN m
Moment capacity of rafter (Fig. 11.6)
= 0.0305 × 5925 = 181 kN m

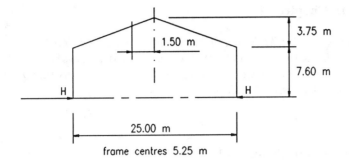

loading	unfactored design	load factor	factored load	total factored load
imposed	0.75 kN/m²	1.6	1.2 kN/m²	158 kN
dead	0.43 kN/m²	1.4	0.6 kN/m²	79 kN
total				237 kN

factored load/m (w) = 237/25 = 9.48 kN/m

Fig. 11.8 Portal frame design example

Moment capacity of leg (Fig. 11.7)
$$= 0.059 \times 5925 = 350 \text{ kN m}$$

Assuming a design strength of 275 N/mm² (grade 43 steel):

S_x required for rafter = 181 × 1000/275 = 658 cm³
S_x required for leg = 350 × 1000/275 = 1270 cm³

Trial sections (*NB* these are first trials and may not be adequate):

Rafter 406 × 140 × 39 kg UB (grade 43 steel);
$S_x = 721$ cm³, $I_x = 12\,500$ cm⁴

Leg 457 × 152 × 60 kg UB (grade 43 steel);
$S_x = 1280$ cm³, $I_x = 25\,500$ cm⁴

It is suggested that the next stage is to check the overall stability of the frame. The main reason for this is that the only way to correct insufficient stability is to change the main member sizes. If any other checks are not satisfied, additional bracing can frequently be used to rectify the situation, without altering the member sizes.

References to Chapter 11

1. Kleinlogel A. (1931) *Mehrstielige Rahmen*. Ungar, New York.
2. British Standards Institution (1990) *Structural use of steelwork in building*. Part 1: *Code of practice for design in simple and continuous construction: hot rolled sections*. BS 5950, BSI, London.

Further reading for Chapter 11

Baker J.F. (1954) *The Steel Skeleton, Vol. 1*, 1st edn. Cambridge University Press.
Baker J.F., Horne M.R. & Heyman J. (1956) *The Steel Skeleton, Vol. 2*. Cambridge University Press.
Horne M.R. & Morris L.S. (1981) *Plastic Design of Low-Rise Frames*. Constrado Monograph, Collins, London.
Neal B.G. (1956) *Plastic Methods of Structural Analysis*. Chapman and Hall.
Weller A.D. (1988) *Portal Frame Design, Introduction to Steelwork Design to BS 5950: Part 1*. Steel Construction Institute, Ascot, Berks.

Chapter 12
Applicable dynamics

by MICHAEL WILLFORD

12.1 Introduction

The dynamic performance of steel structures has traditionally only been an area of interest for special classes of structure, particularly slender wind-sensitive structures (masts, towers and stacks), structures supporting mechanical equipment, offshore structures and earthquake-resistant structures. However, a wider interest in dynamic behaviour has recently come about as a result of the widespread adoption of long span composite floors in buildings, for which vibration criteria must be satisfied.

Dynamic loads in structures may arise from a number of sources including the following:

Forces generated inside a structure	– Machinery
	– Impacts
	– Human activity (walking, dancing, etc.)
External forces	– Wind buffeting and other aerodynamic effects
	– Waves (offshore structures)
	– Impacts from vehicles, etc.
Ground motions	– Earthquakes
	– Ground-borne vibration due to railways, roads, pile driving, etc.

The principal effects of concern are:

Strength	– The structure must be strong enough to resist the peak dynamic forces that arise.
Fatigue	– Fatigue cracks can initiate and propagate when large numbers of cycles of vibration inducing significant stress are experienced, leading to reduction in strength and failure.
Perception	– Human occupants of a building can perceive very low amplitudes of vibration, and, depending on the circumstances, may find vibration objectionable. Certain items of precision equipment are also extremely sensitive to vibration. Perception will generally be the most onerous dynamic criterion in occupied buildings.

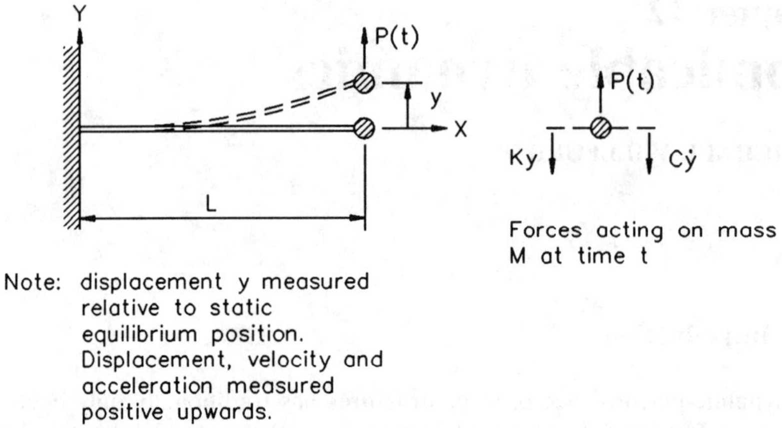

Fig. 12.1 Example of a dynamic system

Note: displacement y measured relative to static equilibrium position. Displacement, velocity and acceleration measured positive upwards.

It is beyond the scope of this book to address all these issues, and references are suggested for more detailed guidance. The following sections are intended to give an overview of the fundamental features of dynamic behaviour and to introduce some of the terminology employed and analysis procedures available. It must be borne in mind that dynamic behaviour is influenced by a larger number of parameters than static behaviour, and that some of the parameters cannot be predicted precisely at the design stage. It is often advisable to investigate the effects of varying initial assumptions to ensure that the most critical situations that may occur have been examined.

12.2 Fundamentals of dynamic behaviour

The principal features of dynamic behaviour may be illustrated by the examination of a very simple dynamic system, a concentrated mass M supported on a light cantilever of flexural rigidity EI and length L as illustrated in Fig. 12.1. The mass is subjected to forces P which vary with time t.

12.2.1 Dynamic equilibrium

One of the basic methods of dynamic analysis is the examination of dynamic equilibrium to formulate an equation of motion. Consider the dynamic equilibrium of the mass illustrated in Fig. 12.1. If at some time t the mass is displaced upwards from its static equilibrium position by y, and has velocity $\dot{y}$ and acceleration $\ddot{y}$ (positive upwards) the mass is in general subjected to the following forces:

Table 12.1

Mass	Force	Displacement	Time
kg	N	m	s
tonnes	kN	m	s

External force $P(t)$
Stiffness force Ky ($K = 3EI/L$ for uniform cantilever)
Damping force $C\dot{y}$ (a dissipative force assumed to act in the opposite direction to the velocity).

The resultant of these forces will cause the mass to accelerate according to Newton's 2nd law of motion. Hence the dynamic equilibrium equation may be written as:

$$M\ddot{y} = P(t) - Ky - C\dot{y}$$

or

$$M\ddot{y} + C\dot{y} + Ky = P(t) \qquad (12.1)$$

Equation (12.1) is the general equation of motion for a single degree of freedom dynamic system (a system whose behaviour can be defined by a single quantity, in this case the deflection of the mass, y). It is clearly most important that the quantities in this equation are defined in dynamically consistent units. Examples of consistent units are given in Table 12.1.

Various solutions to Equation (12.1) can give an insight into the principal features of dynamic behaviour.

12.2.2 Undamped free vibration

In this case it is assumed that the system has been set into motion in some way and is then allowed to vibrate freely in the absence of external forces. It is also assumed initially that there is no damping.

The corresponding equation of motion derived from Equation (12.1) is:

$$M\ddot{y} + Ky = 0 \qquad (12.2)$$

By putting $K/M = \omega_n^2$ it is easily shown that a motion of the form $y = Y \cos \omega_n t$ satisfies this equation. This is known as simple harmonic motion; the mass oscillates about its static equilibrium position with amplitude Y as shown in Fig. 12.2. ω_n is known as the circular frequency, measured in radians per second. There are 2π radians in a complete cycle of vibration and so the vibration frequency, f_n, is $\omega_n/2\pi$ cycles per second (hertz). In the absence of damping or external forces the system will vibrate in this manner indefinitely.

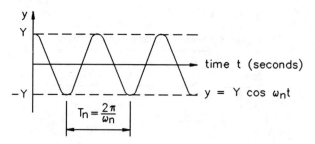

Fig. 12.2 Simple harmonic motion

The amplitude of vibration Y is the peak displacement of the mass relative to its static equilibrium position. Vibration amplitudes are sometimes referred to as root mean square (rms) quantities, where $y_{rms} = \dfrac{1}{T}\sqrt{\left(\int_0^T y^2 dt\right)}$, and T is the total time over which the vibration is considered. For continuous simple harmonic motion $y_{rms} = Y/\sqrt{2}$.

12.2.3 Damped free vibration

Energy is always dissipated to some extent during vibration of real structures. Inclusion of the damping force in the free vibration equation of motion leads to:

$$M\ddot{y} + C\dot{y} + Ky = 0 \tag{12.3}$$

The solution to this equation for a lightly damped system when the mass is initially displaced by Y and then released is:

$$y = Ye^{-\xi\omega_n t} \cos(\omega_d t - \phi) \tag{12.4}$$

The motion takes the form shown in Fig. 12.3(a), and it can be seen that the vibration amplitude decays exponentially with time. The rate of decay is governed by the amount of damping present.

If the damping constant C is sufficiently large then oscillation will be prevented and the motion will be as in Fig. 12.3(b). The minimum damping required to prevent overshoot and oscillation is known as critical damping, and the damping constant for critical damping is given by $C_o = 2\sqrt{(KM)}$, where K and M are the stiffness and mass of the system.

Practical structures are lightly damped and the damping present is often expressed as a proportion of critical. In Equation (12.4)

$$\xi = \text{critical damping ratio} = \frac{C}{C_o}$$

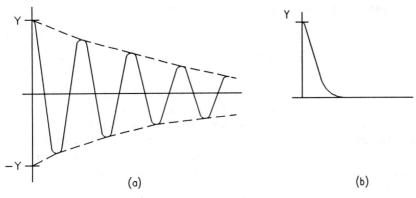

Fig. 12.3 Damped free vibration: (a) lightly damped system, (b) critically damped system

The phase shift angle ϕ is small when the damping is small.
The frequency of damped oscillation is:

$$\omega_d = \omega_n (1 - \xi^2)^{\frac{1}{2}} \text{ radians/s}$$

where ω_n is the undamped natural frequency. With low damping ($\xi \ll 1$) the reduction of natural frequency (increase in natural period) resulting from damping is negligible.

12.2.4 Response to harmonic loads

A load which varies sinusoidally with time at a constant frequency is known as a harmonic load. This form of dynamic load is characteristic of machinery operating at constant speed and many other types of continuous vibration can be approximated to this form.

When such a force of amplitude P and frequency f Hz (or $\omega = 2\pi f$ radians/s) is applied to the simple structure described in previous sections the structure will be caused to vibrate. After some time the motions of the structure will reach a steady-state, that is, vibration of a constant amplitude and frequency will be achieved. The dynamic equation of motion is:

$$M\ddot{y} + C\dot{y} + Ky = P \cos \omega t \tag{12.5}$$

and it can be shown that the steady-state response motion is described by:

$$y = Y \cos(\omega t - \phi) \tag{12.6}$$

Note that the steady-state vibration occurs at the frequency of the harmonic force exciting the motion, not at the natural frequency of the structure.

328   *Applicable dynamics*

The displacement amplitude Y can be shown to be:

$$Y = \frac{P}{K} \frac{1}{\left\{\left[1 - \left(\frac{\omega}{\omega_n}\right)^2\right]^2 + \left(2\xi \frac{\omega}{\omega_n}\right)^2\right\}^{\frac{1}{2}}} \qquad (12.7)$$

P/K is the deflection of the structure under a static force P, and the multiplying term can be regarded as a magnification factor. The variation of magnification factor with frequency ratio and damping ratio is shown in Fig. 12.4.

When the dynamic force is applied at a frequency much lower than the natural frequency of the system ($\omega/\omega_n \ll 1$), the response is quasi-static; the response is governed by the stiffness of the structure and the amplitude is close to the static deflection P/K.

When the dynamic force is applied at a frequency much higher than the natural frequency ($\omega/\omega_n \gg 1$), the response is governed by the mass (inertia) of the structure; the amplitude is less than the static deflection.

When the dynamic force is applied at a frequency close to the natural frequency, the stiffness and inertia forces in the vibrating system are almost equal and opposite at any instant, and the external force is resisted by the damping force. This is the condition known as resonance, when very large dynamic magnification factors are possible. When damping is low the maximum steady-state magnification occurs when $\omega = \omega_n$, and then

$$Y = \frac{P}{K} \frac{1}{2\xi} \qquad (12.8)$$

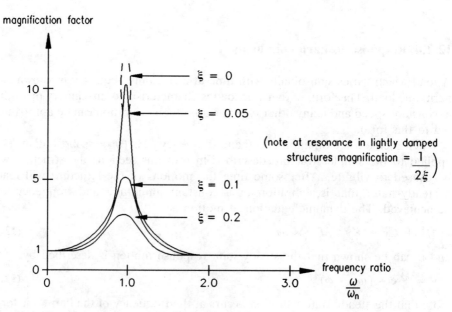

Fig. 12.4   Steady-state response to harmonic loads

Since in many structural systems ξ is of the order of 0.01, magnification factors of the order of 50 may result. The force in the structure is proportional to the displacement so the same magnification factor applies to structural forces.

Human perception of motion is usually related to acceleration levels rather than displacement. The peak acceleration amplitude at steady-state is given by:

$$\hat{a} = \omega^2 Y = (2\pi f)^2 Y \tag{12.9}$$

In practice, there is usually advantage in avoiding the possibility of resonance whenever possible by ensuring that structural frequencies are well away from the frequencies of any known sources of substantial dynamic force.

12.2.5 Response to an impact

Another dynamic loading case of interest is the response of a structure to an impact, say from an object falling on to the structure. A full discussion of impact loading is given in Reference 1, but a simple approximate method is useful for many practical situations when the mass of the impacting object is small compared to the mass of the structure, and the impact duration is short compared to the natural period of the structure. In these cases the effect of the impact can be assessed as an impulse I acting on the structure. The magnitude of I may be calculated as $m \Delta v$, where m is the mass of the falling object and Δv its change in velocity at impact. If there is no rebound Δv can be taken as the approach velocity. For the simple system discussed in previous sections conservation of momentum at impact requires the initial velocity of the structural mass to be I/M. A lightly damped system then displays damped free vibration corresponding to an initial displacement amplitude of approximately:

$$Y = \frac{I}{\omega_n M} = \frac{I}{2\pi f_n M} \tag{12.10}$$

12.2.6 Response to base motion

The previous sections have illustrated the behaviour of a dynamic system with a fixed base subject to applied forces. When the dynamic excitation takes the form of base motion, as for example in an earthquake, the formulation of the equation of motion and the solutions are slightly modified. Detailed treatment of this type of excitation is beyond the scope of this chapter and References 1 and 2 are recommended for discussion of the solutions.

Although not correct in detail, the general form of response indicated by Fig. 12.4 for harmonic applied forces is still relevant. Resonance occurs for those components of the base motion close in frequency to the natural frequency of the structure.

12.2.7 Response to general time-varying loads

Although some dynamic loads (e.g. impacts and harmonic loads) are simple and can be dealt with analytically, many forces and ground motions that occur in practice are complex.

In general, numerical analysis procedures are required for the evaluation of responses to these effects, and these are often available in finite element analysis programs.

The techniques employed include:

- direct step-by-step integration of the equations of motion using small time increments,
- Fourier analysis of the forcing function followed by solution for Fourier components in the frequency domain,
- for random forces, random vibration theory and spectral analysis.

The background and application of these techniques are discussed in References 1 and 3.

12.3 Distributed parameter systems

The dynamic system considered in section 12.2 was simple in that its entire mass was concentrated at one point. This enabled Newton's second law and therefore the equation of motion to be written in terms of a single variable.

Although some practical structures can be represented adequately in this way, in other cases it is not realistic to assume that the mass is concentrated at one point. The classical treatment of distributed systems is illustrated below by analysis of a uniform beam; similar techniques can be used for two-dimensional (plate) structures. The analysis calculates the natural frequency of the beam and, by reference to section 12.2.2, this can be done by ignoring external forces and damping forces.

12.3.1 Dynamic equilibrium

The beam under consideration is shown in Fig. 12.5. For a small element of the beam of length dx the variation of shear force V along the beam results in a net force on the element of $\frac{\partial V}{\partial x} dx$ which must cause the mass $m\, dx$ to accelerate.

Dynamic equilibrium therefore requires:

$$\frac{\partial V}{\partial x} dx = m\, dx\, \frac{\partial^2 y}{\partial t^2} \qquad (12.11)$$

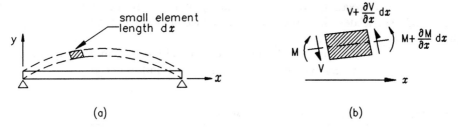

Fig. 12.5 Dynamic equilibrium of vibrating beams. (a) Uniform beam mass/unit length = m, (b) small element mass = $m\,dx$

For a flexural beam the deflection y is a function of the bending moment distribution, with:

$$\frac{\partial^2 y}{\partial x^2} = \frac{M}{EI}$$

Noting also that for moment equilibrium on the element:

$$V = -\frac{\partial M}{\partial x}$$

$$\frac{\partial V}{\partial x} = -\frac{\partial^2 M}{\partial x^2}$$

Substituting these relationships into Equation (12.11) leads to the equation of motion for a flexural beam:

$$\frac{\partial^4 y}{\partial x^4} + \frac{m}{EI}\frac{\partial^2 y}{\partial t^2} = 0 \tag{12.12}$$

It is sometimes convenient to represent an entire framed structure as a shear beam. In this case the deflection y is a function of the shear force with:

$$\frac{\partial y}{\partial x} = \frac{V}{GA_s}$$

The equation of motion for a shear beam is therefore:

$$\frac{\partial^2 y}{\partial x^2} - \frac{m}{GA_s}\frac{\partial^2 y}{\partial t^2} = 0 \tag{12.13}$$

12.3.2 Modes of vibration

The partial differential equations derived above can be solved for given sets of boundary conditions. The solution in each case identifies a family of modes of vibration (eigenvectors) with corresponding natural frequencies (eigenvalues). The mode shape describes the relative displacements of the various parts of the beam at any instant in time.

A number of standard results are given in Tables 12.2 and 12.3, and more comprehensive lists are contained in References 4 and 5.

Table 12.2 Natural frequencies of uniform flexural beams

Length = L (m) (typical units)
Flexural rigidity = EI (N m^2)
Mass/unit length = m (kg/m)

$$f_n = \frac{k_n}{2\pi}\sqrt{\left(\frac{EI}{mL^4}\right)} \text{ Hz: values of } k_n \text{ given below}$$

	Mode	Shape	k_n	Nodal points at $x/L =$
Simply-supported	1		9.87	0 1
	2		39.5	0 0.5 1
	3		88.8	0 0.333 0.667 1
	n	$x = \sin\left(n\pi\frac{x}{L}\right)$	$n^2\pi^2$	0 $\frac{1}{n}$ ... $\frac{n-1}{n}$ 1
Encastré	1		22.4	0 1
	2		61.7	0 0.5 1
	3		121	0 0.359 0.641 1
Cantilever	1		3.52	0
	2		22.0	0 0.774
	3		61.7	0 0.5 0.868

Table 12.3 Natural frequencies for uniform shear cantilevers

	Length = L	(m) (typical units)
	Shear rigidity = GA_s	(N)
	Mass/unit length = m	(kg/m)

$$f_n = \frac{k_n}{2\pi}\sqrt{\left(\frac{GA_s}{mL^2}\right)} \text{ Hz: values of } k_n \text{ given below}$$

Mode	Shape	k_n	Nodal points at x/L =
1		1.57	0
2		4.71	0 0.667
3		7.85	0 0.4 0.8
n	$y = \sin\left((2n-1)\dfrac{\pi x}{2L}\right)$	$(2n-1)\dfrac{\pi}{2}$	0 $\dfrac{2}{(2n-1)}$ $\dfrac{4}{(2n-1)}$ $\dfrac{6}{(2n-1)}$ etc.

12.3.3 Calculation of responses

For linear elastic behaviour, once the mode shapes and frequencies have been established, dynamic responses can be calculated treating each mode as a single degree of freedom system such as the one described in section 12.2.2. In theory distributed systems have an infinite number of modes of vibration, but in practice only a few modes, usually those of lowest frequency, will contribute significantly to the overall response.

It is convenient to describe a mode shape by a displacement parameter ϕ defined (or normalized) such that the maximum value at any point is 1.0. If the mode is excited by dynamic forces resulting in a modal response amplitude of Y_m, then the displacement amplitude at a point i on the structure is:

$$y_i = \phi_i Y_m$$

where ϕ_i is the mode shape value at the point.

The magnitude of Y_m can be calculated for simple dynamic loads treating the mode as a single degree of freedom system having the following properties. These are often referred to as modal or generalized properties:

334 Applicable dynamics

modal mass $\quad M^* = \Sigma M_i \phi_i^2 \quad$ sum over whole structure
modal stiffness $K^* = \omega_n^2 M^*$
modal force $\quad P^* = \Sigma P_i \phi_i \quad$ sum for all loaded points

Note that for mode n of a uniform simply-supported beam of span L and mass/unit length m with $\phi = \sin(n\pi x/L)$, the modal mass M^* is:

$$M^* = m \int_L^0 [\sin(n\pi x/L)]^2 \, dx$$

which is $0.5 \, mL$ (half the total mass of the beam).

The application of these concepts to the problems of floor vibration and wind-induced vibration is described in References 6 and 7.

12.3.4 Approximate methods to determine natural frequency

Approximate methods are useful for estimating the natural frequencies of structures not conforming with one of the special cases for which standard solutions exist, and for checking the predictions of computer analyses when these are used.

One of the most useful approximate methods relates the natural frequency of a system to its static deflection under gravity load, δ. With reference to section 12.2.2 the natural frequency of a single lumped mass system is:

$$f_n = \frac{1}{2\pi} \sqrt{\left(\frac{K}{M}\right)}$$

This may be rewritten, replacing the mass term M by the corresponding *weight* Mg as:

$$f_n = \frac{\sqrt{g}}{2\pi} \sqrt{\left(\frac{K}{Mg}\right)}$$

Since Mg/K is the static deflection under gravity load (δ)

$$f_n = \frac{\sqrt{g}}{2\pi \sqrt{\delta}}$$
$$= 15.76/\sqrt{\delta} \quad \text{when } \delta \text{ is measured in mm} \tag{12.14}$$

This formula is exact for any single lumped mass system.

For distributed parameter systems a similar correspondence is found, although the numerical factor in Equation (12.14) varies from case to case, generally between 16 and 20. For practical purposes a value of 18 will give results of sufficient accuracy.

When applying these formulae the following points should be noted.

(1) The static deflection should be calculated assuming a weight corresponding to the loading for which the frequency is required. This is usually a dead load with an allowance for expected imposed load.

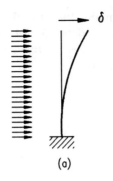

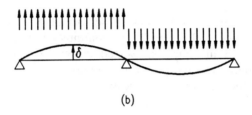

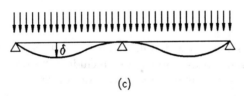

Fig. 12.6 Use of gravity deflections to estimate natural frequencies. (a) For a vertical structure the gravity load is applied horizontally: $f_n \simeq 18/\sqrt{\delta}$. (b) Continuous structure gravity load applied in opposite directions on alternate spans: $f_n \simeq 18/\sqrt{\delta}$. (c) Continuous structure – symmetric mode has higher frequency: $f_n \simeq 18/\sqrt{\delta}$

(2) For horizontal modes of vibration (e.g. lateral vibration of an entire structure) the gravity force must be applied laterally to obtain the appropriate lateral deflection, as shown in Fig. 12.6(a).

(3) The mode shape required must be carefully considered in multi-span structures. In the two-span beam of Fig. 12.6(b) the lowest frequency will correspond to an asymmetrical mode as illustrated; the corresponding δ must be obtained by applying gravity in opposite directions on the two spans. The normal gravity deflection will correspond to the symmetrical mode with a higher natural frequency, Fig. 12.6(c).

336 *Applicable dynamics*

These concepts can be extended to estimating the natural frequencies of primary beam – secondary beam systems. If the static deflection of the primary beam is δ_p and the static deflection of a secondary beam is δ_s (relative to the primary) the combined natural frequency is approximately:

$$f_n = \frac{18}{\sqrt{(\delta_p + \delta_s)}}$$

from which it can be shown that

$$\frac{1}{f_n^2} = \frac{1}{f_p^2} + \frac{1}{f_s^2} \qquad (12.15)$$

where f_p and f_s are the natural frequencies of the primary and secondary beams alone.

Care must always be taken in using these formulae so that a realistic mode shape is implied. There will generally be continuity between adjacent spans at small amplitudes even in simply-supported designs and in many situations a combined mode where both the primaries and secondaries are vibrating together in a 'simply-supported' fashion is not possible.

12.4 Damping

Damping arises from the dissipation of energy during vibration. A number of mechanisms contribute to the dissipation, including material damping, friction at interfaces between components and radiation of energy from the structure's foundations.

Material damping in steel provides a very small amount of dissipation and in most steel structures the majority of the damping arises from friction at bolted connections and frictional interaction with non-structural items, particularly partitions and cladding. Damping is found to increase with increasing amplitude of vibration.

The amount of damping that will occur in any particular structure cannot be calculated or predicted with a high degree of precision, and design values for damping are generally derived from dynamic measurements on structures of a corresponding type.

Damping can be measured by a number of methods, including:

- rate of decay of free vibration following an impact (Fig. 12.3(a))
- forced excitation by mechanical vibrator at varying frequency to establish the shape of the steady-state resonance curve (Fig. 12.4)
- spectral methods relying on analysis of response to ambient random vibration such as wind loading.

Table 12.4

Structure type	Structure damping (% critical)
Unclad welded steel structures (e.g. steel stacks)	0.3%
Unclad bolted steel structures	0.5%
Floor (fitted out), composite and non-composite	1.5% – 3% (may be higher when many partitions on floor)
Clad buildings (lateral sway)	1%

All these methods can run into difficulty when several modes close in frequency are present. One result of this is that on floor structures where there are often several closely spaced modes the apparent damping seen in the initial rate of decay after impact can be substantially higher than the true modal damping.

Damping is usually expressed as a fraction or percentage of critical (ξ), but the logarithmic decrement (δ) is also used. The relationship between the two expressions is $\xi = \delta/2\pi$.

Table 12.4 gives values of modal damping that are suggested for use in calculations when amplitudes are low (e.g. for occupant comfort). Somewhat higher values are appropriate at large amplitudes where local yielding may develop, e.g. in seismic analysis.

12.5 Finite element analysis

Many simple dynamic problems can be solved quickly and adequately by the methods outlined in previous sections. However there are situations where more detailed numerical analysis may be required and finite element analysis is a versatile technique widely available for this purpose. Numerical analysis is often necessary for problems such as:

(1) determination of natural frequencies of complex structures
(2) calculation of responses due to general time-varying loads or ground motions
(3) response spectrum analysis to determine seismic design forces.

12.5.1 Basis of the method

As explained in Chapter 9 the finite element method describes the state of a structure by means of deflections at a finite number of node points. Nodes are connected by elements which represent the stiffness of the structural components.

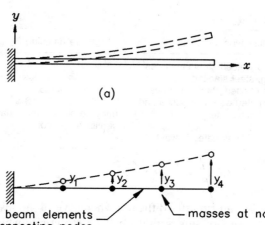
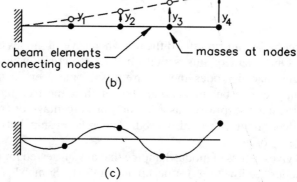

Fig. 12.7 Finite element idealization. (a) Uniform beam in free vibration, (b) finite element representation, (c) fourth mode not accurately represented

In static problems the equilibrium of every degree of freedom at the nodes of the idealization is described by the stiffness equation:

$F = KY$

where F is the vector of applied forces, Y is the vector of displacements for every degree of freedom, and K is the stiffness matrix. Solution of unknown displacements for a known force vector involves inversion of the stiffness matrix.

The extension of the method to dynamic problems can be visualized in simple terms by considering the dynamic equilibrium of a vibrating structure.

Figure 12.7(a) shows the instantaneous deflected shape of a vibrating uniform cantilever, and Fig. 12.7(b) shows a finite element idealization of this condition. The shape is described by the deflections of the nodes, Y, and a mass is associated with each degree of freedom of the idealization.

At the instant considered when the deflection vector is Y the forces at the nodes provided by the stiffness elements must be KY.

These forces may in part be resisting external instantaneous nodal forces P and may in part be causing the mass associated with each node to accelerate. The equations of motions of all the nodes may therefore be written as:

$P - M\ddot{Y} = KY$

or

$$KY + M\ddot{Y} = P$$

where M is the mass matrix, $\ddot{Y}$ is the acceleration vector, and P is the external force vector.

The natural frequencies and mode shapes are obtained by solving the undamped free vibration equations:

$$KY + M\ddot{Y} = 0$$

Assuming a solution of the form $y = Y \cos \omega_n t$, it follows that $\ddot{Y} = -\omega_n^2 Y$ and hence:

$$[K - \omega_n^2 M]Y = 0$$

This is a standard eigenvalue problem of matrix algebra for which various numerical solution techniques exist. The solution provides a set of mode shape vectors Y with corresponding natural frequencies ω_n. The number of modes possible will be equal to the number of degrees of freedom in the solution.

12.5.2 Modelling techniques

Dynamic analysis is more complex than static analysis and care is required so that results of appropriate accuracy are obtained at reasonable cost when using finite element programs. It is often advisable to investigate simple idealizations initially before embarking upon detailed models. As problems and programs vary it is possible to give only broad guidance; individual program manuals must be consulted and experience with the program being used is invaluable. More detailed background is given in Reference 8.

The first stage in any dynamic analysis will invariably be to obtain the natural frequencies and mode shapes of the structure. As can be seen from Fig. 12.7(c) a given finite element model will represent higher modes with decreasing accuracy. If it is only necessary to obtain a first mode frequency accurately then a relatively coarse model, such as that illustrated in Fig. 12.7(b) will be perfectly adequate. In order to obtain an accurate estimate of the fourth mode a greater subdivision of the structure would be necessary, since the distribution of inertia load along the uniform beam in this mode is not well represented by just four masses.

As models increase in size and complexity, the solution of the eigenvalue problem for all the degrees of freedom of the structure becomes prohibitive and unnecessary. It is usual in these cases to select a limited number of 'master' degrees of freedom (which become the effective degrees of freedom in the analysis). These may be selected manually, or automatic procedures are available in most programs. In either case the number and distribution of master must be sufficient to enable the mode shapes required to be defined. It is generally better to retain translational degrees of freedom as masters rather than rotational.

Displacements for the remaining 'slave' nodes are obtained by interpolation. An eigensolution based on N master degrees of freedom will yield N modes of vibration, though only a limited number of these will be of high accuracy. Examination of the mode shapes and the distribution of master degrees of freedom (as in Figs 12.7(b) and 12.7(c)) will indicate which of the higher modes are well represented.

Clearly dynamically consistent units must be used throughout, and these units may not be the same as those used for a static analysis using the same model. A hand check of the first mode frequency using an approximate or empirical method is strongly advisable to ensure that the results are realistic. In addition, there is even more need than with static analysis to view computer analysis results as approximate. It is very difficult to predict natural frequencies of real structures with a high degree of precision unless the real boundary conditions and structural stiffness can be defined with confidence. This is rarely the case and these are uncertainties that finite element analysis cannot resolve.

References to Chapter 12

1. Clough R.W. & Penzien J. (1975) *Dynamics of Structures*. McGraw-Hill.
2. Dowrick D.J. (1988) *Earthquake Resistant Design for Engineers and Architects*, 2nd edn. John Wiley & Sons.
3. Warburton G.B. (1976) *The Dynamical Behaviour of Structures*, 2nd edn. Pergamon Press.
4. Harris C.M. & Crede C.E. (1976) *Shock and Vibration Handbook*. McGraw-Hill.
5. Roark R.J. & Young W.C. (1975) *Formulas for Stress and Strain*, 5th edn. McGraw-Hill.
6. Construction Industry Research & Information Association (CIRIA)/The Steel Construction Institute (SCI) (1989) *Design Guide on the Vibration of Floors*. SCI Publication 076, SCI, Ascot, Berks.
7. Engineering Sciences Data Unit (1976) *The Response of Flexible Structures to Atmospheric Turbulence*. ESDU Data Item 76001, ESDU International, London.
8. Bathe K.J. (1982) *Finite Element Procedures in Engineering Analysis*. Prentice-Hall.

Chapter 13
Local buckling and cross-section classification

by DAVID NETHERCOT

13.1 Introduction

The efficient use of material within a steel member requires those structural properties which most influence its load-carrying capacity to be maximized. This, coupled with the need to make connections between members, has led to the majority of structural sections being thin-walled as illustrated in Fig. 13.1. Moreover, apart from circular tubes, structural steel sections (such as universal beams and columns, cold-formed purlins, built-up box columns and plate girders) normally comprise a series of flat plate elements. Simple considerations of minimum material consumption frequently suggest that some plate elements be made extremely thin but limits must be imposed if certain potentially undesirable structural phenomena are to be avoided. The most important of these in everyday steelwork design is local buckling.

Figure 13.2 shows a short UC section after it has been tested as a column. Considerable distortion of the cross-section is evident with the flanges being deformed out of their original flat shape. The web, on the other hand, appears to be comparatively undeformed. The buckling has therefore been confined to certain plate elements, has not resulted in any overall deformation of the member and its centroidal axis has not deflected. In the particular example of Fig. 13.2, local buckling did not develop significantly until well after the column had sustained its 'squash load' equal to the product of its cross-sectional area times its material strength. Local buckling did not affect the load-carrying capacity because the proportions of the web and flange plates are sufficiently compact. The fact that the local buckling appeared in the flanges before the web is due to these elements being the more slender.

Terms such as *compact* and *slender* are used to describe the proportions of the individual plate elements of structural sections based on their susceptibility to local buckling. The most important governing property is the ratio of plate width to plate thickness, β, often referred to as the 'b/t ratio'. Other factors that have some influence are material strength, the type of stress system to which the plate is subjected, the support conditions provided and whether the section is produced by hot-rolling or welding.

Although the rigorous treatment of plate buckling is a mathematically complex topic,[1] it is possible to design safely and in most cases economically with no direct consideration of the subject. For example, the properties of the majority of

342   *Local buckling and cross-section classification*

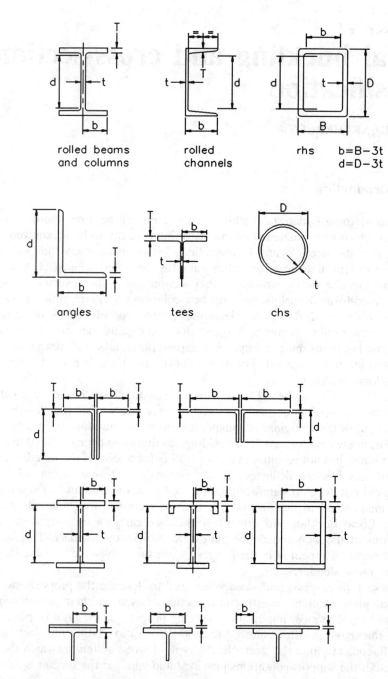

Fig. 13.1 Structural cross sections

Fig. 13.2 Local buckling of column flange

standard hot-rolled sections have been selected to be such that local buckling effects are unlikely to affect significantly their load-carrying capacity when used as beams or columns. Greater care is, however, necessary when using fabricated sections for which the proportions are under the direct control of the designer. Also cold-formed sections are often proportioned such that local buckling effects must be accounted for.

13.2 Cross-sectional dimensions and moment−rotation behaviour

Figure 13.3 illustrates a rectangular box section used as a beam. The plate slenderness ratios for the flanges and webs are b/T and d/t and elastic stress diagrams for both components are also shown. If the beam is subject to equal and opposite end moments M, Fig. 13.4 shows in a qualitative manner different forms of relationship between M and the corresponding rotation θ.

Assuming d/t to be such that local buckling of the webs does not occur, which of the four different forms of response given in Fig. 13.4 applies depends on the compression flange slenderness b/T. The four cases are defined as:

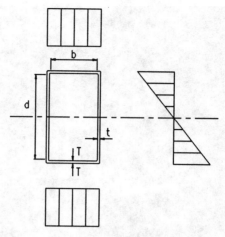

Fig. 13.3 Rectangular hollow section used as a beam

(a) $b/T \leq \beta_1$, full plastic moment capacity M_p is attained and maintained for large rotations and the member is suitable for plastic design – *plastic* cross-section (Class 1).
(b) $\beta_1 < b/T < \beta_2$, full plastic moment capacity M_p is attained but is only maintained for small rotations and the member is suitable for elastic design using its full capacity – *compact* cross-section (Class 2).
(c) $\beta_2 < b/T \leq \beta_3$, full elastic moment capacity M_y (but not M_p) is attained and the member is suitable for elastic design using this limited capacity – *semi-compact* cross-section (Class 3).
(d) $\beta_3 < b/T$, local buckling limits moment capacity to less than M_y – *slender* cross-section (Class 4).

The relationship between moment capacity M_u and compression flange slenderness b/T indicating the various β limits is illustrated diagrammatically in Fig. 13.5. In the figure the value of M_u for a semi-compact section is conservatively taken as the moment corresponding to extreme fibre yield M_y for all values of b/T between β_2 and β_3. This is more convenient for practical calculation than the more correct representation shown in Fig. 13.4 in which a moment between M_y and M_p is indicated. Since the classification of the section as plastic, compact, etc., is based on considerations of the compression flange alone, the assumption concerning the web slenderness d/t is that its classification is the same as or better than that of the flange. For example, if the section is semi-compact, governed by the flange proportions, then the web must be plastic, compact or semi-compact; it cannot be slender.

If the situation is reversed so that the webs are the controlling elements, then the same four categories, based on the same definitions of moment–rotation

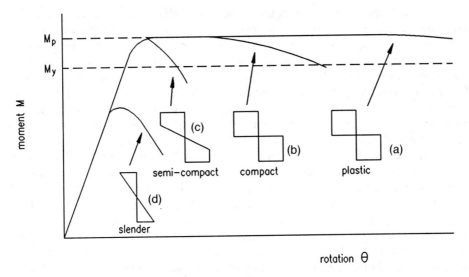

Fig. 13.4 Behaviour in bending of different classes of section

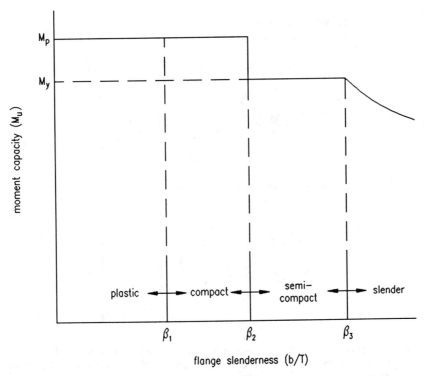

Fig. 13.5 Moment capacity as a function of flange slenderness

behaviour, are now determined by the value of web slenderness d/t. However, the governing values of β_1, β_2 and β_3 change since the web stress distribution differs from the pure compression in the top flange. Since the rectangular fully plastic condition, the triangular elastic condition and any intermediate condition contain less compression, the values of β are larger. Thus section classification also depends upon the type of stress system to which the plate element under consideration is subjected. If, in addition to the moment M, an axial compression F is applied to the member, then for elastic behaviour the pattern of stress in the web is of the form shown in Fig. 13.6(a). The values of σ_1 and σ_2 are dependent on the ratio F/M with σ_2 approaching σ_y, if F is large and M is small. In this case it may be expected that the appropriate β limits will be somewhere between the values for pure compression and pure bending, approaching the former if $\sigma_2 \approx \sigma_y$, and the latter if $\sigma_2 \approx -\sigma_1$. A qualitative indication of this is given in Fig. 13.7 which shows M_u as a function of d/t for three different σ_2/σ_1 ratios corresponding to pure compression, $\sigma_2 = 0$ and pure bending. If the value of d/t is sufficiently small that the web may be classified as compact or plastic, then the stress distribution will adopt the alternative plastic arrangement of Fig. 13.6(b).

For a plate element in a member which is subject to pure compression the load-carrying capacity is not affected by the degree of deformation since the scope for

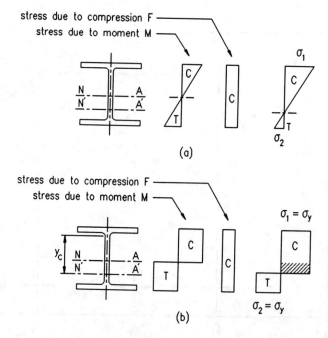

Fig. 13.6 Stress distributions in webs of symmetrical sections subject to combined bending and compression. (a) Semi-compact, elastic stress distribution. (b) Plastic or compact, plastic stress distribution

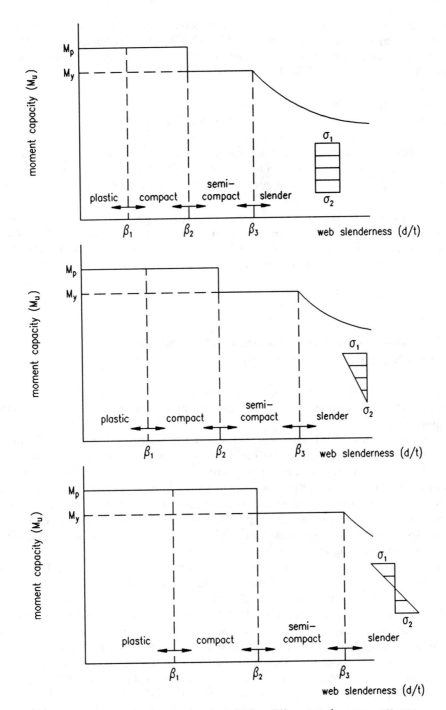

Fig. 13.7 Moment capacity as a function of d/t for different web stress patterns

a change in strain distribution as the member passes from a wholly elastic to a partially plastic state, as illustrated in Fig. 13.4 for pure bending, does not exist. The plastic and compact classifications do not therefore have any meaning; the only decision required is whether or not the member is slender and specific values are only required for β_3.

In the introduction to this chapter several other factors which affect local buckling are listed. These have a corresponding influence on β limits. As an example, the flanges of an I-section receive support along one longitudinal edge only, with the result that their buckling resistance is less than that of the flange of a box section and lower β values may be expected. Similarly the plate elements in members fabricated by welding generally contain a more severe pattern of residual stress, again leading to reduced β values.

One special case is the webs of beams and girders subject to shear. Although β limits for the purpose of section classification are normally provided for designers, the efficient design of plate girder webs may well require these to be exceeded. Special procedures (see Chapter 17) are, however, normally provided for such members.

13.3 Effect of moment–rotation behaviour on approach to design and analysis

The types of member present in a structure must be compatible with the method employed for its design. This is particularly important in the context of section classification.

Taking the most restrictive case first, for a plastically-designed structure, in which plastic hinge action in the members is being relied upon as the means to obtain the required load-carrying capacity, only plastic sections are admissible. Members which contain any plate elements that do not meet the required β_1 limit for the stress condition present are therefore unsuitable. This restriction could be relaxed for those members in a plastically-designed structure not required to participate in plastic hinge action; members other than those in which the plastic hinges corresponding to the collapse mechanism form. However, such an approach could be considered unsound on the basis of the effects of overstrength material, changes in the elastic pattern of moments due to settlement or lack of fit, and so on. Something less than a free choice of member types in structures designed plastically is required and BS 5950: Part 1 therefore generally restricts the method to structures in which only plastic or compact sections are present.

When elastic design − in the sense that an elastically determined set of member forces forms the basis for member selection − is being used, any of compact, semi-compact or plastic sections may be used, providing member strengths are properly determined. This point is discussed more fully in Chapters 14−18 which deal with different types of member. As a simple illustration, however, for members subject to pure bending the available moment capacity M_u must be taken as M_p or M_y. If slender sections are being used the loss of effectiveness due to local buckling will reduce not just their strength but also their stiffness. Moreover, reductions in stiffness are dependent upon load level, becoming greater as stresses increase sufficiently to cause local buckling effects to become more significant. Strictly speaking, such changes should be included when determining member forces. However, any attempt to do this would render design calculations prohibitively difficult and it is therefore usual to make only a very approximate allowance for the effect. It is of most importance for cold-formed sections.

In practice the designer, having decided upon the design approach (essentially either elastic or plastic), should check section classification first using whatever design aids are available to him. Since most hot-rolled sections are at least compact in both grade 43 and grade 50 steel, this will normally be a relatively trivial task. When using cold-formed sections, which will often be slender, sensible use of manufacturer's literature will often eliminate much of the actual calculation. Greater care is required when using sections fabricated from plate, for which the freedom to select dimensions and thus b/T and d/t ratios means that any class is possible.

13.4 Classification table

Part of a typical classification table, extracted from BS 5950: Part 1, is given in Table 13.1. Values of β_1, β_2 and β_3 for flanges, defined as plates supported along one longitudinal edge, and webs, defined as plates supported along both longitudinal edges, under pure compression, pure bending and combined compression and bending are listed. While the third case reduces to the second as the compression component reduces to zero, it does not accord with the first case when the web is wholly subject to uniform compression. The reason for this is that the neutral axis of a member subject to bending and compression in which the web is wholly in compression must lie in the flange, or at least at the web/flange junction, with the result that the tensile strains in the flange provide some degree of stabilizing influence. A slightly higher set of limits than those provided for a plate supported by other elements which are themselves in compression, such as the compression flange of a box beam, is therefore appropriate.

Table 13.1 Extract from table of section classification limits (BS 5950: Part 1)

Type of element	Class of section		
	Plastic (β_1)	Compact (β_2)	Semi-compact (β_3)
Outstand element of compression flange	$b/T \leq 7.5\varepsilon$	$b/T \leq 8.5\varepsilon$	$b/T \leq 13\varepsilon$
Internal element of compression flange	$b/T \leq 23\varepsilon$	$b/T \leq 25\varepsilon$	$b/T \leq 28\varepsilon$
Web with neutral axis at mid-depth	$d/t \leq 79\varepsilon$	$d/t \leq 98\varepsilon$	$d/t \leq 120\varepsilon$
Web, generally	$d/t \leq 79\varepsilon/(0.4 + 0.6\alpha)$	$d/t \leq 98\varepsilon/\alpha$	$d/t = f(\alpha,\varepsilon)$

13.5 Economic factors

When design is restricted to a choice of suitable standard hot-rolled sections, local buckling is not normally a major consideration. For plastically designed structures only plastic sections are suitable; thus the designer's choice is slightly restricted, although no UBs and only 6 UCs in grade 43 steel and 11 UBs and 10 UCs in grade 50 steel are outside the limits of BS 5950: Part 1 when used in pure bending. Although considerably more sections are unsatisfactory if their webs are subject to high compression, the number of sections barred from use in plastically designed portal frames is, in practice, extremely small. Similarly for elastic design no UB is other than semi-compact or better, providing it is not required to carry high compression in the web, while all UCs are at least semi-compact even when carrying their full squash load.

The designer should check the class of any trial section at an early stage. This can be done most efficiently using information of the type given in Reference 2. For webs under combined compression and bending the first check should be for pure compression as this is the more severe; providing the section is satisfactory no additional checks are required; if it does not meet the required limit a decision on whether it is likely to do so under the less severe combined load case must be made.

The economic use of cold-formed sections, including profiled sheeting of the type used as decking and cladding, often requires that members are non-compact. Quite often they contain plate elements that are slender, with the forming process being exploited to provide carefully proportioned shapes. Since cold-formed sections are proprietary products, manufacturers normally provide design literature in which member capacities which allow for the presence of slender plate elements are listed. If rigorous calculations are, however, required, then Parts 5 and 6 of BS 5950 contain the necessary procedures.

When using fabricated sections the opportunity exists for the designer to optimize on the use of material. This leads to a choice between three courses of action:

(1) eliminate all considerations of local buckling by ensuring that the width-to-thickness ratios of every plate element are sufficiently small;
(2) if employing higher width-to-thickness ratios, use stiffeners to reduce plate proportions sufficiently so that the desired strength is achieved;
(3) determine member capacities allowing for reductions due to exceeding the relevant compact or semi-compact limits.

Effectively only the first of these is available if plastic design is being used. For elastic design when the third approach is being employed and the sections are slender, then calculations inevitably are more involved as even the determination of basic cross-sectional capacities requires allowances for local buckling effects through the use of concepts such as the effective width technique.[1]

References to Chapter 13

1. Bulson P.S. (1970) *The Stability of Flat Plates*. Chatto and Windus, London.
2. The Steel Construction Institute (SCI) (1991) *Steelwork Design Guide to BS 5950: Part 1: 1985, Vol. 1: Section Properties, Member Capacities*, 3rd edn. SCI, Ascot, Berks.

Chapter 14
Tension members

by JOHN RIGHINIOTIS

14.1 Introduction

Theoretically, the tension member transmitting a direct tension between two points in a structure is the simplest and most efficient structural element. In many cases this efficiency is seriously impaired by the end connections required to join tension members to other members in the structure. In some situations (for example, in cross-braced panels) the load in the member reverses, usually by the action of wind, and then the member must also act as a strut. Where the load can reverse, the designer often permits the member to buckle, with the load then being taken up by another member.

14.2 Types of tension member

The main types of tension members, their applications and behaviour are:

(a) open and closed single rolled sections such as angles, tees, channels and the structural hollow sections. These are the main sections used for tension members in light trusses and lattice girders for bracing.
(b) compound sections consisting of double angles or channels. At least one axis of symmetry is present and so the eccentricity in the end connection can be minimized. When angles or other shapes are used in this fashion, they should be interconnected at intervals to prevent vibration especially when moving loads are present.
(c) heavy rolled sections and heavy compound sections of built-up H- and box sections. The built-up sections are tied together either at intervals (batten plates) or continuously (lacing or perforated cover plates). Batten plates or lacing do not add any load-carrying capacity to the member but they do serve to provide rigidity and to distribute the load among the main elements. Perforated plates can be considered as part of the tension member.
(d) bars and flats. In the sizes generally used, the stiffness of these members is very low; they may sag under their own weight or that of workmen. Their small cross-sectional dimensions also mean high slenderness values and, as a consequence, they may tend to flutter under wind loads or vibrate under moving loads.

(e) ropes and cables. Further discussion on these types of tension members is included in section 14.7 and Chapter 5, section 5.3.

The main types of tension members are shown in Fig. 14.1.
Typical uses of tension members are:

(a) tension chords and internal ties in trusses and lattice girders in buildings and bridges.
(b) bracing members in buildings.
(c) main cables and deck suspension cables in cable-stayed and suspension bridges.
(d) hangers in suspended structures.

Typical uses of tension members in buildings and bridges are shown in Fig. 14.2.

14.3 Design for axial tension

Rolled sections behave similarly to tensile test specimens under direct tension (Fig. 14.1).
For a straight member subject to direct tension, F:

tensile stress, $f_t = \dfrac{F}{A}$

elongation, $\delta_L = \dfrac{FL}{AE}$ (in the linear elastic range)

load at yield, $P_y = p_y A$ = load at failure (neglecting strain hardening)

For typical stress–strain curves for structural steel and wire rope see Fig. 14.3.

14.3.1 BS 5950: Part 1

The design of axially loaded tension members is given in clause 4.6. The tension capacity is:

$P_t = A_e p_y$

where A_e is the effective area defined in clause 3.3.3. This clause states that the effective area of a member with holes may be taken as K_e times its net area but not more than its gross area. The net area is defined as the gross area less deductions for fastener holes. For members with staggered holes reference should be made to Chapter 2, section 2.5.

The factor K_e, values of which are given below for steels complying with BS 4360, has been introduced as a result of tests which show that the presence of

Tension members

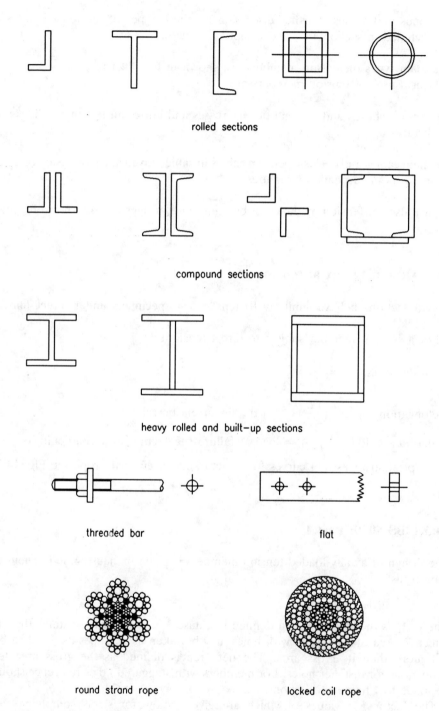

Fig. 14.1 Tension members

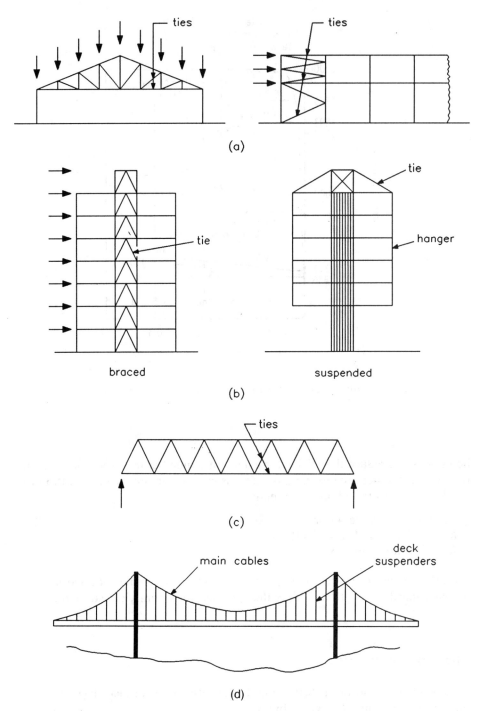

Fig. 14.2 Tension members in buildings and bridges. (a) Single-storey building – roof and truss bracing. (b) Multi-storey building. (c) Bridge truss. (d) Suspension bridge

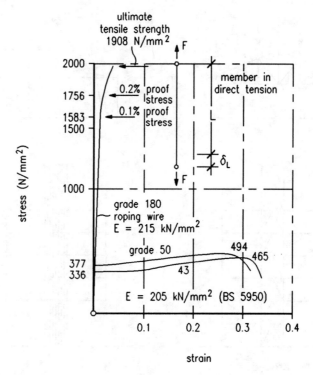

Fig. 14.3 Stress–strain curves for structural steels and wire rope

holes does not reduce the effective capacity of a member in tension provided that the ratio of the net area to the gross area is suitably greater than the ratio of the yield strength to the ultimate strength.

$K_e = 1.2$ for grade 40 or 43
$K_e = 1.1$ for grade 50 or WR50
$K_e = 1.0$ for grade 55

For other steels, $K_e = 0.75 \, U_s/Y_s$ but ≤ 1.2, where U_s is the specified minimum ultimate tensile strength and Y_s is the specified minimum yield strength.

14.3.2 BS 5400: Part 3

The member should be such that the design ultimate axial load does not exceed the tensile resistance P_D, given by:

$$P_D = \frac{\sigma_y A_e}{\gamma_m \gamma_{f_3}}$$

where A_e = effective cross-sectional area
= $k_1 k_2 A_t$
A_t = net cross-sectional area
A = gross area
σ_y = nominal yield stress for steel given in BS 4360
γ_m = partial safety factor.[a] Values are given in Table 2 of BS 5400: Part 3. This takes account of material variation.
γ_{f_3} = partial safety factor.[a] Values are given in clause 4.3.3. This is often called the 'gap factor'.
k_1, k_2 = factors. Values are given in clause 11.

[a] Note that the design ultimate load incorporates other safety factors (γ_{fL}).

14.4 Combined bending and tension

Bending in tension members arises from

(a) eccentric connections
(b) lateral loading
(c) rigid frame action.

If a structural member is subjected to axial tension combined with bending about axes *xx* and *yy* then:

tensile stress, $\quad f_t = F/A$
bending stress *xx* axis, $f_{bx} = M_x/Z_x$
bending stress *yy* axis, $f_{by} = M_y/Z_y$
maximum stress, $\quad f_{max} = f_t + f_{bx} + f_{by}$

The separate stress diagrams are shown in Fig. 14.4(a). The values of the separate actions causing yield are:

tensile load at yield, $\quad P_y = Y_s A$
moment at yield (*xx* axis), $M_{xx} = Y_s Z_x$
moment at yield (*yy* axis), $M_{yy} = Y_s Z_y$

The points defined by P_y, M_{xx} and M_{yy} form an interaction surface. Any point on this surface gives the co-existent values of F, M_x and M_y for which the maximum stress equals the yield stress.

358   Tension members

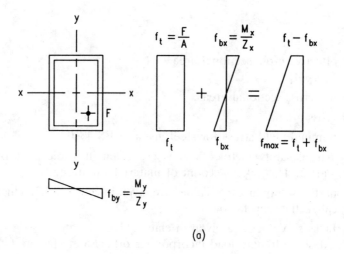

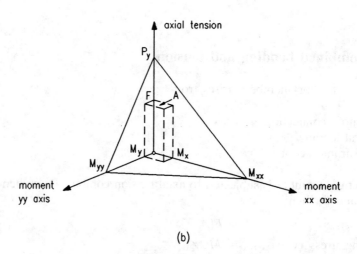

Fig. 14.4 Combined bending and tension – elastic analysis. (a) Stress diagram. (b) Elastic interaction surface

An elastic interaction surface for a rectangular hollow section is shown in Fig. 14.4(b).

The maximum values for axial tension and plastic moment for the same member are:

axial tension, $P_y = Y_s A$
plastic moment (xx axis), $M_{px} = Y_s S_x$
plastic moment (yy axis), $M_{py} = Y_s S_y$

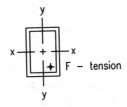

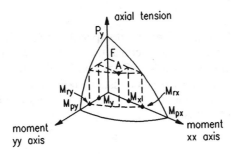

Fig. 14.5 Plastic interaction surface

The difference between the elastic and plastic interaction surfaces represents the additional design strength available if plasticity is taken into account (Fig. 14.5).

14.4.1 BS 5950: Part 1

The design of tension members with moments is given in clause 4.8.2. This states that tension members should be checked for capacity at the points of greatest bending moments and axial loads, usually at the ends. The following relationship should be satisfied:

$$\frac{F}{A_e p_y} + \frac{M_x}{M_{cx}} + \frac{M_y}{M_{cy}} \leq 1$$

where F = applied axial force in the member
A_e = effective area
p_y = design strength
M_x = applied moment about the major axis at the critical region
M_{cx} = moment capacity about the major axis in the absence of axial load
M_y = applied moment about the minor axis
M_{cy} = moment capacity about the minor axis in the absence of axial load.

For plastic or compact sections with low shear load:

$$M_c = p_y S \leq 1.2 \, p_y Z$$

where S = plastic modulus about the relevant axis
Z = elastic modulus about the relevant axis.

Alternatively for greater economy, in plastic or compact cross sections only, the following relationship should be satisfied:

$$\left(\frac{M_x}{M_{rx}}\right)^{z_1} + \left(\frac{M_y}{M_{ry}}\right)^{z_2} \leq 1$$

where M_{rx}, M_{ry} are the reduced moment capacities about their respective axes in the presence of axial load; z_1 is a constant, taken as 2.0 for solid and hollow circular sections, I- and H-sections, 5/3 for solid and hollow rectangular sections and 1.0 for all other cases; and z_2 is a constant, taken as 1.0 for I- and H-sections, 2.0 for solid and hollow circular sections, 5/3 for solid and closed hollow rectangular sections and 1.0 for all other cases.

14.4.2 BS 5400: Part 3

A member subjected to co-existent tension and bending should be such that at all cross sections the following relationship is satisfied:

$$\frac{P}{P_D} + \frac{M_x}{M_{Dxt}} + \frac{M_y}{M_{Dyt}} \leq 1.0$$

where P = axial tensile force
P_D = tensile resistance, see section 14.3.2
M_x, M_y = bending moments at the section about xx and yy axes, respectively
M_{Dxt}, M_{Dyt} = corresponding bending resistances with respect to the extreme tensile fibres.

Additionally, if at any section within the middle third of the length of the member the maximum compressive stress due to bending exceeds the tensile stress due to axial load, the design should be such that:

$$\frac{M_{xmax}}{M_{Dxc}} + \frac{M_{ymax}}{M_{Dyc}} < 1 + \frac{P}{P_D}$$

where M_{xmax}, M_{ymax} = maximum moments anywhere within the middle third
M_{Dxc}, M_{Dyc} = corresponding bending resistances with respect to the extreme compression fibres.

For compact sections the bending resistance is given in clause 9.9.1.2. This is:

$$M_D = \frac{Z_{pe} \, \sigma_{lc}}{\gamma_m \, \gamma_{f_3}}$$

where σ_{lc} = limiting compressive stress (see clause 9.8)
Z_{pe} = plastic modulus of the effective section (see clause 9.4.2).

Clause 9.9.1.3 should be referred to for the resistance of non-compact sections.

14.5 Eccentricity of end connections

Simplified design rules are given in both BS 5950 and BS 5400 for the effects of combined tension and bending caused by the eccentric load introduced into the member by the end connection.

14.5.1 BS 5950: Part 1

Angles, channels and tees may be designed as axially loaded members following the simplified design rules given below:

(a) Single angles connected through one leg, single channels connected only through the web and T-sections connected only through the flange have an effective area given by:

$$A_e = a_1 + \left(\frac{3a_1}{3a_1 + a_2}\right) a_2$$

where a_1 is the net sectional area of the connected leg and a_2 is the sectional area of the unconnected leg.

Double angles connected to one side of a gusset or section may be designed individually as given above.

(b) For back-to-back double angles connected to one side of a gusset or section which is in contact or separated by a distance not exceeding the aggregate thickness of the parts with solid packing pieces or connected by bolts or welding such that the slenderness of the individual component does not exceed 80, then the effective area A_e may be taken as:

$$A_e = a_1 + \left(\frac{5a_1}{5a_1 + a_2}\right) a_2$$

where a_1 is the net sectional area of the connected leg and a_2 is the sectional area of the unconnected leg.

14.5.2 BS 5400: Part 3

The bending moment resulting from an eccentricity of the end connections should be taken into account. For single angles connected by one leg these provisions may be considered to be met if the effective area of the unconnected leg is taken as:

$$\left(\frac{3A_1}{3A_1 + A_2}\right) A_2$$

where A_1 = net area of the connected leg
      A_2 = net area of the unconnected leg.

14.6 Other considerations

14.6.1 Serviceability, fatigue and corrosion

Ropes and bars are not normally used in building construction because they lack stiffness, but they have been used in some cases as hangers in suspended buildings. Very light, thin tension members are susceptible to excessive elongation under direct load as well as lateral deflection under self-weight and lateral loads. Special problems may arise where the members are subjected to vibration or conditions leading to failure by fatigue, such as can occur in bridge deck hangers. Damage through corrosion is undesirable; adequate protective measures must be adopted. All these factors can make the design of tension members a complicated process in some cases.

The light-rolled sections used for tension members in trusses and for bracing are easily damaged during transport. It is customary to specify a minimum size for such members to prevent this happening. For angle ties, a general rule is to make the leg length not less than one-sixtieth of the member length.

To avoid buckling of ties due to reversal of load under wind conditions, BS 5950: Part 1, clause 4.7.3 states that the slenderness value λ for any member acting as a tie but subject to reversal of stress resulting from the action of wind should not exceed 350. Slenderness λ is calculated by dividing the effective length L_E of a member by the radius of gyration about the relevant axis. The nominal effective length L_E can be determined from Table 24 of BS 5950: Part 1.

In a fatigue assessment of a member, the stress range considered is the greatest algebraic difference between principal stresses occurring during any one stress cycle. The above applies when the stresses occur on principal planes of more than 45° apart.

The stresses should be calculated using elastic theory and taking account of all axial, bending and shear stresses occurring under the design loadings. Alternatively the design loadings in BS 5400: Part 10 can be used.

In assessing the fatigue behaviour of a member, the following effects have to be included:

- shear lag, restrained torsion and distortion, transverse stresses and flange curvatures,

- effective width of steel plates,
- stresses in triangulated skeletal structures due to load applications away from the joints, member eccentricities to joints and rigidity of joints,
- cracking of concrete in composite elements.

The following effects however, can be ignored:

- residual stresses,
- eccentricities unnecessarily arising in a standard detail,
- plate buckling.

14.6.2 Stress concentration factors

In cases of geometrical discontinuity, such as a change of cross section or an aperture, the resulting stress concentrations may be determined either by special analysis or by the use of stress concentration factors. Stress concentrations are not usually important in ductile materials but can be the cause of failure due to fatigue or brittle fracture in certain considerations. A hole in a flat member can increase the stresses locally on the net section by a factor which depends on the ratio of hole diameter to net plate width.

When designing against fatigue it is convenient to consider three levels of stress concentration.

(a) stress concentrations from structural action due to the difference between the actual structural behaviour and the static model chosen,
(b) macroscopic stress concentrations due to large scale geometric interruptions to stress flow,
(c) microscopic and local geometry stress concentrations due to imperfections within the weld or the heat affected zone.

For the design of welded details and connections further reference should be made to BS 5400: Part 10. However, for non-standard situations it may be necessary to determine the stress concentration factor directly from a numerical analysis study or from an experimental model.

In many cases the detail under consideration is very likely not to fit neatly into one of the classes. On site, the actual stress range for a particular loading occurrence is likely to be strongly influenced by detailed fit of the joint and overall fit of the structure. Therefore the overall form should be such that load paths are as smooth as possible and unintended load paths should be avoided, particularly where fit could significantly influence behaviour. Discontinuities must be avoided by tapering and appropriate choice of radii.

Typical stress concentration factor graphs are shown in Fig. 14.6.

364 *Tension members*

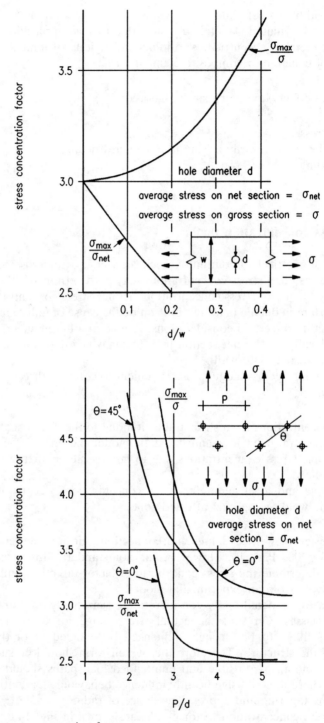

Fig. 14.6 Stress concentration factors

14.6.3 Fabrication and erection

The behaviour of tension members in service depends on the fabrication tolerances and the erection sequence and procedure. Provision must be made so that all tension members are immediately active in resisting applied loads in service. For bars or cables used as tension members tensioning up should be ensured through screwed ends.

Bracing members fabricated from rolled sections should be installed and properly tightened before other connections and column base plates are bolted up, to bring the structure into line and square. To enable the bracing members to be immediately effective and to avoid sagging, they are usually specified slightly shorter than the exact length.

Trial shop assembly is often specified for the member in heavy industrial trusses and bridge members to ensure that the fabrication is accurate and that erection is free from problems.

14.7 Cables

This section is directed mainly towards bridge cables; see Chapter 5, section 5.3 for cables in building structures.

14.7.1 Composition

A cable may be composed of one or more structural ropes, structural strands, locked coil strands or parallel wire strands.

A strand, with the exception of a parallel wire strand, is an assembly of wires formed vertically around a central wire in one or more symmetrical layers. A strand may be used either as an individual load-carrying member, where radius of curvature is not a major requirement, or as a component in the manufacture of structural rope.

A rope is composed of a plurality of strands vertically laid around a core. In contrast to the strand, a rope provides increased curvature capability and is used where curvature of the cable becomes an important consideration. The significant differences between strand and rope are as follows:

- at equal sizes, a rope has lower breaking strength than a strand,
- the modulus of elasticity of a rope is lower than that of a strand,
- a rope has more curvature capability than a strand,
- the wires in a rope are smaller than those in a strand of the same diameter, consequently, a rope for a given size coating is less corrosion-resistant because of the thinner coating on the smaller diameter wires.

14.7.2 Application

Cables used in structural applications, namely for suspension systems in bridges, fall into the following categories:

(a) parallel-bar cables;
(b) parallel-wire cables;
(c) stranded cables (see Fig. 14.1);
(d) locked-coil cables (see Fig. 14.1).

The final choice depends on the properties required by the designer, i.e. modulus of elasticity, ultimate tensile strength, durability. Other criteria include economic and structural detailing, i.e. anchorages, erection, etc.

14.7.3 Parallel-bar cables

Parallel-bar cables are formed of steel rods or bars, parallel to each other in metal ducts, kept in position by polyethylene spacers. The process of tensioning the bar or rods individually is simplified by the capability of the bars to slide longitudinally. Cement grout, injected after erection, makes sure that the duct plays its part in resisting the stresses due to live loads.

Transportation in reels is only possible for the smaller diameters while for the larger sizes delivery is made in straight bars of 15.0–20.0 m in length. Continuity of the bars has to be provided by the use of couplers which considerably reduces the fatigue strength of the stay.

The use of mild steel necessitates larger sections than when using high-strength wires or strands. This leads to a reduction in the stress variation and thus lessens the risk of fatigue failure.

14.7.4 Parallel-wire cables

Parallel wires are used for cable-stayed bridges and pre-stressed concrete. Their fatigue strength is satisfactory, mainly because of their good mechanical properties.

14.7.5 Corrosion protection

Wires in the cables should be protected from corrosion. The most effective protection is obtained by hot galvanizing by steeping or immersing the wires in a bath of melted zinc, automatically controlled to avoid overheating. A wire is described as 'terminally galvanized' or 'galvanized re-drawn' depending on whether the operation has taken place after drawing or in between two wire drawings prior to the wire being brought to the required diameter. For reinforcing bars and cables, the first method is generally adopted. A quantity of zinc in the range of $250-330$ g/m^2 is deposited, providing a protective coating $25-45$ microns thick.

14.7.6 Coating

The coating process, used currently for locked-coil cables, consists of coating the bare wires with an anti-corrosion product with a good bond and long service life. The various substances used generally have a high dropping point so as not to run back towards the lower anchorages. They are usually high viscosity resins or oil-based grease, paraffins or chemical compounds.

14.7.7 Protection of anchorages

The details of the connections between the ducts and the anchorages must prevent any inflow or accumulation of water. The actual details depend on the type of anchorages used, on the protective systems for the cables and on their slope. There are different arrangements intended to ensure water tightness of vital zones.

14.7.8 Protection against accidents

Cables should be protected against various risks of accident, such as vehicle impact, fire, explosion and vandalism. Measures to be taken may be based on the following:

(a) protection of the lower part of the stay, over a height of about 2.0 m, by a steel tube fixed into the deck and fixed into the duct; the tube dimensions (thickness and diameter) must be adequate.
(b) strength of the lower anchorage against vehicle impact.
(c) replacement of protective elements possible without affecting the cables themselves and, as far as possible, without interrupting traffic.

Tension members

Further reading for Chapter 14

Adams P.F., Krentz H.A. & Kulak G.L. (1973) *Canadian Structural Steel Design*. Canadian Institute of Steel Construction, Ontario.

Bresler B., Lin T.Y. & Scalzi J.B. (1968) *Design of Steel Structures*. Wiley, Chichester.

Dowling P.J., Knowles P. & Owens G.W. (1988) *Structural Steel Design*. Butterworths, London.

Horne M.R. (1971) *Plastic Theory of Structures*, 1st edn. Nelson, Walton-on-Thames.

Owens G.W. & Cheal B.D. (1989) *Structural Steelwork Connections*. Butterworths, London.

Timoshenko S.P. & Goodier J.N. (1970) *Theory of Elasticity*. McGraw-Hill, Maidenhead, Berks.

Trahair N.S. (1977) *The Behaviour and Design of Steel Structures*. Chapman and Hall, London.

Troitsky M.S. (1988) *Cable-Stayed Bridges*, 2nd edn. BSP Professional Books, Oxford.

A series of worked examples follows which are relevant to Chapter 14.

The Steel Construction Institute Silwood Park, Ascot, Berks SL5 7QN	Subject **TENSION MEMBERS**		Chapter ref. **14**
	Design code **BS5950**	Made by **JR** Checked by **GWO**	Sheet no. **1**

Problem

BS 5950: Part 1

Design a single angle tie for the member AB shown.

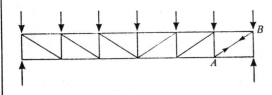

Tensile force in member AB

Dead load	=	*144 kN*
Imposed load	=	*240 kN*
Total	=	*384 kN*

Material: Use Grade 50 steel

Connections: a)   Welded
                       b)   Bolted

Factored load

$F = (1.4 \times 144) + (1.6 \times 240) = 585.6 \text{ kN}$ *Table 2*

Material Grade 50 steel, thickness ≤ 16 mm   *Table 6*

Design strength p_y = 355 N/mm^2

The Steel Construction Institute Silwood Park, Ascot, Berks SL5 7QN	Subject **TENSION MEMBERS**		Chapter ref. **14**
	Design code **BS5950**	Made by **JR** Checked by **GWO**	Sheet no. **2**

a) <u>Welded connections</u>

Try <u>125 × 75 × 10</u> connected by long leg.

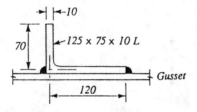

<u>Effective Area, A_e</u> 4.6.3.1

a_1 = 120 × 10 = 1200 mm²
a_2 = 70 × 10 = 700 mm²

$$A_e = a_1 + \left[\frac{3a_1}{3a_1 + a_2}\right] a_2$$

$$\therefore A_e = 1200 + \left[\frac{3 \times 1200}{3 \times 1200 + 700}\right] \times 700$$

= 1786 mm²

<u>Tension capacity</u> P_t = $A_e p_y$

$P_t = \dfrac{1786 \times 355}{10^3}$ = 634 kN > 585.6 kN OK 4.6.1

∴ Use <u>125 × 75 × 10</u> Grade 50

The Steel Construction Institute Silwood Park, Ascot, Berks SL5 7QN	Subject *TENSION MEMBERS*	Chapter ref. **14**
	Design code **BS5950**	Made by **JR** / Checked by **GWO** · Sheet no. **3**

b) <u>Bolted connections</u>

Try <u>150 × 75 × 10</u> connected by long leg

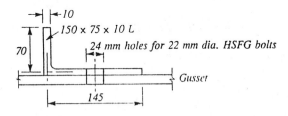

Effective Area, A_e

a_1 = 145 × 10 = 1450 mm²   4.6.3.1

a_2 = 70 × 10 = 700 mm²

$a_{1\ net}$ = 1450 − (24 × 10)

= 1210 mm²

A_e = $a_1 + \left[\dfrac{3a_1}{3a_1 + a_2}\right] \times a_2$

A_e = $1210 + \left[\dfrac{3 \times 1210}{3 \times 1210 + 700}\right] \times 700$

= 1797 mm²

Note: Effective area factor K_e for sections with holes (3.3.3) does not apply in this case.

<u>Tension capacity</u> P_t = $A_e p_y$

                                                               4.6.1

P_t = $\dfrac{1797 \times 355}{10^3}$ = 638 kN > 585.6 kN

∴ Use <u>150 × 75 × 10</u> Grade 50

Worked example

The Steel Construction Institute Silwood Park, Ascot, Berks SL5 7QN	Subject **TENSION MEMBERS**		Chapter ref. **14**
	Design code **BS5950**	Made by **JR** Checked by **GWO**	Sheet no. **4**

Efficiency of single angle tie

Calculation of the efficiency of a 150 × 75 × 10 used as a single angle tension member. Gross area = 21.6 cm²

a)   Bolted through the long leg with bolts in 24 mm dia. holes

From the above calculations, Net area = 1797 mm²

$$\text{Efficiency} = \left[\frac{1797}{21.6 \times 10^2}\right] \times 100 = \underline{83\%}$$

b)   Welded through long leg

a_1 = 145 × 10 = 1450 mm²
a_2 = 70 × 10 = 700 mm²

Effective area:

$$A_e = 1450 + \left[\frac{3 \times 1450}{3 \times 1450 + 700}\right] \times 700$$

$$= 2053 \text{ mm}^2$$

$$\text{Efficiency} = \left[\frac{2053}{21.6 \times 10^2}\right] \times 100 = \underline{95\%}$$

The Steel Construction Institute Silwood Park, Ascot, Berks SL5 7QN	Subject **TENSION MEMBERS**		Chapter ref. **14**
	Design code **BS5950**	Made by **JR** Checked by **GWO**	Sheet no. **5**

Problem

Design a double angle tie for the member AB in the figure on Sheet 1. Assume the use of double angles, connected back to back through a 8 mm gusset. Assume welded connections.

BS 5950: Part 1

Try <u>2 × 75 × 50 × 8</u>, connected through long leg.

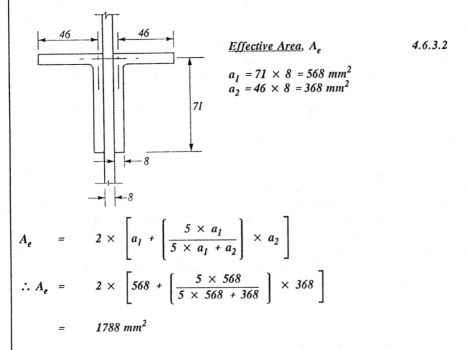

Effective Area, A_e 4.6.3.2

$a_1 = 71 \times 8 = 568 \ mm^2$
$a_2 = 46 \times 8 = 368 \ mm^2$

$$A_e = 2 \times \left[a_1 + \left(\frac{5 \times a_1}{5 \times a_1 + a_2} \right) \times a_2 \right]$$

$$\therefore A_e = 2 \times \left[568 + \left(\frac{5 \times 568}{5 \times 568 + 368} \right) \times 368 \right]$$

$$= 1788 \ mm^2$$

Tension capacity

$P_t = A_e \times p_y$ 4.6.1

$P_t = \dfrac{1788 \times 355}{10^3} = 635 \ kN > 585.6 \ kN \ OK$

$\therefore$ Use <u>2 × 75 × 50 × 8</u> Grade 50   4.6.3.2.(b)

Note: The two angles must be connected at intervals so that their individual slenderness does not exceed 80.

Chapter 15
Columns and struts

by DAVID NETHERCOT

15.1 Introduction

Members subject to compression, referred to as either 'columns' or 'struts', form one of the basic types of load-carrying component. They may be found, for example, as vertical columns in building frames, in the compression chords of a bridge truss or in any position in a space frame.

In many practical situations struts are not subject solely to compression but, depending upon the exact nature of the load path through the structure, are also required to resist some degree of bending. For example, a corner column in a building is normally bent about both axes by the action of the beam loads, a strut in a space frame is not necessarily loaded concentrically, the compression chord of a roof truss may also be required to carry some lateral loads. Thus many compression members are actually designed for combined loading as beam-columns. Notwithstanding this, the ability to determine the compressive resistance of members is of fundamental importance in design, both for the struts loaded only in compression and as one component in the interaction type of approach normally used for beam-column design.

The most significant factor that must be considered in the design of struts is buckling. Depending on the type of member and the particular application under consideration, this may take several forms. One of these, local buckling of individual plate elements in compression, has already been considered in Chapter 13. Much of this chapter is devoted to the consideration of the way in which buckling is handled in strut design.

15.2 Common types of member

Various types of steel section may be used as struts to resist compressive loads; Fig. 15.1 illustrates a number of them. Practical considerations such as the methods to be employed for making connections often influence the choice, especially for light members. Although closed sections such as tubes are theoretically the most efficient, it is normally much easier to make simple site connections, using the minimum of skilled labour or special equipment, to open sections. Typical arrangements include:

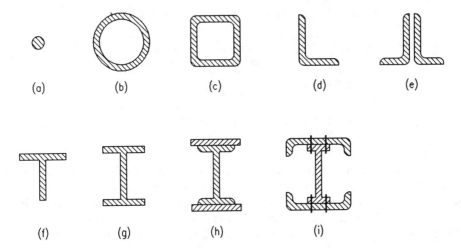

Fig. 15.1 Typical column cross sections

(1) light trusses and bracing – angles (including compound angles back to back) and tees
(2) larger trusses – circular hollow sections, rectangular hollow sections, compound sections and universal columns
(3) frames – universal columns, fabricated sections e.g. reinforced UCs
(4) bridges – box columns
(5) power stations – stiffened box columns.

15.3 Design considerations

The most important property of a strut as far as the determination of its load-carrying capacity is concerned is its slenderness, λ, defined as the ratio of its effective length, L_E, divided by the appropriate radius of gyration, r. Codes of Practice such as BS 5950 place upper limits on λ, as indicated in Table 15.1, so as to avoid the use of flimsy construction, i.e. to ensure that a member which will

Table 15.1 Maximum slenderness values for struts

Condition	BS 5950: Part 1 limits	BS 5950: Part 5
Members in general	180[a]	180
Members resisting self-weight wind loads only	250[a]	250
Members normally in tension but subject to load reversal due to wind	–	350

[a] Check for self-weight deflection if $\lambda > 180$; allow for bending effects in design if this deflection $> L/1000$

Table 15.2 Upper limits on plate slenderness of BS 5400: Part 3

b/t or D/t limit[a]	Unstiffened outstand	Stiffened outstand	CHS
p_y = 355 N/mm^2	12	14	100
p_y = 275 N/mm^2	14	16	114

[a] See Fig. 13.1 for definitions of b; D = outside diameter

ordinarily be subject only to axial load does have some limited resistance to an accidental lateral load, does not rattle, etc. In the case of BS 5400: Part 3, no actual limits are specified; the user is merely provided with a general advisory note to the effect that construction should be suitably robust. By contrast, BS 5950 places upper bounds on local plate slenderness to avoid consideration of local buckling (see Table 15.2).

Strut design will normally require that, once a trial member has been selected and its loading and support conditions determined, attention be given to whichever of the following checks are relevant for the particular application:

(1) Overall flexural buckling – largely controlled by the slenderness ratio, λ, which is a function of member length, cross-sectional shape and the support conditions provided; also influenced by the type of member
(2) Local buckling – controlled by the width-to-thickness ratios of the component plate elements (see Chapter 13); with some care in the original choice of member this need not involve any actual calculation
(3) Buckling of component parts – only relevant for built-up sections such as laced and battened columns; the strength of individual parts must be checked, often by simply limiting distances between points of interconnection
(4) Torsional or torsional-flexural buckling – for cold-formed sections and in extreme cases of unusually shaped heavier open sections, the inherent low torsional stiffness of the member may make this form of buckling more critical than simple flexural buckling.

In principle, local buckling and overall buckling (flexural or torsional) should always be checked. In practice, providing cross sections that at least meet the semi-compact limits for pure compression are used, then no local buckling check is necessary since the cross section will be fully effective.

15.4 Cross-sectional considerations

Since the maximum attainable load-carrying capacity for any structural member is controlled by its local cross-sectional capacity (factors such as buckling may prevent this being achieved in practice), the first step in strut design must involve consideration of local buckling as it influences axial capacity. Only two classes of section are relevant for purely axially compressed members: either the section is not slender, in which case its full capacity, $p_y A$, is available, or it is slender and some allowance in terms of a reduced capacity is required. The distinctions between plastic, compact and semi-compact as described in chapter 13 therefore have no relevance when the type of member under consideration is a strut.

The general approach for a member containing slender plate elements designed in accordance with BS 5950: Part 1 requires the use in all calculations relating to that member, apart from those concerned with components of connections to the member, of a reduced design strength, with the magnitude of the reduction being dependent on the extent to which the semi-compact limits (the boundary between not slender and slender) are exceeded. Effectively this means that the member is assumed to be made of a steel having a lower material strength than is actually the case.

When the cross section is classified as slender because of the proportions of one or more outstand elements, only this reduced strength method is permitted. However, if web proportions control and any outstands present are at least semi-compact, for example, an RHS or a UB with compact flanges, then the following alternatives are available. As noted above, BS 5400: Part 3 does not permit the use of columns with slender outstands.

15.4.1 Columns with slender webs

Reference 1 indicates that most UB and a few RHS, even in grade 43 steel, will be slender according to BS 5950: Part 1 if used as struts. A suitable allowance for the weakening effect of local buckling should therefore be made. Three different procedures are given, outlined in the following example.

Consider a 406 × 140 × 39 UB (the UB with the highest d/t) in grade 43 material. This has a d/t of 57.1 compared with the limit in Table 7 of 39.

(1) Use clause 3.6.3, limiting $d/t = 39 \, (275/p_{yr})^{\frac{1}{2}}$

and reduced design strength, $p_{yr} = (39/57.1)^2 \times 275$
$= 128.3 \text{ N/mm}^2$

For $L_E/r_y = 100$, using Appendix C for $p_{yr} = 128.3 \text{ N/mm}^2$ and $a = 3.5$

$p_c = 89.7 \text{ N/mm}^2$
$P_c = 49.4 \times 87.7/10$
$= 433 \text{ kN}$

(2) Use clause 3.6.4 for Table 8 strength reduction factor
$= 31/(d/t\varepsilon - 8)$

and $p_{yr} = 275 \times 31/(57.1 - 8)$
$= 173.6$ N/mm^2

For $L_E/r_y = 100$, using Appendix C, $p_c = 108.9$ N/mm^2

$P_c = 49.4 \times 108.9/10$
$= 538$ kN

These results may be compared with a P_c value assuming a fully effective section of 697 kN.

(3) *Alternative using footnote to Table 8.*

This permits the effective area method of Part 5 to be employed; a similar approach is adopted in Eurocode 3. It is suitable for the slender webs of UB sections but not for slender flange outstands or for slender webs of unsymmetrical sections such as channels or unequal flange Is.

The method uses a parameter Q defined as (effective area)/(gross area); for each element of the cross section this corresponds to the strength reduction factor of Table 8. The value of P_c is then obtained using the effective area and a p_c value calculated from a slenderness reduced by $Q^{\frac{1}{2}}$. The following illustration follows the Eurocode 3 presentation rather than that of Part 5.

$$A_w = dt$$
$$= 359.7 \times 6.3$$
$$= 2266 \text{ mm}^2$$

Using Table 8, effective web area $= 2266 \times 31/(57.1 - 8)$
$= 1431$ mm^2

Total effective area $= 4940 - (2266 - 1431)$
$= 4105$ mm^2

Q-factor $= 4105/4940$
$= 0.831$

For $L_E/r_y = 100$, effective slenderness $= 100 \, (0.831)^{\frac{1}{2}}$
$= 91.2$

and $p_c = 158$ N/mm^2
$P_c = 4105 \times 158/10^3$
$= 649$ kN

The third method leads to a significantly higher result, being only some 7% below the fully effective figure. The main reason for this is that both the first and second methods apply the same reduction in strength, based on the proportions of the most slender plate element, to the whole of the cross section, whereas the third method makes a reduction only for the plate elements which are slender.

Both BS 5400: Part 3 and BS 5950: Part 5 employ a method for slender sections in compression that uses the concept of effective area. Its implementation for cold-formed sections has been illustrated by the third method of the above example. The bridge code prohibits the use of slender outstands by specifying the absolute upper limits on b/t given in Table 15.2 with no procedure to cover arrangements outside these, and determining the effective area of the cross section by summing the full areas of any outstands and compact webs and the effective areas, $K_c A_c$, of the slender webs. Values of K_c are obtained as a function of b/t from a graph.

It is worth emphasizing that for design using hot-rolled sections the majority of situations may be treated simply by ensuring that the proportions of the cross section lie within the semi-compact limits. Reference 1, in addition to listing flange b/T and web d/t values for all standard sections, also identifies those sections that are slender according to BS 5950: Part 1 when used in either grade 43 or grade 50 steel. Table 15.3 lists these.

Table 15.3 Sections classified as 'slender' when used as struts

Section	Grade 43	Grade 50
Universal Beam	All *except* 914 × 419UB388 610 × 305UB238 610 × 305UB179	All *except* 610 × 305UB238 356 × 171UB 67
		305 × 127UB 48
	533 × 210UB122	305 × 127UB 42
	457 × 191UB 98	254 × 146UB 43
	457 × 191UB 89	254 × 146UB 37
	457 × 152UB 82	203 × 133UB 30
		203 × 133UB 25
	406 × 178UB 74	
		203 × 102UB 23
	356 × 171UB 67	
		178 × 102UB 19
	305 × 165UB 54	
		152 × 89UB 16
	305 × 127UB 48	
	305 × 127UB 42	127 × 76UB 13
	305 × 127UB 37	
	254 × 146UB 43	
	254 × 146UB 37	
	254 × 146UB 31	
	254 × 102UB 28	
	254 × 102UB 25	
	254 × 102UB 22	
	203 × 133UB 30	
	203 × 133UB 25	
	203 × 102UB 23	
	178 × 102UB 19	
	152 × 89UB 16	
	127 × 76UB 13	
Universal Column	none	none

Table 15.3 Continued

Section	Grade 43	Grade 50
Circular Hollow Section	none	none
Rectangular Hollow Section	300 × 200 × 6.3 450 × 250 × 10.0	250 × 250 × 6.3 400 × 400 × 10.0 200 × 100 × 5.0 250 × 150 × 6.3 300 × 200 × 6.3 300 × 200 × 8.0 400 × 200 × 10.0 450 × 250 × 10.0
Rolled Steel Angle	200 × 150 × 15 200 × 150 × 12 200 × 100 × 12 200 × 100 × 10 150 × 90 × 10 125 × 75 × 8 100 × 65 × 7 80 × 60 × 6 200 × 200 × 16 150 × 150 × 12 150 × 150 × 10 120 × 120 × 10 120 × 120 × 8 100 × 100 × 8 90 × 90 × 7 90 × 90 × 6 80 × 80 × 6 70 × 70 × 6 60 × 60 × 5	200 × 150 × 15 200 × 150 × 12 200 × 100 × 15 200 × 100 × 12 200 × 100 × 10 150 × 90 × 10 100 × 75 × 10 125 × 75 × 8 100 × 75 × 8 100 × 65 × 8 100 × 65 × 7 80 × 60 × 6 75 × 50 × 6 65 × 50 × 5 200 × 200 × 18 200 × 200 × 16 150 × 150 × 12 150 × 150 × 10 120 × 120 × 10 120 × 120 × 8 100 × 100 × 8 90 × 90 × 8 90 × 90 × 7 90 × 90 × 6 80 × 80 × 6 70 × 70 × 6 60 × 60 × 5

15.5 Compressive resistance

The axial load-carrying capacity for a single compression member is a function of its slenderness, its material strength, cross-sectional shape and method of manufacture. Using BS 5950: Part 1, the compression resistance, P_c, is given by clause 4.7.4 as:

$$P_c = A_g p_c$$

in which A_g is the gross area and p_c is the compressive strength.

Values of p_c in terms of slenderness λ and material design strength p_y are given in Tables 27a–d. Slenderness is defined as the ratio of the effective length, L_E (taken as the geometrical length L for the present but see section 15.7), to the least radius of gyration, r.

The basis for Table 27 is the set of four column curves shown in Fig. 15.2. These have resulted from a comprehensive series of full-scale tests, supported by detailed numerical studies, on a representative range of cross sections.[2] They are often referred to as the 'European Column Curves'. Four curves are used in recognition of the fact that for the same slenderness certain types of cross section consistently perform better than others as struts. This is largely due to the arrangement of the material but is also influenced by the residual stresses that form as a result of differential cooling after hot rolling. It is catered for in design by using the strut curve selection table given as Table 25 in BS 5950: Part 1.

The first step in column design is therefore to consult Table 25 to see which of Tables 27a–d is appropriate. For example, if the case being checked is a UC liable to buckle about its minor axis, Table 27c should be used. Selection of a trial section fixes r and A_g; the geometrical length wil be defined by the application required, so λ and thus p_c and P_c may be obtained.

The above process should be used for all types of rolled section, including those reinforced by the addition of welded cover plates for which Table 26 should be used to supplement Table 25 when deciding on which of Tables 27a–d to use for determining p_c. When sections are fabricated by welding plates together, however,

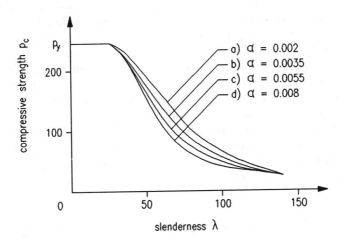

Fig. 15.2 Strut curves of BS 5950

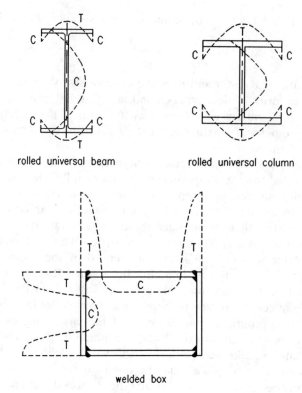

Fig. 15.3 Distribution of residual stresses

the pattern of residual stresses produced by the heating and cooling of the welding process will be rather different from that typically found in a hot-rolled section[3] as illustrated in Fig. 15.3. This tends to produce lower buckling strengths. It is allowed for in BS 5950: Part 1 by the modification to the column curves shown in Fig. 15.4 in which the basic curve of Fig. 15.2 is replotted for a reduced material design strength. This has the effect of reducing the basic curve down to the level of that for the welded member in the important medium slenderness region. Clause 4.7.5 thus requires that when checking the axial resistance of a member that has been fabricated from plate by welding, the value of p_c be obtained using a p_y value of 20 N/mm² less than the actual figure.

BS 5400: Part 3 is intended to cover a wider range of construction than BS 5950: Part 1. Specifically in this context it recognizes the greater likelihood that struts may contain slender elements (other than outstands) and so it defines compression resistance, P_c, as

$P_c = A_e p_c$ (clause 10.6.1.1)

in which A_e is the effective area defined as

$A_e = \Sigma K_c (k_h A_c)$ for members other than CHSs

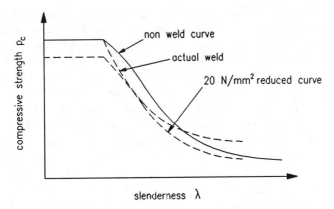

Fig. 15.4 Modified strut curve for welded sections used in BS 5950: Part 1

where K_c allows for loss of effectiveness in slender plate elements, determined from Figure 36, k_h allows for the presence of holes, and A_c is the net area; and

$$A_e = A_c \quad \text{for CHSs for which } \frac{D}{t}\sqrt{\left(\frac{p_y}{355}\right)} \leq 50$$

$$A_e = A_c\left[1.15 - 0.003\frac{D}{t}\sqrt{\left(\frac{p_y}{355}\right)}\right] \quad \text{for CHSs for which } \frac{D}{t}\sqrt{\left(\frac{p_y}{355}\right)} > 50$$

p_c is obtained from a set of curves (Figure 37) similar to Fig. 15.2.

Selection of the appropriate column curve is made using a simpler selection table than that of Part 1 of BS 5950. Essentially it distinguishes between curves on the basis of the ratio of radius of gyration to distance from neutral axis to extreme fibre, apart from heavy sections and hot-finished SHS which are universally allocated to the lowest and highest curves respectively.

15.6 Torsional and flexural-torsional buckling

In addition to the simple flexural buckling described in the previous section, struts may buckle due to either pure twisting about their longitudinal axis or a combination of bending and twisting. The first type of behaviour is only possible for centrally-loaded doubly-symmetrical cross sections for which the centroid and shear centre coincide. The second, rather more general form of response, occurs for centrally-loaded struts such as channels for which the centroid and shear centre do not coincide.

In practice pure torsional buckling of hot-rolled structural sections is highly unlikely, the pure flexural mode normally requiring a lower load, unless the strut is of a somewhat unusual shape so that its torsional and warping stiffnesses are low as for a cruciform section. In such cases a reasonable design approach consists of determining an effective slenderness based on the direct use of the member's elastic critical load for torsional buckling (assuming this to be lower than its elastic critical load for pure flexural buckling) and using this to enter the basic column design curve. This approach is well substantiated for aluminium members for which torsional buckling more commonly controls. Similarly for unsymmetrical sections, use of an effective slenderness based on the member's lowest elastic critical load, corresponding to flexural-torsional buckling in this case, permits the basic column design curve to be retained and used in a more general way. For certain types of section, such as hot-rolled angles, special empirically-based design approaches are provided in BS 5950: Part 1 which recognize the possibility of some torsional influence. They are discussed in detail in section 15.8.

Torsional-flexural buckling is of greater practical significance in the design and use of cold-formed sections. This arises for two reasons:

(1) Because the torsion constant, J, depends on t^3, the use of thin material results in the ratio of torsional to flexural stiffness being much reduced as compared with hot-rolled sections
(2) The forming process leads naturally to a preponderance of singly-symmetrical or unsymmetrical open sections as these can be produced from a single sheet.

Procedures are given in BS 5950: Part 5 for determining the axial strength of singly-symmetrical sections using a factored effective length αL_E, in which α is obtained from

$\alpha = 1$ for $P_{EY} \leq P_{TF}$
$\alpha = P_{EY}/P_{TF}$ for $P_{EY} > P_{TF}$

Formulae for P_{TF} in terms of basic cross-sectional properties are also provided. The use of $\alpha \neq 1$ is only required for those situations illustrated in Fig. 15.5 for which P_{TF} is the lowest elastic critical load.

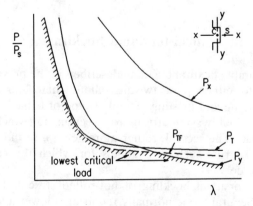

Fig. 15.5 Elastic critical load curves for a member subject to flexural–torsional buckling

15.7 Effective lengths

Basic design information relating column strength to slenderness is normally founded on the concept of a pin-ended member, e.g. Fig. 15.2. Stated more precisely, this means a member whose ends are supported such that they cannot translate relative to one another but are able to rotate freely. Compression members in actual structures are provided with a variety of different support conditions which are likely to be less restrictive in terms of translational restraint, giving fixity in position, with or without more restriction in terms of rotational restraint, giving fixity in direction.

The usual way of treating this topic in design is to use the concept of an effective column length which may be defined as 'the length of an equivalent pin-ended column having the same load-carrying capacity as the member under consideration provided with its actual conditions of support'. This engineering definition of effective length is illustrated in Fig. 15.6 which compares a column strength curve for a member with some degree of rotational end restraint with the basic curve for the same member when pin-ended.

In determining the column slenderness ratio the geometrical length, L, is replaced by the effective length, L_E. Values of effective length factors $k = L_E/L$ for a series of standard cases are provided in BS 5950: Part 1, BS 5400: Part 3, etc.; Fig. 15.7 illustrates typical values. When compared with values given by elastic stability theory,[2] these appear to be high for those cases in which reliance is being placed on externally provided rotational fixity; this is in recognition of the practical difficulties of providing sufficient rotational restraint to approach the condition of full fixity. On the other hand, translational restraints of comparatively

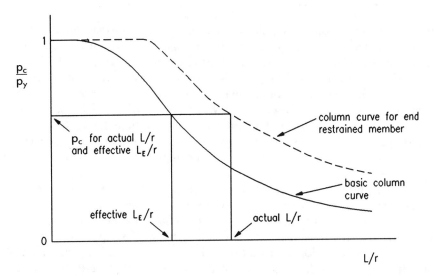

Fig. 15.6 Use of effective length with column curve to allow for end restraint

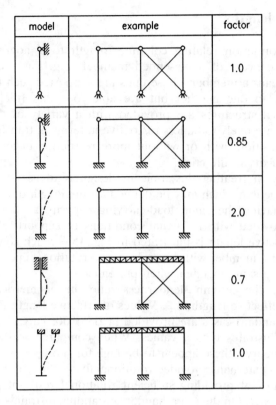

Fig. 15.7 Typical effective length factors for use in strut design

modest stiffness are quite capable of preventing lateral displacements. A certain degree of judgement is required of the designer in deciding which of these standard cases most nearly matches his arrangements. In cases of doubt the safe approach is to use a high approximation, leading to an overestimate of column slenderness and thus an underestimate of strength. The idea of an effective column length may also be used as a device to deal with special types of column, such as compound or tapered members, the idea then being to convert the complex problem into one of an equivalent simple column for which the basic design approach of the relationship between compressive strength and slenderness may be employed.

Of fundamental importance when determining suitable effective lengths is the classification of a column as either a sway case for which translation of one end relative to the other is possible or a non-sway case for which end translation is prevented. For the first case, effective lengths will be at least equal to the geometrical length, tending in theory to infinity for a pin-base column with no restraint at its top, while for the non-sway case, effective lengths will not exceed the geometrical length, decreasing as the degree of rotational fixity increases.

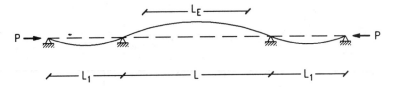

Fig. 15.8 Restraint to critical column segment from adjacent segments

For non-standard cases, it is customary to make reference to published results obtained from elastic stability theory. Providing these relate to cases for which buckling involves the interaction of a group of members with the less critical restraining the more critical, as illustrated in Fig. 15.8, such evidence as is available suggests that the use of effective lengths derived directly from elastic stability theory in conjunction with a column design curve of the type shown in Fig. 15.2 will lead to good approximations of the true load-carrying capacity.

For compression members in rigid-jointed frames the effective length may, in both cases, be directly related to the restraint provided by the surrounding members by using charts presented in terms of the stiffness of these members provided in Appendix E of BS 5950: Part 1. Useful guidance on effective column lengths for a variety of more complex situations is available from several sources.[2-4]

When designing compression members in frames configured on the basis of simple construction, the use of effective column lengths provides a simple means of recognizing that real connections between members will normally provide some degree of rotational end restraint, leading to compressive strengths somewhat in excess of those that would be obtained if columns were treated as pin-ended. If axial load levels and unsupported lengths change within the length of a member that is continuous over several segments such as a building frame column spliced so as to act as a continuous member but carrying decreasing compression with height or a compression chord in a truss, then the less heavily loaded segments will effectively restrain the more critical segments.

Even though the distribution of internal member forces has been made on the assumption of pin joints, some allowance for rotational end restraint when designing the compression members is therefore appropriate. Thus the apparent contradiction of regarding a structure as pin-jointed but using compression member effective lengths that are less than their actual lengths does have a basis founded upon an approximate version of reality. Figure 15.9 presents results obtained from elastic stability theory for columns continuous over a number of storeys which show how the effective length of the critical segment will be reduced if more stable segments (shorter unbraced lengths in this case) are present. A practical equivalent for each case in terms of simple braced frames with pinned beam-to-column connections is also shown. It is also necessary to recognize that practical equivalents of pin joints may also be capable of transferring limited moments. This point is considered explicitly in BS 5950: Part 1 for both building frames and trusses; the effect on the

388   Columns and struts

column	frame	a/L	0	0.1	0.2	0.3	0.4	0.5	0.6	0.7	0.8	0.9	1.0
(column with load N at distance a, length L, EI)	(portal frame with N loads, deflected shape)	L_E/L	1.0	1.11	1.24	1.40	1.56	1.74	1.93	2.16	2.31	2.50	2.70
(column with load N inclined, length L and a, EI)	(frame with a and L dimensions, deflected shape)	L_E/L	2.0	2.07	2.13	2.20	2.27	2.34	2.41	2.48	2.55	2.62	2.70
(simply supported column with N load, a and L)	(braced frame)	L_E/L	0.70	0.72	0.74	0.77	0.79	0.81	0.84	0.87	0.91	0.95	1.0

column	frame	0	0.1	0.2	0.3	0.4	0.5	0.6	0.7	0.8	0.9	1.0	a/L
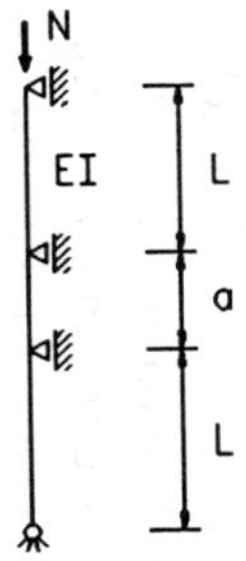	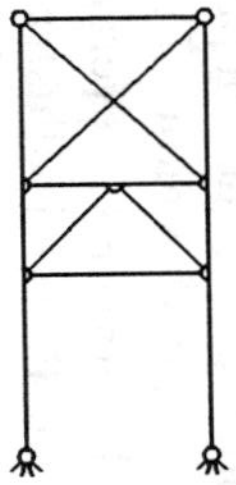	0.70	0.73	0.76	0.79	0.82	0.85	0.88	0.91	0.94	0.97	1.0	L_E/L
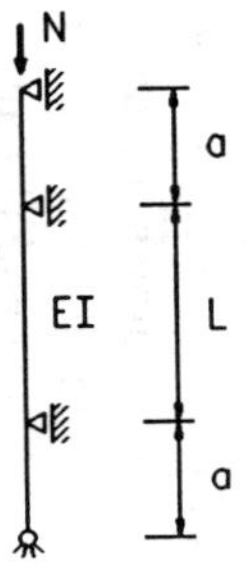	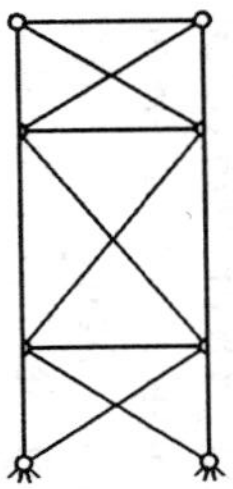	0.50	0.53	0.57	0.61	0.65	0.70	0.75	0.81	0.87	0.93	1.0	L_E/L

Fig. 15.9 Effective length factors for continuous columns based on elastic stability theory

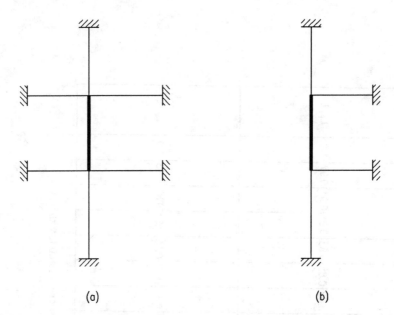

Fig. 15.10 Limited frames. (a) Internal column. (b) External column

design of compression members is considered in detail in Chapter 18 for the former and in Chapter 19 for the latter.

For compression members in rigid-jointed frames the effective length is directly related to the restraint provided by all the surrounding members. Strictly speaking an interaction of all the members in the frame occurs because the real behaviour is one of frame buckling rather than column buckling, but for design purposes it is often sufficient to consider the behaviour of a limited region of the frame. Variants of the 'limited frame' concept are to be found in several Codes of Practice and design guides. That used in BS 5950: Part 1 is illustrated in Fig. 15.10. The limited frame comprises the column under consideration and each immediately adjacent member treated as if its far end were fixed. The effective length of the critical column is then obtained from a chart which is entered with two coefficients k_1 and k_2, the values of which depend on the stiffnesses of the surrounding members K_U, K_{TL}, etc., relative to the stiffness of the column K_C, a concept similar to the well-known moment distribution method. Two distinct cases are considered: columns in non-sway frames and columns in frames that are free to sway. Figures 15.11 (a) and (b) and 15.11 (c) and (d) illustrate both cases as well as giving the associated effective length charts. For the former, the factors will vary between 0.5 and 1.0 depending on the values of k_1 and k_2, while for the latter, the variation will be between 1.0 and ∞. These end points correspond to cases of: rotationally fixed ends with no sway and rotationally free ends with no sway; rotationally fixed ends with free sway and rotationally free ends with free sway.

Effective lengths

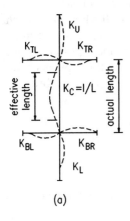
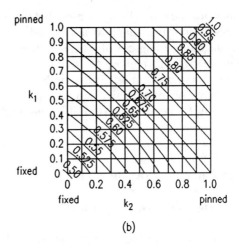
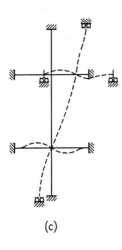
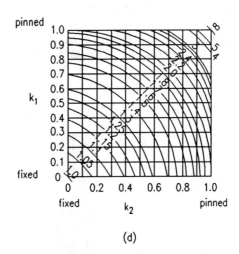

Fig. 15.11 Limited frames and corresponding effective length charts of BS 5950: Part 1. (a) Limited frame and (b) effective length ratios ($k_3 = \infty$), for non-sway frames. (c) Limited frame and (d) effective length ratios (without partial bracing, $k_3 = 0$), for sway frames

For beams not rigidly connected to the column or for situations in which significant plasticity either at a beam end or at either column end would prevent the restraint being transferred into the column, the K (and thus the k) values must be suitably modified. Similarly at column bases, k_2 values, in keeping with the degree of restraint provided, should be used. Guidance is also provided on K

Columns and struts

values for beams, distinguishing between both non-sway and sway cases and beams supporting concrete floors and bare steelwork. A further pair of charts permits modest degrees of partial restraint against sway, as might be provided for example by infill panels, to be allowed for in slightly reducing effective length values for sway frames. Full details of the background to this approach to the determination of effective lengths in rigidly jointed sway, partially braced and non-sway frames may be found in the work of Wood.[4]

15.8 Special types of strut

The design of two types of strut requires that certain additional points be considered:

(1) built-up sections or compound struts (Fig. 15.12), for which the behaviour of the individual components must be taken into account
(2) angles, channels and tees (Fig. 15.13), for which the eccentricity of loading produced by normal forms of end connection must be acknowledged.

In both cases, however, it will often be possible to design this more complex type of member as an equivalent single axially-loaded strut.

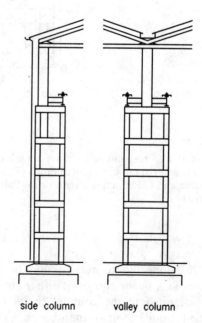

Fig. 15.12 Built-up struts

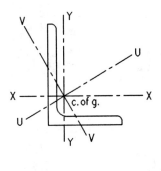

axis X–X } rectangular axes
axis Y–Y

axis U–U   maximum principal axis
axis V–V   minimum principal axis

Fig. 15.13 Geometrical properties of an angle section

15.8.1 Design of compound struts

Individual members may be combined in a variety of ways to produce a more efficient compound section. Figure 15.14 illustrates the most common arrangements. In each case the concept is one of providing a compound member whose overall slenderness will be such that its load-carrying capacity will significantly exceed the sum of the axial resistances of the component members, i.e. for the case of Fig. 15.14(b) the laced strut will be stronger than the four corner angles treated separately.

Thus the design approach of BS 5950: Part 1 is to set conditions which when met permit the compound member to be designed as a single integral member. The following cases are considered explicitly:

(1) Laced struts conforming with the provisions of clause 4.7.8
(2) Battened struts conforming with the provisions of clause 4.7.9
(3) Batten-starred angles conforming with the provisions of clause 4.7.11, which uses much of clause 4.7.9
(4) Batten parallel angle struts conforming with the provisions of clause 4.7.12
(5) Back-to-back struts conforming with the provisions of clause 4.7.13.1 if the components are separated and of clause 4.7.13.2 if the components are in contact.

The detailed rules contained in these clauses are essentially of two types:

(1) Covering construction details such as the arrangements for interconnection in a general 'good practice' manner

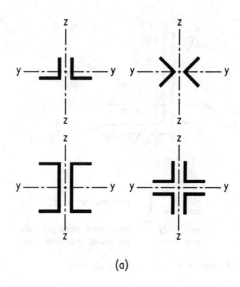

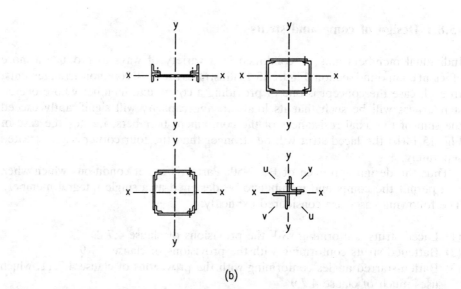

Fig. 15.14 Typical arrangements for compound struts: (a) closely spaced, (b) laced or battened

(2) Quantitative rules for the determination of the overall slenderness, limits necessary for component slenderness, forces for which the interconnections should be designed, etc.

BS 5400: Part 3 also contains specific rules for the design of:

(1) Batten struts (clause 10.8)
(2) Laced struts (clause 10.9)
(3) Struts connected by perforated cover plates (clause 10.10)
(4) Struts consisting of back-to-back components (clause 10.11).

These are somewhat more detailed than those of BS 5950: Part 1, particularly in the matter of determining suitable design forces for the interconnections, i.e. battens, lacings, etc.

15.8.2 Design of angles, channels and tees

Four specific cases are covered in detail by BS 5950: Part 1:

(1) Single angles in clause 4.7.10.2
(2) Double angles in clause 4.7.10.3
(3) Single channels in clause 4.7.10.4
(4) Single tees in clause 4.7.10.5.

In all cases, guidance is provided on the determination of the slenderness to be used when obtaining p_c for each of the more common forms of fastening arrangement. In cases where only a single fastener is used at each end the resulting value of p_c should then be reduced to 80% so as to allow for the combined effects of load eccentricity and lack of rotational end restraint. For ease of use the whole set of slenderness relationships is grouped together in Table 28.

BS 5400: Part 3 only gives specific consideration to single angles. This distinguishes between single bolt and 'other' forms of end connection. In both cases load eccentricity may be ignored but for the former only 80% of the calculated resistance may be used.

15.9 Economic points

Strut design is a relatively straightforward design task involving choice of cross-sectional type, assessment of end restraint and thus effective length, calculation of slenderness, determination of compressive strength and hence checking that the trial section can withstand the design load. Certain subsidiary checks may also be required part way through this process to ensure that the chosen cross section is not slender (or make suitable allowances if it is) or to guard against local failure in compound members. Thus only limited opportunities occur for the designer to use judgement and to make choices on the grounds of economy. Essentially these are restricted to control of the effective length, by introducing intermediate restraints where appropriate, and the original choice of cross section.

However, certain other points relating to columns may well have a bearing on the overall economy of the steel frame or truss. Of particular concern is the need to be able to make connections simply. In a multi-storey frame, the use of heavier UC sections thus may be advantageous, permitting beam-to-column connections to be made without the need for stiffening the flanges or web. Similarly in order to accommodate beams framing into the column web an increase in the size of UC may eliminate the need for special detailing.

While compound angle members were a common feature of early trusses, maintenance costs due both to the surface area requiring painting and the incidence of corrosion caused by the inherent dirt and moisture traps have caused a change to the much greater use of tubular members. If site joints are kept to the minimum tubular trusses can be transported and handled on site in long lengths and a more economic as well as a visually more pleasing structure is likely to result.

References to Chapter 15

1. The Steel Construction Institute (SCI) (1991) *Steelwork Design Guide to BS 5950: Part 1: 1985, Vol. 1: Section Properties, Member Capacities*, 3rd edn. SCI, Ascot, Berks.
2. Ballio G. & Mazzolani F.M. (1983) *Theory and Design of Steel Structures*. Chapman and Hall.
3. Allen H.G. & Bulson P.S. (1980) *Background to Buckling*. McGraw-Hill.
4. Wood R.H. (1974) Effective lengths of columns in multi-storey buildings. *The Structural Engineer*, **52**, Part 1, July, 235–44, Part 2, Aug., 295–302, Part 3, Sept., 341–6.

Further reading for Chapter 15

ECCS (1986) *Behaviour and Design of Steel Plated Structures* (Ed. by P. Dubas & E. Gehri). ECCS Publication No. 44.
Galambos T.V. (Ed.) (1988) *Guide to Stability Design Criteria for Metal Structures*, 4th edn. Wiley Interscience.
Hancock G.J. (1988) *Design of Cold-Formed Steel Structures*. Australian Institute of Steel Construction.
Kirby P.A. & Nethercot D.A. (1985) *Design for Structural Stability*. Collins, London.
Trahair N.S. & Nethercot D.A. (1984) *Bracing Requirements in Thin-Walled Structures, Developments in Thin-Walled Structures – 2* (Ed. by J. Rhodes & A.C. Walker), pp. 92–130. Elsevier Applied Science Publishers, Barking, Essex.

A series of worked examples follows which are relevant to Chapter 15.

The Steel Construction Institute Silwood Park, Ascot, Berks SL5 7QN	Subject COLUMN EXAMPLE 1 ROLLED UNIVERSAL COLUMN	Chapter ref. 15
	Design code BS5950: Part 1	Made by DAN Checked by GWO	Sheet no. 1

Problem

Check the ability of a 203 × 203 × 52 UC in grade 43 steel to withstand an axial compressive load of 1250 kN over an unsupported height of 3.6 m assuming that both ends are held in position but are provided with no restraint in direction. Design to BS 5950: Part 1.

The problem is as shown in the sketch.

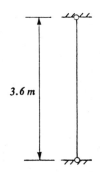

Take effective length	L_E	=	1.0 L	Table 24
	∴ L_E	=	$1.0 \times 3.6 \times 10^3$	
		=	3600 mm	

On the assumption that weak axis flexural buckling will govern determine compressive strength p_c from Table 27c.   Table 25

A	=	66.4 cm²	b/T	=	8.16	Steelwork Design Guide Vol 1
r_y	=	5.16 cm	d/t	=	20.1	
Since T < 16 mm			take p_y	=	275 N/mm²	Table 6

Check section classification for pure compression.   3.5

Need only check section is not slender;   Table 7

for outstand   b/T   ≤   15ε

The Steel Construction Institute Silwood Park, Ascot, Berks SL5 7QN	Subject **COLUMN EXAMPLE 1** **ROLLED UNIVERSAL COLUMN**	Chapter ref. **15**
	Design code *BS5950: Part 1*	Made by **DAN** Checked by **GWO**	Sheet no. **2**

for web $d/t \leq 39\epsilon$

$\epsilon = \sqrt{(275/p_y)} = 1.0$

actual b/T = 8.16, *within limit*

actual d/t = 20.1, *within limit*

∴ *Section is not slender.*

$P_c = A_g p_c$ 4.7.4

$\lambda = L_E/r_y = 3600/51.6 = 70$ 4.7.3

For $\lambda = 70$ and $p_y = 275 \text{ N/mm}^2$ Table 27c

value of $p_c = 202 \text{ N/mm}^2$

∴ $P_c = 6640 \times 202 = 1341 \times 10^3 \text{ N}$
$= 1341 \text{ kN}$

This exceeds required resistance of 1250 kN and section is therefore OK.

∴ <u>Use 203 × 203 × 52 UC</u>

The Steel Construction Institute Silwood Park, Ascot, Berks SL5 7QN	Subject **COLUMN EXAMPLE 2 WELDED BOX**		Chapter ref. **15**
	Design code BS5950: Part 1	Made by DAN Checked by GWO	Sheet no. 1

Problem

Check the ability of a 960 mm square box column fabricated from 30 mm thick grade 50 plate to withstand an axial compressive load of 22000 kN over an unsupported height of 15 m assuming that both ends are held in position but are provided with no restraint in direction. Design to BS 5950: Part 1.

The problem is as shown in the sketches.

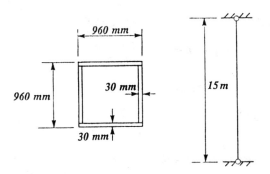

Take effective length	L_E =	1.0 L	Table 24
∴	L_E =	$1.0 \times 15 \times 10^3$	
	=	15000 mm	
Determine p_c from Table 27b			Table 25
Since T > 16 mm, take p_y =		340 N/mm²	Table 6
Check section classification for pure compression.			3.5
Need only check section is not slender;			Table 7
for flange	$b/T \leq 28\epsilon$		
ϵ =	$\sqrt{(275/340)}$ =	0.9	

The Steel Construction Institute Silwood Park, Ascot, Berks SL5 7QN	Subject **COLUMN EXAMPLE 2 WELDED BOX**	Chapter ref. **15**
	Design code BS5950: Part 1	Made by **DAN** · Checked by **GWO** · Sheet no. **2**

actual b/T = $(960 - 2 \times 30)/30$ = 30

limit = 0.9×28 = 25.2

∴ Section is slender.

Strength reduction factor = $\dfrac{21}{30/0.9 - 7}$ = 0.8   Table 8

Use reduced p_y = 0.8×340 = 272 N/mm²

I_y = $(960 \times 960^3 - 900 \times 900^3)/12$ = 1.61×10^{10} mm⁴

A_g = $2 \times 30 (960 + 900)$ = 111600 mm²

r_y = $\sqrt{1.61 \times 10^{10}/111600}$ = 380 mm

λ = L_E/r_y = 15000/380   4.3.7.1
 = 39.5

Since section fabricated by welding, use Table 27b with   4.7.5

p_y = 272 − 20
 = 252 N/mm²

For λ = 39.5 and p_y = 252 N/mm²   Table 27b

value of p_c = 230 N/mm²

P_c = $A_g p_c$ = 111600×230 4.7.4
 = 25.7×10^6 N
 = 25700 kN

This exceeds required resistance of 22000 kN and section is OK.

∴ Use section shown in figure.

The Steel Construction Institute Silwood Park, Ascot, Berks SL5 7QN		Subject **COLUMN EXAMPLE 3 ROLLED UNIVERSAL BEAM**		Chapter ref. **15**	
		Design code **BS5950: Part 1**	Made by **DAN**	Sheet no. **1**	
			Checked by **GWO**		

Problem

A 457 × 191 × 89 UB in Grade 43 steel is to be used as an axially loaded column over a free height of 7.0 m. Both ends will be held in position but not direction for both planes. The possibility exists to provide discrete bracing members capable of preventing deflection in the plane of the flanges only. Investigate the advisability of using such bracing.

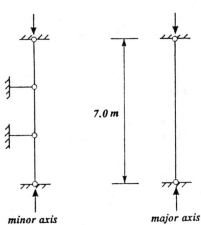

Clearly strength cannot exceed that for major axis failure. Check no. of intermediate (minor-axis) braces needed to achieve this.

Take L_{Ex} = 1.0 L

∴ L_{Ex} = 1.0 × 7.0 × 10³ = 7000 mm

A	=	114 cm²	b/T	=	5.42
r_y	=	4.28 cm	d/t	=	38.5
r_x	=	19.0 cm	p_y	=	265 N/mm² (for T = 17.7 mm)

Steelwork Design Guide Vol 1

Section not slender   Table 7

$P_c = A_g p_c$ 4.7.4

$\lambda_x = L_{Ex}/r_x$ = 7000/190 = 37   4.7.3

Use Table 27a for p_{cx} Table 25

The Steel Construction Institute Silwood Park, Ascot, Berks SL5 7QN	Subject COLUMN EXAMPLE 3 ROLLED UNIVERSAL BEAM		Chapter ref. 15
	Design code BS5950: Part 1	Made by DAN Checked by GWO	Sheet no. 2

For λ = 37 and p_y = 265 N/mm²   Table 27a

value of p_{cx} = 253 N/mm²

P_{cx} = 11400 × 253 = 2.884 × 10⁶ N
= <u>2884 kN</u>

Use Table 27b for p_{cy} & Table 25

For a p_{cy} value of 253 N/mm², value of $\lambda_y \not> 30$ Table 27b

$\therefore L_{Ey}/42.8$ = 30

and $L_{Ey} \not> 1284$ mm

$\therefore$ to achieve a minor axis buckling resistance $\not> 2987$ kN would require 7.0 m height to be provided with 7000/1284 → 5 restraints

L_{Ey}/r_y = (7000/6)/42.8 = 27

p_{cy} = 256 N/mm²   Table 27b

P_{cy} = 11400 × 256 = 2.918 × 10⁶
= <u>2918 kN</u>

Note however that substantial improvements on the basic minor axis resistance for L_{Ey} = 7.0 m of 741 kN may be achieved for rather less restraints as shown below.

no. of restraints	P_{cy} (kN)	P_{cy}/P_{cx}
0	730	0.25
1	1972	0.68
2	2531	0.88
3	2645	0.92

Chapter 16
Beams

by DAVID NETHERCOT

16.1 Common types of beam

Beams are possibly the most fundamental type of member present in a civil engineering structure. Their principal function is the transmission of vertical load by means of flexural (bending) action into, for example, the columns in a rectangular building frame or the abutments in a bridge which support them.

Table 16.1 provides some idea of the different structural forms suitable for use as beams in a steel structure; several of these are illustrated in Fig. 16.1. For

Table 16.1 Typical usage of different forms of beam

Beam type	Span range (m)	Notes
(0) Angles	3–6	Used for roof purlins, sheeting rails, etc., where only light loads have to be carried
(1) Cold-formed sections	4–8	Used for roof purlins, sheeting rails, etc., where only light loads have to be carried
(2) Rolled sections: UBs, UCs, RSJs, RSCs	1–30	Most frequently used type of section; proportions selected to eliminate several possible types of failure
(3) Open web joists	4–40	Prefabricated using angles or tubes as chords and round bar for web diagonals, used in place of rolled sections
(4) Castellated beams	6–60	Used for long spans and/or light loads; depth of UB increased by 50%; web openings may be used for services, etc.
(5) Compound sections e.g. UB + RSC	5–15	Used when a single rolled section would not provide sufficient capacity; often arranged to provide enhanced horizontal bending strength as well
(6) Plate girders	10–100	Made by welding together 3 plates sometimes automatically; web depths up to 3–4 m sometimes need stiffening
(7) Trusses	10–100	Heavier version of (3); may be made from tubes, angles or if spanning large distances, rolled sections
(8) Box girders	15–200	Fabricated from plate, usually stiffened; used for OHT cranes and bridges due to good torsional and transverse stiffness properties

404 *Beams*

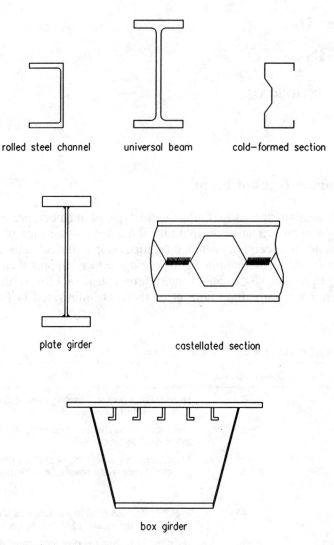

Fig. 16.1 Types of beam cross-section

modest spans, including the majority of those found in buildings, the use of standard hot-rolled sections (normally UBs but possibly UCs if minimizing floor depth is a prime consideration or channels if only light loads need to be supported) will be sufficient. Lightly loaded members such as the purlins supporting the roof of a portal-frame building are frequently selected from the range of proprietary cold-formed sections produced from steel sheet only a few millimetres thick, normally already protected against corrosion by galvanizing, in a variety of highly efficient shapes, advantage being taken of the roll-forming process to produce sections with properties carefully selected for the task they are required to perform. For spans in excess of those that can be achieved sensibly using ready-made sections some form of built-up member is required. Castellated beams, formed by profile cutting of the web and welding to produce a deeper section, typically 50% deeper using the standard UK geometry, are visually attractive but cannot withstand high shear loads unless certain of the castellations are filled in with plate. The range of spans for which UBs may be used can be extended if cover plates are welded to both flanges.

Alternatively a beam fabricated entirely by welding plates together may be employed allowing variations in properties by changes in depth, for example, flange thickness, or, in certain cases where the use of very thin webs is required, stiffening to prevent premature buckling failure is necessary. A full treatment of the specialist aspects of plate-girder design is provided in Chapter 17. If spans are so large that a single member cannot economically be provided, then a truss may be a suitable alternative. In addition to the deep truss fabricated from open hot-rolled sections, SHS or both, used to provide long clear spans in sports halls and supermarkets, smaller prefabricated arrangements using RHS or CHS provide an attractive alternative to the use of standard sections for more modest spans. Truss design is discussed in Chapter 19.

Since the principal requirement of a beam is adequate resistance to vertical bending, a very useful indication of the size of section likely to prove suitable may be obtained through the concept of the span to depth ratio. This is simply the value of the clear span divided by the overall depth. An average figure for a properly designed steel beam is between 15 and 20, perhaps more if a particularly slender form of construction is employed or possibly less if very heavy loadings are present.

When designing beams, attention must be given to a series of issues, in addition to simple vertical bending, that may have some bearing upon the problem. Torsional loading may often be eliminated by careful detailing or its effects reasonably regarded as of negligible importance by a correct appreciation of how the structure actually behaves; in certain instances it should, however, be considered. Section classification (allowance for possible local buckling effects) is more involved than is the case for struts since different elements in the cross-section are subject to different patterns of stress; the flanges in an I-beam in the elastic range will be in approximately uniform tension or compression while the web will contain a stress gradient. The possibility of members being designed for an elastic or a plastic state, including the use of a full plastic design for the complete structure, also affects section classification. Various forms of instability of the beam as a whole or of parts subject to locally high stresses, such as the web over a support, also require attention. Finally certain forms of construction may lend themselves to the appearance of unacceptable vibrations; although this is likely to be affected by the choice of beams, its coverage is left to Chapter 20 on floors.

16.2 Cross-section classification and moment capacity, M_c

The possible influences of local buckling on the ability of a particular cross-section to attain, and where appropriate also to maintain, a certain level of moment are discussed in general terms in Chapter 13. In particular, section 13.2 covers the influence of flange (b/T) and web (d/t) slenderness on moment–rotation behaviour (Fig. 13.4) and moment capacity (Fig. 13.5). When designing beams it is usually sufficient to consider the web and the compression flange separately using the appropriate sets of limits.[1]

In building design when using hot-rolled sections, for which relevant properties are tabulated,[2] it will normally be sufficient to ascertain a section's classification and moment capacity simply by referring to the appropriate table. An example illustrates the point.

Example

Using BS 5950: Part 1[1] determine the section classification and moment capacity for a 533 × 210 UB82 when used as a beam in (1) grade 43 steel and (2) grade 50 steel.

(1) From Reference 2, p. 129, section is 'plastic' and M_{cx} = 566 kN m
Alternatively, from Reference 2, p. 26, b/T = 7.91, d/t = 49.6
Since T = 13.2 mm from Reference 1, Table 6, p_y = 275 N/mm² and $\varepsilon = (275/p_y)^{\frac{1}{2}} = 1.0$
From Reference 1, Table 7, limits for plastic section are $b/t \leq 8.5\varepsilon$, $d/t \leq 79\varepsilon$; actual b/t and d/t are within these limits and section is plastic.
From Reference 1, clause 4.2.5, $M_c = p_y\, S$
$= 275 \times 2060 \times 10^{-3}$
$= 566.5$ kN m

(2) From Reference 2, p. 303, section is compact and M_{cx} = 731 kN m
Alternatively, from Reference 2, p. 26, b/T = 7.91, d/t = 49.6
Since T = 13.2 mm from Reference 1, Table 6, p_y = 355 N/mm² and $\varepsilon = (275/p_y)^{\frac{1}{2}} = 0.88$
From Reference 1, Table 7, limits for plastic section are $b/T \leq 8.5\varepsilon$, $d/t = 79\varepsilon$, which give $b/T \leq 7.48$, $d/t \leq 70$; actual b/T of 7.91 exceeds 7.48 so check compact limits of $b/T \leq 9.5\varepsilon$, which gives $b/T \leq 8.36$; actual b/T of 7.91 meets this and section is compact.
From Reference 1, clause 4.2.5, $M_c = p_y\, S$
$= 355 \times 2060 \times 10^{-3}$
$= 731.3$ kN m

Thus whichever grade of steel is being used, the moment capacity, M_c, will be the maximum attainable value corresponding to the section's full plastic moment capacity, M_p. Provided that grade 43 material is used, the beam is capable of redistributing moments, since the plastic cross section behaves as illustrated in Fig. 13.4(a) and so could be used in a plastically designed frame. Use of the higher strength grade 50 material, while not affecting the beam's ability to attain M_p, precludes the use of plastic design since redistribution of moments cannot occur due to the lack of rotation capacity implied by the behaviour illustrated in Fig. 13.4(b). For a continuous structure designed on the basis of elastic analysis to determine the distribution of internal forces and moments or for simple construction, the question of rotation capacity is not relevant and in both cases moment capacity should be taken as M_p. The fact that the section is plastic for grade 43 material is then of no particular relevance; a more appropriate classification would be to regard the section as compact or better.

Inspection of the relevant tables in Reference 2 shows that when used as beams:

Grade 43 steel
all but 2 UBs are plastic
all but 6 UCs are plastic
of these 6, 3 are compact and 3 are semi-compact
Grade 50 steel
all but 11 UBs are plastic
of these 11, 8 are compact and 3 are semi-compact
all but 10 UCs are plastic
of these 10, 4 are compact and 6 are semi-compact

In those cases of semi-compact sections, moment capacities based on M_y are listed. The full list of non-plastic beam sections is given as Table 16.2.

When using fabricated sections individual checks on the web and compression flange using the actual dimensions of the trial section must be made. It normally proves much simpler if proportions are selected so as to ensure that the section is compact or better since the resulting calculations need not then involve the various complications associated with the use of non-compact sections. This point is particularly noticeable when designing bridge beams to BS 5400: Part 3. When slender sections are used the amount of calculation increases considerably due to the need to consider loss of effectiveness of some parts of the cross section due to local buckling when determining M_c. Probably the most frequent use of slender sections involves cold-formed shapes used for example as roof purlins. Because of their proprietary nature, the manufacturers normally provide design information, much of it based on physical testing, listing such properties as moment capacity. In the absence of design information, reference should be made to Part 5 of BS 5950 for suitable calculation methods.

Table 16.2 Non-plastic UBs and UCs in bending

	Compact	Semi-compact
	Grade 43	
UB	356 × 171 × 44 203 × 133 × 25	
UC	356 × 368 × 153 254 × 254 × 73 203 × 203 × 46	356 × 368 × 129 305 × 305 × 97 152 × 152 × 23
	Grade 50	
UB	838 × 292 × 176 686 × 254 × 125 610 × 305 × 149 610 × 229 × 101 533 × 210 × 82 406 × 178 × 54 406 × 140 × 38 305 × 165 × 40	356 × 171 × 45 254 × 146 × 31 203 × 133 × 25
UC	356 × 368 × 177 305 × 305 × 118 203 × 203 × 52 152 × 152 × 30	356 × 368 × 153 356 × 368 × 129 305 × 305 × 97 254 × 254 × 73 203 × 203 × 46 152 × 152 × 23

Although the part of the web between the flange and the horizontal edge of the castellation in a castellated beam frequently exceeds the compact limit for an outstand, sufficient test data exist to show that this does not appear to influence the moment capacity of such sections. Section classification should therefore be made in the same way as for solid web beams, the value of M_c being obtained using the net modulus value for the section at the centre of a castellation.

16.3 Basic design

One (or more) of a number of distinct limiting conditions may, in theory, control the design of a particular beam as indicated in Table 16.3, but in any particular practical case only a few of them are likely to require full checks. It is therefore convenient to consider the various possibilities in turn, noting the conditions under which each is likely to be important. For convenience the various phenomena are first considered principally within the context of using standard hot-rolled sections i.e. UBs, UCs, RSJs and channels; other types of cross-section are covered in the later parts of this chapter.

Table 16.3 Limiting conditions for beam design

Ultimate	Serviceability
Moment capacity, M_c (including influence of local buckling Shear capacity, P_v Lateral–torsional buckling, M_b Web buckling, P_w Web bearing, P_{yw} Moment–shear interaction Torsional capacity, M_T Bending–torsion interaction	Deflections due to bending (and shear if appropriate) Twist due to torsion Vibration

16.3.1 Moment capacity, M_c

The most basic design requirement for a beam is the provision of adequate in-plane bending strength. This is provided by ensuring that M_c for the selected section exceeds the maximum moment produced by the factored loading. Determination of M_c, which is linked to section classification, is fully covered in section 16.2.

For a statically determinate structural arrangement, simple considerations of statics provide the moment levels produced by the applied loads against which M_c must be checked. For indeterminate arrangements, a suitable method of elastic analysis such as moment distribution or slope deflection is required. The justification for using an elastically obtained distribution of moments with, in the case of compact or plastic cross-sections, a plastic cross-sectional resistance has been fully discussed by Johnson and Buckby.[3]

16.3.2 Effect of shear

Only in cases of high coincident shear and moment, found for example at the internal supports of continuous beams, is the effect of shear likely to have a significant influence on the design of beams.

Shear capacity P_v is normally calculated as the product of a shear strength p_v, often taken for convenience as $0.6p_y$ which is close to the yield stress of steel in shear of $1/\sqrt{3}$ of the uniaxial tensile yield stress, and an appropriate shear area A_v. The process approximates the actual distribution of shear stress in a beam web as well as assuming some degree of plasticity. While suitable for rolled sections, it may not therefore be applicable to plate girders. An alternative design approach, more suited to webs containing large holes or having variations in thickness, is to work from first principles and to limit the maximum shear stress to a suitable value; BS 5950: Part 1 uses $0.7p_y$. In cases where $d/t > 63$, shear buckling limits the effectiveness of the web and reference to Chapter 17 should be made for methods of determining the reduced capacity.

In principle the presence of shear in a section reduces its moment capacity. In practice the reduction may be regarded as negligibly small up to quite large fractions of the shear capacity P_v. For example, BS 5950: Part 1 only requires a reduction in M_c for plastic or compact sections when the applied shear exceeds $0.6P_v$ and permits the full value of M_c to be used for all cases of semi-compact or slender sections.

Figure 16.2 illustrates the application of the BS 5950: Part 1 rule for plastic or compact sections to a typical UB and UC having approximately equal values of plastic section modulus S_x. Evaluation of the formula in cases where $F_v/P_v > 0.6$, first requires that the value of S_v, the plastic modulus of the shear area A_v (equal to tD in this case), be determined. This is readily obtained from the tabulated values of S for rectangles given in Reference 2 corresponding to a linear reduction from S_x to $(S_x - S_v)$.

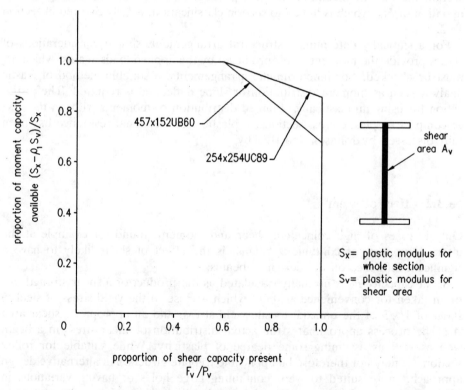

Fig. 16.2 Moment–shear interaction for plastic or compact sections to BS 5950: Part 1

16.3.3 Deflection

When designing according to limit states principles it is customary to check that deflections at working load levels will not be such as to impair the proper function of the structure. For beams, examples of potentially undesirable consequences of excessive serviceability deflections include:

- cracking of plaster ceilings
- allowing crane rails to become misaligned
- causing difficulty in opening large doors.

Although earlier codes of practice specified limits for working load deflections, the tendency with more recent documents[1] is to draw attention to the need for deflection checks and to provide advisory limits to be used only when more specific guidance is not available. Table 5 of BS 5950: Part 1 gives 'recommended limitations' for certain types of beams and crane gantry girders and states that 'Circumstances may arise where greater or lesser values would be more appropriate.' Not surprisingly surveys of current practice[4] reveal large variations in what is considered appropriate for different circumstances.

When checking deflections of steel structures under serviceability loading, the central deflection Δ_{max} of a uniformly-loaded simply-supported beam, assuming linear–elastic behaviour, is given by

$$\Delta_{max} = \frac{5}{384} \frac{WL^3}{EI} 10^{12} \text{(mm)} \tag{16.1}$$

in which W = total load (kN)
E = Young's modulus (N/mm^2)
I = second moment of area (mm^4)
L = span (m).

If Δ_{max} is to be limited to a fraction of L, Equation (16.1) may be rearranged to give

$$I_{rqd} = 0.62 \times 10^{-2} \, \alpha \, WL^2 \tag{16.2}$$

in which α defines the deflection limit as

$$\Delta_{max} = L/\alpha$$

and I_{rqd} is now in cm^4.

Writing $I_{rqd} = KWL^2$, Table 16.4 gives values of K for a range of values of α.

Table 16.4 Minimum I values for uniformly-loaded simply-supported beams for various deflection limits

α	200	240	250	325	360	400	500	600	750	1000
K	1.24	1.49	1.55	2.02	2.23	2.48	3.10	3.72	4.65	6.20

Table 16.5 Equivalent UDL coefficients

a/L	K_Δ	No. of equal loads	b/L	c/L	K
0.5	1.0	2	0.2	0.6	0.91
			0.25	0.5	1.10
0.4	0.86		0.333	0.333	1.3
0.375	0.82	3	0.167	0.333	1.05
			0.2	0.3	1.14
0.333	0.74		0.25	0.25	1.27
		4	0.125	0.25	1.03
0.3	0.68		0.2	0.2	1.21
0.25	0.58	5	0.1	0.2	1.02
			0.167	0.167	1.17
0.2	0.47	6	0.083	0.167	1.01
			0.143	0.143	1.15
0.1	0.24	7	0.071	0.143	1.01
			0.125	0.125	1.12
		8	0.063	0.125	1.01
			0.111	0.111	1.11

$$K_\Delta = 1.6 \frac{a}{L}\left[3 - 4\left(\frac{a}{L}\right)^2\right]$$

a/L	0.01	0.05	0.1	0.15	0.2	0.25	0.3	0.35	0.4	0.45	0.5
K_Δ	0.05	0.24	0.47	0.70	0.91	1.10	1.27	1.41	1.51	1.58	1.60

Since deflection checks are essentially of the 'not greater than' type, some degree of approximation normally is acceptable, particularly if the calculations are reduced as a result. Converting complex load arrangements to a roughly equivalent UDL permits Table 16.4 to be used for a wide range of practical situations. Table 16.5 gives values of the coefficient K by which the actual load arrangement shown should be multiplied in order to obtain an approximately equal maximum deflection.

Tables of deflections for a number of standard cases are provided in the Appendix.

16.3.4 Torsion

Beams subjected to loads which do not act through the point on the cross-section known as the shear centre normally suffer some twisting. Methods for locating the shear centre for a variety of sectional shapes are given in Reference 5. For doubly symmetrical sections such as UBs and UCs it coincides with the centroid while for channels it is situated on the opposite side of the web from the centroid; for rolled channels its location is included in the tables of Reference 2. Figure 16.3 illustrates its position for a number of standard cases.

The effects of torsional loading may often be minimized by careful detailing, particularly when considering how loads are transferred between members. Proper attention to detail can frequently lead to arrangements in which the load transfer is organized in such a way that twisting should not occur.[5] Whenever possible this approach should be followed as the open sections normally used as beams are inherently weak in resisting torsion. In circumstances where beams are required to withstand significant torsional loading, consideration should be given to the use of a torsionally more efficient shape such as a structural hollow section.

16.3.5 Local effects on webs

At points within the length of a beam where vertical loads act, the web is subject to concentrations of stresses, additional to those produced by overall bending. Failure by buckling, rather in the manner of a vertical strut, or by the development of unacceptably high bearing stresses in the relatively thin web material immediately adjacent to the flange, is a possibility. Methods for assessing the likelihood of both types of failure are given in BS 5950: Part 1 and tabulated data to assist in the evaluation of the formulae required are provided for rolled sections in Reference 2. The parallel approach for cold-formed sections is discussed in section 16.7.

In cases where the web is found to be incapable of resisting the required level of load, additional strength may be provided through the use of stiffeners. The design of load-carrying stiffeners (to resist web buckling) and bearing stiffeners is covered in both BS 5950 and BS 5400. However, web stiffeners may be required to resist shear buckling, to provide torsional support at bearings or for other reasons; a full treatment of their design is provided in Chapter 17.

16.3.6 Lateral–torsional buckling

Beams for which none of the conditions listed in Table 16.6 are met (explanation of these requirements is delayed until section 16.3.7 so that the basic ideas and parameters governing lateral–torsional buckling may be presented first) are liable to have their load-carrying capacity governed by the type of failure illustrated in

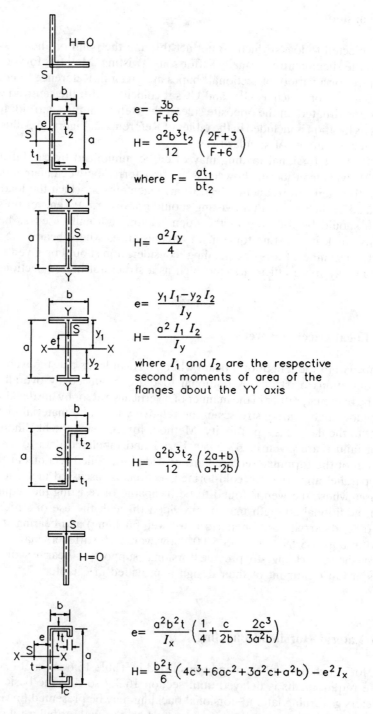

Fig. 16.3 Location of shear centre for standard sections (H is warping constant)

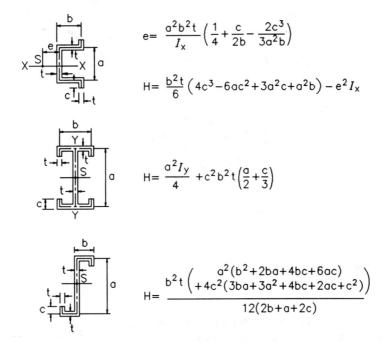

$$e = \frac{a^2 b^2 t}{I_x}\left(\frac{1}{4} + \frac{c}{2b} - \frac{2c^3}{3a^2 b}\right)$$

$$H = \frac{b^2 t}{6}\left(4c^3 - 6ac^2 + 3a^2 c + a^2 b\right) - e^2 I_x$$

$$H = \frac{a^2 I_y}{4} + c^2 b^2 t\left(\frac{a}{2} + \frac{c}{3}\right)$$

$$H = \frac{b^2 t \left(\begin{array}{c} a^2(b^2 + 2ba + 4bc + 6ac) \\ + 4c^2(3ba + 3a^2 + 4bc + 2ac + c^2) \end{array}\right)}{12(2b + a + 2c)}$$

Fig. 16.3 (continued)

Table 16.6 Types of beam not susceptible to lateral–torsional buckling

loading produces bending about the minor axis

beam provided with closely spaced or continuous lateral restraint

closed section

Fig. 16.4. Lateral–torsional instability is normally associated with beams subject to vertical loading buckling out of the plane of the applied loads by deflecting sideways and twisting; behaviour analogous to the flexural buckling of struts. The presence of both lateral and torsional deformations does cause both the governing mathematics and the resulting design treatment to be rather more complex.

The design of a beam taking into account lateral–torsional buckling consists essentially of assessing the maximum moment that can safely be carried from a knowledge of the section's material and geometrical properties, the support conditions provided and the arrangement of the applied loading. Codes of practice, such as BS 5400: Part 3, BS 5950: Parts 1 and 5, include detailed guidance on the

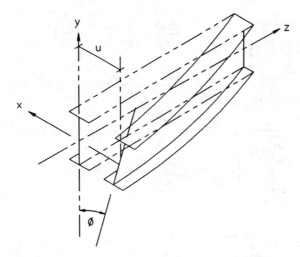

Fig. 16.4 Lateral–torsional buckling

subject. Essentially the basic steps required to check a trial section (using BS 5950: Part I for a UB as an example) are:

(1) assess the beam's effective length L_E from a knowledge of the support conditions provided (clause 4.3.5 and 4.3.6)
(2) determine beam slenderness λ_{LT} using the geometrical parameters u (tabulated in Reference 2), L_E/r_y, v (Table 14 of BS 5950: Part 1) using values of x (tabulated in Reference 2), and n (clause 4.3.6)
(3) obtain corresponding bending strength p_b (Table 11)
(4) calculate buckling resistance moment $M_b = p_b S_x$, where S_x is the plastic section modulus (tabulated in Reference 2).

The central feature in the above process is the determination of a measure of the beam's lateral–torsional buckling strength (p_b) in terms of a parameter (λ_{LT}) which represents those factors which control this strength. Modifications to the basic process permit the method to be used for unequal flanged sections including tees, fabricated Is for which the section properties must be calculated, sections containing slender plate elements, members with properties that vary along their length, closed sections and flats. Various techniques for allowing for the form of the applied loading are also possible; some care is required in their use.

The relationship between p_b and λ_{LT} of BS 5950: Part 1 (and between σ_{li}/σ_{yc} and $\lambda_{LT} \sqrt{(\sigma_{yc}/355)}$ in BS 5400: Part 3) assumes the beam between lateral restraints to be subject to uniform moment. Other patterns, such as a linear moment gradient reducing from a maximum at one end or the parabolic distribution produced by a uniform load, are generally less severe in terms of their effect on lateral stability; a given beam is likely to be able to withstand a larger peak

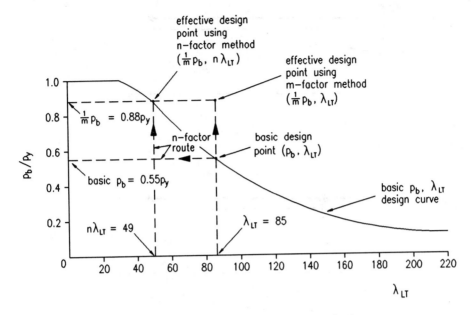

Fig. 16.5 Design modifications using m-factor or n-factor methods

moment before becoming laterally unstable. One means of allowing for this in design is to adjust the beam's slenderness by a factor (n in BS 5950: Part 1), the value of which has been selected so as to ensure that the resulting value of p_b correctly reflects the enhanced strength due to the non-uniform moment loading. An alternative approach consists of basing λ_{LT} on the geometrical and support conditions alone but making allowance for the beneficial effects of non-uniform moment by comparing the resulting value of M_b with a suitably adjusted value of design moment $\overline{M}$. $\overline{M}$ is taken as a factor m times the maximum moment within the beam M_{max}; $m = 1.0$ for uniform moment and $m < 1.0$ for non-uniform moment. Provided that suitably chosen values of m and n are used, both methods can be made to yield identical results; the difference arises simply in the way in which the correction is made, whether on the slenderness axis of the p_b versus λ_{LT} relationship for the n-factor method or on the strength axis for the m-factor method. Figure 16.5 illustrates both concepts, although for the purpose of the figure the m-factor method has been shown as an enhancement of p_b by $1/m$ rather than a reduction in the requirement of checking M_b against $\overline{M} = mM_{max}$. However, BS 5950: Part 1 restricts the use of the m-factor method to beams loaded only at points of effective lateral restraint having moment diagrams of the sort shown as Fig. 16.6(a), while BS 5400: Part 3 only includes the n-factor method. Even a modest allowance for beam self-weight causes the moment diagram of Fig. 16.6(a) to alter to that of Fig. 16.6(b) and so apparently contravenes the requirement concerning loading between points of effective lateral restraint.

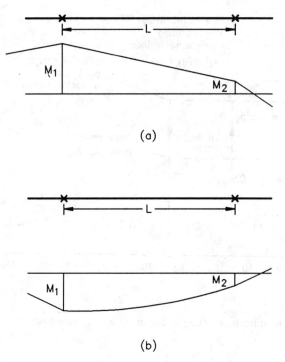

Fig. 16.6 Beam segment between points of effective lateral restraint (a) subject to moment gradient, (b) subject to moment gradient plus a modest UDL

Some degree of judgement is therefore required and it is suggested that where the additional parabolic component of the moment diagram does not exceed 15% of the larger end moment M_1 the *m*-factor method may be used. As a final caution on the use of *m* and *n* factors, it is always safe to set $m = n = 1$ basing design on the uniform moment case. There is no situation for which both $m \neq 1$ and $n \neq 1$ should be used together; only one method should be used for any situation.

When the *m*-factor method is used the buckling check is conducted in terms of a moment $\overline{M}$ less than the maximum moment in the beam segment M_{max}, then a separate check that the capacity of the beam cross-section M_c is at least equal to M_{max} must also be made. In cases where $\overline{M}$ is taken as M_{max}, then the buckling check will be more severe than (or in the case of a stocky beam for which $M_b = M_c$, identical to) the cross-section capacity check.

Allowance for non-uniform moment loading on cantilevers is normally treated somewhat differently. For example, the set of effective length factors given in Table 10 of Reference 1 includes allowances for the variation from the arrangement used as the basis for the strength–slenderness relationship due to both the lateral support conditions and the form of the applied loading. When a cantilever is subdivided by one or more intermediate lateral restraints positioned between its root and tip, then segments other than the tip segment should be treated as

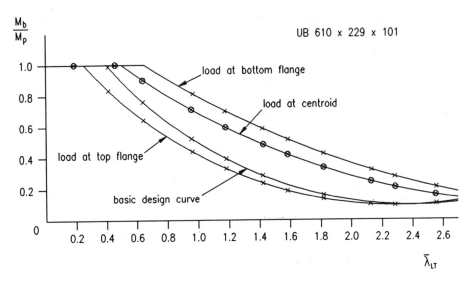

Fig. 16.7 Lateral–torsional buckling of a tip-loaded cantilever

ordinary beam segments when assessing lateral–torsional buckling strength. Similarly a cantilever subject to an end moment such as horizontal wind load acting on a façade, should be regarded as an ordinary beam since it does not have the benefit of non-uniform moment loading.

For more complex arrangements that cannot reasonably be approximated by one of the standard cases covered by correction factors, codes normally permit the direct use of the elastic critical moment M_E. Values of M_E may conveniently be obtained from summaries of research data.[6] For example, BS 5950: Part 1 permits λ_{LT} to be calculated from

$$\lambda_{LT} = \sqrt{(\pi^2 E/p_y)} \sqrt{(M_p/M_E)} \tag{16.3}$$

As an example of the use of this approach Fig. 16.7 shows how significantly higher load-carrying capacities may be obtained for a cantilever with a tip load applied to its bottom flange, a case not specifically covered by BS 5950: Part 1.

16.3.7 Fully restrained beams

The design of beams is considerably simplified if lateral–torsional buckling effects do not have to be considered explicitly; a situation which will occur if one or more of the conditions of Table 16.6 are met.

In these cases the beam's buckling resistance moment M_b may be taken as its moment capacity M_c and, in the absence of any reductions in M_c due to local

420   **Beams**

buckling, high shear or torsion, it should be designed for its full in-plane bending strength. Certain of the conditions corresponding to the case where a beam may be regarded as 'fully restrained' are virtually self-evident but others require either judgement or calculation.

Lateral–torsional buckling cannot occur in beams loaded in their weaker principal plane; under the action of increasing load they will collapse simply by plastic action and excessive in-plane deformation. Much the same is true for rectangular box sections even when bent about their strong axis. Figure 16.8,

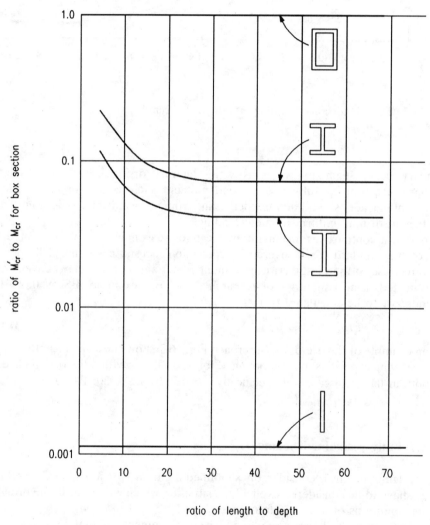

Fig. 16.8 Effect of type of cross-section on theoretical elastic critical moment

which is based on elastic critical load theory analogous to the Euler buckling of struts, shows that typical RHS beams will be of the order of ten times more stable than UB or UC sections of the same area. The limits on λ below which buckling will not affect M_b of Table 38 of BS 5950: Part 1, are sufficiently high ($\lambda = 350$, 225 and 170 for D/B ratios of 2, 3 and 4, and $p_y = 275$ N/mm^2) that only in very rare cases will lateral–torsional buckling be a design consideration.

Situations in which the form of construction employed automatically provides some degree of lateral restraint or for which a bracing system is to be used to enhance a beam's strength require careful consideration. The fundamental requirements of any form of restraint if it is to be capable of increasing the strength of the main member is that it limits the buckling type deformations. An appreciation of exactly how the main member would buckle if unbraced is a prerequisite for the provision of an effective system. Since lateral–torsional buckling involves both lateral deflection and twist as shown in Fig. 16.4, either or both deformations may be addressed. Clauses 4.3.2 and 4.3.3 of BS 5950: Part 1 set out the principles governing the action of bracing designed to provide either lateral restraint or torsional restraint. In common with most approaches to bracing design these clauses assume that the restraints will effectively prevent movement at the braced cross-sections, thereby acting as if they were rigid supports. In practice, bracing will possess a finite stiffness. A more fundamental discussion of the topic, which explains the exact nature of bracing stiffness and bracing strength may be found in References 7 and 8. Noticeably absent from the code clauses is a quantitative definition of 'adequate stiffness', although it has subsequently been suggested that a bracing system that is 25 times stiffer than the braced beam would meet this requirement. Examination of Reference 7 shows that while such a check does cover the majority of cases, it is still possible to provide arrangements in which even much stiffer bracing cannot supply full restraint.

16.4 Lateral bracing

For design to BS 5950: Part 1, unless the engineer is prepared to supplement the code rules with some degree of working from first principles, only restraints capable of acting as rigid supports are acceptable. Despite the absence of a specific stiffness requirement, adherence to the strength requirement together with an awareness that adequate stiffness is also necessary, avoiding obviously very flexible yet strong arrangements, should lead to satisfactory designs. Doubtful cases will merit examination in a more fundamental way.[7,8] Where properly designed restraint systems are used the limits on λ_{LT} for $M_b = M_c$ (or more correctly $p_b = p_y$) are given in Table 16.7.

For beams in plastically-designed structures it is vital that premature failure due to plastic lateral–torsional buckling does not impair the formation of the full plastic collapse mechanism and the attainment of the plastic collapse load. Clause 5.3.5 provides a basic limit on L/r_y to ensure satisfactory behaviour; it is not

Table 16.7 Maximum values of λ_{LT} for which $p_b = p_y$ for rolled sections

p_y (N/mm^2)	Value of λ_{LT} up to which $p_b = p_y$
245	37
265	35
275	34
325	32
340	31
365	30
415	28
430	27
450	26

necessarily compatible with the elastic design rules of section 4 of the code since acceptable behaviour can include the provision of adequate rotation capacity at moments slightly below M_p.

The expression of clause 5.3.5 of BS 5950: Part 1,

$$L_m \leq \frac{38 r_y}{[f_c/130 + (p_y/275)^2 (x/36)^2]^{\frac{1}{2}}} \tag{16.4}$$

makes no allowance for either of two potentially beneficial effects:

(1) moment gradient
(2) restraint against lateral deflection provided by secondary structural members attached to one flange as by the purlins on the top flange of a portal frame rafter.

The first effect may be included in Equation (16.4) by adding the correction term of Brown,[9] the basis of which is the original work on plastic instability of Horne.[10] A method of allowing for both effects when the beam segment being checked is either elastic or partially plastic is given in Appendix G of BS 5950: Part 1; alternatively the effect of intermittent tension flange restraint alone may be allowed for by replacing L_m with an enhanced value L_t obtained from clause 5.5.3.5 of BS 5950: Part 1.

In both cases the presence of a change in cross-section, for example, as produced by the type of haunch usually used in portal frame construction, may be allowed for. When the restraint is such that lateral deflection of the beam's compression flange is prevented at intervals, then Equation (16.4) applies between the points of effective lateral restraint. A discussion of the application of this and other approaches for checking the stability of both rafters and columns in portal frames designed according to the principles of either elastic or plastic theory is given in section 18.7.

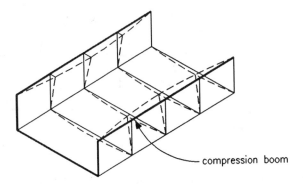

Fig. 16.9 Buckling of main beams of half-through girder

16.5 Bracing action in bridges – U-frame design

The main longitudinal beams in several forms of bridge construction will, by virtue of the structural arrangement employed, receive a significant measure of restraint against lateral–torsional buckling by a device commonly referred to as 'U-frame action'. Figure 16.9 illustrates the original concept based on the half-through girder form of construction. (See Chapter 4 for a discussion of different bridge types.) In a simply-supported span, the top (compression) flanges of the main girders, although laterally unbraced in the sense that no bracing may be attached directly to them, cannot buckle freely in the manner of Fig. 16.4 since their lower flanges are restrained by the deck. Buckling must therefore involve some distortion of the girder web into the mode given in Fig. 16.9. An approximate way of dealing with this is to regard each longitudinal girder as a truss in which the tension chord is fully laterally restrained and the web members, by virtue of their lateral bending stiffness, inhibit lateral movement of the top chord. It is then only a small step to regard this top chord as a strut provided with a series of intermediate elastic spring restraints against buckling in the horizontal plane. The stiffness of each support corresponds to the stiffness of the U-frame comprising the two vertical web stiffeners and the cross-girder and deck shown in Fig. 16.10.

The elastic critical load for the top chord is:

$$P_{cr} = \pi^2 \, EI/L_E^2 \tag{16.5}$$

in which L_E is the effective length of the strut.

If the strut receives continuous support of stiffness $(1/\delta \, L_u)$ per unit length, in which L_u is the distance between U-frames and buckles in a single half-wave, this load will be increased to:

$$P_{cr} = (\pi^2 \, EI/L^2) + (L^2/\pi^2 \delta L_u) \tag{16.6}$$

which gives a minimum value when

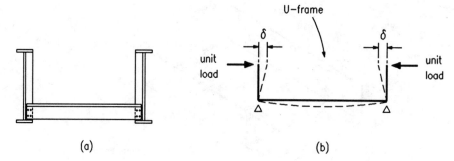

Fig. 16.10 U-frame restraint action. (a) Components of U-frame. (b) U-frame elastic support stiffness.

$$L = \pi(EI\delta L_u)^{0.25} \tag{16.7}$$

giving

$$P_{cr} = 2(EI/\delta L_u)^{0.5} \tag{16.8}$$

or

$$L_E = (\pi/\sqrt{2})(EI\delta L_u)^{0.25} \tag{16.9}$$

In clause 9.6.5 of BS 5400: Part 3, this is rounded up to

$$L_E = 2.5(EI_c L_u \delta)^{0.25}$$

If lateral movement at the ends of the girder is not prevented by sufficiently stiff end U-frames as shown in Fig. 16.11, then the effective length is

$$L_E = \pi(EI \delta L_u)^{0.25} \tag{16.10}$$

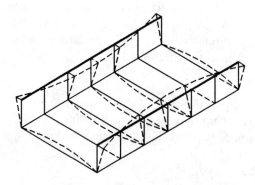

Fig. 16.11 Buckling mode for half-through construction with flexible end frames

In both Equations (16.9) and (16.10) a coefficient K_3 may be inserted, with values less than unity being adopted if some resistance to rotational restraint in plan of the top flange is present.

For unstiffened girders a similar approach is possible with the effective U-frame now comprising a unit length of girder web plus the cross-member. In all cases the assessment of U-frame stiffness via the δ parameter is based on summing the deflections due to bending of the horizontal and vertical components, including any flexibility of the upright to cross-frame connections. Clauses 9.6.5 and 9.6.6 deal respectively with the cases where actual vertical members are either present or absent.

Because the U-frames are required to resist the buckling deformations, they will attract forces which may be estimated as the product of the amplification of the initial lateral deformation of the top chord and the U-frame stiffness as

$$F_u = \left(\frac{1}{1 - \sigma_{fc}/\sigma_{li}}\right) \frac{L_E}{667\delta} \qquad (16.11)$$

in which the assumed initial bow over an effective length of flange (L_E) has been taken as $L_E/667$ and $1/(1 - \sigma_{fc}/\sigma_{li})$, is the amplification which depends in a non-linear fashion on the level of stress σ_{fc} in the flange.

For a frame spacing L_u and a flange critical stress σ_{li} corresponding to a force level of $\pi^2 EI_c/L_u^2$, the maximum possible value of F_u given in clause 9.12.2 of BS 5400: Part 3 is

$$F_u = \left(\frac{1}{1 - \sigma_{fc}/\sigma_{li}}\right) EI_c/16.7 \, L_u^2 \qquad (16.12)$$

Additional forces in the web stiffeners are produced by rotation θ of the ends of the cross beam to give (clause 9.12.2.3 of BS 5400: Part 3):

$$F_c = 3EI_1 \, \theta/d_2^2 \qquad (16.13)$$

However, the code rule assumes that the top flange is restrained in position by neighbouring unloaded U-frames; this is conservative in regions away from the stiff end frames and for approximately uniform bridge loadings. One possible approach consists of redefining θ as the rotation of the base of the stiffener, allowing for fixity reducing the cross-beam rotation and arriving at the expression for F_c of

$$F_c = \frac{\theta_o}{d_2} \left(\frac{1}{d_1/3EI_1 + B/4EI_2 + f}\right) \qquad (16.14)$$

in which θ_o is the cross-beam rotation determined as if it were simply supported. An alternative is to accept that transverse flexibility of the top flange allows the stiffener to deflect and shed some of its moment, leading to

$$F_c = \theta_o d \left(\frac{1}{\delta + L_u^3/15EI_c}\right) \qquad (16.15)$$

16.6 Design for restricted depth

Frequently beam design will be constrained by a need to keep the beam depth to a minimum. This restriction is easy to understand in the context of floor beams in a multi-storey building for which savings in overall floor depth will be multiplied several times over, thereby permitting the inclusion of extra floors within the same overall building height or effecting savings on expensive cladding materials by reducing building height for the same number of floors. Within the floor zone of buildings with large volumes of cabling, ducting and other heavy services, only a fraction of the depth is available for structural purposes.

Such restrictions lead to a number of possible solutions which appear to run contrary to the basic principles of beam design. However, structural designers should remember that the main framing of a typical multi-storey commercial building typically represents less than 10% of the building cost and that factors such as the efficient incorporation of the services and enabling site work to proceed rapidly and easily are likely to be of greater overall economic significance than trimming steel weight.

An obvious solution is the use of universal columns as beams. While not as structurally efficient for carrying loads in simple vertical bending as UB sections, as illustrated by the example of Table 16.8, their design is straightforward. Problems of web bearing and buckling at supports are less likely due to the reduced web d/t ratios. Lateral–torsional buckling considerations are less likely to control the design of laterally unbraced lengths because the wider flanges will provide greater lateral stiffness (L/r_y values are likely to be low). Wider flanges are also advantageous for supporting floor units, particularly the metal decking used frequently as part of a composite floor system.

Table 16.8 Comparison of use of UB and UC for simple beam design

71 kN/m, 6 m span, beams at 3 m spacing

M_{max} = 320 kN m
F_v = 213 kN

457 × 152 × UB 60	254 × 254 × UC 89
M_c = 352 kN m P_v = 600 kN $F_v < 0.6\, P_v$ – no interaction	M_c = 326 kN m P_v = 435 kN $F_v < 0.6\, P_v$ – no interaction

From Equation (16.2) and Table 16.5, assuming deflection limit is L/360 and service load is 47 kN m

I_{rqd} = 2.23 (47 × 3) 6² = 11 319 cm⁴

I_x = 25 500 cm⁴	I_x = 14 300 cm⁴

Difficulties can occur, because of the reduced depth, with deflections, although dead load deflections may be taken out by precambering the beams. This will not assist in limiting deflections in service due to imposed loading, although composite action will provide a much stiffer composite section. Excessive deflection of the floor beams under the weight of wet concrete can significantly increase slab depths at mid-span, leading to a substantially higher dead load. None of these problems need cause undue difficulty providing they are recognized and the proper checks made at a sufficiently early stage in the design.

Another possible source of difficulty arises in making connections between shallow beams and columns or between primary and secondary beams. The reduced web depth can lead to problems in physically accommodating sufficient bolts to carry the necessary end shears. Welding cleats to beams removes some of the dimensional tolerances that assist with erection on site as well as interfering with the smooth flow of work in a fabricator's shop that is equipped with a dedicated saw and drill line for beams. Extending the connection beyond the beam depth by using seating cleats, is one solution, although a requirement to contain the connection within the beam depth may prevent their use.

Beam depths may also be reduced by using moment-resisting beam-to-column connections which provide end fixity to the beams; a fixed end beam carrying a central point load will develop 50% of the peak moment and only 20% of the central deflection of a similar simply-supported beam. Full end fixity is unlikely to be a realistic proposition in normal frames but the replacement of the notionally pinned beam-to-column connection, provided by an arrangement such as web cleats, with a substantial end plate that functions more or less as a rigid connection permits the development of some degree of continuity between beams and columns. These arrangements will need more careful treatment when analyzing the pattern of internal moments and forces in the frame since the principles of simple construction will no longer apply.

An effect similar to the use of UC sections may be achieved if the flanges of a UB of a size that is incapable of carrying the required moment are reinforced by welding plates over part of its length. Additional moment capacity can be provided where it is needed as illustrated in Fig. 16.12; the resulting non-uniform section is

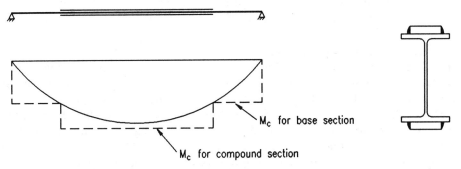

Fig. 16.12 Selective increase of moment capacity by use of a plated UB

stiffer and deflects less. Plating of the flanges will not improve the beam's shear capacity since this is essentially provided by the web and the possibility of shear or indeed local web capacity governing the design must be considered. A further development of this idea is the use of tapered sections fabricated from plate.[11] To be economic tapered sections are likely to contain plate elements that lie outside the limits for compact sections.

Because of the interest in developing longer spans for floors and the need to improve the performance of floor beams, a number of ingenious arrangements have developed in recent years.[12] Since these all utilize the benefits of composite action with the floor slab, they are considered in Chapter 21.

16.7 Cold-formed sections as beams

In situations where a relatively lightly loaded beam is required such as a purlin or sheeting rail spanning between main frames supporting the cladding in a portal frame, it is common practice to use a cold-formed section produced cold from flat steel sheet, typically between about 1 mm and 6 mm in thickness, in a wide range of shapes of the type shown in Fig. 16.13. A particular feature is that normally each section is formed from a single flat bent into the required shape; thus most available sections are not doubly symmetric but channels, zeds and other singly symmetric shapes. The forming process does, however, readily permit the use of quite complex cross-sections, incorporating longitudinal stiffening ribs and lips at the edges of flanges. Since the original coils are usually galvanized, the members do not normally require further protective treatment.

The structural design of cold-formed sections is covered by BS 5950: Part 5, which permits three approaches:

(1) design by calculation using the procedures of the code, section 5, for members in bending
(2) design on the basis of testing using the procedures of section 10 to control the testing; section 10.3 for members in bending
(3) for three commonly used types of member (zed purlins and sheeting rails and lattice joists), design using the simplified set of rules given in section 9.

In practice, option (2) is the most frequently used with all the major suppliers providing design literature, the basis of which is usually extensive testing of their product range, design being often reduced to the selection of a suitable section for a given span, loading and support arrangement using the tables provided.

Most cold-formed section types are the result of considerable development work by their producers. The profiles are therefore highly engineered so as to produce a near optimum performance, a typical example being the ranges of purlins produced by the leading UK suppliers. Because of the combination of the thin material and the comparative freedom provided by the forming process, this

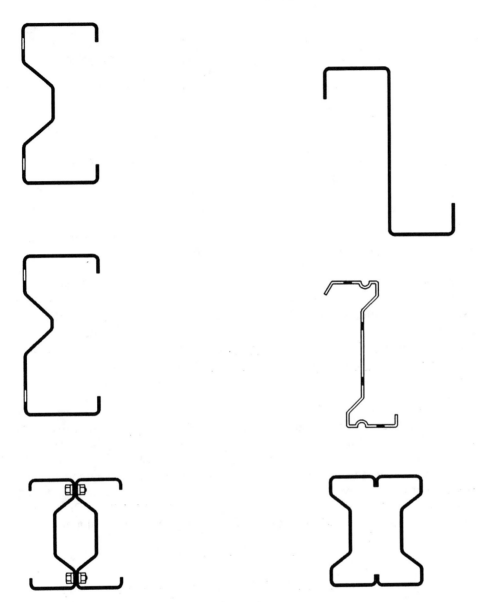

Fig. 16.13 Typical cold-formed section beam shapes

means that most sections will contain plate elements having high width-to-thickness ratios. Local buckling effects, either due to overall bending because the profile is non-compact, or due to the introduction of localized loads, are of greater importance than is usually the case for design using hot-rolled sections. BS 5950: Part 5 therefore gives rather more attention to the treatment of slender cross-sections

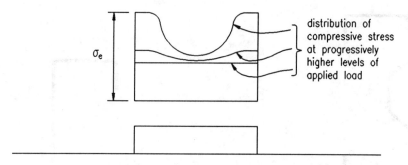

Fig. 16.14 Loss of plating effectiveness at progressively higher compressive stress

than does BS 5950: Part 1. In addition, manufacturers' design data normally exploit the post-buckling strength observed in their development tests.

The approach used to deal with sections containing slender elements in BS 5950: Part 5 is the well accepted effective width technique. This is based on the observation that plates, unlike struts, are able to withstand loads significantly in excess of their initial elastic buckling load, providing some measure of support is available to at least one of their longitudinal edges. Buckling then leads to a redistribution of stress with the regions adjacent to the supported longitudinal edges attracting higher stresses and the other parts of the plate becoming progressively less effective as shown in Fig. 16.14. A simple design representation of the condition of Fig. 16.14 consists of replacing the actual post-buckling stress distribution with the approximation shown in Fig. 16.15. The structural properties of the member (strength and stiffness) are then calculated for this effective cross-section as illustrated in Fig. 16.16. Tabulated information in BS 5950: Part 5 for steel of yield strength 280 N/mm^2 makes the application of this approach simpler in the sense that effective widths may readily be determined, although cross-sectional properties have still to be calculated. The use of manufacturer's literature removes this requirement. For beams, Part 5 also covers the design of reinforcing lips on the usual basis of ensuring that the free edge of a flange supported by a

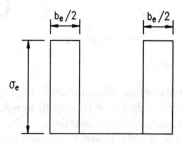

Fig. 16.15 Effective width design approximation

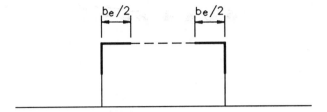

Fig. 16.16 Effective cross-section

single web behaves as if both edges were supported; web crushing under local loads, lateral–torsional buckling and the approximate determination of deflections take into account any loss of plating effectiveness.

For zed purlins or sheeting rails section 9 of BS 5950: Part 5 provides a set of simple empirically based design rules. Although easy to use, these are likely to lead to heavier members for a given loading, span and support arrangement than either of the other permitted procedures. A particular difference of this material is its use of unfactored loads, with the design conditions being expressed directly in member property requirements.

16.8 Beams with web openings

One solution to the problem of accommodating services within a restricted floor depth is to run the services through openings in the floor beams. Since the size of hole necessary in the beam web will then typically represent a significant proportion of the clear web depth, it may be expected that it will have an effect on structural performance. The easiest way of visualizing this is to draw an analogy between a beam with large rectangular web cut-outs and a Vierendeel girder. Figure 16.17 shows how the presence of the web hole enables the beam to deform locally in a similar manner to the shear type deformation of a Vierendeel panel. These deformations, superimposed on the overall bending effects, lead to increased deflection and additional web stresses.

A particular type of web hole is the castellation formed when a UB is cut, turned and rewelded as illustrated in Fig. 16.18. For the normal UK module geometry this leads to a 50% increase in section depth with a regular series of hexagonal holes. Other geometries are possible, including a further increase in depth through the use of plates welded between the two halves of the original beam. Some aspects of the design of castellated beams are covered by the provisions of BS 5950: Part 1, while more detailed guidance is available in a Constrado publication.[13]

Based on research conducted in the USA, a comparatively simple elastic method for the design of beams with web holes, including a fully worked example, is

432 Beams

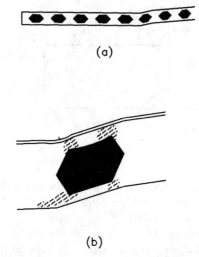

Fig. 16.17 Vierendeel-type action in beam with web openings: (a) overall view, (b) detail of deformed region

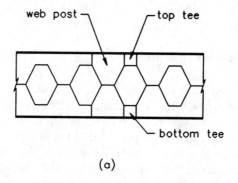

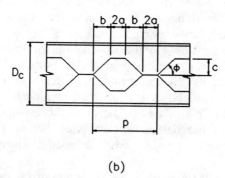

Fig. 16.18 Castellated beam: (a) basic concept, (b) details of normal UK module geometry

available.[14] This uses the concept of an analysis for girder stresses and deflections that neglects the effects of the holes, coupled with checking against suitably modified limiting values. The full list of design checks considered in Reference 14 is:

(1) web shear due to overall bending acting on the reduced web area
(2) web shear due to local Vierendeel bending at the hole
(3) primary bending stresses (little effect since overall bending is resisted principally by the girder flanges)
(4) local bending due to Vierendeel action
(5) local buckling of the tee formed by the compression flange and the web adjoining the web hole
(6) local buckling of the stem of the compression tee due to secondary bending
(7) web crippling under concentrated loads or reactions near a web hole; as a simple guide, Reference 14 suggests that for loads which act at least $(d/2)$ from the edge of a hole this effect may be neglected
(8) shear buckling of the web between holes; as a simple guide, Reference 14 suggests that for a clear distance between holes that exceeds the hole length this effect may be neglected
(9) vertical deflections; as a rough guide, secondary effects in castellated beams may be expected to add about 30% to the deflections calculated for a plain web beam of the same depth $(1.5D)$. Beams with circular holes of diameter $(D/2)$ may be expected to behave similarly, while beams with comparable rectangular holes may be expected to deflect rather more.

As an alternative to the use of elastic methods, significant progress has been made in recent years in devising limit states approaches based on ultimate strength conditions. A CIRIA/SCI design guide [15] dealing with the topic principally from the point of composite beams is now available. If some of the steps in the 24-point design check of Reference 15 are omitted, the method may be applied to non-composite beams, including composite beams under construction. Much of the basis for Reference 15 may be traced back to the work of Redwood and Choo,[16] and the following treatment of bare steel beams is taken from Reference 16.

The governing condition for a stocky web in the vicinity of a hole is taken as excessive plastic deformation near the opening corners and in the web above and below the opening as illustrated in Fig. 16.19. A conservative estimate of web strength may then be obtained from a moment–shear interaction diagram of the type shown as Fig. 16.20. Values of M_0 and V_1 in terms of the plastic moment capacity and plastic shear capacity of the unperforated web are given in Reference 14 for both plain and reinforced holes; M_1 may also be determined in this way. Solution of these equations is tedious, but some rearrangement and simplification are possible so that an explicit solution for the required area of reinforcement may be obtained. However, the whole approach is best programmed for a microcomputer and a program based on the full method of Reference 15 is available from the SCI.

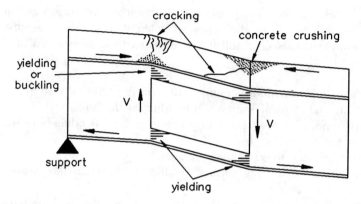

Fig. 16.19 Hole-induced failure

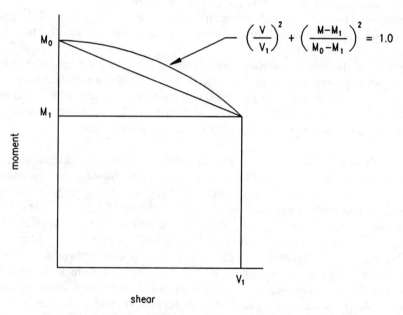

Fig. 16.20 Moment–shear interaction for a stocky web in the vicinity of a hole

$$\left(\frac{V}{V_1}\right)^2 + \left(\frac{M-M_1}{M_0-M_1}\right)^2 = 1.0$$

References to Chapter 16

1. British Standards Institution (1990) Part 1: *Code of practice for design in simple and continuous construction: hot rolled sections*. BS 5950, BSI, London.
2. The Steel Construction Institute (SCI) (1991) *Steelwork Design Guide to BS*

5950: Part 1: 1990, Vol. 1, Section Properties, Member Capacities, 3rd edn. SCI, Ascot, Berks.
3. Johnson R.P. & Buckby R.J. (1979) *Composite Structures of Steel and Concrete, Vol. 2: Bridges with a Commentary on BS 5400: Part 5*, 1st edn. Granada, London. (2nd edn, 1986).
4. Woodcock S.T. & Kitipornchai S. (1987) Survey of deflection limits for portal frames in Australia. *J. Construct. Steel Research*, 7, No. 6, 399–418.
5. Nethercot D.A., Salter P. & Malik A. (1989) *Design of Members Subject to Bending and Torsion*. The Steel Construction Institute, Ascot, Berks (SCI Publication 057).
6. Dux P.F & Kitipornchai S. (1986) Elastic buckling strength of braced beams. *Steel Construction*, (AISC), 20, No. 1, May.
7. Trahair N.S. & Nethercot D.A. (1984) Bracing requirements in thin-walled structures. In *Developments in Thin-Walled Structures – 2* (Ed. by J. Rhodes & A.C. Walker), pp. 93–130. Elsevier Applied Science Publishers, Barking, Essex.
8. Nethercot D.A. & Lawson R.M. (1992) *Lateral stability of steel beams and columns – common cases of restraint*. SCI Publication 093. The Steel Construction Institute, Ascot, Berks.
9. Brown B.A. (1988) The requirements for restraints in plastic design to BS 5950. *Steel Construction Today*, 2, No. 6, Dec., 184–6.
10. Horne M.R. (1964) Safe loads on I-section columns in structures designed by plastic theory. *Proc. Instn. Civ. Engrs*, 29, Sept., 137–50.
11. Raven G.K. (1987) The benefits of tapered beams in the design development of modern commercial buildings. *Steel Construction Today*, 1, No. 1, Feb., 17–25.
12. Owens G.W. (1987) Structural forms for long span commercial building and associated research needs. In *Steel Structures, Advances, Design and Construction* (Ed. by R. Narayanan), pp. 306–319. Elsevier Applied Science Publishers, Barking, Essex.
13. Knowles P.R. (1985) *Design of Castellated Beams for use with BS 5950 and BS 449*. Constrado.
14. Constrado (1977) *Holes in Beam Webs: Allowable Stress Design*. Constrado.
15. Lawson R.M. (1987) *Design for Openings in the Webs of Composite Beams*. CIRIA Special Publication S1 and SCI Publication 068. CIRIA/Steel Construction Institute.
16. Redwood R.G. & Choo S.H. (1987) Design tools for steel beams with web openings. In: Composite Steel Structures, Advances, Design and Construction (Ed. by R. Narayanan), pp. 75–83. Elsevier Applied Science Publishers, Barking, Essex.

A series of worked examples follows which are relevant to Chapter 16.

Worked example

The Steel Construction Institute Silwood Park, Ascot, Berks SL5 7QN	Subject **BEAM EXAMPLE 1** **LATERALLY RESTRAINED** **UNIVERSAL BEAM**	Chapter ref. **16**
	Design code **BS5950: Part 1** Made by **DAN** Checked by **GWO**	Sheet no. **1**

Problem

Select a suitable UB section to function as a simply supported beam carrying a 140 mm thick solid concrete slab together with an imposed load of 7.0 kN/m². Beam span is 7.2 m and beams are spaced at 3.6 m intervals. The slab may be assumed capable of providing continuous lateral restraint to the beam's top flange.

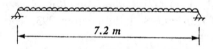

7.2 m

Due to restraint from slab there is no possibility of lateral-torsional buckling, so design beam for:

i)   Moment capacity

ii)   Shear capacity

iii)   Deflection limit

Loading

D.L.	=	$(2.4 \times 9.81 \times 0.14)$	=	3.3 kN/m^2	
I.L.			=	7.0 kN/m^2	
Total serviceability loading			=	10.3 kN/m^2	Table 2

Total load for ultimate limit state

= $1.4 \times 3.3 + 1.6 \times 7.0$ = 15.8 kN/m^2

Design ultimate moment   = $(15.8 \times 3.6) \times 7.2^2/8$
                                           = 369 kNm

Design ultimate shear   = $(15.8 \times 3.6) \times 7.2/2$
                                         = 205 kN

	Subject	Chapter ref.
The Steel Construction Institute Silwood Park, Ascot, Berks SL5 7QN	**BEAM EXAMPLE 1** **LATERALLY RESTRAINED** **UNIVERSAL BEAM**	**16**
	Design code BS5950: Part 1 — Made by **DAN** / Checked by **GWO**	Sheet no. **2**

Assuming use of Grade 43 steel and no material greater than 16 mm thick,   Table 6

take p_y =   275 N/mm^2

Required S_x = $369 \times 10^6 / 275$
           = 1.34×10^6 mm^3 =   1340 cm^3

A $457 \times 152 \times 67$ UB has a value of S_x of 1440 cm^3 Steelwork Design Guide Vol 1

T = $15.0 < 16.0$ mm

$\therefore$ p_y =   275 N/mm^2

Check section classification   3.5.2

Actual b/T =   5.06 d/t =   44.7

ϵ = $(275/p_y)^{1/2}$ =   1   Table 7

Limit on b/T for plastic section   = $8.5 > 5.06$

Limit on d/t for shear   = $63 > 44.7$

$\therefore$ <u>Section is plastic</u>

Actual M_c = $275 \times 1440 \times 10^3$ 4.2.5

            = 396×10^6 Nmm

            =   <u>396 kNm</u> > <u>369 kNm</u>   OK

Vertical shear capacity

P_v = $0.6 \, p_y \, A_v$ 4.2.3

where A_v = $t D$

$\therefore$ P_v = $0.6 \times 275 \times 9.1 \times 457.2 = 686 \times 10^3$ N

          =   <u>686 kN</u> > <u>205 kN</u>   OK

	Subject	Chapter ref.
The Steel Construction Institute Silwood Park, Ascot, Berks SL5 7QN	**BEAM EXAMPLE 1** **LATERALLY RESTRAINED** **UNIVERSAL BEAM**	**16**
	Design code *BS5950: Part 1*	Made by **DAN** Checked by **GWO**	Sheet no. **3**

Check serviceability deflections under imposed load   2.5.1

$$\delta = \frac{5 \times (7.0 \times 3.6) \times 7200^4}{384 \times 205000 \times 32400 \times 10^4}$$

  = *13.3 mm*   =   *span/541*

From Table 5, limit is span/360   ∴   δ *OK*

∴ <u>Use 457 × 152 × 67 UB Grade 43</u>

The Steel Construction Institute Silwood Park, Ascot, Berks SL5 7QN	Subject **BEAM EXAMPLE 2 LATERALLY UNRESTRAINED UNIVERSAL BEAM**	Chapter ref. **16**
	Design code **BS5950: Part 1**	Made by **DAN**	Sheet no. **1**
		Checked by **GWO**	

Problem

For the same loading and support conditions of example 1 select a suitable UB assuming that the member must be designed as laterally unrestrained.

It is not now possible to arrange the calculations in such a way that a direct choice is possible; a guess and check approach must be adopted.

Try $610 \times 229 \times 125$ UB

u	=	0.873	r_y	=	4.98 cm

Steelwork Design Guide Vol 1

x = 34.0 S_x = 3680 cm³

λ = L_E / r_y = 7200/49.8   4.3.7.5
                      = 145

λ / x = 145/34 = 4.26

v = 0.85   Table 14

λ_{LT} = $n\,u\,v\,\lambda$

For simplicity make a safe approximation of $n = 1.0$

∴ λ_{LT} = $1.0 \times 0.873 \times 0.85 \times 145$

           = 108

p_b = 116 N/mm²   Table 11

M_b = $S_x\,p_b$ = $3680 \times 10^3 \times 116$ 4.3.7.3

       = 427×10^6 Nmm

       = <u>427 kNm</u> > <u>369 kNm</u>   OK

The Steel Construction Institute Silwood Park, Ascot, Berks SL5 7QN	Subject **BEAM EXAMPLE 2** **LATERALLY UNRESTRAINED** **UNIVERSAL BEAM**	Chapter ref. **16**
	Design code **BS5950: Part 1** Made by **DAN** Checked by **GWO**	Sheet no. **2**

Since section is larger than before, P_v and δ will also be satisfactory

∴ <u>Adopt 610 × 229 × 125 UB</u>

However for a UDL, n = 0.94 when determining λ_{LT} Table 16

Refer to member capacities section for values of M_b directly, noting grade 43 material.   Steelwork Design Guide Vol 1

Value of M_b required is 369 kNm; interpolating for n = 0.94 & L_E = 7.2 m suggests as possible sections:

533 × 210 × 122 UB M_b OK

610 × 229 × 113 UB M_b OK

Since both are larger than that checked for shear capacity and serviceability deflection in the previous example, either may be adopted.

∴ <u>Adopt 533 × 210 × 122 UB or 610 × 229 × 113 UB Grade 43</u>

The Steel Construction Institute Silwood Park, Ascot, Berks SL5 7QN	Subject **BEAM EXAMPLE 3 LATERALLY UNRESTRAINED UNIVERSAL BEAM**	Chapter ref. **16**
	Design code **BS5950: Part 1**	Made by **DAN** Checked by **GWO**	Sheet no. **1**

Problem

Select a suitable UB section in Grade 43 steel to carry the pair of point loads at the third points transferred by crossbeams as shown in the accompanying sketch. Design to BS 5950: Part 1.

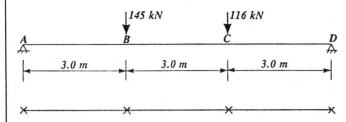

The crossbeams may reasonably be assumed to provide full lateral and torsional restraint at B and C; assume further that ends A and D are similarly restrained. Thus the actual level of transfer of load at B and C (relative to the main beam's centroid) will have no effect, the lateral-torsional buckling aspects of the design being one of considering the 3 segments AB, BC and CD separately.

From statics the BMD and SFD are

BMD

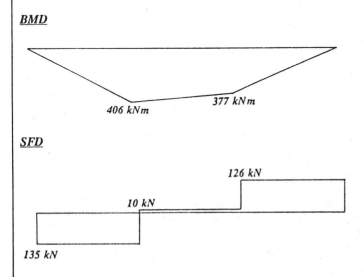

SFD

The Steel Construction Institute Silwood Park, Ascot, Berks SL5 7QN	Subject **BEAM EXAMPLE 3** **LATERALLY UNRESTRAINED** **UNIVERSAL BEAM**	Chapter ref. **16**
Design code **BS5950: Part 1**	Made by **DAN** Checked by **GWO**	Sheet no. **2**

For initial trial section select a UB with $M_{cx} >$ 406 kNm.

A 457 × 152 × 74 UB provides M_{cx} of 429 kNm. Now check lateral-torsional buckling strength for segments AB, BC & CD.

Steelwork Design Guide Vol 1

AB

β = 0/406 = 0.0

m = 0.57   *Table 18*

$\bar{M}$ = 0.57 × 406 = 231.4 kNm

For L_E = 3.0 m, M_b = 288 kNm

∴ 288 > 231.4 kNm OK

Steelwork Design Guide Vol 1

BC

β = 377/406 = 0.93

m = 0.97   *Table 18*

$\bar{M}$ = 0.97 × 406 = 393.8 kNm

But for L_E = 3.0 m, M_b = 288 kNm <u>Not</u> OK

Try 457 × 191 × 82 UB

this provides, M_b of 396 kNm > 393.8 kNm OK

Steelwork Design Guide Vol 1

CD

Satisfactory by inspection OK

A 457 × 191 × 82 UB provides sufficient resistance to lateral-torsional buckling for each segment and thus for the beam as a whole.

		Subject **BEAM EXAMPLE 3** **LATERALLY UNRESTRAINED** **UNIVERSAL BEAM**	Chapter ref. **16**
The Steel Construction Institute Silwood Park, Ascot, Berks SL5 7QN			
		Design code **BS5950: Part 1**	Made by **DAN**	Sheet no. **3**
			Checked by **GWO**	

Check shear capacity; maximum shear at A is 135 kN

P_v = 752 kN > 135 kN OK

*Check bearing and buckling capacity of web at the supports -
required capacity is 135 kN.*

Since C_1 values exceed 135 kN in both cases, section is clearly adequate.

Steelwork
Design Guide
Vol 1

For initial check on serviceability deflections assume an equivalent UDL and factor down all loads by 1.5 to obtain

$$w = \frac{145 + 116}{1.5 \times 9} = 19.3 \text{ kN/m}$$

$$\delta = \frac{5 \times 19.3 \times 9000^4}{384 \times 205000 \times 37100 \times 10^4}$$

= 21.68 mm   =   span/415

Limiting deflection is span/360 ∴ OK

Table 5

Beam is clearly satisfactory for deflection since these (approximate) calculations have used the full load and not just the imposed load.

∴ <u>Adopt 457 × 191 × 82 UB</u>

Chapter 17
Plate girders

by RANGACHARI NARAYANAN

17.1 Introduction

Plate girders are employed to support heavy vertical loads over long spans; the resulting bending moments are larger than the moment capacity of available rolled sections. In its simplest form the plate girder is a built-up beam consisting of two flange plates, fillet welded to a web plate to form an I-section. The primary function of the top and bottom flange plates is to resist the axial compressive and tensile forces caused by the applied bending moments; the main function of the web is to resist the shear. Indeed this partition of structural action is used as the basis for design in some Codes of Practice.

For a given bending moment the required flange areas can be reduced by increasing the distance between them. Thus for an economical design, it is advantageous to increase the distance between flanges. To keep the self-weight of the girder to a minimum the web thickness should be reduced as the depth increases, but this leads to web buckling considerations being more significant in plate girders than in rolled beams.

Plate girders are sometimes used in buildings and are often used in small- to medium-span bridges. They are designed in accordance with the provisions contained in BS 5950: Part 1[1] and BS 5400: Part 3[2] respectively. This chapter explains current practice in designing plate girders for buildings and bridges; references to the relevant clauses in the Codes are made.

17.2 Advantages and disadvantages

The development of highly automated workshops in recent years has reduced the fabrication costs of plate girders very considerably; box girders and trusses still have to be fabricated manually, with consequently high fabrication costs. Optimum use of material is made compared with rolled sections as the girder is fabricated from plates and so the designer has greater freedom to vary the section to respond to a change in the applied forces. Thus variable depth plate girders have been increasingly designed in recent years. Plate girders are aesthetically more pleasing than trusses and are easier to transport and erect than box girders.

There are only a very few limitations in the use of plate girders. Compared with trusses, they are heavier, more difficult to transport and have larger wind resist-

Table 17.1 Recommended span-to-depth ratios for plate girders used in buildings

Applications	Span-to-depth ratio
(1) Constant-depth beams used in simply-supported composite girders, and for simply-supported non-composite girders, with concrete decking	12 to 20
(2) Constant-depth beams used in continuous non-composite girders using concrete decking (*NB* continuous composite girders are rare in buildings)	15 to 20
(3) For simply-supported crane girders (non-composite construction is usual)	10 to 15

ance. The provision of openings for services is also more difficult. The low torsional stiffness of plate girders makes them difficult to use on bridges having small plan radius. Plate girders can sometimes pose problems during erection because of concern for the stability of compression flanges.

17.3 Initial choice of cross-section in buildings

17.3.1 Span-to-depth ratios

Advances in fabrication methods allow the economic manufacture of plate girders of constant or variable depth. Traditionally, constant-depth girders were more common in buildings; however this may change as designers become more inclined to modify the steel structure to accommodate services.[3] Recommended span-to-depth ratios are given in Table 17.1.

17.3.2 Recommended plate thicknesses and proportions

The choice of the thickness of the web is related to the stiffening. Care should be exercised in ensuring that not too thin a web is chosen; some of the material saved has to be put back by way of stiffening and the extra workmanship required is expensive.

The minimum web thickness chosen for plate girders in buildings usually varies from 10 mm for girders up to 1200 mm deep to 20 mm for those 2500 mm deep.

Changes in flange size along the girder are not worthwhile in buildings. For non-composite girders the flange width chosen is usually within a range of 0.3–0.5 times the depth of the section (0.4 is most common). It is important to ensure that the flange size chosen is within the breadth-to-thickness limits for semi-compact sections in order to avoid local buckling of the flange. As a preliminary guide it is suggested that the *overall* flange width-to-thickness ratio be limited to 24. For simply-supported composite girders these rules can still be employed for the preliminary sizing of top flanges. The width of bottom (tension) flanges can be increased by 30%.

17.3.3 Stiffeners

Horizontal web stiffeners are not usually required for plate girders used in buildings. Vertical web stiffeners may be provided close to supports to enhance the resistance to shear near the supports. Intermediate stiffening at locations far away from supports will, in general, be unnecessary due to reduced shear.

The provision of vertical stiffeners enables the web to sustain shear stresses well in excess of its unstiffened elastic critical shear stress. Furthermore, additional strength is obtained as a consequence of *tension field action*, whereby the web panel between the stiffeners resists additional shear loads by developing diagonal tensile stresses.

Where vertical stiffeners are specified, they are spaced such that the panel aspect (width-to-depth) ratio is around 1.2 to 1.6. Where end panels are designed without utilizing the tension field action (see Fig. 17.6), the aspect ratio of the end panels is reduced to 0.6–1.0. Sometimes double stiffeners are employed as bearing stiffeners at the end (see Fig. 17.7). The overhang of the beam beyond the support is generally limited to a maximum of one-eighth of the depth of the girder.

17.4 Design of plate girders used in buildings to BS 5950: Part 1

17.4.1 General

Any cross section of a plate girder will normally be subjected to a combination of shear force and bending moment, present in varying proportions. BS 5950: Part 1 allows one of the following three methods to be used for the determination of capacity (see clause 4.4.4.2):

(1) The moment may be assumed to be resisted by the flanges alone and the web designed for shear only

(2) The moment may be assumed to be resisted by the whole section, the web being designed for combined shear and longitudinal stresses
(3) A proportion of the loading may be assumed to be resisted by Method (1), the remainder being resisted by Method (2).

The separation of moment and shear effects in Method (1) leads to considerable simplification and a better physical appreciation of the load-carrying action. Although slightly conservative it is probably the most commonly used, and is the method described in the following sections.

17.4.2 Web thickness

The slenderness at which web plates become prone to buckling is defined as 63ε in BS 5950: Part 1, where $\varepsilon = (275p_y)^{\frac{1}{2}}$ (see clause 4.4.4.2). When $d/t \leq 63\varepsilon$ (see Fig. 17.1), webs will not buckle and the design principles for plate girders are similar to those for universal beams. For economy, plate girders are designed with thin webs ($d/t > 63\varepsilon$), hence shear buckling *must* be taken into account.

The buckling capacity of slender webs can be increased by the provision of web stiffeners. In general, plate girder webs used in buildings are unstiffened or have transverse stiffeners only.

Minimum web thickness requirements (clause 4.4.2 of BS 5950: Part 1) are largely fixed by serviceability considerations such as adequate stiffness to prevent unsightly buckles developing during normal handling and erection, 'breathing of the web' during loading and unloading in service, etc. The following minimum thickness values are prescribed in BS 5950: Part 1 (see Fig. 17.1):

(1) For unstiffened webs, $t \geq \dfrac{d}{250}$

(where t = thickness of the web, and d = depth of the web)
(2) For transversely stiffened webs,

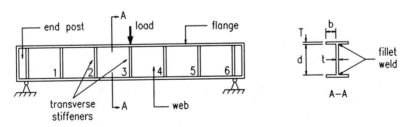

Fig. 17.1 Elevation of a typical plate girder

with stiffener spacing $a > d$, $\quad t \geq \dfrac{d}{250}$

with stiffener spacing $a \leq d$, $\quad t \geq \dfrac{d}{250}\left(\dfrac{a}{d}\right)^{\frac{1}{2}}$.

To prevent the flange buckling into the web, the Code specifies:

(1) For unstiffened webs, $\quad t \geq \dfrac{d}{250}\dfrac{p_{yf}}{345}$

(where p_{yf} is the design stress of the flange material)
(2) For transversely stiffened webs,

with stiffener spacing $a > 1.5d$, $\quad t \geq \dfrac{d}{250}\left(\dfrac{p_{yf}}{345}\right)^{\frac{1}{2}}$

with stiffener spacing $a \leq 1.5d$, $\quad t \geq \dfrac{d}{250}\left(\dfrac{p_{yf}}{455}\right)^{\frac{1}{2}}$.

17.4.3 Flange thickness

Local buckling of the compression flange may also occur if the flange plate is of slender proportions, i.e. if $b/T > 13\varepsilon$. However, flanges of such slender proportions are rare in buildings since there would seldom appear to be a good reason to exceed the specified outstand width in normal design.

17.4.3.1 Moment capacity

Following Method (1), the overall moment capacity, M_c, is calculated from the plastic capacity of the flange plates only:

$$M_c = p_{yf} S_{xf}$$

where p_{yf} is the design stress of the flange material and S_{xf} is the plastic modulus of the flanges only about the section axis.

17.4.4 Shear capacity

Following Method (1), the shear capacity is determined as that of the web plate only, assuming the web to be unaffected by the bending moment.

Thin webs may be designed with or without transverse stiffeners; the procedures are considered individually.

17.4.4.1 Webs without intermediate stiffeners
(clause 4.4.5.3 of BS 5950: Part 1)

The shear capacity of unstiffened webs is limited to their shear buckling resistance,

$$V_{cr} = q_{cr}dt$$

where q_{cr} is the critical shear stress.

Based on the elastic buckling theory, the Code gives the following values for q_{cr}:

(1) When $\lambda_w \leq 0.8$,

$$q_{cr} = 0.6p_{yw}$$

(2) When $0.8 < \lambda_w < 1.25$,

$$q_{cr} = 0.6p_{yw}\,[1 - 0.8(\lambda_w - 0.8)]$$

(3) When $\lambda_w \geq 1.25$,

$$q_{cr} = q_e$$

where λ_w is the equivalent slenderness of the web and equals $(0.6p_{yw}/q_e)^{\frac{1}{2}}$; q_e is the elastic critical shear strength of the panel (in N/mm²) given by:

(a) when $a/d \leq 1$, $\quad q_e = \left[0.75 + \dfrac{1}{(a/d)^2}\right]\left(\dfrac{1000}{d/t}\right)^2$

(b) when $a/d > 1$, $\quad q_e = \left[1 + \dfrac{0.75}{(a/d)^2}\right]\left(\dfrac{1000}{d/t}\right)^2$

where a is the spacing of transverse stiffeners; d is the depth of the web; t is the thickness of the web; p_{yw} is the design strength of the web.

Values of q_{cr} are tabulated in the Code.

17.4.4.2 Webs with intermediate transverse stiffeners
(clause 4.4.5.4 of BS 5950: Part 1)

Webs with intermittent stiffeners are illustrated in Fig. 17.2. BS 5950: Part 1 allows such webs to be designed by limiting their shear capacity to the shear buckling resistance of the panels in a similar way to section 17.4.4.1. Values of q_{cr} are tabulated in the Code for different stiffener spacing-to-web depth ratios, a/d.

However, this approach is very conservative since the introduction of transverse stiffeners increases the shear capacity in two ways.[4] As has already been noted, the buckling resistance V_{cr} is increased as a consequence of reduced a/d values and the development of post-critical strength; secondly, the stiffeners enable the girder to withstand loads considerably in excess of the conventional elastic critical

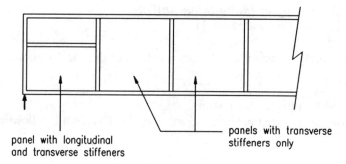

panel with longitudinal
and transverse stiffeners

panels with transverse
stiffeners only

Fig. 17.2 Different stiffener arrangements

strength of the web in shear, because of the development of tension field action. Advantage should be taken of the tension field action to achieve an effective design; the Code permits the designer to do so, should he so wish.

Figure 17.3(a) shows the development of a tension field in a typical girder. Once a web panel has buckled, its capacity to carry additional diagonal compressive stresses is ignored and it is assumed that a new load-carrying mechanism is developed, whereby any additional shear load is carried by an inclined tensile membrane stress field. This tensile force anchors against the top and bottom flanges and against the transverse stiffener on either side of the web panel, as shown. The load-carrying action of the plate girder then becomes similar to that

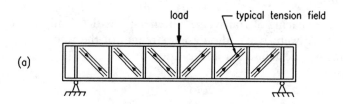

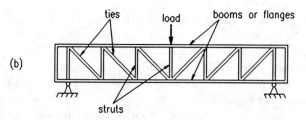

Fig. 17.3 Tension field action (a) in individual sub-panels as a result of transverse stiffeners. (b) Typical N-truss for comparison

of the N-truss in Fig. 17.3(b). The resistance offered by the web plates is analogous to that of the diagonal tie bars in the truss.

The behaviour of a girder under an increasing shear load may thus be divided into the three phases shown in Fig. 17.4. Figure 17.4(a) shows the situation prior to buckling when equal tensile and compressive principal stresses are developed in the plate. Figure 17.4(b) shows the development of a tension field in the post-buckling phase, and Fig. 17.4(c) shows the situation at collapse. As indicated, failure occurs when the web yields and four plastic hinges form in the flanges to allow the development of a sway collapse mechanism.

In the Code the full shear buckling resistance, V_b, is expressed as:

$$V_b = [q_b + q_f (K_f)^{\frac{1}{2}}] dt \quad \text{but} \leq 0.6 p_y dt$$

In this expression q_b represents the *basic* shear strength of the web. It combines the buckling strength q_{cr} and the post-buckling strength derived from that part of the web tension field supported by the transverse stiffeners. Values of the basic shear strength, q_b, are tabulated in the Code for different stiffener spacing ratios (a/d) and web slenderness ratios (d/t), but may be calculated by formulae given below.

The basic shear strength of the web panel, q_b, is given by:

$$q_b = q_{cr} + \frac{y_b}{2\{a/d + [1 + (a/d)^2]^{\frac{1}{2}}\}}$$

where q_{cr} is the elastic critical stress in shear of the web panel ($a \times d$) calculated from elastic buckling theory; y_b is the basic tension field strength given by:

$$y_b = (p_{yw}^2 - 3q_{cr}^2 + \phi_t^2)^{\frac{1}{2}} - \phi_t$$

in which $\phi_t = \dfrac{1.5 q_{cr}}{[1 + (a/d)^2]^{\frac{1}{2}}}$

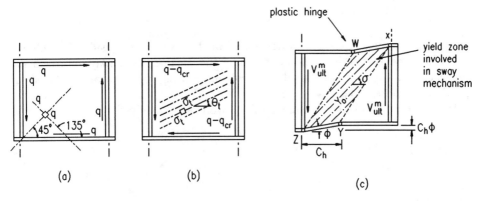

Fig. 17.4 Phases in behaviour up to collapse of a typical panel in shear: (a) unbuckled, (b) post-buckled, (c) collapsed

a is the spacing of transverse stiffeners; d is the depth of the web; t is the thickness of the web; p_{yw} is the design strength of the web.

NB If q_b is greater than $0.57p_{yw}$, it may be taken as equal to but not greater than $0.6p_{yw}$.

The second term in the equation for V_b (i.e. $q_f(K_f)^{\frac{1}{2}}$) represents the contribution made by the flanges to the post-buckling strength. The flanges support that part of the web tension field that pulls against them, as in Fig. 17.4(b), and they finally develop plastic hinges at collapse.

The term q_f helps to evaluate the flange-dependent shear strength, which is again tabulated in the Code for different a/d and d/t ratios. The term K_f is a non-dimensional parameter relating the plastic moment capacity of the flange, M_{pf}, to that of the web, M_{pw}.

The flange-dependent shear strength, q_f, is given by:

$$q_f = \left[4\sqrt{3} \sin\left(\frac{\theta}{2}\right)\left(\frac{y_b}{p_{yw}}\right)^{\frac{1}{2}}\right] \times 0.6p_{yw}$$

where $\theta = \tan^{-1}\left(\frac{d}{a}\right)$.

If the girder is subjected to shear forces only, then M_{pf} is taken as the full plastic moment capacity of the flange:

$$M_{pf} = \frac{2b}{4} T^2 p_{yf}$$

When a girder is subjected to a bending moment as well as shear then the flanges have to carry the axial forces arising from the bending action in addition to the lateral pull exerted by the tension field. These axial forces will reduce the plastic moment capacity of the flange plates, the reduced value M_{pf} being obtained from plasticity theory as:

$$M_{pf} = M_{pf}\left(1 - \frac{f}{p_{yf}}\right)$$

where f is the mean axial stress developed in the flange as a result of bending.

Thus, at sections of high moment, where the mean stress f may approach the flange design strength p_{yf}, the flange will make little contribution to the shear buckling resistance, V_b.

The plastic moment of resistance of the web, M_{pw}, is given by:

$$M_{pw} = \frac{1}{4} d^2 t\, p_{yw}$$

and

$$K_f = \frac{M_{pf}}{4M_{pw}}$$

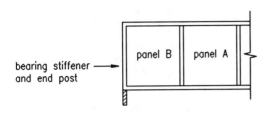

Fig. 17.5 End panel designed using tension field action (single stiffener)

17.4.4.3 Special consideration for end panels
(clause 4.4.5.4.3 of BS 5950: Part 1)

Tension field forces can only develop when an adequate anchorage is provided by the members bounding the panel. This is a particular problem at the ends of a girder where the anchorage has to be provided by the end stiffener alone.

The Code suggests that the anchor force H_q arising from tension field action be calculated as:

$$H_q = 0.75 dt p_y \left(1 - \frac{q_{cr}}{0.6 p_y}\right)^{\frac{1}{2}}$$

The Code then offers two possible approaches to end panel design. In the first (Fig. 17.5), the capacity of the end panel B may be limited to its shear buckling resistance, V_{cr}, so that no tension field develops within it. Panel B and its bounding transverse stiffeners then act as a beam, spanning vertically between the girder flanges to resist the shear and bending moment arising from the tension field pull, H_q, of panel A.

This first approach is conservative and results in a substantial end member. It has the disadvantage of omitting the additional shear capacity that could be derived from tension field action in the end panel, and this in the region close to the support where applied shear forces could well be a maximum. To prevent the calculated shear capacity of panel B being much lower than that of panel A, the intermediate stiffener would have to be positioned close to the support to make B a narrow panel.

To overcome this disadvantage, the second approach offered by the Code (Fig. 17.6) does allow the development of a tension field in the end panel, i.e. panel B is designed in the same way as the internal panels. The disadvantage now is that the single end stiffener by itself, spanning vertically between the flanges, has to resist the tension field pull of panel B. This represents a significant demand on the capacity of the single stiffener.

To overcome this, a double stiffener may be used as in Fig. 17.7 to form a rigid end post in which the two stiffeners and the portion of the web projecting beyond the support now form the vertically spanning beam. The disadvantage in this case

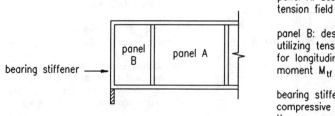

Fig. 17.6 End panel designed without using tension field action

is that adequate space must be available to allow the girder to project beyond its support.

Thus, each of the three alternatives shown in Figs 17.5, 17.6 and 17.7 has certain disadvantages. However, their inclusion in the Code enables the most appropriate to be chosen for each particular design situation. In most cases, the double stiffener solution results in economy, assuming the site conditions permit the girder to project beyond the support.

17.4.4.4 Special considerations for webs with openings

Web openings frequently have to be provided in girders used on building construction for service ducts, etc. When any dimension of such an opening exceeds 10% of the minimum dimensions of the panel in which it is placed, then the Code limits the shear capacity of the panel to its shear buckling resistance, V_{cr}. Any tension field capacity of a perforated web is thus conservatively ignored to avoid design complexity. Also, the panels immediately adjacent to that containing the large openings are designed as end panels according to the special considerations outlined above.

Guidance on design of plate girders with openings is provided in Reference 5.

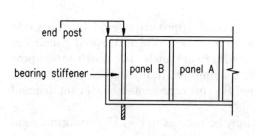

Fig. 17.7 End panel designed using tension field action (double stiffener)

17.4.5 Design of web stiffeners

The provision of adequate transverse stiffeners is essential to the satisfactory performance of the web panels of girders. The three most important types of stiffeners are:

(1) load-carrying
(2) bearing
(3) intermediate transverse.

Stiffeners of each of these types are subjected to compression and should be checked for local buckling.

17.4.5.1 Load-carrying stiffeners

These are provided to prevent local buckling of the web due to concentrated loads or reactions applied through the flange. The Code requires that the stiffeners should be positioned wherever such actions occur, if the buckling resistance of the unstiffened web would otherwise be exceeded.

Load-carrying stiffeners must be checked for both buckling and bearing. When checking for buckling an effective web width of $20t$ on either side of the centreline of the stiffener is considered to act with it to form a cruciform section. The resulting stiffener strut is then assumed to have an effective length of 0.7 times its actual length if the flange is restrained against rotation in the plane of the stiffener. If the flange is not so restrained, the actual length is taken. The buckling resistance of the stiffener strut is then determined as for a normal compression member.

When checking the stiffener for bearing, only that area of the stiffener in contact with the flange is taken into account.

Sometimes it is not possible to provide stiffeners immediately under the externally applied vertical loading, e.g. in the case of a travelling load on a gantry girder or during launching of a girder in construction. In such cases of patch loading (loads applied between the stiffeners), an additional check is required to ensure that the specified compressive strength of the web for edge loading is not exceeded.

17.4.5.2 Bearing stiffeners

These are provided to prevent local crushing of the web due to concentrated loads or reactions applied through the flange. Should such actions not exceed the buckling resistance of the web, then the load-carrying stiffeners of section 17.4.5.1 are not needed but the local bearing resistance can still be exceeded. If so, then bearing stiffeners are introduced designed to provide a sufficient bearing area to resist the loading.

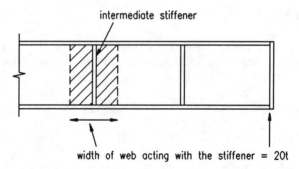

Fig. 17.8 Web acting with stiffener

17.4.5.3 Intermediate transverse stiffeners

The need for intermediate transverse stiffeners and their required spacing is dictated by the rules for shear web design discussed in section 17.4.4.2 (webs with intermediate transverse stiffeners).

The Code requires the stiffeners alone to have a specified minimum second moment of area about the centreline of the web. This ensures that the stiffener is able to remain straight and prevent out-of-plane buckling of the web at the stiffener position.

Designing for tension field action imposes additional loading on the transverse stiffeners. Because of the additional compressive forces induced, the stiffener must be designed as a strut and checked for buckling. As discussed later in section 17.5.1 for load-carrying stiffeners, an effective portion of the web, of width $20t$ on each side of the centreline of the stiffener, is considered to act with the stiffener to form a cruciform section (Fig. 17.8). The buckling resistance, P_q, of the stiffener strut, calculated as for a normal compression member, is then required to be in excess of the stiffener loading, F_q, calculated as:

$$F_q = V - V_s$$

where V is the maximum shear force in the panels adjacent to the stiffener, and V_s is the shear buckling resistance of the web panel designed without using tension field action, i.e. smaller values of V_{cr} for the panels on either side of the stiffener.

17.4.5.4 Combined loads on stiffeners

Stiffeners may, of course, be required to perform more than one of the above functions. For example, an intermediate stiffener may also be subjected to an external load, in which case it should also satisfy the requirements for a load-carrying stiffener.

17.4.6 Gantry girders

Plate girders used to support cranes are specifically excluded from utilizing the extra load capacity provided by tension field action. Apart from this limitation, gantry girders are designed using procedures outlined above.

17.5 Initial choice of cross-section for plate girders used in bridges (see also Chapter 4)

17.5.1 Choice of span

Plate girders are frequently employed to support railway and highway loadings on account of their economic advantages and ease of fabrication and are considered suitable for spans in the region of 25–100 m. Plate girders are considered when rolled sections are not big enough to carry the loads over the chosen span.

Spans are usually fixed by site restrictions and clearances. If there is freedom for the designer, simply-supported spans within the range of 25–45 m will be found to be appropriate; the optimum for continuous spans is about 45 m, as 27 m long *span* girders can be spliced with *pier* girders 18 m long. However, plate girders can be employed for spans of up to about 100 m in continuous construction.

17.5.2 Span-to-depth ratios

For supporting highways, a composite concrete decking, having a concrete thickness in the region of 250 mm is commonly employed. A span-to-depth ratio of 20 for simply-supported spans serves as an initial choice for such girders; where continuous spans are employed, the depth of the girder can be reduced at least by one-third, compared with a simply-supported span.

Through girders of constant depth are rarely employed in continuous construction. For simply-supported plate girders designed for highway loading, a span-to-depth ratio of 20 can serve as an initial choice; for through girders supporting railway loading, larger depths are required, hence a (span-to-depth) ratio of 15 is more appropriate. Variable depth plate girders are more appropriate when spans in excess of 30 m are required.

For continuous girders in composite construction, a (span-to-depth) ratio of 25 at the pier and 40 at mid-span is suitable. For highway bridges provided with an orthotropic deck, the corresponding values are increased to 30 and 60 respectively.

Table 17.2 Guide for selection of deck thickness

Transverse girder spacing (m)	Deck	Thickness (mm)
2.5–3.8	Reinforced concrete slab on permanent formwork	Constant 225–275
3.3–5.5	Haunched slab	Min. 250 up to 350 at haunch
5.0–7	Stringer	Constant 225–275
6.5+	Cross girders	Typically 250

17.5.3 Initial sizing of flange proportions

It is important to keep the flanges as wide as possible consistent with outstand limitations. The outstand should not ordinarily be wider than 12 times its thickness if the flange is fully stressed in compression, or 16 times its thickness if the flange is not fully stressed or is in tension. This ensures that the flange is stable during erection with the minimum number of bracings. For practical reasons, e.g. to accommodate detailing for certain types of permanent formwork, it is desirable to use a minimum flange width of 400 mm. A maximum flange thickness of 65 mm is recommended to avoid heavy welds and the consequent distortion.

Changes in the width of the top flange can be incorporated easily in composite decks to suit design requirements and these do not invite criticism on appearance grounds. On the other hand, changes in widths of bottom flanges are less acceptable visually. In any case it is desirable not to change flange sizes frequently lest the economy achieved by saving in material should be offset by expensive butt welding.

For composite girders having concrete decks it is usually necessary to allow for at least two rows of shear connectors on top flanges of beams; for longer spans three rows may be required at piers where high shear transfer takes place.

The cross sections chosen affect deck thickness and overall structural form. The guide in Table 17.2 is useful for initial selection of deck thickness.

17.5.4 Initial sizing of the webs

The thickness of web chosen for initial sizing is related to stiffening. There is no special advantage in using too slender a web, as the material saved will have to be replaced by stiffening; moreover, the workmanship with a thicker web plate is often superior. Probably the biggest single problem when determining the girder layout for a concrete decked plate girder road bridge is to achieve a solution with

Table 17.3 Initial values of web thickness

Beam depth (mm)	Web thickness (mm)
Up to 1200	10
1200–1800	12
1800–2250	15
2250–3000	20

Consideration should also be given to the economy of providing thicker webs, without any intermediate stiffening.

optimum transverse deck spans and minimum cost of web steel, as this involves the balancing of weight against workmanship.

The following advice is intended as general guidance:

(1) Two or three vertical stiffeners should be provided at, say, 1 m centres close to the bearings; thereafter, their spacing should be increased to 1.5 × girder depth.
(2) Generally speaking, horizontal stiffeners should be avoided. In long-span continuous plate girders they may be necessary in locations close to the piers.
(3) It is desirable to provide vertical stiffeners on one side and horizontal stiffeners on the other side, where possible.
(4) The initial values for web thickness are suggested in Table 17.3, if it is intended to provide web stiffening. Consideration should also be given to the economy of providing thicker webs, without any intermediate stiffening.
(5) Compact girders are more economical for most simply-supported spans and shorter continuous spans; economical plate girders for longer continuous spans will be non-compact.

17.6 Design of steel bridges to BS 5400: Part 3

The Code requires that the global analysis of the structure should be carried out elastically to determine the load effects (i.e. bending moments, shear forces, etc.). The section properties to be used will generally be those of the gross section. (See clauses 7.1 and 7.2 of the Code.[2])

Plastic analysis of the structure (i.e. redistribution of moments due to plastic hinge formation) is *not* allowed under BS 5400: Part 3.

Analysis should be carried out for individual and unfactored load cases. Summation of the load effects in different combinations and with different load factors can then be carried out in tabular form, as required by the design process.

460   Plate girders

17.6.1 Design of beams at the ultimate limit state

In the design of beam cross sections at the ultimate limit state the following need to be considered:

(1) Material strength
(2) Limitations on shape on account of local buckling of individual elements (i.e. webs and flanges)
(3) Effective section (reductions for compression buckling and holes)
(4) Lateral torsional buckling
(5) Web buckling (governed by depth-to-thickness ratio of web and panel size)
(6) Combined effects of bending and shear

17.6.1.1 Material strength (clause 4.3)

The nominal material strength is the yield stress of steel. The partial factor on strength, γ_m, depends on the structural component and behaviour; generally it is 1.05 at the ultimate limit state, but higher values are specified in the Code for compressive stress in bending or buckling.

17.6.1.2 Shape limitations (clause 9.3)

The capacity of a section can be limited by local buckling if the flange outstand-to-thickness ratio is large. The Code limits this ratio so that local buckling will not govern (Chapter 13 discusses local buckling in detail). Similarly, limits are given for outstands in tension to limit local shear-lag effects.

Where the compression region of a web without stiffeners has a very large depth-to-thickness ratio (i.e. > 68) the bending capacity of the section is modified by the requirement that webs are effectively reduced to take account of local buckling. Webs with longitudinal stiffeners are not so reduced.

When calculating the bending resistance of beams, it is important to recognize the contribution made by the webs and make appropriate provision for it in computing the modulus of the 'effective section'. Slender webs without horizontal stiffeners buckle in the compression zone, when subjected to high bending stresses. Based on parametric studies, the bending resistance contributed by the web (associated with a flange) is assessed by reducing the thickness of the web using the formula:

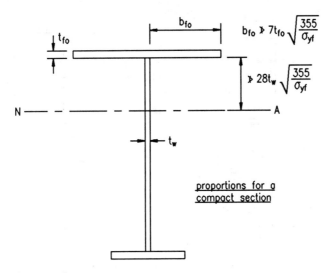

Fig. 17.9 Section classification

$$\frac{t_{we}}{t_w} = 1 \qquad \text{if } \frac{y_c}{t_w}\left(\frac{\sigma_{yw}}{355}\right)^{\frac{1}{2}} \leq 68$$

$$\frac{t_{we}}{t_w} = \left[1.425 - 0.00625\frac{y_c}{t_w}\left(\frac{\sigma_{yw}}{355}\right)^{\frac{1}{2}}\right] \qquad \text{if } 68 < \frac{y_c}{t_w}\left(\frac{\sigma_{yw}}{355}\right)^{\frac{1}{2}} < 228$$

$$\frac{t_{we}}{t_w} = 0 \qquad \text{if } \frac{y_c}{t_w}\left(\frac{\sigma_{yw}}{355}\right)^{\frac{1}{2}} \geq 228$$

where y_c is the depth of the web measured in its plane from the elastic neutral axis of the gross section of the beam to the compressive edge of the web.

The Code further defines cross sections as *compact or non-compact* and a different design approach is required for each classification.[6]

The classification of a section depends on the width-to-thickness ratio of the elements of the cross section considered, as shown in Fig. 17.9 for a compact plate girder.

A compact cross section can develop the full plastic moment capacity of the section (i.e. a rectangular stress block) and *local buckling of the individual elements of the cross section will not occur before this stage is reached* (see Fig. 17.10).

In a non-compact section, however, *local buckling of elements of the cross section may occur before the full moment capacity is reached* and hence the design of such sections is limited to first yield on the extreme fibre (i.e. triangular stress block) (see Fig. 17.11).

Fig. 17.10 Design stresses for compact sections

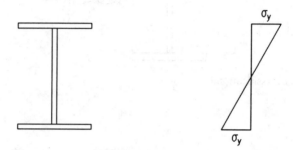

Fig. 17.11 Design stresses for non-compact sections

These classifications can best be illustrated by considering a beam in fully restrained bending. This gives the idealized moment capacity of the cross sections as:

compact sections $\quad M_1 = Z_{pe}\sigma_y/\gamma_m\gamma_{f3}$
non-compact sections $\quad M_1 = Z\sigma_y/\gamma_m\gamma_{f3}$

where σ_y is the yield stress; Z_{pe} is the plastic modulus; Z is the elastic modulus; γ_m is the partial factor on strength; and γ_{f3} is the partial factor on loads, reflecting uncertainty of loads.

The use of the plastic modulus for compact sections does not imply that plastic analysis can be employed. In fact, it is specifically excluded by BS 5400: Part 3. The achievement of a rectangular stress block does not necessarily mean that there has been redistribution of moments along the member.

17.6.1.3 Lateral torsional buckling (clause 9.6)

Where a member has portions of its length with unrestrained elements in compression, lateral torsional buckling may occur (see Fig. 17.12 and section 16.3.6).

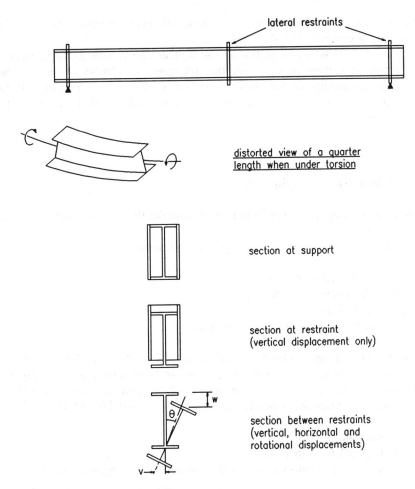

Fig. 17.12 Lateral torsional buckling

The Code deals with these effects by the use of a slenderness parameter λ_{LT} which is a modified form of the ratio L_e/r_y.

$$\lambda_{LT} = \frac{L_e}{r_y} K_4 \eta v$$

where L_e is the effective length for lateral torsional buckling; r_y is the radius of gyration of the whole beam about its yy axis; K_4 is a torsion factor (taken as 0.9 for rolled I or channel sections or 1.0 for all other sections); η is a factor that takes account of moment gradient, i.e. shape of moment diagram; it takes account of the fact that a uniform moment over the unrestrained length will cause buckling more readily than a non-uniform moment; and v is a torsion factor dependent on the shape of the beam.

The process of calculating the value of the limiting compressive stress, σ_{lc}, from the value λ_{LT} is described in the Code.

Portions of beams between restraints can deflect downwards and sideways and rotate. Failure may then occur before the full moment capacity of the section is reached. The possibility of this type of failure is dictated by the unrestrained length of the compression flange, the cross section geometry of the beam and the moment gradient.

17.6.1.4 Shear buckling (clause 9.9.2)

This phenomenon becomes significant in webs with depth-to-thickness ratios greater than about 80.

The actual shear strength of a web panel is dependent on

(1) Yield stress (suitably factored)
(2) Depth-to-thickness ratio of web
(3) Spacing of stiffeners
(4) Conditions of restraint provided by flanges.

The web of a plate girder between stiffeners acts similarly to the diagonal of a Pratt truss (this phenomenon known as tension field action has already been described in section 17.4.4.2). The theory stipulates that the web will resist the applied loading in *three* successively occurring stages (see Fig. 17.4):

Stage 1 A pure shear field
Stage 2 A diagonal tension field
Stage 3 A collapse mechanism due to the formation of hinges in the flanges.

In BS 5400: Part 3, the theory explained in section 17.4.4.2 is slightly modified for the design of steel bridges. In slender web panels, the limit of stage 1, i.e. the pure shear field, is reached when the applied shear stress reaches the elastic critical stress, τ_c. To allow for the effects of residual stresses and initial imperfections in stocky plates, a limiting value, τ_1, is taken less than τ_c when τ_c is greater than 0.8 times the yield stress in shear (τ_y), and equal to τ_y when τ_c exceeds 1.5 times τ_y.

The elastic critical stress for a plate loaded in shear, τ_c, is given by:

$$\tau_c = K \frac{\pi^2 E}{12 (1-\mu^2)} \left(\frac{t_w}{d_{we}}\right)^2$$

where

$$K = 5.34 + \frac{4}{\phi^2} \quad \text{when } \phi \geq 1$$

and $K = 4 + \dfrac{5.34}{\phi^2}$ when $\phi < 1$

$$\phi = \text{aspect ratio} = \dfrac{a}{d_{we}}$$

The criteria outlined previously have been incorporated in computing the value of τ_c:

$$\dfrac{\tau_c}{\tau_y} = \dfrac{904}{\beta^2} \quad \text{when } \beta \geq 33.62$$
$$= 1 \quad \text{when } \beta \leq 24.55$$
$$= 1.54 - 0.022\beta \quad \text{when } 24.55 < \beta < 33.62$$

where $\beta = \dfrac{\lambda}{K^{\frac{1}{2}}}$

and $\lambda = \dfrac{d_{we}}{t_w}\left(\dfrac{\sigma_{yw}}{355}\right)^{\frac{1}{2}}$

In stage 2, a tensile membrane stress field develops in the panel, the direction of which does not necessarily coincide with the diagonal of the panel. The maximum shear capacity is reached in stage 3, when the pure shear stress, τ_1, of stage 1 and the membrane stress, σ_t, of stage 2 cause yielding of the panel according to the von Mises yield criterion *and* plastic hinges are formed in the flanges.

The magnitude of the membrane tensile stress, σ_t, in terms of its *assumed* direction θ and the first stage limiting stress, τ_1, is given by:

$$\dfrac{\sigma_t}{\tau_y} = \left[3 + (2.25 \sin^2 2\theta - 3)\left(\dfrac{\tau_c}{\tau_y}\right)^2\right]^{\frac{1}{2}} - 1.5 \dfrac{\tau_c}{\tau_y} \sin 2\theta$$

Adding the resistance in the three stages, the ultimate shear capacity is obtained as:

$$\dfrac{\tau_u}{\tau_y} = \left[\dfrac{\tau_c}{\tau_y} + 5.264 \sin\theta \left(m_{fw}\dfrac{\sigma_t}{\tau_y}\right)^{\frac{1}{2}} + \dfrac{\sigma_t}{\tau_y}(\cot\theta - \phi)\sin^2\theta\right]$$

when $m_{fw} \leq \dfrac{\phi^2}{4\sqrt{3}}\dfrac{\sigma_t}{\tau_y}\sin^2\theta$

$$\dfrac{\tau_u}{\tau_y} = \left[\dfrac{4\sqrt{3}\, m_{fw}}{\phi} + \dfrac{\sigma_t}{2\tau_y}\sin^2\theta + \dfrac{\tau_c}{\tau_y}\right]$$

when $m_{fw} > \dfrac{\phi^2}{4\sqrt{3}}\dfrac{\sigma_t}{\tau_y}\sin^2\theta$

To provide an added measure of safety in respect of slender webs, the above values are multiplied by a varying correction factor, f, to obtain the limiting shear stress.

$$f = 1 \quad \text{when } \lambda \le 56$$

$$f = \frac{1.15}{1.35} \quad \text{when } 156 \le \lambda$$

$$f = \frac{1.15}{1.15 + 0.002(\lambda - 56)} \quad \text{when } 56 < \lambda < 156$$

The value of $\dfrac{\tau_l}{\tau_y}$ is taken as the lower of $\left(\dfrac{\tau_u}{\tau_y}\right)$, computed as given above, and 1.0.

The term m_{fw} in the above equations is a non-dimensional representation of plastic moment of resistance of the flange (taking the smaller value at the top and bottom flanges, ignoring any concrete).

$$m_{fw} = \frac{M_p}{d_{we}^2 \, t_w \, \sigma_{yw}}$$

$$= \frac{\sigma_{yf} \, b_{fe} \, t_f^2}{2 d_{we}^2 \, t_w \, \sigma_{yw}}$$

Values of $\left(\dfrac{\tau_l}{\tau_y}\right)$ versus λ are plotted in Figures 11–17 of the Code corresponding to various values of m_{fw} and ϕ. These can be used directly by designers.

The term b_{fe} is the width of flange associated with the web and is limited to

$$10 t_f \left(\frac{355}{\sigma_{yf}}\right)^{\frac{1}{2}}$$

Where two flanges are unequal, the value of m_{fw} is conservatively taken as that corresponding to the weaker flange; moreover only a section symmetric about the mid-plane is taken as effective and any portion of the flange plate outside this plane of symmetry is ignored so that complexities due to the torsion of the flange are eliminated.

The above procedure has to be repeated for several values of θ and the highest value of τ_u/τ_y is to be used. From parametric studies it has been established that θ is never less than $(1/3)\cot^{-1}\phi$ or more than $(4/3)\cot^{-1}\phi$.

When tension field action is used, consideration must be given to the anchorage of the tension field forces created in the end panels and special procedures must be adopted for designing the end stiffener.

17.6.1.5 Combined bending and shear (clause 9.9.3)

The Code has simplified the procedure for girders *without* longitudinal stiffeners by allowing shear and bending capacities to be calculated independently and then combined by employing an interaction relationship on the basis given below:

(1) The bending capacity of the whole section is determined with and without contribution from the web (M_1 and M_R respectively).
(2) The shear capacity using the tension field theory discussed above is determined with and without contribution from the flanges (V_D and V_R respectively).
(3) The bending and shear capacities without any contribution from the web and flanges respectively (M_R and V_R) can be mobilized simultaneously.
(4) The pure bending capacity of the whole section, M_D, can be obtained even when there is a coincident shear on the section provided the latter is less than $\tfrac{1}{2}V_R$. Similarly the theoretical design shear capacity, V_D, of the whole beam section is attained even if the coexisting bending moment is not greater than $\tfrac{1}{2}M_R$.

The interaction relationship is linear between this set of values of shear force and bending moment and is shown graphically in Fig. 17.13. In this figure, V_D and M_D represent capacity in shear and bending; V_R is the shear capacity ignoring flange contribution; and M_R is the bending capacity ignoring web contribution, i.e. flange capacity × depth.

The important feature from the designer's point of view is that full values of calculated bending capacity can be utilized in the presence of a moderate magnitude of shear.

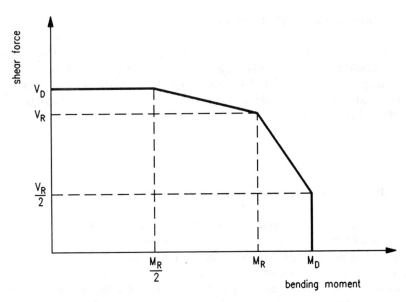

Fig. 17.13 Interaction between shear and bending resistances

For girders with longitudinal stiffeners, account is taken of the combined effects of bending and shear by the comparison of the stresses in the web panels with the relevant critical buckling strength of the panel. The buckling coefficients are based upon large-deflection elastic-plastic finite element studies. An interaction expression is then used which is based on an equivalent stress check. This makes suitable allowance for webs with longitudinal stiffeners and having large depth-to-thickness ratios: no reduction need be made to account for local buckling.

17.6.1.6 Bearing stiffeners (clause 9.14)

BS 5400: Part 3 requires such stiffeners to be provided at supports. The design forces to be applied to the stiffener include the direct reaction from the bearing (less the force component in the flange if the beam soffit is haunched), a force to account for the de-stabilizing effect of the web, forces from any local cross bracing (including forces from restraint system) and a force due to tension field. The derivation of these is far more precise and lengthy than BS 5950: Part 1 requires, largely because plate girders in bridges are usually part of a much more complex structural system than in buildings.

Bearing stiffeners are almost always placed symmetrically about the web centre-line. Stiffeners that are symmetrical about the axis perpendicular to the web, i.e. flats or tees rather than angles, are preferred.

17.6.1.7 Intermediate stiffeners (clause 9.13)

Intermediate stiffeners are usually placed on one side of the web only. Standard flat sizes should be used. BS 5400: Part 3 does not prescribe stiffness criteria for intermediate stiffeners (which have an associated length of web), but it does prescribe design forces for which these stiffeners are designed similar to those outlined for bearing stiffeners.

17.6.2 Design of beams at the serviceability limit state

In practice, due to the proportions of partial safety factors at the ultimate and serviceability limits, the serviceability check on the steel section is automatically satisfied if designs are satisfactory at the ultimate stage.

The design of beams is carried out using an elastic distribution throughout the cross section in a manner generally similar to that used at the ultimate limit but with different partial safety factors, i.e. different values of γ_{fL}, γ_m and γ_{f3}. The determination of effective section must take account of shear lag, where appropriate.

Stresses should be checked at critical points in the steel member to ensure that no permanent deformation due to yielding takes place.[7,8,9] Plastic moment capacity is not considered at the serviceability limit state.

In composite construction, crack widths of concrete deck may often govern the design at serceability limit state. For non-compact sections, because the analysis at the ultimate limit state has been carried out using a triangular stress distribution (i.e. an elastic distribution), the serviceability limit check is always satisfied for bending analysis, as these factors are less than the ultimate limit state. The Code lists the clauses which require checking at both limit states.

17.6.3 Fatigue

Fatigue should be checked with reference to BS 5400: Part 10. Generally it is only bracing connections, stiffeners and shear connectors and their welding to the girders that have to be checked for fatigue. The details of the connection have a significant effect on the fatigue life and by careful detailing fatigue may be 'designed out' of the bridge.

17.6.4 Design format

One of the significant features of BS 5400: Part 3 is that plastic redistribution across the section is now permitted in certain circumstances prior to the attainment of the ultimate limit state. Such redistributions are, of course, much less than those permitted in simple plastic design for buildings, but they can give a considerable enhancement in strength.

For example, design bending strength is based on plastic section modulus for compact sections, and the basic shear strength is based on rules which take account of tension field action, which only develops in the presence of considerable plasticity.

In addition, the interaction between shear strength and bending strength of a cross section is empirically based and also implicitly assumes that plastic strains may occur prior to the attainment of design strength.

Finally, the design strength of compact sections that are built in stages assumes that a redistribution of stresses may take place within the cross-section; thus the ultimate limit state check for the completed structure is simply to ensure that the bending resistance (given by the limiting stress times the plastic section modulus divided by partial factors) is greater than the maximum design moment (obtained by summing all the moments due to the various design loads).

This recognition of the reserve strength beyond first yield is limited to situations where plasticity may be permitted safely. The principal limitations are as follows:

Plate girders

(1) Plastic section moduli may only be used for compact sections, i.e. those that can sustain local compressive yielding without any local buckling.
(2) Where the structure is built in stages and is non-compact, plasticity is permitted implicitly only in considering the interaction between bending and shear strength.
(3) Where longitudinal stiffeners are present, neither plastic section modulus nor plastic bending/shear interaction may be used.
(4) If flanges are curved in elevation, similar restrictions to (3) apply.

References to Chapter 17

1. British Standards Institution (1990) *Structural use of steelwork in building. Part 1: Code of practice for design in simple and continuous construction: hot rolled sections*. BS 5950, BSI, London.
2. British Standards Institution (1982) *Steel, concrete and composite bridges. Part 3: Code of practice for the design of steel bridges*. BS 5400, BSI, London.
3. Owens, G.W. (1989) *Design of Fabricated Composite Beams in Buildings*. The Steel Construction Institute, Ascot, Berks.
4. Evans H.R. (1988) Design of plate girders. In *Introduction to Steelwork Design to BS 5950: Part 1*, pp. 12.1–12.10. The Steel Construction Institute, Ascot, Berks.
5. Lawson R.M., Rackham J.W. (1989) *Design for Openings in Webs of Composite Beams*. The Steel Construction Institute, Ascot, Berks.
6. Chaterjee S. (1981) Design of webs and stiffeners in plate and box girders. In *The Design of Steel Bridges* (Ed. by K.C. Rockey & H.R. Evans), Chapter 11. Granada.
7. Iles D.C. (1989) *Design Guide for Continuous Composite Bridges 1: Compact Sections*. SCI Publication 065, The Steel Construction Institute, Ascot, Berks.
8. Iles D.C. (1989) *Design Guide for Continuous Composite Bridges 2: Non-Compact Sections*. SCI Publication 066, The Steel Construction Institute, Ascot, Berks.
9. Iles D.C. (1991) *Design Guide for Simply Supported Composite Bridges*. The Steel Construction Institute, Ascot, Berks.

Further reading for Chapter 17

References 3, 4, 7, 8 and 9 are recommended wider reading.

A worked example follows which is relevant to Chapter 17.

The Steel Construction Institute Silwood Park, Ascot, Berks SL5 7QN	Subject **PLATE GIRDERS**		Chapter ref. **17**
	Design code **BS5950: Part 1**	Made by **RN** Checked by **GWO**	Sheet no. **1**

The girder shown is fully restrained throughout its length. For the loading shown design a stiffened plate girder in Grade 43 steel. Girder depth is unrestricted.

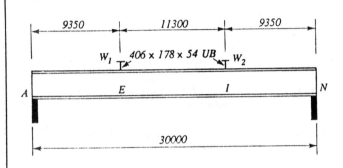

$w = w_d + w_i$

$W_1 = W_{1d} + W_{1i}$

$W_2 = W_{2d} + W_{2i}$

Loading

Dead loads:

UDL	w_d	=	20 kN/m
Point load	W_{1d}	=	200 kN
Point load	W_{2d}	=	200 kN

Imposed loads:

UDL	w_i	=	40 kN/m
Point load	W_{1i}	=	300 kN
Point load	W_{2i}	=	300 kN

The Steel Construction Institute Silwood Park, Ascot, Berks SL5 7QN	Subject **PLATE GIRDERS**	Chapter ref. **17**
	Design code **BS5950: Part 1**	Made by **RN**	Sheet no. **2**
		Checked by **GWO**	

Load factors

Dead load factor γ_{fd} = 1.4

Imposed load factor γ_{fi} = 1.6

2.4.1.1
Table 2

Factored loads

w' = $w_d \times \gamma_{fd} + w_i \times \gamma_{fi}$ = $20 \times 1.4 + 40 \times 1.6$
= 92 kN/m

W_1' = $W_{1d} \times \gamma_{fd} + W_{1i} \times \gamma_{fi}$ = $200 \times 1.4 + 300 \times 1.6$
= 760 kN

W_2' = $W_{2d} \times \gamma_{fd} + W_{2i} \times \gamma_{fi}$ = $200 \times 1.4 + 300 \times 1.6$
= 760 kN

The design shear forces and moments are shown below:

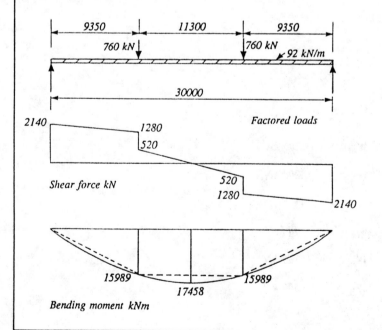

The Steel Construction Institute Silwood Park, Ascot, Berks SL5 7QN	Subject **PLATE GIRDERS**		Chapter ref. **17**
	Design code BS5950: Part 1	Made by **RN** Checked by **GWO**	Sheet no. **3**

Initial sizing of plate girder

The recommended span/depth ratio for simply supported non-composite girders varies between 12 for short span girders and 20 for long span girders. Here the depth will be assumed to be one-fifteenth of the span.

Depth, $\quad D \approx \dfrac{span}{15} \;=\; \dfrac{30000}{15} \;=\; 2000 \; mm$

Single flange area, (assuming $p_y = 255 \; N/mm^2$, see Table 6)

$A_f \approx \dfrac{M_{max}}{D \, p_y} \;=\; \dfrac{17458}{2000 \times 255} \times 10^6 \qquad\qquad 4.4.4.2(a)$

$\therefore A_f \approx 34232 \; mm^2$

For non-composite girders, the flange width chosen is usually within a range of 0.3 to 0.5 of the depth of the section (0.4 is more common). Suggest $700 \times 50 = 35000 \; mm^2$.

Minimum web thickness for plate girders in buildings usually varies between 10 mm for girders up to 1200 mm deep to 20 mm for those of 2500 mm deep. Here the web thickness t, will be chosen as 14 mm.

Try girder with flanges of 700 mm × 50 mm thick and web 2000 mm × 14 mm thick

Section classification

Flange

For T = 50 mm $\hfill$ Table 6

$p_y \;=\; p_{yf} \qquad\qquad p_{yf} \;=\; 255 \; N/mm^2$

$\epsilon \;=\; \left[\dfrac{275}{p_y} \right]^{½} \qquad\qquad \epsilon \;=\; \left[\dfrac{275}{255} \right]^{½} = 1.04 \qquad$ Table 7 Note 3

The Steel Construction Institute Silwood Park, Ascot, Berks SL5 7QN	Subject **PLATE GIRDERS**	Chapter ref. **17**
	Design code BS5950: Part 1	Made by RN Checked by GWO	Sheet no. **4**

Outstand, $b = \dfrac{B - t}{2} = \dfrac{700 - 14}{2} = 343$ mm   Fig 3

$\dfrac{b}{T} = \dfrac{343}{50} = 6.86$

Compact limiting value of b/T for outstand of welded section is $8.5\,\epsilon$ Table 7

$8.5\,\epsilon = 8.5 \times 1.04 = 8.84$

$6.86 < 8.84$

∴   Flange is compact

Web

For $t = 14$ mm   Table 6

$p_y = p_{yw}$ $p_{yw} = 275$ N/mm^2

$\epsilon = \left[\dfrac{275}{p_y}\right]^{1/2}$ $\epsilon = \left[\dfrac{275}{275}\right]^{1/2} = 1.00$ Table 7 Note 3

$\dfrac{d}{t} = \dfrac{2000}{14} = 142.8$

$120\,\epsilon = 120$

$142.8 > 120$ Table 7

∴   the web is slender

since $\dfrac{d}{t} > 63\epsilon$ $142.8 > 63$ Table 7 Note 2

the web must be checked for shear buckling as given in 4.4.4.2

The Steel Construction Institute Silwood Park, Ascot, Berks SL5 7QN	Subject **PLATE GIRDERS**		Chapter ref. **17**
	Design code BS5950: Part 1	Made by RN Checked by GWO	Sheet no. 5

Dimension of webs and flanges

4.4.2

Assuming stiffener spacing, $a > d$

4.4.2.2

$t \geq \dfrac{d}{250}$ $\dfrac{d}{250} = \dfrac{2000}{250} = 8.0\ mm$

4.4.2.2(b)(1)

since $t > \dfrac{d}{250}$ $14 > 8.0$

web is adequate for serviceability

Assuming stiffener spacing, $a > 1.5d$

4.4.2.3

$t \geq \dfrac{d}{250} \times \dfrac{p_{yf}}{345} = 8.0 \times \dfrac{255}{345} = 5.9\ mm$

4.4.2.3(b)(1)

since $t = 14\ mm > 5.9\ mm$

the web is adequate to avoid flange buckling into web

For sections with slender webs (and flanges which are not slender) three methods of design are given. Method (2) will be used in this example. The moment is assumed to be resisted by the flanges alone and the web designed for shear only.

4.4.4.2

Moment capacity of the section ignoring the web

$M_c = p_{yf} \times A_f \times h_s$

$p_{yf} = 255\ N/mm^2$

h_s = distance between centroid of flanges

$h_s = d + T = 2000 + 50 = 2050\ mm$

$A_f = B \times T = 700 \times 50 = 35000\ mm^2$

The Steel Construction Institute Silwood Park, Ascot, Berks SL5 7QN	Subject **PLATE GIRDERS**	Chapter ref. **17**
	Design code **BS5950: Part 1** — Made by **RN** / Checked by **GWO**	Sheet no. **6**

∴ $M_c = 255 \times 35000 \times 2050 \times 10^{-6}$

$M_c = 18296$ kNm

Maximum design moment, $M = 18296$ kNm

Since M_c > Moment at G $\quad 18296 > 17458$ kNm

section is adequate for carrying the moment

<u>Web design for shear only</u>   4.4.5

Webs with intermediate stiffeners may be designed according to 4.4.5.3 or 4.4.5.4.

Although the Code allows three alternative methods for providing stiffeners (see Clauses 4.4.5.4.2 and 4.4.5.4.3), the most commonly employed method consists of designing end panels without accounting for tension field action (Clause 4.4.5.4.2). This method is illustrated in this worked example.

Try the stiffener spacing shown below:

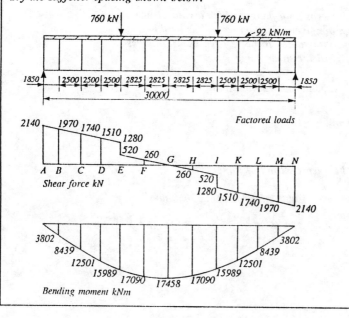

	Subject	Chapter ref.
The Steel Construction Institute Silwood Park, Ascot, Berks SL5 7QN	**PLATE GIRDERS**	**17**
	Design code BS5950: Part 1 Made by **RN** Checked by **GWO**	Sheet no. **7**

<u>Consider end panel AB</u>   4.4.5.4.2

$d \;=\; 2000 \text{ mm}$

$t \;=\; 14 \text{ mm}$

Maximum shear stress in the panel

$f_v \;=\; \dfrac{F_{v,A}}{dt}$

$f_v \;=\; \dfrac{2140 \times 10^3}{2000 \times 14} \;=\; 76.43 \text{ N/mm}^2$

$\dfrac{a}{d} \;=\; \dfrac{1850}{2000} \;=\; 0.925$

$\dfrac{d}{t} \;=\; \dfrac{2000}{14} \;=\; 142.8$

$p_y \;=\; 275 \text{ N/mm}^2$

Critical shear strength, $q_{cr} \;=\; 94.22 \text{ N/mm}^2$ Table 21b

Since $f_v < q_{cr}$ $76.43 < 94.22$

∴   <u>Tension field action need not be utilized for design.</u>

End panel AB should also be checked as a beam spanning between the flanges of the girder capable of resisting a shear force R_{tf} and a moment M_{tf} due to anchor forces.   Fig 5

$R_{tf} \;=\; \dfrac{H_q}{2}$ 4.4.5.4.4

$M_{tf} \;=\; \dfrac{H_q \, d}{10}$

478   Worked example

The Steel Construction Institute Silwood Park, Ascot, Berks SL5 7QN	Subject PLATE GIRDERS		Chapter ref. 17
	Design code BS5950: Part 1	Made by RN Checked by GWO	Sheet no. 8

$H_q = 0.75 \, d \, t \, p_y \left[1 - \dfrac{q_{cr}}{0.6 \, p_y}\right]^{1/2}$

$q_{cr} = 94.22 \text{ N/mm}^2$

$H_q = 0.75 \times 2000 \times 14 \times 275 \times \left[1 - \dfrac{94.22}{0.6 \times 275}\right]^{1/2} \times 10^{-3}$

$H_q = 3782.4 \text{ kN}$

$R_{tf} = \dfrac{3782.4}{2} = 1891.2 \text{ kN}$

Although panel AB and its bounding transverse stiffeners act as a beam, spanning vertically between the girder flanges to resist the shear and bending moment arising from the tension field pull of panel BC, the effect of the bounding transverse stiffeners will be disregarded in the following calculations. In cases where the moment M_{tf} arising from the anchor forces is higher than the moment capacity of the web alone, the moment resistance of the bounding stiffeners should be included in the design.

Shear capacity of the end panel AB

$P_v = 0.6 \, p_{yw} \, A_v$ 4.2.3

$A_v = t \, a$

'a' is the distance between the transverse stiffeners making up the end panel

$a = 1850 \text{ mm}$

$P_v = 0.6 \times 275 \times 14 \times 1850 \times 10^{-3}$

$P_v = 4273.5 \text{ kN}$

The Steel Construction Institute Silwood Park, Ascot, Berks SL5 7QN	Subject **PLATE GIRDERS**	Chapter ref. **17**
	Design code BS5950: Part 1 · Made by RN · Checked by GWO	Sheet no. **9**

Since $R_{tf} < P_v$ 1891.2 < 4273.5

the end panel is capable of carrying the shear force

$$M_{tf} = \frac{H_q \, d}{10}$$

$$M_{tf} = \frac{3782.4 \times 2000}{10} \times 10^{-3} = 756.5 \text{ kNm}$$

<u>Moment capacity of the end panel AB, M_q</u>

$$M_q = \frac{I}{y} p_y$$

$$I = \frac{1}{12} t \, a^3 \quad \text{(Stiffeners \underline{not} included)}$$

$$y = \frac{a}{2} = 925 \qquad a = 1850 \text{ mm}$$

$$I = \frac{1}{12} \times 14 \times 1850^3$$

$$I = 738689.6 \times 10^4 \text{ mm}^4$$

$$y = \frac{1850}{2} = 925 \text{ mm}$$

$$M_q = \frac{738689.6 \times 10^4}{925} \times 275 \times 10^{-6}$$

$$M_q = 2196 \text{ kNm}$$

Since $M_{tf} < M_q$ 756.5 < 2196

the end panel is capable of carrying the bending moment.

The Steel Construction Institute Silwood Park, Ascot, Berks SL5 7QN	Subject **PLATE GIRDERS**		Chapter ref. **17**
	Design code BS5950: Part 1	Made by **RN** Checked by **GWO**	Sheet no. **10**

Consider panel BC

Panel BC will be designed utilizing tension field action   Fig 5

d = 2000 mm

t = 14 mm

Maximum shear stress in the panel

$$f_v = \frac{F_{v,B}}{dt}$$

$$f_v = \frac{1970}{2000 \times 14} \times 10^3 = 70.36 \text{ N/mm}^2$$

$$\frac{a}{d} = \frac{2500}{2000} = 1.25$$

$$\frac{d}{t} = \frac{2000}{14} = 142.8$$

p_y = 275 N/mm²

q_b = 105.65 N/mm²   Table 22b

Since $q_b > f_v$ 105.65 > 70.36

panel BC is safe against shear buckling

Load bearing stiffener at A   Fig 5

Design should be made for a compression force due to bearing plus the compression force due to the moment M_{tf}.   4.4.5.4.2

Design force due to bearing, F_b = 2140 kN

Force due to moment M_{tf}, F_m

$$F_m = \frac{M_{tf}}{a} = \frac{756.5}{1850} \times 10^3 = 409 \text{ kN}$$

The Steel Construction Institute Silwood Park, Ascot, Berks SL5 7QN	Subject **PLATE GIRDERS**	Chapter ref. **17**
	Design code **BS5950: Part 1** Made by **RN** Checked by **GWO**	Sheet no. **11**

Total compression force, F_x = $F_b + F_m$

F_x = $2140 + 409$ = 2549 kN

Area of stiffener in contact with the flange   4.5.4.2

$$A > \frac{0.8 \, F_x}{p_{ys}}$$

$$\frac{0.8 \, F_x}{p_{ys}} = \frac{0.8 \times 2549}{265} \times 10^3 = 7695 \text{ mm}^2$$

Try stiffeners consisting of 2 flats 240 mm × 22 mm thick

For t_s = 22 mm < 40 mm p_{ys} = 265 N/mm²   Table 6

Allow 15 mm to cope for the web/flange weld

A = $225 \times 22 \times 2 = 9900$ mm² > 7695 mm²

∴   Bearing check is OK

Check outstands   4.5.1.2

Outstands from the face of the web should not be greater than $19 \, t_s \, \epsilon$

$$\epsilon = \left[\frac{275}{p_y}\right]^{1/2} = \left[\frac{275}{265}\right]^{1/2} = 1.02 \quad \text{Table 8}$$

Outstand, b_s = 240 mm

$19 \, t_s \, \epsilon$ = $19 \times 22 \times 1.02$ = 425 mm

$b_s < 19 \, t_s \, \epsilon$ 240 < 425   OK

482 Worked example

The Steel Construction Institute Silwood Park, Ascot, Berks SL5 7QN		Subject **PLATE GIRDERS**	Chapter ref. **17**
	Design code *BS5950: Part 1*	Made by **RN** Checked by **GWO**	Sheet no. **12**

Check if $\quad b_s < 13\, t_s\, \epsilon$

$13\, t_s\, \epsilon \quad = \quad 13 \times 22 \times 1.02 \quad = \quad 291 \text{ mm}$

$b_s < 13\, t_s\, \epsilon \qquad 240 < 291 \therefore \text{ Satisfactory}$

<u>Check stiffener for buckling</u> 4.5.1.5

Although the effective section from 4.5.1.5 may be used for providing buckling resistance of stiffeners, here the buckling resistance due to the web will be disregarded. Thus, no moment will occur on the stiffener coming from the eccentricity in the load due to an unsymmetric effective section. But if the required stiffener cross section using this method comes out to be very massive, the effective section should be regarded as a T section and the stiffener designed accordingly.

The effective section of the stiffener is shown below

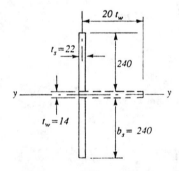

In the above figure:

——————— *shows the effective section that will be used in this example.*

— — — — — *shows the additional area contributed by the web. That section will be disregarded here for simplicity.*

	Subject	Chapter ref.
The Steel Construction Institute Silwood Park, Ascot, Berks SL5 7QN	**PLATE GIRDERS**	**17**
	Design code **BS5950: Part 1**	Made by **RN** Checked by **GWO**	Sheet no. **13**

$I_y = \dfrac{22 \times 494^3}{12} - \dfrac{22 \times 14^3}{12}$

$I_y = 22101 \times 10^4 \text{ mm}^4$

A_E: effective area of the stiffener

$A_E = 240 \times 22 \times 2$

$A_E = 10560 \text{ mm}^2$

$r_y = \left[\dfrac{I}{A}\right]^{1/2}$

$r_y = \left[\dfrac{22101 \times 10^4}{10560}\right]^{1/2}$

$r_y = 144.7 \text{ mm}$

Flange is restrained against rotation in the plane of the stiffener   4.5.1.5(a)

$\therefore L_E = 0.7 L = 0.7 \times 2000 = 1400 \text{ mm}$

$\lambda = \dfrac{L_E}{r_y} = \dfrac{1400}{144.7} = 9.68$

$p_y = 265 - 20 = 245 \text{ N/mm}^2$ when using Table 27   4.7.5

for $p_y = 245$, $\lambda = 9.68$

$p_c = 245 \text{ N/mm}^2$ Table 27c

$\therefore$ Buckling resistance of the stiffener

$P_x = p_c A_E = 245 \times 10560 = 2587.2 \text{ kN}$

Since $F_x < P_x$ $2549 < 2587.2$

stiffener is adequate to resist buckling

The Steel Construction Institute Silwood Park, Ascot, Berks SL5 7QN	Subject **PLATE GIRDERS**	Chapter ref. **17**
	Design code BS5950: Part 1	Made by **RN** Sheet no. **14** Checked by **GWO**

<u>Check stiffener at A as a bearing stiffener</u>   4.5.5

Local capacity of the web

Assume stiff bearing length $b_1 = 0$

$P_{crip} = (b_1 + n_2) \, t \, p_{yw}$ 4.5.3

$n_2 = 2.5 \times 50 \times 2 = 250 \text{ mm}$

$P_{crip} = (0 + 250) \times 14 \times 275 \times 10^{-3} = 962.5 \text{ kN}$

Bearing stiffener should be designed for a force, F_A

$F_A = F_x - P_{crip} = 2549 - 962.5 = 1586.5 \text{ kN}$ 4.5.5

Bearing capacity of the stiffener alone

$P_A = p_{ys} \times A = 265 \times 9900 = 2623.5 \text{ kN}$

Since $F_A < P_A$ $1586.5 < 2623.5$

Stiffener is OK for bearing.

<u>Stiffener at A: Adopt 2 flats 240 mm × 22 mm thick, Grade 43 Steel</u>

<u>Intermediate stiffeners at B,C,D,F & G</u>   4.4.6

Stiffener at B is the most critical one and will be the design criterion

Minimum stiffness   4.4.6.4

$I_s \geq 0.75 \, dt^3$ for $a \geq d\sqrt{2}$

$I_s \geq \dfrac{1.5 \, d^3 t^3}{a^2}$ for $a < d\sqrt{2}$

The Steel Construction Institute Silwood Park, Ascot, Berks SL5 7QN	Subject **PLATE GIRDERS**		Chapter ref. **17**
	Design code BS5950: Part 1	Made by RN Checked by GWO	Sheet no. **15**

$d\sqrt{2}$ = $\sqrt{2} \times 2000$ = 2828 mm

∴ $a < d\sqrt{2}$ 2825 < 2828 (maximum 'a')

Note: Conservatively 't' will be taken as the actual web thickness

$$\frac{1.5 d^3 t^3}{a^2} = \frac{1.5 \times 2000^3 \times 14^3}{2500^2} = 526.8 \times 10^4 \text{ mm}^4$$

(conservatively minimum 'a' is used above)

Try intermediate stiffener 2 flats 80 mm × 13 mm

Note: 13 mm is not a standard thickness but is available

$$I_s = \frac{13 \times 174^3}{12} - \frac{13 \times 14^3}{12} = 570.4 \times 10^4 \text{ mm}^4 \text{ OK}$$

Outstand of the stiffener ≤ $13 t_s \epsilon$ 4.5.1.2

$13 t_s \epsilon$ = 13 × 13 × 1 = 169 mm

Outstand = 80 mm   80 < 169   OK

<u>Buckling check on intermediate stiffeners at B, C, D, F & G</u>   4.4.6.6

Design force on the stiffener

F_q = $V - V_s$ V = 1970 kN

V_s = V_{cr}

V_{cr} = $q_{cr} d t$

$\frac{a}{d}$ = $\frac{2825}{2000}$ = 1.41   ('a' is conservatively taken as the maximum)

The Steel Construction Institute Silwood Park, Ascot, Berks SL5 7QN		Subject **PLATE GIRDERS**		Chapter ref. **17**
		Design code BS5950: Part 1	Made by **RN** Checked by **GWO**	Sheet no. **16**

$\dfrac{d}{t} = \dfrac{2000}{14} = 142.8$

$p_y = 275 \text{ N/mm}^2$

$q_{cr} = 67.89 \text{ N/mm}^2$　　　　　　　　　　　　　　　　　　　　*Table 21b*

$V_{cr} = 67.89 \times 2000 \times 14 \times 10^{-3}$

$V_{cr} = 1900 \text{ kN}$

$F_q = 1970 - 1900 = 70 \text{ kN}$

Buckling resistance of the intermediate stiffener at B, P_q

The effective section is shown below

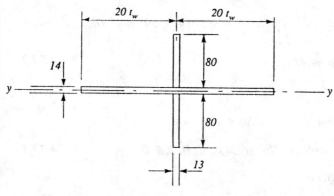

$20 t_w = 20 \times 14 = 280 \text{ mm}$

$I_y = \dfrac{13 \times 174^3}{12} + \dfrac{560 \times 14^3}{12} - \dfrac{13 \times 14^3}{12} = 583.2 \times 10^4 \text{ mm}^4$

$A = 160 \times 13 + 560 \times 14 = 9920 \text{ mm}^2$

$r_y = \left[\dfrac{583.2 \times 10^4}{9920} \right]^{1/2} = 24.25 \text{ mm}$

The Steel Construction Institute Silwood Park, Ascot, Berks SL5 7QN	Subject **PLATE GIRDERS**		Chapter ref. **17**
	Design code **BS5950: Part 1**	Made by **RN** Checked by **GWO**	Sheet no. **17**

L_E = 0.7 × 2000 = 1400 mm

λ = $\dfrac{L_E}{r_y}$ = $\dfrac{1400}{24.25}$ = 57.7

for p_y = 255 N/mm², λ = 57.7   4.7.5

p_c = 192.6 N/mm²   Table 27c

Buckling resistance, P_q = $p_c A$

P_q = 192.6 × 9920 × 10^{-3} = 1910.6 kN

Since $F_q < P_q$ 70 kN < 1910.6 kN

intermediate stiffener at B is safe against buckling.

<u>Intermediate Stiffener at B, C, D, F & G</u>
<u>Adopt 2 flats 80 mm × 13 mm thick, Grade 43 Steel</u>

<u>Stiffener at E</u>

Intermediate stiffener subject to external load

Minimum stiffness calculations: 4.4.6.5

a = 2825 < $d\sqrt{2}$ = 2828

$\therefore I_s \geq \dfrac{1.5\,d^3 t^3}{a^2}$ = 526.8 × 10^4 as calculated for stiffener at B

Try intermediate stiffeners 2 flats 80 mm × 13 mm thick

I_s = 570.4 × 10^4 mm⁴ (from stiffener design at B)

$\therefore$ Satisfactory

The Steel Construction Institute Silwood Park, Ascot, Berks SL5 7QN		Subject **PLATE GIRDERS**		Chapter ref. **17**
		Design code *BS5950: Part 1*	Made by **RN** Checked by **GWO**	Sheet no. **18**

Intermediate stiffener subject to external load 4.4.6.6

$$\frac{F_q - F_x}{P_q} + \frac{F_x}{P_x} + \frac{M_s}{M_{ys}} \leq 1$$

$F_q = V - V_s \qquad V = 1280 \text{ kN}$

$V_s = V_{cr}$

$V_{cr} = q_{cr}\, d\, t$

$\dfrac{a}{d} = \dfrac{2825}{2000} = 1.41$

$\dfrac{d}{t} = \dfrac{2000}{14} = 142.8$

$p_y = 275 \text{ N/mm}^2$

$q_{cr} = 67.89$ Table 21b

$V_{cr} = 67.89 \times 2000 \times 14 \times 10^{-3}$

$V_{cr} = 1900 \text{ kN}$

$F_q = 1280 - 1900 < 0$

F_q is negative

so no tension field action occurs

$\therefore \quad F_q - F_x = 0$ 4.4.6.6

$M_s = 0$

$F_x = 760 \text{ kN}$

The Steel Construction Institute Silwood Park, Ascot, Berks SL5 7QN	Subject **PLATE GIRDERS**	Chapter ref. **17**
	Design code **BS5950: Part 1** Made by **RN** Checked by **GWO**	Sheet no. **19**

Buckling resistance of load carrying stiffener at E, P_x   4.5.1.5

P_x = P_q as calculated for stiffener at B

P_x = 1910.6 kN

Since $F_x < P_x$ 760 < 1910.6

stiffener at E is OK for buckling

Stiffener at E: Adopt flats 80 mm × 13 mm thick, Grade 43 Steel

Web check between stiffeners   4.5.2.2

f_{ed} ≤ p_{ed}

f_{ed} = w'/t = 92/14 = 6.57 N/mm²

When the compression flange is restrained against rotation relative to the web

$$p_{ed} = \left[2.75 + \frac{2}{(a/d)^2}\right] \frac{E}{(d/t)^2}$$

$$p_{ed} = \left[2.75 + \frac{2}{\left(\frac{2825}{2000}\right)^2}\right] \frac{205000}{\left(\frac{2000}{14}\right)^2} \quad \text{(check the largest panel only)}$$

p_{ed} = 37.7 N/mm²

Since $f_{ed} < p_{ed}$ 6.57 < 37.7

the web is OK for all panels.

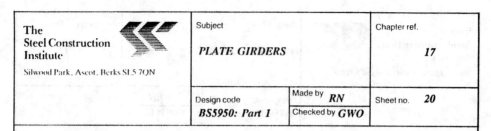

Subject	Chapter ref.
PLATE GIRDERS	**17**

Design code	Made by **RN**	Sheet no. **20**
BS5950: Part 1	Checked by **GWO**	

FINAL GIRDER

All Grade 43 Steel

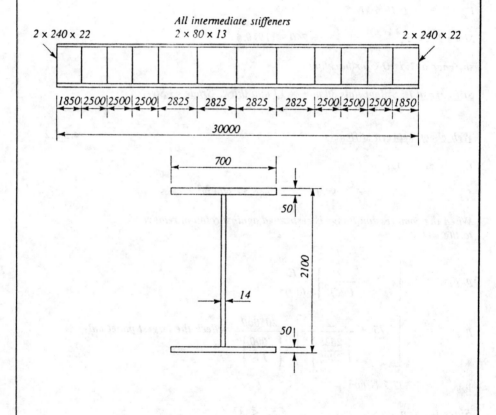

Chapter 18
Members with compression and moments

by DAVID NETHERCOT

18.1 Occurrence of combined loading

Chapters 15 and 16 deal respectively with the design of members subject to compression and bending when these loadings act in isolation. However, in practice a combination of the two effects is frequently present. Figure 18.1 illustrates a number of common examples.

The balance between compression and bending, which may be induced about one or both principal axes, depends on a number of factors, the most important ones being the type of structure, the form of the applied loading, the member's location in the structure and the way in which the connections between the members function.

For building frames designed according to the principles of simple construction, it is customary to regard column moments as being produced only by beam reactions acting through notional eccentricities as illustrated in Fig. 18.2. Thus, column axial load is accumulated down the building but column moments are only ever generated by the floor levels under consideration with the result that they typically contain increasing ratios of compression to moment. Many columns are therefore designed for high axial loads but rather low moments. Corner columns suffer bending about both axes, but may well carry less axial load; edge columns are subject to bending about at least one axis; and internal columns may, if both the beam framing arrangements and the loading are balanced, be designed for axial load only.

Conversely the columns of portal frames are required to carry high moments in the plane of the frame but relatively low axial loads, unless directly supporting cranes. Rafters also attract some small axial load, particularly when wind loading is being considered. Portals employ rigid connections between members permitting the transfer of moments around the frame. Similarly, multi-storey frames designed on the basis of rigid beam-to-column connections are likely to contain columns with large moments.

Members required to carry combined compression and moments are not restricted to rectangular building frames. Although trusses are often designed on the basis that member centrelines intersect at the joints, this is not always possible and the resulting eccentricities induce some moments in the predominantly axially loaded members. Sometimes the transfer of loads from secondary members into the main

492 Members with compression and moments

booms of trusses is arranged so that they do not coincide with joints, leading to beam action between joints being superimposed on the compression produced by overall bending.

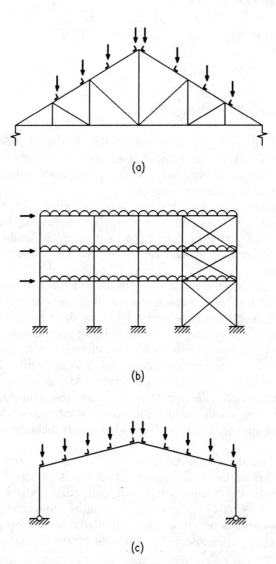

Fig. 18.1 Occurrence of beam-columns in different types of steel frames. (a) Roof truss – top chord members subject to bending from purlin loads and compression due to overall bending. (b) Simple framing – columns subject to bending from eccentric beam reactions and compression due to gravity loading. (c) Portal frame – rafters and columns subject to bending and compression due to frame action

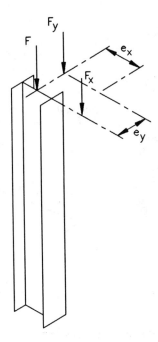

Fig. 18.2 Loading on a beam-column in a 'simply-designed' frame

18.2 Types of response − interaction

It is customary in many parts of the world, but not in the UK, to designate members subject to combined compression and bending as 'beam-columns'. The term is helpful in appreciating member response because the combination of loads produces a combination of effects incorporating aspects of the behaviour of the two extreme examples: a beam carrying only bending loads and a column carrying only compressive load. This then leads naturally to the concept of interaction as the basis for design, an approach in which the proportions of the member's resistance to each component load type are combined using diagrams or formulae. Figure 18.3 illustrates the concept in general terms for cases of two and three separate load components. Any combination is represented by a point on the diagram and an increasing set of loads with fixed ratios between the components corresponds to a straight line starting from the origin. Points that fall inside the boundary given by the design condition are safe, those that fall on the boundary just meet the design condition and those that lie outside the boundary represent an unsafe load combination. For the two load component case if one load type is fixed and the maximum safe value of the other is required, the vertical co-ordinate corresponding to the specified load is first located; projecting horizontally to meet the design boundary the horizontal co-ordinate can be read off.

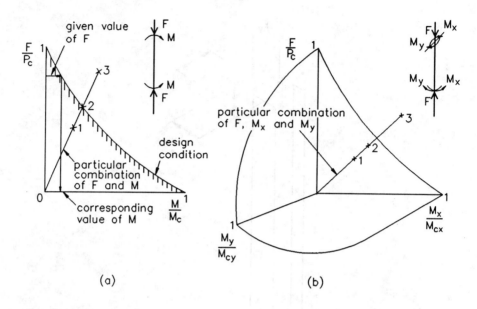

point 1 — safe
point 2 — design condition just satisfied
point 3 — unsafe

Fig. 18.3 Concept of interaction diagrams for combined loading: (a) two-dimensional, (b) three-dimensional

The exact version of Fig. 18.3 appropriate for a design basis depends upon several factors which include the form of the applied loading, the type of response that is possible, the member slenderness and the cross-sectional shape. Design methods for beam-columns must therefore seek to balance the conflicting requirements of rigour, which would try to adjust the form of the design boundary of Fig. 18.3 to reflect the influence of each of these factors, and simplicity. However, some appreciation of the role of each factor is necessary if even the simplest design approach is to be properly appreciated and applied.

The importance of member slenderness may be appreciated readily with reference to the two-dimensional example illustrated in Fig. 18.4. The member is loaded by compression plus equal and opposite end moments and is assumed to respond simply by deflecting in the plane of the applied loading. Under the action of the applied moments bending occurs leading to a lateral deflection v. The moment at any point within the length comprises two components: a constant primary moment M due to the applied end moments plus a secondary moment Fv due to the axial load F acting through the lateral deflection v. Summing the effects of compression and bending gives

$$F/P_c + M_{max}/M_c = 1.0 \qquad (18.1)$$

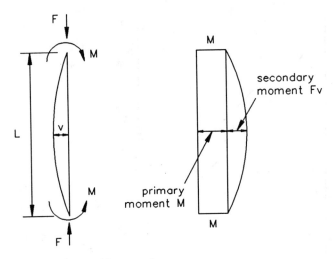

Fig. 18.4 In-plane behaviour of beam-columns

in which P_c and M_c are the resistance as a strut and a beam respectively and M_{max} is the total moment.

Analysis of beam-column problems shows that M_{max} may be closely approximated by

$$M_{max} = M/(1 - F/P_{cr}) \tag{18.2}$$

in which $P_{cr} = \pi^2 EI/L^2$ is the elastic critical load.

Combining these two expressions gives

$$F/P_c + M/(1 - F/P_{cr})M_c = 1.0 \tag{18.3}$$

Figure 18.5 shows how this expression plots in an increasingly concave fashion as member slenderness increases and the amplification of the primary moments M becomes more significant due to reductions in P_{cr}. Clearly if secondary moments are neglected, and the member is designed on the basis of summing the effects of F and M without allowing for their interaction, then an unsafe result is obtained.

If the member can respond only by deforming in the plane of the applied bending, the foregoing is an adequate representation of that response and so can be used as the basis for design. However, more complex forms of response are also possible. Figure 18.6 illustrates an I-section column subject to compression and bending about its major axis. Reference to Chapters 15 and 16 on columns and beams indicates that either the compression or the bending if acting alone would induce failure about the minor axis, in the first case by simple flexural buckling at a load P_{cy} and in the second by lateral–torsional buckling at a load M_b. Both tests and rigorous analysis confirm that the combination of loads will

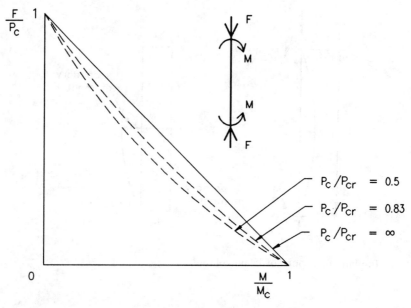

Fig. 18.5 Effect of slenderness on form of interaction according to Equation (18.3)

also produce a minor axis failure. Noting the presence of the amplification effect of the axial load acting through the bending deflections in the plane of the web and not the out-of-plane deformations associated with the eventual failure mode, leads therefore to a modified form of Equation (18.3) as an interaction equation that might form a suitable basis for design

$$F/P_{cy} + M/(1 - F/P_{crx})M_b = 1.0 \tag{18.4}$$

Note that both resistances P_{cy} and M_b relate to out-of-plane failure but that the amplification depends upon the in-plane Euler load. Consideration of the term F/P_{crx}, given that F is limited to P_{cy} which is less than P_{cx} and which is, in turn, less than P_{crx}, suggests that amplification effects are less significant than for the purely in-plane case.

Applied moments are not necessarily restricted to a single plane. For the most general case illustrated in Fig. 18.7, in which compression is accompanied by moments about both principal axes, the member's response is a three-dimensional one involving bending about both axes combined with twisting. This leads to a complex analytical problem which cannot really be solved in such a way that it provides a direct indication of the type of interaction formulae that might be used as a basis for design. A practical view of the problem, however, suggests that some form of combination of the two previous cases might be suitable, providing any proposal were properly checked against data obtained from tests and reliable analyses. This leads to two possibilities: combining the acceptable moments M_x

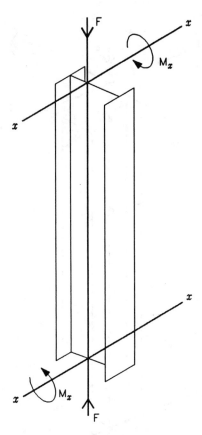

Fig. 18.6 Compression plus major axis bending for an I-section column

and M_y that can safely be combined separately with the axial load F obtained by solving Equations (18.3) and (18.4) to give

$$M_x/M_{ax} + M_y/M_{cy} = 1.0 \tag{18.5}$$

in which M_{ax} and M_{ay} are the solutions of Equations (18·3) and (18.4); or simply adding the minor axis bending effect to Equation (18.4) as

$$F/P_{cy} + M_x/(1 - F/P_{crx})M_b + M_y/(1 - F/P_{cry})M_{cy} = 1.0 \tag{18.6}$$

Although the first of these two approaches does not lead to such a seemingly straightforward end result as the second, it has the advantage that interaction about both axes may be treated separately, and so leads to a more logical treatment of cases for which major axis bending does not lead to a minor axis failure as for a rectangular tube in which $M_b = M_{cx}$, and P_{cx} and P_{cy} are likely to be much closer than is the case for a UB. Similarly for members with different effective lengths for the two planes, for example, due to intermediate bracing

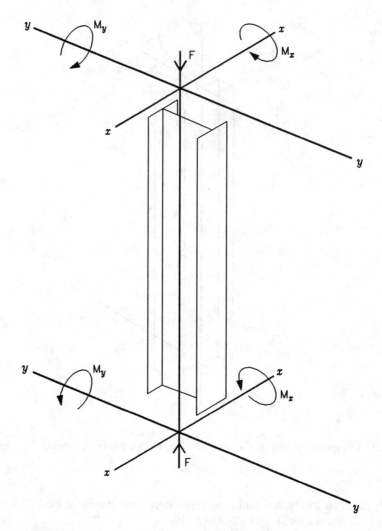

Fig. 18.7 Compression plus major and minor axis bending for an I-section column

acting in the weaker plane only, the ability to treat in-plane and out-of-plane response separately and to combine the weaker with minor axis bending leads to a more rational result.

The foregoing discussion has deliberately been conducted in rather general terms, the main intention being to illustrate those principles on which beam-column design should be based. Collecting them together:

(1) interaction between different load components must be recognized; merely summing the separate components can lead to unsafe results
(2) interaction tends to be more pronounced as member slenderness increases

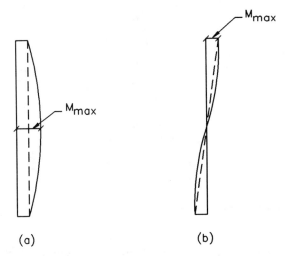

Fig. 18.8 Primary and secondary moments 1: (a) single curvature, (b) double curvature

(3) different forms of response are possible depending on the form of the applied loading.

Having identified these three principles it is comparatively easy to recognize their inclusion in the design procedures of BS 5950 and BS 5400.

18.3 Effect of moment gradient loading

Returning to the comparatively simple in-plane case, Fig. 18.8 illustrates the patterns of primary and secondary moments in a pair of members subject to unequal end moments that produce either single- or double-curvature bending. For the first case, the point of maximum combined moment occurs near mid-height where secondary bending effects are greatest. On the other hand, for double-curvature bending the two individual maxima occur at quite different locations and for the case illustrated, in which the secondary moments have deliberately been shown as small, the point of absolute maximum moment is at the top. Had larger secondary moments been shown, as is the case in Fig. 18.9, then the point of maximum moment moves down slightly but is still far from that of the single-curvature case.

Theoretical and experimental studies of steel beam-columns constrained to respond in-plane and subject to different moment gradients, as represented by the factor of the ratio of the numerically smaller end moment to the numerically larger end moment (β), show clearly that, when all other parameters are held constant, failure loads tend to increase as β is varied from $+1$ (uniform single-curvature bending) to -1 (uniform double-curvature bending). Figure 18.10 illus-

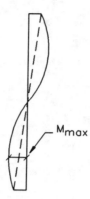

Fig. 18.9 Primary and secondary moments 2

trates the point in the form of a set of interaction curves. Clearly if all beam-column designs were to be based upon the $\beta = +1$ case, safe but rather conservative designs would result.

Figure 18.10 also shows how for high moments the curves for $\beta \neq +1$ tend to merge into a single line corresponding to the condition in which the more heavily stressed end of the member controls design. Reference to Figs 18.8 and 18.9 illustrates the point. The left-hand and lower parts of Fig. 18.10 correspond to situations in which Fig. 18.9 controls, while the right-hand and upper parts represent failure at one end. The two cases are sometimes referred to as 'stability' and 'strength' failure respectively. While this may offend the purists, for in both cases the limiting condition is one of exhausting the cross-sectional capacity, but at different locations within the member length, it does, nonetheless, serve to draw attention to the principal difference. Also shown in Fig. 18.10 is a line corresponding to the cross-sectional interaction; the combinations of F and M corresponding to the full strength of the cross-section. This is the 'strength' limit, representing the case where the primary moment acting in conjunction with the axial load accounts for all the cross-section's capacity.

The substance of Fig. 18.10 can be incorporated within the type of interaction formula approach of section 18.2 through the concept of 'equivalent uniform moment' presented in Chapter 16 in the context of the lateral–torsional buckling of beams; its meaning and use for beam-columns is virtually identical. For moment gradient loading member stability is checked using an equivalent moment $\overline{M} = mM$, as shown in Fig. 18.11. Coincidentally, suitable values of m, based on both test data and rigorous ultimate strength analyses, for the in-plane beam-column case are almost the same as those for laterally unrestrained beams (see section 16.3.6); m may conveniently be represented simply in terms of the moment gradient parameter β.

The situation corresponding to the upper boundary or strength failure of Fig. 18.10 must be checked separately using an appropriate means of determining cross-sectional capacity under F and M. The strength check is superfluous for $\beta = +1$ as

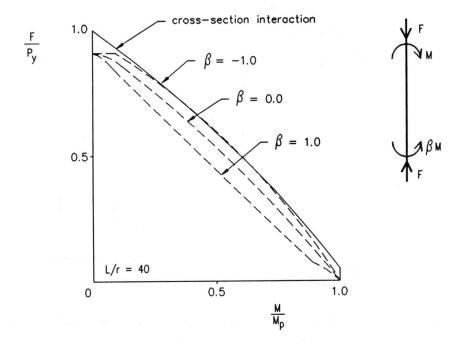
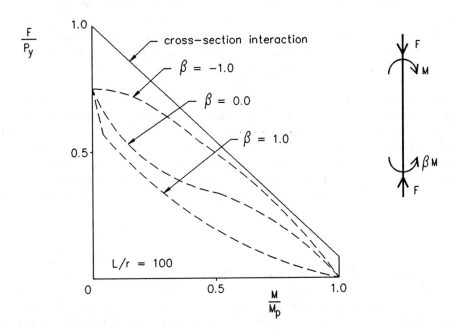

Fig. 18.10 Effect of moment gradient on interaction

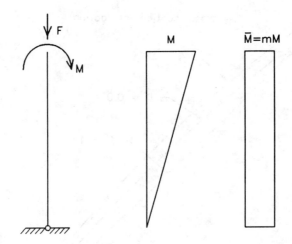

Fig. 18.11 Concept of equivalent uniform moment applied to primary moments on a beam-column

it can never control, while as $\beta \to -1$ and $\overline{M} \to M_c$ it becomes increasingly likely that the strength check will govern. The procedure is:

(1) check stability using an interaction formula in terms of buckling resistance P_c and moment capacity M_c with axial load F and equivalent moment $\overline{M}$,
(2) check strength using an interaction procedure in terms of axial capacity P_s and moment capacity M_c with coincident values of axial load F and maximum applied end moment M_1 (this check is unnecessary if $m = 1.0$; $\overline{M} = M_1$ is used in the stability check).

Consideration of other cases involving out-of-plane failure or moments about both axes shows that the equivalent uniform moment concept may also be applied. For simplicity the same m values are normally used in design, although minor variations for the different cases can be justified. For biaxial bending, two different values, m_x and m_y, for bending about the two principal axes may be appropriate. An exception occurs for a column considered pinned at one end about both axes for which $\beta_x = \beta_y = 0.0$, whatever the sizes of the moments at the top.

18.4 Selection of type of cross-section

Several different design cases and types of response for beam-columns are outlined in section 18.2 of this chapter. Selection of a suitable member for use as a beam-column must take account of the differing requirements of these various factors. In addition to the purely structural aspects, practical requirements such as the need to connect the member to adjacent parts of the structure in a simple and efficient fashion must also be borne in mind. A tubular member may appear to be the best solution for a given set of structural conditions of compressive load, end moments, length, etc., but if site connections are required, very careful thought is necessary to ensure that they can be made simply and economically. On the other hand, if the member is one of a set of similar web members for a truss that can be fabricated entirely in the shop and transported to site as a unit, then simple welded connections should be possible and the best structural solution is probably the best overall solution too.

Generally speaking when site connections, which will normally be bolted, are required, open sections which facilitate the ready use of, for example, cleats or endplates are to be preferred. UCs are designed principally to resist axial load but are also capable of carrying significant moments about both axes. Although buckling in the plane of the flanges, rather than the plane of the web, always controls the pure axial load case, the comparatively wide flanges ensure that the strong-axis moment capacity M_{cx} is not reduced very much by lateral–torsional buckling effects for most practical arrangements. Indeed the condition $M_b = M_{cx}$ will often be satisfied.

In building frames designed according to the principles of simple construction, the columns are unlikely to be required to carry large moments. This arises from the design process by which compressive loads are accumulated down the building but the moments affecting the design of a particular column lift are only those from the floors at the top and bottom of the storey height under consideration. In such cases preliminary member selection may conveniently be made by adding a small percentage to the actual axial load to allow for the presence of the relatively small moments and then choosing an appropriate trial size from the tables of compressive resistance given in Reference 1. For moments about both axes, as in corner columns, a larger percentage to allow for biaxial bending is normally appropriate, while for internal columns in a regular grid with no consideration of pattern loading, the design condition may actually be one of pure axial load.

The natural and most economic way to resist moments in columns is to frame the major beams into the column flanges since even for UCs, M_{cx} will always be comfortably larger than M_{cy}. For structures designed as a series of two-dimensional frames in which the columns are required to carry quite high moments about one axis but relatively low compressive loads, UBs may well be an appropriate choice of member. The example of this arrangement usually quoted is the single-storey portal building, although here the presence of cranes, producing much higher axial loads, the height, leading to large column slenderness, or a combination of the two, may result in UCs being a more suitable choice. UBs used as columns also suffer from the fact that the d/t values for the webs of many sections are non-compact when the applied loading leads to a set of web stresses that have a mean compressive component of more than about 70–100 N/mm^2.

18.5 Basic design procedure

When the distribution of moments and forces throughout the structure has been determined, for example, from a frame analysis in the case of continuous construction or by statics for simple construction, the design of a member subject to compression and bending consists of checking that a trial member satisfies the design conditions being used by ensuring that it falls within the design boundary defined by the type of diagram shown as Fig. 18.3. BS 5950 and BS 5400 therefore contain sets of interaction formulae which approximate such boundaries, use of which will automatically involve the equivalent procedures for the component load cases of strut design and beam design, to define the end points. Where these procedures permit the use of equivalent uniform moments for the stability check, they also require a separate strength check.

BS 5950: Part 1 requires that stability be checked using

$$\frac{F}{A_g p_c} + \frac{\overline{M}_x}{M_b} + \frac{\overline{M}_y}{p_y Z_y} \leq 1 \tag{18.7}$$

In Equation (18.7) the use of $p_y Z_y$, rather than M_{cy}, makes some allowance in the case of plastic and compact sections for the effects of secondary moments as described in section 18.2. For non-compact sections, for which $M_{cy} = p_y Z_y$, no such allowance is made and an unconservative effect is therefore present. Evaluation of Equation (18.7) may be effected quite rapidly if the tabulated values of P_{cy}, P_{cx}, M_b and $p_y Z_y$ given in Reference 1 for all UB, UC, RSJ and SHS are used. In the cases where $\overline{M}$ values are being used and M_x and M_y are not taken as the full values of the maximum moments about the two principal axes, it is essential to check that the most highly stressed cross section is capable of sustaining the coincident compression and moment(s). BS 5950: Part 1 covers this with the expression:

$$\frac{F}{A_g p_{yu}} + \frac{M_x}{M_{cx}} + \frac{M_y}{M_{cy}} \not> 1 \tag{18.8}$$

Clearly when both M_{cx} and M_b values are the same Equation (18.7) is always a more severe check, or in the limit is identical, and only Equation (18.7) need be used. Values of $A_g p_y$, M_{cx} and M_{cy} are also tabulated in Reference 1.

As an alternative to the use of Equations (18.7) and (18.8), BS 5950: Part 1 does, for plastic or compact sections only, permit the use of an alternative pair of interaction formulae. These are both presented in the same form as:

$$\left(\frac{\overline{M}_x}{M_{ax}}\right) + \left(\frac{\overline{M}_y}{M_{ay}}\right) \not> 1 \tag{18.9}$$

$$\left(\frac{M_x}{M_{rx}}\right)^{z_1} + \left(\frac{M_y}{M_{ry}}\right)^{z_2} \not> 1 \tag{18.10}$$

The first controls overall buckling, the second local capacity and in both cases the denominators in the two terms (M_a and M_r) are a measure of the moment that can be carried in the presence of the axial load F. If bending is present about one axis only both checks reduce to a single term on the left-hand side, which corresponds to ensuring that

$M \not> M_a$ for overall buckling
$M \not> M_r$ for local capacity

Evaluation of M_{ax} and M_{ay} is explained in the code, while M_{rx} and M_{ry} may be obtained either from the formulae given in the first part of Reference 1, or, much more conveniently using the tabulated values given in the later parts of that handbook. For fabricated sections, the principles of plastic theory may be applied first to locate the plastic neutral axis for a given combination of F, M_x and M_y and then to calculate M_{rx} and M_{ry}. This is manageable for uniaxial bending — F and M_x or F and M_y — but it is tedious for the full three-dimensional case and some use of approximate results [1] may well be preferable.

18.6 Cross-section classification under compression and bending

It is assumed in the discussion of the use of the BS 5950: Part 1 procedure that the designer has conducted the necessary section classification checks so as to ensure that the appropriate values of M_{cx}, M_{cy}, etc. are used. When the tabulated data of Reference 1 are being employed, any allowances for non-compactness are included in the listed values of M_{cx} and M_{cy} but only if P_{cx} and P_{cy} have been taken from the strut tables rather than the beam-column tables will these contain any reduction. The reason is that for pure compression the stress pattern is known, whereas under combined loading the requirement may be to sustain only a very small axial load; to reduce P_{cx} and P_{cy} on the basis of uniform compression in each plate element of the section is much too severe. For simplicity, section classification may initially be conducted under the most severe conditions of pure axial load; if the result is either plastic or compact nothing is to be gained by conducting additional calculations with the actual pattern of stresses. However, if the result is a non-compact section, possibly when checking the web of a UB, then it is normally advisable for economy of both design time and actual material use to repeat the classification calculations more precisely, as illustrated in the following.

18.6.1 Compactness check under combined loading

When checking a web subject to combined compression and bending for compactness according to BS 5950: Part 1, careful ordering of the calculations will simplify what may well at first sight appear to be a complex process. Rearrangement of the limits of Table 7 of the code into the five categories of Table 18.1[2] is

Table 18.1 Check for compact web under combined loading

Category	d/t limit	F limit	Result/action
1	39ε	Any	Compact
2	49ε	tdp_y	Compact
3	$> 49\varepsilon$	$t^2p_y(98\varepsilon - d/t)$	Compact
4	$> 49\varepsilon$	$> t^2p_y(98\varepsilon - d/t)$	App. H3.2 of BS 5950: Part 1
5	$> 39\varepsilon$	$> tdp_y$	App. H3.2 of BS 5950: Part 1

the first step. In drawing up this table the limitations on axial load F are a measure of the proportion of axial load F to moment M; since the check is being conducted at the compact/semi-compact boundary, the use of plastic (rectangular) stress blocks is appropriate. It is therefore necessary that the compression flange is at least compact.

Since all webs satisfying $d/t < 39\varepsilon$ are compact even if the whole section is under uniform compression, as shown in Fig. 18.12(a), it follows that such webs subject to combined loading are also compact whatever the value of F.

The upper limits for the second category are illustrated in Fig. 18.12(b), which shows the case of a web subject to uniform compression throughout ($y_c = d$, $\alpha = 2$). The flanges are resisting all the moment, leading to a maximum value for F of tdp_y, and, using the condition from Table 7 for 'web generally', a compact limit of

$$d/t \leq 98\varepsilon/\alpha$$
$$\leq 98\varepsilon/2$$
$$\leq 49\varepsilon$$

Increasing the moment so that part of it must be resisted by a portion of the web leads to the condition of Fig. 18.12(c), in which equilibrium requires that:

$C = T$
$F = C'$
C and T together resist M

Block C' must be symmetrical about the geometrical axis, and so y_c may be obtained as:

$$y_c = d/2 + \tfrac{1}{2}(F/tp_y)$$

and since $\alpha = 2y_c/d$, it follows that

$$\alpha = 2/d[d/2 + \tfrac{1}{2}(F/tp_y)]$$
$$= 1 + F/tdp_y$$

From Table 7 for 'web generally', a compact limit is

$$d/t \leq 98\varepsilon/\alpha$$
$$\leq 98\varepsilon/(1 + F/tdp_y)$$

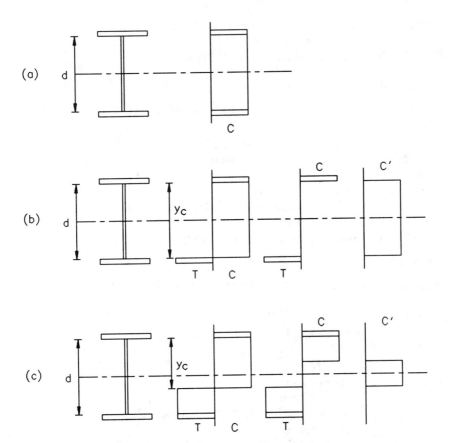

Fig. 18.12 Compactness checks for webs under combined compression and bending: (a) category 1, (b) category 2, (c) category 3

$$\therefore d/t\,(1 + F/tdp_y) \le 98\varepsilon$$
$$d/t + d/t(F/tdp_y) \le 98\varepsilon$$
and $(F/tdp_y)d/t \le 98\varepsilon - d/t$
or $F/t^2 p_y \le 98\varepsilon - d/t$
$$F \le (98\varepsilon - d/t)\,t^2 p_y$$

If F is limited to this value, and d/t does not exceed the absolute limit of 98ε, the position of the neutral axis means that the 'web generally' limit will be met.

If the value of F exceeds this limit (and $d/t > 49\varepsilon$), the web may still be compact if it satisfies the more extensive procedure of Appendix H3.2 of the code, use of which is best demonstrated by means of an example.

Take a 406 × 140 × 39 UB (the UB with the highest d/t) with $\varepsilon = 1$ and $F = 450$ kN and $M = 150$ kN m.

Plastic resistance, $dtp_y = 359.7 \times 6.3 \times 275/10^3$
$= 623$ kN
$F = 450$ kN < 623 and plastic neutral axis is in web

Reduced plastic moment, $M_{pr} = M_p - \dfrac{F^2}{4tp_y}$
$= 721 \times 275/10^3 - 450^2/4 \times 6.3 \times 275$
$= 198.3 - 29.2$
$= 169.1$ kN m

$M = 150$ kN < 169.1 so plastic resistance okay.
Compactness, $d/t = 57.1 > 49$
Check whether $F > (98\varepsilon - d/t)t^2 p_y$
$> (98 - 57.1)6.3^2 \times 275/10^3$
$= 446.4$ kN
$F = 450$ kN > 446.4 so use Appendix H3.2

Local buckling

M_p of web $= 6.3 \times 359.7^2 \times 275/4 \times 10^3 \times 10^3$
$= 56.0$ kN m
M_p of flanges $= 198.3 - 56.0$
$= 142.3$ kN m
M_w $= 150 - 142.3$
$= 7.7$ kN m
$p_{b.cr}$ $= (1630/57.1)^2$
$= 814.9$ N/mm^2
M_{cr} $= 6.3 \times 359.7^2 \times 814.9/4 \times 10^3 \times 10^3$
$= 166.1$ kN m
$p_{c.cr}$ $= (815/57.1)^2$
$= 203.7$ N/mm^2
$P_{c.cr}$ $= dtp_{c.cr}$
$= 359.7 \times 6.3 \times 203.7/10^3$
$= 461.6$ kN

$\dfrac{F_c}{P_{c.cr}} + \left(\dfrac{M_w}{M_{cr}}\right)^2 = \dfrac{450}{461.6} + \left(\dfrac{7.7}{166.1}\right)^2$
$= 0.975 + 0.002$
$= 0.977 < 1$ okay

Despite the limits on d/t and F given by Table 7 being exceeded, the more rigorous check shows the web to be just safe against buckling under a fully plastic stress distribution and so compact. Were F to be increased to the level of $P_{c.cr}$ then the corresponding value of M_w would fall to zero; alternatively if F is taken as the limiting value $(98 - d/t)t^2 p_y$:

$M_{pr} = 198.3 - 446.4^2/4 \times 6.3 \times 275$
$= 198.3 - 23.8$
$= 169.5$ kN m

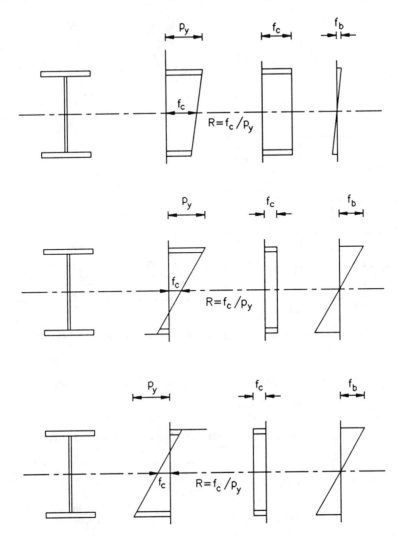

Fig. 18.13 Definition of R when checking semi-compact webs under compression and bending – symmetrical sections

$M_w = 169.5 - 142.3$
  $= 27.2$ kN m

$$\frac{F_c}{P_{c.cr}} + \left(\frac{M_w}{M_{cr}}\right)^2 = \frac{446.4}{461.6} + \left(\frac{27.2}{166.1}\right)^2$$
$$= 0.967 + 0.027$$
$$= 0.994 \approx 1$$

which shows that Appendix H3.2 leads to higher results with the web resisting some of the moment. If the compression exceeds the web capacity and the web

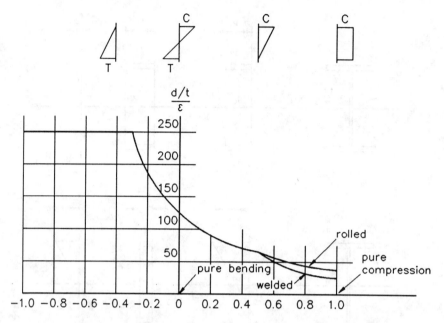

Fig. 18.14 Semi-compact limits for webs subject to compression and bending

Table 18.2 Semi-compact limits of webs subject to compression and bending

	d/t limits	
Range of values for R	Rolled sections	Welded sections
$R < -0.307$	250ε	250ε
$0 \geq R \geq -0.307$	$\dfrac{120\varepsilon}{(1+R)^2}$	$\dfrac{120\varepsilon}{(1+R)^2}$
$0.504 \geq R \geq 0$	—	$\dfrac{120\varepsilon}{1+1.5R}$
$R > 0.504$	—	$(41/R - 13)\varepsilon$
$0.656 \geq R \geq 0$	$\dfrac{120\varepsilon}{1+1.5R}$	—
$R > 0.656$	$(41/R - 2)\varepsilon$	—

slenderness exceeds the limit for 'web subject to compression throughout' i.e. $F > tdp_y$ and $d/t > 39\varepsilon$, Appendix H3.2 should be used since the plastic neutral axis is located in the flange.

Appendix H3.2 may also be used if F exceeds the web capacity ($F > tdp_y$) and d/t exceeds the 'uniform compression throughout the section' limit of 39ε, although in such cases the plastic neutral axis is located part way through the flange.

Web elements which are non-compact should be checked against the semi-compact limit since meeting this will require no reductions in P_c and the simple limitation of M_c (or M_b) to $p_y Z$. BS 5950: Part 1 uses a set of formulae expressed in terms of a property R which is the ratio of the mean longitudinal stress in the web to p_y. From Fig. 18.13 it is clear that for symmetrical sections, R is equal to F/Ap_y. The formulae ensure a smooth transition between the Table 7 values for compression throughout to pure bending as R reduces from unity to zero as shown in Fig. 18.14. Table 18.2 defines the range of R-values for which each formula is applicable.

Taking the UB with the most slender web (a 406 × 140 × 39 UB with $d/t = 57.1$) gives limits on R of 0.66 and 0.52 for grade 43 and 50 material respectively, up to which levels of compression the section will not be slender. Higher figures are appropriate for other UBs; no UC has a sufficiently high d/t value to be other than compact even under pure compression ($d/t \leq 39\varepsilon$).

18.7 Special design methods for members in portal frames

18.7.1 Design requirements

Both the columns and the rafters in the typical pitched roof portal frame represent particular examples of members subject to combined bending and compression. Providing such frames are designed elastically, the methods already described for assessing local cross-sectional capacity and overall buckling resistance may be employed. However, these general approaches fail to take account of some of the special features present in normal portal frame construction, some of which can, when properly allowed for, be shown to enhance buckling resistance significantly.

When plastic design is being employed, the requirements for member stability change somewhat. It is no longer sufficient simply to ensure that members can safely resist the applied moments and thrust; rather for members required to participate in plastic hinge action, the ability to sustain the required moment in the presence of compression during the large rotations necessary for the development of the frame's collapse mechanism is essential. This requirement is essentially the same as that for a 'plastic' cross-section discussed in Chapter 13. The performance requirement for those members in a plastically designed frame actually required to take part in plastic hinge action is therefore equivalent to the most onerous type of response shown in Fig. 13.4. If they cannot achieve this level of performance, for example, because of premature unloading caused by local buckling, then they will prevent the formation of the plastic collapse mechanism assumed as the basis for the design, with the result that the desired load factor will not be attained. Put simply the requirement for member stability in plastically-designed structures is to impose limits on slenderness and axial load level, for example, that ensure stable behaviour while the member is carrying a moment

512 *Members with compression and moments*

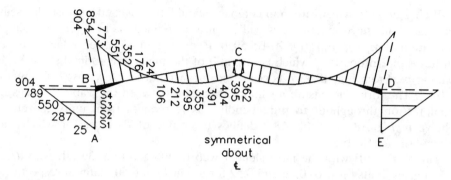

Fig. 18.15 Moment distribution for dead plus imposed load condition

equal to its plastic moment capacity suitably reduced so as to allow for the presence of axial load. For portal frames, advantage may be taken of the special forms of restraint inherent in that form of construction by, for example, purlins and sheeting rails attached to the outside flanges of the rafters and columns respectively.

Figure 18.15 illustrates a typical collapse moment diagram for a single-bay pin-base portal subject to gravity load only (dead load + imposed load), this being the usual governing load case in the UK. The frame is assumed to be typical of UK practice with columns of somewhat heavier section than the rafters and haunches of approximately 10% of the clear span and twice the rafter depth at the eaves. It is further assumed that the purlins and siderails which support the cladding and are attached to the outer flanges of the columns and rafters provide positional restraint to the frame, i.e. prevent lateral movement of the flange, at these points. Four regions in which member stability must be ensured may be identified.

(1) full column height AB
(2) haunch, which should remain elastic throughout its length
(3) eaves region of rafter for which the lower unbraced flange is in compression due to the moments, from end of haunch
(4) apex region of the rafter between top compression flange restraints.

18.7.2 Column stability

Figure 18.16 provides a more detailed view of the column AB, including both the bracing provided by the siderails and the distribution of moment over the column height. Assuming the presence of a plastic hinge immediately below the haunch, the design requirement is to ensure stability up to the formation of the collapse mechanism.

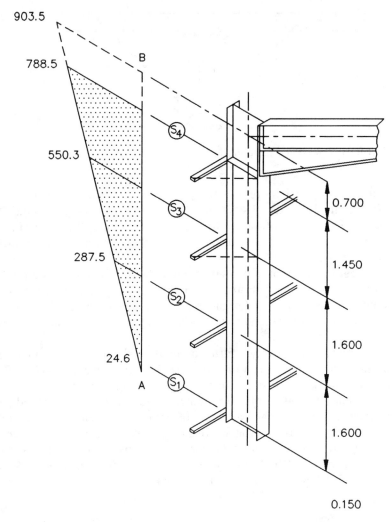

Fig. 18.16 Member stability – column

According to clause 5.3.5 of BS 5950: Part 1, torsional restraint must be provided no more than $D/2$, where D is the overall column depth, measured along the column axis, from the underside of the haunch. This may conveniently be achieved by means of the knee brace arrangement of Fig. 18.17. The simplest means of ensuring adequate stability for the region adjacent to this braced point is to provide another torsional restraint within a distance of not more than L_m, where L_m is obtained from clause 5.3.5 as

$$L_m \leq \frac{38 r_y}{[f_c/130 + (p_y/275)^2 (x/36)^2]^{\frac{1}{2}}} \tag{18.11}$$

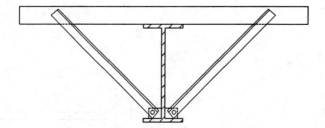

Fig. 18.17 Effective torsional restraints

Noting that the mean axial stress in the column f_c is normally small, that p_y is around 275 N/mm² for grade 43 steel and that x has values between about 20 and 45 for UBs, gives a range of values for L_m/r_y of between 30 and 68. Placing a second torsional restraint at this distance from the first therefore ensures the stability of the upper part of the column.

Below this region the distribution of moment in the column normally ensures that the remainder of the length is elastic. Its stability may therefore be checked using the procedures of section 18.5. Frequently no additional intermediate restraints are necessary, the elastic stability condition being much less onerous than the plastic one.

Equation (18.11) is effectively a fit to the limiting slenderness boundary of the column design charts[3] that were in regular use until the advent of BS 5950: Part 1, based on the work of Horne,[4] which recognized that for lengths of members between torsional restraints subject to moment gradient, longer unbraced lengths could be permitted than for the basic case of uniform moment. Equation (18.11) may therefore be modified to recognize this by means of the coefficients proposed by Brown.[5] Figure 18.18 illustrates the concept and gives the relevant additional formulae. For a 533 × 210 UB82 of grade 43 steel for which $x = 41.6$ and assuming $f_c = 15$ N/mm², the key values become:

$L_m = 31.55 r_y$
$KL_m = 122.7 r_y$
$K_0 KL_m = 90.6 r_y$
$\beta_m = 0.519$

When checking a length for which the appropriate value of β is significantly less than +1.0, use of this modification permits a more relaxed approach to the provision of bracing. Some element of trial and error is involved since the exact value of β to be used is itself dependent upon the location of the restraints.

Neither the elastic nor the plastic stability checks described above take account of the potentially beneficial effect of the tension flange restraint provided by the sheeting rails. This topic has been extensively researched,[6] with many of the findings being distilled into the design procedures of Appendix G of BS 5950: Part 1. Separate procedures are given for both elastic and plastic stability checks. Although significantly more complex than the use of Equation (18.11) or the

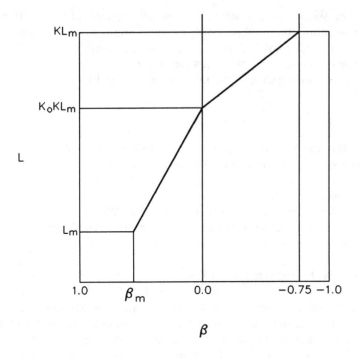

$20 < x \leqslant 30 \quad K = 2.3 + 0.03\,x - x\,f_c/3000$

$30 < x < 50 \quad K = 0.8 + 0.08\,x - (x - 10)f_c/2000$

$K_0 = (180 + x)/300$

Grade 43 steel $\beta_m = 0.44 + x/270 - f_c/200$

Grade 50 steel $\beta_m = 0.47 + x/270 - f_c/250$

Fig. 18.18 Modification to Equation (18.11) to allow for moment gradient

methods of section 18.5, their use is likely to lead to significantly increased allowable unbraced lengths, particularly for the plastic region.

18.7.3 Rafter stability

Stability of the eaves region of the rafter may most easily be ensured by satisfying the conditions of clause 5.5.3.5. If tension flange restraint is not present between

516 *Members with compression and moments*

points of compression flange restraint i.e. widely spaced purlins and a short unbraced length requirement, this simply requires the use of Equation (18.11). However, when the restraint is present in the form illustrated in Fig. 18.19, the distance between compression flange restraints for grade 43 steel and a haunch that doubles the rafter depth may be taken as L_s, given by

$$L_s = \frac{495\, r_y x}{(72x^2 - 10^4)^{\frac{1}{2}}} \tag{18.12}$$

Variants of this expression are given in the code for changes in the grade of steel or haunch depth. Certain other limitations must also be observed:

(1) the rafter must be a UB
(2) the haunch flange must not be smaller than the rafter flange
(3) the distance between tension flange restraints must be stable when checked as a beam using the procedure of section 16.3.6.

Equation (18.12) is less sensitive than Equation (18.11) to changes in x with the result that it gives an average value for L_s/r_y of about 65. It is often regarded as good practice to provide bracing at the toe of the haunch since this region corresponds to major changes in the pattern of force transfer due to the change in the line of action of the compression in the bottom flange. In cases where the use of clause 5.5.3.5 does not give a stable haunch because the length from eaves to

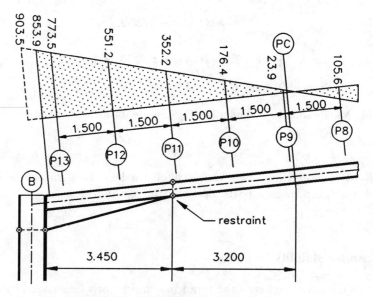

Fig. 18.19 Member stability in haunched rafter region

toe exceeds L_s, Appendix G may be used to obtain a larger value of L_s. If this is still less than the haunch length, then additional compression flange restraints are required.[6]

18.7.4 Bracing

The general requirements of lateral bracing systems have already been referred to in Chapter 16 — sections 16.3, 16.4 and 16.5 in particular. When purlins or siderails are attached directly to a rafter or column compression flange it is usual to assume that adequate bracing stiffness and strength are available without conducting specific calculations. In cases of doubt the ability of the purlin to act as a strut carrying the design bracing force may readily be checked. Definitive guidance on the appropriate magnitude to take for such a force is noticeably lacking in codes of practice. A recent suggestion for members in plastically-designed frames[7] is 2% of the squash load of the compression flange of the column or rafter; $0.02p_yBT$ at every restraint. In order that bracing members possess sufficient stiffness a second requirement that their slenderness be not more than 100 has also been proposed.[7] Both suggestions are largely based on test data. For elastic design the provisions of BS 5950: Part 1 may be followed.

When purlins or siderails are attached to the main member's tension flange, any positional restraint to the compression flange must be transferred through both the bracing to main member interconnection and the webs of the main member. Both effects are allowed for in the work on which the special provisions in BS 5950: Part 1 for tension flange restraint are based.[8] When full torsional restraint is required so that interbrace buckling may be assumed, the arrangement of Fig. 18.17 is often used. The stays may be angles, tubes (providing simple end connections can be arranged) or flats (which are much less effective in compression than in tension). In theory a single member of sufficient size would be adequate but practical considerations such as hole clearance[6] normally dictate the use of pairs of stays. It should also be noted that for angles to the horizontal of more than 45° the effectiveness of the stay is significantly reduced.

Reference 9 discusses several practical means of bracing or otherwise restraining beam-columns.

References to Chapter 18

1. The Steel Construction Institute (SCI) (1991) *Steelwork Design Guide to BS 5950: Part 1: 1990, Vol. 1, Section Properties, Member Capacities*, 3rd edn. SCI, Ascot, Berks.
2. Advisory Desk (1988) *Steel Construction Today*, **2**, Apr., 61–2.
3. Morris L.J. & Randall A.L. (1979) *Plastic Design*. Constrado. (See also *Plastic Design (Supplement)*, Constrado, 1979.)

4. Horne M.R. (1964) Safe loads on I-section columns in structures designed by plastic theory. *Proc. Instn Civ. Engrs*, **29**, Sept., 137–50 and *Discussion*, **32**, Sept. 1965, 125–34.
5. Brown B.A. (1988) The requirements for restraint in plastic design to BS 5950. *Steel Construction Today*, **2**, 184–96.
6. Morris L.J. (1981 & 1983) A commentary on portal frame design. *The Structural Engineer*, **59A**, No. 12, 394–404 and **61A**, No. 6, 181–9.
7. Morris L.J. & Plum D.R. (1988) *Structural Steelwork Design to BS 5950*. Longman, Harlow, Essex.
8. Horne M.R. & Ajmani J.L. (1972) Failure of columns laterally supported on one flange. *The Structural Engineer*, **50**, No. 9, Sept., 355–66.
9. Nethercot D.A. & Lawson R.M. (1992) *Lateral stability of steel beams and columns – common cases of restraint*. SCI publication 093, The Steel Construction Institute.

Further reading for Chapter 18

Chen W.F. & Atsuta T. (1977) *Theory of Beam-Columns, Vols 1 and 2*. McGraw-Hill, New York.
Galambos T.V. (1988) *Guide to Stability Design Criteria for Metal Structures*, 4th edn. Wiley, New York.
Horne M.R. (1979) *Plastic Theory of Structures*, 2nd edn. Pergamon, Oxford.
Horne M.R., Shakir-Khalil H. & Akhtar S. (1967) The stability of tapered and haunched beams. *Proc. Instn Civ. Engrs*, **67**, No. 9, 677–94.
Morris L.J. & Nakane K. (1983) Experimental behaviour of haunched members. In *Instability and Plastic Collapse of Steel Structures* (Ed. by L.J. Morris), pp. 547–59. Granada.

A series of worked examples follows which are relevant to Chapter 18.

The Steel Construction Institute Silwood Park, Ascot, Berks SL5 7QN	Subject BEAM-COLUMN EXAMPLE 1 ROLLED UNIVERSAL COLUMN	Chapter ref. 18
	Design code BS5950: Part 1	Made by DAN Checked by GWO	Sheet no. 1

Problem

Select a suitable UC in grade 43 steel to carry safely a combination of 940 kN in direct compression and a moment about the minor axis of 16 kNm over an unsupported height of 3.6 m.

Problem is one of uniaxial bending producing failure by buckling about the minor axis. Since no information is given on distribution of applied moments make conservative (& simple) assumption of uniform moment ($\beta = 1.0$).

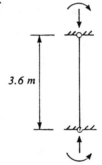

3.6 m

Try 203 × 203 × 60 UC - member capacities suggest P_{cy} of approximately 1400 kN will provide correct sort of margin to carry the moment

Steelwork Design Guide Vol 1

r_y = 5.19 cm Z_y = 199 cm³

A = 75.8 cm² S_y = 303 cm³

λ_y = 3600/51.9 = 69.4   4.7.3

Use Table 27c for p_c Table 25

For p_y = 275 N/mm² and λ = 69.4

value of p_c = 183 N/mm²

P_{cy} = 183 × 7580 = 1387 × 10³ N   4.7.4
= 1387 kN

$$\frac{940}{1387} + \frac{16}{275 \times 199000 \times 10^{-6}} = 0.68 + 0.29$$ 4.8.3.3.1

= 0.97

∴ <u>Adopt 203 × 203 × 60 UC</u>

The Steel Construction Institute Silwood Park, Ascot, Berks SL5 7QN	Subject **BEAM-COLUMN EXAMPLE 1** **ROLLED UNIVERSAL COLUMN**	Chapter ref. **18**
	Design code **BS5950: Part 1**	Made by **DAN** Checked by **GWO**	Sheet no. **2**

The determination of P_{cy} assumed that the section is not slender; similarly the use of Clause 4.8.3.3.1 in the present form presumes that the section is not slender. The actual stress distribution in the flanges will vary linearly due to the minor axis moment component of the load. Since the actual case cannot be more severe than uniform compression, check classification for pure compression.

3.5

Flange limiting b/T $\quad = \quad$ 15 $\hspace{4cm}$ Table 7

Web limiting d/t $\quad = \quad$ 39

Actual b/T $\quad = \quad$ 7.23

Actual d/t $\quad = \quad$ 17.3

∴ <u>section is not slender</u>

The Steel Construction Institute Silwood Park, Ascot, Berks SL5 7QN	Subject BEAM-COLUMN EXAMPLE 2 ROLLED UNIVERSAL BEAM		Chapter ref. 18	
	Design code BS5950: Part 1	Made by DAN Checked by GWO	Sheet no.	1

Problem

Check the suitability of a 533 × 210 × 82 UB in grade 50 steel for use as the column in a portal frame of clear height 5.6 m if the axial compression is 160 kN, the moment at the top of the column is 530 kNm and the base is pinned. The ends of the column are adequately restrained against lateral displacement (i.e. out of the plane) and rotation.

Loading corresponds to compression and major axis moment distributed as shown. Check initially over full height.

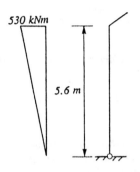

r_y	=	4.38 cm	u	=	0.865	Steelwork Design Guide Vol 1
S_x	=	2060 cm³	x	=	41.6	
A	=	104 cm²				
p_y	=	355 N/mm²				Table 6
λ_y	=	5600/43.8	=	128		4.7.3
λ/x	=	128/41.6	=	3.08		
v	=	0.91				Table 14
λ_{cr}	=	0.865 × 0.91 × 128	=	101		4.3.7.5
p_b	=	139 N/mm²				Table 11
M_b	=	139 × 2060000	=	286 × 10⁶ Nmm		4.3.7.3
			=	286 kNm		

The Steel Construction Institute Silwood Park, Ascot, Berks SL5 7QN	Subject **BEAM-COLUMN EXAMPLE 2 ROLLED UNIVERSAL BEAM**	Chapter ref. **18**
	Design code BS5950: Part 1	Made by **DAN** Sheet no. **2** Checked by **GWO**

Use Table 27b for p_c Table 25

p_c = 153 N/mm²   Table 27b

P_{cy} = 153 × 10400   =   1591 × 10³ N   4.7.4
                             =   1591 kN

For β = 0/530 = 0   take m = 0.57   Table 18

$$\frac{\overline{M}}{M_b} = \frac{0.57 \times 530}{286} = 1.06$$

∴ member has insufficient buckling resistance moment. Check moment capacity

M_{cx} = 355 × 2060000 =   731 × 10⁶ Nmm
                                   =   731 kNm

∴ section capacity OK so increase stability by inserting a brace from a suitable side rail to the compression flange. Estimate suitable location as 1.6 m below top.

For upper part of column

λ_y = 1600/43.8 = 37   4.7.3

λ/x = 37/41.6 = 0.9

v = 0.99   Table 14

λ_{LT} = 0.865 × 0.99 × 37 = 32   4.7.3.5

p_b = 350 N/mm²   Table 11

M_b = 350 × 2060000 =   721 × 10⁶ Nmm   4.3.7.3
                                 =   721 kNm

p_c = 320 N/mm²   Table 27b

P_{cy} = 320 × 10400 =   3328 × 10³ N   4.7.4
                             =   3328 kN

The Steel Construction Institute Silwood Park, Ascot, Berks SL5 7QN	Subject **BEAM-COLUMN EXAMPLE 2 ROLLED UNIVERSAL BEAM**	Chapter ref. **18**
	Design code BS5950: Part 1	Made by DAN Sheet no. 3 Checked by GWO

Table 18

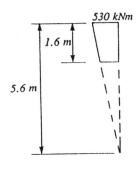

$\beta = \dfrac{(5.6 - 1.6)}{5.6} = 0.72$

$m = 0.85$

$\dfrac{\overline{M}}{M_b} = \dfrac{0.85 \times 530}{721} = 0.62$

$\dfrac{P}{P_c} = \dfrac{160}{3328} = 0.05$

$0.05 + 0.62 = 0.67 \quad OK$ 4.8.3.3.1

Check lower part of column for moment of

$0.72 \times 530 = 382 \ kNm$

$\lambda_y = 4000/43.8 = 91$ 4.7.3

$\lambda/x = 91/41.6 = 2.2$

$v = 0.96$ Table 14

$\lambda_{LT} = 0.865 \times 0.96 \times 91 = 76$ 4.7.3.5

$p_b = 202 \ N/mm^2$

$M_b = 416 \ kNm$ 4.7.3.3

	Subject	Chapter ref.
The Steel Construction Institute Silwood Park, Ascot, Berks SL5 7QN	BEAM-COLUMN EXAMPLE 2 ROLLED UNIVERSAL BEAM	18
	Design code BS5950: Part 1 — Made by DAN / Checked by GWO	Sheet no. 4

p_c = 178 N/mm²   Table 27b

P_{cy} = 1851 kN   4.7.4

$$\frac{\overline{M}}{M_b} = \frac{0.57 \times 382}{416} = 0.52$$

$$\frac{P}{P_{cy}} = \frac{160}{1851} = 0.09$$

0.09 + 0.52 = 0.61 OK   Use 1 brace   4.8.3.3.1

Capacity of cross-section under compression and bending should also be checked at point of maximum coincident values. However, since M_{cx} = 355 × 2060000 × 10^{-6} = 731 kNm and compression is small by inspection, capacity is OK.

As before, use of M_b presumes section is at least compact.

b/T limit = 9.5ε   Table 7

d/t limit (pure compression) = 39ε

Since ε = $(275/355)^{1/2}$ = 0.88   these are:

8.4 and 34.3

Actual b/T = 5.98

Actual d/t = 41.2

∴ d/t greater than limit for pure compression. However, actual loading is principally bending for which limit is 98ε = 86

∴ without performing a rigorous check (by locating plastic neutral axis position etc.) it is clear that section will meet the limit for principally bending.

Section compact

∴   <u>Adopt 533 × 210 × 82 UC</u>

	Subject	Chapter ref.
The Steel Construction Institute Silwood Park, Ascot, Berks SL5 7QN	**BEAM-COLUMN EXAMPLE 3** **RHS IN BIAXIAL BENDING**	18
Design code **BS5950: Part 1**	Made by **DAN** Checked by **GWO**	Sheet no. 1

Problem

Select a suitable RHS in grade 50 material for the top chord of the 26.2 m span truss shown below.

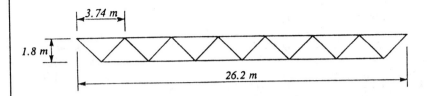

Trusses are spaced at 6 m intervals with purlins at 1.87 m intervals; these may be assumed to prevent lateral deflection of the top chord at these points. Under the action of the applied loading the chord loads in the most severely loaded bay are:

compression   664 kN

vertical moment   24.4 kNm

horizontal moment 19.6 kNm

It is necessary to consider a length between nodes, allowing for the lateral restraint at mid-length under the action of compression plus biaxial bending.

Take L_{Ex} as distance between nodes and L_{Ey} as distance between purlins

L_{Ex} = 3.74 m

L_{Ey} = 1.87 m

Try 150 × 150 × 10 RHS

For L_{Ex}	= 3.74 m	P_{cx}	=	1560 kN			*Steelwork Design Guide Vol 1*
For L_{Ey}	= 1.87 m	P_{cy}	=	1900 kN			
P_z	= 1970 kN	M_{cx}	=	M_{cy}	=	102 kN	

The Steel Construction Institute Silwood Park, Ascot, Berks SL5 7QN	Subject BEAM-COLUMN EXAMPLE 3 RHS IN BIAXIAL BENDING		Chapter ref. 18
	Design code BS5950: Part 1	Made by DAN Checked by GWO	Sheet no. 2

Check local capacity using "more exact" method for plastic section   4.8.3.2(b)

$$\left[\frac{M_x}{M_{rx}}\right]^{5/3} + \left[\frac{M_y}{M_{ry}}\right]^{5/3} \leq 1$$

F/P_z = 664/1970 = 0.337

M_{rx} = M_{ry} = 87 kNm

$$\left[\frac{24.4}{87}\right]^{5/3} + \left[\frac{19.6}{87}\right]^{5/3} = 0.120 + 0.083$$

= 0.203 < 1   local capacity OK

Steelwork Design Guide Vol 1

Check overall buckling using "more exact" method   4.8.3.3.2

$$\left[\frac{m\,M_x}{M_{ax}}\right] + \left[\frac{m\,M_y}{M_{ay}}\right] \leq 1$$

Take m = 1.0 for both axes since exact moment distributions not given; this will be conservative

$$M_{ax} = M_{cx}\,\frac{(1 - F/P_{cx})}{(1 + 1/2 \times F/P_{cx})}$$

$$= 102\,\frac{(1 - 664/1560)}{(1 + 1/2 \times 664/1560)} = 48.3\ kNm$$

For RHS, in-plane case will govern for M_{ay}

$$M_{ay} = M_{cy}\,\frac{(1 - F/P_{cy})}{(1 + 1/2 \times F/P_{cy})}$$

$$= 102\,\frac{(1 - 664/1900)}{(1 + 1/2 \times 664/1900)} = 56.5\ kNm$$

	Subject	Chapter ref.
The Steel Construction Institute Silwood Park, Ascot, Berks SL5 7QN	**BEAM-COLUMN EXAMPLE 3 RHS IN BIAXIAL BENDING**	**18**
	Design code Made by **DAN**	Sheet no. **3**
	BS5950: Part 1 Checked by **GWO**	

$$\left(\frac{24.4}{48.3}\right) + \left(\frac{19.6}{56.5}\right) = 0.505 + 0.347$$

$$= 0.852 < 1 \quad \text{overall buckling OK}$$

∴ <u>Adopt 150 × 150 × 10 RHS</u>

Care is needed if the simplified approach of Clause 4.8.3.3.1 is used since this presumes that F and M_x would produce a lateral-torsional buckling failure. Since $M_b = M_{cx}$ for RHS this will not be the case. One possibility would be to approximate the in-plane interactions for M_x & M_y in the same way using

$$\frac{F}{A_g \, p_c} + \frac{m \, M_x}{p_y \, Z_x} + \frac{m \, M_y}{p_y \, Z_y} \leq 1$$

$$\frac{664}{1560} + \frac{24.4}{85} + \frac{19.6}{85} = 0.426 + 0.287 + 0.231$$

$$= 0.944 \leq 1 \quad OK$$

For this example since m = 1.0 has been used throughout overall buckling will always control.

Chapter 19
Trusses

by PAUL TASOU

19.1 Common types of trusses

19.1.1 Buildings

The most common use of trusses in buildings is to provide support to roofs, floors and such internal loading as services and suspended ceilings. There are many types and forms of trusses; some of the most widely used are shown in Fig. 19.1. The type of truss adopted in design is governed by architectural and client requirements, varied in detail by dimensional and economic factors.

The Pratt truss, Fig. 19.1(a) and (e), has diagonals in tension under normal vertical loading so that the shorter vertical web members are in compression and the longer diagonal web members are in tension. This advantage is partially offset by the fact that the compression chord is more heavily loaded than the tension chord at mid-span under normal vertical loading. It should be noted, however, that for a light-pitched Pratt roof truss wind loads may cause a reversal of load thus putting the longer web members into compression.

The converse of the Pratt truss is the Howe truss (or English truss), Fig. 19.1(b). The Howe truss can be advantageous for very lightly loaded roofs in which reversal of load due to wind will occur. In addition the tension chord is more heavily loaded than the compression chord at mid-span under normal vertical loading. The Fink truss, Fig. 19.1(c), offers greater economy in terms of steel weight for long-span high-pitched roofs as the members are subdivided into shorter elements. There are many ways of arranging and subdividing the chords and web members under the control of the designer.

The mansard truss, Fig. 19.1(d), is a variation of the Fink truss which has the advantage of reducing unusable roof space and so reducing the running costs of the building. The main disadvantage of the mansard truss is that the forces in the top and bottom chords are increased due to the smaller span-to-depth ratio.

However, it must not escape the designer's mind that any savings achieved in steel weight by introducing a greater number of smaller members may, as is often the case, substantially increase fabrication and maintenance costs.

The Warren truss, Fig. 19.1(f), has equal length compression and tension web members, resulting in a net saving in steel weight for smaller spans. The added advantage of the Warren truss is that it avoids the use of web members of differing length and thus reduces fabrication costs. For larger spans the modified

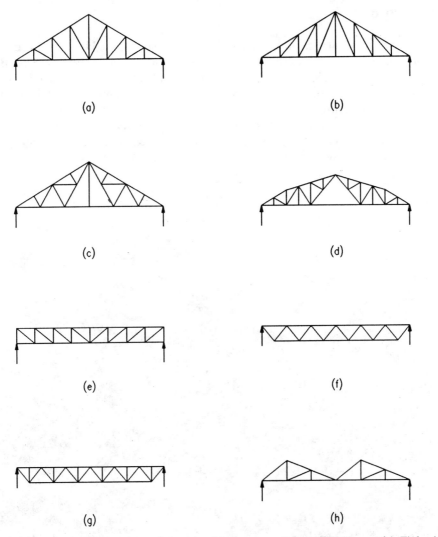

Fig. 19.1 Common types of roof trusses: (a) Pratt – pitched, (b) Howe, (c) Fink, (d) mansard, (e) Pratt – flat, (f) Warren, (g) modified Warren, (h) saw-tooth

Warren truss, Fig. 19.1(g), may be adopted where additional restraint to the chords is required (this also reduces secondary stresses). The modified Warren truss requires more material than the parallel-chord Pratt truss, but this is offset by its symmetry and pleasing appearance. The saw-tooth or butterfly truss, Fig. 19.1(h), is just one of many examples of trusses used in multi-bay buildings, although the other types described above are equally suitable.

530   *Trusses*

19.1.2 Bridges

Trusses are now infrequently used for road bridges in the UK because of high fabrication and maintenance costs. However, the recent award-winning Brinnington railway bridge (Fig. 19.2) demonstrates that they can still be used to create efficient and attractive railway structures. In many parts of the world, particularly the Third World countries where labour costs are low and material costs are high, trusses are often adopted for their economy in steel. Their structural form also lends itself to transportation in small components and piece-small erection, which may be suitable for remote locations.

Fig. 19.2   Brinnington Railway Bridge

Some of the most commonly used trusses suitable for both road and rail bridges are illustrated in Fig. 19.3. Pratt, Howe and Warren trusses. Fig. 19.3(a), (b) and (c), which are discussed in section 19.2.1, are more suitable for short to medium spans. The economic span range of the Pratt and Howe trusses may be extended by subdividing the diagonals and the deck support chord as shown for the Petit truss, Fig. 19.3(h), although this often gives rise to high secondary stresses for short to medium spans. In the case of the Warren truss the unsupported length of the chords may be too great for the economic span and depth range, in such a case the modified Warren truss, Fig. 19.3(d), is more appropriate.

The variable depth type truss such as the Parker, Fig. 19.3(e), offers an aesthetically pleasing structure. With this type of truss its structural function is emphasized and the material is economically distributed at the cost of having expensive fabrication due to the variable length and variable inclination of the web and top chord members.

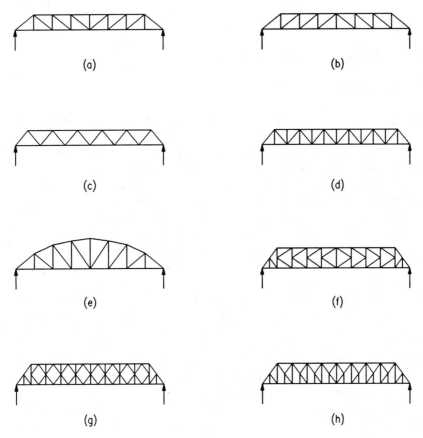

Fig. 19.3 Common types of bridge trusses: (a) Pratt, (b) Howe, (c) Warren, (d) modified Warren, (e) Parker, (f) K, (g) diamond, (h) Petit.

For economy the truss depth is ideally set at a fixed proportion of the span. As the span increases the truss depth and bay width increase accordingly. The bay width is usually fixed by providing truss nodes on the centrelines of the deck cross-beams, thus avoiding high local bending stress in the deck chord. For large-span bridges with an economical spacing for the deck cross-beams, the height of the truss may be as much as four times the bay width. In such a case a subdivided form of truss will be required to avoid very long uneconomical compression web members, or tensile members subjected to load reversal due to moving live loads. The diamond, Petit and K-trusses, Fig. 19.3(g), (h) and (f), are just three types which can be used.

The diamond and Petit trusses have the advantage of having shorter diagonals than the K-truss. The main disadvantage of trusses such as the diamond or Petit, which have intermediate bracing members connected to the chords away from the main joints, is that they give rise to high secondary stresses for short to medium spans due to differential joint deformations caused by moving live loads. The K-truss is far superior in this respect.

19.2 Guidance on overall concept

19.2.1 Buildings

For pitched-roof trusses such as the Pratt, Howe or Fink, Fig. 19.1(a), (b) and (c), the most economical span-to-depth ratio (at apex) is between 4 and 5, with a span range of 6 m to 12 m, the Fink truss being the most economical at the higher end of the span range. Spans of up to 15 m are possible but the unusable roof space becomes excessive and increases the running costs of the building. In such circumstances the span-to-depth ratio may be increased to about 6 to 7, the additional steel weight (increase in initial capital expenditure) being offset by the long-term savings in the running costs. For spans of between 15 m and 30 m, the mansard truss, Fig. 19.1(d), reduces the unusable roof space but retains the pitched appearance and offers an economic structure at span-to-depth ratios of about 7 to 8.

The parallel (or near parallel) chord trusses (also known as lattice girders) such as the Pratt or Warren, Fig. 19.1(e) and (f), have an economic span range of between 6 m and 50 m, with a span-to-depth ratio of between 15 and 25 depending on the intensity of the applied loads. For the top end of the span range the bay width should be such that the web members are inclined at approximately 50° or slightly steeper. For long, deep trusses the bay widths become too large and are often subdivided with secondary web members.

For roof trusses the web member intersection points with the chords should ideally coincide with the secondary transverse roof members (purlins). In practice this is not often the case for economic truss member arrangements, thus resulting in the supporting chord being subject to local bending stresses.

The most economical spacing for roof trusses is a function of the span and load intensity and to a lesser extent the span and spacing of the purlins, but as a general rule the spacing should be between $\frac{1}{4}$ and $\frac{1}{5}$ of the span, which results in a spacing of between 4 m and 10 m for the economic range of truss spans.

For short-span roof trusses between 6 m and 15 m the minimum spacing should be limited to 3–4 m.

19.2.2 Bridges

Road and railway truss bridges can either be underslung (deck at top chord level), through (deck at bottom chord level) or semi-through.

Limits on headroom, navigation height and construction depth will determine whether an underslung or through truss will be the most appropriate. For large-span bridges the through type is often adopted as ample headroom will be available to permit direct lateral restraint to the top compression chord using cross bracing. For short–span bridges however, the underslung type is most appropriate provided the navigation height or construction depth limits are satisfied. Underslung trusses are usually more economical than either through or semi-through trusses as the deck structure performs the dual function of directly supporting the traffic loads and providing lateral restraint to the compression chords. In the case of short-span through trusses the span-to-depth ratio may be uneconomically low if the top chord is restrained by cross-beams, as sufficient traffic headroom must be provided. In such a situation it is more economical to brace the top compression chords by U-frame action.

A span-to-depth ratio of between 6 and 8 for railway bridges and between 10 and 12 for road bridges offers the most economical design. In general terms the proportions should be such that the chords and web members have approximately an equal weight.

The bay widths should be proportioned so that the diagonal members are inclined at approximately 50° or slightly steeper. For large-span trusses subdivision of the bays is necessary to avoid having excessively long web members.

The Pratt, Howe and Warren trusses, Fig. 19.3(a)–(c), have an economic span range of between 40 m and 100 m for railway bridges and up to 150 m for road bridges. For the shorter spans of the range the Warren truss requires less material than either the Pratt or the Howe trusses. For medium spans of the range the Pratt or Howe trusses are both more favourable and by far the most common types. For large spans the modified Warren truss and subdivided Parker (inclined chord) truss are the most economical.

For spans of between 100 m and 250 m the depth of the truss may be up to four times the economic bay width, and in such a case the K-, diamond or Petit (subdivided Pratt or Howe) trusses are more appropriate. For the shorter spans of the range the diamond or Petit trusses, by their nature, are subject to very high secondary stresses. In such a case the K-truss, with primary truss members at all nodes, is more appropriate as joint deflections are uniform, greatly reducing the secondary stresses.

For spans greater than 150 m variable depth trusses are normally adopted for economy.

The spacing of bridge trusses depends on the width of the carriageway for road bridges and the required number of tracks for railway bridges, in addition to considerations regarding lateral strength and rigidity. However, in general the spacing should be limited to between $\frac{1}{18}$ and $\frac{1}{20}$ of the span, with a minimum of 4 m to 5 m for through trusses and approximately $\frac{1}{15}$ of the span, with a minimum of 3 m to 4 m, for underslung trusses.

19.3 Effects of load reversal

For buildings with light pitched roofs, load reversal is often caused by wind suction and internal pressure. Load reversal caused by wind load is of particular importance as light sections normally acting as ties under dead and imposed loads may be severely overstressed or even fail by buckling when required to act as struts. For heavy pitched or flat roofs load reversal is rarely a problem because the dead load usually exceeds the wind uplift forces.

For bridge trusses, load reversal in the component elements may be caused by the erection technique adopted or by moving live loads, particularly in continuous bridges. During the detailed design stage, consideration should be given to the method of erection to ensure stability and adequacy of any member likely to experience load reversal. For short-span simply-supported trusses erected whole, load reversal in the chords and web members is attained if the crane pick-up points during erection are at or near mid-span, Fig. 19.4(a) and (b). For large-span bridges, erection by the cantilever method causes load reversal in the chords and web members. Load reversal caused by moving loads is usually more significant in continuous trusses. A convenient way of overcoming the problem of load reversal in web elements which are likely to buckle is to provide either temporary or permanent counter bracing, Fig. 19.4(c). This will ensure that the web elements are always in tension under all load conditions and avoids the use of heavy compression elements.

19.4 Selection of elements and connections

19.4.1 Elements

For light roof trusses in buildings the individual members are normally chosen from rolled sections for economy; these are illustrated in Fig. 19.5(a). Structural hollow sections are becoming more popular due to their efficiency in compression

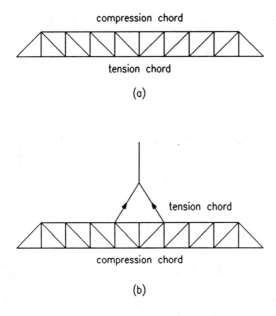

Fig. 19.4 Effects of load reversal. (a) Normal loading; (b) reversal during erection; (c) counter bracing

and their neat and pleasing appearance in the case of exposed trusses. Structural hollow sections, however, have higher fabrication costs and are only suited to welded construction. For larger-span heavily-loaded roof trusses and small span bridge trusses it often becomes necessary to use heavier sections such as rolled universal beams and columns and multiples of the smaller rolled sections such as back-to-back angles and channels, Fig. 19.5(b).

For large-span bridge trusses, compound or fabricated sections are normally necessary, particularly for the chords and the compression web members. Figure 19.5(c) illustrates some typical arrangements often used, although the choice available to the designer is very wide. For heavily-loaded bridge trusses, fabricated box sections offer economy in material due to their high efficiency in compression.

536 Trusses

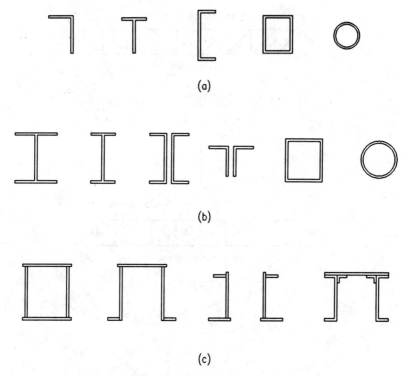

Fig. 19.5 Typical element cross sections. (a) Light building trusses; (b) heavy building trusses and light, small-span bridges; (c) road and railway bridges

The open sections shown assist the connections to the web members of the truss but require lacing or battening and are prone to distortion during fabrication.

Providing suitable access to all members and surfaces for inspection, cleaning and painting should be a primary consideration in deciding on the sections and details to be incorporated in the design. Laced sections are disadvantageous in this respect as access can be severely restricted. In highly corrosive environments welded closed box or circular hollow sections with welded connections are usually used in order to reduce maintenance costs as all exposed surfaces are readily accessible.

19.4.2 Connections

There are basically three types of connections used for connecting truss elements to each other, that is, welding, bolting and riveting. Riveting is rarely used in the UK due to the very high labour costs involved, although it is still widely used in

Third World countries where labour costs are low. Small-span trusses which can be transported whole from the fabrication shop to the site can be entirely welded. In the case of large-span roof trusses which cannot be transported whole, welded sub-components are delivered to site and are either bolted or welded together on site. Generally in steelwork construction bolted site splices are much preferred to welded splices for economy and speed of erection. In light-building roof trusses entirely bolted connections are less favoured than welded connections due to the increased fabrication costs, and usually bolted connections require cumbersome and obtrusive gusset plates. However, bolted connections are more widely used in bridge trusses, particularly medium- to large-span road bridges and railway bridges, due to their improved performance under fatigue loading. In addition, bolted connections may sometimes permit site erection of the individual elements without the need for expensive heavy craneage. Gusset plates are often associated with bolted bridge trusses, their size being dependent on the size of the incoming members and the space available for bolting.

Gusset plates also enable the incoming members to be positioned in such a way that their centroidal axes meet at a single point thus avoiding load eccentricities. Ideally for all types of trusses the connections should be arranged so that the centres of gravity of all incoming members meeting at the joint coincide. If this is not possible the out-of-balance moments caused by the eccentricities must be taken into account in the design.

Some typical joint details are illustrated in Fig. 19.6.

19.5 Guidance on methods of analysis

Loads are generally assumed to be applied at the intersection points of the members, so that they are principally subjected to direct stresses. To simplify the analysis the weights of the truss members are assumed to be apportioned to the top and bottom chord panel points and the truss members are assumed to be pinned at their ends, even though this is usually not the case. Normally chords are continuous and the connections are either welded or contain multiple bolts; such joints tend to restrict relative rotations of the members at the nodes and end moments develop.

Generally, in light building trusses secondary stresses are negligible and are often ignored. Guidance on whether or not to consider secondary stresses in light building trusses is given in BS 5950: Part 1. In accordance with clause 4.10, secondary stresses may be neglected provided that:

- the slenderness of the chord members in the plane of the truss is greater than 50, and
- the slenderness of most of the web members, about the same axis, is greater than 100.

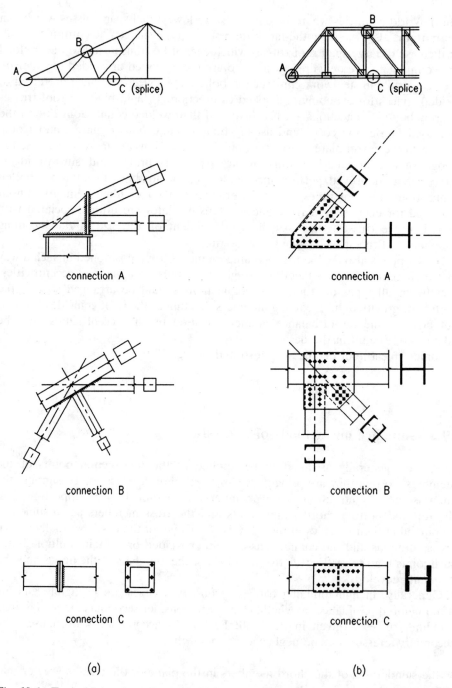

Fig. 19.6 Typical joints in trusses. (a) Welded RHS building roof truss; (b) bolted bridge truss

However in bridge trusses, the secondary stresses can be a significant proportion of the primary stresses and must be taken into account. The British Standard for the design of steel bridges, BS 5400: Part 3, 1982, requires the fixity of the joints to be taken into account although axial deformation of the members may be ignored for the ultimate limit state.

The magnitude of the secondary stresses depends on a number of factors including member layout, joint rigidity, the relative stiffness of the incoming members at the joints and lack of fit.

Manual methods of analysis are often adopted to analyse the stresses, particularly in simple trusses. For simple, statically-determinate trusses, methods of analysis include joint resolution, graphical analysis (Bow's notation or Maxwell diagram) and the method of sections. The last method is particularly useful as the designer can limit the analysis to the critical sections.

Statically-indeterminate trusses are more laborious to analyse manually; methods available include virtual work, least work and the reciprocal theorem with influence lines. For a full discussion on these methods of analysis the reader should refer to textbooks on structural analysis.

Computers are nowadays readily available to designers and provide a useful means of analyzing the most complex of trusses. In addition, joint and member rigidities can easily be incorporated in the modelling thus avoiding laborious hand calculations in determining out-of-balance moments caused by joint deformations. Local stresses caused by loads not applied at the panel points, joint eccentricities and axial deformation should generally be calculated and superimposed on the direct stresses. However, stresses due to axial deformation are normally neglected except for bridge trusses and trusses of major importance.

Careful consideration must be given to the out-of-plane stability of a truss and resistance to lateral loads such as wind loads or eccentric loads causing torsion about their longitudinal axis. An individual truss is very inefficient and generally sufficient bracing must be provided between trusses to prevent instability. In bridges, plan bracing is normally provided between trusses at the chord levels in addition to stiff end portals to prevent lateral instability.

19.6 Detailed design considerations for elements

19.6.1 Design loads

The current British Standards for steel structures in buildings and bridges are both limit-state codes. The magnitude of the partial load factors to be applied is dependent on the load type, the load combination and the limit state (ultimate or serviceability) under consideration.

The following approach may be adopted in deriving the critical load combinations for each truss member:

(1) The member forces and moments are calculated for each, unfactored, load type (dead, superimposed dead, imposed, wind, etc.) using an appropriate method of analysis.
(2) Load combinations are identified and the appropriate load factors for each combination applied for both serviceability and ultimate limit states.
(3) The critical loads in each element and joint are extracted for both limit states.

The above process is long-winded but with experience the designer can often take short cuts in determining the critical load combinations for each element.

In the analysis the member forces and moments due to joint fixity should be calculated and superimposed on the global member forces. For trusses in buildings to BS 5950: Part 1: 1985, the secondary effects due to joint fixity may be ignored provided the slenderness, in the plane of the truss, is greater than 50 for the chord elements and 100 for most of the web members. If this condition is satisfied the members are assumed to be pin jointed in the analysis. Secondary effects due to axial deformations are usually ignored in building trusses. Local effects due to joint eccentricities and where loads are not applied at nodes should be taken into account.

For bridge trusses to BS 5400: Part 3: 1982, the effects of joint rigidity are required to be taken into account. Secondary stresses due to axial deformations may be ignored at the ultimate limit state but should be considered at the serviceability limit state and for fatigue checks. As for building trusses, the local effects due to joint eccentricities and cases where loads are not applied at nodes must be considered in bridge trusses.

19.6.2 Effective length of compression members

For building trusses the fixity of the joints and the rigidity of adjacent members may be taken into account for the purpose of calculating the effective length of compression members. BS 5950: Part 1: 1985 limits the maximum slenderness permissible in struts to the following:

180 for members resisting loads other than wind loads
250 for members resisting self-weight and wind loads
350 for members normally acting as a tie but subject to load reversal under wind action.

In addition, members whose slenderness exceeds 180 must be checked for self-weight deflection. If this exceeds (length/1000) the effect of bending should be taken into account.

The designer should be careful to ensure that he identifies the critical slenderness. For chords, out-of-plane unrestrained lengths do not necessarily relate to the truss nodes and effective length factors are usually unity; in-plane effective length factors may be demonstrated to be less than unity if the restraining actions of tension members and non-critical compression members are mobilized at the ends of the member. Single angle elements, both for the webs and chord, have minimum radii of gyration that do not lie either in, or normal to, the plane of the truss.

For compression members in bridge trusses the effective lengths may either be obtained from Table 11 of BS 5400: Part 3: 1982 or be determined by an elastic critical buckling analysis of the truss.

In the case of simply supported underslung trusses the top compression chord will be effectively restrained laterally throughout its length provided the connection between the chord and the deck is capable of resisting a uniformly distributed lateral force of 2.5% of the maximum force in the chord. The effective length in such a case is taken as zero where friction provides the restraint, or as equal to the spacing of discrete connections where these are provided.

The economic advantages of underslung trusses over through or semi-through trusses is obvious in this respect, due to the dual function of the deck structure.

In the case of unbraced compression chords, that is, chords with no lateral restraints, the provision of U-frames is necessary. The effective length of the compression chord is a function of the stiffness of the chord and the spacing and stiffness of the U-frame members. Clause 12.5 of BS 5400: Part 3: 1982 gives guidance on the calculation of the effective length of compression chords restrained by U-frames.

19.6.3 Detailed design

For building trusses to BS 5950 the members need only be designed at the ultimate limit state for strength, stability and fatigue where applicable, and at the serviceability limit state for deflection and durability.

Compression members in bridge trusses to BS 5400 are designed at both the ultimate and serviceability limit states. Certain compression members, however, are exempt from the serviceability limit state check as defined in clause 12.2.3 of BS 5400: Part 3. Tension members need only be designed at the ultimate limit state.

For guidance on the detailed design of axially loaded members the reader should refer to Chapters 14 and 15, and to Chapter 18 for members subject to combined axial load and bending.

19.7 Factors dictating the economy of trusses

Some of the general factors dictating the economy of trusses relating to truss type, spacing, span-to-depth ratios, pitch, etc. have already been discussed earlier. However, factors such as the location of the structure, contractors' experience and material availability, may have a significant effect on the choice of truss type and details adopted. When designing trusses for overseas locations, particularly the Third World countries or for remote areas with difficult access, the designer should consider the following:

- Material available locally, i.e. weldable or unweldable steel
- Preferred connections, i.e. welded, bolted or riveted
- Maximum size of elements that can be transported to site
- Method of erection, capacity and type of plant available
- Use of rolled, compound or fabricated sections
- Redundancy of structure in case of overloading and lack of maintenance
- Experience of local contractors
- Simple design with maximum repetition.

The designer should always try to maximize the use of local materials, labour and expertise so as to avoid expensive importation of materials and trained manpower. In addition, the relative costs between materials and labour should be reflected in the design. For Third World countries material costs are often high and labour costs are very low, whereas the reverse is true for developed countries.

19.8 Other applications of trusses

Trusses are often used as secondary structures in buildings and bridges in the form of triangulated bracing. Bracing is generally required to resist horizontal loading in buildings or bridges or to prevent deformations and provide torsional rigidity to stiffening girders or box girders.

In buildings, bracing is often required for stability and to transmit horizontal wind loads or crane surges down to foundation level. To avoid the use of heavy compression bracing members, the members are usually arranged so that they always act in tension. Although this requires a high degree of redundancy it is normally more economical than providing compression members. Some examples of wind bracing to single-storey building are illustrated in Fig. 19.7(a). For multi-storey buildings with 'simple' connections, vertical bracing is required on all elevations to stabilize the building. Normally the floor slabs act as horizontal bracing which transmits the lateral wind loads to the vertical bracing. If the steel frame to the building is erected before the floors are constructed, temporary horizontal bracing must be supplied which can be removed once the floors are in place. Bracing at floor levels may be required if the slabs are discontinuous. Although horizontal bracing in buildings can often be hidden within the depth of the floor, vertical bracing can be obtrusive and undesirable, particularly if the building is clad in glass. In such an instance rigidly-jointed frames are adopted in which the wind loading is resisted by bending in the beams and stanchions. However, if the joints to such frames are made with friction-grip bolts, then

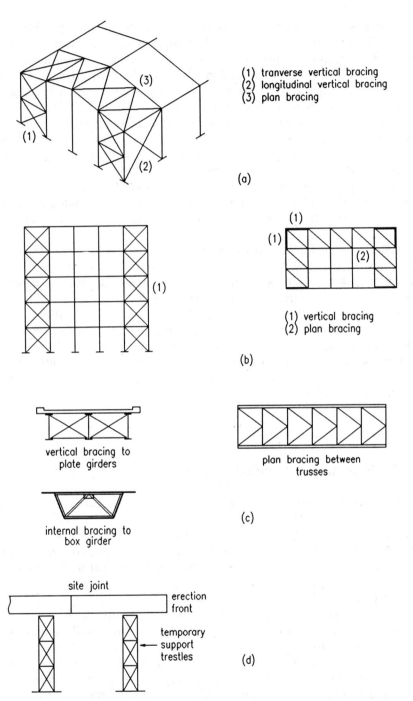

Fig. 19.7 Other applications of trusses. (a) Bracing to single-storey building; (b) bracing to multi-storey building; (c) typical bracing to bridges; (d) other uses

temporary bracing may be required for stability prior to the joints being completed. Figure 19.7(b) illustrates some typical bracing systems to multi-storey buildings.

In bridges, secondary truss or bracing frames are often required for stability and to resist lateral loads due to wind in addition to loads due to road and railway loading such as centrifugal or braking forces. Depending on the type of structure the bracing may be temporary or permanent, usually placed in both horizontal and vertical planes. For trusses, permanent plan bracing is provided at both chord levels for underslung bridges and at deck level for through or semi-through bridges. Where headroom permits, plan bracing is also provided at the top chord level for through trusses thus conveniently providing restraint to the top compression chord and avoiding the need for stiff U-frames. In addition, vertical bracing is also provided between trusses to reduce differential loading and therefore distortion between trusses and to provide added restraint to the compression chord.

In composite steel plate girder and concrete slab decks temporary vertical bracing may be required when the concrete is poured to provide lateral restraint to the plate girder compression flanges. It may be removed once the concrete has gained sufficient strength to act compositely with the steelwork. In stiffening girders or box girders bracing is often provided in place of plated diaphragms to avoid torsional distortion and to maintain the shape of the cross section under service loads. Figure 19.7(c) illustrates some uses of trusses in bridgeworks. Other uses of trusses in bridgeworks include trestling, i.e. triangulated temporary support frames normally used to support medium- to large-span bridges over land during erection, see Fig. 19.7(d).

19.9 Rigid-jointed Vierendeel girders

19.9.1 Use of Vierendeel girders

Vierendeel girders, unlike trusses or lattice girders, are rigidly-jointed open-web girders having only vertical members between the top and bottom chords. The chords are normally parallel or near parallel; some typical forms are shown in Fig. 19.8(a).

The elements in Vierendeel girders are subjected to bending, axial and shear stress, unlike conventional trusses with diagonal web members where the members are primarily design for axial loads. Vierendeel girders are usually more expensive than conventional trusses and their use is limited to instances where diagonal web members are either obtrusive or undesirable. Vierendeel girders in bridges are rare; they are more commonly used in buildings where access for circulation or a large number of services is required within the depth of the girder.

The economic proportions and span lengths are similar to those of the parallel chord trusses already discussed in section 19.2.

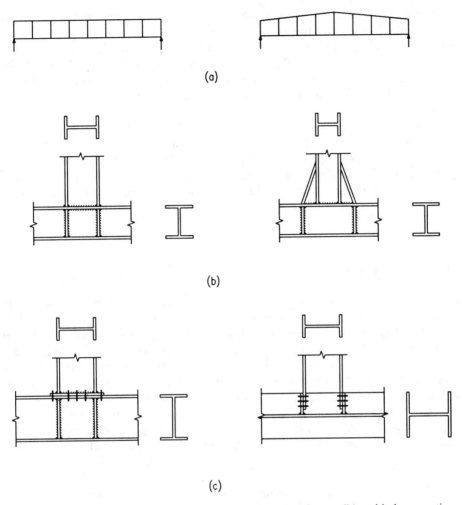

Fig. 19.8 Typical details of Vierendeel girders: (a) typical forms, (b) welded connections, (c) bolted connections

19.9.2 Analysis

Vierendeel girders are statically indeterminate structures but various manual methods of analysis have been developed. The statically determinate method assumes pin joints at the mid-points of the verticals and chords of each panel. The method, however, is only suitable for girders with parallel chords of constant stiffness and when the loads are applied at the node points. Various modified moment distribution methods have been developed for the analysis of Vierendeel girders which allow for inclined chords, chords of different stiffness in the panels and member widening at the node positions.

The use of computers offers the most accurate and efficient way of analyzing Vierendeel girders, particularly those with inclined chords, chords of varying stiffness and when the loading is not applied at the node positions. A further advantage of computer analysis is that joint rotations and deflections are easily calculated.

Plastic theory may be applied to the design of Vierendeel girders in a similar way to its application to other rigid frames such as portal frames. Failure of the structure, as a whole, generally results from local failure of a small number of its members to form a mechanism. Once the failure mode is established the chords and vertical are designed against failure. Computer programs are available for the plastic analysis of plane frameworks including Vierendeel arrangements.

19.9.3 Connections

Vierendeel girders have rigid joints with full fixity and so the connections must be of the type which prevents rotation or slip of the incoming members, such as welded or friction-grip bolted connections. Welded connections are usually the most efficient and compact although undesirable if the connections are required to be made on site. Normally site splices are bolted for economy. For very large Vierendeel girders delivered and erected piecemeal, fully bolted connections are normally used. For member and joint efficiency the ends of the verticals are often splayed. This is of advantage in heavily-loaded girders as the high concentrated local stresses are reduced thus avoiding the need for heavy stiffening.

Some typical joint examples are illustrated in Fig. 19.8(b) and (c).

A series of worked examples follows which are relevant to Chapter 19.

The Steel Construction Institute Silwood Park, Ascot, Berks SL5 7QN	Subject ROOF TRUSS	Chapter ref. 19
	Design code BS5950: Part 1 Made by **RT** Checked by **GWO**	Sheet no. **1**

Problem

Design the roof trusses for an industrial building with two 25 m spans, 120 m long. The roofing is insulated metal sheeting with purlins at node positions. The building is 9 m to the eaves, basic wind speed is 42 m/s and ground roughness is category 3c.

Structural form:

For 25 m span a pitched roof will not be economical as the height at the apex would be in the order of 5.5 m.

Ideal solutions would either be a Mansard truss or a parallel chord Pratt, Howe or Warren truss.

For the purpose of this design example a Mansard truss will be adopted.

Economical span to depth ratio between 7 and 8

For 3.5 m depth $\dfrac{span}{depth} = \dfrac{25}{3.5}$

$\hspace{9em} = 7.14$ acceptable

Truss spacing should be in the region of 1/4 to 1/5th of the span

For 6 m truss centres

spacing/span = 6/25 = 1/4.17   acceptable

A truss spacing of 6 m conveniently suits the length of the building.

548   Worked example

The Steel Construction Institute Silwood Park, Ascot, Berks SL5 7QN		Subject ROOF TRUSS		Chapter ref. 19
		Design code BS5950: Part 1	Made by RT Checked by GWO	Sheet no. 2

Number of bays = 120/6 = **20 No.**

Truss dimensioning:

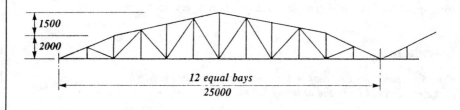

Loading: kN/m²   BS: 648

Dead load   -   Steel sheeting   =   0.075

                          Insulation   =   0.020

                          Fixings   =   0.025

                          Services etc   =   0.100

                          Total   =   0.22 kN/m²

for 6 m bays

Roof dead load = 0.22 × 25 × 6   =   33.0 kN

Allow for purlins   =   11.8 kN

Allow for own weight   =   30.0 kN

                          Total load =   74.8 kN

nodal load   = $\dfrac{74.8}{12}$ =   6.2 kN

	Subject		Chapter ref.
The Steel Construction Institute Silwood Park, Ascot, Berks SL5 7QN	ROOF TRUSS		19
	Design code BS5950: Part 1	Made by **RT** Checked by **GWO**	Sheet no. 3

Imposed load: BS: 6399

No access to roof, $\therefore$ *snow load* = 0.75 kN/m^2

nodal load $= 0.75 \times 6 \times 25/12 =$ 9.4 kN

Wind loading: CP3:
                                                                                                           Chapter V:

Basic wind speed   =   42 m/s   P2

Design wind speed V_s = $V S_1, S_2 S_3$

Topography factor S_1 =   1.0

Statistical factor S_3 =   1.0

Ground roughness category   =   3 c

Roof elevation between 9 m & 12.5 m

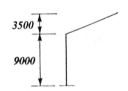

$\therefore S_2 =$ 0.74   CP3:
                                                                                          Chapter V:
                                                                                          P2 Table 3

$V_s = 42 \times 1 \times 0.74 \times 1$ =   31.1 m/s

Dynamic pressure q = kV_s^2

where k =   0.613

                     q = $0.613 \times 31.1^2 \times 10^{-3}$

                               = 0.593 kN/m^2

External pressure coefficients, C_{pe}

Roof slopes approximately   26° *at ends*

                           &   10° *at C (see diagram, sheet 5)*

550   Worked example

The Steel Construction Institute Silwood Park, Ascot, Berks SL5 7QN	Subject **ROOF TRUSS**		Chapter ref. **19**
	Design code **BS5950: Part 1**	Made by **RT** Checked by **GWO**	Sheet no. **4**

Building Dimensions   w = 28 m   (b/N = 0.32)

l = 120 m

h = 9 m

CP3:
Chapter V:
P2 Table 8

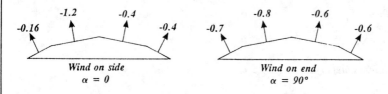

Wind on side   Wind on end
$\alpha = 0$ $\alpha = 90°$

Internal pressure coefficients Cpi =   + 0.2 or − 0.3

Wind force = $q A (Cpe - Cpi)$

Maximum $(Cpe - Cpi)$

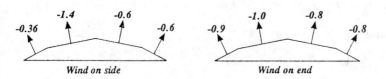

Wind on side   Wind on end

Wind forces:

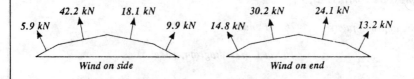

Wind on side   Wind on end

	Subject	Chapter ref.
The Steel Construction Institute Silwood Park, Ascot, Berks SL5 7QN	**ROOF TRUSS**	19
	Design code BS5950: Part 1	Made by **RT** Sheet no. **5** Checked by **GWO**

Member forces (unfactored)

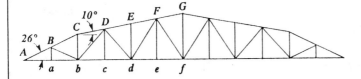

Member	Dead Load kN	Imposed Load kN	Wind on Side kN	Wind on End kN
A-B	-78.8	-118.2	104.0	91.0
B-C	-72.7	-109.1	101.7	85.2
C-D	-67.0	-100.5	96.2	81.5
D-E	-76.8	-115.2	110.6	87.4
E-F	-76.8	-115.2	113.2	89.2
F-G	-66.6	-99.9	102.0	79.1
A-a	72.5	108.8	-106.9	-97.2
a-b	72.5	108.8	-106.9	-97.2
b-c	74.3	111.5	-115.3	-94.4
c-d	74.3	111.5	-115.3	-94.4
d-e	71.9	107.9	-105.9	-82.8
e-f	71.9	107.9	-105.9	-82.8
B-a	0	0	0	0
C-b	12.7	19.1	-15.1	-11.8
D-c	0	0	0	0
E-d	-6.2	-9.3	11.9	8.5
F-e	0	0	0	0
G-f	18.6	27.9	-37.4	-29.4
B-b	-7.4	-11.1	6.3	9.5
b-D	-12.8	-19.2	17.8	8.8
D-d	1.7	2.6	-1.9	-2.8
d-F	6.2	9.3	-14.9	-12.8
F-f	-11.5	-17.3	22.4	17.8

The above forces have been calculated assuming pin joints

		Subject			Chapter ref.
The Steel Construction Institute Silwood Park, Ascot, Berks SL5 7QN		ROOF TRUSS			19
		Design code BS5950: Part 1	Made by RT Checked by GWO	Sheet no.	6

Load factors & combinations:

For Dead + Imposed
$1.4 \times DL + 1.6 \times IL$

For Dead + Wind
$1.0 \times DL + 1.4 \times WL$

For Dead + Imposed + Wind

not critical as wind loads act in opposite direction to Dead and Imposed loads

Member forces (factored)

Member	Dead + Imposed kN	Dead + Wind kN
A-B	-299.4	66.8
B-C	-276.3	69.7
C-D	-254.6	67.7
D-E	-291.8	78.0
E-F	-291.8	81.7
F-G	-253.1	76.2
A-a	275.6	-77.2
a-b	275.6	-77.2
b-c	282.4	-87.1
c-d	282.4	-87.1
d-e	273.3	-76.4
e-f	273.3	-76.4
B-a	0	0
C-b	48.3	-8.4
D-c	0	0
E-d	-23.6	10.5
F-e	0	0
G-f	70.7	-33.8
B-b	-28.1	5.9
b-D	-48.6	12.1
D-d	6.5	-2.2
d-F	23.6	-14.7
F-f	-43.8	19.9

BS 5950:
Part 1
Table 2

The Steel Construction Institute Silwood Park, Ascot, Berks SL5 7QN	Subject ROOF TRUSS		Chapter ref. 19
	Design code BS5950: Part 1	Made by RT Checked by GWO	Sheet no. 7

Secondary stresses:

Provided the slenderness of the chords and (most) web members are greater than 50 and 100 respectively, secondary stresses may be ignored (slenderness in the plane of the truss). This condition is to be checked after preliminary sizing of the members.

BS 5950:
Part 1
4.10

Member Design:

(A) *Top Chord* -   member A-B

Maximum compressive force   =   299.4 kN

Effective length L_E =   1.0L   =   2311 mm   Table 24

Try 90 × 90 × 5 RHS Grade 43

$\lambda = \dfrac{L_E}{r_y} = \dfrac{2311}{34.6}$

= 66.8   (> 50 ignore secondary stresses)   4.10

Maximum slenderness ratio λ =   180 OK   4.7.3.2

p_y = 275 N/mm^2

p_c = 228 N/mm^2 Table 27(a)

Allowable compressive force P_c = $A_g p_c$ 4.7.4

A_g = 16.9 cm^2

P_c = 16.9 × 10^2 × 228 × 10^{-3} =   385 kN > 299.4 kN OK

∴ *Use 90 × 90 × 5 RHS Grade 43*

(B) *Bottom Chord*   -   Member b-c

Maximum tensile force   =   282.4 kN
Maximum compressive force   =   87.1 kN

The Steel Construction Institute Silwood Park, Ascot, Berks SL5 7QN	Subject ROOF TRUSS		Chapter ref. 19
	Design code BS5950: Part 1	Made by RT Checked by GWO	Sheet no. 8

For welded connections allowable tensile force, P_t 4.6.1

P_t = $A_e p_y$ A_e = gross area in this case

Assume the same size section as the top chord

∴ Try 90 × 90 × 5 RHS Grade 43

Allowable tensile force = $16.9 \times 10^2 \times 275 \times 10^{-3}$
 = 464.8 kN > 282.4 kN OK

Under Wind + Dead load the bottom chord goes into compression.

By inspection, the bottom chord will require longitudinal ties. Assume that these ties are at quarter positions i.e. 6.25 m spacing.

∴ L_E = 6.25 × 1.0

λ = $6.25 \times 10^3 / 34.6$ = 181 < 350   4.7.3.2

∴ p_c = 56 N/mm^2 Table 27(a)

p_c = $16.9 \times 10^2 \times 56 \times 10^{-3}$

 = 94.6 kN > 87.1 kN OK

∴ <u>Use 90 × 90 × 5 RHS Grade 43</u> with longitudinal ties at 6.25 m intervals.

(C)   <u>Web members</u>   -   member G - f

Maximum tensile force   = 70.7 kN

Maximum compressive force = 33.8 kN

Try using 40 × 40 × 3.2 RHS

Allowable tensile force P_t = $466 \times 275 \times 10^{-3}$ 4.6.1

 = 128.2 kN > 70.7 kN ∴ OK

The Steel Construction Institute Silwood Park, Ascot, Berks SL5 7QN	Subject ROOF TRUSS		Chapter ref. 19
	Design code BS5950: Part 1	Made by RT Checked by GWO	Sheet no. 9

Effective length L_E = 1.0 L   Table 24

Slenderness λ = $\dfrac{L_E}{r_y}$ = $\dfrac{3500}{15.0}$ = 233.3   4.7.3.1
4.10

(> 100 ∴ ignore secondary stresses)

Max allowable slenderness λ = 350 > 233.3 ∴ OK   4.7.3.2

(in compression due to wind loads)

p_c = 35 N/mm²   Table 27(a)

Allowable compressive force P_c = 466 × 35 × 10⁻³

P_c = 16.3 kN < 33.8 kN

section is overstressed due to wind load reversal

Try using 80 × 40 × 4 RHS

λ = $\dfrac{3500}{15.9}$ = 220

∴ p_c = 39 N/mm²   Table 27(a)

Allowable compressive force P_c = 39 × 888 × 10⁻³

= 34.6 kN > 33.8 kN OK

<u>Use 80 × 40 × 4 RHS</u>

For guidance on the design of connections the reader should refer to Chapter 26.

556 *Worked example*

The Steel Construction Institute Silwood Park, Ascot, Berks SL5 7QN		Subject **ROOF TRUSS**	Chapter ref. **19**
		Design code **BS5950: Part 1**	Made by **RT** Checked by **GWO**	Sheet no. **10**

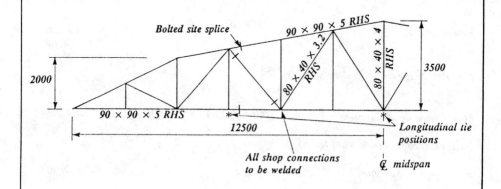

The Steel Construction Institute Silwood Park, Ascot, Berks SL5 7QN	Subject **ROAD BRIDGE**		Chapter ref. **19**
	Design code *BS5400*	Made by **RT** Checked by **GWO**	Sheet no. **1**

Problem

Design the principal structure for a bridge span of 70 m, carrying a 2 lane carriageway 7.5 m wide with 2 footways 1.5 m wide. The loading is to be HA plus 37.5 units of HB. Construction depth below the road level is limited to 2.5 m maximum.

Structural form:

For road bridge span to depth ratio ~ 10 to 12

Try using 6.5 m depth

span/depth $= \dfrac{70}{6.5} =$ 10.8 adequate

Bay width say 5 m

angle of diagonals $=$ arc tan $\dfrac{6.5}{5}$

$=$ 52.4°

$> 50°$ adequate

Cross beams to be provided at each panel point.

From the limitations imposed on the construction depth below road level an underslung truss will not be appropriate. A through truss will be adopted with an unbraced compression chord to meet the unlimited headroom requirement. Restraint to the compression chord to be provided by U-frame action.

A Pratt truss will be appropriate for a span of 70 m although a Warren or Howe truss could also be adopted.

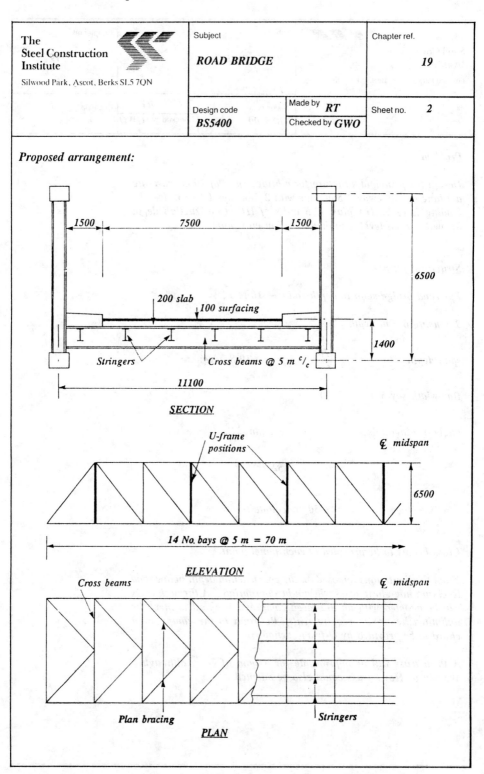

Worked example 559

The Steel Construction Institute Silwood Park, Ascot, Berks SL5 7QN	Subject **ROAD BRIDGE**	Chapter ref. **19**
	Design code **BS5400**	Made by **RT**	Sheet no. **3**
		Checked by **GWO**	

Sections:

The truss is analysed using a plane frame stress analysis program.
The members are estimated using hand analysis.

Try using the following sections:

Top chord

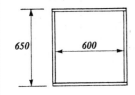

650 × 25 top plate

600 × 25 webs

650 × 25 bottom plate

Bottom chord

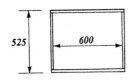

640 × 20 top plate

485 × 20 webs

640 × 20 bottom plate

Diagonal webs

2 No. 381 × 102 C's

Vertical webs

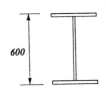

400 × 25 flange

15 web

400 × 25 flange

560   Worked example

The Steel Construction Institute Silwood Park, Ascot, Berks SL5 7QN		Subject *ROAD BRIDGE*		Chapter ref. 19
		Design code *BS5400*	Made by *RT* Checked by *GWO*	Sheet no. 4

Section properties:

Member	Area mm^2	$\bar{y}$ mm	I In Plane mm^4	I Out of Plane mm^4	wt/m kN/m
Top chord	62500	325	4.074 E9	4.074 E9	4.95
Bottom chord	45000	263	2.012 E9	2.738 E9	3.53
Diagonals	14038		0.298 E9	1.072 E9	1.10
Verticals	28250		0.267 E9	1.861 E9	2.16

For the purpose of this example the members will be kept constant throughout the span.

Loading

Loads calculated for each truss at each cross beam location.

(i)   Surfacing = $0.1 \times 7.5 \times 5 \times 0.5 \times 22$ = 41.3 kN

(ii) Parapets = 5×1.0 = 5.0 kN

(iii) Slab   = $0.2 \times 10.5 \times 5 \times 0.5 \times 24$ = 126.0 kN

(iv) Footways = $0.2 \times 1.5 \times 5 \times 24$ = 36.0 kN

(v)   Deck steelwork & bracing = 24 kN

(vi) Own wt.   = $(4.95 + 3.53) \times 5 + 1.1 \times 8.2 + 2.16 \times 6.5$

           = 65.5 + 10% (extras) = 72 kN

(vii) HA UDL

No. of notional lanes 2   Part 2, 3.2.9.3

The Steel Construction Institute Silwood Park, Ascot, Berks SL5 7QN	Subject ROAD BRIDGE		Chapter ref. 19
	Design code BS5400	Made by RT Checked by GWO	Sheet no. 5

For 70 m loaded length, HA UDL per lane   =   22.5 kN/m   BD 14/82
                                                                (Amended)

for spans > 50 m   1 lane 1.2 × HA UDL

                        1 lane 0.8 × HA UDL

Assume lanes equally loaded for purpose of this example

∴ HA UDL   =   22.5 × 5   =   112.5 kN

(viii)   HA KEL: Part 2, 6.2.2

120 kN per lane

Point load   =   120 kN at any position per truss

(ix)   HB Associated HA UDL:

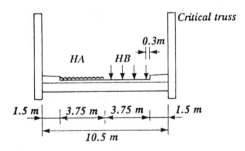

Using static distribution, HA UDL on critical truss

$$= 22.5 \times 5 \times \frac{3.375}{10.5} = 36.2 \text{ kN}$$

The Steel Construction Institute Silwood Park, Ascot, Berks SL5 7QN	Subject ROAD BRIDGE		Chapter ref. 19
	Design code BS5400	Made by **RT** Checked by **GWO**	Sheet no. **6**

(x) **HB load (37.5 units):**   Part 2, 6.3

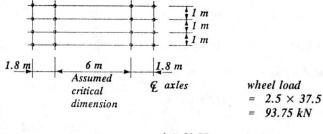

wheel load
= 2.5 × 37.5
= 93.75 kN

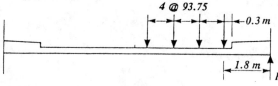

Reaction $R = \dfrac{93.75}{10.5} \times (8.7 + 7.7 + 6.7 + 5.7)$

$ = 257$ kN

Travelling load to be applied

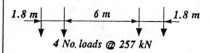

4 No. loads @ 257 kN

(xi) **Footway loading:**   Part 2, 7.1.1

$= 5 \times \left[\dfrac{22.5}{30} \right] \times 1.5 \times 5 \quad = \quad 28.2$ kN

For the purpose of this example the following loads shall be ignored, but would normally be considered in a similar way to the primary loading ie.

Wind; Temperature; Erection; Braking, traction; etc.

The Steel Construction Institute Silwood Park, Ascot, Berks SL5 7QN	Subject **ROAD BRIDGE**		Chapter ref. **19**
	Design code **BS5400**	Made by **RT** Checked by **GWO**	Sheet no. **7**

Design for load combination *Part 2, 4.4*

Load cases:

(a) Steel Dead Load - top nodal points   = 72/2   = 36 kN

Steel Dead Load - bottom nodal points   = 36 + 24   = 60 kN

(b) Concrete dead load - bottom nodal points   = 126 + 36   = 162.0 kN

(c) Superimposed dead - bottom nodal points   = 41.3 + 5   = 46.3 kN

(d) HA UDL + KEL - bottom nodal points   = 112.5 kN

& Travelling load   = 120 kN

(e) HB + Associated HA UDL = bottom nodal points   = 36.2 kN

& 4 No. Travelling loads @ 257 kN

(f) Footway loading - bottom nodal points   = 28.2 kN

Nodal points notation:

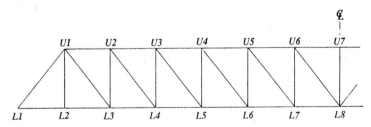

The design will be limited to the following members

U6 - U7   Top chord
L7 - L8   Bottom chord
U1 - L3   Diagonal
U2 - L3   Vertical

Worked example

The Steel Construction Institute Silwood Park, Ascot, Berks SL5 7QN	Subject **ROAD BRIDGE**	Chapter ref. **19**
	Design code **BS5400**	Made by **RT** Sheet no. **8** Checked by **GWO**

From the computer output,

Member		Dead Steel kN	Dead Concrete kN	Super Dead kN	HA kN	HB kN	FTWY kN	
U6 - U7		- 1695	- 3041	- 869	- 2428	- 3113	- 530	*Unfactored axial loads*
L7 - L8		1658	2974	850	2342	3244	518	
U1 - L3		580	1039	297	836	1157	181	
U2 - L3		- 419	- 812	- 232	- 648	- 843	- 141	
(U6 - U7)	U6	32.5	58.3	16.7	60.1	145.8	10.1	
	U7	34.1	60.4	17.5	62.1	87.0	10.7	
(L7 - L8)	L7	15.2	27.2	7.8	33.3	95.8	4.7	*Unfactored bending moments kNm*
	L8	16.2	28.9	8.3	32.7	34.0	5.1	
(U1 - L3)	U1	25.3	45.4	13.0	36.7	48.5	7.9	
	L3	- 24.4	- 43.8	- 12.5	- 35.3	- 50.0	- 7.6	
(U2 - L3)	U2	22.1	39.7	11.4	32.2	44.4	6.9	
	L3	- 20.5	- 36.6	- 10.5	- 29.9	- 41.5	- 6.4	

for load combination 1   Part 2, 4.4

load combination 1a   Part 2, T1

at SLS $1.0DS + 1.0DC + 1.25DL + 1.2HA + 1.0FTWY$

at ULS $1.05DS + 1.2DC + 1.75SDL + 1.5HA + 1.0FTWY$

load combination 1b

at SLS $1.0DS + 1.0DC + 1.25DL + 1.1HB + 1.0FTWY$

at ULS $1.0DS + 1.2DC + 1.75SDL + 1.3HB + 1.5FTWY$

	Subject	Chapter ref.
The Steel Construction Institute Silwood Park, Ascot, Berks SL5 7QN	ROAD BRIDGE	19
	Design code BS5400	Made by RT / Checked by GWO	Sheet no. 9

Factored Axial loads & Bending Moments

Member		Load Combination 1a				Load Combination 1b			
		ULS		SLS		ULS		SLS	
		AXIAL	BM	AXIAL	BM	AXIAL	BM	AXIAL	BM
		kN	kNm	kN	kNm	kN	kNm	kN	kNm
(U6-U7)	U6 U7	-11387	239 248	-9222	193 201	-11792	338 268	-9733	281 222
(L7-L8)	L7 L8	11087	119 123	8980	96 99	11791	194 118	9738	162 98
(U1-L3)	U1 L3	3901	171 -164	3160	138 -133	4151	179 -177	3429	148 -146
(U2-L3)	U2 L3	-3004	149 -138	-2428	121 -112	-3128	159 -147	-2578	131 -122

Load combination (1b) critical for members selected.

The tensile members are designed at the ULS only.

Part 3,
12.2.3

The compression members are designed at the ULS and the SLS to meet the requirements of Clause 12.2.3(a) and (b).

Compression chord effective length:

U-frame restraints are 10 m centres

Effective length l_e = $2.5 k_3 (EI_c \, a \, \delta)^{0.25}$

Ignoring the contribution of stiffness from the concrete slab

I_c = $4.074 \times 10^9 \, mm^4$

a = $10000 \, mm$

E = $205000 \, N/mm^2$

k_3 = 1.0

Worked example

The Steel Construction Institute 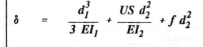 Silwood Park, Ascot, Berks SL5 7QN	Subject **ROAD BRIDGE**		Chapter ref. **19**
	Design code **BS5400**	Made by **RT** Checked by **GWO**	Sheet no. **10**

$$\delta = \frac{d_1^3}{3\,EI_1} + \frac{US\,d_2^2}{EI_2} + f\,d_2^2$$

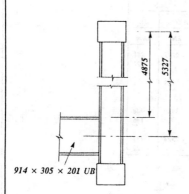

914 × 305 × 201 UB

$d_1 = 4875$ mm

$d_2 = 5327$ mm

$I_1 = 1.861 \times 10^9$ mm^4

$I_2 = 3.255 \times 10^9$ mm^4

$U = 0.5$

$S = 11100$ mm

$f = 0.1 \times 10^{-10}$

$$\delta = \frac{4875^3}{3 \times 2.05 \times 10^5 \times 1.861 \times 10^9} + \frac{0.5 \times 11100 \times 5327^2}{2.05 \times 10^5 \times 3.255 \times 10^9} + 0.1 \times 10^{-10} \times 5327^2$$

$\delta = 6.21 \times 10^{-4}$

$\therefore$ effective length $l_e = 2.5 \times 1.0 \times (2.05 \times 10^5 \times 4.074 \times 10^9 \times 10000 \times 6.21 \times 10^{-4})^{0.25}$

$\underline{l_e = 21216 \text{ mm}}$

Effective length of vertical web members:

For buckling in plane of truss $\quad l_e = 0.7\,l$ $\qquad$ Part 3, T11
For buckling out of plane of truss $\quad l_e = 1.0\,l$

Member design:

The individual members may now be designed. For guidance on the design of axially loaded members with bending, reference should be made to Chapters 14, 15 and 18. Further analysis may be required if the chosen members are found to be over or understressed.

The Steel Construction Institute Silwood Park, Ascot, Berks SL5 7QN	Subject **ROAD BRIDGE**	Chapter ref. **19**
	Design code **BS5400**	Made by **RT** / Checked by **GWO**	Sheet no. **11**

Typical connection detail:

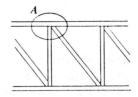

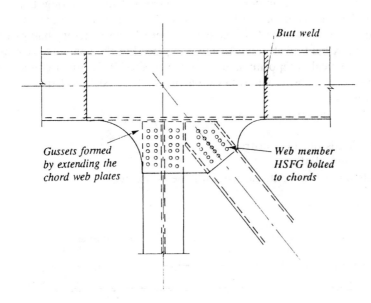

Enlarged Detail A

For guidance on the detailed design of connections the reader should refer to Chapter 26.

Chapter 20
Composite deck slabs

by MARK LAWSON and PETER WICKENS

20.1 Introduction

20.1.1 Form of construction

'Composite floor' is the general term used to denote the composite action of steel beams and concrete or composite slabs that form a structural floor. 'Composite deck slabs', in this context, comprise profiled steel decking (or sheeting) as the permanent formwork to the underside of a concrete slab spanning between support beams. The decking acts compositely with the concrete under service loading. It also supports the loads applied to it before the concrete has gained adequate strength. A light mesh reinforcement is placed in the concrete. A cross-section through a typical composite slab is shown in Fig. 20.1. Shear-connectors are used to develop composite action between the concrete slab and steel beams (see also Chapter 21).

The decking has a number of roles. It

(1) supports the loads during construction
(2) acts as a working platform
(3) develops adequate composite action with the concrete

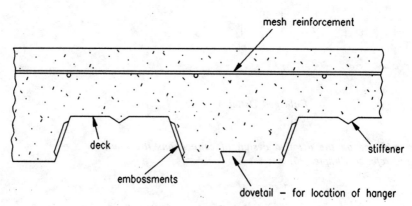

Fig. 20.1 Section through a typical composite deck slab

(4) transfers in-plane loads by diaphragm action to vertical bracing or walls (in conjunction with the concrete topping)
(5) stabilizes the beams against lateral buckling
(6) acts as transverse reinforcement to the composite beams
(7) distributes shrinkage strains preventing serious cracking.

In addition it has a number of advantages over precast or in situ concrete alternatives:

(1) construction periods are reduced
(2) decking is easily handled
(3) attachments (e.g. ceiling hangers) can be made easily
(4) openings can be formed
(5) shear-connectors can be welded through the decking
(6) decking can be cut to length and is less prone to tolerance problems.

The main economy sought in buildings is speed of construction and for this reason slabs and beams are generally designed to be unpropped during the construction stage. Spans of the order of 3 m to 3.6 m between support beams are most common, but can be increased to over 4 m if propping is used. Design of composite slabs is covered by BS 5950: Part 4.[1]

20.2 Deck types

Deck profiles are usually in the range of 38–75 mm height and 150–300 mm trough spacing with sheet thicknesses between 0.8 mm and 1.5 mm, use of the lower thickness being limited by local buckling and the upper thickness by difficulties in rolling. A summary of the different decking profiles marketed for use in composite slabs is shown in Fig. 20.2. There are two well-known generic types: the dovetail profile and the trapezoidal profile with web indentations.

The shape of the profile is controlled by a number of criteria:

(1) the need to maximize the efficiency of the cross-section in bending (both positive and negative moments)
(2) the need to develop adequate composite action with the concrete by use of embossments or indentations or by the shape of the profile itself
(3) the efficient transfer of shear from the beam into the concrete slab (a similar problem to the haunch design in reinforced concrete slabs).

Deeper, thicker profiles are used where there is a need to maximize spans in unpropped construction. Practice in North America has been to use profiles with a trough pitch typically 300 mm and 50 mm height based on a 3 m span.[2] In the UK there is a tendency to design for longer spans with slightly deeper profiles, of typically 60 mm.

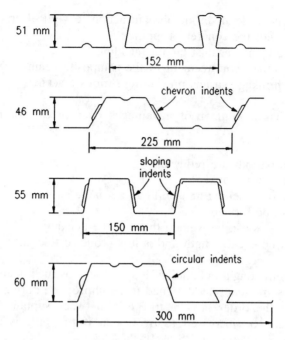

Fig. 20.2 Examples of composite deck profiles

Steel grades are generally Z25 or Z28 (250 N/mm² and 280 N/mm² yield strength respectively). There is normally no benefit in using higher grade steel because serviceability criteria often control the design of the deck (see section 20.5.5).

Steel decking is usually galvanized to a standard of 275 g/m² (roughly 0.04 mm per face) which gives adequate protection for internal use. In North America, unprotected decks have been used but one problem lies in the specification of steel decks for car parks because of potential salt attack through cracks in the concrete.

20.3 Normal and lightweight concretes

One of the principal advantages of steel decking is that it acts as a working platform. The loads that it supports arise mainly during the concreting operation. Lightweight concrete (LWC), therefore, has advantages in terms of its reduced self-weight in comparison to normal-weight concrete (NWC). In the UK the main form of LWC comprises Lytag and sand with a dry density range of 1750–1900 kg/m³. The compressive strength of Lytag LWC is similar to NWC, although there is a slight reduction in shear strength.

Concrete is usually placed by pump, mainly for reasons of speed, but also because it is difficult to 'skip' the concrete with the decking in place above. Indeed, it is usual practice to deck-out a number of floors ahead of the concreting operation. The decking is attached to its supports by shot-fired pins or screws and later by welded shear-connectors.

Unlike traditional reinforced concrete, there is no need to restrict bay sizes during construction because the decking serves to distribute early age and shrinkage strains, thereby eliminating the formation of wide cracks. Typical pumped pours are 500–1000 m². LWC also has a higher tensile strain capacity than NWC, reducing the tendency for cracking. The mesh reinforcement is of a nominal size to control cracking at supports, and to act as 'fire reinforcement'. A142 or A193 are the common sizes specified.

The concrete grade (cube strength in N/mm²) is normally specified as 30 to 40. Pumped concrete also contains additives to aid lubrication. For this reason the slump test is not a good measure of workability and so the 'flow-meter' method is often used. The concrete is usually tamped level by fixing the tamping rails to the support beams. As these beams deflect, the slab level adopts that of the support beams. In propped construction, the slab level deflects further on removal of the props.

20.4 Selection of floor system

Slab depths are normally in the range of 110–150 mm. Clearly, the minimum depth for structural adequacy is usually selected, but the slab depth is affected by insulation requirements in a fire (see section 20.6.3) and serviceability criteria as influenced by the ratio of slab span to depth. As a general rule, span:depth ratios (considering the overall slab depth) of continuous composite slabs should be less than 30 for LWC and 35 for NWC. Longer spans may suffer from excessive deflections.

In designing the composite slab it is difficult to develop the full flexural capacity of the slab determined by the area of the deck acting as effective tensile reinforcement. Failure is normally one of shear-bond rupture resulting in slip between the deck and the concrete. Nevertheless, load capacities of composite slabs with the above proportions are normally adequate for most imposed loads up to 10 kN/m². The shear-bond capacity of composite slabs is determined from tests (see section 20.5.3); this capacity may be enhanced by end-anchorage provided by shear-connectors attached to the support beams.

The concrete slab also acts as the principal compressive element in the design of the composite beams (see Chapter 21). This interaction between the flexural behaviour of the slab and the beams is only important where both the slab and beam span in the same direction.

20.5 Basic design

20.5.1 Construction condition

The decking supports the weight of concrete in the finished slab, excess weight from concrete placement and ponding, the weight of operatives and any impact loads during construction. There is a variety of recommendations for this temporary construction load to be considered in the design of the decking; these are expressed in terms of either a uniformly-distributed load or a single line load transverse to the span being checked. In BS 5950: Part 4,[1] the standard used in the design of the decking and composite slab, these construction loads are specified as 1.5 kN/m^2 or a transverse line load of 2 kN/m^2 (for spans up to 3 m) respectively. Self-weight and construction loads are multiplied by a load factor of 1.4.[1]

However, significant loads can be developed before concreting due to storage of equipment and materials. To avoid premature failure of the deck these loads applied before concreting should not exceed 3 kN/m^2. Similarly, loads applied to the composite slab before it has gained adequate strength for fully composite action should not exceed 1.5 kN/m^2. The definition of 'adequate strength' is considered to be a cube strength of 75% of the specified value, which is often achieved in 5–7 days after concreting.

The design of single-span decking is usually controlled by deflection and ponding of concrete. Continuous decking is generally designed on strength rather than deflection criteria and longer unpropped spans are permissible. The construction load is applied in design as a pattern load even though only one span is likely to be loaded to this extent, in view of the progressive nature of concreting.

In BS 5950[1] a limit on the residual deflection of the soffit of the slab after concreting of the smaller of (span/180) or (slab depth/10) is specified, increased to (span/130) if the effects of ponding are included. Greater deflections are likely to be experienced during concreting.

The design of the decking in bending is dependent on the properties of the profile, and particularly the thin plate elements in compression. Where profiles are stiffened by one or two folds in the compression plate, the section is more efficient in bending (Fig. 20.3).

The design of continuous decking is based, according to code requirements, on an elastic distribution of moment, as a safe lower bound to the collapse strength. Elastic moments are normally greatest at internal supports. For the two equal span case this negative (hogging) moment is equal to that of the positive (sagging) moment of single-span decking. Many profiles with wide troughs are weaker under negative moment than positive, and the effect of the localized reaction at the internal support is to reduce further the negative moment capacity. In design to code requirements, therefore, continuous decking may appear to be weaker than simply-supported decking even though in reality it must be considerably stronger.

In general the capacity of the decking is obtained from tests. However, it is possible to calculate strength conservatively using the following approach.

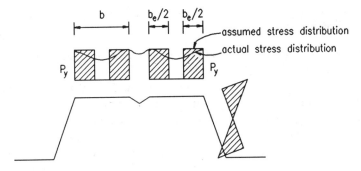

Fig. 20.3 Behaviour of stiffened plate in compression

The compressive strength of a thin plate is presented in terms of an effective width of plate acting at its full yield strength. This effective width b_e is based on a semi-empirical formula[1] considering the post-buckling strength of the plate, and is given by:

$$b_e = \frac{857t}{\sqrt{p_y}} \left(1 - \frac{187t}{b\sqrt{p_y}}\right) \quad \text{but} \quad \not> b \tag{20.1}$$

where t is the plate thickness, b is the plate width, both in mm, and p_y is the design strength of the steel in N/mm^2.

Where a stiffener is introduced, then the dimension b is based on the flat plate width between the stiffener and the web (see Fig. 20.3). However, there is a limiting size of stiffener below which it does not contribute to the increased capacity of the unstiffened plate. In most practical cases the web of the profile is fully effective.

The moment capacity of the section is evaluated using this effective width. In the calculations the design strength of the steel is taken as 0.93 times the specified yield strength. Under positive moment the top plate is in compression, whereas under negative moment the bottom plate is in compression. This suggests that the most efficient profiles are symmetric in shape. For serviceability calculations, the actual stress replaces the design stress in Equation (20.1). For most serviceability calculations, this stress may be taken as 65% of the design stress.

The construction condition generally determines the permissible spans of the slabs; this is the case which has the greatest influence on the economy of the method of construction. Design based on elastic moments in continuous decking is very conservative because at failure there is a significant redistribution of moment from the most highly stressed areas at the supports to the mid-span area, initially as a result of elastic effects because the stiffness of the section changes with applied moment. Beyond the point at which yield takes place, plate collapse and some 'plastic' deformation occurs with increasing deformation.

If the decking were able to deform in an ideal plastic manner, as in Fig. 20.4(a), then the 'plastic' failure load of an end-span would be approximately:

$$w = \frac{8}{L^2}(M_p + 0.46KM_n)$$

where M_p and M_n are the elastic moment capacities of the profile under positive and negative moment respectively. The factor K (less than unity) is introduced because only a proportion of M_n can be developed at failure. The key to this is the post-elastic behaviour of the section illustrated in Fig. 20.4(b).

From tests on different profile shapes it can be shown that the value of K has a lower bound over the normal range of sheet thicknesses and spans. Typically, the reduction in M_n (as reflected in the K value) corresponding to a support rotation of 2–3° to develop M_p is 50%. This method of design is covered in more detail in CIRIA Technical Note 116.[3]

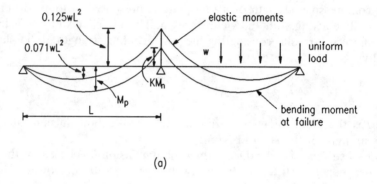

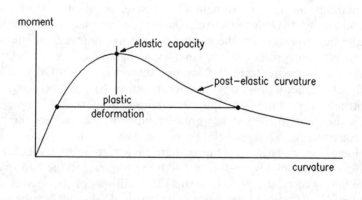

Fig. 20.4 Ultimate behaviour of continuous decking: (a) moment on two-span decking, (b) typical moment–curvature relationship of decking

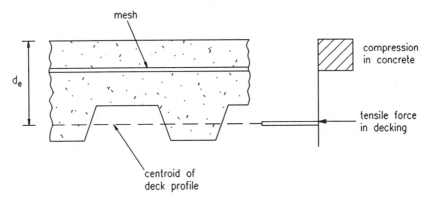

Fig. 20.5 Behaviour of composite deck slab as an equivalent reinforced concrete section

Spans 10% to 15% in excess of those predicted by elastic design are possible while maintaining an adequate factor of safety against failure, and satisfactory performance at working load. It is for this reason that manufacturers often present load–span tables on the basis of tests rather than the elastic method specified in BS 5950: Part 4.[1]

20.5.2 Composite condition

The cross-sectional area of the decking A_p acts as conventional reinforcement at a lever arm determined by the centroid of the profile, as shown in Fig. 20.5. The tensile forces in the decking are developed as a result of the bond between the decking and the concrete, enhanced by some form of mechanical connection, such as by indenting the profile.

If the shear-connectors used to develop composite action between the slab and the steel beam are also included in the design, then the full flexural capacity of the slab can usually be mobilized because of the end anchorage provided. If not, failure is usually by slip between the decking and the concrete, known as incomplete shear connection as indicated in Fig. 20.6. This means that the full flexural capacity of the slab is not developed. Nevertheless, the degree of composite action is sufficiently good that, for all but very heavy imposed loads, permissible spans are controlled by the construction condition. Vertical shear rarely influences the design.

Composite slabs are usually designed as simply-supported elements with no account taken of continuity provided by the reinforcement in the concrete at internal beams, reflecting the behaviour of an end-bay slab. However, internal bays will not fail by incomplete shear connection provided the decking is laid continuously over the supports, an observation which also applies to end anchorage which need only be developed at the edge beams or at sheet discontinuities in a continuous composite slab.

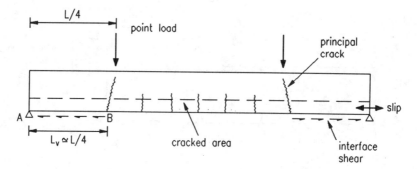

Fig. 20.6 Failure of composite slab by incomplete shear connection

The ultimate capacity of composite slabs, in the absence of end anchorage, is controlled by a combination of friction and chemical bond, followed by mechanical interlock after initial slippage. As a result, the performance of a particular decking system can only be readily assessed by testing. A typical slab under test is shown in Fig. 20.7.

Fig. 20.7 Testing of composite slab to establish composite action

Mechanical interlock is partly dependent on the local plate stiffness and so indentations are best situated close to the more rigid sections of the profile, such as corners or narrow plate elements. Slippage is associated with separation; some profiles, such as the dovetail section, achieve good shear-bond capacity by preventing separation. Different trapezoidal profiles incorporate a wide variety of indent and embossment shapes, as illustrated in Fig. 20.2, which have varying degrees of effectiveness.

If design were to be carried out on elastic principles, permissible bond strengths between the deck profile and the concrete would be of the order of 0.05 N/mm^2 for plain profiles rising to about 0.2 N/mm^2 for some indented profiles. First slip and flexural failure of plain trapezoidal profiles are coincident, whereas there is often a considerable reserve following initial slip in properly designed composite slabs.

If failure of a composite slab that is propped during construction occurs by incomplete shear connection, then the applied load to be considered in the analysis of its shear bond strength is the imposed load plus self-weight loads on depropping. If the slab is not propped, then only the imposed load contributes to the shear bond mode of failure. If flexure governs strength, as in an equivalent reinforced concrete section, then the ultimate strength of the section is independent of the sequence of loading, and therefore the total load is to be considered in both propped and unpropped construction.

20.5.3 Requirements of BS 5950

The method of determining the ultimate shear bond capacity of composite slabs with a particular deck profile is based on load tests. A simply-supported slab of the appropriate proportions is subject to 2 or 4 point loads to simulate a uniformly distributed load (see Fig. 20.7). Crack inducers are cast into the slab so that a predetermined length of the outer portion of the slab can slip under load relative to the deck, without developing the tensile reserve of the concrete. These crack inducers are placed at the desired position for the form of loading (normally quarter span points).

Testing to BS 5950: Part 4 [1] requires that a minimum of six composite slabs are tested covering a range of design parameters. The slabs are first subject to a dynamic loading between 50% and 150% of the desired working load, and then the load is increased statically to failure. The objective of the dynamic part of the test is to identify those cases where there is inherent fragility in the concrete–deck connection. The ultimate load is then recorded.

The problem remains of how to use the test information in design, where the parameters may be different from those tested. This is achieved by using an empirical design formula in terms of experimentally derived constants.[4]

The ultimate vertical shear strength (per unit width) of a composite slab is given by the formula:

$$V_u = 0.8A_c (m_d\, p\, d_e/L_v + k_d \sqrt{f_{cu}}) \tag{20.2}$$

where p is the ratio of the cross-sectional area of the profile to that of the concrete A_c per unit width of slab; f_{cu} is the cube strength of the concrete; d_e is the effective slab depth to the centroid of the profile; and L_v is the shear span length, taken as one quarter of the slab span L.

The empirical constants m_d and k_d are calculated from the slope and intercept, respectively, of the reduced regression line of Fig. 20.8. Tests are carried out at the extremes of the regression line (such as low and high L). The regression line is to be reduced by 15% if fewer than eight slab tests are performed over the range of spans. Extrapolation outside the limits of the tests is not permitted. The constant of 0.8 represents 1 (material factor) and takes account of the potential variability of this form of failure. The slab capacity is then compared to the applied shear forces using the load factors in Part 1 of BS 5950.

Physically, m_d is a broad measure of the effect of mechanical interlock and k_d broadly represents the friction bond. Despite this, k_d can be negative. In reality, it is unlikely that interlock is linearly dependent on profile area but Equation (20.2) may be considered to be reasonably accurate for small variations from the test parameters p, d_e L_v and f_{cu}.

20.5.4 Design tables

Manufacturers normally present design information in terms of load–span tables for the different decks that are marketed. A typical load–span table for a composite slab is shown in Table 20.1. It can be seen that the permissible spans of unpropped slabs are largely unaffected by imposed load because spans are controlled by the construction condition and the composite capacity of slabs of these spans exceeds the tabulated imposed loads. Conversely, permissible spans of propped slabs are considerably greater because design is now based on the composite condition. However, some of these designs may be controlled by deflection (see the span-to-depth ratios in section 20.4).

20.5.5 Serviceability

The two key serviceability aspects relating to the design of composite slabs are avoidance of premature slip and control of deflections. It is not normally considered necessary to control cracking in concrete in heated buildings. Nevertheless, standard mesh reinforcement of area greater than 0.1% of the cross-sectional area of the concrete is placed at about 25 mm from the slab surface. This reinforcement also acts as 'fire-reinforcement' (see section 20.6). However, this minimum amount of

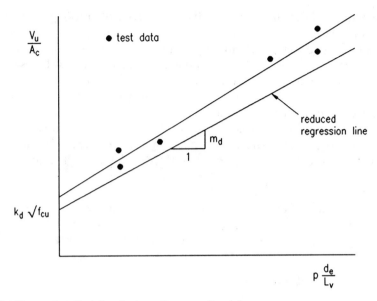

Fig. 20.8 Regression line for design of composite slabs

Table 20.1 Permissible spans (m) according to BS 5950: Part 4 for typical 50 mm deep deck of thickness, t

Support condition	Slab depth (mm)	$t = 0.9$ mm			$t = 1.2$ mm		
		Imposed loading (kN/m²)					
		2.5	5.0	7.5	2.5	5.0	7.5
Single span – no props	100	2.6	2.6	2.6	3.1	3.1	3.1
	150	2.2	2.2	2.2	2.7	2.7	2.7
Multiple span	100	2.7	2.7	2.7	3.2	3.2	3.1
	150	2.3	2.3	2.3	2.8	2.8	2.8
Single span – one prop	100	3.5	3.3	2.8	3.5	3.5	3.1
	150	4.6	4.1	3.5	5.2	4.5	3.8

Maximum span-to-depth ratio limited to 35 for normal weight concrete slab

reinforcement would in theory be insufficient to control cracking at the supports of continuous slabs.

A typical load–deflection relationship of a composite slab is shown in Fig. 20.9. Initial slip occurs well before the ultimate load is reached in a well-designed slab. Nevertheless, it would be inadvisable if first slip occurred at below half of the ultimate load, because this might suggest poor serviceability performance.

The deflection of a composite slab is usually calculated on the assumption that it behaves as a reinforced concrete element with the deck area acting as an equivalent reinforcing bar. The section is assumed to be cracked in mid-span. The neutral axis depth below the upper slab surface is determined from:

$$x_e = d_e \{\sqrt{[(\alpha_e p)^2 + 2\alpha_e p]} - \alpha_e p\} \qquad (20.3)$$

where α_e is the ratio of the elastic moduli of steel to concrete. The selection of an appropriate modular ratio is discussed in Chapter 21 (section 21.7.2). The values p and d_e are defined in section 20.5.3 and Fig. 20.5 respectively.

The cracked second moment of area (in steel units) is therefore given by:

$$I_c = x_e^3/(3\alpha_e) + pd_e(d_e - x_e)^2 + I_s \qquad (20.4)$$

where I_s is the second moment of area of the deck profile per unit width.

In determining deflections it is common practice to use the average value of the cracked and uncracked second moments of area of the section reflecting the fact that only part of the slab is cracked. Shrinkage-induced deflection of slabs is usually ignored.

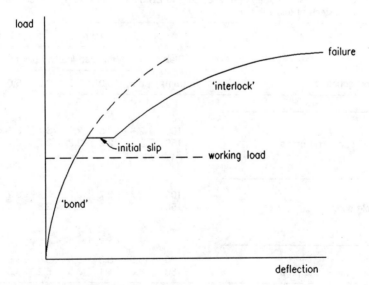

Fig. 20.9 Typical load–deflection behaviour of composite deck slabs

Although deflections are usually calculated assuming that the slab is simply-supported, the effect of continuity of negative (hogging) reinforcement may be estimated by reducing the deflection δ_0 of the simply-supported slab according to:

$$\delta_c = \delta_o [1 - 0.6 (M_1 + M_2)/M_0] \tag{20.5}$$

where M_1 and M_2 are the negative moments at each support (not more than the moment capacity of the reinforced slab) and M_0 is the simply-supported moment for the same loading conditions.

20.6 Fire resistance

20.6.1 Fire tests

Three criteria are imposed by the standard fire-resistance test on floors to BS 476: Part 20.[5] These are:

(1) strength or stability under load
(2) integrity to the passage of smoke or flame
(3) insulation, so that the rise of temperature on the upper surface of the slab is not excessive.

Failure of a floor or beam is deemed to occur when the maximum deflection exceeds (span/20) or when the rate of deflection exceeds a specified amount.[5] To satisfy the insulation criterion the temperature on the upper surface of the slab should not exceed a maximum of 185°C or an average of 140°C above ambient.

In principle, the minimum depth of composite slabs is based on an insulation criterion, and the amount of reinforcement needed in a fire is based on a strength criterion. It would be reasonable to assume that most of the tensile capacity of the deck would be lost in a severe fire and therefore the load-carrying capacity derives mainly from the reinforcement.

There are a number of beneficial factors which make fires in real buildings generally less onerous than in a standard fire test. From a structural point of view, the most important factor to be modelled in a fire test is the effect of structural continuity. Membrane action (or in-plane restraint) is generally ignored because of its indeterminate magnitude. Until recently, few fire tests had included the effect of continuity of the slab and its reinforcement.

Simply-supported composite slabs with nominal reinforcement rarely exceed a fire-resistance period of 30 minutes whereas tests on continuous slabs with the same reinforcement can achieve over 60 minutes fire resistance. Indeed, a series of ten fire tests carried out in the UK showed that 90 minutes fire resistance can be achieved for slabs with standard mesh reinforcement and subject to imposed loads of up to 6.7 kN/m². These imposed load and fire resistance requirements are typical of those specified currently for most commercial buildings.

20.6.2 Fire engineering method

The phrase 'fire engineering method' (FEM) is the term given to the means of calculating the amount of emergency reinforcement needed for a certain fire resistance period. This is described fully in a Steel Construction Institute publication.[6] In principle, the moment of resistance of the section is calculated taking into account the temperatures in the section and the reduced strength of the various elements at elevated temperatures. Temperatures are determined from tables or indicative tests. The plastic capacity of continuous members can be evaluated and compared to the applied moment with load factors of unity. In some areas, BS 5950: Part 8 [7] permits a further 20% reduction in imposed load in a fire. Typically, therefore, the applied moment in a fire would be 50–60% of the ultimate design moment.

The critical element determining the moment of resistance of the section is the reinforcement. Mesh reinforcement is normally well-insulated from the effects of the fire and contributes significantly to the fire resistance of the slab. Where fire resistance periods greater than 90 minutes are required (see below) then additional reinforcing bars or heavier mesh can be introduced. In such cases it may be economic to design the slab as a reinforced concrete ribbed slab to BS 8110,[8] and treat the decking as permanent formwork. Bars of diameter 10 mm or 12 mm are normally placed singly in the troughs at the appropriate cover.

20.6.3 Design recommendations

Simple design tables covering common design cases are given in Reference 9. A minimum slab depth and mesh size are given for different profile types. Data for only two spans (i.e. 3 m and 3.6 m) and one imposed load (6.7 kN/m^2) are presented, but these may be converted to other cases by using the same equivalent moment as the tabulated cases. The data are reproduced in Table 20.2.

20.7 Diaphragm action

The decking serves to transfer lateral loads to vertical bracing of concrete walls. It is normally attached on all four sides to support beams at spacings of not less than 600 mm. However deck–deck seam fasteners are not normally installed. The shear stiffness of the 'diaphragm' is very high and the strength is normally determined by the capacity of the fixings. Typical ultimate strengths of the standard screws or pins are 6 kN/mm sheet thickness. Where through-deck welding of shear-connectors is used the ultimate shear strength of the decking is considerably enhanced.

Table 20.2 Simplified rules for fire resistance of composite deck slabs[9]

(a) Trapezoidal profiles (depth not exceeding 60 mm)

Max. span (m)	Fire rating (h)	Sheet thickness (mm)	Slab depth (mm) NWC	Slab depth (mm) LWC	Mesh size
2.7	1	0.8	130	120	A142
3.0	1	0.9	130	120	A142
	1½	0.9	140	130	A142
3.6	1	1.0	130	120	A193
	1½	1.0	140	130	A193

(b) Dovetail profiles (depth not exceeding 50 mm)

Max. span (m)	Fire rating (h)	Sheet thickness (mm)	Slab depth (mm) NWC	Slab depth (mm) LWC	Mesh size
2.5	1	0.8	100	100	A142
	1½	0.8	110	105	A142
3.0	1	0.9	120	110	A142
	1½	0.9	130	120	A142
3.6	1	1.0	125	120	A193
	1½	1.2	135	125	A193

The steel beams are laterally supported in simple bending by the decking, provided the decking crosses the beams and is attached to them by shear-connectors or pins at spacings not exceeding 600 mm. However, beams running parallel to the decking are laterally supported only at transverse beam connections. All beams are laterally supported under positive moment in their composite state.

20.8 Other constructional features

In the role of the decking as formwork, shuttering at the edge of the slab and at large openings is usually in the form of light cold-formed channels which are attached by pins or screws to the decking. Small openings can be formed by leaving 'boxed out' voids in the slab and the decking cut away once the concrete has gained adequate strength. Additional reinforcement should be placed to redistribute the loads that would otherwise have been resisted by the composite slabs in this zone.

Propping of composite slabs is rarely used, but if so, it is important to ensure that the slab beneath can resist the self-weight and construction load of the slab being cast. This may be critical if the support slab has not gained its full strength for composite action. Timber spreaders should be used to avoid damage to the deck.

References to Chapter 20

1. British Standards Institution (1982) *Structural use of steelwork in building*. Part 4: *Code of practice for design of floors with profiled steel sheeting*. BS 5950, BSI, London (currently being revised).
2. Steel Deck Institute (1989) *Design Manual for Composite Decks, Form Decks and Roof Decks*. Steel Deck Institute (St Louis).
3. Bryan E.R. & Leach P. (1984) *Design of Profiled Sheeting as Permanent Formwork*. Construction Industry Research and Information Association (CIRIA) Technical Note 116.
4. Shuster R.M. (1976) Composite steel-deck concrete floor systems. *Proc. Am. Soc. Civ. Engrs, J. Struct. Div.*, **102**, No. ST5, May, 899–917.
5. British Standards Institution (1987) *Fire tests on building materials and structures*. Part 20: *Method for determination of the fire resistance of elements of construction (general principles)*. BS 476, BSI, London.
6. Steel Construction Institute (SCI) (1988) *Fire resistance of composite floors with steel decking*. SCI, Ascot, Berks.
7. British Standards Institution (1990) *Structural use of steelwork in building*. Part 8: *Code of practice for fire resistant design*. BS 5950. BSI, London.
8. British Standards Institution (1985) *Structural use of concrete*. Part 2: *Code of practice for special circumstances*. BS 8110, BSI, London.
9. Construction Industry Research and Information Association (CIRIA) (1986) *Data Sheet: Fire Resistance of Composite Slabs with Steel Decking*. CIRIA Special Publication 42.

A worked example follows which is relevant to Chapter 20.

The Steel Construction Institute Silwood Park, Ascot, Berks SL5 7QN	Subject **DESIGN OF COMPOSITE SLAB**		Chapter ref. **20**
	Design code **BS5950: Part 4**	Made by **RML** Checked by **GWO**	Sheet no. **1**

Problem

Check the design of a composite slab 125 mm deep, spanning 3 m and using the deck shown below. (Note: this design is carried out to BS 5950: Part 4: 1982 which is currently under review.)

Deck shape :

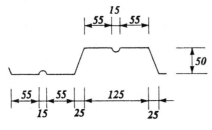

Steel thickness	t	=	1.2 mm (bare thickness of steel)
Steel grade	p_y	=	280 N/mm²
Trough spacing		=	300 mm

Loads

Imposed load	=	5.0 kN/m²
Partitions (imposed load)	=	1.0 kN/m²
Ceiling	=	0.5 kN/m²
Self weight (slab)	=	2.0 kN/m² - see later
Construction load (temporary)	=	1.5 kN/m²

Concrete

Cube strength	f_{cu}	=	30 N/mm²
Density (dry)		=	1800 kg/m³ (lightweight)
Density (wet)		=	1900 kg/m³ (lightweight)

586 Worked example

The Steel Construction Institute	Subject	Chapter ref.
Silwood Park, Ascot, Berks SL5 7QN	DESIGN OF COMPOSITE SLAB	20
	Design code BS5950: Part 4	Made by RML Checked by GWO	Sheet no. 2

Construction condition

Self weight of slab

$$w_d = 1900 \times 9.81 \times 10^{-6} (75 + 50/2) = 1.86 \text{ kN/m}^2$$

Self weight of deck $= 0.12 \text{ kN/m}^2$

Total self weight $\simeq 2.0 \text{ kN/m}^2$

Design moment in construction condition

$$M_u = (2.0 + 1.5) \times 1.4 \times \frac{3.0^2}{8}$$

$$= 5.51 \text{ kNm}$$

Moment resistance of deck in sagging (positive bending)

Assume in the worst case that the deck is laid simply supported. The moment resistance derives from the positive (sagging) elastic moment resistance of the section.

Effective breadth of flat compression plate   Equation 20.1

$$b_e = \frac{857\, t}{\sqrt{p_y}} \left[1 - \frac{187\, t}{b\sqrt{p_y}} \right] \not> b$$

where $b = 55$ mm

$$b_e = \frac{857 \times 1.2}{\sqrt{280}} \left[1 - \frac{187 \times 1.2}{55\sqrt{280}} \right]$$

$$= 46.5 \text{ mm} < b$$

According to BS5950: Part 4 the stiffener area is ignored when calculating the compressive resistance of the plate.

	Subject	Chapter ref.
The Steel Construction Institute Silwood Park, Ascot, Berks SL5 7QN	*DESIGN OF COMPOSITE SLAB*	20
Design code BS5950: Part 4	Made by RML Checked by GWO	Sheet no. 3

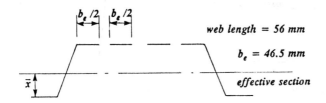

web length = 56 mm

b_e = 46.5 mm

effective section

Neutral axis position

$$\bar{x} = \frac{(46.5 \times 2 + 56) \times 50}{(125 + 2 \times 56 + 2 \times 46.5)} = 22.5 \text{ mm}$$

Second moment of area

$$I = 1.2 \ [46.5 \times 2 \times (48.8 - 22.5)^2 + 125 \times 22.5^2$$

$$+ 2 \times 56 \times \frac{50^2}{12} + 2 \times 56 \times (24.4 - 22.5)^2]$$

$$= 1.2 \ (64.3 + 63.3 + 23.3 + 0.4) \times 10^3$$

$$= 181.6 \times 10^3 \text{ mm}^4 \text{ / trough}$$

$$= 605 \times 10^3 \text{ mm}^4 \text{ / m width}$$

Depth of web in compression = 26.3 mm

d/t = 21.9 - it is not necessary to check for local buckling of web

Elastic section modulus

$$Z_e = \frac{605 \times 10^3}{(49.4 - 22.5)} = 22.5 \times 10^3 \text{ mm}^3 \text{ / m}$$

Elastic moment resistance of section (sagging moment)

$$M_c = Z_e \times 0.93 \ p_y = 22.4 \times 10^3 \times 0.93 \times 280 \times 10^{-6}$$
$$= 5.8 \text{ kNm} > M_u$$

(using a design strength of 0.93 p_y as in BS 5950: Part 4)

Worked example

The Steel Construction Institute Silwood Park, Ascot, Berks SL5 7QN	Subject DESIGN OF COMPOSITE SLAB	Chapter ref. 20
	Design code BS5950: Part 4 — Made by RML / Checked by GWO	Sheet no. 4

Check deflection of slab after construction

Bending moment on deck after construction

$$= 2.0 \times 3.0^2/8 \quad = \quad 2.25 \text{ kNm}$$

Equivalent stress in compression plate

$$\sigma \approx \frac{2.25}{6.3} \times 280 = 100 \text{ N/mm}^2$$

Repeating the calculation of second moment of area using this stress (rather than p_y) in equation 20.1 gives:

$$I = 649 \times 10^3 \text{ mm}^4 / \text{m width}$$

Deflection of soffit of deck (simply supported)

$$\delta_2 = \frac{5 \times 2.0 \times 3000^4}{384 \times 205 \times 10^3 \times 649 \times 10^3}$$

$$= 15.9 \text{ mm} \quad (\text{span}/188)$$

This is less than the limit of span/180 in BS 5950: Part 4. If this limit had been exceeded it would be necessary to include for the effects of ponding of concrete at greater deflections.

Composite condition

Check moment resistance of the composite section as a reinforced concrete slab, assuming full shear connection.

Area of deck $= (125 + 56) \times 2 \times 1.2$

$\qquad = 434 \text{ mm}^2 / \text{trough}$

$A_p = 1448 \text{ mm}^2 / \text{m}$

Tensile resistance $= 0.93 \times 280 \times 1448 \times 10^{-3}$

$\qquad = 377 \text{ kN}$

	Subject	Chapter ref.
The Steel Construction Institute Silwood Park, Ascot, Berks SL5 7QN	*DESIGN OF COMPOSITE SLAB*	20
	Design code BS5950: Part 4	Made by RML Checked by GWO	Sheet no. 5

Neutral axis depth into concrete

$$x_c = \frac{377 \times 10^3}{0.4 \times f_{cu} \times 10^3} = 31.4 \text{ mm}$$

Plastic moment resistance of composite section

$$M_p = 377 \times (125 - 25 - 31.4/2) \times 10^{-3}$$
$$= 31.8 \text{ kNm}$$

Applied moment in composite condition

$$M_u = [\,1.6\,(5.0 + 1.0) + 1.4\,(2.0 + 0.5)\,] \times 3.0^2/8$$
$$= 14.7 \text{ kNm}$$

Therefore, the plastic moment resistance of the section is more than adequate.

Check for incomplete shear connection   *Equation 20.2*

$$V_u = 0.8\,A_c\,(\,m_d \times p \times \frac{d_e}{L_v} + k_d\,\sqrt{f_{cu}}\,)$$

where $p = \dfrac{A_p}{A_c} = \dfrac{1448}{100 \times 1000} = 0.0145$

$d_e = 100 \text{ mm}$

$L_v = 3000/4 = 750 \text{ mm}$

Typical values of empirical constants are:

$m_d = 130$ and $k_d = 0.004$

The Steel Construction Institute  Silwood Park, Ascot, Berks SL5 7QN	Subject **DESIGN OF COMPOSITE SLAB**	Chapter ref. **20**
	Design code BS5950: Part 4 — Made by RML / Checked by GWO	Sheet no. **6**

$$V_u = 0.8 \times 100 \times \times (130 \times 0.0145 \times \frac{100}{750} + 0.004\sqrt{30})$$

$$= 80 (0.251 + 0.022)$$

$$= 21.8 \text{ kN}$$

Applied ultimate shear force

$$V = 4 M_u / L = 4 \times 14.7/3.0 = 19.6 \text{ kN}$$

The slab can resist the applied loads adequately without failure by incomplete shear connection.

Check shear stress on concrete

$$v_c = \frac{19.6 \times 10^3}{100 \times 10^3} = 0.20 \text{ N/mm}^2$$

This is relatively small according to BS8110.

Check deflection of composite slab

Properties of cracked section:

Neutral axis depth below upper surface of slab

$$x_e = d_e (\sqrt{(\alpha_e p)^2 + 2 \alpha_e p} - \alpha_e p) \qquad \text{Equation 20.3}$$

$$\alpha_e = \text{modular ratio} = 15 \text{ for LWC}$$

$$\alpha_e p = 0.0145 \times 15 = 0.218$$

$$x_e = 100 (\sqrt{0.218^2 + 2 \times 0.218} - 0.218)$$

$$= 51.5 \text{ mm}$$

The Steel Construction Institute Silwood Park, Ascot, Berks SL5 7QN	Subject **DESIGN OF COMPOSITE SLAB**		Chapter ref. **20**
	Design code **BS5950: Part 4**	Made by **RML** Checked by **GWO**	Sheet no. **7**

Second moment of area of cracked section ⟶ *Equation 20.4*

$$I_c = \frac{51.5^3}{3 \times 15} + 0.0145 \times 100 \times (100 - 51.5)^2 + 649$$

$$= 3035 + 3411 + 649$$

$$= 7095 \text{ mm}^4 / \text{mm width (in steel units)}$$

Service load on composite slab

$$w_s = 5.0 + 1.0 + 0.5 = 6.5 \text{ kN/m}^2$$

Deflection under service load

$$\delta = \frac{5 \times 6.5 \times 3000^4 \times 10^{-3}}{384 \times 205 \times 10^3 \times 7095}$$

$$= 4.7 \text{ mm (or span/638)}$$

This is satisfactory.

Note:

Mesh reinforcement is required for crack control and fire resistance requirements.

Chapter 21
Composite beams

by MARK LAWSON and PETER WICKENS

21.1 Applications of composite beams

In buildings and bridges steel beams often support concrete slabs. Under load each component acts independently with relative movement or slip occurring at the interface. If the components are connected so that slip is eliminated, or considerably reduced, then the slab and steel beam act together as a composite unit (Fig. 21.1). There is a consequent increase in the strength and stiffness of the composite beam relative to the sum of the components.

The slab may be solid in situ concrete or the composite deck slab considered in Chapter 20. It may also comprise precast concrete units with an in situ concrete topping. In buildings, steel beams are usually of standard UB section, but UC sections are sometimes used where there is need to minimize the beam depth. A typical building under construction is shown in Fig. 21.2. Welded fabricated sections are often used for long-span beams in buildings and bridges.

Design of composite beams in buildings is now covered by BS 5950: Part 3,[1] although guidance was formerly available in an SCI publication.[2] The design of composite beams incorporating composite slabs is affected by the shape and orientation of the decking, as indicated in Fig. 21.3.

One of the advantages of composite construction is smaller construction depths. Services can usually be passed beneath, but there are circumstances where the

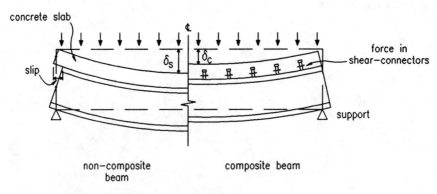

Fig. 21.1 Behaviour of composite and non-composite beams

Fig. 21.2 Composite building under construction showing decking and shear-connectors

beam depth is such that services can be passed through the structure, either by forming large openings, or by special design of the structural system. A good example of this is the stub-girder.[3] The bottom chord is a steel section and the upper chord is the concrete slab. Short steel sections or 'stubs' are introduced to transfer the forces between the chords.

Openings through the beam webs can be provided for services. Typically, these can be up to 70% of the beam depth and can be rectangular or circular in shape. Guidance on the design of composite beams with web openings is given in Reference 4. Examples of the above methods of introducing services within the structure are shown in Fig. 21.4.

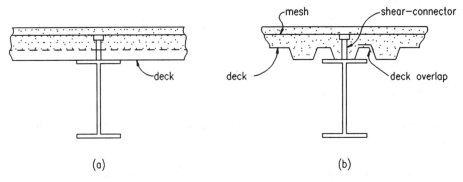

Fig. 21.3 Composite beams incorporating composite deck slabs: (a) deck perpendicular to beam, (b) deck parallel to beam

Composite beams

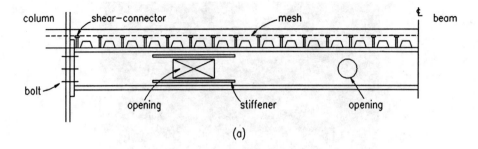

(a)

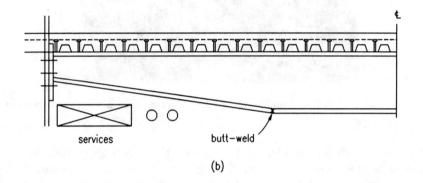

(b)

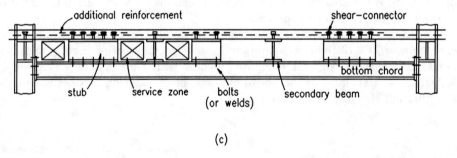

(c)

Fig. 21.4 Different methods of incorporating services within the structural depth

21.2 Economy

Composite beam construction has a number of advantages over non-composite construction:

(1) savings in steel weight are typically 30% to 50% over non-composite beams.
(2) the greater stiffness of the system means that beams can be shallower for the same span, leading to lower storey heights and savings in cladding, etc.

It also shares the advantage of rapid construction.

The main disadvantage is the need to provide shear-connectors at the interface between the steel and concrete. There may also be an apparent increase in complexity of design. However, design tables have been presented to aid selection of member sizes.[2]

The normal method of designing simply-supported beams for strength is by plastic analysis of the cross-section. Full shear connection means that sufficient shear-connectors are provided to develop the full plastic capacity of the section. Beams designed for full shear connection result in the lightest beam size. Where fewer shear-connectors are provided (known as partial shear connection) the beam size is heavier, but the overall design may be more economic.

Partial shear connection is most attractive where the number of shear-connectors is placed in a standard pattern; such as one per deck trough or one per alternate trough where profiled decking is used. In such cases, the resistance of the shear-connectors is a fixed quantity irrespective of the size of the beam or slab.

Conventional elastic design of the section results in heavier beams than with plastic design because it is not possible to develop the full tensile capacity of the steel section. Designs based on elastic principles are to be used where the compressive elements of the section are non-compact or slender, as defined in BS 5950 Part 1. This mainly affects the design of continuous beams (see section 21.6.3).

21.3 Guidance on span-to-depth ratios

Beams are frequently designed to be unpropped during construction. Therefore, the steel beam is sized first to support the self-weight of the slab before the concrete has gained adequate strength for composite action. Beams are assumed to be laterally restrained by the decking in cases where the decking crosses the beams and is directly attached to them. These beams can develop their full flexural capacity.

Where simply-supported unpropped composite beams are sized on the basis of their plastic capacity it is normally found that span-to-depth ratios can be in the range of 18 to 22 before serviceability criteria influence the design. The 'depth' in these cases is defined as the overall depth of the beam and slab. Grade 50 steel is often specified in preference to grade 43 in composite beam design because the stiffness of a composite beam is often three to four times that of the non-composite beam, justifying the use of higher working stresses.

The span-to-depth ratios of continuous composite beams are usually in the range of 22 to 25 for end spans and 25 to 30 for internal spans before serviceability criteria influence the design. Many continuous bridges are designed to satisfy the serviceability limit state.

21.4 Types of shear connection

The modern form of shear-connector is the welded headed stud ranging in diameter from 13–25 mm and from 65–125 mm in height. The most popular size is 19 mm diameter and 100 mm height. When used with steel decking, studs are often welded through the decking using a hand tool connected via a control unit to a power generator. Each stud takes only a few seconds to weld in place. Alternatively, the studs can be welded directly to the steel beams in the factory and the decking butted up to or slotted over the studs.

There are, however, some limitations to through-deck welding: the top flange of the beam must not be painted, the galvanized steel should be less than around 1.25 mm thick and the deck should be clean and free of moisture. The power generator needs 415 V electrical supply, and the maximum cable length between the weld gun and the power control units should be limited to around 70 m to avoid loss of power.

Where precast concrete planks are used, the positions of the shear-connectors are usually such that they project through holes in the slab which are later filled with concrete. Alternatively, a gap is left between the ends of the units sitting on the top flange of the beam on to which the shear connectors are fixed. Reinforcement (usually in the form of looped bars) is provided around the shear-connectors.

There is a range of other forms of welded shear-connector, but most lack practical applications. The 'bar and hoop' and 'channel' welded shear-connectors have been used in bridge construction.[5] Shot-fired shear-connectors are marketed for use in smaller building projects where site power might be a problem. All shear-connectors should be capable of resisting uplift forces; hence the use of headed rather than plain studs.

The number of shear-connectors placed along the beam is usually sufficient to develop the full flexural capacity of the member. However it is possible to reduce the number of shear-connectors in cases where the moment capacity exceeds the applied moment and the shear-connectors have adequate ductility. This is known as 'partial shear connection' and is covered in section 21.7.4.

21.5 Span conditions

In buildings, composite beams are usually designed to be simply-supported, mainly to simplify the design process, to reduce the complexity of the beam-to-column connections, and to minimize the amount of slab reinforcement and shear-connectors that are needed to develop continuity at the ultimate limit state.

However, there are ways in which continuity can be readily introduced, in order to improve the stiffness of composite beams. Figure 21.5 shows how a typical connection detail at an internal column can be modified to develop continuity. The stub girder system also utilizes continuity of the secondary members (see Fig. 21.4(c)). Other methods of continuous design are presented in References 3 and 6.

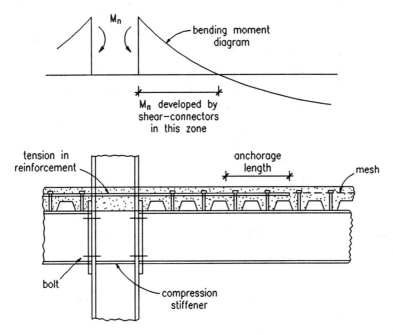

Fig. 21.5 Representation of conditions at internal column of continuous beam

Continuous composite beams may be more economic than simply-supported beams where plastic hinge analysis of the continuous member is carried out, provided the section is plastic according to BS 5950: Part 1. However, where the lower flange or web of the beam is non-compact or slender in the negative (hogging) moment region, then elastic design must be used, both in terms of the distribution of moment along the beam, and also for analysis of the section. Lateral instability of the lower flange is an important design condition, although torsional restraint is developed by the web of the section and the concrete slab.[6]

In bridges, continuity is often desirable for serviceability reasons, both to reduce deflections, and to minimize cracking of the concrete slab, finishes and wearing surface in road bridges. Special features of composite construction appropriate to bridge design are covered in the publication by Johnson and Buckby[7] based on BS 5400: Part 5.[5]

21.6 Analysis of composite section

21.6.1 Elastic analysis

Elastic analysis is employed in establishing the serviceability performance of composite beams, or the strength of beams subject to the effect of instability, for

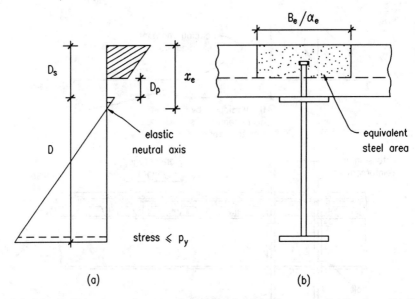

Fig. 21.6 Elastic behaviour of composite beam. (a) Elastic stress distribution. (b) Transformed section

example, in continuous construction, or in beams where the ductility of the shear connection is not adequate.

The important properties of the section are the section modulus and the second moment of area. First it is necessary to determine the centroid (elastic neutral axis) of the transformed section by expressing the area of concrete in steel units by dividing the concrete area within the effective breadth of the slab, B_e, by an appropriate modular ratio (ratio of the elastic modulus of steel to concrete).

In unpropped construction, account is taken of the stresses induced in the non-composite section as well as the stresses in the composite section. In elastic analysis, therefore, the order of loading is important. For elastic conditions to hold, extreme fibre stresses are kept below their design values, and slip at the interface between the concrete and steel should be negligible.

The elastic section properties are evaluated from the transformed section as in Fig. 21.6. The term α_e is the modular ratio. The area of concrete within the profile depth is ignored (this is conservative where the decking troughs lie parallel to the beam). The concrete can usually be assumed to be uncracked under positive moment.

The elastic neutral axis depth, x_e, below the upper surface of the slab is determined from the formula:

$$x_e = \frac{\dfrac{D_s - D_p}{2} + \alpha_e r \left(\dfrac{D}{2} + D_s\right)}{(1 + \alpha_e r)} \qquad (21.1)$$

where $r = A/[(D_s - D_p)B_e]$, D_s is the slab depth, D_p is profile height (see Fig. 21.6) and A is the cross-sectional area of the beam of depth D.

The second moment of area of the uncracked composite section is:

$$I_c = \frac{A(D + D_s + D_p)^2}{4(1 + \alpha_e r)} + \frac{B_e(D_s - D_p)^3}{12\alpha_e} + I \tag{21.2}$$

where I is the second moment of area of the steel section. The section modulus for the steel in tension is:

$$Z_t = I_c/(D + D_s - x_e) \tag{21.3}$$

and for concrete in compression is:

$$Z_c = I_c \alpha_e/x_e \tag{21.4}$$

The composite stiffness can be 3 to 5 times, and the section modulus 1.5 to 2.5 times that of the I-section alone.

21.6.2 Plastic analysis

The ultimate strength of a composite section is determined from its plastic capacity. It is assumed that the strains across the section are sufficiently great that the steel stresses are at yield throughout the section and that the concrete stresses are at their design strength. The plastic stress blocks are therefore rectangular, as opposed to linear in elastic design.

The plastic capacity of the section is independent of the order of loading (i.e. propped or unpropped construction). The plastic capacity is compared to the moment resulting from the total factored loading using the load factors in BS 5950: Part 1.

The plastic neutral axis of the composite section is evaluated assuming stresses of p_y in the steel and $0.45 f_{cu}$ in the concrete. The tensile capacity of the steel is therefore $R_s = p_y A$, where A is the cross-sectional area of the beam. The compressive capacity of the concrete slab depends on the orientation of the decking. Where the decking crosses the beams the depth of concrete contributing to the compressive capacity is $D_s - D_p$ (Fig. 21.6(a)). Clearly, D_p is zero in a solid slab. Where the decking runs parallel to the beams (Fig. 21.6(b)), then the total cross-sectional area of the concrete is used. Taking the first case:

$$R_c = 0.45 f_{cu}(D_s - D_p)B_e \tag{21.5}$$

where B_e is the effective breadth of the slab considered in section 21.7.1.

Three cases of plastic neutral axis depth x_p (measured from the upper surface of the slab) exist. These are presented in Fig. 21.7. It is not necessary to calculate x_p explicitly if the following formulae for the plastic capacity of I-section beams subject to positive (sagging) moment are used. The value R_w is the axial capacity of the web and R_f is the axial capacity of one steel flange (the section is assumed

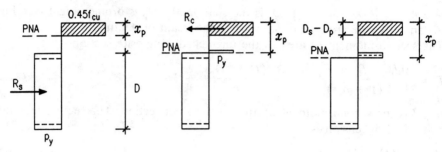

Fig. 21.7 Plastic analysis of composite section under positive moment (PNA: plastic neutral axis)

to be symmetrical). The top flange is considered to be fully restrained by the concrete slab.

The moment capacity, M_{pc}, of the composite beam is given by:

Case 1: $R_c < R_s$ (plastic neutral axis lies in concrete slab):

$$M_{pc} = R_s \left[\frac{D}{2} + D_s - \frac{R_s}{R_c} \left(\frac{D_s - D_p}{2} \right) \right] \quad (21.6)$$

Case 2: $R_s > R_c > R_w$ (plastic neutral axis lies in steel flange):

$$M_{pc} = R_s \frac{D}{2} + R_c \left(\frac{D_s + D_p}{2} \right) - \frac{(R_s - R_c)^2}{R_f} \frac{T}{4} \quad (21.7)$$

NB the last term in this expression is generally small (T is the flange thickness).

Case 3: $R_c < R_w$ (plastic neutral axis lies in web):

$$M_{pc} = M_s + R_c \left(\frac{D_s + D_p + D}{2} \right) - \frac{R_c^2}{R_w} \frac{D}{4} \quad (21.8)$$

where M_s is the plastic moment capacity of the steel section alone. This formula assumes that the web is compact i.e. not subject to the effects of local buckling. For this to be true the depth of the web in compression should not exceed $78t\varepsilon$, where t is the web thickness (ε is defined in BS 5950: Part 1). If the web is non-compact, a method of determining the moment capacity of the section is given in BS 5950: Part 3, Appendix B.[1]

21.6.3 Continuous beams

Bending moments in continuous composite beams can be evaluated from elastic global analysis. However, these result in an overestimate of moments at the supports because cracking of the concrete reduces the stiffness of the section and permits a relaxation of bending moment. A simplified approach is to redistribute

Table 21.1 Maximum redistribution of support moment based on elastic design of continuous beams at the ultimate limit state

Assumed section properties at supports	Classification of compression flange at supports				
				Plastic	
	Slender	Semi-compact	Compact	Generally	Beams (with nominal slab reinforcement
Gross uncracked	10%	20%	30%	40%	50%
Cracked	0%	10%	20%	30%	30%

the support moment based on gross (uncracked) section properties by the amounts given in Table 21.1. Alternatively, moments can be determined using the appropriate cracked and uncracked stiffnesses in a frame analysis. In this case, the permitted redistribution of moment is less.

The section classification is expressed in terms of the proportions of the compression (lower) flange at internal supports. This determines the permitted redistribution of moment. A special category of plastic section is introduced where the section is of uniform shape throughout and nominal reinforcement is placed in the slab which does not contribute to the flexural capacity of the beam. In this case the maximum redistribution of moment under uniform loading is increased to 50%. A simplified elastic approach is to use the design moments in Table 21.2 assuming that:

(1) the unfactored imposed load does not exceed twice the unfactored dead load;
(2) the load is uniformly distributed;
(3) end spans do not exceed 115% of the length of the adjacent span;
(4) adjacent spans do not differ in length by more than 25% of the longer span.

An alternative to the elastic approach is plastic hinge analysis of plastic sections. Conditions on the use of plastic hinge analysis are presented in BS 5950: Part 3,[1] and Eurocode 4 (draft).[8] However, large redistributions of moment may adversely affect serviceability behaviour (see section 21.7.8).

The ultimate capacity of a continuous beam under positive moment is determined as for a simply-supported beam. The effective breadth of the slab is based on the effective span of the beam under positive moment (see section 21.7.1). The number of shear-connectors contributing to the positive moment capacity is ascertained knowing the point of contraflexure.

The negative (hogging) moment capacity of a continuous beam or cantilever should be based on the steel section together with any properly anchored tension reinforcement within the effective breadth of the slab. This poses problems at edge columns, where it may be prudent to neglect the effect of the reinforcement unless particular measures are taken to provide this anchorage. The behaviour of a continuous beam is represented in Fig. 21.5.

Table 21.2 Moment coefficients (multiplied by free moment of WL/8) for elastic design of continuous beams

Location		Slender	Semi-compact	Compact	Plastic Generally	Plastic Beams (with nominal slab reinforcement)
Middle of end span	2 spans	0.71	0.71	0.71	0.75	0.79
	3 or more spans	0.80	0.80	0.80	0.80	0.82
First internal support	2 spans	0.91	0.81	0.71	0.61	0.50
	3 or more spans	0.86	0.76	0.67	0.57	0.48
Middle of internal spans		0.65	0.65	0.65	0.65	0.67
Internal supports (except first)		0.75	0.67	0.58	0.50	0.42
Redistribution		10%	20%	30%	40%	50%

The negative moment capacity is evaluated from plastic analysis of the section:

Case 1: $R_r < R_w$ (plastic neutral axis lies in web):

$$M_{nc} = M_s + R_s \left(\frac{D}{2} + D_r\right) - \frac{R_q^2}{R_w}\frac{D}{4} \tag{21.9}$$

where R_r is the tensile capacity of the reinforcement over width B_e, R_q is the capacity of the shear-connectors between the point of contraflexure and the point of maximum negative moment (see section 21.7.3). D_r is the height of the reinforcement above the top of the beam.

Case 2: $R_r > R_w$ (plastic neutral axis lies in flange):

$$M_{nc} = R_s \frac{D}{2} + R_r D_r - \frac{(R_s - R_r)^2}{R_f}\frac{T}{4} \tag{21.10}$$

NB the last term in this expression is generally small.

The formulae assume that the web and lower flange are compact i.e. not subject to the effects of local buckling. The limiting depth of the web in compression is $78t\varepsilon$ (where ε is defined in Chapter 2) and the limiting breath : thickness ratio of the flange is defined in Table 7 of BS 5950: Part 1.

If these limiting slendernesses are exceeded then the section is designed elastically; often the situation in bridge design. The appropriate effective breadth of slab is used because of the sensitivity of the position of the elastic neutral axis and hence the zone of the web in compression to the tensile force transferred by the reinforcement. The elastic section properties are determined on the assumption that the concrete is cracked and does not contribute to the resistance of the section.

21.7 Basic design

21.7.1 Effective breadths

The structural system of a composite floor or bridge deck is essentially a series of parallel T beams with wide thin flanges. In such a system the contribution of the concrete flange in compression is limited because of the influence of 'shear lag'. The change in longitudinal stress is associated with in-plane shear strains in the flanges.

The ratio of the effective breadth of the slab to the actual breadth (B_e/B) is a function of the type of loading, the support conditions and the cross-section under consideration as illustrated in Fig. 21.8. The effective breadth of slab is therefore not a precise figure but approximations are justified. A common approach in plastic design is to consider the effective breadth as a proportion (typically 20% – 33%) of the beam span. This is because the conditions at failure are different from the elastic conditions used in determining the data in Fig. 21.8, and the plastic bending capacity of a composite section is relatively insensitive to the precise value of effective breadth used.

Eurocode 4 (in draft)[8] and BS 5950: Part 3[1] define the effective breadth as (span/4) (half on each side of the beam) but not exceeding the actual slab breadth

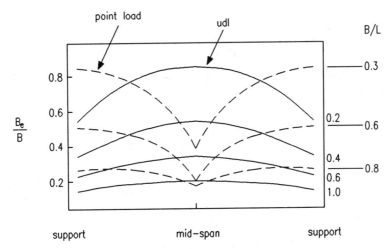

Fig. 21.8 Variation of effective breadth along beam and with loading

considered to act with each beam. Where profiled decking spans in the same direction as the beams, as in Fig. 21.3(b), allowance is made for the combined flexural action of the composite slab and the composite beam by limiting the effective breadth to 80% of the actual breadth.

In building design, the same effective breadth is used for section analysis at both the ultimate and serviceability limit states. In bridge design to BS 5400: Part 5,[5] tabular data of effective breadths are given for elastic design at the serviceability limit state. If plastic design is appropriate, the effective breadth is modified to (span/3).[5]

In the design of continuous beams, the effective breadth of the slab may be based conservatively on the effective span of the beam subject to positive or negative moment. For positive moment, the effective breadth is 0.25 times 0.7 span, and for negative moment, the effective breadth is 0.25 times 0.25 times the sum of the adjacent spans. These effective breadths reduce to 0.175 span and 0.125 span respectively for positive and negative moment regions of equal-span beams.

21.7.2 Modular ratio

The modular ratio is the ratio of the elastic modulus of steel to the creep-modified modulus of concrete which depends on the duration of the load. The short- and long-term modular ratios given in Table 21.3 may be used for all grades of concrete. The effective modular ratio used in design should be related to the proportions of the loading that are considered to be of short- and long-term duration. Typical values used for office buildings are 10 for normal weight and 15 for lightweight concrete.

21.7.3 Shear connection

The shear strength of shear-connectors is established by the push-out test, a standard test using a solid slab. A typical load–slip curve for a welded stud is

Table 21.3 Modular ratios (α_e) of steel to concrete

Type of concrete	Duration of loading		
	Short-term	Long-term	'Office' loading
Normal weight	6	18	10
Lightweight (density > 1750 kg/m^3)	10	25	15

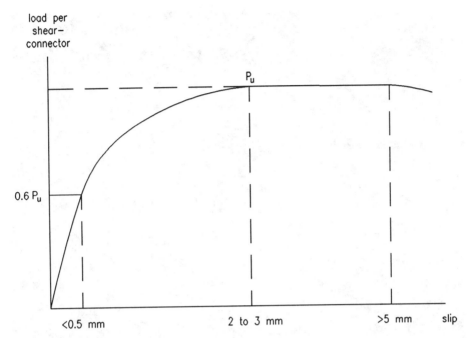

Fig. 21.9 Load–slip relationship for ductile welded shear-connector

shown in Fig. 21.9. The loading portion can be assumed to follow an empirical curve.[9] The strength plateau is reached at a slip of 2–3 mm.

The shear strength of shear-connectors is a function of the concrete strength, connector type and the weld, related to the diameter of the connector. The purpose of the head of the stud is to prevent uplift. The common diameter of stud which can be welded easily on site is 19 mm, supplied in 75 mm, 100 mm or 125 mm heights. The material properties, before forming, are typically:

Ultimate tensile strength   450 N/mm^2
Elongation at failure   15%

Higher tensile strengths (495 N/mm^2) are required in bridge design.[5] Nevertheless, the 'push-out' strength of the shear-connectors is relatively insensitive to the strength of the steel because failure is usually one of the concrete crushing for concrete grades less than 40. Also the weld collar around the base of the shear-connector contributes to increased shear resistance.

The modern method of attaching studs in composite buildings is by through-deck welding. An example of this is shown in Fig. 21.10. The common in situ method of checking the adequacy of the weld is the bend-test, a reasonably easy method of quality control which should be carried out on a proportion of studs (say 1 in 50) and the first 2 to 3 after start up.

Other forms of shear-connector such as the shot-fired connector have been developed (Fig. 21.11). The strength of these types is controlled by the size of the

Fig. 21.10 Welding of shear-connector through steel decking to a beam

pins used. Typical strengths are 30%–40% of the strengths of welded shear-connectors, but they demonstrate greater ductility.

The static strength data for stud shear-connectors are given in Table 21.4 taken from BS 5400: Part 5 and also incorporated in BS 5950: Part 3. As-welded heights are some 5 mm less than the nominal heights. The strengths of shear-connectors in structural lightweight concrete (density > 1750 kg/m^2) are taken as 90% of these values.

In Eurocode 4,[8] the approach is slightly different. Empirical formulae are given based on two failure modes; failure of the concrete and failure of the steel. The upper bound strength is given by shear failure of the shank and therefore there is apparently little advantage in using high-strength concrete.

In BS 5950: Part 3 and Eurocode 4, the design capacity of the shear-connectors

Fig. 21.11 Shot-fired shear-connector

Table 21.4 Characteristic strengths of headed stud shear-connectors in normal weight concrete

Dimensions of stud shear-connectors (mm)			Characteristic strength of concrete (N/mm^2)			
Diameter	Nominal height	As-welded height	25	30	35	40
25	100	95	146	154	161	168
22	100	95	119	126	132	139
19	100	95	95	100	104	109
19	75	70	82	87	91	96
16	75	70	70	74	78	82
13	65	60	44	47	49	52

For concrete of characteristic strength greater than 40 N/mm^2 use the values for 40 N/mm^2

For connectors of heights greater than tabulated use the values for the greatest height tabulated

is taken as 80% of the nominal static strength. Although this may broadly be considered to be a material factor applied to the material strength, it is, more correctly, a factor to ensure that the criteria for plastic design are met (see below). The design capacity of shear-connectors in negative moment regions is taken as 60% of the nominal strength. In BS 5400: Part 5, an additional material factor of 1.1 is introduced to further reduce the design capacity of the shear connectors.

In plastic design, it is important to ensure that the shear-connectors display adequate ductility. It may be expected that shear-connectors maintain their ultimate capacity at displacements of up to 5 mm. A possible exception is where concrete strengths exceed grade 40 as the form of failure may be more brittle.

For beams subject to uniform load, the degree of shear connection that is provided by uniformly-spaced shear-connectors (defined in section 21.7.4) reduces more rapidly than the applied moment away from the point of maximum moment. To ensure that the shear connection is adequate at all points along the beam, the design strength of the shear-connectors is taken as 80% of their static strength. This also partly ensures that flexural failure will occur before shear failure. For beams subject to point loads, it is necessary to design for the appropriate shear connection at each major load point.

In a simple composite beam subject to uniform load the elastic shear flow defining the shear transfer between the slab and the beam is linear, increasing to a maximum at the ends of the beam. Beyond the elastic limit of the connectors there is a transfer of force among the shear-connectors, such that, at failure, each of the shear-connectors is assumed to be subject to equal force, as shown in Fig. 21.12. This is consistent with a relatively high slip between the concrete and the

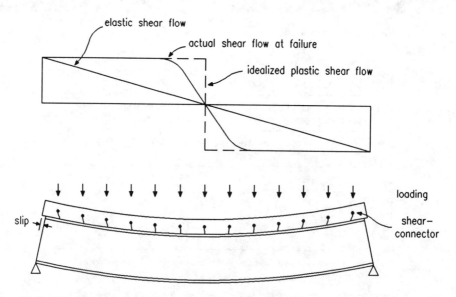

Fig. 21.12 Idealization of forces transferred between concrete and steel

steel. The slip increases as the beam span increases and the degree of shear connection reduces. For this reason BS 5400: Part 5[5] requires that shear-connectors in bridges are spaced in accordance with elastic theory. In building design, shear-connectors are usually spaced uniformly along the beam when the beam is subject to uniform load.

No serviceability limit is put on the force in the shear-connectors, despite the fact that consideration of the elastic shear flow suggests that such forces can be high at working load. This is reflected in the effect of slip on deflections in cases where partial shear connection is used. When designing bridges[5] or structures subject to fatigue loading, a limit of 55% of the design capacity of the shear connectors is appropriate for design at the serviceability limit state.

21.7.4 Partial shear connection

In plastic design of composite beams the longitudinal shear force to be transferred between the concrete and the steel is the lesser of R_c and R_s. The number of shear-connectors placed along the beam between the points of zero and maximum positive moment should be sufficient to transfer this force.

In cases where fewer shear-connectors than the number required for full shear connection are provided it is not possible to develop M_{pc}. If the total capacity of the shear-connectors between the points of zero and maximum moment is R_q (less than the smaller of R_s and R_c), then the stress block method in section 21.6.2 is modified as follows, to determine the moment capacity, M_c:

Case 4: $R_q > R_w$ (plastic neutral axis lies in flange):

$$M_c = R_s \frac{D}{2} + R_q \left[D_s - \frac{R_q}{R_c} \left(\frac{D_s + D_p}{2} \right) \right] - \frac{(R_s - R_q)^2}{R_f} \frac{T}{4} \qquad (21.11)$$

NB the last term in this expression is generally small.

Case 5: $R_q < R_w$ (plastic neutral axis lies in web):

$$M_c = M_s + R_q \left[\frac{D}{2} + D_s - \frac{R_q}{R_c} \left(\frac{D_s - D_p}{2} \right) \right] - \frac{R_q^2}{R_w} \frac{D}{4} \qquad (21.12)$$

The formulae are obtained by replacing R_c by R_q and re-evaluating the neutral axis position. The method is similar to that used in the American method of plastic design[10] which predicts a non-linear increase of moment capacity with degree of shear connection K defined as:

$K = R_q/R_s$ for $R_s < R_c$
or $K = R_q/R_c$ for $R_c < R_s$

An alternative approach, which has proved attractive, is to define the moment capacity in terms of a linear interaction of:

$$M_c = M_s + K(M_{pc} - M_s) \qquad (21.13)$$

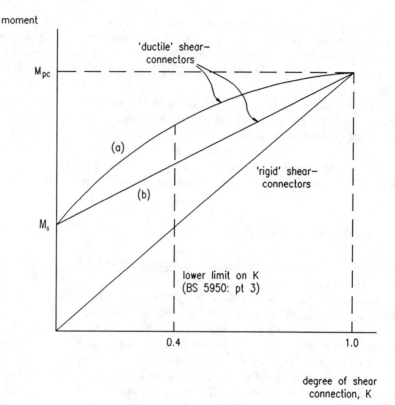

Fig. 21.13 Interaction between moment capacity and degree of shear connection. (a) Stress block method. (b) Linear interaction method

The 'stress block' and 'linear interaction' methods are presented in Fig. 21.13 for a typical beam. It can be seen that there is a significant benefit in the stress block method in the important range of $K = 0.5$ to 0.7.

In using methods based on partial shear connection a lower limit for K of 0.5 is specified in Eurocode 4 (draft).[8] This is to overcome any adverse effects arising from the limited deformation capacity of the shear connectors.

In BS 5950: Part 3, the limiting degree of shear connection increases with span (L in metres) such that:

$$K \leq (L - 6)/10 \geq 0.4 \qquad (21.14)$$

This formula means that beams longer than 16 m span should be designed for full shear connection, and beams of up to 10 m span designed for not less than 40% shear connection, with a linear transition between the two cases. Partial shear connection is also not permitted for beams subject to heavy off-centre point loads.

A further requirement is that the degree of shear connection should be adequate at all points along the beam length. For a beam subject to point loads, it follows that

Basic design 611

Table 21.5 Comparison of designs of simply-supported composite beams

Beam data	Elastic design	Plastic design		
		No connectors (BS 5950: Part 1)	Partial connection	Full shear connection
Full depth (mm)	536	536	435	435
Beam size (mm)	406 × 140	406 × 140	305 × 102	305 × 102
Beam weight (kg/m)	39	46	33	28
Number of 19 mm diameter shear-connectors	50 (every trough)	0	25 (alternate troughs)	50 (every trough)
Deflection (mm) imposed load	7	19	14	13
Deflection (mm) self-weight	14	11	25	30

Beam span: 7.5 m (unpropped)
Beam spacing: 3.0 m
Slab depth: 130 mm
Deck height: 50 mm
Steel grade: 50 ($p_y = 355$ N/mm^2)
Concrete grade: 30 (normal weight)
Imposed load: 5 kN/m^2

the shear-connectors should be distributed in accordance with the shear force diagram.

Comparison of the method of partial shear connection with other methods of design is presented in Table 21.5. Partial shear connection can result in overall economy by reducing the number of shear-connectors at the expense of a slightly heavier beam than that needed for full shear connection. Elastic design is relatively conservative and necessitates the placing of shear-connectors in accordance with the elastic shear flow. Deflections are calculated using the guidance in section 21.7.8.

21.7.5 Influence of deck shape on shear connection

The efficiency of the shear connection between the composite slab and the composite beam may be reduced because of the shape of the deck profile. This is analogous to the design of haunched slabs where the strength of the shear-connectors is strongly dependent on the area of concrete around them. Typically, there should be a 45° projection from the base of the connector to the core of the solid slab to transfer shear smoothly in the concrete without local cracking.

612   Composite beams

The model for the action of a shear-connector placed in the trough of a deck profile is shown in Fig. 21.14. For comparison also shown is the behaviour of a connector in a solid slab. Effectively, the centre of resistance in the former case moves towards the head of the stud and the couple created is partly resisted by bending of the stud but also by tensile and compressive forces in the concrete, encouraging concrete cracking and consequently the strength of the stud is reduced.

A number of tests on different stud heights and profile shapes have been performed. The following formula has been adopted by most standards worldwide. The strength reduction factor (relative to a solid slab) for the case where the decking crosses the beams is:

$$r_p = \frac{0.85}{\sqrt{n}} \frac{b_a}{D_p} \left(\frac{h - D_p}{D_p} \right) \begin{array}{l} \leq 1.0 \quad \text{for } n = 1 \\ \leq 0.8 \quad \text{for } n = 2 \end{array} \tag{21.15}$$

where b_a is the average width of the trough, h is the stud height, and n is the number of studs per trough ($n < 3$). The limit for pairs of studs is given in BS 5950 Part 3 and takes account of less ductile behaviour when $n > 1$. This formula does not apply in cases where the shear-connector does not project at

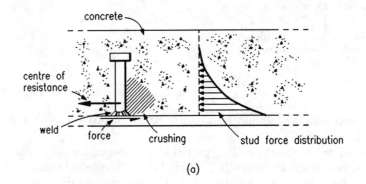

(a)

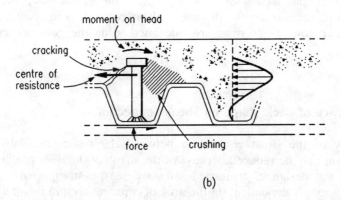

(b)

Fig. 21.14 Model of behaviour of shear-connector. (a) Shear-connector in plain slab. (b) Shear-connector fixed through profile sheeting

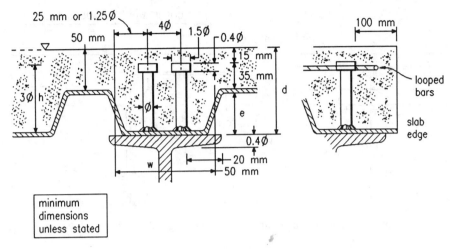

Fig. 21.15 Geometric limits on location of shear-connectors

least 35 mm above the top of the deck. A further limit is that $h \not> 2D_p$ in evaluating r_p.

Where the decking is placed parallel to the beams no reduction is made for the number of connectors but the constant in Equation (21.15) is reduced to 0.6 (instead of 0.85). No reduction is made in the second case when $b_a/D_p > 1.5$.

Further geometric limits on the placing of the shear-connectors are presented in Fig. 21.15. The longitudinal spacing of the shear-connectors is limited to a maximum of 600 mm and a minimum of 100 mm.

21.7.6 Longitudinal shear transfer

In order to transfer the thrust from the shear-connectors into the slab, without splitting, the strength of the slab in longitudinal shear should be checked. The strength is further influenced by the presence of pre-existing cracks along the beam as a result of the bending of the slab over the beam support.

The design recommendations used to check the resistance of the slab to longitudinal shear are based on research into the behaviour of reinforced concrete slabs. The design shear stress which can be transferred is taken as 0.9 N/mm² for normal weight and 0.7 N/mm² for lightweight concrete; this strength is relatively insensitive to the grade of concrete.[11]

It is first necessary to establish potential planes of longitudinal shear failure around the shear-connectors. Typical cases are shown in Fig. 21.16. The top reinforcement is assumed to develop its full tensile capacity, resisted by an equal and opposite compressive force close to the base of the connector. Both top and bottom reinforcement play an important role in preventing splitting of the concrete.

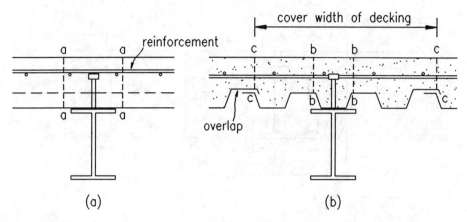

Fig. 21.16 Potential failure planes through slab in longitudinal shear

The shear resistance per unit length of the beam which is equated to the shear force transferred through each shear plane (in the case of normal weight concrete) is:

$$V = 0.9L_s + 0.7A_r f_y \leq 0.15 \, L_s \, f_{cu} \tag{21.16}$$

where L_s is the length of each shear plane considered on a typical cross-section, which may be taken as the mean slab depth in Fig. 21.16(a) or the minimum depth in Fig. 21.16(b). The total area of reinforcement (per unit length) crossing the shear plane is A_r. For an internal beam, the slab shear resistance is therefore $2V$.

The effect of the decking in resisting longitudinal shear is considerable. Where the decking is continuous over the beams, as in Fig. 21.3(a), or rigidly attached by shear-connectors, there is no test evidence of the splitting mode of failure. It is assumed, therefore, that the deck is able to provide an important role as transverse reinforcement. The term $A_r f_y$ may be enhanced to include the contribution of the decking, although it is necessary to ensure that the ends of the deck (butt joints) are properly welded by shear-connectors. In this case, the anchorage provided by each weld may be taken as $4\phi \, t_s p_y$, where ϕ is the stud diameter, t_s is the sheet thickness, and p_y is the strength of the sheet steel. Dividing by the stud spacing gives the equivalent shear resistance per unit length provided by the decking.

Where the decking is laid parallel to the beams, longitudinal shear failure can occur through the sheet-to-sheet overlap close to the line of studs. Failure is assessed assuming the shear force transferred diminishes linearly across the slab to zero at a distance of $B_e/2$. Therefore, the effective shear force that is transferred across the longitudinal overlaps is (1 − cover width of decking/effective breadth) × shear force. The shear resistance at these points excludes the effect of the decking.

To prevent splitting of the slab at edge beams, the distance between the line of shear-connectors and the edge of the slab should not be less than 100 mm. When this distance is between 100 mm and 300 mm, additional reinforcement in the form of U bars located below the head of the shear-connectors is to be provided.[12] No additional reinforcement (other than that for transverse reinforcement) need be provided where the edge of the slab is more than 300 mm from the shear-connectors.

21.7.7 Interaction of shear and moment in composite beams

Vertical shear can cause a reduction in the plastic capacity of a composite beam where high moment and shear co-exist at the same position along the beam (i.e. the beam is subject to one or two point loads). Where the shear force F_v exceeds $0.5P_v$ (where P_v is the lesser of the shear capacity and the shear buckling capacity, determined from Part 1 of BS 5950), the reduced moment capacity is determined from:

$$M_{cv} = M_c - (M_c - M_f)(2F_v/P_v - 1)^2 \qquad (21.17)$$

where M_c is the plastic capacity of the composite section, and M_f is the plastic capacity of the composite section having deducted the shear area (i.e. the web of the section).

The interaction is presented diagrammatically in Fig. 21.17. A quadratic relationship has been used, as opposed to the linear relationship in BS 5950: Part 1, because of its better agreement with test data, and because of the need for greater economy in composite sections which are often more highly stressed in shear than non-composite beams.

21.7.8 Deflections

Deflection limits for beams are specified in BS 5950: Part 1. Composite beams are, by their nature, shallower than non-composite beams and often are used in structures where long spans would otherwise be uneconomic. As spans increase, so traditional deflection limits based on a proportion of the beam span may not be appropriate. The absolute deflection may also be important and pre-cambering may need to be considered for beams longer than 10 m.

Elastic section properties, as described in section 21.6.1, are used in establishing the deflection of composite beams. Uncracked section properties are considered to be appropriate for deflection calculations. The appropriate modular ratio is used, but it is usually found that the section properties are relatively insensitive to the precise value of modular ratio. The effective breadth of the slab is the same as that used in evaluating ultimate strength.

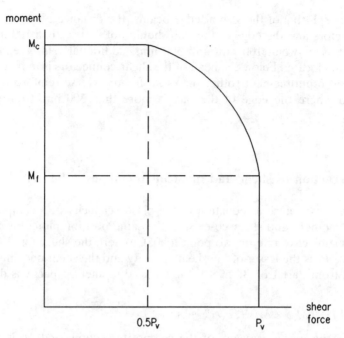

Fig. 21.17 Interaction between moment and shear

The deflection of a simple composite beam at working load, where partial shear connection is used, can be calculated from:[13]

$$\left. \begin{array}{l} \delta'_c = \delta_c + 0.5\,(1 - K)\,(\delta_s - \delta_c) \quad \text{for propped beams} \\ \delta'_c = \delta_c + 0.3\,(1 - K)\,(\delta_s - \delta_c) \quad \text{for unpropped beams} \end{array} \right\} \quad (21.18)$$

where δ_c and δ_s are the deflections of the composite and steel beam respectively at the appropriate serviceability load; K is the degree of shear connection used in the determining of the plastic strength of the beam (section 21.7.4). The difference between the coefficients in these two formulae arises from the different shear-connector forces and hence slip at serviceability loads in the two cases. These formulae are conservative with respect to other guidance.[10]

The effect of continuity in composite beams may be considered as follows. The imposed load deflection at mid-span of a continuous beam under uniform load or symmetric point loads may be determined from the approximate formula:

$$\delta_{cc} = \delta'_c\,(1 - 0.6\,(M_1 + M_2)/M_0) \quad (21.19)$$

where δ'_c is the deflection of the simply-supported composite beam for the same loading conditions; M_0 is the maximum moment in a simply-supported beam subject to the same loads; M_1 and M_2 are the end moments at the adjacent supports of the span of the continuous beam under consideration.

To determine appropriate values of M_1 and M_2, an elastic global analysis is carried out using the flexural stiffness of the uncracked section.

For buildings of normal usage, these support moments are reduced to take into account the effect of pattern loading and concrete cracking. The redistribution of support moment under imposed load should be taken as the same as that used at the ultimate limit state (see Table 21.1), but not less than 30%.

For buildings subject to semi-permanent or variable loads (e.g. warehouses), there is a possibility of alternating plasticity under repeated loading leading to greater imposed load deflections. This also affects the design of continuous beams designed by plastic hinge analysis, where the effective redistribution of support moment exceeds 50%. In such cases a more detailed analysis should be carried out considering these effects (commonly referred to as 'shakedown') as follows:

(1) evaluate the support moments based on elastic analysis of the continuous beam under a first loading cycle of dead load and 80% imposed load (or 100% for semi-permanent load);
(2) evaluate the excess moment where the above support moment exceeds the plastic capacity of the section under negative moment;
(3) the net support moments based on elastic analysis of the continuous beam under imposed load are to be reduced by 30% (or 50% for semi-permanent loads) and further reduced by the above excess moment;
(4) these support moments are input into Equation (21.19) to determine the imposed load deflection.

21.7.9 Vibration

Shallower beams imply greater flexibility and although the in-service performance of composite beams and floors in existing buildings is good, the designer may be concerned about the susceptibility of the structure to vibration-induced oscillations. The parameter commonly associated with this effect is the natural frequency of the floor or beams. The damping of the vibration by a bare steel-composite structure is often low. However, when the building is occupied, damping increases considerably.

The lower the natural frequency, the more the structure may respond dynamically to occupant-induced vibration. A limit of 4 Hz (cycles per second) is a commonly accepted lower bound to the natural frequency of each element of the structure. Clearly, vibrating machinery or external vibration effects pose particular problems and in such cases it is often necessary to isolate the source of the vibration.

In practice, the mass of the structure is normally such that the exciting force is very small in comparison leading to the conclusion that long-span structures may respond less than light short-span structures. Guidance is given in Reference 14 and Chapter 12.

21.7.10 Shrinkage, cracking and temperature

It is not normally necessary to check crack widths in composite floors in heated buildings, even where the beams are designed as simply-supported, provided that the slab is reinforced as recommended in BS 5950: Part 4 or BS 8110 as appropriate. In such cases crack widths may be outside the limits given in BS 8110, but experience shows that no durability problems arise.

In other cases additional reinforcement over the beam supports may be required to control cracking and the relevant clauses in BS 8110: Part 2[15] and BS 5400: Part 5[5] should be followed. This is particularly important where hard finishes are used.

Questions of long-term shrinkage and temperature-induced effects often arise in long-span continuous composite beams, as they cause additional negative (hogging) moments and deflections. In buildings these effects are generally neglected, but in bridges they can be important.[5,7]

The curvature of a composite section resulting from a free shrinkage (or temperature induced) strain ε_s in the slab is:

$$K_s = \frac{\varepsilon_s (D + D_s + D_p) A}{2 (1 + \alpha_e r) I_c} \tag{21.20}$$

where I_c and r are defined in section 21.6.1 and α_e is the appropriate modular ratio for the duration of the action considered. The free shrinkage strain may be taken to vary between 100×10^{-6} in external applications and 300×10^{-6} in dry heated buildings. A creep reduction factor is used in BS 5400: Part 5[5] when considering shrinkage strains. This can reduce the effective strain by up to 50%. The central deflection of a simply-supported beam resulting from shrinkage strain is then $0.125 K_s L^2$.

References to Chapter 21

1. British Standards Institution (1990) *Structural use of steelwork in building*. Part 3, Section 3.1: *Code of practice for design of composite beams*. BS 5950, BSI, London.
2. The Steel Construction Institute (SCI) (1989) *Design of Composite Slabs and Beams with Steel Decking*. SCI, Ascot, Berks.
3. Chien E.Y.L. & Ritchie J.K. (1984) *Design and Construction of Composite Floor Systems*. Canadian Institute of Steel Construction.
4. Lawson R.M. (1988) *Design for Openings in the Webs of Composite Beams*. The Steel Construction Institute, Ascot, Berks.
5. British Standards Institution (1979) *Steel, concrete and composite bridges*. Part 5: *Code of practice for design of composite bridges*. BS 5400, BSI, London.

6. Brett P.R., Nethercot D.A. & Owens G.W. (1987) Continuous construction in steel for roofs and composite floors. *The Structural Engineer*, **65A**, No. 10, Oct.
7. Johnson R.P. & Buckby R.J. (1986) *Composite Structures of Steel and Concrete, Vol. 2: Bridges*, 2nd edn. Collins.
8. Commission of the European Communities (1985) *Eurocode 4: Common unified rules for composite steel and concrete structures*. Report EUR 9886 EN.
9. Yam L.C.P. & Chapman J.C. (1968) The inelastic behaviour of simply supported composite beams of steel and concrete. *Proc. Instn Civ. Engrs*, **41**, Dec., 651–83.
10. American Institute of Steel Construction (1986) *Manual of Steel Construction: Load and Resistance Factor Design*. AISC, Chicago.
11. Johnson R.P. (1975 & 1986) *Composite Structures of Steel and Concrete. Vol. 1: Beams. Vol. 2: Bridges*, 2nd edn. Granada.
12. Johnson R.P. & Oehlers D.J. (1982) Design for longitudinal shear in composite L beams. *Proc. Instn Civ. Engrs*, **73**, Part 2, March, 147–70.
13. Johnson R.P. & May I.M. (1975) Partial interaction design of composite beams. *The Structural Engineer*, **53**, No. 8, Aug., 305–11.
14. Wyatt T.A. (1989) *Design Guide on the Vibration of Floors*. The Steel Construction Institute (SCI)/CIRIA.
15. British Standards Institution (1985) *Structural use of concrete. Part 2: Code of practice for special circumstances*. BS 8110, BSI, London.

A series of worked examples follows which are relevant to Chapter 21.

620   Worked example

The Steel Construction Institute Silwood Park, Ascot, Berks SL5 7QN	Subject **DESIGN OF SIMPLY SUPPORTED COMPOSITE BEAM**	Chapter ref. **21**
	Design code **BS5950: Part 3**	Made by **RML** Checked by **GWO**	Sheet no. **1**

Problem

Design a composite floor with beams at 3 m centres spanning 12 m. The composite slab is 130 mm deep. The floor is to resist an imposed load of 5.0 kN/m², partition loading of 1.0 kN/m² and a ceiling load of 0.5 kN/m². The floor is to be unpropped during construction.

Deck:

Profile height	D_p	=	50 mm
Trough spacing		=	300 mm
Trough width (average)		=	150 mm

Shear connectors:

19 mm diameter

95 mm as-welded length

Concrete:

Compressive strength f_{cu} = 30 N/mm²

Density (lightweight concrete) is 1800 kg/m³ (dry)
(no extra allowance is made for wet weight - assumed to be included in construction load)

Loads:

Imposed load	=	5.0 kN/m²	
Partitions (imposed load)	=	1.0 kN/m²	
Ceiling	=	0.5 kN/m²	
Self weight (slab)	=	2.00 kN/m²	} see later
Self weight (beam)	=	0.67 kN/m	}
Construction load (temporary)	=	0.5 kN/m²	

Finishes and other loads are neglected.

The Steel Construction Institute Silwood Park, Ascot, Berks SL5 7QN		Subject DESIGN OF SIMPLY SUPPORTED COMPOSITE BEAM	Chapter ref. 21
		Design code BS5950: Part 3	Made by RML Checked by GWO	Sheet no. 2

Initial selection of beam size

Choose beam depth so that ratio of span/overall depth is about 20
Choose 457 × 191 × 67 kg/m beam steel grade 50
$p_y = 355$ N/mm² as flange thickness is less than 16 mm
Alternatively, use "Design of Composite Slabs and Beams with Steel Decking".

SCI Publication

Construction condition

Self weight of slab (symmetric deck)
Self weight $= 105 \times 1800 \times 9.81 \times 10^{-6} = 1.85$ kN/m²
Self weight of deck is taken as 0.14 kN/m²
Total slab weight is 2.0 kN/m²

Design load in construction condition

$= (1.6 \times 0.5 + 1.4 \times 2.0) \times 3.0 + 0.67 \times 1.4$
$= 11.74$ kN/m

Design moment in construction condition

$= 11.74 \times 12.0^2/8 = 211.0$ kNm

Moment resistance of steel section

$M_s = 1470 \times 10^3 \times 355 \times 10^{-6}$
$= 521.9$ kNm > 211.0 kNm

(the beam is assumed to be laterally restrained by the decking)

Check stress due to self weight

Moment $= (2.0 \times 3.0 + 0.67) \times 12^2/8$
$= 120.1$ kNm

Bending stress $= \dfrac{120.1 \times 10^6}{1300 \times 10^3} = 92$ N/mm²

Deflection of soffit of beam after construction

$\delta = \dfrac{5}{384} \times \dfrac{(2.0 \times 3.0 + 0.67) \times 12000^4}{205 \times 10^3 \times 29400 \times 10^4}$

Worked example

The Steel Construction Institute Silwood Park, Ascot, Berks SL5 7QN	Subject **DESIGN OF SIMPLY SUPPORTED COMPOSITE BEAM**		Chapter ref. **21**	
	Design code **BS5950: Part 3**	Made by **RML** Checked by **GWO**	Sheet no.	**3**

Composite condition - ultimate load

Factored load on composite beam

w_u = $(5.0 \times 1.6 + 1.0 \times 1.6 + 0.5 \times 1.4 + 2.0 \times 1.4) \times 3.0 + 0.67 \times 1.4$

= 40.2 kN/m

Design moment on composite beam

M_u = $\dfrac{40.2 \times 12^2}{8}$ = 723.6 kNm

Effective breadth of slab

B_e = 12000/4 = 3000 mm ≯ b = 3000 mm

Resistance of slab in compression

R_c = $0.45 f_{cu} B_e (D_s - D_p)$

= $0.45 \times 30 \times 3000 \times (130 - 50) \times 10^{-3}$

= 3240 kN

Resistance of steel section in tension

R_s = $A p_y$

= $8540 \times 355 \times 10^{-3}$

= 3032 kN

As $R_c > R_s$ plastic neutral axis lies in concrete.

Moment resistance of composite beam (equation 21.6) for full shear connection is:

M_{pc} = $R_s \left[\dfrac{D}{2} + D_s - \dfrac{R_s}{R_c} \dfrac{(D_s - D_p)}{2} \right]$ for $R_s \leq R_c$

= $3032 \left\{ \dfrac{453.6}{2} + 130 - \dfrac{3032}{3240} \times \dfrac{80}{2} \right\} \times 10^{-3}$

= 968.3 kNm > M_u

The Steel Construction Institute Silwood Park, Ascot, Berks SL5 7QN	Subject **DESIGN OF SIMPLY SUPPORTED COMPOSITE BEAM**		Chapter ref. **21**
	Design code **BS5950: Part 3**	Made by **RML** Checked by **GWO**	Sheet no. **4**

Degree of shear connection

*Use two shear connectors per trough
(i.e. 2 per 300 mm spacing)*

Number of shear connectors in half span $= \dfrac{6000}{150} = 40$

Resistance of shear connector:

    From Table 21.4, characteristic strength $= 100$ kN

    Design strength $= 0.8 \times 100 = 80$ kN

    For lightweight concrete multiply by 0.9

    Design strength (LWC) $= 0.9 \times 80 = 72$ kN

Reduction factor for profile shape (equation 21.15)

$$r = \dfrac{0.85}{\sqrt{2}} \times \dfrac{150}{50} \times \dfrac{(95-50)}{50} = 1.62 > 0.8$$

Therefore use $r = 0.8$ *for pairs of shear connectors as required in BS 5950: Part 3 Clause 5.4.7.2.*

Total resistance of shear connectors

$$R_q = 0.8 \times 72 \times 40 = 2304 \text{ kN}$$

As $R_q < R_s$ *and* $R_q < R_c$
redesign beam for partial shear connection.

The Steel Construction Institute Silwood Park, Ascot, Berks SL5 7QN	Subject DESIGN OF SIMPLY SUPPORTED COMPOSITE BEAM		Chapter ref. 21
	Design code BS5950: Part 3	Made by RML Checked by GWO	Sheet no. 5

Moment resistance of composite beam with partial shear connection (equation 21.11)

Tensile resistance of web

R_w = $8.5 \times (453.6 - 2 \times 12.7) \times 355 \times 10^{-3}$

   = 1292 kN

As $R_q > R_w$, plastic neutral axis lies in steel flange

$$M_c = R_s \frac{D}{2} + R_q \left[D_s - \frac{R_q}{R_c} \frac{(D_s - D_p)}{2} \right] - \frac{(R_s - R_q)^2}{R_f} \frac{T}{4}$$

(normally the final term can be ignored)

$$M_c = 3032 \times \frac{453.6}{2} \times 10^{-3} + 2304 \times \left[130 - \frac{2304}{3240} \times \frac{80}{2} \right] \times 10^{-3}$$

$$- \frac{(3032 - 2304)^2}{(3032 - 1292) \times 0.5} \times \frac{12.7}{4} \times 10^{-3}$$

   = 687.6 + 234.0 − 1.9 = 919.7 kNm > 723.6 kNm

Degree of shear connection provided

K = R_q / R_s = 2304 / 3032 = 0.76

Check possibility of providing fewer shear connectors. Limit on degree of shear connection is:

K ≥ $\frac{L - 6}{10}$; but K ≥ 0.4

For L = 12 m   K ≥ 0.6

One shear connector every trough (i.e. 300 mm spacing) corresponds to K = 0.47 < 0.6

which does not provide adequate shear connection.

	Subject	Chapter ref.
The Steel Construction Institute Silwood Park, Ascot, Berks SL5 7QN	DESIGN OF SIMPLY SUPPORTED COMPOSITE BEAM	21
	Design code BS5950: Part 3	Made by RML Sheet no. 6 Checked by GWO

Composite condition - service load

Determine elastic section properties

Modular ratio $\alpha_e = 15$ for LWC (see Table 21.3)

Elastic neutral axis depth (equation 21.1):

$$x_e = \frac{\frac{D_s - D_p}{2} + \alpha_e \, r \left[\frac{D}{2} + D_s\right]}{(1 + \alpha_e \, r)}$$

where $r = \dfrac{A}{(D_s - D_p) B_e} = \dfrac{8540}{80 \times 3000} = 0.0355$

$$x_e = \frac{\frac{80}{2} + 0.0355 \times 15 \times \left[\frac{453.6}{2} + 130\right]}{(1 + 0.0355 \times 15)}$$

$ = 150.1$ mm below top of slab

Second moment of area (equation 21.2):

$$I_c = \frac{A (D + D_s + D_p)^2}{4 (1 + \alpha_e \, r)} + \frac{B_e (D_s - D_p)^3}{12 \, \alpha_e} + I_s$$

$$= \frac{8540 (453.6 + 130 + 50)^2}{4 (1 + 0.0355 \times 15)} + \frac{3000 \times 80^3}{12 \times 15} + 294 \times 10^6$$

$= 861.8 \times 10^6$ mm^4

Elastic section modulus - steel flange (equation 21.3)

$$Z_t = \frac{861.8 \times 10^6}{(453.6 + 130 - 150.1)} = 1.98 \times 10^6 \text{ mm}^3$$

Elastic section modulus - concrete (equation 21.4)

$$Z_c = \frac{861.8 \times 10^6}{150.1} \times 15 = 86.1 \times 10^6 \text{ mm}^3$$

626   Worked example

The Steel Construction Institute Silwood Park, Ascot, Berks SL5 7QN	Subject **DESIGN OF SIMPLY SUPPORTED COMPOSITE BEAM**	Chapter ref. **21**
	Design code **BS5950: Part 3**	Made by **RML** / Sheet no. **7** Checked by **GWO**

Service load on composite section

$$= (5.0 + 1.0 + 0.5) \times 3.0 = 19.5 \text{ kN/m}$$

Service moment

$$= 19.5 \times 12^2/8 = 351 \text{ kNm}$$

Stress in steel $= \dfrac{351 \times 10^6}{1.98 \times 10^6} = 177 \text{ N/mm}^2$

Additional stress on non-composite section $= 92 \text{ N/mm}^2$

Total serviceability stress $= 177 + 92$
$$= 269 \text{ N/mm}^2 < p_y$$

Stress in concrete $= \dfrac{351 \times 10^6}{86.1 \times 10^6}$

$$= 4.1 \text{ N/mm}^2 < 0.5 f_{cu}$$

Deflection - imposed load

Imposed load $= (5.0 + 1.0) \times 3.0 = 18.0 \text{ kN/m}$

Deflection $\delta_c = \dfrac{5}{384} \times \dfrac{18.0 \times 12000^4 \times 10^{-3}}{205 \times 10^3 \times 861.8 \times 10^6}$

$$\delta_c = 27.5 \text{ mm}$$

Additional deflection from partial shear-connection (equation 21.18)

$$\delta_s = \dfrac{861.8}{294.0} \times 27.5 = 80.6 \text{ mm}$$

$$\delta_c^1 = \delta_c + 0.3 (1 - K)(\delta_s - \delta_c)$$
$$= 27.5 + 0.3 (1 - 0.76)(80.6 - 27.5)$$
$$= 31.3 \text{ mm} \quad (\text{span}/383)$$

Deflection is the limiting criterion.
It would be possible to show that the imposed load could be increased to 6 kN/m² for the same beam size.

The Steel Construction Institute Silwood Park, Ascot, Berks SL5 7QN	Subject **DESIGN OF SIMPLY SUPPORTED COMPOSITE BEAM**		Chapter ref. **21**
	Design code **BS5950: Part 3**	Made by **RML** Checked by **GWO**	Sheet no. **8**

Check natural frequency of beam

Consider weight of floor in dynamic calculations to include: self weight of slab and beam + 10% imposed load + ceiling load but excluding partitions

$$= (2.0 + 0.1 \times 5.0 + 0.5) \times 3.0 + 0.67 = 9.67 \text{ kN/m}$$

Accurate calculations show that the ratio of the dynamic to static stiffness of a composite beam used in deflection calculations is 1.1 to 1.15. Consider deflection reduction of 1.1 for the same section properties as calculated under imposed load.

Deflection of composite beam, when subject to instantaneously applied self weight (as above)

$$\delta_{sw} = \frac{5 \times 9.67 \times 12^4 \times 10^3}{384 \times 205 \times 10^6 \times 1.1 \times 861.8 \times 10^{-6}}$$

$$= 13.4 \text{ mm}$$

Natural frequency of beam

$$f \simeq 18/\sqrt{\delta_{sw}} = 18/\sqrt{13.4} = 4.91 \text{ Hz} \quad \underline{Satisfactory}$$

(A detailed check to Reference 21.14 demonstrates that this floor is satisfactory for general office use).

Worked example

The Steel Construction Institute Silwood Park, Ascot, Berks SL5 7QN	Subject **DESIGN OF CONTINUOUS COMPOSITE BEAM**	Chapter ref. **21**
	Design code BS5950: Parts 1 & 3	Made by **RML** Checked by **GWO**	Sheet no. **1**

Use same design data as for simply - supported composite beams.

Choose 406 × 178 × 60 kg/m UB grade 50 - N.B. Section is plastic. (span/depth ratio = 22, and 12% weight saving).

Construction condition

Design load in construction condition

w_u = (1.6 × 0.5 + 1.4 × 2.0) × 3.0 + 0.60 × 1.4 = 11.6 kN/m

Check stability of bottom flange under worst combination of loading

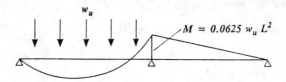

$M = 0.0625\, w_u\, L^2$

M = $0.0625 \times 11.6 \times 12^2$ = 104.4 kNm

Consider effective length as $0.85\, L = 0.85 \times 12000 = 10200$ mm

Slenderness λ = $10200/39.7$ = 257

Slenderness correction factor for shape of bending moment diagram (linear) is obtained from Table 20 of BS 5950: Part 1.

From SCI publication 'Steelwork Design: Guide to BS 5950: Part 1: Volume 1' (page 29)

u = 0.882 x = 33.9

Assume top flange is laterally restrained by the decking. Use design formula for effective slenderness in BS 5950: Part 1 (Appendix G3.3).

$$\lambda_{LT} = n\, u\, v_t\, \lambda \quad \text{where } v_t = \left[\frac{\frac{4a}{d}}{1 + \left(\frac{2a}{d}\right)^2 + \frac{1}{20}\left(\frac{\lambda}{x}\right)^2} \right]^{0.5}$$

and a = distance of restraint above shear centre of beam (= $d/2$ in this case)
  d = clear web depth

The Steel Construction Institute Silwood Park, Ascot, Berks SL5 7QN	Subject **DESIGN OF CONTINUOUS COMPOSITE BEAM**		Chapter ref. **21**
	Design code BS5950: Parts 1 & 3	Made by RML Checked by GWO	Sheet no. **2**

Hence:

$$v_t = \left[\frac{1}{1 + \frac{1}{40}\left[\frac{\lambda}{x}\right]^2}\right]^{0.5} = \left[\frac{1}{1 + \frac{1}{40}\left[\frac{257}{33.9}\right]^2}\right]^{0.5} = 0.64$$

λ_{LT} = 0.77 × 0.882 × 0.64 × 257 = 112

From Table 11 of BS 5950: Part 1, bending strength

p_b = 122 N/mm^2

Buckling resistance moment

M_b = 122 × 1194 × 10^3 × 10^{-6}

= 145.6 kNm > 104.4 kNm i.e. stable during construction

Maximum negative (hogging) moment

= 0.125 w_u L^2 = 208.8 kNm

Plastic moment resistance of steel section

= 355 × 1194 × 10^3 × 10^{-6}

= 423.9 kNm > 208.8 kNm

Hence, the steel beam can support the loads during construction.

Moment in mid-span of two span beam after construction

= 0.0703 × (2.0 × 3.0 + 0.6) × 12^2 = 66.8 kNm

Bending stress (elastic)

$$= \frac{66.8 \times 10^6}{1058 \times 10^3} = 63 \text{ N/mm}^2$$

Deflection of underside of beam after construction

$$\delta_o = \frac{2.1 \times (2.0 \times 3.0 + 0.6) \times 12000^4}{384 \times 205 \times 10^3 \times 21508 \times 10^4}$$

= 17.0 mm   (span/706)

630   Worked example

The Steel Construction Institute Silwood Park, Ascot, Berks SL5 7QN		Subject **DESIGN OF CONTINUOUS COMPOSITE BEAM**	Chapter ref. **21**
	Design code *BS5950: Parts 1 & 3*	Made by **RML** Checked by **GWO**	Sheet no. **3**

Composite condition - ultimate loading

Factored loading on composite beam

w_u = $[(5.0 + 1.0) \times 1.6 + (0.5 + 2.0) \times 1.4] \times 3.0 + 0.6 \times 1.4$

   = 40.1 kN/m

Design moment on composite beam

Ratio of imposed load: dead load

$$= \frac{5.0 + 1.0}{2.0 + 0.5 + 0.6/3.0} = 2.22 > 2$$

Therefore, the use of Table 21.2 is not permitted. However, calculate moments from pattern load cases as follows:

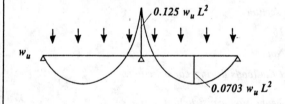

No redistribution

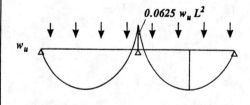

50% redistribution

$(0.0703 + 0.032) w_u L^2$
$= 0.103 w_u L^2$
(accurate value $= 0.096 w_u L^2$)

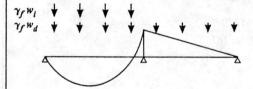

Imposed load on one span; dead load on both spans

The Steel Construction Institute Silwood Park, Ascot, Berks SL5 7QN	Subject **DESIGN OF CONTINUOUS COMPOSITE BEAM**		Chapter ref. **21**
	Design code BS5950: Parts 1 & 3	Made by **RML** Checked by **GWO**	Sheet no. **4**

The section is 'plastic' when grade 50 steel is used. According to Table 21.1, 50% redistribution of the negative (hogging) moment can be made provided the slab has nominal reinforcement.

Negative moment $= 0.0625 \times 40.1 \times 12^2$

$\qquad = 360.9 \ kNm$

Positive moment $= 0.096 \times 40.1 \times 12^2$

$\qquad = 554.3 \ kNm \quad (50\% \ redistribution)$

Alternatively, single span loading gives:

Positive (sagging) moment

$= \{0.096 \times 1.6 \times (5.0 + 1.0) \times 3.0 + 0.070 \times 1.4 \times [(2.0 + 0.5) \times 3.0 + 0.6]\} \times 12^2$

$= 512.4 \ kNm$

Plastic moment resistance of steel section

$= 355 \times 1194 \times 10^3 \times 10^{-6}$

$= 423.9 \ kNm > 360.9 \ kNm$

This exceeds the design moment. It may be shown that 42% redistribution of moment is adequate. Check moment resistance of composite section.

<u>Moment resistance of composite section</u>

Effective breadth of slab (for internal span)

$B_e \quad = \quad 0.7 \times 12000/4 = 2100 \ mm \not> beam \ spacing$

(note 0.7 has been used for internal span, rather than 0.8 for external span)

Worked example

The Steel Construction Institute Silwood Park, Ascot, Berks SL5 7QN	Subject **DESIGN OF CONTINUOUS COMPOSITE BEAM**		Chapter ref. **21**
	Design code BS5950: Parts 1 & 3	Made by **RML** Checked by **GWO**	Sheet no. **5**

Resistance of slab in compression

$$R_c = 0.45 f_{cu} B_e (D_s - D_p)$$

$$= 0.45 \times 30 \times 2100 \times (130 - 50) \times 10^{-3}$$

$$= 2268 \text{ kN}$$

Resistance of steel section in tension

$$R_s = A p_y = 7600 \times 355 \times 10^{-3}$$

$$= 2698 \text{ kN}$$

As $R_s > R_c$, plastic neutral axis lies in steel flange

Plastic moment resistance of composite section

$$M_{pc} = R_s \frac{D}{2} + R_c \frac{(D_s + D_p)}{2} - \left[\frac{(R_s - R_c)^2}{R_f} \times \frac{T}{4} \right]$$

where $R_f = 12.8 \times 177.8 \times 355 \times 10^{-3}$

$$= 808 \text{ kN}$$

The final term is small and may be neglected

$$M_{pc} = \left[2698 \times \frac{406.4}{2} + 2268 \times \frac{(130 + 50)}{2} - \frac{(2698 - 2268)^2}{808} \times \frac{12.8}{4} \right] \times 10^{-3}$$

$$= 548.2 + 204.1 - 0.7$$

$$= 751.6 \text{ kNm} > 554.3 \text{ kNm}$$

As there is sufficient reserve in the composite section, consider reducing the amount of shear connectors.

The Steel Construction Institute	Subject	Chapter ref.
Silwood Park, Ascot, Berks SL5 7QN	DESIGN OF CONTINUOUS COMPOSITE BEAM	21
	Design code BS5950: Parts 1 & 3	Made by RML Sheet no. 6 Checked by GWO

Partial shear connection

Zone of beam subject to positive (sagging) moment

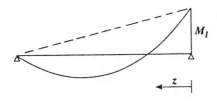

$M = 0$; when $z = L \left[\dfrac{2 M_1}{w_u L^2} \right]$

If $\quad M_1 \quad = \quad 0.0625 \, w_u \, L^2 \qquad$ *(after redistribution)*

$\quad\quad z \quad = \quad L/8$

Zone of beam from point of zero to max. positive moment $\simeq 0.43 \, L$

Use two shear connectors every 300 mm

Number of shear connectors $= \dfrac{0.43 \times 12000 \times 2}{300} = 34$

Resistance of one shear connector in LWC

(see simple beam calculations) $= \quad 72 \text{ kN}$

$R_q \quad = \quad 0.8 \times 72 \times 34 \quad = \quad 1958 \text{ kN}$

Degree of shear connection ($R_c < R_s$)

$K \quad = \quad 1958/2268 \quad = \quad 0.86$

Minimum shear connection

$K \quad > \quad \dfrac{L - 6}{10} \quad = \quad 0.6 < 0.86 \qquad$ OK

The Steel Construction Institute Silwood Park, Ascot, Berks SL5 7QN	Subject *DESIGN OF CONTINUOUS COMPOSITE BEAM*	Chapter ref. 21
	Design code BS5950: Parts 1 & 3	Made by RML Checked by GWO	Sheet no. 7

As $R_q > R_w$; *plastic neutral axis lies in steel flange*

Plastic moment resistance of section

$$M_c = R_s \frac{D}{2} + R_q \left[D_s - \frac{R_q}{R_c} \frac{(D_s - D_p)}{2} \right]$$

$$M_c = \left\{ 2698 \times \frac{406.4}{2} + 1958 \left[130 - \frac{1958}{2268} \frac{(130 - 50)}{2} \right] \right\} \times 10^{-3}$$

$$= 548.2 + 186.9$$

$$= 735.1 \text{ kNm} > 554.3 \text{ kNm}$$

<u>Vertical shear resistance</u>

Shear resistance (to BS 5950: Part 1, Clause 4.2.6)

$$P_v = 0.6 \times 355 \times 406.4 \times 7.8 \times 10^{-3}$$

$$= 675.2 \text{ kN}$$

Applied shear force at internal support

$$F_v = w_u \frac{L}{2} + \frac{M_1}{L}$$

$$= 1.25 \left[\frac{w_u L}{2} \right]$$

$$= 1.25 \times 40.1 \times 12/2 = 300.7 \text{ kN}$$

$$F_v = 0.45 P_v < 0.5 P_v$$

So no reduction in moment resistance is necessary.

The Steel Construction Institute Silwood Park, Ascot, Berks SL5 7QN	Subject **DESIGN OF CONTINUOUS COMPOSITE BEAM**		Chapter ref. **21**
	Design code BS5950: Parts 1 & 3	Made by RML Checked by GWO	Sheet no. **8**

Composite condition - service loading

Determine elastic section properties under positive moment

Elastic neutral axis depth

$$x_e = \frac{\frac{D_s - D_p}{2} + \alpha_e r \left[\frac{D}{2} + D_s\right]}{1 + \alpha_e r}$$

Take α_e = 15 for LWC

$$r = \frac{A}{(D_s - D_p) B_e} + \frac{7600}{(130 - 50) \, 2100} = 0.045$$

$$x_e = \frac{\left[\frac{80}{2} + 0.045 \times 15 \times \left[\frac{406.4}{2} + 130\right]\right]}{(1 + 0.045 \times 15)}$$

= 158.1 mm

Second moment of area of section (equation 21.2)

$$I_c = \frac{A(D + D_s + D_p)^2}{4(1 + \alpha_e r)} + \frac{B_e (D_s - D_p)^3}{12 \alpha_e} + I_x$$

$$= \left[\frac{7600 \, (406.4 + 130 + 50)^2}{4 \, (1 + 0.045 \times 15)} + \frac{2100 \times (130 - 50)^3}{12 \times 15} + 215.1\right] \times 10^6$$

= (390.0 + 6.0 + 215.1) × 10^6

= 611.1 × 10^6 mm^4

Elastic section modulus - steel flange (equation 21.3)

$$Z_t = \frac{611.1 \times 10^6}{(406.4 + 130 - 158.1)}$$

= 1.62 × 10^6 mm^3

The Steel Construction Institute Silwood Park, Ascot, Berks SL5 7QN		Subject *DESIGN OF CONTINUOUS COMPOSITE BEAM*		Chapter ref. 21
		Design code BS5950: Parts 1 & 3	Made by RML Checked by GWO	Sheet no. 9

Service load on composite section

$$= (5.0 + 1.0 + 0.5) \times 3 = 19.5 \text{ kN/m}$$

Mid span moment for serviceability conditions should take into account 'shakedown effects' as degree of moment redistribution at ultimate loads exceeds 30%.

Shakedown load = Dead load + 80% Imposed load

$$w_s = [2.0 + 0.5 + 0.8 \times (5.0 + 1.0)] \times 3 + 0.6$$
$$= 22.5 \text{ kN/m}$$

Shakedown moment (elastic negative (hogging) moment)

$$= 0.125 \, w_s \, L^2 = 0.125 \times 22.5 \times 12^2$$
$$= 405 \text{ kNm}$$

This is less than the plastic moment resistance of the beam (= 423.9 kNm), so no plastification occurs at working load.

Service moment under imposed loading on all spans (allowing for 30% redistribution of negative moment)

$$= (0.0703 + 0.15 \times 0.125) \times 19.5 \times 12^2$$
$$= 250.0 \text{ kNm}$$

Steel stress (lower flange)

$$= \frac{250.0 \times 10^6}{1.62 \times 10^6} = 154 \text{ N/mm}^2$$

Additional stress in steel beam arising from self weight of floor

$$= 63 \text{ N/mm}^2$$

Total stress = $63 + 154$ = $217 \text{ N/mm}^2 < p_y$

Concrete stress

$$= \frac{250 \times 10^6}{611.1 \times 10^6} \times \frac{158.1}{15} = 4.3 \text{ N/mm}^2 < 0.5 f_{cu}$$

The Steel Construction Institute Silwood Park, Ascot, Berks SL5 7QN	Subject **DESIGN OF CONTINUOUS COMPOSITE BEAM**		Chapter ref. **21**
	Design code BS5950: Parts 1 & 3	Made by **RML** Checked by **GWO**	Sheet no. **10**

Deflection - imposed loading

Imposed loading $\quad w_i \quad = \quad (5.0 + 1.0) \times 3 \quad = \quad 18.0$ kN/m

Free moment $\quad M_o \quad = \quad w_i L^2/8 \quad = \quad 18.0 \times 12^2/8$

$\quad\quad\quad\quad\quad\quad\quad\quad = \quad 324$ kNm

Negative (hogging) moment $\quad = \quad 324$ kNm for a two span beam

According to section 21.7.8, reduce negative moment by 30% to represent the effects of pattern loading

$M_1 \quad = \quad 0.7 \times 324 \quad = \quad 226.8$ kNm

$M_2 \quad = \quad 0$

Deflection of simply supported beam

$$\delta_o = \frac{5 \times 18.0 \times 12^4 \times 10^3}{384 \times 205 \times 10^6 \times 611.1 \times 10^{-6}}$$

$\quad\quad = \quad 38.8$ mm

Allowing for partial shear connection $K = 0.86$

Deflection of steel beam $\delta_s = \dfrac{611.1}{215.1} \times 38.8 = 110.2$ mm

Modified deflection $\delta_o' \quad = \quad \delta_o + 0.3 (1 - K) \delta_s$

$\quad\quad\quad\quad\quad\quad\quad = \quad 38.8 + 0.3 \times (1 - 0.86) \times 110.2$

$\quad\quad\quad\quad\quad\quad\quad = \quad 43.4$ mm

Deflection of continuous beam

$$\delta_c = \delta_o \left[1 - 0.6 \frac{(M_1 + M_2)}{M_o} \right] = 43.4 \left[1 - 0.6 \times \frac{226.8}{324} \right]$$

$\quad\quad = \quad 25.2$ mm $\quad$ (span/476) <u>Satisfactory</u>

The Steel Construction Institute Silwood Park, Ascot, Berks SL5 7QN	Subject **DESIGN OF CONTINUOUS COMPOSITE BEAM**	Chapter ref. **21**
	Design code BS5950: Parts 1 & 3	Made by **RML** Checked by **GWO**	Sheet no. **11**

Stability of bottom flange

Check the stability of the bottom flange in the negative (hogging) moment region. The worst case for instability is when only one span is loaded:

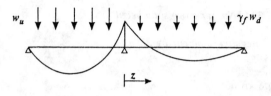

Negative moment = $0.0625 \, (w_u + \gamma_f w_d) \, L^2$

$\gamma_f w_d$ = $1.4 \, [\,(2.0 + 0.5)3 + 0.6\,] = 11.3 \text{ kN/m}$

w_u = 40.1 kN/m

Negative moment = $0.0625 \, (40.1 + 11.3) \, 12^2$

= 462.6 kNm

This exceeds the plastic resistance moment of the beam, $M_p = 423.9 \text{ kNm}$

Use this moment M_p at the support

$M = 0$ when $z = L \left[\dfrac{2 \times M_p}{\gamma_f w_d L^2} \right]$

$z = 12 \left[\dfrac{2 \times 423.9}{11.3 \times 12^2} \right] = 6.3 \text{ m}$

It is necessary to introduce an additional lateral restraint close to the support such that $\lambda_{LT} < 30$ for grade 50 steel.

The Steel Construction Institute Silwood Park, Ascot, Berks SL5 7QN	Subject **DESIGN OF CONTINUOUS COMPOSITE BEAM**		Chapter ref. **21**
	Design code BS5950: Parts 1 & 3	Made by **RML** Checked by **GWO**	Sheet no. **12**

Worked example   639

The distortional restraint provided by the web has been neglected. However, it is reasonable to take:

$n = 0.95; \quad u = 0.88; \quad v_t = 0.9$

such that $\lambda_{max} = \dfrac{30}{0.95 \times 0.88 \times 0.9} = 39.9$

and $L_{max} = 39.9 \times 39.7 = 1582 \text{ mm}$

(say 1500 mm)

Check the stability of the remaining portion of the beam

Moment at the above position

$\approx \dfrac{(6.2 - 1.5)}{6.2} \times 423.9 = 321.3 \text{ kNm}$

From Appendix G3.6, it is possible to evaluate an effective value of n as follows:

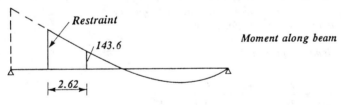

$n = \left[\dfrac{1}{12} \left(\dfrac{321.3}{423.0} + 3 \times \dfrac{143.6}{423.9} \right) \right]^{0.5} = 0.38$ BS 5950:
Part 1
G.3.6.1.

$\lambda = \dfrac{10500}{39.7} = 264$

$v_t = \dfrac{1}{\left[1 + \dfrac{1}{40} \left(\dfrac{\lambda}{x} \right)^2 \right]^{0.5}}$

The Steel Construction Institute Silwood Park, Ascot, Berks SL5 7QN	Subject **DESIGN OF CONTINUOUS COMPOSITE BEAM**		Chapter ref. **21**
	Design code BS5950: Parts 1 & 3	Made by RML Checked by GWO	Sheet no. **13**

$$v_t = \left[1 + \frac{1}{40}\left[\frac{264}{33.9}\right]^2\right]^{-0.5} = 0.63$$

$$\lambda_{LT} = 0.38 \times 0.88 \times 0.63 \times 264$$

$$= 56$$

Buckling resistance moment

$$M_b = 271 \times 1194 \times 10^3 \times 10^{-6}$$

$$= 323.6 \text{ kNm} > 321.3 \text{ kNm}$$

BS 5950:
Part 1
4.3.7.3

So adequate restraint is provided by a single lateral restraint 1.5 m from the internal support. This is done by introducing a web stiffener at this point.

<u>Check natural frequency of beam</u>

To calculate the natural frequency of a continuous beam it is necessary to consider the mode shape of vibration. Because of the influence of the asymmetric inertial forces, the natural frequency is taken as the same as for a simply supported beam.

Considering the same loads as previously:

$$\delta_{sw} = \frac{5 \times 9.67 \times 12^4 \times 10^3}{384 \times 205 \times 10^6 \times 1.1 \times 611.1 \times 10^{-6}}$$

$$= 18.9 \text{ mm}$$

Natural frequency $= 18/\sqrt{\delta_{sw}} = 18/\sqrt{18.9} = 4.1 \text{ Hz}$

However, when carrying out the full Response Factor analysis of Reference (21.14), account can be taken of the effective mass of both spans, thus reducing the amplitude of the vibration from a given excitation. This reduces the sensitivity of the floor to vibration.

The Steel Construction Institute Silwood Park, Ascot, Berks SL5 7QN	Subject ***DESIGN OF CONTINUOUS COMPOSITE BEAM***	Chapter ref. **21**
	Design code ***BS5950: Parts 1 & 3***	Made by **RML**	Sheet no. **14**
		Checked by **GWO**	

Conclusion:

The design using a 406 × 178 × 60 kg/m UB is adequate. However, reducing the section to 406 × 178 × 54 kg/m would not result in an acceptable design because this section is 'compact' in grade 50 steel. Consequently, only 40% redistribution of support moment is permitted and design is limited by the moment resistance of the steel section at the supports.

Chapter 22
Composite columns

by MARK LAWSON and PETER WICKENS

22.1 Introduction

22.1.1 Form of construction

Composite columns comprise steel sections with a concrete encasement or core. Encased columns usually consist of standard I-beam or column sections with a rectangular or square concrete section case around them to form a solid composite section. Additional reinforcement is placed in the concrete cover around the steel section in order to prevent spalling under axial stress.

Concrete-filled columns consist of circular, square or rectangular hollow sections filled with concrete. Additional reinforcement is not normally required except in columns of large section. Examples of these different types of composite columns are shown in Fig. 22.1.

The steel structure is normally constructed in advance of the formation of the composite section. The minimum size of the steel element is therefore controlled by the construction condition. For practical reasons it is normally necessary to fill the tubular sections shortly after their erection but I-sections can be encased much later. However, there are practical problems in concreting around a column with the floors in place. Alternatively, columns can be pre-encased in the factory and in situ concrete used to fill the zone around the beam–column connections.

22.1.2 Advantages of composite columns

In many cases of design, concrete or some other protective material is needed around steel columns for reasons of fire-resistance and durability. It would seem appropriate, therefore, to develop composite action between the steel and concrete, thereby taking advantage of the inherent compressive strength of the concrete, increasing the capacity of the section and leading to considerable savings in the cost of the steelwork. Even ignoring this composite action, the slenderness of the steel column in lateral buckling is reduced, thereby increasing the compressive stress that can be resisted by the steel section.

Much research has gone into the behaviour of concrete-filled tubular sections. Architecturally, tubular columns have many attractive features; concrete filling

Introduction

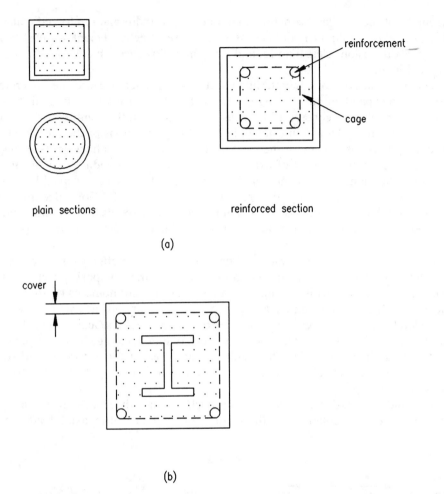

Fig. 22.1 Examples of composite columns: (a) RHS and CHS concrete-filled section, (b) concrete-encased section

has no visual effect on their appeal. The advantages from a structural point of view are, first, the triaxial confinement of the concrete within the section, and second, the fire-resistance of the column which largely depends on the residual capacity of the concrete core.

22.1.3 Principles of design

The axial load capacity of a stocky column, defined as a column that is not subject to the effects of instability, is determined by adding the ultimate compressive strengths of the steel and concrete components. In traditional reinforced concrete design a 'short' column is defined as one whose effective height to least cross-sectional dimension is less than 12, and a nominal allowance is made for eccentricity of axial load.

The axial capacity of concrete-filled sections is greater because the concrete is not able to expand laterally (Poisson's ratio effect) under load and triaxial stresses are developed in the concrete. This causes an increase in the compressive strength of the concrete by an amount dependent on the proportions of the cross-section. The hoop tensions created in the steel have a small adverse affect on its strength.

The effect of eccentricity of axial load is to develop a bending moment in the section. The moment resistance of the section (in the absence of axial load) can be calculated considering plastic stress blocks (see Fig. 22.2). Formulae are given in BS 5400: Part 5, Appendix C.[1] The interaction between axial load and bending moment can be considered in terms of a simplified interaction formula (section 22.2.2).

Slender columns require a more refined treatment. The effective slenderness of a column is determined from the proportions of the composite section. The second moment of area is obtained by adding the second moments of area of the steel and concrete (divided by an appropriate modular ratio). This represents a considerable increase over the properties of steel alone. The axial stress that the section can resist is then determined from the column buckling curves for the steel section under consideration. The resulting axial stress, relative to the yield stress of the steel, is effectively a capacity reduction factor to be applied to the 'stocky column' capacity.

It should be noted that the above approach assumes that loads are not applied laterally over the column length. Concentric and eccentric axial loads cause

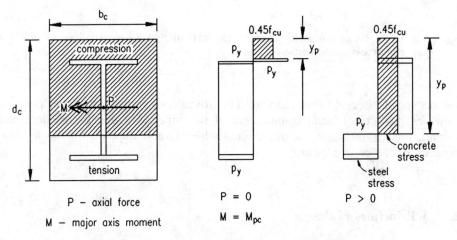

Fig. 22.2 Plastic stress blocks acting on cross section of composite column subject to moment and axial force

relatively low interface shear stresses between the steel and the concrete. Lateral loads cause greater shear stesses and may necessitate the introduction of shear-connectors (as for composite beams – see Chapter 21, section 21.4).

22.1.4 Cased strut method

The traditional method of designing cased columns, presented in BS 449: 1969,[2] can be very conservative but is readily accepted as a method of simple design.

The minimum width of the concrete casing, according to BS 449, should be the flange width b of the steel section plus 100 mm. At least four reinforcing bars are to be located in the concrete at a cover not exceeding 50 mm, to which 5 mm diameter steel stirrups are attached at spacings not exceeding 200 mm. The normal aggregate size is 10 mm and the minimum concrete grade is 21 N/mm^2.

To establish the axial capacity of the cased column (or strut), the radius of gyration of the solid section is taken as 0.2 $(b + 100)$ mm, or alternatively that of the major axis of the steel section. The cross section excludes any concrete cover in excess of 75 mm from the overall dimension of the steel section. The net area of concrete is replaced by an equivalent area of steel by dividing by a modular ratio of 30.

Therefore, knowing the permissible axial stress, as a function of the effective slenderness of the cased column, and the equivalent area of steel, the axial capacity of the section can be easily calculated. However, an onerous limit on the use of this method is that axial load on the cased column should not exceed twice that permitted on the uncased section.

22.2 Design of encased composite columns

22.2.1 Axial load capacity

The design of composite columns is described in BS 5400: Part 5[1] and in Euro-code 4 (draft).[3] It is based on the method developed by Basu and Sommerville[4] and modified by Virdi and Dowling.[5] The maximum compressive strength (squash load) of a stocky column is:

$$P_u = 0.45 f_{cu} A_c + A_s p_y + 0.87 A_r f_y \tag{22.1}$$

where A_c is the cross-sectional area of concrete
A_s is the cross-sectional area of the steel section
A_r is the cross-sectional area of the reinforcement
f_{cu} is the cube strength of concrete
p_y is the design strength of the steel
f_y is the yield strength of the reinforcement.

In BS 5400: Part 5 a factor of 0.91 (corresponding to a material factor of 1.1) is used to modify the term $A_s p_y$. This formula is restricted to concrete contribution factors, defined as $0.45 f_{cu} A_c / P_u$, between 0.15 and 0.8. In order for any reinforcement to contribute to the axial load capacity of the column, shear links of not less than 5 mm diameter should be provided at not more than 150 mm spacing. The value of A_r should not exceed $0.03 A_c$.

Limits on the proportion of the cross section[3] are that the concrete casing should provide a minimum cover to the steel section of 40 mm, and that the dimensions of the concrete section used in determining P_u should not exceed $1.8 \times b$ or $1.6 \times D$ (where b is the flange width, and D is the depth of the steel section.)

The relative slenderness of a composite column of length L is defined by the slenderness factor:

$$\bar{\lambda} = \frac{L}{\pi} \sqrt{\left(\frac{P_u}{E_s \Sigma I}\right)} \tag{22.2}$$

where ΣI is the combined second moments of area of the concrete, steel section and reinforcement expressed in steel units. To do this the second moment of area of the concrete is divided by the modular ratio $E_s/(450 f_{cu})$, where E_s is the elastic modulus of steel in N/mm^2.

The slenderness factor $\bar{\lambda}$, converted to an effective slenderness λ by multiplying $\bar{\lambda}$ by $\pi \sqrt{(E_s/p_y)}$, can then be used to determine the axial stress that may be resisted by the steel section. The selection of the appropriate column design curve in BS 5950: Part 1 depends on the steel section used (i.e. Tables 27(a)–(d) in BS 5950: Part 1).

The column design stress p_c, as determined from λ, divided by p_y is then used as a capacity reduction factor K_1 to be applied to all the components of P_u in Equation (22.1). Hence, this method can be used to determine the axial capacity of a composite column $K_1 P_u$ in the absence of applied moment. This method can be used for slenderness factors $\bar{\lambda}$ less than 2.0, as proposed in Eurocode 4;[3] for greater slendernesses second order effects are underestimated. Columns of slenderness factor less than 0.2 may be taken as 'stocky' (i.e. $K_1 = 1.0$).

In BS 5400: Part 5,[1] short columns are defined as those where the ratio of the length L to the least lateral dimension b_c of the composite column does not exceed 12. Where short columns are not designed to be subject to significant applied moment, the axial capacity may be taken as $0.85 K_1 P_u$ to allow for slight eccentricity of load.

Slender columns are those where L/b_c exceeds 12. In design to BS 5400: Part 5,[1] the analysis of the member should consider a minimum additional moment given by the axial load times an eccentricity of $0.03 b_c$. This, together with any further applied moments, may be treated as in the following section.

22.2.2 Combined axial load and bending moments

The moment resistance of a composite column may be determined by establishing plastic stress blocks defining the strengths of the portions of the section under tension and compression. The plastic neutral axis depth y_p is defined as below the extreme edge of the concrete in compression. Three cases of neutral axis position exist: in the concrete, through the steel flange, and in the web of the section. The position depends on the relative proportions of steel and concrete (see Fig. 22.2). Commonly, y_p lies within the steel flange (i.e. $y_p \approx (d_c - D)/2$) for major axis bending of a composite section. In this case:

$$A_s p_y \geq \left(\frac{d_c - D}{2}\right) \times (0.45 f_{cu} b_c) \geq A_w p_y + A_f (0.45 f_{cu}) \tag{22.3}$$

The moment capacity of the composite section is then given by:

$$M_{pc} = 0.5 A_s p_y (d_c - y_p) + 0.5 A_r (0.87 f_y)(d_c - 2 d_r) \tag{22.4}$$

where d_c is the depth and b_c is the breadth of the concrete section, d_r is the cover to the reinforcement, D is the depth of the steel section, A_w is the cross-sectional area of the web, A_f is the cross-sectional area of the flange, and A_r is the cross-sectional area of any additional reinforcement.

Other cases are defined in BS 5400: Part 5, Appendix C.[1]

Because the plastic stress block method slightly overestimates the strength of the section, the moment resistance M_{pc} is to be multiplied by 0.9 for design purposes.

In the presence of axial load the plastic neutral axis depth increases. For small to medium axial loads the plastic neutral axis remains within the steel web, but for higher axial loads most of the section is in compression. A typical interaction diagram representing the variation of moment capacity with axial load is shown in Fig. 22.3. An interesting phenomenon is that there is a slight increase in moment resistance with increasing axial load (compression), and there is a certain axial load where the moment resistance of the section equals that in the absence of axial load (i.e. M_{pc} from Equation (22.4)).

For simple design, the possibility exists of defining the interaction between moment and axial force in terms of three intercepts A, B and C on the moment and axial load axes, and also point D which corresponds to the axial load at which the moment capacity remains unchanged. Therefore, a trilinear relationship AD, DC, CB closely models the real interaction diagram. It is normal practice to ignore the beneficial effect of axial load as it cannot always be assumed to be coincident with the applied moment. The curvature of the interaction diagram at higher axial loads can also be ignored without much loss of economy. The method developed by Basu and Sommerville[4] empirically follows the shape of the interaction diagram using coefficients K_1, K_2 and K_3 (see Fig. 22.3).

The value of axial load P at which the moment resistance remains unchanged (point D) may be evaluated as follows.

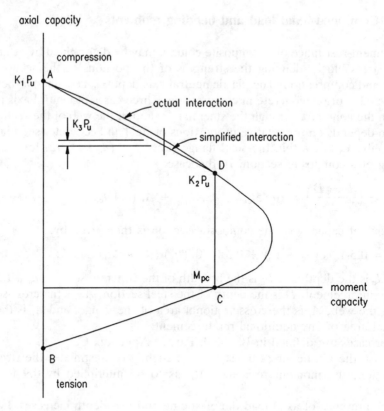

Fig. 22.3 Interaction between moment and axial force

Determine the neutral axis depth y_p for P of zero, corresponding to equal tension and compression forces across the composite section. Redefine a neutral axis depth, $y'_p = d_c - y_p$, which corresponds to an axis symmetrically placed with respect to y_p around the centre of the section. The net axial resistance of the section with neutral axis y'_p corresponds to no change in moment capacity because the net moment effect of the section contained between depths y_p and y'_p is zero as illustrated in Fig. 22.2.

It may be shown that the axial resistance of the section (termed P_0), corresponding to depth y'_p, is, in fact, the axial resistance of the concrete section ignoring the contribution of the steel member and the reinforcement. Hence,

$$P_0 = 0.45 f_{cu} d_c b_c \tag{22.5}$$

Dividing by P_u from Equation (22.1) gives a non-dimensional ordinate on the moment/axial-force diagram, corresponding to the moment resistance of the composite section. This ordinate is also equivalent to the concrete contribution factor for major axis bending, and can also be used for minor axis bending. For stocky columns it also corresponds to the appropriate value of K_2 in the Basu and Sommerville approach.[4]

For slender columns account is to be taken of the moments arising from eccentricity of axial load in addition to the applied moments. The Basu and Sommerville method[4] has been codified in BS 5400: Part 5.[1] Because the method uses empirical formulae for K_1, K_2 and K_3 as a function of the slenderness of the column, it appears to be relatively complicated, but it may be simplified by taking K_3 as zero so that the moment resistance at any value of axial load is given by:

$$M = \left(\frac{K_1 - P/P_u}{K_1 - K_2}\right)M_{pc} \le 0.9M_{pc} \tag{22.6}$$

In design to BS 5400: Part 5,[1] slender columns subject to major axis moment are treated as subject to biaxial moment by including an additional minor axis moment of $0.03Pb_c$, where P is the axial load on the column. An interaction formula given for combining major and minor axis effects is not considered further here.

Provision should be made for the smooth transfer of force between the concrete and steel in cases where the section is subject to high moment. No mechanical shear connection need be provided where the shear stress at the interface between the concrete and the steel is less than 0.6 N/mm² for encased columns.

22.2.3 Method of Eurocode 4

The method proposed in Eurocode 4 differs in the treatment of slender columns subject to moment. The moment/axial-force diagram may be determined for a stocky column as above, and a straight line drawn from the origin to intersect the point on the diagram corresponding to an axial force of K_1P_u. The assumption is made that the moments to the left of this line are utilized as second-order effects under axial load, and moments to the right are available to resist the applied moment. These second-order effects only apply to columns with moments about their major axis.

The simple bilinear interaction diagram may therefore be modified by subtracting the moment to the left of this line (see Fig. 22.4(a)). The method also permits treatment of columns subject to non-uniform moment, but it is conservative to ignore this beneficial effect and to design for the maximum moment.

This approach may be followed independently for major and minor axis bending. A simple method of combining these effects is by a linear interaction of the two directions of bending such that:

$$\frac{M_x}{\mu_x M_{pc,x}} + \frac{M_y}{\mu_y M_{pc,y}} \le 1.0 \quad \text{for } P \le K_1 P_u \tag{22.7}$$

and $M_x \le 0.9M_{pc,x}$ and $M_y \le 0.9M_{pc,y}$, where the subscripts x and y represent the bending actions in the two directions. The values μ_x and μ_y represent the reduction

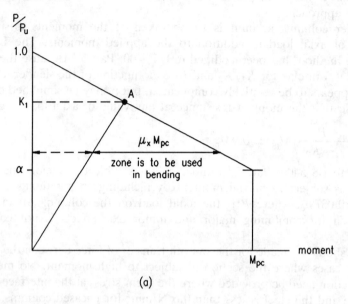

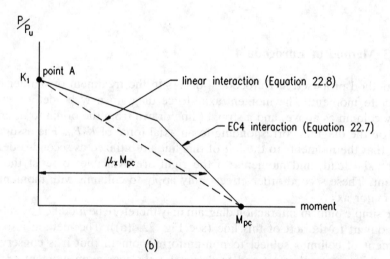

Fig. 22.4 Simplified treatment of slender composite columns in Eurocode 4: (a) moment/axial-force interaction for stocky column, (b) moment/axial-force interaction for slender column

factors on moment resistance considering the effect of axial load. Imperfections are only considered in the weak direction of the column when determining K_1. The values μ_x and μ_y for slender columns may be determined from relationships of the form of Fig. 22.4(b) (taking K_1 as unity when determining μ_y).

For low values of concrete contribution factor, a simple linear interaction formula of the following form may be used for slender columns:

$$\frac{P}{K_1 P_u} + \frac{M_x}{M_{pc,x}} + \frac{M_y}{M_{pc,y}} \leq 1.0 \qquad (22.8)$$

This is very conservative, but at least overcomes the need to determine interaction charts and can be used for scheme design.

22.3 Design of concrete-filled tubes

22.3.1 Axial load capacity

The axial capacity of a concrete-filled rectangular or circular section is enhanced by the confining effect of the steel section on the concrete which depends in magnitude on the shape of the section and the length of the column. Buckling tends to reduce the benefit of confinement on the squash load as the column slenderness increases. To account for this, modification factors are introduced. In circular sections it is possible to develop the cylinder strength $(0.83f_{cu})$ of the concrete.

The 'squash' load capacity of a circular concrete-filled column is:

$$P_u = C_1 p_y A_s + 0.87 f_y A_r + (0.83 f_{cu}/\gamma_{mc}) A_c \{1 + C_2(t/\phi) [p_y/(0.83 f_{cu})]\} \quad (22.9)$$

where t is the thickness and ϕ is the diameter of the tubular section, and γ_{mc} is the material factor for concrete (= 1.5). The terms C_1 and C_2 are coefficients which are a function of the slenderness factor $\bar{\lambda}$ of the column, defined in Equation (22.2); C_1 is less than unity because of the effect of hoop tensions created in the steel. The values of C_1 and C_2 are presented in Table 22.1.

The method derives from research[6,7] carried out by CIDECT and CIRIA and has been incorporated into Eurocode 4.[3] Limits to the use of the method are that the term $A_s p_y$ should represent between 20% and 90% of P_u, and to avoid local buckling, $\phi \leq 85t\varepsilon$ (where ε is $\sqrt{(275/p_y)}$).

The effect of slenderness on the axial strength of a concrete-filled column may be treated as in section 22.2.1. The slenderness factor $\bar{\lambda}$ may be determined from Equation (22.2) as a function of P_u. This involves iteration as P_u is partly dependent on $\bar{\lambda}$. As a reasonable approximation, $\bar{\lambda}$ may be determined assuming

Table 22.1 Values of C_1 and C_2 defining the behaviour of concrete-filled circular hollow sections

$\bar{\lambda}$	0	0.1	0.2	0.3	0.4	≥ 0.5
C_1	0.75	0.80	0.85	0.90	0.95	1.00
C_2	4.90	3.22	1.88	0.88	0.22	0.00

that $C_1 = 1.0$ and $C_2 = 0.0$. Having evaluated $\bar{\lambda}$, the strength modification factor K_1 can be determined from the column buckling curve given in Table 27(a) of BS 5950: Part 1.

No design formulae are readily available for the axial strength of rectangular filled sections in section 22.2.1. A limit on the proportions of rectangular sections is that $b \leq 52t\varepsilon$ and $d \leq 52t\varepsilon$.

22.3.2 Combined axial load and bending moment

The effect of eccentric axial load or applied moment may be treated conservatively by evaluating the moment capacity of the section taking the concrete strength as $0.45f_{cu}$ and ignoring the benefits of confinement of the concrete (i.e. $C_1 = 1.0$ and $C_2 = 0.0$). Formulae for M_{pc} are given in Reference 7. The interface shear stress is limited to 0.4 N/mm².[3]

The reduced moment capacity of a concrete-filled section subject to axial load may be determined from the method in section 22.2.2. Design tables for both circular and rectangular hollow sections are given in Reference 7 and are based on the Basu and Sommerville approach.[4]

22.3.3 Fire-resistant design

The fire-resistance of encased composite columns may be treated in the same way as reinforced concrete columns. The steel is insulated by an appropriate concrete cover, as given in BS 8110: Part 2.[8] To preserve the cover in the event of a fire, light reinforcement to the steel section is required. In such cases, two-hour fire-resistance can usually be achieved.

The fire-resistance of concrete-filled columns has been the subject of extensive research.[9] During a fire, sufficient redistribution of stress occurs between the hot steel section and the relatively cool concrete cover, such that a minimum of 30 minutes fire-resistance can be achieved.

The fire-resistance of a concrete-filled hollow section may be defined by redefining the load ratio in fire conditions as:[10]

$$\eta = \frac{P_f}{K_1 P_u} \tag{22.10}$$

where $\quad P_u = 0.83 f_{cu} A_c + f_y A_r \tag{22.11}$

The axial load P_f is determined for load factors of unity. The concrete strength of $0.83f_{cu}$ is the cylinder strength of concrete which can be developed in both circular and rectangular sections in fire. The material factors for concrete and for reinforcement are taken as 1.0 in this formula.

Table 22.2 Fire-resistance of concrete-filled hollow sections as a function of load ratio[10]

Fire-resistance (min)	Load ratio, η	
	Plain concrete	5% fibre concrete
30	—	—
60	0.51	0.67
90	0.40	0.53
120	0.36	0.49

The term K_1 is determined as in section 22.2.1 or as in Table 10 of BS 5950 Part 8.[10] It excludes the contribution of the steel hollow section and includes the enhanced strength of the concrete. The reinforcing bars are assumed to be fully effective, provided they are located at the appropriate cover.[8] A method of including the effect of applied moment is given in BS 5950: Part 8.[10]

Fire-resistance periods that can be achieved for plain (or bar-reinforced) or fibre-reinforced concrete-filled hollow sections are presented in Table 22.2.

Where enhanced periods of fire-resistance or greater load-carrying capacity are required, external fire protection may be provided. Normally, the effect of concrete filling offers a considerable reduction in the thickness of fire protection material that would be required for an unfilled hollow section.[9]

An important practical requirement for the use of concrete-filled sections is the provision of vent holes at the top and bottom of each column. These are to prevent dangerous build-up of steam pressure inside the columns in the event of a fire. They also permit seepage of any excess moisture in the concrete after construction. Two 12 mm diameter holes placed diametrically opposite each other at the top and bottom of each storey height have been used in testing and proved to be adequate.

References to Chapter 22

1. British Standards Institution (1979) *Steel, concrete and composite bridges.* Part 5: *Code of practice for design of composite bridges.* BS 5400, BSI, London.
2. British Standards Institution (1969) *Specification for the use of structural steelwork in building.* BS 449, BSI, London.
3. Commission of the European Communities (1985) *Eurocode 4: Common unified rules for composite steel and concrete structures.* Report EUR 9886 EN.

4. Basu A.K. & Sommerville W. (1969) Derivation of formulae for the design of rectangular composite columns. *Proc. Instn Civ. Engrs*, supplementary volume, Paper 7206S, 233–80.
5. Virdi K.S. & Dowling P.J. (1973) The ultimate strength of composite columns in biaxial bending. *Proc. Instn Civ. Engrs*, **55**, Part 2, March, 251–72.
6. Sen H.K. & Chapman J.C. (1970) *Ultimate Load Tables for Concrete-Filled Tubular Steel Columns*. Construction Industry Research and Information Association (CIRIA), Technical Note 13.
7. British Steel plc (1986) *Design Manual SHS Concrete Filled Columns*. BS Tubes Division, Corby.
8. British Standards Institution (1985) *Structural use of concrete*. Part 2: *Code of practice for special circumstances*. BS 8110, BSI, London.
9. British Steel plc (1985) *Recent developments in the assessment of the fire-resistance of structural hollow sections*. BS Tubes Division, Corby.
10. British Standards Institution (1990) *Structural use of steelwork in building*. Part 8: *Code of practice for fire-resistant design*. BS 5950, BSI, London.

A worked example follows which is relevant to Chapter 22.

Worked example

The Steel Construction Institute Silwood Park, Ascot, Berks SL5 7QN		Subject **DESIGN OF COMPOSITE COLUMN**		Chapter ref. **22**	
		Design code *EUROCODE 4* *(1990 DRAFT)*	Made by RML Checked by GWO	Sheet no.	*1*

Problem

Determine the envelope of design resistance to axial load, major axis moment and minor axis moment of the composite column shown below. Its effective length is 4 m.

Column size	400 mm square
Steel section	254 × 254 × 89 kg/m UC
Grade 50 steel	p_y = 355 N/mm²
Concrete grade	f_{cu} = 30 N/mm² (normal weight)
Reinforcement	4 No. T12 bars
	f_y = 460 N/mm²

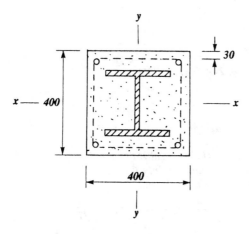

Cross-section through column

Note: *EC3 does not limit the design strength for steel for* $t \le 40$ *mm.*

The Steel Construction Institute Silwood Park, Ascot, Berks SL5 7QN	Subject **DESIGN OF COMPOSITE COLUMN**	Chapter ref. **22**
	Design code EUROCODE 4 (1990 DRAFT)	Made by **RML** Checked by **GWO**	Sheet no. **2**

Column section properties

Area of section $\quad A_s \quad = \quad 11400 \text{ mm}^2$

Area of reinforcement $\quad A_r \quad = \quad 4 \times \pi \times \dfrac{12^2}{4} \quad = \quad 452 \text{ mm}^2$

Area of concrete $\quad A_c \quad = \quad 400^2 - 11400 - 452$
$ = \quad 148.1 \times 10^3 \text{ mm}^2$

Squash load (compressive resistance) of stocky column - equation 22.1

$P_u \quad = \quad A_s p_y + A_r \times 0.87 f_y + A_c \times 0.45 f_{cu}$

$ \quad = \quad (11.4 \times 10^3 \times 355 + 452 \times 0.87 \times 460$
$ + 148.1 \times 10^3 \times 0.45 \times 30) \times 10^{-3}$

$ \quad = \quad 6228 \text{ kN}$

Tensile resistance of column

$P_t \quad = \quad A_s p_y + A_r \times 0.87 f_y$

$ \quad = \quad (11.4 \times 10^3 \times 355 + 452 \times 0.87 \times 460) \times 10^{-3}$

$ \quad = \quad 4228 \text{ kN}$

Moment resistance of column - major axis

Determine position of plastic neutral axis (PNA) parallel to x-x axis of column. Assume that PNA lies at a distance Y into the web of the steel section. Concrete tensile strength is neglected.

Equating tension and compression, it follows that the forces in flanges and reinforcement cancel out.

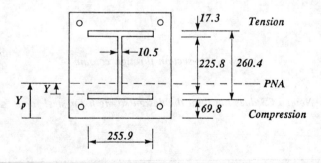

The Steel Construction Institute Silwood Park, Ascot, Berks SL5 7QN	Subject **DESIGN OF COMPOSITE COLUMN**	Chapter ref. **22**
	Design code EUROCODE 4 (1990 DRAFT)	Made by **RML** Sheet no. **3** Checked by **GWO**

Therefore

tensile force in web depth $(225.8 - 2Y)$ = *compressive force in concrete*

$$(225.8 - 2Y) \times 10.5 \times 355 = [(Y + 69.8 + 17.3) \times 400 - \frac{452}{2} - 17.3 \times 225.9 - Y \times 10.5] \times 0.45 \times 30$$

$$841.7 \times 10^3 - 7.46 \times 10^3 Y = 5.4 \times 10^3 Y + 470.3 \times 10^3 - 3.05 \times 10^3 - 52.7 \times 10^3 - 0.14 \times 10^3 Y$$

or $\quad (5.4 + 7.46 - 0.14) Y = 841.7 - 470.3 + 3.05 + 52.7$

$\therefore \quad Y = 33.6 \; mm$

Hence plastic neutral axis depth from edge of slab is

$Y_p = 33.6 + 69.8 + 17.3 = 120.7 \; mm$

Plastic moment of resistance about x-x axis:
Take moments about PNA at depth Y_p

$$M_{px} = 17.3 \times 255.9 \times 355 \times 10^{-6} \times (260.4 - 17.3)$$
$$+ \frac{452}{2} \times 0.87 \times 460 \times 10^{-6} \times (400 - 30 \times 2 - 12)$$
$$+ 10.5 \times \frac{Y^2}{2} \times 355 \times 10^{-6} + 10.5 \times \frac{(225.8 - Y)^2}{2} \times 355 \times 10^{-6}$$
$$+ (5.4 Y + 470.3 - 3.05 - 52.7 - 0.14 Y) \times 10^{-3} \times Y_p /2$$

$M_{px} = 382.0 + 29.7 + 2.1 + 68.8 + 35.7 \; for \; Y = 33.6 \; mm$

$M_{px} = 518.3 \; kNm$

$0.9 \; M_{px} = 466.5 \; kNm$

Compressive resistance of column with co-existing moment (equation 22.5)

This is equivalent to the resistance of the concrete in compression (see text)

$P_{o,x} = 148.1 \times 10^3 \times 0.45 \times 30 \times 10^{-3} = 1999 \; kN$

Design of Composite Column

Chapter ref. 22

Design code: EUROCODE 4 (1990 DRAFT)
Made by: RML
Checked by: GWO
Sheet no. 4

Moment resistance of column - minor axis

Assume PNA at distance X from edge of web.

Equating tensile and compressive forces.

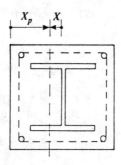

If $A_w \, p_y > 0.45 \, f_{cu} \, (b^2 - A_f - \dfrac{A_r}{2})$

then PNA lies in web

$10.5 \times 225.8 \times 355 \times 10^{-3} > 0.45 \times 30 \times (400^2 / 2 - 17.3 \times 255.9 - 452 / 2)$

$841.7 < 1017.2$

So PNA lies just outside web

Therefore distance X is obtained from:

$(2X + 10.5) \times 2 \times 17.3 \times 355 \times 10^{-3} + 10.5 \times 225.8 \times 355 \times 10^{-3}$

$= 0.45 \times 30 \times [\, (200 - \dfrac{10.5}{2} - X) \times 400 - \dfrac{452}{2} - \left(\dfrac{255.9}{2} - \dfrac{10.5}{2} - X \right) \times 17.3 \,]$

$24.6X + 129.0 + 841.7 \quad = \quad 1051.6 - 5.4X - 3.1 - 28.7 + 0.2X$

$X (24.6 + 5.4 - 0.2) \quad = \quad 1051.6 - 841.7 - 129.0 - 3.1 - 28.7$

$X = 1.7 \text{ mm}$

Therefore $X_p \quad = \quad 200 - 10.5/2 - 1.7 \quad = \quad 193.0 \text{ mm}$

The Steel Construction Institute Silwood Park, Ascot, Berks SL5 7QN	Subject **DESIGN OF COMPOSITE COLUMN**		Chapter ref. 22
	Design code EUROCODE 4 (1990 DRAFT)	Made by RML Checked by GWO	Sheet no. 5

Plastic moment resistance about y-y axis

Take moments about X_p

$$M_{py} = \frac{452}{2} \times 0.87 \times 460 \times 10^{-6} \times (400 - 30 \times 2 - 12)$$

$$+ 10.5 \times 225.8 \times (X + \frac{10.5}{2}) \times 355 \times 10^{-6}$$

$$+ \left[\left(\frac{255.9}{2} - X - \frac{10.5}{2} \right)^2 + \left(\frac{255.9}{2} + X + \frac{10.5}{2} \right)^2 \right] \times 17.3 \times 355 \times 10^{-6}$$

$$+ (1051.6 - 5.4X - 3.1 - 28.7 + 0.2X) \frac{X_p}{2} \times 10^{-3}$$

M_{py} = 29.7 + 5.8 + 201.6 + 97.6

= 334.7 kNm

$0.9\, M_{py}$ = 301.2 kNm

Compressive resistance of column subject to co-existing moment of M_{py}

$P_{o,y}$ = $P_{o,x}$ = 1999 kN

660   Worked example

The Steel Construction Institute Silwood Park, Ascot, Berks SL5 7QN	Subject **DESIGN OF COMPOSITE COLUMN**		Chapter ref. 22
	Design code EUROCODE 4 (1990 DRAFT)	Made by RML Checked by GWO	Sheet no. 6

Effect of column slenderness

Modular ratio α_e = $\dfrac{E_s}{450 f_{cu}}$ = $\dfrac{205 \times 10^3}{450 \times 30}$ = 15.2

Second moment of area of column - major axis bending

$$I_{xx} \approx \left[I_{xs} + \dfrac{452}{4} \times (400 - 60 - 12)^2 \right] \left[1 - \dfrac{1}{\alpha_e} \right] + \dfrac{400^4}{12} \times \dfrac{1}{\alpha_e}$$

I_{xs} = 143.0×10^6 mm^4

I_{xx} = 285.3×10^6 mm^4

Second moment of area of column - minor axis bending

$$I_{yy} = \left[I_{ys} + \dfrac{452}{4} \times (400 - 60 - 12)^2 \right] \left[1 - \dfrac{1}{\alpha_e} \right] + \dfrac{400^4}{12} \times \dfrac{1}{\alpha_e}$$

I_{ys} = 48.5×10^6 mm^4

I_{yy} = 197.0×10^6 mm^4

P_u = 6228 kN squash load - see calculation sheet no.2

Slenderness factor

$$\bar{\lambda} = \dfrac{L}{\pi} \left[\dfrac{P_u}{E_s \, \Sigma \, I_{yy}} \right]^{0.5}$$

$$= \dfrac{4000}{\pi} \left[\dfrac{6228 \times 10^3}{205 \times 10^3 \times 197.0 \times 10^6} \right]^{0.5} = 0.5$$

λ_{eff} = $\bar{\lambda} \pi (E_s/p_y)^{0.5}$ = $0.5 \pi (205 \times 10^3/355)^{0.5} \simeq 38$

From BS 5950: Part 1 Table 27(c)

Axial strength p_c = 308 N/mm^2

		Chapter ref.	
The Steel Construction Institute Silwood Park, Ascot, Berks SL5 7QN	Subject **DESIGN OF COMPOSITE COLUMN**	22	
	Design code EUROCODE 4 (1990 DRAFT)	Made by RML Checked by GWO	Sheet no. 7

Factor K_1 representing reduced axial strength of column is
$308/355 = 0.87$

Hence $K_1 P_u = 0.87 \times 6228 = 5418$ kN

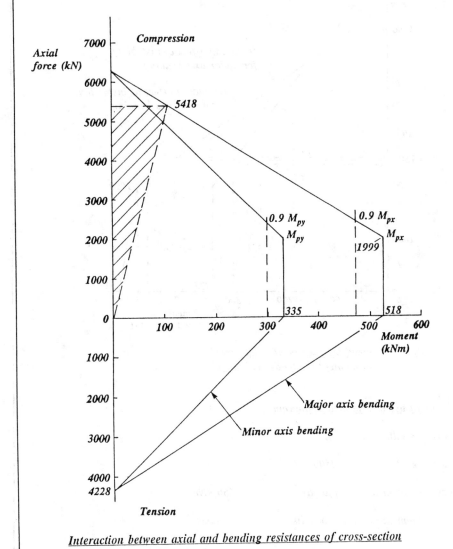

Interaction between axial and bending resistances of cross-section

The Steel Construction Institute Silwood Park, Ascot, Berks SL5 7QN	Subject **DESIGN OF COMPOSITE COLUMN**		Chapter ref. 22
	Design code EUROCODE 4 (1990 DRAFT)	Made by RML Checked by GWO	Sheet no. 8

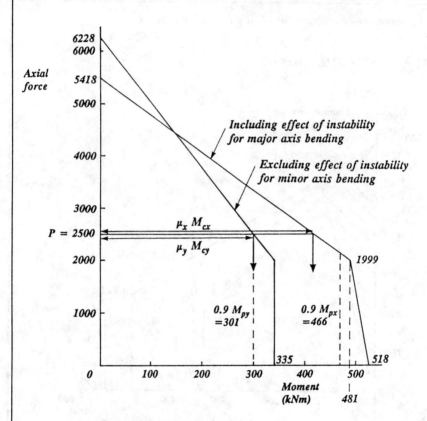

Reduced interaction of axial force and moment in slender column

Example of use of interaction diagram

Assume the following design loads

Axial force P = 2500 kN

Bending moment about major axis = 150 kNm

Bending moment about minor axis = 75 kNm

The Steel Construction Institute	Subject	Chapter ref.
Silwood Park, Ascot, Berks SL5 7QN	DESIGN OF COMPOSITE COLUMN	22
Design code EUROCODE 4 (1990 DRAFT)	Made by RML Checked by GWO	Sheet no. 9

Additional moment arising from potential eccentricity of vertical load. Take eccentricity as $0.03\, b_c$.

Moment $M_x = 150 + 2500 \times 0.03 \times 0.4 = 180\ kNm$

$M_y = 75 + 2500 \times 0.03 \times 0.4 = 105\ kNm$

Simple interaction (equation 22.8)

$$\frac{P}{K_1\, P_u} + \frac{M_x}{M_{px}} + \frac{M_y}{M_{py}} \leq 1.0$$

$$\frac{2500}{5418} + \frac{180}{518.3} + \frac{105}{334.7} = 1.12 > 1.0$$

Consider EC4 interaction (equation 22.7)

Major axis bending

At $P = 2500\ kN$, resistance moment $= \mu_y\, M_{px}$

$$\mu_y\, M_{px} = \frac{5418 - 2500}{5418 - 1999} \times 488 = 416\ kNm$$

Minor axis bending $= \mu_y\, M_{py}$

$$\mu_y\, M_{py} = \frac{6228 - 2500}{6228 - 1999} \times 335 = 295\ kNm$$

$$\frac{M_x}{\mu_y\, M_{px}} + \frac{M_y}{\mu_y\, M_{py}} = \frac{180}{416} + \frac{105}{295} = 0.79 < 1.0$$

Therefore, the design is satisfactory for biaxial bending.

Chapter 23
Bolts

by HUBERT BARBER

23.1 Types of bolt

23.1.1 Grade 4.6 and 8.8 bolts

The most frequently used bolts in structural connections are grade 4.6 black bolts and 8.8 high-strength bolts, the manufacture of which is governed by British Standards as follows:

BS 4190: *Specification for ISO metric black hexagon bolts, screws and nuts*[1]
BS 3692: *Specification for ISO metric precision hexagon bolts, screws and nuts.*[2]

In practice many bolts are supplied that are made from BS 3692 material to the sizes and tolerances of BS 4190.

Bolts of other strength grades are also available: for example, grade 4.6, 4.8, 6.9 to BS 4190, and grade 4.6 through to 14.9 to BS 3692.

The strength grade designation, which is in accordance with ISO/R 898/1, is defined in both these British Standards. The system consists of two figures: the first is one-tenth of the minimum tensile strength in kilograms force per square millimetre and the second is one-tenth of the ratio between the minimum yield and minimum tensile strength expressed as a percentage. The two figures are separated by a decimal point.

In the case of higher strength bolts which have a less easily defined yield strength the stress at permanent set limit $R_{0.2}$ is used.

From this it can be seen that the minimum ultimate strength of a grade 4.6 bolt is 40 kgf/mm^2 (392 N/mm^2) and that its minimum yield stress is $0.6 \times 40 = 24$ kgf/mm^2 (235 N/mm^2). With higher grade bolts the permanent set limit will occur at 80% or 90%, depending on the classification.

There is a similar designation system for nuts. This is a single figure being one-tenth of the ultimate strength of the steel. Hence a grade 8 nut has an ultimate strength of 80 kgf/mm^2 (785 N/mm^2) and would normally be used with a grade 8.8 bolt. It is permissible however to use a nut of higher grade than the bolt to minimize the risk of thread stripping.

23.1.2 High-strength friction-grip bolts

High-strength friction-grip bolts are manufactured to the requirements of BS 4395.[3] This British Standard covers three distinctly different types of fastener.

Part 1 – General grade. The most commonly used type in general structural steelwork. The strength grade is about that of 8.8 bolts up to 24 mm diameter. For larger sizes the minimum ultimate strength is reduced to approximately 73 kgf/mm^2 (715 N/mm^2) and the yield to approximately 77% of that figure. Their use is governed by BS 4604: Part 1[5].

Part 2 – Higher grade. These are made from 10.9 grade and although a higher tensile load can be applied to them there is a much reduced margin between the yield load and the ultimate strength. It is therefore not permissible to use them for connections in which there is applied tension. The 'part turn' method of tightening is not permitted by BS 4604: Part 2 which requires in Table 2 of that Standard that the prestressing is within strictly controlled limits; given as from 0.85 to 1.15 of the proof load.

Part 3 – Higher grade (waisted shank). These are made from higher grade steel quality as for Part 2 bolts, that is 10.9 grade, but as a consequence of the reduced diameter of the waisted shank the ultimate strengths are somewhat less than those of general grade bolts, except for diameters of 27 mm and over, which are higher than those of Part 1 bolts, for the reasons given in the paragraph 'Part 1 – General grade'. As the area of the waisted shank is less than that of the threaded portion the combined torsional and tensile stresses during tightening do not become critical. The 'part turn' method of tightening, or any method utilizing direct measurement of the shank tension, is therefore allowed. BS 4604: Part 3 however expressly proscribes the torque control method for waisted shank bolts.

23.2 Methods of tightening and their application

The usual methods of tightening of friction-grip bolts are outlined in section 33.7.5. Further details are given in References 6 and 7.

23.3 Geometric considerations

23.3.1 Hole sizes

Ordinary bolts should be used in holes having a suitable clearance in order to facilitate insertion. For bolts up to and including diameters of 24 mm the clearance should be 2 mm, and above 24 mm should be 3 mm.

666 Bolts

Normal clearance holes as given for ordinary bolts are usually used for HSFG fasteners but it is permissible to use oversize, short or long slotted holes provided standard hardened washers are used over the holes in the outer plies.

Oversize and short slotted holes may be used in all plies but long slotted holes may only be used in one single ply in any connection. If a long slotted hole occurs in an outer ply it should be covered with a washer plate longer than the slot and at least 8 mm thick.

Maximum dimensions of holes are given in Table 35 of BS 5950 and in the Appendix *Connection design: table of hole sizes*.

The assessment of the slip resistance is affected when oversize or slotted holes are used. The constant K_s (clause 6.4.2.1 of BS 5950), which is 1.0 for bolts in clearance holes, is reduced to 0.85 and 0.6 as given in paragraph 23.5.2.

23.3.2 Spacing of fasteners, end and edge distances

Spacing is covered fully in Section 6.2 of BS 5950, summarized as follows:

Minimum requirements (see Fig. 23.1)
Centres of fasteners   2.5d
Edge distances
    to rolled, sawn, planed or machine flame cut edge   1.25D
    to sheared or hand flame cut edge   1.4D
End distances
    in all cases $1.4D^a$

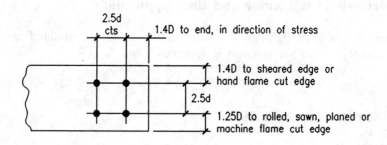

Fig. 23.1 Minimum dimensions (see Table in Appendix)

[a] The full bearing value of a bolt through a connected part cannot be developed if the end distance is less than $2D$. For end distances between $1.4D$ and $2D$ the bearing value is proportionally reduced. The minimum end distance for the full bearing value in the case of parallel shank HSFG bolts is $3d$ (see below).

Maximum requirements (see Fig. 23.2)
Centres of fasteners in the direction of stress $14t$

where t is the thickness of the thinner element.

Edge distances
Distance to the nearest line of fasteners from the edge of an unstiffened part $11t\varepsilon$

In corrosive environments
Centres of fasteners in any direction $16t$ or 200 mm
Edge distances $4t + 40$ mm

d = fastener diameter
D = hole diameter
t = thickness of the thinner outside ply
$\varepsilon = \left[\dfrac{275}{p_y}\right]^{\frac{1}{2}}$ where p_y is the design strength of the element concerned.

In the case of high-strength parallel-shank friction-grip bolts the slip resistance is a serviceability criterion and the connection could slip into bearing between working and failure load. It is necessary therefore to check the bearing strength of the joint. The end distance to attain the full bearing strength is increased to $3d$.

This does not apply to waisted shank fasteners, which cannot act satisfactorily in bearing. Accordingly the allowable slip resistance is reduced so that slip becomes the failure condition.

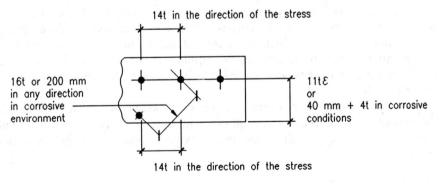

Fig. 23.2   Maximum dimensions (see Table in Appendix)

23.3.3 Back marks and cross centres

The back mark is the distance from the back of an angle or channel web to the centre of a hole through the leg or flange. This dimension is determined so as to allow the tightening of a bolt with a standard podger spanner, to be as near as possible to the centroidal axis and to allow the required edge distance. Recommended back marks and diameters are given in the tables for channels and for angles (Appendix *Connection design: Back marks in channel flanges* and *Back marks in angles*).

The distances between centres of holes (cross centres) in the flanges of joists, universal beams and universal columns are similarly determined after consideration of accessibility and edge distances.

Recommended cross centres and diameters are given in the Appendix *Connection design: Cross centres through flanges*.

23.4 Methods of analysis of bolt groups

23.4.1 Introduction

Any group of bolts may be required to resist an applied load acting through the centroid of the group either in or out of plane producing shear or tension respectively. The load may also be applied eccentrically producing additionally torsional shear or bending tension. Examples are given in Fig. 23.3.

23.4.2 Bolt groups loaded in shear

British and Australian practice is to distribute the torsional shear due to eccentricity elastically in proportion to the distance of each bolt from the centroid of the group. This is referred to as the polar inertia method.

(In some countries, notably Canada and in some cases the USA, the instantaneous centre method is used. This is a redistribution system, developed by Crawford and Kulak,[4] in which the assumed centre of rotation is continually adjusted until the three basic equations of equilibrium are satisfied. The method is a limit-state concept and has been shown to be less conservative than the traditional elastic methods.)

Consider first a single line of bolts subject to a torsional moment, Fig. 23.4(a). If the area of each bolt is a, the second moment of area of a typical bolt is ay^2 and the total is $\Sigma ay^2 = a\Sigma y^2$. The stress in the extreme bolt due to the eccentricity then becomes

$$\frac{My_1}{I} = \frac{My_1}{a\Sigma y^2}$$

Methods of analysis of bolt groups 669

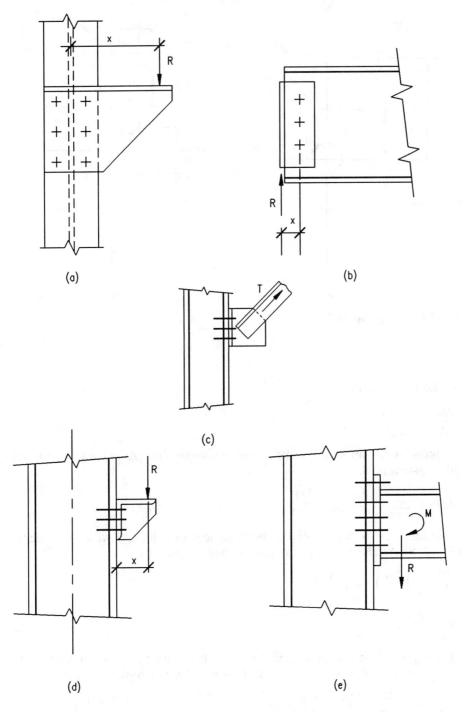

Fig. 23.3 Bolt groups

Bolts

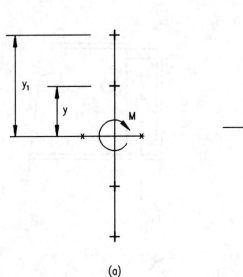

Fig. 23.4 Bolt group analysis

and the force per bolt

$$\frac{May_1}{a\Sigma y^2} = \frac{My_1}{\Sigma y^2}$$

The polar inertia about the centroid for any single line group containing n bolts with constant pitch p is

$$I_0 = \sum_{J=0}^{J=(n-1)} \left[\frac{(n-1-2J)}{2}p\right]^2$$

Consider next a double line of bolts subject to a load R with eccentricity x (Fig. 23.4(b)). The radius to the nearest bolt is given by

$$r = \sqrt{\left[\left(\frac{p}{2}\right)^2 + \left(\frac{c}{2}\right)^2\right]}$$

$$r^2 = \left(\frac{p}{2}\right)^2 + \left(\frac{c}{2}\right)^2$$

I_0 for a typical bolt is then ar^2 and it follows that I_0 for the whole group becomes $I_{xx} + I_{yy}$ which, if there are m vertical rows and n horizontal rows, is

$$I_{00} = m \sum_{J=0}^{J=(n-1)} \left[\frac{(n-1-2J)}{2}p\right]^2 + n \sum_{J=0}^{J=(m-1)} \left[\frac{(m-1-2J)}{2}c\right]^2$$

where c is the cross centres between the vertical lines. The distance to the extreme bolt is

$$r = \sqrt{\left\{\left[\frac{(n-1)}{2}p\right]^2 + \left[\frac{(m-1)}{2}c\right]^2\right\}}$$

The force in the extreme bolt due to the moment is

$$f_m = \frac{R x r}{I_{00}}$$

The force in each bolt due to the shear (assumed equally divided between all bolts) is

$$f_v = \frac{R}{mn}$$

The combined force per bolt is the resultant of these two, see Fig. 23.5.

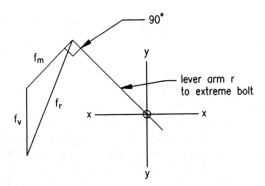

Fig. 23.5 Resultant force

The resultant bolt force is then checked against the bolt strength in single shear, double shear or bearing as is appropriate. In the case of bearing, however, it should be remembered that the full strength cannot be achieved if the end distance measured along the line of the resultant is less than twice the diameter of the bolt. In this case the bearing strength is reduced in proportion.

23.4.3 Bolt groups loaded in tension and out of plane moment

The connection types illustrated in Fig. 23.3(c), (d) and (e) are dealt with in Chapter 26.

23.5 Design strengths

23.5.1 General

Ordinary bolts have to resist forces in shear and bearing or tension or combinations of these. HSFG bolts resist shear by developing friction between the plies; they may also resist external tension. The load capacity of a joint may be affected if it is excessively long, has a large grip length or is subjected to prying action.

23.5.2 Shear

When bolts are loaded in shear some of the threads of the bolt may lie within the shear plane. The shear strength is then assessed on the tensile area of the bolt. If however it is known with certainty that no part of the threaded section of the bolt will fall within the shear plane the full shank area may be used in determining the strength.

Until an HSFG bolted shear connection slips shear force is resisted by friction. The slip resistance, P_{sL}, is given by

$$P_{sL} = 1.1 \, K_s \, \mu \, P_0$$

where K_s = 1.0 for clearance holes;
= 0.85 for oversized holes, short slotted holes or long slotted holes loaded perpendicular to the slot direction;
= 0.6 for long slotted holes loaded parallel to the slot direction
μ = slip factor, determined from tests in accordance with BS 4604, must not exceed 0.55; but for BS 4395: Part 1 fasteners in connections with untreated surfaces in contact in accordance with BS 4604, μ may be taken as 0.45 without further tests
P_0 = the minimum shank tension.

This expression gives the value for P_{sL} for bolts to BS 4395: Part 1 (general grade) taking μ, the slip factor, as 0.45, the permitted value for untreated surfaces meeting the requirements of BS 4604: Part 1.[5]

In the case of higher grade parallel and waisted shank fasteners it is inappropriate to use the value 0.45 for μ that is given for general grade bolts in BS 5950. The actual value of the slip factor, μ, should always be established by tests in accordance with BS 4604: Part 2 or Part 3[5] as appropriate. BS 5950 however specifies in upper limit of 0.55. For waisted shank fasteners $P_{sL} = 0.9 K_s \mu P_0$, where K_s, μ and P_0 are as previously defined for parallel shank fasteners.

It is also required by BS 4604: Part 2[5] that the shank tension should be between 0.85 and 1.15 of the proof load so as to avoid distress in the threads during tightening. The minimum shank tension P_0 then becomes 0.85 proof load.

23.5.3 Bearing

Ordinary bolts

In cases where the strength is assessed from the bearing of the bolt on one of the connected parts it is necessary to limit the bearing strength in accordance with the end distance requirements.

For end distances between the minimum (1.4D) and the minimum value for full bearing (2.0D) the bearing value is proportional to the end distance.

HSFG bolts

After slip has occurred a parallel shank HSFG bolt acts in shear or bearing.

When the end distance is at least three times the nominal diameter of the bolt the bearing strengths, p_{bg}, for steel to BS 4360 are as follows:

Grade 43 – 825 N/mm^2
Grade 50 – 1065 N/mm^2
Grade 55 – 1210 N/mm^2

This is a condition which can only apply after joint slip has occurred, and the joint is beyond the serviceability limit.

23.5.4 Tension

When bolts are loaded in tension additional axial forces are sometimes induced due to prying action. The strength values given in Table 32 of BS 5950 have been reduced to make allowance for these forces. Further discussion of prying action is given in Chapter 26.

It is permissible to subject HSFG bolts complying with Part 1 or Part 3 of BS 4395 to externally applied tension. Fasteners to Part 2 of that standard however should not be required to carry external tension.

The maximum tension, P_t, which may be applied to Part 1 and Part 3 fasteners is 0.9P_0, where P_0 is the minimum shank tension given in the Appendix and specified in BS 4604.

The values of P_t are given in the Appendix *Connection design: Bolt capacities in tension*.

23.5.5 Combined shear and tension

Ordinary bolts which are subject to both tension and shear should satisfy the relationship

$$\frac{F_s}{P_s} + \frac{F_t}{P_t} \leq 1.4$$

where F_s is the applied shear, F_t is the applied tension, P_s is the shear capacity, and P_t is the tensile capacity, in addition to the other specified requirements of section 6.3 of BS 5950. The shear capacity, P_s, may be based on the tension stress area or on the full shank area as appropriate (see section 23.5.2).

This expression allows that when either the shear or tensile capacity is fully taken up (i.e. F_s/P_s or $F_t/P_t = 1.0$), the other may be stressed to 40% of its capacity.

HSFG bolts complying with Part 1 or Part 3 of BS 4395, subject to both shear and externally applied tension, should satisfy the relationship

$$\frac{F_s}{P_{sL}} + 0.8 \frac{F_t}{P_t} \leq 1$$

where F_s is the applied shear, F_t is the applied external tension, P_t is the tension capacity given in the Appendix *Connection design: Bolts to BS 4395, long joints*, and P_{sL} is the lesser appropriate slip resistance.

23.5.6 Long joints and large grips

When the joint length in a splice or end connection, L_j, defined as the distance between the first and last bolt on either side of the joint is greater than 500 mm, the strength of the joint is reduced by the factor $(5500 - L_j)/5000$, where L_j is the joint length as defined above (see Fig. 23.6). Similarly when the grip length, T_g, the total thickness of the connected plies, is greater than five times the nominal diameter of the bolts, the strength of the joint is reduced by the factor $8d/(3d + T_g)$, where d is the nominal bolt diameter and T_g is as defined above (see Fig. 23.7).

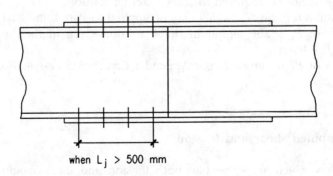

Fig. 23.6 Long joint

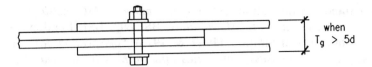

Fig. 23.7 Large grip

In cases when both the above conditions apply it is only necessary to apply the factor producing the greater reduction.

The large grip reduction does not apply to HSFG bolts except when parallel shank bolts are being used in bearing for considerations beyond serviceability.

'Long joint' connections, in the case of parallel shank HSFG bolted joints, are defined by the same criterion and the joint capacity due to slip is reduced by the same factor, $(5500 - L_j)/5000$.

$$P_{sL} = 0.6P_0 \frac{(5500 - L_j)}{5000}$$

as given in clause 6.4.2.3 of BS 5950.

23.6 Tables of strengths

23.6.1 Bolt strengths

Tables 32 and 33 of BS 5950 give respectively the strength of bolts in clearance holes and the bearing strength of connected parts to BS 4360 for ordinary bolts in clearance holes. The values are given in the Appendix *Connection design: Bolt strengths* for connected parts of steel to BS 4360.

23.6.2 Bolt capacities

Tables of bolt capacities for various grades and applications are given in the Appendix *Connection design*.

References to Chapter 23

1. British Standards Institution (1967) *Specification for ISO metric black hexagon bolts, screws and nuts.* BS 4190, BSI, London.

2. British Standards Institution (1967) *Specification for ISO metric precision hexagon bolts, screws and nuts. Metric units.* BS 3692, BSI, London.
3. British Standards Institution (1969 & 1973) *Specification for high strength friction grip bolts and associated nuts and washers for structural engineering. Parts 1 and 2* (1969) and *Part 3* (1973) BS 4395, BSI, London.
4. Crawford S.F. & Kulak G.L. (1971) Eccentrically loaded bolted connections. *J. Struct. Div., ASCE*, **97**, No. ST3, March, 765–83.
5. British Standards Institution (1970 & 1973) *Specification for the use of high strength friction grip bolts in structural steelwork. Metric series. Parts 1 and 2* (1970) and *Part 3* (1973). BS 4604, BSI, London.
6. Owens G.W. & Cheal B.D. (1989) *Structural Steelwork Connections.* Butterworths, London.
7. Boston R.M. and Park J.W. (1978) *Structural Fasteners and their Application.* BCSA Publication 4/78.

Further reading for Chapter 23

British Constructional Steelwork Association (1979) *Metric Practice for Structural Steelwork.* BCSA Publication No. 5

Douty R.T. & McGuire W. (1965) High strength bolted moment connections. *J. Struct. Div., ASCE*, **91**, ST2, April.

Kulak G.L., Fisher J.W. & Struik J.H.A. (1987) *Design Criteria for Bolted and Riveted Joints*, 2nd edn. John Wiley & Sons.

Chapter 24
Welds

by HUBERT BARBER

24.1 Full penetration and partial penetration welds

24.1.1 Butt welds

Butt welds are defined as those which join parts together essentially within the thickness of the parent metal. When the parts connected are thin it may be possible to achieve full penetration without preparation, depending on the welding process used. For thicker joints, edge preparation is undertaken in almost all butt-welded joints. In lightly-loaded joints partial penetration welds, nevertheless involving some preparation, are often adequate.

Partial penetration butt welds

In partial penetration butt joints welded from one or both sides the throat thickness should be taken as the design depth of penetration. In the case of V or bevel butt welds the depth of penetration should be taken as the depth of preparation minus 3 mm. Except when it can be shown that greater penetration can be consistently achieved, the depth of penetration on one side of a J or U weld should be taken as the depth of preparation.

The specified preparation should not be less than $2\sqrt{t}$, where t is the thickness in millimetres of the thinner part joined.

In the design of partial penetration butt joints the eccentricity of the joints should be allowed for in calculating the stress, and the joint should be suitably restrained against rotation where appropriate.

The above paragraphs summarize the requirements of clauses 6.6.6.2 and 6.6.6.3 of BS 5950.

Incomplete penetration butt welds welded from one side only should not be subjected to bending so as to put the root in tension. The use of such welds to resist repeating or alternating forces should also be avoided. If it is unavoidable, the design strengths should be suitably adjusted. Reference should be made to BS 5400: Part 10.

Full penetration butt welds

Provided that the weld is made with suitable welding consumables, the strength of a full penetration butt weld is equal to that of the parent metal.

Butt weld preparation

The importance of the choice of the most appropriate form of preparation cannot be overemphasized as it often influences the integrity of the weld, the control of cracking and the control of distortion. The shape of the weld preparation in relation to the metal thickness also has an important influence on accessibility for the electrode. These matters are discussed in detail in Appendix B of BS 5135.[1] Table 15 of that British Standard gives full profiles of some commonly used preparation, of which thirteen are tabulated together with limiting dimensions and tolerances in Fig. 24.1.

When parts of unequal cross section are butt jointed with the centroidal axes out of line with each other, local bending will result and must be taken into account in assessing the strength of the joint. If it becomes desirable to taper the thicker member, the slope should not exceed one in four.

24.1.2 Fillet welds

Fillet welds may be made between surfaces where the fusion faces form an angle of between 120° and 60°. It is acceptable for the two legs to be of unequal lengths but the throat thickness should be appropriately calculated.

When fillet welds are used through holes or slots in one or more of the connected parts (plug welds) the holes should be of adequate size to allow welding access. The appropriate dimensions are given in Appendix A of BS 5135 and are summarized in the following paragraph.

The diameter of the hole to be not less than 3 × (thickness of part connected) or 25 mm. The ends of the slots should be radiused at 1.5 × (thickness of the plate) or 12 mm. The distance between the edge of the part and the edge of the slot should be not less than twice the plate thickness or 25 mm. Holes and slots should not be completely filled with weld metal in a single operation as there is a risk of cracking during cooling. If it is specifically required that the hole be filled it should be done as a separate pass after the slag has been removed from the first weld and it has been inspected.

24.2 Geometric considerations

24.2.1 Effective throats

The throat thickness of fillet welds is given in Table 24.1. The factors given in the table are approximately equal to the cosine of the half angle between the fusion faces for welds of equal leg length, and when multiplied by the leg length will give the perpendicular distance between the root and a line joining the intersections of

Geometric considerations

Weld type	Shape of preparation	Dimensions in mm					
		T	α	G	R	r	L
Open square butt (without backing)	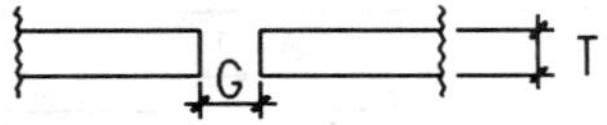	3 to 6		3			
Open square butt (with backing)	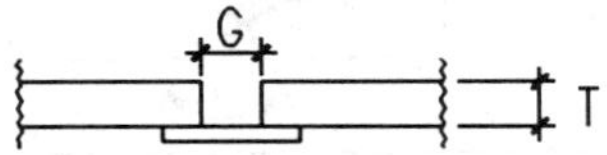	3 to 5 5 to 8 8 to 16		6 8 10			
Single V (without backing) Weld from one side or both	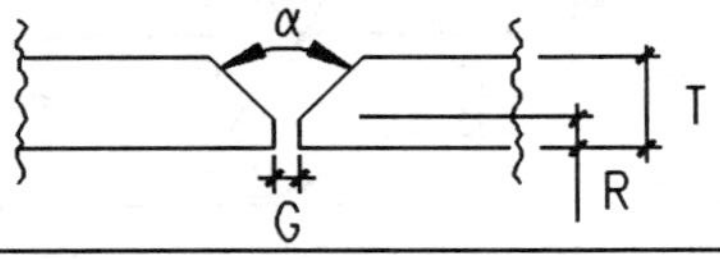	5 to 12 over 12	60 60	2 2	1 2		
Single V (with backing; backing temporary or permanent)	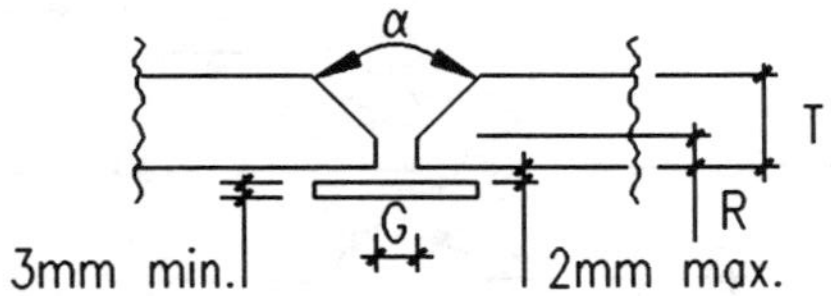	>10 >10	45 20	6 10	0 0		Single run Double run
Double V welded from both sides	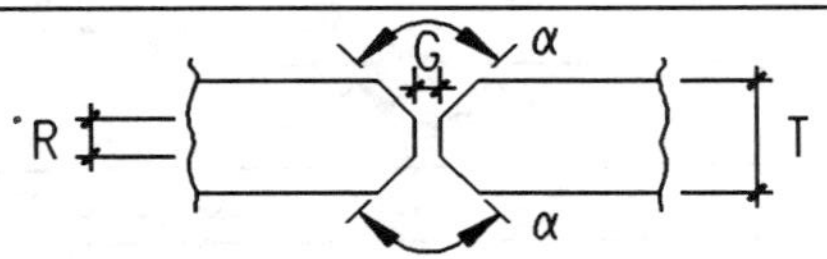	>12	60	3	2		

Fig. 24.1 Butt weld preparation from BS 5153, Table 5

Welds

Weld type	Shape of preparation	Dimensions in mm					
		T	α	G	R	r	L
Double V asymmetric, welded from both sides		>12	α = 60 β = 60	3	2		
Single U		>20	20	0	5	5	
Double U		>40	20	0	5	5	
Double U Asymmetric		>30	20	0	5	5	6

Geometric considerations 681

Weld type	Shape of preparation	Dimensions in mm					
		T	α	G	R	r	L
Single J, welded both sides	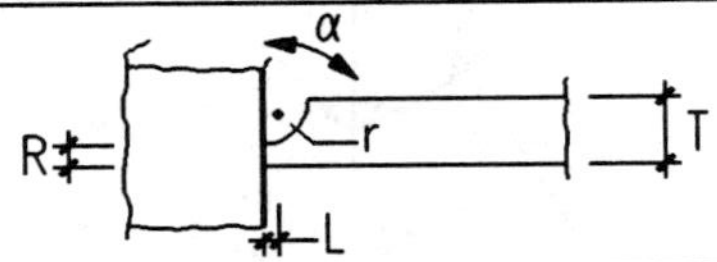	>20	20	0	5	5	5
Double J, welded from both sides	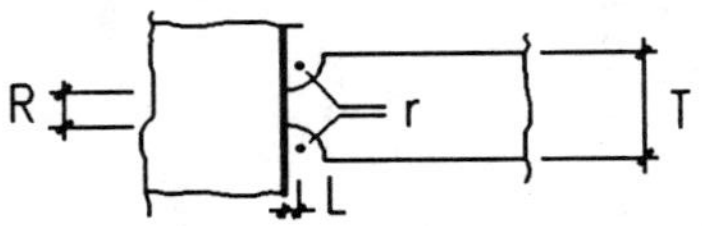	>40	20	0	5	5	5
Single bevel welded both sides	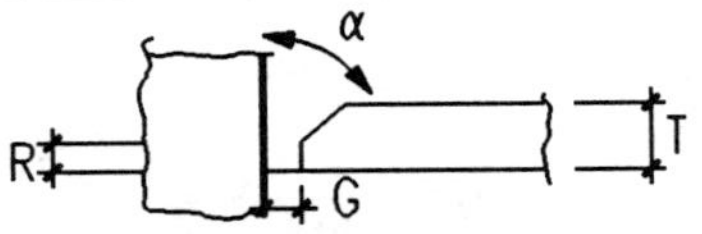	5 to 12	45	3	1		
Double bevel welded both sides	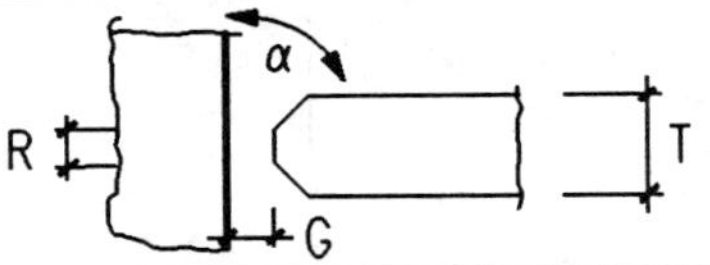	>12	45	3	2		

Fig. 24.1 Continued

682 Welds

Table 24.1 Throat thickness of fillet welds (from BS 5135, Table 14)

Angle between fusion faces (degrees)	Factor (to be applied to the leg lengths)
60 to 90	0.7
91 to 100	0.65
101 to 106	0.6
107 to 113	0.55
114 to 120	0.5

the weld with the fusion faces. This by definition is the throat thickness. For welds with an angle less than 90° between the fusion faces, for which the defined distance would be greater than that for a right angle weld, there is an upper limit to the factor of 0.7. Similarly in the case of welds of unequal leg lengths the assumed throat thickness is not to exceed 0.7 multiplied by the shorter of the two legs (Fig. 24.2).

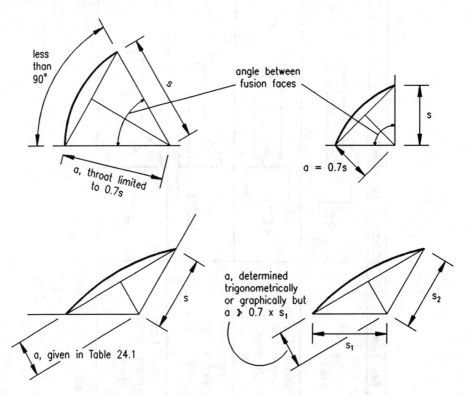

Fig. 24.2 Throat thickness of fillet welds

24.2.2 Effective lengths

The effective length of a fillet weld is the actual length less twice the leg length to allow for the starting and stopping of the weld. It should not be less than four times the leg length. When a fillet weld terminates at the end or edge of a plate it should be returned continuously round the corner for a distance of twice the leg length.

Intermittent fillet welds are laid in short lengths with gaps between. They should not be used in fatigue situations or where capillary action could lead to the formation of rust pockets. The effective length of each run within a length is calculated in accordance with the general requirements for fillet welds.

24.2.3 Spacing limitations

The longitudinal spacing between effective lengths of weld along any edge of an element should not exceed 300 mm or $16t$ for compression elements, where t is the thickness of the thinner part joined.

24.3 Methods of analysis of weld groups

24.3.1 Introduction

Any weld group may be required to resist an applied load acting through the centroid of the group either in or out of plane producing shear or tension respectively. The load may also be applied eccentrically producing in addition bending tension or torsional shear. Examples are given in Fig. 24.3.

24.3.2 Weld groups loaded in shear

British and Australian practice is to distribute the torsional shear due to eccentricity elastically in proportion to the distance of each element of the weld from the centroid of the group. This is referred to as the polar inertia method.

In some countries, notably Canada and in some cases the USA, the instantaneous centre method, referred to in Chapter 23 for bolt groups, is also used for weld groups.

684 *Welds*

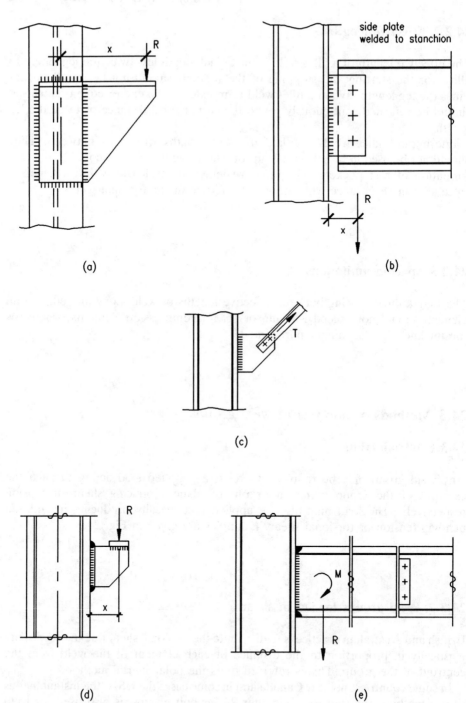

Fig. 24.3 Weld groups loaded eccentrically

The polar inertia method

Consider the four-sided weld group shown in Fig. 24.4(a). Assume the throat thickness is unity.

$$I_{xx} = \frac{2b^3}{12} + 2a\left(\frac{b}{2}\right)^2$$

$$I_{yy} = \frac{2a^3}{12} + 2b\left(\frac{a}{2}\right)^2$$

$$I_{00} = I_{xx} + I_{yy}$$

$$I_{00} = \frac{b^3 + 3ab^2 + 3ba^2 + a^3}{6}$$

Distance r to extreme fibre Fig. 24.4(b)

$$r = \frac{1}{2}\sqrt{(a^2 + b^2)}$$

$$Z_{00} = \frac{b^3 + 3ab^2 + 3ba^2 + a^3}{3\sqrt{(a^2 + b^2)}}$$

By similar reasoning to that given in Chapter 23 for bolts, f_m, the force vector per unit length from the moment, is

$$f_m = \frac{R x r}{Z_{00}}$$

f_v, the shear, is assumed uniformly distributed around the weld group

$$f_v = \frac{R}{2a + 2b}$$

The resultant force vector per unit length, f_r, may be determined as shown in Fig. 24.4(c) either graphically or trigonometrically

$$\tan \alpha = b/a$$

$$f_r = \sqrt{[(f_m \cos \alpha + f_v)^2 + (f_m \sin \alpha)^2]}$$

The value f_r can then be compared with the strength of the weld proposed from the three tables in the Appendix *Capacities of fillet welds* as appropriate.

24.3.3 Weld groups loaded in shear and tension/compression

Weld groups loaded in shear and bending (tension/compression), such as crane brackets or portal knee joints, are dealt with in Chapter 26.

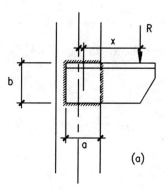

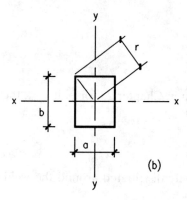

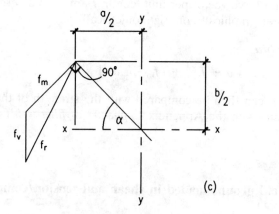

Fig. 24.4 Weld groups loaded in shear

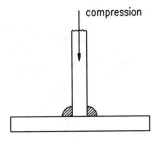

Fig. 24.5 Weld in compression

24.4 Design strengths

24.4.1 General

In fillet-welded joints which are subject to compression forces as shown in Fig. 24.5, it should not be assumed, unless provision is made to ensure this, that the parent metal surfaces are in bearing contact. In such cases the fillet weld should be designed to carry the whole of the load.

Single-sided fillet welds should not be used in cases where there is a moment about the longitudinal axis: see Fig. 24.6 and clause 6.6.2.4 of BS 5950. Ideally they should not be used to transmit tension.

A further limitation for fillet welds with unequal leg lengths, deep penetration fillet welds or partial penetration butt welds with superimposed fillet welds, is that the shear and tension on the fusion line should not exceed $0.7p_y$ and $1.0p_y$ respectively (clause 6.6.5.5 of BS 5950).

24.4.2 Strength

The design strengths of fillet welds when electrodes complying with BS 639[2] have been used and the parent metal complies with BS EN 10025 are given in Table 24.2 which is reproduced from BS 5950.

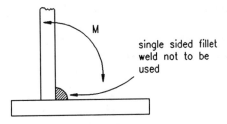

Fig. 24.6 Weld with moment

Table 24.2 Design strength of fillet welds

Grade steel to BS 4360	Design strength, P_w			Other types[b] (N/mm^2)
	Electrode strength to BS 639			
	E43 (N/mm^2)	E51 (N/mm^2)	E51[a] (N/mm^2)	
40 or 43	215	215	—	$0.5U_e$ but $\leq 0.54U_s$
WR50 and 50	215	255	—	
55	—	255	275	

[a] Only applies to electrodes having minimum tensile strength of 550 N/mm^2 and a minimum yield strength of 450 N/mm^2
[b] U_e = minimum tensile strength of electrode based on all weld tensile tests to BS 709[4]
U_s = minimum ultimate tensile strength of the steel

Three tables in the Appendix *Capacities of fillet welds* show fillet weld capacities using appropriate strengths of 215 N/mm^2, 255 N/mm^2 and 275 N/mm^2 respectively.

References to Chapter 24

1. British Standards Institution (1984) *Specification for arc welding of carbon and carbon manganese steels.* BS 5135, BSI, London.
2. British Standards Institution (1986) *Specification for covered carbon and carbon manganese steel electrodes for manual metal arc welding.* BS 639, BSI, London.
3. British Standards Institution (1990) *Specification for weldable structural steels.* BS EN 10025, BSI, London.
4. British Standards Institution (1983) *Methods of destructive testing of fusion welded joints and weld metal in steel.* BS 709, BSI, London.

Further reading for Chapter 24

Blodgett O.S. (1966) *Design of Welded Structures.* James F. Lincoln Arc Welding Foundation, Cleveland.
Owens G.W. & Cheal B.D. (1989) *Structural Steelwork Connections.* Butterworths, London.
Pratt J.L. (1989) *Introduction to the Welding of Structural Steelwork.* The Steel Construction Institute, Ascot, Berks.

Chapter 25
Plate and stiffener elements in connections

by BRIAN CHEAL

25.1 Dispersion of load through plates and flanges

Where loads (or reactions) are applied to the flanges of beams, columns or girders the web adjacent to the flange must be checked for its local bearing capacity. The effective length of web to be used for checking the bearing capacity is obtained by assuming a dispersion of the load through the plates and flanges. Generally the dispersion to find the *stiff bearing length* is taken at an angle of 45° through solid material as shown in Fig. 25.1(a) and (b).[1] The dispersion depends upon the local bending resistance of the plate and so the dispersion can only occur when there is some restraint to balance the bending moment. For example, in Fig. 25.1(c) where the loose pack is not symmetrical about the point of application of the load, the 45° dispersion should not be taken through the pack.

In the case of a flange which is integral with or is connected to the web a greater angle of dispersion at a slope of 1:2.5 to the plane of the flange is allowed.[1] The dispersion of 1:2.5, taken to the web-to-flange connection, has been verified by tests with loads applied to columns, remote from the column ends. It can also be established by calculation, assuming that at failure the web

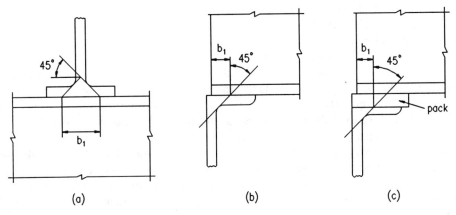

Fig. 25.1 Dispersion of load, b_1 = stiff bearing length

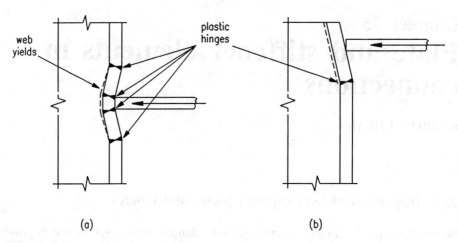

(a) (b)

Fig. 25.2 Failure mechanism of flange

crushes and four plastic hinges form in the flange (Fig. 25.2(a)). The theoretical formula that is obtained is given in the draft Eurocode 3:[2]

$$P = (b_1 + n)t\, p_{yw}/\gamma_{M1}$$

in which the length due to dispersion (n) is given as (Fig. 25.3):

$$n = 2T(B/t)^{\frac{1}{2}} (p_{yf}/p_{yw})^{\frac{1}{2}} [1 - (f_a/p_{yf})^2]^{\frac{1}{2}}$$

where P = crushing resistance of the web,
 b_1 = length of stiff bearing,
 t = thickness of web,
 B = width of flange, but not greater than $25T$,
 T = thickness of flange,
 p_{yf} = yield stress of flange,
 p_{yw} = yield stress of web,
 f_a = longitudinal stress in the flange,
 γ_{M1} = partial safety factor.

Conventionally for rolled sections dispersion has been taken to the K-line, i.e. through a distance equal to the flange thickness (T) plus the flange to web root radius (r). To adjust the formula to this practice the minimum value of $T(B/t)^{\frac{1}{2}}$ in terms of ($T + r$) for rolled sections is substituted in the equation, i.e. $T(B/t)^{\frac{1}{2}} = 2.5(T + r)$, hence the 1:2.5 dispersion.

If the load is applied near the end of a beam or column (Fig. 25.2(b)) only one plastic hinge can be assumed to form in the flange, compared to the two hinges each side of the load in the columns tested; the allowable dispersion will be less than 1:2.5 and until further research is carried out the standard dispersion of 45° should be taken.

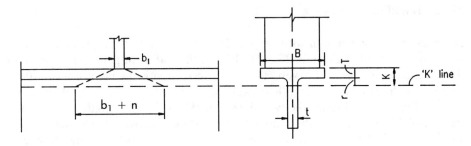

Fig. 25.3 Dispersion of load through flange

Compared with British and American sections European sections have a relatively larger root radius and the simplification of a 1:2.5 dispersion does not apply.

Where there are axial stresses in a flat plate the plastic bending moment capacity is reduced. As the dispersion depends upon the bending capacity of the flange, when there are relatively high longitudinal stresses in the flange caused by axial load and bending moments the angle of dispersion will be significantly reduced. To allow for this effect the 1:2.5 spread should be modified to 1:2.5$\sqrt{\mu}$, where $\mu = 1 - (f_a/p_y)^2$ and f_a = average longitudinal stress in the flange. This modification is included in the formula in the draft Eurocode 3.[2]

25.2 Stiffeners

25.2.1 General

At connections to beams and columns web stiffeners are provided where compressive forces applied through a flange by loads or reactions exceed the buckling resistance of the unstiffened web, or where the compressive or tensile forces applied through the flange exceed the local capacity of the web at its connection with the flange. They may also be provided to stiffen the flange where it is inadequate in bending, e.g. in a bolted tension connection.

25.2.2 Outstand of stiffeners

For flat stiffeners the outstand from the face of the web should not exceed $19t_s\varepsilon$, where t_s is the thickness of stiffener and $\varepsilon = (275/p_y)^{\frac{1}{2}}$; but a maximum of $13t_s\varepsilon$ should be used for the design of the stiffeners. The limit of $13t_s\varepsilon$ corresponds to a semi-compact section in Table 7 of BS 5950: Part 1.

25.2.3 Buckling resistance

The buckling resistance of a stiffener should be based on the compressive strength of a strut using Table 27(c) in BS 5950: Part 1, the radius of gyration being taken about the axis parallel to the web. The effective section is the area of the stiffener together with an effective length of web on each side of the centreline of the stiffener limited to 20 times the web thickness. The effective length of the strut is taken as $0.7L$ (where L is the length of the stiffener) provided that the flange is restrained against rotation in the plane of the stiffener and that the stiffener is fitted or welded to the flange so that the restraint is in fact applied to the stiffener. Where the stiffener is not so restrained the effective length is taken as L. These effective lengths assume that the flange through which the load or reaction is applied is effectively restrained against lateral movement relative to the other flange. If this is not so the stiffener should be designed as part of the compression member applying the load, and the design of the connection should include the moments due to strut action (see BS 5950: Part 1, clause 6.1.7.2 and Appendix C.3).

25.2.4 Local bearing

Web stiffeners are often *sniped* (the internal corners are cut) to clear the root radius or the web-to-flange welds. Since this reduces the effective bearing area of the stiffeners, it is necessary to check the local bearing capacity.

For the design of stiffeners where the web alone would be inadequate in bearing, and for the design of tension stiffeners, BS 5950: Part 1 allows the stiffeners to be designed to carry the applied load or reaction less the load capacity of the web without stiffeners.

In a compression detail, when stiffeners are used, a greater proportion of the load is initially carried by the stiffeners and it may well be advisable to limit the dispersion to 45° when calculating the portion of load carried by the web. This reservation follows from the stiffener outstand limitation being for a semi-compact section which by definition has limited ductility. Alternatively, the full 1:2.5 dispersion can be used with the compact criteria limitation $b \not> 8.5 t_s \varepsilon$ applied to the stiffener outstand. In the case of a tensile load the potential problem is that the fillet welds attaching the stiffeners to the flange have limited ductility. To overcome this problem, for a pair of symmetrical fillet welds, the leg length of the fillet welds should be not less than $0.85 t_s$ for grade 43 steel and $1.0 t_s$ for grade 50 steel, or the welds can be designed for a net load based on the 45° dispersion. BS 5950: Part 1 also requires that the area of the stiffener in contact with the flange should be greater than $0.8 F_x / p_{ys}$, where F_x is the external load or reaction and p_{ys} is the design strength of the stiffener.

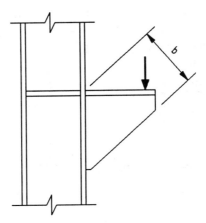

Fig. 25.4 Column bracket

25.2.5 Bracket stiffeners

The outside edge of the welded column bracket, shown in Fig. 25.4, is in compression and provided that a reasonably conservative approach is used in calculating the extreme fibre stress, semi-compact criteria for the outstand are appropriate (i.e. $b \not> 13t_s\varepsilon$). With relatively light loads, this may require an unnecessarily thick stiffener. To overcome the problem, the simple design approach of substituting f_a for p_y in the formula for ε, so that $b \not> 13t_s (275/f_a)^{\frac{1}{2}}$, where $f_a = 1.5 \times$ the extreme fibre compressive stress (at factored loading), can be used.

25.3 Plates loaded perpendicular to their plane

25.3.1 Prying forces

Where bolts are used to carry tensile forces they almost invariably connect together plates or flanges which flex in bending and are as a result subject to prying action. This prying action is best illustrated by considering a tee-connection (Fig. 25.5). When the flange of the tee is relatively flexible it bends under the applied load and the flanges outside the bolts are pressed against the supporting plate. The reactions generated at the points of contact are referred to as prying forces. For equilibrium, the total force in the bolts must equal the applied force plus the prying forces. Only when the flange is very stiff, with no plastic deformation so that its flexural deformation is small relative to the elongation of the bolts, will the prying forces be insignificant.

694   *Plate and stiffener elements in connections*

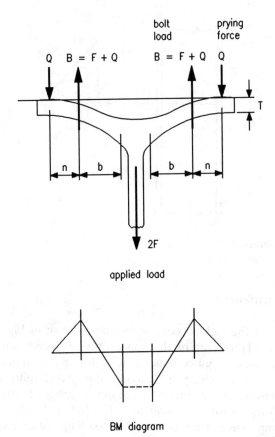

Fig. 25.5 Tee-connection prying forces

A number of semi-empirical formulae for calculating the prying forces have been published. A study of these formulae showed that although for a particular narrow range of connections they gave similar values for the prying forces, for other proportions of connection they gave very different values. Nowadays, the design procedure generally used assumes that a plastic hinge forms at the root and if required at the bolt line allowing the designer to select any suitable value for the prying forces. This design procedure is adopted in CIRIA Technical Note 98, BS 5400: Part 3 and the draft Eurocode 3.[3,4,2]

With this simple design procedure, due to the ductility of the plate in bending, the design will be adequate as far as strength is concerned. However, the ductility of the bolts may be limited, or plastic strain of the bolts may lead to an unacceptable loss of preload in the case of HSFG bolts or to a loose nut with ordinary bolts. A limitation is placed on the distance n from the centreline of the bolt to the point at which the prying force, Q, is assumed to act, so that the bolt will not be required to stretch by an unacceptable amount to accommodate the elastic curvature of the plate. The following rule has been proposed for determining the distance n:[5]

n = distance from the centreline of the bolt to the edge of the plate or supporting plate if this is less

$$\text{but} \not> 1.1T \sqrt{\left(\frac{\beta p_0}{p_y}\right)}$$

where p_0 = minimum proof stress of bolt,
p_y = yield stress of steel in the plate,
β = 1.0 for preloaded bolts
2.0 for non-preloaded bolts,
T = thickness of flange of tee.

For practical design, where HSFG (or grade 8.8) bolts are used, with grade 43 or grade 50 plate, the limitation on n may be taken as $n \not> 1.5T$ for preloaded bolts and $n \not> 2.0T$ for non-preloaded bolts.

It is sometimes convenient for the designer to increase the plate thickness, so that the moment at the root can be increased with a corresponding decrease in the moment at the bolt line and a reduction in the prying force. This is another situation where it is possible that the bolt could be required to stretch by an unacceptable amount to accommodate the plate flexure assumed in the design. To safeguard against the possibility it has been suggested that the prying force, Q, should not be taken as less than that calculated from the following formula:[5]

$$Q = \frac{b}{2n}\left(F - \frac{\beta \gamma p_0 w T^4}{27nb^2}\right)$$

where b = distance from centreline of bolt to toe of fillet weld (or weld reinforcement) for a welded tee, or to half the root radius for a rolled section,
n = distance from centreline of bolt to point where the prying force is assumed to act (as defined above),
w = effective length of flange per bolt,
T = thickness of flange of tee,
F = applied load per bolt,
p_0 = minimum proof stress of bolt,
β = 1.0 for preloaded bolts
2.0 for non-preloaded bolts,
γ = 1.0 for working load design
1.5 for limit state design (with factored loads).

When $\frac{\beta \gamma p_0 w T^4}{27nb^2}$ is greater than F, the prying force derived by the above formula should be taken as zero (i.e. $Q = 0$). The units of $\frac{\beta \gamma p_0 w T^4}{27nb^2}$ must be consistent with the units of F.

With minimum plate thickness design, when plastic hinges are assumed at the root and the bolt line, the above check for the minimum value of Q is not required since for equilibrium

$$Q = b/(2n).$$

In the formulae for prying force, Q, and the limitation of the distance n, preloaded bolts may be considered as 'non-preloaded' bolts if they are not required to contribute to the shear strength of the connection and fatigue, vibration and stiffness considerations do not apply.

BS 5950: Part 1 has two different rules for the design of bolts in tension.[1] If the bolts are used as ordinary bolts (i.e. they are not pretensioned to a specific preload) the allowable tensile strength, given in Table 32, is $0.58U_f$ but $\leq 0.83Y_f$ and in clause 6.3.6.2 it is specifically stated that prying action need not be taken into account. Ignoring the prying action is inconsistent with the general recommendations for the design of connections in clause 6.1.1 and can be shown to be unsafe as a general procedure. If the bolts are installed as 'friction grip fasteners' the allowable tensile capacity given in clause 6.4.4.2 is $0.9P_0$ and the implication is that prying action should be included in determining the bolt load, although this is not specifically stated.

Because of the danger of an unsafe design, it is recommended that, when it can occur, prying action should always be taken into account. Since the bolt load is increased by the prying force it is reasonable to use a higher tension strength, p_t, than that in Table 32. A suitable approach is to use the formulae below from clause 14.5.3.3 of BS 5400: Part 3,[4] taking $\gamma_m\gamma_{f3}$ equal to 1.1 instead of 1.32 to bring it into line with BS 5950.

$$p_t = \frac{\sigma_t}{\gamma_m\gamma_{f3}}$$

where σ_t is the lesser of $0.7 \times$ minimum ultimate tensile stress, and either the yield stress or the stress at permanent set of 0.2% as appropriate.

That is,

$$p_t = 0.64U_f \quad \text{but} \leq 0.91Y_f$$

where Y_f is the specified minimum yield strength of the fastener, and U_f is the specified minimum ultimate tensile strength of the fastener.

It is generally considered that a higher safety factor should be used for bolts in tension than for members in tension. The reasons for this are, the importance of connections at the final collapse of a structure, the relatively lower ductility of the bolts and the fact that the actual distribution of loads in the bolts (in for example an end plate connection) will not be the same as the simplified distribution assumed for the design. In the above proposal the higher margin of safety is obtained by limiting σ_t to $0.7U_f$ and taking $\gamma_m\gamma_{f3}$ equal to 1.1.

For a grade 8.8 bolt, the tension capacity obtained by taking $p_t = 0.64U_f$ is only 2% less than the tension capacity of $0.9P_0$ which would apply if it were to be designed as a preloaded friction grip fastener in accordance with clause 6.4.4.2 of BS 5950: Part 1.

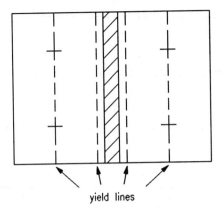

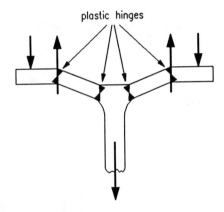

Fig. 25.6 Tee-connection yield lines

25.3.2 Short and long flanges

In a short tee-connection, provided that the capacity of the bolts is more than adequate, the maximum design capacity of the flange will be reached when parallel yield lines form at the connection of the flange to the supporting web and on the line of the bolts (Fig. 25.6). Alternatively, if the cross-section is considered, this can be stated in the terms used for plastic analysis, that the maximum design capacity will be reached when plastic hinges form at the flange-to-web connection and on the line of the bolts.

When the flange of the tee-connection is long (e.g. where it is part of the flange of a column and the detail is remote from the end of the column) the yield lines do not run to the end of the column but turn towards the edges of the flange (Fig. 25.7). The capacity of the flange can be assessed using yield line analysis. Five of the many possible yield line patterns are shown in Fig. 25.8. For a single bolt there is a mirror image of the yield line pattern on the right-hand side of the bolt; but if there is an adjacent bolt, there is a yield line linking the bolts with a parallel yield line adjacent to the web (Fig. 25.7).

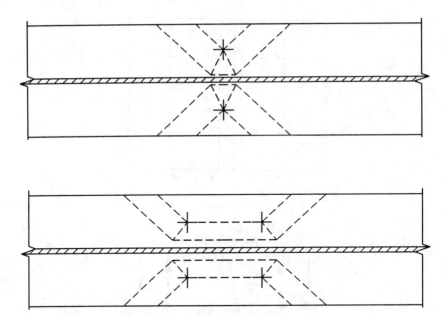

Fig. 25.7 Long flange yield lines

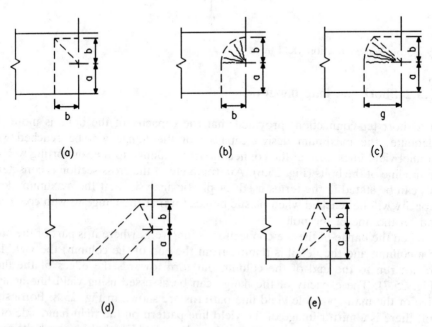

Fig. 25.8 Yield line patterns

Figure 25.8(a) shows a simple pattern with a 45° diagonal yield line; Fig. 25.8(b) is a modification of (a) with a fan yield line in the corner; Fig. 25.8(c) has a partial fan yield line at the corner (the dimension g being calculated to correspond to the minimum capacity of the pattern) and Fig. 25.8(d) and (e) are other patterns that have been proposed.[5,6,7]

The problem with the use of yield line analysis is that it is necessary to find the pattern which corresponds to, or is acceptably close to, the minimum capacity. It also requires a good working knowledge of the geometry of intersecting inclined planes (and cones where 'fan' yield lines are used). In view of this it is normal practice to use formulae based on yield line analysis. The most straightforward concept is to consider the formulae as giving the equivalent effective length of parallel yield lines so that the flange can be designed as an equivalent 'short flange' tee-connection. The equivalent lengths for the patterns in Fig. 25.8(a)–(e) are plotted in Fig. 25.9. The equivalent effective length derived from the yield line pattern in Fig. 25.8(d) is used for design since it corresponds to the minimum capacity for the practical range of values of a/b.

The equivalent length cannot be greater than π since this corresponds to another mode of failure which is a circular fan around the hole. If the effective value of a is limited to $1.4b$, in the formula corresponding to the pattern in

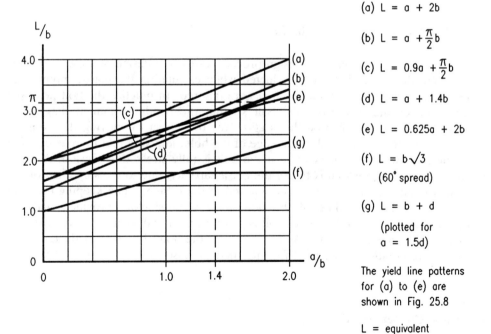

(a) $L = a + 2b$

(b) $L = a + \frac{\pi}{2} b$

(c) $L = 0.9a + \frac{\pi}{2} b$

(d) $L = a + 1.4b$

(e) $L = 0.625a + 2b$

(f) $L = b\sqrt{3}$
    (60° spread)

(g) $L = b + d$
    (plotted for
    $a = 1.5d$)

The yield line patterns for (a) to (e) are shown in Fig. 25.8

L = equivalent effective length

Fig. 25.9 Comparison of yield line patterns (graphs of L/b versus a/b)

Fig. 25.8(d), this covers the circular fan failure mode and also limits the deformation that can occur, before the full yield line pattern is formed, if large values of a relative to b are used in the formula.

The 60° maximum angle of spread used in BS 5400: Part 3 is shown as line (f) in Fig. 25.9. It can be seen that it is a good approximation for lower values of a/b (bearing in mind that a should not normally be less than $b/3$) and although conservative for higher values it has the advantage of providing a relatively stiffer connection. The line for an effective length taken as the bolt diameter plus a 45° spread is shown as (g) in Fig. 25.9. This formula is used when formation of the full yield line pattern is not acceptable (e.g. when the connected tee or extended end plate is designed with yield lines at the bolt holes and for compatibility of the deformations and prying forces the column flange is designed as a short tee).

The prying forces, for a long flange tee calculated using the equivalent effective length, are higher than the theoretical values. However, as with many other instances in the design of structures where the theoretical analysis is simplified for practical design, it is considered that the advantages of a faster simpler approach to design outweigh the disadvantages. It is worth noting that one advantage of determining the prying force using the equivalent effective length is that it overcomes the difficult problem of determining how far the detail needs to be from the end of the column for the full 'long flange' pattern to develop.

25.3.3 Allowance for bolt holes

In some of the formulae and design procedures for the design of tee-connections and end plates, the holes are deducted when calculating the plastic moment of resistance; whereas in others the holes are ignored. In practice the bolt head (or nut) spreads the load from the bolt so that the bending moment diagram is modified. Some design procedures allow for this effect by moving the point of application of the bolt load to the edge of the hole, implying that there is bending in the bolt, which is not possible if the bolt is fully stressed in tension. A more straightforward approach is to use the extra capacity, due to the spreading of the load, to allow the bolt holes to be ignored in the design, while keeping the load applied on the centreline of the bolt. This is satisfactory provided that the effective length, w, of flange per bolt is greater than the distance, b, from the bolt line to the hinge at the root. When $w < b/2$, the hole should be deducted, with linear interpolation being used for intermediate cases.

25.3.4 Corner details

When the plate in a tension connection is supported on two adjacent sides, the yield line pattern for 'minimum capacity' incorporates fan yield lines (Fig. 25.10).

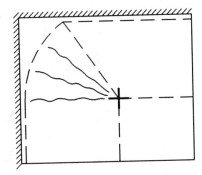

Fig. 25.10 Corner yield line pattern

A simple design approach, which allows for the various patterns that could occur in the corner, is to assume a yield line pattern with a diagonal line from the bolt to the corner; but to take only 70% of the lines in the corner as being effective (Fig. 25.11(a)). It can be shown that the diagonal yield line is equivalent to the projected yield lines on the perpendicular axes, so that the pattern with the diagonal line can be split into two parallel line patterns (Fig. 25.11(b)). This is convenient, because it not only simplifies the design of the flange plate, but it enables the bolt load to be divided between the supporting sides for the design of the supporting plates and welds.

For the collapse pattern associated with the yield lines, the load is divided between the supported sides in proportion to w/b. However, if the plate is thicker or its yield stress higher than the minimum required it can behave elastically and so change the distribution of the load. Normally, ductility takes care of this; but where fillet welds, which have limited ductility, transfer the load it is prudent to design them for the greater of the loads obtained by assuming that the distribution is proportional to w/b (plastic) or w/b^3 (elastic).

It is a useful design point that if a simple pattern with a 45° diagonal yield line is assumed for the long flange case, with the corner yield lines only 70% effective, the equivalent effective length of $a + 1.4b$ is obtained.

25.3.5 Summary

(1) For minimum plate thickness assume plastic hinges (yield lines) at the web root and the bolt line (Fig. 25.6).
(2) Where desired (to justify a smaller size of bolt), a smaller prying force can be assumed; but it is not advisable to assume a prying force, Q, less than that calculated using the formula in section 25.3.1.

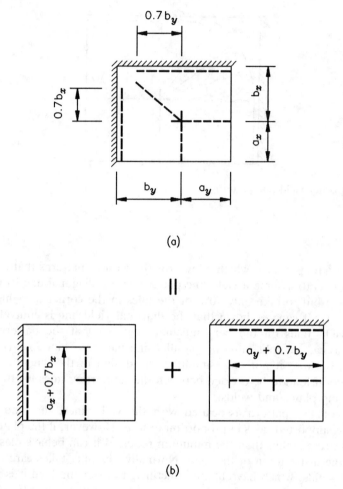

Fig. 25.11 Corner yield lines – equivalent pattern

(3) Take the distance, n, from the centreline of the bolt to the point at which the prying force, Q, is assumed to act as the least of:
  (a) the distance from the centreline of the bolt to the edge of the plate
  (b) the distance from the centreline of the bolt to the edge of the supporting plate
  (c) $1.5T$ for preloaded bolts or $2.0T$ for non-preloaded bolts (and preloaded bolts where some loss of preload would be acceptable); where T is the plate thickness.
(4) Use the plastic modulus of the plate $wT^2/4$. The bolt holes may be ignored provided that the effective length, w, of flange per bolt is greater than the distance, b, from the bolt line to the hinge at the root.

(5) Calculate the effective length of plate associated with each bolt by taking, on each side of the bolt, the least of:
    (a) distance from the centreline of the bolt to the end of the plate
    (b) 0.5 × distance to the adjacent bolt
    (c) $a + 1.4b$ (but $\not> 2.8b$)
        where a = distance from centreline of bolt to edge of plate
              b = distance from centreline of bolt to toe of fillet weld (or weld reinforcement) for a welded detail, or half the root radius for a rolled section.
        If stiffness is important, limit spread to 60° (i.e. $1.73b$) instead of ($a + 1.4b$).
        For a fillet welded detail (which has limited ductility), when calculating the effective length of fillet weld, limit spread to $d + b$; where d is the bolt diameter.
(6) For the purposes of design, when a plate is supported on two adjacent sides it may be considered as two independent 'cantilever' plates; but only 70% of the yield lines in the corner should be taken as being effective (see section 25.3.4).
    For the design of the supporting 'plates' distribute the load in proportion to w/b.
    For the design of fillet welds (which have limited ductility) design for the greater of the loads obtained by distribution in proportion to w/b or w/b^3 and assume that the load is applied to the weld on the line of the bolt.

25.4 Plates loaded in-plane

25.4.1 Deductions for holes

Where holes are staggered the effective net section through the line of the holes is taken as the area of the gross cross section less a deduction of the area of the holes in any (zig-zag) section less $s^2t/4g$ for each gauge space in the line of holes (Fig. 25.12);[1] where s is the bolt pitch in the direction of stress, t is the thickness of the material and g is the bolt gauge transverse to the pitch.

25.4.2 Gusset plates

The design of gusset plates can be carried out by developing separate rules for each of the possible modes of failure (e.g. in Fig. 25.13, failure on line A−F or failure by tearing out of the section G−C−D−H, etc.). A simpler procedure is normally adopted in which dispersion of the load is assumed and only transverse sections are checked in tension or compression (with the addition of bending

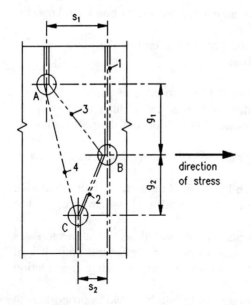

Deduction for holes equals greatest of:

(1) Area of hole B

(2) Area of holes $B+C- \dfrac{s_2^2 t}{4g_2}$

(3) Area of holes $A+B+C- \dfrac{s_1^2 t}{4g_1} - \dfrac{s_2^2 t}{4g_2}$

(4) Area of holes $A+C- \dfrac{(s_1-s_2)^2 t}{4(g_1+g_2)}$

Fig. 25.12 Deduction factors

where appropriate). It is considered that 30° is a satisfactory maximum angle of dispersion.

The ratio of D−E to D−J is 0.577 (i.e. tan 30°); if this is compared with the relationship between shear strength and tensile strength the reason why the check on B−E also covers the other modes of failure is apparent.

25.4.3 Notched beams

Although not absolutely clear in BS 5950: Part 1, it is normal practice to ignore bolt holes when checking the shear capacity of beams and channels. Other codes, such as the AISC specification, require a check allowing for the effect of the fastener holes.[8] In view of this, for a web without flanges (including beams and channels with one or both flanges notched) it is prudent to check against ultimate rupture of the detail:

shear capacity, $P_v = 0.6 p_y \times 0.9 A_{ve}$

where $A_{ve} = K_e A_{vn}$ but $\not> A_{vg}$

A_{ve}, A_{vn} and A_{vg} are the effective, net and gross areas and K_e is the factor defined in BS 5950: Part 1, clause 3.3.3.

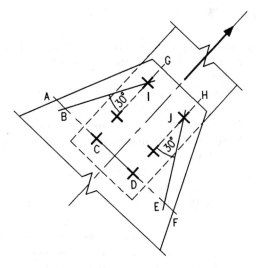

Fig. 25.13 Gusset plate load dispersion

Where, for example, the connection is concentrated toward the top of a notched beam a 'block' failure must also be considered; e.g. a tear-out failure along A–B–C in Fig. 25.14. In the light of test results, some allowance for the eccentricity of the reaction is made by taking a triangular stress distribution on B–C:[9]

shear capacity, P_v = shear capacity of A–B + tensile capacity of B–C
$= 0.6p_y \times 0.9A_{\text{ve(AB)}} + 0.5p_y A_{\text{ve(BC)}}$

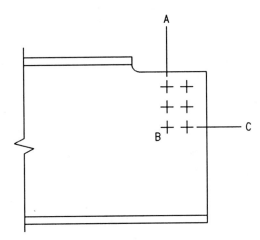

Fig. 25.14 Notched beam tear-out failure

References to Chapter 25

1. British Standards Institution (1990) *Structural use of steelwork in building*. Part 1: *Code of practice for design in simple and continuous construction: hot rolled sections*. BS 5950, BSI, London.
2. Commission of the European Communities (1988) *Design of steel structures*. Part 1: *General rules and rules for buildings*. Eurocode 3 (final draft), Dec.
3. Cheal B.D. (1980) *Design Guidance Notes for Friction Grip Bolted Connections*. CIRIA Technical Note 98. Construction Industry Research and Information Association, London, May.
4. British Standards Institution (1982) *Steel, concrete and composite bridges*. Part 3: *Code of practice for design of steel bridges*. BS 5400, BSI, London.
5. Owens G.W. & Cheal B.D. (1989) *Structural Steelwork Connections*. Butterworths, London.
6. Cheal B.D. (1983) *Guidance Notes for the Design of Bolted End Plate Connections*. Interim report on project RP306. Construction Industry Research and Information Association (CIRIA), London, Oct.
7. Zoetmeijer P. (1974) A design method for the tension side of statically loaded bolted beam-to-column connections. *Heron*, **20**, No. 1, Delft.
8. American Institute of Steel Construction (1978) *Specification for the Design, Fabrication and Erection of Structural Steel for Buildings*. AISC, Chicago.
9. Ricles J.M. & Yura J.A. (1983) Strength of double row bolted web connections. *J. Struct. Engng. ASCE*, **109**, No. 1

Chapter 26
Design of connections

by BRIAN CHEAL

26.1 Conceptual design of connections

26.1.1 General

The design of the individual elements, the bolts, welds, plates and stiffeners, is described in Chapters 23, 24 and 25. For most connections the main problem to be dealt with is the determination of the distribution of forces, moments and stresses in the constituent elements. Other matters that need to be considered are the overall stiffness of the connection and the practical aspects of fabrication, erection and inspection.

26.1.2 Load paths

In some simple details the distribution of forces in the elements of the connection is fairly obvious; but in more complex cases, such as the bolted end plates used for rigid beam-to-column connections, it is more difficult to ascertain the distribution of forces. The problem can be overcome by considering the load paths by which the applied moments and forces are carried through the connection. The principle is best illustrated by an example: the moment applied to the beam-to-column connection in Fig. 26.1 is replaced by a tensile force in the top flange and a compressive force in the bottom flange. The load paths taken by these forces through the various elements of the connection are illustrated in Fig. 26.2. The force in the tension flange is transmitted through

(1) the flange to end plate fillet welds;
(2) the end plate in bending;
(3) the bolts in tension;
(4) the column flange in bending;
(5) the column flange to web fillet welds and the column web in tension and shear.

The force in the compression flange is transmitted through the end plate and column flange in bearing into the column stiffeners from whence it is transmitted by the fillet welds into the column web.

708 *Design of connections*

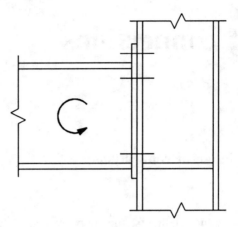

Fig. 26.1 Beam-to-column connection

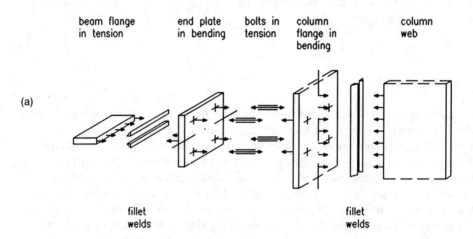

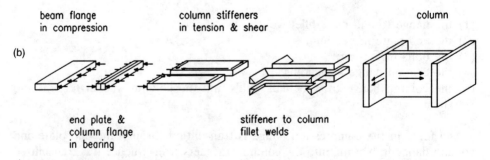

Fig. 26.2 Beam-to-column connection – load paths (a) at top flange, (b) at bottom flange

Conceptual design of connections

Every element in the chain that forms the load path must be adequate to carry the force being transmitted. In practice some of the elements in the load paths will be seen to be adequate by inspection and specific calculations will not be required. Nevertheless, the designer must be sure that there are no weak links in his detail.

26.1.3 Relative flexibility and hard spots

In the connection shown in Fig. 26.3(a), the applied moment is transmitted by compression in the lower part of the beam and by the bolts in tension in the upper part. The compression path is in direct bearing and is relatively much stiffer than the tension paths which include the end plate and the column flange in bending. Due to this difference in stiffness, the connection tends to rotate about the bottom flange; this assumption is usually made when choosing the load paths. At the stage of loading when the end plate and column flange are elastic they may be considered as springs. In practice the spring stiffnesses for the bolt positions vary with bolt spacing, are affected by the presence of the top flange (and column stiffeners if used) and by the partial formation of yield lines, and are further influenced by other variables such as lack of fit and variations in the preload of the bolts. For design these effects are generally ignored; the spring stiffnesses are assumed to be the same at each bolt and the tension in each bolt is then proportional to its distance from the point of rotation (Fig. 26.3(b)). There is one generally accepted modification to the distribution of bolt forces indicated in Fig. 26.3(b), which is used for the extended end plate detail where there is no

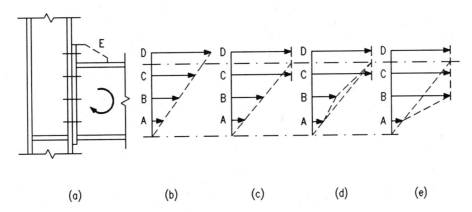

Fig. 26.3 Distribution of bolt loads

stiffener (E) above the tension flange (Fig. 26.3(a)). In this case equal forces are assumed in the bolts above and immediately below the flange as shown in Fig. 26.3(c). The reason for this modification is that the end plate below the flange (which is stiffened by the web of the beam) is stiffer than the cantilever portion of the end plate above the flange and tends to attract more load. The additional load due to stiffness is balanced against the effect of distance from the point of rotation.

Although the tension load paths are each associated with the pair of bolts at a given level, the strength of each load path is governed by the weakest link in the chain (i.e. bolts, welds, flange or end plate in bending or webs in tension). In some cases a small variation from the triangular distribution of forces may be reasonable. For example, if load path B is governed by the tension capacity of the web, the distribution in Fig. 26.3(d) can be used. The justification for this is that the web in tension has sufficient ductility to yield and allow the other load paths to reach their design capacity. On the other hand, if load path B is governed by the capacity of the fillet welds, the modification of the triangular distribution is not permissible because end fillet welds have very limited ductility. Compared with that shown in Fig. 26.3(d), the distribution shown in Fig. 26.3(e) is not generally acceptable because additional deformation at load paths C and D is required to allow the capacity assumed for load path B to be generated. This would reduce the overall stiffness of the connection.

26.1.4 Ductility

It is clear that the stresses obtained by the load path analysis are unlikely to be the actual stresses that occur in the real connection. In fact, due to variables such as lack of fit, if two nominally identical connections are fabricated and tested the stresses are unlikely to be the same. What is required from the design of a connection is that the resulting detail will have adequate strength and stiffness or ductility. Stiffness is required for a fixed connection (and possibly ductility, although using a connection as a plastic hinge in the frame analysis is not recommended). Ductility is required for a simply-supported connection.

In a fixed connection, such as that shown in Fig. 26.3(a), provided that reasonable load paths are assumed, the ductility of the steel should accommodate any 'inaccuracy' in the load distribution used for analysis. For example, if one of the load paths is stiffer than implied by the triangular distribution, the load path will initially carry a greater proportion of load; but when the weakest element of the load path is fully stressed it will deform plastically and as the applied load is increased the distribution between the load paths will move towards that assumed in the design.

The elements in the load path will not have the same capacity for plastic deformation; this has to be kept in mind while carrying out the design. It is useful to consider the relative deformation capacities of the elements in the load path:

Plates in bending	Relatively large elastic deformation. Large deformation capacity when plastic hinges form.
Plates in tension (also plates in compression with 'plastic' cross-sections)	Small elastic deformation. Reasonable plastic deformation.
Butt welds	Deformation occurs in plate adjacent to weld.
Bolts	Some plastic deformation (but deformation capacity dependent upon number of threads in the grip length).
Fillet welds	End fillet welds have very little deformation capacity.

Due to the limited deformation capacity of end fillet welds care should be taken to avoid the possibility of their being subject to a load much greater than the design load. For example, in Fig. 26.1 the column flange could be designed as a long flange tee-connection using the full plastic collapse load. However, if the thickness of the flange is greater than the minimum required and/or the yield stress of the flange is greater than the minimum, the full yield line pattern will not form and there will be a higher load to be carried by the fillet welds near the bolts. To allow for this possibility it is reasonable to take a lesser spread of load when designing the fillet welds, for example, a maximum spread of the bolt diameter plus 45° from the centreline of the bolt.

26.1.5 Compatibility of design assumptions

Care is required to ensure that the assumptions made for the design of the various elements in the connection are compatible. For example, in a cover plate splice, sharing the load between ordinary bolts (in shear and bearing) and fillet welds is not acceptable. The reason for this is that the deformation characteristics of ordinary bolts and fillet welds are incompatible. The fillet welds are stiff compared with the bolts in clearance holes, and as the load is applied the welds will initially carry most of the load. Then, due to their limited ductility, the fillet welds will be liable to break before the bolts can take up their share of the load.

Another example of the need for compatibility is in the design of bolted tension connections where the two tees have their webs at right angles to each other. This is best understood by first of all considering the interaction of two short tees (Fig. 26.4). The shaded areas in Fig. 26.4(b) and (c) indicate where the two tees are in contact (ignoring elastic curvature) if they are attached to solid reaction plates and the crosses indicate where the prying forces required to generate the

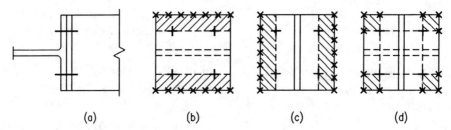

Fig. 26.4 Interaction of two short tees

yield lines through the bolts occur. Figure 26.4(d) shows the contact areas and the position of the prying forces for the two tees in Fig. 26.4(a), which is satisfactory. Now consider the interaction of a short tee with a long flange tee (Fig. 26.5). Using the same approach as for the two short tees, it can be seen that there are no common shaded areas and that, although prying forces can occur in suitable positions to generate the long flange yield line pattern, there are no prying forces to generate any yield lines through the bolt on the short tee. As shown in Fig. 26.5(e), the short tee is able to nest around the deformed shape of the long tee. The problem can be overcome either by designing the column flange as a long tee and the outstands of the short tee as simple cantilevers or by designing both parts as short tees. This is achieved by limiting the effective length of the tees, using a maximum spread of the bolt diameter plus 45°.

The above design provisions also apply to an extended end plate beam-to-column connection (Fig. 26.1), where the end plate at the top flange is in effect a short tee with its web (the beam flange) at right angles to the column web.

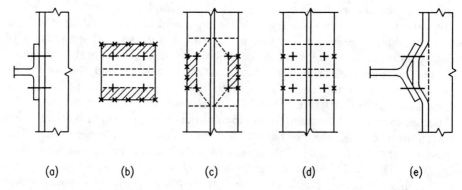

Fig. 26.5 Interaction of short and long tee

26.1.6 Stiffness

One other aspect to be considered is the stiffness of the connection. This has to be acceptable with respect to the deflection of the structure, the overall and local stability, the analysis and the assumed distribution of forces and moments.

26.1.7 Summary

Important points that must be kept in mind when designing connections are listed below:

(1) The design of the connection should follow the assumptions made in the analysis of the frame or truss. Note that the forces and moments are related to the line diagram used in the analysis.
(2) If fixed ended members are assumed in the analysis (e.g. in a lattice girder) the connections should be designed for the end moments as well as the axial forces. (Preferably analysis and design should assume pinned joints.)
(3) Check whether any additional moment capacity is required to ensure stability (e.g. at a splice in the middle of a strut or beam).
(4) Check that the load combinations taken in the analysis give the maximum forces for the connection design.
(5) When choosing an analytical model of the connection, consider the effect of hard spots (in bearing) and the relatively more flexible plates (in bending) on the distribution of forces.
(6) The load paths through the connection must be complete and in equilibrium and should be as direct as possible.
(7) The elements in the connection must not only have adequate strength, but they must also have adequate ductility for the assumptions/simplifications made in the design to be satisfactory.
(8) Wherever practicable, the centroidal axis of the splice material should coincide with the centroidal axis of the element joined.
(9) Generally in lattice structures centroidal axes should intersect at the node points. Where they do not intersect the effect of the eccentricity should be allowed for in the design.
(10) The load should not be shared between welds and fasteners of different stiffnesses.
(11) The effects of lack of fit on the behaviour of the connection. Also the effect of the introduction of packs and shims. In particular beware of the effect of lack of fit on the fatigue life of the connection.
(12) Avoid severe stress concentrations, particularly where fatigue could be a problem.
(13) Consider the economics and ease of fabrication, erection and inspection. Allow adequate clearance for installation and tightening of fasteners (particularly HSFG bolts), welding, inspection and maintenance.

26.2 Types of connection

26.2.1 Column splices

26.2.1.1 General

As well as providing continuity of strength, column splices are required to provide adequate continuity of stiffness about both axes. If a splice is near to a point of lateral restraint, and the strut is designed as pinned at that point, the splice is designed simply for the axial load and applied moments (if any). However, if the splice is away from a point of lateral restraint, or end fixity or continuity is assumed in calculating the effective length, the additional moments that can be induced by strut action must be taken into account. The procedure for calculating the additional moment is given in Appendix C.3 of BS 5950: Part 1.[1]

Where the web and flanges are spliced, the axial load can be shared between the web and the flanges in proportion to their areas. However, in bolted connections it is usual to assume that the bending moments are carried solely by the flanges.

26.2.1.2 Bolted cover plate splices

The column ends are usually prepared for full contact, in which case compression forces may be transmitted in bearing. Alternatively the cover plates can be designed to carry all the loads and moments and full bearing need not be attained.

When the splice is near to the mid-point of the 'effective length' of the strut, the size of the flange cover plates should be not less than that of the associated flanges (of the lighter section where a change of section occurs) with respect to area and section modulus.

A typical splice is shown in Fig. 26.6(a). If the sections are of the same serial size but of different rolling weights, packs are required to make up the differences in flange thickness. If the two sections are of different serial size, some means, such as the division plate in Fig. 26.6(b), must be provided to transfer the bearing load between the flanges which are not in line.

26.2.1.3 Welded splices (primarily used for shop connections)

Where the sections are of the same serial size, a full strength butt weld can be used (Fig. 26.7(a)). However, it is normally easier and cheaper to prepare the ends for full bearing and use partial penetration welds. Where the flanges do not line up, some provision must be made for transferring the flange forces (Fig. 26.7(b)).

Types of connection 715

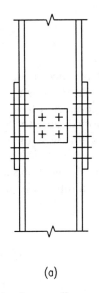
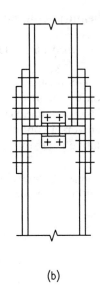

(a)   (b)

Fig. 26.6 Bolted column splices

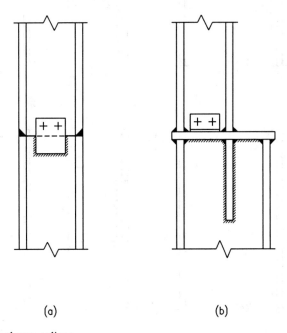

(a)   (b)

Fig. 26.7 Welded column splices

26.2.2 Simple beam-to-column connections

26.2.2.1 Bolted seating cleats (Fig. 26.8(a))

If the seating cleat is unstiffened, it is accepted practice to design the cleat-to-column connection for shear only. This assumes that there is a second cleat near the top flange of the beam and that the beam is connected to a column flange or to the web of a universal column. The detail is not suitable for the connection of a beam to a relatively flexible element such as the web of a plate girder or universal beam.

If the seating cleat is stiffened, the cleat-to-column connection should be designed for the eccentric moment as well as the shear. For a fully stiffened cleat, the load

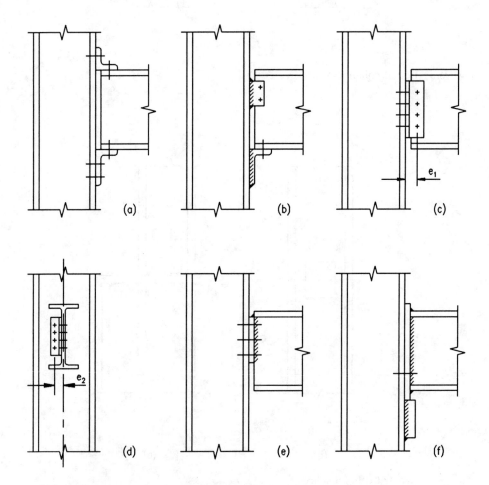

Fig. 26.8 Simple beam-to-column connections

is assumed to be applied at the toe of the horizontal leg. Since the horizontal leg of the top cleat is relatively flexible, the top cleat cannot be assumed to contribute to the shear resistance of the connection.

An alternative detail for the top cleat is to connect it to the top of the web. This alternative should be considered where the top flange cleat would interfere with the floor construction.

26.2.2.2 Shop-welded seating cleats (Fig. 26.8(b))

This is a variation of the bolted detail where the cleats are shop-welded to the column and drilling of the column is avoided.

On the minor axis, where the seating cleat is connected to the column web, it is difficult to make sound welds if the edges of the angle cleat are close to the column flanges.

26.2.2.3 Bolted web cleats (Fig. 26.8(c))

Normally the cleats are bolted and are in pairs. The bolt group connecting the cleats to the beam web should be designed for the eccentricity e_1. Conventionally, for a pair of cleats, the effects of eccentricity are ignored when designing the bolts connecting the cleats to the column flange. In practice, the cleat to column flange bolts are rarely critical, design is almost always governed by bolt bearing on the web of the beam.

If only one cleat is used, the eccentricity e_2 must be taken into account when designing the bolts in the cleat-to-column connection (Fig. 26.8(d)). It is also prudent to use HSFG bolts to limit the beam rotation.

26.2.2.4 Welded web cleats and fin plates

A variation of the bolted web cleat detail is to weld the cleats to the beams, or where a single cleat is used, to weld the cleat to the column. A further variation is to use a flat (fin plate) shop-welded to the column and site-bolted to the beam. Care is required to ensure that the flat is welded on perpendicular to the web or flange. In the web the flat is 'protected' by the column flanges, but if a flat is used on the flange, extra care is required to avoid damage during transportation. For the connection to the column web, the flange of the minor axis beam usually has to be notched to allow the beam to be erected.

With cleats welded to the column, the designer has to ensure that the end connection is able to accommodate the end rotation of the simply-supported beam. To achieve this:

(1) use ordinary (not preloaded HSFG) bolts and ensure that deformation of the plate in bearing occurs before shear deformation of the bolts, for example, with 8.8 bolts make the cleat (or beam web) thickness less than $0.5d$ for grade 43 steel and $0.4d$ for grade 50 steel, where d is the diameter of the bolt.
(2) make the edge distance greater than twice the bolt diameter
(3) ensure that the welds attaching the cleat to the column have a greater resistance than the moment that can be applied by the bolts; this is important because of the limited ductility of fillet welds. In the case of a welded flat this can be achieved by making (leg lengths of welds) $> 0.8t$, so that the deformation will take place in the flat rather than the welds.

26.2.2.5 Welded end plates (Fig. 26.8(e))

Welded end plates are economical and widely utilised when welded fabrication is to be used for the beam rather than drilling and bolting. Care in detailing is required to ensure that there is sufficient ductility for the simply-supported beam to take up its end rotation without damage to the detail. To achieve this the bolts must be a sensible distance away from the web and flange of the beam and the end plate must not be too thick. Reference 3 gives more details on practical limits for end plate proportions which ensure that they have adequate deformation capacity.

The local shear capacity of the web of the beam must always be checked. Due to their lack of ductility, the welds attaching the end plate to the beam web must not be the weakest link. The rules in section 26.2.2.4 may be used to achieve this.

26.2.2.6 Shear plates (Fig. 26.8(f))

When designing this type of connection, consideration should be given to the effect of tolerances and the possible use of packs, since the bearing (in the direction of the span of the beam) is relatively narrow.

Nominal bolts are used near the point of rotation but the requirements of lateral forces (wind/restraint/stability) must be considered.

If the shear plates are only welded on three sides, because welding along the top edge requires plate preparation and machining after welding, the possible distortion of the shear plates must be considered. Bolts or welding may be required to control this.

There will be a tendency for the structure to spread with this detail, due to the elongation of the bottom flanges of the beams. The effect should be considered for large structures; to overcome the problem the shear plate can be moved up to the neutral axis, but this introduces the complication of changing the thickness of the end plate where it sits on the shear plate.

Shear plates are normally only used for large end reactions. Their advantage is that the number of site bolts is reduced to those required to tie the structure together and the eccentricity on the column is less than with a stiffened bracket.

26.2.2.7 *General*

The choice between welded and bolted details will depend largely upon the availability of the appropriate equipment in the fabricating shop. For example, if the shop is equipped with numerically controlled cutting and drilling machines there will be a bias towards bolted shop details. In all the simple beam-to-column connections the site connection is bolted.

Another consideration in the choice of detail is that it is more economical if only shop drilling or shop welding is carried out on a particular beam or column.

Erection should always be considered and an advantage, particularly with heavier beams, of the seating cleat is that the beam can be lowered on to the cleat during erection.

26.2.3 Rigid beam-to-column connections

26.2.3.1 *Site-welded connections* (Fig. 26.9(a))

Site-welded connections are rarely used in this country but are frequently used in the USA and other countries.

The flange-to-column welds are usually down-hand welds using backing strips. Column stiffeners are usually required.

A variation of the site-welded connection is shown in Fig. 26.9(b). The web cleat, which is designed to carry the shear, is shop-welded to the column and site-bolted with HSFG bolts. The flanges are then site-welded.

26.2.3.2 *Welded end plates site-bolted to the columns* (Fig. 26.9(c)–(f))

The end plate can either be stopped at the level of the top flange of the beam or be extended beyond the flange to accommodate a row of bolts; the extended end plate is the most economic of the two alternatives. For beams subjected to only 70% or so of their yield moment capacity, the connection is usually idealized for design purposes as a tee-connection at the tension flange and a concentrated reaction at the compression flange. The detail without the extended end plate is only able to carry a relatively small proportion of the beam capacity (assuming that excessively thick end plates and large bolts are not being used).

This type of connection is used for the knee connections of portal frames. Normally it is haunched so that its moment capacity is equal to or greater than the beam capacity. Although end plates with all the tension bolts within the depth of the beam are frequently used, the extended end plate is really the better detail and is generally the more economic. However, it is important to check that the extended end plate and column can be fitted into the overall arrangement of the structure, with respect to such items as the purlins, eaves beams, gutters and curved sheeting at the eaves.

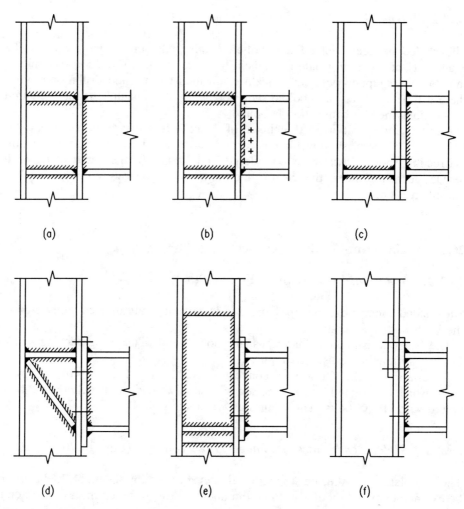

Fig. 26.9 Rigid beam-to-column connections

26.2.3.3 Column stiffening

Where rigid beam-to-column connections are used, some kind of column stiffening is normally required.

Horizontal stiffeners are used where the column web is overstressed by the concentrated load from the beam flanges (Fig. 26.9(a)).

There is often a high shear in the column over the depth of the beam and where this exceeds the shear capacity of the column web, stiffening is required. Normally diagonal stiffeners (Fig. 26.9(d)) are used but the web plate shown in Fig. 26.9(e) is another possibility.

The reason for the high shear in the external columns of rigid frames is illustrated in Fig. 26.10, where the bending moment diagram for the column is superimposed on a sketch of the column detail. The full line is the bending moment diagram obtained from a centreline analysis of the structure. The broken line shows how the diagram is modified locally by the application of the beam moment, which for simplicity is assumed to be applied by forces in the beam flanges. It can be seen that the shear in the column (which is the slope of the bending moment diagram) is much greater over the depth of the beam. The important point is that, whereas shear rarely governs the design of hot rolled section beams, the shear capacity can be critical at the joints in rigid frames.

Wherever possible the angle of diagonal stiffeners should be between 30° and 60°. If the column depth is considerably less than the beam depth, 'K' stiffening may be used.

If the web is adequate but the column flange at the tension connection is inadequate in bending it is possible that the use of horizontal stiffeners would

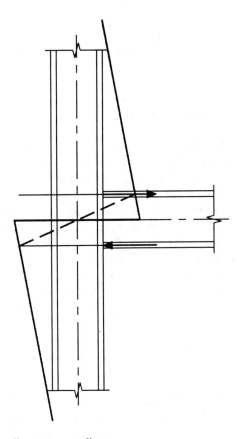

Fig. 26.10 Column bending moment diagram

enable the flange to be justified. Backing plates (Fig. 26.9(f)) are another method of strengthening the column flange to carry the bolt loads, but it should be noted that the use of backing plates increases the prying force on the bolts.

26.2.4 Beam and plate girder splices

26.2.4.1 Butt-welded splices

Connecting the flanges and web with full strength butt welds is the simplest beam splice from the point of view of design in that a full strength connection is obtained and no calculations are required. However, welded splices are of limited use due to the following considerations:

(1) careful programming is necessary to avoid delays due to the time required for welding (which is more weather-dependent than bolting) and weld inspection
(2) provision must be made for support and alignment while the welding is carried out
(3) allowance has to be made for longitudinal shrinkage due to the welding and care is required in the welding sequence to avoid girder distortion and local buckling of the web.

26.2.4.2 Bolted cover plate splices

For plate girders and the heavier beams, double cover plates are normally used so that the bolts are in double shear and the minimum number of bolts is required (Fig. 26.11(a)). It is also a preferred detail because the cover plates are balanced about the flange. However, for smaller beams with flanges less than 200 mm wide, single cover plates on the outside of the flanges are acceptable (Fig. 26.11(b)).

The splice should be situated away from the point of maximum bending moment and normally preloaded HSFG bolts are used to avoid slip in the joint, which could lead to an unacceptable deflection of the beam.

For small beams it is normal practice to design the flanges to carry all of the bending moment. For larger beams and plate girders, the web splice should be designed to carry the portion of bending moment associated with the web as well as the shear.

26.2.4.3 End plate splices

In pitched portal frames, end plate splices are normally used for the site joint at the apex (Fig. 26.12(a)). As the bending moment at this point is usually near to the capacity of the beam an extended end plate connection is required; usually

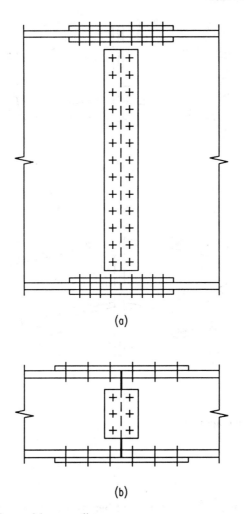

Fig. 26.11 Plate girder and beam splices

with the extended portion stiffened. Sometimes it is necessary to stiffen the extended portion with a beam cutting forming a mini-haunch detail (Fig. 26.12(b)).

26.2.5 Beam-to-beam connections

There are two general forms of beam-to-beam connection, bolted web cleats and welded end plates, and the comments in sections 26.2.2.3 and 26.2.2.5 on the similar beam-to-column connections will apply.

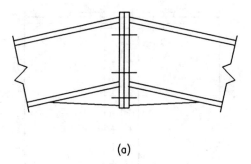

(a)

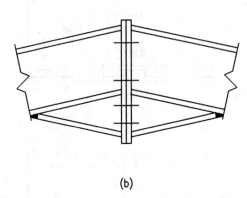

(b)

Fig. 26.12 Portal frame apex joints

Where the top flanges of the connected beams are at the same level, the flange of the supported beam is notched (Fig. 26.13(a)) and the web has to be checked allowing for the effect of the notch (see section 25.4.3).

The top of the web of the notch, which is in compression, has to be checked for local buckling of the unrestrained web. Provided the beam is laterally restrained (e.g. by a floor slab), to ensure that the web at the top of the notch does not buckle, it is recommended that the length of the notch should not exceed[3]

the depth of the beam (D) for grade 43 and for grade 50 where $D/t_w < 50$
half the depth of the beam for grade 50 where $D/t_w > 50$

where t_w is the web thickness of the supported beam. For beams which are not laterally restrained a more detailed investigation is required (see Reference 3). If necessary the web is stiffened with horizontal stiffeners (Fig. 26.13 (b)).

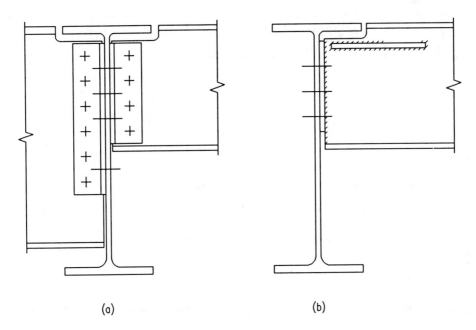

Fig. 26.13 Beam-to-beam connections

References to Chapter 26

1. British Standards Institution (1990) *Structural use of steelwork in building.* Part 1: *Code of practice for design in simple and continuous construction: hot rolled sections.* BS 5950, BSI, London.
2. Anderson D. (1991) *Wind Moment Design for Unbraced Frames.* The Steel Construction Institute, Ascot, Berks.
3. The Steel Construction Institute/British Constructional Steelwork Association (1991) *Joints in Simple Construction. Vol. 1: Design Methods.* SCI/BCSA.

Wider reading for Chapter 26

Owens G.W. & Cheal B.D. (1989) *Structural Steelwork Connections.* Butterworth, London.

A series of worked examples follows which are relevant to Chapter 26.

Worked example

The Steel Construction Institute Silwood Park, Ascot, Berks SL5 7QN		Subject *LIST OF DESIGN EXAMPLES*		Chapter ref. 26
		Design code	Made by **BDC**	Sheet no. *1*
			Checked by **GWO**	

		Code
1.	Warren girder. Bolted gusset plate connection.	BS 5950
2.	Pitched roof truss. Welded node connection.	BS 5950
3.	N - girder. Welded node connection.	BS 5950
4.	Universal column. Bolted cover plate splice. Splice near point of lateral restraint.	BS 5950
5.	Universal column. Bolted cover plate splice. Splice remote from point of lateral restraint.	BS 5950
6.	Beam to column connection. Bolted web cleats.	BS 5950
7.	Pitched portal frame. Extended end plate eaves connection.	BS 5950
8.	Pitched portal frame. Apex connection.	BS 5950
9.	Plate girder. Bolted cover plate splice.	BS 5400

	Subject	Chapter ref.
The Steel Construction Institute Silwood Park, Ascot, Berks SL5 7QN	**No.1 WARREN GIRDER. BOLTED GUSSET PLATE CONNECTION.**	**26**
	Design code **BS5950**	Made by **BDC** Checked by **GWO**	Sheet no. **1**

The connection is for a node in the bottom boom of a Warren girder truss.

Forces are due to factored loads

All steel grade 43

M20 general grade HSFG bolts

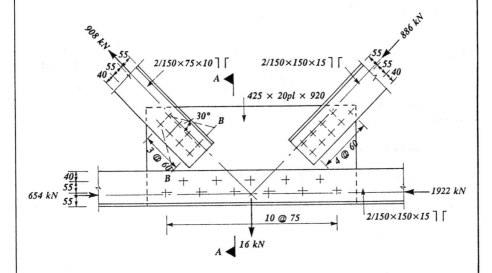

Bolts

Use M20 general grade HSFG bolts in double "shear" ($\mu = 0.45$) (see notes at end of example)

Slip resistance per bolt $= 2 \times 1.1 \, K_s \, \mu \, P_o$ 6.4.2.1

$\qquad = 2 \times 1.1 \times 1.0 \times 0.45 \times 144$

$\qquad = 142.6 \text{ kN}$

Bearing capacity $= d \, t \, p_{bg} < \dfrac{1}{3} e \, t \, p_{bg}$ 6.4.2.2

728 Worked example

The Steel Construction Institute Silwood Park, Ascot, Berks SL5 7QN	Subject No.1 WARREN GIRDER. BOLTED GUSSET PLATE CONNECTION.	Chapter ref. 26
	Design code BS5950	Made by **BDC**	Sheet no. **2**
		Checked by **GWO**	

Minimum thickness of gusset plate

$$= \frac{142.6 \times 10^3}{20 \times 825} = 8.6 \text{ mm}$$

Try 20 mm thick gusset plate

Minimum end distance

$$= \frac{142.6 \times 10^3 \times 3}{20 \times 825} = 25.9 \text{ mm}$$

Also, minimum end distance $= 1.4 D$ 6.2.3
                                                             Table 31

$$= 1.4 (20 + 2) = 30.8 \text{ mm}$$

The 20 mm plate is necessary to ensure adequate local tension capacity (see later).

Bearing thickness of angle members

$$= 2 \times 10 = 20 > 8.6 \text{ mm} \quad OK$$

<u>*Tension diagonal*</u> *(2/150 × 75 × 10 angles)*

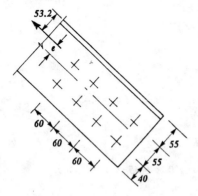

$\Sigma (x^2 + y^2) = 8 \times 27.5^2 + 4 \times 30^2 + 4 \times 90^2$ [23.4.2]

$= 42050 \text{ mm}^2$

The Steel Construction Institute Silwood Park, Ascot, Berks SL5 7QN	Subject **No.1 WARREN GIRDER.** **BOLTED GUSSET PLATE CONNECTION.**	Chapter ref. **26**
	Design code **BS5950**	Made by **BDC** / Sheet no. **3** Checked by **GWO**

Moment = $R\,e$ = $908\,(55 + 27.5 - 53.2)$

$\phantom{\text{Moment}\,=\,}$ = 26604 Nm

(See notes at end of example for discussion of line of action of force)

Maximum resultant shear on bolt

$= \left[\left(\dfrac{R}{n} + \dfrac{R\,e\,x_1}{\Sigma\,(x^2 + y^2)} \right)^2 + \left(\dfrac{R\,e\,y_1}{\Sigma\,(x^2 + y^2)} \right)^2 \right]^{0.5}$

$= \left[\left(\dfrac{908}{8} + \dfrac{26604 \times 27.5}{42050} \right)^2 + \left(\dfrac{26604 \times 90}{42050} \right)^2 \right]^{0.5}$

$= \left[(113.5 + 17.4)^2 + (56.9)^2 \right]^{0.5}$

= 142.7 kN   =   slip resistance OK

<u>Compression diagonal</u> (2/150 × 150 ×15 angles)

Centroid is 42.5 mm from heel of angle.

By inspection 8 bolts are inadequate.

Try 9 bolts.

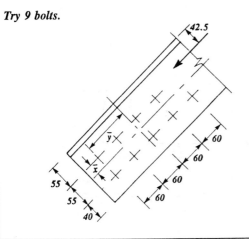

730   Worked example

The Steel Construction Institute Silwood Park, Ascot, Berks SL5 7QN	Subject **No.1 WARREN GIRDER. BOLTED GUSSET PLATE CONNECTION.**	Chapter ref. **26**
	Design code **BS5950**	Made by **BDC** Checked by **GWO**	Sheet no. **4**

$\bar{x} = \dfrac{5 \times 55}{9} = 30.6$ mm

$\bar{y} = \dfrac{2 \times 60 + 2 \times 120 + 2 \times 180 + 1 \times 240}{9} = 106.7$ mm

$\Sigma (x^2 + y^2) = 5 \times 24.4^2 + 4 \times 30.6^2 + 2 \times 106.7^2$

$\qquad\qquad\qquad + 2 \times 46.7^2 + 2 \times 13.3^2 + 2 \times 73.3^2$

$\qquad\qquad\qquad + 1 \times 133.3^2$

$\qquad\qquad = 62722$ mm^2

Moment $= 886 (55 + 30.6 - 42.5) = 38187$ Nm

Maximum resultant shear on bolt

$= \left[\left(\dfrac{886}{9} + \dfrac{38187 \times 30.6}{62722} \right)^2 + \left(\dfrac{38187 \times 133.3}{62722} \right)^2 \right]^{0.5}$

$= [(98.4 + 18.6)^2 + (81.2)^2]^{0.5}$

$= 142.4$ kN < slip resistance OK

<u>Bottom boom</u> (2/150 × 150 × 15 angles)

Force to be transmitted to gusset plate

$= 1922 - 654 = 1268$ kN

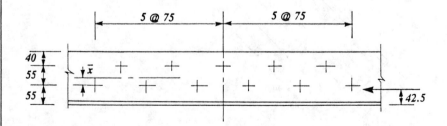

The Steel Construction Institute Silwood Park, Ascot, Berks SL5 7QN	Subject **No.1 WARREN GIRDER. BOLTED GUSSET PLATE CONNECTION.**	Chapter ref. **26**
	Design code **BS5950** Made by **BDC** Checked by **GWO**	Sheet no. **5**

$$\bar{x} = \frac{5 \times 55}{11} = 25.0 \text{ mm}$$

$$\Sigma(x^2 + y^2) = 6 \times 25.0^2 + 5 \times 30.0^2 + 2 \times 75^2$$
$$+ 2 \times 150^2 + 2 \times 225^2 + 2 \times 300^2$$
$$+ 2 \times 375^2$$
$$= 627000 \text{ mm}^2$$

Moment = $1268 (55 + 25.0 - 42.5)$

= 47550 Nm

Maximum resultant shear on bolt

$$= \left[\left(\frac{1268}{11} + \frac{47550 \times 25.0}{627000} \right)^2 + \left(\frac{47550 \times 373}{627000} \right)^2 \right]^{0.5}$$

$$= [(115.3 + 1.9)^2 + (28.3)^2]^{0.5}$$

$$= 120.6 \text{ kN} < \text{slip resistance} \quad OK$$

<u>Gusset plate</u>

Check section A-A

Shear = $\dfrac{908}{\sqrt{2}}$ = 642 kN

Shear capacity = $0.6 \, p_y \times 0.9 \, t \, d$ 4.2.3

= $0.6 \times 265 \times 0.9 \times 20 \times 425 \times 10^{-3}$

= 1216 kN > 642 kN   OK

732   *Worked example*

		Subject **No.1 WARREN GIRDER.** **BOLTED GUSSET PLATE** **CONNECTION.**	Chapter ref. 26
The Steel Construction Institute Silwood Park, Ascot, Berks SL5 7QN		Design code **BS5950**	Made by **BDC** Checked by **GWO**	Sheet no. **6**

Check section B-B

Take 30° spread from extreme bolts   [25.4.2]

Effective width   = $2 \times 180 \tan 30° + 55 - 2 \times 22$

                    = 219 mm

Tension capacity = $p_y \, t \, d$ 4.6.1

                    = $265 \times 20 \times 219 \times 10^{-3}$

                    = 1161 kN

                    > 908 kN OK

HSFG bolts have been used in this example and should be used in practice where the potential additional deflection of the truss due to slippage of the bolts is unacceptable. They should also be used if there is a possibility of vibration affecting the structure. If the effect of bolt slippage is acceptable, the use of ordinary bolts would lead to greater economy.

In this example the centroidal axes of the members have been used for setting out the detail. In the case of bolted framing of angles and tees the code allows the setting out lines of the bolts to be used, instead of the centroidal axes, if this is desired. With 2 lines of bolts, the bolt line nearest the outstand leg of the angle or the flange of the tee should be used for the setting out.   6.1.2

Note that it is necessary to allow for the bolt holes in the design of the tension member.   4.6.1

	Subject	Chapter ref.
The Steel Construction Institute Silwood Park, Ascot, Berks SL5 7QN	**No.2 PITCHED ROOF TRUSS.** **WELDED NODE CONNECTION.**	**26**
	Design code **BS5950**	Made by **BDC**	Sheet no. **1**
		Checked by **GWO**	

The connections are for nodes C and D of the roof truss shown in the sketch.

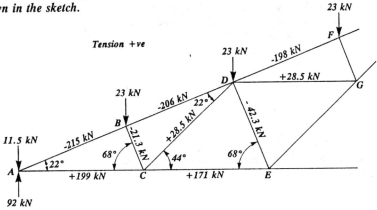

Forces are due to factored loads.

<u>Member sizes</u> (Grade 43)

Member ABDF	146 × 127 × 22 Structural Tee
Member ACE	133 × 102 × 13 Structural Tee
Members BC, CD & DG	50 × 50 × 6 Angle
Member DE	60 × 60 × 6 Angle

<u>Node C</u>

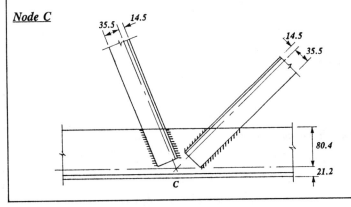

734   *Worked example*

The Steel Construction Institute Silwood Park, Ascot, Berks SL5 7QN	Subject No.2 PITCHED ROOF TRUSS. WELDED NODE CONNECTION.		Chapter ref. 26
	Design code BS5950	Made by BDC Checked by GWO	Sheet no. 2

Try 4 mm fillet welds. (Due to the radius on the toe of the angle, the leg of the fillet weld needs to be 2 mm less than the thickness of the angle. Therefore, 4 mm is the maximum reasonable size of fillet weld on a 6 mm thick angle.)

Capacity of 4 mm fillet weld   6.6.5

$= 4 \times 0.7 \times 215 \times 10^{-3} = 0.602 \text{ kN/mm}$

Member CD

Assume fillet welds to toe and heel of angle.
Line of action of load is on centroid of member; therefore, by moments about heel of angle:

Load on fillet weld to toe

$= 28.5 \times \dfrac{14.5}{50} = 8.3 \text{ kN}$

Effective length of weld required

$= \dfrac{8.3}{0.602} = 14 \text{ mm}$

Length of weld required with end allowance   6.6.5.2

$= 14 + 2 \times 4 = 22 \text{ mm}$

Load on fillet weld to heel

$= 28.5 - 8.3 = 20.2 \text{ kN}$

Total length of weld required

$= \dfrac{20.2}{0.602} + 2 \times 4 = 42 \text{ mm}$

Available lengths of weld at toe and heel are adequate.

		Subject	Chapter ref.
The Steel Construction Institute Silwood Park, Ascot, Berks SL5 7QN		**No.2 PITCHED ROOF TRUSS. WELDED NODE CONNECTION.**	26
		Design code **BS5950**	Made by **BDC** Checked by **GWO**	Sheet no. **3**

Member BC

4 mm fillet welds to toe and heel of angle. OK by inspection.

Member ACE

Shear between inclined members

= 28.5 sin 44° = 19.8 kN

Shear capacity   4.2.3

= $0.6 \, p_y \times (0.9 \, d \, t)$ = $0.6 \times 275 \times 0.9 \times 101.6 \times 5.8 \times 10^{-3}$

= 87.5 kN   >   19.8 kN   OK

Node D

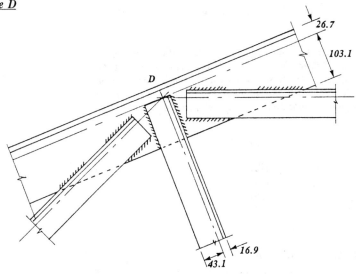

Members CD and DG

From previous calculation for member CD at node C, use 4 mm fillet welds to toe and heel of angle (intermittent fillet welds would be adequate on the heel of the angle).

736   *Worked example*

The Steel Construction Institute Silwood Park, Ascot, Berks SL5 7QN	Subject *No.2 PITCHED ROOF TRUSS.* *WELDED NODE CONNECTION.*		Chapter ref. 26
	Design code *BS5950*	Made by *BDC* Checked by *GWO*	Sheet no. *4*

Member DE

Assume fillet welds to toe and heel of angle.

Load on fillet weld to heel

$= 42.3 \times \dfrac{43.1}{60} = 30.4 \, kN$

Length of 4 mm fillet weld required with end allowance

$= \dfrac{30.4}{0.602} + 2 \times 4 = 58 \, mm <$ *available length at heel OK*

Weld at toe, 4 mm fillet weld. OK by inspection.

By inspection, shear on BDF is not critical, being less than that calculated for ACE.

The single angles are all welded to the same face of the stalk of the tees, for ease of fabrication. This arrangement also has the advantage of reducing the torsional couple due to the eccentricity of the centroids. For example, at node C, if the angles were on opposite faces there would be an out-of-balance couple:

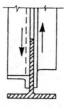

However, with the single sided arrangement, there is a higher out-of-balance moment on plan. In this example the forces are relatively small and the members should be able to accommodate the secondary stresses due to this out-of-balance moment.

The truss is detailed so that the member centroids intersect at the node point; therefore, there are no secondary moments to consider in the plane of the truss.

6.1.2

	Subject		Chapter ref.
The Steel Construction Institute Silwood Park, Ascot, Berks SL5 7QN	No.3 N - GIRDER. WELDED NODE CONNECTION.		26
	Design code BS5950	Made by BDC Checked by GWO	Sheet no. 1

The connection is for a node in the bottom boom of an N - girder truss.

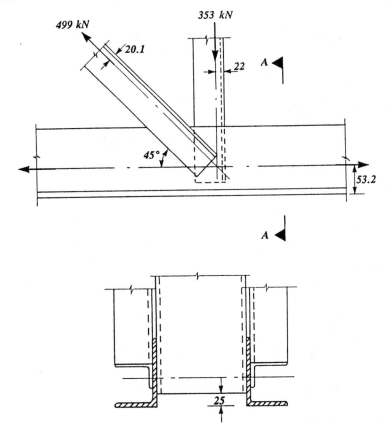

Section A-A

Forces are due to factored loads.

<u>*Member sizes*</u>

Diagonal   2 No 70 × 70 × 8 angles
Vertical   178 × 76 × 20.84 kg/m channel
Bottom boom   2 No 150 × 75 × 10 angles

Worked example

The Steel Construction Institute Silwood Park, Ascot, Berks SL5 7QN	Subject **No.3 N - GIRDER.** **WELDED NODE CONNECTION.**		Chapter ref. **26**	
	Design code **BS5950**	Made by **BDC** Checked by **GWO**	Sheet no.	**2**

Vertical member

Load is applied at centroid of channel.

Load for each weld group $\quad = \quad \dfrac{353}{2} \quad = \quad 176.5\ kN$

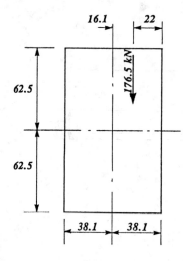

Total length of weld
= 2(125 + 76.2)
= 402.4 mm

$$I_{oo} = 2\left[\frac{125^3}{12} + 76.2 \times 62.5^2 + \frac{76.2^3}{12} + 125 \times 38.1^2\right] \qquad [24.3.2]$$

$$= 1.357 \times 10^6\ mm^3$$

Maximum resultant shear on weld

$$= \left[\left(\frac{R}{2(a+b)} + \frac{R_e \times \frac{a}{2}}{I_{oo}}\right)^2 + \left(\frac{R_e \times \frac{b}{2}}{I_{oo}}\right)^2\right]^{0.5}$$

$$= \left[\left(\frac{176.5}{2(76.2+125)} + \frac{176.5 \times 16.1 \times 38.1}{1.357 \times 10^6}\right)^2 + \left(\frac{176.5 \times 16.1 \times 62.5}{1.357 \times 10^6}\right)^2\right]^{0.5}$$

	Subject	Chapter ref.
The Steel Construction Institute Silwood Park, Ascot, Berks SL5 7QN	**No.3 N - GIRDER.** **WELDED NODE CONNECTION.**	26
	Design code **BS5950**	Made by **BDC** Checked by **GWO**	Sheet no. **3**

$= [(0.439 + 0.080)^2 + (0.131)^2]^{0.5}$

$= 0.535 \text{ kN/mm}$

Use 5 mm fillet welds (as required for diagonal member).

Diagonal member

Load is applied at centroid of angles.

Load for each weld group $= \dfrac{499}{2} = 249.5 \text{ kN}$

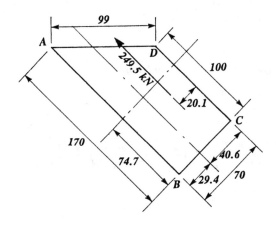

Length of weld $= 170 + 70 + 100 + 99 = 439 \text{ mm}$

By moments about AB

$\bar{x} = \dfrac{70 \times 35 + 100 \times 70 + 99 \times 35}{439} = 29.4 \text{ mm}$

By moments about BC

$\bar{y} = \dfrac{170 \times 85 + 99 \times 135 + 100 \times 50}{439} = 74.7 \text{ mm}$

740 Worked example

The Steel Construction Institute Silwood Park, Ascot, Berks SL5 7QN	Subject No.3 N - GIRDER. WELDED NODE CONNECTION.		Chapter ref. 26
	Design code BS5950	Made by BDC Checked by GWO	Sheet no. 4

$I_{oo} = \dfrac{170^3}{12} + 170 \left[\dfrac{170}{2} - 74.7\right]^2 + 70 \times 74.7^2$

$+ \dfrac{100^3}{12} + 100 \left[74.7 - \dfrac{100}{2}\right]^2 + 99 \times \dfrac{70^2}{12}$

$+ 99 (135 - 74.7)^2 + 170 \times 29.4^2$

$+ \dfrac{70^3}{12} + 70 \left[\dfrac{70}{2} - 29.4\right]^2 + 100 \times 40.6^2$

$+ 99 \times \dfrac{70^2}{12} + 99 \times \left[\dfrac{70}{2} - 29.4\right]^2$

$= 1.749 \times 10^6 \text{ mm}^3$

Eccentricity of load $= 40.6 - 20.1 = 20.5$ mm

Maximum resultant shear on weld

$= \left[\left[\dfrac{249.5}{439} + \dfrac{249.5 \times 20.5 \times 40.6}{1.749 \times 10^6}\right]^2 + \left[\dfrac{249.5 \times 20.5 \times 74.7}{1.749 \times 10^6}\right]^2\right]^{0.5}$

$= [(0.568 + 0.119)^2 + (0.218)^2]^{0.5}$

$= 0.721$ kN/mm

Use 5 mm fillet welds.

Weld all round (angles & channels) with 5 mm fillet welds.

Capacity $= 5 \times 0.7 \times 215 \times 10^{-3} = 0.753$ kN/mm 6.6.5

The Steel Construction Institute Silwood Park, Ascot, Berks SL5 7QN	Subject No.4 UNIVERSAL COLUMN. BOLTED COVER PLATE SPLICE. SPLICE NEAR POINT OF LATERAL RESTRAINT.		Chapter ref. 26	
	Design code BS5950	Made by BDC	Sheet no. 1	
		Checked by GWO		

The splice is between a 203 × 203 × 52 UC upper column and a 203 × 203 × 71 UC lower column. Both sections are grade 43.

The splice is to carry a factored axial load of 653 kN.

The splice is near to a point of lateral restraint and moments due to strut action are not considered.

In this example it is assumed that all of the load is carried by the splice. If the ends of the members were prepared for full contact in bearing, only a nominal splice would be required for erection and continuity.

6.1.7.2
[26.2.1.2]

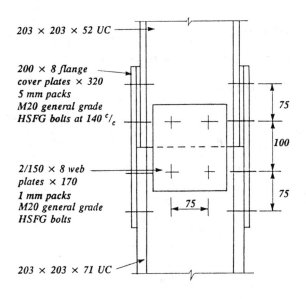

203 × 203 × 52 UC

200 × 8 flange
cover plates × 320
5 mm packs
M20 general grade
HSFG bolts at 140 c/c

2/150 × 8 web
plates × 170
1 mm packs
M20 general grade
HSFG bolts

203 × 203 × 71 UC

Share the load between the web and the flanges in proportion to their areas.

[26.2.1.1]

Area of 203 × 203 × 52 UC   =   66.4 cm^2

Area of web   = $(206.2 - 2 \times 12.5) \times 8.0 \times 10^{-2}$

                        =   14.5 cm^2

742 *Worked example*

The Steel Construction Institute Silwood Park, Ascot, Berks SL5 7QN	Subject *No.4 UNIVERSAL COLUMN. BOLTED COVER PLATE SPLICE. SPLICE NEAR POINT OF LATERAL RESTRAINT.*	Chapter ref. 26
	Design code *BS5950*	Made by *BDC* / Checked by *GWO* / Sheet no. 2

Portion of load carried by web

$$= 653 \times \frac{14.5}{66.4} = 142.6 \text{ kN}$$

Portion of load carried by each flange

$$= \frac{1}{2}(653 - 142.6) = 255.2 \text{ kN}$$

Try M20 general grade HSFG bolts.

<u>*Web splice*</u>

Slip resistance per bolt (in double shear) 6.4.2.1

$$= 2 \times 1.1 \, K_s \, \mu \, P_o = 2 \times 1.1 \times 1.0 \times 0.45 \times 144$$

$$= 142.6 \text{ kN}$$

Bearing resistance per bolt 6.4.2.2

$$= d \, t \, p_{bg} = 20 \times 8.0 \times 825 \times 10^{-3}$$

$$= 132 \text{ kN}$$

Bearing resistance governs.

Number of bolts required $= \dfrac{\text{web load}}{\text{resistance per bolt}}$

$$= \frac{142.6}{132} = 1.08$$

Use 2 No. M20 general grade HSFG bolts for web splice.

Note that end failure check ($P_{bg} \leq \frac{1}{3} e \, t \, p_{bg}$) is not required because bearing acts away from the end. 6.4.2.2

But "edge" distance $\not< 1.4 (20 + 2) = 30.8$ mm 6.2.3
(could be $\not< 1.25 \, D = 27.5$ mm if machine flame cut or sawn). Table 31

The Steel Construction Institute Silwood Park, Ascot, Berks SL5 7QN	Subject No.4 UNIVERSAL COLUMN. BOLTED COVER PLATE SPLICE. SPLICE NEAR POINT OF LATERAL RESTRAINT.	Chapter ref. 26
	Design code BS5950	Made by BDC Checked by GWO	Sheet no. 3

Flange splices

Slip resistance per bolt (in single shear)   6.4.2.1

    = $1.1 \times 1.0 \times 0.45 \times 144$ =   71.3 kN

Bearing resistance per bolt on flange   6.4.2.2

    = $20 \times 12.5 \times 825 \times 10^{-3}$

    =   206.3 kN

Slip resistance governs.

Number of bolts required   = $\dfrac{255.2}{71.3}$ =   3.6

Use 4 No. M20 general grade HSFG bolts for each flange splice.

Edge distance $\leq 1.25 D$ = $1.25 (20 + 2)$ =   27.5 mm   Table 31

Flange splice plates

Use 200 mm wide cover plates, bolts at 140 mm cross centres and 75 mm pitch (100 mm pitch across joint).

Minimum thickness of splice plate

    = $\dfrac{255.2 \times 10^{3}}{275(200 - 140 + 2 \times 75 \tan 30°)}$

    =   6.3 mm

(The effective width of plate is obtained by assuming a maximum dispersion angle of 30°. No deduction is made for holes since the plates are in compression.)   [25.4.2] 4.7.4

The Steel Construction Institute Silwood Park, Ascot, Berks SL5 7QN	Subject **No.4 UNIVERSAL COLUMN. BOLTED COVER PLATE SPLICE. SPLICE NEAR POINT OF LATERAL RESTRAINT.**	Chapter ref. **26**
	Design code **BS5950**	Made by **BDC** Checked by **GWO**	Sheet no. **4**

Also thickness $\not< \dfrac{pitch}{14} = \dfrac{100}{14} = 7.1$ mm   6.2.2

Use 8 mm thick × 200 mm flange cover plates.

(Bearing resistance on bolt OK by inspection)

Web splice plates

Use 150 mm × 8 mm web cover plates.

OK by inspection

Packs will be required to make up the difference in size between the upper and lower column sections. The faying surfaces of the packs must provide a slip factor equal to or greater than that used in the design.

	Subject **No.5 UNIVERSAL COLUMN. BOLTED COVER PLATE SPLICE. SPLICE REMOTE FROM POINT OF LATERAL RESTRAINT.**	Chapter ref. 26
The Steel Construction Institute Silwood Park, Ascot, Berks SL5 7QN	Design code **BS5950**	Made by **BDC** / Sheet no. **1** Checked by **GWO**

The splice is between a 203 × 203 × 52 UC upper column and a 203 × 203 × 71 UC lower column. Both sections are grade 43.

The splice is to carry a factored axial load of 653 kN.

The splice is assumed to be mid-way between points of lateral restraint and the moments due to strut action are considered in the design. (Effective length of strut = 6 m.)

As in the previous example all the load is assumed to be carried by the splice.

6.1.7.2

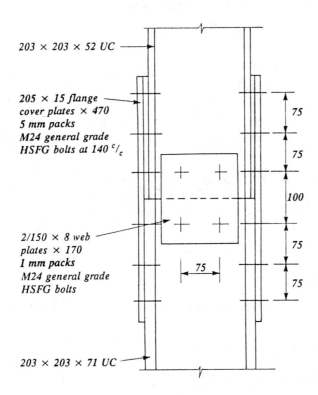

203 × 203 × 52 UC

205 × 15 flange cover plates × 470
5 mm packs
M24 general grade
HSFG bolts at 140 $^c/_c$

2/150 × 8 web plates × 170
1 mm packs
M24 general grade
HSFG bolts

75
75
100
75
75
75

203 × 203 × 71 UC

746 Worked example

The Steel Construction Institute Silwood Park, Ascot, Berks SL5 7QN	Subject No.5 UNIVERSAL COLUMN. BOLTED COVER PLATE SPLICE. SPLICE REMOTE FROM POINT OF LATERAL RESTRAINT.	Chapter ref. 26
	Design code BS5950	Made by BDC Checked by GWO	Sheet no. 2

Moment at splice due to strut action

Appendix C
[26.2.1.1]
C.2

$$\lambda = \frac{600}{5.16} = 116.3$$

$$\lambda_o = 0.2 \left[\frac{\pi^2 E}{p_y} \right]^{0.5} = 0.2 \left[\frac{\pi^2 \times 205 \times 10^3}{275} \right]^{0.5} = 17.2$$

Use strut table 27(c) $a = 5.5$ Table 25

$\eta = 0.001\, a\, (\lambda - \lambda_o) = 0.001 \times 5.5\, (116.3 - 17.2)$ C.2

$ = 0.545$

$$p_E = \frac{\pi^2 E}{\lambda^2} = \frac{\pi^2 \times 205 \times 10^3}{116.3^2} = 149.6\ N/mm^2 \quad\quad C.1$$

$$f_c = \frac{653 \times 10^3}{66.4 \times 10^2} = 98.3\ N/mm^2 \quad\quad C.3$$

$$M_{max} = \frac{\eta f_c S}{\left[1 - \dfrac{f_c}{p_E}\right]} = \frac{0.545 \times 98.3 \times 264 \times 10^{-3}}{\left[1 - \dfrac{98.3}{149.6}\right]} \quad\quad C.3$$

$\phantom{M_{max}} = 41.2\ kNm$

Distribution of axial load

Share the axial load between the web and the flanges in proportion to their areas.   [26.2.1.1]

Area of $203 \times 203 \times 52$ UC $= 66.4\ cm^2$

Area of web $= (206.2 - 2 \times 12.5) \times 8.0 \times 10^{-2}$

$ = 14.5\ cm^2$

The Steel Construction Institute Silwood Park, Ascot, Berks SL5 7QN	Subject **No.5 UNIVERSAL COLUMN. BOLTED COVER PLATE SPLICE. SPLICE REMOTE FROM POINT OF LATERAL RESTRAINT.**		Chapter ref. **26**	
	Design code *BS5950*	Made by **BDC** Checked by **GWO**	Sheet no.	**3**

Portion of load carried by web

$$= 653 \times \frac{14.5}{66.4} = 142.6 \text{ kN}$$

Portion of load carried by each flange

$$= \tfrac{1}{2}(653 - 142.6) = 255.2 \text{ kN}$$

Try M24 general grade HSFG bolts.

<u>*Flange splices*</u>

Slip resistance per bolt (in single shear) 6.4.2.1

$$= 1.1 \, k_s \, \mu \, P_o = 1.1 \times 1.0 \times 0.45 \times 207$$

$$= 102.5 \text{ kN}$$

Bearing resistance per bolt 6.4.2.2

$$= d \, t \, p_{bg} = 24 \times 12.5 \times 825 \times 10^{-3}$$

$$= 247.5 \text{ kN}$$

Slip resistance governs.

Assume 2 lines of bolts at 140 mm cross centres in each flange.

Axial load per line of bolts $= \dfrac{255.2}{2}$

$$= 127.6 \text{ kN}$$

Load per line of bolts due to strut action moment [26.2.1.1]

$$= \frac{41.2 \times 10^3}{2 \times 140} = 147.1 \text{ kN}$$

748   *Worked example*

The Steel Construction Institute Silwood Park, Ascot, Berks SL5 7QN	Subject *No.5 UNIVERSAL COLUMN. BOLTED COVER PLATE SPLICE. SPLICE REMOTE FROM POINT OF LATERAL RESTRAINT.*	Chapter ref. *26*
	Design code *BS5950*	Made by *BDC* Sheet no. *4* Checked by *GWO*

Total load per line of bolts $\quad = \quad 127.6 + 147.1$

$\qquad\qquad\qquad\qquad\qquad\quad = \quad 274.7$ *kN*

Number of bolts required in each line

$\qquad = \quad \dfrac{274.7}{102.5} \quad = \quad 2.7$

Use 3 bolts in each line, 2 lines of 3 M24 general grade HSFG bolts for each flange splice.

Use 205 × 15 flange cover plates.

Note that the size of the flange splice plates should not be less than that of the column flange in respect to area and modulus.   *[26.2.1.2]*

<u>*Web splice*</u>

By inspection use 2 / M24 general grade HSFG bolts.

Packs will be required to make up the difference in size between the upper and lower column sections. The faying surfaces of the packs must provide a slip factor equal to or greater than that used in the design.

	Subject	Chapter ref.
The Steel Construction Institute Silwood Park, Ascot, Berks SL5 7QN	**No.6 BEAM TO COLUMN CONNECTION. BOLTED WEB CLEATS.**	26
	Design code BS5950	Made by BDC Checked by GWO	Sheet no. 1

The connection is between a 356 × 171 × 45 UB (Grade 43) beam and a 254 × 254 × 73 UC (Grade 43) column.

The end reaction of the simply supported beam due to factored loads is 185 kN

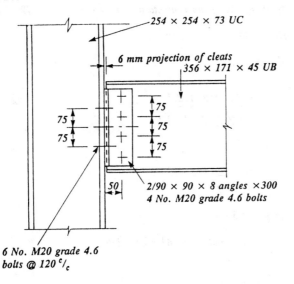

Try 2/90 × 90 × 8 angle cleats × 300 mm and M20 Grade 4.6 bolts.

Connection of cleats to web of beam

Shear capacity of bolt in double shear   6.3.2

$$= 2 p_s A_t = 2 \times 160 \times 245 \times 10^{-3} = 78.4 \text{ kN}$$

Bearing capacity of bolt on web of beam   6.3.3

$$= p_{bb} \, d \, t = 460 \times 20 \times 6.9 \times 10^{-3} = 63.5 \text{ kN}$$ Table 32 (Amd. 1)

(bearing capacity on cleats is greater)

The Steel Construction Institute Silwood Park, Ascot, Berks SL5 7QN	Subject No.6 BEAM TO COLUMN CONNECTION. BOLTED WEB CLEATS.	Chapter ref. 26
	Design code BS5950	Made by BDC Checked by GWO	Sheet no. 2

Try 4 bolts at 75 mm vertical pitch and 50 mm from heel of cleats.

Vertical shear force per bolt $= \dfrac{185}{4} = 46.3 \text{ kN}$

Moment due to eccentricity of reaction from centroid of bolt group [26.2.2.3]

$M_e = P e = 185 \times 50 = 9250 \text{ Nm}$

Horizontal shear due to eccentricity

$= \dfrac{M_e \, r_i}{\Sigma \, (r_i^2)} = \dfrac{9250 \times 112.5}{2(37.5^2 + 112.5^2)}$ [23.4.2]

$= 37.0 \text{ kN}$

Resultant shear force $= \sqrt{(46.3)^2 + (37.0)^2}$

$= 59.3 \text{ kN}$

$<$ bolt capacity $(= 63.5 \text{ kN})$ OK

Connection of cleats to column flange

Shear capacity of bolt in single shear   6.3.2

$= p_s A_t = 160 \times 245 \times 10^{-3} = 39.2 \text{ kN}$

Bearing capacity of bolt on cleat   6.3.3

$= p_{bb} \, d \, t = 460 \times 20 \times 8 \times 10^{-3} = 73.6 \text{ kN}$ Table 32 (Amd. 1)

Try 6 bolts in 2 columns at 75 mm vertical pitch and 120 mm cross centres.

Vertical shear force per bolt $= \dfrac{185}{6} = 30.8 \text{ kN}$

	Subject	Chapter ref.
The Steel Construction Institute Silwood Park, Ascot, Berks SL5 7QN	No.6 BEAM TO COLUMN CONNECTION. BOLTED WEB CLEATS.	26
Design code BS5950	Made by BDC Checked by GWO	Sheet no. 3

Considering one cleat, horizontal shear force on bottom bolt due to eccentricity (assuming a centre of pressure 25 mm below the top of cleat) [26.2.2.3]

$$= \frac{P}{2} \times e \times \frac{r_i}{\Sigma (r_i^2)}$$

$$= \frac{185}{2} \times \frac{1}{2} \times (120 - 6.9) \times \frac{200}{(50^2 + 125^2 + 200^2)}$$

$$= 18.0 \text{ kN}$$

Resultant shear force $= \sqrt{(30.8)^2 + (18.0)^2}$

$= 35.7 \text{ kN}$

$<$ bolt capacity $(= 39.2 \text{ kN})$ OK

Use 2 / 90 × 90 × 8 angles × 300 long

4 No. M20 Grade 4.6 bolts

In the design of the bolts connecting the cleats to the column flange, the horizontal shear due to eccentricity has been included so that the "load path" check is complete. In traditional practice, the eccentricity effect is included in the design of bolts in the beam web, but (for a pair of cleats) is ignored in the design of the bolts in the column flange. Where there are 4 or more rows of bolts, this is reasonable since the effect of eccentricity is relatively small. For 3 rows and less it is probably advisable to check that the "load path" is complete. In this example, where grade 4.6 bolts are used, the horizontal shear could be ignored since the cleat can carry the 18.0 kN force in bending. [26.1.2]

Note that if a similar detail is used for a beam to beam connection (and the top flange is notched) the beam web should be checked for "block shear" failure. [25.4.3]

Worked example

The Steel Construction Institute Silwood Park, Ascot, Berks SL5 7QN	Subject **No.7 PITCHED PORTAL FRAME. EXTENDED END PLATE EAVES CONNECTION.**		Chapter ref. **26**
	Design code **BS5950**	Made by **BDC** Checked by **GWO**	Sheet no. **1**

The connection is between a haunched
406 × 178 × 60 UB rafter (Grade 50)
457 × 191 × 74 UB column (Grade 43)

The factored moment and forces at the end of the rafter are:

Bending moment	427 kNm
Axial force	75 kN (compression)
Shear	136 kN

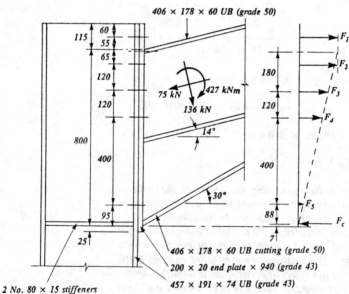

[26.2.3.2]

2 No. 80 × 15 stiffeners
M20 general grade HSFG bolts at 80 c/c (used as ordinary bolts)

[26.1.2]

Distribution of applied loading

Assuming bolt loads as in the sketch.   [26.1.3]

By moments about bottom flange

$$427 \times 10^3 - (75 \cos 14° - 136 \sin 14°) \times \left[800 - \frac{406.4}{2 \cos 14°} - \frac{12.8}{2 \cos 30°} \right]$$

The Steel Construction Institute Silwood Park, Ascot, Berks SL5 7QN	Subject **No.7 PITCHED PORTAL FRAME. EXTENDED END PLATE EAVES CONNECTION.**	Chapter ref. **26**
	Design code **BS5950**	Made by **BDC** / Sheet no. **2** Checked by **GWO**

$$= (2F_1 + 2F_2) \times 788 + 2F_3 \times 608 + 2F_4 \times 488 + 2F_5 \times 88$$

$$= \frac{2F_1}{788}[2 \times 788^2 + 608^2 + 488^2 + 88^2]$$

$F_1 = F_2 = 85.6 \text{ kN}$

$ F_3 = 66.1 \text{ kN}$

$ F_4 = 53.0 \text{ kN}$

$ F_5 = 9.6 \text{ kN}$

Reaction at bottom flange (of haunch)

$= F_c = 2(85.6 + 85.6 + 66.1 + 53.0 + 9.6)$
$ + 75 \cos 14° - 136 \sin 14°$

$ = 639.7 \text{ kN}$

Column flange

Check capacity of flange in bending to resist bolt loads [25.3.5]

Distance from centreline of bolt to centreline of root fillet

$= b = \dfrac{80}{2} - \dfrac{9.1}{2} - \dfrac{10.2}{2} = 30.4 \text{ mm}$

Distance from centreline of bolt to prying force

$= n =$ edge distance but $\not> 2T$ [25.3.1]

$ = 2 \times 14.5 = 29.0 \text{ mm}$

Effective length of flange per bolt

$= w = 2(b + d) = 2(30.4 + 20) = 100.8 \text{ mm}$ [25.3.2]

(assuming M20 bolts)

	Subject	Chapter ref.
The Steel Construction Institute Silwood Park, Ascot, Berks SL5 7QN	No.7 PITCHED PORTAL FRAME. EXTENDED END PLATE EAVES CONNECTION.	26
	Design code BS5950	Made by BDC / Checked by GWO	Sheet no. 3

$w > b$, therefore bolt holes can be ignored. [25.3.3]

A maximum spread of $(b + d)$ is used for checking the flange capacity so that a plastic hinge can be assumed at the bolt line for the design of the end plate. [25.3.2] [26.1.5]

Consider flange for maximum bolt load (F_1) with equal moments at bolt line and root:

$$M = \frac{F_1 \, b}{2} = \frac{85.6 \times 30.4}{2} = 1301.1 \; Nm$$

Moment capacity $= p_y \times \dfrac{w \, T^2}{4}$

$$= \frac{275 \times 100.8 \times 14.5^2 \times 10^{-3}}{4}$$

$= 1457 \; Nm > M$ OK

Prying force $= \dfrac{M}{n} = \dfrac{1301.1}{29.0} = 44.9 \; kN$ [25.3.1]

Bolts

Tension capacity of M20 general grade HSFG bolt, taken as $0.64 \; U_f$ [25.3.1]

$= 0.64 \times 827 \times 245 \times 10^{-3}$

$= 129.7 \; kN$

Bolt load

$= F_1 + prying \; force$

$= 85.6 + 44.9 \; = \; 130.5 \; kN > $ bolt capacity

The Steel Construction Institute Silwood Park, Ascot, Berks SL5 7QN	Subject **No.7 PITCHED PORTAL FRAME. EXTENDED END PLATE EAVES CONNECTION.**	Chapter ref. **26**
	Design code **BS5950**	Made by **BDC** Sheet no. **4** Checked by **GWO**

Check with reduced prying force.

Maximum allowable prying force (Q_1)

$$= 129.7 - 85.6 = 44.1 \text{ kN}$$

'Serviceability check' for minimum prying force

$$Q = \frac{b}{2n}\left[F - \frac{\beta \gamma p_o w T^4}{27 n b^2}\right] \qquad [25.3.1]$$

$$= \frac{30.4}{2 \times 29.0}\left[85.6 - \frac{2 \times 1.5 \times 0.587 \times 100.8 \times 14.5^4}{27 \times 29 \times 30.4^2}\right]$$

$$= 0.524\,[\,85.6 - 10.8\,] = 39.2 \text{ kN} < 44.1 \text{ kN OK}$$

Moment at root of flange, with reduced prying force

$$= F_1 b - Q_1 n$$
$$= 85.6 \times 30.4 - 44.1 \times 29.0$$
$$= 1323 \text{ Nm} < \text{moment capacity}$$

M20 bolts are OK

Use M20 general grade HSFG bolts (as ordinary bolts).

By inspection the bolt group is also satisfactory for shear load.

General grade HSFG bolts are used as 'ordinary' bolts (i.e. they are not pre-tensioned to the proof load). They are 'black' bolts (defined by a British Standard) which are suitable for use in tension. Although it is not assumed in the design that they are pre-tensioned in accordance with BS 4604: Part 1, the bolts in the connection should be uniformly tightened to a reasonably high torque to provide the detail with adequate stiffness.

Worked example

The Steel Construction Institute Silwood Park, Ascot, Berks SL5 7QN	Subject No.7 PITCHED PORTAL FRAME. EXTENDED END PLATE EAVES CONNECTION.		Chapter ref. 26
	Design code BS5950	Made by BDC Checked by GWO	Sheet no. 5

If the column flange (or the bolts) had not been adequate, a further check could have been made using a greater effective length (in this example, $w = 120$ mm) based on the least of the end distance, half the bolt pitch and $(a + 1.4 b)$ each side of the bolt. A thicker end plate would then be required based on a zero bending moment at the bolt line.

<u>Column web</u> (at upper bolts)

Tension capacity of 120 mm length of web

$= \quad 275 \times 120 \times 9.1 \times 10^{-3}$ 4.6.1

$= \quad 300.3 \text{ kN} > 2F_1$ OK

<u>Column web</u> (at bottom flange of haunch)   [26.2.3.3]

Assuming 20 mm thick end plate, local bearing capacity of web for reaction $(F_c) = (b_1 + n_2) \, t \, p_{yw}$ [25.1]
4.5.3

$= \quad (12.8 + 2 \times 20 + 2 \times 2.5 \times (14.5 + 10.2)) \times 9.1 \times 275 \times 10^{-3}$ 4.5.1.3

$= \quad 441.2 \text{ kN} < F_c$ stiffeners required

<u>Stiffeners</u>   [25.2]

Area required in contact with the flange

$= \quad \dfrac{0.8 \, F_c}{p_{ys}} = \dfrac{0.8 \times 639.7 \times 10^3}{275}$ 4.5.4.2

$= \quad 1861 \text{ mm}^2$

Try 2 No. 80×15 stiffeners (with 15 mm snipes).

	Subject	Chapter ref.
The Steel Construction Institute Silwood Park, Ascot, Berks SL5 7ON	No.7 PITCHED PORTAL FRAME. EXTENDED END PLATE EAVES CONNECTION.	26
	Design code BS5950	Made by **BDC** Checked by **GWO**	Sheet no. **6**

Area in contact $= 2(80 - 15) \times 15$

$= 1950 \text{ mm}^2$ OK

Use 2 No. 80 × 15 stiffeners.

Buckling resistance   4.5.1.5

Area of effective section

$= 2 \times 80 \times 15 + 2 \times 20 \times 9.1^2 = 5712 \text{ mm}^2$

$I = \dfrac{15 \times 169.1^3}{12} = 6.04 \times 10^6 \text{ mm}^4$

$r = \left[\dfrac{6.04 \times 10^6}{5712}\right]^{0.5} = 32.5 \text{ mm}$

$\dfrac{L}{r} = \dfrac{0.7 \times (457.2 - 2 \times 14.5)}{32.5} = 9.2$

$p_c = 275 \text{ N/mm}^2$ Table 27(c)

Buckling resistance $= 5712 \times 275 \times 10^{-3}$

$= 1571 \text{ kN} > F_c$ OK

Connection of stiffeners to column web

Effective length of weld

$= 457.2 - 2 \times (14.5 + 15 + 6)$ 6.6.5.2

$= 386.2 \text{ mm}$

Load on welds $= \dfrac{639.7}{4 \times 386.2} = 0.414 \text{ kN/mm}$

Weld stiffeners to column with 6 mm fillet welds.

The Steel Construction Institute Silwood Park, Ascot, Berks SL5 7QN	Subject No.7 PITCHED PORTAL FRAME. EXTENDED END PLATE EAVES CONNECTION.	Chapter ref. 26
Design code BS5950	Made by **BDC** Checked by **GWO**	Sheet no. 7

<u>Shear in column web</u> [26.2.3.3]

Maximum shear in column web (immediately above bottom flange of haunch)

$= 2(85.6 + 85.6 + 66.1 + 53.0 + 9.6)$

$= 599.8 \text{ kN}$

Shear capacity $= 0.6 \, p_y \, t \, d$ 4.2.3

$= 0.6 \times 275 \times 9.1 \times 457.2 \times 10^{-3}$

$= 686.5 \text{ kN} > 599.8 \text{ kN} \quad OK$

<u>End plate</u>

By inspection, design will be governed by extended portion above the top flange. [25.3.5]

Try 200 mm wide plate × 20 mm thick.

(Assume 8 mm fillet welds to flange)

Distance from centreline of bolt to toe of fillet weld

$= b = 55 - 8 = 47 \text{ mm}$

Distance from centreline of bolt to prying force

$= n =$ edge distance but $\not> 2T$ [25.3.1]

$= 2 \times 20 = 40 \text{ mm}$

Effective length of end plate per bolt [25.3.2]

$= w = ½ \times 200 = 100 \text{ mm}$ [25.3.3]

(In this case ½ × flange width governs 'w')

The Steel Construction Institute Silwood Park, Ascot, Berks SL5 7QN	Subject *No.7 PITCHED PORTAL FRAME. EXTENDED END PLATE EAVES CONNECTION.*	Chapter ref. 26
	Design code BS5950	Made by BDC Sheet no. 8 Checked by GWO

With equal moments at bolt line and root

$$M = \frac{F_1 \, b}{2} = \frac{85.6 \times 47}{2} = 2012 \text{ Nm}$$

Moment capacity $= p_y \times \dfrac{w \, T^2}{4}$

Minimum thickness $= \left[\dfrac{4 \times 2012 \times 10^3}{265 \times 100} \right]^{0.5}$

$= 17.4$ mm   Use 20 mm

Prying force $= \dfrac{M}{n} = \dfrac{2012}{40} = 50.3$ kN   [25.3.1]

As with the check on the column flange,
applied load + prying force > bolt capacity

maximum allowable prying force (Q_1)

$= 129.7 - 85.6 = 44.1$ kN

Serviceability check for minimum (Q)

$$Q = \frac{47.0}{2 \times 40.0} \left[85.6 - \frac{2 \times 1.5 \times 0.587 \times 100 \times 20^4}{27 \times 40.0 \times 47.0^2} \right] \quad [25.3.1]$$

$= 0.588 \, [\, 85.6 - 11.8 \,] = 43.4$ kN < 44.1 kN   OK

Check moment at root with reduced prying force

$= F_1 \, b - Q_1 \, n$

$= 85.6 \times 47 - 44.1 \times 40$

$= 2259$ Nm

Worked example

The Steel Construction Institute Silwood Park, Ascot, Berks SL5 7QN	Subject **No.7 PITCHED PORTAL FRAME. EXTENDED END PLATE EAVES CONNECTION.**	Chapter ref. **26**
	Design code **BS5950**	Made by **BDC** / Checked by **GWO** — Sheet no. **9**

Moment capacity of end plate

$$= \frac{265 \times 100 \times 20^2 \times 10^{-3}}{4}$$

$= 2650 \; Nm >$ moment at root   OK

Use 200×20 end plate $\times 940$ Grade 43.

Beam flanges

Capacity of flange $\quad = \quad p_y \, A$

$$= 355 \times 177.8 \times 12.8 \times 10^{-3}$$

$$= 807.9 \; kN$$

Tension in top flange $= \dfrac{2(F_1 + F_2)}{\cos 14°}$

$$= \frac{4 \times 85.6}{\cos 14°}$$

$$= 352.9 \; kN < 807.9 \; kN \quad OK$$

Compression in bottom (haunch) flange

$$= \frac{F_c}{\cos 30°}$$

$$= \frac{639.7}{\cos 30°}$$

$$= 738.7 \; kN < 807.9 \; kN \quad OK$$

Welds

Top flange to end plate

Load from bolts above flange $\quad = \quad 2F_1 \quad = \quad 2 \times 85.6 \quad = \quad 171.2 \; kN$

The Steel Construction Institute Silwood Park, Ascot, Berks SL5 7QN	Subject No.7 PITCHED PORTAL FRAME. EXTENDED END PLATE EAVES CONNECTION.		Chapter ref. 26	
	Design code BS5950	Made by BDC Checked by GWO	Sheet no.	10

Load on weld $= \dfrac{171.2}{177.8} = 0.963$ kN/mm

Capacity of 8 mm fillet weld 6.6.5

$= 8 \times 0.7 \times 215 \times 10^{-3} = 1.20$ kN/mm OK

∴ weld top flange to end plate with 8 mm fillet welds.

<u>Web to end plate</u>

Bolt below top flange, $F_2 = 85.6$ kN

Centreline of bolt to toe of flange weld

$= b_x = 65 - 12.8 - 8 = 44.2$ mm

Centreline of bolt to toe of web weld

$= b_y = \dfrac{80}{2} - \dfrac{7.8}{2} - 6 = 30.1$ mm

Effective length of plate at flange

$= w_x = \dfrac{200 - 80}{2} + 0.7 \times 30.1 = 81.1$ mm

Effective length of plate at web

$= w_g = \dfrac{120}{2} + 0.7 \times 44.2 = 90.9$ m

Bolt load supported by web; plastic distribution of load [25.3.4]
(proportional to $\dfrac{w}{b}$)

$= \left[\dfrac{\dfrac{90.9}{30.1}}{\dfrac{81.1}{44.2} + \dfrac{90.9}{30.1}} \right] \times 85.6 = 53.2$ kN

762   Worked example

The Steel Construction Institute Silwood Park, Ascot, Berks SL5 7QN	Subject **No.7 PITCHED PORTAL FRAME. EXTENDED END PLATE EAVES CONNECTION.**	Chapter ref. **26**
	Design code **BS5950**	Made by **BDC** / Sheet no. **11** Checked by **GWO**

Bolt load supported by web; elastic distribution of load (proportional to $\dfrac{w}{b^3}$)

$$= \left[\dfrac{\dfrac{90.9}{30.1^3}}{\dfrac{81.1}{44.2^3} + \dfrac{90.9}{30.1^3}} \right] \times 85.6 \quad = \quad 66.8 \text{ kN}$$

Take higher load (66.8 kN) for design of weld.

Take equal length of weld either side of bolt line, to avoid eccentricity calculations.

Load on weld $= \dfrac{66.8}{2 \times 44.2} = 0.756$ kN/mm

Use 6 mm fillet welds.

Weld web (and middle flange) to end plate with 6 mm fillet welds.

<u>Bottom (haunch) flange to end plate</u>

Horizontal component of flange force

$= F_c = 639.7$ kN

Vertical component of flange force

$= F_c \tan 30° = 639.7 \tan 30° = 369.3$ kN

Fit end plate to flange and take horizontal component in direct compression and vertical component in shear on fillet welds

load on weld $= \dfrac{369.3}{2 \times 177.8} = 1.04$ kN/mm

Weld bottom (haunch) flange to end plate with 8 mm fillet welds on inside and 10 mm on outside to allow for 120° angle.

6.6.5.3
[24.2.1]

The Steel Construction Institute		Subject	Chapter ref.
Silwood Park, Ascot, Berks SL5 7QN		No.7 PITCHED PORTAL FRAME. EXTENDED END PLATE EAVES CONNECTION.	26
		Design code **BS5950**	Made by **BDC** / Checked by **GWO** / Sheet no. **12**

Web of rafter

Load on web (at second row of bolts taking plastic distribution of load)

$\quad = \quad 2 \times 53.2 \quad = \quad 106.4 \ kN$

Taking effective web as 'b_x' each side of bolt line, capacity of web at second row of bolts $\quad = \quad p_y \ 2 \ b_x \ t$

$\quad = \quad 355 \times 2 \times 44.2 \times 7.8 \times 10^{-3}$

$\quad = \quad 244.8 \ kN$

$\quad > \quad 106.4 \ kN \quad OK$

The strengths of the other parts of the detail also have to be considered. For example, the shear capacity of the end plate to rafter weld and the adequacy of the column flange/end plate/welds/beam web to carry the loads in the third row of bolts. In this example, these are seen to be adequate by inspection (by comparison with the calculations for other parts of the detail) and no further calculations are required.

	Subject	Chapter ref.
The Steel Construction Institute Silwood Park, Ascot, Berks SL5 7QN	No.8 PITCHED PORTAL FRAME. APEX CONNECTION.	26
	Design code BS5950	Made by BDC Checked by GWO	Sheet no. 1

The bolted apex connection is for a 406 × 178 × 60 UB rafter (Grade 50).

The factored moment and axial force in the rafter are:

Bending moment   348 kNm
Axial force   57 kN

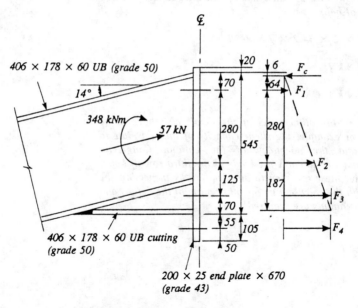

M24 general grade bolts at 100 c/c
(used as 'ordinary' bolts)

Distribution of applied loading   [26.1.2]

By moments about top flange   [26.1.3]

$$348 \times 10^3 - 57 \left[\frac{406.4}{2} - \frac{12.8}{2} \right] = 2F_1 \times 64 + 2F_2 \times 344 + (2F_3 + 2F_4) \times 531$$

	Subject	Chapter ref.
The Steel Construction Institute Silwood Park, Ascot, Berks SL5 7QN	No.8 PITCHED PORTAL FRAME. APEX CONNECTION.	26
	Design code BS5950	Made by BDC Checked by GWO	Sheet no. 2

$$= \frac{2F_4}{531} [2 \times 531^2 + 344^2 + 64^2]$$

$$F_1 = 15.7 \text{ kN}$$

$$F_2 = 84.4 \text{ kN}$$

$$F_3 = F_4 = 130.3 \text{ kN}$$

Reaction at top flange

$$= F_c = 2 (15.7 + 84.4 + 130.3 + 130.3) + 57 \cos 14°$$

$$= 776.7 \text{ kN}$$

Beam flanges

Capacity of flange $= p_y A$

$$= 355 \times 177.8 \times 12.8 \times 10^{-3}$$

$$= 807.9 \text{ kN}$$

Compression in top flange

$$= \frac{F_c}{\cos 14°} = \frac{776.7}{\cos 14°} = 800.5 \text{ kN} < 807.9 \text{ kN} \quad OK$$

Tension in bottom (haunch) flange $< 2 (F_3 + F_4)$

$$= 4 \times 130.3 = 521.2 \text{ kN} < 807.9 \text{ kN} \quad OK$$

(In practice part of F_3 will be carried by the web and the middle flange.)

The Steel Construction Institute Silwood Park, Ascot, Berks SL5 7QN	Subject No.8 PITCHED PORTAL FRAME. APEX CONNECTION.		Chapter ref. 26
	Design code BS5950	Made by BDC Checked by GWO	Sheet no. 3

End plate [25.3.5]

By inspection, design will be governed by extended portion at bottom of detail.

Try 200 mm wide plate.

(Assume 10 mm fillet welds to bottom flange)

Distance from centreline of bolt to toe of fillet weld

$= b = 55 - 10 = 45$ mm

Effective length of end plate per bolt [25.3.2]

$= w = \frac{1}{2} \times 200 = 100$ mm [25.3.3]

(In this case ½ × flange width governs 'w')

With equal moments at bolt line and root

$M = \dfrac{F_4 \, b}{2} = \dfrac{130.3 \times 45}{2} = 2932$ Nm

Moment capacity $= p_y \times \dfrac{w \, T^2}{4}$

Minimum thickness $= \left[\dfrac{4 \times 2932 \times 10^3}{265 \times 100}\right]^{0.5}$

$= 21.0$ mm

Use 200 × 25 end plate × 670 Grade 43.

Take advantage of 'excess' plate thickness to reduce prying force (and bolt size).

Moment capacity $= 265 \times \dfrac{100 \times 25^2}{4 \times 10^3}$

$= 4141$ Nm

The Steel Construction Institute Silwood Park, Ascot, Berks SL5 7QN	Subject **No.8 PITCHED PORTAL FRAME. APEX CONNECTION.**	Chapter ref. **26**
	Design code **BS5950**	Made by **BDC** Sheet no. **4** Checked by **GWO**

Moment required at bolt line

$= M_b = F_4 b - M_{root} = 130.3 \times 45 - 4141 = 1723 \text{ Nm}$

Distance from centreline of bolt to prying force   [25.3.1]

$= n =$ edge distance but $\ngtr 2T$

$= 50 \text{ mm}$

Prying force $= \dfrac{M_b}{n} = \dfrac{1723}{50} = 34.5 \text{ kN}$

'Serviceability' check for minimum prying force

[25.3.1]

$$Q = \dfrac{b}{2n}\left[F - \dfrac{\beta \gamma p_o w T^4}{27 n b^2}\right]$$

$$= \dfrac{45}{2 \times 50}\left[130.3 - \dfrac{2 \times 1.5 \times 0.587 \times 100 \times 25^4}{27 \times 50 \times 45^2}\right]$$

$= 0.45 [130.3 - 25.2] = 47.3 \text{ kN} > 34.5 \text{ kN}$

Design bolt for $(F + Q) = 130.3 + 47.3$

$= 177.6 \text{ kN}$

Bolts

Tension capacity of M24 general grade HSFG bolt, taken as 0.64 U_f   [25.3.1]

$= 0.64 \times 827 \times 353 \times 10^{-3}$

$= 186.8 \text{ kN} > 177.6 \text{ kN} \quad OK$

Use M24 general grade HSFG bolts (as ordinary bolts).

768   Worked example

The Steel Construction Institute Silwood Park, Ascot, Berks SL5 7QN	Subject *No.8 PITCHED PORTAL FRAME. APEX CONNECTION.*	Chapter ref 26
	Design code *BS5950*	Made by **BDC** / Checked by **GWO** — Sheet no. **5**

Welds

Bottom (haunch) flange to end plate

Load from bolts below flange = $2F_4$

= 2×130.3 = 260.6 kN

Load on weld = $\dfrac{260.6}{177.8}$ = 1.466 kN/mm

Capacity of 10 mm fillet weld

= $10 \times 0.7 \times 215 \times 10^{-3}$ = 1.505 kN/mm   OK

Weld bottom (haunch) flange to end plate with 10 mm fillet welds.

Distribution of load (F_3) from bolt between lower flange and haunch

(Assume 6 mm fillet welds, except for 10 mm fillet weld to bottom haunch flange.)

Centreline of bolt to toe of haunch flange weld

= b_x = 57 − 10 = 47 mm

Centreline of bolt to toe of web weld

= b_y = $\dfrac{100}{2} - \dfrac{7.8}{2} - 6$ = 40 mm

Centreline of bolt to toe of weld to 'middle' flange

= b_z = 56 − 6 = 50 mm

Effective length of plate at haunch flange

= w_x = $\dfrac{200 - 100}{2} + 0.7 \times 40$ = 78 mm

The Steel Construction Institute Silwood Park, Ascot, Berks SL5 7QN	Subject **No.8 PITCHED PORTAL FRAME. APEX CONNECTION.**	Chapter ref. **26**
	Design code **BS5950**	Made by **BDC** Checked by **GWO**	Sheet no. **6**

Effective length of plate at web

$= w_y = 0.7 \times 47 + 0.7 \times 50 = 68 \text{ mm}$

Effective length of plate at 'middle' flange

$= w_z = w_x = 78 \text{ mm}$

(a) *Plastic distribution of load (proportional to $\frac{w}{b}$)*   [25.3.4]

    Load supported by haunch flange

$$= \left[\frac{\frac{78}{47}}{\frac{78}{47} + \frac{68}{40} + \frac{78}{50}} \right] \times 130.3 = 44.0 \text{ kN}$$

    Load supported by web $= 45.0 \text{ kN}$

    Load supported by middle flange $= 41.3 \text{ kN}$

(b) *Elastic distribution of load (proportional to $\frac{w}{b^3}$)*

    Load supported by haunch flange

$$= \left[\frac{\frac{78}{47^3}}{\frac{78}{47^3} + \frac{68}{40^3} + \frac{78}{50^3}} \right] \times 130.3 = 40.2 \text{ kN}$$

    Load supported by web $= 56.8 \text{ kN}$

    Load supported by middle flange $= 33.4 \text{ kN}$

Worked example

The Steel Construction Institute Silwood Park, Ascot, Berks SL5 7QN	Subject **No.8 PITCHED PORTAL FRAME. APEX CONNECTION.**		Chapter ref. **26**	
	Design code **BS5950**	Made by **BDC** Checked by **GWO**	Sheet no.	**7**

<u>Welds</u> *(taking higher of (a) and (b) above)*

<u>End plate to web</u>

Load on weld $= \dfrac{56.8}{2 \times 47}$

$= 0.604 \text{ kN/mm}$

Use 6 mm fillet welds.

<u>End plate to middle flange</u>

Take equal lengths of weld either side of bolt line to avoid eccentricity calculations.

Load on weld $= \dfrac{41.3}{2 \times \frac{1}{2}(177.8 - 100)}$

$= 0.531 \text{ kN/mm}$

Use 6 mm fillet welds.

<u>Top flange to end plate</u>

Horizontal component of flange force

$= F_c = 776.7 \text{ kN}$

Vertical component of flange force

$= F_c \tan 14° = 776.7 \tan 14° = 193.7 \text{ kN}$

Fit end plate to flange and take horizontal component in direct compression and vertical component in shear on fillet welds.

Load on weld $= \dfrac{193.7}{2 \times 177.8} = 0.545 \text{ kN/mm}$

Weld end plate to rafter with 6 mm fillet welds (except for 10 mm fillet welds on bottom flange).

Worked example 771

The Steel Construction Institute Silwood Park, Ascot, Berks SL5 7QN	Subject *No.8 PITCHED PORTAL FRAME. APEX CONNECTION.*	Chapter ref. 26
	Design code *BS5950*	Made by **BDC** Sheet no. **8** Checked by **GWO**

Connection of haunch flange to rafter flange

Use full strength butt weld for flange connection.

If haunch flange were fully stressed at this point,

flange force   = $p_y A$ =   *807.9 kN*

Component of force perpendicular to rafter

     =   *807.9 sin 14°*   =   *195.4 kN*

Assume stiff bearing length = flange thickness and take 1 in 2.5   [25.1]
spread to one side only in consideration of the flange being fully
stressed.

Local "bearing" capacity of web   4.5.3

     = $[12.8 + 2.5 \times (12.8 + 10.2)] \times 7.8 \times 355 \times 10^{-3}$

     =   *194.7 kN*

     ≃   *195.4 kN Stiffeners not required*

Other parts of the detail (e.g. the web of the haunch) are satisfactory by inspection.

If the apex detail is part of the same frame as the eaves detail in example 9, it may be desirable to standardize the bolt size and design both details using M24 general grade HSFG bolts.

772 Worked example

The Steel Construction Institute Silwood Park, Ascot, Berks SL5 7QN	Subject **No.9 PLATE GIRDER.** **BOLTED COVER PLATE** **SPLICE.**	Chapter ref. 26
Design code **BS5400**	Made by **BDC** Checked by **GWO**	Sheet no. 1

The connection is for a 1500 mm × 600 mm plate girder, with 40 mm thick flanges and a 15 mm thick web (Grade 43).

The factored loading for the ultimate limit state is:

Bending moment	*4783 kNm*
Axial tensile load	*113 kN*
Shear	*653 kN*

The loading for the serviceability limit state is:

Bending moment	*3565 kNm*
Axial tensile load	*85 kN*
Shear	*496 kN*

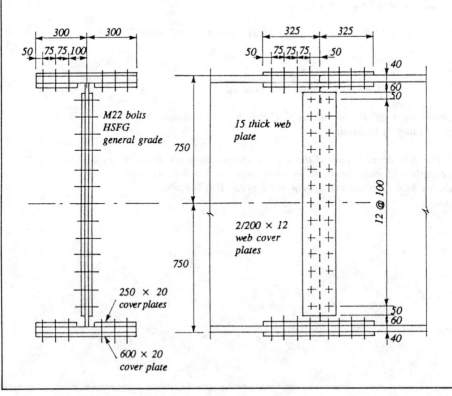

	Subject	Chapter ref.
The Steel Construction Institute Silwood Park, Ascot, Berks SL5 7QN	No.9 PLATE GIRDER. BOLTED COVER PLATE SPLICE.	26
Design code BS5400	Made by BDC Checked by GWO	Sheet no. 2

Properties of plate girder

$I = 2.916 \times 10^6 \text{ cm}^4$
$A = 693 \text{ cm}^2$

Web splice

Try a single line of 13 bolts at 100 mm vertical pitch, 50 mm from centre of joint.

(a) <u>Ultimate limit state</u>

Bending stress in web of girder at level of top and bottom bolts due to moment

$$= f_b = \frac{M_y}{I} = \frac{4783 \times 10^6 \times 600}{2.916 \times 10^6 \times 10^4} = 98.4 \text{ N/mm}^2$$

Horizontal shear force on bolt due to moment

$= f_b \times \text{bolt pitch} \times t$

$= 98.4 \times 100 \times 15 \times 10^{-3} = 147.6 \text{ kN}$

Average stress on girder due to axial load

$$= f_a = \frac{F_a}{A} = \frac{113 \times 10^3}{693 \times 10^2} = 1.63 \text{ N/mm}^2$$

Horizontal shear force on bolt due to axial load

$= f_a \times \text{bolt pitch} \times t$

$= 1.63 \times 100 \times 15 \times 10^{-3} = 2.4 \text{ kN}$

Vertical shear force on bolt due to shear

$$= \frac{V}{n} = \frac{653}{13} = 50.2 \text{ kN}$$

774 Worked example

The Steel Construction Institute Silwood Park, Ascot, Berks SL5 7QN		Subject No.9 PLATE GIRDER. BOLTED COVER PLATE SPLICE.	Chapter ref. 26
	Design code BS5400	Made by BDC Checked by GWO	Sheet no. 3

Horizontal shear force on bolt due to moment due to eccentricity [23.4.2]

$$= \frac{P \times e \times r_i}{\Sigma_n r_i^2}$$

$$= \frac{653 \times 50 \times 600}{2(100^2 + 200^2 + 300^2 + 400^2 + 500^2 + 600^2)}$$

$$= 10.8 \text{ kN}$$

Resultant shear force (= vector sum of vertical and horizontal shear forces)

$$= \sqrt{50.2^2 + (147.6 + 2.4 + 10.8)^2}$$

$$= 168.5 \text{ kN}$$

Try M22 general grade HSFG bolts in "double shear" with 12 mm thick cover plates.

Friction capacity per bolt ($\mu = 0.45$)   14.5.4.2

$$= \frac{k_h F_v \mu N}{\gamma_m \gamma_{f_3}} = \frac{1.0 \times 177 \times 0.45 \times 2}{1.3 \times 1.1}$$

$$= 111.4 \text{ kN}$$

Shear capacity per bolt   14.5.3.4

$$= \frac{\sigma_q n A_{eq}}{\gamma_m \gamma_{f_3} \sqrt{2}} = \frac{635 \times 10^{-3} \times 2 \times 303}{1.1 \times 1.1 \times \sqrt{2}}$$

$$= 224.9 \text{ kN}$$

Bearing capacity on web   14.5.3.6
(edge distance = 2 × hole diameter)

$$= \frac{k_1 k_2 k_3 k_4 \sigma_y}{\gamma_m \gamma_{f_3}} \times d_{bolt} \times t$$

The Steel Construction Institute Silwood Park, Ascot, Berks SL5 7QN	Subject **No.9 PLATE GIRDER. BOLTED COVER PLATE SPLICE.**	Chapter ref. **26**
	Design code **BS5400**	Made by **BDC** / Sheet no. **4** Checked by **GWO**

$$= \frac{1.0 \times 1.97 \times 1.2 \times 1.5 \times 275 \times 10^{-3} \times 22 \times 15}{1.05 \times 1.1}$$

$= 278.6 \text{ kN}$

Bearing capacity on cover plates   14.5.3.6
(edge distance = 2.08 × hole diameter)

$$= \frac{1.0 \times 2.01 \times 0.95 \times 1.5 \times 275 \times 10^{-3} \times 2 \times 22 \times 12}{1.05 \times 1.1}$$

$= 360.1 \text{ kN}$

Shear capacity governs design.   14.5.4.1.1

Bolt capacity > resultant force on bolt
(224.9 kN)   (168.5 kN)

but, since capacity is based on shear strength, serviceability   Table 1
check is required.   14.5.4.1.2

(b) <u>*Serviceability limit state*</u>

Resultant shear force =   125.9 kN
(using the same procedure as for the ultimate limit state)

Friction capacity per bolt ($\mu = 0.45$)   14.5.4.2

$$= \frac{k_h F_v \mu N}{\gamma_m \gamma_{f_3}} = \frac{1.0 \times 177 \times 0.45 \times 2}{1.2 \times 1.0}$$

$= 132.8 \text{ kN} >$ resultant shear force   OK

Note that in this example the serviceability check governs the design.

<u>*Flange splices*</u>

Assume that the area of web associated with the web splice extends half a bolt pitch beyond the top and bottom bolts.

The Steel Construction Institute Silwood Park, Ascot, Berks SL5 7QN	Subject No.9 PLATE GIRDER. BOLTED COVER PLATE SPLICE.		Chapter ref. 26	
	Design code BS5400	Made by BDC Checked by GWO	Sheet no.	5

Second moment of area of portion of web where its share of the applied moment and axial load is carried by the web splice

$$= I'_w = \frac{t\,d^3}{12} = \frac{15 \times 1300^3}{12 \times 10^4}$$

$$= 0.275 \times 10^6 \text{ cm}^4$$

Area of the same portion of web

$$= A'_w = t\,d = 15 \times 1300 \times 10^{-2}$$

$$= 195 \text{ cm}^2$$

Proportion of applied moment carried by web splice

$$= \frac{I'_w}{I_{girder}} = \frac{0.275 \times 10^6}{2.916 \times 10^6} = 0.094$$

Proportion of axial load carried by web splice

$$= \frac{A'_w}{A_{girder}} = \frac{195}{693} = 0.281$$

(a) <u>Ultimate limit state</u>

Force to be carried by flange splice

$$= \frac{4783 \times 10^3}{(1500 - 40)} \times (1 - 0.094) + \frac{113}{2}(1 - 0.281)$$

$$= 2968 + 41 = 3009 \text{ kN}$$

Try M22 general grade HSFG bolts in "double shear" with 20 mm thick cover plates.

By inspection (from web splice design) shear capacity governs (224.9 kN per bolt).

The Steel Construction Institute Silwood Park, Ascot, Berks SL5 7QN	Subject **No.9 PLATE GIRDER.** **BOLTED COVER PLATE SPLICE.**	Chapter ref. **26**
	Design code **BS5400**	Made by **BDC** Sheet no. **6** Checked by **GWO**

No. of bolts required $= \dfrac{3009}{224.9} = 13.4$

(b) *Serviceability limit state*

Force to be carried by flange splice   =   2243 kN
(using the same procedure as for the ultimate limit state)

Friction capacity per bolt   =   132.8 kN

No. of bolts required $= \dfrac{2243}{132.8} = 16.9$

As with the web connection, the serviceability check governs the design.

3 rows of 6 bolts would satisfy the design, but if there is any lack of fit the row of bolts next to the joint may be only partly effective as far as friction capacity is concerned. Therefore, since the slip (serviceability limit state) is critical, use 4 rows of 6 bolts each side of joint for flange splices M22 general grade HSFG bolts.

Splice plates

Effective area required for cover plates

$= \dfrac{P \gamma_m \gamma_{f_3}}{\sigma_y \times 0.8} = \dfrac{3009 \times 10^3 \times 1.05 \times 1.1}{265 \times 0.8}$ 14.4.3.2

$\qquad = 16393 \ mm^2$

Inner splice plates, try 2 No. 250 × 20 mm

Effective area $= k_1 k_2 A_t$ (but $\leq A$)   11.3.2
$\qquad\qquad = 2 \times 1.0 \times 1.2 \ (250 - 3 \times 24) \times 20$
$\qquad\qquad = 8544 \ mm^2$
$\qquad\qquad > \dfrac{16393}{2}$ OK

Use 2 No. 250 × 20 mm

The Steel Construction Institute Silwood Park, Ascot, Berks SL5 7QN	Subject No.9 PLATE GIRDER. BOLTED COVER PLATE SPLICE.	Chapter ref. 26
	Design code BS5400	Made by BDC Checked by GWO	Sheet no. 7

Use 600 × 20 mm outer splice plates to flanges.

Flanges

Moment at end of splice

$= 4783 + 653 \ (3 \times 75 + 50) \times 10^{-3} = 4963 \ kNm$

Average stress in tension flange

$$= \frac{4963 \times 10^6 \times (1500 - 40) \times 0.5}{2.916 \times 10^6 \times 10^4} = 124.2 \ N/mm^2$$

Flange force $= 124.2 \times 600 \times 40 \times 10^{-3} = 2981 \ kN$

Effective area of tension flange 11.3.2

$= k_1 \ k_2 \ A_t \ (but \leq A)$

$= 1.0 \times 1.2 \ (600 - 6 \times 24) \times 40 = 21888 \ mm^2$

Flange capacity $= \dfrac{\sigma_y \ A_e}{\gamma_m \ \gamma_{f_3}}$ 11.5.1

$$= \frac{265 \times 21888 \times 10^{-3}}{1.05 \times 1.1}$$

$= 5022 \ kN > flange \ force \quad OK$

Chapter 27
Foundations and holding-down systems

by HUBERT BARBER

27.1 Foundations

27.1.1 Types of foundation

Pad foundations are used primarily to support the major structural elements in either sheds or multi-storey buildings. The pad foundations to major elements may be either mass concrete or reinforced concrete, the latter when either heavy loads or very poor ground conditions are present. They may be used in the context of cladding to support intermediate posts carrying sheeting rails, in which case the load is almost all from wind forces and is horizontal.

Strip foundations are used in steel-framed buildings to support external masonry or brickwork cladding and masonry internal partitions. In some cases the ground floor is thickened at these locations to provide a foundation but care should be taken with respect to the appropriate depth for clay or frost heave and for compatibility between such foundations and those of the main frame.

Piled foundations, either driven, bored or cast in place, are used on sites where ground conditions are poor or for buildings or structures in which differential settlement is critical. They may also be required in circumstances where heavy concentrations of load occur. In general when piled foundations are used the whole of the construction should be supported on piles. The ground floor slab, ground floor cladding and internal partitions should be carried by ground beams between the pile cap locations. If it is necessary for reasons of economy to support the ground floor independently, provision should be made for differential settlement by the inclusion of suitable movement joints.

Ground improvement techniques are appropriate for some types of poor ground. The most usual techniques are vibro-compaction or vibro-replacement but dynamic compaction can also be useful for improvement of large isolated sites. Ground improvement specialists or specialist consultants should be approached as economy will be the most important factor in the decision.

Typical foundation layouts are shown in Fig. 27.1.

780 Foundations and holding-down systems

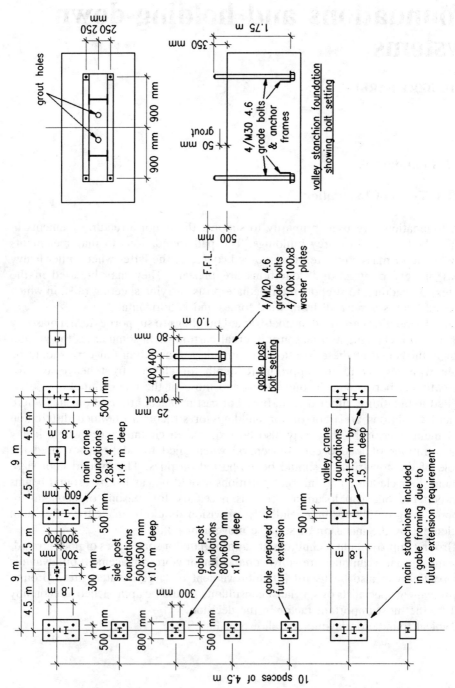

Fig. 27.1 Part plan of typical two-bay crane shed

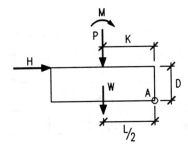

Fig. 27.2 Stability of foundation

27.1.2 Design of foundations

In order to assess the distribution of pressure under a foundation it is necessary to make a reasonable estimate of the weight of the foundation. In addition to distributing the forces to the ground the foundation block is also required to provide stability in cases where overturning moments are present.

Referring to Fig. 27.2, loads P, H and M are factored as appropriate while W, the foundation mass, is factored by 1.0, being a restoring moment. Moments about A give

$$M + HD - PK - \frac{WL}{2} \leq 0$$

From this a minimum value of W for stability is produced.

The minimum value for D for a mass concrete foundation is established by 45° dispersal from the edge of the baseplate shown in Fig. 27.3. Shallower foundations can be used if they are suitably reinforced.

The distribution of pressure under the foundation is then assessed as follows.

Case 1

See Fig. 27.4(a):

$$f_g = \frac{P + W}{LB} \pm \frac{(M + DH)6}{BL^2}$$

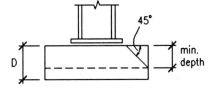

Fig. 27.3 Thickness of foundation

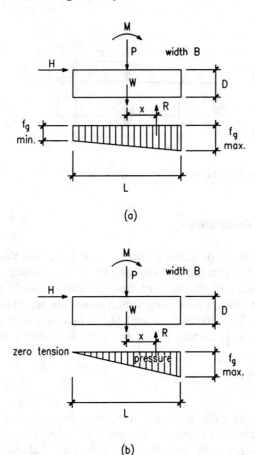

Fig. 27.4 Ground pressure – Case 1

It is necessary for f_{gmax} to be less than the stipulated ground bearing capacity for the foundation to be satisfactory.

For f_{gmin} to be zero (Fig. 27.4(b)):

$$\frac{P + W}{LB} - \frac{(M + DH)6}{BL^2} = 0$$

Replacing the forces by the resultant acting at eccentricity x

$$\frac{R}{LB} - \frac{6Rx}{BL^2} = 0$$

from which

$$x = \frac{L}{6}$$

This is the limiting condition for the application of Case 1.

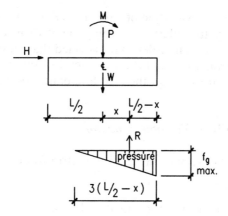

Fig. 27.5 Ground pressure – Case 2

Case 2

This occurs when f_{gmin} is negative. As no tension can exist between the soil and the underside of the concrete base a compressive stress wedge is formed at the compression side of the foundation. The summation of the stress under the block must equal the resultant of the applied loads. When $x > L/6$, the length of the triangular stress wedge is three times the edge distance $(L/2 - x)$ in order that the resultant acts at the centroid of the wedge. The theory proposes that $3(L/2 - x)$ is the length of surface contact between the foundation and the ground (Fig. 27.5).

$$\frac{f_{gmax}}{2}\left(\frac{L}{2} - x\right) 3B = P + W$$

$$f_{gmax} = \frac{2(P + W)}{3B\left(\dfrac{L}{2} - x\right)}$$

When f_{gmax} exceeds the stipulated ground bearing pressure, dimensions B or L or both may be increased within the limits of economy, after which piling or ground improvement techniques can be investigated.

27.1.3 Sub-soil bearing pressure

Foundation bearing pressure should always be determined on the basis of experimental results and field assessments taken during the soil investigation. Almost all sites have considerable variation of strength and quality of the sub-strata and it is therefore necessary to undertake a comprehensive investigation producing a large number of test results in order to be satisfied that reasonable average values are obtained for the various parameters. A factor of safety of 3 is usually applied to the bearing strength to obtain the safe foundation bearing pressure.

27.1.3.1 Clayey soils – Terzaghi's method

(1) Foundation long in relation to width i.e. strip footing.

$$q = cN_c + \gamma z N_q + 0.5\gamma B N_\gamma$$

where q = bearing capacity (ultimate) (kN/m²)
c = cohesion (kN/m²)
γ = bulk density of the soil (kN/m³)
z = depth of foundation (m)
B = breadth of foundation (m)

and N_c, N_q and N_γ are constants dependent upon the angle of cohesion. These constants are given in graph form in soil engineering references; Table 27.1 gives approximate values as guidance only.

Table 27.1 Terzaghi's constants for clayey soils

ϕ	N_c	N_q	N_γ
0	5.7	1.0	0
10	10	3.0	2.0
20	18	8.0	6.5
30	36	21.0	20
35	60	50.0	45.0

(2) Square and circular foundations to isolated piers or columns.

The bearing capacity of foundations rectangular, square or circular is higher than that of strip footings. Terzaghi's expression is adjusted as follows:

$$q = 1.3cN_c + \gamma z N_q + 0.3\gamma B N_\gamma$$

while Skempton applies a factor $(1 + 0.2B/L)$ to the Terzaghi strip footing calculation where B is the breadth and L the length. In the case of a square the enhancement factor is then 1.2.

27.1.3.2 Sandy or cohesionless soils

The appropriate site test for cohesionless soils is the standard penetration test (SPT) in which the number of blows of a standard weight is recorded for unit penetration of a standard cylindrical implement. According to Meyerhof the ultimate bearing capacity is given, in kN/m², by

$$q = 10.7 \times NB \left(1 + \frac{z}{B}\right)$$

where N is the number of blows per metre and q, z and B are as before.

Cohesionless soils subject to flooding will suffer a reduction of capacity at water table level:

capacity when flooded $= K \times$ unflooded capacity

where $K = (\gamma - 9.8)/\gamma$ and γ is the soil bulk density given for convenience in kN/m³.

27.2 Connection of the steelwork

27.2.1 Fixed and pinned bases

The function of a column baseplate is to distribute the column forces to the concrete foundation. In general a plain or slab base is used for pinned conditions or when there is very little tension between the plate and the concrete. A gusseted base is used occasionally to spread very heavy loads but more generally for conditions of large moment in relation to the vertical applied loads, the principal function of the gusset being to allow the holding-down bolt lever arm to be increased to give maximum efficiency while keeping the baseplate thickness to an acceptable minimum. Gusseted or built-up bases give an ideal solution for compound or twin crane stanchions in industrial shed buildings.

Fixed bases are used primarily in low-rise construction either in portal buildings specifically designed as 'fixed base' or in industrial sheds in which the main columns cantilever from the foundations. They are also used, though less frequently, in multi-storey rigid-frame construction. In each of these cases it is assumed by definition that no angular rotation takes place and although this is unlikely to be achieved it is generally accepted that sufficient rigidity can be obtained to justify the assumption.

Pinned bases are those in which it is assumed that there is no restraint against angular rotation. Although this is also difficult to achieve it is accepted that sufficient flexibility can be introduced by minimizing the size of the foundation and similarly reducing the anchorage system. Pinned bases are used in portal and in multi-storey construction.

Typical pinned and fixed bases are shown in Fig. 27.6.

786 *Foundations and holding-down systems*

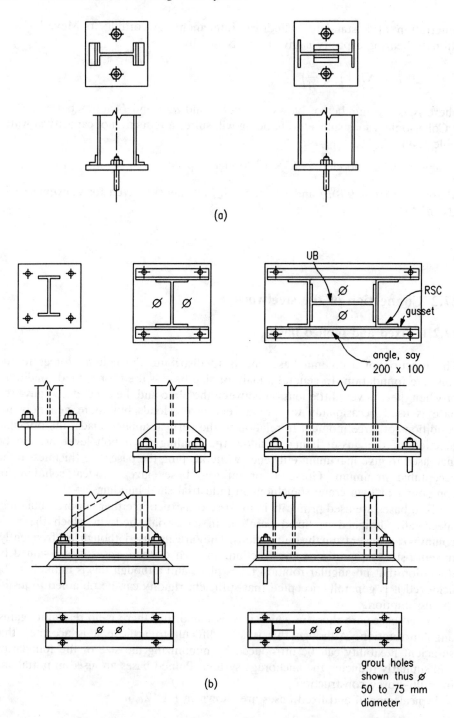

Fig. 27.6 Typical stanchion bases: (a) pinned bases, (b) gusseted and partially fixed bases

27.2.2 Baseplate design

27.2.2.1 Plain bases

(1) The effective area method

The empirical design method given in clause 4.13.2.2 of BS 5950 is to be used with caution in cases where the projections a and b are small and when the depth/width ratio of the column is high.

The effective area method given below will ensure adequacy in this respect. The plate thickness should never be less than the column flange thickness. The effective area method assumes that the whole of the axial load is spread on to an area of bedding material determined by dispersal through the plate to a boundary dimensioned K from the section profile as shown in Fig. 27.7(a). The resulting area is assumed to give an even distribution of pressure and the plate is assumed to cantilever the distance K.

The effective area is therefore given by

$$A_e = 2(B + 2K)(2K + T) + (D - 2T - 2K)(2K + t)$$

where B and D are the column dimensions, T is the flange thickness and t is the web thickness.

then $\dfrac{W}{A_e} \not> $ the bearing strength of the bedding

The resulting quadratic is solved for K and the moment on the cantilever outstand, K, is given by

$$M = \frac{W}{A_e} \times \frac{K^2}{2}$$

then $t_p = \left(\dfrac{6M}{p_{yp}}\right)^{\frac{1}{2}}$

where $p_{yp} = 270$ N/mm^2

(see the first worked example at the end of this chapter).

The method is also applied to tubular columns.

The dispersal dimension K taken radially on either side of the tube wall gives an annular contact area between the plate and the bedding material, as shown in Fig. 27.8(a) Then

$$A_e = (2K + t)(D - t)\pi$$

where D is the tube diameter and $(W/A_e) \not> $ the bearing strength of the bedding.

After solving for K, M and t are determined in the same way as for rolled sections (see the second worked example at the end of this chapter).

If K is greater than $D/2$

Foundations and holding-down systems

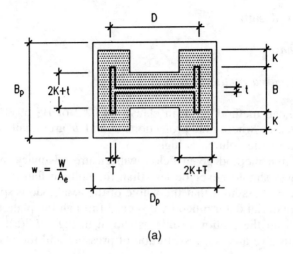

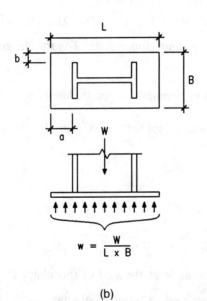

Fig. 27.7 Slab base design (a) the Effective Area Method (b) the Empirical Design Method (clause 4.13.2 or BS 5950)

$$A_e = (D + 2K)^2 \frac{\pi}{4}$$

as shown in Fig. 27.8(b).

Similarly after solving for K, M and t are obtained in the same way as for rolled sections (see the third worked example at the end of this chapter).

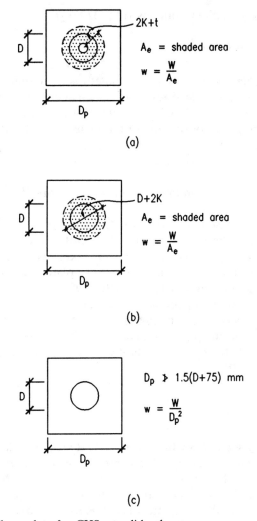

Fig. 27.8 Square base plate for CHS or solid column

(2) The empirical design method

In this method, given in clause 4.13.2 of BS 5950, it is assumed that the load from the column is evenly distributed at the underside of the plate. An empirical formula then defines the required baseplate thickness:

(a) for columns of I, H, channel, box or RHS (Fig. 27.7(b))

$$t = \left[\frac{2.5w}{P_{yp}}(a^2 - 0.3b^2)\right]^{\frac{1}{2}}$$

(b) for circular solid or hollow columns (Fig. 27.8(c))

$$t = \left[\frac{wD_p}{2.4p_{yp}}(D_p - 0.9D)\right]^{\frac{1}{2}}$$

where p_{yp} is the design strength of the plate, t is the required plate thickness, w is the pressure on the underside of the plate, assuming uniform distribution; a and b are the greater and lesser projections of the plates beyond the rolled section; D_p is the length of the side of the plate, or its diameter; and D is the diameter of the column.

NB $\quad D_p \not< 1.5(D + 75)$ mm.

Typical solutions are included in the first and second worked examples at the end of this chapter.

In the event that the baseplate is oversized, that is, greater than is required for the bearing strength of the bedding material, in order to accommodate holding-down bolts, etc., the additional area may be taken as ineffective. The plate thickness may therefore be determined to suit the plate dimensions required for the bedding strength (see clause 4.13.2.1 of BS 5950).

In situations where the baseplate is of irregular shape or when the forces are eccentric giving non-uniform bearing pressures, the bending moments on the plate should be computed and the plate designed accordingly. The moment should not exceed 1.2 $p_{yp}Z_p$, where Z_p is the elastic modulus and $p_{yp} \not> 270$ N/mm² (Table 3.1.1 of BS 5950).

27.2.2.2 Gusseted bases

In a stiffened or gusseted base the moment in the gusset due to the bearing pressure under the effective area of the baseplate or due to the tensile forces in the holding-down bolts should not exceed $p_{yg}Z_g$, where Z_g is the elastic modulus of the gusset and p_{yg} is the design strength of the gusset ($p_{yg} \not> 270$ N/mm²). When the effective area of the baseplate is less than its gross area, the connections of the gusset should be checked for the effects of a nominal distribution of bearing pressure on the gross area as well as for the distribution used in the design.

27.2.2.3 Beam bearing plates

Bearing plates at beam seatings are required to distribute the beam reaction to the masonry support at stress levels within the capacity of the masonry and to ensure that the web-crushing capacity of the beam is not exceeded. The distribution of bearing stresses under the plate is extremely complex although simplifying assumptions are usually made in appropriate cases.

The bending of the plate, shown in Fig. 27.9 in the direction transverse to the beam, will depend on the stiffness of the beam flange and the fixing of the flange

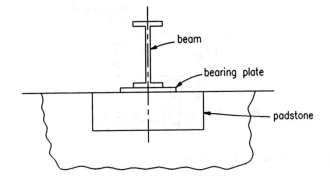

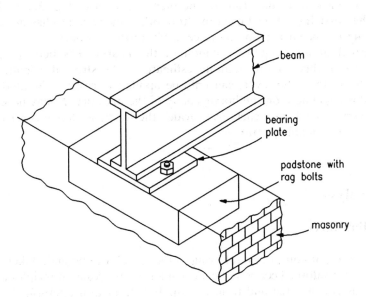

Fig. 27.9 Beam bearing

to the plate. It is usual to assume that the position of maximum bending is the outside edge of the root of the web and that the plate carries the whole of the bending.

In the longitudinal direction, shown in Fig. 27.10, the deflection and rotation of the beam due to its loading will cause a concentration of bearing at the front edge and, depending upon the load from above the bearing, a possible lifting of the back edge of the plate. It is often assumed, therefore, that the distribution will be either trapezoidal or triangular; possibly the triangle may not reach the back of

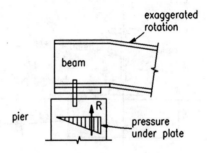

Fig. 27.10 Pressure under bearing plate (1)

the bearing. If it is expected that the front edge concentration will be high the plate is set back from the front of the pier as shown in Fig. 27.11. This is to reduce the possibility of spalling at the front of the pier but also has the advantage of applying the beam reaction more centrally to the masonry.

A method of assessing the rotation of the bearing has been proposed by Lothers. From this a more accurate estimate of the stress distribution can be made. The method, however, can only be applied in cases of isolated masonry piers and is dependent on the homogeneity of the masonry. It may be justified in cases of very heavy beam reactions provided the workmanship in constructing the pier can be reasonably guaranteed.

27.3 Analysis

27.3.1 Bolt forces

The calculation of bolt forces, referring to Fig. 27.12, can be undertaken using the analogy of a reinforced concrete beam. Under elastic design principles a notional neutral axis is computed and from this the bolt lever arm established.

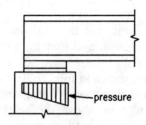

Fig. 27.11 Pressure under bearing plate (2)

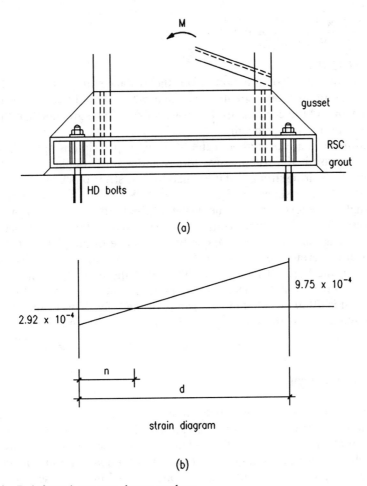

Fig. 27.12 Reinforced concrete beam analogy

From Figure 2.1 of BS 8110:

concrete strain at a stress of $0.4f_{cu} = 0.8 \times 10^{-4} \left(\dfrac{f_{cu}}{\gamma_m}\right)^{\frac{1}{2}} = 2.92 \times 10^{-4}$

Assuming $f_{cu} = 20$ N/mm² (grout) and $\gamma_m = 1.5$:

$$\text{steel strain} = \frac{195 \text{ N/mm}^2}{200 \text{ kN/mm}^2} = 9.75 \times 10^{-4}$$

taking the bolt strength as 195 N/mm² (Table 32 of BS 5950) and E, the modulus of elasticity, as 200 kN/mm². Then

$$n = \frac{2.92 \times 10^{-4} d}{(2.92 + 9.75)10^{-4}} = 0.23d$$

(see Fig. 27.12 (b)).

However, the 'give' of the bolts over the distance of protrusion above the concrete or grout surface is greater than that which would occur internally in a concrete beam and the crushing of the grout is likely to be greater than that which would occur in a concrete beam.

These difficulties are overcome if the load factor principle is applied. The area required to transmit the compressive forces under the baseplate is calculated at the appropriate bearing strength of the concrete. The stress block may be assumed to be partly parabolic as shown in Figure 2.1 of BS 8110 or it may be approximated as a rectangle. The lever arm for the design of the bolts is then from the centroid of this stress block to the bolt position. The centroid of the stress block is often less than the edge distance from the compression edge of the plate to the holding-down bolt: it is therefore often assumed that the lever arm for the bolts is equal to the centres of the bolts. It is also very likely that the point of application of the compressive forces will be near to the holding-down bolts due to the extra stiffening that is often included in the vicinity of the bolts. This is illustrated in the typical design given as the fourth worked example at the end of this chapter.

27.3.2 Bolt anchorage

Anchorage of the holding-down bolts into the concrete foundation should be sufficient to cater for any uplift forces and to provide for any shears applied to the bolts. Attention is particularly directed to the last paragraph of clause 6.7 of BS 5950 which states that rag bolts and indented foundation bolts should not be used to resist uplift forces. The elastic elongation of indented screwed rods or bolts under tension causes the breakdown of the grout surrounding the bolt. This is even more critical in the case of resinous grouts.

The failure mode of bolts pulled from a concrete block is shown in Fig. 27.13(a); a reasonable approximation is shown in Fig. 27.13(b). The surface

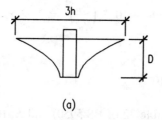
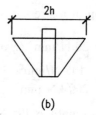

Fig. 27.13 Typical pull-out from concrete block

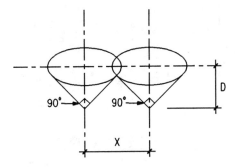

Fig. 27.14 Conical pull-out

area of the conical pull-out (Fig. 27.14) is $4.44D^2$, where D is the depth of embedment. The factored tensile capacity of an M20 (4.6) bolt is $245 \times 195 \times 10^{-3} = 47.7$ kN. For an M20 HD bolt 450 mm long with an embedment of say 350 mm

the conical surface is $4.44 \times 350^2 = 544 \times 10^3$ mm^2

the surface stress is $\dfrac{47.7 \text{ kN}}{544 \times 10^3} = 0.09$ N/mm^2

As holding-down bolts usually act in pairs the conical pull-outs often overlap depending on the depth of embedment. The BCSA have tabulated the surface areas including those which overlap (Table 27.2).

27.4 Holding-down systems

27.4.1 Holding-down bolts

The most generally used holding-down bolts are of grade 4.6 although 8.8 grade are also available (see Table 27.3). They are usually supplied □□○×: square head, square shoulder, round shank, hexagon nut. Each bolt must be provided with an anchor washer (square hole to match the shoulder) or an appropriate anchor frame to embed in the concrete in circumstances of high uplift forces. In such cases the anchor frame may be composed of angles or channels. Typical frames are shown in Fig. 27.15. When long anchors are required a rod threaded at both ends may be used, and in exceptional circumstances when prestressing is required a high tensile rod (usually Macalloy bar) is adopted. In both these cases provision should be made to prevent rotation of the rod during tightening which may result in the embedded nut being slackened.

Corrosion of holding-down bolts to a significant extent has been reported in some instances. This usually occurs between the level of the concrete block and

Table 27.2 Embedded lengths of holding-down bolts based on conical pull-out

Depth, D(mm)	Distance between centres, X (mm)										
	75	100	125	150	200	225	300	450	600	750	1000
	Effective conical surface area (allowing for overlap) (cm^2)										
75	402.1	445.1	479.9	499.8	499.8	499.8	499.8	499.8	499.8	499.8	499.8
100	651.3	714.9	773.2	824.5	888.6	888.6	888.6	888.6	888.6	888.6	888.6
125	955.3	1038.1	1117	1191	1316	1362	1388	1388	1388	1388	1388
150	1315	1416	1514	1608	1780	1855	1999	1999	1999	1999	1999
200	2199	2337	2473	2605	2859	2979	3298	3554	3554	3554	3554
225	2724	2880	3034	3186	3479	3619	4006	4498	4498	4498	4498
300	4633	4843	5052	5258	5664	5862	6434	7420	7997	7997	7997
450	9950	10267	10583	10897	11521	11831	12743	14476	16022	17277	17994
600	17266	17689	18112	18533	19373	19790	21032	23448	25735	27837	30715
750	26581	27111	27640	28168	29221	29746	31312	34392	37371	40211	44507
1000	46550	47256	47962	48667	50076	50779	52882	57048	61141	65134	71486

Values to the right of the heavy zig-zag line are for two non-intersecting cones

Holding-down systems

Table 27.3 Holding-down bolts; tension capacity per pair of bolts

Nominal diameter (mm)	Tensile stress area (mm^2)[b]	Bolts grade 4.6@ 195 N/mm^2 (kN)	Bolts grade 8.8@ 450 N/mm^2 (kN)
M16	314	61.22	141.30
M20	490	95.54	220.50
M22[a]	606	118.16	272.70
M24	706	137.66	317.70
M27[a]	918	179.00	413.10
M30	1122	218.78	504.90

[a] Non-preferred size
[b] Tensile stress areas are taken from BS 4190 and BS 3692

the underside of the steel baseplate, in aggressive chemical environments or at sites where moisture ingress to this level is recurrent. Fine concrete grout, well mixed, well placed and well compacted will provide the best protection against corrosion but in cases where this is not adequate for the prevailing conditions, an allowance may be made in the sizing of the bolts or by specifying a higher grade bolt which provides a larger factor of safety against tensile failure in the event that some corrosion does occur.

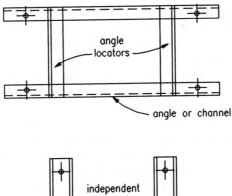

Fig. 27.15 Typical anchorages for holding-down bolts

27.4.2 Grouting

The casting-in of the holding-down bolts with adequate provision for adjustment requires that they are positioned in the concrete surrounded by a tube, conical or cylindrical, or a polystyrene former. After removal of the tube or former, which should be delayed until the last possible time before the erection of the columns, the available lateral movement of the bolts should be between three and four times the bolt diameter. In cases where open tubes are used they should be provided with a cap or cover to prevent the ingress of water, rubbish and mud. After erection, lining, levelling and plumbing of the frame the grout voids around the bolts should be cleaned out by compressed air immediately prior to grouting. The bolt grouting and baseplate filling should be done as two separate operations to allow shrinkage to take place. During the levelling and plumbing operations wedges and packings are driven into the grout space. Before final grouting these should be removed, otherwise, after shrinkage of the grout filling material, they will become hard spots, preventing the even distribution of the compressive forces to the concrete base.

27.4.3 Bedding

Bedding materials are required to perform a number of functions, one of which is the provision of the corrosion protection referred to earlier in section 27.4.1. In accordance with BS 5950, steel baseplates are designed for a compression under the plate of $0.4 f_{cu}$, where f_{cu} is the cube strength of the filling. Design to BS 8110 and more recent design advice[1] recommend the use of a design strength of $0.6 f_{cu}$. The bedding material therefore transmits high vertical stresses including those resulting from the applied moment. The third function is to transmit the horizontal forces or shears resulting from wind or crane surge. It is clear therefore that the bedding material is a structural medium and should be specified, controlled and supervised accordingly.

For heavily-loaded columns or those carrying large moments resulting in high compressive forces the bedding should be fine concrete using a maximum aggregate of 10 mm size. The usual mix is $1:1\frac{1}{4}:2$ with a water–cement ratio of between 0.4 and 0.45 (this is not suitable for filling the bolt tubes as it is too stiff; a pure cement water mix has suitable flow properties and is usually used). It also has high shrinkage properties and should be allowed to set fully before continuing with the bedding. A cement mortar mix is often used for moderately-loaded columns. A suitable mix would be $1:2\frac{1}{2}$. Weaker filling than this should only be used for lightly-loaded columns where the erection packs are left in position and transfer all the load to the foundation. In cases when erection packs are left in place it may be necessary to re-examine the baseplate design as load transference will be directly through the steel packs.

In order to facilitate the compaction of the bedding material, holes are cut in the baseplate of the order of 50 mm diameter or more, near to the centre of the plate in order to allow the escape of air pockets and to ensure that the bedding reaches to centre.

References

1. The Steel Construction Institute/British Constructional Steelwork Association (1991) *Joints in Simple Construction*, Vol. 1. SCI/BCSA.

Further reading for Chapter 27

British Constructional Steelwork Association/The Concrete Society/Constructional Steel Research and Development Council (1980) *Holding-Down Systems for Steel Stanchions*.
British Standards Institution (1985) *Structural use of concrete*. Part 1: *Code of practice for design and construction*. BS 8110, BSI, London.
British Standards Institution (1990) *Structural use of steelwork in building*. Part 1: *Code of practice for design in simple and continuous construction: hot rolled sections*. BS 5950, BSI, London.
Capper P.L. & Cassie W.F. (1969) *The Mechanics of Engineering Soils*, 5th edn. E. & F.N. Spon Ltd.
Capper P.L., Cassie W.F. & Geddes J.W. (1968) *Problems of Engineering Soils*. Spon Civil Engineering Series.
Lothers J.E. (1972) *Design in Structural Steel*. Prentice-Hall.
Pounder C.C. (1940) *The Design of Flat Plates*. Association of Engineering and Shipbuilding Draughtsmen.
Skempton A.W. & McDonald D.H. (1956) *The allowable settlement of buildings*. Proc. Instn Civ. Engrs, Part 3, 5 Dec.
Skempton A.W. & Bjerrum L. (1957) A contribution to settlement analysis of foundations on clay. Géotechnique, **7**, No. 4, Dec.
Terzaghi K. & Peck R.B. (1967) *Soil Mechanics in Engineering Practice*. John Wiley, London.
Tomlinson M.J. (1975) *Foundation Design and Construction*. Pitman, London.

A series of worked examples follows which are relevant to Chapter 27.

The Steel Construction Institute Silwood Park, Ascot, Berks SL5 7QN	Subject **FOUNDATION EXAMPLE 1**	Chapter ref. **27**
	Design code **BS5950: Part 1**	Made by **HB** Checked by **GWO**	Sheet no. **1**

Problem

Design a simple base plate for a 254 × 254 × 73 UC to carry a factored axial load of 1000 kN

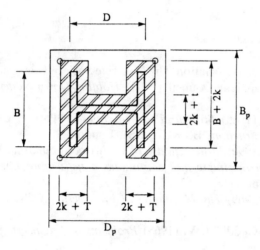

Design by the Effective Area Method using Reference 27.1

Bearing strength of concrete = $0.6 f_{cu}$ - take f_{cu} as 40 N/mm²

Area required (in mm²) = $\dfrac{1000 \times 10^3}{0.6 \times 40}$ = 41667 mm²

Bearing area = hatched area

= $2(B + 2k)(2k + T) + (D - 2T - 2k)(2k + t)$

∴ $2(254 + 2k)(2k + 14.2) + (254 - 28.4 - 2k)(2k + 8.6)$ = 41667

∴ $4k^2 + 1506.8 k - 32513$ = 0

$k = \dfrac{-1506.8 \pm \sqrt{1506.8^2 + 520208}}{8}$ = 20.46 mm

The Steel Construction Institute Silwood Park, Ascot, Berks SL5 7QN		FOUNDATION EXAMPLE 1		27
		Design code BS5950: Part 1	Made by HB Checked by GWO	Sheet no. 2

$$t_p = \left[\frac{6 \times 24 \times 20.46^2}{2 \times 270}\right]^{0.5} = 10.57$$

<u>Use a base plate 300 × 300 × 15</u>

<u>Using the empirical method:</u>

Area required   =   41667 i.e. 204 × 204 plate

Assume projection of 25 mm each dimension

Use 304 × 304 plate

$$w = \frac{1000 \times 10^3}{304^2} = 10.82 \text{ N/mm}^2$$

$$t_p = \left[10.82 \times \frac{2.5}{270}(25^2 - 0.3 \times 25^2)\right]^{0.5}$$

$$t_p = 6.62 \text{ mm}$$

This is inadequate by the effective area method. This demonstrates the necessity for caution when the projections are small.

Worked example

The Steel Construction Institute Silwood Park, Ascot, Berks SL5 7QN	Subject **FOUNDATION EXAMPLE 2**		Chapter ref. **27**
	Design code BS5950: Part 1	Made by **HB** Checked by **GWO**	Sheet no. **1**

Problem

Design a simple base plate for a 219 × 6.3 CHS to carry a factored axial load of 1010 kN

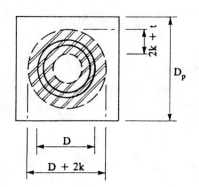

219 × 6.3 CHS

Factored axial load = 1010 kN

Assumed bedding material
f_{cu} = 40 N/mm²

Design by the Effective Area Method using Reference 27.1

Area required $= \dfrac{1010 \times 10^3}{0.6 \times 40} = 42083 \text{ mm}^2$

Area of shaded annulus $= (2k + t)(D - t)\pi = 42083$

$(2k + 6.3)(219 - 6.3) = 13395$, hence $k = 28.3$ mm

$t_p = \left[\dfrac{6 \times 24 \times 28.3^2}{2 \times 270} \right]^{0.5} = 14.62 \text{ mm}$ ∴ Use 280 × 280 × 15 plate

Using the empirical method:

$w = \dfrac{1010 \times 10^3}{280^2} = 12.9 \text{ N/mm}^2$

$t = \left[\dfrac{12.9}{2.4 \times 270} \times 280(280 - 0.9 \times 219) \right]^{0.5} = 21.5 \text{ mm}$

Attention is drawn to Clause 4.13.2.1 of BS 5950

Worked example

The Steel Construction Institute Silwood Park, Ascot, Berks SL5 7QN	Subject *FOUNDATION EXAMPLE 3*	Chapter ref. 27
	Design code *BS5950: Part 1*	Made by HB Sheet no. 1 Checked by GWO

Problem

Design a simple base plate for a 273 × 25 CHS to carry a factored axial load of 6340 kN

Strength of bedding material = 40 N/mm²

$$\text{Area required} = \frac{6340 \times 10^3}{0.6 \times 40} = 264167 \text{ mm}^2$$

Design by the Effective Area Method using Reference 27.1

$(2k + 25)(273 - 25)\pi = 264167$

from which $k = 157$ mm

When $k > \dfrac{D - 2t}{2}$

the bearing area is a circle of $(D + 2k)$ diameter.

Then:

$(D + 2k)^2 \dfrac{\pi}{4} = 264167$

$(273 + 2k)^2 = 336348$

$4k^2 + 1092k + 74529 = 336348$

$k = \dfrac{-1092 \pm [1092^2 - 16 \times (-261819)]^{0.5}}{8}$

$ = 153$

$t_p = \left[\dfrac{6 \times 24 \times 153^2}{2 \times 270} \right]^{0.5}$

$ = 79$ mm ∴ Use 600 × 600 × 80 plate

804   Worked example

The Steel Construction Institute Silwood Park, Ascot, Berks SL5 7QN	Subject **FOUNDATION EXAMPLE 4**		Chapter ref. **27**	
	Design code BS5950: Part 1	Made by **HB** Checked by **GWO**	Sheet no.	**1**

Problem
Design a built-up base for the valley stanchion of a double bay crane shed that is shown below. The stanchion comprises twin 406 × 178 UB.

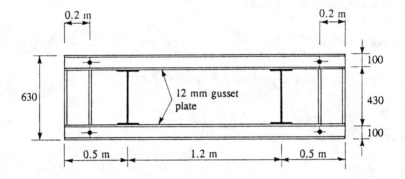

Taking moments about the tensile bolt, with n, the trial neutral axis as 0.4 m depth, and taking

$f_c = 0.4 f_{cu} = 8 \text{ N/mm}^2$

Loading

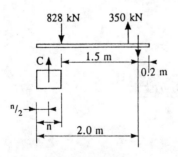

$828 \times 1.5 - 350 \times 0.3 = C(2.0 - 0.2)$

$C = \dfrac{828 \times 1.5 - 350 \times 0.3}{1.8} = 632 \text{ kN}$

The Steel Construction Institute Silwood Park, Ascot, Berks SL5 7QN	Subject **FOUNDATION EXAMPLE 4**		Chapter ref. **27**
	Design code **BS5950: Part 1**	Made by **HB**	Sheet no. **2**
		Checked by **GWO**	

$$f_c = \frac{632 \times 10^3}{0.4 \times 0.6 \times 10^6} = 2.63 \text{ N/mm}^2$$

taking n as 0.2 m, C becomes 598 kN and f_c is 4.98 N/mm²

T is then: 598 + 350 − 828   =   120 kN

to check, take moments about C

$$\frac{828 \times 0.4 - 350 \times 1.6}{1.9} = 120.4 \text{ kN}$$

The relative stiffness of the base plate and channels will determine the point of application of the compressive force. As an alternative therefore assume the lever arm to be equal to the bolt centres and the centre of compression at the bolt line with appropriate stiffening added at this point.

$$C = \frac{828 \times 1.5 - 350 \times 0.3}{1.7} = 669 \text{ kN}$$

T   =   669 + 350 − 828   =   191 kN

$$n = \frac{669 \times 10^3}{630 \times 8 \text{ N/mm}^2} = 133 \text{ mm}$$

This is the minimum value of n for concrete strength of 20 N/mm².

Design of channels & gusset

M   =   669 × 300 / 10³   =   200.7 kNm

Use 2/ 229 × 89 × 32.76 RSCS , M_{cx} = 95 kNm

These are satisfactory by inspection since the gussets and base plate acting compositely would also make a contribution.

The internal stiffener and base plate would similarly be designed as a composite member taking the maximum outstand given in Table 7 of BS 5950.

The Steel Construction Institute Silwood Park, Ascot, Berks SL5 7QN		Subject FOUNDATION EXAMPLE 4	Chapter ref. 27
		Design code BS5950: Part 1	Made by HB Checked by GWO	Sheet no. 3

The base plate panel between the stiffeners should be checked using the Pounder expressions given in Chapter 30 as follows - the panel is shown below.

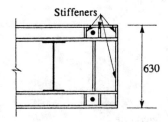

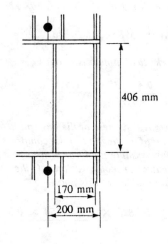

Base plate

$$\frac{L}{B} = \frac{406}{170} = 2.38$$

$$K = \frac{2.38^4}{2.38^4 + 1} = 0.97$$

K_{me}, the Pounder expression for moment, in the centre of the long edge, when all four edges are encastre, is given below:

$$K_{me} = K\left[1 + \frac{11}{35}(1-k) + \frac{79}{141}(1-k)^2\right]$$

$$= 0.9796$$

w_{me}, the ultimate load intensity is given by:

$$w_{me} = \frac{275 \times 12 \times 1.2 \times t^2}{6 \times 170^2 \,[\,K_{me}\,]}$$

$$f_c = \frac{669 \times 10^3}{600 \times 200} = 5.58 \text{ N/mm}^2, \text{ cf. } f_c = 4.98 \text{ N/mm}^2 \text{ for } n = 0.2 \text{ m}$$

the plate thickness of 16 mm is therefore satisfactory.

Chapter 28
Bearings and joints

by STEPHEN MATTHEWS

28.1 Introduction

28.1.1 Movement

All structures move to some extent. Movements may be permanent and irreversible or short-term and possibly reversible. The effects can be significant in terms of the behaviour of the structure, its performance during its lifetime, and the continued integrity of the materials from which it is built.

Movements can arise from a variety of sources:

(1) environmental: thermal, humidity, wind-induced.
(2) material properties: creep, shrinkage.
(3) loading: axial and flexural strains, impact, braking, traction, centrifugal forces.
(4) external sources: tilt, settlement, subsidence, seismic loads.
(5) use of the building: heating, cold storage.
(6) others: requirements for moving or lifting bridges, allowances for jacking procedures, during or after construction.

In general it is necessary to consider the behaviour of the structure at each point in terms of its possible movement in each of three principal directions, together with any associated rotations. The movements of a structure are not in themselves detrimental; the problems arise where movements are restrained, either by the way in which the structure is connected to the ground, or by surrounding elements such as claddings, adjacent buildings, or other fixed or more rigid items. If provision is not made for such movements and associated forces it is possible that they will lead to, or contribute towards, deterioration in one or more elements. Deterioration in this context can range from, for example, cracking or disturbance of the finishes on a building to buckling or failure of primary structural elements due to large forces developed through inadvertent restraint.

28.1.2 Design philosophies

In catering for movement of a structure, one of three methods can be adopted:

(1) Design the structure to withstand all the forces developed by restraint of movement. This is possible with smaller structures (small-span bridges) or structures which are comparatively flexible (portal frames, in the plane of the frame). The method will avoid joints but may require the use of additional material in construction.
(2) Subdivide the structure into smaller structurally stable units, each of which then becomes essentially a structure in its own right, able to move independently of the surrounding units. This principle is ideal for controlling those factors such as thermal movement which are related to the size of the overall structure. In many cases, the need for bearings as discrete elements can be eliminated. The disadvantage lies in the need to provide joints between the various units of the structure capable of accommodating all the anticipated relative movements between the units, while at the same time fulfilling all the other requirements, i.e. visual, practical, etc. It is, however, generally possible to achieve a balance by subdividing the structure so that the movements at the joints between units are kept relatively small, permitting the joints to be simple and economical (possibly at the expense of larger numbers of joints).
(3) Subdivide the structure into fewer but larger sections, and make provision for a smaller number of joints, each with larger movement capacity, and thus possibly more complex than those that would be used at (2). Examples are to be found in bridges where use of the least number of road deck joints is preferable both in terms of riding quality, and also in the minimization of long-term maintenance requirements.

The need to restrict strains on elements and thus to protect finishes will lead to the adoption of the second of the above methods for design of building structures. Bridges, for reasons cited above, are more frequently designed adopting the third method.

28.2 Bearings

28.2.1 Criteria for design and selection

28.2.1.1 Form of the unit

Choice of form depends on several criteria:

(1) *Physical size limitations.* The space available in the structure for the bearing. As bearings are subject to more wear than other parts of the structure they may have a shorter life and consequently this space should include allowance for access, inspection, maintenance and possible replacement.
(2) *Bearing pressure.* The allowable bearing pressure on the materials above and below the bearing will dictate the minimum size of the top and bottom faces of the bearing unit.

(3) *Loading*. The magnitude of the design load to be withstood by the bearing in each of the three principal directions will govern the form and type of the bearing. For each direction the maximum and minimum load should be considered at ultimate limit state, serviceability limit state or working load depending on the requirements of the design. In each case co-existent load and movement effects should be considered, together with a check for the existence of any load combinations which would act so as to separate the components of the bearing (e.g. uplift). For bearings carrying both horizontal and vertical loads it is common that the design of the bearing requires a *minimum* vertical load to be present to ensure satisfactory performance under horizontal loads.

(4) *Rotations*. The magnitude of the maximum anticipated rotations in the three principal directions should be considered. For certain types of bearing (e.g. elastomeric bearings) there exists an interaction between maximum load-carrying capacity and rotation/translation capacity, so that it may be necessary to consider co-existent effects under loading (3) and movement (5).

(5) *Movements*. Provision for maximum calculated movements can affect the size of the moving parts of the bearing and thus the overall size of the unit. As with rotations, the design of certain types of bearing is sensitive to the interaction of movement and loading requirements.

(6) *Stiffness (vertical, rotational or translational)*. Certain structures may be sensitive to the deformation which occurs within the bearing during its support of the loads. The various types of bearing have different stiffness characteristics so that an appropriate form can be selected.

(7) *Dynamic considerations*. Any particularly onerous dynamic loadings on the structure will have to be considered. Certain types of bearings (e.g. elastomeric bearings) have damping characteristics which may be desirable in particular instances, such as vibration of footbridges or machine foundations.

(8) *Connections to structure*. The form of connection of the bearing to the structure requires careful consideration of the materials involved and the need for installation, maintenance and replacement of the bearing. In addition, bearings are frequently at a position in the structure where different forms of construction meet, perhaps constructed by different contractors. In this case, it is necessary to ensure that surrounding construction is properly detailed so that design requirements for load transfer are achieved.

(9) *Use of proprietary bearings*. Many types of bearings are commercially available. These range from items which are available 'off the shelf' to more specialized units which may be designed and proven, but which are only produced to order. It is often appropriate for bearings to be individually designed to meet a particular need in situations where proprietary types may not be suitable. In these instances the engineer has the option of designing the units using available literature (see References to Chapter 28) and perhaps incorporating standard bearings from a manufacturer as components of a completed assembly or alternatively engaging a recognized manufacturer to design and produce the item as a special bearing. For straightforward

applications such as may be required on a short single-span bridge, it may be worthwhile investigating the relative costs of a simple fabricated bearing compared with the equivalent proprietary unit. Bearings (particularly 'special' bearings) can prove to be a large item of expenditure in a structure and an estimate of the costs involved should be made early in the design stage.

(10) *Summary of design requirements*. Before selecting a particular bearing it is suggested that a summary of all relevant parameters is prepared. This can then be used if necessary for submission to the bearing manufacturers for examination and recommendations as to particular bearing types. A typical format for such a sheet is given in Table 9 of BS 5400: Section 9.1.[1]

28.2.1.2 Materials

Generally materials fall into three groups:

(1) those able to withstand high localized contact pressures e.g. steel.
(2) those able to withstand lower contact pressures but having a low coefficient of friction; these slide easily in a direction perpendicular to the direction of the pressure and thus accommodate translational movement, e.g. polytetrafluoroethylene (PTFE).
(3) those able to withstand contact pressure and also to accommodate translational or rotational movements by deformation of the material (e.g. elastomers). Certain of these materials may be confined within a steel cylinder in order to increase their compressive resistance.

(a) Mild or high-yield steel

The coefficient of friction of steel on steel is of the order of 0.3 to 0.5, unless continuously lubricated; in order to provide for movement alternative arrangements are usually necessary. Traditionally this has been through the use of single or multiple rollers or knuckles. Rollers will permit translation in one direction and, if a single roller is used, rotation about an axis perpendicular to that direction. Knuckles permit rotation about one axis only. Rotation in two directions may be achieved using spherical-shaped bearing surfaces.

The allowable pressures between surfaces for steel on steel contact depend upon the radii of the two surfaces and the hardness and ultimate tensile strength of the material used. BS 5400: Section 9.1[1] gives expressions for design load effects in such cases. As load-carrying requirements increase, the use of steels with greater hardness is dictated. This can be achieved by use of high-grade alloy steels of various compositions. For design purposes, Table 2 of BS 5400: Section 9.1[1] gives indicative values of coefficients of friction of between 0.01 and 0.05 for steel roller bearings.

(b) Stainless steel

Stainless steel is frequently used in strip or plate form to provide a smooth path for sliding surfaces. It is important to utilize a material for the sliding surface which will not deteriorate and adversely affect the coefficient of friction assumed for design of the structure. A typical arrangement is a polished austenitic stainless steel surface sliding against dimpled PTFE.

(c) Polytetrafluoroethylene (PTFE)

PTFE has good chemical resistance and very low coefficients of static and dynamic friction. Unfortunately, pure PTFE has a low compressive strength, high thermal expansion and very low thermal conductivity. As a consequence it is frequently used in conjunction with 'filler' materials which improve these detrimental effects without significantly affecting the coefficient of friction.

The coefficient of friction varies with the bearing stress acting upon it. BS 5400: Section 9.1[1] gives the relationship shown in Fig. 28.1 for continuously lubricated pure PTFE sliding on stainless steel.

Lubrication of the pure PTFE is commonly achieved by means of silicone grease confined in dimples which are rolled on to the surface of the material. References 1 and 2 give further guidance on the restrictions on shape, thickness and containment on the PTFE and stainless steel components.

In preliminary design and assessment of forces on structures using PTFE sliding bearings, a figure of 0.06 is usually assumed for the coefficient of friction and the value checked later when the bearing selection is complete.

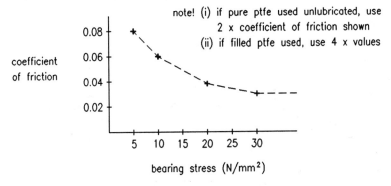

Fig. 28.1 Coefficient of friction for continuously lubricated pure PTFE

(d) Phosphor bronze

For particular applications, such as bearing guides, phosphor bronze may be used. BS 5400: Section 9.1 suggests a coefficient of friction of 0.35 for phosphor bronze sliding on steel or cast iron.

(e) Elastomers

An elastomer is either a natural rubber or a man-made material which has rubber-like characteristics. Elastomers are used frequently in bearings; they either constitute the bulk of the bearing itself or act as a medium for permitting rotation to take place (see sections 28.2.2.2 and 28.2.2.3(7)).

Elastomers are principally characterized by their hardness which is measured in several ways, the most common of which is the international rubber hardness (IRHD). This ranges on a scale from very soft at 0 to very hard at 100. Those elastomers used in bearings which are to comply with BS 5400: Part 9 have hardnesses in the range 45 IRHD to 75 IRHD.

The tensile capacity of most elastomers is considerable. As an illustration BS 5400: Part 9 specifies a minimum tensile elongation at failure of between 300% and 450% depending on IRHD.

When considering the behaviour of a block of elastomer under vertical compression it is assumed that the material is securely bonded to top and bottom loading plates. In this case (which is representative of most bearing situations) the vertical behaviour is related to the material's ability to bulge on the four non-loaded faces and is expressed in terms of the *shape factor* for the block, which is the ratio of the loaded area to the force free surface area (see Fig. 28.2).

$$S = \frac{LB}{2t(L+B)} \quad \text{for a rectangle}$$

$$S = \frac{D}{4t} \quad \text{for a circle}$$

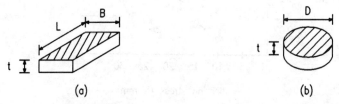

Fig. 28.2 Elastomeric bearing dimensions for (a) a rectangular block, (b) a circular block

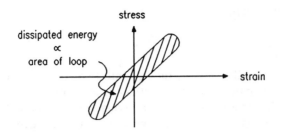

Fig. 28.3 Energy loss in elastomer

The vertical stiffness of the block is significantly reduced if the loaded faces can slip laterally.

The shear stiffness of a block of elastomer bonded to top and bottom plates is more or less linear and independent of shape factor. A detailed treatment of these effects is given in References 3 and 1.

An important property of elastomers is related to the fact that the strain in the material tends to lag behind the stress which causes it. As a consequence, some of the energy input during deformation is dissipated within the bearing as heat. A typical plot of stress versus strain for an elastomer is shown in Fig. 28.3. The energy lost as heat in one loading cycle is represented by the area of the loop.

This effect, known as 'hysteresis', has two implications:

(1) it can result in a build-up of heat in the bearing under dynamic loading conditions,
(2) if appropriately sized it can be used to act as a form of damping device to the structure.

The sensitivity of the elastomer to dynamic loading depends upon both the frequency of the applied stress and also temperature, as elastomers exhibit hardening at low temperatures.

Elastomers are prone to creep and are sensitive to attack by atmospheric oxygen and ozone, petrol and radiation from nuclear sources. They are not suitable for operation in temperatures above 120°C.

A more detailed discussion of these aspects is given by Long,[4] and limits are given for their control in practice by BS 5400.[1]

(f) Concrete

The concept of the use of a small, highly contained block of concrete as a hinge has been employed in the form of the Freysinnet hinge or the Mesnager hinge. Details of these are given in Reference 5.

28.2.2 Types of bearing

28.2.2.1 General

Structural bearings can be broadly divided into two types: elastomeric and mechanical. Elastomeric bearings comprise blocks of elastomeric material reinforced as necessary with other materials. They can be made to accommodate movement by shearing of the elastomer block. Mechanical bearings are generally formed from metal and employ sliding surfaces of PTFE to achieve any necessary movement capabilities. It should be noted that the above divisions are not exclusive as, for example, some bearings are commercially available which employ elastomer blocks with sliding surfaces which cater for larger movements than could be accommodated by shearing of the block, but which do not merit the larger expense of a mechanical bearing.

In the following sections the principal types of bearing are briefly described.

28.2.2.2 Elastomeric bearings

Elastomeric bearings rely for their operation on the interaction between vertical load, rotation and translation. As a consequence, design of most elastomeric bearings must be carefully checked. Large proprietary ranges are available and although load tables of the various capacities are published by manufacturers, it is prudent to ask the supplier to confirm that the selected bearing is suitable for the load/movement conditions under which it will be used. The basis of design of these bearings is related to controlling strains and stresses in the elastomer and any reinforcing material and ensuring that the bearing does not deform excessively, become unstable, lift off, or slip under the anticipated design effects. Further guidance on this subject is given in BS 5400: Part 9.1[1] and by Long.[4] The latter reference also gives a detailed discussion of the properties of elastomers.

Bearing types are:

(1) *Rubber pad or strip bearings*. As the name implies, these bearings consist simply of a block or strip of elastomer. They have the advantage of being inexpensive and simple although their load-carrying and movement capability is limited.
(2) *Fabric-reinforced bearings*. In order to increase the capabilities of the simple pad bearing, use is made of fabric (e.g. compressed cotton duck) to reinforce the elastomer. Movement in these bearings is usually provided by use of a PTFE surface bonded to the top of the block and sliding against a stainless steel plate attached to the underside of the superstructure. In this manner, the elastomer is used to provide rotational capability only, rather than rotation and movement as in the case of other elastomeric bearings.
(3) *Elastomeric-laminated bearings*. This type of bearing consists of a block of elastomeric material reinforced with steel plates to which the elastomer is also

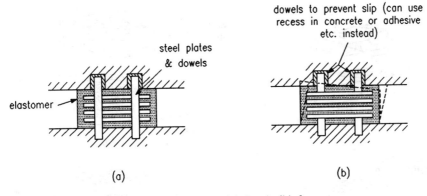

Fig. 28.4 Two types of elastomeric bearing: (a) fixed, (b) free

bonded. The characteristics of the bearings can be varied considerably by alteration of the size, shape and disposition of the layers as well as the usual parameters of bearing plan area and thickness. Generally, these bearings are either 'fixed' for translation by means of a steel dowel passing through the bearing layers (Fig. 28.4(a)) or 'free' bearings which permit translation and rotation by deformation of the bearing (see Fig. 28.4(b)). This type of bearing is capable of carrying quite substantial loadings and movements and has the benefit of being cheaper than mechanical bearings.

28.2.2.3 Mechanical bearings

(1) *Roller* (Fig. 28.5(a)). The earlier and more traditional forms of bearing comprised single or multiple steel rollers sandwiched between upper and lower steel plates. Single rollers will allow for longitudinal movement and rotation about the axis of the roller, while at the same time carrying comparatively high vertical loads, but will not permit transverse rotation or movement. Bearings of very large capacity have been produced by use of special alloy steels to form the contact surfaces. Note that bearings which utilize multiple rollers will not allow rotation about an axis parallel to the axis of the rollers. Rollers are sometimes used enclosed in an oil bath or grease box to exclude deleterious matter. Other forms of bearing have, to a large extent, supplanted the use of rollers for the most common applications.

(2) *Rocker* (Fig. 28.5(b)). Rocker bearings will not permit translational movement. The bearings may be cylindrical or spherical on one surface with the other surface flat or curved. In the cylindrical form there is no provision for transverse rotation which may have consequences for design of the structures above and below the unit. Rocker bearings usually incorporate a pin or shear key between the two surfaces to maintain relative position.

816 Bearings and joints

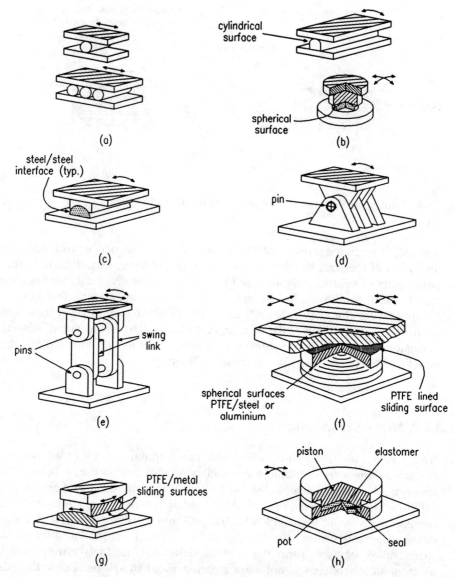

Fig. 28.5 Mechanical bearings: (a) single/multiple roller, (b) cylindrical/spherical rocker, (c) cylindrical knuckle, (d) knuckle leaf, (e) swing link, (f) spherical — with sliding top plate, (g) cylindrical PTFE bearings combined to form 'anticlastic' bearing, (h) 'pot' bearing which can have sliding top plate similar to (f)

(3) *Knuckle bearings* (Fig. 28.5(c)). These are similar to rocker bearings.
(4) *Leaf bearings* (Fig. 28.5(d)). These are formed of leaves of steel with a common pin. They will carry large vertical loads and permit large rotations

about the axis of the pin but not transversely. They have the benefit that they can be designed to resist uplift. It should be noted, however, that they are unlikely to be anything other than produced to order and that there may be other means of controlling comparatively small uplifts (e.g. 'pot' type bearing with separate vertical restraints). Leaf bearings have been used in suspension bridges to form the swing link bearings which are necessary to cater for large movements and uplifts (Fig. 28.5(e)).

(5) *Spherical (PTFE, circular)* (Fig. 28.5(f)). These comprise a spherical lower surface which is lined with PTFE and a matched upper spherical surface of aluminium or stainless steel. This arrangement allows considerable rotation capacity in all directions. Horizontal translation is frequently achieved using another (flat) sliding surface above the upper part of the bearing.

An important consideration with spherical bearings is that in order to withstand any horizontal loads it is necessary to have a minimum co-existent vertical load to prevent instability.

Spherical bearings are capable of carrying high vertical loads and also permit higher rotations than many other types.

(6) *Cylindrical (PTFE) 'anticlastic' bearings* (Fig. 28.5(g)). These are similar in concept to rocker bearings but instead of using (for example) steel on steel bearing surfaces they have enlarged bearing areas which are coated with PTFE on one surface and stainless steel or aluminium on the other. This produces a bearing with high rotation capabilities about an axis as well as high load-carrying capacity. One unit can be combined with another similar arrangement to provide rotation about an axis at right angles to the first and also with a sliding plate arrangement to provide translational capability.

(7) *Disc or 'pot'* (Fig. 28.5(h)). These are often of similar proportions to spherical bearings but instead of a sliding spherical surface being used to provide rotation capability, a disc of elastomeric material is used, confined in a cylindrical pot. Loading is applied to the surface of the disc via a closely fitting steel piston. Under these conditions, the confined elastomer is in a near fluid state, and permits rotation in all directions without significant resistance. Sliding is achieved by means of a PTFE/sliding surface above the piston, in a similar manner to spherical bearings. Disc bearings are popular for many applications, as they tend to be cheaper than spherical bearings but can carry higher loadings than elastomeric-laminated bearings of comparable plan area. They have rotation capabilities intermediate between spherical and laminated bearings.

(8) *Fabricated*. Fabricated bearings have become less popular largely through the availability of a wide range of proprietary units. They are used for footbridges and temporary works applications. There is, however, no reason why properly designed fabricated bearings should not be used to support a structure, particularly say, for a fixed bearing where there is no requirement for sliding surfaces. Guidance on design of bearings is given in References 1–8.

(9) *Special*. Special bearings will always be required for particular locations. Perhaps the most common demands are for:

(a) bearings which will resist horizontal loads only in order to restrain the structure in the horizontal plane, but without providing any vertical support (see section 28.2.4.2(3))
(b) bearings which will withstand uplift under certain loading conditions.

Uplift bearings can be special versions of normal proprietary bearing types, or can use a proprietary bearing set in a subframe which controls the tendency to uplift within prescribed limits adopted in consultation with the bearing manufacturer.

28.2.3 Use of bearings

28.2.3.1 General

The parameters which dictate the form of the bearing as a unit are discussed in section 28.2.1. It is also necessary to consider the action of the bearing in the broader concept of the behaviour of the two elements of structure which the unit connects.

28.2.3.2 Fixings

Various forms of fixings are utilized to connect bearings to the structure. These include:

(1) shear studs, usually in conjunction with a subsidiary plate which is tapped to receive the bearing fixing bolts,
(2) square or cylindrical dowels, tapped to receive the bearing fixing bolts,
(3) direct bolting of the bearing to the structure.

In all forms it is desirable to allow for tolerances in the processes of installation and possible need for replacement of the bearing. The system shown in Fig. 28.6

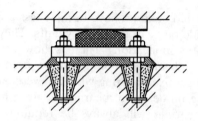

Fig. 28.6 Bearing fixing

allows for support of the bearing during fine adjustment, but requires large jacking capability (possibly more than a continuous structure could accommodate) to remove it.

In bearings subjected to dynamic loadings such as machine foundations, it is necessary to ensure that the fixings are vibration-proof.

28.2.3.3 Effect on the structure

The elements of the structure above and below the bearing are affected by the type of bearing, which can be classified as:

(1) *fixed* – not permitting movement in any horizontal direction,
(2) *guided* – movement, constrained by guides of some form, to be in one horizontal direction only,
(3) *free* – movement permitted in all horizontal directions,
(4) *elastomeric* which may be laminated or not. These bearings can be 'fixed' by means of steel dowels passing through them but are more often used 'free' in all directions and their capability to generate forces when shearing takes place is utilized to withstand horizontal loadings. If the whole structure is supported on such bearings it effectively 'floats', with all horizontal loads shared by all bearings. (See also section 28.2.4.2.)

If the bearings are fixed or guided, the neighbouring structure must be designed for the forces arising from the restraints. Even when the bearing is free in a particular direction and movement is permitted, some forces are developed – either from friction effects at the movement interfaces of a sliding mechanical bearing, or from shearing deformation in the case of an elastomeric bearing (see Fig. 28.7(a)).

In addition to forces developed laterally, the effects of the eccentricities produced by the movement must be allowed for, and also the rotation capability of the bearing in the transverse direction (see Figs 28.7(b), (c) and (d)). It is possible to control the extent of the additional eccentricity effects on a steel superstructure by use of a sliding bearing inverted which transfers the eccentricity to the substructure where it may be more easily accommodated. In this case however care should be taken to protect the sliding surfaces against falling dust, debris, etc. by use of a flexible skirt enclosure.

28.2.3.4 Installation

Bearings must be correctly installed into the structure. The procedure will depend upon the form of the structure above and below the bearing, and the type of bearing, but in general care should be taken not to load the bearing significantly before bedding materials between the bearing and the structure have fully cured,

820 Bearings and joints

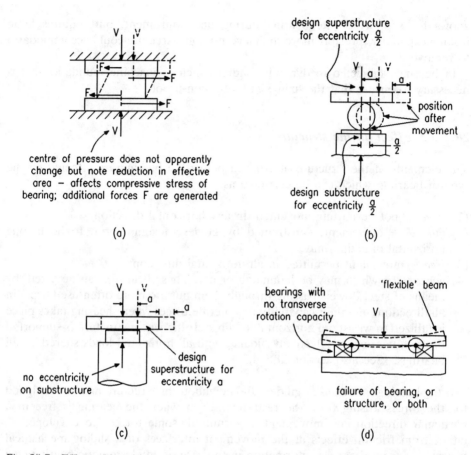

Fig. 28.7 Effect of bearing on structure: (a) elastomeric, (b) roller, (c) sliding, (d) need for transverse rotation capacity

or to load the bearing in a manner for which it has not been designed. A common cause of the latter effect is the incorporation of temporary packs or shims into the bedding in such a manner that they subsequently form hard spots under the bearing. (Note that there is, however, no fault in principle with the concept of temporary packing provided that it is responsibly carried out.)

Factory-assembled bearings are usually provided with transit straps to prevent inadvertent dismantling of the unit. When a significant irreversible movement is anticipated at a bearing, due to shrinkage or prestressing for example, an allowance for this movement may be pre-set in the factory. To allow for departure of the actual structure temperature, when the bearings are set, from the mean temperature assumed in design, it may be necessary to make alterations of the relative positions of the fixed and moving elements of the bearing. For bridges, Lee[5] suggests times when this may be most conveniently carried out — when the bridge temperature is approximately equivalent to the air shade temperature. This can be taken as:

(1) concrete bridges: 0900 BST +/− 1 hour each day,
(2) steel bridges: at about 0400 to 0600 BST each day during the summer, and at any time on 'average' days during the winter.

Further guidance on bridge temperatures is given in Reference 9.

Frequently bearings are incorporated into the structure using a bedding layer above and below the unit, typically of 25 mm thickness which allows some tolerance in fixing of the bearing, and will also permit final adjustment of the levels of the structure above during construction. The form of the bedding may be 'dry pack', trowelable or pourable material. Epoxy resin, sand/cement, sand/epoxy, or sand/polyester compounds are commonly used. The same material may also be used for filling the spaces around fixing devices once final positioning has been carried out.

28.2.4 Assemblies of bearings

28.2.4.1 General

The selection and use of bearings of various types has been discussed in terms of the individual units. The behaviour of the structure or substructure as a whole will now be considered, and the use of the four principal forms of bearing to control movement illustrated.

28.2.4.2 Structures straight in plan

As an example, the movement of a typical bridge deck will be considered in the horizontal plane, although the principles involved can equally be applied in other directions.

The four forms of bearing commonly available are given in section 28.2.3.3.

Consider the bridge deck shown in plan in Fig. 28.8.

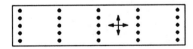

Fig. 28.8 Straight bridge deck

822   Bearings and joints

The deck vertical loading arises from dead and live loads, from which maximum and minimum values of bearing loads can be derived at each position. Longitudinal loading on the deck will arise from wind loads, braking and traction of vehicles, and also from the manner in which the chosen restraint system accommodates movements. Forces from similar effects are generated in the transverse direction also.

Thermal expansion and contraction of the deck is frequently the predominant reversible movement. This can normally be considered to act radially from a particular fixed point on the deck. The options for bearing arrangement are many, but three typical layouts are given in Fig. 28.9(a), (b) and (c). It should be remembered that whether a bearing accommodates horizontal movement through PTFE/sliding or by shearing of an elastomer block a horizontal force (due to friction or shear respectively) will be generated, and the bearing system should be arranged so that wherever possible these forces cancel one another out, and so minimize the net horizontal force to be resisted by the substructure.

(1) In Figure 28.9(a) all the bearings are elastomeric with *no* fixed bearings. The horizontal loads in both directions are shared between all bearings and the structure 'floats'.

Thus all substructures will be loaded when horizontal loads or expansion/contraction occur. This system is economic, but is limited by the maximum capabilities of the bearings in rotation, load, and movement.

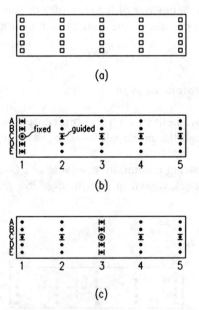

Fig. 28.9 Typical bearing layouts for straight bridge decks: (a) all elastomeric, (b) all mechanical (i), (c) all mechanical (ii)

(2) In Fig. 28.9(b) all the bearings are mechanical (typically spherical or pot bearings). Line 'C' provides fixity in the transverse direction. Line 1 provides fixity in the longitudinal direction. All longitudinal forces from external sources are taken at abutment 1, together with longitudinal forces arising from friction at bearings on piers 2 to 5. At each pier transversely lateral loads are taken by 'C' line bearings. Friction forces due to transverse expansion, etc. will tend to cancel one another out.

(3) The arrangement in Fig. 28.9(c) is better for longitudinal effects than that in Fig. 28.9(b) as friction forces in this direction tend to cancel one another out. It has the disadvantage that external loads are transmitted to an intermediate pier rather than an abutment. The forces due to movement of the deck are minimized in both horizontal directions. Occasionally, the line of fixed or guided bearings with both horizontal and vertical capability such as at 'C' may be replaced by two lines, one with bearings with vertical capability only, and one with bearings with horizontal capability only.

28.2.4.3 Structures curved in plan

If the structure shown in Fig. 28.9(b) is curved in plan, then any expansion or contraction movements longitudinally are accompanied by lateral movements also. This effect can be controlled in two ways:

(1) set the bearing guides to permit radial expansion from a fixed point on the structure,
(2) set the bearing guides tangential to the plan curvature, and so constrain the structure to follow this line when it moves (see Fig. 28.10).

In radially-guided structures the accuracy of setting out and alignment becomes more critical as the distance from the fixed point increases. In tangentially-guided structures, the structure is constrained to move along a particular path, and the horizontal forces developed in so doing must be taken into account in the design of the structure and supports.

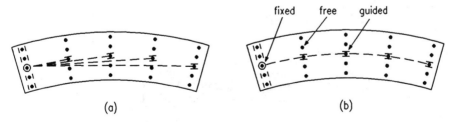

Fig. 28.10 Curved bridge deck: (a) radially-guided, (b) tangentially-guided

824 Bearings and joints

It should be noted that frequently bearings which are nominally 'guided' are manufactured with a gap tolerance at the guides. The actual value of this tolerance should be checked with the manufacturer of the particular bearing, but a value of 0.5 mm is typical. This tolerance can have a significant effect on the permissible accuracy of setting out of radially-guided structures, and the magnitude of the forces developed in tangentially-guided structures. (Reference 6.)

28.2.4.4 Structures with fixed bearings and flexible supports

An alternative to the use of systems of guided bearings is to provide fixed bearings at more than one (possibly all) supports. In this case the supporting structures (e.g. bridge piers) have to be designed to flex and accommodate the necessary movements. They also have to cater for the forces developed by these movements in addition to any other design loading effects. This arrangement may be appropriate when it is required to share horizontal load effects over several supports, but it should be noted that replacement of the bearings may be more difficult owing to horizontal loads which may be locked into the bearing/support arrangement.

28.2.4.5 Other considerations

(1) *Wedging action*. It is possible to utilize a form of 'wedging action' to resist horizontal loadings by setting two (usually elastomeric) bearings on planes inclined to one another as shown in Fig. 28.11. Equally, it is also possible to develop the action inadvertently by errors in bearing setting out, and thus attract more loading than that for which the unit is designed. (See References 3, 4 and 6.)
(2) *Shock transmission units (STUs)*. Although not strictly bearings, these units can be utilized in conjunction with bearings to distribute certain components of loading to other parts of the structure. The units typically consist of a cylinder filled with putty-like material which is acted on by a piston with a hole in it through which the putty can flow. Slow, steadily applied forces such as thermal expansion forces will cause the putty to flow from one side of the piston to the other, and allow dissipation of the force through movement.

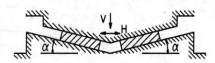

Fig. 28.11 Wedging action

Rapidly applied forces such as seismic loads, braking loads, or wind gusts are too fast to allow the flow to occur, and the unit therefore effectively transmits this 'shock' loading without significant movement. A description of the use of these devices is given in Reference 8.

28.3 Joints

28.3.1 General

The form of joints in a structure will vary to suit particular requirements at each position. The basic parameters to be considered in derivation of a joint detail are discussed below, although they are not all appropriate to every situation. Despite the fact that significant differences exist in the final application, many of the factors involved in joint design are common to both buildings and bridges. Joint detailing and construction is considerably facilitated by the many forms of proprietary sealants, gaskets, and fillers that are now commercially available for use as components, as well as complete prefabricated units which may be used in particular applications. The manufacturers of these products will generally be able to supply technical information on their products, and also to give guidance as to the suitability of items for use in particular applications.

28.3.2 Basic criteria

28.3.2.1 Form of the structure

The form of the structure, and the location and orientation of the joint within the structure, will dictate to a large extent the arrangement of the detail. The basic categories of joint are:

(1) *Wall joints.* These may be vertical (e.g. expansion joint in a building or a bridge substructure); or horizontal (e.g. joint between preformed cladding units on a building façade).
(2) *Floor/roof joints.* Examples are expansion joints in a building, or road deck joints in a bridge.
(3) *Internal/external joints.* This type of joint needs to be weather-proofed.

28.3.2.2 Material to be joined, and method of fixing

The material either side of the joint may be steel or aluminium cladding, concrete, brickwork, blockwork, or various forms of surfacing. The detail of the joint will vary considerably with the properties of the material and the method of fixing to be used. It is important to note that this may affect the stage of construction at which the joint is formed, e.g. PVC waterstops will need to be positioned before concreting of the walls on either side of them takes place. Where it is anticipated that the joint may need repair or replacement during the life of the structure (e.g. expansion joints on heavily trafficked bridges) the fixings of the joint should allow for easy removal and reinstatement.

28.3.2.3 Weather-resistance

It is important to consider the degree of weather-resistance required for a joint. In this respect, joints can be classified (in a somewhat over-simplified form) into three types (see Fig. 28.12(a), (b) and (c)):

(a) 'Closed' joint, with a filler material and exterior sealant,
(b) 'Closed' joint, with a compressible gasket and exterior sealant,
(c) 'Open' joint, with a flexible membrane seal, and arrangements to drain rain-water, etc. from the inside surfaces of the joint which are 'open' to the weather.

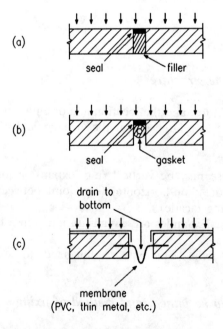

Fig. 28.12 Weather-resistance of joints: (a) closed, with filler, (b) closed, with gasket, (c) open, with membrane

Where appropriate, arrangements should be made at joints in buildings for continuity or sealing of insulation and vapour barriers, etc. to prevent formation of condensation, or loss of heat.

In structures where there is likely to be water in contact with the structural envelope, e.g. structures buried in ground which has a high water table, or water-retaining structures, flexible waterbars are usually incorporated at construction and movement joints. These have the effect of interrupting the path along which any water present has to travel.

Figure 28.13(a), (b) and (c) are typical of wall details in reinforced concrete or brickwork construction. A typical joint detail for steel cladding in a building is shown in Fig. 28.13(d), and Fig. 28.13(e) shows a detail suitable for a building roof joint, or small bridge deck movement joint.

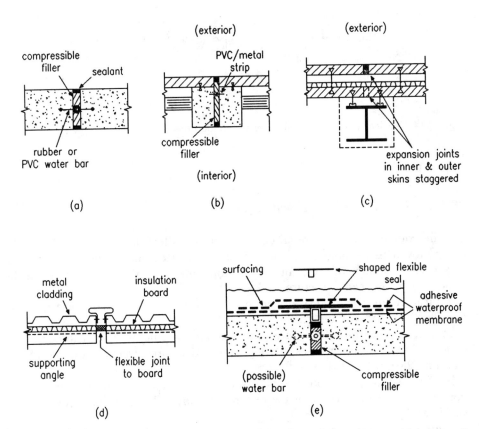

Fig. 28.13 Typical joint details: (a) concrete walls, (b) concrete columns, (c) brick walls, (d) cladding, (e) roof or bridge deck

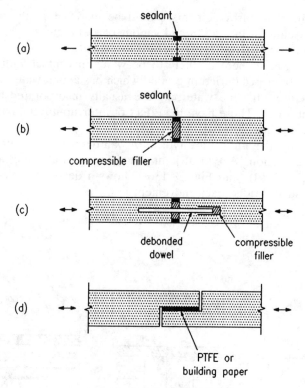

Fig. 28.14 Typical joint details allowing for moderate movements

28.3.2.4 Direction of movement required

The direction of movement required at a joint will affect the form of the joint. All likely movements should be evaluated, as restraint of unanticipated movements may result in failure of the joint, or development of large restraint forces in the joint and the adjacent structure.

Typical broad classifications of joint by movement requirements are given in Fig. 28.14(a), (b), (c) and (d).

In order to illustrate the nature of additional movements which can occur in a structure, two particular cases relating to a bridge superstructure are considered (see Fig. 28.15(a) and (b)).

In Fig. 28.15(a), the bridge deck is set to allow radial expansion from a fixed point. It will be observed that the expansion joint will require both longitudinal *and* lateral movement capability.

Figure 28.15(b) represents a cross-section throughout the end of the last span of a bridge. In (i) the distance from the actual point of rotation to the joint is small,

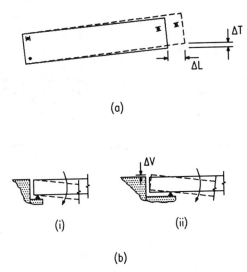

Fig. 28.15 Radial and rotational movement. (a) Radial expansion from a fixed point, (b) End rotation

and little vertical movement of the joint is necessary when flexure of the beam causes rotation. In (ii) the effect on the joint of a large overhang is demonstrated: a much larger vertical movement is induced. Further discussion of these and other related effects is given in References 3–6.

28.3.2.5 Magnitude of movement required

The magnitude of movement to be allowed for has a significant effect on the type of joint. Joints can be broadly classified into three groups on this basis:

(1) *0–25 mm movement*: Common for buildings, where there is a tendency towards the use of larger numbers of joints, each with a small movement. The restricted amount of movement considerably facilitates joint construction as the discontinuity in the structure is smaller. Similar joints are used in bridge work for small-span bridges, or so-called 'buried joints' in the road surface. Typical examples of joints suitable for movements within this range are given in Fig. 28.14. They may be preformed units, but are more likely to be assemblies of proprietary components arranged to suit the particular requirements of the detail and its location.
(2) *25–300 mm movement*: Joints capable of movements of this extent are more common in bridge work, and are usually prefabricated units, although a simple plate arrangement can be used in some cases (Fig. 28.16(a)). Two basic types of prefabricated units are common. Figure 28.16(b) shows the first

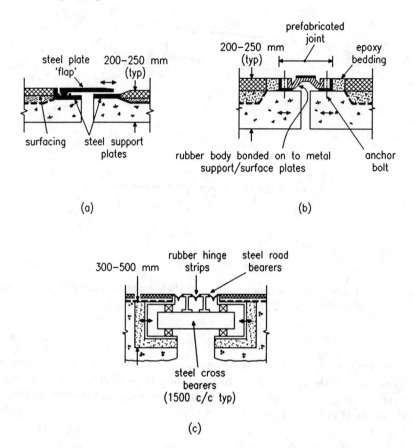

Fig. 28.16 Joints allowing for larger movements: (a) (b) 200–250 mm, (c) 300–500 mm

type, which uses an elastomeric material in its construction to accommodate the movement. Figure 28.16(c) shows the second type, which is suited to the larger movement ranges. This type uses an arrangement of mechanical components supported on stub beams to form a moving joint. Detailing of the structure must allow for the space required by these units.

(3) *>300 mm movements*: Joints for movements of this magnitude are unusual and will be tailored to particular requirements, e.g. expansion joints for suspension bridge decks.

28.3.2.6 Required performance of the joint surface

The nature of effects acting on the surface of the joint should be considered. In building structures, this will frequently be limited to environmental effects on exterior walls or roof, but consideration of the size of the joint gap is necessary in, say, a floor or roof joint where excessive gaps could prove hazardous to pedestrians, or could lead to jamming of the joint through ingress of debris, etc.

Joints on bridge decks are subject to heavy localized effects from wheel loads and corrosive effects from de-icing salts and spilt chemicals. As a result, the practicalities of maintenance and perhaps eventual replacement of such joints should be considered. The resistance to skidding of the surface of such joints is also important.

28.3.2.7 *Load generation at joints*

Frequently the loads generated by compression or extension of movement joints are insignificant in terms of overall structural behaviour. In some cases, however, the loads developed may be large enough to affect the design of other elements of the structure. A particular example is a large expansion joint of the elastomeric type in a bridge, where horizontal forces developed in moving the joint should be considered in design or selection of the bearing which is to form the fixed point of the deck.

28.4 Bearings and joints – other considerations

In design and detailing on bearings and joints, it should be remembered that the positions where they exist may be positions of concentrated load application and/or significant discontinuity in the structure. Care should be taken in design to establish and preclude all likely means of deterioration and failure. This should extend to adequate supervision to ensure correct installation, as many problems have been attributed to substandard workmanship applied to an otherwise competent design.

The need for maintenance of all joints should be assessed, and adequate allowance made where necessary for inspection, servicing, and facility of replacement.

References to Chapter 28

1. British Standards Institution (1983) *Steel, concrete and composite bridges.* Part 9: Section 9.1: *Code of practice for design of bridge bearings.* BS 5400, BSI, London.
2. Kaushke W. & Baigent M. (1986) *Improvements in the Long Term Durability of Bearings in Bridges.* ACI Congress, San Antonio, USA, Sept.

3. Baigent M. *The Design and Application of Structural Bearings in Bridges*. Glacier Metal Co. Ltd.
4. Long J. (1974) *Bearings in Structural Engineering*. Newnes-Butterworths.
5. Lee D.J. (1971) *The Theory and Practice of Bearings and Expansion Joints in Bridges*. Cement and Concrete Association.
6. Nicol T. & Baigent M. *The Importance of Accurate Installation of Structural Bearings and Expansion Joints*. Glacier Metal Co. Ltd.
7. Wallace A.A.C. (1988) *Design: Bearings and Deck Joints*. ECCS/BCSA International Symposium on Steel Bridges, Feb.
8. Pritchard B. & Hayward A.C.G. (1988) Upgrading of the viaducts for the Docklands Light Railway. *Symposium on Repair and Maintenance of Bridges*, June. Construction Marketing Ltd.
9. Emerson M. *Bridge Temperatures and Movements in the British Isles*. Transport and Road Research Laboratory Report LR 228. (*See also* TRRL reports LR 382 (W. Black); LR 491 & LR 532 (M. Taylor); LR 696, LR 744, LR 748, LR 765 (M. Emerson *et al.*))

Chapter 29
Steel piles

by DENNIS WAITE

29.1 Bearing piles

29.1.1 Uses

The very high strength to weight ratio and toughness of steel makes the material particularly suitable for certain piling applications:

(1) where piles need to be handled, pitched and driven in long lengths, e.g. maritime structures, the ability to be transported and handled without concern for overstressing due to self-weight is an important consideration.
(2) the ability to withstand high driving stresses enables steel piles to be installed without damage through difficult strata such as surface deposits of slag or old foundations and subsurface layers of boulders.
(3) energy-absorbing structures such as jetties and dolphins are particularly suited to the use of steel by virtue of its ability to withstand large elastic deformation.

29.1.2 Types of pile

Steel bearing piles are available in three basic profiles (see Fig. 29.1) and in two basic qualities of steel, grade 430A and grade 510A to BS EN 10025.

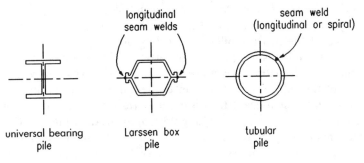

Fig. 29.1 Types of bearing pile

Universal bearing piles

These are H-sections produced on a universal rolling mill – hence their name. They are essentially the same as universal column sections, except that they have uniform thickness throughout the section (see Table 29.1).

They are used principally for land-sited structures where the full length of the pile is embedded, e.g. foundations for bridges and buildings.

They are not suitable for conditions where long lengths of the pile shaft are unsupported by surrounding soils, because buckling failure about the minor axis will occur at relatively low axial loads.

Larssen box piles

These are hollow multi-sided sections made by placing together two Larssen sheet piles in the form of a box profile and longitudinally welding the seams (see Table 29.2).

They are particularly suitable for marine structures, such as jetties and dolphins, where part of the pile shaft is exposed above sea bed level and the pile functions as a free-standing column.

Tubular piles

These are made by forming steel plate into tubes. They are available in a large range of diameters and wall thicknesses.

As with Larssen box piles they are most suitable for marine structures, especially where sited in deep water, e.g. berthing jetties for deep draught vessels.

29.1.3 Design

Loading

Although the superstructures which the bearing piles are required to support are designed by methods using factored loading, it is important that the values used for the design of the bearing piles are working loads.

In most cases the loads are principally vertical, with relatively minor horizontal components. Small horizontal loads are commonly resisted by passive soil resistance on the side of the pile cap, rather than by bending in the pile itself. Where the horizontal loads are significant they are better resisted by the inclusion of raking piles which provide a foundation with greatly improved stiffness in the horizontal plane. The induced uplift forces resulting from the use of raking piles can be resisted by the weight of the superstructure or by tension in the associated piles (see Fig. 29.2).

Table 29.1 Dimensions and properties of universal bearing piles

Serial size (mm)	Mass (kg/m)	Depth of section D (mm)	Width of section B (mm)	Thickness web and flange T (mm)	Root radius r (mm)	Depth between fillets d (mm)	Ratio $\frac{D}{T}$	Area of section (cm²)	Second moment of area Axis X-X (cm⁴)	Second moment of area Axis Y-Y (cm⁴)	Radius of gyration Axis X-X (cm)	Radius of gyration Axis Y-Y (cm)	Section modulus Axis X-X (cm³)	Section modulus Axis Y-Y (cm³)
356 × 368	174	361.5	378.1	20.4	15.2	290	17.7	222.2	51134	18444	15.2	9.11	2829	976
	152	356.4	375.5	17.9	15.2	290	19.9	193.6	43916	15799	15.1	9.03	2464	841
	133	351.9	373.3	15.6	15.2	290	22.5	169.0	37840	13576	15.0	8.96	2150	727
	109	346.4	370.5	12.9	15.2	290	26.9	138.4	30515	10901	14.8	8.87	1762	588
305 × 305	223	338.0	325.4	30.5	15.2	247	11.1	285.0	52817	17570	13.6	7.80	3126	1080
	186	328.4	320.5	25.7	15.2	247	12.8	237.0	42625	14108	13.4	7.70	2597	881
	149	318.5	315.5	20.6	15.2	247	15.4	190.0	33040	10869	13.2	7.60	2075	689
	126	312.4	312.5	17.8	15.2	247	17.6	161.3	27526	9013	13.1	7.50	1760	576
	110	307.9	310.3	15.4	15.2	247	20.0	140.4	23580	7689	13.0	7.40	1532	496
	95	303.8	308.3	13.4	15.2	247	22.7	121.0	20111	6529	12.9	7.30	1324	424
	88	301.7	307.2	12.3	15.2	247	24.5	112.0	18402	5959	12.8	7.30	1220	388
	79	299.2	306.0	11.1	15.2	247	27.0	100.4	16400	5292	12.8	7.26	1096	346
254 × 254	85	254.3	259.7	14.3	12.7	200	17.8	108.1	12264	4188	10.7	6.22	965	323
	71	249.9	257.5	12.1	12.7	200	20.6	91.0	10153	3451	10.6	6.10	813	268
	63	246.9	256.0	10.6	12.7	200	23.3	79.7	8775	2971	10.5	6.11	711	232
203 × 203	54	203.9	207.2	11.3	10.2	161	18.0	68.4	4987	1683	8.54	4.95	489	162
	45	200.2	205.4	9.5	10.2	161	21.1	57.0	4079	1539	8.46	4.90	408	133

836 Steel piles

Table 29.2 Dimensions and properties of Larssen box piles

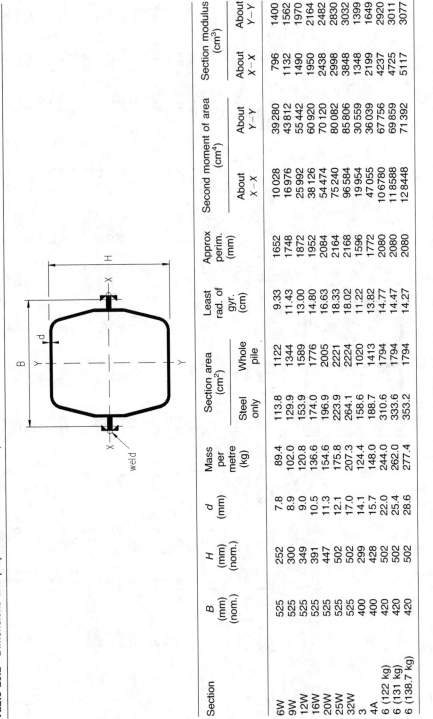

Section	B (mm) (nom.)	H (mm) (nom.)	d (mm)	Mass per metre (kg)	Section area (cm²) Steel only	Section area (cm²) Whole pile	Least rad. of gyr. (cm)	Approx perim. (mm)	Second moment of area (cm⁴) About X–X	Second moment of area (cm⁴) About Y–Y	Section modulus (cm³) About X–X	Section modulus (cm³) About Y–Y
6W	525	252	7.8	89.4	113.8	1122	9.33	1652	10 028	39 280	796	1400
9W	525	300	8.9	102.0	129.9	1344	11.43	1748	16 976	43 812	1132	1562
12W	525	349	9.0	120.8	153.9	1589	13.00	1872	25 992	55 442	1490	1970
16W	525	391	10.5	136.6	174.0	1776	14.80	1952	38 126	60 920	1950	2164
20W	525	447	11.3	154.6	196.9	2005	16.63	2084	54 474	70 120	2438	2482
25W	525	502	12.1	175.8	223.9	2221	18.33	2164	75 240	80 082	2998	2830
32W	525	502	17.0	207.3	264.1	2224	18.02	2168	96 584	85 806	3848	3032
3	400	299	14.1	124.4	158.6	1020	11.22	1596	19 954	30 559	1348	1399
4A	400	428	15.7	148.0	188.7	1413	13.82	1772	47 055	36 039	2199	1649
6 (122 kg)	420	502	22.0	244.0	310.6	1794	14.77	2080	106 780	67 756	4237	2920
6 (131 kg)	420	502	25.4	262.0	333.6	1794	14.47	2080	118 588	69 859	4725	3011
6 (138.7 kg)	420	502	28.6	277.4	353.2	1794	14.27	2080	128 448	71 392	5117	3077

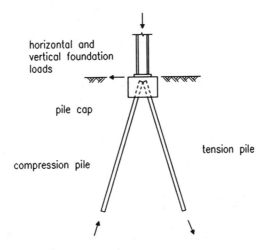

Fig. 29.2 Raking pile foundation

Design stresses

Steel bearing piles are commonly designed on a working stress of 33% of yield stress. A driven pile is usually driven to a resistance of twice the working load to demonstrate an adequate factor of safety. Thus the average stress in the shaft is 66% of yield during driving. The remaining stress capacity up to yield is available to cater for the very high impact stresses which occur in the pile shaft immediately under the hammer. Steel, being an elastic material, shortens elastically under each blow of the hammer and so the high head stresses dissipate rapidly and affect only the top one or two centimetres of the shaft metal. Figure 29.3 shows the typical stress distribution.

Pile capacity

In general, steel bearing piles are most cost effective if they can be designed to develop resistance through end bearing on rock, dense granular soils or very hard clay. In other soils it is difficult to utilize fully the strength of the steel without overstressing the soil.

When achieving end bearing on rock, it is usual to employ grade 50A steel, giving a considerable increase in capacity (37%), for a small cost increase (approximately 10%) compared with grade 43A steel.

Because the allowable axial stress is limited by the reserve of strength required for driving stresses, the pile is capable of carrying significant bending stress in addition to axial stress. The total combined stress should not exceed 0.5 yield stress.

Steel piles

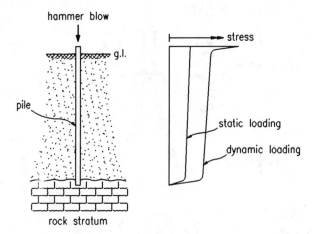

Fig. 29.3 Pile stresses

Alternatively, when the pile is likely to be subject to corrosion (see Table 29.10) and has to support only axial loading, the spare capacity in the static condition can be used to compensate for loss of metal thickness. This represents an increase in useful life of 67%.

In circumstances where the pile is driven through soft overburden to end resistance on rock the allowable design stress can be increased from 33% to 50% of yield stress since the high stresses at the head of the pile shaft will only be generated for a few blows of the hammer while the pile is being seated on the rock.

Unless any part of a pile passes through particularly weak soil, or is exposed above soil level, it will receive full lateral support and can be designed as a short column with no allowance for buckling failure.

Where a pile is exposed above soil level it must be designed as a free-standing column taking full account of slenderness ratios for the exposed part of the shaft.

A similar allowance must be made for those parts of any pile shaft which pass through particularly weak soils. Table 29.3 shows the degree of lateral restraint which may be assumed from various types of soil.

Table 29.3 Lateral support for various soil types

Soil type	Lateral support
Granular	Full
Cohesive:	
$c = 0\text{--}12.5$ kN/m^2	None
$c = 12.5\text{--}20$ kN/m^2	Partial
$c > 20$ kN/m^2	Full

Pile length

There are many different ways of estimating the length of pile which is necessary to provide a given carrying capacity. The following formulae for granular soils and cohesive soils are commonly used but others are equally suitable.

(1) *Granular soils*

   Meyerhof – based on standard penetration test (SPT).

   End resistance $400 \times N \times A$ kN ultimate
   Shaft friction $2 \times N \times a$ kN ultimate

   where
   N = blows/300 mm (SPT)
   A = end area of pile (m^2)
   a = shaft area of pile (m^2)
   For submerged soils, $N = 0.67$ (blows/300 mm)
   For H-piles,

   a = enclosing rectangle $2 \times (B + D) \times$ length
   $A = B \times D$ (firm soils)
   $= 0.5(B \times D)$ (soft soils)

(2) *Cohesive soils*

   End resistance $9 \times c \times A$ kN ultimate
   Shaft resistance $0.7 \times c \times a$ kN ultimate

   where cohesion c is given in Table 29.3.

29.1.4 Installation

Support and guidance

Adequate positional and directional restraint during pitching and driving are essential if piles are to be driven within acceptable tolerance of head location and verticality/rake. Normal tolerance limits are 75 mm in any direction in plan at the pile head, 1:100 verticality for vertical piles and 1:75 for raking piles.

Splices

Site butt splicing is frequently necessary when the required length of pile is too great to be delivered to site in one piece, or when an extension to the designed length is found to be necessary because the required resistance is not available at the anticipated depth. In both cases the butt weld is made after the first length of pile is driven until its upper end is just above ground level, the extension piece then being positioned and welded on. Acceptance standards for the site welds are given in the British Steel Plc publication *Welding Quality Acceptance Standards for Site Butt Welds in Steel Bearing Piles*.

29.1.5 Evaluation of capacity

The safe load capacity of each pile must be demonstrated. The only certain way to do this is by static test loading. Unfortunately, to do this for every pile is neither practical nor economical for most projects.

Alternative, though less accurate, ways of evaluating piles are by:

(1) calibration of the pile-driver by means of a test load
(2) dynamic formula
(3) stress wave analysis.

Pile head settlement is most accurately determined by static test load. Dynamic formulae give no indication of this value. Stress wave analysis will give an indication; the technique is rapidly improving.

Static test loading

The tests can be performed with kentledge loading or reaction anchors as shown in Fig. 29.4. Correct interpretation of the results of static test loading is essential.

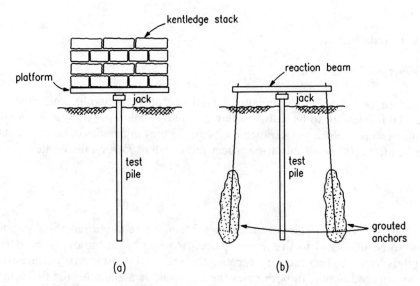

Fig. 29.4 Static load testing. (a) Kentledge test load. (b) Reaction test load

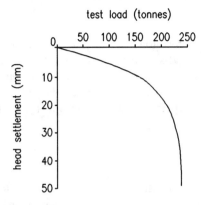

Fig. 29.5 Load–settlement plot

A plot of load/head settlement (Fig. 29.5) provides an understanding of pile head performance, which is an essential requirement for ensuring compatibility between superstructure and foundation, but it reveals only part of the picture of overall pile performance.

It is also very important to know the true limits of pile load capacity. Head settlement does not necessarily indicate toe settlement, which is the true indicator of pile failure. From the diagram in Fig. 29.5, a further diagram can be constructed by allowing for the effects of elastic shortening of the pile shaft, which will show load/toe settlement, Fig. 29.6.

Elastic shortening can be significant in steel piles, especially those which are long and fully stressed in grade 50A quality. Such shortening is of no consequence provided it is anticipated and allowed for in design and evaluation.

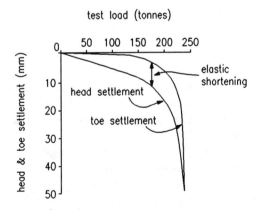

Fig. 29.6 Corrected load–settlement plot

Steel piles

Pile-driver calibration

To overcome the need to test load every pile, careful records taken during the driving of a subsequently satisfactorily test-loaded pile can be used to evaluate later piles. This system is reliable when the hammer, pile size, pile length, and geology remain unchanged for subsequent piles. However, as soon as any one of these elements is changed further test loading and recalibration is required.

Dynamic formulae

These are an attempt to relate the energy output of the impact hammer and the pile penetration per blow of the hammer to the static resistance of the pile. There are a number of such formulae available but all are to some degree unreliable. The resistance of the pile and soil under the influence of dynamic hammer loading may be very different to that under static loading.

Dynamic formulae are therefore useful in indicating relative, rather than absolute, values. Where there are relatively few piles in the project and test loading is not economic, dynamic formulae may be used, but with a suitably enhanced factor of safety.

The use of dynamic formulae also requires careful supervision of site operations. Drop hammers must be as truly 'free fall' as possible — ram guide and cable friction minimized, and drum inertia allowed for. Diesel hammers must not be allowed to overheat as this leads to marked reduction in energy output. Dolly material must be in good condition and pile heads undamaged at the time of taking readings.

Finally, and most important, redrive checks must be carried out to ensure that the resistance of the pile does not change after driving is completed. This is most likely to occur if pore water pressures in the soil are altered by the action of driving the piles. A significant change of pile resistance between the end of the initial drive and the commencement of the redrive means that dynamic formulae are completely unreliable for those circumstances.

Stress wave analysis

This is a relatively recent development which is gaining acceptance through the advent of direct on-site computer analysis of pile gauge readings. It provides information on pile toe resistance, pile shaft resistance, pile shaft integrity, and actual energy input to the pile head.

This form of evaluation provides valuable information as the pile is driven, which is an ideal arrangement. However, it still relies on relating dynamic resistance to static resistance and care must to taken to ensure the results are not misleading. Redrive checks are essential and a cross check by static loading may be advisable at the start of a large contract.

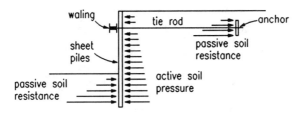

Fig. 29.7 Sheet pile retaining wall

29.2 Sheet piles

29.2.1 Uses

There are various types of earth-retaining structures. Where a differential surface level is to be established with a vertical interface, a retaining wall is used. Steel sheet piling is one form of retaining wall construction. Sheet piles resist soil and water pressures by functioning as a beam spanning vertically between points of support, as shown in Fig. 29.7.

Steel sheet piles are also used for the construction of temporary coffer-dams; the piles enable deep excavations to be made to facilitate construction below ground and water level of other permanent works. On completion of construction the sheet piles are usually extracted for re-use on other projects. Coffer-dams may consist of cantilever sheet pile walls when the excavated depth is small, or, for greater depths, the sheet piles are supported by one or more levels of internal supporting frames (see Fig. 29.8).

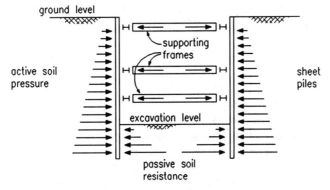

Fig. 29.8 Sheet pile coffer-dam

Steel piles

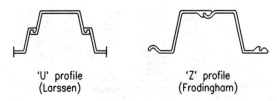

'U' profile (Larssen)   'Z' profile (Frodingham)

Fig. 29.9 Types of sheet pile

29.2.2 Types of piles

The section profile may be U or Z shape, as shown in Fig. 29.9. Details are given in Table 29.4(a) and (b). The essential difference between these sections and ordinary structural beams is their ability to interlock with each other to form a continuous membrane in the soil. Where the depth of excavation is small, trench sheet sections (Table 29.4(c)) may be used as an alternative to sheet piles.

29.2.3 Design

Soil pressures

Active soil pressure and passive soil resistance, at any given depth, are both a function of the effective vertical pressure at that depth and of the strength of the stratum being considered. The active pressure (p_a) is always equal to, or less than, the vertical overburden pressure and the passive resistance (p_b) is always equal to, or greater than, the overburden pressure.

For granular soils (sands, gravels, coarse silts, etc.), the expressions depicting the two conditions are:

$$p_a = \gamma h \times k_a$$
$$p_p = \gamma h \times k_p$$

where γ = effective density of the soil,
       h = depth of overburden,
      k_a = coefficient of active pressure,
      k_p = coefficient of passive resistance.

The expression for cohesive soils (clays) has two alternative forms, depending upon the geological history of the soil mass (normally, or over-consolidated) and upon the life requirement of the proposed earth-retaining structure (short-term temporary works, or long-term permanent works):

Table 29.4(a) Dimensions and properties of Larssen sheet piles

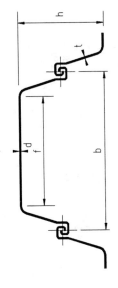

Section	b (mm) (nom.)	h (mm) (nom.)	d (mm)	t (mm) (nom.)	f flat of pan (mm)	Sectional area (cm² per metre of wall)	Mass (kg per linear metre)	Mass (kg per m² of wall)	Combined second moment of area (cm⁴ per metre)	Section modulus (cm³ per metre)
6W	525	212	7.8	6.4	331	108	44.7	85.1	6 459	610
9W	525	260	8.9	6.4	343	124	51.0	97.1	11 726	902
12W	525	306	9.0	8.5	343	147	60.4	115.1	18 345	1199
16W	525	348	10.5	8.6	341	166	68.3	130.1	27 857	1601
20W	525	400	11.3	9.2	333	188	77.3	147.2	40 180	2009
25W	525	454	12.1	10.5	317	213	87.9	167.4	56 727	2499
32W	525	454	17.0	10.5	317	252	103.6	197.4	73 003	3216
3	400	248	14.1	8.5	253	198	62.2	155.5	16 980	1360
4A	400	381	15.7	9.6	219	236	74.0	185.1	44 916	2360
6	420	440	22.0	14.0	248	370	122.0	290.5	92 452	4200
6	420	440	25.4	14.0	251	397	131.0	311.8	102 861	4675
6	420	440	28.6	14.0	251	421	138.7	330.2	111 450	5066

846   Steel piles

Table 29.4(b) Dimensions and properties of Frodingham sheet piles

Section	b (mm) (nom.)	h (mm) (nom.)	d (mm)	t (mm) (nom.)	f₁ (mm) (nom.)	f₂ (mm) (nom.)	Sectional area (cm² per metre of wall)	Mass (kg per linear metre)	Mass (kg per m² of wall)	Second moment of area (cm⁴ per metre)	Section modulus (cm³ per metre)
1BXN	476	143	12.7	12.7	78	123	166.5	62.1	130.4	4 919	688
1N	483	170	9.0	9.0	105	137	126.0	47.8	99.1	6 048	713
2N	483	235	9.7	8.4	97	149	143.0	54.2	112.3	13 513	1150
3N	483	283	11.7	8.9	89	145	175.0	66.2	137.1	23 885	1688
3NA	483	305	9.7	9.5	96	146	165.0	62.6	129.8	25 687	1690
4N	483	330	14.0	10.4	77	127	218.0	82.4	170.8	39 831	2414
5	425	311	17.0	11.9	89	118	302.0	100.8	236.9	49 262	3168

Table 29.4(c) Dimensions and properties of trench sheets

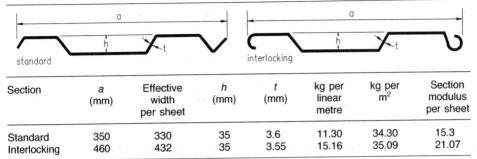

Section	a (mm)	Effective width per sheet	h (mm)	t (mm)	kg per linear metre	kg per m²	Section modulus per sheet
Standard	350	330	35	3.6	11.30	34.30	15.3
Interlocking	460	432	35	3.55	15.16	35.09	21.07

(1) Normally consolidated clays with cohesion, c, up to 50 kN/m², temporary and permanent works

$$p_a = \gamma h - k_{ac} c \qquad p_p = \gamma h + k_{pc} c$$

where k_{ac} = coefficient of active pressure,
k_{pc} = coefficient of passive resistance.

(2) Over-consolidated clays with cohesion, c, greater than 50 kN/m² temporary works

$$p_a = \gamma h - k_{ac} c \qquad p_p = \gamma h + k_{pc} c$$

permanent works

$$p_a = \gamma h \times k_a \qquad \phi'(C' = 0)$$
$$p_p = \gamma h \times k_p \qquad \phi'(C' = 0)$$

Values of k_a and k_p are modified by the effects of wall friction (δ) which is taken as 0.5ϕ for steel piles. Values of k_{ac} and k_{pc} are modified by the effects of wall adhesion (c_w) which, for active pressures, is taken as $c_w = c$, up to a maximum of 50 kN/m², and for passive pressures, is taken as $c_w = 0.5c$, up to a maximum of 25 kN/m². Tables 29.5 to 29.8 show values of k_a, k_{ac}, k_p and k_{pc}.

Active pressure and passive resistance are plotted as a net pressure diagram as shown in Fig. 29.10. From this diagram the required pile length, anchor loads (if any) and pile bending are determined.

There are two basic types of sheet pile structures:

(1) cantilever wall
(2) anchored wall.

The anchored wall can be designed such that the support mobilized from the soil at the bottom of the wall is sufficient to provide either 'fixed earth' support or 'free earth' support to the piles, the difference depending on the depth to which

Table 29.5 Coefficients of active pressure for granular soils, k_a

Values of δ	Values of ϕ				
	25°	30°	35°	40°	45°
0	0.41	0.33	0.27	0.22	0.17
10°	0.37	0.31	0.25	0.20	0.16
20°	0.34	0.28	0.23	0.19	0.15
30°	—	0.26	0.21	0.17	0.14

Table 29.6 Coefficients of passive resistance for granular soils, k_p

Values of δ	Values of ϕ			
	25°	30°	35°	40°
0	2.5	3.0	3.7	4.6
10°	3.1	4.0	4.8	6.5
20°	3.7	4.9	6.0	8.8
30°	—	5.8	7.3	11.4

Table 29.7 Coefficients of active pressure for cohesive soils, k_{ac}

	Values of δ	Values of c_w/c	Values of ϕ					
			0°	5°	10°	15°	20°	25°
k_a	0	All	1.00	0.85	0.70	0.59	0.48	0.40
	ϕ	values	1.00	0.78	0.64	0.50	0.40	0.32
k_{ac}	0	0	2.00	1.83	1.68	1.54	1.40	1.29
	0	1	2.83	2.60	2.38	2.16	1.96	1.76
	ϕ	$\frac{1}{2}$	2.45	2.10	1.82	1.55	1.32	1.15
	ϕ	1	2.83	2.47	2.13	1.85	1.59	1.41

Table 29.8 Coefficients of passive resistance for cohesive soils, k_{pc}

	Values of δ	Values of c_w/c	Values of ϕ					
			0°	5°	10°	15°	20°	25°
k_p	0	All	1.0	1.2	1.4	1.7	2.1	2.5
	ϕ	values	1.0	1.3	1.6	2.2	2.9	3.9
k_{pc}	0	0	2.0	2.2	2.4	2.6	2.8	3.1
	0	$\frac{1}{2}$	2.4	2.6	2.9	3.2	3.5	3.8
	0	1	2.6	2.9	3.2	3.6	4.0	4.4
	ϕ	$\frac{1}{2}$	2.4	2.8	3.3	3.8	4.5	5.5
	ϕ	1	2.6	2.9	3.4	3.9	4.7	5.7

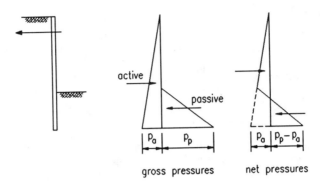

Fig. 29.10 Pressure diagrams

the piles are to penetrate. Figure 29.11 illustrates the pressure and bending moment for the various conditions.

Anchorages comprise a series of tie rods and passive anchor units, or pressure-grouted anchor tendons, at regular intervals along the length of the wall. Load is transferred from the piles to the anchors by means of walings, usually pairs of RSCs back to back with space between them to accommodate the tie rod.

When passive anchor units are used, the distance of the anchor unit from the front retaining wall must be carefully determined to ensure that the zone of unstable soil supported by the wall does not encroach upon the passive soils providing the resistance to the anchor unit. Figure 29.12 illustrates the requirement. Grouted anchors are normally designed and installed by specialist sub-contractors.

Walings are designed as beams spanning between the anchorages (taking advantage of continuity when the wales are continuous over several anchors). Adequate protection against corrosion is essential. Tie rods are selected from manufacturers' published safe load tables. Anchorages may be either steel sheet piles or concrete and either isolated or continuous.

Coffer-dams

In the design of multi-frame coffer-dams the designer must be sure to check every condition of partial excavation which can occur during construction: the temporary condition of partial excavation in preparation for erection of each succeeding frame is often more critical than the fully excavated and framed condition (Figs 29.8 and 29.13).

The number and position of supporting frames is a function of the size of sheet pile, or vice versa. The position of the frames must also be considered with respect to the construction sequence of the permanent works which are built within the coffer-dam. Compatibility of the temporary and permanent works is vital if the project is to be executed efficiently and economically. For particularly

850   *Steel piles*

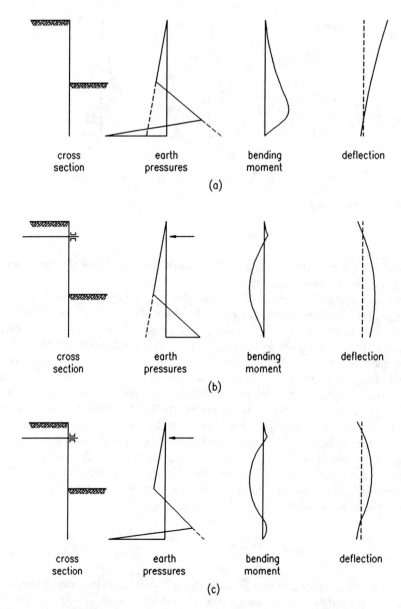

Fig. 29.11 Pressures, bending moments and deflections in retaining walls: (a) cantilever, (b) anchored, free earth support, (c) anchored, fixed earth support

complex permanent works it may preferable to form the coffer-dam as a circular structure which uses circular ring beam wales (in reality, ring struts), thus entirely eliminating cross strutting and creating an unobstructed construction space for the permanent works.

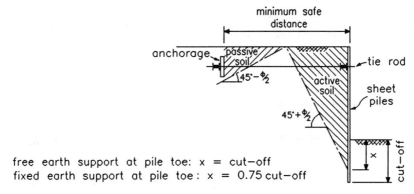

Fig. 29.12 Minimum safe distance of anchorage

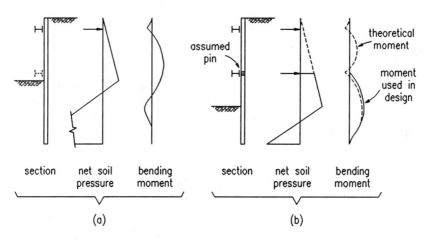

Fig. 29.13 Partial and fully excavated conditions. (a) Temporary condition, (b) final condition

When sheet piles are used for coffer-dams, the length of pile below formation level, the 'cut-off', is often critical to the stability of the bottom of the coffer-dam in terms of controlling the ingress of ground water, which can destabilize the passive zone, when working in permeable soils. Figure 29.14 shows how the overburden in the passive zone is reduced by the pressure of ground water flowing under the toes of the piles.

A flow net investigation, as shown in Fig. 29.14, will quickly demonstrate this and provide a check on the adequacy of the cut-off in general for the control of ground water flow.

852   Steel piles

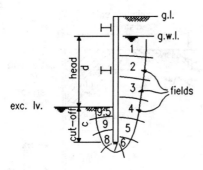

w = water density
ɣ = soil density

no. of field = N(9.5)

diff. head press = d×w

head loss/field = d×w/N

flow head at pile toe
(btm. of field no.8) = 2.5d×w/N

total passive overburden
at pile toe = (ɣ+w)×c

effective passive o/burden
at toe = (ɣ+w)×c − 2.5d×w/N

Fig. 29.14 Ground water flow net

Driving considerations

The size of sheet pile section is frequently determined by the requirement that it must be capable of being driven to the design penetration in the ground without deformation. Table 29.9 gives some guidance to enable a pile of suitable driving properties to be selected.

Table 29.9 Selection of pile size to suit driving conditions

Dominant SPT value (N)	Minimum wall modulus (cm³/m of wall)	
0–10	450	
11–20		450
21–25	850	
26–30		850
31–35	1300	
36–40		1300
41–45	2300	
46–50		2300
51–60	3000	
61–70		3000
71–80	4200	
81–140		4200

NB This table is based upon sheet pile sections of approximately 500 mm interlock centres installed with panel driving techniques. Wider sections and those installed by methods giving less control will require greater moduli.

Durability

The method of achieving the desired working life of the structure can make a very large difference to both first cost and maintenance costs. A steel structure deteriorates with the passage of time due to the effects of corrosion. However the rate of deterioration varies with the degree of aggressiveness of the environment to which the structure is exposed.

In sheet pile retaining walls, especially those in maritime locations, there are several different environmental zones (see Fig. 29.15(a)) which are, in order of decreasing aggressiveness:

(1) high tide and splash zone,
(2) low water zone,
(3) intertidal zone,
(4) atmospheric zone,
(5) submerged zone,
(6) embedded zone.

Recent research has evaluated the relative aggressiveness of each of these zones and typical rates of corrosion can now be ascribed to them. Values are given in Table 29.10. In many cases the effects of corrosion are such that no special measures are necessary since significant corrosion occurs only in locations where stresses are low and considerable loss of metal thickness is permissible (Fig. 29.15(c)).

If corrosion is still deemed to be a problem, its effects can be countered by using grade 50A steels in designs based on grade 43A stresses. This expedient has the benefit of increasing the permissible working stress by a precisely known amount and hence the working life by the same proportion. It also has the secondary benefit of providing a pile section with improved driving capability.

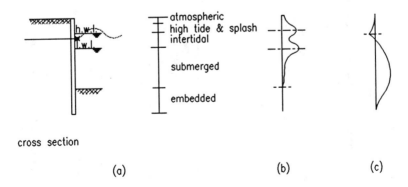

Fig. 29.15 Environmental zones. (a) Zones, (b) Corrosion rate, (c) Bending moment

Steel piles

Table 29.10 Effective life of sheet piling

Section type			Environments			
Frodingham	Larssen	Permissible loss in thickness (mm) at position of maximum stress in pile	Atmospheric corrosion	Splash and low water zones	Sea water immersion and tidal zone	Underground
			Corrosion rates (mm/year)			
			0.05 (mean)	0.09 (mean)	0.05 (mean)	0.03 (max.)
			Effective life (years)			
	6W, 9W	3.5	70	[a]	[a]	117
	12W, 16W	4.5	90	50	90	120+
	202, 3, 4A	5.0	100	55	100	120+
1N		5.2	104	58	104	120+
2N		5.4	108	60	108	120+
3NA		5.5	110	60	110	120+
3N		5.9	118	65	118	120+
	25W, 32W	6.0	120	67	120	120+
	All No. 6	7.0	120+	78	120+	120+
1BXN		7.2	120+	80	120+	120+
4N		7.4	120+	82	120+	120+
5		8.9	120+	99	120+	120+

[a] Not suitable for marine environments

Alternatively the piles can be protected against the effects of corrosion by the application of paint coatings, but even with the best surface preparation and paints, these will only give an increase of life of 15 to 20 years before maintenance becomes necessary.

Further reading for Chapter 29

British Standards Institution (1943) *Code of practice for earth retaining structures.* CP2, BSI, London. (To be replaced by BS 8002: *Code of practice for earth retaining structures.*)
British Standards Institution (1984–8) *Code of practice for maritime structures.* BS 6349, BSI, London.
British Steel, General Steels (1988) *Piling Handbook*, 6th edn. British Steel plc.
Henry F.D.C. (1985) *The Design and Construction of Engineering Foundations.* Chapman and Hall.
Tomlinson M.J. (1969) *Foundation Design and Construction.* Pitman, London.

A worked example follows which is relevant to Chapter 29.

Worked example

The Steel Construction Institute Silwood Park, Ascot, Berks SL5 7QN	Subject QUAY FACILITIES	Chapter ref. 29
Design code BS8002*	Made by DW Checked by GWO	Sheet no. 1

Problem

It is required to provide quay wall berthing facilities, with rail mounted loading/unloading cranage.

Documents used in this design example are:

* BS 8002 Code of practice for earth retaining structures (in course of preparation)

   BS 8004 Code of practice for foundations (used in the design of bearing piles)

   British Steel Piling Handbook

N.B. The following calculations are based on recommendations given in the British Steel Piling Handbook. A number of alternative methods are available for the analysis of sheet pile walls. Further details of these methods are outlined in the SCI Publication "A Comparison of Design Methods for Propped Sheet Pile Walls". Having decided on the method to be adopted, careful consideration must be given to the choice of appropriate soil parameters and Factors of Safety.

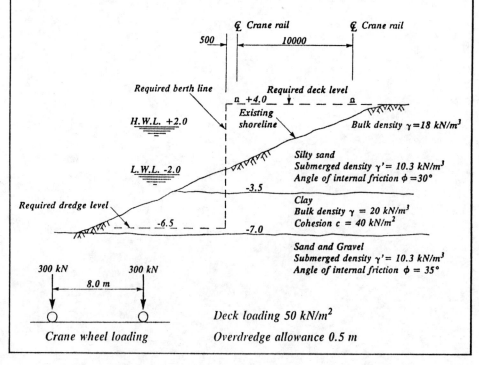

Worked example 857

The Steel Construction Institute Silwood Park, Ascot, Berks SL5 7QN	Subject **QUAY FACILITIES**	Chapter ref. **29**
	Design code **BS8002***	Made by **DW** / Checked by **GWO** — Sheet no. **2**

For construction of the quay wall it is proposed to use an anchored steel sheet pile wall. Fill behind wall to be dredged sand and gravel.

Crane rails will be mounted on ground beams carried on piled foundations, and on the cope beam at the head of the sheet pile wall.

Sheet pile quay wall

B.S. Piling Handbook

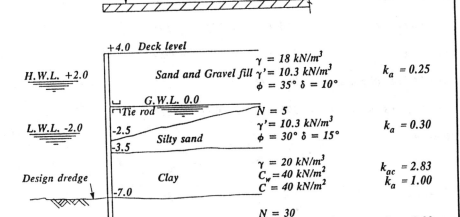

Ground water level will be related to tide level but, due to the low permeability of the sheet pile wall, will always lag behind the rising or falling tide.

Hence, assume ground water level to be at mid-tide level (+ 0.0)

Worst tidal condition for wall design is low water (− 2.0)

Worked example

The Steel Construction Institute Silwood Park, Ascot, Berks SL5 7QN	Subject **QUAY FACILITIES**		Chapter ref. **29**
	Design code **BS8002***	Made by **DW** Checked by **GWO**	Sheet no. **3**

Overburden pressures

Active: kN/m^2

At +4.0 in fill	= 50	=	50
At +0.0 in fill	= 50 + (4 × 18)	=	122
At −2.0 in fill	= 122 + (2 × 10.3)	=	143
At −2.5 in fill & sand	= 143 + (0.5 × 10.3)	=	148
At −3.5 in sand	= 148 + (1 × 10.3)	=	158
At −3.5 in clay	= 158 + (3.5 × 9.8)	=	192
At −7.0 in clay	= 192 + (3.5 × 20)	=	262
At −7.0 in sand/gravel	= 262 − (7 × 9.8)	=	193
At −10.5 in sand/gravel	= 193 + (3.5 × 10.3)	=	229

Passive: kN/m^2

At −7.0	= 0	=	0
At −10.5	= 3.5 × 10.3	=	36

Active soil & water pressures kN/m^2

At +4.0 in fill	= 50 × 0.25	=	13
At +0.0 in fill	= 122 × 0.25	=	31
At −2.0 in fill	= (143 × 0.25) + (2 × 9.8)	=	55
At −2.5 in fill	= (148 × 0.25) + (2.5 × 9.8)	=	62
At −2.5 in fill & sand	= (148 × 0.30) + (2.5 × 9.8)	=	69
At −3.5 in sand	= (158 × 0.30) + (3.5 × 9.8)	=	82
At −3.5 in clay	= (192 × 1.0) − (2.83 × 40)	=	79

	Subject		Chapter ref.
The Steel Construction Institute Silwood Park, Ascot, Berks SL5 7QN	QUAY FACILITIES		29
	Design code BS8002*	Made by DW Checked by GWO	Sheet no. 4

                                                                              kN/m^2

At -7.0 in clay   = $(262 \times 1.0) - (2.83 \times 40)$ = 149

At -7.0 in sand/gravel = $(193 \times 0.23) + (7 \times 9.8)$ = 113

At -10.5 in sand/gravel = $(229 \times 0.23) + (10.5 \times 9.8)$ = 156

<u>Passive soil & water pressures</u> kN/m^2

At -2.0 in water   = 0 = 0

At -2.5 in water   = 0.5×9.8 = 5

At -3.5 in water   = 1.5×9.8 = 15

At -7.0 in water   = 5×9.8 = 49

At -10.5 in sand/gravel = $(36 \times 6) + (8.5 \times 9.8)$ = 299

<u>Net soil & water pressures</u> kN/m^2

At $+4.0$ in fill   = 13

At $+0.0$ in fill   = 31

At -2.0 in fill   = 55

At -2.5 in fill   = $62 - 5$ = 57

At -2.5 in silty sand   = $69 - 5$ = 64

At -3.5 in silty sand   = $82 - 15$ = 67

At -3.5 in clay   = $79 - 15$ = 64

At -7.0 in clay   = $149 - 49$ = 100

At -7.0 in sand/gravel = $113 - 49$ = 64

At -10.5 in sand/gravel = $156 - 299$ = -143

The Steel Construction Institute Silwood Park, Ascot, Berks SL5 7QN	Subject **QUAY FACILITIES**	Chapter ref. **29**
	Design code **BS8002***	Made by **DW** / Checked by **GWO**	Sheet no. **5**

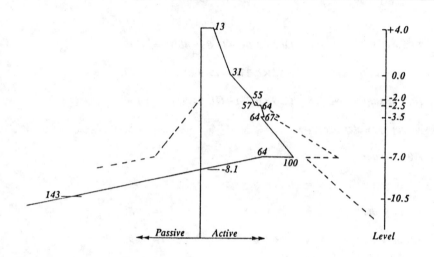

Assume anchor level is at 0.0 level

Moment of net active pressure about 0.0

13×4	=	52	$\times$ -2	=	-104
$18 \times 4 \times 0.5$	=	36	$\times$ -1.33	=	-48
31×2	=	62	$\times$ 1.0	=	62
$24 \times 2 \times 0.5$	=	24	$\times$ 1.33	=	32
55×0.5	=	28	$\times$ 2.25	=	63
$2 \times 0.5 \times 0.5$	=	1	$\times$ 2.33	=	2
64×1	=	64	$\times$ 3.0	=	192
$3 \times 1.0 \times 0.5$	=	2	$\times$ 3.17	=	6
64×3.5	=	224	$\times$ 5.25	=	1176
$36 \times 3.5 \times 0.5$	=	63	$\times$ 5.83	=	367
$64 \times 1.1 \times 0.5$	=	35	$\times$ 7.37	=	258
Total active pressure	=	591 kN			2006 kNm

Design for "free earth support"

Penetration required below -8.1 *level for factor of safety 1 against failure by forward movement of pile toe* $= x$

Slope of pressure line below $-8.1 = 10.3 \times (6 - 0.23) = 59.4 \text{ kN/m}^2/\text{m}$

The Steel Construction Institute Silwood Park, Ascot, Berks SL5 7QN	Subject QUAY FACILITIES		Chapter ref. 29
	Design code BS8002*	Made by DW Checked by GWO	Sheet no. 6

$\therefore 59.4 \times x \times x \times 0.5 \times (0.67x + 8.1) \quad = 2006$

$$19.9x^3 + 240.6x^2 \quad = 2006$$

$$x \quad = 2.6 \text{ m}$$

Pile toe level for factor of safety 1 $\quad = -10.7$

For factor of safety 2

$$19.9x^3 + 240.6x^2 \quad = 2006 \times 2$$

$$x \quad = 3.6 \text{ m}$$

Required actual pile toe level $\quad = -8.1 - 3.6 = -11.7$ Pile toe at -11.7

To find required anchor force at 0.0:

Moment of net active and passive pressures about and above -10.7 level

centre of active pressure occurs at

$+0.0 - \left[\dfrac{2008}{592}\right] \quad = -3.4 \text{ level}$

$592 \times (10.7 - 3.4) \quad = \quad 4322$

$-59.4 \times 2.6^2 \times 0.5 \times 0.87 \quad = \quad \underline{-175}$

$\qquad\qquad\qquad\qquad\qquad\qquad 4147 \text{ kNm}$

Anchor force $= \dfrac{4147}{10.7} = 388 \text{ kN/m}$ Anchor load $= 388 \text{ kN/m}$

Depth of zero shear below $-3.5 = y$

$388 - 270 \quad = 64x + (10.3 \times x^2 \times 0.5)$

$x \quad = 1.7 \text{ m}$

862   Worked example

The Steel Construction Institute Silwood Park, Ascot, Berks SL5 7QN	Subject QUAY FACILITIES		Chapter ref. 29
	Design code BS8002*	Made by DW Checked by GWO	Sheet no. 7

Level of zero shear $\quad = -5.2\ m$

Moments about and above -5.2 *level*

388×5.2	$= 2018$
-52×7.2	$= -374$
-36×6.53	$= -235$
-62×4.2	$= -260$
-24×3.87	$= -93$
-28×2.95	$= -83$
-1×2.87	$= -3$
-64×2.2	$= -141$
$-64 \times 1.7 \times 0.85$	$= -92$
$-10.3 \times 1.7^2 \times 0.5 \times 0.57$	$= -8$

$\therefore$ *max. bending moment in pile* $= \overline{729\ kNm}$

Axial load in sheet piles from crane

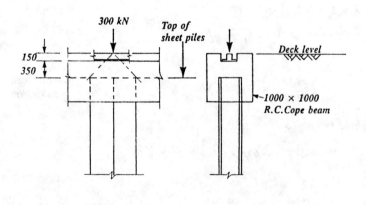

		Subject	Chapter ref.
The Steel Construction Institute Silwood Park, Ascot, Berks SL5 7QN		QUAY FACILITIES	29
	Design code BS8002*	Made by DW Checked by GWO	Sheet no. 8

load per metre run of wall

$$= \frac{300 \times 1.25}{(0.35 + 0.15) \times 2} = 375 \text{ kN}$$

Cope self-weight $= \underline{24}$

399 kN

SHEET PILE LOAD SUMMARY:

Max. axial = 399 kN/m

Max. bending moment = 729 kNm/m

Bearing piles supporting rear crane rail

Assume piles at 5 m centres

Max. axial load/pile = 300 crane

75 impact

120 ground beam SW

$\underline{495 \text{ kN}}$

<u>Anchorage</u>

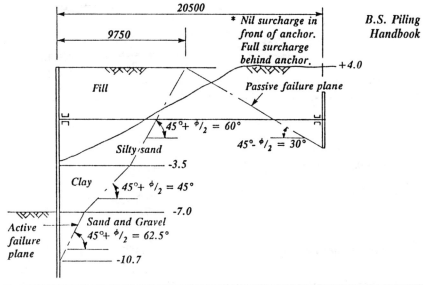

B.S. Piling Handbook

The Steel Construction Institute Silwood Park, Ascot, Berks SL5 7QN	Subject **QUAY FACILITIES**	Chapter ref. **29**
	Design code **BS8002***	Made by **DW** Checked by **GWO**	Sheet no. **9**

NB wall friction is not taken into account on anchorage soil pressures.

p_a behind anchor at +4.0 = 50 × 0.33   = 17 kN/m²

p_a behind anchor at 0.0 = [50 + (4 × 18)] × 0.33   = 40 kN/m²

p_a behind anchor at −3.5 = [122 +(3.5×10.3)]×0.33   = 52 kN/m²

p_p in front of anchor at +4.0 = 0   = 0 kN/m²

p_p in front of anchor at 0.0 = (4 × 18) × 3   = 216 kN/m²

p_p in front of anchor at −3.5 = [72 +(3.5×10.3)]× 3   = 324 kN/m²

Net p_p in front of anchor at +4.0 = 0 − 17   = −17 kN/m²

Net p_p in front of anchor at 0.0 = 216 − 40   = 176 kN/m²

Net p_p in front of anchor at −3.5 = 324 − 52   = 272 kN/m²

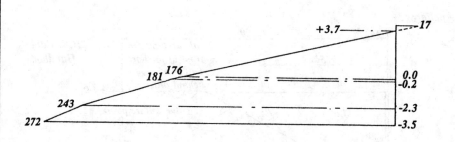

p_p net down to 0.0 = 176 × 3.7 × 0.5   = 325

p_p net 0.0 to −3.5 = (176 + 272) × 0.5 × 3.5   = 784

                                                            1109 kN/m

Assume continuous anchor wall

The Steel Construction Institute Silwood Park, Ascot, Berks SL5 7QN	Subject **QUAY FACILITIES**	Chapter ref. **29**
	Design code *BS8002**	Made by *DW* Sheet no. **10** Checked by *GWO*

For factor of safety 2, required resistance/metre

$$= 388 \times 2 \quad = 776 \text{ kN}$$

Required depth of anchor wall $= -3.5 + \dfrac{(1109 - 776)}{260 \text{ say}} = -2.3 \text{ m}$

∴ *anchor wall toe at* -2.3 *m*

Moment of passive resistance about and above -2.3 *m*

$176 \times 3.7 \times 0.5 \times 3.53 \quad = 1149$
$176 \times 2.3 \times 1.15 \quad\quad\quad = 466$
$67 \times 2.3 \times 0.5 \times 0.77 \quad = \underline{59}$
$\quad\quad\quad\quad\quad\quad\quad\quad\quad\quad\quad 1674 \text{ kNm}$

For balanced anchor, tie level at anchor

tie level at anchor $\quad = -2.3 + \dfrac{1674}{776} \quad = -0.14$ *Allow* -0.2 *m*

∴ *tie level at anchor* $= -0.2$ *m*

For anchor location see calculation sheet 8   **Anchor wall 20.5 m behind front wall**

(Passive and active failure planes must not intersect)

Top of anchor piles $= -0.2 + \left[\dfrac{3.7 + 0.2}{2}\right] = +1.75$ *say* $+1.8$ **Anchor wall head at +1.8**

Bending moment on anchor piles $= \dfrac{176 \times 3.7}{2} \times 1.43 \quad = 466 \text{ kNm}$

(at 2 × working load)

$\quad\quad\quad\quad\quad\quad\quad\quad\quad\quad 179 \times 0.2 \times 0.1 \quad\quad = \underline{4}$
$\quad\quad\quad\quad\quad\quad\quad\quad\quad\quad\quad\quad\quad\quad\quad\quad\quad\quad 470 \text{ kNm}$

Bending moment at working load $= \dfrac{470}{2} = 235 \text{ kNm/m}$

The Steel Construction Institute Silwood Park, Ascot, Berks SL5 7QN	Subject **QUAY FACILITIES**	Chapter ref. **29**
	Design code **BS8002***	Made by **DW** — Sheet no. **11** Checked by **GWO**

Design of members

Front wall sheet piles

Max. bending moment = 729 kNm/m

Axial load = 399 kN/m

Try Larssen No. 6, ~ 25.4 mm thick, ~ BS 4360 Grade 50A

$$f_{bc} = \frac{729 \times 1000}{4675} = 156 \text{ N/mm}^2$$

$$f_c = \frac{399 \times 10}{397} = 10 \text{ N/mm}^2$$

$$\frac{\ell}{r_{xx}} = \frac{10.7 \times 100}{16.1} = 66$$

$$p_c = 158 \times \frac{180}{230} = 123.6 \text{ N/mm}^2$$

$$p_{bc} = 180 \text{ N/mm}^2$$

$$\frac{156}{180} + \frac{10}{123.6} = 0.95 \text{ OK}$$

Use Larssen No. 6, ~ 25.4 mm thick, ~ Grade 50A, × 15.7 m long

Anchor wall sheet piles

Max. bending moment = 235 kNm/m

Required modulus = $\dfrac{235 \times 1000}{140}$ = 1679 cm^3/m

(BS 4360 Grade 43A)

Use Larssen 20W Grade 43A × 4.3 m long

The Steel Construction Institute Silwood Park, Ascot, Berks SL5 7QN		Subject QUAY FACILITIES	Chapter ref. 29
		Design code BS8002*	Made by DW Checked by GWO	Sheet no. 12

Bearing piles - *centres to match tie rods*

Load = 495 kN

Piles are fully embedded, full lateral support from soil, effective length is zero.

p_c = 30% yield stress = 85 N/mm² ~ Grade 43A

Area required = $\dfrac{495 \times 1000}{85 \times 100}$ = 58.2 cm²

<u>Use 254 × 254 × 63 UBP BS 4360 Grade 43A</u>

Pile length (using Meyerhof)

End resistance in sand and gravel

= 400 × 30 × 0.67 × 0.25² = 502 kN

Shaft resistance in silty sand

= 2 × 5 × 0.25 × 4 × 7.5 = 75 kN

Shaft resistance in clay

= 0.7 × 40 × 0.25 × 4 × 3.5 = 98 kN

Shaft resistance in sand and gravel

= 2 × 30 × 0.67 × 0.25 × 4 = 40 kN/m

For factor of safety 2, penetration required below −7.8 level

= $\dfrac{(495 \times 2) - (502 + 75 + 98)}{40}$ = 7.9 m   Pile toe level
                                                                    −15.0

Pile length = 18.5 m

868   Worked example

The Steel Construction Institute Silwood Park, Ascot, Berks SL5 7QN	Subject QUAY FACILITIES	Chapter ref. 29
	Design code BS8002*	Made by DW Checked by GWO	Sheet no. 13

Walings - front wall and anchor wall

Assume tie rods at 6 pile centres (2.52 m)

Bending moment in wales = $\dfrac{388 \times 2.52^2}{12}$ = 205 kNm

Required modulus = $\dfrac{205 \times 1000}{165}$ = 1242 cm^3

Use 2/381 × 102 × 55.1 RSC's - Grade 43A

Web shear = $\dfrac{397 \times 1.26}{2}$ = 250 kN OK

Tie bars

Load per tie = 388 × 2.52 = 978 kN

Use 2/75ϕ bars BS 4360 Grade 50B (rolled threads) × 21.1 m long, - inc. 1/turnbuckle & 1/coupler

Bearing plates 400 × 200 × 30 Grade 43A

2/75ϕ
Grade 50B
@ 200c/c

Anchor bolts

Load per bolt = 388 × 0.42 × 2 = 326 kN

Use 60ϕ bolt BS 4360 Grade 50B (rolled threads) × 0.55 long

60ϕ
Grade 50B

Washer plates   180 × 180 × 30   Grade 43A

Bearing plates   180 × 180 × 30   Grade 43A

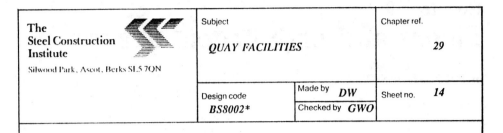

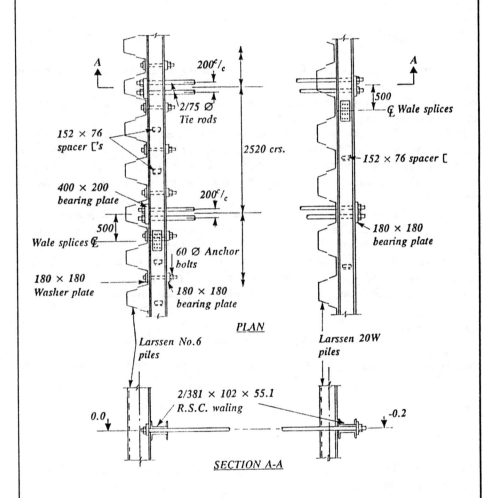

PLAN

SECTION A-A

TIE ROD & WALING DETAILS

Chapter 30
Floors and orthotropic decks

by DICK STAINSBY

30.1 Steel plate floors

Steel plate used for floors, walkways or for staircase treads normally has a raised pattern of the non-slip type such as 'Durbar plate' supplied by British Steel. Where a plated area will have only occasional use, such as walkways for maintenance access, then plain plate may be used.

If properly designed, both plain and non-slip plates have a dual function in carrying floor loads and in acting as a horizontal diaphragm in place of a separate bracing system. For example, a plate may be connected to the top flange of crane gantry girders to act as a surge girder and also as a means of crane or maintenance access.

When steel plate floors are in situations where moisture or high humidity is present, then the plates should be continuously welded to the support members to avoid corrosion problems. In dry areas connections may be fastened by bolting with countersunk bolts, intermittent welds, or clips. When clips are used they must be arranged so that they cannot move out of position.

Plain plate is commonly available in grades 430B and 510B to BS EN 10 025.

Table 30.1 Plain plate

Thickness (mm)	3	4.5	6	8	10	12.5	15
Mass (kg/m^2)	23.6	35.3	47.1	62.8	78.5	98.1	118

Durbar plate

The pattern of Durbar plate is shown in Fig. 30.1. The pattern is raised 1.5 mm to 2.2 mm above the plate surface. When ordering, the thickness should be specified as 'thickness on plain'.

Table 30.2 Durbar plate

Thickness on plain (mm)	3	4.5	6	8	10	12.5
Mass (kg/m^2)	24.9	36.9	48.7	64.4	80.1	99.7

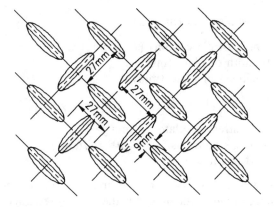

Fig. 30.1 Durbar plate

Durbar-pattern plate is free draining and can be matched width to length without affecting appearance.

Table 30.3 Recommended standard sizes for Durbar plate

Width (mm)	600	1000	1000	1250	1500	1750	1830
Length (m)	2.0, 2.5, 3.0		Any between 1.8 and 12.0				

Durbar plate is commonly available in grade 430B and 510B to BS EN 10025, also 'commercial quality' with a UTS not less than 355 N/mm².

30.1.1 Design of plates simply supported on four edges

This condition is for plates resting on four edges or attached by clips or tack welds only to supporting members. Plates supported on two edges, but with stiffeners in the other direction, can also be considered as simply supported on four edges.

(1) Uniform distributed loading
    The design may be made using Pounder's plate theory.

$$w = \frac{4 p_y t^2}{3kB^2 \left[1 + \frac{14}{75}(1-k) + \frac{20}{57}(1-k)^2\right]}$$

$$d = \frac{m^2 - 1}{m^2} \times \frac{5kwB^4}{32\gamma_f E t^3} \left[1 + \frac{37}{175}(1-k) + \frac{79}{201}(1-k)^2\right]$$

where w = uniformly distributed load on plate (ultimate) (N/mm²)
d = maximum deflection (mm), occurring at serviceability

$\dfrac{1}{m}$ = Poisson's ratio ($m = 3$)
L = length of plate (mm) ($L > B$)
B = breadth of plate (mm)
t = thickness of plate (mm)
$k = \left(\dfrac{L^4}{L^4 + B^4}\right)$
p_y = yield stress of plate (N/mm²)
E = Young's modulus 205×10^3 (N/mm²)
γ_f = load factor

This formula, which is the basis of the first two load tables in the Appendix *Floor plate design tables*, assumes that there is no resistance to uplift in plate corners. Where such resistance is provided, higher design strength may be obtained by using References 1 or 2.

(2) Centre point loads (concentrated over a small circle of radius r)
The design may be made using formulae developed by Roark.[1]
NB There is a small amount of uplift (negative load) on the corners of the plate when design is made using the following formula. Fastening at each corner must be able to resist an uplift of 6% of the centre point load.

$$W = \dfrac{0.67 \pi p_y t^2}{\left(1 + \dfrac{1}{m}\right) \ln \dfrac{2B}{\pi r} + \beta}$$

$$d = \dfrac{\alpha W B^2}{\gamma_f E t^3}$$

where W = concentrated load (ultimate) (N)
d = maximum deflection (mm), occurring at serviceability
L = length of plate (mm) ($L > B$)
B = breadth of plate (mm)
t = thickness of plate (mm)
p_y = yield stress of plate (N/mm²)
E = Young's modulus 205×10^3 (N/mm²)
$\dfrac{1}{m}$ = Poisson's ratio ($m = 3$)
r = radius of contact for a load concentrated on a small area (mm)
ln = natural logarithm of ($2B/\pi r$)
γ_f = load factor

Table 30.4 Values of β and α for centre point loads on simply supported plates

L/B	1.0	1.2	1.4	1.6	1.8	2.0	∞
β	0.435	0.650	0.789	0.875	0.972	0.958	1.0
α	0.1267	0.1478	0.1621	0.1715	0.1770	0.1805	0.1851

30.1.2 Design of plates fixed on four edges

The plates must be secured by continuous bolting or welding which will achieve the fixed condition and prevent uplift which would otherwise occur at the plate corners.

(1) Uniformly distributed loading
   The design may be made using Pounder's plate theory.

$$w = \frac{2p_y t^2}{kB^2 \left[1 + \frac{11}{35}(1-k) + \frac{79}{141}(1-k)^2\right]}$$

$$d = \frac{m^2 - 1}{m^2} \times \frac{kwB^4}{32\gamma_f E t^3}\left(1 + \frac{47}{210}(1-k) + \frac{200}{517}(1-k)^2\right)$$

where L = length of plate (mm) ($L > B$)
   B = breadth of plate (mm)
   t = thickness of plate (mm)
   $k = \left(\frac{L^4}{L^4 + B^4}\right)$
   p_y = yield stress of plate (N/mm^2)
   w = uniformly distributed load on plate (ultimate) (N/mm^2)
   E = Young's modulus 205×10^3 (N/mm^2)
   $\frac{1}{m}$ = Poisson's ratio ($m = 3$)
   d = maximum deflection (mm), occurring at serviceability
   γ_f = load factor

This formula is the basis of the third load table in the Appendix *Floor plate design tables*.

(2) Centre point loads (concentrated over a small circle of radius r)
   The design may be made using formulae developed by Roark.[1]

$$W = \frac{0.67\pi p_y t^2}{\left(1 + \frac{1}{m}\right)\ln\frac{2B}{\pi r} + \beta_1} \quad \text{OR} \quad \frac{p_y t^2}{\beta_2}$$

whichever is the lesser, where

$$d = \frac{\alpha W B^2}{\gamma_f E t^3}$$

W = concentrated load (ultimate) (N)
d = maximum deflection (mm), occurring at serviceability
L = length of plate (mm) ($L > B$)
B = breadth of plate (mm)
t = thickness of plate (mm)
p_y = yield stress of plate (N/mm^2)

874 Floors and orthotropic decks

Table 30.5 Values of β_1, β_2 and α for centre point loads on fixed plates

L/B	1.0	1.2	1.4	1.6	1.8	2.0	∞
β_1	−0.238	−0.078	0.011	0.053	0.068	0.067	0.067
β_2	0.7542	0.8940	0.9624	0.9906	1.000	1.004	1.008
α	0.0611	0.0706	0.0754	0.0777	0.0786	0.0788	0.0791

E = Young's modulus 205×10^3 (N/mm^2)

$\dfrac{1}{m}$ = Poisson's ratio ($m = 3$)

r = radius of contact for a load concentrated on a small area (mm)

ln = natural logarithm of $\dfrac{2B}{\pi r}$

γ_f = load factor

30.1.3 Design criteria

Imposed loading

Walkway loading is specified in British Standard BS 6399: Part 1 *Loading for buildings*. Table 30.6 is an extract from table 10 dealing with workshops and factories.

Horizontal loading applied to handrailing should be taken as the following applied at 1.1 m above the walkway (irrespective of the handrail height):

General duty − regular two-way pedestrian traffic   0.36 kN/m
Heavy duty − high-density pedestrian traffic: escape routes 0.74 kN/m
Potentially crowded areas over 3 m wide   3.00 kN/m

(See also BS 5395: Part 3, *Code of practice for industrial type stairs, permanent ladders and walkways* and BS 6180, *Code of practice for protective barriers in and above buildings*.)

Load factors

When designing to BS 5950: Part 1, the load factors in Table 30.7 should be used.

Table 30.6 Extract from table 10 of BS 6399: Part 1

Floor area usage	Intensity of distribution (kN/m^2)	Concentrated load (kN)
Factories, workshops and similar buildings	5.0	4.5
Access hallways, stairs and footbridges	4.0	4.5
Factories, workshops and similar buildings	5.0	4.5
Cat walks	–	1 kN @ 1 m centres

Table 30.7 Load factors

Loading	Factor
Dead load	1.4
Vertical load	1.6
Horizontal load	1.6
Vertical & horizontal combined	1.4

Deflection

Deflection of floor plating under the action of dead and imposed loads should not exceed $B/100$, where B is the minimum spanning dimension of the plate.

30.2 Open-grid flooring

Open-grid proprietary systems are often the economical solution for industrial flooring and walkways particularly where ventilation or light must be available through the flooring. They are generally available in spans up to 2.0 metres, but can be supplied to span 4.0 metres. The design allows spanning in one direction only. Stair treads are also available. (See Fig. 30.2.)

Open-grid systems are usually manufactured in grade 430 steel, with painted or galvanized finish. Aluminium and stainless steel types are also available. Load-carrying capacities and fixing details are readily available from suppliers' catalogues.

Open-grid flooring is manufactured in accordance with BS 4592: *Industrial type metal flooring, walkways and stair treads*. It has the following parts:

BS 4592: Part 1: 1987, *Specification for open bar gratings*
BS 4592: Part 2: 1987, *Specification for expanded metal grating panels*
BS 4592: Part 3: 1987, *Specification for cold formed planks*.

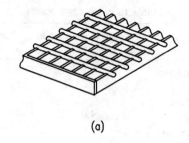

(a)

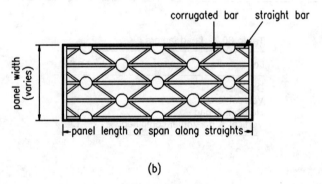

(b)

Fig. 30.2 Open grid flooring. (a) Open bar; (b) diamond pattern; (c) expanded metal; (d) Q-grating

Names and addresses of suppliers of open-grid flooring

William Sharp Ltd, Bescot Crescent, Walsall WS1 4NE. Tel: 0922 27531, Fax 0922 23894
Steelway-Fensecure (Glynwed Engineering Ltd), Queensgate Works, Bilston Road, Wolverhampton, West Midlands, WV2 2NJ. Tel: 0902 451733, Fax 0902 452256
Lionweld Kennedy Ltd, Marsh Road, Middlesbrough, Cleveland TS1 5JS. Tel: 0642 245151, Fax 0642 224710
Expanded Metal Co. Ltd, Longhill Industrial Estate (North), Hartlepool, Cleveland TS25 1PR. Tel: 0429 867388, Fax 0429 866795

30.3 Orthotropic decks

Orthotropic decks first came into regular use for long-span bridges in the 1950s. They replaced the earlier battle deck system where steel plates were supported on an independent grillage of steel plates.

Orthotropic decks 877

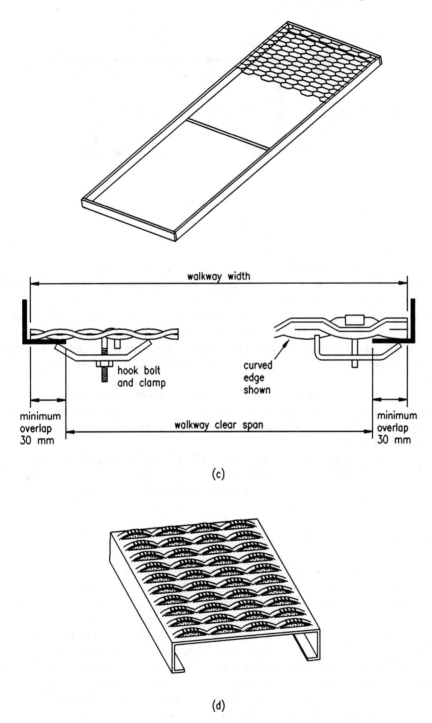

Fig. 30.2 continued

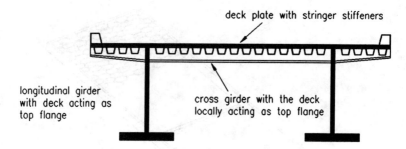

Fig. 30.3 Section through orthotropic deck

In the orthotropic system a stiffened plate deck is integral with the support members which together form the primary members of the bridge. A typical orthotropic bridge system is shown in Fig. 30.3.

All steel orthotropic decks are used when the dead weight of the bridge must be reduced to a minimum. They are therefore ideally suited to long-span bridges with single spans exceeding 120 metres, and are usually incorporated in suspension and cable-stayed bridges. Other applications are for bridges with moving or lifting spans.

It should be noted that for short-span bridges, steel/concrete composite decks are generally more economic than orthotropic decks. A composite deck of depth 250 mm will be three to four times the mass of an orthotropic deck but may be only one-third of the cost.

A typical orthotropic deck consists of a 12–15 mm plate stiffened with longitudinal stringers spaced at about 300 mm, intersecting with deeper transverse cross girders. Stringers are of the closed or open type.

Closed stringers

The closed, torsionally stiff, stringers are formed as a 'vee' or a 'trough' and are rolled or pressed from plate of 6–10 mm thickness. They provide the lightest overall weight since they distribute heavy wheel loads most effectively between the members. Welding to the floor plates has been made with fillet or butt welds, but fatigue cracking of welds on closed section stringers has occurred on some bridges. This has been attributed to lack of penetration on the inaccessible side of the weld. Full penetration butt welds are now generally adopted but they are expensive and difficult to achieve. Closed stringers may be made in short lengths to fit between the cross girders or preferably longer lengths, slotted through to achieve a better fatigue life (Fig. 30.4).

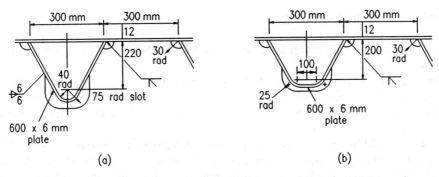

Fig. 30.4 Typical closed-section stringers: (a) vee trough; (b) trapezoidal trough

Open stringers

Open-section, torsionally weak, stringers may be formed from flats, bulb flats or angles. A slightly greater dead weight results because the transverse distributive properties are less efficient than with closed stringers. However, the advantage is that welds to the floor plate can be made from both sides and take the form of fillet welds. Transverse bending stresses between the stringer and floor plate are far less severe than with closed section stiffeners, such that this form is far less fatigue sensitive. Again, the best fatigue performance is achieved when the stringers are slotted through the cross girders. Open-section stringers are suitable for many bridges including lifting or moving structures, railway bridges and footbridges (Fig. 30.5).

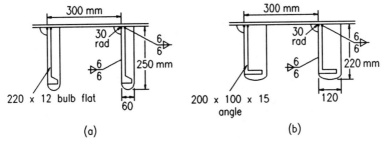

Fig. 30.5 Typical open-section stringers: (a) bulb flat; (b) angle

Floors and orthotropic decks

Loading, analysis and design

For loading to bridges reference should be made to BS 5400: Part 2: *Specification for loads*.

An analysis to take account of the distribution of point loads and partially distributed loads is best achieved with a bridge deck computer program, but hand methods of creating an 'influence surface' for moment and shears are available (see References 3 and 4).

Design should be made in accordance with BS 5400: Part 3: *Code of practice for design of steel bridges*. Fatigue considerations for bridgework are generally covered by BS 5400: Part 10: *Code of practice for fatigue*. It does not specifically deal with orthotropic decks for which it is necessary to consult research papers.[5]

References to Chapter 30

1. Young W.C. (1965) *Roark's Formulas for Stress and Strain*, 6th edn. McGraw-Hill, New York.
2. Timoshenko S. & Weinowsky-Krieger S. (1959) *Theory of Plates and Shells*, 2nd edn. McGraw-Hill, New York.
3. Pucher A. (1973) *Influence Surfaces of Elastic Plates*, 4th edn. Springer-Verlag, New York.
4. Dowling P. & Bawa A.S. (1975) Influence surfaces for orthotropic decks. *Proc. Instn Civ. Engrs*, **59**, Mar., 149–68.
5. Cunningham J.R. (1982) *Steel Bridge Decks, Fatigue Performance of Joints Between Longitudinal Stiffeners*. LR 1066, Transport and Road Research Laboratory, Crowthorne, Berks.

Chapter 31
Tolerances

by COLIN TAYLOR

31.1 Introduction

31.1.1 Why set tolerances?

Compared to other structural materials, steel (and aluminium) structures can be made economically to much closer tolerances. Compared to mechanical parts, however, it is neither economic nor necessary to achieve extreme accuracy.

There are a number of distinct reasons why tolerances may need to be considered. It is important to be quite clear which actually apply in any given case, particularly when deciding the values to be specified, or when deciding the actions to be taken in cases of non-compliance.

Table 31.1 Reasons for specifying tolerances

Structural safety	Dimensions (particularly of cross-sections, straightness, etc.) associated with structural resistance and safety of the structure.
Assembly requirements	Tolerances necessary to enable fabricated parts to be put together.
Fit-up	Requirements for fixing non-structural components, such as cladding panels, to the structure.
Interference	Tolerances to ensure that the structure does not foul with walls, door or window openings or service runs, etc.
Clearances	Clearances necessary between structures and moving parts, such as overhead travelling cranes, elevators, etc. or for rail tracks, and also between the structure and fixed or moving plant items.
Site boundaries	Boundaries of sites to be respected for legal reasons. Besides plan position, this can include limits on the inclination of outer faces of tall buildings.
Serviceability	Floors must be sufficiently flat and even, and crane gantry tracks etc. must be accurately aligned, to enable the structure to fulfil its function.
Appearance	The appearance of a building may impose limits on verticality, straightness, flatness and alignment, though generally the tolerance limits required for other reasons will already be sufficient.

882 *Tolerances*

The various reasons for specifying tolerances are outlined in Table 31.1. In all cases no closer tolerances than are actually needed should normally be specified, because while additional accuracy may be achievable, it generally increases the costs disproportionately.

31.1.2 Terminology

'Tolerance' as a general term means a permitted range of values. Other terms which need definition are given in Table 31.2.

Table 31.2 Definitions – deviations and tolerances

Deviation	The difference between a specified value and the actual measured value, expressed vectorially (i.e. as a positive or negative value).
Permitted deviation	The vectorial limit specified for a particular deviation.
Tolerance range	The sum of the absolute values of the permitted deviations each side of a specified value.
Tolerance limits	The permitted deviations each side of a specified value, e.g. ± 3.5 mm or $+5$ mm -0 mm.

31.1.3 Classes of tolerance

Table 31.3 defines the three classes of tolerances which are recognized in Eurocode 3.

Table 31.3 Classes of tolerances

Normal tolerances	Those which are generally necessary for all buildings. They include those normally required for structural safety, together with normal structural assembly tolerances.
Particular tolerances	Tolerances which are closer than normal tolerances, but which apply only to *certain* components or only to *certain* dimensions. They may be necessary in specific cases for reasons of fit-up or interference or in order to respect clearances or boundaries.
Special tolerances	Tolerances which are closer than normal tolerances, and which apply to a *complete* structure or project. They may be necessary in specific cases for reasons of serviceability or appearance, or possibly for special structural reasons (such as dynamic or cyclic loading or critical design criteria), or for special assembly requirements (such as interchangeability or speed of assembly).

It is important to draw attention to any particular or special tolerances when calling for tenders, as they usually have cost implications. Where nothing is stated, fabricators will automatically assume that only normal tolerances are required.

31.1.4 Types of tolerances

For structural steel there are three types of dimensional tolerance:

(1) *Manufacturing tolerances*, such as plate thickness and the dimensions of rolled sections.
(2) *Fabrication tolerances*, applicable in the workshops.
(3) *Erection tolerances*, relevant to work on site.

Manufacturing tolerances are specified in standards such as BS 4, BS 4360, BS 4848, BS EN 10 029 and various other BS EN standards currently in draft form. Only fabrication and erection tolerances will be covered here.

31.2 Standards

31.2.1 Relevant documents

The standards covering tolerances applicable to building steelwork are:

(1) BS 5950 *Structural use of steelwork in building*.
   Part 2: *Specification for materials fabrication and erection: hot rolled sections*.
   Part 7: *Specification for materials and workmanship: cold formed sections and sheeting*.
(2) *National structural steelwork specification for building construction* NSSS, 2nd edition.
(3) Draft CEN TC 135 standard *Execution of steel structures: Part 1: General rules and rules for buildings*.
(4) BS 5606 *Guide to accuracy in building*.

31.2.2 BS 5950 *Structural use of steelwork in building*

The specification of tolerances for building steelwork was first introduced into British Standards in BS 5950: Part 2: 1985. Only a limited amount of information could be agreed for inclusion at that time. This has since been supplemented in respect of cold formed steelwork by Part 7.

31.2.3 *National structural steelwork specification* (NSSS)

The limitations of the tolerances specified in BS 5950: Part 2 have been extended by an extensive coverage of tolerances in the *National structural steelwork specification for building construction*, 2nd edition.

31.2.4 Draft CEN TC 135 standard *Execution of steel structures*

As part of the harmonization of construction standards in Europe, a draft has been produced for Part 1: *General rules and rules for buildings*. When the final version of this is issued as a European Standard (with a BS EN number) it will supersede BS 5950: Parts 2 and 7.

This document includes comprehensive recommendations for both erection and manufacturing tolerances. To a large extent these recommendations are consistent with BS 5950: Part 2 and the NSSS. However, some of them are more detailed.

31.2.5 BS 5606 *Guide to accuracy in building*

BS 5606 is concerned with buildings generally and is not specific to steelwork. The 1990 version has been rewritten as a Guide, following difficulties due to incorrect application of the previous (1978) version which was in the form of a Code.

BS 5606 is not intended as a document to be simply called up in a contract specification. It is primarily addressed to designers to explain the need for them to include means for adjustment, rather than to call for unattainable accuracy of construction. Provided that this advice is heeded, its tables of 'normal' accuracy can then be included in specifications, except where they conflict with overriding structural requirements. This can in fact happen, so it is important to remember that the requirements of BS 5950, or its BS EN successor, must always take precedence over BS 5606.

BS 5606 introduces the idea of 'characteristic accuracy', the concept that any construction process will inevitably lead to deviations from the target dimensions, and its objective is to advise designers on how to avoid resulting problems on site by appropriate detailing. The emphasis in BS 5606 is on the practical tolerances which will normally be achieved by good workmanship and proper site supervision. This can only be improved upon by adopting intrinsically more accurate techniques, which are likely to incur greater costs. These affect the fit-up, the boundary dimensions, the finishes and the interference problems. Data are given on the normal tolerances (to be expected and catered for in detailed design) under two headings:

(1) Site construction (table 1 of BS 5606).
(2) Manufacture (table 2 of BS 5606).

Unfortunately many of the values for site construction of steelwork are only estimated. No specific consideration is given in BS 5606 to dimensional tolerances necessary to comply with the assumptions inherent in structural design procedures, which may in fact be more stringent. It does however recognize that special accuracy may be necessary for particular details, joints and interfaces.

Another important point mentioned in BS 5606 is the need to specify methods of monitoring compliance, including methods of measurement. It has to be recognized that methods of measurement are also subject to deviations; for the methods necessary for monitoring site dimensions, these measurement deviations may in fact be quite significant compared to the permitted deviations of the structure itself.

31.3 Implications of tolerances

31.3.1 Member sizes

31.3.1.1 Encasement

The tolerances on cross-sectional dimensions have to be allowed for when encasing steel columns or other members, whether for appearance, fire resistance or structural reasons. It should not be forgotten that the permitted deviations represent a further variation over and above the difference between the serial size and the nominal size.

For example, a 356 × 406 × 235 UC has a nominal size of 381 mm deep by 395 mm wide, but with tolerances to BS 4 may actually measure 401 mm wide by 387 mm deep one side, and have a depth of 381 mm the other side. The same is true of continental sections. A 400 × 400 × 237 HD also has a nominal size of 381 mm deep by 395 mm wide, but with tolerances to Euronorm 34 may actually measure 398 mm wide by 389 mm deep one side, and have a depth of 380 mm the other side.

31.3.1.2 Fabrication

Variations of cross-sectional dimensions (with permitted deviations) may also need to be allowed for, either in detailing the workmanship drawings or in the fabrication process itself, if problems are to be avoided during erection on site.

The most obvious case is a splice between two components of the same nominal size, where packs may be needed before the flange splice plates fit properly, unless the components are carefully matched. Similarly variations in the depths of adjacent crane girders or runway beams may necessitate the provision of packs, unless the members are carefully matched.

Less obviously, if the sizes of columns vary, the lengths of beams connected between them will need some form of adjustment, even if the columns are accurately located and the beams are exactly to length.

31.3.2 Attachment of non-structural components

It is good practice to ensure that all other items attached to the steel frame have adequate provision for adjustment in their fixings to cater for the effects of all steelwork tolerances, plus an allowance for deviations in their own dimensions. Where necessary, further allowances may be needed to cater for structural movements under load and for differential expansion due to temperature changes.

Where possible, the number of fixing points should be limited to three or four, only one of which should be positive with all the others having slotted holes or other means of adjustment.

31.3.3 Building envelope

It must be appreciated that erection tolerances, including variation in the position of the site grid lines, will affect the exact location of the external building envelope relative to other buildings or to site boundaries, and there may be legal constraints to be respected which will have to be taken into account at the planning and preliminary stages of design.

These effects also need to be taken into account where a building is intended to have provision for future extension or where the project is an extension of an existing building, in which case deviations in the actual dimensions have to be catered for at the interface.

In the case of tall multi-storey buildings, the building envelope deviates increasingly with height compared to the location at ground level, even though permitted deviations for column lean generally reduce with height. Unless there are step-backs or other features with a similar effect, it may be necessary to impose particular tolerance limits on the outward deviations of the columns.

31.3.4 Lift shafts for elevators

The deviations from verticality that can be tolerated in the construction of guides for 'lifts' or elevators are commonly more stringent than those for the construction of the building in which they operate. In low-rise buildings sufficient adjustment can be provided in association with the clearances, but in tall buildings it becomes necessary either to impose 'special' tolerances on column verticality or else to impose 'particular' tolerances on those columns bounding the lift shaft.

In agreeing the limits to be observed with the lift supplier, it should not be overlooked that the horizontal deflections of the building due to wind load also have implications for the verticality of the lift shafts.

31.4 Fabrication tolerances

31.4.1 Scope of fabrication tolerances

The description 'fabrication tolerances' is used here to include tolerances for all normal workshop operations except welding. It thus covers tolerances for:

(1) cross sections, other than rolled sections,
(2) member length, straightness and squareness,
(3) webs, stiffened plates and stiffeners,
(4) holes, edges and notches,
(5) bolted joints and splices,
(6) column baseplates and cap plates.

However tolerances for cross sections of rolled sections and for thicknesses of plates and flats are treated as manufacturing tolerances. Welding tolerances (including tolerances on weld preparations and fit-up and sizes of permitted weld defects) are treated elsewhere.

31.4.2 Relation to erection tolerances

An overriding requirement for accuracy of fabrication must always be to ensure that it is possible to erect the steelwork within the specified erection tolerances.

Due to the wide variety of steel structures and the even wider variety of their components, any recommended tolerances must always be specified in a very general way. Even if it were possible to specify fabrication tolerances in such a way that their cumulative effect would always permit the specified erection tolerances to be satisfied, the resulting permitted deviations would be so small as to be unreasonably expensive, if not impossible, to achieve.

Fortunately in most cases it is possible to rely on the inherent improbability of all unfavourable extreme deviations occurring together. Also the usually accepted values for fabrication tolerances do make some limited allowances for the need to avoid cumulative effects developing on site. They are tolerances that have been shown by experience to be workable, provided that simple means of adjustment are incorporated where the effects of a number of deviations could otherwise become cumulative. For example, beams with bolted end cleats usually have sufficient adjustment available due to hole clearances, but where a line of beams all have end plate connections, provision for packing at intervals may be advisable, unless other measures are taken to ensure that the beams are not all systematically over-length or under-length by the normal permitted deviation. Other possible means for adjustment include threaded rods and slotted holes.

Where it can be seen from the drawings that the fabrication tolerances could easily accumulate in such a way as to create a serious problem in erection, either closer tolerances or means of adjustment should be considered; however the co-incident occurrence of all extreme deviations is highly improbable, and judgement should be exercised both on the need for providing means of adjustment and on the range of adjustment to be incorporated.

31.4.3 Full contact bearing

31.4.3.1 Application

The requirements for contact surfaces in joints which are required to transmit compression by 'full contact bearing' probably cause more trouble than any other item in a fabrication specification, largely due to misapprehension of what is actually intended to be achieved.

First it is necessary to be clear about the kind of joint to which the requirements for full contact bearing should be applied. Figure 31.1(a) shows the normal case, where the profile of a member is required to be in full contact bearing on a baseplate or cap plate or division plate. The stress on the contact area equals the stress in the member, thus full contact is needed to transmit this stress from the member into the plate. Only that part of the plate in contact with the member need satisfy the full contact bearing criteria, though it may be easier to prepare the whole plate.

Figure 31.1(b) shows two end plates in simple bearing. The potential contact area is substantially larger than the cross-sectional area of the member, thus full contact bearing is not necessary. All that is needed is for the end plates to be square to the axis of the member. Another common case of simple bearing is shown in Fig. 31.1(c).

By contrast, the case shown in Fig. 31.1(d) is one where if full contact bearing is needed, it is also necessary to take special measures to ensure that the profiles of the two members align accurately, otherwise the area in contact may be significantly less than the area required to transmit the load. Particular tolerances should be specified in such cases, based on the maximum local reduction of area that can be accepted according to the design calculations. Alternatively a division plate could be introduced; if the stresses are high this may well prove to be the most practical solution.

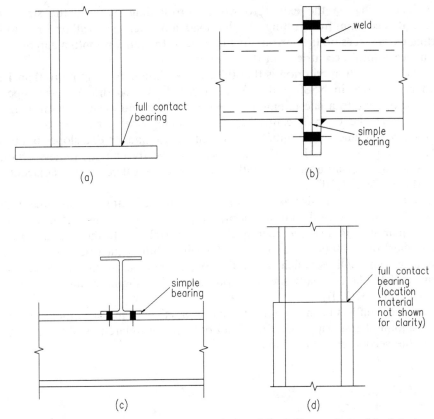

Fig. 31.1 Types of member-to-member bearing: (a) profile to plate, (b) plate to plate, (c) flange to flange, (d) profile to profile (accurate alignment necessary)

31.4.3.2 Requirements

Where full contact bearing is required, there are in fact three different criteria involved:

(1) Squareness.
(2) Flatness.
(3) Smoothness.

31.4.3.3 Squareness

If the ends of a length of column are not square to its axis, then after erection either the column will not be vertical or else there may be tapered gaps at the joints, depending on the extent to which surrounding parts of the structure prevent the column from tilting. Under load any such gap will try to close, exerting extra forces on the surrounding members. In addition, both a gap or a tilt will induce a local eccentricity in the column.

A practical erection criterion is that the column should not lean more than 1 in x (where x is 600 in NSSS and 500 in the CEN TC 135 draft). This slope is measured relative to a line joining the centres of each end of the column length, referred to as the over-all centreline. The column is also allowed a 'lack of straightness' tolerance of (length/1000), which corresponds to end slopes of about 1/300 (see Fig. 31.2(a)). It is thus necessary to specify end squareness criteria relative to the over-all centreline, rather than to the local centreline adjacent to the end (see Fig. 31.2(b)).

There is generally a design assumption that the line of action of the force in the column does not change direction at a braced joint by more than 1/250, requiring an end squareness in a simple bearing connection (relative to the over-all axis of the member) of 1/500 (see Fig. 31.2(c)). However, full contact bearing generally arises at column splices which are not at braced points, so an end squareness tolerance of 1/1000 is usually specified, producing a maximum change of slope of 1/500 (see Fig. 31.2(d)).

Once a column has been erected, it is more practical to measure the remaining gaps in a joint. These gaps are affected not only by the squareness of the ends but also by the second criterion, flatness.

31.4.3.4 Flatness

Ends have to be reasonably flat (as distinct from curved or grossly uneven) to enable the load to be transferred properly. Following a history of arguments over appropriate specifications, the American Institute of Steel Construction (AISC) commissioned some tests, which are the basis for their current specifications.

It was found that a surprisingly high tolerance was quite acceptable, and that beyond its limit (or to compensate for end squareness deviations) the use of localized packs or shims was acceptable. Basically similar rules are now beginning to appear in other specifications including the CEN TC 135 draft standard (see section 31.5.6 in relation to erection tolerances). This is an essentially simple and effective method of correcting excessive gaps on site (see also section 31.5.6). However, inserting shims into column joints is not a matter to be undertaken lightly. It is normally more economic to avoid the need for shimming by working to close fabrication tolerances in joints where full contact bearing is required.

31.4.3.5 Smoothness

In the light of the findings of the flatness tests, it can be appreciated that if absolute local flatness is not in fact needed, absolute smoothness is irrelevant also.

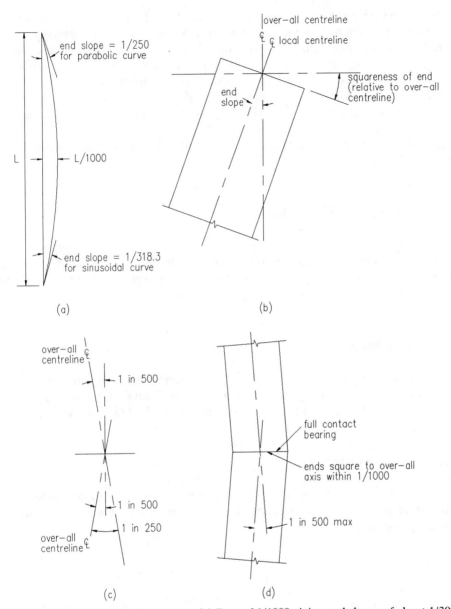

Fig. 31.2 Squareness of column ends. (a) Bow of 1/1000 giving end slopes of about 1/300. (b) Squareness of end measured relative to over-all centreline. (c) Change of direction at a braced joint. (d) End squareness at full contact bearing splice

The best description of the smoothness that is needed, is the smoothness of a surface produced by a good-quality modern saw in proper working order. This degree of smoothness is indeed very good.

Where sawing is not possible, ending machines (i.e. special end-milling machines) can be used for correcting the squareness (or flatness) of ends of built-up (fabricated) columns, such as box columns or other welded-up constructions. Where baseplates are not flat and are too thick to be pressed flat, either they are milled locally in the contact zone or else planing machines are used.

However it cannot be overemphasized that the normal preparation for a rolled section column required to transmit compression by full contact in bearing is by saw cutting square to the axis of the member.

It is, of course, unnecessary to flatten the undersides of baseplates supported on concrete foundations.

31.4.4 Other compression joints

Compression joints, transferring compression through end plates in simple bearing, also need to have their ends square to the axis. If after the members have been firmly drawn together, a gap remains which would introduce eccentricity into the joint, it should be shimmed.

31.4.5 Lap joints

Steel packs should be used where necessary to limit the maximum step between adjacent surfaces in a lap joint (see Fig. 31.3) to 2 mm with ordinary bolts or 1 mm (before tightening the bolts) where preloaded HSFG bolts are used.

31.4.6 Beam end plates

Where the length of a beam with end plates is too short to fit between the supporting columns, or other supporting members, packs should be supplied to make up the difference.

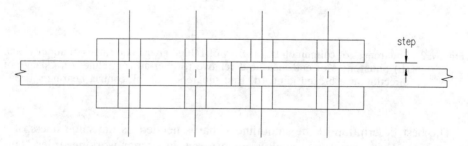

Fig. 31.3 Maximum step between adjacent surfaces

Gaps arising from distortion caused by welding, as shown in Fig. 31.4, need not be packed if the members can be firmly drawn together. However, they may need to be filled or sealed to avoid corrosion where the steelwork is external or is exposed to an aggressive internal environment.

31.4.7 Values for fabrication tolerances

The values for fabrication tolerances currently given in the NSSS are reproduced for convenience in Table 31.4. Each of the specified criteria should be considered and satisfied separately. The cumulative effect of several permitted deviations should not be considered as overriding the specific criteria.

These values represent current practice at the time of writing and are taken from the second edition of the NSSS. However, it is likely that this will be updated, particularly when the CEN Standard comes into force. In any case, the latest relevant standards should always be specified.

The clause numbers referred to in Table 31.4 are clause numbers in the NSSS which should be referred to for further information.

31.5 Erection tolerances

31.5.1 Importance of erection tolerances

Erection tolerances potentially have a significant effect on structural behaviour. There are four matters to be considered:

(1) overall position,
(2) fixing bolts,
(3) internal accuracy,
(4) external envelope.

31.5.2 Erection − positional tolerance

31.5.2.1 Setting out

The position in plan, level and orientation can only be defined relative to some fixed references, such as the national grid and the Ordnance datum level. From the national system, it is usual to set subsidiary site datum points, and often a site datum level, and then refer the accuracy of the structure to these.

For any site the use of a grid of established column lines together with an established site level is strongly recommended. For a large site it is virtually

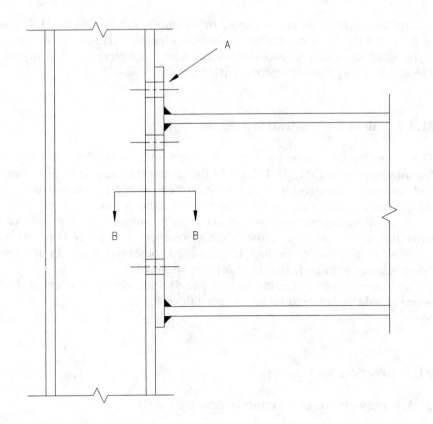
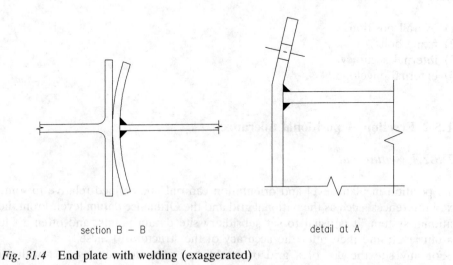

section B – B   detail at A

Fig. 31.4 End plate with welding (exaggerated)

Table 31.4 (Extract from National Structural Steelwork Specification)

WORKMANSHIP - ACCURACY OF FABRICATION

7.1 PERMITTED DEVIATIONS

Permitted deviations in cross section, length, straightness, flatness, cutting, holing and position of fittings shall be as specified in clauses 7.2 to 7.5 below (see clause 3.4.6(i)).

Notes: Deviations marked
* *are in accordance with BS 5950 Part 2*
** *are modified BS 5950 Part 2 values*
‡ *are not included in BS 5950 Part 2*

7.2 PERMITTED DEVIATIONS IN ROLLED COMPONENTS AFTER FABRICATION
(Including Structural Hollow Sections)

7.2.1 Cross Section after Fabrication*

7.2.2 Squareness of Ends Not Prepared for Bearing**

See also clause 4.3.3 (i).

7.2.3 Squareness of Ends Prepared for Bearing **

Prepare ends with respect to the longitudinal axis of the member. See also clause 4.3.3 (ii) and (iii).

7.2.4 Straightness on Both Axes **

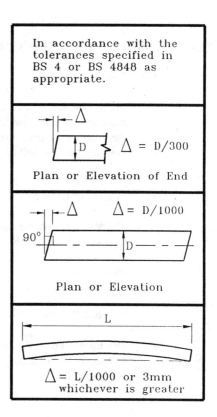

In accordance with the tolerances specified in BS 4 or BS 4848 as appropriate.

$\Delta = D/300$

Plan or Elevation of End

$\Delta = D/1000$

Plan or Elevation

$\Delta = L/1000$ or 3mm whichever is greater

Table 31.4 *(contd)*

7.2.5 Length **

Length after cutting, measured on the centre line of the section or on the corner of angles.

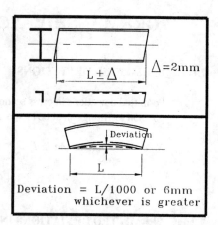

$L \pm \Delta$ $\Delta = 2\text{mm}$

7.2.6 Curved or Cambered **

Deviation from intended curve or camber at mid-length of curved portion when measured with web horizontal.

Deviation $= L/1000$ or 6mm whichever is greater

7.3 PERMITTED DEVIATIONS FOR ELEMENTS OF FABRICATED MEMBERS

7.3.1 Position of Fittings ‡

Fittings and components whose location is critical to the force path in the structure, the deviation from the intended position shall not exceed Δ.

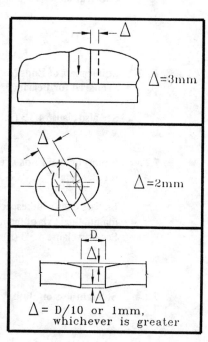

$\Delta = 3\text{mm}$

7.3.2 Position of Holes ‡

The deviation from the intended position of an isolated hole, also a group of holes, relative to each other shall not exceed Δ.

$\Delta = 2\text{mm}$

7.3.3 Punched Holes ‡

The distortion caused by a punched hole shall not exceed Δ.
(see clause 4.6.3.)

$\Delta = D/10$ or 1mm, whichever is greater

Table 31.4 *(contd)*

7.3.4 Sheared or Cropped Edges of Plates or Angles ‡

The deviation from a 90° edge shall not exceed Δ.

7.3.5 Flatness **

Where bearing is specified, the flatness shall be such that when measured against a straight edge not exceeding one metre long, which is laid against the full bearing surface in any direction, the gap does not exceed Δ.

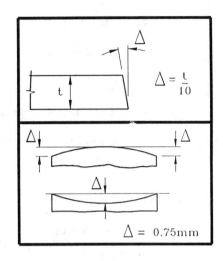

$\Delta = \dfrac{t}{10}$

$\Delta = 0.75$ mm

7.4 PERMITTED DEVIATIONS IN PLATE GIRDER SECTIONS

7.4.1 Depth *

Depth on centre line.

7.4.2 Flange Width *

Width of B_w or B_n

7.4.3 Squareness of Section **

Out of Squareness of Flanges.

7.4.4 Web Eccentricity *

Intended position of web from one edge of flange.

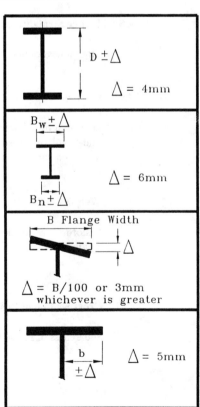

$D \pm \Delta$

$\Delta = 4$ mm

$B_w \pm \Delta$

$B_n \pm \Delta$

$\Delta = 6$ mm

B Flange Width

$\Delta = B/100$ or 3mm whichever is greater

$\Delta = 5$ mm

Table 31.4 (contd)

7.4.5	**Flanges** **	
	Out of flatness.	
7.4.6	**Top Flange of Crane Girder** ‡	
	Out of flatness where the Rail seats.	
7.4.7	**Length** *	
	Length on centre line.	
7.4.8	**Flange Straightness** *	
	Straightness of individual flanges.	
7.4.9	**Camber** **	
	Deviation from intended camber at mid-length of curved portion when measured with web horizontal.	
7.4.10	**Web Distortion** *	
	Distortion on web depth or gauge length.	
7.4.11	**Cross Section at Bearings** *	
	Squareness of Flanges to Web.	

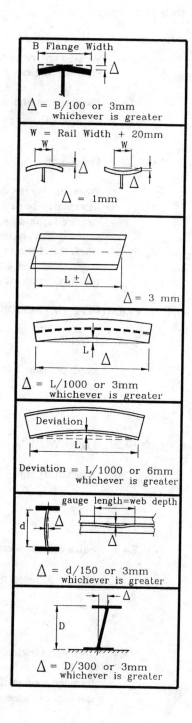

Table 31.4 (contd)

7.4.12 Web Stiffeners ‡

Straightness of stiffener out of plane with Web after welding.

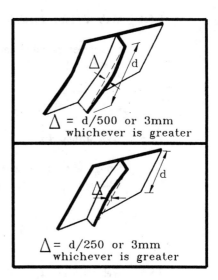

$\Delta = d/500$ or 3mm whichever is greater

7.4.13 Web Stiffeners ‡

Straightness of stiffener in plane with Web after welding.

$\Delta = d/250$ or 3mm whichever is greater

7.5 PERMITTED DEVIATIONS IN BOX SECTIONS

7.5.1 Plate Widths ‡

Width of B_f or B_w

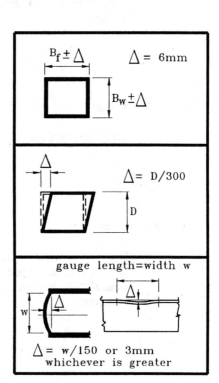

$B_f \pm \Delta$ $\Delta = 6$mm

$B_w \pm \Delta$

7.5.2 Squareness **

Squareness at diaphragm positions.

$\Delta = D/300$

7.5.3 Plate Distortion *

Distortion on width or gauge length.

gauge length = width w

$\Delta = w/150$ or 3mm whichever is greater

Table 31.4 *(contd)*

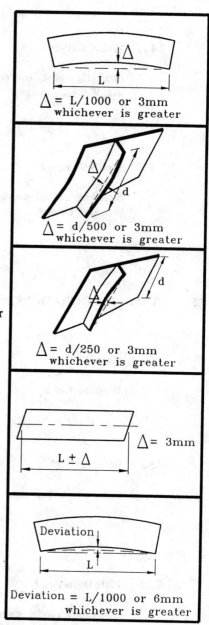

7.5.4 **Web or Flange Straightness** ‡

Straightness of individual web or flanges.

7.5.5 **Web Stiffeners** ‡

Straightness in plane with Plate after welding.

7.5.6 **Web Stiffeners** ‡

Straightness out of plane to Plate after welding.

7.5.7 **Length** *

Length on centre line.

7.5.8 **Camber** **

Deviation from intended camber at mid-length of curved portion when measured with the uncambered side horizontal.

indispensable. To help appreciate this, consider what happens on the site of a steel structure.

31.5.2.2 Site practice

Normal site practice is for the supporting concrete foundations, and other supporting structures, to be prepared in advance of steel erection, generally by an organization separate from the steel erector. Depending on the system of holding-down bolts or other fixings to be used, this may involve casting-in of holding-down bolts, preparation of pockets in the concrete and preparation of surfaces to receive fixings to the steelwork.

Even with care, the standard of accuracy achievable is limited and the concrete requires time to harden to a sufficient strength for steel erection to proceed. Once all the foundations etc. are available for steel erection (or at least a sufficient proportion of them on a large site), it is prudent to survey them to review their accuracy.

31.5.2.3 Established column lines and established site level

From this survey it is convenient to introduce a grid of established column lines (ECL) and an established site level (ESL) of the foundations and other supporting structures in such a way that the positions and levels of steel columns etc. can readily be related to the site grid and site level.

The established column lines are defined as that grid of site grid lines that best represents the actual mean positions of the installed foundations and fixings. Similarly the established site level is defined as that level which best represents the actual mean level of the installed foundations. Of course it should also be verified that the deviation of the ECL grid and the ESL from those specified are within the relevant permitted deviations.

31.5.3 Erection − fixing bolts

31.5.3.1 Types of fixing bolts

Fixing bolts include both holding-down bolts for columns and various types of fixing bolts used to locate or to support other members, such as beams or brackets carried by walls or concrete members.

Holding-down bolts and other fixing bolts are either:

(1) fixed in position,
(2) adjustable, in sleeves or pockets.

31.5.3.2 Fixed bolts

Fixed bolts used to be solidly cast in, an operation requiring care and the use of jigs or templates to achieve accurately. However they are now also commonly produced by placing resin-grouted bolts in holes drilled in the concrete after casting. It may also be possible to use expanding bolts.

However fixed bolts are achieved, they need to be positioned accurately, as the only adjustment possible is in the steelwork, so relatively close tolerances are normally specified.

31.5.3.3 Adjustable bolts

Adjustable bolts are placed in tubes or in tapered trapezoidal or conical holes cast in the concrete, so that a degree of movement of the threaded end of the bolt is possible, while the other end is held in place by a steel washer or other anchoring device embedded in the concrete.

This alternative permits the use of more easily achieved tolerances for the bolts, while using relatively simple details for the steelwork. Adjustment of the bolt necessitates its axis deviating from the vertical to some extent, and the holes in the steelwork need to be large enough to allow for this, particularly if the baseplate is thick. The use of loose plate washers is recommended to span oversize holes if necessary. If required they can be welded in place after the bolts are tightened, but this should not normally be necessary. 'Particular' tolerances need to be worked out for each case, depending on the details, including the length of the bolts, because this affects their slope.

31.5.3.4 Length of bolts

The level of the top of an HD bolt is also important to ensure that the nuts can be fitted properly after erection. To provide the necessary tolerances for the fixing of the bolts, they should be longer than theoretically required, long threaded lengths should be provided and the nominal level for the top should be above the theoretical position.

Similar considerations apply to the lengths of fixing bolts located horizontally.

31.5.4 Erection − internal accuracy

In terms of structural performance, the main erection tolerance is verticality of columns; positions of beams etc. on brackets may also be important. Levels of beams, particularly of one end relative to the other end and of one beam relative to the next one, are important in terms of serviceability.

Otherwise the internal accuracy of one part of the structure relative to another is largely a matter of assembly tolerances, provided that these do not cause any problem of fit-up, interference or clearances. Where the structural accuracy resulting from the assembly tolerances is liable to infringe any of these limits, 'particular' tolerances should be specified.

The necessary tolerances are specified in relation to readily identifiable points and levels. For columns and other vertical members, the reference points are conveniently defined as the actual centre of the member at each end of the fabricated piece. For beams and other horizontal members the reference points are more conveniently defined by the actual centre of the top surface at each end. Either the column system or the beam system should be used for any other cases, and the relevant system should be indicated on the erection drawings. The tolerances are then defined by the permitted deviations of these reference points from the established column lines ECL and established floor level EFL.

The concept of an ECL grid and an established site level ESL have already been explained in section 31.5.2.3. The established floor level EFL is defined as that level which best represents the actual mean level of the as-built floor levels. The EFL must not deviate from the specified floor level (relative to the ESL) by more than the permitted deviation for height of columns.

The reference points for each beam must then be within the permitted deviation from the EFL. In addition the difference in level of each end of a beam and the difference in level between adjacent beams must also be within their respective limits.

In the case of columns, the permitted deviations at each level form an 'envelope' within which the column must lie at all levels. In addition, the permitted inclination of each column within a storey height is limited, but except where columns are fabricated as individual storey-height pieces, the overall envelope normally governs.

31.5.5 Erection – external envelope

Generally the same erection tolerances for verticality apply to external columns as to internal columns. When the envelope of extreme permitted deviations is plotted from the extreme position of the base (allowing for the permitted deviation of the ECL from the theoretical position as well as the permitted deviation of the column base from the ECL), it may be found that this is unacceptable in terms of site boundaries or building lines, especially for a tall multi-storey building. If so, 'particular' tolerances should be specified.

Fit-up problems with cladding could also occur if alternate adjacent columns at the periphery were allowed too large a deviation alternately in and out from the theoretical line of the building face. Even if the fit-up problems could be overcome, the visual appearance might be affected. Again, 'particular' tolerances should be specified if necessary.

31.5.6 Shimming full contact bearing splices

As mentioned in section 31.4.3.4 in relation to fabrication tolerances, tests commissioned by AISC, and used as the basis for several modern standards, showed that shims can be used to reduce gaps in full contact bearings to within the specified tolerances. Shimmed gaps up to 6.35 mm were tested, so it is not prudent to permit shimming for gaps exceeding 6 mm; gaps larger than this should be corrected by other means.

Gaps which would otherwise remain over the specified tolerance when the members are in their final alignment, should be shimmed. As the tests were on flat shims, it is acceptable to use flat shims in practice. In the tests the shims were of mild steel, and this is permitted in the AISC specification and in the CEN TC 135 draft.

The shims should be inserted such that no remaining gap exceeds the specified permitted deviation. Short lengths of shim are appropriate in a variety of thicknesses in steps not exceeding the permitted deviation. No more than three layers of shims should be used at any point, and preferably only one or two. The shims (and the lengths of columns) may be held in place by means of a partial penetration butt weld extending over the shims (see Fig. 31.5(a)).

In bolted compression splices, bolted 'finger' shims (shaped as indicated in Fig. 31.5(b)) can be used.

In some cases shims can be driven in, but if so they need to be fairly robust (usually over 2 mm thick), so shims of various thicknesses are needed throughout

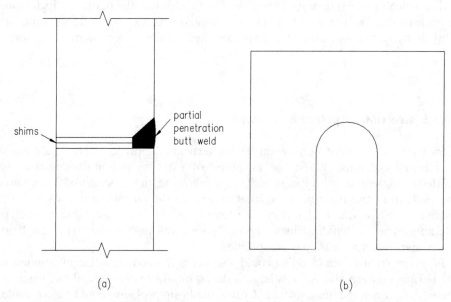

Fig. 31.5 Shims for full contact bearing. (a) Shims with partial penetration butt weld. (b) Finger shim

Table 31.5 (Extract from National Structural Steelwork Specification)

WORKMANSHIP
ACCURACY OF ERECTED STEELWORK

9.1 FOUNDATIONS

The Steelwork Contractor shall inspect the prepared foundations and holding down bolts for position and level not less than seven days before erection of steelwork starts. He shall then inform the Employer if he finds any discrepancies which are outside the deviations specified in clause 9.4 requesting that remedial work be carried out before erection commences.

9.2 STEELWORK

Permitted maximum deviations in erected steelwork shall be as specified in clause 9.5. All measurements shall be taken in calm weather, and due note is to be taken of temperature effects on the structure. (see clause 8.6.2.).

Note: Deviations marked * are in accordance with BS 5950 Part 2
     **are modified BS 5950 Part 2 values
     ‡ are not included in BS 5950 Part 2

9.3 INFORMATION FOR OTHER CONTRACTORS

The Engineer shall advise contractors engaged in operations following steel erection of the deviations acceptable in this document in fabrication and erection, so that they can provide the necessary clearances and adjustments.

9.4 PERMITTED DEVIATIONS FOR FOUNDATIONS, WALLS AND FOUNDATION BOLTS

9.4.1 Foundation Level ‡

Deviation from exact level.

9.4.2 Vertical Wall ‡
Deviation from exact position at steelwork support point.

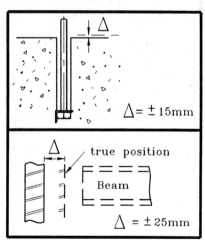

Table 31.5 *(contd)*

9.4.3 **Pre-set Foundation Bolt or Bolt Groups when Prepared for Adjustment ***

Deviation from the exact location and level and minimum movement in pocket.

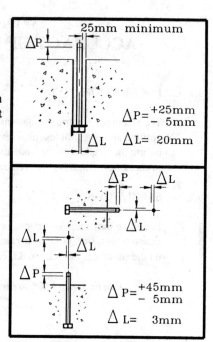

9.4.4 **Pre-set Foundation Bolt or Bolt Groups when Not Prepared for Adjustment ***

Deviation from the exact location, level and protrusion.

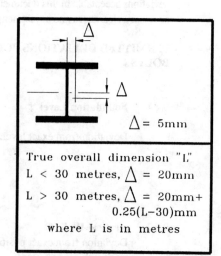

9.5 **PERMITTED DEVIATIONS OF ERECTED COMPONENTS**

9.5.1 **Position at Base of First Column Erected ***

Deviation of section centreline from the specified position.

Δ = 5mm

9.5.2 **Overall Plan Dimensions ‡**

Deviation in length or width.

True overall dimension "L"
L < 30 metres, Δ = 20mm
L > 30 metres, Δ = 20mm + 0.25(L−30)mm
where L is in metres

Table 31.5 *(contd)*

9.5.3	**Single Storey Columns Plumb**** Deviation of top relative to base, excluding portal frame columns, on main axes. *See clauses 1.5 (iv) and 3.4.6 (iii) regarding pre-setting continuous frames.*	$\Delta = \pm H/600$ or 5mm whichever is greater Max $= \pm 25$mm
9.5.4	**Multi-storey Columns Plumb**** Deviation in each storey and maximum deviation relative to base.	$\Delta h = h/600$ or 5mm whichever is greater $\Delta H = 50$mm maximum
9.5.5	**Gap Between Bearing Surfaces ‡** (See clauses 4.3.3 (iii), 6.2.1 and 7.2.3).	$\Delta = (D/1000) + 1$mm
9.5.6	**Alignment of Adjacent Perimeter Columns ‡** Deviation relative to the next column in line.	$\Delta = 5$mm

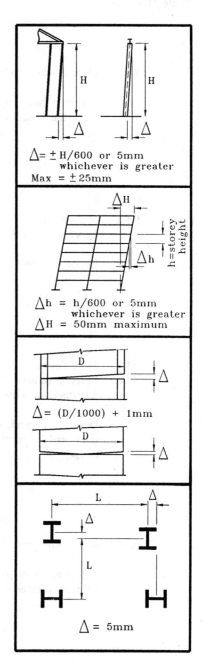

Table 31.5 (contd)

9.5.7 **Floor Beams Level*** Deviation from specified level at supporting stanchion.	
9.5.8 **Floor Beams.** **Level at Each End of Same Beam** ‡ Deviation in level.	
9.5.9 **Floor Beams.** **Level of Adjacent Beams within a distance of 5 metres*** Deviation from relative levels (measured on centreline of top flange).	
9.5.10 **Beams Alignment** ‡ Deviation relative to an adjacent beam above or below.	

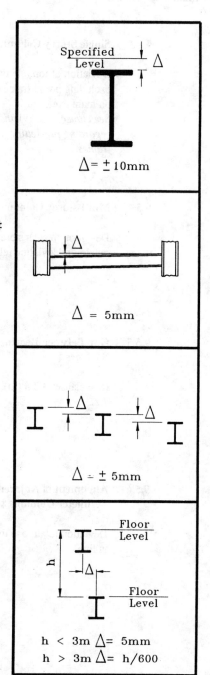

Table 31.5 *(contd)*

9.5.11	Crane Gantry Columns Plumb ‡ Deviation of cap relative to base.	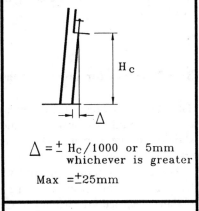 $\Delta = \pm H_C/1000$ or 5mm whichever is greater Max $= \pm 25$mm
9.5.12	Crane Gantries Gauge of Rail Tracks ‡ Deviation from true gauge.	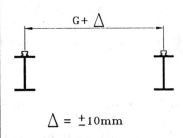 $\Delta = \pm 10$mm
9.5.13	Joints in Gantry Crane Rails ‡	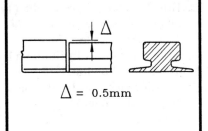 $\Delta = 0.5$mm

the joint. Driven shims are best limited to vertical joints e.g. between a beam end plate and a column. More commonly the joint must be jacked or wedged open (or else the upper portion lifted by a crane) so that the shims can be inserted. Tapered shims are particularly difficult to insert; as they are not necessary they are best avoided.

31.5.7 Values for erection tolerances

The values for erection tolerances are given in Table 31.5. Each of the specified criteria should be considered and satisfied separately. The permitted deviations should not be considered as cumulative, except to the extent that they are specified relative to points or lines that also have permitted deviations.

These values represent current practice at the time of writing and are taken from the second edition of the NSSS. However, it is likely that this will be updated, particularly when the CEN Standard comes into force. In any case, the latest relevant standards should always be used.

The clause numbers referred to in Table 31.5 are clause numbers in the NSSS which should be referred to for further information.

Further reading for Chapter 31

1. British Standards Institution (1980) *Specification for hot rolled sections*. BS 4: Part 1: 1980, BSI, London.
2. British Standards Institution (1990) *Specification for hot rolled products of non-alloy structural steels and their technical delivery conditions*. BS EN 10 025 1990, BSI, London.
3. British Standards Institution (1986) *Equal and unequal angles*. BS 4848: Part 4: 1986, BSI, London.
4. British Standards Institution (1990) *Guide to accuracy in building*. BS 5606 1990, BSI, London.
5. British Standards Institution (1985) *Specification for materials fabrication and erection: hot rolled sections*. BS 5950: Part 2: 1985, BSI, London.
6. British Standards Institution (1991) *Specification for tolerances and dimensions, shape and mass for hot rolled steel plates 3 mm thick and above*. BS EN 10 029 1991, BSI, London.
7. CEN TC 135 Standard: *Execution of steel structures* (in draft at time of publication).
8. The British Constructional Steelwork Association (1991) *National structural steelwork specification for building structures*. Second edition, BCSA/SCI Publication No.203/91.

Chapter 32
Fabrication

by DAVID DIBB-FULLER

32.1 Introduction

The steel-framed building derives most of its competitive advantage from the virtues of prefabricated components which can be assembled speedily on site. Additional economies can be significant provided the designer seeks through his design to minimize the value added costs of fabrication. This is proper 'value engineering' of the product and is applied to perhaps the most influential sector of the delivered-to-site cost of structural steelwork.

This chapter explains the processes of fabrication and links them with design decisions. It is increasingly more important for the designer to understand the skills and techniques available from different fabricators so that he can adapt his design to keep overall costs down. The choice of fabricator can then be made on the basis of ability to conform in production engineering terms to design assumptions made much earlier. It is unacceptable to allow the fabricator or designer to undertake the production engineering element of design in perfect isolation; the dialogue between designers and fabricators must be a continuous one. The effects that fabrication and assembly have on design assumptions and vice versa and, in particular, the achievable fit-up of components and permissible limits of tolerance of the raw steel as received from the mill and their effect on component shape after fabrication are described.

Design must be viewed as a complete process covering strength and stiffness as well as production engineering. The most economical structures feature these principles in full measure.

32.2 Economy of fabrication

Structural form has a significant effect on the delivered-to-site cost of steelwork. This is due to a number of factors additional to the cost of raw steel from steel suppliers. Some forms will prove to be more costly from some fabricators than others; they tend to attract work by aligning their methods to specific market sectors. A few years ago the industrial building market was taken over by the introduction of the portal-framed structure. The fabricators in this sector adopted high-volume, low-cost production and deliberately excluded themselves from other areas of the market. By the use of pre-engineered standards they were able to maximize repetition and minimize input from both design and drawing activities: a classic example of a combination of design and production engineering. Other fabricators specialize in tubular structures, lightweight sections and heavy sections. Fabricators who claim skill in all forms of structural steelwork rarely exhibit good economy in all of them and can prejudice their performance when carrying out work for which they are, in reality, poorly equipped.

32.2.1 Fabrication as a cost consideration

Figures 32.1 to 32.4 give an indication of the proportional costs associated with the fabrication and erection of structural steelwork. Actual costs in monetary terms have not been included as they will change with the demand level of the market. The proportions of cost will vary a little from fabricator to fabricator; those shown represent a reasonable average.

The cost headings are:

- raw steel
- fabrication ⎫
- painting   ⎬ fabrication shop activities
- transport
- erection
- site painting.

Raw steel cost covers the average cost of rolled steel in grade 43A delivered to the fabricator by the British Steel Corporation. No allowance has been made for extra costs arising from the use of stockholders' steel or the extremes of section variation.

Fabrication cost covers a number of different sized jobs and incorporates cleaning of the raw steel by shot blasting, preparation of small parts (cleats and plates and connection components), assembly of the components into complete structural members ready for shop-applied paint treatments. It also includes an allowance for consumables.

Paint cost covers the shop application of 75 microns of primer by spray immediately after fabrication. No allowance has been made for blast cleaning of areas affected by welding. A coverage allowance of 28 m^2/tonne at the rate of 3 m^2/litre has been made which represents the likely consumption for rolled section beams and columns. This allowance has been adjusted for various specific work types as described in the accompanying text.

Transport cost is based on 18 tonne loads per trailer which delivers finished products to sites within a fifty mile radius of the fabrication shop. Transport costs will rise for loads of less than 18 tonnes or when oversize components need special police escort or permission.

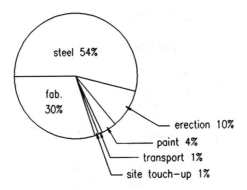

Fig. 32.1 Cost breakdown: portal-framed industrial buildings

The erection cost is average for the work type and includes preliminaries for high-rise multi-storey work. Erection costs are largely beyond the control of the smaller fabricators or fabricators who do not employ their own erection crews; in certain cases, this cost element can be more influential than fabrication.

The cost of site painting has been included but it only covers the average cost for touching-up damage to the primer coat. Other site-applied protection systems vary enormously in cost and so have not been considered.

Portal-framed industrial buildings

The breakdown shown in Fig. 32.1 follows the assumptions given above. There is no adjustment in either the shop painting or the transport cost as this type of work fits the basic parameters well. It can be seen that design economies come principally from the weight of the structure combined with efficient fabrication processes.

Simple beam and column structures

The breakdown shown in Fig. 32.2 incorporates the following limitations:

(1) the maximum height of the building is three storeys,
(2) erection is carried out using mobile cranes on the ground floor slab.

It can be seen that again tonnage and fabrication efficiencies are the dominant criteria; 83% of the costs arise from these elements with a slightly greater emphasis on tonnage than was the case with portal frames.

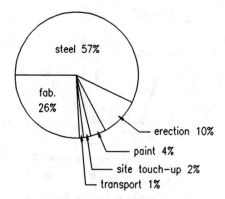

Fig. 32.2 Cost breakdown: simple beam and column structures

High-rise multi-storey

The breakdown shown in Fig. 32.3 incorporates the following adjustments:

(1) the steel grade has been taken as grade 50B with an allowance for cambering and straightening,
(2) the paint system generally is shot blast and 75 micron primer, 10% of steel coated with 100 micron primer,
(3) no allowance has been made for any concrete encased beams or stanchions,
(4) transport includes for off-site stockpiling, bundling and out-of-hours delivery to site (city centre sites often incur these costs).

This sector of the market has a very different cost profile to those already shown. Raw steel still dominates but erection charges have now overtaken the fabrication element. This type of steelwork lends itself particularly well to automated fabrication techniques featuring drilling lines.

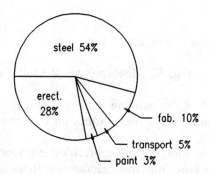

Fig. 32.3 Cost breakdown: high-rise multi-storey

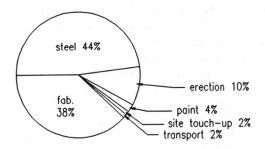

Fig. 32.4 Cost breakdown: lattice structures

Lattice structures

This is perhaps the most difficult of the sectors on which to carry out an analysis as lattice structures vary enormously in size, complexity and element make-up. The costs shown in Fig. 32.4 are indicative and incorporate the following:

(1) angle booms and lacings,
(2) welded joints without gusset plates,
(3) transportable lengths and widths with full-depth splices only.

Here fabrication is virtually the same cost as the raw steel. The designer should work very closely with the fabricator to ensure simplicity of assembly for this form of structure.

32.2.2 Production engineering

Production engineering is concerned with the processes of material procurement, planning, manufacture and delivery. Some of these processes are entirely the province of the production engineer, others can be materially affected by those who specify the manufactured items.

In the context of structural steelwork these items are usually specified by the designer, in more detail than a pure performance specification, in terms of material quality, standard of workmanship and tolerances. However, the production engineer can contribute significantly to the cost effectiveness of the product when given the opportunity and the designer can assist also by considering the production engineering requirements.

Materials are specified largely by direct reference to British Standards which are wide in scope and, if used indiscriminately, can cause difficulties to the production engineer. Keeping the specification brief and relevant along with a restricted range of plate or section sizes will be of benefit. Small quantities with large variations make material supply costly and prone to error. Choosing sections without regard to rolling programmes can delay manufacture and choice of uncommon section sizes may incur stockist charges. Some sections in the range are rolled infrequently and often to a minimum production run.

Planning the flow of work varies between fabricators but as a general rule the workpiece will travel through the factory in accordance with the work being carried out on it. Designing a beam which requires multiple different fabrication inputs will increase handling costs and reduce factory output. Whenever possible, fabrication should be kept to a minimum number of differing operations, e.g. keep welding away from beams which can be fabricated with bolted connections and vice versa. Mixing these operations increases costs despite the apparent material savings.

Economic delivery is linked to marshalling and transport efficiencies by ensuring that components are kept compact and that brackets are robust enough to withstand the inevitable handling mistakes which can damage flimsy attachments. Keeping the number of components required to a minimum certainly helps costs and has enormous impact on designing for production.

32.2.3 Design for production

This is part of production engineering but focusses on the manufacturing operations themselves. The creation of production and design standards is more appropriate than standardizing complete buildings. Any manufacturing operation benefits from repetition yet in structural steelwork recently there has been a trend towards lower production runs and increasing complexity. Even within this environment, standards can be set up. What is needed is a dialogue between the designer and the production engineer. Each can understand the needs of the other and a consensus can be reached on what will be of benefit to the job in hand.

32.2.4 Standards

Typical examples of production standards are:

- Limiting hole sizes simplifies bolt ordering and avoids tool changes for drills and punches.
- Standard hole patterns benefit those fabricators using computer-controlled punching and drilling lines. Keeping cross centres of holes standard has a similar effect as it increases repeatability of hole formation.
- The use of angle cleats of the same section throughout a job increases the potential for computerized production, particularly when the range of different cleat hole patterns is decreased.

- Keeping plate widths and thicknesses constant creates production standards which favour computer-controlled machinery, if material grades are also standardized.
- Where prefabrication primer paints are specified, keep welding to a minimum as the paint has to be removed local to the weld area to avoid porosity problems; the steel has then to be recycled through the spray booth.
- If specifying weld testing, consider how this is to be carried out and be sympathetic to the production cycle. Try to design connections for which testing can be achieved reliably rather than awkward details with difficult access or impossible geometries for ultrasonic readouts.

Typical examples of design standards are:

- The rationalization of end connection details which can increase repeatability and assist in creating longer production runs of components. Designing each connection to minimize material input may appear beneficial, but in fact it increases fabrication costs by lowering production runs and raising wastage rates. A 25% increase in connection material may represent only a 1% increase in overall job tonnage but save money by the resulting rationalization of components. A saving of course is also realized in design time and drawing costs.
- Maintaining uniform serial sizes which can result in increased repeatability of component assembly. The trade-off is weight versus workshop costs. In particular, the decision to splice a column on the grounds of material saving can incur increased costs for the column as a whole. A full-strength splice is very costly and rarely proves to be the economic solution.

It can be seen that design standards tend to rationalize components and details, thus setting the scene for production standards.

32.3 Welding

Fusion welding processes are used to join structural steel components together. These processes can be carried out either in the workshop or on site though it is generally accepted that site welding should not be used as a primary source of fabrication. Welding should only be undertaken by welders who are certificated to the appropriate level required by British Standards or other recognized authority.

Welding processes are classified by the protection medium used to shield the weld metal from the atmosphere. This protection ensures that a sound weld is produced free from contamination-induced cracks and porosity. The forms of protection used during the fabrication of structural steelwork are:

(1) chemical;
(2) gas-shielded.

Chemical protection covers many processes, with manual metal-arc (MMA) and submerged-arc welding being those used most commonly.

Gas shielding is achieved in a multitude of ways, the most common being a mixture of carbon dioxide and argon. This system is commonly referred to as MIG welding (metal inert gas) though strictly it is MIG/MAG (metal active gas). MIG welding is gradually superseding MMA welding for reasons of productivity and cost effectiveness. MIG welding can be operated either manually or with some degree of mechanization.

32.3.1 Shop welding processes

Stick welding is the traditional form of manual metal-arc welding. The consumable electrode is in stick form and usually covered in a flux which is the chemical protection medium.

The MMA process uses an electrode holder attached via cables to an electrical power source, either a.c. or d.c. (Fig. 32.5). The welding operation is carried out by hand with the rate of deposition controlled by the welder. Stick electrodes are available in diameters from 1.5 to 8 mm with various flux formulations. Joint access is good due to the compactness of the equipment and the ability to use short rods in particularly tight situations.

Figure 32.6 gives access requirements for MMA welding.

Automated methods normally employ submerged-arc techniques in which a heap of flux material laid along the joint receives the bare electrode during the welding operation. The flux buries the arc, acting as a chemical shield. Some automatic processes, however, use flux-coated electrodes. Automatic submerged-arc welding gives high quality results but is limited to uniform welds with long weld runs such as flange-to-web welds on plate girders. Access requirements for submerged-arc welding depend upon the particular equipment available; each

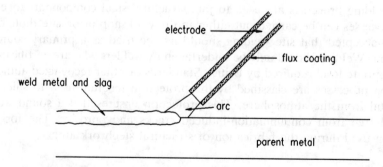

Fig. 32.5 Manual metal-arc welding

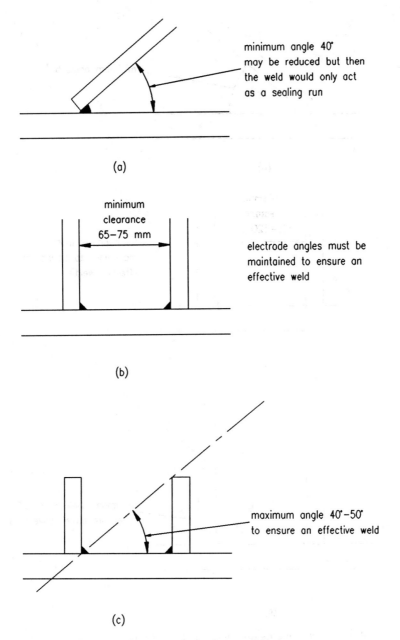

Fig. 32.6 Access requirements for manual metal-arc welding

joint should be considered with detailed knowledge of the welding gun carriage being employed (Fig. 32.7).

MIG welding can be semi-automatic or fully automatic. The electrode is fed through a water-cooled welding head from which it picks up its current and which

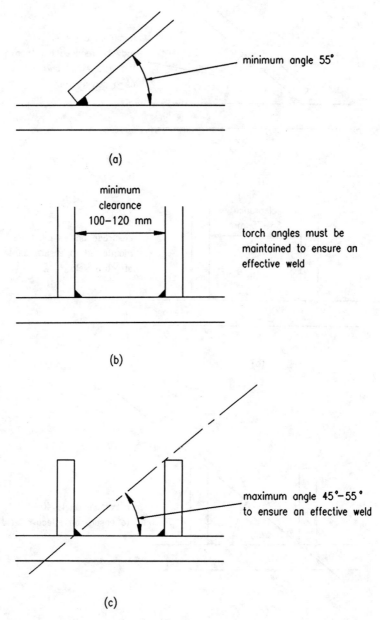

Fig. 32.7 Access requirements for submerged-arc welding

acts as a hood to direct the carbon dioxide into the arc area (Fig. 32.8). The electrode is fed at a constant rate selected to give the required current and the arc is controlled by the power source. The operator does not control the rate of deposition as it is semi-automatic. The process is confined to welding in the flat or

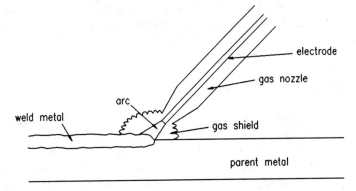

Fig. 32.8 Metal active (or inert) gas welding

horizontal–vertical position. Access is not as good as manual metal-arc. Figure 32.9 gives access requirements for MAG welding.

Other forms of welding may be used in specific circumstances. They are described here for the sake of completeness.

Electro-slag welding

This is a specialist form of welding used to join plates together at their edges. The weld is formed by the flow of electric current through a molten flux (slag). The joint is contained between sliding shoes and the electrode is fed automatically into the slag where it forms a pool of weld metal. As the pool increases in depth, the

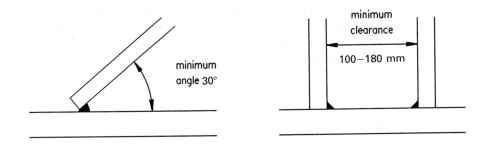

due to the range of welding gun carriages available
each joint should be considered separately

Fig. 32.9 Access requirements for metal active gas welding

shoes are moved up the joint leaving behind a solid joint. Welds of as much as 100 mm thickness can be successfully undertaken by the use of this process. Weld thicknesses of less than 25 mm are not suitable for this process.

Tungsten inert gas welding (TIG)

Use is made of a tungsten electrode in a process somewhat similar to MIG welding. The weld metal in this case is shielded by an inert gas, argon, nitrogen or helium, although argon is the most widely used. The other significant difference arises from the fact that the electrode is not consumed and therefore the operator has to control the feed of filler metal, a very skilled operation. This form of welding is confined to high-quality, high-integrity applications within the aircraft, nuclear and chemical plant industries which use the more exotic metals.

32.3.2 Access requirements

The different welding processes make differing demands upon the space needed for access to the joint. Proper access is a prerequisite for a sound weld; poor access may give rise to incomplete or defective welds particularly where the angle of attack of the electrode is too shallow. MMA welding gives the best flexibility for access due to the compactness of the electrode. MIG welding is more restricted due to the need for clearance around the gas nozzle. All mechanized welding processes are extremely limited because the machinery involved restricts these processes to long welds between plates with the weld laid in the flat position.

Deposition rates are dependent upon welding current, electrode size and number of electrodes (only applicable to mechanized processes such as submerged-arc and electro-slag). Because of the short set-up times for MMA and MIG welding, relatively short lengths of weld can be formed economically. In this respect, there is no rule as to a minimum length, although due allowance must be made for start and stop lengths which reduce the effective length of a weld. Obviously, mechanized techniques are only viable for long runs and probably limited to large weld sections of at least 3 m in length.

32.3.3 Weld defects

Welding can create defects in either the weld metal or the parent metal. Some defects are metallurgical while others are a result of post-welding tension forces caused by cooling contraction. Poor weld preparation, weld geometry and poor technique are the primary source of defects.

Weld preparation induced defects include the following:

- undercut
- hydrogen cracking
- porosity
- lack of fusion
- slag inclusion
- root concavity
- excessive penetration
- excessive reinforcement
- lack of penetration
- lamellar tearing.

Poor process technique induced defects include the following:

- undercut
- overlap
- porosity
- lack of side wall fusion
- incomplete penetration
- crater pipes
- worm holes
- lamellar tearing
- hot cracking
- heat affected zone (HAZ) cracking.

The notching effect of undercuts and overlaps will be to weaken joints in members subject to fatigue loading patterns. Porosity is an undesirable defect which often gives cause for concern but in itself does not impair strength. Excessive porosity points to the possible existence of other defects associated with poor preparation or technique. HAZ cracking seriously undermines the structural strength of the joint as do hot cracking and lamellar tearing. Lack of side wall fusion can be tolerated provided it occurs infrequently along the weld and is of limited size. Ultrasonic testing will normally reveal these defects and give a good indication of their extent. Incomplete penetration points to poor technique or poor weld design both of which must be corrected. Back gouging root runs will often cure the problem and should be carried out and regarded as good welding practice. Slag inclusion can occur when multipass welds are used due to incomplete clearance of slag from earlier weld runs. When this defect occurs, it should be judged in a similar manner to lack of side wall fusion.

In all cases, the designer should weigh the existence of defects against the required performance of the joint. Critical highly-stressed joints may require very much higher standards of weld integrity than other joints. The lower characteristic strength of welds in some part recognizes that defects do occur and that it is not economic to demand perfect results for all weld situations. Visual inspection of a weld is important as the size and physical appearance of welds is some indication that the joint is sound.

32.4 Bolting

Shop bolting may form part of the fabrication process. There are implications arising from the use of shop bolting which need to be appreciated by the designer so that he can ensure that a cost penalty does not occur.

At one time it was common practice to assemble components in the workshop using bolts or rivets. With the increased implementation of welding, this practice has declined due to the costs associated with bolted fabrications. Instead of a simple run of fillet weld, holes need to be formed and bolts introduced, increasing total man hours and cost. In many respects the ease with which welding can be undertaken has diverted designers' attention from the use of shop bolting. Today, fabricators are looking at increased automation to keep costs down and machines have been developed which considerably speed up hole forming. The designer should look to maximize the benefits of automation by appreciating the place of shop bolting in modern fabrication.

32.4.1 Shop bolting

There are three ways in which shop bolting can be used:

(1) permanently-bolted assemblies;
(2) tack-bolted assemblies;
(3) trial assembly.

There is still a demand for structural members to be bolted arising from a requirement to avoid welding because of the service conditions of the member under consideration. These may be low temperature criteria, the need to avoid welding stresses or the requirement for the component to be taken apart during service (e.g. bolted-on crane rails). For lattice structures, the designer should specify the bolting bearing in mind the effect of hole clearances around bolt shanks. HSFG bolts will not give problems but other bolts in clearance holes will allow a 'shake-out' which can cause significant additional displacement at joints. Typically, a truss with bolted connections may deflect due to the take-up of lack of fit in clearance holes to such an extent that it loses its theoretical camber. The use of close-tolerance bolts in holes of the same diameter reduces this effect without incurring the cost of turned and fitted bolts in reamed holes.

Tack bolting is used to position members prior to welding, combining the effective use of CNC punching or drilling with the speed and convenience of welding. The tack-bolted assemblies are self-aligning and so do not require expensive jigging. The tack bolts are sacrificial as they cannot be considered as contributing to the strength of a welded joint. Tack bolting is used for lattice work where angles, channels or tees are used as the booms and lacings.

Large and complex assemblies which are to be bolted together on site may be trial assembled in the fabrication shop. This increases fabrication costs but may pay for itself many times over by ensuring that the steel delivered to site will fit. Restricting trial assembly to highly repetitive items or items critical to the site programme is to be recommended.

32.4.2 Types of bolt

The choice of which type of bolt to use may not necessarily be made on the basis of strength alone but may be influenced by the actual situation in which the bolt is used, e.g. in non-slip connections.

There are four basic types of bolts. They are structural bolts, friction-grip bolts, close-tolerance bolts and turned-barrel bolts.

Structural bolts

Bolts with low material strength and wide manufacturing tolerance were until recently known as 'black bolts' because of their appearance. Now they are called 'structural bolts', normally have a protective coating which gives them a bright appearance and are available in a range of tensile strengths.

Friction-grip bolts

Friction-grip bolts are normally supplied with a protective coating and are differentiated by material grade. Friction-grip bolts are used in connections which resist shear by clamping action in contrast to structural bolts which resist shear and bearing. When considering the use of friction-grip bolts during the fabrication process, adequate means of access must be provided so that the bolt can be properly tensioned.

Close-tolerance bolts

This type of bolt differs from a structural bolt in that it is manufactured to closer tolerances. To gain the full benefit of close-tolerance bolts, they should be used in close-fitting holes produced by reaming which adds considerably to the expense of fabrication. Where a limited slip connection is required, close-tolerance bolts can be used in holes of the same nominal diameter as the bolt but not reamed; this gives a connection which is subject to far less slip than would be the case for structural bolts in clearance holes.

Turned-barrel bolts

The turned-barrel bolt is a close-tolerance bolt with a shank which is of a greater diameter than the threaded area. Typically, a 20 mm diameter turned-barrel bolt has a 22 mm shank. These bolts are used in reamed holes and the virtue of the increased shank diameter is that it ensures that the threads are not damaged during the insertion of the bolt.

32.4.3 Hole forming

Most fabrication shops have a range of machines which can form holes in structural steelwork. There is a growing tendency towards the use of computer-controlled machines with which a number of fabrication activities, typically punching, marking, shearing and clipping, can be carried out on the one piece of equipment. Numerically-controlled drilling lines feature machines which carry out the drilling operation only although there are some which incorporate plate-burning facilities.

Despite the introduction of computer numerically controlled (CNC) machinery, use is still made of equipment which requires human operation. The traditional drilling machine is the radial drill, a manually-operated machine which drills individual holes in structural steelwork. Given the enormous difference in output between the two generations of machine, commercial pressures will ensure that the older type is removed from service. Therefore, only the features of modern equipment will be outlined here.

To gain the best output, CNC machines are incorporated in conveyor lines. There is an infeed line which takes the unprocessed raw steel and an outfeed line which distributes the finished product either to the despatch area or further along the fabrication cycle and into an assembly area. The infeed conveyor for punching and drilling machines that handle angles, flats, small channels and joists will normally be configured to handle 12 m long bars. There will be a marking unit which carries a set of marking dyes that stamp an identification mark on to the steel if required. There may be a number of hydraulic punch presses each suitable for accepting up to three punching tools and it is normal that the tool holders are quick-change units. Typically, hydraulic presses have 1000 kN capacity. The punch presses can form differently-shaped holes so that angles can be produced with slots. Some machines produce the slot by a series of circular punching operations, others have a single slot shaped dye.

Once the material has passed through the punch presses, it is then positioned for shearing. The hydraulic shear has a capacity of between 3000 and 5000 kN. Machines which incorporate drilling facilities can have these positioned next to the punch presses prior to the hydraulic shear. The output of the machine can be as high as a thousand holes per hour. Obviously, these machines can be obtained with different capacities so that if the work of the fabricator is at the light end of the section range, he does not need to purchase machines with large capacity.

For larger sections, where drilling and cutting are required, there are two basic machines available. One is a drilling and plate-cutting system which is used for heavy plates; the other machine is of a larger capacity and is used for drilling rolled sections of all sizes. Modern drills normally have three axis numeric control and air-cooled drills. These machines have sensors which can detect the position of the web to ensure that hole patterns are symmetrical about the web centreline.

Where components only need a small number of holes, mobile drills are used. These are normally magnetic limpet drills hand operated by the fabricator. The holes are formed with rotary-broach drill bits which have a central guide drill surrounded by a cylindrical cutter.

The fabricator will always carry out hole-forming operations prior to any further fabrication as the presence of any stiffeners or cleats on a bar would severely disrupt the input of NC machines.

32.5 Cutting and machining

The fabrication processes of cutting and machining are tending towards greater automation. Many traditional methods are falling into disuse due to increased labour costs and low levels of productivity.

Universal beams, columns and the larger angles, tees and channels are normally cut to length by saws. Hollow sections are also treated in this way. Small sections of joist, channel, angle and flat are cut to length by shearing either as a separate operation or as part of a punching and cropping operation which is computer controlled. Large plate sections may be sheared but this involves specialist plant and equipment which is not available to all fabricators.

Machining can take a number of forms but commonly is the process of bringing surfaces into a closer tolerance of flatness and squareness than can be achieved by cold sawing or burning. Machining can be applied to the ends of rolled sections and the surfaces of plates. In the past, specifications often called for the ends of columns, which were transmitting loads through bearing, to be machined to ensure close contact. This process was known as 'ending'. Nowadays a good quality saw cut is accepted for column end bearing.

32.5.1 Cutting and shaping techniques

Flame cutting or burning

This technique produces a cut by the use of a cutting torch. The process may be either manual or machine controlled. Manual cutting produces a rough edge profile which can be very jagged and may need further treatment to improve its appearance. Edges of plates cut manually require greater edge distances to holes

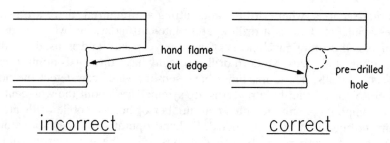

Fig. 32.10 Hand flame cutting

for this reason. Notches or holes with square corners should not be hand cut unless the corner is first radiused by a drilled hole, Fig. 32.10.

When flame cutting is carried out under machine control, the cut edge is smooth and therefore the restrictions on edge distances to holes are relaxed. Notches which are machine cut need radiusing either by a predrilled hole or by the control of the cutting machine. Unradiused notches are to be avoided due to their stress-raising characteristics. Notching or coping of beam ends can now be carried out by computer-controlled machines which burn the webs first and create the radiused corner then cut the flanges with a cut which coincides with the web cut. Numerically-controlled coping machines can cut top and bottom notches simultaneously and the notches can be of different dimensions.

Castellated beams are formed by cutting a web pattern along a UB or UC section followed by realignment and rewelding. The web cutting is always machine controlled to ensure accurate fit-up and an edge which is suitable for subsequent welding.

Arc plasma cutting

Plates may be cut by arc plasma techniques. In this case the cutting energy is produced electrically by heating a gas in an electric arc produced between the tungsten electrode and the workpiece which ionizes the gas, enabling it to conduct an electric current. The high-velocity plasma jet melts the metal of the workpiece and blows it away. The cut so produced is very clean and its quality can be improved by using a water injection arc plasma torch. Plasma cutting can be used on thicknesses up to about 150 mm but the process is then substantially reduced in speed.

Shearing and cropping

Sections can be cut to length or width by cropping or shearing using hydraulic shears. Heavy sections or long plates can be shaped and cut to length by specialist plate shears. These are large and very expensive machines normally to be found in the workshops of those fabricators who specialize in plate girder work for bridges, power stations and other heavy steelwork fabrications. For the more commonplace range of smaller plates and sections, there is a range of equipment available which is suitable for cutting to length or shaping operations. These machines feature a range of shearing knives which can accept the differing section shapes. Shearing can be adjusted so that angled cuts may be made across a section. This is particularly useful for lacings of latticed structures. One version of the shear is a 'notcher' which can cut shaped notches. Special dies are made to suit the notch dimensions and it is possible to obtain dies to cut the ends of hollow sections in preparation for welding together.

Cold sawing

When, because of either specification or size, a section cannot be cut to length by cropping or shearing, then it is normally sawn. All saws for structural applications are mechanical and feature some degree of computer control. Sawing is normally carried out after steel is shot blasted as the saw can be easily incorporated within the conveyor systems associated with shot blast plants.

There are three forms of mechanical saw, circular, band and hack. The circular saw has the blade rotating in a vertical plane which can cut either downwards or upwards, though the former is the more common. The blade is a large milling wheel approximately 5 mm thick. The diameter of the saw blade determines its capacity in terms of the maximum size of section which can be cut. Normally, fabricators will have saws capable of dealing with the largest sections produced by British Steel. For increased productivity, sections may be nested or stacked together and cut simultaneously. Some circular saws allow the blade to move transversely across the workpiece, which is useful for wide plates. Most circular saws can make raking cuts across a section of any angle though some saws are restricted to single side movement. The preferred axis of cut is across the $Y-Y$ axis of the section. In the case of beams, channels and columns, this is with the web horizontal and the flange toes upwards. Depending on the control exercised, a circular saw can make a cut within the flatness and squareness tolerances necessary for end bearing of members. Tighter tolerances will require 'ending'.

Band saws have less capacity in terms of section size which can be cut. Generally, sections greater than 600 mm × 600 mm cannot be sawn using band saws. The saw blade is a continuous metal band edged with cutting teeth which is driven by an electric motor. The speed of cut is adjustable to suit the workpiece. Band saws can make mitre cuts and can cut through stacked sections. Cutting accuracy is dependent on machine set-up but should produce results similar to circular saws.

Hack saws are as the name implies mechanically driven reciprocating saws. They have normal format blades carried in a heavy duty hack saw frame. Hack saws have more limited cutting capacity than band saws and have the capability to produce mitre cuts.

All saws feature computer-controlled positioning carriages for accurate length set-up; most also have computer sensing for angle cuts. The highest output saws are circular saws.

Machining

In metalworking terms, machining covers a very wide range of processes. In regard to structural steelwork, machining is limited. At its simplest it covers grinding of surfaces or edges to improve appearance, remove notches or in preparation for welding. Edge preparation for welding may be carried out by feeding material through a grinding machine or by hand grinding using a hand-held tool. Local tidying up of plate edges after burning by hand is normally done with hand-held tools.

The larger machining operations are confined to surface preparation of the ends of members or surfaces of plates which need to butt closely together. Grinding the surface of baseplates or splice plates to columns is often carried out to ensure the flatness of the plate surface. This is either performed prior to the delivery of plates to the fabricator or it is carried out by him. This process is called 'flash grinding'. The preparation of column ends is a specialist operation called 'ending'. An ending machine is a milling bed often mounted in a long trench or pit. This machine can be traversed along guide rails so that a number of columns can be processed one after the other. The columns are laid flat and adjusted so that they are normal to the milling axis. The milling machine then trims the column ends to produce a flat ground surface. This surface is the bearing surface between column lengths at splices or at the baseplate. 'Ending' and 'flash grinding' should be coincident and carried out after straightening the column length.

Gouging

The gouging process is the removal of metal at the underside of butt welds. Gouging techniques are also used for the removal of defective material or welds. The various forms of gouging are flame gouging, air-arc gouging, oxygen-arc gouging and metal-arc gouging.

Flame gouging is an oxy fuel gas cutting process and uses the same torch but with a different nozzle. It is important that a proper gouging nozzle is adopted so that the gouging profile is correct.

Air-arc gouging uses the same equipment as manual metal arc welding. The process differs in that the electrode is made of a bonded mixture of carbon and graphite encased in a layer of copper and jets of compressed air are emitted from the specially designed electrode holder. These air jets blow away the parent metal from beneath the arc.

A special electrode holder is used in the oxygen-arc gouging process. A special tubular coated steel electrode controls the release of a supply of oxygen. Oxygen is released once the arc is established and when the gouge is being made the oxygen flow is increased to a maximum.

Standard manual metal arc welding equipment is used for metal-arc gouging. The electrodes, however, are specially designed for cutting or gouging. This process relies on the metal being forced out of the cut by the arc and not blown away as in the other processes.

32.5.2 Surface preparation

Structural sections from the rolling mills may require surface cleaning prior to fabrication and painting. Hand preparation, such as wire brushing, does not normally conform to the requirements of modern paint or surface protection systems.

Blast cleaning is the accepted way of carrying out surface preparation. It involves blasting dry steelwork with either shot or grit at high velocity to remove rust, oil, paint, mill scale and any other surface contaminants. The most productive form of blast cleaning plant has special equipment comprising infeed conveyor, drying oven, blast chamber, spray chamber and outfeed conveyor.

Steel is loaded on to the infeed conveyor either as separate bars or side by side depending on the blast chamber passage opening. It then travels through the drying oven which ensures that any surface moisture is removed prior to blasting. The blast chamber receives the steel and passes it over racks and between the blasting turbines, which are impellers fed centrally with either shot or grit. The material is thrown out from the edge of the turbine and impacts against the steelwork. The speed of the turbine, its location and aperture determine the velocity and direction of the blasting medium. While many blast chambers operate in a satisfactory manner with four turbines, a better result is achieved on larger sections when six turbines are present (Fig. 32.11).

The blasting medium is retrieved from within the blast chamber and recycled until it is exhausted. Shot provides a good medium for steel which is to be painted; grit can cause excessive wear of the blast chamber and results in a surface which is more pitted, giving a reduced paint thickness over high points.

The steel leaves the blast chamber and immediately enters the spray chamber. Depending upon the requirements of the specification, the steel is then sprayed with a primer paint which protects it from flash rusting. The use of prefabrication primers should be discussed with the fabricator and paint system supplier as requirements may vary. For most structural applications in building work, there is little need to specify a prefabrication primer. Once the steel has passed through the spray chamber, it is fed on to the outfeed conveyor and continues to its next process centre.

The other form of blast cleaning is called 'vacu blasting', a manual method where the blasting medium is blown out of a hand-held nozzle under the pressure of compressed air. This process is normally performed in a special sealed cabinet

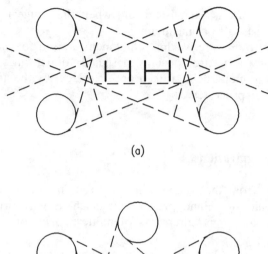

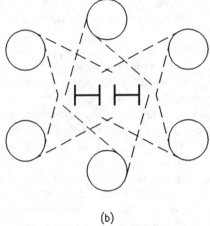

Fig. 32.11 Blast chambers: (a) four turbine, (b) six turbine

with the operator wearing protective clothing and breathing equipment. It relies heavily on the skill of the operator and does not give production rates approaching those of mechanical blasting plants.

Local areas of damage caused by welding can be cleaned by vacu blasting or by needle gunning. The needle gun has a set of hardened needles which are propelled and withdrawn rapidly. The action of the needles on the steel effectively removes weld slag and provides a good surface for painting. Needle gunning is not suitable for large areas or for heavily contaminated surfaces.

Surface preparation is intended to provide a uniform substrate for even paint application. It is important to consider the edges of plates where paint thickness is often low. A slight edge chamfer produced by grinding will improve paint film thickness performance.

32.5.3 Cambering, straightening and bending

Each of these operations has a different purpose. Cambering is done to compensate for anticipated deflections of beams or trusses under permanent loads, dead loads or superimposed dead loads such as finishes. Straightening is part of the fabrication process aimed at bringing sections back within straightness tolerances. Bending is to form the section to a shape which is outside normal cambering limits.

Rolled sections are normally bought in by the fabricator and then sent to specialist firms for cambering. The steel section is cold cambered by being passed between rolls. The rolls are adjusted on each pass until the required camber is achieved. The following cambers can be produced:

- circular profile (specify radius)
- parabolic profile (specify equation)
- specified offsets (tabulate co-ordinates)
- reverse cambers (circular, parabolic, offset or composite).

Bending is also carried out by specialists, the process being identical to that for cambering. Curves can be formed on either the $X-X$ or $Y-Y$ axes.

Straightening can be carried out by the fabricator or by the cambering specialists. Sections are often straightened by the fabricator by applying heat local to the flange or web. This is a skilled operation which achieves excellent results.

32.6 Handling and routing of steel

The fabrication process involves the movement of steel around the workshop from one process activity to the next. Clearly, all these movements should be planned so that the maximum throughput of work is achieved and costs are minimized. Even with the highly automated facilities available today, planning is essential as work centres may become overloaded and cause delays to other activities. In this respect, the work flow has to be balanced between work centres.

The routing of steel through a fabrication workshop is planned on the basis of the work content required on the workpiece. The workpiece will consist of one or more main components which may have smaller items attached (cleats, tabs and brackets). The main components are normally obtained from the steel mills though some fabricators obtain steel from stockholders. The steel is stored in a stockyard local to the workshop. The control of steel stock is an important management function which can balance the cash flow of the fabricator and provide clear identification of the steel for traceability purposes. Stock control systems are commonly computerized and should allow the fabricator to identify:

(1) steel on order but not yet received,
(2) steel which is in the stockyard and its location,

(3) allocation of steel to particular contracts,
(4) levels of unallocated steel (free stock).

Prior to the fabrication cycle, an estimate is made of the work content of a workpiece. This assessment may be crudely based upon tonnage or more commonly based upon time study data of previous similar work. The work content assessment is a production control function which will identify:

(1) weight of finished product,
(2) content of workpiece (parts list),
(3) work centre activities required (sawing, drilling, welding, etc.),
(4) labour content,
(5) machine centre utilization.

Production control systems are now often computerized and can produce a detailed map of the movement of the workpiece and its component parts within the fabrication workshop. Computer systems can identify any potential overloadings of work centres so that the production controller can specify an alternative path.

Generally, the cycle of work centre activity is as shown in Table 32.1 (the exact sequence of events may, however, vary between fabricators).

32.6.1 Lifting equipment in fabrication workshops

Not all steel can be moved around a workshop on conveyor lines. It is not practical or cost effective to do so and would restrict the flexibility for planning workshop activities. Much of the steel will be moved by cranes of one form or another. Craneage capacity can be a restriction on the work which a fabricator can undertake. Restricted headroom is another annoying problem particularly when considering the fabrication of deep roof trusses. Knowing the fabricator's limitations in these regards is useful when making a selection.

There are various types of crane in common use in most workshops, they are:

- electric overhead travelling (EOT)
- goliath
- semi-goliath
- jib
- gantry.

EOT cranes are the main lifting vehicles in the larger fabrication workshops. Their capacity varies from fabricator to fabricator but rarely exceeds 30 tonnes SWL. EOT cranes run on rails supported on crane beams carried off either the main building stanchions or a separate gantry and are used for moving main

Table 32.1 Sequence of activities in fabricating shops

Cleaning and cutting to length	Shot blasting Pre-fabrication priming Sawing or burning
Manufacture of attachments	Preparation of: cleats brackets stiffening plates end plates baseplates
Preparation of main components	Pre-assembly: drilling punching coping cropping notching cutting of openings
Assembly of workpiece	Marshalling of components Setting-out of components Shop bolting Welding Local drilling Local cutting Cleaning up
Quality control (assembly)	Check assembly – dimensional – NDT – visual inspection Sign-off
Surface treatment	Painting Metal spraying } often by other Galvanizing } specialists
Quality control (final product)	Conformity with specification Sign-off
Transportation	Marshalling Loading Despatch

components and finished workpieces between work centres. Heavy components need turning over for welding and this is achieved by the use of EOT cranes.

Goliath and semi-goliath cranes are heavy-duty lifting devices which tend to be used on a more local basis within the workshop than EOT cranes. The goliath is a free-standing gantry supported on rails laid along the workshop floor. Semi-goliath cranes are ground rail supported on one side and gantry rail supported on the other. The lifting capacity of goliath-type cranes is normally less than that of an EOT crane though greater in some workshops where there are EOT crane support restrictions (Fig. 32.12).

Jib cranes are light-duty (1–3 tonne capacity) cranes mounted from building side stanchions and operating on a radial arm with a reach of up to 5 m. They are useful for lifting or turning the lighter components (Fig. 32.13).

936 Fabrication

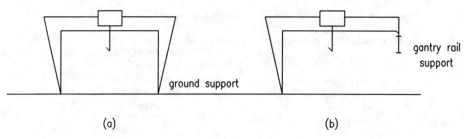

Fig. 32.12 Goliath cranes: (a) goliath, (b) semi-goliath

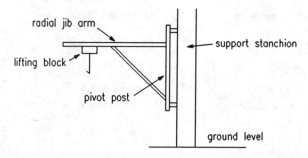

Fig. 32.13 Jib crane

Gantry cranes are purpose-built structures with limited lifting capacity (3–10 tonne). The lifting block travels along a gantry beam and is either underslung or top mounted. Gantry cranes are useful lifting devices where no building can be used to support an EOT crane and some gantry cranes feature EOT cranes riding on rails along side gantries. Stockyards normally are serviced by gantry cranes (Fig. 32.14).

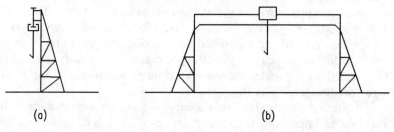

Fig. 32.14 Gantry cranes: (a) single underslung, (b) EOT

32.6.2 Conveyor systems

The use of factory automation is increasing the use of roller conveyor systems for moving main components. The roller conveyors can be designed so that they move steel between automated work centres either laterally across the workshop or longitudinally along it. The modern conveyor systems are computer controlled and bring steel accurately into position for the work centre activity to be performed. An example of a roller conveyor system is shown in Fig. 32.15.

Roller conveyors are suitable for steel which does not need turning and does not have cleats or brackets attached to it. Assembly work is not performed on conveyors but the assembly work benches are fed with components from conveyors or cranes.

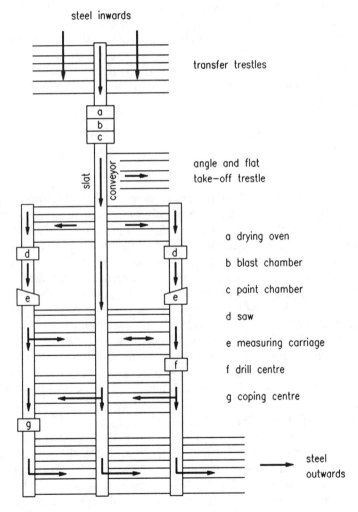

Fig. 32.15 Roller conveyor system

32.6.3 Handling aids

Most steelwork assemblies can be picked up easily using chains or strops but the designer should be alert to the possibility of having to provide for temporary lifting brackets suitably stiffened. This may prove necessary where it is important that a component is lifted in a particular way either for stability reasons or for reasons of strength. Lifting brackets need not be permanent of course but should be used and identified for the benefit of the fabricator who has to lift the component in his workshop.

The designer should also be cautious in his provision of welded-on brackets. Flimsy outstands *will* get damaged either in the workshop, during transport or on site. Flimsy brackets should be supplied separately for bolting on at site.

Some components by nature of their design or geometry may require to be lifted in a specific way using strongbacks or lifting beams. The designer must make the fabricator aware of this requirement to avoid possible accidental damage or injury to workshop staff; steel is a very unforgiving material.

The fabricator will normally organize transport of steel from the workshop to site. There are loading restrictions on vehicles in terms of weight, width, height and length.

32.7 Quality management

The key activities within any quality management system are quality assurance and quality control. These activities are fully embodied in many industry wide quality standards. While it is not a prerequisite that a manufacturer is registered under a standard based scheme, for appropriate levels of quality management to prevail, it does, however, allow specifiers to easily assess the relevance of the system under review. As far as design and manufacture of structural steelwork is concerned, there are a number of standards which apply. These standards are the basis for the formulation and implementation of appropriate quality management systems. British Standard 5750 provides an excellent basis for a system to be applied in structural steelwork manufacturing and design. The various parts of the standard give the opportunity for fabricators to develop quality management systems which cover design, manufacture and the erection of structural steelwork. The use of systems based on BS 5750 will allow progression to the International System Standard ISO 9000.

While the use of these standards is commended to those setting up quality management systems, it does not imply any formal recognition that compliance with the standard has been met. Formal recognition can be obtained by application for assessment of the quality management system by an independent body. The British Standards Institution will perform this assessment and provide registration to those companies which can successfully demonstrate compliance with the standard. Other registration schemes in the steelwork industry are run by Lloyd's and Steelwork Construction Quality Assurance Ltd. Within the quality management system there must be written procedures to be followed which will give the basis for adequate QA and QC. A brief description of some of these relevant to the fabrication of structural steelwork are given below.

32.7.1 Traceability

It is necessary to demonstrate that both materials and manufacturing processes can be traced by a clear audit trail. This aspect of quality management allows the specifier to determine the source of material for the product he receives and the origin of workmanship. This is particularly important when a defect is discovered. In terms of fabrication, the materials should have the support of documentation from the steelmaker in the form of mill certificates. Suppliers of other materials or consumable items such as bolts, welding electrodes, etc., will also be required to give details which show appropriate manufacturing data. The fabrication process involves workmanship and evidence should be maintained which traces the source of workmanship on each item.

32.7.2 Inspection

This is a prime area of quality control. It is important that quality inspectors are independent of workshop management, that is they must be made responsible to a higher level of management in the company which is not concerned with the production of fabricated steel in the workshop. They may report to the quality manager of the company who in turn may report to the managing director. It is also important that all inspection activities are part of a predefined quality plan for the job in question. There should be no opportunity for the level or extent of inspection to be decided upon an *ad hoc* basis. The quality plan will define levels of inspection required and when the inspection is to take place, typically:

(1) inspection of incoming materials or components,
(2) inspection of fabricated assemblies during and/or after fabrication,
(3) inspection of finished products prior to despatch,
(4) calibration of measuring equipment,
(5) certification of welders.

32.7.3 Defect feedback

At some stage or another, there will be occasions when defects are discovered in the fabricated item. This may be during inspection or on site. It is important that the defect is reported and the quality management system must cater for this. The report should be formal and it must identify the defect and possible cause. This report is then acted upon to prevent recurrence of the defect. The audit trail of traceability and inspection should highlight the actual source of the defect and this may be corrected by defined and planned actions.

32.7.4 Corrective action

All defects will cause corrective action to be taken. This action may take the form of revised procedures but more commonly will involve a process change. The most important corrective action is training as many defects stem from a lack of understanding on the part of someone in the production cycle. An important part of the prevention of recurrence of defects is trend analysis. Defects are recorded and trends studied which may highlight underlying problems which can be tackled. Occasionally, a finished product does not fully comply with the client's requirements and an approach may be made to determine if the product is suitable. While this process is going on the offending items must be put into quarantine areas which clearly distinguish them from others which do conform to requirements.

Further reading for Chapter 32

British Standards Institution (1984) *Specification for arc welding of carbon and carbon manganese steels*. BS 5135, BSI, London.
British Steel plc *SHS Welding*. BSC Publication.
(1988) CNC burning technology for structural members. *The Fabricator*, Oct.
British Constructional Steelwork Association (1980) *Structural Steelwork Fabrication, Vol. 1*. BCSA Publication No. 7/80.
Dewsnap H. (1987) Submerged-arc welding of plate girders. *Metal Construction*, Oct.
Hicks J.G. (1987) *Welded Joint Design*, 2nd edn. Blackwell Scientific Publications, Oxford.
Pratt J.L. (1989) *Introduction to the Welding of Structural Steelwork*, 3rd edn. SCI Publication 014, The Steel Construction Institute, Ascot, Berks.
Taggart R. (1986) Structural Steelwork Fabrication. *The Structural Engineer*, **64A**, No. 8, Aug.

Chapter 33
Erection

by HARRY ARCH

33.1 Introduction

If a structure is to be erected, stage by stage, safely, within the shortest possible time and at the lowest possible cost, then positive action is required from all the parties concerned. The first step is for the designer to have clearly in mind the means by which it is to be erected. The second step is to convey these ideas to the fabricators and the erectors through an initial method statement.

Notes on the drawing and in the documentation to indicate where erection should start in order to establish initial stability are essential, as is a statement indicating the need for any temporary support or stiffening in cases where instability would result from omission or early removal of the temporary support or stiffening. The method which is decided upon should conform with the intention of the designer of the structure wherever possible. If the eventual decision is for an alternative method to be used, then that new method must be fully documented in order to satisfy all the parties involved that it is acceptable. The agreed method statement is used, together with other, more detailed, planning documents, to monitor the work as it progresses.

The number of pieces to be lifted into place, their average height, and the weight, size and final position of the maximum lift determine the capacity of the cranes deployed on site. Attempts should be made to optimize the choice by arranging that every lift matches the capacity of the crane at that particular jib length and radius. Every time a hook goes up with less than its rated load the crane is being under-utilized.

One option which has the dual advantages of reducing the number of small lifts and man hours at risk to be spent at height, is to consider sub-assembly at ground level. To be fully effective the connections for a sub-assembled unit must be considered at the design stage in order that all the splices can be easily made with entry from one direction, from above, as the sub-assembly is lowered into place.

Another option is to lift components in bundles for distribution and fixing from a high-level dispersal point.

Whether sub-assembly techniques are being considered or not, it is vital that the positioning of splices is fully considered. How is safe access to be arranged; can a welded connection be arranged to reduce the amount of positional welding? Could a bolted joint be substituted to produce a more economical connection, and one which is less susceptible to the influence of weather, less sensitive to operator skills and easier to monitor and inspect using less sophisticated methods and equipment?

Finally, the method eventually selected must take account of the possibility that other contractors may have to be on site at the same time, using the same access routes and possibly even the same cranes. They may have to obtain access through parts of the structure left open by the omission of important structural members. Steps must be taken to ensure that the structure remains stable throughout such incomplete phases particularly where loads are to be imposed on the structure when components are missing. Members which are redundant in the completed frame can be vital during the erection phases.

Any temporary works which are needed to ensure the stability of the incomplete frame must receive full design attention, including foundations, connections, the permanent member to which they are connected, and the eventual removal method and timing within the overall programme.

33.2 The method statement

Guidance Note GS 28 Parts 1 and 2, produced by the Health and Safety Executive, contains full details of the contents of a method statement. The main items are:

(1) Arrangements for scheme management, including co-ordination and the responsibilities and authority of supervisory personnel at all levels
(2) Erection sequences, noting the scheduled starting position, or positions if phased construction is required
(3) Methods of ensuring stability at all times of individual members (including columns) and sub-assemblies, as well as the partially erected structure
(4) The detailed method of erecting the structure, the erection scheme, devised to ensure that activities such as lifting, unslinging, initial connecting, alignment and final connecting can be carried out safely
(5) Provisions to prevent falls from height, including safe means of access and safe places of work, special platforms and walkways, arranging to complete permanent walkways early, mobile towers, aerial platforms, slung, suspended or other scaffolds, secured ladders, safety harnesses and safety nets
(6) Protection from falls of materials, tools and debris by the provision of barriers such as screens, fans and nets
(7) The provision of suitable plant (including cranes), tools and equipment of sufficient strength and quantity
(8) Contingency arrangements, for example, against a breakdown of essential plant, or if components are delivered out of sequence
(9) Arrangements for delivery, stacking, storing, movement on site, on-site fabrication or pre-assembly and the siting of offices and mess rooms
(10) Details of site features, layout and access, with notes on how they may affect proposed arrangements and methods of working.

These items enable the requirements of the Health and Safety at Work Act of 1974 to be satisfied. A complex job may require a series of complex method statements, each covering an aspect of the work.

The statement, or statements, should form a single document capable of modification if necessary, indexed for easy reference. It should follow a logical sequence, be completely unambiguous, refer to components or items of work using their drawing identification marks and exact location, and be dated and marked with modification letters, as is a drawing, in order to prevent the use of an out-of-date statement.

Repetitive or 'standard' jobs can be covered by standard sheets which should identify any critical stages of the work that they detail.

33.3 Programme

33.3.1 Introduction

Construction planning begins with an understanding of the purpose of the project, and with a full and detailed evaluation of the site including ground conditions, the location of the site and what plant is available in the area. In the case of a building the initial site evaluation should consider whether there will be space around the building for cranes to work, whether they will have to climb with the building's own framework or whether tower cranes may be utilised.

The management team charged with responsibility for the initial design proposals must not produce their documents without taking account of all the contexts in which the project will have to be built as well as those in which it will operate.

Most aspects of construction work are seriously affected by weather; earth-moving by wet weather, concreting by cold weather, and steel erection by high winds. Considerations of extreme temperature and wind conditions can, for example, affect decisions about the use of site welding in major connections. Initial programming of the construction phases of a project must take account of these factors otherwise a programme can easily become meaningless but still have contractual implications.

It seldom happens that an erector has a free hand in the determination of the period of time within which his work on site must be completed. The client or his representatives will have integrated the steel erection phase with his overall programme for the total project. As shown in Fig. 33.1 there will be interlocking work processes to consider, starting with foundation completion and ending with following trades on floors and façades, etc. There may also be periods when access must be given to other contractors who have to work within the erection area, for instance, to install the boiler drums in a power station frame. A further particular case concerns railway possession periods.

Precast floor panels which the designer may have relied upon to give resistance to wind loading in the completed building have to be programmed too, as do the temporary bracings which will be needed before the floor panels can take up the

944   *Erection*

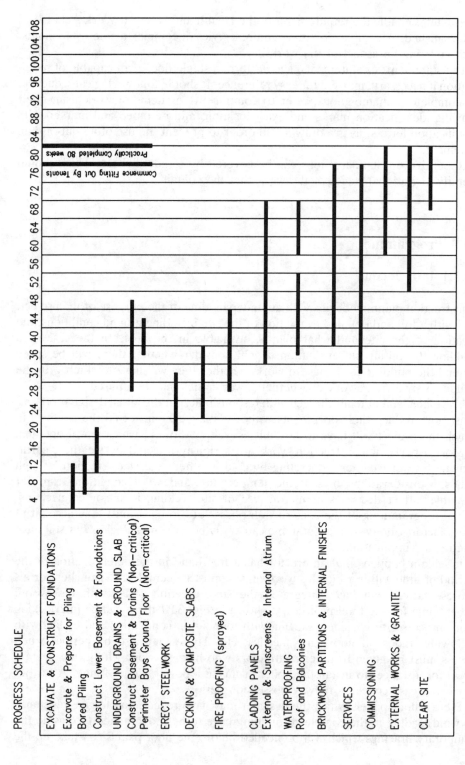

Fig. 33.1 Construction programme for typical multi-storey building (in weeks)

load. All of the foregoing conditions, and more, form the constraints within which the erectors' programme must be fitted.

33.3.2 Controlling the flow of materials to site

A job can only proceed smoothly on site with a regular flow of the materials required for its construction. The preparation of a delivery schedule based on a compromise between expected outputs, possibly from more than one fabricator, and the demands of the site programme is an early priority for the management team who must also consider the need to ensure early stability of the frame.

The almost inevitable lack of correlation between these demands often makes the provision of a stockyard on a site a necessity. A stockyard is additionally a surge tank which can absorb, for example, variations in transport or shipping times due to outside influences which, without a buffer stock of material, would leave the erection squads unemployed. As much as four weeks' erection work may have to be provided by the stockyard capacity.

It may be found that the demands of the site and its sequenced delivery requirements will be beyond the capability of any one fabricator. Hence another early role of management is to plan the schedule for all the material requirements. Typically this will involve a purchasing department using specifications provided by the designers and detailers, working to the demands of the programme and the original cost estimate for the supply of each item required within the financial limit, to define delivery requirements.

The day-by-day monitoring of the actual performance of the suppliers in terms of the quality of the work, its adherence to the specifications of workmanship and tolerance and adherence to programme is a management responsibility. A typical delivery demand and monitoring chart is shown in Fig. 33.2 from which it can be seen that the job is short of bolts and side framing, main beams and wall beams are critical, but the site is clogged with unwanted columns, secondary beams, side cladding and gutters.

In addition it is always helpful as a visual check for all concerned in the office and eventually on site, to mark up a general arrangement drawing as each marked piece is received on site and stored, or taken to the erection front. The bolt lists prepared by the drawing office are similarly broken down into phased requirements and used to order and monitor deliveries into the site store.

33.3.3 Cash flow and material monitoring

There are two reasons for making a delivery schedule and then monitoring it. The first is to ensure that the site is not flooded with unwanted components while being at the same time unable to erect anything through the lack of an essential

946 *Erection*

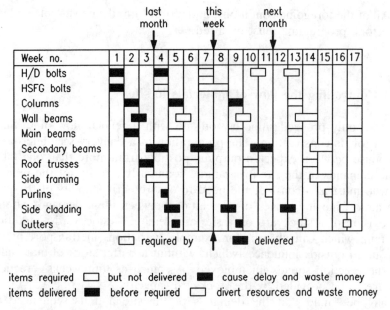

Fig. 33.2 Site delivery schedule

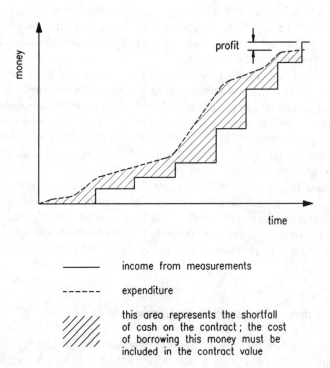

Fig. 33.3 Cash flow diagram

part. The second, which is often more important especially on a large job, is the determination of the cash flow out of, and into, the project account.

Cash flow is determined by a 'cash flow diagram' which plots cash requirements and availability month by month throughout the expected duration of the job, and compares the two (see Fig. 33.3).

The cost of the design, of the steel bolts, labour involved in fabrication, transport and erection are all items which produce the graph of expenditure. The other line charts the income expected from the stage payments which are made as work progresses.

As far as a client is concerned these payments may typically be after the completion of significant phases of construction. His income line will not begin until the whole is completed and his revenue starts to flow. The contractor's cash flow will be generated by stage payments as the job progresses.

The amounts by which expenditure exceeds income from month to month have to be financed; the cost of this finance is an important factor in the build-up of the cost of the project. There may additionally be considerations of possible fluctuations in exchange rates to be taken into account where part of the work has to be carried out in a different country.

33.4 Cranes and craneage

33.4.1 Introduction

On a large job cranes of varying size and capacity are provided to perform the various tasks: a heavy one for the main columns and a smaller one for the side framing posts and angles for instance. The number of pieces to be lifted in any one working period determines the number of cranes required to achieve the desired rate of working. In practice the choice of crane type and capacity is a compromise, with efforts being made to get as near to the optimum as possible.

In order to decide on a crane layout it is necessary to prepare a sketch or drawing of each critical lifting position showing, on a large scale, the position of the crane on the ground, the location of the hook on plan, the clearance between the jib and its load, and between the jib and the existing structure as the piece is being landed in its final position. These drawings will enable a check to be made of the match of each crane to its load, of the ability of the ground under the crane to sustain the crane with its load, and clearance to lift the component and place it in position. Of particular importance is the clearance needed by the tail of the crane as it slews. A series of these drawings will rapidly confirm whether a particular crane is suitable. If there is an anomalous heavy lift in the series, the designer can be asked whether a splice position can be moved to reduce weight.

Reports on overturned cranes and crippled jibs bear adequate testimony to the need to plan and to the extent to which it is neglected. The Code of Practice CP 3010 *The safe use of cranes* indicates clearly the dangers of improper use.

948   *Erection*

There are two ways in which the ingenuity of the erection planner can produce an economic scheme; first in his choice of crane and second in the organization of getting the material under the hook and up into place.

33.4.2 Types of crane

Cranes used on steel erection sites are of two main types: those described as 'mobile' which arrive on site ready to be used, needing only to have their jibs extended to the length required, and those that are delivered to site piece-small which have to be assembled before they can be used. They are fully described and their safe use is defined in CP 3010 *The safe use of cranes*.

Mobile cranes

This group includes truck, or wheel-mounted, and tracked cranes. Truck-mounted cranes are able to travel on the public roads under their own power, but with their jibs shortened. Mobile cranes mounted on crawler tracks are not permitted to travel on the road under their own power. A truck-mounted crane is a relevant

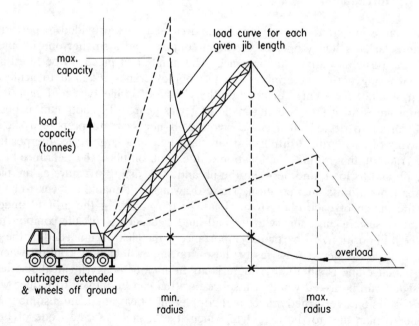

Fig. 33.4 Safe load diagram for a mobile crane

choice if a crane is only required on site for a short time; it can come and go relatively easily. However if a crane is to be on site for longer, the high transportation costs of moving a crawler crane around the country can be justified. The real advantage of a crawler crane is that its weight plus the reactions from the load being lifted are spread more widely by its tracks but it does not have the added stability of a truck-mounted crane with its outriggers set.

Both types are typically part of the erection fleet owned by steel erection and plant hire companies. The decision to hire from a plant hire company is often determined by the geography of the site – is it close to a plant hire company's yard or is it close to the contractor's own plant depot; is the crane to be used for short or long duration; and is the lift for which the crane is required one of a series or simply a 'one-off'?

Capacities of mobile cranes normally range up to 50 tonnes but exceptional cranes can lift up to 300 tonnes. Figures 33.4 and 33.5 provide performance data for a typical 25-tonne crane. It is most important to note that the rated capacity of a mobile crane and the load which can be safely lifted are two very different things. The maker's safe loading diagram for the crane must always be carefully consulted to see what load can be lifted with the jib length and the radius proposed. It is essential that a mobile crane is used on level ground so that no side loads are imposed on the jib structure.

main boom capacities (tonnes) – through full 360° circle slew – with outriggers fully extended							
	boom length						
radius in metres	10.07 m fully retracted	10.07 m to 12.50 m	12.50 m to 15.00 m	15.00 m to 17.50 m	17.50 m to 20.00 m	20.00 m to 22.50 m	22.50 m to 24.57 m
3.0 m	25.40	20.70	20.10	20.10			
3.5 m	22.00	20.00	19.00	18.80	16.00		
4.0 m	19.50	18.00	17.80	17.60	15.50		
4.5 m	17.00	16.80	16.70	16.50	14.90	12.70	
5.0 m	15.30	15.30	15.30	15.00	13.90	12.30	10.40
6.0 m	13.00	12.80	12.40	12.40	12.20	11.60	9.80
7.0 m	10.50	10.50	10.50	10.50	10.50	10.50	9.40
8.0 m		8.30	8.30	8.30	8.30	8.30	8.30
10.0 m		5.35	5.35	5.35	5.35	5.35	5.35
12.0 m			3.85	3.85	3.85	3.85	3.85
14.0 m				2.80	2.80	2.80	2.80
16.0 m					2.15	2.15	2.15
18.0 m						1.70	1.70
20.0 m						1.30	1.30
22.0 m							0.90

Fig. 33.5 Safe load charge for a 25-tonne capacity crane

Mobility for non-mobile cranes

Non-mobile cranes can be made mobile, but only normally by mounting them on rails. This has two advantages: the positioning of the crane can be more easily dictated and controlled, and the loads transmitted by the rails to the ground are in a precisely known location. Many a crane has collapsed because of insufficient support underneath but most rail-mounted crane failures have occurred from overloading.

Where the crane is to work over complex plant foundations the rail can be carried on a beam supported, if necessary, on piles especially driven for the purpose. If the rail is supported only by a beam on sleepers directly on the ground the load can be properly distributed by suitable spreaders. In either case the conditions will have been properly considered and designed for, instead of the often rather uninformed trust which is put in the capability of the ground to support a mobile crane and its outriggers.

Non-mobile cranes

Non-mobile cranes are generally larger either in their capacity to reach to great heights, or in their capability of lifting their rated loads at a useful radius. It is in both of these respects that the mobile crane is deficient.

There are two main types of non-mobile cranes (see Fig. 33.6), the tower crane and the derrick; with each their size precludes the possibility of the crane arriving on site other than in pieces. So the disadvantage of a non-mobile crane is that it has to be assembled on site. Having been assembled it must receive a structural test in addition to the winch and stability tests required for all cranes before they can be put into service.

Derrick cranes, once very common, are still occasionally used where their particular advantages are relevant. The big advantage of a scotch derrick is its simplicity. It can be built using relatively unskilled labour, is easy to maintain and is less susceptible to breakdown than its more complicated sisters. Its disadvantages are that it takes up a large ground area, requires two standard gauge railway tracks for travelling and needs a great deal of ballast which has to be loaded on to the ends of its 'lying' and 'raking' legs. It also requires a secondary derrick on a lower gabbard, or supporting framework, for its erection.

Where a tower crane is available with sufficient height and lifting capacity, it has several advantages (see Fig. 33.7).

(1) It requires only two rails for it to be 'mobile', and these two rails, although at a wide gauge, take up less ground space than a derrick.
(2) It carries most of its ballast at the top of the tower on the slewing jib/counter balance structure, and so very much less ballast is needed at the bottom – indeed in some cases there is no need for any ballast at all at the tower base or portal.

Fig. 33.6 Types of non-mobile cranes: tower cranes and derricks

(3) Because the jib of a tower crane is often horizontal, with the luffing of a derrick jib replaced by a travelling crab, the crane can work much closer to the structure and can reach over to positions inaccessible to a luffing jib crane.

952 *Erection*

Fig. 33.7 A tower crane tied to the erected structure

(4) A tower crane is 'self-erecting' in the sense that after initial assembly at or near ground level the telescoping tower eliminates the need for secondary cranes.
(5) As shown in Fig. 33.7 it can be tied into the structure it is erecting, thus permitting its use at heights beyond its free standing capacity.

Climbing cranes

The wide spread of the reaction points needed to stabilize either a scotch derrick, or its first cousin the guy derrick, becomes an advantage if the crane is to be mounted on a structure. In a guy derrick, the fixed legs of a scotch derrick are replaced by a number of guy ropes arranged radially around the top of the mast, from the bottom of which the jib is mounted in the same way as that of a scotch derrick (see Fig. 33.8).

The guy derrick was for many years the standard method of erecting medium- or high-rise frames where the crane must climb with the structure. The hoisting winch can be remote from the crane structure in a guy derrick while a tower crane or scotch derrick carries its winch with it. The climbing process for the guy derrick is therefore much simplified and the winch power supplies need to be moved less often.

Once a suitable means of climbing a tower crane had been developed it became the standard solution for tall building construction. By arranging connection points at various levels of floor steelwork it is possible to give the tower both support and stability. This, together with the telescoping movements of the inner and outer portions of the tower, enables the crane's own equipment to be used to power the climbing mechanism.

Provided the designer of the framework has been aware of and has made provision in his design for the loads, vibrations and connection points from the crane, there is no limit to how far it can climb. Intermediate load transfer platforms, each with its own crane, can be provided if the capacity of the winch drum on the crane is unable to provide the full length of rope required for the height of the overall lift.

Cranes for the stockyard

Stockyard cranes have to work hard. The tonnage in the job has to be handled twice in the same period of time, often with many fewer cranes. It is therefore important that cranes be selected and sited carefully to ensure maximum efficiency.

It is usual to arrange for stockyard cranes to be given a short length of track to enable them to cover a greater area than would be possible under a fixed crane. Tracks use up valuable area in a stockyard so the choice often falls on the goliath type crane (see Fig. 33.9). Running on only a single rail at each end it uses the minimum possible of that valuable space; furthermore only part of its coverage needs to be dedicated to actual transference of steel from and to the trucks, wagons and trailers. It also does not suffer the reduction in capacity that is incurred by jib cranes.

Where a contract involves the construction of a piece of industrial plant or a power house, for example, where all the lifts are concentrated in a relatively small area, a goliath type crane can also be used. The crane runs on tracks laid outside the limits of the structure and has sufficient headroom to lift items over the top of

Fig. 33.8 A guy derrick used as a climbing crane

the unit being built. In exceptional cases such a goliath may be almost in the category of a special lifting device (see Fig. 33.10). One penalty of such an arrangement is that only one hook is available for all the lifts and programming must take this into account.

Fig. 33.9 A goliath crane serving a stockyard

33.4.3 Other solutions

If there is no suitable crane, or if there is no working place around or inside the building where a crane may be placed, then consideration must be given to the need for a special mounting device for a standard crane, or even a special lifting device to do the work of a crane, designed to be supported on the growing framework. In either event this is an example of a situation where the closest collaboration between the designer and the erector members of the management team is of paramount importance, and where a lack of adequate communications can lead to trouble.

Once the decision to consider the use of a special lifting device has been made, a new range of options becomes available. The most important of these is the possibility of sub-assembling larger and heavier components so reducing the number of man hours worked at height. This is particularly true in the case of bridgework, where substantial sums which would otherwise have to be expended on other temporary supports and stiffening can possibly be saved. The big disadvantage is that the special lifting device being considered is often so special that is is unlikely to be of use on another job and so the whole cost has to be written off on the one job for which it has been initially designed. (See Fig. 33.11.)

At the cost of only a slight increase in time, and of the labour force needed, it is possible to eliminate the not inconsiderable cost of bringing a crane on to a site where the frame is single-storey. With the help of a winch, powered by either compressed air or an internal combustion engine, and some blocks and tackle, a

956 *Erection*

Fig. 33.10 A super goliath covering a complete project

light lattice guyed pole can be used to give very economic erection (see Fig. 33.12). In this instance the pole is carried in a cradle of wires attached at points on the tower. These connection points need to be carefully designed to be sure that they will carry the load and the tower structure will not be crippled.

Fig. 33.11 A purpose-made lifting beam for cantilever bridge erection

It is vital that any pole is used in as near a vertical position as possible as the capacity drops off severely as the droop increases. This requires careful planning and the employment of a gang of men experienced in the use of the method.

In a different context, pairs of heavier poles provided with a cat head to support the top block of the tackle can be used inside existing buildings to erect the components of, say, an overhead travelling crane, or to lift in a replacement girder (Fig. 33.13). The arrangements for a pole and its appurtenances take up much less floor space in a working bay than a mobile crane, which additionally needs a wide access route and space to manoeuvre itself into position. The method is particularly useful where headroom is restricted.

33.4.4 Crane layout

Having decided on the type, size and number of cranes that are required to carry out the work, each has a designated range of positions relative to the work it is to perform. These positions are then co-ordinated into an overall plan which enables each crane to work without interfering with its neighbours, and at the same time enables each to work in a position where adequate support can safely be provided

Fig. 33.12 An erection pole used to build a transmission tower

Fig. 33.13 Two poles connected at their tops and guyed, used to erect a 160 tonne girder

(see Fig. 33.14). This plan will then form the basis of the erection method statement documentation.

A major factor in planning craneage is to ensure that access is available and adequate to enable the necessary quantity and size of components to be moved. On large green field developments these movements may often have to take place along common access roads used by all contractors and along routes which may be subject to weight or size restrictions. On a tight urban site the access may be no more than a narrow one-way street subject to major traffic congestion.

33.5 Use of sub-assemblies

The crane layout shown in Fig. 33.14 makes use of a minimum amount of sub-assembly in the main rafter members of the portal frames, using the area behind the cranes on the erection front. If it has been decided to make use of much larger sub-assemblies on a larger frame, plans are drawn up to incorporate this into the design of the steelwork. The plans must also be incorporated into the erection

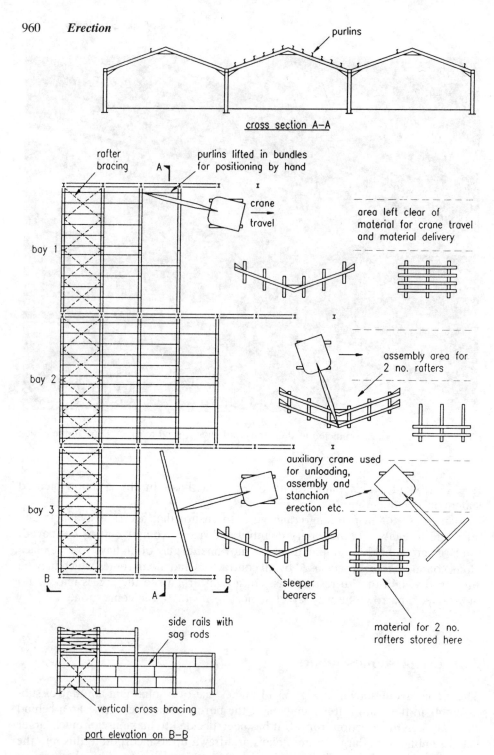

Fig. 33.14 Erection plan for multi-bay portal frame

method statement to show where and how the sub-assemblies are to be put together, transported and erected.

If the decision to sub-assemble has been made there will remain the further question of deciding whether an area in the stockyard is to be dedicated for this purpose, or whether the work will be done behind the cranes on the erection front so that the assembly can be lifted straight off the ground and into position as the crane is moved back.

The most common components to be assembled behind the crane are roof trusses or lattice girders which, because of their size, are almost always delivered to site 'piece small'. However, there is not always space for this to be done and other places have to be made available, possibly in the stockyard.

Where the potential sub-assembly area, and even the stockyard itself, is remote from the erection site, very careful investigations are needed before a decision to sub-assemble can be agreed. Local transportation size restrictions on the route to the erection cranes may rule it out.

There are three factors which affect the practicability and economy of sub-assembling a unit on the ground: (1) the weight of the eventual assembly including any lifting beams needed, (2) the degree to which the unit is capable of being temporarily stiffened without unduly increasing its weight, (3) the bulk of the unit – will it be possible to lift it to the height needed without fouling the crane jib?

It is often necessary to make a drawing or model of the crane jib and the unit at the point of highest lift before the last point can be checked satisfactorily.

Often a jig is used to represent the connection points of the unit to the building. A unit assembled on such a jig will fit when it is lifted, one built without a jig is very unlikely to fit. Sub-assembly is only worthwhile if the unit can be lifted and bolted into place almost as easily as a single beam. The object is to avoid carrying out operations at height which can easily be done at ground level.

33.6 Safety

33.6.1 Introduction

Everyone who is concerned in any way with the design and construction of any type of structure must constantly be reminding himself of the fact that while it is being erected it is not as stable and secure as it will be in its finished state. Most structures that have fallen, have fallen during erection. The men who have been killed or injured in an accident on a construction site, have suffered because of a lack of thought on someone's part somewhere along the way leading up to the event. BS 5531 and the Health and Safety Executive Guidance Note GS 28 are both aimed at improving the safety record during the erection process.

Accidents are caused by people. Accidents happen to people. People can prevent accidents. A weakness in the structure or in the overloading of a crane are both brought about by people, not by any fault in the inanimate objects which failed. The designer, the planner, the erector, the man with the spanner, are all the people who with forethought and care could have prevented an accident happening.

The man who hits his own thumb with his hammer is a victim not only of his own clumsiness but also of the man who put him in the awkward working position that led to the bruised thumb, and of the designer who produced the awkward design of the detail that had to be hammered. If he subsequently falls and is fatally injured the responsibility pattern is the same as if he had just bruised his thumb.

Research into the causes of accidents has shown that management has the greatest contribution to make to the cause of an accident, followed by the victim of the accident, and lastly by his workmates. Management is taken here in a broad sense. No one involved can escape from their responsibility.

If the limits and tolerances of dimension, of workmanship, of temperature, of the wind, of the load factor designated for the particular component, all conspire together at their limit, only a small 'nudge' is needed to trigger a succession of events leading to a disaster. The 'domino effect' is not however an inevitability. The better the planning, the better the design, the more thought that has been devoted to real understanding of what is going on in a structure as it is slowly built up from its component parts, the more likely it is that the next domino will refuse to fall, and the dangerous occurrence will have been contained.

What can be done to build in a stop to prevent such a succession? The first step is to consider each potential hazard and be sure that something is done to cope with it. An example of this is in the design of the holding-down bolts. These may eventually only serve as locators, but when the column is first erected and before any bracing is fixed they are all that prevents that column from falling (see Fig. 33.15). It is subject to loading from the wind and from knocks and bumps as the connecting steelwork is attached. A concrete column in a similar situation has to be propped in four directions with temporary adjustable props each anchored at each end. Steel columns are much lighter and much simpler to place on end, but they do depend on their holding-down bolts for their first stability.

33.6.2 The safety of the workforce

This is the aspect of site safety subject to statutory regulation, and to inspection by the Health and Safety Executive, in a way that the structural safety of a project is not.

Regulations lay down minimum acceptable standards for the width of working platforms, the height of guardrails, the fixing of ladders, and so on. They refer to use of safety belts and safety nets. They lay down the frequency with which a shackle or chain sling must be tested and the records that must be kept to show that this was done. Reference should be made to the appropriate regulation for the details of these requirements.

Fig. 33.15 Initial column erection

By its very nature the erection of a structural frame is a process involving a certain amount of risk. The work is carried out at height above the ground, and until it has progressed to a certain point there is nothing to which a safe working platform can be attached; the process of erecting a safe platform can be as hazardous as the erection process itself. One solution is to provide mobile access equipment if ground conditions permit.

Different access platforms are appropriate in different circumstances. One advantage of modern composite floor construction is that the decking can quickly provide a safe working platform only requiring the addition of a handrail. Figure 33.16 shows a safe platform for the erection of bare steel work – a prefabricated platform slung over a convenient beam. In this case weather protection may be added for site welding.

There is a special onus on designers and planners to ensure that no platforms have to be put up for work that ought to have been done either in the fabrication shops or on the ground before the component concerned was lifted into place.

A prime consideration at planning stage is to see if a working platform can be eliminated altogether. Can the operation be carried out at ground level before the component is erected? If not, can the platform be designed so that it is assembled on the component while it is still on the ground? It is impracticable to have to consider the provision of a safe working platform in order to be able to safely erect the main safe working platform.

Fig. 33.16 A prefabricated working platform slung over a convenient beam

The object of safety procedures is to ensure that everything possible is done to eliminate the risk of an accident injuring or killing anyone working on the site.

In summary, the methods of achieving safety are:

(1) Communication

Communication of the details of safety procedures to all concerned, with their implementation through the display of the regulations themselves in the form of abstracts, by issuing safety procedure documents and by running of training courses. Individuals must be aware of the location of particularly hazardous areas and the type of protection they have been given, the types of protective clothing and equipment that are available and how to obtain them, the restrictions in force on the site regarding the use of scaffolding or certain items of plant and any access restrictions to certain areas. They should be encouraged to tell someone in authority if they see a potential hazard developing before it causes an accident. (See Fig. 33.17).

(2) Equipment

Making the necessary equipment available on the site and maintaining it in good order. Equipment includes safety helmets, ladders and working platforms, safety belts and properly selected tools.

(3) Avoidance of working at height

Organization of the task so that the least amount of work possible has to be done at height:

(a) by the use of sub-assembly techniques,
(b) by the fixing of ladders and working platforms on to the steelwork on the ground before it is lifted into place,
(c) by the early provision of horizontal access walkways,
(d) by the provision of temporary staircases or hoists where appropriate.

All of these measures enable some of the hazards of working at height to be reduced by conferring on that work some of the advantages of ground level working.

(4) Fixing of portable equipment

Ensuring that portable equipment such as gas bottles and welding plant is firmly anchored while it is being used. The horizontal pull on a gas pipe or a welding cable being used at height is considerable, and easily able to dislodge the plant from a working platform, causing it to fall, possibly taking the operator with it. Care should also be taken to ensure that there are no flammable materials below on which sparks could fall.

(5) Design

Thoughtful design makes an important contribution to safety. The positioning of a splice so that it is just above rather than just below a floor level will reduce the risks in completing the splice on site. The arrangement of the splice so that the entry of the next component can be effected readily and without a fight, will reduce the risk of completing the splice up in the air.

1. EMPLOYEE JOINING THE COMPANY

- Discuss and obtain details of person's previous safety training experience.
  Details should be recorded.

- Show and explain company safety policy, organizations and arrangements.
  Supply copy or abbreviated version as necessary, OR confirm where it can be seen.

- Show and describe safety rules and procedures.
  Supply copy as necessary; OR confirm where details can be seen.

- Provide name and location of Safety Adviser.
  Details of Safety Supervisor/Safety Representatives (where appropriate) to be provided on site.

- Explain the necessity for personal protection.
  Highlight specific processes as necessary.

- Inform employees of their responsibility for health and safety at work.
  Safeguard/co-operation and mis-use of equipment.

- Explain the necessity for authorization in the use of plant machinery and equipment.
  Obtain details of previous experience – record. Arrange for training as necessary.

- Discuss importance of reporting dangerous incidents and defective plant and equipment.
  Explain system of reporting.

- Explain company procedures for grievances, disputes on safety matters.
  Whom to contact.

- Explain method of communicating safety information within the company.
  Notice boards, Safety Bulletins, Safety Newsletters – as applicable.

Fig. 33.17 Safety check lists

Lifting cleats and connections for heavy and complex components should be designed and incorporated in the shop fabrications as should fixing cleats, brackets or holes for working platforms and for safety belts or safety net anchorages. They will then be incorporated as part of the fabrication, rather than having to be provided by the erector while working at height. Access to a level should be provided by attaching the ladder, and the working platform that may be required, to the member at ground level prior to lifting. Ideally these connections should be designed so that they can be dismantled after the

2. EMPLOYEE'S FIRST VISIT TO SITE

A tour of the site should be made during the induction process to aid in the identification and location of key personnel and siting of equipment, fire and first aid points, etc.

- Confirm location of hazard areas.
  Fragile roofs, excavations, electricity, etc.

- Provide detail of first aid arrangements.
  Describe procedure for receiving first aid. Contact point/person.

- Confirm company policy for use of helmets.
  Areas where risk of head injury exist.

- Describe fire fighting arrangements.
  Confirm fire alarm points. Allocate fire assembly points. Ascertain knowledge of fire extinguishers – training as necessary.

- Describe accident procedure.
  Treatment – reporting. Accident book. Absence from work.

- Confirm protective clothing and equipment. Availability/use/issue.
  Issue relevant items. Record. Instruct in use. Storage – loss arrangements.

- Site tidiness.
  Storage/disposal of materials. Safe access, vermin, disease.

- Confirm authority to operate.
  Issue permit as necessary. Driver's licence. Site rules for plant and transport. Arrange training.

- Scaffolding – means of access.
  Highlight basic safety with work on scaffolding. Non use of incomplete scaffold. Access – to work place.

- Reporting defects. Malpractice. Safe systems of work.
  Stress need to report defects and malfunctions in plant and equipment. Horseplay. Safe practice.

- Locate and provide details of welfare facilities.
  Canteen opening times. Designated area where smoking is permitted. Storage of clothing, equipment and tools.

erector has left the platform and descended the ladder. He should not have to come down an unfixed ladder or stand on an unfixed platform as he removes these items after use.

Thought given to all these points at drawing board stage will reduce the risk inherent in a hastily improvised solution, often reached without proper consideration, and by unqualified personnel.

33.6.3 The safety of the structure

The stability of the structure itself, which is not prescribed by statutory regulations in the way that the safety of those working on it has been protected, is clearly the most important product of the care which has been lavished on its design.

Where a collapse of a partly-built structure occurs, the loss of life is generally heavy. One of the main indictments of the design process is that where there is a collapse the investigation and inquests which follow show that there have been lapses in the understanding of the behaviour of the incomplete structure, lapses in the detailed consideration of each and every temporary condition, and, most important of all, lapses in the communication of information to all involved. In the bridge collapse shown in Fig. 33.18, the temporary loading condition was not considered properly. At the point of failure of the bottom flange, the splice shown in Fig. 33.19 (in tension in the finished structure) was not adequate to take the compression imposed during erection.

Fig. 33.18 Bridge failure during cantilever erection

Fig. 33.19 Splice detail that failed in compression

A designer must communicate his plan for building the structure to those who will actually have to do the building; nothing could be worse than the designer who has not even considered how his creation is to be turned into reality.

Columns

The effective length, and even design condition of a column can change during the construction of a building. Each condition must be checked to ensure that the column is adequate for each one and the risks inherent at each stage must be assessed and provided for.

Plate girders and box girders

The checks which are applied to the webs and stiffeners of a plate girder during its design normally take account only of conditions at points where stiffeners are located and at points where loads are applied to the girder. It may not be obvious to the designer of a bridge girder that it may be subjected to a rolling load when the bottom flange is rolled out over the piers, subjecting the girder to a compressive load which it is not required to carry in its permanent position.

Splices

The effects of stress reversals are most severe on splice details which almost inevitably involve some degree of eccentricity that can trigger a collapse if the condition, transient though it is, has not been considered in design (see Fig. 32.19).

Bracings

Bracing is built into all types of structures to give them a capability to withstand horizontal forces produced by wind, temperature and the movements of cranes and other plant in and on the building.

Erection cranes carried on the structure during its erection produce vibrations and also loads where no load is intended when the building is complete. Reactions to cranes slewing, luffing and hoisting are carried by the framework supporting the crane down to the ground; these loads and vibrations must be considered and the structure's ability to carry them assessed at the outset.

Temporary bracings required at some stages of the work must have properly designed connections and be specifically referred to in the erection method statement. Early or unauthorized removal of temporary bracings is a common cause of collapse in a partially completed frame.

Having considered the need for installing temporary bracings and the need to postpone fixing permanent bracings, consideration should be given to the overall economy of retaining the temporary bracings and perhaps leaving out the permanent bracings. It is a costly and potentially dangerous business to go back into a structure solely in order to take out temporary members, or to insert components that have had to be left out temporarily.

Temperatures and wind effects

On a partly erected and unclad building frame the effects of temperature on the framework can exceed the effects of the wind. A tall framework will lean away from the sun in much the same way as a plant leans towards it as the sun moves round from east to west.

This must be recognized in any attempt to check the plumb of a building and should only be done on a cloudy day or after the whole structure has been allowed to reach a uniform temperature, say at night and then only when the temperature is at or near the design mean figure. Tightening the bolts in the bracing when a building is at non-uniform temperature can lock in the error and make it very difficult to correct later.

Wind effects can bring a building down if it is not adequately braced and guyed. The wind can have two effects: (1) pressure exerted on anything in its path (2) vibrations in a member obstructing its path. The problem is compounded by the variability of the direction and speed of a wind, and by the variability of the aerodynamic shape of the structure as each new piece is added. Care must be taken to ensure that they have been properly considered at each stage of erection for potentially problem structures, for example, bridges erected by cantilevering. Bracings, guy ropes and damping weights may all have to be considered as methods of changing critical frequencies of vibration and of limiting movements as the job progresses.

33.6.4 The safe use of cranes

Mention has already been made of the UK Statutory Regulations. These lay down not only requirements for safe access and safe working but also a series of test requirements for cranes and other lifting appliances.

It is the responsibility of management to ensure that plant put on to a site has a sufficient capacity to do the job for which it is intended, and that it remains in good condition during the course of the project. Shackles and slings must have test certificates showing when they were last tested. Cranes must be tested to an overload after they have been assembled. The crane test is to ensure that the winch capacity, as well as the resistance to overturning and the integrity of the structure, is adequate.

British Standards lay down the various requirements for safe working. Lists of those standards, and the necessary forms to enable each of the tests to be recorded, must be provided by management, often in the form of a 'site pack' which the site agent must then display and bring into use as each test is carried out. It is the site agent's responsibility to ensure that these requirements are fulfilled. He may also be required to produce them from time to time for inspection by the factory inspector during one of his periodic visits to the site.

A crane which has been tested and used safely in many locations might overturn at its next location and use (see Fig. 33.20). The cause is often found to be that an inadequate foundation has been provided under the tracks or outriggers of the crane, i.e. adequate support under the tracks or outriggers is an absolute requirement. It is equally important that the crane should work on level ground, as an overload can easily be imposed, either directly or as a sideways twist to the jib, if the ground is not level.

33.6.5 Temporary supports and temporary conditions

The design of the structure will have had a considerable amount of effort devoted to it. The design of the temporary works on which that structure may have to depend while it is being built may not have had so much attention. The number of collapses which have taken place after an initial failure in the temporary supports,

Fig. 33.20 Crane overturning

suggests that temporary works seldom get the design attention they deserve (see Fig. 33.21). For example a temporary support may only be designed to take a vertical load. In practice the structure it is intended to support may move due to changes in temperature and wind loading – imposing significant additional horizontal loads.

Fig. 33.21 Falsework collapse

Consideration starts at the foundations. Settlement in a trestle foundation can profoundly affect the stress distribution in the girder work that it supports. Settlement under a crane outrigger from a load applied only momentarily can lead to the collapse of the crane and its load.

The Code of Practice BS 5957 for falsework, which includes all temporary works, trestling, guy wires, etc., as well as temporary works associated with earth works, deals with a wide range of falsework types and should be carefully read and observed. Particular attention should be paid to the paragraphs dealing with communication, co-ordination and supervision since failure in any of these can lead to a failure of the falsework itself.

Re-used steelwork showing signs of severe corrosion must not be used for temporary falsework carrying critical loading, but in other situations re-used steel should be measured to be sure that it is adequate.

During construction a structure will move as its parts take up their design load. Connections to temporary supports have to be capable of absorbing these movements. Unless the design allows for these movements, eccentricities can result which will trigger a collapse. The cross-heads at the tops of bridge trestles are typically members which have a bad record of failure from this cause. They are often called upon to resist wind-induced loads, vibrations and temperature-induced movements in the structure, in addition to their more obvious direct loading, and so must receive a special design study.

Very tall buildings and chimneys as well as bridges can be affected by wind-induced vibrations as can working platforms and those who have to work with them. The force of the wind can make welding impossible without adequate shelter; the fixings for a working platform must be able to take the load of the wind blowing on the area of shelter which the welder has to have for his work.

The rule that one should 'put' before one 'takes' is a good one. Too many examples exist of a collapse following the removal of guy wires before the bracing was fitted, or before column bases designed to be 'fixed' had been grouted and fixed in reality. What is needed here is a clear communication down to the foreman and the workforce of exactly which sequence of working must be followed. Supervision alone is unlikely to be looking in the right direction at the precise time the deed is done. The only way to be sure is to issue a clear directive coupled with an explanation of why the instruction is being given and to have a skilled workforce who know what they are doing.

Attention has already been drawn to the need for an organization chart. A second chart showing who needs to know what, and why, and when should also be produced. If the lines of communication and the patterns of responsibility up and down the various management levels and between the various organizations on the chart are to be effective there must be a conscious effort on the part of everyone, firstly to understand why the links are there, and secondly to ensure that the necessary information flows swiftly up and down the lines. Assumptions can be dangerous things: 'It is for the want of thinking that most men are undone.'

33.7 Site practice

33.7.1 Setting-out tolerances

Accuracy of setting out is a very important part of maintaining an acceptable control of tolerances and achieving a structure which is acceptable to all the following trades. (Chapter 31 discusses the topic of tolerances generally for structural steelwork.) It is concerned with the establishment of a grid of lines and the transfer of those lines to the foundations and columns, and with the establishment of accurate levels.

The site should be provided with a dumpy or quick-set level and with a simple theodolite. The advantage of a theodolite over a simpler box square is that the transit of the instrument can later be used for checking the plumb of the columns and the telescope for reading offsets to check their alignment.

Time spent on laying down and marking the cross-centre lines of the column bases on individual boards at each base and subsequently on to the concrete of the base is time very well spent. Lines should be drawn on chalked areas in such a way that the mark made is reasonably permanent and will not be washed away in the first shower of rain. The base can then be quickly and accurately set in both directions and to the correct level on the pre-set pads. The value of this work is that it enables errors in the concrete work to be identified at a time when corrective measures can be undertaken. It also allows much argument, claim and counter claim, to be avoided. The accurate setting of grouted packer plates on which to land the column serves the same purpose. It will highlight concrete which is high and holding-down bolts which are low. It makes sense to discover these errors as early as possible.

Whether the column base is a socket type with shear connectors to the concrete, or if it has a plate or slab base plate, or even if it sits on a fabricated grillage, the same rule holds true. It is far easier to set, adjust and make true a pile of packers and a line than it is to start heaving the whole framework around after it has been erected. Final bolting can be done right away, without having to leave the connections loose to give the framework freedom to be moved into plumb and line as a later operation.

33.7.2 Slinging and lifting

Components, whether they are on transport or are lying in the stockyard, should always be landed on timber packers. The packers should be strong enough to support the weight of steel placed above them and thick enough to enable a sling to be slipped between each component.

The aim when lifting a component for transport only is to have it hang horizontally. This means that it is necessary to estimate its centre of gravity; easy for a simple beam but less easy for a complex component. The first lift should be made very slowly in order to check how it will behave, and also to check that the slings are properly bedded (see Fig. 33.22).

Most steelwork arrives on site with some or all of its paint treatment. The damage which slinging and handling can do must be put right and it is therefore very important to try to minimize damage to the paint. The same measures which achieve this saving also ensure that the load will not slip as it is being lifted and that the slings (chain or wire) are not themselves damaged as they bend sharply around the corners. Softwood packers should be used to ease these sharp corners.

Packers to prevent slipping are even more necessary if the piece being erected does not end up horizontal. The aim should always be to arrange the slinging in such a way that the piece hangs at the same attitude as that which it will assume in its erected position. Pieces being lifted are usually controlled by a light hand line affixed to one end. This hand line is there to control the swing of the piece in the wind, and not to pull it into level. Wherever possible non-metallic slings should be used. Their use will reduce damage to paintwork; they are less likely to slip than chain or wire slings.

In extreme cases two pieces may have to be erected simultaneously using two cranes. Staff working back at the office should have identified such possibilities

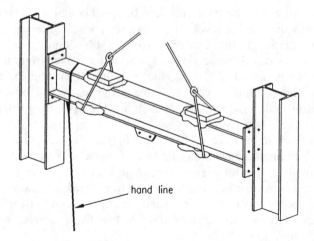

Fig. 33.22 Correct slinging practice

and placed them in the site erection method statement. It is too late to find out about it when the erection is attempted with only one crane, or with no plan made of how to pull back the head of the column.

Reference was made earlier to the need to consider the stiffness of large assemblies such as roof trusses as they are lifted from a horizontal position on the ground. Reference has also been made to the need to build assemblies in a jig to represent the various points at which connection has to be made in the main framework. Another jig for lifting can be useful particularly if there are many similar lifts to be made. This can be made to combine the need to stiffen with the need to connect to stiff points in the sub-frame, and the need to have the sub-frame hang in the correct attitude on the crane hook. The weight of any such stiffening and of any jig must of course be taken into account in allocating a crane to make the lift.

Some situations require that temporary stiffening be left in position after the initial erection until the permanent connections are made. This should also have been foreseen and sufficient stiffeners and lifting devices provided to avoid an unnecessary bottleneck caused by a shortage of a device for erection of the next sub-frame.

Where a particularly awkward or heavy lift has to be made a much quicker and safer job of slinging and lifting can be made if cleats for the slings have been incorporated in the fabrication. Each trial lift made after the first one wastes time until the piece hangs true. The drawing office knows exactly where the centre of gravity is.

A chart giving details of standard hand signals is illustrated in Fig. 33.23. Their use is essential when a banksman has to be used to control the rear end of the

Fig. 33.23 Standard hand signals for lifting

transport bringing the component to the hook as it is reversed. The banksman is needed to relay the signal from the one man directing the movements of the crane if he is necessarily in such a position as to be out of sight of the crane driver. A clear system of signals should be agreed for the hand-over of control of the crane from the man on the ground to the man up on the steel who will control the actual landing of the component. A banksman may also be needed up on the steel if the crane driver cannot see clearly the top man who is giving the control instructions. It is vital that there is no confusion over who is giving instructions to the crane driver.

33.7.3 Lining, levelling and plumbing

If, in spite of having taken care over the initial levelling of packers and of carefully positioning the column bases to the pre-set lines on the foundations, the structure still needs adjustment during lining, levelling and plumbing, then something must be wrong.

The first thing to check is that no error has been made in the erection. If nothing is out of place then the drawings and the fabrication need to be checked. If there is an error it will be necessary to take careful note of all the circumstances and advise all concerned immediately.

If the steelwork has just been thrown together with no great regard for accurate positioning it will be necessary to provide equipment and manpower to carry out the lining, levelling and plumbing. The equipment will include jacks, tirfor-type wire pullers with wires, straining screws, wedges, piano wires, and heavy plumb bobs with damping arrangements, all in addition to the level and simple theodolite which are needed for the basic check.

It is necessary to do the lining and levelling check before final bolting up is done. In practice this means that it must be done immediately following erection since it is a waste of time to have to slacken off bolts already tightened in order to be able to move the steelwork about. There is no way that erection can go ahead until the braced bay is bolted up and that cannot be done until it is level and plumb.

A supply of steel landing wedges and slip plates is needed to adjust the levels. They must be positioned in pairs opposite to each other on two sides of the base plate of the columns to be levelled. If the wedge is placed on one side only the column will be supported eccentrically and can in fact be brought down, especially if more than one column had been lifted on a series of wedges all driven in from the same direction. Alternatively toe jacks can be used in pairs to lift the columns. The engineer may direct that temporary packings be removed after the grouting medium has set. A temporary bench-mark should be established in the vicinity of the columns in a position where it cannot be disturbed. Its value should be agreed with the client's representative. The level will then be used to check the final setting with the seating packers inserted, and the column landed back on them and bolted down with the holding-down bolts. The column can then be moved about in plan on these packers to bring it into line and bay length from its neighbours.

The position of a line, offset from the column centreline by an amount to clear the columns, should be marked and agreed. This line is then used either to string and strain a piano wire or to set and sight a theodolite telescope. Offsets from it will then give the amounts by which movements must be made.

Running dimensions from the previous column will give required movements longitudinally. Care should however be taken to watch the plumb of the columns in this direction in view of the tendency of steelwork to 'grow'. Regard should be paid to column-to-column dimensions rather than to running dimensions from the end of the building.

Having got the bases of the columns in their correct position and level, it is possible to check the plumb. The comments already made regarding temperature effects are relevant here too. It is a complete waste of time to check the plumb of a building which is not all at the same temperature, and in the case of a long building if the temperature is not at the standard which in the UK is normally 10°C.

The fact that the column bases are level does not mean that a level check is not needed on the various floors in a tall building. It is important however that the levels of any one floor are only checked for variations from a plane rather than on the basis of running vertical measurements from the base which are affected by temperature and the variable shortening effect on the lower columns as weight is added to them.

Plumb can be most readily checked with a theodolite using its vertical axis and reading against a rule held to zero on the column centreline. This eliminates the effect of rolling errors and is a check against the same centreline used in the fabrication shops. If a theodolite cannot be used a heavy plumb bob hung on a piano wire and provided with a simple damping arrangement, such as a bucket of water into which the bob is submerged, is a second best method. Measurements are made from the wire to the centrelines in the same way as before. The disadvantage of using a plumb wire is that all the operatives have to climb on the steel to take and then to check the readings. Optical or laser plumbing units are available; these are particularly useful for checking multi-storey frames.

Adjustments, and in extreme cases provision for holding the framework in its correct position, are normally only necessary if the frame is not self-stable. If it will eventually depend say on concrete diaphragm panels for stability, consideration should have been given at the design stage to one of two alternatives. Either the concrete panels should be erected with the steel frame or else temporary bracing should have been provided as part of the original planning. Any bracing must be positioned so that it can be left in place until after the concrete panel has been placed and fixed so as to be able to carry out its function.

If diagonal wires have to be used they should be tied off to the frame at node points rather than in mid-beam, in order to avoid bending members, and their ends fixed using timber packers in the same way as is done when the pieces are slung for lifting. Turnbuckles or tirfor-type pullers provide the effort necessary to tension the wires. The sequence of placing, tensioning and ultimately removing these temporary arrangements should form part of the method statement prepared for the particular situation.

Once the components of the building have been pulled into position the final bolting up can be completed. In the case of a stiff-jointed frame, after the joints have been fully bolted or welded it is impossible to make any further adjustments.

33.7.4 Site welding

Special consideration must be given to the planning of the erection of a structural steel frame where site welding has been specified. Site welding has to be carried out in conditions of weather and in positions which make it more difficult and expensive than welding carried out in workshop conditions. In a workshop the work can be positioned to give optimum conditions for the deposition of the weld metal, while most welds on site are positional. Most site welding is carried out by manual metal arc and electrodes since this is more flexible in use and is more able to lay good weld fillets in the vertical up and overhead positions.

Some means must be provided for temporarily aligning adjacent components which are to be welded together, and of holding them in position until they are welded (see Fig. 33.24). The methods adopted to cope with the need for alignment may have to carry the weight of the components, and in some cases a substantial load from the growing structure.

Safe means of access and of working must be provided for the welder and his equipment. The working platform may also have to incorporate weather protection since wind and rain and cold can all adversely affect the quality of the weld being produced.

The design of the weld preparation on the components to be joined must take into account the position of these components in the structure. The erection method statement and the weld procedure statement for each joint must take all these factors into account. BS 5135 gives an outline of the many factors which must be defined in a weld procedure statement. Reference should also be made to BS 4870.

Provision must be made for the necessary run-off coupon plates for butt welds and for their subsequent removal. Run-off plates are required in order to ensure that the full quality of the weld metal being deposited is maintained along the full design length of the weld (see Fig. 33.25).

Also provision must be made in the initial setting and positioning of the components in order that the weld shrinkage which will take place as the joint cools will not result in loss of the required dimensional tolerances across the joints.

Fig. 33.24 Temporary web splice plates supporting and aligning joint for site welding

33.7.5 Site bolting

Wherever possible site splices should be designed to be bolted. This process is less affected by adverse weather than welding, uses simple equipment and presents much less complex problems regarding access for carrying out the work and for its subsequent inspection.

Impact wrenches and powered nut runners speed the tightening of bolts. When small quantities of HSFG bolts are to be used, manual torque wrenches are available. Where large quantities of friction-grip bolts have to be tightened, an impact wrench is an essential tool. Where access is difficult, wrenches are made which run the nut up and are then used as manual ratchet wrenches to finally

Fig. 33.25 Run-off plates for a butt weld

tighten the bolt. A large impact wrench able to tighten 33 mm bolts weighs 8.3 kg (18 lb) and consumes 21 l/s of air (44 ft^3/min). The drive to the nut is through a square shaft and an interchangeable socket sized to fit the nut being tightened. The sockets are retained on the drive shaft by a pin which is in turn retained by a rubber ring. A range of sockets and adaptors is available enabling the usefulness of the wrenches to be increased.

The podger spanner is the tool that is associated most closely with a steel erector. The pointed end is used for the initial lining of holes before the bolt is inserted and tightened with the jaws of the spanner.

Friction-grip bolts must be tightened first to bring the plates together and thereafter be brought up to the required preload either by applying a further part turn of the nut or by using a wrench calibrated to indicate that the required torque has been reached. The manual wrench has a break action which gives the indication by means of a spring without endangering the erector by a complete collapse. Alternatively load-indicating washers may be used to demonstrate that the bolt has been tightened to the requisite tension.

33.8 Special structures

While it is a mistake to consider that any structure is not special in some way, there are, nevertheless, a number of structures which, by their complexity, require special consideration when planning their erection and therefore their design. The length or height of the complete structure may bring it into this category, or the depth of individual members, or the fact that it is designed to move.

Temperature differentials over the depth of bridge box girders will produce changes in the camber of the girder. Temperature differentials over the width of a structure will produce changes in verticality. Temperature changes will affect the verticality of the columns at each end of a long single-storey factory building, or

Fig. 33.26 Humber Bridge during erection

of a bridge. Some of these effects can be, and commonly are, accommodated by the provision of expansion joints, but others must be recognized in the planning and execution of the work.

The building of a suspension bridge and a cable-stayed bridge are perhaps the best examples where the movements and changes to the shape of the structure as it grows are most apparent. A radio telescope is the best example of a special structure which is designed to move and yet must maintain very close tolerances as the extremities of the structure are reached. Other structures move as they grow and their temporary supports can fail as a result. This is too often the result of a lack of appreciation of construction movements, vibrations from wind, or local loads from erection plant. But, with a suspension bridge (Fig. 33.26), the obvious complexities demand special attention and so it can truly be said that it is where a structure gives the appearance of simplicity that problems are most likely to arise.

Chapter 34
Fire protection and fire engineering

by JEF ROBINSON

34.1 Introduction

Fire safety must be regarded as a major priority at the earliest stage as it can have a major impact on the design of a building and its structural form. Nevertheless, it should not stifle aesthetic or functional freedom; 'fire engineering' techniques are now available which permit a more rational treatment of fire development and fire protection in buildings.

The strength of all materials reduces as their temperature increases. Steel is no exception. It is essential that the structure should not weaken in fire to the extent that collapse occurs prematurely, while the occupants are seeking to make their way to safety. For this reason it is necessary to provide a minimum degree of 'fire resistance' to the building structure. Additionally, a measure of 'property protection' is implied in the current Building Regulations, although in principle the only concern of the Regulations is safety of life.

There are two basic ways to provide fire resistance. First, to design the structure using the ordinary temperature properties of the material and then to insulate the members so that the temperature of the structure remains sufficiently low, or secondly, to take into account the high temperature properties of the material, in which case no insulation may be necessary.

The first method, the 'fire-protection' approach, is the most common at present but is a prescriptive method. Recent research has shown that the failure temperature of a member depends upon load, temperature gradient, dimensions and stress distribution. By quantifying these effects a more rational 'fire-resistance' approach is being developed which is the basis of the recently published BS 5950 Part 8: *Code of practice for fire resistant design*.[1]

34.2 Standards and Building Regulations

34.2.1 Building Regulations

All buildings in the UK are required to comply with the Building Regulations[2] which are concerned with safety of life. The provisions of Approved Document B, for England and Wales, of the Regulations, dealing with structural fire precautions, are aimed at reducing the danger to people who are in or around a building when a fire occurs, by containing the fire and ensuring the stability of the structure for sufficient time to allow the occupants to reach safety. Generally, in Scotland and Northern Ireland the provisions are similar. Approved Document B requires that adequate provision for fire safety be provided either by fulfilling the requirements laid down or by suitable alternative methods.

The fire-resistance requirements of Document B apply only to structural elements used in:

(a) buildings, or parts of buildings, of more than one storey,
(b) single-storey buildings built close to a property boundary.

The degree of fire resistance required of a structural member is governed by the building function (office, shop, factory, etc.), by the building height and by the compartment size in which the member is located.

Fire resistance provisions are expressed in units of time: ½, 1, 2, 3 and 4 hours. It is important to realise that these times are not allowable escape times for building occupants or even survival times for the structure. They are simply a convenient way of grading different categories of buildings by fire load, from those in which a fire is likely to be relatively small, such as low-rise offices, to those in which a fire might result in a major conflagration, such as a library. Fire-resistance requirements for structural elements are given in Reference 2.

34.2.2 BS 5950: Part 8

BS 5950: Part 8[1] permits two methods of assessing the fire resistance of bare steel members. The first, the 'load ratio' method consists of comparing the 'design temperature', which is defined as the temperature reached by an unprotected member in the required fire-resistance time, with the 'limiting temperature', which is the temperature at which it will fail. The load ratio is defined as:

$$\text{load ratio} = \frac{\text{load applied at the fire limit state}}{\text{load causing the member to fail under normal conditions}}$$

If the limiting temperature exceeds the design temperature no protection is necessary. The method permits designers to make use of reduced loads and higher strength steels to achieve improved fire-resistance times in unprotected sections.

The second method, which is applicable to beams only, gives benefits when members are partially exposed and when the temperature distribution is known. It consists of comparing the calculated moment capacity at the required fire-resistance time with the applied moment. When the moment capacity exceeds the applied moment no protection is necessary. This method of design is used for unusual structural forms such as 'shelf-angle' floor beams. Some examples of the use of the moment capacity method are given in the handbook to BS 5950: Part 8.

Limiting temperatures for various structural members are presented in the Appendix *Limiting temperatures*. These 'failure' temperatures are independent of the form or amount of fire protection. Beams supporting concrete floors fail at a much higher limiting temperature than columns, for example.

The rate of heating of a given section is related to its 'section factor' which is the ratio of the surface perimeter exposed to radiation and convection and the mass, which is directly related to cross-sectional area.

$$\text{section factor} = \frac{H_p}{A} = \frac{\text{perimeter of the section exposed to fire (m)}}{\text{cross-sectional area of the member (m}^2\text{)}}$$

A member with a low H_p/A value will heat up at a slower rate than one with a high H_p/A value and will require less insulation (fire protection) to achieve the same fire-resistance rating. Standard tables are available listing H_p/A ratios for structural sections (see the Appendix *Section factors for UBs, UCs, CHSs and RHSs*). These factors are calculated as indicated in Fig. 34.1.

Sections at the heavy end of the structural range have such low H_p/A ratios, and therefore such slow heating rates, that failure does not occur within $\frac{1}{2}$ hour under standard BS 476 heating conditions even when they are unprotected.

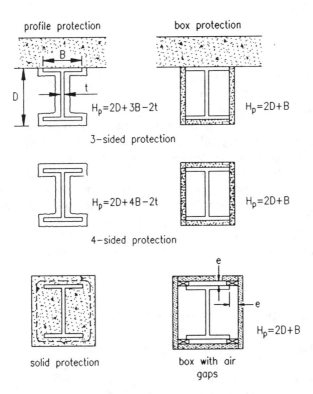

Fig. 34.1 Different forms of fire protection to I-section members

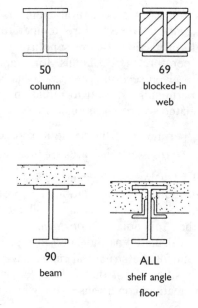

Fig. 34.2 Maximum ratios of H_p/A (m^{-1}) of exposed steel to give 30 minutes fire resistance.

Limiting section factors for various structural elements are given in Fig. 34.2 (corresponding to a load ratio of 0.5).

Manufacturers of fire-protection products now give guidance on the required thickness of fire protection depending on the section factor of the member. The example of a table for a typical spray applied (profile protection) shown in the Appendix *Minimum thickness of spray protection* is taken from Reference 3.

Most fire-protection materials are assessed at a single limiting temperature of 550°C. However, the limiting temperatures of beams supporting concrete floors can be considerably higher, leading potentially to the use of less fire protection to the section. This is the basis of the approach in BS 5950: Part 8, under which, when fully implemented, manufacturers will provide thermal properties of their materials to enable design to any limiting temperature to be carried out.

34.3 Structural performance in fire

34.3.1 Strength of steel at elevated temperatures

Steel begins to lose strength at about 200°C and continues to lose strength at an increasing rate up to a temperature of about 750°C when the rate of strength loss

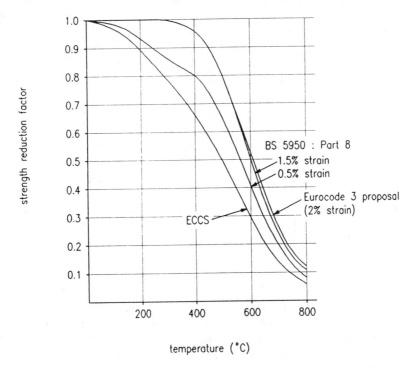

Fig. 34.3 Strength reduction factor for grade 43 steel at elevated temperatures

flattens off. This relationship is shown in Fig. 34.3. An important parameter is the strain at which the strength is assessed. It is reasonable to take a higher strain limit than in normal design, partly because there is a gradual increase of strength with strain at elevated temperatures, and partly because much higher deflections are attained in fire tests than in normal structural tests.

The relevant strain limit in BS 5950: Part 8 is 1.5%, which may be used for beam sections with 'robust' fire-protection materials, or for unprotected sections. A lower strain limit of 0.5% is used for more 'brittle' fire-protection materials, or for columns (in all cases). The 2% strain limit proposed in Eurocode 3, Part 10 gives results close to the 1.5% limit.

34.3.2 Performance of beams

Beams supporting concrete slabs behave better than uniformly heated sections, for which the material performance is the dominant factor. The concrete slab causes the top flange to be significantly cooler than the bottom flange and therefore the plastic neutral axis of the section rises close to the top flange (see Fig. 34.4).

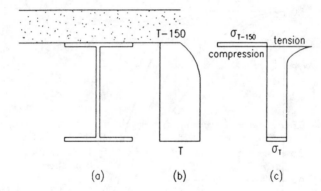

Fig. 34.4 Temperature and stress variation in I-beam supporting concrete slab when limiting temperature is reached. (a) Section through beam and slab. (b) Temperature variation. (c) Stress variation.

The section resistance is therefore determined by the strength of steel at 1.5% strain, but in this case, more of the web is effective in resisting tension.

The limiting temperatures of beams supporting concrete slabs are shown in Fig. 34.5 with the test results for a range of beam sizes and load ratios. These limiting temperatures are increased by about 60°C relative to a uniformly heated section.

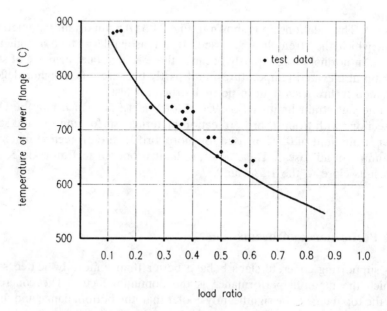

Fig. 34.5 Limiting temperatures for I-section beams supporting concrete floors

34.3.3 Performance of columns

Steel columns fail at a cross-section strain of less than 1%, and the 0.5% strain limit is taken as being appropriate when determining the limiting temperature of the members. This is applicable for columns up to the slenderness of 70 normally encountered in buildings.

The fire resistance of columns can be increased by partial protection in the form of concrete blocks or bricks, either by building into a wall or by fitting blocks between the flanges. A minimum fire resistance of 30 minutes can usually be achieved in this way.

Concrete-filled CHS or RHS columns can be used to provide typically 60 minutes fire resistance without additional fire protection. Intumescent coatings and reinforcement in the concrete can be used to provide enhanced fire resistance of these columns.

34.3.4 Performance of composite slabs

Modern steel frames often involve the use of composite slabs comprising steel decking acting compositely with the concrete floor (see Chapter 21). A large number of fire tests have now been carried out to justify the use of 90 minutes fire resistance for composite slabs with standard mesh reinforcement and no additional fire protection. Guidance is given in *Handbook to BS 5950: Part 8*.[4]

34.4 Methods of protection

Basic information on methods of protection is summarized in the Appendix.

34.4.1 Spray-applied protection

Sprays are the cheapest method with costs commonly in the range of £6 to £16/ square metre applied (1990 prices), depending on the fire rating required and the size of the job. As implied, spray protection is applied around the exposed perimeter of the member and therefore the relevant section factors are for profile protection (see the Appendix *Limiting temperatures*). Application is fast and it is easy to protect complex shapes or connections. However, they are applied wet, which can create problems in winter conditions, they can be messy, and the appearance is often poor. For this reason they are generally used in hidden areas such as on beams above suspended ceilings, or in plant rooms.

34.4.2 Board protection

Boards tend to be more expensive, commonly in the range £9 to £30/square metre applied (1990 prices), because of the higher labour content in fixing. The price depends on the rating required and the surface finish chosen but tends to be less sensitive to job size. Board systems form a box around the section and therefore have a reduced heated perimeter in comparison to spray systems (see the Appendix *Limiting temperatures* and Fig. 34.1). They are dry fixed by gluing, stapling or screwing, so there is less interference with other trades on site, and the box appearance is often more suitable for frame elements, such as free-standing columns, which will be in view.

34.4.3 Intumescent coatings

Intumescent coatings achieve insulation in a totally different way; the insulating layer is formed only by the action of heat when the fire breaks out.

The coating is applied as a thin layer, perhaps as thin as 1 mm, but it contains a compound in its formulation which releases a gas when heat is applied. This gas inflates the coating into a thick carbonaceous foam, which provides heat insulation to the steel underneath. The coatings are available in a range of colours and may be used for aesthetic reasons on visible steelwork.

Two types of intumescent coating are currently available. The first, which is water resistant, has a maximum rating of 2 hours, but is expensive, costing up to £60/square metre (1990 prices). The second type has maximum ratings up to $1\frac{1}{2}$ hours but is not so resistant to moisture and so is not recommended for wet applications such as swimming pools but is satisfactory in dry buildings. Costs range from about £10 to £20/square metre (1990 prices).

34.4.4 Concrete encasement

In comparison with these lightweight materials, protection by in situ concrete costs about £17 to £30/square metre of steelwork encased (1990 prices). However, concrete is often perceived as a 'traditional' fire-protection material and indeed data on its use is given in the Building Research Establishment publication *Guidelines for the Construction of Fire Resisting Structural Elements*.[5]

34.5 Fire testing

Fire testing requirements are based largely on the *Fire Grading of Buildings* report of 1946.[6] The fire-resistance periods refer to the time in a standard fire, defined by BS 476: Parts 20 and 21,[7] that an element of structure (a column, beam, compartment wall, etc.) should maintain:

(1) stability − it should not collapse under load at the fire limit state,
(2) integrity − it should not crack or otherwise allow the passage of flame to an adjoining compartment,
(3) insulation − it should not allow passage of heat by conduction which might induce ignition in an adjoining compartment.

The 'standard' fire time−temperature relationship defined in BS 476 is presented in Fig. 34.6. Most attention (and cost) is directed at satisfying the stability (or strength) criterion.

In a fire test columns are exposed to fire on all four sides and axially loaded vertically, whereas beams are loaded horizontally in bending and are exposed on three sides, the upper flange being in contact with the floor slab, which also acts as the furnace roof.

The stability limit is deemed to have been reached for columns when 'run-away' deflection occurs. For beams this is more accurately specified when the deflection rate reaches $L^2/9000d$ (where L = beam span, d = beam depth), for deflections exceeding $L/30$. An upper limit on the total deflection is set at $L/20$.

Indicative (i.e. thermal) and loaded fire tests are also carried out on beams and columns to assess the performance of fire-protection materials. Interpolation from test results in Reference 3 is by an empirical method, but in BS 5950: Part 8 it is possible to use the thermal properties of the materials to assess the required protection. Some materials also benefit from stored moisture giving rise to the dwell in the temperature response shown in Fig. 34.7.

34.6 Fire engineering

There are many special building forms which can take advantage of a rational approach, called 'fire engineering',[8,9] which takes account of the temperatures developed in a real fire, as opposed to a standard fire (see Fig. 34.6).

Essentially the fire engineering design method can be divided into two main steps:

(1) Determination of the fire load
  The fire load of a compartment is the maximum heat that can theoretically be generated by the combustible items of contents and structure i.e. weight × calorific value per unit weight.
  Fire load is usually expressed in relation to floor area, sometimes as MJ/m^2 or $Mcal/m^2$ but more often converted to an equivalent weight of wood and expressed as 'kg wood/m^2' (1 kg wood ≡ 18 MJ). Standard data tables giving fire loads of different materials are available.[8]

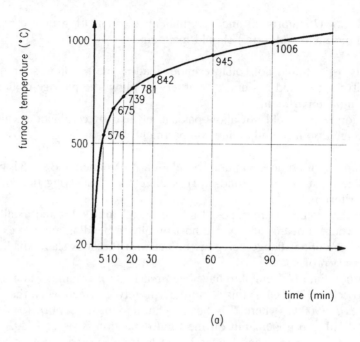

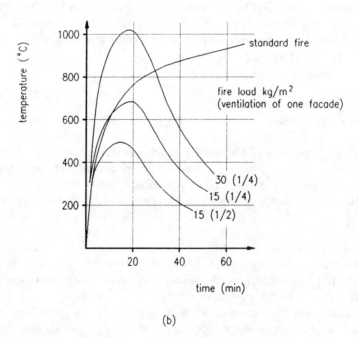

Fig. 34.6 Temperature–time curves: (a) standard ISO fire test, (b) standard and natural fires in small compartments

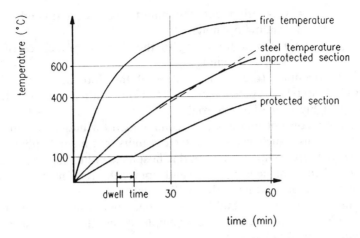

Fig. 34.7 Rate of heating in unprotected and protected sections

Examples of typical fire loads (wood equivalent) are:

schools	15 kg/m²
hospital wards	20 kg/m²
hotels	25 kg/m²
offices	35 kg/m²
department stores	35 kg/m²
textile warehouses	>200 kg/m²

(2) Prediction of maximum compartment temperature

The heat that is retained in the burning compartment depends upon the thermal characteristics of the wall, floor and ceiling materials and the degree of ventilation. Sheet steel walls will dissipate heat by conduction and radiation while blockwork will retain heat in the compartment and lead to higher temperatures.

It is assumed that window glass breaks in fire conditions and calculations take into account the size and position of such ventilation. Openings close to the ceiling level of a compartment (or disintegrating roofing materials) will tend to dissipate heat whereas openings close to the floor will provide oxygen to feed the fire.

Since a great deal of data have been gathered over the years on the performance of materials in the standard fire test, methods have been sought to relate real fire conditions to standard fire performance in order that the existing data can be used in fire engineering. The 'time equivalent' or equivalent required fire resistance is given by:

$$T_{eq} = CWQ_f$$

where Q_f = fire load density in MJ/m² i.e. the amount of combustible material per unit area of compartment floor,

W = ventilation factor relating to the area and height of door and window openings,

C = a constant relating to the thermal properties of the walls, floor and ceiling.

For large openings (greater than 10% of the floor area), the fire is 'fuel controlled' and W may be taken as 1.5. Conservatively, C may be taken as 0.09 for highly insulating materials.

It is not a concept that can be recommended for buildings that are subject to change of use, such as advance factory units, but many buildings are 'fixed' in terms of their occupancy (car parks, hospitals, swimming pools, etc.) and in such cases fire engineering is a valid approach. It is most appropriate for buildings of large volume with low fire load.

In the UK a number of buildings have been built using unprotected steel on fire engineering principles, one example being the north stand at Ibrox football ground in Glasgow.

Other examples of the use of calculation methods in determining structural response in fire are external steel in framed buildings,[10] and portal frames with fire-resistant boundary wall.[11]

References to Chapter 34

1. British Standards Institution (1990) *Structural use of steelwork in building. Part 8: Code of practice for fire resistant design.* BS 5950, BSI, London.
2. The Building Regulations (a) Approved document B (1991), Fire safety, Department of the Environment and the Welsh Office, HMSO.
   (b) Technical standards for compliance with the building standards (Scotland) regulations (1990) HMSO.
   (c) The Building Regulations (Northern Ireland) (1990) Department of the Environment, HMSO.
3. The Steel Construction Institute/Association of Structural Fire Protection Contractors and Manufacturers (1989) *Fire Protection of Structural Steel in Buildings*, 2nd edn. SCI/ASFPCM.
4. Lawson R.M. & Newman G.M. (1990) *Fire Resistant Design of Steel Structures – A Handbook to BS 5950: Part 8*. The Steel Construction Institute, Ascot, Berks.
5. Morris W.A., Read R.E.H. & Cooke G.N.E. (1988) *Guidelines for the Construction of Fire Resisting Structural Elements*. Building Research Establishment, Garston, Watford.
6. Ministry of Public Buildings and Works (1946) *Fire Grading of Buildings, Part 1, General Principles and Structural Precautions*. Post War Building Studies No. 20, HMSO.
7. British Standards Institution (1987) *Fire tests on building materials and structures. Part 20: Method of determination of the fire resistance of elements of*

construction (general principles). Part 21: *Method for determination of the fire resistance of load-bearing elements of construction*. BS 476, BSI, London.
8. Report of CIB Workshop 14 (1983) Design guide: structural fire safety. *Fire Safety Journal*, **6**, No. 1.
9. European Convention for Constructional Steelwork (1981) *European Recommendations for the Fire Safety of Steel Structures*. ECCS Technical Committee 3. Elsevier Applied Science Publishers, Barking, Essex. (Also *Design Manual*, 1985).
10. Law M. & O'Brien T.P. (1981) *Fire and Steel Construction – Fire Safety of Bare External Structural Steel*. Constrado.
11. Newman G.M. (1990) *Fire and Steel Construction – The Behaviour of Steel Portal Frames in Boundary Conditions*, 2nd edn. The Steel Construction Institute, Ascot, Berks.

Further reading for Chapter 34

References 4, 5 and 10 are recommended further reading.
EUROFER (1990) *Steel and Fire Safety – A Global Approach*. Eurofer, Brussels. (Available from The Steel Construction Institute, Ascot, Berks.)

Chapter 35
Corrosion resistance

by KEN JOHNSON

35.1 The corrosion process

35.1.1 General corrosion

Most corrosion of steel can be considered as an electrochemical process which occurs in stages. Initial attack occurs at anodic areas on the surface, where ferrous ions go into solution. Electrons are released from the anode and move through the metallic structure to the adjacent cathodic sites on the surface where they combine with oxygen and water to form hydroxyl ions. These react with the ferrous ions from the anode to produce ferrous hydroxide which itself is further oxidized in air to produce hydrated ferric oxide; red rust (Fig. 35.1).

The sum of these reactions is described by the following equation:

$$4Fe + 3O_2 + 2H_2O = 2Fe_2O_3H_2O$$
(iron/steel) + (oxygen) + (water) = rust

Two important points emerge:

(1) for iron or steel to corrode it is necessary to have the simultaneous presence of water and oxygen; in the absence of either, corrosion does not occur.
(2) all corrosion occurs at the anode; no corrosion occurs at the cathode.

However, after a period of time, polarization effects such as the growth of corrosion products on the surface cause the corrosion process to be stifled. New, reactive anodic sites may then be formed thereby allowing further corrosion. Over long periods the loss of metal is reasonably uniform over the surface and so this case is usually described as 'general corrosion'.

35.1.2 Other forms of corrosion

Various types of localized corrosion can also occur:

(1) *Pitting corrosion.* In some circumstances the attack on the original anodic area is not stifled and continues deep into the metal, forming a corrosion pit.

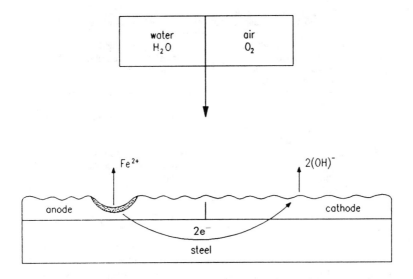

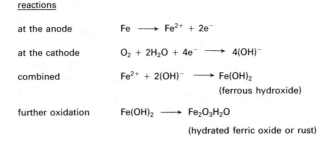

Fig. 35.1 Diagrammatic representation of the corrosion of steel

Pitting more often occurs with mild steels immersed in water or buried in soil rather than those exposed in air. Pitting can also occur on stainless steels in certain environments.
(2) *Crevice corrosion.* Crevices can be formed by design-detailing, welding, surface debris, etc. Available oxygen in the crevice is quickly used by the corrosion process and, because of limited access, cannot be replaced. The entrance to the crevice becomes cathodic, since it can satisfy the oxygen-demanding cathode reaction. The tip of the crevice becomes a localized anode and high corrosion rates occur at this point.
(3) *Bimetallic corrosion.* When two dissimilar metals are joined together in an electrolyte an electrical current passes between them and corrosion occurs on

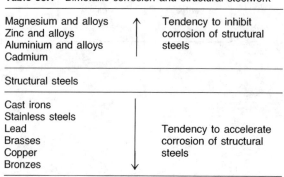

Table 35.1 Bimetallic corrosion and structural steelwork

Magnesium and alloys / Zinc and alloys / Aluminium and alloys / Cadmium	↑	Tendency to inhibit corrosion of structural steels
Structural steels		
Cast irons / Stainless steels / Lead / Brasses / Copper / Bronzes	↓	Tendency to accelerate corrosion of structural steels

the anodic metal. Some metals (e.g. nickel and copper) cause steel to corrode preferentially whereas other metals corrode preferentially themselves, thereby protecting the steel. The tendency of dissimilar metals to bimetallic corrosion is partly dependent upon their respective positions in the galvanic series (Table 35.1): the further apart the two metals are in the series the greater the tendency. Other aspects which influence bimetallic corrosion are the nature of the electrolyte and the respective surface areas of the anodic and cathodic metals. Bimetallic corrosion is most serious for immersed or buried structures but should also be considered for steel in the atmosphere.

(4) *Stress corrosion cracking*. This occurs under the simultaneous influence of a static tensile stress (or residual stress), which may be well below the yield strength of the steel, and a specific corrosive environment. Cracking occurs, which may lead to mechanical failure. This type of corrosion is not common with ferrous metals though some stainless steels are susceptible in chloride environments and mild steels can exhibit stress-corrosion cracking in the presence of nitrates or in highly alkaline solutions.

(5) *Bacterial corrosion*. This can occur in soils and water as a result of microbiological activity. The most commonly encountered is that arising from the presence of sulphate-reducing bacteria. These reduce sulphates in the soil to sulphides and cause corrosion under anaerobic conditions (i.e. in the absence of oxygen). They are characterized by black corrosion products having the distinctive 'rotten-egg' smell of sulphide. Bacterial corrosion is most commonly encountered in pipelines and other buried structures and is rarely, if ever, found in driven steel piles.

35.1.3 Corrosion rates

The principal factors that determine the rate of corrosion of steel in air are:

(1) the type and amount of pollution, e.g. sulphur dioxide, chlorides, dust
(2) the 'time of wetness', i.e. the proportion of total time during which the surface is wet, due to rainfall, condensation, etc.
(3) the temperature; though in the UK this is of less importance.

Within a given local environment corrosion rates can vary markedly. For example, steel may corrode more on a particular side of a building because it is in the shade and so remains wet for longer periods. Prevailing winds may carry airborne contaminants (e.g. chlorides, SO_2) predominantly on to one face of a structure. It is therefore the 'microclimate' immediately surrounding the structure which determines corrosion rates for practical purposes.

35.2 Effect of the environment

Steel is most commonly used in the following environments:

(1) *Rural atmospheric* – essentially inland, unpolluted environments; steel corrosion rates tend to be low, usually less than 50 microns per annum (*NB* 1 micron = 0.001 mm)
(2) *Industrial atmospheric* – inland, polluted environments; corrosion rates are usually between 50–100 microns per annum, dependent upon level of SO_2
(3) *Marine atmospheric* – in the UK a two-kilometre strip around the coast is broadly considered as being in a marine environment; corrosion rates are usually between 50–100 microns per annum, largely dependent upon proximity to the sea
(4) *Marine/industrial atmospheric* – polluted coastal environments which produce the highest corrosion rates e.g. between 50–150 microns/annum
(5) *Sea-water immersion* – in tidal waters four vertical zones are usually encountered:

  (a) the splash zone, immediately above the high-tide level, is usually the most corrosive zone with a mean corrosion rate of about 100 microns/annum
  (b) the tidal zone, between high-tide and low-tide levels, is often covered with marine growths and exhibits low corrosion rates e.g. 50 microns/annum
  (c) the low-water zone, a narrow band just below the low-water level, exhibits corrosion rates similar to the splash-zone
  (d) the permanent immersion zone, from the low-water level down to bed level, exhibits low corrosion rates e.g. 50 microns/annum

(6) *Fresh-water immersion* – corrosion rates are lower in fresh water than in salt water e.g. 30–50 microns/annum
(7) *Soils* – the corrosion process is complex and very variable; various methods are used to assess the corrosivity of soils:

(a) resistivity; generally high-resistance soils are least corrosive
(b) redox potential; to assess the soil's capability of anaerobic bacterial corrosion
(c) pH; highly acidic soils (e.g. pH less than 3.0) can be corrosive
(d) water content; corrosion depends upon the presence of moisture in the soil, the position of the water-table has an important bearing.

Long buried steel structures, e.g. pipelines, are most susceptible to corrosion. Steel piles driven into undisturbed soils are much less susceptible due to the low availability of oxygen.

35.3 Design and corrosion

Design can have an important bearing on the corrosion of steel structures. The prevention of corrosion should therefore be taken into account during the design stage of a project. The main points to be considered are:

(1) Entrapment of moisture and dirt:

   (a) avoid sharp edges, sharp corners, cavities, crevices
   (b) welded joints are preferable to bolted joints
   (c) avoid or seal lap joints
   (d) edge-seal HSFG faying surfaces
   (e) provide drainage holes for water, where necessary
   (f) seal box sections
   (g) provide free circulation of air around the structure.

(2) Contact with other materials:

   (a) avoid bimetallic connections or insulate the contact surfaces
   (b) provide adequate depth of cover and quality of concrete (see BS 8110)
   (c) separate steel and timber by the use of coatings or sheet plastics.

(3) Coating application; design should ensure that the selected protective coatings can be applied efficiently:

   (a) radius edges and corners
   (b) provide vent-holes and drain-holes for items to be hot-dip galvanized
   (c) provide adequate access for metal spraying, paint spraying, etc.

(4) General factors:

   (a) large flat surfaces are easier to protect than more complicated shapes
   (b) ideally, locate load-bearing members in the least corrosive locations
   (c) provide access for subsequent maintenance
   (d) provide lifting lugs or brackets where possible to reduce damage during handling and erection.

BS 5493, Appendix A,[1] provides a detailed account of designing for the prevention of corrosion.

35.4 Surface preparation

Structural steel is a hot-rolled product. Sections leave the last rolling pass at about 1000°C and as they cool the steel surface reacts with oxygen in the atmosphere to produce mill-scale, a complex oxide which appears as a blue-grey tenacious scale completely covering the surface of the as-rolled steel section. Unfortunately, mill-scale is unstable. On weathering, water penetrates fissures in the scale and rusting of the steel surface occurs. The mill-scale loses adhesion and begins to shed. Mill-scale is therefore an unsatisfactory base and needs to be removed before protective coatings are applied.

As mill-scale sheds, further rusting occurs. Rust is a hydrated oxide of iron which forms at ambient temperatures, producing a layer on the surface which is itself an unsatisfactory base and also needs to be removed before protective coatings are applied.

Surface preparation of steel is therefore principally concerned with removal of mill-scale and rust. Various methods of surface preparation are available:

(1) *Manual preparation*

The simplest form of surface preparation. It involves chipping, scraping and brushing with hand-held implements. Although the method is not very effective (only about 30% removal of rust and scale can be achieved) it is nevertheless often used, usually for economic reasons. The degree of cleaning achieved can be specified by reference to the St series of photographic standards included in BS 7079 Part A1: 1989 (ISO 8501−1: 1988).

(2) *Mechanical preparation*

Similar to manual preparation but utilizes power-driven tools, e.g. rotary wire brushing. A marginal improvement in efficiency can be achieved (up to 35%), and the same photographic standards can be used. Care must be taken to avoid confusing burnished scale with clean steel, both of which have a similar appearance.

The above methods are used on site, usually after a weathering period to promote loosening of mill-scale. A suitable primer must then be applied, which is tolerant of poor surface preparation. Many modern primers are quite unsuitable for such surfaces and, indeed, the old tried and trusted red lead in oil primers (e.g. BS 2523 Type B) cannot be bettered for manually cleaned surfaces.

(3) *Flame cleaning*

Not used extensively in the UK. An oxy-gas flame is applied to the surface. Differential thermal expansion and steam generated behind the mill-scale serve to loosen the mill-scale layer, which can then be removed by mechanical scraping.

(4) *Acid pickling*

The steel is immersed in a bath of suitably inhibited acids which dissolve or remove mill-scale and rust but do not appreciably attack the exposed steel surface. It can be 100% effective.

Acid pickling is always used on structural steel intended for hot-dip galvanizing but is now rarely used as a pre-treatment before painting.

(5) *Blast-cleaning*

Abrasive particles are projected at high speed on to the steel surface. The abrasive can consist of either spherical particles, described as 'shot', or angular particles, described as 'grit'.

The abrasive is projected towards the surface either in a jet of compressed air or by centrifugal impeller wheel. The particles impinge on the steel surface, removing scale and rust, producing a rough, clean surface. The size and shape of the surface roughness produced is largely dependent upon the size and shape of the abrasive used; angular grits produce angular surface profiles, round shots produce a rounded profile.

Grit-blast abrasives can be either metallic (e.g. chilled iron grit) or non-metallic (e.g. slag grit). The latter are used only once and are referred to as 'expendable'. They are used exclusively for site work. Metallic grits are expensive and are used only where they can be recycled.

Grit blasting is always used for metal-sprayed coatings, where adhesion is at least partly dependent upon mechanical keying. It is also used for some paint coatings, particularly on site and for primers where adhesion may be a problem (e.g. zinc silicates).

Shot-blast abrasives are always metallic, usually cast steel shot, and are used particularly on shot-blast plants, utilizing impeller wheels and abrasive recycling. They are the preferred abrasive for paints, particularly for thin film coatings (e.g. prefabrication primers).

Blast-cleaned surfaces are normally specified in terms of surface cleanliness and surface roughness. A number of standards have been used in the past, including BS 4232 and Swedish Standard SIS 055900. These are now superseded by ISO 8501−1: 1988 (BS 7079: Part A1: 1989) which utilises photographic replicas of four grades of surface cleanliness after blast-cleaning: Sa1, Sa2, Sa2.5 and Sa3.

Surface roughness of blast-cleaned surfaces is defined in ISO 8503−1: 1988 and Parts 2, 3 and 4 of this standard describe methods of measuring surface roughness.

(6) *Wet-blasting*

A further variation on the blast-cleaning process. In this process a small amount of water is entrained in the abrasive/compressed air stream. This is particularly useful in washing from the surface soluble iron salts that are formed in the rust by atmospheric pollutants (e.g. chlorides and sulphates) during weathering. These are often located deep in corrosion pits on the steel surface and cannot be removed by conventional dry blast-cleaning methods. Wet-blasting has proved to be particularly useful on offshore structures and prior to maintenance painting of structures in heavily-polluted environments.

35.5 Metallic coatings

There are four commonly used methods of applying metal coating to steel surfaces: hot-dip galvanizing, metal spraying, electroplating and sherardizing. The latter two processes are not used in structural steelwork but are used for fittings, fasteners and other small items.

In general the corrosion protection afforded by metallic coatings is largely dependent upon the choice of coating metal and its thickness and is not greatly influenced by the method of application.

35.5.1 Hot-dip galvanizing

The most common method of applying a metal coating to structural steel is by galvanizing. Hot-dipping processes are used for the application of other metals, e.g. aluminium, lead/tin (terne), but these are not used on structural steelwork.

The galvanizing process involves the following stages:

(1) The steel is cleaned of all rust and scale by acid pickling. This may be preceded by blast-cleaning to remove scale and roughen the surface but such surfaces are always subsequently pickled in hydrochloric acid producing ferric chloride on the steel surface which acts as a flux for the hot-dip process.
(2) The cleaned and fluxed steel is dipped into a bath of molten zinc at a temperature of about 450°C at which the steel reacts with the molten zinc to form a series of zinc/iron alloys on its surface.
(3) As the steel workpiece is removed from the bath a layer of relatively pure zinc is deposited on top of the alloy layers.

As the zinc solidifies it assumes a crystalline metallic lustre, usually referred to as 'spangling'. The thickness of the galvanized coating is influenced by various factors:

(1) The size of the workpiece; thicker, heavier sections tend to produce heavier coatings.
(2) The steel surface; blast-cleaned surfaces tend to produce heavier coatings.
(3) The steel composition. Silicon, particularly, can have a marked effect on the coating weight deposited. The thickness of the coating varies with the silicon content of the steel and bath immersion time. These thick coatings sometimes have a dull dark-grey appearance and can be brittle and less adherent.

Since this is a bath-dipping process there is obviously some limitation on the size of components which can be galvanized. Double-dipping can often be used when the length or width of the workpiece exceeds the size of the bath.

Some aspects of design need to take the galvanizing process into account:

(1) Filling, venting and draining. Holes not less than 10 mm diameter must be provided in hollow articles (e.g. tubes, rectangular hollow sections) to allow rapid access for molten zinc, venting of hot gases and subsequent draining of zinc.
(2) Distortion of fabricated steelwork can be caused by differential thermal expansion and contraction and by relief of residual stresses.

The specification of hot-dip galvanized coatings for structural steelwork is covered by BS 729 which requires, for sections not less than 5 mm thick, a minimum mean zinc coating weight of 610 g/m^2, equivalent to a coating thickness of 85 microns. The coating will usually protect the underlying steel for up to 30 years in a clean rural environment and between 10 and 25 years in marine and urban situations. However, in heavily-polluted environments durability can be reduced to less than 10 years.

For most applications galvanizing does not need to be painted. Where, for reasons of extra durability or decorative effect, there is a need to paint, then special primers are normally required such as calcium plumbate and etch primers.

35.5.2 Metal spray coatings

An alternative method of applying a metallic coating to structural steelwork is by metal-spraying of either zinc or aluminium. The metal, in powder or wire form, is fed through a special spray-gun containing a heat source which can be either an oxy-gas flame or an electric-arc. Molten globules of the metal are blown by a compressed air jet on to the previously blast-cleaned steel surface. No alloying occurs and the coating which is produced consists of overlapping platelets of metal and is porous. The pores are subsequently sealed, either by applying a thin organic coating which soaks into the surface, or by allowing the metal coating to weather, when corrosion products block the pores.

The adhesion of sprayed metal coatings to steel surfaces is considered to be essentially mechanical in nature. It is therefore necessary to apply the coating to a clean roughened surface for which blast-cleaning with a coarse grit abrasive is normally specified, usually chilled-iron grit, but for steels with a hardness exceeding 360 HV, alumina or silicon carbide grits may be necessary.

Coating thickness varies from 100−250 microns for aluminium to 75−400 microns for zinc. The metals perform similarly in most environments but aluminium is more durable in highly industrial environments.

Metal spray coatings can be applied in the shops or at site and there is no limitation on the size of the workpiece, as there is with hot-dip galvanizing. Since the steel surface remains cool there are no distortion problems. However, metal spraying is considerably more expensive than hot-dip galvanizing.

For many applications metal-spray coatings are further protected by the subsequent application of paint coatings. A sealer is first applied which fills the pores in the metal spray coating and provides a smooth surface for application of the paint coating.

The protection of structural steelwork against atmospheric corrosion by metal-sprayed aluminium or zinc coatings is covered in BS 2569: Part 1:

35.6 Paint coatings

Painting is the principal method of protecting structural steelwork from corrosion.

35.6.1 Composition of paints and film-formation

Paints are made by mixing and blending three main components:

(1) Pigments; finely ground inorganic or organic powders which provide colour, opacity, film-cohesion and sometimes corrosion-inhibition.
(2) Binders; usually resins or oils but can be inorganic compounds such as soluble silicates. The binder is the film-forming component in the paint.
(3) Solvents; used to dissolve the binder and to facilitate application of the paint. Solvents are usually organic liquids or water.

Paints are applied to steel surfaces by many methods but in all cases they produce a 'wet film'. The thickness of the 'wet film' can be measured, before the solvent evaporates, using a comb-gauge.

As the solvent evaporates, film-formation occurs, leaving the binder and pigments on the surface as a 'dry film'. The thickness of the 'dry film' can be measured, usually with a magnetic induction gauge.

The relationship between the applied 'wet film' thickness and the final 'dry film' thickness (d.f.t.) is determined by the percentage volume solids of the paint, i.e.

$$\text{d.f.t.} = \text{'wet film' thickness} \times \% \text{ vol. solids}$$

In general the corrosion protection afforded by a paint film is directly proportional to its dry film thickness.

35.6.2 Classification of paints

Since, in the broadest terms, a paint consists of a particular pigment, dispersed in a particular binder, dissolved in a particular solvent, the number of generic types of paint is limited. The most common methods of classifying paints are either by their pigmentation or by their binder-type.

Primers for steel are usually classified according to the main corrosion-inhibitive pigments used in their formulation, e.g. zinc phosphate, zinc chromate, red-lead, metallic-zinc. Each of these inhibitive pigments can be incorporated into a range of binder resins, e.g. zinc phosphate alkyd primers, zinc phosphate epoxy primers, zinc phosphate chlorinated-rubber primers.

Intermediate coats and finishing coats are usually classified according to their binders, e.g. vinyl finishes, urethane finishes.

35.6.3 Painting systems

Paints are usually applied one coat on top of another, each coat having a specific function or purpose.

The primer is applied directly on to the cleaned steel surface. Its purpose is to wet the surface and to provide good adhesion for subsequently applied coats. Primers for steel surfaces are also usually required to provide corrosion inhibition.

The intermediate coats (or undercoats) are applied to build the total film thickness of the system. This may involve the application of several coats.

The finishing coats provide the first-line defence against the environment and also determine the final appearance in terms of gloss, colour, etc.

The various superimposed coats within a painting system have, of course, to be compatible with one another. They may be all of the same generic type or may be different, e.g. chlor-rubber-based intermediate coats may be applied on to an epoxy primer. However, as a first precaution, all paints within a system should normally be obtained from the same manufacturer.

35.6.4 Main generic types of paint and their properties

(1) *Air-drying paints*, e.g. oil-based, alkyd, epoxy-ester, dry and form a film by an oxidative process which involves absorption of oxygen from the atmosphere. They are therefore limited to relatively thin films. Once the film has formed it has limited solvent resistance and usually poor chemical resistance.
(2) *One-pack chemical-resistant paints*, e.g. chlor-rubbers, vinyls, form a film by solvent evaporation and no oxidative process is involved. They can be applied as moderately thick films, although retention of solvent in the film can be a

problem at the upper end of the range. The film formed remains relatively soft, and has poor solvent resistance but good chemical resistance.

Bituminous paints also dry by solvent-evaporation. They are essentially solutions of either asphaltic bitumen or coal-tar pitch in organic solvents.

(3) *Two-pack chemical-resistant paints*, e.g. epoxy, urethane, are supplied as two separate components, usually referred to as the base and the curing agent. When the two components are mixed, immediately before use, a chemical reaction begins. These materials therefore have a limited 'pot-life' by which the mixed coating must be applied. The polymerization reaction continues after the paint has been applied and after the solvent has evaporated to produce a densely cross-linked film which can be very hard and has good solvent and chemical resistance.

Liquid resins of low viscosity can be used in the formulation thereby avoiding the need for a solvent. Such coatings are referred to as 'solventless' or 'solvent-free' and can be applied as very thick films.

A summary of the main generic types of paint and their properties is shown in Table 35.2.

35.6.5 Blast-primers (also referred to as prefabrication primers, shop-primers, weldable primers, temporary primers, holding primers, etc.)

These primers are used on structural steelwork, immediately after blast-cleaning, to hold the reactive blast-cleaned surface in a rust-free condition until final painting can be undertaken. They are mainly applied to steel plates and sections before fabrication. The main requirements of a blast-primer are as follows:

(1) The primer should be capable of airless-spray application to produce a very thin even coating. Dry-film thickness is usually limited to 15–30 microns. Below 15 microns the peaks of the blast profile are not protected and 'rust-rashing' occurs on weathering. Above 30 microns the primer affects the quality of the weld and produces excessive weld-fume.
(2) The primer must dry very quickly. Priming is often done in-line with automatic blast-cleaning plant which may be handling plates or sections at a pass-rate of 1–3 metres/minute. The interval between priming and handling is usually of the order of 1–10 minutes and hence the primer film must dry within this time.
(3) Normal fabrication procedures e.g. welding, gas-cutting, must not be significantly impeded by the coating.
(4) The primer should not cause excessive weld-porosity. (*NB* at the present time there are no blast-primers available which are compatible in this respect with the submerged-arc welding process and in this case the primer has to be removed before welding.)

Table 35.2 Main generic types of paint and their properties

	Cost	Tolerance of poor surface preparation	Chemical resistance	Solvent resistance	Over-coatability after ageing	Other comments
Bituminous	Low	Good	Moderate	Poor	Good with coatings of same type	Limited to black and dark colours Thermoplastic
Oil based	Low	Good	Poor	Poor	Good	Cannot be overcoated with paints based on strong solvent
Alkyd, epoxy-ester, etc.	Low–medium	Moderate	Poor	Poor–moderate	Good	Good decorative properties
Chlor-rubber	Medium	Poor	Good	Poor	Good	High-build films remain soft and are susceptible to 'sticking'
Vinyl	High	Poor	Good	Poor	Good	
Epoxy	Medium–high	V. poor	V. good	Good	Poor	Very susceptible to chalking in UV
Urethane	High	V. poor	V. good	Good	Poor	Better decorative properties than epoxies
Inorganic silicate	High	V. poor	Moderate	Good	Moderate	May require special surface preparation

(5) Weld-fumes emitted by the primer must not exceed the appropriate occupational exposure limits. Proprietary primers are tested and certificated by the North of England Industrial Health Service.
(6) The primer coating should provide adequate protection. It should be noted that many manufacturers make misleading claims about the durability of their blast-primers and suggested exposure periods of 6–12 months are not uncommon. In practice, such claims are never met except in the least arduous conditions, e.g. indoor storage. In aggressive conditions, durability can be measured in weeks rather than months.
(7) The primed surface, after weathering, should require the minimum of preparation for subsequent painting and must be compatible with the intended painting system.

Many proprietary blast-primers are available but they can be classified under the following main generic types:

(1) *Etch primers* are based on polyvinyl butyral resin reinforced with a phenolic resin to uprate water resistance. The curing agent contains phosphoric acid or alkyl phosphates and can be added as a separate component immediately before application. Alternatively the primer can be supplied in a single-pack form.

In general, two-pack etch primers provide better durability than the one-pack types which are usually preferred because of their handling convenience, particularly on automatic spraying plants.

(2) *Epoxy primers* are two-pack materials utilizing epoxy resins and usually either polyamide or polyamine curing agents. They are pigmented with a variety of inhibitive and non-inhibitive pigments. Zinc phosphate epoxy primers are the most frequently encountered and give the best durability within the group.

(3) *Zinc epoxy primers* can be subdivided into zinc-rich and reduced-zinc types. Zinc-rich primers produce films which contains about 90% by weight of metallic zinc powder. These primers provide the highest order of protection of all blast-primers.

Reduced-zinc primers are formulated with metallic zinc content as low as 55% by weight on the dry film, the remainder of the pigmentation usually being made up with siliceous extenders. This reduces the cost of the primer, avoids possible difficulties with intercoat adhesion in marine environments, but slightly reduces the standard of protection that can be achieved.

Zinc epoxy primers all produce zinc oxide fumes during welding and gas-cutting which can cause a health-hazard. When exposed in either marine or highly industrial environments, zinc epoxy primers are prone to the formation of insoluble white zinc corrosion products which must be removed from the surface before subsequent overcoating.

(4) *Zinc silicate primers* can be based upon either ethyl silicate or inorganic silicates e.g. sodium or potassium. Only the ethyl silicate primers are suitable as blast-primers.

Ethyl zinc silicate primers produce a level of protection which is comparable with the zinc-rich epoxy types and they suffer from the same drawbacks, e.g. formation of zinc salts and production of zinc oxide fumes during welding. They are however more expensive and usually are less convenient to use.

35.7 Application of paints

The method of application and the conditions under which paints are applied have a significant effect on the quality and durability of the coating.

35.7.1 Methods of application

The standard methods used for applying paints to structural steelwork are brush, roller, conventional air-spray, and airless-spray, although other methods, e.g. dipping, can be used.

(1) Brush. The simplest and also the slowest and therefore most expensive method. Nevertheless it has certain advantages over the other methods, e.g. better wetting of the surface; can be used in restricted spaces; useful for small areas; less wastage; less contamination of surroundings; can be used for application of certain toxic materials like lead-based primers which cannot be sprayed.
(2) Roller. Much quicker than brushing; useful for large flat areas; demands suitable rheological properties of the paint.
(3) Air-spray. The paint is atomized at the gun-nozzle by jets of compressed air; application rates are quicker than for brushing or rolling; paint wastage by overspray is high.
(4) Airless-spray. The paint is atomized at the gun-nozzle by very high hydraulic pressures; application rates are higher than for air-spray and overspray wastage is greatly reduced.

Airless-spraying has become the most commonly used method of applying paint coatings to structural steelwork under controlled shop-conditions. Brush and roller application are more commonly used for site-application, though spraying methods are also used.

35.7.2 Conditions for application

The principal conditions which affect the application of paint coatings are temperature and humidity. These can be more easily controlled under shop-conditions than on site.

(1) Temperature. Air temperature and steel temperature affect solvent evaporation, brushing and spraying properties, drying and curing times, pot-life of two-pack materials, etc. Heating, if required, should only be by indirect methods.
(2) Humidity. Paints should not be applied when there is condensation present on the steel surface or the relative humidity of the atmosphere is such that it will affect the application or drying of the coating. Normal practice is to measure the steel temperature with a contact thermometer and to ensure that it is maintained at at least 3°C above dew-point.

35.8 Weathering steels

Weathering steels are high-strength, low-alloy weldable structural steels which possess excellent weathering resistance in many atmospheric conditions. They contain up to 3% of alloying elements, e.g. chromium, copper, nickel, phosphorus. On exposure to air, under suitable conditions, they form an adherent protective oxide coating. This acts as a protective film which, with time and under appropriate conditions, causes the corrosion rate to reduce until it reaches a low terminal level. Conventional coatings are therefore not usually necessary since the steel provides its own protection.

Weathering steels have properties comparable with those of grade 50 steels to BS 4360. They carry a material cost premium of about 25% over mild steel but in many cases this can be substantially reduced if advantage is taken of the greater corrosion resistance and 30% greater yield strength than mild steel.

BS 4360 *Specification for weldable structural steels* includes weathering steels, designated WR50 grades. Corten, one of the best known proprietary weathering steels, is also manufactured by British Steel, under licence to the United States Steel Corporation.

35.8.1 Formation of the protective oxide layer

During the early part of their life weathering steels corrode in a similar manner and at a similar rate to mild steels. As the protective oxide layer develops, the corrosion rate falls to a low terminal level (Fig. 35.2). The time required for a weathering steel to form a stable protective coating depends upon its orientation, the degree of atmospheric pollution and the frequency with which the surface is wetted and dried. The steel should be blast-cleaned, to remove mill-scale, before exposure in order to provide a sound uniform surface for the formation of the oxide coatings. Under appropriate conditions a stable oxide patina will be established in about two years, changing in colour from brown through to almost black, although the ultimate colour will depend upon the conditions of the site.

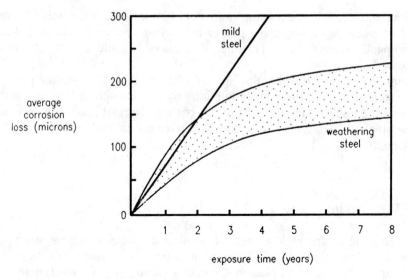

Fig. 35.2 Corrosion rates of mild and weathering steel in the UK

35.8.2 Precautions and limitations

During fabrication and erection care must be taken in handling and storage of bare weathering steel sections and sub-assemblies. Unsightly gouges, scratches, dents, etc. should be avoided. Scale and discoloration produced by welding should be removed from all exposed surfaces. Staining by concrete, mortar, oil, paint, marking crayons, etc. should be avoided. If such staining occurs additional cleaning is necessary to remove these blemishes.

Drainage of corrosion products can be expected during the first two or three years' exposure which can stain or streak adjacent materials, e.g. concrete piers. Provision should be made to divert this from vulnerable surfaces.

A continually wet surface will cause weathering steels to corrode at an unacceptably high rate. Detailing should therefore avoid crevices, or other water-retention areas. Drainage and ventilation should be provided. All interior surfaces, e.g. box girders, and faying surfaces should be painted or sealed to prevent entry of moisture. Bare weathering steel is not a completely maintenance-free material and structures should be periodically inspected to ensure that all joints and surfaces are performing satisfactorily.

Dissimilar metals such as stainless steel, anodized aluminium, copper, bronze and brass can generally be used adjacent to weathering steels providing the coupled area does not act as a crevice which might collect and hold water and debris. Zinc- and cadmium-coated items are not suitable for use in contact with weathering steels.

Weathering steels should not be used in the bare condition in certain environments, e.g. where

(1) the steel surface would be exposed to recurrent wetting by salt water or salt water spray (the Department of Transport Standard BD/7/81 prohibits weathering steel bridges in highly marine environments or in situations where the steel would be subjected to de-icing salt spray).
(2) high concentrations of strong chemical or industrial fume are present.
(3) the steel is buried in soil as it will not provide corrosion resistance greater than mild steel. If, for example, columns are taken below ground level conventional methods of protection such as concrete encasement or a high-quality coating are generally suitable and the protection must extend for a small distance above the ground level.
(4) the steel is immersed in water as it will not provide corrosion resistance greater than mild steel.

Bare weathering steels are not intended for interior architectural applications where the controlled atmosphere would restrict the development of the protective oxide layer.

35.8.3 Welding and bolted connections

Weathering steels can be welded by all the usual methods, manual metal arc, gas shielded, submerged arc, and electrical resistance, including spot-welding. Mild steel electrodes can be used for ordinary single-pass welds for steel less than 12 mm thick since diffusion of alloyed base-metal into the filler-metal will result in a weld-bead having corrosion resistance and colour conversion similar to that of the weathering steel.

For multi-pass welds or where the weathering appearance of welded areas is particularly important, electrodes containing 2.5% − 3.5% nickel should be used.

Since any interruption in the surface of the structure can cause the oxide coating to develop unevenly, welds may be dressed flush.

For structural joints where high-strength bolts are required, ASTM A325, Type 3 bolts (Corten X) must be used. Where lower-strength bolts are satisfactory these may be in Corten A or stainless steel. Galvanized, sherardized or electroplated nuts and bolts are not suitable for use in weathering steel structures since, in time, the coatings will be consumed leaving an exposed fastener which is less corrosion resistant than the surrounding weathering steel.

35.8.4 Painting of weathering steels

The painting requirements for weathering steels do not differ from those for mild steel. They require the same surface preparation and the same painting systems may be used. Claims have been made that in many circumstances the durability of painting systems is greater on weathering steel than on mild steel.

35.9 The protective treatment specification

35.9.1 Factors affecting choice

For a given structure the following will be largely predetermined:

(1) The expected life of the structure and the feasibility of maintenance
(2) The environment/s to which the steelwork will be subjected
(3) The size and shape of the structural members
(4) The shop-treatment facilities which are available to the fabricator and/or his coatings sub-contractor
(5) The site conditions which will determine whether the steelwork can be treated after erection
(6) The money which is available to provide protection.

These facts, and possibly others, have to be considered before making decisions on:

- the types of coating to be used,
- the method of surface preparation,
- the method/s of application,
- the number of coats and the thickness of each coat.

In general, each case has to be decided on its own merits. However, the following points may be of assistance in making these decisions:

(1) Protection requirements are minimal inside dry, heated buildings. Hidden steelwork in such situations requires no protection at all.
(2) The durability of painting systems is increased several times over by using blast-cleaning rather than manual surface preparation.
(3) Shot-blasting is preferred for most painting systems.
(4) Grit-blasting is essential for metal-spraying and some primers, e.g. zinc silicates.
(5) If blast-cleaning is to be used, two alternative process routes are available, i.e.

   (a) blast/prime/fabricate/repair damage
   (b) fabricate/blast/prime

   The former is usually cheaper but requires the use of a weldable prefabrication primer.

(6) Prefabrication primers have to be applied to blast-cleaned surfaces as thin films, usually 25 microns maximum. Their durability is therefore limited and further shop-coating is often desirable.
(7) Manual preparation methods are dependent upon weathering to loosen the mill-scale. These methods are therefore not usually appropriate for shop-treatments. On site an adequate weathering period must be allowed.
(8) Many modern primers based on synthetic resins are not compatible with manually prepared steel surfaces since they have a low tolerance for rust and scale.
(9) Many oil-based and alkyd-based primers cannot be overcoated with finishing coats which contain strong solvents, e.g. chlorinated rubbers, epoxies, bituminous coatings, etc.
(10) Two-pack epoxies have poor resistance to UV radiation and are highly susceptible to 'chalking'. Overcoating problems can arise with two-pack epoxies unless they are overcoated before the prior coat is fully cured. This is particularly relevant when an epoxy system is to be applied partly in the shops and partly on site.
(11) Steelwork which is to be encased in concrete does not normally require any other protection, given an adequate depth of concrete cover (British Standard BS 8110).
(12) Perimeter steelwork hidden in cavity walls falls into two categories:

   (a) Where an air gap (40 mm min.) exists between the steelwork and the outer brick or stone leaf, then adequate protection can be achieved by applying relatively simple painting systems.
   (b) Where the steelwork is in direct contact with the outer leaf, or is embedded in it, then the steel should be hot-dip galvanized and painted.

(13) Where fire-protection systems are to be applied to the steelwork, consideration must be given to the question of compatibility between the corrosion-protection and the fire-protection systems.
(14) New hot-dip galvanized surfaces can be difficult to paint and, unless special primers are used, adhesion problems can arise. Weathering the zinc surface before painting reduces this problem.
(15) Metal spraying produces a porous coating which can be sealed by applying either a thin etch-primer or a bituminous solution. Further painting is then optional.
(16) Particular attention should be paid to the treatment of weld areas. Flux residues and weld-spatter should be removed before application of coatings. In general the objective should be to achieve the same standard of surface preparation and coating on the weld area as on the general surface.
(17) Black-bolted joints require protection of the contact surfaces. This is normally restricted to the priming coat, which can be applied either in the shops or on site before the joint is assembled.
(18) For high-strength friction-grip bolted joints, the faying surfaces must be free of any contaminant or coating which would reduce the slip-factor required

on the joint. Some metal-spray coatings and some inorganic zinc silicate primers can be used but virtually all organic coatings adversely affect the slip-factor.

35.9.2 Writing the specification

The specification is intended to provide clear and precise instructions to the contractor on what is to be done and how it is to be done. It should be written in a logical sequence, starting with surface preparation, going through each paint-coat to be applied and finally dealing with specific areas e.g. welds. It should also be as brief as possible, consistent with providing all the necessary information. The most important items of a specification are as follows:

(1) The method of surface preparation and the standard required, which can often be specified by reference to an appropriate standard, e.g. ISO 8501−1, Sa2−3
(2) The maximum interval between surface preparation and subsequent priming.
(3) The types of paint to be used, supported by standards where these exist.
(4) The method/s of application to be used.
(5) The number of coats to be applied and the interval between coats.
(6) The wet and dry film thickness for each coat
(7) Where each coat is to be applied (e.g. shop or site) and the application conditions that are required, in terms of temperature, humidity, etc.
(8) Details for treatment of welds, connections, etc.
(9) Rectification procedures for damage, etc.

Possibly the most convenient method of presenting this information is in tabular form and an example is shown in Table 35.3.

Reference to Chapter 35

1. British Standards Institution (1977) *Code of practice for protective coating of iron and steel structures against corrosion*. BS 5493, BSI, London.

Further reading for Chapter 35

See Appendix *Basic data on corrosion* for extracts from British Steel Publications *Corrosion Protection Guides*.

Table 35.3 Protective coating specification for structural steelwork (example)

Contract no. – 1234/56/R	Surface preparation	Before fabrication or After fabrication
Client – J. Bloggs	Shot-blast ☐	Maximum interval before overcoating – 4 hours
Project title – Docklands Warehouse	Grit-blast ☐	Level of inspection – Periodic
Date – 27.6.90	Other ☐	Other comments –
	Standard Sa 2.5	

	1st coat	2nd coat	3rd coat	4th coat	
	Shop or ~~Site~~	Shop or ~~Site~~	~~Shop~~ or Site	~~Shop~~ or Site	
Brand name of paint	Excote 2	Excote HB ZP	Excote HB CR	Excote CR	Treatment of welds: Blast clean Sa 2.5 and full system.
Generic type	Zinc-rich epoxy	Zinc phosphate epoxy	HB chlor rubber	Chlor rubber finish	
Reference number	1/597/P	5/643/u	3/124/u	4/510/F	Treatment of bolted connections: HSFG faying surfaces left bare. Edges sealed on site.
Specification	BS 4652 Type 3	N/A	N/A	N/A	
Supplier	A.N. Other	A.N. Other	A.N. Other	A.N. Other	
Method of application	Airless spray	Airless spray	Airless spray or brush	Airless spray or brush	
Wet film thickness	N/A	150 μm	250 μm	75 μm	Treatment of damaged areas: Abrade smooth or local blast cleaning. Full system.
Dry film thickness	20 μm	75 μm	75 μm	25 μm	
Spreading rate	10 m²/l	5 m²/l	2.5 m²/l	10 m²/l	
Minimum temperature	10°C	5°C	5°C	5°C	
Maximum relative humidity	95%	95%	95%	95%	Other comments: Site finish colour – BS 4800, 12 B 15.
Overcoating period, minimum	4 hours	24 hours	48 hours	N/A	
Overcoating period, maximum	7 days	7 days	N/A	N/A	
Inspection required	Full	Full	Random	Random	

Appendix

Steel technology
Properties of steel   1023

Design theory
Bending moment, shear and deflection   1026
Bending moment and reaction   1051
Influence lines   1054
Second moments of area   1066
Plane sections   1074
Plastic moduli   1077
Formulae for rigid frames   1080

Element design
Notes on section dimensions and properties   1098
Dimensions and properties   1106
Extracts from BS 5950: Part 1   1150

Connection design
Bolt data   1164
Weld data   1186

Other Elements
Piling information   1194
Floor plates   1204

Construction
Fire resistance   1206
Corrosion resistance   1230

Miscellaneous
Conversion tables   1234
British Standards for steelwork   1248

Elastic properties of steel

Modulus of elasticity (Young's modulus) $E = 205 \text{ kN/mm}^2$
Poisson's ratio $v = 0.30$
Coefficient of linear thermal expansion $\alpha = 12 \times 10^{-6}$ per °C

Selection of BS 4360: 1990 and BS EN 10025: 1990 Weldable structural steel plate

Grade	Chemical composition (ladle), (max.)						Supply condition	UTS (N/mm²) (min.)	Mechanical properties				Charpy (J)
	C (%)	Mn (%)	Si (%)	P (%)	S (%)	Nb (%)	V (%)			Min. yield strength (N/mm²) for thickness in mm				
										<16	16–40	40–63	63–100	
Fe430A	0.25	1.60	0.50	0.050	0.050				410	275	265	255	235	—
Fe430B[b]	0.21			0.045	0.045				410	275	265	255	235	27@ +20°C
Fe430C[b]	0.18			0.040	0.040				410	275	265	255	235	27@ 0°C
Fe430D1/D2	0.18			0.035	0.035				410	275	265	255	235	27@ −20°C
43EE	0.16	1.50	0.50	0.040	0.030			Normalized	430	275	265	245		27@ −50°C
Fe510B[b]	0.24	1.60	0.55	0.045	0.045				490	355	345	335	315	27@ +20°C
Fe510C[b]	0.22	1.60	0.55	0.040	0.040				490	355	345	335	315	27@ 0°C
Fe510D1/D2	0.22	1.60	0.55	0.035	0.035				490	355	345	335	315	27@ −20°C
Fe510DD1/DD2	0.22	1.60	0.55	0.035	0.035				490	355	345	335	315	40@ −20°C
50EE	0.18	1.50	0.50	0.040	0.030	0.003/0.10		Normalized	490	355	345	340	325	27@ −50°C
50F	0.16	1.50	0.50	0.025	0.025	0.003/0.008		QT	490	390	390	—	—	27@ −60°C
55C	0.22	1.60	0.60	0.040	0.040	0.003/0.10		Normalized	550	450	430[c]	—	—	27@ 0°C
55EE	0.22	1.60	0.50	0.040	0.030	0.003/0.10		Normalized	550	450	430[c]	400	—	27@ −50°C
55F	0.16	1.50	0.50	0.025	0.025	0.003/0.008		QT	550	450	430[c]	—	—	27@ −60°C
WR50B[a]	0.19	1.25	0.65	0.040	0.050		0.10	As-rolled	480	345	345	340[d]	—	27@ 0°C
WR50C[a]	0.22	1.45	0.65	0.040	0.050		0.10	or normalized	480	345	345	340[d]	—	27@ −15°C

[a] Cr 0.50/0.65, Cu 0.25/0.40, weather resisting grades
[b] N 0.009
[c] limit 25 mm
[d] limit 50 mm

The figures in this table are indicative only, and apply to plate material. There are many detailed additional requirements and differences in BS 4360. The figures in the table should under no circumstances be used for specification purposes. For full details reference should be made to BS 4360: 1990 and BS EN 10 025: 1990.

Selection of BS 970: Part 1: 1991 Austenitic stainless steels

Grade	Chemical composition (max.)								Mechanical properties		
	C (%)	Mn (%)	Si (%)	P (%)	S (%)	Cr (%)	Mo (%)	Ni (%)	Others (%)	UTS (N/mm^2)	0.2% proof stress (N/mm^2)	1% proof stress (N/mm^2)	HB max.
304S11	0.030	2.0	1.0	0.045	0.030	19		12		480	180	215	183
304S15	0.06	2.0	1.0	0.045	0.030	19		11		480	195	230	183
304S31	0.07	2.0	1.0	0.045	0.030	19		11		490	195	230	183
321S31	0.08	2.0	1.0	0.045	0.030	19		12	Ti 5C	510	200	235	183
347S31	0.08	2.0	1.0	0.045	0.030	19		12	Nb 10C	510	205	240	183
316S11	0.030	2.0	1.0	0.045	0.030	18.5	2.5	14		490	190	225	183
316S31	0.07	2.0	1.0	0.045	0.030	18.5	2.5	13.5		510	205	240	183
310S31	0.15	2.0	1.5	0.045	0.030	26		22		510	205	240	207
320S31	0.08	2.0	1.0	0.045	0.030	18.5	2.5	14	Ti 5C	510	210	245	183
303S31	0.12	2.0	1.0	0.06	0.35	19	1.0	10		510	190	225	183
325S31	0.12	2.0	1.0	0.045	0.35	19		11	Ti 5C	510	200	235	183

The figures in this table are indicative only and should under no circumstances be used for specification purposes. For full details reference should be made to BS 970: Part 1: 1991.

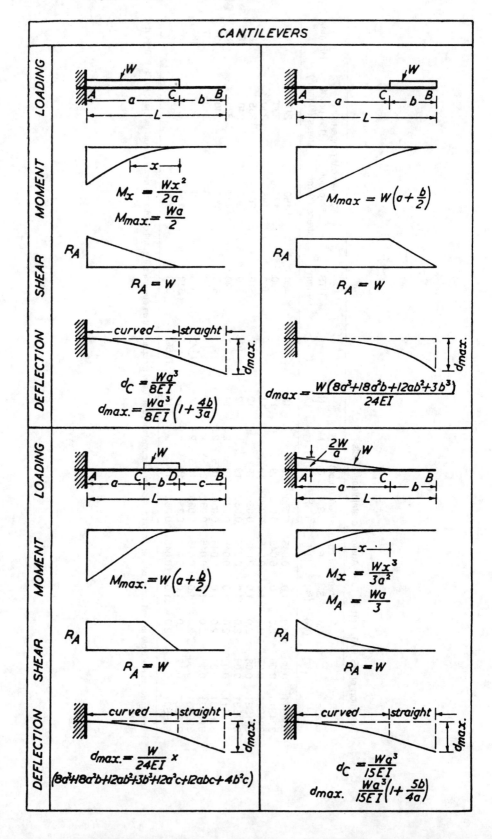

Bending moment, shear and deflection

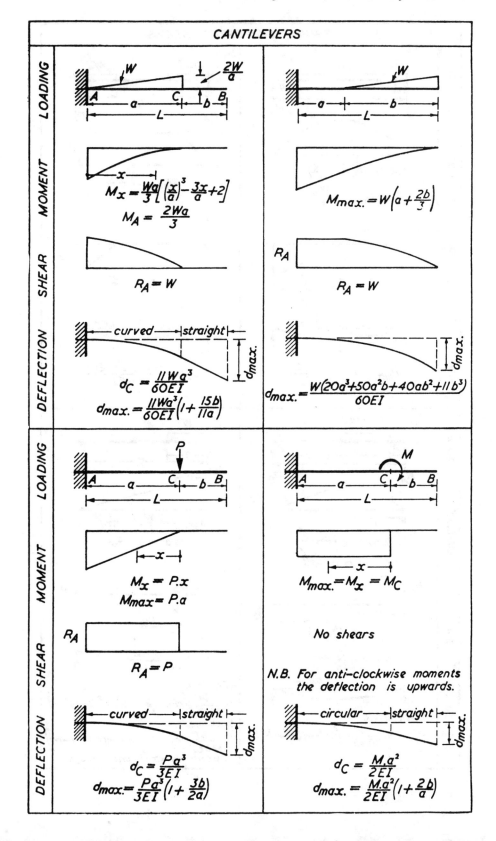

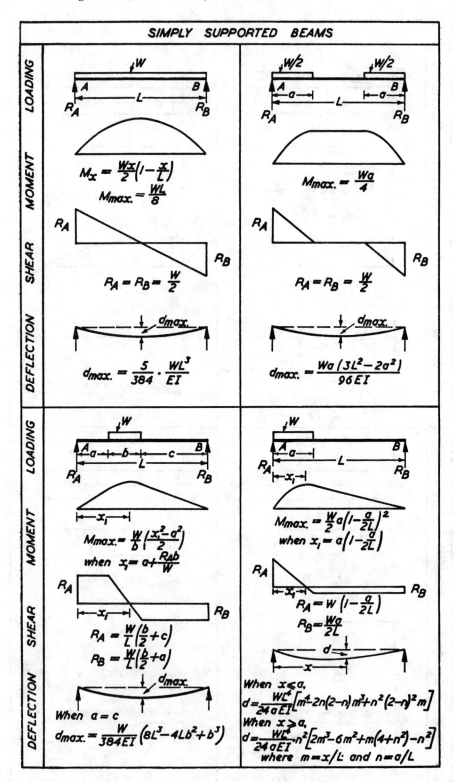

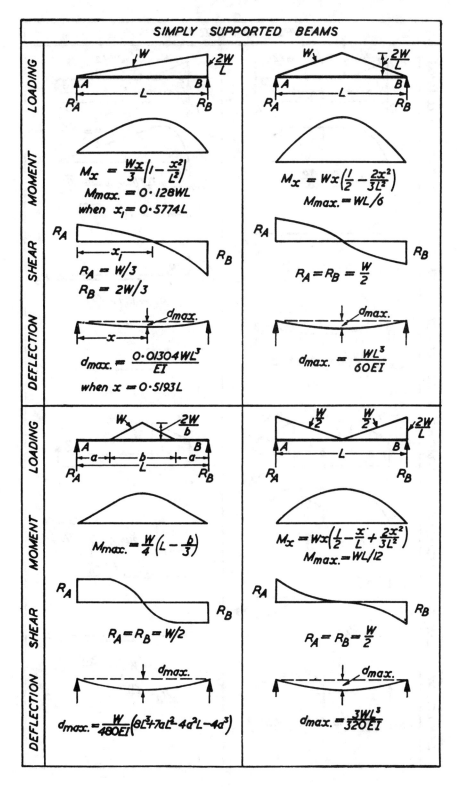

SIMPLY SUPPORTED BEAMS

LOADING (top-left): Triangular loads of $W/2$ at each end, W/a between, spans a-b-a, length L, reactions R_A, R_B.

MOMENT:
$$M_{max.} = \frac{Wa}{6}$$

SHEAR:
$$R_A = R_B = W/2$$

DEFLECTION:
$$d_{max.} = \frac{Wa}{240EI}(18a^2 + 20ab + 5b^2)$$

LOADING (top-right): Triangular load W at left rising to $2W/a$, spans a-b, length L.

MOMENT:
$$m = a/L$$
$$M_{max.} = \frac{Wa}{3}\left(1 - m + \frac{2m}{3}\sqrt{\frac{m}{3}}\right)$$
when $x = a\left(1 - \sqrt{\frac{m}{3}}\right)$

SHEAR:
$$R_A = W\left(1 - \frac{m}{3}\right)$$
$$R_B = \frac{Wm}{3}$$

LOADING (bottom-left): Triangular loads of $W/2$ at each end with peak W/a inward, spans a-b-a, length L.

MOMENT:
$$M_{max.} = \frac{Wa}{3}$$

SHEAR:
$$R_A = R_B = W/2$$

DEFLECTION:
$$d_{max.} = \frac{Wa}{120EI}(16a^2 + 20ab + 5b^2)$$

LOADING (bottom-right): Triangular load with W at left, peak $2W/a$, spans a-b, length L.

MOMENT:
$$M_{max.} = \frac{2Wa}{3}\left(1 - \frac{2m}{3}\right)^{3/2}$$
when $x = a\sqrt{1 - \frac{2m}{3}}$

SHEAR:
$$R_A = W\left(1 - \frac{2m}{3}\right)$$
$$R_B = \frac{2Wm}{3}$$

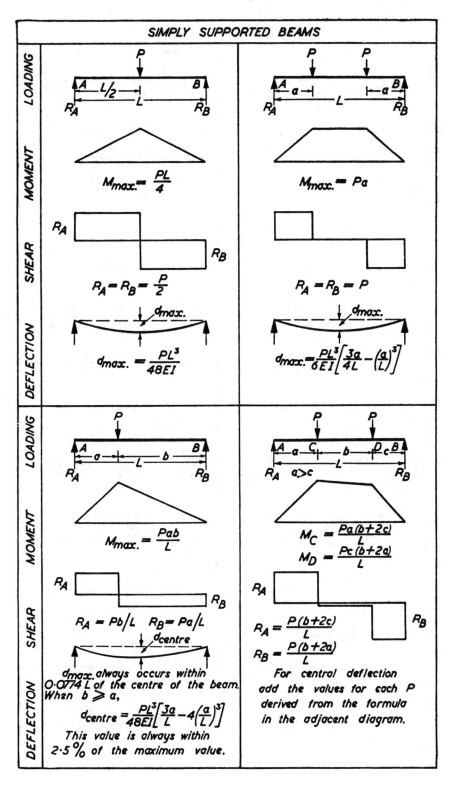

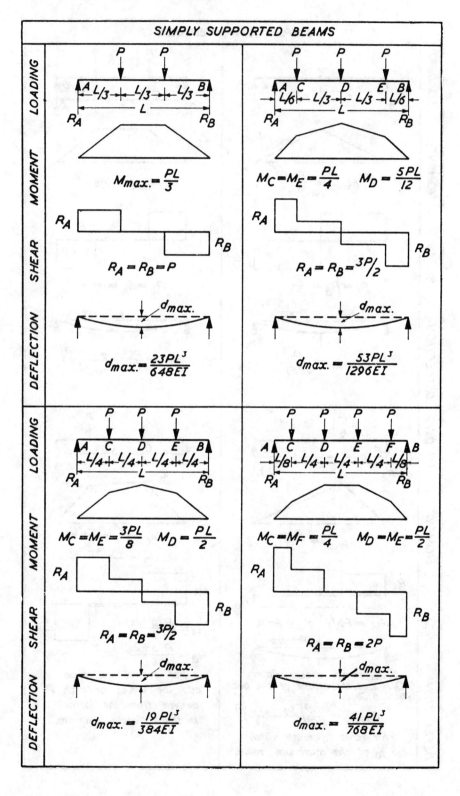

Bending moment, shear and deflection

SIMPLY SUPPORTED BEAMS

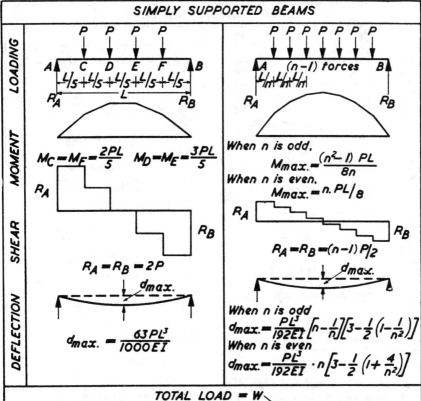

$M_C = M_F = \dfrac{2PL}{5} \quad M_D = M_E = \dfrac{3PL}{5}$

$R_A = R_B = 2P$

$d_{max.} = \dfrac{63 PL^3}{1000 EI}$

When n is odd,
$M_{max.} = \dfrac{(n^2 - 1) PL}{8n}$

When n is even,
$M_{max.} = n \cdot PL/8$

$R_A = R_B = (n-1)P/2$

When n is odd
$d_{max.} = \dfrac{PL^3}{192 EI}\left[n - \dfrac{1}{n}\right]\left[3 - \dfrac{1}{2}\left(1 - \dfrac{1}{n^2}\right)\right]$

When n is even
$d_{max.} = \dfrac{PL^3}{192 EI} \cdot n\left[3 - \dfrac{1}{2}\left(1 + \dfrac{4}{n^2}\right)\right]$

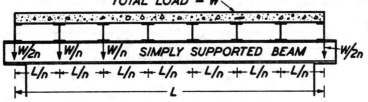

When $n > 10$, consider the load uniformly distributed

The reaction at the supports $= W/2$, but the maximum S.F. at the ends of the beam $= \dfrac{W(n-1)}{2n} = A \cdot W$

The value of the maximum bending moment $= C \cdot WL$

The value of the deflection at the centre of the span $= k \cdot \dfrac{WL^3}{EI}$

Value of n	A	C	k
2	0.2500	0.1250	0.0105
3	0.3333	0.1111	0.0118
4	0.3750	0.1250	0.0124
5	0.4000	0.1200	0.0126
6	0.4167	0.1250	0.0127
7	0.4286	0.1224	0.0128
8	0.4375	0.1250	0.0128
9	0.4444	0.1236	0.0129
10	0.4500	0.1250	0.0129

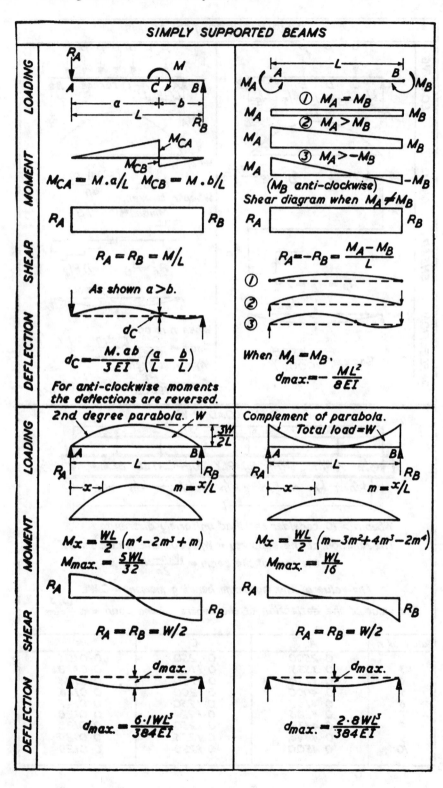

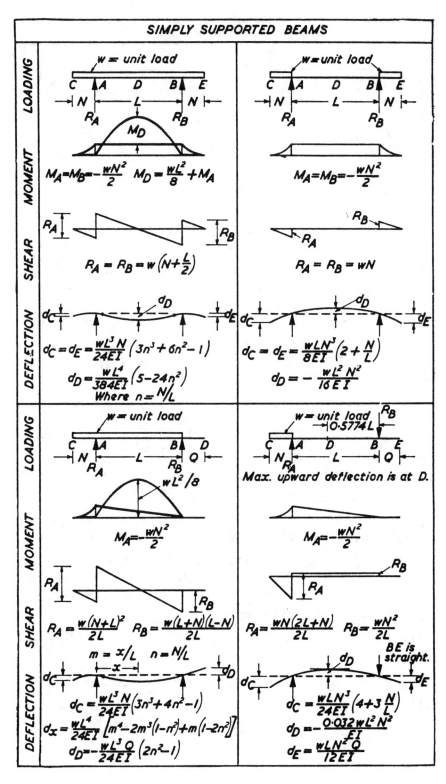

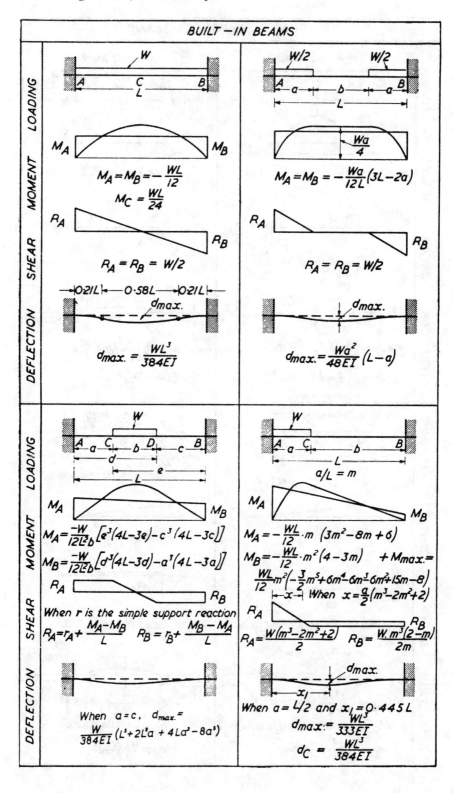

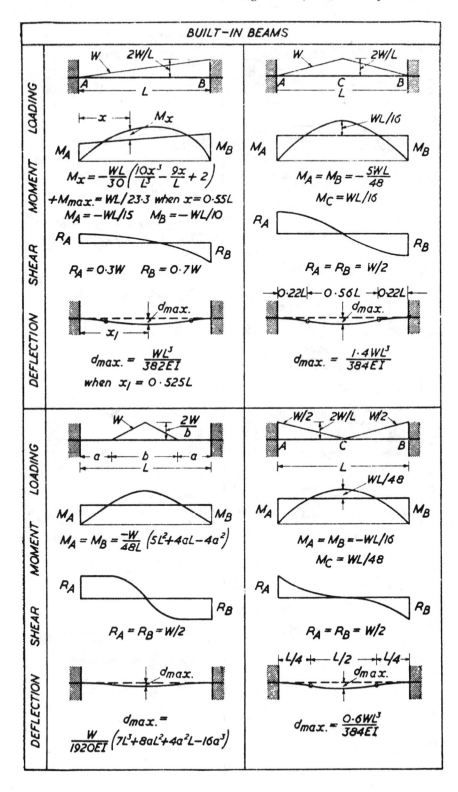

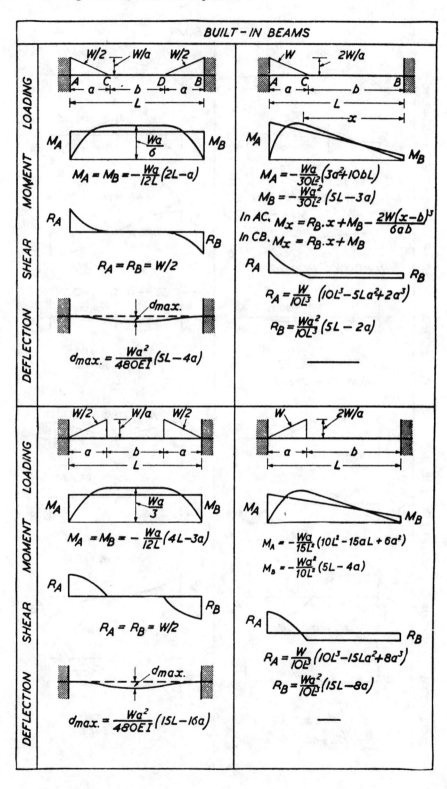

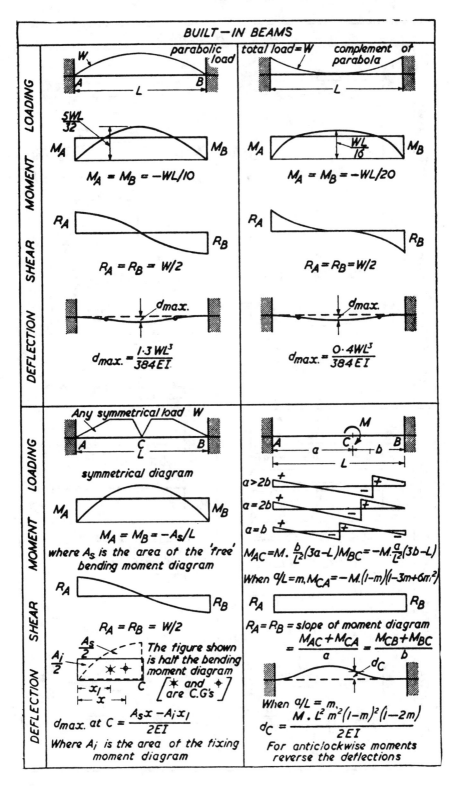

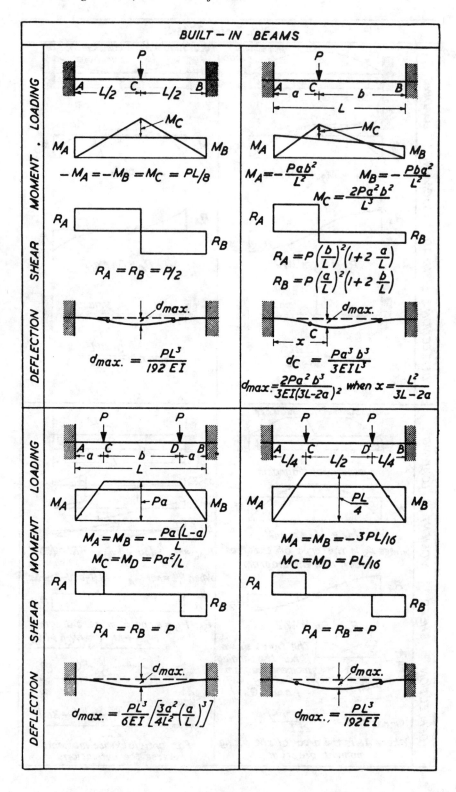

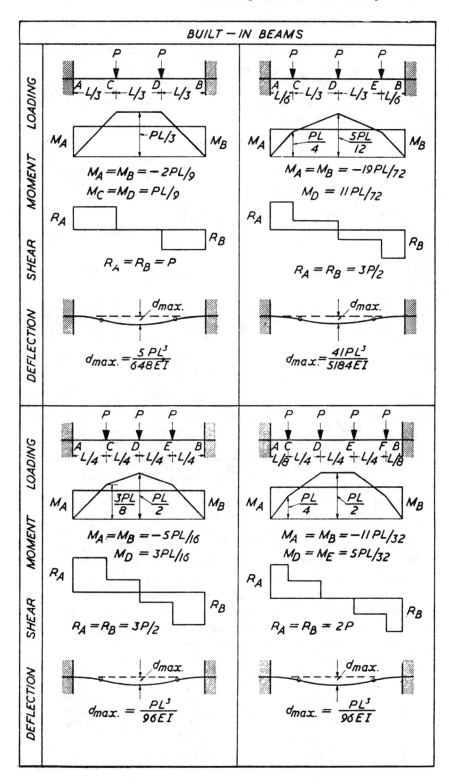

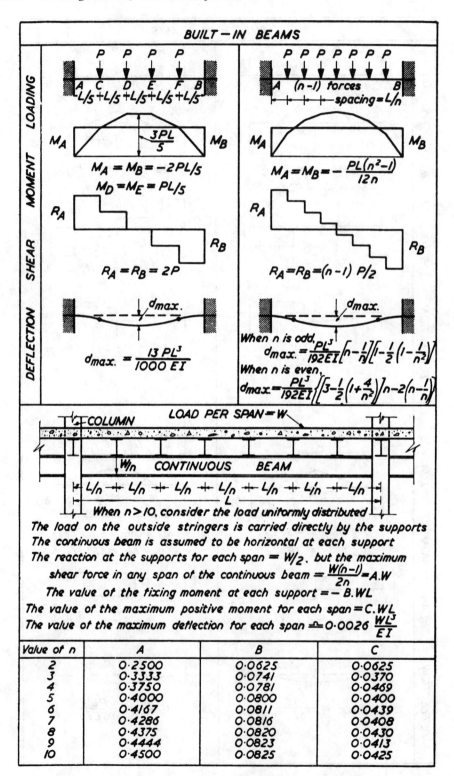

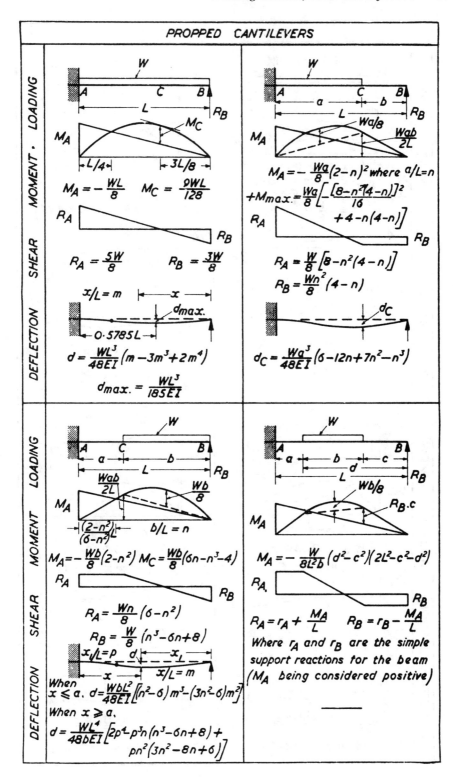

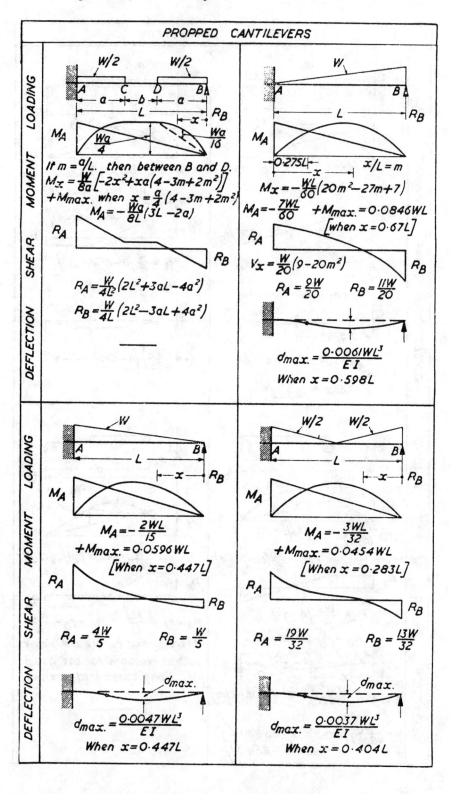

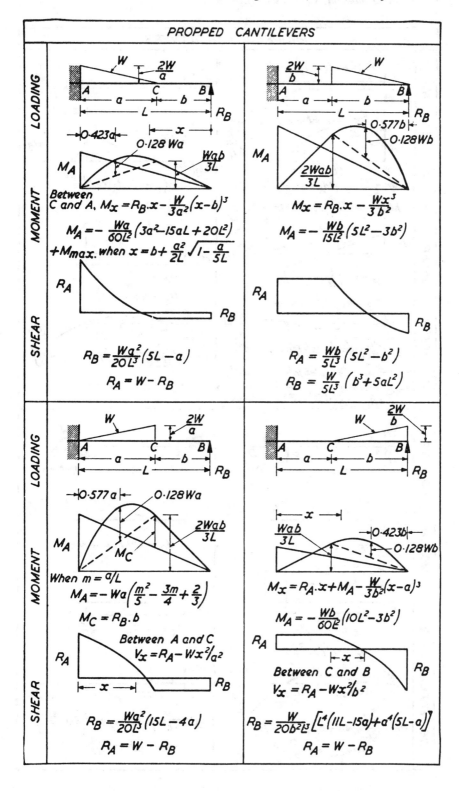

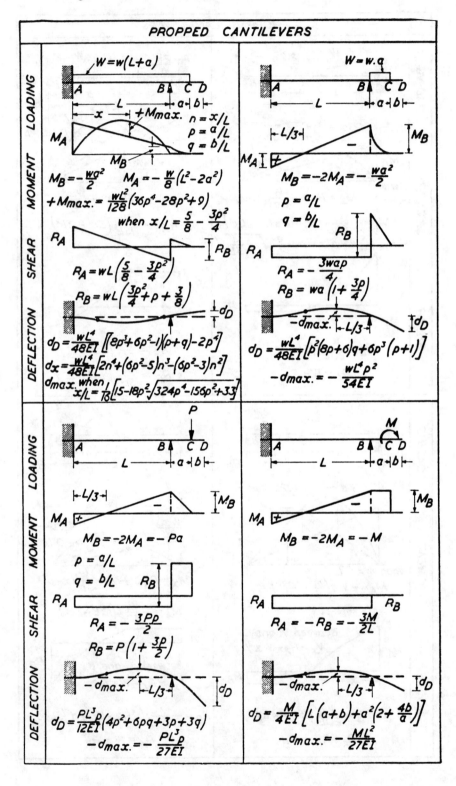

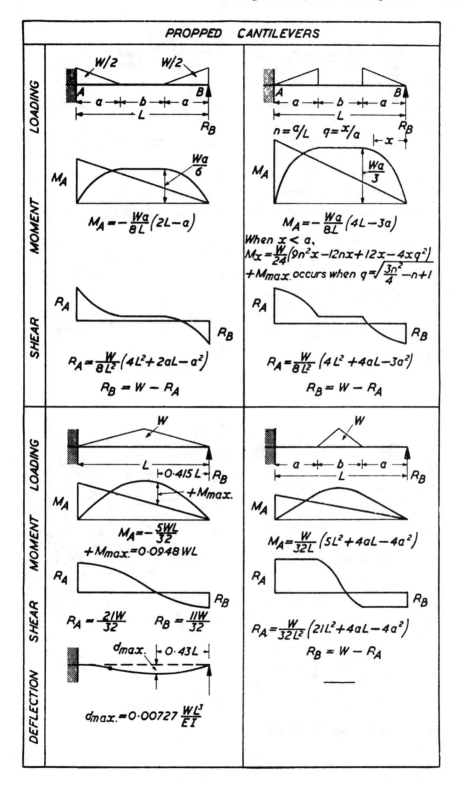

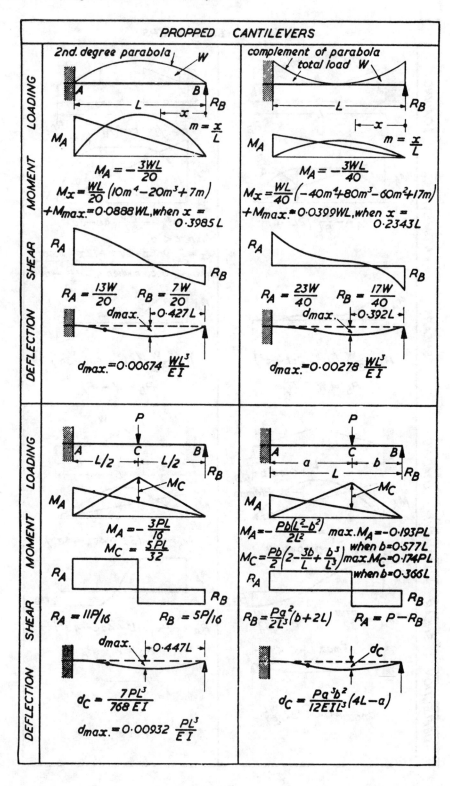

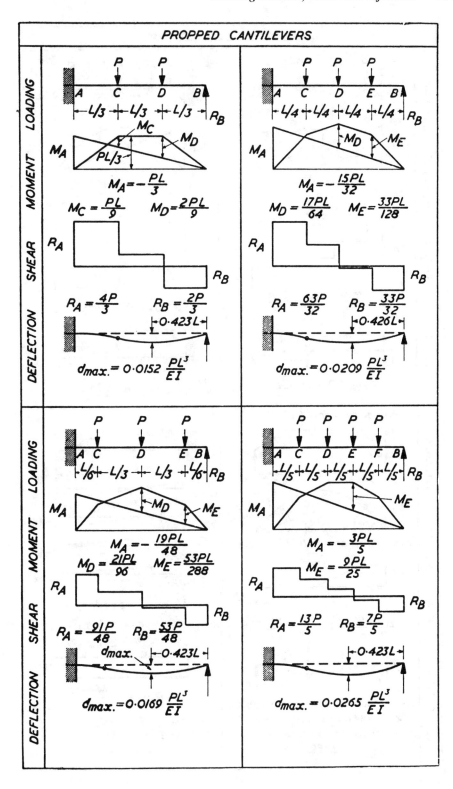

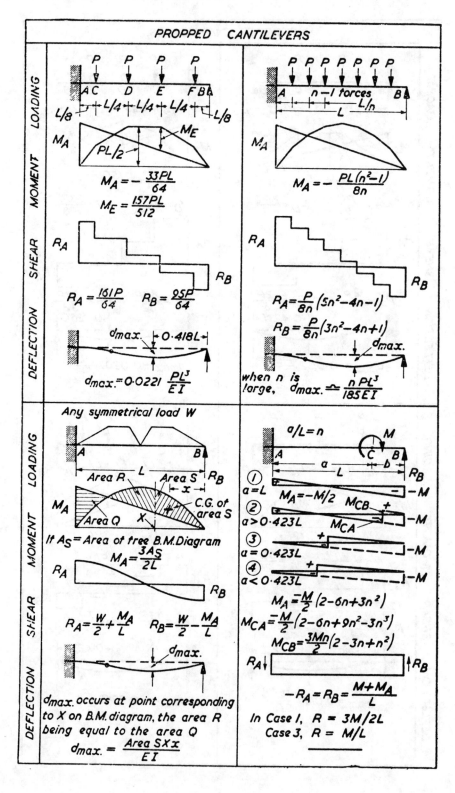

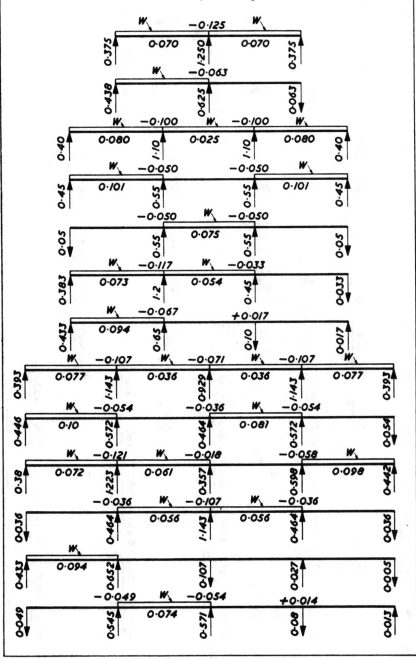

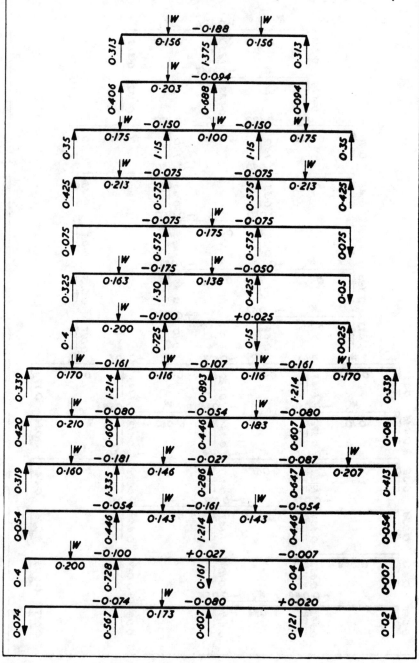

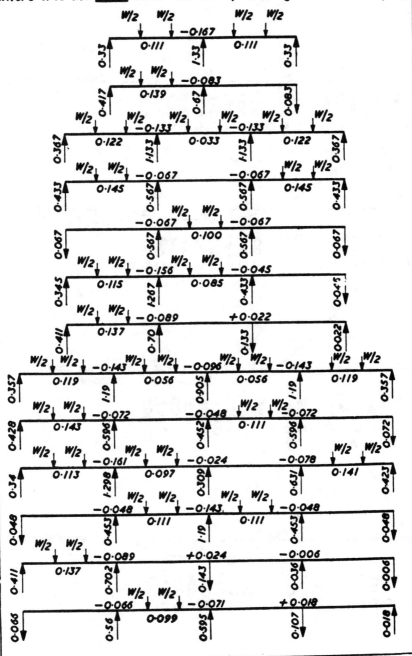

Influence lines for bending moments — two-span beam

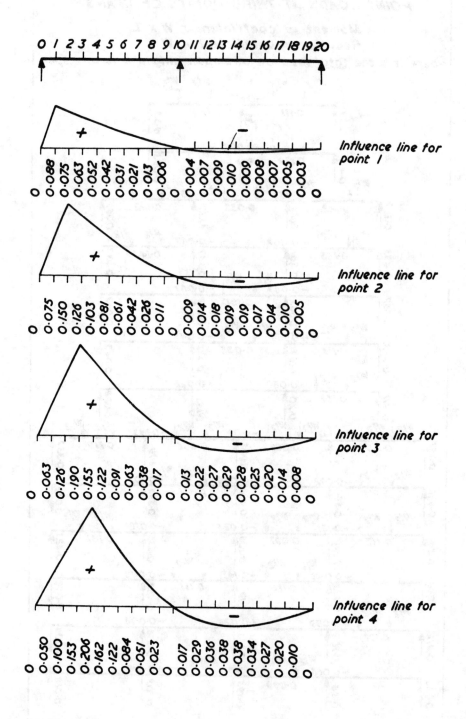

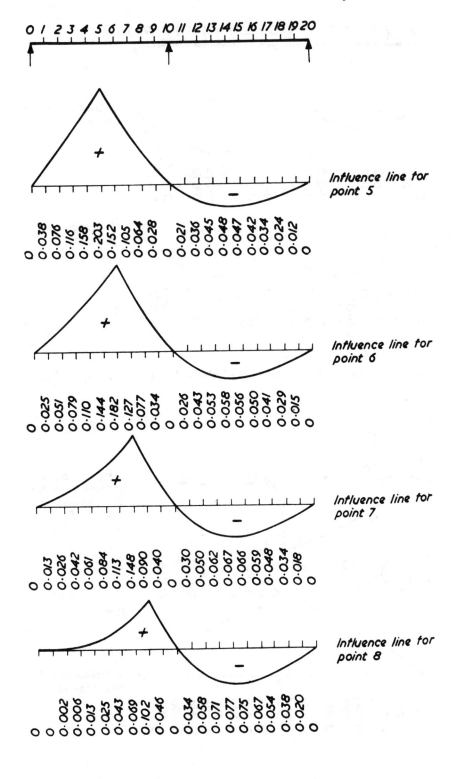

Influence lines

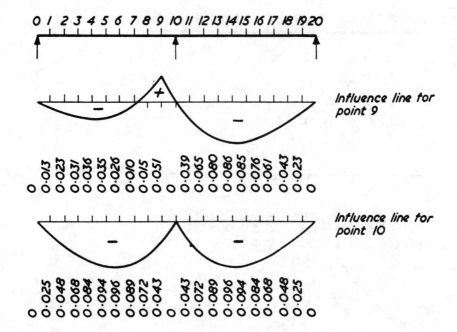

Influence line for point 9

Influence line for point 10

Influence lines for reactions and shear forces – two-span beam

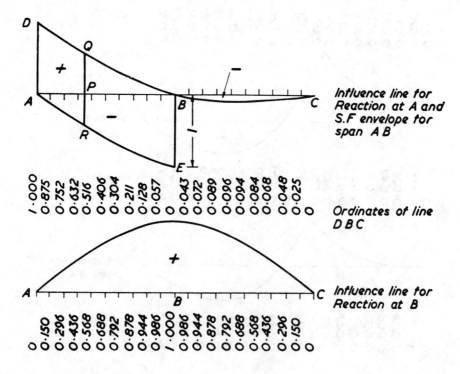

Influence line for Reaction at A and S.F envelope for span AB

Ordinates of line DBC

Influence line for Reaction at B

Influence lines for bending moments — three-span beam

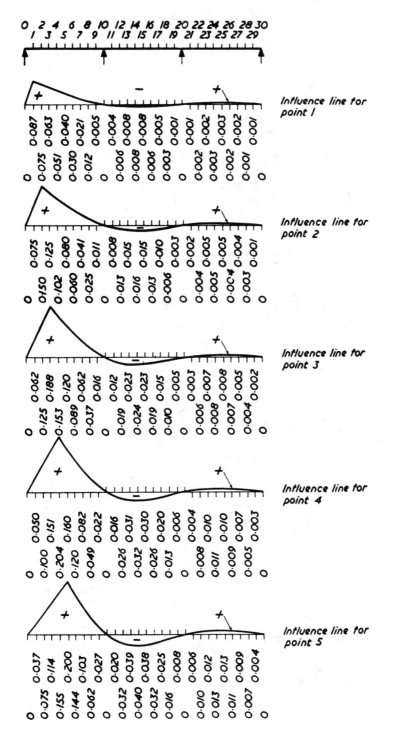

Influence lines for bending moments — three-span beam

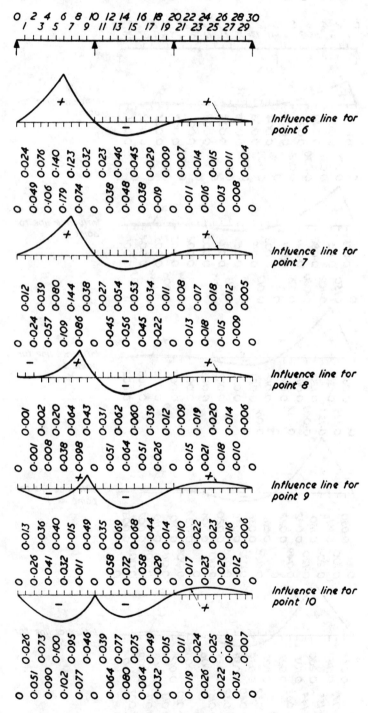

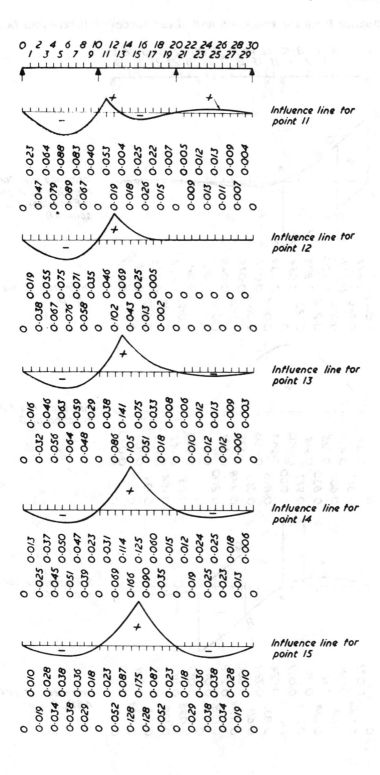

Influence lines for reactions and shear forces — three-span beam

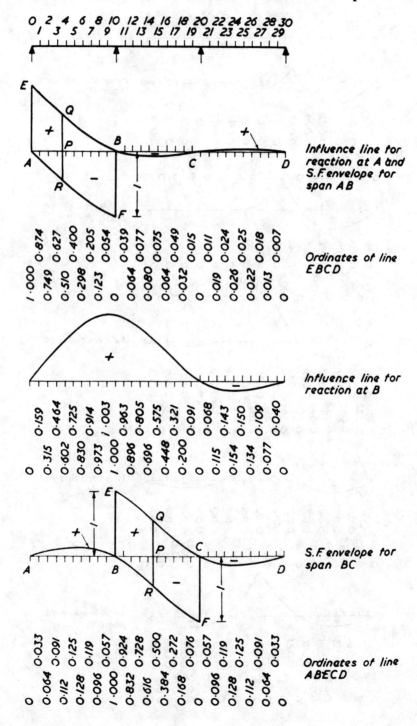

Influence lines for bending moments – four-span beam

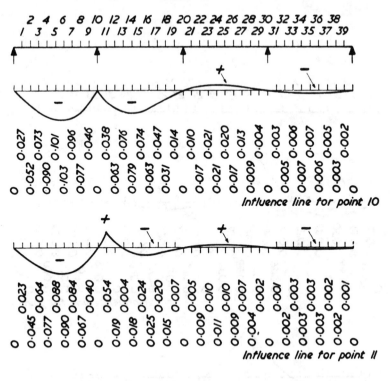

Influence line for point 10

Influence line for point 11

Influence line for point 12

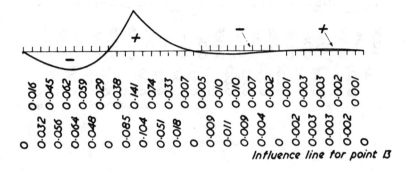

Influence line for point 13

Influence lines for bending moments – four-span beam

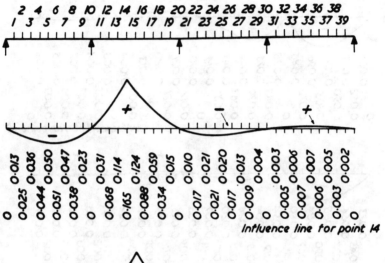

Influence line for point 14

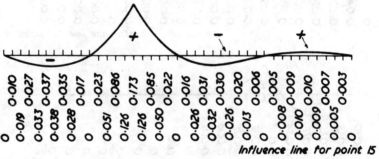

Influence line for point 15

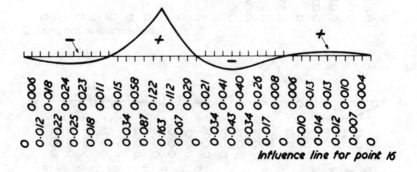

Influence line for point 16

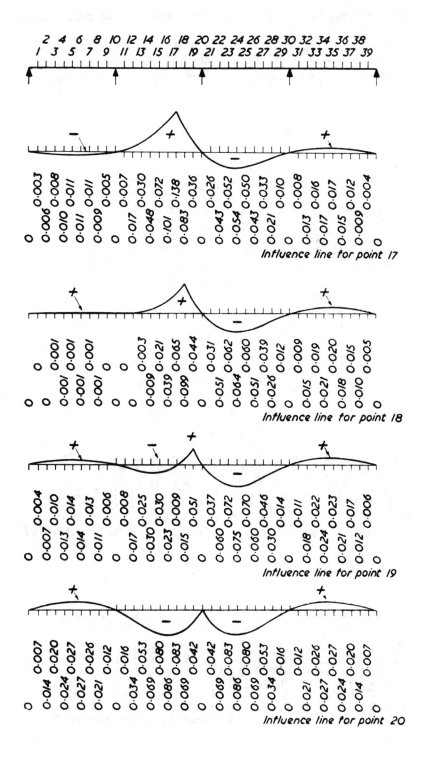

Influence lines for reactions and shear forces – four-span beam

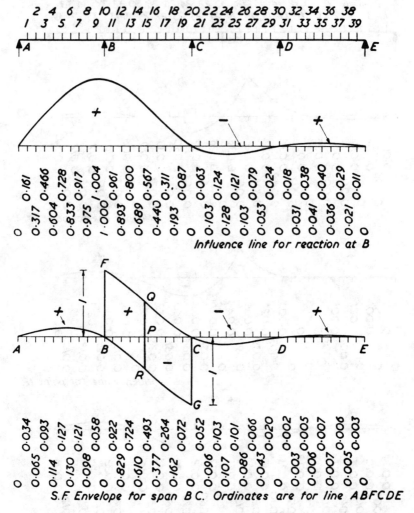

SECOND MOMENTS OF AREA (cm⁴) OF TWO FLANGES
per millimetre of width

Distance d_w mm	THICKNESS OF EACH FLANGE IN MILLIMETRES									
	10	12	15	18	20	22	25	28	30	32
1000	510.1	614.5	772.7	932.8	1041	1149	1314	1480	1592	1705
1100	616.1	742.0	932.5	1125	1255	1385	1582	1782	1916	2051
1200	732.1	881.4	1107	1335	1489	1643	1876	2112	2270	2429
1300	858.1	1033	1297	1564	1743	1923	2195	2469	2654	2839
1400	994.1	1196	1502	1810	2017	2224	2539	2855	3068	3282
1500	1140	1372	1721	2074	2311	2548	2907	3269	3512	3756
1600	1296	1559	1956	2356	2625	2894	3301	3711	3986	4262
1700	1462	1759	2206	2656	2959	3262	3720	4181	4490	4800
1800	1638	1970	2471	2975	3313	3652	4164	4679	5024	5371
1900	1824	2193	2750	3311	3687	4064	4632	5204	5588	5973
2000	2020	2429	3045	3665	4081	4498	5126	5758	6182	6607
2100	2226	2676	3355	4037	4495	4953	5645	6340	6806	7273
2200	2442	2936	3680	4428	4929	5431	6189	6950	7460	7971
2300	2668	3207	4019	4836	5383	5931	6757	7588	8144	8702
2400	2904	3491	4374	5262	5857	6453	7351	8254	8858	9464
2500	3150	3786	4744	5706	6351	6997	7970	8947	9602	10258
2600	3406	4094	5129	6169	6865	7563	8614	9669	10376	11084
2700	3672	4413	5528	6649	7399	8150	9282	10419	11180	11943
2800	3948	4744	5943	7147	7953	8760	9976	11197	12014	12833
2900	4234	5088	6373	7663	8527	9392	10695	12003	12878	13755
3000	4530	5443	6818	8198	9121	10046	11439	12837	13772	14709
3100	4836	5811	7277	8750	9735	10722	12207	13699	14696	15696
3200	5152	6190	7752	9320	10369	11420	13001	14588	15650	16714
3300	5478	6582	8242	9908	11023	12139	13820	15506	16634	17764
3400	5814	6985	8747	10515	11697	12881	14664	16452	17648	18846
3500	6160	7401	9266	11139	12391	13645	15532	17426	18692	19961
3600	6516	7828	9801	11781	13105	14431	16426	18428	19766	21107
3700	6882	8267	10351	12441	13839	15239	17345	19458	20870	22285
3800	7258	8719	10916	13120	14593	16069	18289	20515	22004	23495
3900	7644	9182	11495	13816	15367	16920	19257	21601	23168	24738
4000	8040	9658	12090	14530	16161	17794	20251	22715	24362	26012
4100	8446	10145	12700	15262	16975	18690	21270	23857	25586	27318
4200	8862	10645	13325	16012	17809	19608	22314	25027	26840	28656
4300	9288	11156	13964	16781	18663	20548	23382	26225	28124	30027
4400	9724	11679	14619	17567	19537	21510	24476	27450	29438	31429
4500	10170	12215	15289	18371	20431	22494	25595	28704	30782	32863
4600	10626	12762	15974	19193	21345	23499	26739	29986	32156	34329
4700	11092	13322	16673	20034	22279	24527	27907	31296	33560	35827
4800	11568	13893	17388	20892	23233	25577	29101	32634	34994	37358
4900	12054	14477	18118	21768	24207	26649	30320	34000	36458	38920
5000	12550	15072	18863	22662	25201	27743	31564	35393	37952	40514

SECOND MOMENTS OF AREA (cm⁴) OF TWO FLANGES
per millimetre of width

THICKNESS OF EACH FLANGE IN MILLIMETRES										Distance
35	38	40	45	50	55	60	65	70	75	d_w mm
1875	2048	2164	2459	2758	3064	3374	3691	4013	4341	1000
2255	2461	2600	2951	3308	3671	4040	4416	4797	5184	1100
2670	2913	3076	3489	3908	4334	4766	5205	5651	6103	1200
3120	3402	3592	4072	4558	5052	5552	6060	6575	7097	1300
3604	3930	4148	4700	5258	5825	6398	6980	7569	8166	1400
4124	4495	4744	5372	6008	6652	7304	7965	8633	9309	1500
4679	5099	5380	6090	6808	7535	8270	9014	9767	10528	1600
5269	5740	6056	6853	7658	8473	9296	10129	10971	11822	1700
5893	6420	6772	7661	8558	9466	10382	11309	12245	13191	1800
6553	7137	7528	8513	9508	10513	11528	12554	13589	14634	1900
7248	7892	8324	9411	10508	11616	12734	13863	15003	16153	2000
7978	8686	9160	10354	11558	12774	14000	15238	16487	17747	2100
8742	9517	10036	11342	12658	13987	15326	16678	18041	19416	2200
9542	10387	10952	12374	13808	15254	16712	18183	19665	21159	2300
10377	11294	11908	13452	15008	16577	18158	19752	21359	22978	2400
11247	12240	12904	14575	16258	17955	19664	21387	23123	24872	2500
12151	13223	13940	15743	17558	19388	21230	23087	24957	26841	2600
13091	14245	15016	16955	18908	20875	22856	24852	26861	28884	2700
14066	15304	16132	18213	20308	22418	24542	26681	28835	31003	2800
15076	16401	17288	19516	21758	24016	26288	28576	30879	33197	2900
16120	17537	18484	20864	23258	25669	28094	30536	32993	35466	3000
17200	18710	19720	22256	24808	27376	29960	32561	35177	37809	3100
18315	19922	20996	23694	26408	29139	31886	34650	37431	40228	3200
19465	21171	22312	25177	28058	30957	33872	36805	39755	42722	3300
20649	22459	23668	26705	29758	32830	35918	39025	42149	45291	3400
21869	23784	25064	28277	31508	34757	38024	41310	44613	47934	3500
23124	25147	26500	29895	33308	36740	40190	43659	47147	50653	3600
24414	26549	27976	31558	35158	38778	42416	46074	49751	53447	3700
25738	27988	29492	33266	37058	40871	44702	48554	52425	56316	3800
27098	29466	31048	35018	39008	43018	47048	51099	55169	59259	3900
28493	30981	32644	36816	41008	45221	49454	53708	57983	62278	4000
29923	32535	34280	38659	43058	47479	51920	56383	60867	65372	4100
31387	34126	35956	40547	45158	49792	54446	59123	63821	68541	4200
32887	35756	37672	42479	47308	52159	57032	61928	66845	71784	4300
34422	37423	39428	44457	49508	54582	59678	64797	69939	75103	4400
35992	39128	41224	46480	51758	57060	62384	67732	73103	78497	4500
37596	40872	43060	48548	54058	59593	65150	70732	76337	81966	4600
39236	42653	44936	50660	56408	62180	67976	73797	79641	85509	4700
40911	44473	46852	52818	58808	64823	70862	76926	83015	89128	4800
42621	46330	48808	55021	61258	67521	73808	80121	86459	92822	4900
44365	48226	50804	57269	63758	70274	76814	83381	89973	96591	5000

SECOND MOMENTS OF AREA (cm⁴) OF RECTANGULAR PLATES
about axis x–x

Depth d_w mm	THICKNESS t MILLIMETRES					
	3	4	5	6	8	10
25	.391	.521	.651	.781	1.04	1.30
50	3.13	4.17	5.21	6.25	8.33	10.4
75	10.5	14.1	17.6	21.1	28.1	35.2
100	25.0	33.3	41.7	50.0	66.7	83.3
125	48.8	65.1	81.4	97.7	130	163
150	84.4	113	141	169	225	281
175	134	179	223	268	357	447
200	200	267	333	400	533	667
225	285	380	475	570	759	949
250	391	521	651	781	1042	1302
275	520	693	867	1040	1386	1733
300	675	900	1125	1350	1800	2250
325	858	1144	1430	1716	2289	2861
350	1072	1429	1786	2144	2858	3573
375	1318	1758	2197	2637	3516	4395
400	1600	2133	2667	3200	4267	5333
425	1919	2559	3199	3838	5118	6397
450	2278	3038	3797	4556	6075	7594
475	2679	3572	4465	5359	7145	8931
500	3125	4167	5208	6250	8333	10417
525	3618	4823	6029	7235	9647	12059
550	4159	5546	6932	8319	11092	13865
575	4753	6337	7921	9505	12674	15842
600	5400	7200	9000	10800	14400	18000
625	6104	8138	10173	12207	16276	20345
650	6866	9154	11443	13731	18308	22885
675	7689	10252	12814	15377	20503	25629
700	8575	11433	14292	17150	22867	28583
725	9527	12703	15878	19054	25405	31757
750	10547	14063	17578	21094	28125	35156
775	11637	15516	19395	23274	31032	38790
800	12800	17067	21333	25600	34133	42667
825	14038	18717	23396	28076	37434	46793
850	15353	20471	25589	30706	40942	51177
875	16748	22331	27913	33496	44661	55827
900	18225	24300	30375	36450	48600	60750

SECOND MOMENTS OF AREA (cm⁴) OF RECTANGULAR PLATES
about axis x—x

THICKNESS t MILLIMETRES						Depth
12	15	18	20	22	25	d_w mm
1.56	1.95	2.34	2.60	2.86	3.26	25
12.5	15.6	18.8	20.8	22.9	26.0	50
42.2	52.7	63.3	70.3	77.3	87.9	75
100	125	150	167	183	208	100
195	244	293	326	358	407	125
338	422	506	563	619	703	150
536	670	804	893	983	1117	175
800	1000	1200	1333	1467	1667	200
1139	1424	1709	1898	2088	2373	225
1563	1953	2344	2604	2865	3255	250
2080	2600	3120	3466	3813	4333	275
2700	3375	4050	4500	4950	5625	300
3433	4291	5149	5721	6293	7152	325
4288	5359	6431	7146	7860	8932	350
5273	6592	7910	8789	9668	10986	375
6400	8000	9600	10667	11733	13333	400
7677	9596	11515	12794	14074	15993	425
9113	11391	13669	15188	16706	18984	450
10717	13396	16076	17862	19648	22327	475
12500	15625	18750	20833	22917	26042	500
14470	18088	21705	24117	26529	30146	525
16638	20797	24956	27729	30502	34661	550
19011	23764	28516	31685	34853	39606	575
21600	27000	32400	36000	39600	45000	600
24414	30518	36621	40690	44759	50863	625
27463	34328	41194	45771	50348	57214	650
30755	38443	46132	51258	56384	64072	675
34300	42875	51450	57167	62883	71458	700
38108	47635	57162	63513	69864	79391	725
42188	52734	63281	70313	77344	87891	750
46548	58186	69823	77581	85339	96976	775
51200	64000	76800	85333	93867	106667	800
56152	70189	84227	93586	102945	116982	825
61413	76766	92119	102354	112590	127943	850
66992	83740	100488	111654	122819	139567	875
72900	91125	109350	121500	133650	151875	900

SECOND MOMENTS OF AREA (cm^4)
OF RECTANGULAR PLATES
about axis x–x

Depth d_w mm	THICKNESS t MILLIMETRES					
	3	4	5	6	8	10
1000	25000	33333	41667	50000	66667	83333
1100	33275	44367	55458	66550	88733	110917
1200	43200	57600	72000	86400	115200	144000
1300	54925	73233	91542	109850	146467	183083
1400	68600	91467	114333	137200	182933	228667
1500	84375	112500	140625	168750	225000	281250
1600	102400	136533	170667	204800	273067	341333
1700	122825	163767	204708	245650	327533	409417
1800	145800	194400	243000	291600	388800	486000
1900	171475	228633	285792	342950	457267	571583
2000	200000	266667	333333	400000	533333	666667
2100	231525	308700	385875	463050	617400	771750
2200	266200	354933	443667	532400	709867	887333
2300	304175	405567	506958	608350	811133	1013917
2400	345600	460800	576000	691200	921600	1152000
2500	390625	520833	651042	781250	1041667	1302083
2600	439400	585867	732333	878800	1171733	1464667
2700	492075	656100	820125	984150	1312200	1640250
2800	548800	731733	914667	1097600	1463467	1829333
2900	609725	812967	1016208	1219450	1625933	2032417
3000	675000	900000	1125000	1350000	1800000	2250000
3100	744775	993033	1241292	1489550	1986067	2482583
3200	819200	1092267	1365333	1638400	2184533	2730667
3300	898425	1197900	1497375	1796850	2395800	2994750
3400	982600	1310133	1637667	1965200	2620267	3275333
3500	1071875	1429167	1786458	2143750	2858333	3572917
3600	1166400	1555200	1944000	2332800	3110400	3888000
3700	1266325	1688433	2110542	2532650	3376867	4221083
3800	1371800	1829067	2286333	2743600	3658133	4572667
3900	1482975	1977300	2471625	2965950	3954600	4943250
4000	1600000	2133333	2666667	3200000	4266667	5333333
4100	1723025	2297367	2871708	3446050	4594733	5743417
4200	1852200	2469600	3087000	3704400	4939200	6174000
4300	1987675	2650233	3312792	3975350	5300467	6625583
4400	2129600	2839467	3549333	4259200	5678933	7098667
4500	2278125	3037500	3796875	4556250	6075000	7593750
4600	2433400	3244533	4055667	4866800	6489067	8111333
4700	2595575	3460767	4325958	5191150	6921533	8651917
4800	2764800	3686400	4608000	5529600	7372800	9216000
4900	2941225	3921633	4902042	5882450	7843267	9804083
5000	3125000	4166667	5208333	6250000	8333333	10416667

SECOND MOMENTS OF AREA (cm⁴) OF RECTANGULAR PLATES
about axis x—x

| THICKNESS t MILLIMETRES ||||||| Depth |
12	15	18	20	22	25	d_w mm
100000	125000	150000	166667	183333	208333	1000
133100	166375	199650	221833	244017	277292	1100
172800	216000	259200	288000	316800	360000	1200
219700	274625	329550	366167	402783	457708	1300
274400	343000	411600	457333	503067	571667	1400
337500	421875	506250	562500	618750	703125	1500
409600	512000	614400	682667	750933	853333	1600
491300	614125	736950	818833	900717	1023542	1700
583200	729000	874800	972000	1069200	1215000	1800
685900	857375	1028850	1143167	1257483	1428958	1900
800000	1000000	1200000	1333333	1466667	1666667	2000
926100	1157625	1389150	1543500	1697850	1929375	2100
1064800	1331000	1597200	1774667	1952133	2218333	2200
1216700	1520875	1825050	2027833	2230617	2534792	2300
1382400	1728000	2073600	2304000	2534400	2880000	2400
1562500	1953125	2343750	2604167	2864583	3255208	2500
1757600	2197000	2636400	2929333	3222267	3661667	2600
1968300	2460375	2952450	3280500	3608550	4100625	2700
2195200	2744000	3292800	3658667	4024533	4573333	2800
2438900	3048625	3658350	4064833	4471317	5081042	2900
2700000	3375000	4050000	4500000	4950000	5625000	3000
2979100	3723875	4468650	4965167	5461683	6206458	3100
3276800	4096000	4915200	5461333	6007467	6826667	3200
3593700	4492125	5390550	5989500	6588450	7486875	3300
3930400	4913000	5895600	6550667	7205733	8188333	3400
4287500	5359375	6431250	7145833	7860417	8932292	3500
4665600	5832000	6998400	7776000	8553600	9720000	3600
5065300	6331625	7597950	8442167	9286383	10552708	3700
5487200	6859000	8230800	9145333	10059867	11431667	3800
5931900	7414875	8897850	9886500	10875150	12358125	3900
6400000	8000000	9600000	10666667	11733333	13333333	4000
6892100	8615125	10338150	11486833	12635517	14358542	4100
7408800	9261000	11113200	12348000	13582800	15435000	4200
7950700	9938375	11926050	13251167	14576283	16563958	4300
8518400	10648000	12777600	14197333	15617067	17746667	4400
9112500	11390625	13668750	15187500	16706250	18984375	4500
9733600	12167000	14600400	16222667	17844933	20278333	4600
10382300	12977875	15573450	17303833	19034217	21629792	4700
11059200	13824000	16588800	18432000	20275200	23040000	4800
11764900	14706125	17647350	19608167	21568983	24510208	4900
12500000	15625000	18750000	20833333	22916667	26041667	5000

Second moments of area

SECOND MOMENT OF A PAIR OF UNIT AREAS
about axis x—x

Distance d_u mm	0	5	10	15	20	25	30	35	40	45
500	1250	1275	1301	1326	1352	1378	1405	1431	1458	1485
550	1513	1540	1568	1596	1625	1653	1682	1711	1741	1770
600	1800	1830	1861	1891	1922	1953	1985	2016	2048	2080
650	2113	2145	2178	2211	2245	2278	2312	2346	2381	2415
700	2450	2485	2521	2556	2592	2628	2665	2701	2738	2775
750	2813	2850	2888	2926	2965	3003	3042	3081	3121	3160
800	3200	3240	3281	3321	3362	3403	3445	3486	3528	3570
850	3613	3655	3698	3741	3785	3828	3872	3916	3961	4005
900	4050	4095	4141	4186	4232	4278	4325	4371	4418	4465
950	4513	4560	4608	4656	4705	4753	4802	4851	4901	4950
1000	5000	5050	5101	5151	5202	5253	5305	5356	5408	5460
1050	5513	5565	5618	5671	5725	5778	5832	5886	5941	5995
1100	6050	6105	6161	6216	6272	6328	6385	6441	6498	6555
1150	6613	6670	6728	6786	6845	6903	6962	7021	7081	7140
1200	7200	7260	7321	7381	7442	7503	7565	7626	7688	7750
1250	7813	7875	7938	8001	8065	8128	8192	8256	8321	8385
1300	8450	8515	8581	8646	8712	8778	8845	8911	8978	9045
1350	9113	9180	9248	9316	9385	9453	9522	9591	9661	9730
1400	9800	9870	9941	10011	10082	10153	10225	10296	10368	10440
1450	10513	10585	10658	10731	10805	10878	10952	11026	11101	11175
1500	11250	11325	11401	11476	11552	11628	11705	11781	11858	11935
1550	12013	12090	12168	12246	12325	12403	12482	12561	12641	12720
1600	12800	12880	12961	13041	13122	13203	13285	13366	13448	13530
1650	13613	13695	13778	13861	13945	14028	14112	14196	14281	14365
1700	14450	14535	14621	14706	14792	14878	14965	15051	15138	15225
1750	15313	15400	15488	15576	15665	15753	15842	15931	16021	16110
1800	16200	16290	16381	16471	16562	16653	16745	16836	16928	17020
1850	17113	17205	17298	17391	17485	17578	17672	17766	17861	17955
1900	18050	18145	18241	18336	18432	18528	18625	18721	18818	18915
1950	19013	19110	19208	19306	19405	19503	19602	19701	19801	19900
2000	20000	20100	20201	20301	20402	20503	20605	20706	20808	20910
2050	21013	21115	21218	21321	21425	21528	21632	21736	21841	21945
2100	22050	22155	22261	22366	22472	22578	22685	22791	22898	23005
2150	23113	23220	23328	23436	23545	23653	23762	23871	23981	24090
2200	24200	24310	24421	24531	24642	24753	24865	24976	25088	25200
2250	25313	25425	25538	25651	25765	25878	25992	26106	26221	26335
2300	26450	26565	26681	26796	26912	27028	27145	27261	27378	27495
2350	27613	27730	27848	27966	28085	28203	28322	28441	28561	28680
2400	28800	28920	29041	29161	29282	29403	29525	29646	29768	29890
2450	30013	30135	30258	30381	30505	30628	30752	30876	31001	31125
2500	31250	31375	31501	31626	31752	31878	32005	32131	32258	32385
2550	32513	32640	32768	32896	33025	33153	33282	33411	33541	33670
2600	33800	33930	34061	34191	34322	34453	34585	34716	34848	34980
2650	35113	35245	35378	35511	35645	35778	35912	36046	36181	36315
2700	36450	36585	36721	36856	36992	37128	37265	37401	37538	37675

Second moments are tabulated in cm^4 and are for unit areas of 1 cm^2 each.

SECOND MOMENT OF A PAIR OF UNIT AREAS
about axis x—x

Distance d_u mm	0	5	10	15	20	25	30	35	40	45
2750	37813	37950	38088	38226	38365	38503	38642	38781	38921	39060
2800	39200	39340	39481	39621	39762	39903	40045	40186	40328	40470
2850	40613	40755	40898	41041	41185	41328	41472	41616	41761	41905
2900	42050	42195	42341	42486	42632	42778	42925	43071	43218	43365
2950	43513	43660	43808	43956	44105	44253	44402	44551	44701	44850
3000	45000	45150	45301	45451	45602	45753	45905	46056	46208	46360
3050	46513	46665	46818	46971	47125	47278	47432	47586	47741	47895
3100	48050	48205	48361	48516	48672	48828	48985	49141	49298	49455
3150	49613	49770	49928	50086	50245	50403	50562	50721	50881	51040
3200	51200	51360	51521	51681	51842	52003	52165	52326	52488	52650
3250	52813	52975	53138	53301	53465	53628	53792	53956	54121	54285
3300	54450	54615	54781	54946	55112	55278	55445	55611	55778	55945
3350	56113	56280	56448	56616	56785	56953	57122	57291	57461	57630
3400	57800	57970	58141	58311	58482	58653	58825	58996	59168	59340
3450	59513	59685	59858	60031	60205	60378	60552	60726	60901	61075
3500	61250	61425	61601	61776	61952	62128	62305	62481	62658	62835
3550	63013	63190	63368	63546	63725	63903	64082	64261	64441	64620
3600	64800	64980	65161	65341	65522	65703	65885	66066	66248	66430
3650	66613	66795	66978	67161	67345	67528	67712	67896	68081	68265
3700	68450	68635	68821	69006	69192	69378	69565	69751	69938	70125
3750	70313	70500	70688	70876	71065	71253	71442	71631	71821	72010
3800	72200	72390	72581	72771	72962	73153	73345	73536	73728	73920
3850	74113	74305	74498	74691	74885	75078	75272	75466	75661	75855
3900	76050	76245	76441	76636	76832	77028	77225	77421	77618	77815
3950	78013	78210	78408	78606	78805	79003	79202	79401	79601	79800
4000	80000	80200	80401	80601	80802	81003	81205	81406	81608	81810
4050	82013	82215	82418	82621	82825	83028	83232	83436	83641	83845
4100	84050	84255	84461	84666	84872	85078	85285	85491	85698	85905
4150	86113	86320	86528	86736	86945	87153	87362	87571	87781	87990
4200	88200	88410	88621	88831	89042	89253	89465	89676	89888	90100
4250	90313	90525	90738	90951	91165	91378	91592	91806	92021	92235
4300	92450	92665	92881	93096	93312	93528	93745	93961	94178	94395
4350	94613	94830	95048	95266	95485	95703	95922	96141	96361	96580
4400	96800	97020	97241	97461	97682	97903	98125	98346	98568	98790
4450	99013	99235	99458	99681	99905	100128	100352	100576	100801	101025
4500	101250	101475	101701	101926	102152	102378	102605	102831	103058	103285
4550	103513	103740	103968	104196	104425	104653	104882	105111	105341	105570
4600	105800	106030	106261	106491	106722	106953	107185	107416	107648	107880
4650	108113	108345	108578	108811	109045	109278	109512	109746	109981	110215
4700	110450	110685	110921	111156	111392	111628	111865	112101	112338	112575
4750	112813	113050	113288	113526	113765	114003	114242	114481	114721	114960
4800	115200	115440	115681	115921	116162	116403	116645	116886	117128	117370
4850	117613	117855	118098	118341	118585	118828	119072	119316	119561	119805
4900	120050	120295	120541	120786	121032	121278	121525	121771	122018	122265
4950	122513	122760	123008	123256	123505	123753	124002	124251	124501	124750

Second moments are tabulated in cm^4 and are for unit areas of 1 cm^2 each.

GEOMETRICAL PROPERTIES OF PLANE SECTIONS

Section		Area	Position of Centroid	Moments of Inertia	Section Moduli
TRIANGLE		$A = \dfrac{bh}{2}$	$e_x = \dfrac{h}{3}$	$I_{XX} = bh^3/36$ $I_{YY} = hb^3/48$ $I_{aa} = bh^3/4$ $I_{bb} = bh^3/12$	Z_{XX} base $= bh^2/12$ apex $= bh^2/24$ $Z_{YY} = bh^2/24$
RECTANGLE		$A = bd$	$e_x = \dfrac{h}{2}$	$I_{XX} = bd^3/12$ $I_{YY} = db^3/12$ $I_{bb} = bd^3/3$	$Z_{XX} = bd^2/6$ $Z_{YY} = db^2/6$
RECTANGLE	axis on diagonal	$A = bd$	$e_x = \dfrac{bd}{\sqrt{b^2+d^2}}$	$I_{XX} = \dfrac{b^3 d^3}{6(b^2+d^2)}$	$Z_{XX} = \dfrac{b^2 d^2}{6\sqrt{b^2+d^2}}$
RECTANGLE	axis through C.G.	$A = bd$	$e_x = \dfrac{b\sin\theta + d\cos\theta}{2}$	$I_{XX} = \dfrac{bd(b^2\sin^2\theta + d^2\cos^2\theta)}{12}$	$Z_{XX} = \dfrac{bd(b^2\sin^2\theta + d^2\cos^2\theta)}{6(b\sin\theta + d\cos\theta)}$
SQUARE		$A = s^2$	$e_x = \dfrac{s}{2}$ $e_v = \dfrac{s}{\sqrt{2}}$	$I_{XX} = I_{YY} = s^4/12$ $I_{bb} = s^4/3$ $I_{VV} = s^4/12$	$Z_{XX} = Z_{YY} = \dfrac{s^3}{6}$ $Z_{VV} = \dfrac{s^3}{6\sqrt{2}}$
TRAPEZIUM		$A = \dfrac{d(a+b)}{2}$	$e_{x_1} = \dfrac{d(2a+b)}{3(a+b)}$	$I_{XX} = \dfrac{d^3(a^2+4ab+b^2)}{36(a+b)}$ $I_{YY} = \dfrac{d(a^3+a^2b+ab^2+b^3)}{48}$	$Z_{XX} = \dfrac{I_{XX}}{d-e_x}$ (two values) $Z_{YY} = \dfrac{2 I_{YY}}{b}$
DIAMOND		$A = \dfrac{bd}{2}$	$e_x = \dfrac{d}{2}$	$I_{XX} = \dfrac{bd^3}{48}$ $I_{YY} = \dfrac{db^3}{48}$	$Z_{XX} = \dfrac{bd^2}{24}$ $Z_{YY} = \dfrac{db^2}{24}$
HEXAGON		$A = 0.866 d^2$	$e_x = 0.866 s$ $= d/2$	$I_{XX} = I_{YY} = I_{VV}$ $= 0.0601 d^4$	$Z_{XX} = 0.1203 d^3$ $Z_{YY} = Z_{VV}$ $= 0.1042 d^3$

GEOMETRICAL PROPERTIES OF PLANE SECTIONS

Section		Area	Position of Centroid	Moments of Inertia	Section Moduli
OCTAGON		$A = 0.8284 d^2$ $s = 0.4142 d$	$e_x = \dfrac{d}{2}$ $e_v = 0.541 d$	$I_{XX} = I_{YY} = I_{VV}$ $= 0.0547 d^4$	$Z_{XX} = Z_{YY}$ $= 0.1095 d^3$ $Z_{VV} = 0.1011 d^3$
POLYGON	Regular figure, n sides	$A = \dfrac{n s^2 \cot\theta}{4}$ $A = n r^2 \tan\theta$ $A = \dfrac{n R^2 \sin 2\theta}{2}$	$e = r$ or R depending on the axis and value of n	$I_1 = I_2$ $= \dfrac{A(6R^2 - s^2)}{24}$ $= \dfrac{A(12 r^2 + s^2)}{48}$	$Z = \dfrac{I}{e}$
CIRCLE		$A = \pi r^2$ $A = 0.7854 d^2$	$e = r = \dfrac{d}{2}$	$I = \dfrac{\pi d^4}{64}$ $I = 0.7854 r^4$	$Z = \dfrac{\pi d^3}{32}$ $Z = 0.7854 r^3$
SEMI-CIRCLE		$A = 1.5708 r^2$	$e_x = 0.424 r$	$I_{XX} = 0.1098 r^4$ $I_{YY} = 0.3927 r^4$	Z_{XX} base $= 0.2587 r^3$ crown $= 0.1907 r^3$ $Z_{YY} = 0.3927 r^3$
SEGMENT		$A = \dfrac{r^2}{2}\left(\dfrac{\pi\theta°}{180°} - \sin\theta\right)$	$e_0 = \dfrac{c^3}{12 A}$ $e_x = e_0 - r\cos\dfrac{\theta}{2}$	$I_{XX} = \dfrac{r^4}{16}\left(\dfrac{\pi\theta°}{90°} - \sin 2\theta\right)$ $- \dfrac{20 r^4 (1-\cos\theta)^3}{\pi\theta° - 180°\sin\theta}$ $I_{YY} = \dfrac{r^4}{48}\left(\dfrac{\pi\theta°}{30°} - 8\sin\theta + \sin 2\theta\right)$	Z_{XX} base $= I_{XX}/e_x$ crown $= \dfrac{I_{XX}}{b - e_x}$ $Z_{YY} = \dfrac{2 I_{YY}}{c}$
SECTOR		$A = \dfrac{\theta°}{360°}\pi r^2$	$e_x = \dfrac{2}{3} r \dfrac{c}{a}$ $e_x = \dfrac{r^2 c}{3 A}$	$I_{XX} = I_0 - \dfrac{360°}{8\pi}\sin\dfrac{2\theta}{2}\cdot\dfrac{4 r^4}{3}$ $I_{YY} = \dfrac{r^4}{8}\left(\dfrac{\pi\theta°}{180°} - \sin\theta\right)$ $I_0 = \dfrac{r^4}{8}\left(\dfrac{\pi\theta°}{180°} + \sin\theta\right)$	Z_{XX} centre $= I_{XX}/e_x$ crown $= \dfrac{I_{XX}}{r - e_x}$ $Z_{YY} = \dfrac{2 I_{YY}}{c}$
QUADRANT		$A = \dfrac{\pi r^2}{4}$	$e_x = 0.424 r$ $e_v = 0.6 r$ $e_u = 0.707 r$	$I_{XX} = I_{YY} = 0.0549 r^4$ $I_{bb} = 0.1963 r^4$ $I_{UU} = 0.0714 r^4$ $I_{VV} = 0.0384 r^4$	Minimum Values $Z_{XX} = Z_{YY}$ $= 0.0953 r^3$ $Z_{UU} = 0.1009 r^3$ $Z_{VV} = 0.064 r^3$
COMPLEMENT		$A = 0.2146 r^2$	$e_x = 0.777 r$ $e_v = 1.098 r$ $e_u = 0.707 r$ $e_a = 0.316 r$ $e_b = 0.391 r$	$I_{XX} = I_{YY} = 0.0076 r^4$ $I_{UU} = 0.012 r^4$ $I_{VV} = 0.0031 r^4$	Minimum Values $Z_{XX} = Z_{YY}$ $= 0.0097 r^3$ $Z_{UU} = 0.017 r^3$ $Z_{VV} = 0.0079 r^3$

GEOMETRICAL PROPERTIES OF PLANE SECTIONS

Section		Area	Position of Centroid	Moments of Inertia	Section Moduli
ELLIPSE		$A = \pi ab$	$e_x = a$ $e_y = b$	$I_{XX} = 0.7854 ba^3$ $I_{YY} = 0.7854 ab^3$	$Z_{XX} = 0.7854 bd^2$ $Z_{YY} = 0.7854 ab^2$
SEMI-ELLIPSE		$A = \dfrac{\pi ab}{2}$	$e_x = 0.424a$ $e_y = b$	$I_{XX} = 0.1098 ba^3$ $I_{YY} = 0.3927 ab^3$ $I_{base} = 0.3927 ba^3$	Z_{XX} – base $= 0.2587 ba^2$ Z_{XX} – crown $= 0.1907 ba^2$ $Z_{YY} = 0.3927 ab^2$
¼ ELLIPSE		$A = 0.7854 ab$	$e_x = 0.424a$ $e_y = 0.424b$	$I_{XX} = 0.0549 ba^3$ $I_{YY} = 0.0549 ab^3$ $I_{b_1 a_1} = 0.1963 ba^3$ $I_{b_1 c_1} = 0.1963 ab^3$	Z_{XX} – base $= 0.1293 ba^2$ Z_{XX} – crown $= 0.0953 ba^2$ Z_{YY} – base $= 0.1293 ab^2$ Z_{YY} – crown $= 0.0953 ab^2$
COMPLEMENT		$A = 0.2146 ab$	$e_x = 0.777a$ $e_y = 0.777b$	$I_{XX} = 0.0076 ba^3$ $I_{YY} = 0.0076 ab^3$	Z_{XX} – base $= 0.0338 ba^2$ Z_{XX} – apex $= 0.0097 ba^2$ Z_{YY} – base $= 0.0338 ab^2$ Z_{YY} – apex $= 0.0097 ab^2$
FULL PARABOLA		$A = \dfrac{4ab}{3}$	$e_x = \dfrac{2a}{5}$ $e_y = b$	$I_{XX} = 0.0914 ba^3$ $I_{YY} = 0.2666 ab^3$ $I_{base} = 0.3048 ba^3$	Z_{XX} – base $= 0.2286 ba^2$ Z_{XX} – crown $= 0.1524 ba^2$ $Z_{YY} = 0.2666 ab^2$
SEMI-PARABOLA		$A = \dfrac{2ab}{3}$	$e_x = \dfrac{2a}{5}$ $e_y = \dfrac{3b}{8}$	$I_{XX} = 0.0457 ba^3$ $I_{YY} = 0.0396 ab^3$ $I_{b_1 a_1} = 0.1524 ba^3$ $I_{b_1 c_1} = 0.1333 ab^3$	Z_{XX} – base $= 0.1143 ba^2$ Z_{XX} – crown $= 0.076 ba^2$ Z_{YY} – base $= 0.1055 ab^2$ Z_{YY} – crown $= 0.0633 ab^2$
COMPLEMENT		$A = \dfrac{ab}{3}$	$e_x = \dfrac{7a}{10}$ $e_y = \dfrac{3b}{4}$	$I_{XX} = 0.0176 ba^3$ $I_{YY} = 0.0125 ab^3$ $I_{a_1 b_1} = 0.181 ba^3$ $I_{b_1 c_1} = 0.2 ab^3$	Z_{XX} – base $= 0.0587 ba^2$ Z_{XX} – apex $= 0.0252 ba^2$ Z_{YY} – base $= 0.05 ab^2$ Z_{YY} – apex $= 0.0167 ab^2$
FILLET		$A = \dfrac{s^2}{6}$	$e_u = e_v = \dfrac{4s}{5}$	$I_{UU} = I_{VV} = 0.00524 s^4$ $I_{ab} = 0.1119 a^4$	$Z_{UU} = Z_{VV}$ base $= 0.0262 a^3$ apex $= 0.0066 a^3$

PLASTIC MODULUS OF TWO FLANGES

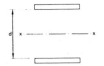

Dist d mm	Plastic Modulus Sxx(cm³) For Thickness t(mm)												
	15	20	25	30	35	40	45	50	55	60	65	70	75
1000	15.2	20.4	25.6	30.9	36.2	41.6	47.0	52.5	58.0	63.6	69.2	74.9	80.6
1100	16.7	22.4	28.1	33.9	39.7	45.6	51.5	57.5	63.5	69.6	75.7	81.9	88.1
1200	18.2	24.4	30.6	36.9	43.2	49.6	56.0	62.5	69.0	75.6	82.2	88.9	95.6
1300	19.7	26.4	33.1	39.9	46.7	53.6	60.5	67.5	74.5	81.6	88.7	95.9	103
1400	21.2	28.4	35.6	42.9	50.2	57.6	65.0	72.5	80.0	87.6	95.2	103	111
1500	22.7	30.4	38.1	45.9	53.7	61.6	69.5	77.5	85.5	93.6	102	110	118
1600	24.2	32.4	40.6	48.9	57.2	65.6	74.0	82.5	91.0	99.6	108	117	126
1700	25.7	34.4	43.1	51.9	60.7	69.6	78.5	87.5	96.5	106	115	124	133
1800	27.2	36.4	45.6	54.9	64.2	73.6	83.0	92.5	102	112	121	131	141
1900	28.7	38.4	48.1	57.9	67.7	77.6	87.5	97.5	108	118	128	138	148
2000	30.2	40.4	50.6	60.9	71.2	81.6	92.0	102	113	124	134	145	156
2100	31.7	42.4	53.1	63.9	74.7	85.6	96.5	107	119	130	141	152	163
2200	33.2	44.4	55.6	66.9	78.2	89.6	101	112	124	136	147	159	171
2300	34.7	46.4	58.1	69.9	81.7	93.6	106	117	130	142	154	166	178
2400	36.2	48.4	60.6	72.9	85.2	97.6	110	122	135	148	160	173	186
2500	37.7	50.4	63.1	75.9	88.7	102	115	127	141	154	167	180	193
2600	39.2	52.4	65.6	78.9	92.2	106	119	132	146	160	173	187	201
2700	40.7	54.4	68.1	81.9	95.7	110	124	137	152	166	180	194	208
2800	42.2	56.4	70.6	84.9	99.2	114	128	142	157	172	186	201	216
2900	43.7	58.4	73.1	87.9	103	118	133	147	163	178	193	208	223
3000	45.2	60.4	75.6	90.9	106	122	137	152	168	184	199	215	231
3100	46.7	62.4	78.1	93.9	110	126	142	157	174	190	206	222	238
3200	48.2	64.4	80.6	96.9	113	130	146	162	179	196	212	229	246
3300	49.7	66.4	83.1	99.9	117	134	151	167	185	202	219	236	253
3400	51.2	68.4	85.6	103	120	138	155	172	190	208	225	243	261
3500	52.7	70.4	88.1	106	124	142	160	177	196	214	232	250	268
3600	54.2	72.4	90.6	109	127	146	164	182	201	220	238	257	276
3700	55.7	74.4	93.1	112	131	150	169	187	207	226	245	264	283
3800	57.2	76.4	95.6	115	134	154	173	192	212	232	251	271	291
3900	58.7	78.4	98.1	118	138	158	178	197	218	238	258	278	298
4000	60.2	80.4	101	121	141	162	182	202	223	244	264	285	306
4100	61.7	82.4	103	124	145	166	187	207	229	250	271	292	313
4200	63.2	84.4	106	127	148	170	191	212	234	256	277	299	321
4300	64.7	86.4	108	130	152	174	196	217	240	262	284	306	328
4400	66.2	88.4	111	133	155	178	200	222	245	268	290	313	336
4500	67.7	90.4	113	136	159	182	205	227	251	274	297	320	343
4600	69.2	92.4	116	139	162	186	209	232	256	280	303	327	351
4700	70.7	94.4	118	142	166	190	214	237	262	286	310	334	358
4800	72.2	96.4	121	145	169	194	218	242	267	292	316	341	366
4900	73.7	98.4	123	148	173	198	223	247	273	298	323	348	373
5000	75.2	100.0	126	151	176	202	227	252	278	304	329	355	381

PLASTIC MODULUS OF RECTANGLES

Depth d mm	Plastic Modulus Sxx(cm^3) For Thickness t(mm)									
	5	6	7	8	9	10	12.5	15	20	25
25	0.78	0.93	1.09	1.25	1.41	1.56	1.95	2.34	3.13	3.91
50	3.13	3.75	4.37	5.00	5.62	6.25	7.81	9.37	12.5	15.6
75	7.03	8.44	9.84	11.3	12.7	14.1	17.6	21.1	28.1	35.2
100	12.5	15.0	17.5	20.0	22.5	25.0	31.2	37.5	50.0	62.5
125	19.5	23.4	27.3	31.2	35.2	39.1	48.8	58.6	78.1	97.7
150	28.1	33.8	39.4	45.0	50.6	56.2	70.3	84.4	112	141
175	38.3	45.9	53.6	61.2	68.9	76.6	95.7	115	153	191
200	50.0	60.0	70.0	80.0	90.0	100.0	125	150	200	250
225	63.3	75.9	88.6	101	114	127	158	190	253	316
250	78.1	93.7	109	125	141	156	195	234	312	391
275	94.5	113	132	151	170	189	236	284	378	473
300	112	135	158	180	203	225	281	338	450	563
325	132	158	185	211	238	264	330	396	528	660
350	153	184	214	245	276	306	383	459	613	766
375	176	211	246	281	316	352	439	527	703	879
400	200	240	280	320	360	400	500	600	800	1000
425	226	271	316	361	406	452	564	677	903	1130
450	253	304	354	405	456	506	633	759	1010	1270
475	282	338	395	451	508	564	705	846	1130	1410
500	312	375	437	500	562	625	781	937	1250	1560
525	345	413	482	551	620	689	861	1030	1380	1720
550	378	454	529	605	681	756	945	1130	1510	1890
575	413	496	579	661	744	827	1030	1240	1650	2070
600	450	540	630	720	810	900	1130	1350	1800	2250
625	488	586	684	781	879	977	1220	1460	1950	2440
650	528	634	739	845	951	1060	1320	1580	2110	2640
675	570	683	797	911	1030	1140	1420	1710	2280	2850
700	613	735	858	980	1100	1230	1530	1840	2450	3060
725	657	788	920	1050	1180	1310	1640	1970	2630	3290
750	703	844	984	1120	1270	1410	1760	2110	2810	3520
775	751	901	1050	1200	1350	1500	1880	2250	3000	3750
800	800	960	1120	1280	1440	1600	2000	2400	3200	4000
825	851	1020	1190	1360	1530	1700	2130	2550	3400	4250
850	903	1080	1260	1440	1630	1810	2260	2710	3610	4520
875	957	1150	1340	1530	1720	1910	2390	2870	3830	4790
900	1010	1210	1420	1620	1820	2020	2530	3040	4050	5060

PLASTIC MODULUS OF RECTANGLES

Depth d mm	Plastic Modulus Sxx(cm^3) For Thickness t(mm)									
	5	6	7	8	9	10	12.5	15	20	25
1000	1250	1500	1750	2000	2250	2500	3120	3750	5000	6250
1100	1510	1810	2120	2420	2720	3020	3780	4540	6050	7560
1200	1800	2160	2520	2880	3240	3600	4500	5400	7200	9000
1300	2110	2530	2960	3380	3800	4220	5280	6340	8450	10600
1400	2450	2940	3430	3920	4410	4900	6130	7350	9800	12300
1500	2810	3370	3940	4500	5060	5620	7030	8440	11200	14100
1600	3200	3840	4480	5120	5760	6400	8000	9600	12800	16000
1700	3610	4330	5060	5780	6500	7220	9030	10800	14400	18100
1800	4050	4860	5670	6480	7290	8100	10100	12100	16200	20200
1900	4510	5410	6320	7220	8120	9020	11300	13500	18000	22600
2000	5000	6000	7000	8000	9000	10000	12500	15000	20000	25000
2100	5510	6620	7720	8820	9920	11000	13800	16500	22100	27600
2200	6050	7260	8470	9680	10900	12100	15100	18100	24200	30200
2300	6610	7930	9260	10600	11900	13200	16500	19800	26500	33100
2400	7200	8640	10100	11500	13000	14400	18000	21600	28800	36000
2500	7810	9370	10900	12500	14100	15600	19500	23400	31200	39100
2600	8450	10100	11800	13500	15200	16900	21100	25400	33800	42200
2700	9110	10900	12800	14600	16400	18200	22800	27300	36400	45600
2800	9800	11800	13700	15700	17600	19600	24500	29400	39200	49000
2900	10500	12600	14700	16800	18900	21000	26300	31500	42000	52600
3000	11200	13500	15700	18000	20200	22500	28100	33700	45000	56200
3100	12000	14400	16800	19200	21600	24000	30000	36000	48000	60100
3200	12800	15400	17900	20500	23000	25600	32000	38400	51200	64000
3300	13600	16300	19100	21800	24500	27200	34000	40800	54400	68100
3400	14400	17300	20200	23100	26000	28900	36100	43300	57800	72200
3500	15300	18400	21400	24500	27600	30600	38300	45900	61200	76600
3600	16200	19400	22700	25900	29200	32400	40500	48600	64800	81000
3700	17100	20500	24000	27400	30800	34200	42800	51300	68400	85600
3800	18000	21700	25300	28900	32500	36100	45100	54100	72200	90200
3900	19000	22800	26600	30400	34200	38000	47500	57000	76000	95100
4000	20000	24000	28000	32000	36000	40000	50000	60000	80000	100000
4100	21000	25200	29400	33600	37800	42000	52500	63000	84000	105000
4200	22100	26500	30900	35300	39700	44100	55100	66200	88200	110000
4300	23100	27700	32400	37000	41600	46200	57800	69300	92400	116000
4400	24200	29000	33900	38700	43600	48400	60500	72600	96800	121000
4500	25300	30400	35400	40500	45600	50600	63300	75900	101000	127000
4600	26500	31700	37000	42300	47600	52900	66100	79300	106000	132000
4700	27600	33100	38700	44200	49700	55200	69000	82800	110000	138000
4800	28800	34600	40300	46100	51800	57600	72000	86400	115000	144000
4900	30000	36000	42000	48000	54000	60000	75000	90000	120000	150000
5000	31200	37500	43700	50000	56200	62500	78100	93700	125000	156000

Frame I

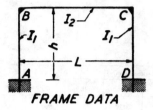

FRAME DATA

Coefficients:

$$k = \frac{I_2}{I_1} \cdot \frac{h}{L}$$

$$N_1 = k + 2 \qquad N_2 = 6k + 1$$

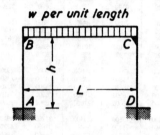

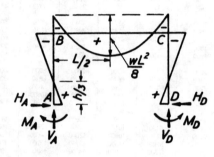

$$M_A = M_D = \frac{wL^2}{12N_1} \qquad M_B = M_C = -\frac{wL^2}{6N_1} = -2M_A$$

$$M_{max} = \frac{wL^2}{8} + M_B \qquad V_A = V_D = \frac{wL}{2} \qquad H_A = H_D = \frac{3M_A}{h}$$

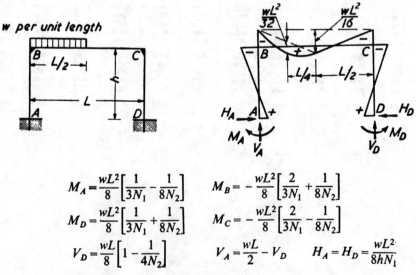

$$M_A = \frac{wL^2}{8}\left[\frac{1}{3N_1} - \frac{1}{8N_2}\right] \qquad M_B = -\frac{wL^2}{8}\left[\frac{2}{3N_1} + \frac{1}{8N_2}\right]$$

$$M_D = \frac{wL^2}{8}\left[\frac{1}{3N_1} + \frac{1}{8N_2}\right] \qquad M_C = -\frac{wL^2}{8}\left[\frac{2}{3N_1} - \frac{1}{8N_2}\right]$$

$$V_D = \frac{wL}{8}\left[1 - \frac{1}{4N_2}\right] \qquad V_A = \frac{wL}{2} - V_D \qquad H_A = H_D = \frac{wL^2}{8hN_1}$$

Extract: 'Kleinlogel, Rahmenformeln' 11. Auflage Berlin—Verlag von Wilhelm Ernst & Sohn.

Formulae for rigid frames

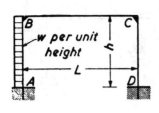 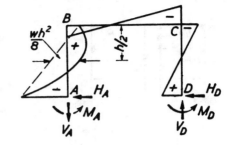

$$M_A = \frac{wh^2}{4}\left[-\frac{k+3}{6N_1} - \frac{4k+1}{N_2}\right] \qquad M_B = \frac{wh^2}{4}\left[-\frac{k}{6N_1} + \frac{2k}{N_2}\right]$$

$$M_D = \frac{wh^2}{4}\left[-\frac{k+3}{6N_1} + \frac{4k+1}{N_2}\right] \qquad M_C = \frac{wh^2}{4}\left[-\frac{k}{6N_1} - \frac{2k}{N_2}\right]$$

$$H_D = \frac{wh(2k+3)}{8N_1} \qquad H_A = -(wh - H_D) \qquad V_A = -V_D = -\frac{wh^2 k}{LN_2}$$

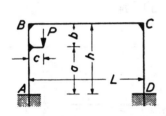 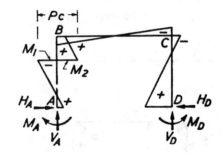

Constants: $a_1 = \frac{a}{h} \qquad b_1 = \frac{b}{h}$

$$X_1 = \frac{Pc}{2N_1}[1 + 2b_1 k - 3b_1^2(k+1)] \qquad X_2 = \frac{Pcka_1(3a_1 - 2)}{2N_1}$$

$$X_3 = \frac{3Pcka_1}{N_2}$$

$$M_A = +X_1 - \left(\frac{Pc}{2} - X_3\right) \qquad M_B = +X_2 + X_3$$

$$M_D = +X_1 + \left(\frac{Pc}{2} - X_3\right) \qquad M_C = +X_2 - X_3$$

$$H_A = H_D = \frac{Pc}{2h} + \frac{X_1 - X_2}{h} \qquad V_D = \frac{2X_3}{L} \qquad V_A = P - V_D$$

$$M_1 = M_A - H_A a \qquad M_2 = M_B + H_D b$$

Extract: 'Kleinlogel, Rahmenformeln' 11. Auflage Berlin—Verlag von Wilhelm Ernst & Sohn.

Formulae for rigid frames

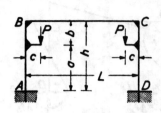 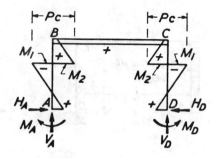

$$\text{Constants: } a_1 = \frac{a}{h} \quad b_1 = \frac{b}{h}$$

$$X_1 = \frac{Pc}{2N_1}[1 + 2b_1 k - 3b_1^2(k+1)] \quad X_2 = \frac{Pcka_1(3a_1 - 2)}{2N_1}$$

$$M_A = M_D = \frac{Pc}{N_1}[1 + 2b_1 k - 3b_1^2(k+1)] = 2X_1$$

$$M_B = M_C = \frac{Pcka_1(3a_1 - 2)}{N_1} = 2X_2$$

$$V_A = V_D = P \quad H_A = H_D = \frac{Pc + M_A - M_B}{h}$$

$$M_1 = M_A - H_A a \quad M_2 = M_B + H_D b$$

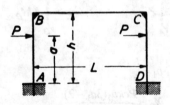 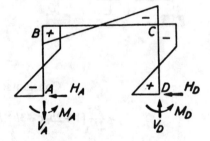

$$\text{Constants: } a_1 = \frac{a}{h} \quad X_1 = \frac{3Paa_1 k}{N_2}$$

$$M_A = -Pa + X_1 \quad M_B = X_1$$

$$M_D = +Pa - X_1 \quad M_C = -X_1$$

$$V_A = -V_D = -\frac{2X_1}{L} \quad H_A = -H_D = -P$$

Extract: 'Kleinlogel, Rahmenformeln' 11. Auflage Berlin—Verlag von Wilhelm Ernst & Sohn.

Formulae for rigid frames

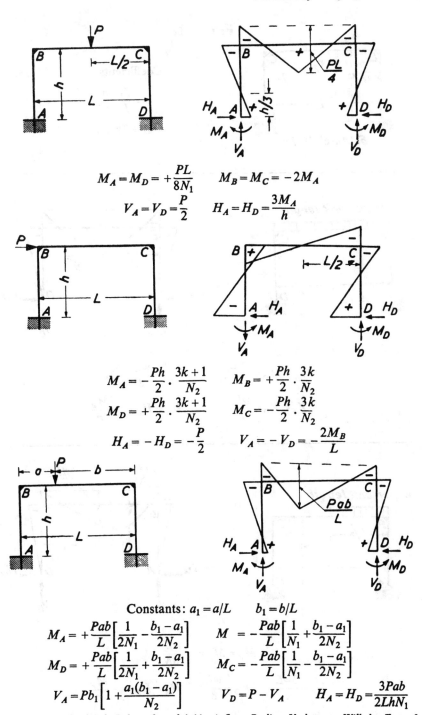

$$M_A = M_D = +\frac{PL}{8N_1} \qquad M_B = M_C = -2M_A$$

$$V_A = V_D = \frac{P}{2} \qquad H_A = H_D = \frac{3M_A}{h}$$

$$M_A = -\frac{Ph}{2} \cdot \frac{3k+1}{N_2} \qquad M_B = +\frac{Ph}{2} \cdot \frac{3k}{N_2}$$

$$M_D = +\frac{Ph}{2} \cdot \frac{3k+1}{N_2} \qquad M_C = -\frac{Ph}{2} \cdot \frac{3k}{N_2}$$

$$H_A = -H_D = -\frac{P}{2} \qquad V_A = -V_D = -\frac{2M_B}{L}$$

Constants: $a_1 = a/L \qquad b_1 = b/L$

$$M_A = +\frac{Pab}{L}\left[\frac{1}{2N_1} - \frac{b_1 - a_1}{2N_2}\right] \qquad M = -\frac{Pab}{L}\left[\frac{1}{N_1} + \frac{b_1 - a_1}{2N_2}\right]$$

$$M_D = +\frac{Pab}{L}\left[\frac{1}{2N_1} + \frac{b_1 - a_1}{2N_2}\right] \qquad M_C = -\frac{Pab}{L}\left[\frac{1}{N_1} - \frac{b_1 - a_1}{2N_2}\right]$$

$$V_A = Pb_1\left[1 + \frac{a_1(b_1 - a_1)}{N_2}\right] \qquad V_D = P - V_A \qquad H_A = H_D = \frac{3Pab}{2LhN_1}$$

Extract: 'Kleinlogel, Rahmenformeln' 11. Auflage Berlin—Verlag von Wilhelm Ernst & Sohn.

Frame II

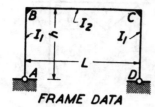

FRAME DATA

Coefficients:

$$k = \frac{I_2}{I_1} \cdot \frac{h}{L}$$

$$N = 2k + 3$$

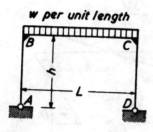

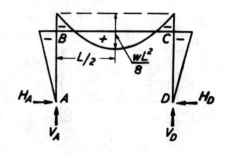

$$M_B = M_C = -\frac{wL^2}{4N} \qquad M_{\max} = \frac{wL^2}{8} + M_B$$

$$V_A = V_D = \frac{wL}{2} \qquad H_A = H_D = -\frac{M_B}{h}$$

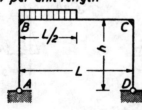

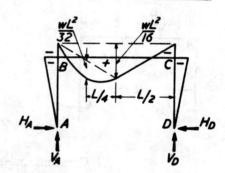

$$M_B = M_C = -\frac{wL^2}{8N}$$

$$V_A = \frac{3wL}{8} \qquad V_D = \frac{wL}{8} \qquad H_A = H_D = -\frac{M_B}{h}$$

Extract: 'Kleinlogel, Rahmenformeln' 11. Auflage Berlin—Verlag von Wilhelm Ernst & Sohn.

Formulae for rigid frames

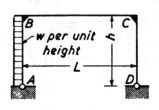 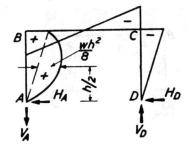

$$M_B = \frac{wh^2}{4}\left[-\frac{k}{2N} + 1\right] \qquad H_D = -\frac{M_C}{h}$$

$$M_C = \frac{wh^2}{4}\left[-\frac{k}{2N} - 1\right] \qquad H_A = -(wh - H_D)$$

$$V_A = -V_D = -\frac{wh^2}{2L}$$

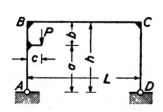 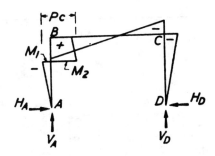

$$\text{Constant: } a_1 = \frac{a}{h}$$

$$M_B = \frac{Pc}{2}\left[\frac{(3a_1^2 - 1)k}{N} + 1\right]$$

$$M_C = \frac{Pc}{2}\left[\frac{(3a_1^2 - 1)k}{N} - 1\right] \qquad H_A = H_D = -\frac{M_C}{h}$$

$$V_D = \frac{Pc}{L} \qquad V_A = P - V_D$$

$$M_1 = -H_A a \qquad M_2 = Pc - H_A a$$

Extract: 'Kleinlogel, Rahmenformeln' 11. Auflage Berlin—Verlag von Wilhelm Ernst & Sohn.

1086 Formulae for rigid frames

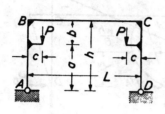 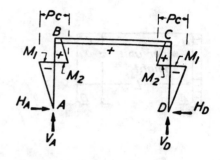

Constant: $a_1 = \dfrac{a}{h}$

$$M_B = M_C = \dfrac{Pc(3a_1^2 - 1)k}{N}$$

$$H_A = H_D = \dfrac{Pc - M_B}{h} \qquad V_A = V_D = P$$

$$M_1 = -H_A a \qquad M_2 = Pc - H_A a$$

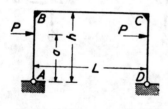 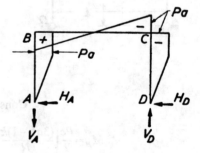

$$M_B = -M_C = Pa \qquad H_A = H_D = P$$

$$V_A = -V_D = -\dfrac{2Pa}{L}$$

Moment at loads $= \pm Pa$

Extract: 'Kleinlogel, Rahmenformeln' 11. Auflage Berlin—Verlag von Wilhelm Ernst & Sohn.

Formulae for rigid frames

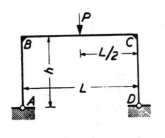 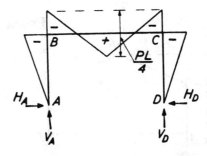

$$M_B = M_C = -\frac{3PL}{8N} \quad V_A = V_D = \frac{P}{2} \quad H_A = H_D = -\frac{M_B}{h}$$

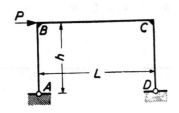 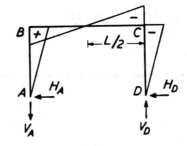

$$M_B = -M_C = +\frac{Ph}{2}$$

$$V_A = -V_D = -\frac{Ph}{L} \quad H_A = -H_D = -\frac{P}{2}$$

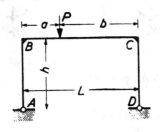 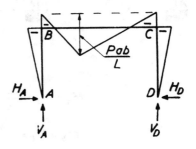

$$M_B = M_C = -\frac{Pab}{L} \cdot \frac{3}{2N}$$

$$V_A = \frac{Pb}{L} \quad V_D = \frac{Pa}{L} \quad H_A = H_D = -\frac{M_B}{h}$$

Extract: 'Kleinlogel, Rahmenformeln' 11. Auflage Berlin—Verlag von Wilhelm Ernst & Sohn.

Frame III

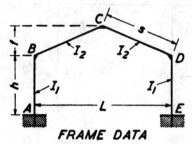

FRAME DATA

Coefficients:
$$k = \frac{I_2}{I_1} \cdot \frac{h}{s} \qquad \phi = \frac{f}{h}$$
$$m = 1 + \phi$$
$$B = 3k + 2 \qquad C = 1 + 2m$$

$$K_1 = 2(k + 1 + m + m^2) \qquad K_2 = 2(k + \phi^2)$$
$$R = \phi C - k \qquad N_1 = K_1 K_2 - R^2 \qquad N_2 = 3k + B$$

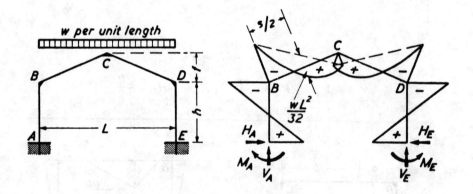

$$M_A = M_E = \frac{wL^2}{16} \cdot \frac{k(8 + 15\phi) + \phi(6 - \phi)}{N_1}$$

$$M_B = M_D = -\frac{wL^2}{16} \cdot \frac{k(16 + 15\phi) + \phi^2}{N_1}$$

$$M_C = \frac{wL^2}{8} - \phi M_A + m M_B$$

$$V_A = V_E = \frac{wL}{2} \qquad H_A = H_E = \frac{M_A - M_B}{h}$$

Extract: 'Kleinlogel, Rahmenformeln' 11. Auflage Berlin—Verlag von Wilhelm Ernst & Sohn.

Formulae for rigid frames

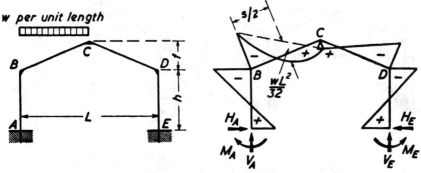

Constants: $*X_1 = \dfrac{wL^2}{32} \cdot \dfrac{k(8+15\phi)+\phi(6-\phi)}{N_1}$

$*X_2 = \dfrac{wL^2}{32} \cdot \dfrac{k(16+15\phi)+\phi^2}{N_1}$ $X_3 = \dfrac{wL^2}{32N_2}$

$M_A = +X_1 - X_3$ $M_B = -X_2 - X_3$ $M_E = +X_1 + X_3$ $M_D = -X_2 + X_3$

$*M_C = \dfrac{wL^2}{16} - \phi X_1 - mX_2$

$V_E = \dfrac{wL}{8} - \dfrac{2X_3}{L}$ $V_A = \dfrac{wL}{2} - V_E$ $H_A = H_E = \dfrac{X_1 + X_2}{h}$

* Note that X_1, $-X_2$ and M_C are respectively half the values of $M_A (=M_E)$, $M_B (=M_D)$ and M_C from the previous set of formulæ where the whole span was loaded.

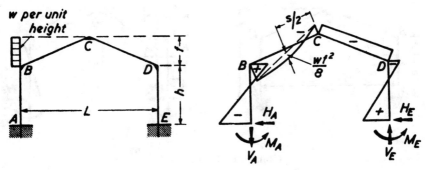

Constants: $X_1 = \dfrac{wf^2}{8} \cdot \dfrac{k(9\phi+4)+\phi(6+\phi)}{N_1}$

$X_2 = \dfrac{wf^2}{8} \cdot \dfrac{k(8+9\phi)-\phi^2}{N_1}$ $X_3 = \dfrac{wfh}{8} \cdot \dfrac{4B+\phi}{N_2}$

$M_A = -X_1 - X_3$ $M_B = +X_2 + \left(\dfrac{wfh}{2} - X_3\right)$

$M_E = -X_1 + X_3$ $M_D = +X_2 - \left(\dfrac{wfh}{2} - X_3\right)$

$M_C = -\dfrac{wf^2}{4} + \phi X_1 + mX_2$

$V_A = -V_E = -\dfrac{wfh(2+\phi)}{2L} + \dfrac{2X_3}{L}$ $H_E = \dfrac{wf}{2} - \dfrac{X_1+X_2}{h}$ $H_A = -(wf - H_E)$

Formulae for rigid frames

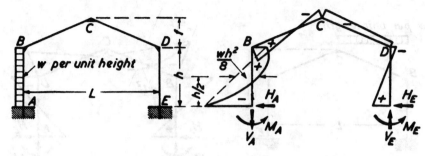

Constants: $X_1 = \dfrac{wh^2}{8} \cdot \dfrac{k(k+6) + k\phi(15+16\phi) + 6\phi^2}{N_1}$

$X_2 = \dfrac{wh^2 k(9\phi + 8\phi^2 - k)}{8N_1}$ $\qquad X_3 = \dfrac{wh^2(2k+1)}{2N_2}$

$M_A = -X_1 - X_3 \qquad M_B = +X_2 + \left(\dfrac{wh^2}{4} - X_3\right)$

$M_E = -X_1 + X_3 \qquad M_D = +X_2 - \left(\dfrac{wh^2}{4} - X_3\right)$

$M_C = -\dfrac{whf}{4} + \phi X_1 + m X_2$

$V_A = -V_E = -\dfrac{wh^2}{2L} + \dfrac{2X_3}{L} \qquad H_E = \dfrac{wh}{4} - \dfrac{X_1 + X_2}{h} \qquad H_A = -(wh - H_E)$

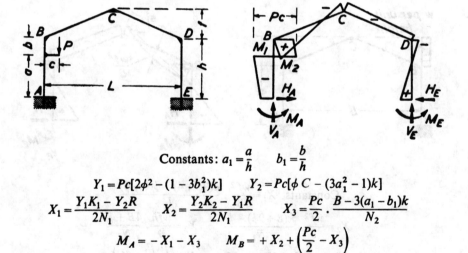

Constants: $a_1 = \dfrac{a}{h} \qquad b_1 = \dfrac{b}{h}$

$Y_1 = Pc[2\phi^2 - (1 - 3b_1^2)k] \qquad Y_2 = Pc[\phi C - (3a_1^2 - 1)k]$

$X_1 = \dfrac{Y_1 K_1 - Y_2 R}{2N_1} \qquad X_2 = \dfrac{Y_2 K_2 - Y_1 R}{2N_1} \qquad X_3 = \dfrac{Pc}{2} \cdot \dfrac{B - 3(a_1 - b_1)k}{N_2}$

$M_A = -X_1 - X_3 \qquad M_B = +X_2 + \left(\dfrac{Pc}{2} - X_3\right)$

$M_E = -X_1 + X_3 \qquad M_D = +X_2 - \left(\dfrac{Pc}{2} - X_3\right) \qquad M_C = -\dfrac{\phi Pc}{2} + \phi X_1 + m X_2$

$M_1 = M_A - H_A a \qquad M_2 = M_B + H_E b$

$V_E = \dfrac{Pc - 2X_3}{L} \qquad V_A = P - V_E \qquad H_A = H_E = \dfrac{Pc}{2h} - \dfrac{X_1 + X_2}{h}$

Extract: 'Kleinlogel, Rahmenformeln' 11. Auflage Berlin—Verlag von Wilhelm Ernst & Sohn.

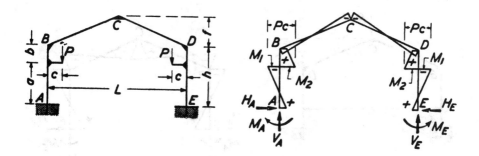

Constants: $a_1 = \dfrac{a}{h}$ $\quad b_1 = \dfrac{b}{h}$

$$Y_1 = Pc[2\phi^2 - (1 - 3b_1^2)k]$$
$$Y_2 = Pc[\phi C + (3a_1^2 - 1)k]$$
$$M_A = M_E = \frac{Y_2 R - Y_1 K_1}{N_1} \qquad M_B = M_D = \frac{Y_2 K_2 - Y_1 R}{N_1}$$
$$M_C = -\phi(Pc + M_A) + mM_B$$
$$V_A = V_D = P \qquad H_A = H_E = \frac{Pc + M_A - M_B}{h}$$
$$M_1 = M_A - H_A a \qquad M_2 = M_B + H_E b$$

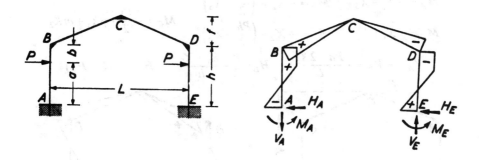

Constant: $X_1 = \dfrac{Pa(B + 3b_1 k)}{N_2}$

$$M_A = -M_E = -X_1 \qquad M_B = -M_D = Pa - X_1 \qquad M_C = 0$$
$$V_A = -V_E = -2\left[\frac{Pa - X_1}{L}\right] \qquad H_A = -H_E = -P$$

Extract: 'Kleinlogel, Rahmenformeln' 11. Auflage Berlin—Verlag von Wilhelm Ernst & Sohn.

Formulae for rigid frames

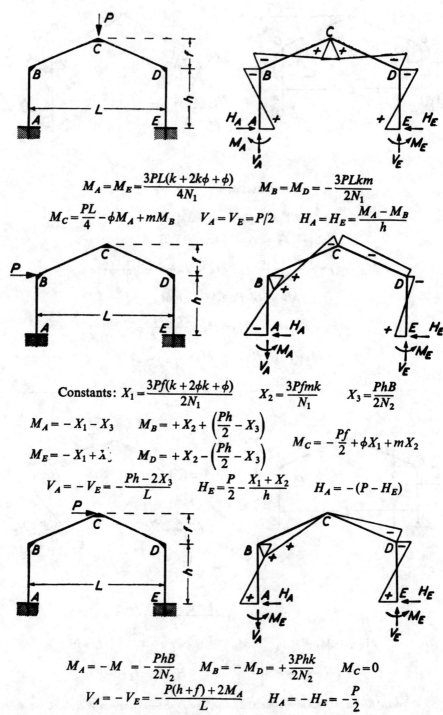

$$M_A = M_E = \frac{3PL(k + 2k\phi + \phi)}{4N_1} \qquad M_B = M_D = -\frac{3PLkm}{2N_1}$$

$$M_C = \frac{PL}{4} - \phi M_A + m M_B \qquad V_A = V_E = P/2 \qquad H_A = H_E = \frac{M_A - M_B}{h}$$

Constants: $X_1 = \dfrac{3Pf(k + 2\phi k + \phi)}{2N_1} \qquad X_2 = \dfrac{3Pfmk}{N_1} \qquad X_3 = \dfrac{PhB}{2N_2}$

$$M_A = -X_1 - X_3 \qquad M_B = +X_2 + \left(\frac{Ph}{2} - X_3\right)$$

$$M_E = -X_1 + X_3 \qquad M_D = +X_2 - \left(\frac{Ph}{2} - X_3\right) \qquad M_C = -\frac{Pf}{2} + \phi X_1 + m X_2$$

$$V_A = -V_E = -\frac{Ph - 2X_3}{L} \qquad H_E = \frac{P}{2} - \frac{X_1 + X_2}{h} \qquad H_A = -(P - H_E)$$

$$M_A = -M_E = -\frac{PhB}{2N_2} \qquad M_B = -M_D = +\frac{3Phk}{2N_2} \qquad M_C = 0$$

$$V_A = -V_E = -\frac{P(h + f) + 2M_A}{L} \qquad H_A = -H_E = -\frac{P}{2}$$

Extract: 'Kleinlogel, Rahmenformeln' 11. Auflage Berlin—Verlag von Wilhelm Ernst & Sohn.

Frame IV

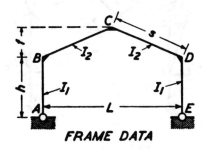

FRAME DATA

Coefficients:

$$k = \frac{I_2}{I_1} \cdot \frac{h}{s}$$

$$\phi = \frac{f}{h}$$

$$m = 1 + \phi$$

$$B = 2(k+1) + m \qquad C = 1 + 2m \qquad N = B + mC$$

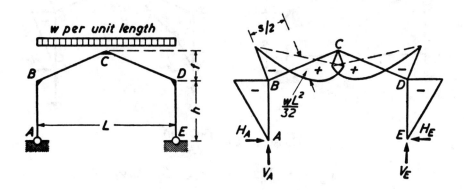

$$M_B = M_D = -\frac{wL^2(3+5m)}{16N} \qquad M_C = \frac{wL^2}{8} + mM_B$$

$$H_A = H_E = -\frac{M_B}{h} \qquad V_A = V_E = \frac{wL}{2}$$

Extract: 'Kleinlogel, Rahmenformeln' 11. Auflage Berlin—Verlag von Wilhelm Ernst & Sohn.

Formulae for rigid frames

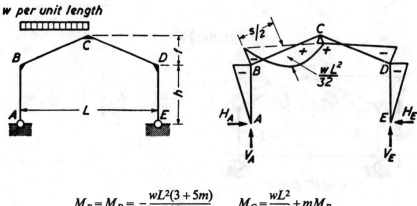

$$M_B = M_D = -\frac{wL^2(3+5m)}{32N} \qquad M_C = \frac{wL^2}{16} + mM_B$$

$$H_A = H_E = -\frac{M_B}{h} \qquad V_A = \frac{3wL}{8} \qquad V_E = \frac{wL}{8}$$

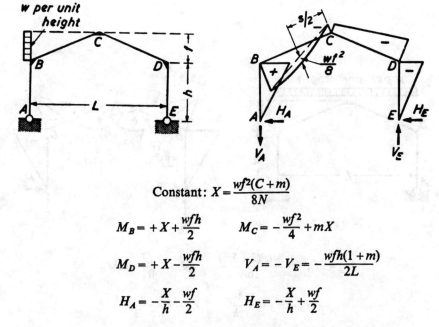

$$\text{Constant: } X = \frac{wf^2(C+m)}{8N}$$

$$M_B = +X + \frac{wfh}{2} \qquad M_C = -\frac{wf^2}{4} + mX$$

$$M_D = +X - \frac{wfh}{2} \qquad V_A = -V_E = -\frac{wfh(1+m)}{2L}$$

$$H_A = -\frac{X}{h} - \frac{wf}{2} \qquad H_E = -\frac{X}{h} + \frac{wf}{2}$$

Extract: 'Kleinlogel, Rahmenformeln' 11. Auflage Berlin—Verlag von Wilhelm Ernst & Sohn.

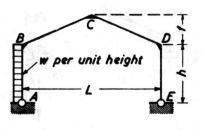
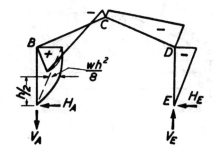

$$M_D = -\frac{wh^2}{8} \cdot \frac{2(B+C)+k}{N} \qquad M_B = \frac{wh^2}{2} + M_D$$

$$M_C = \frac{wh^2}{4} + mM_D$$

$$V_A = -V_E = -\frac{wh^2}{2L} \qquad H_E = -\frac{M_D}{h} \qquad H_A = -(wh - H_E)$$

$$\text{Constants: } a_1 = \frac{a}{h} \qquad X = \frac{Pc}{2} \cdot \frac{B + C - k(3a_1^2 - 1)}{N}$$

$$M_B = Pc - X \qquad M_D = -X \qquad M_C = \frac{Pc}{2} - mX$$

$$M_1 = -a_1 X \qquad M_2 = Pc - a_1 X$$

$$V_E = \frac{Pc}{L} \qquad V_A = P - V_E \qquad H_A = H_E = \frac{X}{h}$$

Extract: 'Kleinlogel, Rahmenformeln' 11. Auflage Berlin—Verlag von Wilhelm Ernst & Sohn.

Formulae for rigid frames

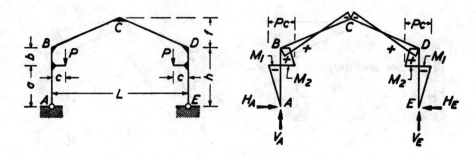

Constant: $a_1 = \dfrac{a}{h}$

$$M_B = M_D = Pc \cdot \dfrac{\phi C + k(3a_1^2 - 1)}{N} \qquad M_C = -\phi Pc + mM_B$$

$$H_A = H_E = \dfrac{Pc - M_B}{h} \qquad V_A = V_E = P$$

$$M_1 = -a_1(Pc - M_B) \qquad M_2 = (1 - a_1)Pc + a_1 M_B$$

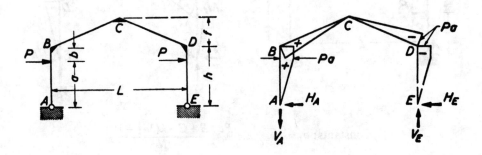

$$M_B = -M_D = Pa \qquad M_C = 0$$

$$H_A = -H_E = -P \qquad V_A = -V_E = -\dfrac{2Pa}{L}$$

Moment at loads $= \pm Pa$

Extract: 'Kleinlogel, Rahmenformeln' 11. Auflage Berlin—Verlag von Wilhelm Ernst & Sohn.

Formulae for rigid frames   1097

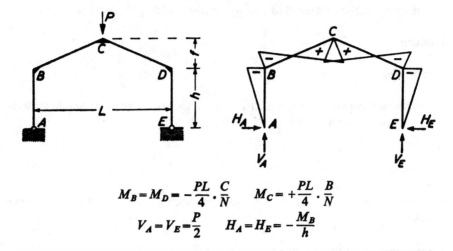

$$M_B = M_D = -\frac{PL}{4} \cdot \frac{C}{N} \qquad M_C = +\frac{PL}{4} \cdot \frac{B}{N}$$

$$V_A = V_E = \frac{P}{2} \qquad H_A = H_E = -\frac{M_B}{h}$$

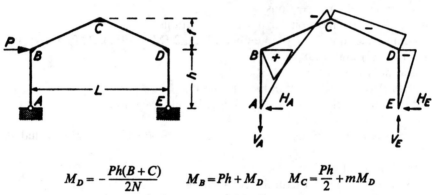

$$M_D = -\frac{Ph(B+C)}{2N} \qquad M_B = Ph + M_D \qquad M_C = \frac{Ph}{2} + mM_D$$

$$V_A = -V_E = -\frac{Ph}{L} \qquad H_E = -\frac{M_D}{h} \qquad H_A = -(P - H_E)$$

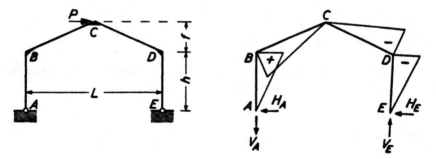

$$M_B = -M_D = +\frac{Ph}{2} \qquad M_C = 0 \qquad V_A = -V_E = -\frac{Phm}{L} \qquad H_A = -H_E = -\frac{P}{2}$$

Extract: 'Kleinlogel, Rahmenformeln' 11. Auflage Berlin—Verlag von Wilhelm Ernst & Sohn.

Explanatory notes on section dimensions and properties

1 General

1.1 Material

The structural components referred to in this handbook are of weldable structural steels to BS 4360: 1990 and BS EN 10 025: 1990.

1.2 Sections

The structural sections referred to in this handbook are produced to the following specifications:

Universal beams, columns, joists and channels:
  BS 4: Structural steel sections
  Part 1: 1980 Specification for hot rolled sections
  Including Amendments 1 (September 1981), 2 (May 1982), 3 (February 1986), 4 (June 1987) and 5 (June 1988)
Structural hollow sections:
  BS 4848: Hot rolled structural steel sections
  Part 2: 1991 Specification for hot-finished hollow sections

Some other hollow sections currently in regular production are also included.

Equal and unequal angles:
  BS 4848: Hot rolled structural steel sections
  Part 4: 1972 Equal and unequal angles
  Including Amendments 1 (February 1974), 2 (October 1982) and 3 (February 1986)

1.3 Dimensional units

The dimensions of sections are given in millimetres (mm). For tolerances on shape, reference should be made to the above standards.

1.4 Property units

Generally the centimetre (cm) is used for the calculated properties but for surface areas and for the warping constant (H), the metre (m) and the decimetre (dm) respectively are used.

NB 1 dm = 0.1 m = 100 mm
    1 dm^6 = 1 × 10^{-6} m^6 = 1 × 10^{12} mm^6

1.5 Mass and force units

The units used are the kilogram (kg), the newton (N) and the metre per second per second (m/s^2) so that 1 N = 1 kg × 1 m/s^2. For convenience a standard value of the acceleration due to gravity has been generally accepted as 9.80665 m/s^2. Thus the force exerted by 1 kg under the action of gravity is 9.80665 N and the force exerted by 1 tonne (1000 kg) is 9.80665 kilonewtons (kN).

2 Dimensions of sections

2.1 Masses

The masses per metre have been calculated assuming that the density of steel is 7850 kg/m^3. The values listed are subject to a rolling margin of $2\frac{1}{2}$%.

In all cases including compound sections the tabulated masses are for the steel section alone and no allowance has been made for connecting material or fittings.

2.2 Ratios for local buckling

The ratios of the flange outstand to thickness (b/T) and the web depth to thickness (d/t) are given for I, H and channel sections. The ratios of the outside diameter to thickness (D/t) are given for circular hollow sections. The ratios d/t and b/t are also given for square and rectangular hollow sections. They have been calculated using the dimensions specified in Figure 3 of BS 5950: Part 1 and are for use when element and section class are being checked to the limits given in Table 7 of BS 5950: Part 1.

2.3 Dimensions for detailing

The dimensions C, N and n have the meanings given in the figures at the head of the tables and have been calculated as follows.

2.3.1 Universal beams, columns and bearing piles

$$N = \left[\frac{B-t}{2}\right] + 10 \text{ mm to the nearest 2 mm above}$$

$$n = \left[\frac{D-d}{2}\right] \text{ mm to the nearest 2 mm above}$$

$$C = \frac{t}{2} + 2 \text{ mm to the nearest mm}$$

2.3.2 Joists

$$N = \left[\frac{B-t}{2}\right] + 6 \text{ mm to the nearest 2 mm above}$$

$$n = \left[\frac{D-d}{2}\right] \text{ mm to the nearest 2 mm above}$$

$$C = \frac{t}{2} + 2 \text{ mm to the nearest mm}$$

NB Flanges of BS 4 joists have an 8° taper.

2.3.3 Channels

$$N = (B - t) + 6 \text{ mm to the nearest 2 mm above}$$

$$n = \left[\frac{D-d}{2}\right] \text{ mm to the nearest 2 mm above}$$

$$C = t + 2 \text{ mm to the nearest mm}$$

NB All channels are rolled with a 5° taper on the inside of the flanges.

2.3.4 Castellated sections

Generally the depth of the castellated section D_c is given by

$$D_c = D + D_s/2$$

where D is the actual depth of the original section, and
D_s is the serial depth of the original section, except as noted below.

The value of D_c has now been calculated from the metric dimensions of D and D_s.

The exceptions to the above are columns in the 356 × 406 series and the column core section for which D_s = 381 mm and 406.4 mm respectively.

3 Section properties

3.1 General

All section properties have been accurately calculated and rounded to three significant figures except those above 1000 which have been rounded to four significant figures. The section properties have been calculated from the metric dimensions given in BS 4 and BS 4848: Parts 2 and 4 as appropriate. In BS 4: Part 1: 1980, the section properties are calculated from the original dimensions in inch units. For angles, BS 4848: Part 4: 1972 repeats the values from ISO 657−1 which are calculated assuming the toe radius equals half the root radius. The section properties in this publication are therefore more accurate, but differ slightly in some cases from those listed in the current version of the relevant British Standards.

3.2 Sections other than hollow sections

3.2.1 Second moment of area (I)

The second moment of area of the sections, often referred to as moment of inertia, has been calculated taking into account all radii and fillets of the sections.

3.2.2 Radius of gyration (r)

The radius of gyration is a parameter used in buckling calculations and is derived from

$$r = \left[\frac{I}{A}\right]^{\frac{1}{2}}$$

where A is the cross-sectional area.

For castellated sections, the radius of gyration is that at the net section as required for use in design to BS 5950.

3.2.3 Elastic modulus (Z)

The elastic modulus, or section modulus, is a parameter used to calculate the moment capacity from the design strength of the stress at the extreme fibre from a known moment. It is derived from

$$Z = \frac{I}{y}$$

where y is the distance to the extreme fibre of the section from the neutral axis.

For castellated sections, the elastic moduli given are those at the net section. The elastic moduli of the tee are calculated at the outer face of the flange and toe of the tee formed at the net section.

For channels, the elastic modulus about the minor $(y-y)$ axis is given at the toe of the section

i.e. $\quad y = B - C_y$

where B is the width of the section, and
$\quad\quad\;\; C_y$ is the distance from the centroid to the outer face of the web.

For angles, the elastic moduli about both axes are given at the toes of the section

i.e. $\quad y_x = A - c_x \quad y_y = B - c_y$

where A is the leg length perpendicular to the $x-x$ axis,
$\quad\quad\;\; B$ is the leg length perpendicular to the $y-y$ axis,
$\quad\quad\;\; c_x$ and c_y are the distance of the centre of gravity from the back of the angle, referred to the $x-x$ axis and $y-y$ axis respectively.

3.2.4 Buckling parameter (u) and torsional index (x)

The buckling parameter and torsional index are parameters used in buckling calculations and have been derived as follows:

(1) For flanged sections symmetrical about the minor axis

$$u = \left[\frac{4S_x^2\,\gamma}{A^2\,h^2}\right]^{\frac{1}{4}}$$

$$x = 0.566h\left[\frac{A}{J}\right]^{\frac{1}{2}}$$

(2) For flanged sections symmetrical about the major axis

$$u = \left[\frac{I_y\,S_x^2\,\gamma}{A^2\,H^2}\right]^{\frac{1}{4}}$$

$$x = 1.132\left[\frac{AH}{I_y J}\right]^{\frac{1}{2}}$$

where S_x is the plastic modulus about the major axis,

$$\gamma = \left[1 - \frac{I_y}{I_x}\right],$$

I_x is the second moment of area about the major axis,
I_y is the second moment of area about the minor axis,
A is the cross-sectional area,

h is the distance between shear centres of flanges (for T sections, h is the distance between the shear centre of the flange and the toe of the web),
H is the warping constant, and
J is the torsion constant.

NB For bisymmetric sections either of the above expressions for u and x may be used.

3.2.5 Warping constant (H) and torsion constant (J)

The warping constants and torsion constants are complex expressions which may be derived from a number of different sources.

3.2.6 Plastic modulus (S)

The full and reduced plastic moduli under axial load about both principal axes are tabulated for sections other than angle sections. Values for angle sections are excluded as BS 5950 requires that they be designed using the elastic modulus.

When a section is loaded to full plasticity by a combination of bending and axial thrust about the major axis the neutral axis shifts and may be in either the web or the tension flange depending on the relative value of bending and thrust. Different formulae giving the reduced plastic modulus under combined loading have to be used. Further details for calculating the reduced plastic modulus can be obtained from *The Plastic Properties of Rolled Sections* published by the British Welding Research Association (now The Welding Institute). They incorporate the term n where

$$n = \frac{F}{A_g \, p_y}$$

where F is the factored axial load,
A_g is the gross cross-sectional area, and
p_y is the design strength of the steel.

For each section there is a 'change' value of n.

If the value of n calculated is less than the change value the neutral axis is in the web and the formula for lower values of n must be used. If n is greater than the change value the neutral axis lies in the tension flange and the formula for higher values of n must be used. The same principles apply to I-sections loaded axially and bent about the minor axis, lower and higher values of n indicating that the neutral axis lies inside or outside the web respectively.

The reduced plastic modulus of channels bending about the minor axis depends on whether the stresses induced by the loading are the same or of opposite kind towards the back of the channel, any moment due to eccentricity of load being assessed relative to the centroidal axis. Where the stresses are of the same kind they cause an initial rise of minor axis moment capacity when lower values of n apply.

3.3 Hollow sections

3.3.1 Common properties

For second moment of area, radius of gyration and elastic modulus see notes 3.2.1, 3.2.2 and 3.2.3.

3.3.2 Torsion constant (J)

For circular hollow sections

$$J = 2I$$

For square and rectangular hollow sections

$$J = \frac{t^3 h}{3} + 2k A_h$$

where I is the second moment of area,
        t is the thickness of section,
        h is the mean perimeter $= 2\,[(B - t) + (D - t)] - 2R_c\,(4 - \pi)$,
        A_h is the area enclosed by mean perimeter $= (B - t)(D - t) - R_c^2\,(4 - \pi)$,
        $k\ =\ \dfrac{2 A_h t}{h}$,
        B is the breadth of section,
        D is the depth of section, and
        R_c is the average of internal and external corner radii.

3.3.3 Torsion modulus constant (C)

For circular hollow sections

$$C = 2Z$$

For square and rectangular hollow sections

$$C = \frac{J}{t + \dfrac{k}{t}}$$

where Z is the elastic modulus and J, t and k are as defined in note 3.3.2.

3.3.4 Plastic modulus of hollow sections

The full plastic modulus only is given in the tables. When a member is subject to axial loads as well as bending stresses, the plastic modulus needs modification to take account of the reduction in plastic moment of resistance.

The following expressions give the values of the reduced plastic moduli to be used.

For circular hollow sections

$$S_r = S \cos\left(\frac{n\pi}{2}\right)$$

For square and rectangular hollow sections,
bending about major axis

$$S_{rx} = S_x - \frac{A^2 n^2}{8t} \quad \text{for } n \leq \frac{2t(D - 2t)}{A}$$

i.e. PNA in the webs

$$S_{rx} = \frac{A^2}{4(B - t)}(1 - n)\left[\frac{2D(B - t)}{A} + n - 1\right] \quad \text{for } n > \frac{2t(D - 2t)}{A}$$

i.e. PNA in the flanges

bending about minor axis

$$S_{ry} = S_y - \frac{A^2 n^2}{8t} \quad \text{for } n \leq \frac{2t(B - 2t)}{A}$$

i.e. PNA in the webs

$$S_{ry} = \frac{A^2}{4(D - t)}(1 - n)\left[\frac{2B(D - t)}{A} + n - 1\right] \quad \text{for } n > \frac{2t(B - 2t)}{A}$$

i.e. PNA in the flanges

where S, S_x, S_y are the full plastic moduli about the relevant axes,

$$n = \frac{F}{A p_y},$$

- F is the factored axial load,
- A is the gross cross-sectional area,
- p_y is the design strength of the steel,
- D, B and t are as defined in note 3.3.2, and
- PNA is plastic neutral axis.

1106 Dimensions and properties

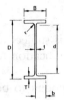

UNIVERSAL BEAMS

DIMENSIONS

Designation		Depth Of Section D	Width Of Section B	Thickness		Root Radius	Depth Between Fillets	Ratios For Local Buckling		Dimensions For Detailing	Notch		Surface Area	
Serial Size	Mass Per Metre			Web t	Flange T	r	d	Flange b/T	Web d/t	End Clearance C	N	n	Per Metre	Per Tonne
mm	kg	mm	mm	mm	mm	mm	mm			mm	mm	mm	m²	m²
914x419	388	920.5	420.5	21.5	36.6	24.1	799.1	5.74	37.2	13	210	62	3.44	8.86
	343	911.4	418.5	19.4	32.0	24.1	799.2	6.54	41.2	12	210	58	3.42	9.96
914x305	289	926.6	307.8	19.6	32.0	19.1	824.4	4.81	42.1	12	156	52	3.01	10.4
	253	918.5	305.5	17.3	27.9	19.1	824.5	5.47	47.7	11	156	48	2.99	11.8
	224	910.3	304.1	15.9	23.9	19.1	824.3	6.36	51.8	10	156	44	2.97	13.3
	201	903.0	303.4	15.2	20.2	19.1	824.4	7.51	54.2	10	156	40	2.96	14.7
838x292	226	850.9	293.8	16.1	26.8	17.8	761.7	5.48	47.3	10	150	46	2.81	12.5
	194	840.7	292.4	14.7	21.7	17.8	761.7	6.74	51.8	9	150	40	2.79	14.4
	176	834.9	291.6	14.0	18.8	17.8	761.7	7.76	54.4	9	150	38	2.78	15.8
762x267	197	769.6	268.0	15.6	25.4	16.5	685.8	5.28	44.0	10	138	42	2.55	13.0
	173	762.0	266.7	14.3	21.6	16.5	685.8	6.17	48.0	9	138	40	2.53	14.6
	147	753.9	265.3	12.9	17.5	16.5	685.9	7.58	53.2	8	138	34	2.51	17.1
686x254	170	692.9	255.8	14.5	23.7	15.2	615.1	5.40	42.4	9	132	40	2.35	13.8
	152	687.6	254.5	13.2	21.0	15.2	615.2	6.06	46.6	9	132	38	2.34	15.4
	140	683.5	253.7	12.4	19.0	15.2	615.1	6.68	49.6	8	132	36	2.33	16.6
	125	677.9	253.0	11.7	16.2	15.2	615.1	7.81	52.6	8	132	32	2.32	18.5
610x305	238	633.0	311.5	18.6	31.4	16.5	537.2	4.96	28.9	11	158	48	2.45	10.3
	179	617.5	307.0	14.1	23.6	16.5	537.3	6.50	38.1	9	158	42	2.41	13.4
	149	609.6	304.8	11.9	19.7	16.5	537.2	7.74	45.1	8	158	38	2.39	16.0
610x229	140	617.0	230.1	13.1	22.1	12.7	547.4	5.21	41.8	9	120	36	2.11	15.0
	125	611.9	229.0	11.9	19.6	12.7	547.3	5.84	46.0	8	120	34	2.09	16.8
	113	607.3	228.2	11.2	17.3	12.7	547.3	6.60	48.9	8	120	30	2.08	18.4
	101	602.2	227.6	10.6	14.8	12.7	547.2	7.69	51.6	7	120	28	2.07	20.5
533x210	122	544.6	211.9	12.8	21.3	12.7	476.6	4.97	37.2	8	110	34	1.89	15.5
	109	539.5	210.7	11.6	18.8	12.7	476.5	5.60	41.1	8	110	32	1.88	17.2
	101	536.7	210.1	10.9	17.4	12.7	476.5	6.04	43.7	7	110	32	1.87	18.5
	92	533.1	209.3	10.2	15.6	12.7	476.5	6.71	46.7	7	110	30	1.86	20.2
	82	528.3	208.7	9.6	13.2	12.7	476.5	7.91	49.6	7	110	26	1.85	22.6
457x191	98	467.4	192.8	11.4	19.6	10.2	407.8	4.92	35.8	8	102	30	1.67	17.0
	89	463.6	192.0	10.6	17.7	10.2	407.8	5.42	38.5	7	102	28	1.66	18.6
	82	460.2	191.3	9.9	16.0	10.2	407.8	5.98	41.2	7	102	28	1.65	20.1
	74	457.2	190.5	9.1	14.5	10.2	407.8	6.57	44.8	7	102	26	1.64	22.2
	67	453.6	189.9	8.5	12.7	10.2	407.8	7.48	48.0	6	102	24	1.63	24.4
457x152	82	465.1	153.5	10.7	18.9	10.2	406.9	4.06	38.0	7	82	30	1.51	18.4
	74	461.3	152.7	9.9	17.0	10.2	406.9	4.49	41.1	7	82	28	1.50	20.2
	67	457.2	151.9	9.1	15.0	10.2	406.8	5.06	44.7	7	82	26	1.49	22.2
	60	454.7	152.9	8.0	13.3	10.2	407.7	5.75	51.0	6	84	24	1.49	24.8
	52	449.8	152.4	7.6	10.9	10.2	407.6	6.99	53.6	6	84	22	1.48	28.4

FOR EXPLANATION OF TABLES SEE NOTE 2

Dimensions and properties

UNIVERSAL BEAMS

PROPERTIES

Designation		Second Moment Of Area		Radius Of Gyration		Elastic Modulus		Plastic Modulus		Buckling Parameter	Torsional Index	Warping Constant	Torsional Constant	Area of Section
Serial Size mm	Mass Per Metre kg	Axis x-x cm⁴	Axis y-y cm⁴	Axis x-x cm	Axis y-y cm	Axis x-x cm³	Axis y-y cm³	Axis x-x cm³	Axis y-y cm³	u	x	H dm⁶	J cm⁴	A cm²
914x419	388	719300	45440	38.1	9.58	15630	2161	17670	3342	0.884	26.7	88.8	1739	495
	343	625200	39160	37.8	9.46	13720	1871	15470	2890	0.883	30.1	75.7	1193	437
914x305	289	504800	15610	37.0	6.50	10900	1015	12590	1603	0.866	31.9	31.2	930	369
	253	436400	13300	36.8	6.42	9503	871	10940	1371	0.866	36.2	26.4	626	323
	224	376300	11240	36.3	6.27	8268	739	9533	1163	0.861	41.3	22.1	422	286
	201	325900	9433	35.6	6.06	7217	622	8372	983	0.853	46.7	18.4	294	257
838x292	226	339700	11360	34.3	6.27	7985	773	9155	1212	0.870	35.0	19.3	514	289
	194	279200	9066	33.6	6.06	6641	620	7640	974	0.862	41.6	15.2	306	247
	176	246000	7791	33.1	5.90	5892	534	6806	841	0.856	46.5	13.0	221	224
762x267	197	239800	8175	30.9	5.71	6232	610	7164	959	0.869	33.2	11.3	404	251
	173	205200	6850	30.5	5.58	5385	514	6195	807	0.864	38.1	9.39	267	220
	147	168800	5462	30.0	5.39	4478	412	5169	648	0.857	45.1	7.40	160	188
686x254	170	170300	6630	28.0	5.53	4916	518	5631	811	0.872	31.8	7.42	308	217
	152	150400	5784	27.8	5.46	4375	455	5001	710	0.871	35.5	6.43	220	194
	140	136300	5183	27.6	5.39	3987	409	4558	638	0.868	38.7	5.72	169	178
	125	118000	4383	27.2	5.24	3481	346	3994	542	0.862	43.9	4.80	116	159
610x305	238	207700	15850	26.1	7.22	6564	1018	7462	1576	0.886	21.1	14.3	790	304
	179	151500	11400	25.8	7.08	4907	742	5515	1143	0.886	27.5	10.0	340	228
	149	124700	9308	25.6	6.99	4093	611	4575	938	0.886	32.5	8.10	201	190
610x229	140	111700	4499	25.0	5.03	3619	391	4139	611	0.875	30.6	3.98	216	178
	125	98500	3932	24.9	4.97	3219	343	3673	535	0.873	34.1	3.45	154	159
	113	87380	3434	24.6	4.88	2878	301	3287	470	0.869	37.9	2.99	112	144
	101	75820	2915	24.2	4.75	2518	256	2887	401	0.863	42.9	2.51	77.6	129
533x210	122	76180	3388	22.1	4.66	2798	320	3203	500	0.876	27.6	2.32	179	156
	109	66800	2939	21.9	4.60	2476	279	2827	435	0.875	30.9	1.99	126	139
	101	61650	2696	21.8	4.57	2297	257	2619	400	0.874	33.1	1.82	102	129
	92	55330	2389	21.7	4.50	2076	228	2366	356	0.871	36.4	1.60	76.3	118
	82	47520	2004	21.3	4.38	1799	192	2058	300	0.864	41.6	1.33	51.5	105
457x191	98	45770	2347	19.1	4.33	1959	243	2234	379	0.881	25.8	1.18	121	125
	89	41140	2093	19.0	4.28	1775	218	2020	339	0.879	28.2	1.04	91.3	114
	82	37090	1871	18.8	4.23	1612	196	1832	304	0.877	30.9	0.923	69.2	105
	74	33430	1674	18.7	4.20	1462	176	1659	273	0.876	33.8	0.820	52.2	95.1
	67	29410	1452	18.5	4.12	1297	153	1472	237	0.872	37.9	0.706	37.1	85.5
457x152	82	36250	1144	18.6	3.31	1559	149	1802	236	0.872	27.3	0.570	89.5	105
	74	32470	1013	18.5	3.26	1408	133	1624	209	0.870	30.0	0.500	66.8	95.1
	67	28600	879	18.3	3.21	1251	116	1442	183	0.867	33.5	0.430	47.6	85.3
	60	25450	795	18.3	3.24	1119	104	1283	163	0.869	37.6	0.387	33.5	75.8
	52	21370	645	17.9	3.11	950	84.6	1096	133	0.859	43.9	0.311	21.4	66.6

FOR EXPLANATION OF TABLES SEE NOTE 3

1108 Dimensions and properties

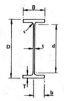

UNIVERSAL BEAMS

DIMENSIONS

Designation		Depth Of Section D	Width Of Section B	Thickness		Root Radius	Depth Between Fillets	Ratios For Local Buckling		Dimensions For Detailing			Surface Area	
Serial Size	Mass Per Metre			Web t	Flange T	r	d	Flange b/T	Web d/t	End Clearance C	Notch N	Notch n	Per Metre	Per Tonne
mm	kg	mm	mm	mm	mm	mm	mm			mm	mm	mm	m²	m²
406x178	74	412.8	179.7	9.7	16.0	10.2	360.4	5.62	37.2	7	96	28	1.51	20.4
	67	409.4	178.8	8.8	14.3	10.2	360.4	6.25	41.0	6	96	26	1.50	22.4
	60	406.4	177.8	7.8	12.8	10.2	360.4	6.95	46.2	6	96	24	1.49	24.8
	54	402.6	177.6	7.6	10.9	10.2	360.4	8.15	47.4	6	96	22	1.48	27.5
406x140	46	402.3	142.4	6.9	11.2	10.2	359.5	6.36	52.1	5	78	22	1.34	29.2
	39	397.3	141.8	6.3	8.6	10.2	359.7	8.24	57.1	5	78	20	1.33	34.1
356x171	67	364.0	173.2	9.1	15.7	10.2	312.2	5.52	34.3	7	94	26	1.39	20.7
	57	358.6	172.1	8.0	13.0	10.2	312.2	6.62	39.0	6	94	24	1.37	24.1
	51	355.6	171.5	7.3	11.5	10.2	312.2	7.46	42.8	6	94	22	1.37	26.8
	45	352.0	171.0	6.9	9.7	10.2	312.2	8.81	45.2	5	94	20	1.36	30.1
356x127	39	352.8	126.0	6.5	10.7	10.2	311.0	5.89	47.8	5	70	22	1.18	30.2
	33	348.5	125.4	5.9	8.5	10.2	311.1	7.38	52.7	5	70	20	1.17	35.4
305x165	54	310.9	166.8	7.7	13.7	8.9	265.7	6.09	34.5	6	90	24	1.26	23.3
	46	307.1	165.7	6.7	11.8	8.9	265.7	7.02	39.7	5	90	22	1.25	27.1
	40	303.8	165.1	6.1	10.2	8.9	265.6	8.09	43.5	5	90	20	1.24	31.0
305x127	48	310.4	125.2	8.9	14.0	8.9	264.6	4.47	29.7	6	70	24	1.09	22.7
	42	306.6	124.3	8.0	12.1	8.9	264.6	5.14	33.1	6	70	22	1.08	25.7
	37	303.8	123.5	7.2	10.7	8.9	264.6	5.77	36.8	6	70	20	1.07	29.0
305x102	33	312.7	102.4	6.6	10.8	7.6	275.9	4.74	41.8	5	58	20	1.01	30.6
	28	308.9	101.9	6.1	8.9	7.6	275.9	5.72	45.2	5	58	18	1.00	35.7
	25	304.8	101.6	5.8	6.8	7.6	276.0	7.47	47.6	5	58	16	0.991	39.7
254x146	43	259.6	147.3	7.3	12.7	7.6	219.0	5.80	30.0	6	82	22	1.08	25.1
	37	256.0	146.4	6.4	10.9	7.6	219.0	6.72	34.2	5	80	20	1.07	29.0
	31	251.5	146.1	6.1	8.6	7.6	219.1	8.49	35.9	5	82	18	1.06	34.3
254x102	28	260.4	102.1	6.4	10.0	7.6	225.2	5.11	35.2	5	58	18	0.903	32.3
	25	257.0	101.9	6.1	8.4	7.6	225.0	6.07	36.9	5	58	16	0.896	35.9
	22	254.0	101.6	5.8	6.8	7.6	225.2	7.47	38.8	5	58	16	0.890	40.4
203x133	30	206.8	133.8	6.3	9.6	7.6	172.4	6.97	27.4	5	74	18	0.923	30.8
	25	203.2	133.4	5.8	7.8	7.6	172.4	8.55	29.7	5	74	16	0.915	36.6
203x102	23	203.2	101.6	5.2	9.3	7.6	169.4	5.46	32.6	5	60	18	0.789	34.3
178x102	19	177.8	101.6	4.7	7.9	7.6	146.8	6.43	31.2	4	60	16	0.740	38.9
152x89	16	152.4	88.9	4.6	7.7	7.6	121.8	5.77	26.5	4	54	16	0.638	39.9
127x76	13	127.0	76.2	4.2	7.6	7.6	96.6	5.01	23.0	4	46	16	0.537	41.3

FOR EXPLANATION OF TABLES SEE NOTE 2

Dimensions and properties

UNIVERSAL BEAMS

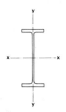

PROPERTIES

Designation		Second Moment Of Area		Radius Of Gyration		Elastic Modulus		Plastic Modulus		Buckling Parameter	Torsional Index	Warping Constant	Torsional Constant	Area of Section
Serial Size mm	Mass Per Metre kg	Axis x-x cm^4	Axis y-y cm^4	Axis x-x cm	Axis y-y cm	Axis x-x cm^3	Axis y-y cm^3	Axis x-x cm^3	Axis y-y cm^3	u	x	H dm^6	J cm^4	A cm^2
406x178	74	27430	1551	17.0	4.03	1329	173	1509	268	0.880	27.5	0.610	63.7	95.3
	67	24330	1365	16.9	3.99	1189	153	1346	237	0.880	30.5	0.533	46.1	85.5
	60	21540	1201	16.8	3.97	1060	135	1195	209	0.881	33.8	0.465	33.0	76.1
	54	18670	1019	16.5	3.86	927	115	1051	178	0.872	38.4	0.391	22.9	68.6
406x140	46	15670	540	16.3	3.03	779	75.9	889	119	0.870	38.8	0.207	19.2	59.0
	39	12410	410	15.9	2.89	625	57.8	718	90.7	0.859	47.6	0.155	10.5	49.2
356x171	67	19540	1362	15.1	3.99	1073	157	1213	243	0.886	24.4	0.413	55.7	85.5
	57	16060	1106	14.9	3.91	896	129	1009	198	0.883	28.9	0.330	33.1	72.2
	51	14160	968	14.8	3.87	796	113	895	174	0.882	32.2	0.287	23.6	64.6
	45	12080	810	14.6	3.77	686	94.7	773	146	0.875	37.0	0.237	15.7	57.0
356x127	39	10100	358	14.3	2.69	573	56.8	654	88.9	0.872	35.2	0.105	14.9	49.4
	33	8192	280	14.0	2.59	470	44.7	539	70.2	0.864	42.3	0.0810	8.65	41.8
305x165	54	11690	1061	13.1	3.94	752	127	843	195	0.891	23.7	0.234	34.3	68.2
	46	9935	896	13.0	3.90	647	108	722	166	0.891	27.2	0.195	22.2	58.8
	40	8551	766	12.9	3.85	563	92.8	626	142	0.888	31.0	0.165	14.9	51.6
305x127	48	9507	460	12.5	2.75	613	73.5	706	116	0.874	23.3	0.101	31.5	60.9
	42	8159	389	12.4	2.70	532	62.6	612	98.4	0.872	26.5	0.0843	21.1	53.4
	37	7162	337	12.3	2.67	471	54.6	540	85.6	0.871	29.6	0.0724	14.9	47.4
305x102	33	6501	194	12.5	2.15	416	37.9	481	60.0	0.866	31.6	0.0442	12.2	41.8
	28	5439	158	12.2	2.08	352	30.9	408	49.2	0.859	36.9	0.0355	7.69	36.4
	25	4364	119	11.8	1.96	286	23.5	336	37.8	0.844	44.1	0.0265	4.57	31.2
254x146	43	6554	677	10.9	3.51	505	92.0	568	141	0.890	21.1	0.103	24.0	55.0
	37	5547	571	10.8	3.47	433	78.0	485	119	0.889	24.3	0.0857	15.4	47.4
	31	4428	448	10.5	3.35	352	61.3	395	94.2	0.879	29.5	0.0660	8.65	39.9
254x102	28	4013	178	10.5	2.21	308	34.9	354	54.8	0.873	27.4	0.0279	9.68	36.3
	25	3420	149	10.3	2.15	266	29.2	307	46.1	0.865	31.3	0.0230	6.52	32.3
	22	2853	119	10.0	2.06	225	23.5	260	37.3	0.854	36.1	0.0182	4.23	28.3
203x133	30	2888	384	8.72	3.18	279	57.4	313	88.0	0.882	21.5	0.0373	10.2	38.0
	25	2349	309	8.54	3.10	231	46.3	259	71.2	0.876	25.5	0.0295	6.05	32.2
203x102	23	2091	163	8.49	2.37	206	32.1	232	49.5	0.890	22.5	0.0153	6.87	29.0
178x102	19	1357	138	7.49	2.39	153	27.2	171	41.9	0.889	22.6	0.00998	4.37	24.2
152x89	16	838	90.4	6.40	2.10	110	20.3	124	31.4	0.889	19.5	0.00473	3.61	20.5
127x76	13	477	56.2	5.33	1.83	75.1	14.7	85.0	22.7	0.893	16.2	0.00200	2.92	16.8

FOR EXPLANATION OF TABLES SEE NOTE 3

UNIVERSAL BEAMS

PLASTIC MODULI-MAJOR AXIS / PLASTIC MODULI-MINOR AXIS

Designation		Plastic Modulus Axis x-x	Reduced Value Of Modulus Under Axial Load					Plastic Modulus Axis y-y	Reduced Value Of Modulus Under Axial Load				
Serial Size	Mass Per Metre		Lower Values Of n		Change Formula	Higher Values Of n			Lower Values Of n		Change Formula	Higher Values Of n	
mm	kg	cm³	K1	K2	At n =	K3	K4	cm³	K1	K2	At n =	K3	K4
914x419	388	17670	17670	28490	0.367	1463	14.6	3342	3342	665	0.400	8096	0.266
	343	15470	15470	24630	0.375	1148	16.3	2890	2890	524	0.404	7192	0.250
914x305	289	12590	12590	17390	0.457	1111	14.4	1603	1603	368	0.492	5153	0.0864
	253	10940	10940	15060	0.462	857	16.3	1371	1371	284	0.492	4498	0.0777
	224	9533	9533	12830	0.479	675	18.3	1163	1163	224	0.507	4085	0.0421
	201	8372	8372	10850	0.510	549	20.1	983	983	183	0.534	3877	-0.0185
838x292	226	9155	9155	12930	0.444	712	16.2	1212	1212	245	0.475	3750	0.112
	194	7640	7640	10360	0.474	525	18.8	974	974	181	0.501	3361	0.0526
	176	6806	6806	8959	0.498	434	20.5	841	841	150	0.522	3173	0.00571
762x267	197	7164	7164	10070	0.447	589	15.4	959	959	204	0.479	2984	0.107
	173	6195	6195	8488	0.466	458	17.3	807	807	159	0.495	2696	0.0694
	147	5169	5169	6845	0.493	336	20.1	648	648	117	0.517	2398	0.0156
686x254	170	5631	5631	8106	0.431	461	15.3	811	811	170	0.463	2398	0.139
	152	5001	5001	7135	0.438	372	16.9	710	710	137	0.468	2159	0.124
	140	4558	4558	6419	0.448	316	18.3	638	638	116	0.475	2009	0.105
	125	3994	3994	5435	0.473	254	20.3	542	542	93.8	0.497	1869	0.0552
610x305	238	7462	7462	12420	0.348	744	11.9	1576	1576	365	0.387	3590	0.304
	179	5515	5515	9189	0.353	424	15.6	1143	1143	210	0.382	2657	0.295
	149	4575	4575	7607	0.356	299	18.4	938	938	148	0.381	2209	0.289
610x229	140	4139	4139	6055	0.421	346	14.9	611	611	129	0.454	1745	0.159
	125	3673	3673	5331	0.427	278	16.5	535	535	104	0.457	1568	0.146
	113	3287	3287	4660	0.443	230	18.1	470	470	85.9	0.471	1456	0.114
	101	2887	2887	3952	0.468	185	20.0	401	401	69.6	0.493	1358	0.0643
533x210	122	3203	3203	4748	0.411	288	13.7	500	500	112	0.447	1382	0.178
	109	2827	2827	4154	0.419	230	15.3	435	435	89.3	0.451	1237	0.164
	101	2619	2619	3829	0.423	200	16.4	400	400	77.8	0.453	1154	0.155
	92	2366	2366	3406	0.433	167	17.8	356	356	65.2	0.461	1067	0.134
	82	2058	2058	2953	0.459	132	19.9	300	300	51.8	0.485	986	0.0818
457x191	98	2234	2234	3442	0.389	204	13.3	379	379	84.0	0.425	977	0.222
	89	2020	2020	3079	0.397	170	14.5	339	339	70.4	0.430	897	0.207
	82	1832	1832	2758	0.405	143	15.8	304	304	59.3	0.436	828	0.191
	74	1659	1659	2485	0.409	119	17.2	273	273	49.5	0.437	755	0.182
	67	1472	1472	2151	0.425	96.9	19.0	237	237	40.3	0.451	693	0.151
457x152	82	1802	1802	2558	0.436	179	12.6	236	236	58.9	0.476	701	0.129
	74	1624	1624	2285	0.444	149	13.8	209	209	49.0	0.480	642	0.113
	67	1442	1442	2001	0.455	120	15.2	183	183	39.8	0.488	583	0.0915
	60	1283	1283	1796	0.451	94.6	17.2	163	163	31.6	0.480	517	0.0991
	52	1096	1096	1461	0.487	73.6	19.4	133	133	24.7	0.513	483	0.0266

$n = F/(A_g . p_y)$. Where F is the factored axial load and A_g is the gross cross sectional area.
For lower values of n reduced plastic modulus$(S_r) = K1 - K2n^2$
For higher values of n reduced plastic modulus$(S_r) = K3(1-n)(K4+n)$
FOR EXPLANATION OF TABLES SEE NOTE 3

Dimensions and properties

UNIVERSAL BEAMS

PLASTIC MODULI-MAJOR AXIS / PLASTIC MODULI-MINOR AXIS

Designation		Plastic Modulus Axis x-x cm³	Reduced Value Of Modulus Under Axial Load					Plastic Modulus Axis y-y cm³	Reduced Value Of Modulus Under Axial Load				
Serial Size mm	Mass Per Metre kg		Lower Values Of n		Change Formula	Higher Values Of n			Lower Values Of n		Change Formula	Higher Values Of n	
			K1	K2	At n =	K3	K4		K1	K2	At n =	K3	K4
406x178	74	1509	1509	2342	0.387	127	14.5	268	268	55.0	0.420	688	0.227
	67	1346	1346	2079	0.391	103	16.0	237	237	44.7	0.421	617	0.219
	60	1195	1195	1857	0.390	82.0	17.9	209	209	35.6	0.416	544	0.222
	54	1051	1051	1546	0.421	66.7	19.7	178	178	29.2	0.446	514	0.158
406x140	46	889	889	1261	0.443	61.7	18.2	119	119	21.6	0.470	367	0.114
	39	718	718	962	0.485	43.4	21.5	90.7	90.7	15.2	0.508	328	0.0305
356x171	67	1213	1213	2010	0.353	106	13.7	243	243	50.3	0.387	563	0.295
	57	1009	1009	1631	0.367	76.3	16.0	198	198	36.4	0.397	482	0.266
	51	895	895	1430	0.375	61.4	17.7	174	174	29.4	0.402	434	0.251
	45	773	773	1178	0.402	48.1	19.9	146	146	23.1	0.426	397	0.198
356x127	39	654	654	939	0.435	49.0	16.8	88.9	88.9	17.3	0.464	267	0.132
	33	539	539	739	0.467	35.4	19.5	70.2	70.2	12.5	0.492	236	0.0676
305x165	54	843	843	1511	0.319	70.0	14.1	195	195	37.4	0.351	412	0.362
	46	722	722	1289	0.323	52.4	16.2	166	166	28.1	0.350	353	0.356
	40	626	626	1093	0.334	40.7	18.3	142	142	22.0	0.359	314	0.333*
305x127	48	706	706	1041	0.412	74.3	11.7	116	116	29.8	0.454	318	0.178
	42	612	612	890	0.422	57.6	13.2	98.4	98.4	23.2	0.460	281	0.157
	37	540	540	782	0.427	45.9	14.7	85.6	85.6	18.5	0.461	250	0.146
305x102	33	481	481	663	0.458	43.0	14.2	60.0	60.0	14.0	0.493	193	0.0845
	28	408	408	543	0.487	32.8	16.1	49.2	49.2	10.7	0.518	176	0.0275
	25	336	336	420	0.540	24.3	18.5	37.8	37.8	7.99	0.567	167	-.0788
254x146	43	568	568	1036	0.310	51.5	12.9	141	141	29.1	0.345	290	0.380
	37	485	485	878	0.316	38.5	14.7	119	119	21.9	0.346	250	0.369
	31	395	395	653	0.357	27.5	17.3	94.2	94.2	15.8	0.384	223	0.286
254x102	28	354	354	515	0.423	32.5	13.5	54.8	54.8	12.7	0.459	157	0.156
	25	307	307	427	0.453	25.8	15.1	46.1	46.1	10.1	0.486	146	0.0956
	22	260	260	344	0.492	19.9	17.0	37.3	37.3	7.8	0.521	137	0.0173
203x133	30	313	313	573	0.310	27.2	13.5	88.0	88.0	17.5	0.343	181	0.381
	25	259	259	447	0.337	19.6	15.7	71.2	71.2	12.7	0.366	158	0.327
203x102	23	232	232	404	0.330	20.9	13.1	49.5	49.5	10.3	0.364	107	0.341
178x102	19	171	171	311	0.314	14.5	13.8	41.9	41.9	8.21	0.346	86.9	0.374
152x89	16	124	124	228	0.306	12.0	12.0	31.4	31.4	6.89	0.342	63.4	0.390
127x76	13	85.0	85.0	167	0.278	9.38	10.4	22.7	22.7	5.54	0.318	42.5	0.447

n = F/(Ag.py). Where F is the factored axial load and Ag is the gross cross sectional area.
For lower values of n reduced plastic modulus(Sr) = K1-K2n²
For higher values of n reduced plastic modulus(Sr) = K3(1-n)(K4+n)
FOR EXPLANATION OF TABLES SEE NOTE 3

Dimensions and properties

UNIVERSAL COLUMNS

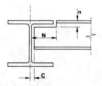

DIMENSIONS

Designation		Depth Of Section D	Width Of Section B	Thickness		Root Radius	Depth Between Fillets	Ratios For Local Buckling		Dimensions For Detailing	Notch		Surface Area	
Serial Size	Mass Per Metre			Web	Flange			Flange	Web	End Clearance			Per Metre	Per Tonne
mm	kg	mm	mm	t mm	T mm	r mm	d mm	b/T	d/t	C mm	N mm	n mm	m²	m²
356x406	634	474.7	424.1	47.6	77.0	15.2	290.3	2.75	6.10	26	200	94	2.52	3.98
	551	455.7	418.5	42.0	67.5	15.2	290.3	3.10	6.91	23	200	84	2.48	4.49
	467	436.6	412.4	35.9	58.0	15.2	290.2	3.56	8.08	20	200	74	2.42	5.19
	393	419.1	407.0	30.6	49.2	15.2	290.3	4.14	9.49	17	200	66	2.38	6.05
	340	406.4	403.0	26.5	42.9	15.2	290.2	4.70	11.0	15	200	60	2.35	6.90
	287	393.7	399.0	22.6	36.5	15.2	290.3	5.47	12.8	13	200	52	2.31	8.06
	235	381.0	395.0	18.5	30.2	15.2	290.2	6.54	15.7	11	200	46	2.28	9.70
COLCORE	477	427.0	424.4	48.0	53.2	15.2	290.2	3.99	6.05	26	200	70	2.43	5.09
356x368	202	374.7	374.4	16.8	27.0	15.2	290.3	6.93	17.3	10	190	44	2.19	10.8
	177	368.3	372.1	14.5	23.8	15.2	290.3	7.82	20.0	9	190	40	2.17	12.3
	153	362.0	370.2	12.6	20.7	15.2	290.2	8.94	23.0	8	190	36	2.15	14.1
	129	355.6	368.3	10.7	17.5	15.2	290.2	10.5	27.1	7	190	34	2.14	16.6
305x305	283	365.3	321.8	26.9	44.1	15.2	246.7	3.65	9.17	15	158	60	1.94	6.85
	240	352.6	317.9	23.0	37.7	15.2	246.8	4.22	10.7	14	158	54	1.90	7.94
	198	339.9	314.1	19.2	31.4	15.2	246.7	5.00	12.8	12	158	48	1.87	9.45
	158	327.2	310.6	15.7	25.0	15.2	246.8	6.21	15.7	10	158	42	1.84	11.6
	137	320.5	308.7	13.8	21.7	15.2	246.7	7.11	17.9	9	158	38	1.82	13.3
	118	314.5	306.8	11.9	18.7	15.2	246.7	8.20	20.7	8	158	34	1.81	15.3
	97	307.8	304.8	9.9	15.4	15.2	246.6	9.90	24.9	7	158	32	1.79	18.4
254x254	167	289.1	264.5	19.2	31.7	12.7	200.3	4.17	10.4	12	134	46	1.58	9.44
	132	276.4	261.0	15.6	25.3	12.7	200.4	5.16	12.8	10	134	38	1.54	11.7
	107	266.7	258.3	13.0	20.5	12.7	200.3	6.30	15.4	9	134	34	1.52	14.2
	89	260.4	255.9	10.5	17.3	12.7	200.4	7.40	19.1	7	134	30	1.50	16.9
	73	254.0	254.0	8.6	14.2	12.7	200.2	8.94	23.3	6	134	28	1.48	20.3
203x203	86	222.3	208.8	13.0	20.5	10.2	160.9	5.09	12.4	9	108	32	1.24	14.4
	71	215.9	206.2	10.3	17.3	10.2	160.9	5.96	15.6	7	108	28	1.22	17.2
	60	209.6	205.2	9.3	14.2	10.2	160.8	7.23	17.3	7	108	26	1.20	20.1
	52	206.2	203.9	8.0	12.5	10.2	160.8	8.16	20.1	6	108	24	1.19	23.0
	46	203.2	203.2	7.3	11.0	10.2	160.8	9.24	22.0	6	108	22	1.19	25.8
152x152	37	161.8	154.4	8.1	11.5	7.6	123.6	6.71	15.3	6	84	20	0.912	24.6
	30	157.5	152.9	6.6	9.4	7.6	123.5	8.13	18.7	5	84	18	0.900	30.0
	23	152.4	152.4	6.1	6.8	7.6	123.6	11.2	20.3	5	84	16	0.889	38.7

FOR EXPLANATION OF TABLES SEE NOTE 2

Dimensions and properties 1113

UNIVERSAL COLUMNS

PROPERTIES

Designation		Second Moment Of Area		Radius Of Gyration		Elastic Modulus		Plastic Modulus		Buckling Parameter	Torsional Index	Warping Constant	Torsional Constant	Area of Section
Serial Size mm	Mass Per Metre kg	Axis x-x cm^4	Axis y-y cm^4	Axis x-x cm	Axis y-y cm	Axis x-x cm^3	Axis y-y cm^3	Axis x-x cm^3	Axis y-y cm^3	u	x	H dm^6	J cm^4	A cm^2
356x406	634	275000	98190	18.5	11.0	11590	4631	14240	7112	0.843	5.46	38.8	13730	808
	551	227000	82670	18.0	10.9	9964	3951	12080	6057	0.841	6.06	31.1	9232	702
	467	183100	67930	17.5	10.7	8388	3295	10010	5040	0.839	6.86	24.3	5817	595
	393	146700	55370	17.1	10.5	7001	2721	8225	4154	0.837	7.87	18.9	3545	501
	340	122500	46850	16.8	10.4	6029	2325	6997	3543	0.836	8.85	15.5	2340	433
	287	99930	38680	16.5	10.3	5077	1939	5814	2949	0.835	10.2	12.3	1441	366
	235	79150	31040	16.2	10.2	4155	1572	4691	2386	0.834	12.1	9.55	813	300
COLCORE	477	172500	68090	16.9	10.6	8078	3209	9704	4981	0.815	6.90	23.8	5705	607
356x368	202	66330	23630	16.0	9.57	3541	1262	3978	1917	0.843	13.4	7.14	561	258
	177	57110	20450	15.9	9.52	3101	1099	3455	1667	0.844	15.0	6.07	382	226
	153	48640	17510	15.8	9.46	2687	946	2970	1433	0.844	17.0	5.10	252	196
	129	40300	14580	15.6	9.39	2266	792	2485	1198	0.843	19.8	4.17	154	165
305x305	283	78800	24540	14.8	8.25	4314	1525	5101	2337	0.855	7.65	6.33	2034	360
	240	64150	20220	14.5	8.14	3639	1272	4243	1945	0.854	8.74	5.01	1270	305
	198	50860	16240	14.2	8.02	2993	1034	3438	1577	0.854	10.2	3.86	735	252
	158	38690	12500	13.9	7.89	2365	805	2675	1225	0.852	12.5	2.85	376	201
	137	32770	10650	13.7	7.82	2045	690	2293	1049	0.851	14.2	2.38	249	174
	118	27610	9006	13.6	7.76	1756	587	1952	892	0.851	16.2	1.97	160	150
	97	22200	7272	13.4	7.68	1443	477	1589	724	0.850	19.3	1.55	91.1	123
254x254	167	29920	9792	11.9	6.79	2070	740	2418	1131	0.852	8.49	1.62	625	212
	132	22550	7506	11.6	6.67	1632	575	1872	877	0.850	10.3	1.18	321	169
	107	17500	5894	11.3	6.57	1312	456	1484	695	0.848	12.4	0.893	173	137
	89	14280	4835	11.2	6.52	1097	378	1225	574	0.849	14.5	0.714	103	114
	73	11370	3880	11.1	6.46	895	306	990	463	0.849	17.3	0.558	57.5	92.9
203x203	86	9461	3114	9.27	5.32	851	298	979	455	0.849	10.2	0.317	138	110
	71	7634	2530	9.16	5.28	707	245	801	373	0.852	11.9	0.249	81.0	90.9
	60	6103	2047	8.96	5.19	582	199	654	303	0.847	14.1	0.195	46.9	76.0
	52	5254	1767	8.90	5.16	510	173	567	263	0.848	15.8	0.166	31.9	66.4
	46	4565	1539	8.81	5.12	449	151	497	230	0.846	17.7	0.142	22.2	58.8
152x152	37	2213	706	6.84	3.87	274	91.5	309	140	0.848	13.3	0.0399	19.3	47.3
	30	1748	560	6.75	3.82	222	73.3	248	112	0.848	16.0	0.0307	10.6	38.4
	23	1258	402	6.51	3.68	165	52.7	184	80.5	0.837	20.5	0.0213	4.82	29.7

FOR EXPLANATION OF TABLES SEE NOTE 3

UNIVERSAL COLUMNS

Designation		Plastic Modulus Axis x-x	Reduced Value Of Modulus Under Axial Load (PLASTIC MODULI-MAJOR AXIS)					Plastic Modulus Axis y-y	Reduced Value Of Modulus Under Axial Load (PLASTIC MODULI-MINOR AXIS)				
Serial Size	Mass Per Metre		Lower Values Of n		Change Formula	Higher Values Of n			Lower Values Of n		Change Formula	Higher Values Of n	
mm	kg	cm³	K1	K2	At n =	K3	K4	cm³	K1	K2	At n =	K3	K4
356x406	634	14240	14240	34270	0.189	3847	3.98	7112	7112	3436	0.280	10520	0.623
	551	12080	12080	29300	0.192	2942	4.43	6057	6057	2701	0.273	9047	0.617
	467	10010	10010	24690	0.193	2150	5.04	5040	5040	2030	0.263	7574	0.615
	393	8225	8225	20470	0.196	1540	5.81	4154	4154	1495	0.256	6301	0.609
	340	6997	6997	17660	0.196	1163	6.56	3543	3543	1152	0.249	5391	0.609
	287	5814	5814	14800	0.198	839	7.58	2949	2949	849	0.243	4517	0.605
	235	4691	4691	12150	0.197	570	9.02	2386	2386	590	0.235	3659	0.606
COLCORE	477	9704	9704	19220	0.253	2175	4.96	4981	4981	2160	0.337	8586	0.495
356x368	202	3978	3978	9908	0.208	446	9.85	1917	1917	444	0.244	3021	0.584
	177	3455	3455	8775	0.206	343	11.1	1667	1667	345	0.237	2613	0.589
	153	2970	2970	7594	0.206	260	12.6	1433	1433	264	0.233	2252	0.588
	129	2485	2485	6376	0.207	186	14.8	1198	1198	192	0.230	1890	0.586
305x305	283	5101	5101	12070	0.206	1010	5.52	2337	2337	889	0.273	3626	0.588
	240	4243	4243	10140	0.208	735	6.33	1945	1945	661	0.266	3040	0.585
	198	3438	3438	8298	0.210	508	7.44	1577	1577	469	0.259	2485	0.580
	158	2675	2675	6421	0.216	326	9.09	1225	1225	308	0.256	1964	0.569
	137	2293	2293	5497	0.219	247	10.3	1049	1049	237	0.254	1696	0.563
	118	1952	1952	4708	0.220	184	11.8	892	892	178	0.250	1447	0.562
	97	1589	1589	3838	0.222	126	14.1	724	724	123	0.247	1183	0.557
254x254	167	2418	2418	5875	0.203	427	6.19	1131	1131	390	0.261	1749	0.594
	132	1872	1872	4560	0.208	273	7.54	877	877	257	0.256	1376	0.584
	107	1484	1484	3590	0.214	181	9.05	695	695	175	0.254	1108	0.573
	89	1225	1225	3074	0.208	127	10.7	574	574	124	0.241	904	0.585
	73	990	990	2510	0.208	85.6	12.8	463	463	85.0	0.235	731	0.584
203x203	86	979	979	2330	0.213	145	7.42	455	455	136	0.263	723	0.574
	71	801	801	2006	0.205	101	8.76	373	373	95.7	0.245	582	0.591
	60	654	654	1554	0.221	70.7	10.3	303	303	68.9	0.256	493	0.559
	52	567	567	1376	0.218	54.3	11.6	263	263	53.4	0.249	425	0.565
	46	497	497	1185	0.224	42.9	12.9	230	230	42.6	0.252	378	0.553
152x152	37	309	309	689	0.237	36.3	9.53	140	140	34.5	0.277	236	0.526
	30	248	248	558	0.238	24.2	11.5	112	112	23.4	0.271	189	0.525
	23	184	184	361	0.284	14.6	14.5	80.5	80.5	14.5	0.313	154	0.432

$n = F/(A_g \cdot p_y)$. Where F is the factored axial load and A_g is the gross cross sectional area.
For lower values of n reduced plastic modulus$(S_r) = K1 - K2n^2$
For higher values of n reduced plastic modulus$(S_r) = K3(1-n)(K4+n)$
FOR EXPLANATION OF TABLES SEE NOTE 3

Dimensions and properties

JOISTS

DIMENSIONS

Designation		Depth Of Section D	Width Of Section B	Thickness		Root Radius r1	Toe Radius r2	Depth Between Fillets d	Ratios For Local Buckling		Dimensions For Detailing			Surface Area	
Serial Size mm	Mass Per Metre kg	mm	mm	Web t mm	Flange T mm	mm	mm	mm	Flange b/T	Web d/t	End Clearance C mm	Notch N mm	Notch n mm	Per Metre m²	Per Tonne m²
254x203	81.85	254.0	203.2	10.2	19.9	19.6	9.7	166.6	5.11	16.3	7	104	44	1.21	14.8
254x114	37.20	254.0	114.3	7.6	12.8	12.4	6.1	199.3	4.46	26.2	6	60	28	0.899	24.2
203x152	52.09	203.2	152.4	8.9	16.5	15.5	7.6	133.2	4.62	15.0	6	78	36	0.932	17.9
152x127	37.20	152.4	127.0	10.4	13.2	13.5	6.6	94.3	4.81	9.07	7	66	30	0.737	19.8
127x114	29.76	127.0	114.3	10.2	11.5	9.9	4.8	79.5	4.97	7.79	7	60	24	0.646	21.7
	26.79	127.0	114.3	7.4	11.4	9.9	5.0	79.5	5.01	10.7	6	60	24	0.650	24.3
114x114	26.79	114.3	114.3	9.5	10.7	14.2	3.2	60.8	5.34	6.40	7	60	28	0.618	23.1
102x102	23.07	101.6	101.6	9.5	10.3	11.1	3.2	55.2	4.93	5.81	7	54	24	0.549	23.8
102x44	7.44♦	101.6	44.5	4.3	6.1	6.9	3.3	74.6	3.65	17.3	4	28	14	0.350	47.0
89x89	19.35	88.9	88.9	9.5	9.9	11.1	3.2	44.2	4.49	4.65	7	46	24	0.476	24.6
76x76	14.67♦	76.2	80.0	8.9	8.4	9.4	4.6	38.1	4.76	4.28	6	42	20	0.419	28.5
	12.65	76.2	76.2	5.1	8.4	9.4	4.6	38.1	4.54	7.47	5	42	20	0.411	32.5

♦ Check availability of section.
FOR EXPLANATION OF TABLES SEE NOTE 2

JOISTS

PROPERTIES

Designation		Second Moment Of Area		Radius Of Gyration		Elastic Modulus		Plastic Modulus		Buckling Parameter	Torsional Index	Warping Constant	Torsional Constant	Area of Section
Serial Size mm	Mass Per Metre kg	Axis x-x cm⁴	Axis y-y cm⁴	Axis x-x cm	Axis y-y cm	Axis x-x cm³	Axis y-y cm³	Axis x-x cm³	Axis y-y cm³	u	x	H dm⁶	J cm⁴	A cm²
254x203	81.85	12020	2280	10.7	4.67	947	224	1077	371	0.890	11.0	0.312	152	105
254x114	37.20	5082	269	10.4	2.39	400	47.1	459	79.1	0.885	18.7	0.0392	25.2	47.3
203x152	52.09	4798	816	8.49	3.50	472	107	541	176	0.891	10.7	0.0711	64.8	66.6
152x127	37.20	1818	378	6.19	2.82	239	59.6	279	99.8	0.866	9.33	0.0183	33.9	47.5
127x114	29.76	979	242	5.12	2.54	154	42.3	181	70.8	0.853	8.76	0.00807	20.8	37.4
	26.79	946	236	5.26	2.63	149	41.3	172	68.2	0.868	9.32	0.00788	16.9	34.2
114x114	26.79	736	224	4.62	2.55	129	39.2	151	65.8	0.841	7.92	0.00601	18.9	34.5
102x102	23.07	486	154	4.07	2.29	95.6	30.3	113	50.6	0.836	7.43	0.00321	14.2	29.3
102x44	7.44	153	7.82	4.01	0.907	30.1	3.51	35.4	6.03	0.872	14.9	0.000178	1.25	9.50
89x89	19.35	307	101	3.51	2.02	69.0	22.8	82.7	38.0	0.830	6.57	0.00158	11.5	24.9
76x76	14.67	172	60.9	3.00	1.78	45.2	15.2	54.2	25.8	0.820	6.42	0.000700	6.83	19.1
	12.65	158	51.8	3.12	1.79	41.5	13.6	48.7	22.4	0.852	7.22	0.000595	4.59	16.2

FOR EXPLANATION OF TABLES SEE NOTE 3

JOISTS

PLASTIC MODULI-MAJOR AXIS / PLASTIC MODULI-MINOR AXIS

Designation		Plastic Modulus Axis x-x	Reduced Value Of Modulus Under Axial Load					Plastic Modulus Axis y-y	Reduced Value Of Modulus Under Axial Load				
			Lower Values Of n		Change Formula	Higher Values Of n			Lower Values Of n		Change Formula	Higher Values Of n	
Serial Size mm	Mass Per Metre kg	cm³	K1	K2	At n =	K3	K4	cm³	K1	K2	At n =	K3	K4
254x203	81.85	1077	1077	2677	0.205	143	8.25	371	371	107	0.248	573	0.597
254x114	37.20	459	459	737	0.363	51.8	10.6	79.1	79.1	22.1	0.408	185	0.282
203x152	52.09	541	541	1245	0.224	76.9	7.80	176	176	54.5	0.272	284	0.561
152x127	37.20	279	279	542	0.270	47.2	6.67	99.8	99.8	37.0	0.334	178	0.471
127x114	29.76	181	181	342	0.279	32.1	6.38	70.8	70.8	27.5	0.347	129	0.454
	26.79	172	172	396	0.221	27.1	7.02	68.2	68.2	23.1	0.275	109	0.567
114x114	26.79	151	151	313	0.247	28.1	6.00	65.8	65.8	26.0	0.315	111	0.517
102x102	23.07	113	113	226	0.255	22.4	5.64	50.6	50.6	21.2	0.329	86.6	0.502
102x44	7.44	35.4	35.4	52.5	0.398	5.38	7.97	6.03	6.03	2.22	0.460	15.3	0.214
89x89	19.35	82.7	82.7	163	0.256	18.4	5.01	38.0	38.0	17.4	0.340	64.7	0.503
76x76	14.67	54.2	54.2	103	0.267	12.2	4.98	25.8	25.8	12.0	0.354	45.0	0.480
	12.65	48.7	48.7	129	0.181	9.26	5.68	22.4	22.4	8.66	0.239	32.4	0.649

n = F/(Ag.py). Where F is the factored axial load and Ag is the gross cross sectional area.
For lower values of n reduced plastic modulus(Sr) = K1-K2n²
For higher values of n reduced plastic modulus(Sr) = K3(1-n)(K4+n)
FOR EXPLANATION OF TABLES SEE NOTE 3

UNIVERSAL BEARING PILES

DIMENSIONS

Designation		Depth Of Section D	Width Of Section B	Thickness		Root Radius	Depth Between Fillets	Ratios For Local Buckling		Dimensions For Detailing			Surface Area	
				Web	Flange			Flange	Web	End Clearance	Notch		Per Metre	Per Tonne
Serial Size mm	Mass Per Metre kg	mm	mm	t mm	T mm	r mm	d mm	b/T	d/t	C mm	N mm	n mm	m²	m²
356x368	174	361.5	378.1	20.4	20.4	15.2	290.3	9.27	14.2	12	190	36	2.17	12.5
	152	356.4	375.5	17.9	17.9	15.2	290.2	10.5	16.2	11	190	34	2.15	14.2
	133	351.9	373.3	15.6	15.6	15.2	290.3	12.0	18.6	10	190	32	2.14	16.1
	109	346.4	370.5	12.9	12.9	15.2	290.2	14.4	22.5	8	190	30	2.12	19.5
305x305	223	338.0	325.4	30.5	30.5	15.2	246.6	5.33	8.09	17	158	46	1.89	8.48
	186	328.3	320.5	25.6	25.6	15.2	246.7	6.26	9.64	15	158	42	1.86	10.0
	149	318.5	315.6	20.7	20.7	15.2	246.7	7.62	11.9	12	158	36	1.83	12.3
	126	312.4	312.5	17.7	17.7	15.2	246.6	8.83	13.9	11	158	34	1.81	14.4
	110	307.9	310.3	15.4	15.4	15.2	246.7	10.1	16.0	10	158	32	1.80	16.4
	95	303.8	308.3	13.4	13.4	15.2	246.6	11.5	18.4	9	158	30	1.79	18.8
	88	301.7	307.2	12.3	12.3	15.2	246.7	12.5	20.1	8	158	28	1.78	20.2
	79	299.2	306.0	11.1	11.1	15.2	246.6	13.8	22.2	8	158	28	1.77	22.5
254x254	85	254.3	259.7	14.3	14.3	12.7	200.3	9.08	14.0	9	134	28	1.50	17.6
	71	249.9	257.5	12.1	12.1	12.7	200.3	10.6	16.6	8	134	26	1.48	20.9
	63	246.9	256.0	10.6	10.6	12.7	200.3	12.1	18.9	7	134	24	1.47	23.4
203x203	54	203.9	207.2	11.3	11.3	10.2	160.9	9.17	14.2	8	108	22	1.20	22.2
	45	200.2	205.4	9.5	9.5	10.2	160.8	10.8	16.7	7	108	20	1.19	26.3

FOR EXPLANATION OF TABLES SEE NOTE 2

UNIVERSAL BEARING PILES

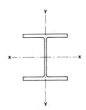

PROPERTIES

Designation		Second Moment Of Area		Radius Of Gyration		Elastic Modulus		Plastic Modulus		Buckling Parameter	Torsional Index	Warping Constant	Torsional Constant	Area of Section
Serial Size mm	Mass Per Metre kg	Axis x-x cm^4	Axis y-y cm^4	Axis x-x cm	Axis y-y cm	Axis x-x cm^3	Axis y-y cm^3	Axis x-x cm^3	Axis y-y cm^3	u	x	H dm^6	J cm^4	A cm^2
356x368	174	51020	18400	15.2	9.11	2823	974	3187	1494	0.821	15.8	5.35	332	222
	152	43950	15810	15.1	9.03	2466	842	2766	1290	0.821	17.8	4.53	224	194
	133	37730	13540	15.0	8.96	2144	725	2391	1109	0.822	20.2	3.83	149	168
	109	30620	10940	14.8	8.87	1768	591	1957	901	0.823	24.1	3.04	85.2	139
305x305	223	52840	17590	13.6	7.86	3127	1081	3664	1683	0.826	9.51	4.16	955	285
	186	42580	14090	13.4	7.71	2594	879	3002	1363	0.827	11.1	3.23	562	237
	149	33050	10870	13.2	7.56	2075	689	2370	1063	0.828	13.5	2.41	296	190
	126	27540	9019	13.1	7.47	1763	577	1996	888	0.829	15.6	1.96	186	162
	110	23550	7680	13.0	7.40	1530	495	1720	760	0.830	17.7	1.64	123	140
	95	20170	6552	12.9	7.34	1328	425	1484	651	0.830	20.1	1.38	81.7	122
	88	18380	5949	12.8	7.30	1218	387	1356	593	0.831	21.7	1.25	63.6	112
	79	16430	5306	12.8	7.26	1098	347	1218	530	0.831	23.8	1.10	47.2	101
254x254	85	12250	4181	10.7	6.22	963	322	1089	495	0.826	15.7	0.602	81.2	108
	71	10140	3448	10.6	6.15	812	268	910	411	0.827	18.2	0.487	49.5	91.0
	63	8764	2967	10.5	6.11	710	232	791	355	0.827	20.6	0.414	33.6	79.6
203x203	54	4979	1678	8.54	4.96	488	162	552	249	0.827	15.9	0.156	32.1	68.2
	45	4092	1374	8.46	4.90	409	134	458	205	0.828	18.6	0.125	19.2	57.1

FOR EXPLANATION OF TABLES SEE NOTE 3

PLASTIC MODULI-MAJOR AXIS / PLASTIC MODULI-MINOR AXIS

Designation		Plastic Modulus Axis x-x cm^3	Reduced Value Of Modulus Under Axial Load					Plastic Modulus Axis y-y cm^3	Reduced Value Of Modulus Under Axial Load				
			Lower Values Of n		Change Formula	Higher Values Of n			Lower Values Of n		Change Formula	Higher Values Of n	
Serial Size mm	Mass Per Metre kg		K1	K2	At n =	K3	K4		K1	K2	At n =	K3	K4
356x368	174	3187	3187	6022	0.294	326	11.3	1494	1494	340	0.333	2932	0.412
	152	2766	2766	5246	0.295	251	12.7	1290	1290	263	0.329	2545	0.410
	133	2391	2391	4549	0.296	191	14.5	1109	1109	202	0.326	2197	0.408
	109	1957	1957	3741	0.297	132	17.3	901	901	139	0.322	1794	0.407
305x305	223	3664	3664	6656	0.296	625	6.70	1683	1683	601	0.362	3257	0.410
	186	3002	3002	5486	0.298	439	7.85	1363	1363	428	0.355	2674	0.404
	149	2370	2370	4360	0.301	287	9.53	1063	1063	283	0.347	2112	0.399
	126	1996	1996	3690	0.302	210	11.0	888	888	209	0.342	1778	0.396
	110	1720	1720	3192	0.303	160	12.5	760	760	160	0.338	1530	0.394
	95	1484	1484	2764	0.304	121	14.2	651	651	122	0.334	1317	0.393
	88	1356	1356	2533	0.304	103	15.4	593	593	103	0.332	1202	0.392
	79	1218	1218	2282	0.304	84.1	16.9	530	530	84.7	0.330	1076	0.392
254x254	85	1089	1089	2037	0.298	113	11.2	495	495	115	0.337	980	0.405
	71	910	910	1711	0.299	81.2	13.0	411	411	82.9	0.332	818	0.403
	63	791	791	1494	0.300	62.6	14.7	355	355	64.1	0.329	709	0.402
203x203	54	552	552	1029	0.299	56.5	11.3	249	249	57.0	0.338	495	0.402
	45	458	458	859	0.300	40.1	13.5	205	205	40.8	0.333	410	0.400

n = F'/(Ag.py). Where F is the factored axial load and Ag is the gross cross sectional area.
For lower values of n reduced plastic modulus(Sr) = K1-K2n^2
For higher values of n reduced plastic modulus(Sr) = K3(1-n)(K4+n)
FOR EXPLANATION OF TABLES SEE NOTE 3

CIRCULAR HOLLOW SECTIONS

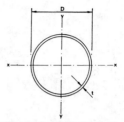

DIMENSIONS AND PROPERTIES

Designation		Mass Per Metre	Area Of Section	Ratio For Local Buckling	Second Moment Of Area	Radius Of Gyration	Elastic Modulus	Plastic Modulus	Torsional Constants		Surface Area Per Metre
Outside Diameter D mm	Thickness t mm	kg	A cm^2	D/t	I cm^4	r cm	Z cm^3	S cm^3	J cm^4	C cm^3	m^2
21.3	3.2}	1.43	1.82	6.66	0.768	0.650	0.722	1.06	1.54	1.44	0.0669
26.9	3.2}	1.87	2.38	8.41	1.70	0.846	1.27	1.81	3.40	2.54	0.0845
33.7	2.6}	1.99	2.54	13.0	3.09	1.10	1.84	2.52	6.18	3.68	0.106
	3.2}	2.41	3.07	10.5	3.60	1.08	2.14	2.99	7.20	4.28	0.106
	4.0}	2.93	3.73	8.43	4.19	1.06	2.49	3.55	8.38	4.98	0.106
42.4	2.6}	2.55	3.25	16.3	6.46	1.41	3.05	4.12	12.9	6.10	0.133
	3.2}	3.09	3.94	13.2	7.62	1.39	3.59	4.93	15.2	7.18	0.133
	4.0}	3.79	4.83	10.6	8.99	1.36	4.24	5.92	18.0	8.48	0.133
48.3	3.2	3.56	4.53	15.1	11.6	1.60	4.80	6.52	23.2	9.60	0.152
	4.0	4.37	5.57	12.1	13.8	1.57	5.70	7.87	27.6	11.4	0.152
	5.0	5.34	6.80	9.66	16.2	1.54	6.69	9.42	32.4	13.4	0.152
60.3	3.2	4.51	5.74	18.8	23.5	2.02	7.78	10.4	47.0	15.6	0.189
	4.0	5.55	7.07	15.1	28.2	2.00	9.34	12.7	56.4	18.7	0.189
	5.0	6.82	8.69	12.1	33.5	1.96	11.1	15.3	67.0	22.2	0.189
76.1	3.2	5.75	7.33	23.8	48.8	2.58	12.8	17.0	97.6	25.6	0.239
	4.0	7.11	9.06	19.0	59.1	2.55	15.5	20.8	118	31.0	0.239
	5.0	8.77	11.2	15.2	70.9	2.52	18.6	25.3	142	37.2	0.239
88.9	3.2	6.76	8.62	27.8	79.2	3.03	17.8	23.5	158	35.6	0.279
	4.0	8.38	10.7	22.2	96.3	3.00	21.7	28.9	193	43.4	0.279
	5.0	10.3	13.2	17.8	116	2.97	26.2	35.2	232	52.4	0.279
114.3	3.6	9.83	12.5	31.8	192	3.92	33.6	44.1	384	67.2	0.359
	5.0	13.5	17.2	22.9	257	3.87	45.0	59.8	514	90.0	0.359
	6.3	16.8	21.4	18.1	313	3.82	54.7	73.6	626	109	0.359
139.7	5.0	16.6	21.2	27.9	481	4.77	68.8	90.8	962	138	0.439
	6.3	20.7	26.4	22.2	589	4.72	84.3	112	1178	169	0.439
	8.0	26.0	33.1	17.5	720	4.66	103	139	1440	206	0.439
	10.0	32.0	40.7	14.0	862	4.60	123	169	1724	246	0.439
168.3	5.0	20.1	25.7	33.7	856	5.78	102	133	1712	204	0.529
	6.3	25.2	32.1	26.7	1053	5.73	125	165	2106	250	0.529
	8.0	31.6	40.3	21.0	1297	5.67	154	206	2594	308	0.529
	10.0	39.0	49.7	16.8	1564	5.61	186	251	3128	372	0.529
193.7	5.0	23.3	29.6	38.7	1320	6.67	136	178	2640	272	0.609
	6.3	29.1	37.1	30.7	1630	6.63	168	221	3260	336	0.609
	8.0	36.6	46.7	24.2	2016	6.57	208	276	4032	416	0.609
	10.0	45.3	57.7	19.4	2442	6.50	252	338	4884	504	0.609
	12.5	55.9	71.2	15.5	2934	6.42	303	411	5868	606	0.609
	16.0‡◆	70.1	89.3	12.1	3554	6.31	367	507	7108	734	0.609
219.1	5.0	26.4	33.6	43.8	1928	7.57	176	229	3856	352	0.688
	6.3	33.1	42.1	34.8	2386	7.53	218	285	4772	436	0.688
	8.0	41.6	53.1	27.4	2960	7.47	270	357	5920	540	0.688
	10.0	51.6	65.7	21.9	3598	7.40	328	438	7196	656	0.688
	12.5	63.7	81.1	17.5	4345	7.32	397	534	8690	794	0.688
	16.0‡◆	80.1	102	13.7	5297	7.20	483	661	10590	966	0.688
	20.0‡◆	98.2	125	11.0	6261	7.07	572	795	12520	1144	0.688

} Sections marked thus are rolled in grade 43C only
‡ Sections marked thus are seamless and rolled in grade 50B only
◆ Check availability of section
FOR EXPLANATION OF TABLES SEE NOTES 2 AND 3

Dimensions and properties 1119

CIRCULAR HOLLOW SECTIONS

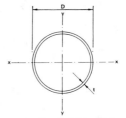

DIMENSIONS AND PROPERTIES

Designation		Mass Per Metre	Area Of Section	Ratio For Local Buckling	Second Moment Of Area	Radius Of Gyration	Elastic Modulus	Plastic Modulus	Torsional Constants		Surface Area Per Metre
Outside Diameter D mm	Thickness t mm	kg	A cm^2	D/t	I cm^4	r cm	Z cm^3	S cm^3	J cm^4	C cm^3	m^2
244.5	6.3	37.0	47.1	38.8	3346	8.42	274	358	6692	548	0.768
	8.0	46.7	59.4	30.6	4160	8.37	340	448	8320	680	0.768
	10.0	57.8	73.7	24.5	5073	8.30	415	550	10150	830	0.768
	12.5	71.5	91.1	19.6	6147	8.21	503	673	12290	1006	0.768
	16.0	90.2	115	15.3	7533	8.10	616	837	15070	1232	0.768
	20.0‡♦	111	141	12.2	8957	7.97	733	1011	17910	1466	0.768
	25.0+‡♦	135	172	9.78	10520	7.81	860	1210	21040	1720	0.768
273.0	6.3	41.4	52.8	43.3	4696	9.43	344	448	9392	688	0.858
	8.0	52.3	66.6	34.1	5852	9.37	429	562	11700	858	0.858
	10.0	64.9	82.6	27.3	7154	9.31	524	692	14310	1048	0.858
	12.5	80.3	102	21.8	8697	9.22	637	849	17390	1274	0.858
	16.0	101	129	17.1	10710	9.10	784	1058	21420	1568	0.858
	20.0‡♦	125	159	13.6	12800	8.97	938	1283	25600	1876	0.858
	25.0‡♦	153	195	10.9	15130	8.81	1108	1543	30260	2216	0.858
323.9	6.3	49.3	62.9	51.4	7929	11.2	490	636	15860	980	1.02
	8.0	62.3	79.4	40.5	9910	11.2	612	799	19820	1224	1.02
	10.0	77.4	98.6	32.4	12160	11.1	751	986	24320	1502	1.02
	12.5	96.0	122	25.9	14850	11.0	917	1213	29700	1834	1.02
	16.0	121	155	20.2	18390	10.9	1136	1518	36780	2272	1.02
	20.0‡♦	150	191	16.2	22140	10.8	1367	1850	44280	2734	1.02
	25.0‡♦	184	235	13.0	26400	10.6	1630	2239	52800	3260	1.02
355.6	8.0	68.6	87.4	44.5	13200	12.3	742	967	26400	1484	1.12
	10.0	85.2	109	35.6	16220	12.2	912	1195	32440	1824	1.12
	12.5	106	135	28.4	19850	12.1	1117	1472	39700	2234	1.12
	16.0	134	171	22.2	24660	12.0	1387	1847	49320	2774	1.12
	20.0‡♦	166	211	17.8	29790	11.9	1676	2255	59580	3352	1.12
	25.0‡♦	204	260	14.2	35680	11.7	2007	2738	71360	4014	1.12
406.4	10.0	97.8	125	40.6	24480	14.0	1205	1572	48960	2410	1.28
	12.5	121	155	32.5	30030	13.9	1478	1940	60060	2956	1.28
	16.0	154	196	25.4	37450	13.8	1843	2440	74900	3686	1.28
	20.0‡♦	191	243	20.3	45430	13.7	2236	2989	90860	4472	1.28
	25.0‡♦	235	300	16.3	54700	13.5	2692	3642	109400	5384	1.28
	32.0‡♦	295	376	12.7	66430	13.3	3269	4497	132900	6538	1.28
457.0	10.0	110	140	45.7	35090	15.8	1536	1998	70180	3072	1.44
	12.5	137	175	36.6	43140	15.7	1888	2470	86280	3776	1.44
	16.0	174	222	28.6	53960	15.6	2361	3113	107900	4722	1.44
	20.0‡♦	216	275	22.9	65680	15.5	2874	3822	131400	5748	1.44
	25.0‡♦	266	339	18.3	79420	15.3	3475	4671	158800	6950	1.44
	32.0‡♦	335	427	14.3	97010	15.1	4246	5791	194000	8492	1.44
	40.0‡♦	411	524	11.4	114900	14.8	5031	6977	229800	10060	1.44
508.0	10.0	123	156	50.8	48520	17.6	1910	2480	97040	3820	1.60
	12.5	153	195	40.6	59760	17.5	2353	3070	119500	4706	1.60
	16.0	194	247	31.7	74910	17.4	2949	3874	149800	5898	1.60
	20.0‡♦	241	307	25.4	91430	17.3	3600	4766	182900	7200	1.60
	25.0‡♦	298	379	20.3	110900	17.1	4367	5837	221800	8734	1.60
	32.0‡♦	376	479	15.9	136100	16.9	5360	7261	272200	10720	1.60
	40.0‡♦	462	588	12.7	162200	16.6	6385	8782	324400	12770	1.60
	50.0‡♦	565	719	10.2	190900	16.3	7515	10530	381800	15030	1.60

\+ Sections marked thus are not included in BS4848: Part 2
‡ Sections marked thus are seamless and rolled in grade 50B only
♦ Check availability of section
FOR EXPLANATION OF TABLES SEE NOTES 2 AND 3

SQUARE HOLLOW SECTIONS

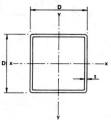

DIMENSIONS AND PROPERTIES

Designation		Mass Per Metre	Area Of Section	Ratio For Local Buckling	Second Moment Of Area	Radius Of Gyration	Elastic Modulus	Plastic Modulus	Torsional Constants		Surface Area Per Metre
Size D D mm	Thickness t mm	kg	A cm^2	d/t	I cm^4	r cm	Z cm^3	S cm^3	J cm^4	C cm^3	m^2
20x20	2.0	1.12	1.42	7.00	0.759	0.731	0.759	0.951	1.22	1.07	0.0757
	2.5	1.35	1.72	5.00	0.865	0.709	0.865	1.12	1.41	1.21	0.0746
25x25	2.0	1.43	1.82	9.50	1.59	0.935	1.27	1.56	2.52	1.81	0.0957
	2.5	1.74	2.22	7.00	1.85	0.914	1.48	1.86	2.97	2.09	0.0946
	3.0	2.04	2.60	5.33	2.06	0.892	1.65	2.12	3.36	2.31	0.0936
	3.2	2.15	2.74	4.81	2.14	0.883	1.71	2.21	3.49	2.38	0.0931
30x30	2.5	2.14	2.72	9.00	3.40	1.12	2.27	2.79	5.40	3.22	0.115
	3.0	2.51	3.20	7.00	3.84	1.10	2.56	3.21	6.17	3.61	0.114
	3.2	2.65	3.38	6.38	4.00	1.09	2.67	3.37	6.45	3.75	0.113
40x40	2.5	2.92	3.72	13.0	8.67	1.53	4.33	5.21	13.6	6.23	0.155
	3.0	3.45	4.40	10.3	9.96	1.51	4.98	6.07	15.7	7.11	0.154
	3.2	3.66	4.66	9.50	10.4	1.50	5.22	6.40	16.5	7.43	0.153
	4.0	4.46	5.68	7.00	12.1	1.46	6.07	7.61	19.5	8.56	0.151
	5.0	5.40	6.88	5.00	13.8	1.42	6.92	8.92	22.6	9.65	0.149
50x50	2.5	3.71	4.72	17.0	17.7	1.94	7.07	8.38	27.4	10.2	0.195
	3.0	4.39	5.60	13.7	20.5	1.91	8.20	9.83	32.0	11.8	0.194
	3.2	4.66	5.94	12.6	21.6	1.91	8.62	10.4	33.8	12.4	0.193
	4.0	5.72	7.28	9.50	25.5	1.87	10.2	12.5	40.4	14.5	0.191
	5.0	6.97	8.88	7.00	29.6	1.83	11.9	14.9	47.6	16.7	0.189
	6.3	8.49	10.8	4.94	33.9	1.77	13.6	17.5	55.3	18.9	0.186
60x60	3.0	5.34	6.80	17.0	36.6	2.32	12.2	14.5	56.9	17.7	0.234
	3.2	5.67	7.22	15.7	38.7	2.31	12.9	15.3	60.1	18.6	0.233
	4.0	6.97	8.88	12.0	46.1	2.28	15.4	18.6	72.4	22.1	0.231
	5.0	8.54	10.9	9.00	54.4	2.24	18.1	22.3	86.3	25.8	0.229
	6.3	10.5	13.3	6.52	63.4	2.18	21.1	26.6	102	29.7	0.226
	8.0	12.8	16.3	4.50	72.4	2.11	24.1	31.4	119	33.5	0.223
70x70	3.0	6.28	8.00	20.3	59.6	2.73	17.0	20.0	92.1	24.8	0.274
	3.6	7.46	9.50	16.4	69.5	2.70	19.9	23.6	108	28.7	0.272
	5.0	10.1	12.9	11.0	90.1	2.64	25.7	31.2	142	36.8	0.269
	6.3	12.5	15.9	8.11	106	2.59	30.4	37.6	169	43.0	0.266
	8.0	15.3	19.5	5.75	123	2.51	35.3	45.0	200	49.4	0.263
80x80	3.0	7.22	9.20	23.7	90.6	3.14	22.7	26.5	139	33.1	0.314
	3.6	8.59	10.9	19.2	106	3.11	26.5	31.3	164	38.5	0.312
	5.0	11.7	14.9	13.0	139	3.05	34.7	41.7	217	49.8	0.309
	6.3	14.4	18.4	9.70	165	3.00	41.3	50.5	261	58.8	0.306
	8.0	17.8	22.7	7.00	194	2.92	48.6	60.9	312	68.5	0.303
90x90	3.6	9.27	12.4	22.0	154	3.52	34.1	40.0	237	49.7	0.352
	5.0	13.3	16.9	15.0	202	3.46	45.0	53.6	315	64.9	0.349
	6.3	16.4	20.9	11.3	242	3.41	53.9	65.3	381	77.1	0.346
	8.0	20.4	25.9	8.25	288	3.33	64.0	79.2	459	90.7	0.343
100x100	4.0	12.0	15.3	22.0	234	3.91	46.8	54.9	361	68.2	0.391
	5.0	14.8	18.9	17.0	283	3.87	56.6	67.1	439	81.9	0.389
	6.3	18.4	23.4	12.9	341	3.81	68.2	82.0	533	97.9	0.386
	8.0	22.9	29.1	9.50	408	3.74	81.5	99.9	646	116	0.383
	10.0	27.9	35.5	7.00	474	3.65	94.9	119	761	134	0.379

FOR EXPLANATION OF TABLES SEE NOTES 2 AND 3

Dimensions and properties 1121

SQUARE HOLLOW SECTIONS

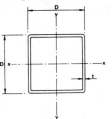

DIMENSIONS AND PROPERTIES

Designation		Mass Per Metre	Area Of Section	Ratio For Local Buckling	Second Moment Of Area	Radius Of Gyration	Elastic Modulus	Plastic Modulus	Torsional Constants		Surface Area Per Metre
Size D D mm	Thickness t mm	kg	A cm²	d/t	I cm⁴	r cm	Z cm³	S cm³	J cm⁴	C cm³	m²
120x120	5.0	18.0	22.9	21.0	503	4.69	83.8	98.4	775	122	0.469
	6.3	22.3	28.5	16.0	610	4.63	102	121	949	147	0.466
	8.0	27.9	35.5	12.0	738	4.56	123	149	1159	176	0.463
	10.0	34.2	43.5	9.00	870	4.47	145	178	1381	206	0.459
	12.5	41.6	53.0	6.60	1009	4.36	168	212	1624	237	0.453
140x140	5.0+	21.1	26.9	25.0	814	5.50	116	136	1251	170	0.549
	6.3+	26.3	33.5	19.2	994	5.45	142	168	1538	206	0.546
	8.0+	32.9	41.9	14.5	1212	5.38	173	207	1889	249	0.543
	10.0+	40.4	51.5	11.0	1441	5.29	206	250	2269	294	0.539
	12.5+	49.5	63.0	8.20	1691	5.18	242	299	2695	342	0.533
150x150	5.0	22.7	28.9	27.0	1009	5.91	135	157	1548	197	0.589
	6.3	28.3	36.0	20.8	1236	5.86	165	194	1907	240	0.586
	8.0	35.4	45.1	15.7	1510	5.78	201	240	2348	291	0.583
	10.0	43.6	55.5	12.0	1803	5.70	240	290	2829	345	0.579
	12.5	53.4	68.0	9.00	2125	5.59	283	348	3372	403	0.573
	16.0	66.4	84.5	6.38	2500	5.44	333	421	4029	468	0.566
160x160	5.0+	24.2	30.9	29.0	1234	6.32	154	179	1890	226	0.629
	6.0+	28.9	36.8	23.7	1450	6.28	181	212	2230	264	0.627
	6.3+	30.3	38.5	22.4	1513	6.27	189	222	2330	276	0.626
	8.0+	37.9	48.3	17.0	1853	6.19	232	275	2875	335	0.623
	10.0+	46.7	59.5	13.0	2219	6.11	277	333	3473	399	0.619
	12.5+	57.3	73.0	9.80	2627	6.00	328	402	4154	468	0.613
180x180	6.3	34.2	43.6	25.6	2186	7.08	243	283	3357	355	0.706
	8.0	43.0	54.7	19.5	2689	7.01	299	352	4156	434	0.703
	10.0	53.0	67.5	15.0	3237	6.92	360	429	5041	519	0.699
	12.5	65.2	83.0	11.4	3856	6.82	428	519	6062	613	0.693
	16.0	81.4	104	8.25	4607	6.66	512	634	7339	725	0.686
200x200	5.0+	30.5	38.9	37.0	2460	7.95	246	284	3752	362	0.789
	6.3	38.2	48.6	28.7	3033	7.90	303	353	4647	444	0.786
	8.0	48.0	61.1	22.0	3744	7.83	374	439	5770	545	0.783
	10.0	59.3	75.5	17.0	4525	7.74	452	536	7020	655	0.779
	12.5	73.0	93.0	13.0	5419	7.63	542	651	8479	779	0.773
	16.0	91.5	117	9.50	6524	7.48	652	799	10330	929	0.766
250x250	6.3	48.1	61.2	36.7	6049	9.94	484	559	9228	712	0.986
	8.0	60.5	77.1	28.2	7510	9.87	601	699	11510	880	0.983
	10.0	75.0	95.5	22.0	9141	9.78	731	858	14090	1065	0.979
	12.5	92.6	118	17.0	11050	9.68	884	1048	17140	1279	0.973
	16.0	117	149	12.6	13480	9.53	1078	1298	21110	1548	0.966
300x300	6.3+	57.9	73.8	44.6	10600	12.0	706	812	16120	1043	1.19
	8.0	73.1	93.1	34.5	13210	11.9	881	1018	20170	1294	1.18
	10.0	90.7	116	27.0	16150	11.8	1077	1254	24780	1575	1.18
	12.5	112	143	21.0	19630	11.7	1309	1538	30290	1905	1.17
	16.0	142	181	15.7	24160	11.6	1610	1916	37570	2327	1.17
350x350	8.0	85.7	109	40.7	21240	14.0	1214	1398	32350	1789	1.38
	10.0	106	136	32.0	26050	13.9	1489	1725	39840	2186	1.38
	12.5	132	168	25.0	31810	13.8	1817	2122	48870	2655	1.37
	16.0	167	213	18.9	39370	13.6	2250	2655	60900	3265	1.37
400x400	10.0	122	156	37.0	39350	15.9	1968	2272	60030	2896	1.58
	12.5	152	193	29.0	48190	15.8	2409	2800	73820	3530	1.57
	16.0	192	245	22.0	59910	15.7	2995	3514	92310	4363	1.57
	20.0‡	237	302	17.0	72390	15.5	3620	4292	112300	5240	1.56

+ Sections marked thus are not included in BS4848: Part 2
‡ Sections marked thus are rolled in grade 50C only
FOR EXPLANATIONS OF TABLES SEE NOTES 2 AND 3

Dimensions and properties

RECTANGULAR HOLLOW SECTIONS

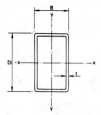

DIMENSIONS AND PROPERTIES

Designation		Mass Per Metre	Area Of Section	Ratios for Local Buckling		Second Moment Of Area		Radius Of Gyration		Elastic Modulus		Plastic Modulus		Torsional Constants		Surface Area Per Metre
Size D B mm	Thickness t mm	kg	cm²	d/t	b/t	Axis x-x cm⁴	Axis y-y cm⁴	Axis x-x cm	Axis y-y cm	Axis x-x cm³	Axis y-y cm³	Axis x-x cm³	Axis y-y cm³	J cm⁴	C cm³	m²
50x25	2.5	2.72	3.47	17.0	7.00	10.6	3.44	1.75	0.996	4.25	2.75	5.41	3.26	8.41	4.62	0.145
	3.0	3.22	4.10	13.7	5.33	12.2	3.89	1.73	0.975	4.88	3.11	6.30	3.77	9.64	5.21	0.144
	3.2	3.41	4.34	12.6	4.81	12.8	4.05	1.72	0.966	5.11	3.24	6.64	3.96	10.1	5.42	0.143
50x30	2.5	2.92	3.72	17.0	9.00	12.0	5.30	1.80	1.19	4.81	3.53	6.01	4.16	11.7	5.74	0.155
	3.0	3.45	4.40	13.7	7.00	13.9	6.04	1.78	1.17	5.54	4.03	7.01	4.83	13.5	6.52	0.154
	3.2	3.66	4.66	12.6	6.38	14.5	6.31	1.77	1.16	5.82	4.21	7.39	5.08	14.2	6.81	0.153
	4.0	4.46	5.68	9.50	4.50	17.0	7.25	1.73	1.13	6.80	4.83	8.81	6.01	16.6	7.79	0.151
	5.0	5.40	6.88	7.00	3.00	19.5	8.13	1.68	1.09	7.79	5.42	10.4	6.98	19.0	8.71	0.149
60x40	2.5	3.71	4.72	21.0	13.0	23.1	12.2	2.21	1.61	7.71	6.10	9.43	7.09	25.0	9.74	0.195
	3.0	4.39	5.60	17.0	10.3	26.9	14.1	2.19	1.59	8.96	7.04	11.1	8.29	29.2	11.2	0.194
	3.2	4.66	5.94	15.7	9.50	28.3	14.8	2.18	1.58	9.44	7.39	11.7	8.75	30.8	11.8	0.193
	4.0	5.72	7.28	12.0	7.00	33.6	17.3	2.15	1.54	11.2	8.67	14.1	10.5	36.6	13.7	0.191
	5.0	6.97	8.88	9.00	5.00	39.2	20.0	2.10	1.50	13.1	10.0	16.8	12.4	43.0	15.8	0.189
	6.3	8.49	10.8	6.52	3.35	45.1	22.6	2.04	1.45	15.0	11.3	19.9	14.6	49.7	17.7	0.186
80x40	3.0	5.34	6.80	23.7	10.3	55.0	18.2	2.85	1.64	13.8	9.10	17.3	10.5	43.7	15.3	0.234
	3.2	5.67	7.22	22.0	9.50	58.1	19.1	2.84	1.63	14.5	9.56	18.3	11.1	46.1	16.1	0.233
	4.0	6.97	8.88	17.0	7.00	69.6	22.6	2.80	1.59	17.4	11.3	22.2	13.4	55.1	18.9	0.231
	5.0	8.54	10.9	13.0	5.00	82.4	26.2	2.75	1.55	20.6	13.1	26.7	15.9	65.0	21.9	0.229
	6.3	10.5	13.3	9.70	3.35	96.5	29.8	2.69	1.50	24.1	14.9	31.9	18.8	75.8	24.9	0.226
	8.0	12.8	16.3	7.00	2.00	111	33.1	2.61	1.42	27.7	16.6	37.8	21.8	86.3	27.6	0.223
90x50	3.0	6.28	8.00	27.0	13.7	85.4	33.8	3.27	2.05	19.0	13.5	23.4	15.5	76.4	22.4	0.274
	3.6	7.46	9.50	22.0	10.9	99.8	39.1	3.24	2.03	22.2	15.6	27.6	18.1	89.3	25.9	0.272
	5.0	10.1	12.9	15.0	7.00	130	50.0	3.18	1.97	28.9	20.0	36.6	23.9	116	32.9	0.269
	6.3	12.5	15.9	11.3	4.94	154	58.1	3.12	1.91	34.2	23.3	44.2	28.5	138	38.2	0.266
	8.0	15.3	19.5	8.25	3.25	180	66.3	3.04	1.84	40.0	26.5	53.0	33.7	161	43.4	0.263
100x50	3.0	6.75	8.60	30.3	13.7	111	37.1	3.59	2.08	22.2	14.8	27.6	16.9	88.3	25.0	0.294
	3.2	7.18	9.14	28.2	12.6	117	39.1	3.58	2.07	23.5	15.6	29.2	17.9	93.3	26.4	0.293
	4.0	8.86	11.3	22.0	9.50	142	46.7	3.55	2.03	28.4	18.7	35.7	21.7	113	31.4	0.291
	5.0	10.9	13.9	17.0	7.00	170	55.1	3.50	1.99	34.0	22.0	43.3	26.1	135	37.0	0.289
	6.3	13.4	17.1	12.9	4.94	202	64.2	3.44	1.94	40.5	25.7	52.5	31.3	160	43.0	0.286
	8.0	16.6	21.1	9.50	3.25	238	73.5	3.36	1.86	47.6	29.4	63.1	37.1	187	49.1	0.283
100x60	3.0	7.22	9.20	30.3	17.0	125	56.2	3.69	2.47	25.0	18.7	30.5	21.3	121	30.7	0.314
	3.6	8.59	10.9	24.8	13.7	147	65.4	3.66	2.45	29.3	21.8	36.0	25.1	142	35.6	0.312
	5.0	11.7	14.9	17.0	9.00	192	84.7	3.60	2.39	38.5	28.2	48.1	33.3	187	45.9	0.309
	6.3	14.4	18.4	12.9	6.52	230	99.9	3.54	2.33	46.0	33.3	58.4	40.2	224	53.9	0.306
	8.0	17.8	22.7	9.50	4.50	272	116	3.46	2.26	54.4	38.7	70.5	48.1	266	62.4	0.303
120x60	3.6	9.72	12.4	30.3	13.7	230	76.9	4.31	2.49	38.3	25.6	47.6	29.2	183	43.3	0.352
	5.0	13.3	16.9	21.0	9.00	304	99.9	4.24	2.43	50.7	33.3	63.9	38.8	242	56.0	0.349
	6.3	16.4	20.9	16.0	6.52	366	118	4.18	2.38	61.0	39.4	78.0	46.9	290	66.0	0.346
	8.0	20.4	25.9	12.0	4.50	437	138	4.10	2.31	72.8	45.9	94.8	56.4	344	76.8	0.343
120x80	5.0	14.8	18.9	21.0	13.0	370	195	4.43	3.21	61.7	48.8	75.4	56.7	401	77.9	0.389
	6.3	18.4	23.4	16.0	9.70	447	234	4.37	3.16	74.6	58.4	92.3	69.1	486	93.0	0.386
	8.0	22.9	29.1	12.0	7.00	537	278	4.29	3.09	89.5	69.4	113	83.9	586	110	0.383
	10.0	27.9	35.5	9.00	5.00	628	320	4.20	3.00	105	80.0	134	99.4	688	126	0.379

FOR EXPLANATION OF TABLES SEE NOTES 2 AND 3

Dimensions and properties

RECTANGULAR HOLLOW SECTIONS

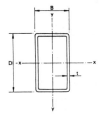

DIMENSIONS AND PROPERTIES

Designation		Mass Per Metre	Area Of Section	Ratios for Local Buckling		Second Moment Of Area		Radius Of Gyration		Elastic Modulus		Plastic Modulus		Torsional Constants		Surface Area Per Metre
Size D B mm	Thickness t mm	kg	cm²	d/t	b/t	Axis x-x cm⁴	Axis y-y cm⁴	Axis x-x cm	Axis y-y cm	Axis x-x cm³	Axis y-y cm³	Axis x-x cm³	Axis y-y cm³	J cm⁴	C cm³	m²
150x100	5.0	18.7	23.9	27.0	17.0	747	396	5.59	4.07	99.5	79.1	121	90.8	806	127	0.489
	6.3	23.3	29.7	20.8	12.9	910	479	5.53	4.02	121	95.9	148	111	985	153	0.486
	8.0	29.1	37.1	15.7	9.50	1106	577	5.46	3.94	147	115	183	137	1202	184	0.483
	10.0	35.7	45.5	12.0	7.00	1312	678	5.37	3.86	175	136	220	164	1431	215	0.479
	12.5	43.6	55.5	9.00	5.00	1532	781	5.25	3.75	204	156	263	194	1680	246	0.473
160x80	5.0	18.0	22.9	29.0	13.0	753	251	5.74	3.31	94.1	62.8	117	71.7	599	106	0.469
	6.3	22.3	28.5	22.4	9.70	917	302	5.68	3.26	115	75.6	144	87.7	729	127	0.466
	8.0	27.9	35.5	17.0	7.00	1113	361	5.60	3.19	139	90.2	177	107	882	151	0.463
	10.0	34.2	43.5	13.0	5.00	1318	419	5.50	3.10	165	105	213	127	1041	175	0.459
	12.5	41.6	53.0	9.80	3.40	1536	476	5.38	3.00	192	119	254	150	1206	199	0.453
200x100	5.0	22.7	28.9	37.0	17.0	1509	509	7.23	4.20	151	102	186	115	1202	172	0.589
	6.3	28.3	36.0	28.7	12.9	1851	618	7.17	4.14	185	124	231	141	1473	208	0.586
	8.0	35.4	45.1	22.0	9.50	2269	747	7.09	4.07	227	149	286	174	1802	251	0.583
	10.0	43.6	55.5	17.0	7.00	2718	881	7.00	3.98	272	176	346	209	2154	296	0.579
	12.5	53.4	68.0	13.0	5.00	3218	1022	6.88	3.88	322	204	417	249	2541	342	0.573
	16.0	66.4	84.5	9.50	3.25	3808	1175	6.71	3.73	381	235	505	297	2988	393	0.566
200x120	5.0+	24.2	30.9	37.0	21.0	1699	767	7.42	4.98	170	128	206	144	1646	210	0.629
	6.0+	28.9	36.8	30.3	17.0	2000	899	7.37	4.94	200	150	244	171	1940	245	0.627
	6.3+	30.3	38.5	28.7	16.0	2087	937	7.36	4.93	209	156	255	178	2025	256	0.626
	8.0+	37.9	48.3	22.0	12.0	2564	1140	7.28	4.86	256	190	316	220	2491	310	0.623
	10.0+	46.7	59.5	17.0	9.00	3079	1356	7.19	4.77	308	226	384	266	2997	367	0.619
	12.5+	57.3	73.0	13.0	6.60	3658	1589	7.08	4.67	366	265	464	319	3567	429	0.613
250x150	5.0+	30.5	38.9	47.0	27.0	3382	1535	9.33	6.28	271	205	326	229	3275	337	0.789
	6.3	38.2	48.6	36.7	20.8	4178	1886	9.27	6.23	334	252	405	284	4049	413	0.786
	8.0	48.0	61.2	28.2	15.7	5167	2317	9.19	6.16	413	309	505	353	5014	506	0.783
	10.0	59.3	75.5	22.0	12.0	6259	2784	9.10	6.07	501	371	618	430	6082	606	0.779
	12.5	73.0	93.0	17.0	9.00	7518	3310	8.99	5.97	601	441	751	520	7317	717	0.773
	16.0	91.5	117	12.6	6.38	9089	3943	8.83	5.82	727	526	924	635	8863	851	0.766
300x200	6.3	48.1	61.2	44.6	28.7	7880	4216	11.3	8.30	525	422	627	475	8468	681	0.986
	8.0	60.5	77.1	34.5	22.0	9798	5219	11.3	8.23	653	522	785	593	10550	840	0.983
	10.0	75.0	95.5	27.0	17.0	11940	6331	11.2	8.14	796	633	964	726	12890	1016	0.979
	12.5	92.6	118	21.0	13.0	14460	7619	11.1	8.04	964	762	1179	886	15650	1217	0.973
	16.0	117	149	15.7	9.50	17700	9239	10.9	7.89	1180	924	1462	1094	19230	1469	0.966
400x200	8.0	73.1	93.1	47.0	22.0	19710	6695	14.5	8.48	985	669	1210	746	15720	1135	1.18
	10.0	90.7	116	37.0	17.0	24140	8138	14.5	8.39	1207	814	1492	916	19240	1377	1.18
	12.5	112	143	29.0	13.0	29410	9820	14.3	8.29	1471	982	1831	1120	23410	1657	1.17
	16.0	142	181	22.0	9.50	36300	11950	14.2	8.14	1815	1195	2285	1388	28840	2011	1.17
450x250	8.0+	85.7	109	53.2	28.2	30270	12200	16.7	10.6	1345	976	1630	1086	27060	1629	1.38
	10.0	106	136	42.0	22.0	37180	14900	16.6	10.5	1653	1192	2013	1338	33250	1986	1.38
	12.5	132	168	33.0	17.0	45470	18100	16.5	10.4	2021	1448	2478	1642	40670	2407	1.37
	16.0	167	213	25.1	12.6	56420	22250	16.3	10.2	2508	1780	3103	2047	50480	2948	1.37
500x200	8.0+	85.7	109	59.5	22.0	34270	8170	17.7	8.65	1371	817	1716	900	21100	1430	1.38
	10.0+	106	136	47.0	17.0	42110	9945	17.6	8.57	1684	994	2119	1106	25840	1738	1.38
	12.5+	132	168	37.0	13.0	51510	12020	17.5	8.46	2060	1202	2609	1354	31480	2097	1.37
	16.0+	167	213	28.2	9.50	63930	14670	17.3	8.31	2557	1467	3267	1683	38830	2554	1.37
500x300	10.0	122	156	47.0	27.0	54120	24560	18.7	12.6	2165	1638	2609	1834	52400	2696	1.58
	12.5	152	193	37.0	21.0	66360	29970	18.5	12.5	2655	1998	3218	2257	64310	3282	1.57
	16.0	192	245	28.2	15.7	82670	37080	18.4	12.3	3307	2472	4042	2825	80220	4046	1.57
	20.0‡	237	302	22.0	12.0	100100	44550	18.2	12.1	4006	2970	4942	3442	97310	4845	1.56

+ Sections marked thus are not included in BS4848: Part 2
‡ Sections marked thus are rolled in grade 50C only
FOR EXPLANATION OF TABLES SEE NOTES 2 AND 3

Dimensions and properties

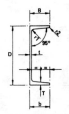

CHANNELS

DIMENSIONS

Designation		Depth Of Section D	Width Of Section B	Thickness		Root Radius	Toe Radius	Depth Between Fillets	Ratios For Local Buckling		Dimensions For Detailing			Surface Area	
Nominal Size	Mass Per Metre			Web	Flange				Flange	Web	End Clearance	Notch		Per Metre	Per Tonne
				t	T	r1	r2	d	b/T	d/t	C	N	n		
mm	kg	mm	mm	mm	mm	mm	mm	mm			mm	mm	mm	m²	m²
432x102	65.54	431.8	101.6	12.2	16.8	15.2	4.8	362.5	6.05	29.7	14	96	36	1.22	18.6
381x102	55.10	381.0	101.6	10.4	16.3	15.2	4.8	312.6	6.23	30.1	12	98	36	1.12	20.3
305x102	46.18	304.8	101.6	10.2	14.8	15.2	4.8	239.3	6.86	23.5	12	98	34	0.966	20.9
305x89	41.69	304.8	88.9	10.2	13.7	13.7	3.2	245.4	6.49	24.1	12	86	30	0.920	22.1
254x89	35.74	254.0	88.9	9.1	13.6	13.7	3.2	194.7	6.54	21.4	11	86	30	0.820	23.0
254x76	28.29	254.0	76.2	8.1	10.9	12.2	3.2	203.9	6.99	25.2	10	76	26	0.774	27.4
229x89	32.76	228.6	88.9	8.6	13.3	13.7	3.2	169.9	6.68	19.8	11	88	30	0.770	23.5
229x76	26.06	228.6	76.2	7.6	11.2	12.2	3.2	177.8	6.80	23.4	10	76	26	0.725	27.8
203x89	29.78	203.2	88.9	8.1	12.9	13.7	3.2	145.2	6.89	17.9	10	88	30	0.720	24.2
203x76	23.82	203.2	76.2	7.1	11.2	12.2	3.2	152.4	6.80	21.5	9	76	26	0.675	28.3
178x89	26.81	177.8	88.9	7.6	12.3	13.7	3.2	121.0	7.23	15.9	10	88	30	0.671	25.0
178x76	20.84	177.8	76.2	6.6	10.3	12.2	3.2	128.8	7.40	19.5	9	76	26	0.625	30.0
152x89	23.84	152.4	88.9	7.1	11.6	13.7	3.2	96.9	7.66	13.6	9	88	28	0.621	26.0
152x76	17.88	152.4	76.2	6.4	9.0	12.2	2.4	105.9	8.47	16.5	8	76	24	0.575	32.2
127x64	14.90	127.0	63.5	6.4	9.2	10.7	2.4	84.0	6.90	13.1	8	64	22	0.476	32.0
102x51	10.42♦	101.6	50.8	6.1	7.6	9.1	2.4	65.8	6.68	10.8	8	52	18	0.379	36.3
76x38	6.70♦	76.2	38.1	5.1	6.8	7.6	2.4	45.8	5.60	8.98	7	40	16	0.282	42.1

♦ Check availability of section
FOR EXPLANATION OF TABLES SEE NOTE 2

Dimensions and properties 1125

CHANNELS

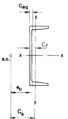

PROPERTIES

Designation		Second Moment Of Area		Radius Of Gyration		Elastic Modulus		Plastic Modulus		Buckling Parameter	Torsional Index	Warping Constant	Torsional Constant	Area of Section
Nominal Size mm	Mass Per Metre kg	Axis x-x cm^4	Axis y-y cm^4	Axis x-x cm	Axis y-y cm	Axis x-x cm^3	Axis y-y cm^3	Axis x-x cm^3	Axis y-y cm^3	u	x	H dm^6	J cm^4	A cm^2
432x102	65.54	21370	627	16.0	2.74	990	79.9	1205	153	0.875	24.5	0.216	61.5	83.4
381x102	55.10	14870	579	14.6	2.87	781	75.7	931	144	0.895	22.6	0.153	46.4	70.1
305x102	46.18	8208	499	11.8	2.91	539	66.5	638	128	0.899	18.8	0.0842	35.9	58.9
305x89	41.69	7078	326	11.5	2.47	464	48.6	559	92.9	0.887	20.3	0.0552	28.1	53.3
254x89	35.74	4445	302	9.89	2.58	350	46.7	414	89.6	0.906	17.0	0.0347	23.2	45.4
254x76	28.29	3355	162	9.67	2.12	264	28.1	316	53.9	0.886	21.2	0.0194	12.3	35.9
229x89	32.76	3383	285	9.01	2.61	296	44.8	348	86.3	0.912	15.5	0.0263	20.6	41.6
229x76	26.06	2615	159	8.87	2.19	229	28.3	271	54.5	0.901	18.7	0.0151	11.6	33.2
203x89	29.78	2492	265	8.11	2.64	245	42.4	287	81.7	0.916	14.0	0.0192	18.1	37.9
203x76	23.82	1955	152	8.02	2.24	192	27.7	226	53.5	0.911	16.5	0.0112	10.6	30.4
178x89	26.81	1753	241	7.17	2.66	197	39.3	230	75.4	0.915	12.6	0.0134	15.3	34.1
178x76	20.84	1338	134	7.10	2.25	151	24.8	176	48.1	0.911	15.3	0.00765	8.26	26.6
152x89	23.84	1168	216	6.20	2.66	153	35.8	178	68.3	0.909	11.2	0.00882	12.7	30.4
152x76	17.88	852	114	6.11	2.23	112	21.0	130	41.2	0.901	14.4	0.00486	6.05	22.8
127x64	14.90	482	67.2	5.04	1.88	76.0	15.2	89.4	29.3	0.909	11.7	0.00188	5.00	19.0
102x51	10.42	207	29.0	3.95	1.48	40.8	8.14	48.7	15.7	0.900	10.8	0.000512	2.58	13.3
76x38	6.70	74.3	10.7	2.95	1.12	19.5	4.09	23.5	7.78	0.908	9.06	0.000101	1.26	8.56

FOR EXPLANATION OF TABLES SEE NOTE 3

Dimensions and properties

CHANNELS

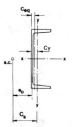

PLASTIC MODULI-MAJOR AXIS

Designation		Area of Section	Dimension				Plastic Modulus Axis x-x	Reduced Value Of Modulus Under Axial Load				
Nominal Size	Mass Per Metre		eo	Cs	Cy	Ceq		Lower Values Of n		Change Formula	Higher Values Of n	
mm	kg	cm^2	cm	cm	cm	cm	cm^3	K1	K2	At n =	K3	K4
432x102	65.54	83.4	3.13	4.83	2.31	0.966	1205	1205	1427	0.579	177	9.17
381x102	55.10	70.1	3.44	5.44	2.52	0.920	931	931	1181	0.514	125	9.65
305x102	46.18	58.9	3.59	5.73	2.65	0.966	638	638	849	0.473	89.1	9.07
305x89	41.69	53.3	2.93	4.60	2.18	0.874	559	559	696	0.527	83.0	8.79
254x89	35.74	45.4	3.20	5.16	2.42	0.894	414	414	567	0.450	60.5	8.55
254x76	28.29	35.9	2.55	4.00	1.85	0.707	316	316	398	0.519	44.4	9.28
229x89	32.76	41.6	3.32	5.42	2.53	1.10	348	348	504	0.413	50.9	8.35
229x76	26.06	33.2	2.70	4.32	2.00	0.727	271	271	363	0.467	37.9	9.01
203x89	29.78	37.9	3.44	5.68	2.65	1.40	287	287	444	0.374	42.3	8.10
203x76	23.82	30.4	2.84	4.62	2.14	0.892	226	226	325	0.418	31.7	8.73
178x89	26.81	34.1	3.55	5.93	2.76	1.70	230	230	383	0.336	34.5	7.81
178x76	20.84	26.6	2.93	4.80	2.20	1.09	176	176	267	0.386	24.5	8.65
152x89	23.84	30.4	3.65	6.16	2.87	1.99	178	178	326	0.296	27.5	7.42
152x76	17.88	22.8	2.96	4.85	2.21	1.15	130	130	203	0.371	18.3	8.49
127x64	14.90	19.0	2.46	4.08	1.94	1.09	89.4	89.4	141	0.360	15.0	7.07
102x51	10.42	13.3	1.89	3.10	1.51	0.759	48.7	48.7	72.1	0.390	9.09	6.41
76x38	6.70	8.56	1.43	2.37	1.19	0.673	23.5	23.5	35.9	0.366	5.01	5.51

eo is the distance from the centre of the web to the shear centre.
Cs is the distance from the centroidal axis to the shear centre.
Cy is the distance from the back of the web to the centroidal axis.
Ceq is the distance from the back of the web to the equal area axis.
n = F/(Ag.py). Where F is the factored axial load and Ag is the gross cross sectional area.
For lower values of n reduced plastic modulus(Sr) = K1-K2n^2
For higher values of n reduced plastic modulus(Sr) = K3(1-n)(K4 + n)
FOR EXPLANATION OF TABLES SEE NOTE 3

CHANNELS

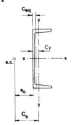

PLASTIC MODULI-MINOR AXIS

Designation		Dimension C_y	Plastic Modulus Axis y-y	Reduced Value Of Plastic Modulus Under Axial Load About Centroidal Axis											
Nominal Size	Mass Per Metre			Axial load and moment inducing stresses of same kind towards back of web						Change Formula At n =	Axial load and moment inducing stresses of opposite kind towards back of web					
				Lower Values Of n			Higher Values Of n				Lower Values Of n			Higher Values Of n		
mm	kg	cm	cm^3	K1	K2	K3	K1	K2	K3		K1	K2	K3	K1	K2	K3
432x102	65.54	2.31	153	153	40.4	2.79	125	453	0.720	0.261	153	40.4	2.79	153	40.4	2.79
381x102	55.10	2.52	144	144	32.3	3.47	139	329	0.580	0.130	144	32.3	3.47	144	32.3	3.47
305x102	46.18	2.65	128	128	28.4	3.51	127	251	0.490	0.0528	128	28.4	3.51	128	28.4	3.51
305x89	41.69	2.18	92.9	92.6	23.1	3.00	87.1	223	0.610	0.166	92.6	23.1	3.00	92.6	23.1	3.00
254x89	35.74	2.42	89.6	89.6	20.4	3.39	89.5	165	0.460	0.0204	89.6	20.4	3.39	89.6	20.4	3.39
254x76	28.29	1.85	53.9	54.1	12.8	3.24	51.7	126	0.590	0.146	54.1	12.8	3.24	54.1	12.8	3.24
229x89	32.76	2.53	86.3	86.4	89.9	0.670	86.2	153	0.440	0.0538	86.4	89.9	0.670	86.6	19.0	3.55
229x76	26.06	2.00	54.5	54.2	12.1	3.50	54.0	105	0.490	0.0494	54.2	12.1	3.50	54.2	12.1	3.50
203x89	29.78	2.65	81.7	81.6	85.2	0.560	80.6	143	0.440	0.129	81.6	85.2	0.560	82.7	17.7	3.67
203x76	23.82	2.14	53.5	53.3	54.4	0.700	53.2	95.1	0.440	0.0474	53.3	54.4	0.700	53.4	11.3	3.72
178x89	26.81	2.76	75.4	75.4	77.1	0.470	73.1	133	0.450	0.207	75.4	77.1	0.470	78.0	16.4	3.74
178x76	20.84	2.20	48.1	48.1	48.7	0.610	47.6	85.9	0.450	0.115	48.1	48.7	0.610	48.6	9.91	3.90
152x89	23.84	2.87	68.3	68.1	67.1	0.400	63.5	123	0.480	0.286	68.1	67.1	0.400	72.4	15.1	3.75
152x76	17.88	2.21	41.2	41.3	40.0	0.600	40.5	74.9	0.460	0.150	41.3	40.0	0.600	42.0	8.51	3.92
127x64	14.90	1.94	29.3	29.3	29.4	0.540	28.8	51.4	0.440	0.150	29.3	29.4	0.540	29.8	7.09	3.20
102x51	10.42	1.51	15.7	15.7	15.2	0.660	15.7	27.2	0.430	0.0671	15.7	15.2	0.660	15.8	4.34	2.63
76x38	6.70	1.19	7.78	7.76	7.60	0.580	7.71	13.2	0.420	0.0927	7.76	7.60	0.580	7.80	2.39	2.26

C_y is the distance from the back of the web to the centroidal axis.
$n = F/(A_g.p_y)$. Where F is the factored axial load and A_g is the gross cross sectional area.
For axial load and moment inducing stresses of same kind towards back of web, reduced plastic modulus$(S_r) = K1 + K2n(K3-n)$
For axial load and moment inducing stresses of opposite kind towards back of web, reduced plastic modulus$(S_r) = K1 - K2n(K3 + n)$
FOR EXPLANATION OF TABLES SEE NOTE 3

COMPOUND STRUTS
TWO CHANNELS LACED

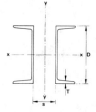

GROSS DIMENSIONS AND SECTION PROPERTIES

Composed Of Two Channels mm	Total Mass Per Metre kg	Total Area cm²	Space Between Webs s mm	Second Moment Of Area		Radius Of Gyration		Elastic Modulus		Plastic Modulus	
				Axis x-x cm⁴	Axis y-y cm⁴	Axis x-x cm	Axis y-y cm	Axis x-x cm³	Axis y-y cm³	Axis x-x cm³	Axis y-y cm³
432x102	131	167	270	42750	42990	16.0	16.0	1980	1817	2411	2639
381x102	110	140	240	29740	30690	14.6	14.8	1561	1385	1862	2035
305x102	92.4	118	180	16420	16990	11.8	12.0	1077	887	1276	1372
305x89	83.4	107	190	14160	15190	11.5	11.9	929	826	1117	1244
254x89	71.5	90.9	150	8891	9546	9.89	10.2	700	582	828	901
254x76	56.6	71.8	160	6710	7297	9.67	10.1	528	467	632	708
229x89	65.5	83.3	130	6766	7364	9.01	9.40	592	478	696	752
229x76	52.1	66.4	140	5231	5704	8.87	9.26	458	390	542	598
203x89	59.6	75.9	110	4985	5566	8.11	8.57	491	387	573	618
203x76	47.6	60.8	120	3910	4330	8.02	8.44	385	318	452	495
178x89	53.6	68.3	80	3506	3600	7.17	7.26	394	279	459	461
178x76	41.7	53.1	100	2676	3024	7.10	7.55	301	240	351	383
152x89	47.7	60.8	60	2337	2527	6.20	6.45	307	213	356	357
152x76	35.8	45.6	70	1703	1714	6.11	6.13	224	154	260	260
127x64	29.8	38.0	55	965	970	5.04	5.05	152	107	179	178
102x51	20.8♦	26.5	45	415	433	3.95	4.04	81.6	59.1	97.3	99.8
76x38	13.4♦	17.1	30	149	146	2.95	2.92	39.0	27.4	46.9	46.1

♦ Check availability of section

**COMPOUND STRUTS
TWO CHANNELS BACK TO BACK**

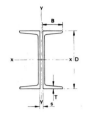

GROSS DIMENSIONS AND SECTION PROPERTIES

Composed Of Two Channels mm	Total Mass Per Metre kg	Total Area cm^2	Properties About Axis x-x				Radius of Gyration ryy About Axis y-y in cm				
			Ixx cm^4	rxx cm	Zxx cm^3	Sxx cm^3	Space between webs in mm				
							0	8	10	12	15
432x102	131	167	42750	16.0	1980	2411	3.59	3.86	3.93	4.00	4.11
381x102	110	140	29740	14.6	1561	1862	3.82	4.09	4.17	4.24	4.35
305x102	92.4	118	16420	11.8	1077	1276	3.94	4.22	4.29	4.37	4.48
305x89	83.4	107	14160	11.5	929	1117	3.30	3.57	3.65	3.72	3.84
254x89	71.5	90.9	8891	9.89	700	828	3.54	3.82	3.90	3.97	4.09
254x76	56.6	71.8	6710	9.67	528	632	2.82	3.10	3.17	3.25	3.36
229x89	65.5	83.3	6766	9.01	592	696	3.64	3.93	4.00	4.08	4.20
229x76	52.1	66.4	5231	8.87	458	542	2.97	3.25	3.32	3.40	3.52
203x89	59.6	75.9	4985	8.11	491	573	3.74	4.03	4.11	4.19	4.30
203x76	47.6	60.8	3910	8.02	385	452	3.09	3.38	3.46	3.54	3.65
178x89	53.6	68.3	3506	7.17	394	459	3.83	4.13	4.20	4.28	4.40
178x76	41.7	53.1	2676	7.10	301	351	3.15	3.44	3.52	3.59	3.71
152x89	47.7	60.8	2337	6.20	307	356	3.91	4.22	4.29	4.37	4.49
152x76	35.8	45.6	1703	6.11	224	260	3.14	3.43	3.51	3.59	3.71
127x64	29.8	38.0	965	5.04	152	179	2.70	3.00	3.08	3.16	3.28
102x51	20.8♦	26.5	415	3.95	81.6	97.3	2.12	2.42	2.50	2.58	2.70
76x38	13.4♦	17.1	149	2.95	39.0	46.9	1.64	1.95	2.03	2.11	2.24

♦ Check availability of section
Properties about y-y axis:
Iyy = (Total area) $(ryy)^2$
Zyy = Iyy/(B + 0.5s)
where s is the space between webs

1130 Dimensions and properties

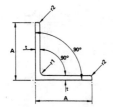

EQUAL ANGLES

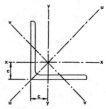

DIMENSIONS AND PROPERTIES

Designation		Mass Per Metre	Radius		Area Of Section	Distance Of Centre Of Gravity	Second Moment Of Area			Radius Of Gyration			Elastic Modulus
Size A A mm	Thickness t mm	kg	Root r1 mm	Toe r2 mm	cm²	cx and cy cm	Axis x-x, y-y cm⁴	Axis u-u cm⁴	Axis v-v cm⁴	Axis x-x, y-y cm	Axis u-u cm	Axis v-v cm	Axis x-x, y-y cm³
250x250	35	128	20.0	4.8	164	7.51	9305	14720	3886	7.54	9.49	4.88	532
	32	118	20.0	4.8	151	7.40	8650	13710	3592	7.58	9.54	4.89	491
	28	104	20.0	4.8	133	7.25	7741	12290	3194	7.63	9.61	4.90	436
	25	93.6	20.0	4.8	120	7.14	7030	11170	2890	7.67	9.67	4.92	394
200x200	24	71.1	18.0	4.8	90.8	5.85	3356	5322	1391	6.08	7.65	3.91	237
	20	59.9	18.0	4.8	76.6	5.70	2877	4569	1185	6.13	7.72	3.93	201
	18	54.2	18.0	4.8	69.4	5.62	2627	4174	1080	6.15	7.76	3.95	183
	16	48.5	18.0	4.8	62.0	5.54	2369	3765	973	6.18	7.79	3.96	164
150x150	18	40.1	16.0	4.8	51.2	4.38	1060	1680	440	4.55	5.73	2.93	99.8
	15	33.8	16.0	4.8	43.2	4.26	909	1442	375	4.59	5.78	2.95	84.6
	12	27.3	16.0	4.8	35.0	4.14	748	1187	308	4.62	5.82	2.97	68.9
	10	23.0	16.0	4.8	29.5	4.06	635	1008	262	4.64	5.85	2.99	58.0
120x120	15	26.6	13.0	4.8	34.0	3.52	448	710	186	3.63	4.57	2.34	52.8
	12	21.6	13.0	4.8	27.6	3.41	371	589	153	3.66	4.62	2.35	43.1
	10	18.2	13.0	4.8	23.3	3.32	316	502	130	3.69	4.65	2.37	36.4
	8	14.7	13.0	4.8	18.8	3.24	259	411	107	3.71	4.67	2.38	29.5
100x100	15	21.9	12.0	4.8	28.0	3.02	250	395	105	2.99	3.76	1.94	35.8
	12	17.8	12.0	4.8	22.8	2.91	208	330	86.5	3.02	3.81	1.95	29.4
	10 +	15.0	12.0	4.8	19.2	2.83	178	283	73.7	3.05	3.84	1.96	24.8
	8	12.2	12.0	4.8	15.6	2.75	146	232	60.5	3.07	3.86	1.97	20.2
90x90	12	15.9	11.0	4.8	20.3	2.66	149	235	62.0	2.70	3.40	1.75	23.5
	10	13.4	11.0	4.8	17.2	2.58	128	202	52.9	2.73	3.43	1.76	19.9
	8	10.9	11.0	4.8	13.9	2.50	105	167	43.4	2.75	3.46	1.77	16.2
	7 ♦	9.61	11.0	4.8	12.3	2.46	93.2	148	38.6	2.76	3.47	1.77	14.3
	6	8.30	11.0	4.8	10.6	2.41	81.0	128	33.6	2.76	3.48	1.78	12.3
80x80	10	11.9	10.0	4.8	15.1	2.34	87.6	139	36.4	2.41	3.03	1.55	15.5
	8	9.63	10.0	4.8	12.3	2.26	72.4	115	29.9	2.43	3.06	1.56	12.6
	6	7.34	10.0	4.8	9.36	2.17	56.0	88.7	23.2	2.45	3.08	1.57	9.60

+ Section not included in BS4848: Part 4
♦ Check availability of section
FOR EXPLANATION OF TABLES SEE NOTES 2 AND 3

Dimensions and properties

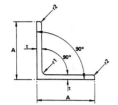

EQUAL ANGLES

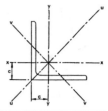

DIMENSIONS AND PROPERTIES

Designation		Mass Per Metre	Radius		Area Of Section	Distance Of Centre Of Gravity	Second Moment Of Area			Radius Of Gyration			Elastic Modulus
Size A A mm	Thickness t mm	kg	Root r1 mm	Toe r2 mm	cm²	cx and cy cm	Axis x-x, y-y cm⁴	Axis u-u cm⁴	Axis v-v cm⁴	Axis x-x, y-y cm	Axis u-u cm	Axis v-v cm	Axis x-x, y-y cm³
70x70	10	10.3	9.0	2.4	13.1	2.10	58.0	91.6	24.4	2.10	2.64	1.36	11.8
	8	8.36	9.0	2.4	10.7	2.02	48.3	76.5	20.1	2.12	2.67	1.37	9.70
	6	6.38	9.0	2.4	8.19	1.94	37.7	59.8	15.6	2.15	2.70	1.38	7.45
60x60	10	8.69	8.0	2.4	11.1	1.85	35.3	55.6	15.0	1.78	2.24	1.16	8.51
	8	7.09	8.0	2.4	9.07	1.78	29.6	46.7	12.4	1.80	2.27	1.17	7.00
	6	5.42	8.0	2.4	6.95	1.70	23.2	36.8	9.64	1.83	2.30	1.18	5.39
	5	4.57	8.0	2.4	5.86	1.65	19.8	31.4	8.23	1.84	2.31	1.18	4.56
50x50	8	5.82	7.0	2.4	7.44	1.53	16.5	25.9	6.96	1.49	1.87	0.968	4.74
	6	4.47	7.0	2.4	5.72	1.45	13.0	20.6	5.43	1.51	1.90	0.974	3.67
	5	3.77	7.0	2.4	4.83	1.41	11.1	17.7	4.63	1.52	1.91	0.979	3.11
	4	3.06	7.0	2.4	3.92	1.37	9.16	14.5	3.82	1.53	1.92	0.987	2.52
	3	2.33	7.0	2.4	2.99	1.32	7.06	11.1	2.97	1.54	1.93	0.996	1.92
45x45	6	4.00	7.0	2.4	5.12	1.33	9.30	14.7	3.90	1.35	1.69	0.872	2.93
	5	3.38	7.0	2.4	4.33	1.29	7.99	12.6	3.33	1.36	1.71	0.877	2.49
	4	2.74	7.0	2.4	3.52	1.24	6.58	10.4	2.75	1.37	1.72	0.883	2.02
	3	2.09	7.0	2.4	2.69	1.20	5.08	8.03	2.14	1.37	1.73	0.892	1.54
40x40	6	3.52	6.0	2.4	4.49	1.20	6.37	10.1	2.68	1.19	1.50	0.773	2.28
	5	2.97	6.0	2.4	3.80	1.17	5.48	8.68	2.29	1.20	1.51	0.776	1.93
	4	2.42	6.0	2.4	3.09	1.12	4.53	7.18	1.89	1.21	1.52	0.781	1.58
	3	1.84	6.0	2.4	2.36	1.08	3.51	5.55	1.47	1.22	1.53	0.788	1.20
30x30	5	2.18	5.0	2.4	2.78	0.919	2.17	3.42	0.919	0.883	1.11	0.575	1.04
	4	1.78	5.0	2.4	2.27	0.879	1.81	2.86	0.756	0.893	1.12	0.577	0.852
	3	1.36	5.0	2.4	1.74	0.836	1.41	2.23	0.588	0.900	1.13	0.581	0.652
25x25	5	1.77	3.5	2.4	2.25	0.796	1.19	1.87	0.515	0.728	0.912	0.478	0.701
	4	1.45	3.5	2.4	1.84	0.758	1.00	1.58	0.421	0.737	0.926	0.478	0.574
	3	1.11	3.5	2.4	1.41	0.718	0.784	1.24	0.325	0.745	0.939	0.480	0.440

FOR EXPLANATION OF TABLES SEE NOTES 2 AND 3

1132 Dimensions and properties

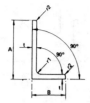

UNEQUAL ANGLES

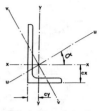

DIMENSIONS AND PROPERTIES

Designation		Mass Per Metre	Radius		Area Of Section	Distance Centre Of Gravity		Second Moment Of Area				Radius Of Gyration				Elastic Modulus		Angle x-x Axis to u-u Axis Tan α
Size A B mm	Thickness t mm	kg	Root r1 mm	Toe r2 mm	cm²	cx cm	cy cm	Axis x-x cm⁴	Axis y-y cm⁴	Axis u-u cm⁴	Axis v-v cm⁴	Axis x-x cm	Axis y-y cm	Axis u-u cm	Axis v-v cm	Axis x-x cm³	Axis y-y cm³	
200x150	18	47.1	15.0	4.8	60.1	6.34	3.86	2390	1155	2922	623	6.30	4.38	6.97	3.22	175	104	0.549
	15	39.6	15.0	4.8	50.6	6.22	3.75	2037	989	2495	531	6.34	4.42	7.02	3.24	148	87.8	0.551
	12	32.0	15.0	4.8	40.9	6.10	3.63	1667	812	2045	435	6.38	4.45	7.07	3.26	120	71.4	0.553
200x100	15	33.7	15.0	4.8	43.1	7.17	2.23	1772	303	1879	197	6.41	2.65	6.60	2.13	138	39.0	0.260
	12	27.3	15.0	4.8	34.9	7.04	2.11	1454	252	1544	162	6.45	2.68	6.65	2.15	112	31.9	0.263
	10	23.0	15.0	4.8	29.4	6.95	2.03	1233	215	1310	138	6.48	2.70	6.68	2.17	94.5	26.9	0.265
150x90	15	26.6	12.0	4.8	34.0	5.21	2.24	764	207	844	127	4.74	2.47	4.99	1.93	78.0	30.6	0.354
	12	21.6	12.0	4.8	27.6	5.09	2.12	630	172	698	104	4.78	2.50	5.03	1.95	63.6	25.0	0.359
	10	18.2	12.0	4.8	23.2	5.00	2.04	536	147	595	89.1	4.81	2.52	5.06	1.96	53.6	21.2	0.361
150x75	15	24.8	11.0	4.8	31.7	5.53	1.81	715	120	756	79.2	4.75	1.95	4.89	1.58	75.5	21.1	0.254
	12	20.2	11.0	4.8	25.7	5.41	1.70	591	100	626	65.2	4.79	1.98	4.93	1.59	61.6	17.3	0.259
	10	17.0	11.0	4.8	21.7	5.32	1.62	503	86.3	534	55.7	4.82	2.00	4.96	1.60	52.0	14.7	0.262
137x102	9.5+	17.3	11.0	4.8	22.0	4.23	2.50	415	198	506	107	4.35	3.00	4.80	2.20	43.8	25.7	0.543
	7.9+	14.5	11.0	4.8	18.4	4.16	2.44	351	168	428	90.4	4.36	3.02	4.82	2.22	36.8	21.6	0.544
	6.4+	11.7	11.0	4.8	15.0	4.09	2.37	289	138	352	74.7	4.38	3.03	4.84	2.23	30.0	17.6	0.544
125x75	12	17.8	11.0	4.8	22.7	4.31	1.84	355	96.0	392	58.8	3.95	2.06	4.16	1.61	43.4	17.0	0.354
	10	15.0	11.0	4.8	19.2	4.23	1.76	303	82.5	336	50.2	3.98	2.08	4.19	1.62	36.7	14.4	0.358
	8	12.2	11.0	4.8	15.5	4.14	1.68	249	68.1	275	41.2	4.00	2.10	4.21	1.63	29.7	11.7	0.360
	6.5+	9.98	11.0	4.8	12.7	4.07	1.62	206	56.6	228	34.3	4.02	2.11	4.23	1.64	24.4	9.62	0.361
100x75	12	15.4	10.0	4.8	19.7	3.27	2.03	189	90.3	230	49.5	3.10	2.14	3.42	1.59	28.1	16.5	0.540
	10	13.0	10.0	4.8	16.6	3.19	1.95	162	77.7	198	42.2	3.12	2.16	3.45	1.59	23.8	14.0	0.544
	8	10.6	10.0	4.8	13.5	3.10	1.87	133	64.2	163	34.7	3.14	2.18	3.48	1.61	19.3	11.4	0.547
100x65	10	12.3	10.0	4.8	15.6	3.36	1.63	154	51.1	175	30.2	3.14	1.81	3.35	1.39	23.2	10.5	0.410
	8	9.94	10.0	4.8	12.7	3.28	1.56	127	42.3	144	24.9	3.17	1.83	3.38	1.40	18.9	8.56	0.414
	7	8.77	10.0	4.8	11.2	3.23	1.51	113	37.7	128	22.1	3.18	1.84	3.39	1.41	16.6	7.56	0.415
80x60	8	8.34	8.0	4.8	10.6	2.55	1.56	65.8	31.5	80.2	17.1	2.49	1.72	2.75	1.27	12.1	7.09	0.544
	7	7.36	8.0	4.8	9.35	2.50	1.52	58.5	28.1	71.4	15.2	2.50	1.73	2.76	1.27	10.6	6.20	0.545
	6	6.37	8.0	4.8	8.08	2.46	1.48	50.9	24.5	62.2	13.2	2.51	1.74	2.77	1.28	9.19	5.41	0.546
75x50	8	7.39	7.0	2.4	9.44	2.53	1.29	52.4	18.6	60.1	10.9	2.36	1.40	2.52	1.07	10.5	5.00	0.430
	6	5.65	7.0	2.4	7.22	2.44	1.21	40.9	14.6	47.1	8.48	2.38	1.42	2.55	1.08	8.10	3.87	0.436
65x50	8	6.75	6.0	2.4	8.61	2.12	1.37	34.9	17.8	43.1	9.62	2.01	1.44	2.24	1.06	7.97	4.92	0.569
	6	5.16	6.0	2.4	6.59	2.04	1.30	27.4	14.1	34.0	7.49	2.04	1.46	2.27	1.07	6.14	3.80	0.575
	5	4.35	6.0	2.4	5.55	2.00	1.26	23.3	12.0	29.0	6.38	2.05	1.47	2.28	1.07	5.18	3.21	0.577
60x30	6	3.99	6.0	2.4	5.09	2.20	0.724	18.3	3.05	19.3	2.01	1.90	0.774	1.95	0.629	4.82	1.34	0.252
	5	3.37	6.0	2.4	4.30	2.16	0.684	15.7	2.64	16.6	1.72	1.91	0.783	1.96	0.632	4.08	1.14	0.256
40x25	4	1.93	4.0	2.4	2.45	1.36	0.621	3.86	1.15	4.32	0.692	1.26	0.685	1.33	0.532	1.46	0.612	0.380

+ Sections marked thus are not included in BS4848: Part 4
FOR EXPLANATION OF TABLES SEE NOTES 2 AND 3

Dimensions and properties 1133

**COMPOUND EQUAL ANGLES
LEGS BACK TO BACK**

GROSS DIMENSIONS AND SECTION PROPERTIES

Composed Of Two Angles		Total Mass Per Metre	Distance nx	Total Area	Properties About Axis x-x			Radius Of Gyration ryy About Axis y-y in cm				
					Ixx	rxx	Zxx	Back To Back Spacing, s, in mm				
A A mm	t mm	kg	cm	cm²	cm⁴	cm	cm³	0	8	10	12	15
250x250	35	256	17.5	327	18610	7.54	1064	10.6	10.9	11.0	11.1	11.2
	32	236	17.6	301	17300	7.58	983	10.6	10.9	10.9	11.0	11.1
	28	208	17.8	266	15480	7.63	872	10.5	10.8	10.9	10.9	11.1
	25	187	17.9	239	14060	7.67	787	10.5	10.8	10.8	10.9	11.0
200x200	24	142	14.1	182	6713	6.08	474	8.44	8.72	8.79	8.86	8.97
	20	120	14.3	153	5754	6.13	402	8.37	8.65	8.72	8.79	8.90
	18	108	14.4	139	5253	6.15	365	8.34	8.61	8.68	8.75	8.86
	16	97.0	14.5	124	4737	6.18	328	8.30	8.57	8.64	8.71	8.82
150x150	18	80.2	10.6	102	2120	4.55	200	6.31	6.60	6.67	6.74	6.86
	15	67.6	10.7	86.4	1817	4.59	169	6.26	6.54	6.61	6.68	6.79
	12	54.6	10.9	70.0	1496	4.62	138	6.21	6.48	6.55	6.62	6.73
	10	46.0	10.9	58.9	1270	4.64	116	6.17	6.44	6.51	6.58	6.68
120x120	15	53.2	8.48	68.0	896	3.63	106	5.06	5.34	5.42	5.49	5.61
	12	43.2	8.59	55.2	742	3.66	86.3	5.00	5.28	5.36	5.43	5.54
	10	36.4	8.68	46.5	632	3.69	72.9	4.96	5.24	5.31	5.38	5.50
	8	29.4	8.76	37.6	517	3.71	59.1	4.92	5.19	5.27	5.34	5.45
100x100	15	43.8	6.98	55.9	500	2.99	71.7	4.25	4.55	4.62	4.70	4.82
	12	35.6	7.09	45.5	416	3.02	58.7	4.20	4.48	4.56	4.63	4.75
	10+	30.0	7.17	38.4	356	3.05	49.7	4.16	4.44	4.51	4.59	4.70
	8	24.4	7.25	31.1	293	3.07	40.4	4.12	4.39	4.47	4.54	4.65
90x90	12	31.8	6.34	40.6	297	2.70	46.9	3.80	4.09	4.16	4.24	4.35
	10	26.8	6.42	34.3	255	2.73	39.8	3.76	4.04	4.12	4.19	4.31
	8	21.8	6.50	27.8	210	2.75	32.3	3.72	4.00	4.07	4.14	4.26
	7♦	19.2	6.54	24.5	186	2.76	28.5	3.69	3.97	4.04	4.12	4.23
	6	16.6	6.59	21.2	162	2.76	24.6	3.67	3.94	4.02	4.09	4.20
80x80	10	23.8	5.66	30.2	175	2.41	31.0	3.36	3.65	3.72	3.80	3.92
	8	19.3	5.74	24.6	145	2.43	25.2	3.31	3.60	3.67	3.75	3.87
	6	14.7	5.83	18.7	112	2.45	19.2	3.27	3.55	3.62	3.70	3.81
70x70	10	20.6	4.90	26.3	116	2.10	23.7	2.97	3.27	3.34	3.42	3.54
	8	16.7	4.98	21.4	96.6	2.12	19.4	2.93	3.22	3.30	3.38	3.49
	6	12.8	5.06	16.4	75.4	2.15	14.9	2.89	3.18	3.25	3.33	3.44
60x60	10	17.4	4.15	22.2	70.6	1.78	17.0	2.57	2.87	2.95	3.03	3.15
	8	14.2	4.22	18.1	59.1	1.80	14.0	2.53	2.83	2.91	2.98	3.11
	6	10.8	4.30	13.9	46.4	1.83	10.8	2.49	2.78	2.86	2.94	3.06
	5	9.14	4.35	11.7	39.6	1.84	9.11	2.47	2.76	2.83	2.91	3.03

+ Section not included in BS4848: Part 4
♦ Check availability of section
Properties about y-y axis:
Iyy = (Total area) $(ryy)^2$
Zyy = Iyy/(0.5Bo)

COMPOUND UNEQUAL ANGLES
SHORT LEGS BACK TO BACK

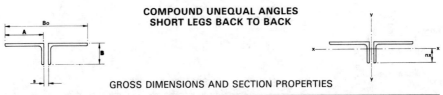

GROSS DIMENSIONS AND SECTION PROPERTIES

Composed Of Two Angles			Total Mass Per Metre	Distance nx	Total Area	Properties About Axis x-x			Radius Of Gyration ryy About Axis y-y in cm				
A B mm		t mm	kg	cm	cm^2	Ixx cm^4	rxx cm	Zxx cm^3	Back To Back Spacing, s, in mm				
									0	8	10	12	15
200x150		18	94.2	11.1	120	2310	4.38	207	8.94	9.23	9.30	9.38	9.49
		15	79.2	11.3	101	1977	4.42	176	8.88	9.17	9.24	9.32	9.43
		12	64.0	11.4	81.9	1625	4.45	143	8.83	9.11	9.18	9.25	9.36
200x100		15	67.4	7.77	86.3	607	2.65	78.0	9.62	9.92	10.0	10.1	10.2
		12	54.6	7.89	69.9	503	2.68	63.8	9.55	9.85	9.92	10.0	10.1
		10	46.0	7.97	58.8	430	2.70	53.9	9.50	9.79	9.87	9.95	10.1
150x90		15	53.2	6.76	67.9	413	2.47	61.1	7.05	7.35	7.42	7.50	7.62
		12	43.2	6.88	55.1	344	2.50	50.1	6.98	7.28	7.35	7.43	7.55
		10	36.4	6.96	46.4	295	2.52	42.4	6.94	7.23	7.31	7.38	7.50
150x75		15	49.6	5.69	63.3	240	1.95	42.2	7.29	7.60	7.68	7.76	7.88
		12	40.4	5.80	51.4	201	1.98	34.6	7.23	7.53	7.61	7.69	7.80
		10	34.0	5.88	43.3	173	2.00	29.3	7.18	7.48	7.56	7.64	7.75
137x102		9.5+	34.6	7.70	43.9	395	3.00	51.3	6.07	6.35	6.42	6.50	6.61
		7.9+	19.0	7.76	36.8	335	3.02	43.2	6.03	6.31	6.39	6.46	6.57
		6.4+	23.4	7.83	30.1	276	3.03	35.3	5.99	6.27	6.35	6.42	6.53
125x75		12	35.6	5.66	45.4	192	2.06	33.9	5.85	6.15	6.23	6.31	6.42
		10	30.0	5.74	38.3	165	2.08	28.8	5.81	6.10	6.18	6.26	6.37
		8	24.4	5.82	31.0	136	2.10	23.4	5.76	6.05	6.13	6.20	6.32
		6.5+	20.0	5.88	25.5	113	2.11	19.2	5.72	6.01	6.08	6.16	6.27
100x75		12	30.8	5.47	39.4	181	2.14	33.0	4.50	4.80	4.88	4.96	5.07
		10	26.0	5.55	33.2	155	2.16	28.0	4.46	4.76	4.83	4.91	5.03
		8	21.2	5.63	27.0	128	2.18	22.8	4.42	4.71	4.78	4.86	4.98
100x65		10	24.6	4.87	31.2	102	1.81	21.0	4.60	4.90	4.98	5.06	5.17
		8	19.9	4.94	25.4	84.7	1.83	17.1	4.56	4.85	4.93	5.00	5.12
		7	17.5	4.99	22.4	75.4	1.84	15.1	4.53	4.82	4.90	4.98	5.09
80x60		8	16.7	4.44	21.2	63.0	1.72	14.2	3.56	3.86	3.93	4.01	4.13
		7	14.7	4.48	18.7	56.1	1.73	12.5	3.54	3.83	3.91	3.99	4.10
		6	12.7	4.52	16.2	49.0	1.74	10.8	3.51	3.81	3.88	3.96	4.08
75x50		8	14.8	3.71	18.9	37.1	1.40	10.0	3.45	3.76	3.83	3.91	4.04
		6	11.3	3.79	14.4	29.3	1.42	7.73	3.41	3.71	3.79	3.87	3.98
65x50		8	13.5	3.63	17.2	35.7	1.44	9.83	2.92	3.22	3.30	3.38	3.50
		6	10.3	3.70	13.2	28.1	1.46	7.60	2.88	3.18	3.26	3.33	3.45
		5	8.70	3.74	11.1	24.1	1.47	6.43	2.86	3.15	3.23	3.31	3.43
60x30		6	7.98	2.28	10.2	6.11	0.774	2.68	2.90	3.22	3.30	3.38	3.51
		5	6.74	2.32	8.61	5.27	0.783	2.28	2.88	3.19	3.27	3.35	3.48

+ Sections marked thus are not included in BS4848: Part 4

Properties about y-y axis:
Iyy = (Total area) (ryy)^2
Zyy = Iyy/(0.5Bo)

COMPOUND UNEQUAL ANGLES
LONG LEGS BACK TO BACK

GROSS DIMENSIONS AND SECTION PROPERTIES

Composed Of Two Angles		Total Mass Per Metre	Distance nx	Total Area	Properties About Axis x-x			Radius Of Gyration ryy About Axis y-y in cm				
					Ixx	rxx	Zxx	Back To Back Spacing, s, in mm				
A B mm	t mm	kg	cm	cm²	cm⁴	cm	cm³	0	8	10	12	15
200x150	18	94.2	13.7	120	4780	6.30	350	5.84	6.11	6.18	6.25	6.36
	15	79.2	13.8	101	4075	6.34	296	5.79	6.06	6.13	6.20	6.30
	12	64.0	13.9	81.9	3335	6.38	240	5.74	6.00	6.07	6.14	6.25
200x100	15	67.4	12.8	86.3	3545	6.41	276	3.46	3.73	3.81	3.88	3.99
	12	54.6	13.0	69.9	2909	6.45	224	3.41	3.67	3.74	3.81	3.92
	10	46.0	13.1	58.8	2466	6.48	189	3.38	3.63	3.70	3.77	3.88
150x90	15	53.2	9.79	67.9	1528	4.74	156	3.33	3.61	3.69	3.76	3.87
	12	43.2	9.91	55.1	1261	4.78	127	3.28	3.55	3.62	3.70	3.81
	10	36.4	10.0	46.4	1073	4.81	107	3.24	3.51	3.58	3.65	3.76
150x75	15	49.6	9.47	63.3	1430	4.75	151	2.66	2.95	3.02	3.10	3.22
	12	40.4	9.59	51.4	1181	4.79	123	2.60	2.88	2.96	3.03	3.15
	10	34.0	9.68	43.3	1006	4.82	104	2.57	2.84	2.91	2.98	3.10
137x102	9.5+	34.6	9.47	43.9	830	4.35	87.6	3.91	4.17	4.24	4.31	4.42
	7.9+	29.0	9.54	36.8	702	4.36	73.6	3.88	4.14	4.21	4.28	4.39
	6.4+	23.4	9.61	30.1	577	4.38	60.1	3.85	4.11	4.17	4.24	4.35
125x75	12	35.6	8.19	45.4	710	3.95	86.8	2.76	3.04	3.12	3.19	3.31
	10	30.0	8.27	38.3	606	3.98	73.3	2.72	3.00	3.07	3.15	3.26
	8	24.4	8.36	31.0	497	4.00	59.5	2.69	2.95	3.03	3.10	3.21
	6.5+	20.0	8.43	25.5	411	4.02	48.8	2.66	2.92	2.99	3.06	3.17
100x75	12	30.8	6.73	39.4	378	3.10	56.2	2.95	3.24	3.32	3.39	3.51
	10	26.0	6.81	33.2	324	3.12	47.6	2.92	3.20	3.27	3.35	3.46
	8	21.2	6.90	27.0	267	3.14	38.7	2.88	3.15	3.23	3.30	3.42
100x65	10	24.6	6.64	31.2	309	3.14	46.5	2.44	2.72	2.80	2.88	2.99
	8	19.9	6.72	25.4	254	3.17	37.8	2.40	2.68	2.75	2.83	2.94
	7	17.5	6.77	22.4	225	3.18	33.3	2.38	2.65	2.73	2.80	2.92
80x60	8	16.7	5.45	21.2	132	2.49	24.1	2.32	2.61	2.69	2.76	2.88
	7	14.7	5.50	18.7	117	2.50	21.3	2.30	2.59	2.66	2.74	2.86
	6	12.7	5.54	16.2	102	2.51	18.4	2.28	2.56	2.63	2.71	2.83
75x50	8	14.8	4.97	18.9	105	2.36	21.1	1.91	2.20	2.27	2.35	2.48
	6	11.3	5.06	14.4	81.9	2.38	16.2	1.87	2.15	2.23	2.31	2.43
65x50	8	13.5	4.38	17.2	69.8	2.01	15.9	1.99	2.28	2.36	2.44	2.57
	6	10.3	4.46	13.2	54.8	2.04	12.3	1.95	2.24	2.32	2.40	2.52
	5	8.70	4.50	11.1	46.7	2.05	10.4	1.94	2.22	2.29	2.37	2.49
60x30	6	7.98	3.80	10.2	36.6	1.90	9.64	1.06	1.37	1.45	1.54	1.67
	5	6.74	3.84	8.61	31.3	1.91	8.16	1.04	1.34	1.42	1.51	1.64

+ Sections marked thus are not included in BS4848: Part 4
Properties about y-y axis:
Iyy = (Total area) (ryy)²
Zyy = Iyy/(0.5Bo)

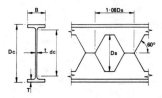

CASTELLATED UNIVERSAL BEAMS

DIMENSIONS AND PROPERTIES

Designation			Depth Of Section Dc mm	Width Of Section B mm	Thickness		Depth Between Fillets dc mm	Pitch 1.08 x Ds mm	Net Second Moment Of Area		Net Radius Of Gyration	
Original Serial Size mm	Castellated Serial Size mm	Mass Per Metre kg			Web t mm	Flange T mm			Axis x-x cm^4	Axis y-y cm^4	Axis x-x cm	Axis y-y cm
914x419	1371x419	388	1377.5	420.5	21.5	36.6	1256.1	987.1	1666000	45400	64.8	10.7
		343	1368.4	418.5	19.4	32.0	1256.2	987.1	1452000	39130	64.5	10.6
914x305	1371x305	289	1383.6	307.8	19.6	32.0	1281.4	987.1	1164000	15580	64.5	7.47
		253	1375.5	305.5	17.3	27.9	1281.5	987.1	1009000	13280	64.3	7.38
		224	1367.3	304.1	15.9	23.9	1281.3	987.1	872600	11220	64.0	7.26
		201	1360.0	303.4	15.2	20.2	1281.4	987.1	757900	9419	63.6	7.09
838x292	1257x292	226	1269.9	293.8	16.1	26.8	1180.7	905.0	780900	11350	59.4	7.16
		194	1259.7	292.4	14.7	21.7	1180.7	905.0	644500	9055	59.0	6.99
		176	1253.9	291.6	14.0	18.8	1180.7	905.0	569400	7781	58.7	6.86
762x267	1143x267	197	1150.6	268.0	15.6	25.4	1066.8	823.0	553400	8163	53.8	6.53
		173	1143.0	266.7	14.3	21.6	1066.8	823.0	475000	6840	53.5	6.42
		147	1134.9	265.3	12.9	17.5	1066.9	823.0	392300	5455	53.2	6.27
686x254	1029x254	170	1035.9	255.8	14.5	23.7	958.1	740.9	393100	6622	48.5	6.30
		152	1030.6	254.5	13.2	21.0	958.2	740.9	348000	5777	48.4	6.23
		140	1026.5	253.7	12.4	19.0	958.1	740.9	315900	5178	48.2	6.17
		125	1020.9	253.0	11.7	16.2	958.1	740.9	274400	4378	47.9	6.06
610x305	915x305	238	938.0	311.5	18.6	31.4	842.2	658.8	475300	15840	43.8	8.00
		179	922.5	307.0	14.1	23.6	842.3	658.8	349300	11390	43.5	7.85
		149	914.6	304.8	11.9	19.7	842.2	658.8	288800	9303	43.3	7.77
610x229	915x229	140	922.0	230.1	13.1	22.1	852.4	658.8	257600	4494	43.2	5.70
		125	916.9	229.0	11.9	19.6	852.3	658.8	227900	3928	43.0	5.65
		113	912.3	228.2	11.2	17.3	852.3	658.8	202700	3431	42.9	5.58
		101	907.2	227.6	10.6	14.8	852.2	658.8	176400	2912	42.6	5.48
533x210	800x210	122	811.1	211.9	12.8	21.3	743.1	575.6	175100	3383	37.9	5.27
		109	806.0	210.7	11.6	18.8	743.0	575.6	154000	2935	37.8	5.22
		101	803.2	210.1	10.9	17.4	743.0	575.6	142400	2693	37.7	5.19
		92	799.6	209.3	10.2	15.6	743.0	575.6	128100	2387	37.6	5.13
		82	794.8	208.7	9.6	13.2	743.0	575.6	110400	2002	37.4	5.03
457x191	686x191	98	695.9	192.8	11.4	19.6	636.3	493.6	105200	2344	32.6	4.86
		89	692.1	192.0	10.6	17.7	636.3	493.6	94840	2091	32.5	4.82
		82	688.7	191.3	9.9	16.0	636.3	493.6	85710	1869	32.4	4.78
		74	685.7	190.5	9.1	14.5	636.3	493.6	77420	1672	32.3	4.74
		67	682.1	189.9	8.5	12.7	636.3	493.6	68290	1451	32.1	4.69
457x152	686x152	82	693.6	153.5	10.7	18.9	635.4	493.6	83640	1142	32.3	3.77
		74	689.8	152.7	9.9	17.0	635.4	493.6	75110	1011	32.2	3.73
		67	685.7	151.9	9.1	15.0	635.3	493.6	66340	878	32.1	3.69
		60	683.2	152.9	8.0	13.3	636.2	493.6	59090	794	32.0	3.71
		52	678.3	152.4	7.6	10.9	636.1	493.6	49820	644	31.8	3.62

FOR EXPLANATION OF TABLES SEE NOTES 2 AND 3

CASTELLATED UNIVERSAL BEAMS

PROPERTIES (Continued)

Designation			Net Elastic Modulus		Elastic Modulus Of Tee		Net Plastic Modulus		Net Buckling Parameter	Net Torsional Index	Net Warping Constant	Net Torsional Constant	Net Area
Original Serial Size mm	Castellated Serial Size mm	Mass Per Metre kg	Axis x-x cm³	Axis y-y cm³	Flange cm³	Toe cm³	Axis x-x cm³	Axis y-y cm³	u	x	H dm⁶	J cm⁴	An cm²
914x419	1371x419	388	24180	2159	1376	314	25610	3289	0.975	37.9	204	1587	397
		343	21230	1870	1251	275	22420	2847	0.974	42.9	175	1082	349
914x305	1371x305	289	16820	1013	1079	289	17960	1559	0.972	44.8	71.2	816	280
		253	14670	869	964	249	15610	1336	0.972	50.9	60.3	547	244
		224	12760	738	863	223	13570	1135	0.971	58.4	50.6	361	213
		201	11150	621	773	207	11860	957	0.969	67.0	42.3	240	187
838x292	1257x292	226	12300	772	811	204	13080	1184	0.972	49.0	43.8	455	221
		194	10230	619	704	179	10880	951	0.970	59.0	34.7	261	185
		176	9082	534	639	166	9655	821	0.969	66.4	29.7	183	165
762x267	1143x267	197	9619	609	619	159	10240	935	0.972	46.7	25.8	356	191
		173	8312	513	553	141	8836	788	0.971	53.9	21.5	230	166
		147	6914	411	477	123	7344	632	0.970	64.6	17.0	133	139
686x254	1029x254	170	7590	518	480	120	8070	793	0.973	44.8	17.0	273	167
		152	6753	454	438	107	7165	695	0.973	50.1	14.7	193	149
		140	6155	408	407	98.6	6524	625	0.972	54.8	13.1	147	136
		125	5375	346	367	90.4	5697	531	0.971	62.8	11.0	97.9	119
610x305	915x305	238	10130	1017	548	134	10800	1549	0.973	30.0	32.5	724	247
		179	7573	742	437	94.6	8003	1128	0.974	39.2	23.0	311	185
		149	6316	610	380	77.0	6647	927	0.974	46.4	18.6	184	154
610x229	915x229	140	5589	391	347	86.1	5941	598	0.973	43.0	9.10	193	138
		125	4971	343	317	76.4	5273	525	0.973	48.1	7.91	137	123
		113	4443	301	291	70.2	4709	460	0.972	53.8	6.87	97.8	110
		101	3890	256	263	64.5	4121	392	0.971	61.5	5.80	65.5	97.1
533x210	800x210	122	4317	319	263	66.4	4598	489	0.973	38.9	5.28	161	122
		109	3821	279	239	58.6	4059	426	0.973	43.7	4.55	112	108
		101	3545	256	226	54.2	3760	392	0.973	46.9	4.16	90.2	100
		92	3203	228	209	49.7	3394	349	0.972	51.7	3.67	66.8	90.7
		82	2777	192	188	45.4	2941	294	0.971	59.5	3.06	43.6	79.1
457x191	686x191	98	3024	243	177	43.7	3219	371	0.974	36.4	2.68	110	99.2
		89	2741	218	165	39.6	2910	333	0.974	39.9	2.38	82.2	90.0
		82	2489	195	154	36.2	2639	298	0.973	43.8	2.11	61.8	81.9
		74	2258	176	142	32.7	2389	268	0.973	48.0	1.88	46.5	74.3
		67	2002	153	130	29.8	2116	233	0.973	54.1	1.63	32.5	66.1
457x152	686x152	82	2412	149	146	39.6	2578	229	0.973	38.2	1.30	80.2	80.2
		74	2178	132	135	35.8	2323	204	0.973	42.1	1.14	59.4	72.5
		67	1935	116	123	32.0	2060	178	0.972	47.1	0.987	41.9	64.5
		60	1730	104	112	27.9	1836	159	0.973	52.8	0.890	29.6	57.5
		52	1469	84.5	99.2	25.5	1560	130	0.971	62.4	0.717	18.0	49.3

The values of the elastic moduli of the Tee are the elastic modulus at the flange and toe of the Tee formed at the net section.
FOR EXPLANATION OF TABLES SEE NOTE 3

1138 Dimensions and properties

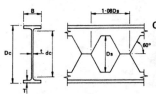

CASTELLATED UNIVERSAL BEAMS

DIMENSIONS AND PROPERTIES

Designation			Depth Of Section	Width Of Section	Thickness		Depth Between Fillets	Pitch 1.08 x Ds	Net Second Moment Of Area		Net Radius Of Gyration	
Original Serial Size mm	Castellated Serial Size mm	Mass Per Metre kg	Dc mm	B mm	Web t mm	Flange T mm	dc mm	mm	Axis x-x cm⁴	Axis y-y cm⁴	Axis x-x cm	Axis y-y cm
406x178	609x178	74	615.8	179.7	9.7	16.0	563.4	438.5	63150	1549	28.9	4.53
		67	612.4	178.8	8.8	14.3	563.4	438.5	56170	1364	28.8	4.49
		60	609.4	177.8	7.8	12.8	563.4	438.5	49840	1200	28.8	4.46
		54	605.6	177.6	7.6	10.9	563.4	438.5	43350	1019	28.6	4.38
406x140	609x140	46	605.3	142.4	6.9	11.2	562.5	438.5	36440	540	28.5	3.46
		39	600.3	141.8	6.3	8.6	562.7	438.5	28990	409	28.2	3.35
356x171	534x171	67	542.0	173.2	9.1	15.7	490.2	384.5	44920	1361	25.5	4.43
		57	536.6	172.1	8.0	13.0	490.2	384.5	37110	1105	25.3	4.37
		51	533.6	171.5	7.3	11.5	490.2	384.5	32810	968	25.2	4.33
		45	530.0	171.0	6.9	9.7	490.2	384.5	28090	809	25.1	4.25
356x127	534x127	39	530.8	126.0	6.5	10.7	489.0	384.5	23520	357	24.9	3.07
		33	526.5	125.4	5.9	8.5	489.1	384.5	19160	280	24.8	2.99
305x165	458x165	54	463.4	166.8	7.7	13.7	418.2	329.4	26930	1060	21.8	4.33
		46	459.6	165.7	6.7	11.8	418.2	329.4	22970	895	21.7	4.29
		40	456.3	165.1	6.1	10.2	418.1	329.4	19840	765	21.6	4.25
305x127	458x127	48	462.9	125.2	8.9	14.0	417.1	329.4	21980	459	21.6	3.12
		42	459.1	124.3	8.0	12.1	417.1	329.4	18940	388	21.5	3.07
		37	456.3	123.5	7.2	10.7	417.1	329.4	16670	337	21.4	3.04
305x102	458x102	33	465.2	102.4	6.6	10.8	428.4	329.4	14900	194	21.7	2.47
		28	461.4	101.9	6.1	8.9	428.4	329.4	12520	157	21.5	2.41
		25	457.3	101.6	5.8	6.8	428.5	329.4	10100	119	21.3	2.31
254x146	381x146	43	386.6	147.3	7.3	12.7	346.0	274.3	15110	677	18.2	3.85
		37	383.0	146.4	6.4	10.9	346.0	274.3	12850	570	18.1	3.81
		31	378.5	146.1	6.1	8.6	346.1	274.3	10320	447	17.9	3.73
254x102	381x102	28	387.4	102.1	6.4	10.0	352.2	274.3	9208	178	18.1	2.51
		25	384.0	101.9	6.1	8.4	352.0	274.3	7887	148	17.9	2.46
		22	381.0	101.6	5.8	6.8	352.2	274.3	6607	119	17.8	2.39
203x133	305x133	30	308.3	133.8	6.3	9.6	273.9	219.2	6663	384	14.5	3.48
		25	304.7	133.4	5.8	7.8	273.9	219.2	5452	309	14.4	3.43
203x102	305x102	23	304.7	101.6	5.2	9.3	270.9	219.2	4875	163	14.3	2.62
178x102	267x102	19	266.8	101.6	4.7	7.9	235.8	192.2	3165	138	12.6	2.63
152x89	229x89	16	228.4	88.9	4.6	7.7	197.8	164.2	1958	90.3	10.7	2.31
127x76	191x76	13	190.5	76.2	4.2	7.6	160.1	137.2	1123	56.2	8.92	2.00

FOR EXPLANATION OF TABLES SEE NOTES 2 AND 3

Dimensions and properties 1139

CASTELLATED UNIVERSAL BEAMS

PROPERTIES (Continued)

Designation			Net Elastic Modulus		Elastic Modulus Of Tee		Net Plastic Modulus		Net Buckling Parameter	Net Torsional Index	Net Warping Constant	Net Torsional Constant	Net Area
Original Serial Size mm	Castellated Serial Size mm	Mass Per Metre kg	Axis x-x cm³	Axis y-y cm³	Flange x-x cm³	Toe x-x cm³	Axis x-x cm³	Axis y-y cm³	u	x	H dm⁶	J cm⁴	An cm²
406x178	609x178	74	2051	172	122	28.9	2177	263	0.974	38.9	1.39	57.5	75.6
		67	1835	153	112	25.6	1942	233	0.974	43.2	1.22	41.5	67.7
		60	1636	135	102	22.3	1727	206	0.974	48.0	1.07	29.8	60.3
		54	1432	115	93.7	21.0	1511	175	0.972	55.0	0.901	19.9	53.1
406x140	609x140	46	1204	75.8	78.5	18.8	1275	116	0.973	54.7	0.476	17.0	45.0
		39	966	57.7	66.1	16.4	1023	88.7	0.971	68.0	0.358	8.83	36.4
356x171	534x171	67	1657	157	92.7	21.4	1759	239	0.974	34.7	0.942	51.2	69.3
		57	1383	128	81.6	17.9	1462	196	0.974	41.2	0.758	30.1	58.0
		51	1230	113	74.9	16.0	1297	172	0.974	46.0	0.659	21.3	51.6
		45	1060	94.6	67.9	14.6	1117	144	0.972	53.2	0.548	13.7	44.7
356x127	534x127	39	886	56.7	55.9	13.5	939	87.1	0.974	49.7	0.242	13.3	37.8
		33	728	44.6	48.1	11.8	771	68.6	0.972	60.1	0.188	7.43	31.3
305x165	458x165	54	1162	127	61.6	13.3	1229	193	0.974	33.8	0.536	32.0	56.5
		46	1000	108	55.0	11.1	1053	164	0.974	38.8	0.449	20.7	48.6
		40	870	92.7	50.1	9.78	914	141	0.974	44.4	0.381	13.7	42.3
305x127	458x127	48	950	73.3	54.2	14.6	1015	113	0.973	33.1	0.231	27.9	47.3
		42	825	62.4	48.7	12.7	879	96.0	0.973	37.7	0.194	18.5	41.2
		37	731	54.5	44.2	11.1	776	83.7	0.973	42.2	0.167	13.0	36.5
305x102	458x102	33	641	37.8	41.9	11.3	685	58.4	0.971	44.2	0.100	10.7	31.8
		28	543	30.9	36.8	10.0	580	47.7	0.969	52.1	0.0805	6.53	27.1
		25	442	23.5	31.0	9.07	473	36.5	0.966	63.8	0.0605	3.57	22.4
254x146	381x146	43	782	91.9	39.2	8.76	829	139	0.973	30.2	0.237	22.4	45.7
		37	671	77.9	35.1	7.35	708	118	0.973	34.9	0.197	14.3	39.3
		31	545	61.2	31.5	6.56	574	93.0	0.972	42.8	0.153	7.69	32.2
254x102	381x102	28	475	34.8	29.4	7.62	507	53.5	0.972	38.7	0.0633	8.57	28.2
		25	411	29.1	26.5	6.94	438	44.9	0.970	44.6	0.0524	5.56	24.5
		22	347	23.5	23.4	6.32	370	36.3	0.968	52.4	0.0417	3.41	20.9
203x133	305x133	30	432	57.3	22.0	4.79	457	87.0	0.970	31.1	0.0856	9.36	31.6
		25	358	46.3	19.7	4.14	377	70.4	0.969	37.1	0.0681	5.39	26.3
203x102	305x102	23	320	32.0	16.1	3.66	339	48.8	0.975	32.2	0.0355	6.40	23.7
178x102	267x102	19	237	27.2	11.8	2.55	251	41.4	0.974	32.5	0.0232	4.06	20.0
152x89	229x89	16	171	20.3	7.94	1.83	182	30.9	0.973	28.1	0.0110	3.36	17.0
127x76	191x76	13	118	14.7	4.74	1.17	126	22.5	0.974	23.4	0.00470	2.76	14.1

The values of the elastic moduli of the Tee are the elastic modulus at the flange and toe of the Tee formed at the net section.
FOR EXPLANATION OF TABLES SEE NOTE 3

Dimensions and properties

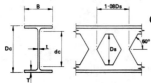

CASTELLATED UNIVERSAL COLUMNS

DIMENSIONS AND PROPERTIES

Designation			Depth Of Section	Width Of Section	Thickness		Depth Between Fillets	Pitch 1.08 x Ds	Net Second Moment Of Area		Net Radius Of Gyration	
Original Serial Size mm	Castellated Serial Size mm	Mass Per Metre kg	Dc mm	B mm	Web t mm	Flange T mm	dc mm	mm	Axis x-x cm⁴	Axis y-y cm⁴	Axis x-x cm	Axis y-y cm
356x406	546x406	634	665.2	424.1	47.6	77.0	480.8	411.5	600500	98020	28.9	11.7
		551	646.2	418.5	42.0	67.5	480.8	411.5	503800	82550	28.5	11.5
		467	627.1	412.4	35.9	58.0	480.7	411.5	413300	67860	28.0	11.3
		393	609.6	407.0	30.6	49.2	480.8	411.5	336500	55320	27.6	11.2
		340	596.9	403.0	26.5	42.9	480.7	411.5	284400	46820	27.3	11.1
		287	584.2	399.0	22.6	36.5	480.8	411.5	234800	38660	27.0	10.9
		235	571.5	395.0	18.5	30.2	480.7	411.5	188300	31030	26.7	10.8
COLCORE	559x424	477	630.2	424.4	48.0	53.2	493.4	438.9	408900	67900	28.3	11.5
356x368	534x368	202	552.7	374.4	16.8	27.0	468.3	384.5	152100	23630	25.8	10.2
		177	546.3	372.1	14.5	23.8	468.3	384.5	131700	20440	25.7	10.1
		153	540.0	370.2	12.6	20.7	468.2	384.5	112900	17510	25.5	10.1
		129	533.6	368.3	10.7	17.5	468.2	384.5	94100	14570	25.4	9.99
305x305	458x305	283	517.8	321.8	26.9	44.1	399.2	329.4	172000	24520	23.2	8.76
		240	505.1	317.9	23.0	37.7	399.3	329.4	141900	20200	22.9	8.64
		198	492.4	314.1	19.2	31.4	399.2	329.4	114000	16230	22.6	8.53
		158	479.7	310.6	15.7	25.0	399.3	329.4	87910	12490	22.3	8.40
		137	473.0	308.7	13.8	21.7	399.2	329.4	75010	10640	22.1	8.34
		118	467.0	306.8	11.9	18.7	399.2	329.4	63620	9004	22.0	8.27
		97	460.3	304.8	9.9	15.4	399.1	329.4	51560	7271	21.8	8.20
254x254	381x254	167	416.1	264.5	19.2	31.7	327.3	274.3	66900	9785	18.9	7.21
		132	403.4	261.0	15.6	25.3	327.4	274.3	51270	7502	18.6	7.10
		107	393.7	258.3	13.0	20.5	327.3	274.3	40310	5891	18.3	7.00
		89	387.4	255.9	10.5	17.3	327.4	274.3	33170	4834	18.2	6.94
		73	381.0	254.0	8.6	14.2	327.2	274.3	26660	3880	18.0	6.88
203x203	305x203	86	323.8	208.8	13.0	20.5	262.4	219.2	21430	3112	14.9	5.67
		71	317.4	206.2	10.3	17.3	262.4	219.2	17480	2529	14.7	5.61
		60	311.1	205.2	9.3	14.2	262.3	219.2	14130	2046	14.6	5.54
		52	307.7	203.9	8.0	12.5	262.3	219.2	12230	1767	14.5	5.51
		46	304.7	203.2	7.3	11.0	262.3	219.2	10680	1539	14.4	5.47
152x152	228x152	37	237.8	154.4	8.1	11.5	199.6	164.2	5038	706	11.1	4.14
		30	233.5	152.9	6.6	9.4	199.5	164.2	4018	560	11.0	4.10
		23	228.4	152.4	6.1	6.8	199.6	164.2	2926	401	10.8	4.00

FOR EXPLANATION OF TABLES SEE NOTES 2 AND 3

Dimensions and properties

CASTELLATED UNIVERSAL COLUMNS

PROPERTIES (Continued)

Designation			Net Elastic Modulus		Elastic Modulus Of Tee		Net Plastic Modulus		Net Buckling Parameter	Net Torsional Index	Net Warping Constant	Net Torsional Constant	Net Area
Original Serial Size mm	Castellated Serial Size mm	Mass Per Metre kg	Axis x-x cm³	Axis y-y cm³	Flange x-x cm³	Toe x-x cm³	Axis x-x cm³	Axis y-y cm³	u	x	H dm⁶	J cm⁴	An cm²
356x406	546x406	634	18050	4623	707	325	20640	7004	0.946	7.81	84.8	13040	717
		551	15590	3945	573	245	17620	5973	0.946	8.72	69.1	8762	622
		467	13180	3291	453	177	14700	4978	0.947	9.95	54.9	5523	527
		393	11040	2719	359	127	12160	4109	0.947	11.5	43.4	3363	442
		340	9528	2324	297	96.5	10400	3510	0.947	13.0	35.9	2221	382
		287	8037	1938	244	71.4	8683	2925	0.948	15.1	29.0	1367	323
		235	6588	1571	196	50.6	7044	2370	0.948	17.9	22.7	773	265
COLCORE	559x424	477	12980	3200	427	178	14390	4864	0.945	10.5	56.5	4956	510
356x368	534x368	202	5502	1262	198	46.9	5875	1905	0.949	19.5	16.3	533	228
		177	4822	1099	175	37.9	5118	1657	0.949	21.9	14.0	364	200
		153	4180	946	155	30.8	4412	1426	0.950	24.9	11.8	240	173
		129	3527	791	136	24.4	3701	1193	0.950	29.2	9.70	147	146
305x305	458x305	283	6643	1524	256	90.6	7380	2309	0.951	10.9	13.8	1935	319
		240	5617	1271	210	67.6	6170	1924	0.951	12.5	11.0	1208	270
		198	4630	1033	170	48.9	5028	1563	0.951	14.7	8.62	699	223
		158	3665	804	137	34.3	3932	1216	0.952	18.1	6.46	357	177
		137	3172	690	120	27.8	3381	1042	0.952	20.6	5.42	235	153
		118	2725	587	106	22.3	2886	886	0.952	23.6	4.52	152	132
		97	2240	477	90.7	17.1	2357	720	0.953	28.2	3.60	86.2	108
254x254	381x254	167	3216	740	115	37.8	3535	1120	0.951	12.2	3.61	595	188
		132	2542	575	88.9	25.5	2755	869	0.951	15.0	2.68	304	149
		107	2048	456	72.2	18.3	2195	689	0.951	18.1	2.05	164	120
		89	1712	378	60.1	13.5	1820	570	0.952	21.2	1.66	98.0	100
		73	1399	305	50.4	10.1	1476	461	0.952	25.4	1.30	54.8	82.0
203x203	305x203	86	1324	298	47.4	13.8	1437	451	0.951	14.8	0.716	130	96.9
		71	1101	245	38.3	9.84	1183	371	0.952	17.3	0.569	77.3	80.5
		60	908	199	33.2	7.76	968	301	0.951	20.6	0.451	44.2	66.6
		52	795	173	29.3	6.29	842	262	0.952	23.2	0.385	30.1	58.2
		46	701	151	26.9	5.40	740	229	0.952	26.1	0.332	20.9	51.4
152x152	228x152	37	424	91.4	16.9	4.17	454	138	0.951	19.4	0.0904	17.9	41.1
		30	344	73.3	14.1	3.08	365	111	0.952	23.3	0.0703	9.87	33.4
		23	256	52.7	12.3	2.48	270	79.8	0.951	30.5	0.0493	4.24	25.1

The values of the elastic moduli of the Tee are the elastic modulus at the flange and toe of the Tee formed at the net section.
FOR EXPLANATION OF TABLES SEE NOTE 3

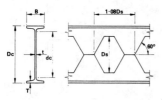

CASTELLATED JOISTS

DIMENSIONS AND PROPERTIES

Designation			Depth Of Section	Width Of Section	Thickness		Depth Between Fillets	Pitch 1.08 x Ds	Net Second Moment Of Area		Net Radius Of Gyration	
Original Serial Size mm	Castellated Serial Size mm	Mass Per Metre kg	Dc mm	B mm	Web t mm	Flange T mm	dc mm	mm	Axis x-x cm^4	Axis y-y cm^4	Axis x-x cm	Axis y-y cm
254x203	381x203	81.85	381.0	203.2	10.2	19.9	293.6	274.3	28700	2279	17.7	4.99
254x114	381x114	37.20	381.0	114.3	7.6	12.8	326.3	274.3	11910	269	17.8	2.67
203x152	305x152	52.09	304.7	152.4	8.9	16.5	234.7	219.2	11460	815	14.1	3.76
152x127	228x127	37.20	228.4	127.0	10.4	13.2	170.3	164.2	4354	378	10.5	3.09
127x114	191x114	29.76	190.5	114.3	10.2	11.5	143.0	137.2	2352	241	8.73	2.80
		26.79	190.5	114.3	7.4	11.4	143.0	137.2	2274	236	8.78	2.83
127x76	191x76	16.37	190.5	76.2	5.6	9.6	150.0	137.2	1361	60.8	8.81	1.86
114x114	171x114	26.79	171.3	114.3	9.5	10.7	117.8	123.1	1777	224	7.82	2.77
102x102	153x102	23.07	152.6	101.6	9.5	10.3	106.2	110.2	1180	154	6.94	2.51
102x44	153x44	7.44♦	152.6	44.5	4.3	6.1	125.6	110.2	362	7.78	7.04	1.03
89x89	134x89	19.35	133.4	88.9	9.5	9.9	88.7	96.1	749	101	6.02	2.21
76x76	114x76	14.67♦	114.2	80.0	8.9	8.4	76.1	82.1	419	60.7	5.16	1.96
		12.65	114.2	76.2	5.1	8.4	76.1	82.1	386	51.7	5.19	1.90

♦ Check availability of section
FOR EXPLANATION OF TABLES SEE NOTES 2 AND 3

CASTELLATED JOISTS

PROPERTIES (Continued)

Designation			Net Elastic Modulus		Elastic Modulus Of Tee		Net Plastic Modulus		Net Buckling Parameter	Net Torsional Index	Net Warping Constant	Net Torsional Constant	Net Area
Original Serial Size mm	Castellated Serial Size mm	Mass Per Metre kg	Axis x-x cm^3	Axis y-y cm^3	Flange x-x cm^3	Toe x-x cm^3	Axis x-x cm^3	Axis y-y cm^3	u	x	H dm^6	J cm^4	A_n cm^2
254x203	381x203	81.85	1506	224	47.6	13.3	1617	367	0.969	16.1	0.743	147	91.6
254x114	381x114	37.20	625	47.0	30.9	8.19	668	77.3	0.975	26.5	0.0911	23.3	37.7
203x152	305x152	52.09	752	107	24.4	7.18	810	174	0.970	15.7	0.169	62.4	57.5
152x127	228x127	37.20	381	59.5	13.3	4.50	414	97.7	0.964	13.8	0.0437	31.0	39.6
127x114	191x114	29.76	247	42.2	8.56	3.00	269	69.2	0.960	13.1	0.0193	18.5	30.9
		26.79	239	41.2	7.51	2.41	258	67.3	0.962	13.8	0.0189	16.0	29.5
127x76	191x76	16.37	143	15.9	5.27	1.60	154	25.9	0.974	17.0	0.00497	6.35	17.5
114x114	171x114	26.79	207	39.1	7.07	2.59	226	64.5	0.953	11.8	0.0144	17.2	29.0
102x102	153x102	23.07	155	30.2	5.09	1.97	169	49.5	0.953	11.2	0.00778	12.7	24.5
102x44	153x44	7.44	47.5	3.50	2.21	0.711	51.2	5.79	0.973	21.2	0.000418	1.12	7.31
89x89	134x89	19.35	112	22.7	3.79	1.57	124	37.0	0.951	9.92	0.00385	10.2	20.6
76x76	114x76	14.67	73.3	15.2	2.49	1.06	81.0	25.0	0.948	9.76	0.00170	5.94	15.8
		12.65	67.5	13.6	2.04	0.790	74.1	22.2	0.954	10.8	0.00145	4.43	14.3

The values of the elastic moduli of the Tee are the elastic modulus at the flange and toe of the Tee formed at the net section.
FOR EXPLANATIONS OF TABLES SEE NOTE 3

Dimensions and properties

STRUCTURAL TEES CUT FROM UNIVERSAL BEAMS

DIMENSIONS

Designation		Cut From		Width Of Section	Depth Of Section	Thickness		Root Radius	Ratios for Local Buckling		Dimension Cx
Serial Size mm	Mass Per Metre kg	Serial Size mm	Mass Per Metre kg	B mm	d mm	Web t mm	Flange T mm	r mm	d/t	b/T	cm
419x457	194	914x419	388	420.5	460.2	21.5	36.6	24.1	21.4	5.74	10.3
	172		344	418.5	455.7	19.4	32.0	24.1	23.5	6.54	10.2
305x457	145	914x305	290	307.8	463.3	19.6	32.0	19.1	23.6	4.81	12.2
	127		254	305.5	459.2	17.3	27.9	19.1	26.5	5.47	12.0
	112		224	304.1	455.2	15.9	23.9	19.1	28.6	6.36	12.1
	101		202	303.4	451.5	15.2	20.2	19.1	29.7	7.51	12.6
292x419	113	838x292	226	293.8	425.4	16.1	26.8	17.8	26.4	5.48	10.8
	97		194	292.4	420.4	14.7	21.7	17.8	28.6	6.74	11.1
	88		176	291.6	417.4	14.0	18.8	17.8	29.8	7.76	11.4
267x381	99	762x267	198	268.0	384.8	15.6	25.4	16.5	24.7	5.28	9.9
	87		174	266.7	381.0	14.3	21.6	16.5	26.6	6.17	10.0
	74		148	265.3	376.9	12.9	17.5	16.5	29.2	7.58	10.2
254x343	85	686x254	170	255.8	346.5	14.5	23.7	15.2	23.9	5.40	8.7
	76		152	254.5	343.8	13.2	21.0	15.2	26.0	6.06	8.6
	70		140	253.7	341.8	12.4	19.0	15.2	27.6	6.68	8.6
	63		126	253.0	339.0	11.7	16.2	15.2	29.0	7.81	8.9
305x305	119	610x305	238	311.5	316.5	18.6	31.4	16.5	17.0	4.96	7.1
	90		180	307.0	308.7	14.1	23.6	16.5	21.9	6.50	6.6
	75		150	304.8	304.8	11.9	19.7	16.5	25.6	7.74	6.4
229x305	70	610x229	140	230.1	308.5	13.1	22.1	12.7	23.5	5.21	7.6
	63		126	229.0	305.9	11.9	19.6	12.7	25.7	5.84	7.5
	57		114	228.2	303.7	11.2	17.3	12.7	27.1	6.60	7.6
	51		102	227.6	301.1	10.6	14.8	12.7	28.4	7.69	7.8
210x267	61	533x210	122	211.9	272.3	12.8	21.3	12.7	21.3	4.97	6.7
	55		110	210.7	269.7	11.6	18.8	12.7	23.2	5.60	6.6
	51		102	210.1	268.4	10.9	17.4	12.7	24.6	6.04	6.6
	46		92	209.3	266.6	10.2	15.6	12.7	26.1	6.71	6.6
	41		82	208.7	264.2	9.6	13.2	12.7	27.5	7.91	6.8
191x229	49	457x191	98	192.8	233.7	11.4	19.6	10.2	20.5	4.92	5.5
	45		90	192.0	231.8	10.6	17.7	10.2	21.9	5.42	5.5
	41		82	191.3	230.1	9.9	16.0	10.2	23.2	5.98	5.5
	37		74	190.5	228.6	9.1	14.5	10.2	25.1	6.57	5.4
	34		68	189.9	226.8	8.5	12.7	10.2	26.7	7.48	5.5
152x229	41	457x152	82	153.5	232.5	10.7	18.9	10.2	21.7	4.06	6.0
	37		74	152.7	230.6	9.9	17.0	10.2	23.3	4.49	6.0
	34		68	151.9	228.6	9.1	15.0	10.2	25.1	5.06	6.0
	30		60	152.9	227.3	8.0	13.3	10.2	28.4	5.75	5.8
	26		52	152.4	224.9	7.6	10.9	10.2	29.6	6.99	6.0

FOR EXPLANATION OF TABLES SEE NOTE 2

Dimensions and properties 1145

STRUCTURAL TEES
CUT FROM UNIVERSAL BEAMS

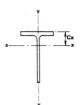

PROPERTIES

Designation		Second Moment Of Area		Radius Of Gyration		Elastic Modulus			Plastic Modulus		Buckling Parameter	Torsional Index	Torsional Constant	Area
Serial Size mm	Mass Per Metre kg	Axis x-x cm^4	Axis y-y cm^4	Axis x-x cm	Axis y-y cm	Axis x-x Flange cm^3	Axis x-x Toe cm^3	Axis y-y cm^3	Axis x-x cm^3	Axis y-y cm^3	u	x	J cm^4	A cm^2
419x457	194	44180	22720	13.4	9.58	4280	1238	1081	2190	1671	0.528	13.4	867	247
	172	38960	19580	13.3	9.46	3823	1101	936	1942	1445	0.534	15.1	595	219
305x457	145	37720	7807	14.3	6.50	3085	1106	507	1980	802	0.654	16.0	464	185
	127	32720	6651	14.2	6.42	2721	965	435	1728	685	0.655	18.1	312	161
	112	29080	5618	14.3	6.27	2395	871	369	1566	582	0.667	20.7	211	143
	101	26490	4716	14.4	6.06	2110	813	311	1474	492	0.687	23.4	146	128
292x419	113	24620	5680	13.1	6.27	2276	776	387	1384	606	0.639	17.5	256	144
	97	21350	4533	13.2	6.06	1926	690	310	1238	487	0.659	20.9	152	123
	88	19560	3895	13.2	5.90	1723	644	267	1164	421	0.675	23.3	110	112
267x381	99	17510	4087	11.8	5.71	1770	613	305	1093	479	0.641	16.6	202	125
	87	15480	3425	11.9	5.58	1550	550	257	986	404	0.653	19.1	133	110
	74	13320	2731	11.9	5.39	1307	484	206	874	324	0.671	22.6	79.9	94.0
254x343	85	12060	3315	10.5	5.53	1389	464	259	826	406	0.623	15.9	153	108
	76	10760	2892	10.5	5.46	1250	418	227	743	355	0.627	17.8	110	97.0
	70	9918	2592	10.5	5.39	1149	388	204	692	319	0.633	19.4	84.1	89.2
	63	8983	2191	10.6	5.24	1015	359	173	643	271	0.651	22.0	57.9	79.7
305x305	119	12300	7926	8.99	7.22	1731	501	509	895	788	0.483	10.6	394	152
	90	8934	5699	8.86	7.08	1344	369	371	651	571	0.482	13.8	169	114
	75	7372	4654	8.80	6.99	1145	307	305	538	469	0.483	16.3	100	95.1
229x305	70	7740	2250	9.32	5.03	1017	333	196	592	305	0.613	15.3	108	89.1
	63	6892	1966	9.30	4.97	915	299	172	531	268	0.617	17.1	76.9	79.6
	57	6305	1717	9.34	4.88	828	277	150	493	235	0.628	19.0	55.9	72.2
	51	5723	1457	9.40	4.75	733	257	128	459	200	0.646	21.5	38.7	64.7
210x267	61	5196	1694	8.16	4.66	777	253	160	450	250	0.602	13.8	89.4	78.0
	55	4603	1469	8.14	4.60	697	226	139	401	218	0.606	15.5	63.0	69.4
	51	4278	1348	8.14	4.57	652	211	128	374	200	0.608	16.6	50.7	64.6
	46	3916	1195	8.15	4.50	595	195	114	346	178	0.615	18.2	38.0	58.9
	41	3531	1002	8.21	4.38	523	180	96.0	321	150	0.634	20.8	25.7	52.3
191x229	49	2974	1174	6.89	4.33	537	167	122	296	189	0.573	12.9	60.5	62.6
	45	2709	1046	6.89	4.28	493	153	109	272	169	0.578	14.1	45.5	57.1
	41	2480	935	6.89	4.23	453	141	97.8	250	152	0.584	15.5	34.5	52.3
	37	2248	837	6.88	4.20	415	129	87.9	228	136	0.586	16.9	26.0	47.6
	34	2039	726	6.91	4.12	373	118	76.5	210	119	0.598	19.0	18.5	42.8
152x229	41	2612	572	7.07	3.31	433	152	74.5	271	118	0.641	13.7	44.6	52.3
	37	2367	506	7.06	3.26	395	139	66.3	248	105	0.645	15.0	33.3	47.6
	34	2122	440	7.05	3.21	356	126	57.9	225	91.3	0.651	16.8	23.7	42.7
	30	1865	397	7.01	3.24	321	110	52.0	197	81.4	0.646	18.8	16.7	37.9
	26	1672	322	7.08	3.11	277	102	42.3	183	66.7	0.671	22.0	10.7	33.3

Properties have been calculated assuming that there is no loss of material due to splitting.
– Indicates that no values of u and x are given as there is no possibility of lateral torsional buckling due to bending about the x-x axis because the second moment of area about y-y axis exceeds the second moment of area about x-x axis.
The value of H is virtually zero and has not therefore been given.
FOR EXPLANATION OF TABLES SEE NOTE 3

Dimensions and properties

STRUCTURAL TEES CUT FROM UNIVERSAL BEAMS

DIMENSIONS

Designation		Cut From		Width Of Section	Depth Of Section	Thickness		Root Radius	Ratios for Local Buckling		Dimension Cx
Serial Size	Mass Per Metre	Serial Size	Mass Per Metre	B	d	Web t	Flange T	r	d/t	b/T	
mm	kg	mm	kg	mm	mm	mm	mm	mm			cm
178x203	37	406x178	74	179.7	206.4	9.7	16.0	10.2	21.3	5.62	4.8
	34		68	178.8	204.7	8.8	14.3	10.2	23.3	6.25	4.7
	30		60	177.8	203.2	7.8	12.8	10.2	26.1	6.95	4.6
	27		54	177.6	201.3	7.6	10.9	10.2	26.5	8.15	4.8
140x203	23	406x140	46	142.4	201.2	6.9	11.2	10.2	29.2	6.36	5.0
	20		40	141.8	198.6	6.3	8.6	10.2	31.5	8.24	5.3
171x178	34	356x171	68	173.2	182.0	9.1	15.7	10.2	20.0	5.52	4.0
	29		58	172.1	179.3	8.0	13.0	10.2	22.4	6.62	4.0
	26		52	171.5	177.8	7.3	11.5	10.2	24.4	7.46	3.9
	23		46	171.0	176.0	6.9	9.7	10.2	25.5	8.81	4.0
127x178	20	356x127	40	126.0	176.4	6.5	10.7	10.2	27.1	5.89	4.4
	17		34	125.4	174.2	5.9	8.5	10.2	29.5	7.38	4.5
165x152	27	305x165	54	166.8	155.4	7.7	13.7	8.9	20.2	6.09	3.2
	23		46	165.7	153.5	6.7	11.8	8.9	22.9	7.02	3.1
	20		40	165.1	151.9	6.1	10.2	8.9	24.9	8.09	3.1
127x152	24	305x127	48	125.2	155.2	8.9	14.0	8.9	17.4	4.47	3.9
	21		42	124.3	153.3	8.0	12.1	8.9	19.2	5.14	3.9
	19		38	123.5	151.9	7.2	10.7	8.9	21.1	5.77	3.8
102x153	17	305x102	34	102.4	156.3	6.6	10.8	7.6	23.7	4.74	4.1
	14		28	101.9	154.4	6.1	8.9	7.6	25.3	5.72	4.2
	13		26	101.6	152.4	5.8	6.8	7.6	26.3	7.47	4.5
146x127	22	254x146	44	147.3	129.8	7.3	12.7	7.6	17.8	5.80	2.7
	19		38	146.4	128.0	6.4	10.9	7.6	20.0	6.72	2.6
	16		32	146.1	125.7	6.1	8.6	7.6	20.6	8.49	2.7
102x127	14	254x102	28	102.1	130.2	6.4	10.0	7.6	20.3	5.11	3.3
	13		26	101.9	128.5	6.1	8.4	7.6	21.1	6.07	3.3
	11		22	101.6	127.0	5.8	6.8	7.6	21.9	7.47	3.5
133x102	15	203x133	30	133.8	103.4	6.3	9.6	7.6	16.4	6.97	2.1
	13		26	133.4	101.6	5.8	7.8	7.6	17.5	8.55	2.1
102x102	12	203x102	24	101.6	101.6	5.2	9.3	7.6	19.5	5.46	2.2
102x89	10	178x102	20	101.6	88.9	4.7	7.9	7.6	18.9	6.43	1.8
89x76	8	152x89	16	88.9	76.2	4.6	7.7	7.6	16.6	5.77	1.6
76x64	7	127x76	14	76.2	63.5	4.2	7.6	7.6	15.1	5.01	1.3

FOR EXPLANATION OF TABLES SEE NOTE 2

STRUCTURAL TEES CUT FROM UNIVERSAL BEAMS

PROPERTIES

Designation		Second Moment Of Area		Radius Of Gyration		Elastic Modulus			Plastic Modulus		Buckling Parameter	Torsional Index	Torsional Constant	Area
Serial Size mm	Mass Per Metre kg	Axis x-x cm^4	Axis y-y cm^4	Axis x-x cm	Axis y-y cm	Axis x-x		Axis y-y cm^3	Axis x-x cm^3	Axis y-y cm^3	u	x	J cm^4	A cm^2
						Flange cm^3	Toe cm^3							
178x203	37	1766	775	6.09	4.03	367	112	86.3	198	134	0.559	13.8	31.7	47.7
	34	1575	682	6.07	3.99	333	100	76.3	177	118	0.561	15.2	23.0	42.8
	30	1384	600	6.03	3.97	300	88.1	67.5	155	104	0.559	16.9	16.5	38.1
	27	1283	510	6.12	3.86	267	83.7	57.4	148	89.0	0.585	19.2	11.4	34.3
140x203	23	1130	270	6.19	3.03	224	75.0	37.9	133	59.3	0.635	19.4	9.60	29.5
	20	964	205	6.26	2.89	183	66.1	28.9	119	45.4	0.665	23.8	5.24	24.6
171x178	34	1162	681	5.21	3.99	289	81.9	78.6	145	121	0.501	12.2	27.8	42.8
	29	983	553	5.22	3.91	248	70.4	64.3	124	99.2	0.513	14.5	16.5	36.1
	26	879	484	5.22	3.87	224	63.4	56.5	112	87.0	0.519	16.1	11.8	32.3
	23	795	405	5.28	3.77	197	58.6	47.3	103	73.1	0.544	18.5	7.81	28.5
127x178	20	718	179	5.39	2.69	163	54.2	28.4	96.5	44.5	0.629	17.6	7.44	24.7
	17	617	140	5.43	2.59	137	47.8	22.3	85.6	35.1	0.651	21.2	4.31	20.9
165x152	27	633	530	4.31	3.94	199	51.2	63.6	91.0	97.6	0.380	11.9	17.1	34.1
	23	539	448	4.28	3.90	175	43.9	54.1	77.4	82.8	0.383	13.6	11.1	29.4
	20	477	383	4.30	3.85	156	39.3	46.4	69.0	71.0	0.402	15.5	7.42	25.8
127x152	24	654	230	4.64	2.75	167	56.4	36.7	101	57.9	0.599	11.7	15.7	30.4
	21	571	194	4.62	2.70	148	49.8	31.3	88.7	49.2	0.606	13.3	10.5	26.7
	19	504	169	4.61	2.67	133	44.2	27.3	78.8	42.8	0.608	14.8	7.44	23.7
102x153	17	487	97.1	4.82	2.15	118	42.3	19.0	75.8	30.0	0.656	15.8	6.08	20.9
	14	427	78.8	4.84	2.08	101	38.1	15.5	68.6	24.6	0.674	18.5	3.83	18.2
	13	373	59.7	4.89	1.96	83.5	34.7	11.8	63.1	18.9	0.705	22.1	2.27	15.6
146x127	22	348	339	3.56	3.51	131	33.7	46.0	60.3	70.6	0.239	10.6	12.0	27.5
	19	297	285	3.54	3.47	115	29.0	39.0	51.5	59.7	0.262	12.2	7.71	23.7
	16	263	224	3.63	3.35	97.9	26.6	30.6	46.8	47.1	0.386	14.8	4.31	20.0
102x127	14	281	89.0	3.93	2.21	85.8	28.8	17.4	51.3	27.4	0.610	13.7	4.82	18.2
	13	253	74.4	3.96	2.15	75.6	26.6	14.6	47.6	23.0	0.632	15.7	3.25	16.1
	11	226	59.7	4.00	2.06	64.9	24.5	11.7	44.3	18.7	0.660	18.1	2.11	14.1
133x102	15	153	192	2.84	3.18	72.9	18.5	28.7	33.1	44.0	–	–	5.09	19.0
	13	133	155	2.87	3.10	62.8	16.5	23.2	29.2	35.6	–	–	3.01	16.1
102x102	12	117	81.4	2.84	2.37	54.3	14.6	16.0	26.1	24.7	0.453	11.3	3.43	14.5
102x89	10	72.4	69.2	2.45	2.39	40.1	10.2	13.6	18.2	20.9	0.275	11.3	2.18	12.1
89x76	8	44.1	45.2	2.07	2.10	28.1	7.29	10.2	13.1	15.7	–	–	1.80	10.2
76x64	7	23.3	28.1	1.67	1.83	18.1	4.59	7.37	8.47	11.4	–	–	1.45	8.39

Properties have been calculated assuming that there is no loss of material due to splitting.
– Indicates that no values of u and x are given as there is no possibility of lateral torsional buckling due to bending about the x-x axis because the second moment of area about y-y axis exceeds the second moment of area about x-x axis.
The value of H is virtually zero and has not therefore been given.
FOR EXPLANATION OF TABLES SEE NOTE 3

Dimensions and properties

STRUCTURAL TEES CUT FROM UNIVERSAL COLUMNS

DIMENSIONS

Designation		Cut From		Width Of Section	Depth Of Section	Thickness		Root Radius	Ratios for Local Buckling		Dimension Cx
Serial Size	Mass Per Metre	Serial Size	Mass Per Metre	B	d	Web t	Flange T	r	d/t	b/T	
mm	kg	mm	kg	mm	mm	mm	mm	mm			cm
406x178	317	356x406	634	424.1	237.4	47.6	77.0	15.2	4.99	2.75	6.1
	276		552	418.5	227.8	42.0	67.5	15.2	5.42	3.10	5.6
	234		468	412.4	218.3	35.9	58.0	15.2	6.08	3.56	5.0
	197		394	407.0	209.6	30.6	49.2	15.2	6.85	4.14	4.5
	170		340	403.0	203.2	26.5	42.9	15.2	7.67	4.70	4.2
	144		288	399.0	196.9	22.6	36.5	15.2	8.71	5.47	3.8
	118		236	395.0	190.5	18.5	30.2	15.2	10.3	6.54	3.4
406x178	239	COLCORE	478	424.4	213.5	48.0	53.2	15.2	4.45	3.99	5.4
368x178	101	356x368	202	374.4	187.3	16.8	27.0	15.2	11.1	6.93	3.3
	89		178	372.1	184.2	14.5	23.8	15.2	12.7	7.82	3.1
	77		154	370.2	181.0	12.6	20.7	15.2	14.4	8.94	2.9
	65		130	368.3	177.8	10.7	17.5	15.2	16.6	10.5	2.7
305x152	142	305x305	284	321.8	182.6	26.9	44.1	15.2	6.79	3.65	4.1
	120		240	317.9	176.3	23.0	37.7	15.2	7.67	4.22	3.7
	99		198	314.1	169.9	19.2	31.4	15.2	8.85	5.00	3.4
	79		158	310.6	163.6	15.7	25.0	15.2	10.4	6.21	3.0
	69		138	308.7	160.3	13.8	21.7	15.2	11.6	7.11	2.9
	59		118	306.8	157.2	11.9	18.7	15.2	13.2	8.20	2.7
	49		98	304.8	153.9	9.9	15.4	15.2	15.5	9.90	2.5
254x127	84	254x254	168	264.5	144.5	19.2	31.7	12.7	7.53	4.17	3.1
	66		132	261.0	138.2	15.6	25.3	12.7	8.86	5.16	2.7
	54		108	258.3	133.4	13.0	20.5	12.7	10.3	6.30	2.5
	45		90	255.9	130.2	10.5	17.3	12.7	12.4	7.40	2.2
	37		74	254.0	127.0	8.6	14.2	12.7	14.8	8.94	2.1
203x102	43	203x203	86	208.8	111.1	13.0	20.5	10.2	8.55	5.09	2.2
	36		72	206.2	108.0	10.3	17.3	10.2	10.5	5.96	2.0
	30		60	205.2	104.8	9.3	14.2	10.2	11.3	7.23	1.9
	26		52	203.9	103.1	8.0	12.5	10.2	12.9	8.16	1.8
	23		46	203.2	101.6	7.3	11.0	10.2	13.9	9.24	1.7
152x76	19	152x152	38	154.4	80.9	8.1	11.5	7.6	9.99	6.71	1.5
	15		30	152.9	78.7	6.6	9.4	7.6	11.9	8.13	1.4
	12		24	152.4	76.2	6.1	6.8	7.6	12.5	11.2	1.4

FOR EXPLANATION OF TABLES SEE NOTE 2

Dimensions and properties

STRUCTURAL TEES
CUT FROM UNIVERSAL COLUMNS

PROPERTIES

Designation		Second Moment Of Area		Radius Of Gyration		Elastic Modulus			Plastic Modulus		Torsional Constant	Area
						Axis x-x		Axis				
Serial Size mm	Mass Per Metre kg	Axis x-x cm⁴	Axis y-y cm⁴	Axis x-x cm	Axis y-y cm	Flange cm³	Toe cm³	y-y cm³	Axis x-x cm³	Axis y-y cm³	J cm⁴	A cm²
406x178	317	11970	49100	5.44	11.0	1962	679	2315	1504	3556	6810	404
	276	9570	41330	5.22	10.9	1718	556	1975	1219	3029	4583	351
	234	7431	33970	5.00	10.7	1480	442	1647	957	2520	2891	298
	197	5786	27680	4.81	10.5	1278	352	1360	748	2077	1763	250
	170	4695	23430	4.66	10.4	1131	290	1163	608	1772	1165	216
	144	3752	19340	4.53	10.3	991	236	969	483	1475	718	183
	118	2881	15520	4.38	10.2	846	184	786	368	1193	405	150
406x178	239	8720	34040	5.36	10.6	1623	546	1604	1089	2491	2797	304
368x178	101	2503	11820	4.40	9.57	754	162	631	317	959	279	129
	89	2103	10220	4.32	9.52	678	137	550	264	833	191	113
	77	1769	8756	4.25	9.46	606	117	473	221	717	126	97.8
	65	1455	7288	4.20	9.39	532	96.7	396	180	599	76.8	82.6
305x152	142	3286	12270	4.27	8.25	800	232	763	488	1168	1012	180
	120	2609	10110	4.13	8.14	698	188	636	388	972	632	153
	99	2018	8119	4.00	8.02	598	148	517	299	789	366	126
	79	1526	6248	3.90	7.89	502	115	402	224	613	188	100
	69	1288	5324	3.85	7.82	450	97.8	345	188	525	124	87.1
	59	1071	4503	3.78	7.76	399	82.2	294	155	446	79.9	74.8
	49	857	3636	3.73	7.68	343	66.5	239	123	362	45.5	61.6
254x127	84	1198	4896	3.36	6.79	390	105	370	219	566	311	106
	66	886	3753	3.24	6.67	326	79.8	288	161	438	160	84.3
	54	686	2947	3.17	6.57	278	63.1	228	124	347	86.3	68.3
	45	533	2417	3.06	6.52	238	49.5	189	95.6	287	51.3	56.8
	37	417	1940	3.00	6.46	203	39.2	153	74.0	232	28.7	46.5
203x102	43	381	1557	2.63	5.32	171	42.8	149	86.1	228	68.6	55.0
	36	289	1265	2.52	5.28	145	32.8	123	65.2	187	40.4	45.5
	30	242	1023	2.52	5.19	129	28.2	99.7	54.0	152	23.4	38.0
	26	203	884	2.47	5.16	115	23.7	86.7	45.0	132	15.9	33.2
	23	179	770	2.47	5.12	105	21.2	75.7	39.5	115	11.1	29.4
152x76	19	94.4	353	2.00	3.87	61.1	14.4	45.7	27.5	69.8	9.60	23.6
	15	73.1	280	1.95	3.82	51.6	11.3	36.7	21.2	55.8	5.28	19.2
	12	61.1	201	2.03	3.68	42.6	9.88	26.3	17.7	40.2	2.39	14.8

Properties have been calculated assuming that there is no loss of material due to splitting.
Values of u and x have not been given as there is no possibility of lateral torsional buckling due to bending about the x-x axis, because the second moment of area about y-y axis exceeds the second moment of area about x-x axis.
The value of H is virtually zero and has not therefore been given.
FOR EXPLANATION OF TABLES SEE NOTE 3

Extracts from BS 5950: Part 1

BS 5950: Part 1: 1990 Section two

Table 5. Deflection limits other than for pitched roof portal frames

(a) *Deflection on beams due to unfactored imposed load*	
Cantilevers	Length/180
Beams carrying plaster or other brittle finish	Span/360
All other beams	Span/200
Purlins and sheeting rails	See clause 4.12.2
(b) *Horizontal deflection of columns other than portal frames due to unfactored imposed and wind loads*	
Tops of columns in single-storey buildings	Height/300
In each storey of a building with more than one storey	Height of storey under consideration/300
(c) *Crane gantry girders*	
Vertical deflection due to static wheel loads	Span/600
Horizontal deflection (calculated on the top flange properties alone) due to crane surge	Span/500

NOTE 1. On low-pitched and flat roofs the possibility of ponding needs consideration.
NOTE 2. For limiting deflections in runway beams refer to BS 2853.

BS 5950: Part 1: 1990: Section three

Table 6. Design strengths, p_y, for steel to BS 4360

BS 4360 Grade	Thickness, less than or equal to	Sections, plates and hollow sections p_y
	mm	N/mm²
43	16	275
	40	265
	63	255
	80	245
	100	235
50	16	355
	40	345
	63	335
	80	325
	100	315
55	16	450
	25	430
	40	415
	63	400

BS 5950: Part 1: 1990: Section three

Table 7. Limiting width to thickness ratios
(Elements which exceed these limits are to be taken as class 4, slender cross sections.)

Type of element	Type of section	Class of section		
		(1) Plastic	(2) Compact	(3) Semi-compact
Outstand element of compression flange	Built-up by welding	$\frac{b}{T} \leq 7.5\varepsilon$	$\frac{b}{T} \leq 8.5\varepsilon$	$\frac{b}{T} \leq 13\varepsilon$
	Rolled sections	$\frac{b}{T} \leq 8.5\varepsilon$	$\frac{b}{T} \leq 9.5\varepsilon$	$\frac{b}{T} \leq 15\varepsilon$
Internal element of compression flange	Built-up by welding	$\frac{b}{T} \leq 23\varepsilon$	$\frac{b}{T} \leq 25\varepsilon$	$\frac{b}{T} \leq 28\varepsilon$
	Rolled sections	$\frac{b}{T} \leq 26\varepsilon$	$\frac{b}{T} \leq 32\varepsilon$	$\frac{b}{T} \leq 39\varepsilon$
Web, with neutral axis at mid-depth	All sections	$\frac{d}{t} \leq 79\varepsilon$	$\frac{d}{t} \leq 98\varepsilon$	$\frac{d}{t} \leq 120\varepsilon$
Web, generally	All sections	$\frac{d}{t} \leq \frac{79\varepsilon}{0.4 + 0.6\alpha}$	$\frac{d}{t} \leq \frac{98\varepsilon}{\alpha}$	See clause 3.5.4
Web, where whole section is subject to compression	Built-up by welding	$\frac{d}{t} \leq 28\varepsilon$	$\frac{d}{t} \leq 28\varepsilon$	$\frac{d}{t} \leq 28\varepsilon$
	Rolled sections	$\frac{d}{t} \leq 39\varepsilon$	$\frac{d}{t} \leq 39\varepsilon$	$\frac{d}{t} \leq 39\varepsilon$
Legs of single angle and double angle members with components separated	Rolled angle sections	$\frac{b}{T} \leq 8.5\varepsilon$ and $\frac{d}{T} \leq 8.5\varepsilon$	$\frac{b}{T} \leq 9.5\varepsilon$ and $\frac{d}{T} \leq 9.5\varepsilon$	$\frac{b}{T}$ and $\frac{d}{T} \leq 15\varepsilon$ and $\frac{b+d}{T} \leq 23\varepsilon$
Outstand legs of double angle members with angles in contact back-to-back	Rolled angle sections	$\frac{b}{T} \leq 8.5\varepsilon$	$\frac{b}{T} \leq 9.5\varepsilon$	$\frac{b}{T} \leq 15\varepsilon$
Stems of T-sections	T-section	$\frac{d}{t} \leq 8.5\varepsilon$	$\frac{d}{t} \leq 9.5\varepsilon$	$\frac{d}{t} \leq 19\varepsilon$
Circular tube subject to moment or axial compression	CHS or built-up by welding	$\frac{D}{t} \leq 40\varepsilon^2$	$\frac{D}{t} \leq 57\varepsilon^2$	$\frac{D}{t} \leq 80\varepsilon^2$

NOTE 1. Dimensions b, D, d, T, t are as defined in figure 3. of BS 5950: part 1.
$$\alpha = \frac{2y_c}{d}$$
where y_c is the distance from the plastic neutral axis to the edge of the web connected to the compression flange. But if $\alpha > 2$ the section should be taken as having compression throughout.
NOTE 2. Check webs for shear buckling in accordance with clause 4.4 when $d/t > 63\varepsilon$.

NOTE 3. $\varepsilon = \left(\frac{275}{p_y}\right)^{1/2}$

BS 5950: Part 1: 1990: Section four

Table 11. Bending strength p_b (in N/mm^2) for rolled sections

λ_{LT} \ p_y	245	265	275	325	340	355	415	430	450
30	245	265	275	325	340	355	408	421	438
35	245	265	273	316	328	341	390	402	418
40	238	254	262	302	313	325	371	382	397
45	227	242	250	287	298	309	350	361	374
50	217	231	238	272	282	292	329	338	350
55	206	219	226	257	266	274	307	315	325
60	195	207	213	241	249	257	285	292	300
65	185	196	201	225	232	239	263	269	276
70	174	184	188	210	216	222	242	247	253
75	164	172	176	195	200	205	223	226	231
80	154	161	165	181	186	190	204	208	212
85	144	151	154	168	172	175	188	190	194
90	135	141	144	156	159	162	173	175	178
95	126	131	134	144	147	150	159	161	163
100	118	123	125	134	137	139	147	148	150
105	111	115	117	125	127	129	136	137	139
110	104	107	109	116	118	120	126	127	128
115	97	101	102	108	110	111	117	118	119
120	91	94	96	101	103	104	108	109	111
125	86	89	90	95	96	97	101	102	103
130	81	83	84	89	90	91	94	95	96
135	76	78	79	83	84	85	88	89	90
140	72	74	75	78	79	80	83	84	84
145	68	70	71	74	75	75	78	79	79
150	64	66	67	70	70	71	73	74	75
155	61	62	63	66	66	67	69	70	70
160	58	59	60	62	63	63	65	66	66
165	55	56	57	59	60	60	62	62	63
170	52	53	54	56	56	57	59	59	59
175	50	51	51	53	54	54	56	56	56
180	47	48	49	51	51	51	53	53	53
185	45	46	46	48	49	49	50	50	51
190	43	44	44	46	46	47	48	48	48
195	41	42	42	44	44	44	46	46	46
200	39	40	40	42	42	42	43	44	44
210	36	37	37	38	39	39	40	40	40
220	33	34	34	35	35	36	36	37	37
230	31	31	31	32	33	33	33	34	34
240	29	29	29	30	30	30	31	31	31
250	27	27	27	28	28	28	29	29	29

Table 12. Bending strength p_b (in N/mm²) for welded sections

λ_{LT} \ p_y	245	265	275	325	340	355	415	430	450
30	245	265	275	325	340	355	401	412	427
35	245	265	272	307	317	328	368	378	391
40	231	244	250	282	292	301	337	346	358
45	212	224	230	259	268	276	308	316	327
50	196	207	212	238	246	253	282	288	297
55	180	190	195	219	225	232	257	263	275
60	167	176	180	201	207	212	245	253	264
65	154	162	166	188	196	204	235	242	251
70	142	150	155	182	189	196	224	230	238
75	135	145	151	175	182	188	212	218	225
80	131	141	146	168	174	179	201	205	211
85	127	136	140	160	165	171	188	190	194
90	123	131	135	152	157	162	173	175	178
95	118	125	129	144	147	150	159	161	163
100	113	120	123	134	137	139	147	148	150
105	109	115	117	125	127	129	136	137	139
110	104	107	109	116	118	120	126	127	128
115	97	101	102	108	110	111	117	118	119
120	91	94	96	101	103	104	108	109	111
125	86	89	90	95	96	97	101	102	103
130	81	83	84	89	90	91	94	95	96
135	76	78	79	83	84	85	88	89	90
140	72	74	75	78	79	80	83	84	84
145	68	70	71	74	75	75	78	79	79
150	64	66	67	70	70	71	73	74	75
155	61	62	63	66	66	67	69	70	70
160	58	59	60	62	63	63	65	66	66
165	55	56	57	59	60	60	62	62	63
170	52	53	54	56	56	57	59	59	59
175	50	51	51	53	54	54	56	56	56
180	47	48	49	51	51	51	53	53	53
185	45	46	46	48	49	49	50	50	51
190	43	44	44	46	46	47	48	48	48
195	41	42	42	44	44	44	46	46	46
200	39	40	40	42	42	42	43	44	44
210	36	37	37	38	39	39	40	40	40
220	33	34	34	35	35	36	36	37	37
230	31	31	31	32	33	33	33	34	34
240	29	29	29	30	30	30	31	31	31
250	27	27	27	28	28	28	29	29	29

BS 5950: Part 1: 1990: Section four

Table 25. Strut table selection

Type of section	Thickness (see note 1)	Axis of buckling	
		$x-x$	$y-y$
Hot-rolled structural hollow section		27(a)	27(a)
Rolled I-section (or as shown in table 26(a))		27(a)	27(b)
Rolled H-section (or as shown in table 26(a))	up to 40 mm over 40 mm	27(b) 27(c)	27(c) 27(d)
Welded plate I or H-section (see note 2 and clause 4.7.5) (or as shown in table 26(c))	up to 40 mm over 40 mm	27(b) 27(b)	27(c) 27(d)
Rolled I or H-section with welded flange cover plates (as shown in table 26(b))	up to 40 mm over 40 mm	27(b) 27(c)	27(a) 27(b)
Welded box section (see note 3 and clause 4.7.5)	up to 40 mm over 40 mm	27(b) 27(c)	27(b) 27(c)
Round, square or flat bar	up to 40 mm over 40 mm	27(b) 27(c)	27(b) 27(c)
Rolled angle Rolled channel or T-section Two rolled sections laced or battened Two rolled sections back-to-back Compound rolled sections		Buckling about any axis 27(c)	

NOTE 1. For thicknesses between 40 mm and 50 mm the value of p_c may be taken as the average of the values for thicknesses up to 40 mm and over 40 mm.
NOTE 2. For welded plate I or H-sections where it can be guaranteed that the edges of the flanges will only be flame-cut, table 27(b) may be used for buckling about the $y-y$ axis for flanges up to 40 mm thick, and table 27(c) for flanges over 40 mm thick.
NOTE 3. 'Welded box section' includes any box section fabricated from plates or rolled sections, provided that all longitudinal welds are near the corners of the section. Box sections with welded longitudinal stiffeners are *not* included in this category.

BS 5950: Part 1: 1990: Section four

Table 27(a). Compressive strength p_c (in N/mm^2) for struts

λ \ p_y	225	245	255	265	275	305	320	325	335	340	355	395	410	415	430	450
15	225	245	255	265	275	305	320	325	335	340	355	394	409	414	429	448
20	225	244	254	264	273	303	317	322	332	337	351	390	405	410	424	444
25	222	241	251	261	270	299	314	318	328	333	347	386	400	405	419	438
30	220	239	248	258	267	296	310	315	324	329	343	381	395	399	414	432
35	217	236	245	254	264	292	306	310	320	324	338	375	389	393	407	425
40	214	233	242	251	260	287	301	305	315	319	333	368	382	386	399	417
42	213	231	240	249	258	285	299	303	312	317	330	365	378	383	396	413
44	212	230	239	248	257	283	297	301	310	314	327	362	375	379	392	409
46	210	228	237	246	255	281	294	299	307	312	325	359	371	375	388	404
48	209	227	236	244	253	279	292	296	305	309	322	355	367	371	383	399
50	208	225	234	242	251	277	289	293	302	306	318	351	363	367	379	394
52	206	223	232	241	249	274	286	291	299	303	315	346	358	362	373	388
54	205	222	230	238	247	271	283	287	295	299	311	342	353	356	367	381
56	203	220	228	236	244	268	280	284	292	296	307	336	347	350	361	374
58	201	218	226	234	242	265	277	281	288	292	303	331	341	344	354	366
60	200	216	224	232	239	262	273	277	284	288	298	325	335	337	347	358
62	198	214	221	229	236	259	269	273	280	283	293	318	328	330	339	349
64	196	211	219	226	234	255	265	268	275	278	288	311	320	322	331	340
66	194	209	216	223	230	251	261	264	270	273	282	304	312	314	322	330
68	192	206	213	220	227	247	256	259	265	268	276	296	304	306	313	320
70	189	204	210	217	224	242	251	254	259	262	270	288	295	297	303	310
72	187	201	207	214	220	237	246	248	253	256	264	280	287	288	294	299
74	184	198	204	210	216	233	240	243	247	250	256	272	278	279	284	289
76	182	194	200	206	212	227	235	237	241	243	249	264	269	270	275	279
78	179	191	197	202	208	222	229	231	235	237	242	255	260	261	265	269
80	176	188	193	198	203	217	223	225	229	230	235	247	251	252	256	259
82	173	184	189	194	199	211	217	219	222	224	228	239	243	243	247	250
84	170	181	185	190	194	206	211	213	216	217	221	231	234	235	238	240
86	167	177	181	186	190	200	205	207	209	211	214	223	226	226	229	231
88	164	173	177	181	185	195	199	200	203	204	208	215	218	218	221	223
90	161	169	173	177	180	189	193	195	197	198	201	208	211	211	213	215
92	158	166	169	173	176	184	188	189	191	192	194	201	203	203	206	207
94	154	162	165	168	171	179	182	183	185	186	188	194	196	196	198	200
96	151	158	161	164	166	173	176	177	179	180	182	187	190	189	191	192
98	147	154	157	159	162	168	171	172	173	174	176	181	183	183	185	186
100	144	150	153	155	157	163	166	167	168	169	171	175	176	177	178	179
102	141	146	149	151	153	158	161	161	163	163	165	169	170	171	172	173
104	137	142	145	147	149	154	156	156	158	158	160	163	165	165	166	167
106	134	139	141	143	145	149	151	152	153	153	155	158	159	159	160	161
108	131	135	137	139	141	145	146	147	148	149	150	153	154	154	155	156

BS 5950: Part 1: 1990: Section four

Table 27(a). *(concluded)*

λ \ p_y	225	245	255	265	275	305	320	325	335	340	355	395	410	415	430	450
110	127	132	133	135	137	140	142	143	144	144	145	148	149	149	150	151
112	124	128	130	131	133	136	138	138	139	140	141	143	144	144	145	146
114	121	125	126	128	129	132	134	134	135	135	136	139	140	140	141	141
116	118	121	123	124	125	129	130	130	131	131	132	135	135	135	136	137
118	115	118	120	121	122	125	126	126	127	127	128	130	131	131	132	133
120	112	115	116	118	119	121	122	123	123	124	125	127	127	127	128	129
122	109	112	113	114	115	118	119	119	120	120	121	123	123	123	124	125
124	107	109	110	111	112	115	116	116	116	117	117	119	120	120	120	121
126	104	106	107	108	109	111	112	113	113	113	114	116	116	116	117	117
128	101	104	105	105	106	108	109	109	110	110	111	112	113	113	113	114
130	99	101	102	103	103	105	106	106	107	107	108	109	110	110	110	111
135	93	95	96	96	97	98	99	99	100	100	101	102	102	102	103	103
140	87	89	90	90	91	92	93	93	93	94	94	95	95	96	96	96
145	82	84	84	85	85	86	87	87	87	88	88	89	89	89	90	90
150	78	79	79	80	80	81	82	82	82	82	83	83	84	84	84	89
155	73	74	75	75	75	76	77	77	77	77	78	78	79	79	79	79
160	69	70	70	71	71	72	72	72	73	73	73	74	74	74	74	74
165	65	66	67	67	68	68	68	69	69	69	69	70	70	70	70	70
170	62	63	63	63	64	64	65	65	65	65	65	66	66	66	66	66
175	59	59	60	60	60	61	61	61	61	61	62	62	62	62	62	63
180	56	56	57	57	57	58	58	58	58	58	58	59	59	59	59	59
185	53	54	54	54	54	55	55	55	55	55	55	56	56	56	56	56
190	51	51	51	51	52	52	52	52	52	53	53	53	53	53	53	53
195	48	49	49	49	49	50	50	50	50	50	50	51	51	51	51	51
200	46	46	47	47	47	47	47	47	48	48	48	48	48	48	48	48
210	42	42	42	42	43	43	43	43	43	43	43	44	44	44	44	44
220	38	39	39	39	39	39	39	39	40	40	40	40	40	40	40	40
230	35	35	36	36	36	36	36	36	36	36	36	37	37	37	37	37
240	33	33	33	33	33	33	33	33	33	33	33	34	34	34	34	34
250	30	30	30	30	30	31	31	31	31	31	31	31	31	31	31	31
260	28	28	28	28	28	28	29	29	29	29	29	29	29	29	29	29
270	26	26	26	26	26	26	27	27	27	27	27	27	27	27	27	27
280	24	24	24	24	24	25	25	25	25	25	25	25	25	25	25	25
290	23	23	23	23	23	23	23	23	23	23	23	23	23	23	23	23
300	21	21	21	21	21	21	22	22	22	22	22	22	22	22	22	22
310	20	20	20	20	20	20	20	20	20	20	20	20	20	20	20	21
320	19	19	19	19	19	19	19	19	19	19	19	19	19	19	19	19
330	18	18	18	18	18	18	18	18	18	18	18	18	18	18	18	18
340	17	17	17	17	17	17	17	17	17	17	17	17	17	17	17	17
350	16	16	16	16	16	16	16	16	16	16	16	16	16	16	16	16

Extracts from BS 5950: Part 1

BS 5950: Part 1: 1990: Section four

Table 27(b). Compressive strength p_c (in N/mm^2) for struts

λ \ p_y	225	245	255	265	275	305	320	325	335	340	355	395	410	415	430	450
15	225	245	255	265	275	305	320	325	335	340	355	394	409	413	428	447
20	224	243	253	263	272	301	315	320	330	334	349	387	401	406	420	439
25	220	239	248	258	267	295	309	314	323	328	342	379	393	397	411	430
30	216	234	243	253	262	289	303	307	316	321	335	371	384	389	402	420
35	211	229	238	247	256	283	296	300	309	313	327	361	374	379	392	409
40	207	224	233	241	250	276	288	293	301	305	318	351	364	368	380	396
42	205	222	231	239	248	273	285	289	298	302	314	347	359	363	375	391
44	203	220	228	237	245	270	282	286	294	298	310	342	354	358	369	385
46	201	218	226	234	242	267	279	283	291	294	306	337	349	352	364	379
48	199	215	223	231	239	263	275	279	287	291	302	332	343	347	358	372
50	197	213	221	229	237	260	271	275	283	286	298	327	337	341	351	365
52	195	210	218	226	234	256	267	271	278	282	293	321	331	334	349	358
54	192	208	215	223	230	253	263	267	274	278	288	315	325	328	337	350
56	190	205	213	220	227	249	259	263	269	273	283	309	318	321	330	342
58	188	202	210	217	224	245	255	258	265	268	278	302	311	314	322	333
60	185	200	207	214	221	241	250	254	260	263	272	295	304	306	314	325
62	183	197	204	210	217	236	246	249	255	258	266	288	296	299	306	316
64	180	194	200	207	213	232	241	244	249	252	261	281	289	291	298	307
66	178	191	197	203	210	227	236	239	244	247	255	274	281	283	289	298
68	175	188	194	200	206	223	231	233	239	241	249	267	273	275	281	288
70	172	185	190	196	202	218	226	228	233	235	242	259	265	267	272	279
72	169	181	187	193	198	213	220	223	227	230	236	252	257	259	264	270
74	167	178	183	189	194	208	215	217	222	224	230	244	249	251	255	261
76	164	175	180	185	190	204	210	212	216	218	223	237	241	243	247	252
78	161	171	176	181	186	199	205	206	210	212	217	230	234	235	239	244
80	158	168	172	177	181	194	199	201	204	206	211	222	226	227	231	235
82	155	164	169	173	177	189	194	196	199	200	205	215	219	220	223	227
84	152	161	165	169	173	184	189	190	193	195	199	209	212	213	216	219
86	149	157	161	165	169	179	183	185	188	189	193	202	205	206	208	212
88	146	154	158	161	165	174	178	180	182	183	187	195	198	199	201	204
90	143	150	154	157	161	169	173	175	177	178	181	189	192	192	195	197
92	139	147	150	153	156	165	168	170	172	173	176	183	185	186	188	191
94	136	143	147	150	152	160	164	165	167	168	171	177	179	180	182	184
96	133	140	143	146	148	156	159	160	162	163	165	171	173	174	176	178
98	130	137	139	142	145	151	154	155	157	158	160	166	168	168	170	172
100	127	133	136	138	141	147	150	151	152	153	155	161	162	163	164	166
102	124	130	132	135	137	143	146	146	148	149	151	156	157	158	159	161
104	122	127	129	131	133	139	141	142	144	144	146	151	152	153	154	156
106	119	124	126	128	130	135	137	138	139	140	142	146	148	148	149	151
108	116	121	123	125	126	131	133	134	135	136	138	142	143	143	144	146

BS 5950: Part 1: 1990: Section four

Table 27(b). *(concluded)*

λ \ p_y	225	245	255	265	275	305	320	325	335	340	355	395	410	415	430	450
110	113	118	120	121	123	128	130	130	131	132	134	137	139	139	140	141
112	111	115	117	118	120	124	126	127	128	128	130	133	134	135	136	137
114	108	112	114	115	117	121	123	123	124	125	126	129	130	131	132	133
116	105	109	111	112	114	117	119	120	121	121	122	125	126	127	128	129
118	103	106	108	109	111	114	116	116	117	118	119	122	123	123	124	125
120	100	104	105	107	108	111	113	113	114	114	116	118	119	119	120	121
122	98	101	103	104	105	108	110	110	111	111	112	115	116	116	117	118
124	96	99	100	101	102	105	107	107	108	108	109	112	112	113	113	114
126	94	96	97	99	100	103	104	104	105	105	106	109	109	110	110	111
128	91	94	95	96	97	100	101	101	102	102	103	106	106	106	107	108
130	89	92	93	94	95	97	98	99	99	100	101	103	103	104	104	105
135	84	86	87	88	89	91	92	93	93	93	94	96	96	97	97	98
140	79	81	82	83	84	86	87	87	87	88	88	90	90	91	91	92
145	75	77	78	78	79	81	81	82	82	82	83	84	85	85	85	86
150	71	72	73	74	74	76	77	77	77	77	78	79	80	80	80	81
155	67	69	69	70	70	72	72	72	73	73	73	75	75	75	75	76
160	64	65	66	66	66	68	68	69	69	69	69	70	71	71	71	71
165	60	61	62	63	63	64	65	65	65	65	66	66	67	67	67	67
170	57	58	59	59	60	61	61	61	61	62	62	63	63	63	63	64
175	55	56	56	56	57	58	58	58	58	58	59	59	60	60	60	60
180	52	53	53	54	54	55	55	55	55	55	56	56	57	57	57	57
185	49	50	51	51	51	52	52	52	53	53	53	54	54	54	54	54
190	47	48	48	48	49	49	50	50	50	50	51	50	51	51	51	52
195	45	46	46	46	47	47	47	48	48	48	48	48	49	49	49	49
200	43	44	44	44	44	45	45	45	46	46	46	46	46	46	47	47
210	39	40	40	40	41	41	41	41	42	42	42	42	42	42	42	43
220	36	37	37	37	37	38	38	38	38	38	38	39	39	39	39	39
230	33	34	34	34	34	35	35	35	35	35	35	35	35	35	36	36
240	31	31	31	31	32	32	32	32	32	32	32	33	33	33	33	33
250	29	29	29	29	29	29	30	30	30	30	30	30	30	30	30	30
260	27	27	27	27	27	27	28	28	28	28	28	28	28	28	28	28
270	25	25	25	25	25	26	26	26	26	26	26	26	26	26	26	26
280	23	23	23	24	24	24	24	24	24	24	24	24	24	24	24	24
290	22	22	22	22	22	22	22	22	22	22	23	23	23	23	23	23
300	20	21	21	21	21	21	21	21	21	21	21	21	21	21	21	21
310	19	19	19	19	19	20	20	20	20	20	20	20	20	20	20	20
320	18	18	18	18	18	18	18	18	18	18	19	19	19	19	19	19
330	17	17	17	17	17	17	17	17	18	18	18	18	18	18	18	18
340	16	16	16	16	16	16	16	17	17	17	17	17	17	17	17	17
350	15	15	15	15	15	15	15	16	16	16	16	16	16	16	16	16

BS 5950: Part 1: 1990: Section four

Table 27(c). Compressive strength p_c (in N/mm^2) for struts

λ \ p_y	225	245	255	265	275	305	320	325	335	340	355	395	410	415	430	450
15	225	245	255	265	275	305	320	325	335	340	355	393	408	413	427	446
20	224	242	252	261	271	299	312	317	326	331	345	382	396	401	414	433
25	217	235	245	254	263	290	303	308	317	321	335	370	384	388	402	419
30	211	228	237	246	255	281	294	298	307	311	324	358	371	375	388	405
35	204	221	230	238	247	272	284	288	296	300	313	345	357	361	374	389
40	198	214	222	230	238	262	274	278	285	289	301	332	343	347	358	373
42	195	211	219	227	235	258	269	273	281	285	296	326	337	340	351	366
44	193	208	216	224	231	254	265	269	276	280	291	320	330	334	344	358
46	190	205	213	220	228	250	261	264	271	275	286	314	324	327	337	351
48	187	202	209	217	224	246	256	260	267	270	280	307	317	321	330	343
50	184	199	206	213	220	241	252	255	262	265	275	301	310	314	323	335
52	181	196	203	210	217	237	247	250	257	260	270	294	303	306	315	327
54	179	193	199	206	213	232	242	245	252	255	264	288	296	299	308	319
56	176	189	196	202	209	228	237	240	246	249	258	281	289	292	300	310
58	173	186	192	199	205	223	232	235	241	244	252	274	282	284	292	302
60	170	183	189	195	201	219	227	230	236	238	247	267	274	277	284	293
62	167	179	185	191	197	214	222	225	230	233	241	260	267	269	276	285
64	164	176	182	188	193	210	217	220	225	227	235	253	260	262	268	276
66	161	173	178	184	189	205	212	215	220	222	229	246	252	254	260	268
68	158	169	175	180	185	200	207	210	214	216	223	239	245	247	252	259
70	155	166	171	176	181	195	202	204	209	211	217	232	238	239	244	251
72	152	163	168	172	177	191	197	199	203	205	211	226	231	232	237	243
74	149	159	164	169	173	186	192	194	198	200	205	219	223	225	229	235
76	146	156	160	165	169	181	187	189	193	194	200	212	217	218	222	227
78	143	152	157	161	165	177	182	184	187	189	194	206	210	211	215	220
80	140	149	153	157	161	172	177	179	182	184	188	200	203	205	208	213
82	137	146	150	154	157	168	173	174	177	179	183	193	197	198	201	205
84	134	142	146	150	154	163	168	169	172	174	178	187	191	192	195	199
86	132	139	143	146	150	159	163	165	168	169	173	182	185	186	189	192
88	129	136	139	143	146	155	159	160	163	164	168	176	179	180	183	186
90	126	133	136	139	142	151	155	156	158	159	163	171	173	174	177	180
92	123	130	133	136	139	147	150	152	154	155	158	165	168	169	171	174
94	120	127	130	133	135	143	146	147	149	150	153	160	163	163	166	168
96	118	124	127	129	132	139	142	143	145	146	149	155	158	158	160	163
98	115	121	123	126	129	135	138	139	141	142	145	151	153	154	155	158
100	112	118	120	123	125	132	134	135	137	138	140	146	148	149	151	153
102	110	115	118	120	122	128	131	132	133	134	136	142	144	144	146	148
104	107	112	115	117	119	125	127	128	130	130	133	138	140	140	142	143
106	105	110	112	114	116	121	124	125	126	127	129	134	135	136	137	139
108	102	107	109	111	113	118	120	121	123	123	125	130	131	132	133	135

BS 5950: Part 1: 1990: Section four

Table 27(c). *(concluded)*

λ \ p_y	225	245	255	265	275	305	320	325	335	340	355	395	410	415	430	450
110	100	104	106	108	110	115	117	118	119	120	122	126	127	128	129	131
112	98	102	104	106	107	112	114	115	116	117	118	122	124	124	125	127
114	96	99	101	103	105	109	111	112	113	113	115	119	120	121	122	123
116	93	97	99	101	102	106	108	109	110	110	112	116	117	117	118	120
118	91	95	96	98	100	104	105	106	107	107	109	112	114	114	115	116
120	89	93	94	96	97	101	103	103	104	105	106	109	110	111	112	113
122	87	91	92	93	95	98	100	100	101	102	103	106	107	108	109	110
124	85	88	90	91	92	96	97	98	99	99	100	103	104	105	106	107
126	83	86	88	89	90	94	95	95	96	97	98	101	102	102	103	104
128	82	84	86	87	88	91	93	93	94	94	95	98	99	99	100	101
130	80	82	84	85	86	89	90	91	91	92	93	95	96	97	97	98
135	75	78	79	80	81	84	85	85	86	86	87	89	90	90	91	92
140	71	74	75	76	76	79	80	80	81	81	82	84	85	85	85	86
145	68	70	70	71	72	74	75	76	76	76	77	79	80	80	80	81
150	64	66	67	68	68	70	71	71	72	72	73	74	75	75	76	76
155	61	63	63	64	65	66	67	67	68	68	69	70	71	71	71	72
160	58	59	60	61	61	63	64	64	64	64	65	66	67	67	67	68
165	55	56	57	58	58	60	60	60	61	61	61	63	63	63	64	64
170	52	54	54	55	55	57	57	57	58	58	58	59	60	60	60	61
175	50	51	52	52	53	54	54	54	55	55	55	56	57	57	57	58
180	48	49	49	50	50	51	52	52	52	52	53	54	54	54	54	55
185	46	46	47	47	48	49	49	49	50	50	50	51	51	51	52	52
190	43	44	45	45	46	46	47	47	47	47	48	48	49	49	49	49
195	42	42	43	43	43	44	45	45	45	45	46	46	46	47	47	47
200	40	41	41	41	42	42	43	43	43	43	43	44	44	44	45	45
210	36	37	38	38	38	39	39	39	39	39	40	40	41	41	41	41
220	34	34	35	35	35	36	36	36	36	36	36	37	37	37	37	38
230	31	32	32	32	32	33	33	33	33	33	34	34	34	34	34	35
240	29	29	29	30	30	30	31	31	31	31	31	32	32	32	32	32
250	27	27	27	28	28	28	28	28	29	29	29	29	29	29	29	29
260	25	25	26	26	26	26	26	27	27	27	27	27	27	27	27	27
270	23	24	24	24	24	24	24	25	25	25	25	25	25	25	25	25
280	22	22	22	22	23	23	23	23	23	23	23	23	24	24	24	24
290	21	21	21	21	21	21	21	21	22	22	22	22	22	22	22	22
300	19	20	20	20	20	20	20	20	20	20	20	21	21	21	21	21
310	18	18	18	18	19	19	19	19	19	19	19	19	19	19	19	20
320	17	17	17	17	18	18	18	18	18	18	18	18	18	18	18	18
330	16	16	16	17	17	17	17	17	17	17	17	17	17	17	17	17
340	15	15	15	16	16	16	16	16	16	16	16	16	16	16	16	16
350	14	15	15	15	15	15	15	15	15	15	15	15	15	15	15	15

BS 5950: Part 1: 1990: Section four

Table 27(d). Compressive strength p_c (in N/mm^2) for struts

λ \ p_y	225	245	255	265	275	305	320	325	335	340	355	395	410	415	430	450
15	225	245	255	265	275	305	320	325	335	340	355	393	407	411	425	444
20	223	241	250	259	269	296	309	314	323	327	341	376	390	394	408	426
25	214	231	240	249	257	283	296	301	309	313	326	360	373	377	390	407
30	205	222	230	238	247	271	283	287	296	300	312	344	356	360	372	388
35	196	212	220	228	236	259	271	274	282	286	297	327	339	342	353	368
40	188	203	210	218	225	247	258	261	268	272	283	310	321	324	334	348
42	184	199	206	214	221	242	252	256	263	266	277	304	314	317	327	340
44	181	195	202	209	216	237	247	251	257	261	271	297	306	309	319	331
46	178	192	199	205	212	232	242	245	252	255	265	290	299	302	311	323
48	174	188	195	201	208	227	237	240	246	249	259	283	291	294	303	314
50	171	184	191	197	204	222	232	235	241	244	253	276	284	287	295	306
52	168	181	187	193	199	217	226	229	235	238	246	269	277	279	287	298
54	165	177	183	189	195	213	221	224	229	232	240	262	269	272	279	289
56	161	173	179	185	191	208	216	219	224	227	234	255	262	264	271	281
58	158	170	175	181	187	203	211	213	218	221	229	248	255	257	264	272
60	155	166	172	177	183	198	206	208	213	215	223	241	247	250	256	264
62	152	163	168	173	178	193	201	203	208	210	217	234	240	242	248	256
64	149	159	164	169	174	189	196	198	202	204	211	227	233	235	241	248
66	145	156	160	165	170	184	191	193	197	199	205	221	226	228	234	240
68	142	152	157	162	166	179	186	188	192	194	200	214	220	221	226	233
70	139	149	153	158	162	175	181	183	187	189	194	208	213	215	219	225
72	136	145	150	154	158	170	176	178	182	183	189	202	207	208	213	218
74	133	142	146	150	155	166	171	173	177	178	183	196	200	202	206	211
76	130	139	143	147	151	162	167	169	172	173	178	190	194	195	199	204
78	127	136	139	143	147	157	162	164	167	169	173	184	188	189	193	198
80	125	132	136	140	143	153	158	160	163	164	168	179	182	184	187	191
82	122	129	133	136	140	149	154	155	158	159	163	173	177	178	181	185
84	119	126	130	133	136	145	150	151	154	155	159	168	171	172	176	179
86	117	123	127	130	133	142	146	147	149	151	154	163	166	167	170	174
88	114	120	123	127	130	138	142	143	145	146	150	158	161	162	165	168
90	111	118	121	123	126	134	138	139	141	142	146	154	156	157	160	163
92	109	115	118	120	123	131	134	135	137	138	142	149	152	152	155	158
94	106	112	115	117	120	127	131	132	134	135	138	145	147	148	150	153
96	104	109	112	115	117	124	127	128	130	131	134	140	143	143	146	148
98	101	107	109	112	114	121	124	125	126	127	130	136	138	139	141	144
100	99	104	107	109	111	117	120	121	123	124	126	132	134	135	137	139
102	97	102	104	106	108	114	117	118	120	121	123	129	131	131	133	135
104	95	99	102	104	106	111	114	115	116	117	120	125	127	127	129	131
106	93	97	99	101	103	109	111	112	113	114	116	121	123	124	125	127
108	90	95	97	99	101	106	108	109	110	111	113	118	120	120	122	124

BS 5950: Part 1: 1990: Section four

Table 27(d). *(concluded)*

λ \ p_y	225	245	255	265	275	305	320	325	335	340	355	395	410	415	430	450
110	88	93	95	96	98	103	105	106	108	108	110	115	116	117	118	120
112	86	90	92	94	96	101	103	103	105	105	107	112	113	114	115	117
114	84	88	90	92	94	98	100	101	102	103	104	109	110	110	112	113
116	83	86	88	90	91	96	98	98	99	100	102	106	107	107	109	110
118	81	84	86	88	89	93	95	96	97	97	99	103	104	105	106	107
120	79	83	84	86	87	91	93	94	94	95	96	100	102	102	103	104
122	77	81	82	84	85	89	91	91	92	93	94	97	99	99	100	102
124	76	79	81	82	83	87	88	89	90	90	92	95	96	96	97	99
126	74	77	79	80	81	84	86	87	88	88	89	93	94	94	95	96
128	72	75	77	78	79	83	84	85	85	86	87	90	92	92	93	94
130	71	74	75	76	77	81	82	83	83	84	85	88	89	89	90	91
135	67	70	71	72	73	76	77	78	79	79	80	83	84	84	85	86
140	64	66	67	68	69	72	73	74	74	74	75	78	79	79	80	81
145	60	63	64	65	66	68	69	70	70	70	71	73	74	74	75	76
150	58	59	61	61	62	64	65	66	66	66	67	69	70	70	71	71
155	55	57	58	58	59	61	62	62	63	63	64	65	66	66	67	67
160	52	54	55	55	56	58	59	59	59	60	60	62	63	63	63	64
165	50	51	52	53	53	55	56	56	56	57	57	59	59	59	60	60
170	47	49	50	50	51	52	53	53	54	54	54	56	56	56	57	57
175	45	47	47	48	48	50	50	51	51	51	52	53	54	54	54	54
180	43	45	45	46	46	47	48	48	48	49	49	50	51	51	51	52
185	42	42	43	43	44	45	46	46	46	46	47	48	48	48	49	49
190	40	41	41	42	42	43	44	44	44	44	45	46	46	46	47	47
195	38	39	40	40	40	41	42	42	42	42	43	44	44	44	44	45
200	36	37	38	38	39	40	40	40	40	41	41	42	42	42	42	43
210	34	34	35	35	35	36	37	37	37	37	37	38	39	39	39	39
220	31	32	32	32	33	33	34	34	34	34	34	35	35	35	36	36
230	29	29	30	30	30	31	31	31	31	32	32	32	33	33	33	33
240	27	27	27	28	28	28	29	29	29	29	29	30	30	30	30	30
250	25	25	26	26	26	27	27	27	27	27	27	28	28	28	28	28
260	23	24	24	24	24	25	25	25	25	25	25	26	26	26	26	26
270	22	22	23	23	23	23	23	24	24	24	24	24	24	24	24	24
280	20	21	21	21	21	22	22	22	22	22	22	23	23	23	23	23
290	19	20	20	20	20	20	21	21	21	21	21	21	21	21	21	21
300	18	18	19	19	19	19	19	19	19	19	20	20	20	20	20	20
310	17	17	18	18	18	18	18	18	18	18	18	19	19	19	19	19
320	16	16	17	17	17	17	17	17	17	17	17	18	18	18	18	18
330	15	15	16	16	16	16	16	16	16	16	16	17	17	17	17	17
340	14	15	15	15	15	15	15	15	15	15	15	16	16	16	16	16
350	14	14	14	14	14	14	14	14	14	14	15	15	15	15	15	15

Bolt data

Hole sizes — for ordinary bolts and friction grip connections

Nominal diameter (mm)	Clearance hole diameter[b] (mm)	Oversize hole diameter[a] (mm)	Short slotted holes[a] (mm)		Long slotted holes[a] (mm)	
			Narrow dimension	Slot dimension	Narrow dimension	Maximum dimension
M12[a]	14	17	14	18	14	30
M16	18	21	18	22	18	40
M20	22	25	22	26	22	50
M22	24	27	24	28	24	55
M24	26	30	26	32	26	60
M27	30	35	30	37	30	67
M30	33	38	33	40	33	75

[a] Hardened washers to be used
[b] In cases where there are more than three plies in joint the holes in the inner plies should be one millimetre larger than those in the outer plies

Bolt strengths

	Bolt grade		Steel to BS 4360		
	4.6	8.8	43	50	55
Shear strength, p_s (N/mm^2)	160	375			
Bearing strength, p_{bb} (N/mm^2)	460	1035[a]	460	550	650
Tension strength, p_t (N/mm^2)	195	450			

[a] The bearing value of the connected part is critical

Bolt capacities in tension

Nominal diameter (mm)	Tensile stress area[b] (mm^2)	Bolts grade 4.6 @ 195 N/mm^2 (kN)	Bolts grade 8.8 @ 450 N/mm^2 (kN)
M12	84.3	16.43	37.93
M16	157	30.61	70.65
M20	245	47.77	110.25
M22[a]	303	59.08	136.35
M24	353	68.83	158.85
M27[a]	459	89.50	206.55
M30	561	109.39	252.45

[a] Non-preferred sizes
[b] Tensile stress areas are taken from BS 4190 and BS 3692

Spacing, end and edge distances — minimum values (see Fig. 23.1)

Nominal diameter of fastener (mm)	Diameter of clearance hole (mm)	Minimum spacing (mm)	Edge distance to rolled, sawn, planed, or machine flame cut edge (mm)	Edge distance to sheared edge or hand flame cut edge and end distance (mm)
M12	14	30	18	20
M16	18	40	23	26
M20	22	50	28	31
M22[a]	24	55	30	34
M24	26	60	33	37
M27[a]	30	68	38	42
M30	33	75	42	47

[a] Non-preferred size

Maximum centres of fasteners

Thickness of element (mm)	Spacing in the direction of stress (mm)	Spacing in any direction in corrosive environments (mm)
5	70	80
6	84	96
7	98	112
8	112	128
9	126	144
10	140	160
11	154	176
12	168	192
13	182	200
14	196	200
15	210	200

Bolt data

Maximum edge distances (1)

BS 4360 grade	Thickness less than or equal to (mm)	p_y (N/mm^2)	$\varepsilon = \left(\dfrac{275}{p_y}\right)^{1/2}$	$11t\varepsilon^a$
43 A, B & C	16 40 63 80 100	275 265 255 245 235	1.0 1.02 1.04 1.06 1.08	$11t$ $11.21t$ $11.44t$ $11.65t$ $11.90t$
50 B & C	16 40 63 80 100	355 345 335 325 315	0.88 0.89 0.91 0.92 0.93	$9.68t$ $9.79t$ $10.0t$ $10.12t$ $10.28t$
55 C	16 25 40 63	450 430 415 400	0.78 0.80 0.81 0.83	$8.58t$ $8.8t$ $8.95t$ $9.13t$

[a] This rule does not apply to fasteners interconnecting the components of back-to-back tension members
This table is expanded in the next table (Maximum edge distances (2))

Maximum edge distances (2)

Thickness of element t (mm)	Corrosive environment 40 mm + 4t (mm)	Steel grade 43 A, B & C $\varepsilon = 11t\varepsilon$ (mm)	Steel grade 50 B & C $\varepsilon = 11t\varepsilon$ (mm)	Steel grade 55 C $\varepsilon = 11t\varepsilon$ (mm)
5	60[a]	55[a]	48[a]	42[a]
6	64[a]	66	58[a]	51[a]
7	68[a]	77	67[a]	60[a]
8	72[a]	88	77	68[a]
9	76	99	87	77
10	80	110	96	85
11	84	121	106	94
12	88	132	116	103
13	92	143	125	111
14	96	154	135	120
15	100	165	145	128
16	104	176	154	137
20	120	224	196	176
25	140	280	245	220
30	160	336	294	267
35	180	392	343	312
40	200	448	392	356
45	220	515	445	411
50	240	572	501	457
55	260	629	551	502
60	280	686	601	548
65	300	757	657	
70	320	816	708	
75	340	874	759	

[a] Use the lesser values for the appropriate grade of steel

Bolt data

Back marks in channel flanges

RSC	Nominal flange width (mm)	Back mark (mm)	Edge dist. (mm)	Recommended diameter (mm)
	102	55	47	24
	89	55	34	20
	76	45	31	20
	64	35	29	16
	51	30	21	10
	38	22	—	—

Back marks in angles

Nominal leg (mm)	S_1 (mm)	S_2 (mm)	S_3 (mm)	S_4 (mm)	S_5 (mm)	S_6 (mm)	Nominal leg (mm)	S_1 (mm)
250							75	45(20)
200		75(30)	75(30)	55(20)	55(20)	55(20)	70	40(20)
150		55(20)	55(20)				65	35(20)
125		45(20)	50(20)				60	35(16)
120		45(16)	50(16)				50	28(12)
100	55(24)						45	25
90	50(24)						40	23
80	45(20)						30	20
							25	15

Maximum recommended bolt sizes are given in brackets

This table is reproduced from BCSA Publication No. 5/79, *Metric Practice for Structural Steelwork*, 3rd edn, 1979.

Cross centres through flanges

Flange width (mm)	Minimum for accessibility (mm)	Maximum for edge dist. (mm)	S_1 (mm)	S_2 (mm)	S_3 (mm)	S_4 (mm)
Joists						
44	27 (5)	30	30			
64	38 (10)	39	40			
76	48 (10)	51	48			
89	54 (12)	59	56			
102	60 (16)	62	60			
114	66 (16)	74	70			
127	72 (20)	77	75			
152	75 (20)	102	90			
203	91 (24)	143	140			
UCs						
152	65 (24)	92	90			
203	75 (24)	143	140			
254	87 (24)	194	140			
305	100 (24)	245	140	120 (24)	60 (24)	240 (24)
368	88 (24)	308	140	140 (24)	75 (24)	290 (24)
406	120 (24)	346	140	140 (24)	75 (24)	290 (24)
UBs						
102	50 (16)	62	54			
127	62 (20)	77	70			
133	57 (20)	83	70			
140	69 (24)	80	70			
146	64 (24)	86	70			
152	73 (24)	92	90			
165	67 (24)	105	90			
171	72 (24)	111	90			
178	72 (24)	118	90			
191	74 (24)	131	90			
210	80 (24)	150	140			
229	80 (24)	169	140			
254	87 (24)	194	140			
267	91 (24)	207	140	90 (20)	50 (20)	190 (20)
292	94 (24)	232	140	100 (24)	60 (24)	220 (24)
305	100 (24)	245	140	120 (24)	60 (24)	240 (24)
419	112 (24)	359	140	140 (24)	75 (24)	290 (24)

Maximum bolt diameters for dimensions shown are given in brackets

Grade 4.6 bolts – threads in shear plane
Single shear and bearing – grade 43 steel

Nominal diameter (mm)	Tensile area (mm^2)	P_s @ 160 N/mm^2 (kN)	Bearing[b] @ 460 N/mm^2			t (mm) for $P_s = P_{bb}$
			Plate thickness, t (mm)			
			5	6		
			(kN)			
M12	84.3	13.5	—	—		2.44
M16	157	25.1	—	—		3.41
M20	245	39.2	—	—		4.26
M22[a]	303	48.5	—	—		4.79
M24	353	56.5	55.2	—		5.12
M27[a]	459	73.4	62.1	—		5.91
M30	561	89.8	69.0	82.8		6.51

[a] Non-preferred size
[b] For end distance $>2d$

Grade 4.6 bolts – threads in shear plane
Double shear and bearing – grade 43 steel

Nominal diameter (mm)	Tensile area (mm^2)	P_s @ 2 × 160 N/mm^2 (kN)	Bearing[b] @ 460 N/mm^2								t (mm) for $P_s = P_{bb}$
			Plate thickness, t (mm)								
			5	6	7	8	9	10	11	12	13	
			(kN)									
M12	84.3	27.0	—	—	—	—	—	—	—	—	—	4.89
M16	157	50.2	36.8	44.2	—	—	—	—	—	—	—	6.82
M20	245	78.4	46.0	55.2	64.4	73.6	—	—	—	—	—	8.52
M22[a]	303	97.0	50.6	60.7	70.8	80.9	91.1	—	—	—	—	9.58
M24	353	113	55.2	66.2	77.3	88.3	99.4	110.4	—	—	—	10.23
M27[a]	459	146.8	62.1	74.5	86.9	99.4	111.8	124.2	136.6	—	—	11.82
M30	561	179.6	69.0	82.8	96.6	110.4	124.2	137.9	151.7	165.5	179.4	13.01

[a] Non-preferred size
[b] For end distance $>2d$

Grade 4.6 bolts – threads excluded from shear plane
Single shear and bearing – grade 43 steel

Nominal diameter (mm)	Shank area (mm²)	P_s @ 160 N/mm² (kN)	Bearing[b] @ 460 N/mm²				t (mm) for $P_s = P_{bb}$
			Plate thickness, t (mm)				
			5	6	7	8	
			(kN)				
M12	113.1	18.10	—	—	—	—	3.28
M16	201.1	32.18	—	—	—	—	4.37
M20	314.2	50.27	46.0	—	—	—	5.46
M22[a]	380.1	60.82	50.6	60.7	—	—	6.01
M24	452.39	72.38	55.2	66.2	—	—	6.56
M27[a]	572.56	91.61	61.1	74.5	86.9	—	7.38
M30	706.86	113.1	69.0	82.8	96.6	110.4	8.20

[a] Non-preferred size
[b] End distance >2d

Grade 4.6 bolts – threads excluded from shear plane
Double shear and bearing – grade 43 steel

Nominal diameter (mm)	Shank area (mm²)	P_s @ 2 × 160 N/mm² (kN)	Bearing[b] @ 460 N/mm²									t (mm) for $P_s = P_{bb}$
			Plate thickness, t (mm)									
			5	6	7	8	9	10	11	12	15	
							(kN)					
M12	113.1	36.20	27.6	33.1	—	—	—	—	—	—	—	6.56
M16	201.1	64.36	36.8	44.2	51.5	58.9	—	—	—	—	—	8.74
M20	314.2	100.54	46.0	55.2	64.4	73.6	82.8	92.0	—	—	—	10.93
M22[a]	380.1	121.64	50.6	60.7	70.8	80.9	91.1	101.2	111.3	121.4	—	12.02
M24	452.39	144.76	55.2	66.2	77.3	88.3	99.4	110.4	121.4	132.5	—	13.11
M27[a]	572.56	183.22	61.1	74.5	86.9	99.4	111.8	124.2	136.6	149.0	—	14.75
M30	706.86	226.20	69.0	82.8	96.6	110.4	124.2	137.9	151.8	165.5	206.9	16.39

[a] Non-preferred size
[b] End distance >2d

Grade 8.8 bolts – threads in shear plane
Single shear and bearing – grade 43 steel

Nominal diameter (mm)	Tensile area (mm²)	P_s @ 375 N/mm² (kN)	Bearing[b] @ 460 N/mm² Plate thickness, t (mm) (kN)											t (mm) for $P_s = P_{bb}$
			5	6	7	8	9	10	11	12	13	14	15	
M12	84.3	31.6	27.6	—	—	—	—	—	—	—	—	—	—	5.73
M16	157	58.9	36.8	44.2	51.5	58.9	—	—	—	—	—	—	—	8.00
M20	245	91.9	46.0	55.2	64.4	73.6	82.8	—	—	—	—	—	—	9.99
M22[a]	303	113.6	50.6	60.7	70.8	80.9	91.1	101.2	111.3	—	—	—	—	11.23
M24	353	132.4	55.2	66.2	77.3	88.3	99.4	110.4	121.4	—	—	—	—	11.99
M27[a]	459	172.1	62.1	74.5	86.9	99.4	111.8	124.2	136.6	149.0	161.5	193.1	—	13.86
M30	561	210.4	69.0	82.8	96.6	110.4	124.2	137.9	151.7	165.5	179.3	—	206.9	15.25

[a] Non-preferred size
[b] End distance $>2d$

Grade 8.8 bolts – threads in shear plane
Double shear and bearing – grade 43 steel

Nominal diameter (mm)	P_s @ 2 × 375 N/mm² (kN)	Bearing[b] @ 460 N/mm² Plate thickness, t (mm) (kN)											t (mm) for $P_s = P_{bb}$
		5	6	7	8	9	10	12	15	18	20	25	30	
M12	63.2	27.6	33.1	38.6	44.1	49.6	55.2	—	—	—	—	—	—	11.4
M16	117.7	36.8	44.2	51.5	58.9	66.3	73.6	88.3	110.4	—	—	—	—	16.0
M20	183.7	46.0	55.2	64.4	73.6	82.8	92.0	110.4	138.0	165.5	—	—	—	19.9
M22[a]	227.3	50.6	60.7	70.8	80.9	91.1	101.2	121.4	151.8	182.1	202.4	—	—	22.4
M24	264.8	55.2	66.2	77.3	88.3	99.4	110.4	132.5	165.6	198.8	220.8	—	—	23.9
M27[a]	344.3	61.1	74.5	86.9	99.4	111.8	124.2	149.0	186.3	223.4	248.4	310.5	—	27.7
M30	420.8	69.0	82.8	96.6	110.4	124.2	137.9	165.5	206.9	248.4	276.0	344.9	413.9	30.5

[a] Non-preferred size
[b] End distance $>2d$

Grade 8.8 bolts – threads in shear plane
Single shear and bearing – grade 50 steel

Nominal diameter (mm)	Tensile area (mm^2)	P_s @ 375 N/mm^2 (kN)	Bearing[b] @ 550 N/mm^2 Plate thickness, t (mm) (kN)								t (mm) for $P_s = P_{bb}$
			5	6	7	8	9	10	11	12	
M12	84.3	31.6	—	—	—	—	—	—	—	—	4.79
M16	157	58.9	44.0	52.8	77.0	88.0	—	—	—	—	6.69
M20	245	91.9	55.0	66.0	84.7	96.8	—	—	—	—	8.35
M22[a]	303	113.6	60.5	72.6	92.4	105.6	108.9	—	—	—	9.39
M24	353	132.4	66.0	79.2	104.0	118.8	118.8	132.0	—	—	10.03
M27[a]	459	172.1	74.3	89.1	104.0	118.8	133.7	148.5	163.4	—	11.59
M30	561	210.4	82.5	99.0	115.5	132.0	148.5	165.0	181.5	198.0	12.75

[a] Non-preferred size
[b] End distance >2d

Grade 8.8 bolts – threads in shear plane
Double shear and bearing – grade 50 steel

Nominal diameter (mm)	P_s @ 2 × 375 N/mm^2 (kN)	Bearing[b] @ 550 N/mm^2 Plate thickness, t (mm) (kN)									t (mm) for $P_s = P_{bb}$
		5	6	7	8	9	10	12	15	18	20	25	
M12	63.2	33.0	39.6	46.2	52.8	59.4	—	—	—	—	—	—	9.5
M16	117.7	44.0	52.8	61.6	70.4	79.2	88.0	105.6	165.0	—	—	—	13.3
M20	183.7	55.0	66.0	77.0	88.0	99.0	110.0	132.0	165.0	—	—	—	16.7
M22[a]	227.3	60.5	72.6	84.7	96.8	108.9	121.0	145.2	181.5	217.8	—	—	18.7
M24	264.8	66.0	79.2	92.4	105.6	118.8	132.0	158.4	198.0	237.6	264.0	—	20.0
M27[a]	344.3	74.3	89.1	104.0	118.8	133.7	148.5	178.2	222.8	267.3	297.0	—	23.1
M30	420.8	82.5	99.0	115.0	132.0	148.5	165.0	198.0	247.5	297.0	330.0	412.5	25.5

[a] Non-preferred size
[b] End distance >2d

Bolt data

Grade 8.8 bolts – threads in shear plane
Single shear and bearing – grade 55 steel

Nominal diameter (mm)	Tensile area (mm²)	P_s @ 375 N/mm² (kN)	Bearing[b] @ 650 N/mm² Plate thickness, t (mm) (kN)						t (mm) for $P_s = P_{bb}$
			5	6	7	8	9	10	
M12	84.3	31.6	–	–	–	–	–	–	4.0
M16	157	58.9	52.0	–	–	–	–	–	5.6
M20	245	91.9	65.0	78.0	91.0	–	–	–	7.1
M22[a]	303	113.6	71.5	85.8	100.1	–	–	–	7.9
M24	353	132.4	78.0	93.6	109.1	124.7	–	–	8.5
M27[a]	459	172.1	87.7	105.3	122.8	140.4	157.9	–	9.8
M30	561	210.4	97.5	117.0	136.5	156.0	175.5	195.0	10.8

[a] Non-preferred size
[b] End distance >2d

Grade 8.8 bolts – threads in shear plane
Double shear and bearing – grade 55 steel

Nominal diameter (mm)	P_s @ 2 × 375 N/mm² (kN)	Bearing[b] @ 650 N/mm² Plate thickness, t (mm) (kN)									t (mm) for $P_s = P_{bb}$
		5	6	7	8	9	10	12	15	18	20	
M12	63.2	39.0	46.8	54.6	62.4	–	–	–	–	–	–	8.1
M16	117.7	52.0	62.4	72.8	83.2	93.6	104.0	–	–	–	–	11.3
M20	183.7	65.0	78.0	91.0	103.9	116.9	129.9	155.9	–	–	–	14.1
M22[a]	227.3	71.5	85.8	100.1	114.3	128.6	142.9	171.5	214.5	–	–	15.9
M24	264.8	78.0	93.6	109.1	124.7	140.3	155.9	187.1	234.0	–	–	16.9
M27[a]	344.3	87.7	105.3	122.8	140.4	157.9	175.5	210.6	263.2	315.8	–	19.6
M30	420.8	97.5	117.0	136.5	156.0	175.5	195.0	234.0	292.5	315.0	390.0	21.5

[a] Non-preferred size
[b] End distance >2d

Grade 8.8 bolts – threads excluded from shear plane
Single shear and bearing – grade 43 steel

Nominal diameter (mm)	Shank area (mm²)	P_s @ 375 N/mm² (kN)	Bearing[b] @ 460 N/mm²									t (mm) for $P_s = P_{bb}$
			Plate thickness, t (mm)									
			5	6	7	8	9 (kN)	10	12	15	18	
M12	113.1	42.41	27.6	33.1	38.6	—	—	—	—	—	—	7.68
M16	201.1	75.41	36.8	44.1	51.5	58.9	66.3	73.6	—	—	—	10.25
M20	314.2	117.81	46.0	55.2	64.4	73.6	82.8	92.0	110.4	—	—	12.81
M22[a]	380.1	142.54	50.6	60.7	70.8	80.9	91.1	101.2	121.4	—	—	14.08
M24	452.39	169.64	55.2	66.2	77.3	88.3	99.4	110.4	132.5	165.6	—	15.37
M27[a]	572.56	214.71	62.1	74.5	86.9	99.4	111.8	124.2	149.0	186.3	—	17.29
M30	706.86	265.07	69.0	82.8	96.6	110.4	124.2	137.9	165.5	206.9	248.4	19.21

[a] Non-preferred size
[b] End distance >2d

Grade 8.8 bolts – threads excluded from shear plane
Double shear and bearing – grade 43 steel

Nominal diameter (mm)	Shank area (mm²)	P_s @ 2 × 375 N/mm² (kN)	Bearing[b] @ 460 N/mm²									t (mm) for $P_s = P_{bb}$
			Plate thickness, t (mm)									
			5	10	12	15	18 (kN)	20	25	30	35	37	
M12	113.1	84.8	27.6	55.2	66.2	82.8	—	—	—	—	—	—	15.36
M16	201.1	150.8	36.8	73.6	88.3	110.4	132.4	147.2	—	—	—	—	20.50
M20	314.2	235.6	46.0	92.0	110.4	138.0	165.6	184.0	230.0	—	—	—	25.62
M22[a]	380.1	285.1	50.6	101.2	121.4	151.8	182.1	202.4	253.0	—	—	—	28.16
M24	452.39	339.2	55.2	110.4	132.5	165.6	198.7	220.8	276.0	331.2	—	—	30.74
M27[a]	522.56	429.4	62.1	124.2	149.0	186.3	223.4	248.4	310.5	372.6	483.0	—	34.58
M30	706.86	530.1	69.0	137.9	165.5	206.9	248.4	276.0	344.9	413.9	—	510	38.42

[a] Non-preferred size
[b] End distance >2d

Bolt data

Grade 8.8 bolts – threads excluded from shear plane
Single shear and bearing – grade 50 steel

Nominal diameter (mm)	Shank area (mm²)	P_s @ 375 N/mm² (kN)	Bearing[b] @ 550 N/mm² Plate thickness, t (mm) (kN)							t (mm) for $P_s = P_{bb}$
			5	6	7	8	9	10	15	
M12	113.1	42.41	33	39.6	—	—	—	—	—	6.43
M16	201.1	75.41	44.0	52.8	61.6	70.4	—	—	—	8.57
M20	314.2	117.81	55.0	66.0	77.0	88.0	99.0	110.0	—	10.71
M22[a]	380.1	142.58	60.5	72.6	84.7	96.8	108.9	121.0	—	11.78
M24	452.39	169.64	66.0	79.2	92.4	105.6	118.9	132.0	158.4	12.85
M27[a]	572.56	214.71	74.3	89.1	104.0	118.8	133.7	148.5	178.2	14.46
M30	706.86	265.07	82.5	99.0	115.5	132.0	148.5	165.0	198.0	16.06
									247.5	

[a] Non-preferred size
[b] End distance $>2d$

Grade 8.8 bolts – threads excluded from shear plane
Double shear and bearing – grade 50 steel

Nominal diameter (mm)	Shank area (mm²)	P_s @ 2 × 375 N/mm² (kN)	Bearing[b] @ 550 N/mm² Plate thickness, t (mm) (kN)							t (mm) for $P_s = P_{bb}$
			5	10	12	15	18	20	25	30	
M12	113.1	84.8	33.0	66.0	79.2	—	—	—	—	—	12.86
M16	201.1	150.8	44.0	88.0	105.6	132.0	—	—	—	—	17.14
M20	314.2	235.6	55.0	110.0	132.0	165.0	198.0	220.0	—	—	21.42
M22[a]	380.1	285.1	60.5	121.0	145.2	181.5	217.8	242.0	—	—	23.56
M24	452.39	339.2	66.0	132.0	158.4	198.0	237.6	264.0	300.0	—	25.70
M27[a]	522.56	429.4	74.3	148.5	178.2	222.8	267.3	297.0	337.5	—	28.92
M30	706.86	530.1	82.5	165.0	198.0	247.5	297.0	330.0	412.5	495.0	32.12

[a] Non-preferred size
[b] End distance $>2d$

Grade 8.8 bolts – threads excluded from shear plane
Single shear and bearing – grade 55 steel

Nominal diameter (mm)	Shank area (mm²)	P_s @ 375 N/mm² (kN)	Bearing[b] @ 650 N/mm²							t (mm) for $P_s = P_{bb}$
			Plate thickness, t (mm)							
			5	6	7	8 (kN)	9	10	12	
M12	113.1	42.41	39.0	—	—	—	—	—	—	5.44
M16	201.1	75.41	52.0	62.4	72.8	—	—	—	—	7.25
M20	314.2	117.81	65.0	78.0	91.0	103.9	116.9	—	—	9.06
M22[a]	380.1	142.58	71.5	85.8	100.1	114.3	128.6	—	—	9.97
M24	452.39	169.64	78.0	93.6	109.1	124.7	140.3	155.9	—	10.87
M27[a]	572.56	214.71	87.7	105.3	122.8	140.4	157.9	175.5	210.6	12.24
M30	706.86	265.07	97.5	117.0	136.5	156.0	175.5	195.0	234.0	13.59

[a] Non-preferred size
[b] End distance >2d

Grade 8.8 bolts – threads excluded from shear plane
Double shear and bearing – grade 55 steel

Nominal diameter (mm)	Shank area (mm²)	P_s @ 2 × 375 N/mm² (kN)	Bearing[b] @ 650 N/mm²							t (mm) for $P_s = P_{bb}$
			Plate thickness, t (mm)							
			5	10	12	15 (kN)	18	20	25	
M12	113.1	84.8	39.0	78.0	—	—	—	—	—	10.88
M16	201.1	150.8	52.0	104.0	124.8	—	—	—	—	14.50
M20	314.2	235.6	65.0	129.9	155.9	195.0	234.0	—	—	18.12
M22[a]	380.1	285.1	71.5	142.9	171.5	214.5	257.4	—	—	19.94
M24	452.39	339.2	78.0	155.9	187.1	234.0	280.8	312.0	—	21.74
M27[a]	522.56	429.4	87.7	175.5	210.6	263.2	315.9	351.0	—	24.48
M30	706.86	530.1	97.5	195.0	234.0	292.5	351.0	390.0	487.5	27.18

[a] Non-preferred size
[b] End distance >2d

Bolt data

Bolts to BS 4395: Part 1
Slip resistance

Nominal diameter (mm)	Proof load (kN)	P_{sL} (kN) when $\mu = 0.45$		
		Clearance holes $K_s = 1.0$	Short slot holes, etc. $K_s = 0.85$	Long slot holes $K_s = 0.6$
M12[a]	49.4	24.5	20.7	14.6
M16	92.1	45.5	38.7	27.3
M20	144	71.2	60.5	42.7
M22	177	87.6	74.4	52.5
M24	207	102.4	87.1	61.4
M27	234	115.8	98.5	69.5
M30	286	141.5	120.3	84.9

[a] Non-preferred size

Bolts to BS 4395: Part 2
Shear connections only – external tension not permitted
Slip resistance

Nominal diameter (mm)	Proof load (kN)	Minimum shank tension (kN)	P_{sL} (kN) when $0.45 < \mu < 0.55$					
			Clearance holes $K_s = 1.0$		Short slot holes, etc. $K_s = 0.85$		Long slot holes $K_s = 0.6$	
			$\mu = 0.45$	$\mu = 0.55$	$\mu = 0.45$	$\mu = 0.55$	$\mu = 0.45$	$\mu = 0.55$
M16	122.2	103.8	51.4	62.8	43.7	53.4	30.8	37.7
M20	190.4	161.8	80.1	97.9	68.0	83.2	48.0	58.7
M22	235.5	200.1	99.0	121.1	84.2	102.9	59.4	72.6
M24	274.6	233.4	115.5	141.2	98.2	120.0	69.3	84.7
M27	356.0	302.6	149.7	183.0	127.3	155.6	89.8	109.8
M30	435.0	369.7	183.0	223.6	155.5	190.1	109.8	134.2

Bolts to BS 4395
Maximum external tension capacity

Nominal diameter (mm)	Part 1 (bolts general grade parallel shank)	
	P_o (kN)	P_t (kN)
M16	92.1	82.8
M20	144	129.6
M22	177	159.3
M24	207	186.3
M27	234	210.6
M30	286	257.4

Bolts to BS 4395
Bearing strength
Parallel shank fastener
Steel to BS 4360

Nominal diameter (mm)	Plate thickness and steel grade					
	Grade 43 steel			Grade 50		Grade 55
	6^b	8	10	6^b	8	6^b
M12^a	—	—	—	—	—	—
M16	79.2	—	—	102.4	—	116.1
M20	99.0	132.0	—	127.8	—	145.1
M22	108.9	145.2	—	140.5	—	159.7
M24	118.8	158.4	198.0	153.3	204.5	174.5
M27	133.6	178.2	222.8	172.5	230.0	196.0
M30	148.5	198.0	247.5	191.7	255.6	217.8

These values will only apply in double shear joints in the case of Part 1 fasteners but may need checking against single shear in the case of Part 2 fasteners

^{a} Non-preferred size
^{b} BS 4604 recommends minimum plate thickness of $d/2$ or 10 mm thickness whichever is the lesser

Bolts to BS 4395
Long joints

For fasteners where the slip resistance given in the previous tables is assessed using a slip factor of 0.45, the reduction due to long joints does not become effective until L_j is 1375 mm. When the slip factor is 0.55 the reduction is effective for all joints over 500 mm long.

Parallel shank fasteners
(taking $K_s = 1.0$, $\mu = 0.55$)

Nominal diameter (mm)	For joints where $L_j = 500$ mm			
	Part 1 bolts general grade		Part 2 bolts higher grade	
	P_0 Proof load (kN)	P_{sL} (kN)	P_0 Minimum shank tension (kN)	P_{sL} (kN)
M16	92.1	55.2	103.8	62.2
M20	144	86.4	161.8	97.0
M22	177	106.2	200.1	120.0
M24	207	124.2	233.4	140.0
M27	234	140.4	302.6	181.5
M30	286	171.6	369.7	221.8

For joints longer than 500 mm, columns 3 and 5 should be multiplied by the factor $(5500 - L_j)/5000$

Bolt data

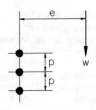

Bolt groups
One row of fasteners; fasteners in the plane of the force

No. of fasteners in vertical row	Pitch, p (mm)	Values of Z_{xx} (cm^3) for diameter of bolt, D(mm)						
		12	16	20	22	24	27	30
2	70	6.0	11.4	18.0	22.5	26.6	35.1	43.8
3		11.9	22.3	35.0	43.4	50.8	66.5	81.9
4		19.7	36.9	57.8	71.6	83.6	109.1	133.9
5		29.6	55.2	86.3	106.9	124.7	162.5	199.2
6		41.4	77.2	120.6	149.3	174.1	226.7	277.6
7		55.1	102.8	160.6	198.7	231.7	301.6	369.2
8		70.9	132.1	206.3	255.3	297.6	387.3	473.8
9		88.6	165.1	257.8	318.9	371.7	483.6	591.6
2	100	8.5	16.0	25.1	31.2	36.6	48.0	59.3
3		16.9	31.6	49.5	61.3	71.6	93.4	114.6
4		28.2	52.5	82.1	101.6	118.5	154.4	189.1
5		42.2	78.7	122.9	152.1	177.3	230.8	282.5
6		59.1	110.1	171.9	212.7	247.9	322.6	394.6
7		78.7	146.7	229.0	283.3	330.2	429.6	525.5
8		101.2	188.5	294.4	364.1	424.3	552.0	675.0
9		126.5	235.6	367.9	455.0	530.2	689.7	843.3

Bolt groups
Two rows of fasteners; fasteners in the plane of the force

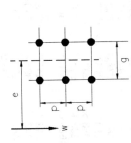

No. of fasteners in vertical row	Pitch, p (mm)	Gauge, q (mm)	Values of Z_{xx} (cm³) for diameter of bolt, D(mm)						
			12	16	20	22	24	27	30
2	70	60	15.7	29.5	46.5	57.9	67.9	89.2	110.3
3			27.8	52.0	81.7	101.3	118.4	154.8	190.4
4			43.5	81.3	127.2	157.6	184.1	240.0	294.5
5			63.1	117.8	184.2	228.0	266.1	346.7	424.8
6			86.7	161.6	252.6	312.7	364.6	474.8	581.4
7			114.2	212.8	332.5	411.5	479.7	624.4	764.2
8			145.6	271.4	423.9	524.4	611.3	795.6	973.4
9			181.0	337.3	526.7	651.6	759.5	988.2	1208.8
2	100	60	19.8	37.1	58.2	72.2	84.6	110.6	136.3
3			36.8	68.6	107.4	133.1	155.4	202.6	248.5
4			59.2	110.4	172.5	213.5	249.1	324.4	397.3
5			87.2	162.6	254.0	314.3	366.4	476.9	583.7
6			120.9	225.3	351.8	435.3	507.4	660.2	807.7
7			160.2	298.5	466.0	576.5	671.9	874.2	1069.1
8			205.1	382.2	596.6	738.0	860.1	1118.8	1368.1
9			255.7	476.3	743.5	919.7	1071.8	1394.0	1704.5

Bolt data

Bolt groups
Four rows of fasteners; fasteners in the plane of the force

No. of fasteners in vertical row	Pitch, p (mm)	S₂ (mm)	S₃ (mm)	S₄ (mm)	Values of Z_{xx} (cm³) for diameter of bolt, D(mm)						
					12	16	20	22	24	27	30
2	70	120	60	240	55.3	103.2	161.3	199.8	233.1	303.7	372.2
3					89.5	166.9	261.0	323.0	376.8	490.8	601.3
4					128.1	238.9	373.4	462.2	539.0	701.9	859.4
5					172.1	320.8	501.3	620.4	723.4	941.6	1152.6
6					222.3	414.3	647.2	800.8	933.6	1215.1	1486.9
7					279.3	520.6	813.0	1005.9	1172.5	1525.7	1866.6
8					343.6	640.2	999.8	1236.9	1441.7	1875.7	2294.4
9					415.3	773.8	1208.2	1494.7	1742.0	2266.3	2771.8
2	70	120	85	290	61.3	114.3	178.7	221.2	258.0	336.1	411.6
3					98.0	182.8	285.6	353.5	412.3	536.8	657.3
4					139.1	259.3	405.1	501.3	584.6	761.0	931.5
5					185.2	345.1	539.1	667.1	777.8	1012.3	1238.8
6					237.1	441.8	690.0	853.7	995.2	1295.1	1584.5
7					295.4	550.4	859.6	1063.5	1239.6	1612.9	1973.0
8					360.7	672.0	1049.3	1298.1	1512.9	1968.3	2407.5
9					433.1	807.0	1259.9	1558.6	1816.5	2363.1	2890.0
2	70	140	50	240	58.8	109.7	171.5	212.4	247.8	322.8	395.5
3					94.2	175.7	274.7	340.0	396.7	516.6	632.8
4					133.6	249.2	389.4	481.9	562.0	731.8	896.0
5					178.1	331.9	518.6	641.7	748.2	974.0	1192.1
6					228.5	425.9	665.2	823.1	959.6	1248.8	1528.1
7					285.7	532.4	831.4	1028.7	1199.1	1560.3	1908.8
8					350.0	652.2	1018.4	1260.0	1468.6	1910.7	2337.2
9					421.8	785.9	1227.1	1518.0	1769.1	2301.5	2814.8

Bolt data   1183

2	70	140	75	290	64.3	119.8	187.3	231.8	270.3	352.1	431.1
3					102.1	190.4	297.5	368.2	429.4	559.1	684.5
4					144.0	268.4	419.4	518.9	605.1	787.7	964.1
5					190.6	355.3	554.9	686.7	800.5	1041.9	1275.0
6					242.8	452.5	706.8	874.5	1019.4	1326.6	1623.0
7					301.4	561.6	877.0	1085.1	1264.8	1645.6	2013.0
8					366.8	683.4	1067.2	1320.2	1538.7	2001.8	2448.5
9					439.4	818.6	1278.1	1581.1	1842.7	2397.1	2931.6
2	100	120	60	240	59.8	111.5	174.4	215.8	251.8	328.0	401.9
3					101.6	189.5	296.1	366.4	427.4	556.4	681.3
4					151.1	281.7	440.1	544.5	634.9	826.3	1011.3
5					209.8	391.0	610.6	755.5	880.6	1145.9	1401.9
6					278.6	519.2	810.6	1002.9	1168.8	1520.7	1860.0
7					358.1	667.2	1041.7	1288.6	1501.7	1953.5	2389.0
8					448.6	835.6	1304.5	1613.6	1880.4	2445.9	2990.8
9					550.0	1024.6	1599.4	1978.4	2305.4	2998.5	3666.1
2	100	120	85	290	65.2	121.6	190.1	235.3	274.4	357.3	437.6
3					109.1	203.4	317.9	393.3	458.7	597.0	730.8
4					160.6	299.3	467.4	578.3	674.2	877.4	1073.6
5					220.7	411.2	642.2	794.5	926.1	1205.0	1474.0
6					290.5	541.3	845.2	1045.6	1218.6	1585.4	1939.0
7					370.7	690.7	1078.3	1333.9	1554.5	2022.1	2472.8
8					461.7	860.0	1342.6	1660.7	1935.3	2517.2	3077.9
9					563.5	1049.7	1638.6	2026.8	2361.7	3071.7	3755.7
2	100	140	50	240	63.2	117.8	184.2	228.0	265.9	346.4	424.3
3					105.8	197.3	308.3	381.6	445.0	579.4	709.3
4					155.7	290.2	453.3	560.9	654.0	851.2	1041.6
5					214.5	399.7	624.3	772.3	900.3	1171.5	1433.1
6					283.4	528.0	824.4	1019.9	1188.7	1546.5	1891.5
7					362.9	676.1	1055.5	1305.7	1521.6	1979.4	2420.6
8					453.3	844.5	1318.3	1630.7	1900.2	2471.7	3022.3
9					554.8	1033.4	1613.2	1995.4	2325.1	3024.2	3697.6
2	100	140	75	290	68.1	127.0	198.4	245.6	286.4	372.9	456.6
3					112.9	210.4	328.7	406.8	474.3	617.4	755.6
4					164.8	307.1	479.6	593.5	691.8	900.3	1101.6
5					225.1	419.5	655.1	810.5	944.7	1229.1	1503.6
6					295.1	549.8	858.4	1062.0	1237.7	1610.2	1969.3
7					375.4	699.3	1091.7	1350.5	1573.8	2047.2	2503.5
8					466.3	868.7	1356.0	1677.4	1954.6	2542.4	3108.7
9					568.2	1058.3	1652.0	2043.4	2381.1	3097.0	3786.5

Bolt data

Bolt groups
One row of fasteners; fasteners not in the plane of the force

No. of fasteners in vertical row	Pitch, p (mm)	Values of Z_{xx} (cm³) for diameter of bolt, D (mm)						
		12	16	20	22	24	27	30
2	70	13.3	24.8	38.8	48.0	55.9	72.8	89.1
3		25.5	47.5	74.2	91.8	107.0	139.2	170.3
4		41.7	77.6	121.2	149.9	174.6	227.2	277.8
5		61.8	115.0	179.5	222.1	258.8	336.5	411.4
6		85.8	159.8	249.4	308.4	359.4	467.3	571.3
7		113.7	211.8	330.6	408.9	476.4	619.6	757.3
8		145.6	271.2	423.3	523.5	610.0	793.2	969.6
9		181.5	338.0	527.4	652.3	760.0	988.3	1208.0
2	100	15.4	28.7	44.8	55.4	64.6	84.1	102.9
3		31.4	58.5	91.3	112.9	131.6	171.2	209.3
4		53.0	98.8	154.2	190.7	222.2	288.9	353.2
5		80.3	149.5	233.4	288.6	336.3	437.3	534.6
6		113.2	210.8	328.9	406.8	474.0	616.3	753.4
7		151.6	282.4	440.8	545.2	635.1	825.9	1009.5
8		195.8	364.6	569.0	703.7	819.9	1066.1	1303.1
9		245.5	457.2	713.5	882.5	1028.1	1336.9	1634.1

Centre of rotation is assumed 60 mm below the bottom bolt
The tabulated values are conservative when the centre of rotation is located more than 60 mm below the bottom bolt. The tabulated values are unconservative when the centre of rotation is located less than 60 mm below the bottom line.

Bolt groups
Two rows of fasteners; fasteners not in the plane of the force

No. of fasteners in vertical row	Pitch, p (mm)	Values of Z_{xx} (cm³) for diameter of bolt, D (mm)						
		12	16	20	22	24	27	30
2	70	26.6	49.6	77.5	95.9	111.8	145.6	178.2
3		51.0	95.1	148.5	183.7	214.1	278.5	340.6
4		83.3	155.2	242.3	299.7	349.3	454.3	555.5
5		123.5	230.1	359.1	444.1	517.5	673.1	822.9
6		171.5	319.5	498.7	616.8	718.7	934.7	1142.6
7		227.5	423.7	661.2	817.8	952.8	1239.1	1514.7
8		291.2	542.5	846.6	1047.1	1219.9	1586.4	1939.2
9		362.9	675.9	1054.8	1304.6	1520.0	1976.5	2416.0
2	100	30.8	57.4	89.6	110.9	129.3	168.2	205.8
3		62.8	117.0	182.6	225.9	263.2	342.4	418.6
4		106.1	197.5	308.3	381.4	444.4	577.9	706.5
5		160.6	299.1	466.7	577.3	672.6	874.7	1069.2
6		226.3	421.5	657.8	813.6	947.9	1232.6	1506.7
7		303.3	564.9	881.6	1090.3	1270.3	1651.8	2019.1
8		391.5	729.2	1138.0	1407.4	1639.7	2132.2	2606.2
9		491.0	914.5	1427.1	1765.0	2056.3	2673.8	3268.2

Centre of rotation is assumed 60 mm below the bottom bolts
The tabulated values are conservative when the centre of rotation is located more than 60 mm below the bottom bolts. The tabulated values are unconservative when the centre of rotation is located less than 60 mm below the bottom bolts.

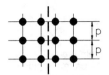

Bolt groups
Four rows of fasteners; fasteners not in the plane of the force

No. of fasteners in vertical row	Pitch, p (mm)	Values of Z_{xx} (cm³) for diameter of bolt, D(mm)						
		12	16	20	22	24	27	30
2	70	53.2	99.2	155.0	191.8	223.7	291.1	356.3
3		102.1	190.2	296.9	367.3	428.1	557.0	681.2
4		166.7	310.5	484.7	599.5	698.6	908.7	1111.1
5		247.0	460.1	718.2	888.3	1035.0	1346.1	1645.7
6		343.1	639.1	997.4	1233.7	1437.4	1869.3	2285.2
7		454.9	847.3	1322.4	1635.6	1905.7	2478.2	3029.4
8		582.5	1084.9	1693.2	2094.1	2439.9	3172.8	3878.3
9		725.8	1351.8	2109.7	2609.2	3039.9	3953.1	4832.0
2	100	61.6	114.8	179.2	221.8	258.5	336.4	411.5
3		125.6	234.0	365.2	451.8	526.5	684.8	837.3
4		212.1	395.1	616.7	762.7	888.7	1155.8	1413.0
5		321.1	598.1	933.5	1154.5	1345.2	1749.3	2138.4
6		452.6	843.0	1315.6	1627.2	1895.8	2465.3	3013.5
7		606.6	1129.8	1763.1	2180.6	2540.6	3303.7	4038.2
8		783.1	1458.4	2276.0	2814.9	3279.5	4264.5	5212.5
9		982.0	1828.9	2854.2	3529.9	4112.6	5347.7	6536.4

Centre of rotation is assumed 60 mm below the bottom bolts
The tabulated values are conservative when the centre of rotation is located more than 60 mm below the bottom bolts. The tabulated values are unconservative when the centre of rotation is located less than 60 mm below the bottom bolts.

Weld data

Weld groups
Welds in the plane of the force

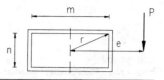

Values of Z_p (cm³) for 1 mm throat thickness

Values of n (mm)	Values of m (mm)										
	50	75	100	125	150	175	200	225	250	275	300
50	4.7	7.2	10.1	13.3	16.9	20.9	25.3	30.1	35.3	40.9	47.0
75	7.2	10.6	14.3	18.3	22.6	27.4	32.5	37.9	43.8	50.1	56.8
100	10.1	14.3	18.9	23.7	28.9	34.4	40.2	46.5	53.1	60.1	67.5
125	13.3	18.3	23.7	29.5	35.5	41.8	48.5	55.5	62.9	70.6	78.7
150	16.9	22.6	28.9	35.5	42.4	49.6	57.2	65.0	73.2	81.7	90.6
175	20.9	27.4	34.4	41.8	49.6	57.7	66.1	74.8	83.9	93.2	102.9
200	25.3	32.5	40.2	48.5	57.2	66.1	75.4	85.0	94.9	105.1	115.6
225	30.1	37.9	46.5	55.5	65.0	74.8	85.0	95.5	106.2	117.3	128.6
250	35.3	43.8	53.1	62.9	73.2	83.9	94.9	106.2	117.9	129.8	142.0
275	40.9	50.1	60.1	70.6	81.7	93.2	105.1	117.3	129.8	142.6	155.7
300	47.0	56.8	67.5	78.7	90.6	102.9	115.6	128.6	142.0	155.7	169.7
325	53.5	64.0	75.3	87.2	99.8	112.9	126.4	140.3	154.5	169.1	184.0
350	60.3	71.5	83.4	96.1	109.4	123.3	137.6	152.3	167.4	182.8	198.6
375	67.6	79.4	92.0	105.4	119.4	134.0	149.1	164.6	180.6	196.9	213.5
400	75.4	87.8	101.1	115.1	129.8	145.1	161.0	177.3	194.1	211.2	228.7
425	83.5	96.5	110.5	125.2	140.6	156.7	173.3	190.4	207.9	225.9	244.2
450	92.0	105.7	120.3	135.7	151.8	168.5	185.9	203.8	222.1	240.9	260.0
475	101.0	115.3	130.5	146.6	163.4	180.8	198.9	217.5	236.6	256.2	276.2
500	110.4	125.3	141.2	157.9	175.4	193.5	212.3	231.7	251.6	271.9	292.7
525	120.2	135.8	152.3	169.6	187.8	206.6	226.1	246.2	266.8	288.0	309.5
550	130.4	146.6	163.8	181.8	200.6	220.1	240.3	261.1	282.5	304.4	326.8
575	141.0	157.9	175.7	194.3	213.8	234.0	254.9	276.4	298.5	321.2	344.3
600	152.0	169.5	188.0	207.3	227.4	248.3	269.8	292.1	314.9	338.3	362.2

Weld groups
Welds in the plane of the force

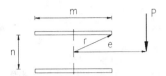

Values of Z_p (cm³) for 1 mm throat thickness

Values of n (mm)	Values of m (mm)										
	50	75	100	125	150	175	200	225	250	275	300
50	2.4	3.6	5.2	7.2	9.5	12.2	15.4	18.9	22.9	27.3	32.1
75	3.6	5.3	7.2	9.3	11.7	14.6	17.8	21.3	25.3	29.7	34.6
100	4.8	7.1	9.4	11.9	14.6	17.5	20.9	24.6	28.6	33.1	37.9
125	6.1	9.0	11.8	14.7	17.8	21.0	24.6	28.4	32.6	37.2	42.1
150	7.4	10.9	14.3	17.7	21.2	24.8	28.7	32.8	37.2	41.9	47.0
175	8.6	12.8	16.8	20.8	24.8	28.9	33.1	37.5	42.2	47.1	52.4
200	9.9	14.7	19.4	24.0	28.5	33.1	37.7	42.5	47.5	52.7	58.2
225	11.2	16.6	21.9	27.1	32.2	37.3	42.5	47.7	53.1	58.7	64.5
250	12.4	18.5	24.5	30.3	36.0	41.7	47.4	53.1	58.9	64.9	71.1
275	13.7	20.4	27.0	33.4	39.8	46.1	52.3	58.6	64.9	71.3	77.9
300	14.9	22.3	29.5	36.6	43.6	50.5	57.3	64.1	71.0	77.8	84.9
325	16.2	24.2	32.0	39.8	47.4	54.9	62.3	69.7	77.1	84.5	92.0
350	17.4	26.1	34.6	43.0	51.2	59.3	67.4	75.4	83.3	91.3	99.2
375	18.7	27.9	37.1	46.1	55.0	63.8	72.5	81.0	89.6	98.1	106.6
400	19.9	29.8	39.6	49.3	58.8	68.2	77.5	86.7	95.8	104.9	114.0
425	21.2	31.7	42.1	52.4	62.6	72.7	82.6	92.4	102.1	111.8	121.5
450	22.5	33.6	44.7	55.6	66.4	77.1	87.7	98.1	108.5	118.7	129.0
475	23.7	35.5	47.2	58.7	70.2	81.5	92.7	103.8	114.8	125.7	136.5
500	25.0	37.4	49.7	61.9	74.0	86.0	97.8	109.5	121.1	132.6	144.1
525	26.2	39.2	52.2	65.0	77.8	90.4	102.9	115.2	127.5	139.6	151.6
550	27.5	41.1	54.7	68.2	81.6	94.8	107.9	120.9	133.8	146.6	159.2
575	28.7	43.0	57.2	71.3	85.4	99.2	113.0	126.6	140.1	153.5	166.8
600	30.0	44.9	59.7	74.5	89.1	103.7	118.1	132.3	146.5	160.5	174.4

Weld data

Weld groups
Welds in the plane of the force

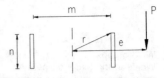

Values of Z_p (cm³) for 1 mm throat thickness

Values of n (mm)	Values of m (mm)										
	50	75	100	125	150	175	200	225	250	275	300
50	2.4	3.6	4.8	6.1	7.4	8.6	9.9	11.2	12.4	13.7	14.9
75	3.6	5.3	7.1	9.0	10.9	12.8	14.7	16.6	18.5	20.4	22.3
100	5.2	7.2	9.4	11.8	14.3	16.8	19.4	21.9	24.5	27.0	29.5
125	7.2	9.3	11.9	14.7	17.7	20.8	24.0	27.1	30.3	33.4	36.6
150	9.5	11.7	14.6	17.8	21.2	24.8	28.5	32.2	36.0	39.8	43.6
175	12.2	14.6	17.5	21.0	24.8	28.9	33.1	37.3	41.7	46.1	50.5
200	15.4	17.8	20.9	24.6	28.7	33.1	37.7	42.5	47.4	52.3	57.3
225	18.9	21.3	24.6	28.4	32.8	37.5	42.5	47.7	53.1	58.6	64.1
250	22.9	25.3	28.6	32.6	37.2	42.2	47.5	53.1	58.9	64.9	71.0
275	27.3	29.7	33.1	37.2	41.9	47.1	52.7	58.7	64.9	71.3	77.8
300	32.1	34.6	37.9	42.1	47.0	52.4	58.2	64.5	71.1	77.9	84.9
325	37.3	39.8	43.2	47.4	52.4	58.0	64.1	70.6	77.4	84.6	92.0
350	42.9	45.4	48.9	53.2	58.2	63.9	70.2	76.9	84.1	91.6	99.3
375	48.9	51.5	55.0	59.3	64.4	70.2	76.7	83.6	91.0	98.8	106.9
400	55.4	57.9	61.4	65.8	71.0	76.9	83.5	90.6	98.2	106.3	114.7
425	62.3	64.8	68.3	72.8	78.0	84.0	90.7	98.0	105.8	114.0	122.7
450	69.6	72.1	75.7	80.1	85.4	91.5	98.2	105.7	113.6	122.1	131.0
475	77.3	79.8	83.4	87.8	93.2	99.3	106.2	113.7	121.9	130.5	139.7
500	85.4	88.0	91.5	96.0	101.4	107.6	114.5	122.2	130.4	139.3	148.6
525	94.0	96.5	100.1	104.6	110.0	116.2	123.2	131.0	139.4	148.4	157.9
550	102.9	105.5	109.0	113.6	119.0	125.3	132.4	140.2	148.7	157.8	167.5
575	112.3	114.9	118.4	123.0	128.4	134.7	141.9	149.8	158.4	167.6	177.5
600	122.1	124.7	128.2	132.8	138.2	144.6	151.8	159.8	168.5	177.8	187.8

Weld groups
Welds in the plane of the force

Values of Z_p (cm³) for 1 mm of throat thickness	
m or n (mm)	Z_p (cm³)
50	0.4
75	0.9
100	1.7
125	2.6
150	3.8
175	5.1
200	6.7
225	8.4
250	10.4
275	12.6
300	15.0
325	17.6
350	20.4
375	23.4
400	26.7
425	30.1
450	33.8
475	37.6
500	41.7
525	45.9
550	50.4
575	55.1
600	60.0

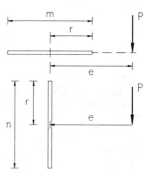

Capacities of fillet welds

(1) Angle between fusion faces 60° to 90° using BS 639 electrodes grade E43, with BS 4360 steel of grade 40, 43, WR50 or 50 (or BS EN 10 025 equivalent) or electrodes grade E51 with grade 40 or 43 steel

Leg length, S (mm)	Throat thickness $a = 0.7S$ (mm)	Capacity @ 215 N/mm^2 (kN/mm)
3	2.1	0.451
4	2.8	0.602
5	3.5	0.752
6	4.2	0.903
8	5.6	1.204
10	7.0	1.505
12	8.4	1.806
15	10.5	2.257
18	12.6	2.719
20	14.0	3.01
22	15.4	3.311
25	17.5	3.762

(2) Angle between fusion faces 60° to 90° using BS 639 electrodes grade E51, with BS 4360 steel of grade WR50, 50 or 55 (or BS EN 10 025 equivalent)

Leg length, S (mm)	Throat thickness $a = 0.7S$ (mm)	Capacity @ 255 N/mm^2 (kN/mm)
3	2.1	0.535
4	2.8	0.714
5	3.5	0.892
6	4.2	1.071
8	5.6	1.428
10	7.0	1.785
12	8.4	2.142
15	10.5	2.677
18	12.6	3.213
20	14.0	3.570
22	15.4	3.927
25	17.5	4.462

(3) Angle between fusion faces 60° to 90° using BS 639 electrodes grade E51[a], with BS 4360 steel of grade 55 (or BS EN 10 025 equivalent)

Leg length, S (mm)	Throat thickness $a = 0.7S$ (mm)	Capacity @ 275 N/mm^2 (kN/mm)[a]
3	2.1	0.577
4	2.8	0.77
5	3.5	0.962
6	4.2	1.155
8	5.6	1.540
10	7.0	1.925
12	8.4	2.310
15	10.5	2.887
18	12.6	3.465
20	14.0	3.850
22	15.4	4.230
25	17.5	4.812

[a] Only applies to electrodes having a minimum tensile strength of 550 N/mm^2 and a minimum yield strength of 450 N/mm^2

Weld groups
Welds not in the plane of the force

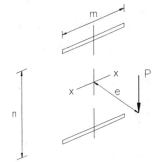

Values of Z_{xx} (cm^3) for 1 mm throat thickness

Values of n (mm)	Values of m (mm)										
	50	75	100	125	150	175	200	225	250	275	300
50	2.5	3.8	5.0	6.3	7.5	8.8	10.0	11.3	12.5	13.8	15.0
75	3.8	5.6	7.5	9.4	11.3	13.1	15.0	16.9	18.8	20.6	22.5
100	5.0	7.5	10.0	12.5	15.0	17.5	20.0	22.5	25.0	27.5	30.0
125	6.3	9.4	12.5	15.6	18.8	21.9	25.0	28.1	31.3	34.4	37.5
150	7.5	11.3	15.0	18.8	22.5	26.3	30.0	33.8	37.5	41.3	45.0
175	8.8	13.1	17.5	21.9	26.3	30.6	35.0	39.4	43.8	48.1	52.5
200	10.0	15.0	20.0	25.0	30.0	35.0	40.0	45.0	50.0	55.0	60.0
225	11.3	16.9	22.5	28.1	33.8	39.4	45.0	50.6	56.3	61.9	67.5
250	12.5	18.8	25.0	31.3	37.5	43.8	50.0	56.3	62.5	68.8	75.0
275	13.8	20.6	27.5	34.4	41.3	48.1	55.0	61.9	68.8	75.6	82.5
300	15.0	22.5	30.0	37.5	45.0	52.5	60.0	67.5	75.0	82.5	90.0
325	16.3	24.4	32.5	40.6	48.8	56.9	65.0	73.1	81.3	89.4	97.5
350	17.5	26.3	35.0	43.8	52.5	61.3	70.0	78.8	87.5	96.3	105.0
375	18.8	28.1	37.5	46.9	56.3	65.6	75.0	84.4	93.8	103.1	112.5
400	20.0	30.0	40.0	50.0	60.0	70.0	80.0	90.0	100.0	110.0	120.0
425	21.3	31.9	42.5	53.1	63.8	74.4	85.0	95.6	106.3	116.9	127.5
450	22.5	33.8	45.0	56.3	67.5	78.8	90.0	101.3	112.5	123.8	135.0
475	23.8	35.6	47.5	59.4	71.3	83.1	95.0	106.9	118.8	130.6	142.5
500	25.0	37.5	50.0	62.5	75.0	87.5	100.0	112.5	125.0	137.5	150.0
525	26.3	39.4	52.5	65.6	78.8	91.9	105.0	118.1	131.3	144.4	157.5
550	27.5	41.3	55.0	68.8	82.5	96.3	110.0	123.8	137.5	151.3	165.0
575	28.8	43.1	57.5	71.9	86.3	100.6	115.0	129.4	143.8	158.1	172.5
600	30.0	45.0	60.0	75.0	90.0	105.0	120.0	135.0	150.0	165.0	180.0

Weld data

Weld groups
Welds not in the plane of the force

Values of Z_{xx} (cm³) for 1 mm throat thickness

n (mm)	Z_{xx} (cm³)
50	0.8
75	1.9
100	3.3
125	5.2
150	7.5
175	10.2
200	13.3
225	16.9
250	20.8
275	25.2
300	30.0
325	35.2
350	40.8
375	46.9
400	53.3
425	60.2
450	67.5
475	75.2
500	83.3
525	91.9
550	100.8
575	110.2
600	120.0

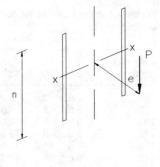

Weld data

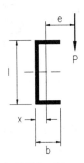

Weld groups
Welds in the plane of the force

	Values of Z_p (cm³) for 1 mm throat thickness									
Values of n (mm)					Values of m (mm)					
	50	75	100	125	150	175	200	225	250	275	300
50	2.8	4.3	6.2	8.5	11.2	14.3	17.9	21.9	26.2	31.0	36.2
75	4.3	6.2	8.4	11.0	13.9	17.3	21.1	25.2	29.8	34.8	40.3
100	6.2	8.5	11.0	13.9	17.1	20.7	24.7	29.2	34.0	39.2	44.9
125	8.3	11.0	14.0	17.2	20.7	24.6	28.9	33.6	38.7	44.1	50.0
150	10.6	13.9	17.3	20.9	24.8	29.0	33.5	38.5	43.8	49.6	55.7
175	13.2	17.0	20.8	24.9	29.1	33.7	38.6	43.8	49.4	55.4	61.8
200	16.0	20.3	24.7	29.1	33.8	38.8	44.0	49.6	55.5	61.8	68.4
225	19.0	23.9	28.8	33.7	38.8	44.2	49.8	55.7	61.9	68.5	75.5
250	22.2	27.7	33.1	38.5	44.1	49.9	55.9	62.2	68.8	75.7	82.9
275	25.6	31.7	37.7	43.6	49.7	55.9	62.3	69.0	75.9	83.2	90.8
300	29.2	35.9	42.5	49.0	55.5	62.2	69.0	76.1	83.4	91.1	99.0
325	33.1	40.4	47.5	54.5	61.6	68.7	76.0	83.5	91.3	99.3	107.6
350	37.1	45.0	52.7	60.3	67.9	75.5	83.3	91.3	99.4	107.8	116.5
375	41.4	49.9	58.2	66.3	74.5	82.6	90.9	99.3	107.8	116.7	125.7
400	45.9	55.0	63.8	72.6	81.2	89.9	98.7	107.5	116.6	125.8	135.3
425	50.6	60.3	69.7	79.0	88.3	97.5	106.7	116.1	125.6	135.3	145.1
450	55.4	65.8	75.8	85.7	95.5	105.3	115.0	124.9	134.8	145.0	155.3
475	60.6	71.5	82.2	92.6	103.0	113.3	123.6	133.9	144.4	155.0	165.7
500	65.9	77.4	88.7	99.8	110.7	121.5	132.4	143.2	154.2	165.2	176.5
525	71.4	83.6	95.4	107.1	118.6	130.0	141.4	152.8	164.2	175.8	187.5
550	77.1	89.9	102.4	114.6	126.7	138.7	150.6	162.5	174.5	186.5	198.7
575	83.0	96.5	109.6	122.4	135.1	147.6	160.1	172.6	185.0	197.6	210.2
600	89.2	103.2	116.9	130.4	143.7	156.8	169.8	182.8	195.8	208.9	222.0

1194 Piling information

FRODINGHAM STEEL SHEET PILING
Dimensions and Properties

Section	b mm (nom)	h mm (nom)	d mm	t mm (nom)	f1 mm (nom)	f2 mm (nom)	Sectional Area sq. cm. per metre of wall	Mass kg per linear metre	Mass kg per sq. m. of wall	Moment of Inertia cm⁴ per metre	Section Modulus cm³ per metre	Section
1BXN	476	143	12·7	12·7	78	123	166·5	62·1	130·4	4919	688	1BXN
1N	483	170	9·0	9·0	105	137	126·0	47·8	99·1	6048	713	1N
2N	483	235	9·7	8·4	97	149	143·0	54·2	112·3	13513	1150	2N
3N	483	283	11·7	8·9	89	145	175·0	66·2	137·1	23885	1688	3N
3NA	483	305	9·7	9·5	96	146	165·0	62·6	129·8	25687	1690	3NA
4N	483	330	14·0	10·4	77	127	218·0	82·4	170·8	39831	2414	4N
5	425	311	17·0	11·9	89	118	302·0	100·8	236·9	49262	3168	5

SPECIAL SECTIONS
Sections may be "rolled up" (thickened) or "rolled down" (thinned) by special arrangement to increase or decrease the thickness of both webs and flanges by a maximum of 0·8mm.

Piling information

High Modulus Piles with Frodingham 1BXN Wings – Dimensions and Properties

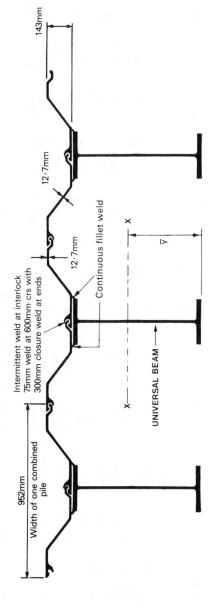

Universal Beam		Crs. of u.b.'s mm (nom)	Mass		Area of Section cm^2	$\bar{y}$ cm	Moment of Inertia xx		Section Modulus xx		Rad of gyration xx cm
serial size mm	mass kg/m		kg/m	kg/m^2			cm^4/m wall	$cm^4/single$ pile	cm^3/m wall	$cm^3/single$ pile	
533 × 210	101	952	225	237	287	45·6	155841	148361	3420	3256	22·7
610 × 229	113	952	237	249	302	50·0	208203	198209	4164	3964	25·6
*610 × 305	149	952	270	284	344	47·4	263788	251126	5564	5297	27·0
686 × 254	152	952	276	290	352	53·1	320324	304949	6036	5746	29·4
762 × 267	173	952	297	312	378	57·0	418126	398056	7331	6979	32·4
762 × 267	197	952	321	337	409	56·2	468674	446178	8346	7945	33·0
*838 × 292	194	952	316	332	403	61·1	542137	516115	8869	8443	35·8
*914 × 305	224	952	344	362	438	64·0	693250	659974	10836	10316	38·8
*914 × 305	253	952	372	391	474	63·0	772405	735330	12261	11672	39·4
*914 × 305	289	952	407	428	519	61·9	860627	819317	13902	13235	39·7
*914 × 419	388	952	469	492	597	56·4	1048950	998600	18599	17706	41·0

*Denotes beam section with top flange reduced to 280mm to facilitate fabrication.
Note: Centre line of common interlock on sheet pile does not coincide with centre line of Universal Beam.

Piling information

High Modulus Piles with Frodingham 3N Wings – Dimensions and Properties

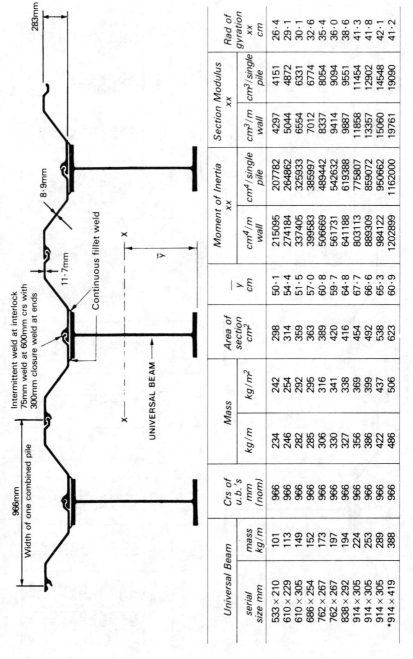

Universal Beam		Crs of u.b.'s mm (nom)	Mass		Area of section cm^2	$\bar{y}$ cm	Moment of Inertia xx		Section Modulus xx		Rad of gyration xx cm
serial size mm	mass kg/m		kg/m	kg/m^2			cm^4/m wall	cm^4/single pile	cm^3/m wall	cm^3/single pile	
533 × 210	101	966	234	242	298	50·1	215095	207782	4297	4151	26·4
610 × 229	113	966	246	254	314	54·4	274184	264862	5044	4872	29·1
610 × 305	149	966	282	292	359	51·5	337405	325933	6554	6331	30·1
686 × 254	152	966	285	295	363	57·0	399583	385997	7012	6774	32·6
762 × 267	173	966	306	316	389	60·8	506669	489442	8337	8054	35·4
762 × 267	197	966	330	341	420	59·7	561731	542632	9414	9094	36·0
838 × 292	194	966	327	338	416	64·8	641188	619388	9887	9551	38·6
914 × 305	224	966	356	369	454	67·7	803113	775807	11858	11454	41·3
914 × 305	253	966	386	399	492	66·6	889309	859072	13357	12902	41·8
914 × 305	289	966	422	437	538	65·3	984122	950662	15060	14548	42·1
*914 × 419	388	966	486	506	623	60·9	1202899	1162000	19761	19090	41·2

*Denotes beam section with one flange reduced to 310mm to facilitate fabrication.

Note: Centre line of common interlock on sheet pile does not coincide with centre line of Universal Beam.

High Modulus Piles with Frodingham 4N Wings – Dimensions and Properties

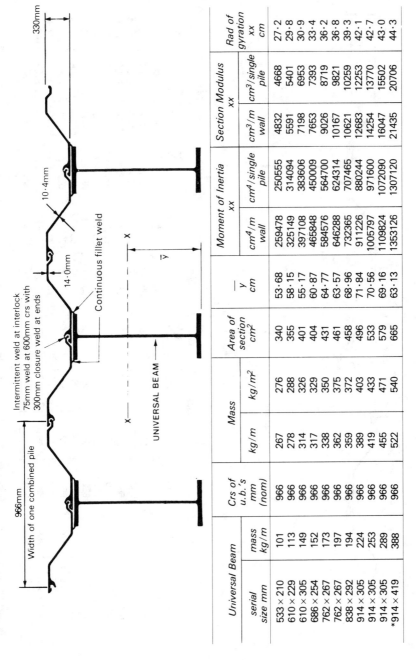

Universal Beam		Crs of u.b.'s mm (nom)	Mass		Area of section cm²	$\bar{y}$ cm	Moment of Inertia xx		Section Modulus xx		Rad of gyration xx cm
serial size mm	mass kg/m		kg/m	kg/m²			cm⁴/m wall	cm⁴/single pile	cm³/m wall	cm³/single pile	
533 × 210	101	966	267	276	340	53·68	259478	250555	4832	4668	27·2
610 × 229	113	966	278	288	355	58·15	325149	314094	5591	5401	29·8
610 × 305	149	966	314	326	401	55·17	397108	383606	7198	6953	30·9
686 × 254	152	966	317	329	404	60·87	465848	450009	7653	7393	33·4
762 × 267	173	966	338	350	431	64·77	584576	564700	9026	8719	36·2
762 × 267	197	966	362	375	461	63·57	646288	624314	10167	9821	36·8
838 × 292	194	966	359	372	458	68·96	732365	707465	10621	10259	39·3
914 × 305	224	966	389	403	496	71·84	911226	880244	12683	12253	42·1
914 × 305	253	966	419	433	533	70·56	1005797	971600	14254	13770	42·7
914 × 305	289	966	455	471	579	69·16	1109824	1072090	16047	15502	43·0
*914 × 419	388	966	522	540	665	63·13	1353126	1307120	21435	20706	44·3

*Denotes beam section with one flange reduced to 310mm to facilitate fabrication.
Note: Centre line of common interlock on sheet pile does not coincide with centre line of Universal Beam.

Piling information

High Modulus Piles with Frodingham 5 Wings – Dimensions and Properties

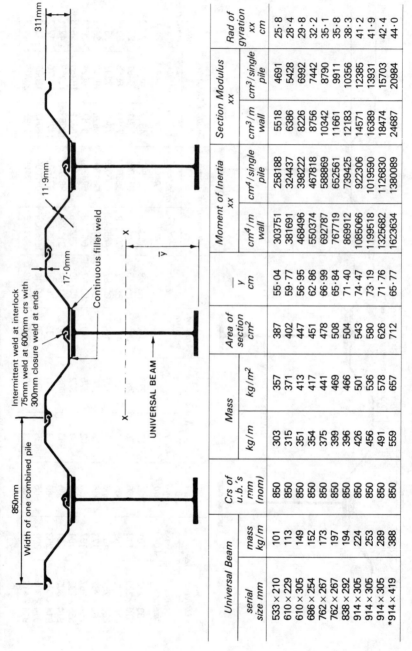

Universal Beam			Mass		Area of section cm²	$\bar{y}$ cm	Moment of Inertia xx		Section Modulus xx		Rad of gyration xx cm
serial size mm	mass kg/m	Crs of u.b.'s mm (nom)	kg/m	kg/m²			cm⁴/m wall	cm⁴/single pile	cm³/m wall	cm³/single pile	
533 × 210	101	850	303	357	387	55·04	303751	258188	5518	4691	25·8
610 × 229	113	850	315	371	402	59·77	381691	324437	6386	5428	28·4
610 × 305	149	850	351	413	447	56·95	468496	398222	8226	6992	29·8
686 × 254	152	850	354	417	451	62·86	550374	467818	8756	7442	32·2
762 × 267	173	850	375	441	478	66·99	692787	588869	10342	8790	35·1
762 × 267	197	850	399	469	508	65·84	767719	652561	11661	9911	35·8
838 × 292	194	850	396	466	504	71·40	869912	739425	12183	10356	38·3
914 × 305	224	850	426	501	543	74·47	1085066	922306	14571	12385	41·2
914 × 305	253	850	456	536	580	73·19	1199518	1019590	16389	13931	41·9
914 × 305	289	850	491	578	626	71·76	1325682	1126830	18474	15703	42·4
*914 × 419	388	850	559	657	712	65·77	1623634	1380089	24687	20984	44·0

*Denotes beam section with one flange reduced to 310mm to facilitate fabrication.
Note: Centre line of common interlock on sheet pile does not coincide with centre line of Universal Beam.

Frodingham Double Box Piles—Dimensions and Properties

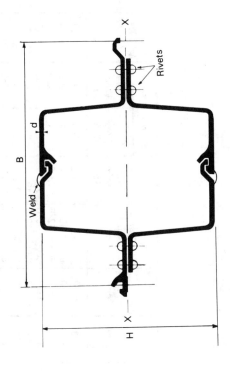

Section	B mm (nominal)	H mm (nominal)	d mm	Mass per metre kg	Section Area cm² Steel only	Section Area cm² Whole Pile	Min. Rad. of Gyr'n cm	Approx. Perimeter mm	Moment of Inertia cm⁴ X–X	Section Modulus cm³ X–X
1BXN	953	286	12·7	231·3	290	1496	9·00	2450	25515	1785
1N	965	340	9·0	179·5	225	1717	10·95	2520	29279	1722
2N	965	470	9·7	201·8	255	2384	15·24	2790	64305	2735
3N	965	565	11·7	246·3	310	2856	18·34	2970	113415	4015
3NA	966	610	9·7	235·4	298	3035	19·69	3100	123861	4064
4N	965	660	14·0	305·1	384	3350	21·34	3200	191660	5805
5	850	622	17·0	367·8	464	2853	20·12	3230	208430	6700

Frodingham Plated Box Piles—Dimensions and Properties

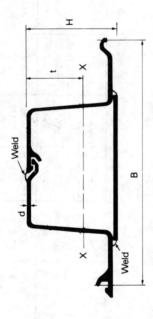

Section	B mm (nom.)	H mm (nom.)	t mm	d mm	*Plate Size mm	Mass per metre kg	Section Area cm² Steel Only	Section Area cm² Whole Pile	Min. Rad. of Gyr'n cm	Approx Perimeter mm	Moment of Inertia cm⁴ X–X	Section Modulus cm³ X–X
1BXN	953	158	103	12.7	711 × 15.0	208.0	265	864	5.74	2210	8705	845
1N	965	180	117	9.0	686 × 10.0	149.4	191	935	7.04	2240	9407	804
2N	965	247.5	164	9.7	660 × 12.5	173.3	221	1290	9.75	2390	20970	1279
3N	965	298	196	11.7	660 × 15.0	210.2	268	1555	11.73	2470	36840	1880
3NA	965	317.5	206	9.7	660 × 12.5	190.1	242	1613	12.60	2550	38528	1870
4N	965	345	220	14.0	660 × 15.0	242.8	310	1819	13.77	2600	58564	2662
5	850	331	204	17.0	533 × 20.0	285.4	364	1561	13.13	2510	62650	3071

*Standard plate thickness have been used.

Frodingham Straight Web Sections—Dimensions and Properties

Section modulus of a single pile = 91·5 cm³

Section	B mm (nominal)	t mm	MASS		Standard Junctions per lin. metre	Minimum Ultimate Strength of Interlock tonnes per metre		
			Per lineal metre of Pile kg	Per square metre of Wall kg		Grade 43A Steel	ASTM–A328	Grade 50A Steel
SW-1	413	9·5	55·3	134·0	83·6kg/m	285	299	384
SW-1A	413	12·7	63·8	154·5	96·3kg/m	285	299	384

LARSSEN STEEL SHEET PILING
Dimensions and Properties

Section	b mm (nominal)	h mm (nominal)	d mm	t mm (nominal)	f Flat of Pan mm	Sectional Area cm² per metre of wall	Mass kg per linear metre	Mass kg per sq. metre of wall	Combined Moment of Inertia cm⁴ per metre	Section Modulus cm³ per metre
6W	525	212	7·8	6·4	331	108	44·7	85·1	6459	610
9W	525	260	8·9	6·4	343	124	51·0	97·1	11726	902
12W	525	306	9·0	8·5	343	147	60·4	115·1	18345	1199
16W	525	348	10·5	8·6	341	166	68·3	130·1	27857	1601
20W	525	400	11·3	9·2	333	188	77·3	147·2	40180	2009
25W	525	454	12·1	10·5	317	213	87·9	167·4	56727	2499
32W	525	454	17·0	10·5	317	252	103·6	197·4	73003	3216
3	400	248	14·1	8·5	253	198	62·2	155·5	16980	1360
4A	400	381	15·7	9·6	219	236	74·0	185·1	44916	2360
6	420	440	22·0	14·0	248	370	122·0	290·5	92452	4200
6	420	440	25·4	14·0	251	397	131·0	311·8	102861	4675
6	420	440	28·6	14·0	251	421	138·7	330·2	111450	5066

Larssen Box Piles—Dimensions and Properties

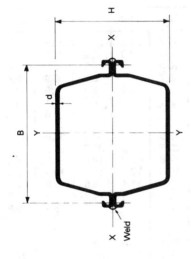

Section	B mm nom.	H mm nom.	d mm	Mass per metre kg	Section Area cm² Steel only	Section Area cm² Whole Pile	Least Rad. of Gyr. cm	Approx Perim. mm	Moment of Inertia cm⁴ About X–X	Moment of Inertia cm⁴ About Y–Y	Section Modulus cm³ About X–X	Section Modulus cm³ About Y–Y
6W	525	252	7·8	89·4	113·8	1122	9·33	1652	10028	39280	796	1400
9W	525	300	8·9	102·0	129·9	1344	11·43	1748	16976	43812	1132	1562
12W	525	349	9·0	120·8	153·9	1589	13·00	1872	25992	55442	1490	1970
16W	525	391	10·5	136·6	174·0	1776	14·80	1952	38126	60920	1950	2164
20W	525	447	11·3	154·6	196·9	2005	16·63	2084	54474	70120	2438	2482
25W	525	502	12·1	175·8	223·9	2221	18·33	2164	75240	80082	2998	2830
32W	525	502	17·0	207·3	264·1	2224	18·02	2168	96584	85806	3848	3032
3	400	296	14·1	124·4	158·6	1020	11·22	1596	19954	30559	1348	1399
4A	400	428	15·7	148·0	188·7	1413	13·82	1772	47055	36039	2199	1649
6 (122kg)	420	502	22·0	244·0	310·6	1794	14·77	2080	106780	67756	4237	2920
6 (131kg)	420	502	25·4	262·0	333·6	1794	14·47	2080	118588	69859	4725	3011
6 (138·7kg)	420	502	28·6	277·4	353·2	1794	14·27	2080	128448	71392	5117	3077

The piling information on pp 1194–1203 is extracted from *Piling Handbook* 6th Edn, published by British Steel General Steels, 1988.

Floor plates

Ultimate load capacity (kN/m^2) for floor plates simply supported on two edges stressed to 275 N/mm^2

Thickness on plain mm	Span (mm)							
	600	800	1000	1200	1400	1600	1800	2000
4.5	20.48	11.62	7.45	5.17	3.80	2.95	2.28	1.87
6.0	36.77	20.68	13.28	9.20	6.73	5.20	4.07	3.30
8.0	65.40	36.87	23.48	16.38	11.97	9.23	7.23	5.93
10.0	102.03	57.42	36.67	25.55	18.70	14.45	11.30	9.25
12.5	159.70	89.85	57.40	39.98	29.27	22.62	17.68	14.50

Stiffeners should be used for spans in excess of 1100 mm to avoid excessive deflections.

Ultimate load capacity (kN/m^2) for floor plates simply supported on all four edges stressed to 275 N/mm^2

Thickness on plain mm	Breadth B mm	Length (mm)							
		600	800	1000	1200	1400	1600	1800	2000
4.5	600	34.9	25.5	22.7	21.7	21.2	21.0	20.8	20.8
	800		19.6	15.1	13.4	12.6	12.2	12.0	11.8
	1000			12.6	10.0	8.8	8.3	7.9	7.7
	1200				8.7	7.1	6.3	5.9	5.6
	1400					6.4	5.3	4.8	4.4
	1600						4.9	4.1	3.7
	1800							3.8	3.3
6.0	600	62.1	45.3	40.4	38.5	37.7	37.3	37.0	36.9
	800		34.9	26.8	23.7	22.3	21.7	21.3	21.1
	1000			22.4	17.8	15.8	14.8	14.2	13.9
	1200				15.5	12.7	11.3	10.6	10.1
	1400					11.4	9.5	8.5	7.9
	1600						8.7	7.4	6.7
	1800							6.9	5.9
8.0	600	110	80.6	71.1	68.4	67.0	66.2	65.8	65.6
	800		62.1	47.7	42.2	39.7	38.5	37.8	37.4
	1000			39.7	31.7	28.1	26.2	25.2	24.6
	1200				27.6	22.6	20.1	18.8	17.9
	1400					20.3	17.0	15.2	14.1
	1600						15.5	13.3	11.9
	1800							12.3	10.6
10.0	600	172*	126*	112*	107*	105*	103*	103*	103*
	800		97.0	74.5	65.9	62.1	60.1	59.1	58.5
	1000			62.1	49.5	43.9	41.0	39.4	38.5
	1200				43.1	35.4	31.5	29.3	28.0
	1400					31.7	26.6	23.8	22.1
	1600						24.3	20.7	18.6
	1800							19.2	16.6
12.5	600	269*	197*	175*	167*	163*	162*	161*	160*
	800		152	116*	103*	97.0*	94.0*	92.3*	91.4*
	1000			97.0	77.4	68.5	64.1	61.6	60.1
	1200				67.4	55.3	49.2	45.8	43.8
	1400					49.5	41.5	37.1	34.5
	1600						37.9	32.4	29.1
	1800							29.9	25.9

Floor plates

Ultimate load capacity (kN/m²) for floor plates fixed on all four edges stressed to 275 N/mm²

Thickness on plain mm	Breadth B mm	Length (mm)							
		600	800	1000	1200	1400	1600	1800	2000
4.5	600	47.7*	36.8*	33.5*	32.2*	31.6*	31.4*	31.2*	31.1*
	800		26.8	21.5*	19.5*	18.6*	18.1*	17.9*	17.7*
	1000			17.2*	14.2*	12.9*	12.2*	11.8*	11.6*
	1200				11.9	10.1	9.1	8.6	8.3
	1400					8.7	7.5	6.9	6.5
	1600						6.7	5.8	5.3
	1800							5.3	4.7
6.0	600	84.8*	65.4*	59.5*	57.3*	56.2*	55.7*	55.5*	55.3*
	800		47.7*	38.3*	34.7*	33.1*	32.2*	31.7*	31.5*
	1000			30.5*	25.3*	22.9*	21.7*	21.0*	20.6*
	1200				21.2*	18.0*	16.3*	15.4*	14.9*
	1400					15.6*	13.4*	12.3*	11.6
	1600						11.9	10.4	9.5
	1800							9.4	8.3
8.0	600	151*	116*	106*	102*	100*	99.1*	98.6*	98.3*
	800		68.1*	61.7*	58.8*	57.3*	56.4*	55.9*	
	1000			54.3*	44.9*	40.7*	38.6*	37.4*	36.7*
	1200				37.7*	31.9*	29.0*	27.4*	26.5*
	1400					27.7*	23.9*	21.8*	20.6*
	1600						21.2*	18.6*	17.0*
	1800							16.9*	14.8*
10.0	600	236*	182*	165*	159*	156*	155*	154*	154*
	800		132*	106*	96.4*	91.8*	89.5*	88.2*	87.4*
	1000			84.8*	70.2*	63.7*	60.3*	58.4*	57.3*
	1200				58.9*	49.9*	45.4*	42.9*	41.3*
	1400					43.3*	37.3*	34.1*	32.2*
	1600						33.1*	29.0*	26.6*
	1800							26.2*	23.2*
12.5	600	368*	284*	258*	249*	244*	242*	241*	240*
	800		207*	166*	151*	144*	140*	138*	137*
	1000			132*	110*	99.5*	94.2*	91.2*	89.5*
	1200				92.0*	77.9*	70.9*	67.0*	64.6*
	1400					67.6*	58.3*	53.3*	50.3*
	1600						51.8*	45.3*	41.6*
	1800							40.9*	36.2*

Note on tables:
Values without an asterisk cause deflection greater than B/100 at serviceability, assuming that the only dead load present is due to self-weight.
Values obtained using Pounder's formula allowing corners to lift.

STEELWORK IN FIRE
INFORMATION SHEET

M01

This series of information sheets is intended to illustrate methods of achieving fire resistance in steel structures. It should not be used for design without consulting detailed design guidance referenced below.

SPRAYED PROTECTION

UP TO 4 HRS

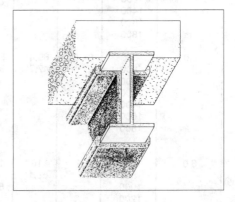

METHOD
Fire protective insulation can be applied by spraying to any type of steel member. Most products can achieve up to 4 hour rating.

PRINCIPLE
Insulation reduces the heating rate of a steel member so that its limiting temperature is not exceeded during the required fire resistance period. The protection material thickness necessary depends on the section factor (Hp/A) of the member and the fire rating required.

ADVANTAGES
a) Low cost
b) Rapid application
c) Easy to cover complex details
d) Often applied to non-primed steelwork
e) Some products may be suitable for external use

LIMITATIONS (check with manufacturer)
a) Appearance may be inadequate for visible members
b) Overspray may need masking or shielding
c) Primer, if used, must be compatible

FOR MORE DETAILED INFORMATION SEE:-
"Fire protection of Structural Steel in Building" published jointly by:
ASFPCM - (0252 336318) and
The Steel Construction Institute - (0344 23345)

Sheet Code
ISF/M01
January 1991

PROTECTION THICKNESS

Thickness recommendations given in "Fire Protection of Structural Steel in Building" have normally been derived from fire tests on orthodox H or I rolled sections. For other sections the recommended thickness for a given section factor and fire rating should be modified as follows:

CASTELLATED SECTIONS

The thickness of fire protection material on a castellated section should be 20% greater than that required for the section from which it was cut.

HOLLOW SECTIONS

For spray applied fire protection materials the recommended thickness (t) should be increased as follows

For section factor (Hp/A) less than 250
modified thickness = t [1 + (Hp/A)/1000]

For section factor (Hp/A) 250 or over
modified thickness = 1.25 x t

STEELWORK IN FIRE
INFORMATION SHEET

M02

This series of information sheets is intended to illustrate methods of achieving fire resistance in steel structures. It should not be used for design without consulting detailed design guidance referenced below.

BOARD PROTECTION

UP TO 4 HRS

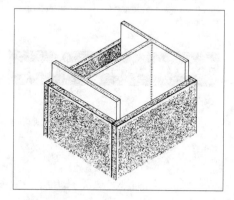

METHOD
Fire protective insulation can be applied by fixing boards to any type of steel member. Most products can achieve up to 4 hour rating. Fixing methods vary.

PRINCIPLE
Insulation reduces the heating rate of a steel member so that its limiting temperature is not exceeded during the required fire resistance period. The protection board thickness necessary depends on the section factor of the member (Hp/A) and the fire rating required.

ADVANTAGES
a) Boxed appearance suitable for visible members
b) Clean dry fixing
c) Factory manufactured, guaranteed thickness
d) Often applied to non-primed steelwork
e) Some products may be suitable for external use

LIMITATIONS (check with manufacturer)
a) Require fitting around complex details
b) May be more expensive and slower to fix than sprays

FOR MORE DETAILED INFORMATION SEE:-
"Fire protection of Structural Steel in Building"
published jointly by:
ASFPCM - (0252 336318) and
The Steel Construction Institute - (0344 23345)

Sheet Code
ISF/M02
January 1991

PROTECTION THICKNESS

Thickness recommendations given in "Fire Protection of Structural Steel in Building" have normally been derived from fire tests on orthodox H or I rolled sections. For other sections the recommended thickness for a given section factor and fire rating should be modified as follows:

CASTELLATED SECTIONS

The thickness of fire protection material on a castellated section should be 20% greater than that required for the section from which it was cut.

1210   *Fire resistance*

STEELWORK IN FIRE
INFORMATION SHEET   M03

This series of information sheets is intended to illustrate methods of achieving fire resistance in steel structures. It should not be used for design without consulting detailed design guidance referenced below

INTUMESCENT COATINGS

UP TO 2 HRS

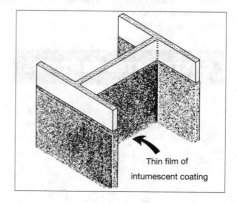

Thin film of intumescent coating

METHOD

Most intumescent coatings can be applied by spray, brush or roller and can achieve up to 1 hour fire resistance on fully exposed steel members. Some thick film products can achieve up to 2 hours fire resistance on some section sizes.

PRINCIPLE

Insulation is created by swelling of the coating at elevated temperatures to generate a foam like char. This reduces the heating rate so that the limiting temperature of the steel member is not exceeded during the required fire resistance period. The coating thickness necessary depends on the section factor (Hp/A) and the fire rating required.

ADVANTAGES
a) Decorative finish
b) Rapid application
c) Easy to cover complex details
d) Easy post protection fixings to steelwork eg service hangers

LIMITATIONS (check with manufacturer)
a) May be suitable for dry internal environments only
b) May be more expensive than sprayed insulation
c) May require blast cleaned surface and compatible primer

FOR MORE DETAILED INFORMATION SEE:-
"Fire Protection of Structural Steel in Building"
published jointly by:
ASFPCM - (0252 336318) and
The Steel Construction Institute - Tel: 0344 23345

Sheet Code
ISF/M03
January 1991

PROTECTION THICKNESS

Thickness recommendations given in "Fire Protection of Structural Steel in Building" have normally been derived from fire tests on orthodox H or I rolled sections. For other sections the recommended thickness for a given section factor and fire rating should be modified as follows:

CASTELLATED SECTIONS

The thickness of fire protection material on a castellated section should be 20% greater than that required for the section from which it was cut.

HOLLOW SECTIONS

For intumescent materials applied to hollow sections the manufacturers should have carried out separate tests and appraisal.

STEELWORK IN FIRE
INFORMATION SHEET

M04

This series of information sheets is intended to illustrate methods of achieving fire resistance in steel structures. It should not be used for design without consulting detailed design guidance referenced below

BLOCK - FILLED COLUMNS

30 MINUTES

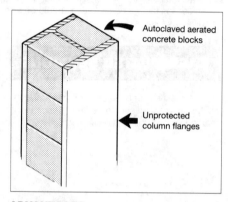

Autoclaved aerated concrete blocks

Unprotected column flanges

METHOD
Unprotected universal sections with section factors up to $69m^{-1}$ (see overleaf) can attain 30 minutes fire resistance by fitting autoclaved aerated concrete blocks between the flanges tied to the web at approximately 1m intervals.

PRINCIPLE
Partial exposure of steel members affects fire resistance in two ways-
Firstly the reduction of exposed surface area reduces the rate of heating by radiation and thus increases the time to reach failure temperature.
Secondly, if fire exposure creates both hot and cold regions in the cross section, plastic yielding occurs in the hot region and load is transferred to the stronger cooler region. Thus a non-uniformly heated section has a higher fire resistance than one heated evenly.

ADVANTAGES
a) Reduced cost - compared with total encasement with insulation
b) More slender finished columns occupy less floor space
c) Good durability - high resistance to impact and abrasion damage

LIMITATIONS
With unprotected steel the method is limited to 30 minute fire rating.
When higher ratings are required exposed steel must be treated with the full insulation or intumescent coating thickness recommended for the higher rating.
This method should not be used when the blockwork also forms a separating wall. In this case the column will be heated on one side only and thermal bowing may cause the wall to crack or collapse. In such cases the flange(s) should be protected. Alternatively, if the limit of wall deformation is known, the bowing can be calculated to ensure no integrity failure.

FOR MORE DETAILED INFORMATION SEE:-
BRE Digest 317, Building Research Advisory Service,
Fire Research Station, Borehamwood,
Herts WD6 2BL
Telephone 081 953 6177

Sheet Code
ISF/M04
January 1991

METHODS OF ACHIEVING 30 MINUTES FIRE RESISTANCE

COLUMN SECTIONS - AXIALLY LOADED [1] FREE STANDING

SERIAL SIZE mm	MASS/METRE kg	PROTECTION METHOD RECOMMENDED
305 x 406	393 and over	No fire protection required
356 x 406	340 and under	Block filling with autoclaved aerated concrete blocks
305 x 305	All weights	
254 x 254	All weights	
203 x 203	52 and over	
203 x 203	46[2]	
152 x 203	All weights	Apply fire protection material as per manufacturers' recomendations

BEAM SECTIONS ACTING AS PORTAL FRAME STANCHIONS [1]

914 x 419	All weights	No fire protection required
914 x 305	289	
*610 x 305	238	
*914 x 305	253 and under	Block filling with autoclaved aerated concrete blocks
838 x 292	All weights	
762 x 267	All weights	
686 x 254	All weights	
*610 x 305	179 and under	
610 x 229	All weights	
533 x 210	All weights	
457 x 191	All weights	
457 x 152	60 and over	
406 x 178	60 and over	
356 x 171	57 and over	
305 x 165	54	
305 x 127	48	
254 x 146	43	
Other beam sizes		Apply fire protection material as per manufacturers' recommendations

Notes:
1) This table applies to sections designed to BS 5950:Part 1:1990 provided the load factor (yf) does not exceed 1.5
2) To achieve 30 min fire resistance, a 203 × 203 × 46 kg/m column with blocked in webs should be loaded only up to 80% of the maximum allowable per BS 449: Part 2: 1969 or BS 5950: Part 1: 1990
*3) The table revises BRE Digest 317 (1986) in accordance with BS 5950: Part 8: 1990

STEELWORK IN FIRE
INFORMATION SHEET

M05

This series of information sheets is intended to illustrate methods of achieving fire resistance in steel structures. It should not be used for design without consulting detailed design guidance referenced below

CONCRETE FILLED HOLLOW COLUMNS

UP TO 2 HRS

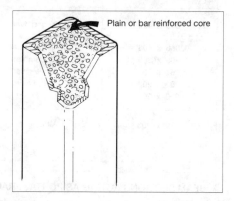
Plain or bar reinforced core

METHOD
Unprotected square or rectangular hollow sections can attain up to 120 minutes fire resistance by filling with plain, fibre reinforced or bar reinforced concrete.

PRINCIPLE
Heat flows through the steel wall into the concrete core which being a poor conductor heats up slowly.
As the temperature increases the steel yield strength reduces and the load is progressively transferred into the concrete core.
The steel acts as a restraint to the concrete preventing spalling and hence the rate of degradation of the concrete.

ADVANTAGES
a) Steel acts as a permanent shuttering
b) More slender finished columns occupy less floor space
c) Good durability - high resistance to impact and abrasion damage

LIMITATIONS
a) A minimum column size of 140mm x 140mm or 100mm x 200mm is required for plain or fibre reinforced sections.
b) A minimum column size of 200mm x 200mm or 150mm x 250mm is required for bar reinforced sections.
c) CHS columns are not included due to insufficient data at present.

FOR MORE DETAILED INFORMATION SEE:-
BS 5950 Part 8
Concrete filled column design manual TD 296 from
British Steel General Steels Welded Tubes (0536 402121)

Sheet Code
ISF/M05
January 1991

Fire resistance

CONCRETE FILLED RECTANGULAR HOLLOW SECTIONS

The fire resistance of externally unprotected concrete filled hollow sections is dependent on three main variables.
- The concrete strength selected
- The ratio of axial load and moment
- The addition of fibre or bar reinforcement

CONCRETE STRENGTH

The core capacity and hence its fire resistance is directly related to the concrete strength selected.

AXIAL LOAD AND MOMENT

Plain concrete does not perform well in tension and when subject to axial load and moment it is necessary to produce a resultant compressive stress in the core.

REINFORCEMENT

Fibre reinforcement will enhance the core axial capacity yet retain the advantage of filling into a section without obstructions.
Bar reinforcement will enhance the moment capacity.

COMBINED PROTECTION

As an alternative the concrete filled section can be designed for full factored loads and provided with external fire protection.
The thickness of the external fire protection is assessed as for an unfilled section, and, due to the effect of the core, the thickness can be reduced.

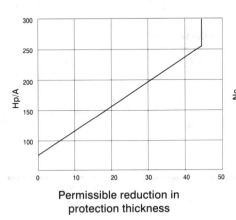

Permissible reduction in protection thickness

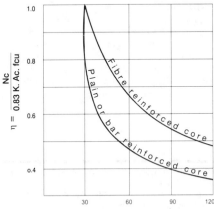

$\eta = \dfrac{N_c}{0.83\, K.\, A_c.\, f_{cu}}$

Fire rating (minutes)

STEELWORK IN FIRE

M06

INFORMATION SHEET

This series of information sheets is intended to illustrate methods of achieving fire resistance in steel structures. It should not be used for design without consulting detailed design guidance referenced below.

COMPOSITE SLABS WITH PROFILED METAL DECK WITH UNFILLED VOIDS

UP TO 2 HRS

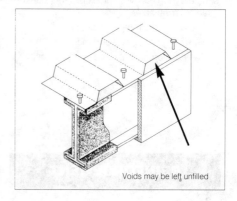

Voids may be left unfilled

METHOD
In composite construction using profiled metal deck floors it is unnecessary to fill the deck voids above the top flange for any fire resistance period using dovetail deck, or up to 90 minutes using trapezoidal deck

PRINCIPLE
In a composite beam/slab member the neutral axis in bending lies in, or close to, the beam top flange. Thus the top flange makes little significant contribution to the structural behaviour of the total composite system and its temperature can be allowed to increase with little detriment to performance in fire.

ADVANTAGES
a) Saving in time on site
b) Saving in cost for filling voids
c) It is unnecessary to build up the full thickness of protection on toes of upper flange
d) Void filling is unnecessary when using dovetail deck

LIMITATIONS
Voids must be filled where:-
a) Trapezoidal deck is used for fire ratings over 90 minutes
b) Trapezoidal deck is used in non-composite construction
c) Any type of deck crosses a fire separating wall

FOR MORE DETAILED INFORMATION SEE:-
Technical Report 087
"Fire resistance of composite beams"
The Steel Construction Institute
(0344 23345)

Sheet Code
ISF/M06
January 1991

COMPOSITE BEAMS - UNFILLED VOIDS

TRAPEZOIDAL DECK

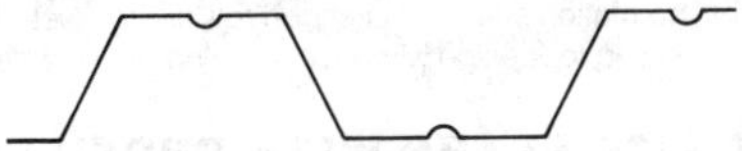

Construction	Fire Protection On Beam	Fire Resistance (minutes)		
		Up to 60	90	Over 90
Composite Beams	BOARD or SPRAY	No increase in thickness*	Increase thickness by 10% (or use thickness* appropriate to beam Hp/A + 15% whichever is less)	Fill voids
	INTUMESCENT	Increase thickness* by 20% (or use thickness* appropriate to beam Hp/A + 30 % whichever is less)	Increase thickness* by 30% (or use thickness* appropriate to beam Hp/A + 50 % whichever is less)	Fill voids
Non-Composite Beams	All types	Fill Voids		

DOVETAIL DECK

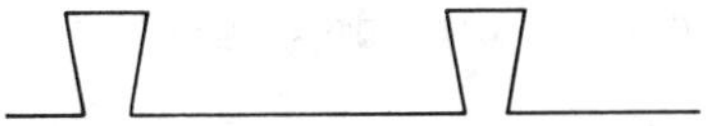

Construction	Fire Protection On Beam	Fire Resistance (minutes)
Composite or Non-composite Beams	All types	Voids may be left unfilled for all fire resistance periods.

* Thickness is the board, spray or intumescent thickness given for 30, 60 or 90 minutes rating in "Fire Protection for Structural Steel in Buildings" published by ASFPCM (0252 336318) and The Steel Construction Institute (0344 23345)

STEELWORK IN FIRE
INFORMATION SHEET

M07

This series of information sheets is intended to illustrate methods of achieving fire resistance in steel structures. It should not be used for design without consulting detailed design guidance referenced below.

COMPOSITE SLABS WITH PROFILED METAL DECK — UP TO 2 HRS

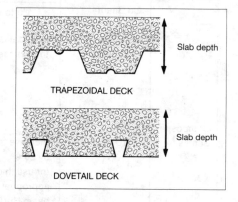

METHOD
Fire resistance of composite slabs up to 90 mins can be achieved using normal A142 mesh reinforcement. This can be increased to 120 mins if heavier mesh is used and the slab depth increased (see overleaf).

Other cases outside the limits overleaf can be evaluated by the "Fire Engineering Method" (See below)

PRINCIPLE
Mesh reinforcement, which is not designed to act structurally under normal conditions, makes a significant contribution to structural continuity in fire.

ADVANTAGES
a) Standard mesh, without additional reinforcing bars, may be used.
b) No fire protection is required on the deck soffit.

LIMITATIONS
a) Applies only to slabs designed to BS5950 Part 4
b) Mesh overlaps should exceed 50 times bar diameters
c) Mesh bar ductility should exceed 12% elongation in tension (to BS 4449)
d) Mesh should lie between 20 & 45mm from slab upper surface
e) Imposed load should not exceed 6.7 kN/m² (including finishes)

FOR MORE DETAILED INFORMATION SEE:-
SCI Publication 059
"Fire resistance of composite floors with steel decking."
The Steel Construction Institute - (0344 23345) and
CIRIA Special publication 42 CIRIA (071 222 8891)

Sheet Code
ISF/M07
January 1991

FIRE RESISTANT COMPOSITE SLABS

TRAPEZOIDAL DECK (60mm max)

Maximum Span (m)	Fire Rating (h)	Sheet thickness	Slab depth (mm) NWC[2]	Slab depth (mm) LWC[3]	Mesh Size
2.7	1	0.8	130	120	A142
3.0	1	0.9	130	120	A142
	1½	0.9	140	130	A142
	2	0.9	155	140	A193
3.6	1	1.0	130	120	A193
	1½	1.2	140	130	A193
	2	1.2	155	140	A252

DOVETAIL DECK (51mm max)

Maximum Span (m)	Fire Rating (h)	Sheet thickness	Slab depth (mm) NWC[2]	Slab depth (mm) LWC[3]	Mesh Size
2.5	1	0.8	100	100	A142
	1½	0.8	110	105	A142
3.0	1	0.9	120	110	A142
	1½	0.9	130	120	A142
	2	0.9	140	130	A193
3.6	1	1.0	125	120	A193
	1½	1.2	135	125	A193
	2	1.2	145	130	A252

1) Imposed load not exceeding 5kN/m² (+ 1.7kN/m² ceiling and services)
2) NWC = Normal weight concrete
3) LWC = Light weight concrete

NOTE: Minimum slab depths given in BS 5950 part 8 are to satisfy the insulation criterion only. Figures given in the table above incorporate a strength criterion also and thus may exceed the minimum depth given in the code.

1220   *Fire resistance*

STEELWORK IN FIRE
INFORMATION SHEET

M08

This series of information sheets is intended to illustrate methods of achieving fire resistance in steel structures. It should not be used for design without consulting detailed design guidance referenced below.

BEAM - COLUMN CONNECTIONS

30 MINUTES

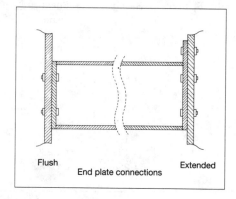

Flush   Extended
End plate connections

METHOD
Most beams can achieve 30 minutes fire resistance without additional protection by mobilising the moment capacity of the connections to reduce the effective load ratio of the beam at the fire limit state (see overleaf).

PRINCIPLE
It is known that so - called "simple" connections can resist moments but this is usually ignored in design. At the fire limit state large rotations are permitted and moments may be transferred to adjacent members. Reduction of moment in the beam increases the limiting temperature and thus the fire resistance of the beam.

ADVANTAGES
a) Elimination of fire protection for 30 minutes rating of most beams.
b) Reduction of fire protection thickness for higher ratings.

LIMITATIONS
a) Applies only to end plate connections designed to be "simple" in normal design
b) Does not apply to continuous frames (eg portal frames)
c) Bolt sizes are M16, 20 or 24 in grades 4.6 or 8.8
d) Applies only to internal beams with span variation less than 25%
e) Special considerations are required for edge columns or for beam-beam connections which do not align through the web

FOR MORE DETAILED INFORMATION SEE:-
SCI Technical Report 086
"Enhancement of fire resistance by beam column connections" published by
The Steel Construction Institute (0344 23345)

Sheet Code
ISF/M08
January 1991

INCREASE IN LIMITING TEMPERATURE

Connections provide an increase in limiting temperature over that of a simply supported beam. The limiting temperature of the simply supported beam is a function of the load ratio. The increase in limiting temperatures due to the connection has been determined conservatively as;

For flush end plates $T°C = 135\, Mc/Mp$

For extended end plates $T°C = 160\, Mc/Mp$

Valid over a range of $0.15 < Mc/Mp < 0.7$

Where Mc = "cold" moment capacity of the connection (obtained by calculation)
      Mp = "cold" moment capacity of the beam

MAXIMUM SECTION FACTORS FOR 30 MINUTES FIRE RESISTANCE OF INTERNAL NON - COMPOSITE BEAMS.

CONNECTION TYPE	APPLIED LOAD RATIO OF SIMPLE BEAM	EFFECTIVE LOAD RATIO	LIMITING TEMPERATURE OF BEAM (°C)	MAXIMUM SECTION FACTOR (m^{-1}) OF BEAM
Extended end plates Moment capacity > 0.6 Mp	0.40 0.45 0,50 0.55	0.20 0.23 0.25 0.28	780 770 755 735	230 210 180 120
Flush end plates Moments capacity > 0.3 Mp	0.35 0.40 0.45 0.50 0.55	0.24 0.27 0.30 0.34 0.37	765 745 725 710 695	200 150 110 105 95

Notes:
1) Beam supporting concrete slabs
2) Internal beam connected to internal column by connections of the above type
3) Load ratio takes into account reduced load factors at limit state
4) Capacity of bolt group in shear is taken as 50% of their design capacity when determining their normal loadcarrying capacity

1222   *Fire resistance*

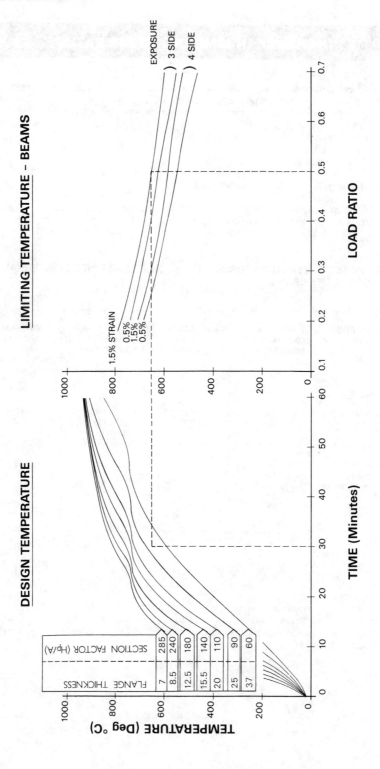

Limiting temperatures for design of protected and unprotected members

Case No.	Members in compression	Load ratio					
		0.7	0.6	0.5	0.4	0.3	0.2
(1)	Uniform heating on all faces; braced members in simple construction: Slenderness ratio ≤ 70	510	540	580	615	655	710
(2)	Slenderness ratio ≤ 180	460	510	545	590	635	635
	Members in bending						
(3)	Supporting concrete or composite deck floors: Unprotected members, or protected members complying with clause 2.3 (a) or (b)	590	620	650	680	725	780
(4)	Other protected members	540	585	625	655	700	745
	Not supporting concrete floors:						
(5)	Unprotected members, or protected members complying with Clause 2.3 (a) or (b)	520	555	585	620	660	745
(6)	Other protected members	460	510	545	590	635	690
	Members in tension						
(7)	All cases	460	510	545	590	635	690

Universal beams

Designation		Depth of section D	Width of section B	Thickness		Area of section	Section factor H_p/A			
Serial size	Mass per metre			Web t	Flange T		Profile 3 sides	Profile 4 sides	Box 3 sides	Box 4 sides
mm	kg	mm	mm	mm	mm	cm²	m⁻¹	m⁻¹	m⁻¹	m⁻¹
914×419	388	920.5	420.5	21.5	36.6	494.4	60	70	45	55
	343	911.4	418.5	19.4	32.0	437.4	70	80	50	60
914×305	289	926.6	307.8	19.6	32.0	368.8	75	80	60	65
	253	918.5	305.5	17.3	27.9	322.8	85	95	65	75
	224	910.3	304.1	15.9	23.9	285.2	95	105	75	85
	201	903	303.4	15.2	20.2	256.4	105	115	80	95
838×292	226	850.9	293.8	16.1	26.8	288.7	85	95	70	80
	194	840.7	292.4	14.7	21.7	247.1	100	115	80	90
	176	834.9	291.6	14	18.8	224.1	110	125	90	100
762×267	197	769.6	268	15.6	25.4	250.7	90	100	70	85
	173	762	266.7	14.3	21.6	220.4	105	115	80	95
	147	753.9	265.3	12.9	17.5	188.0	120	135	95	110
686×254	170	692.9	255.8	14.5	23.7	216.5	95	110	75	90
	152	687.6	254.5	13.2	21.0	193.8	110	120	85	95
	140	683.5	253.7	12.4	19.0	178.6	115	130	90	105
	125	677.9	253	11.7	16.2	159.6	130	145	100	115
610×305	238	633	311.5	18.6	31.4	303.7	70	80	50	60
	179	617.5	307	14.1	23.6	227.9	90	105	70	80
	149	609.6	304.8	11.9	19.7	190.1	110	125	80	95
610×229	140	617	230.1	13.1	22.1	178.3	105	120	80	95
	125	611.9	229	11.9	19.6	159.5	115	130	90	105
	113	607.3	228.2	11.2	17.3	144.4	130	145	100	115
	101	602.2	227.6	10.6	14.8	129.1	145	160	110	130
533×210	122	544.6	211.9	12.8	21.3	155.7	110	120	85	95
	109	539.5	210.7	11.6	18.8	138.5	120	135	95	110
	101	536.7	210.1	10.9	17.4	129.7	130	145	100	115
	92	533.1	209.3	10.2	15.6	117.7	140	160	110	125
	82	528.3	208.7	9.6	13.2	104.4	155	175	120	140
457×191	98	467.4	192.8	11.4	19.6	125.2	120	135	90	105
	89	463.6	192	10.6	17.7	113.9	130	145	100	115
	82	460.2	191.3	9.9	16.0	104.5	140	160	105	125
	74	457.2	190.5	9.1	14.5	94.98	155	175	115	135
	67	453.6	189.9	8.5	12.7	85.44	170	190	130	150
457×152	82	465.1	153.5	10.7	18.9	104.4	130	145	105	120
	74	461.3	152.7	9.9	17.0	94.99	140	155	115	130
	67	457.2	151.9	9.1	15.0	85.41	155	175	125	145
	60	454.7	152.9	8.0	13.3	75.93	175	195	140	160
	52	449.8	152.4	7.6	10.9	66.49	200	220	160	180
406×178	74	412.8	179.7	9.7	16.0	94.95	140	160	105	125
	67	409.4	178.8	8.8	14.3	85.49	155	175	115	140
	60	406.4	177.8	7.8	12.8	76.01	175	195	130	155
	54	402.6	177.6	7.6	10.9	68.42	190	215	145	170
406×140	46	402.3	142.4	6.9	11.2	58.96	205	230	160	185
	39	397.3	141.8	6.3	8.6	49.40	240	270	190	220
356×171	67	364	173.2	9.1	15.7	85.42	140	160	105	125
	57	358.6	172.1	8	13.0	72.18	165	190	125	145
	51	355.6	171.5	7.3	11.5	64.58	185	210	135	165
	45	352	171	6.9	9.7	56.96	210	240	155	185
356×127	39	352.8	126	6.5	10.7	49.40	215	240	170	195
	33	248.5	125.4	5.9	8.5	41.83	250	280	195	225
305×165	54	310.9	166.8	7.7	13.7	68.38	160	185	115	140
	46	307.1	165.7	6.7	11.8	58.90	185	210	130	160
	40	303.8	165.1	6.1	10.2	51.50	210	240	150	180
305×127	48	310.4	125.2	9.9	14.0	60.83	160	180	125	145
	42	306.6	124.3	8	12.1	53.18	180	205	140	160
	37	303.8	123.5	7.2	10.7	47.47	200	225	155	180
305×102	33	312.7	102.4	6.6	10.8	41.77	215	240	175	200
	28	308.9	101.9	6.1	8.9	36.30	245	275	200	225
	25	304.8	101.6	5.8	6.8	31.39	285	315	225	260
254×146	43	259.6	147.3	7.3	12.7	55.10	170	195	120	150
	37	256	146.4	6.4	10.9	47.45	195	225	140	170
	31	251.5	146.1	6.1	8.6	40.00	230	265	160	200
254×102	28	260.4	102.1	6.4	10.0	36.19	220	250	170	200
	25	257	101.9	6.1	8.4	32.17	245	280	190	225
	22	254	101.6	5.8	6.8	28.42	275	315	215	250
203×133	30	206.8	133.8	6.3	9.6	38.00	210	245	145	180
	25	203.2	133.4	5.8	7.8	32.31	240	285	165	210
203×102	23	203.2	101.6	5.2	9.3	29	235	270	175	210
178×102	19	177.8	101.6	4.7	7.9	24.2	265	305	190	230
152×89	16	152.4	88.9	4.6	7.7	20.5	270	310	190	235
127×76	13	127	76.2	4.2	7.6	16.8	275	320	195	240

Fire resistance

Universal columns — Section factor H_p/A

Designation		Depth of section D	Width of section B	Thickness		Area of section	Profile 3 sides	Profile 4 sides	Box 3 sides	Box 4 sides
Serial size	Mass per metre			Web t	Flange T					
mm	kg	mm	mm	mm	mm	cm²	m⁻¹	m⁻¹	m⁻¹	m⁻¹
356 × 406	634	474.7	424.1	47.6	77.0	808.1	25	30	15	20
	551	455.7	418.5	42.0	67.5	701.8	30	35	20	25
	467	436.6	412.4	35.9	58.0	595.5	35	40	20	30
	393	419.1	407.0	30.6	49.2	500.9	40	45	25	35
	340	406.4	403.0	26.5	42.9	432.7	45	55	30	35
	287	393.7	399.0	22.6	36.5	366.0	50	65	30	45
	235	381.0	395.0	18.5	30.2	299.8	65	75	40	50
356 × 368	202	374.7	374.4	16.8	27.0	257.9	70	85	45	60
	177	368.3	372.1	14.5	23.8	225.7	80	95	50	65
	153	362.0	370.2	12.6	20.7	195.2	90	110	55	75
	129	355.6	368.3	10.7	17.5	164.9	105	130	65	90
305 × 305	283	365.3	321.8	26.9	44.1	360.4	45	55	30	40
	240	352.6	317.9	23.0	37.7	305.6	50	60	35	45
	198	339.9	314.1	19.2	31.4	252.3	60	75	40	50
	158	327.2	310.6	15.7	25.0	201.2	75	90	50	65
	137	320.5	308.7	13.8	21.7	174.6	85	105	55	70
	118	314.5	306.8	11.9	18.7	149.8	100	120	60	85
	97	307.8	304.8	9.9	15.4	123.3	120	145	75	100
254 × 254	167	289.1	264.5	19.2	31.7	212.4	60	75	40	50
	132	276.4	261.0	15.6	25.3	167.7	75	90	50	65
	107	266.7	258.3	13.0	20.5	136.6	90	110	60	75
	89	260.4	255.9	10.5	17.3	114.0	110	130	70	90
	73	254.0	254.0	8.6	14.2	92.9	130	160	80	110
203 × 203	86	222.3	208.8	13.0	20.5	110.1	95	110	60	80
	71	215.9	206.2	10.3	17.3	91.1	110	135	70	95
	60	209.6	205.2	9.3	14.2	75.8	130	160	80	110
	52	206.2	203.9	8.0	12.5	66.4	150	180	95	125
	46	203.2	203.2	7.3	11.0	58.8	165	200	105	140
152 × 152	37	161.8	154.4	8.1	11.5	47.4	160	190	100	135
	30	157.5	152.9	6.6	9.4	38.2	195	235	120	160
	23	152.4	152.4	6.1	6.8	29.8	245	300	155	205

Fire resistance

Circular hollow sections

Section factor H_p/A — Profile or Box

Designation Outside diameter D (mm)	Thickness t (mm)	Mass per metre (kg)	Area of section (cm²)	H_p/A (m⁻¹)	Designation Outside diameter D (mm)	Thickness t (mm)	Mass per metre (kg)	Area of section (cm²)	H_p/A (m⁻¹)
21.3	3.2	1.43	1.82	370	244.5	6.3	37.0	47.1	165
26.9	3.2	1.87	2.38	355		8.0	46.7	59.4	130
33.7	2.6	1.99	2.54	415		10.0	57.8	73.7	105
	3.2	2.41	3.07	345		12.5	71.5	91.1	85
	4.0	2.93	3.73	285		16.0	90.2	115	65
42.4	2.6	2.55	3.25	410		20.0	111	141	55
	3.2	3.09	3.94	340	273.0	6.3	41.4	52.8	160
	4.0	3.79	4.83	275		8.0	52.3	66.6	130
48.3	3.2	3.56	4.53	335		10.0	64.9	82.6	105
	4.0	4.37	5.57	270		12.5	80.3	102	85
	5.0	5.34	6.80	225		16.0	101	129	65
60.3	3.2	4.51	5.74	330		20.0	125	159	55
	4.0	5.55	7.07	270		25.0	153	195	45
	5.0	6.82	8.69	220	323.9	6.3	49.3	62.9	160
76.1	3.2	5.75	7.33	325		8.0	62.3	79.4	130
	4.0	7.11	9.06	265		10.0	77.4	98.6	105
	5.0	8.77	11.2	215		12.5	96.0	122	85
88.9	3.2	6.76	8.62	325		16.0	121	155	65
	4.0	8.38	10.70	260		20.0	150	191	55
	5.0	10.3	13.2	210		25.0	184	235	45
114.3	3.6	9.83	12.5	285	355.6	8.0	68.6	87.4	130
	5.0	13.5	17.2	210		10.0	85.2	109	100
	6.3	16.8	21.4	170		12.5	106	135	85
139.7	5.0	16.6	21.2	205		16.0	134	171	65
	6.3	20.7	26.4	165		20.0	166	211	55
	8.0	26.0	33.1	135		25.0	204	260	45
	10.0	32.0	40.7	110	406.4	10.0	97.8	125	100
168.3	5.0	20.1	25.7	205		12.5	121	155	80
	6.3	25.2	37.1	165		16.0	154	196	65
	8.0	31.6	40.3	130		20.0	191	243	55
	10.0	39.0	49.7	105		25.0	235	300	45
193.7	5.0	23.3	29.6	205		32.0	295	376	35
	6.3	29.1	37.1	165	457.0	10.0	110	140	105
	8.0	36.6	46.7	130		12.5	137	175	80
	10.0	45.3	57.7	105		16.0	174	222	65
	12.5	55.9	71.2	85		20.0	216	275	50
	16.0	70.1	89.3	70		25.0	266	339	40
219.1	5.0	26.4	33.6	205		32.0	335	427	35
	6.3	33.1	42.1	165		40.0	411	524	25
	8.0	41.6	53.1	130	508.0	10.0	123	156	100
	10.0	51.6	65.7	105		12.5	153	195	80
	12.5	63.7	81.1	85		16.0	194	247	65
	16.0	80.1	102	65					
	20.0	98.2	125	55					

Rectangular hollow sections

Designation Size D×B (mm)	Thickness t (mm)	Mass per metre (kg)	Area of section (cm²)	Section factor Hₚ/A 3 sides (m⁻¹)	3 sides (m⁻¹)	4 sides (m⁻¹)
50×25	2.5	2.72	3.47	360	290	430
	3.0	3.22	4.10	305	245	365
	3.2	3.41	4.34	290	230	345
50×30	2.5	2.92	3.72	350	295	430
	3.0	3.45	4.40	295	250	365
	3.2	3.66	4.66	280	235	345
	4.0	4.46	5.68	230	195	280
	5.0	5.40	6.88	190	160	235
60×40	2.5	3.71	4.72	340	295	425
	3.0	4.39	5.60	285	250	355
	3.2	4.66	5.94	270	235	335
	4.0	5.72	7.28	220	190	275
	5.0	6.97	8.88	180	160	225
	6.3	8.49	10.8	150	130	185
80×40	3.0	5.34	6.80	295	235	355
	3.2	5.67	7.22	275	220	330
	4.0	6.97	8.88	225	180	270
	5.0	8.54	10.9	185	145	220
	6.3	10.5	13.3	150	120	180
	8.0	12.8	16.3	125	100	145
90×50	3.0	6.28	8.00	290	240	350
	3.6	7.46	9.50	240	200	295
	5.0	10.1	12.9	180	145	215
	6.3	12.5	15.9	145	120	175
	8.0	15.3	19.5	120	95	145
100×50	3.0	6.75	8.60	290	235	350
	3.2	7.18	9.14	275	220	330
	4.0	8.86	11.3	220	175	265
	5.0	10.9	13.9	180	145	215
	6.3	13.4	17.1	145	115	175
	8.0	16.6	21.1	120	95	140
100×60	3.0	7.22	9.20	285	240	350
	3.6	8.59	10.9	240	200	295
	5.0	11.7	14.9	175	150	215
	6.3	14.4	18.4	140	120	175
	8.0	17.8	22.7	115	95	140
120×60	3.6	9.72	12.4	240	195	290
	5.0	13.3	16.9	180	140	215
	6.3	16.4	20.9	145	115	170
	8.0	20.4	25.9	115	95	140
120×80	5.0	14.8	18.9	170	150	210
	6.3	18.4	23.4	135	120	170
	8.0	22.9	29.1	110	95	135
	10.0	27.9	35.5	90	80	115
150×100	5.0	18.7	23.9	165	145	210
	6.3	23.8	29.7	135	120	170
	8.0	29.1	37.1	110	95	135
	10.0	35.7	45.5	90	75	110
	12.5	43.6	55.5	70	65	90
160×80	5.0	18.0	22.9	175	140	210
	6.3	22.3	28.5	140	110	170
	8.0	27.9	35.5	115	90	135
	10.0	34.2	43.5	90	75	110
	12.5	41.6	53.0	75	60	90
200×100	5.0	22.7	28.9	175	140	210
	6.3	28.3	36.0	140	110	165
	8.0	35.4	45.1	110	90	135
	10.0	43.6	55.5	90	70	110
	12.5	53.4	68.0	75	60	90
	16.0	66.4	84.5	60	45	70
250×150	6.3	38.2	48.6	135	115	165
	8.0	48.0	61.1	105	90	130
	10.0	59.3	75.5	85	75	105
	12.5	73.0	93.0	70	60	85
	16.0	91.5	117	55	45	70
300×200	6.3	48.1	61.2	130	115	165
	8.0	60.5	77.1	105	90	130
	10.0	75.0	95.5	85	75	105
	12.5	92.6	118	70	60	85
	16.0	117	149	55	45	65
400×200	10.0	90.7	116	85	70	105
	12.5	112	143	70	55	85
	16.0	142	181	55	45	65
450×250	10.0	106	136	85	70	105
	12.5	132	168	70	55	85
	16.0	167	213	55	45	65

Fire resistance

Rectangular hollow sections (square)

Designation		Mass per metre	Area of section	Section factor H_p/A	
				3 sides	4 sides
Size D × D (mm)	Thickness t (mm)	kg	cm²	m⁻¹	m⁻¹
20×20	2.0	1.12	1.42	425	565
	2.5	1.35	1.72	350	465
25×25	2.0	1.43	1.82	410	550
	2.5	1.74	2.22	340	450
	3.0	2.04	2.60	290	385
	3.2	2.15	2.74	275	365
30×30	2.5	2.14	2.72	330	440
	3.0	2.51	3.20	280	375
	3.2	2.65	3.38	265	355
40×40	2.5	2.92	3.72	325	430
	3.0	3.45	4.40	275	365
	3.2	3.66	4.66	260	345
	4.0	4.46	5.68	210	280
	5.0	5.40	6.88	175	235
50×50	2.5	3.71	4.72	320	425
	3.0	4.39	5.60	270	355
	3.2	4.66	5.94	255	335
	4.0	5.72	7.28	205	275
	5.0	6.97	8.88	170	225
	6.3	8.49	10.8	140	185
60×60	3.0	5.34	6.80	265	355
	3.2	5.67	7.22	250	330
	4.0	6.97	8.88	205	270
	5.0	8.54	10.9	165	220
	6.3	10.5	13.3	135	180
	8.0	12.8	16.3	110	145
70×70	3.0	6.28	8.00	260	350
	3.6	7.46	9.50	220	295
	5.0	10.1	12.9	165	215
	6.3	12.5	15.9	130	175
	8.0	15.3	19.5	110	145
80×80	3.0	7.22	9.20	260	350
	3.6	8.59	10.9	220	295
	5.0	11.7	14.9	160	215
	6.3	14.4	18.4	130	175
	8.0	17.8	22.7	105	140
90×90	3.6	9.72	12.4	220	290
	5.0	13.3	16.9	160	215
	6.3	16.4	20.9	130	170
	8.0	20.4	25.9	105	140
100×100	4.0	12.0	15.3	195	260
	5.0	14.8	18.9	160	210
	6.3	18.4	23.4	130	170
	8.0	22.9	29.1	105	135
	10.0	27.9	35.5	85	115
120×120	5.0	18.0	22.9	155	210
	6.3	22.3	28.5	125	170
	8.0	27.9	35.5	100	135
	10.0	34.2	43.5	85	110
	12.5	41.6	53.0	70	90
140×140	5.0	21.1	26.9	155	210
	6.3	26.3	33.5	125	165
	8.0	32.9	41.9	100	135
	10.0	40.4	51.5	80	110
	12.5	49.5	63.0	65	90
150×150	5.0	22.7	28.9	155	210
	6.3	28.3	36.0	125	165
	8.0	35.4	45.1	100	135
	10.0	43.6	55.5	80	110
	12.5	53.4	68.0	65	90
	16.0	66.4	84.5	55	70
180×180	6.3	34.2	43.6	125	165
	8.0	43.0	54.7	100	130
	10.0	53.0	67.5	80	105
	12.5	65.2	83.0	65	85
	16.0	81.4	104	50	70
200×200	6.3	38.2	48.6	125	165
	8.0	48.0	61.1	100	130
	10.0	59.3	75.5	80	105
	12.5	73.0	93.0	65	85
	16.0	91.5	117	50	70
250×250	6.3	48.1	61.2	125	165
	8.0	60.5	77.1	95	130
	10.0	75.0	95.5	80	105
	12.5	92.6	118	65	85
	16.0	117	149	50	65
300×300	10.0	90.7	116	80	105
	12.5	112	143	65	85
	16.0	142	181	50	65
350×350	10.0	106	136	75	105
	12.5	132	168	65	85
	16.0	167	213	50	65
400×400	10.0	122	156	75	105
	12.5	152	193	60	85
	16.0	192	245	50	65

Minimum thickness of a typical spray protection to an I-section

H_p/A up to	Dry thickness (mm) to provide fire resistance of					
	$\frac{1}{2}$ hour	1 hour	$1\frac{1}{2}$ hours	2 hours	3 hours	4 hours
30	10	10	10	10	15	25
50	10	10	10	14	21	29
70	10	10	12	17	27	36
90	10	10	14	20	31	42
110	10	10	16	22	34	47
130	10	10	17	24	37	51
150	10	11	18	25	40	54
170	10	12	19	27	42	57
190	10	12	20	28	44	59
210	10	13	21	29	45	
230	10	13	21	30	47	
250	10	14	22	31	48	
270	10	14	23	31	49	
290	10	14	23	32	50	
310	10	14	23	33	51	

Linear interpolation is permissible between values of H_p/A

Basic data on corrosion

INTERIOR ENVIRONMENTS

environment	corrosion risk	examples*
Normal	Negligible	Offices Shops Industrial Production/Assembly Warehousing Hospital wards Schools Hotels
Occasional Condensation	Low	Unheated Buildings Vehicle Depots Sports Halls
Frequent Condensation	Significant	Food Processing Plants/Kitchens Laundries Breweries Dairies
Frequent Condensation with Corrosive Atmosphere	High	Chemical Processing Plant Dye Works Swimming Pools Paper Manufacture Boat Yards over seawater Foundries/Smelters

* Buildings may contain areas where different conditions apply, e.g. hospitals will contain kitchen and laundry areas

EXTERIOR ENVIRONMENTS

Environment	Description	Steel Corrosion Risk
Normal inland	Most rural and urban areas with low sulphur dioxide, acid, alkali and salt pollution. (Some apparently non-industrial areas may be polluted from distant sources according to prevailing wind and topography)	Low
Polluted inland	Areas of high airborne sulphur dioxide, or other pollution from industrial or domestic sources.	Significant
Normal coastal	As normal inland with high airborne salt levels. (This salt contaminated zone may extend to a maximum of 3 km from the coast)	High
Polluted coastal	As polluted inland with high airborne salt levels. (This salt contaminated zone may extend to a maximum of 3 km from the coast)	Very high
	Aggressive industrial environments such as steelwork adjacent to acid plants, salt storage depots, electroplating shops, chemical works etc. Buried or immersed steelwork Sea water splash zones.	Variable

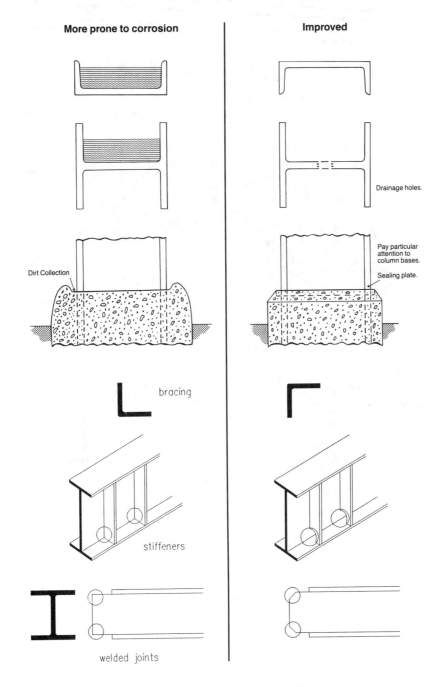

Methods of presenting bi-metallic corrosion.

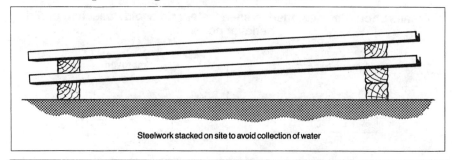

Steelwork stacked on site to avoid collection of water

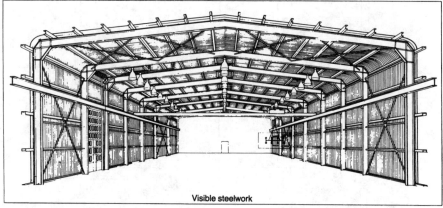

Visible steelwork

Hidden steelwork – hollow encased

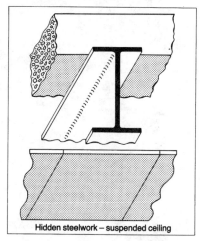

Hidden steelwork – suspended ceiling

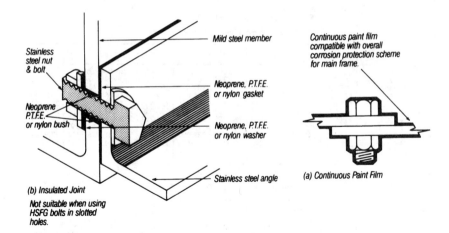

(a) Continuous Paint Film

(b) Insulated Joint
Not suitable when using HSFG bolts in slotted holes.

The above basic data on corrosion are extracted from *Corrosion Protection Guides for Steelwork* in: (i) Building Interiors (issue 3), (ii) Exterior Environments (issue 2) and (iii) Building Refurbishment, published by British Steel General Steels.

SECOND MOMENTS OF AREA
Inches⁴ Units to Centimetres⁴ Units

in.⁴	0	·1	·2	·3	·4	·5	·6	·7	·8	·9
	cm.⁴	cm.⁴	cm.⁴	cm.⁴	cm.⁴	cm.⁴	cm.⁴	cm.⁴	cm.⁴	cm.⁴
1	41·62314256	45·785	49·948	54·110	58·272	62·435	66·597	70·759	74·922	79·084
2	83·24628	87·409	91·571	95·733	99·895	104·058	108·220	112·382	116·545	120·707
3	124·86942	129·032	133·194	137·356	141·519	145·681	149·843	154·006	158·168	162·330
4	166·49257	170·655	174·817	178·979	183·142	187·304	191·466	195·629	199·791	203·953
5	208·11571	212·278	216·440	220·603	224·765	228·927	233·089	237·252	241·414	245·576
6	249·73886	253·901	258·063	262·226	266·388	270·550	274·713	278·875	283·037	287·200
7	291·36200	295·524	299·687	303·849	308·011	312·173	316·336	320·498	324·660	328·883
8	332·98514	337·147	341·310	345·472	349·634	353·797	357·959	362·121	366·284	370·446
9	374·60829	378·771	382·933	387·095	391·257	395·420	399·582	403·744	407·907	412·069
10	416·23143									

Centimetres⁴ Units to Inches⁴ Units

cm.⁴	0	·1	·2	·3	·4	·5	·6	·7	·8	·9
	in.⁴	in.⁴	in.⁴	in.⁴	in.⁴	in.⁴	in.⁴	in.⁴	in.⁴	in.⁴
1	0·02402510	0·02643	0·02883	0·03123	0·03363	0·03604	0·03844	0·04084	0·04324	0·04565
2	0·04805019	0·05045	0·05285	0·05526	0·05766	0·06006	0·06246	0·06487	0·06727	0·06967
3	0·07207529	0·07448	0·07688	0·07928	0·08168	0·08409	0·08649	0·08889	0·09129	0·09370
4	0·09610038	0·09850	0·10091	0·10331	0·10571	0·10811	0·11052	0·11292	0·11532	0·11772
5	0·12012548	0·12253	0·12493	0·12733	0·12974	0·13214	0·13454	0·13694	0·13935	0·14175
6	0·14415058	0·14655	0·14895	0·15136	0·15376	0·15616	0·15856	0·16097	0·16337	0·16577
7	0·16817567	0·17058	0·17298	0·17538	0·17779	0·18019	0·18259	0·18499	0·18740	0·18980
8	0·19220077	0·19460	0·19701	0·19941	0·20181	0·20421	0·20662	0·20902	0·21142	0·21382
9	0·21622586	0·21863	0·22103	0·22343	0·22584	0·22824	0·23064	0·23304	0·23545	0·23785
10	0·24025096									

MODULI OF SECTION
Inches³ Units to Centimetres³ Units

in.³	0	·1	·2	·3	·4	·5	·6	·7	·8	·9
	cm.³	cm.³	cm.³	cm.³	cm.³	cm.³	cm.³	cm.³	cm.³	cm.³
1	16·387064	18·026	19·664	21·303	22·942	24·580	26·219	27·858	29·497	31·135
2	32·77413	34·413	36·051	37·690	39·329	40·967	42·606	44·245	45·884	47·522
3	49·16119	50·800	52·438	54·077	55·716	57·355	58·993	60·632	62·271	63·909
4	65·54826	67·187	68·826	70·464	72·103	73·742	75·380	77·019	78·658	80·297
5	81·93532	83·574	85·213	86·851	88·490	90·129	91·767	93·406	95·045	96·684
6	98·32238	99·961	101·600	103·238	104·877	106·516	108·154	109·793	111·432	113·071
7	114·70945	116·348	117·987	119·625	121·624	122·903	124·542	126·180	127·819	129·458
8	131·09651	132·735	134·374	136·012	137·651	139·290	140·929	142·567	144·206	145·845
9	147·48358	149·122	150·761	152·400	154·038	155·677	157·316	158·954	160·593	162·232
10	163·87064									

Centimetres³ Units to Inches³ Units

cm.³	0	·1	·2	·3	·4	·5	·6	·7	·8	·9
	in.³	in.³	in.³	in.³	in.³	in.³	in.³	in.³	in.³	in.³
1	0·061024	0·06713	0·07323	0·07933	0·08543	0·09154	0·09764	0·10374	0·10984	0·11594
2	0·122047	0·12815	0·13425	0·14035	0·14646	0·15256	0·15866	0·16476	0·17087	0·17697
3	0·183071	0·18917	0·19527	0·20138	0·20748	0·21358	0·21968	0·22579	0·23189	0·23799
4	0·244095	0·25020	0·25630	0·26240	0·26850	0·27461	0·28071	0·28681	0·29291	0·29902
5	0·305119	0·31122	0·31732	0·32342	0·32953	0·33563	0·34173	0·34783	0·35394	0·36004
6	0·366142	0·37224	0·37835	0·38445	0·39055	0·39665	0·40276	0·40886	0·41496	0·42106
7	0·427166	0·43327	0·43937	0·44547	0·45158	0·45768	0·46378	0·46988	0·47599	0·48209
8	0·488190	0·49429	0·50040	0·50650	0·51260	0·51870	0·52480	0·53091	0·53701	0·54311
9	0·549213	0·55532	0·56142	0·56752	0·57362	0·57973	0·58583	0·59193	0·59803	0·60414
10	0·610237									

Based on 1 inch = 25·4 millimetres

Extract from B.S. 350:1944

WEIGHTS PER UNIT LENGTH

POUNDS PER FOOT TO KILOGRAMS PER METRE

Based on 1 inch = 25·4 millimetres ; 1 pound = 0·45359243 kilograms

Pounds per ft.	0	1	2	3	4	5	6	7	8	9
	kg./m.	kg./m.	kg./m.	kg./m.	kg./m.	kg./m.	kg./m.	kg./m.	kg./m.	kg./m.
	—	1·4882	2·9763	4·4645	5·9527	7·4408	8·9290	10·417	11·905	13·393
10	14·882	16·370	17·858	19·346	20·834	22·322	23·811	25·299	26·787	28·275
20	29·763	31·251	32·740	34·228	35·716	37·204	38·692	40·180	41·669	43·157
30	44·645	46·133	47·621	49·109	50·598	52·086	53·574	55·062	56·550	58·038
40	59·527	61·015	62·503	63·991	65·479	66·967	68·456	69·944	71·432	72·920
50	74·408	75·896	77·385	78·873	80·361	81·849	83·337	84·825	86·314	87·802
60	89·290	90·778	92·266	93·754	95·242	96·731	98·219	99·707	101·195	102·683
70	104·171	105·660	107·148	108·636	110·124	111·612	113·100	114·589	116·077	117·565
80	119·053	120·541	122·029	123·518	125·006	126·494	127·982	129·470	130·958	132·447
90	133·935	135·423	136·911	138·399	139·887	141·376	142·864	144·352	145·840	147·328
100	148·816									

KILOGRAMS PER METRE TO POUNDS PER FOOT

Based on 1 inch = 25·4 millimetres ; 1 pound = 0·45359243 kilograms

Kg. per m.	0	1	2	3	4	5	6	7	8	9
	lb./ft.	lb./ft.	lb./ft.	lb./ft.	lb./ft.	lb./ft.	lb./ft.	lb./ft.	lb./ft.	lb./ft.
	—	0·67197	1·3439	2·0159	2·6879	3·3598	4·0318	4·7038	5·3758	6·0477
10	6·7197	7·3917	8·0636	8·7356	9·4076	10·0795	10·7515	11·4235	12·0954	12·7674
20	13·4394	14·1113	14·7833	15·4553	16·1273	16·7992	17·4712	18·1432	18·8151	19·4871
30	20·1591	20·8310	21·5030	22·1750	22·8469	23·5189	24·1909	24·8629	25·5348	26·2068
40	26·8788	27·5507	28·2227	28·8947	29·5666	30·2386	30·9106	31·5825	32·2545	32·9265
50	33·5984	34·2704	34·9424	35·6144	36·2863	36·9583	37·6303	38·3022	38·9742	39·6462
60	40·3181	40·9901	41·6621	42·3340	43·0060	43·6780	44·3500	45·0219	45·6939	46·3659
70	47·0378	47·7098	48·3818	49·0537	49·7257	50·398	51·070	51·742	52·414	53·086
80	53·758	54·429	55·101	55·773	56·445	57·117	57·789	58·461	59·133	59·805
90	60·477	61·149	61·821	62·493	63·165	63·837	64·509	65·181	65·853	66·525
100	67·197	67·869	68·541	69·213	69·885	70·557	71·229	71·901	72·573	73·245
110	73·917	74·589	75·261	75·932	76·604	77·276	77·948	78·620	79·292	79·964
120	80·636	81·308	81·980	82·652	83·324	83·996	84·668	85·340	86·012	86·684
130	87·356	88·028	88·700	89·372	90·044	90·716	91·388	92·060	92·732	93·404
140	94·076	94·748	95·420	96·092	96·764	97·436	98·107	98·779	99·451	100·123

Extract from B.S. 350 : 1944

WEIGHTS

POUNDS TO KILOGRAMS

Based on 1 pound = 0·45359243 kilograms

lb.	0	1	2	3	4	5	6	7	8	9
	kg.	kg.	kg.	kg.	kg.	kg.	kg.	kg.	kg.	kg.
	—	—	0·45359	0·90718	1·36078	1·81437	2·26796	2·72155	3·17515	3·62874	4·08233
10	4·53592	4·98952	5·4431	5·8967	6·3503	6·8039	7·2575	7·7111	8·1647	8·6183
20	9·0718	9·5254	9·9790	10·4326	10·8862	11·3398	11·7934	12·2470	12·7006	13·1542
30	13·6078	14·0614	14·5150	14·9686	15·4221	15·8757	16·3293	16·7829	17·2365	17·6901
40	18·1437	18·5973	19·0509	19·5045	19·9581	20·4117	20·8653	21·3188	21·7724	22·2260
50	22·6796	23·1332	23·5868	24·0404	24·4940	24·9476	25·4012	25·8548	26·3084	26·7620
60	27·2155	27·6691	28·1227	28·5763	29·0299	29·4835	29·9371	30·3907	30·8443	31·2979
70	31·7515	32·2051	32·6587	33·1122	33·5658	34·0194	34·4730	34·9266	35·3802	35·8338
80	36·2874	36·7410	37·1946	37·6482	38·1018	38·5554	39·0089	39·4625	39·9161	40·3697
90	40·8233	41·2769	41·7305	42·1841	42·6377	43·0913	43·5449	43·9985	44·4521	44·9057
100	45·3592	45·8128	46·2664	46·7200	47·1736	47·6272	48·0808	48·5344	48·9880	49·4416
10	49·8952	50·349	50·802	51·256	51·710	52·163	52·617	53·070	53·524	53·977
20	54·431	54·885	55·338	55·792	56·245	56·699	57·153	57·606	58·060	58·513
30	58·967	59·421	59·874	60·328	60·781	61·235	61·689	62·142	62·596	63·049
40	63·503	63·957	64·410	64·864	65·317	65·771	66·224	66·678	67·132	67·585
50	68·039	68·492	68·946	69·400	69·853	70·307	70·760	71·214	71·668	72·121
60	72·575	73·028	73·482	73·936	74·389	74·843	75·296	75·750	76·204	76·657
70	77·111	77·564	78·018	78·471	78·925	79·379	79·832	80·286	80·739	81·193
80	81·647	82·100	82·554	83·007	83·461	83·915	84·368	84·822	85·275	85·729
90	86·183	86·636	87·090	87·543	87·997	88·451	88·904	89·358	89·811	90·265
200	90·718	91·172	91·626	92·079	92·533	92·986	93·440	93·894	94·347	94·801
10	95·254	95·708	96·162	96·615	97·069	97·522	97·976	98·430	98·883	99·337
20	99·790	100·244	100·698	101·151	101·605	102·058	102·512	102·965	103·419	103·873
30	104·326	104·780	105·233	105·687	106·141	106·594	107·048	107·501	107·955	108·409
40	108·862	109·316	109·769	110·223	110·677	111·130	111·584	112·037	112·491	112·945
50	113·398	113·852	114·305	114·759	115·212	115·666	116·120	116·573	117·027	117·480
60	117·934	118·388	118·841	119·295	119·748	120·202	120·656	121·109	121·563	122·016
70	122·470	122·924	123·377	123·831	124·284	124·738	125·192	125·645	126·099	126·552
80	127·006	127·459	127·913	128·367	128·820	129·274	129·727	130·181	130·635	131·088
90	131·542	131·995	132·449	132·903	133·356	133·810	134·263	134·717	135·171	135·624
300	136·078	136·531	136·985	137·439	137·892	138·346	138·799	139·253	139·706	140·160
10	140·614	141·067	141·521	141·974	142·428	142·882	143·335	143·789	144·242	144·696
20	145·150	145·603	146·057	146·510	146·964	147·418	147·871	148·325	148·778	149·232
30	149·686	150·139	150·593	151·046	151·500	151·953	152·407	152·861	153·314	153·768
40	154·221	154·675	155·129	155·582	156·036	156·489	156·943	157·397	157·850	158·304
50	158·757	159·211	159·665	160·118	160·572	161·025	161·479	161·932	162·386	162·840
60	163·293	163·747	164·200	164·654	165·108	165·561	166·015	166·468	166·922	167·376
70	167·829	168·283	168·736	169·190	169·644	170·097	170·551	171·004	171·458	171·912
80	172·365	172·819	173·272	173·726	174·179	174·633	175·087	175·540	175·994	176·447
90	176·901	177·355	177·808	178·262	178·715	179·169	179·623	180·076	180·530	180·983
400	181·437	181·891	182·344	182·798	183·251	183·705	184·159	184·612	185·066	185·519
10	185·973	186·426	186·880	187·334	187·787	188·241	188·694	189·148	189·602	190·055
20	190·509	190·962	191·416	191·870	192·323	192·777	193·230	193·684	194·138	194·591
30	195·045	195·498	195·952	196·406	196·859	197·313	197·766	198·220	198·673	199·127
40	199·581	200·034	200·488	200·941	201·395	201·849	202·302	202·756	203·209	203·663
50	204·117	204·570	205·024	205·477	205·931	206·385	206·838	207·292	207·745	208·199
60	208·653	209·106	209·560	210·013	210·467	210·920	211·374	211·828	212·281	212·735
70	213·188	213·642	214·096	214·549	215·003	215·456	215·910	216·364	216·817	217·271
80	217·724	218·178	218·632	219·085	219·539	219·992	220·446	220·900	221·353	221·807
90	222·260	222·714	223·167	223·621	224·075	224·528	224·982	225·435	225·889	226·343

Extract from B.S. 350: 1944

WEIGHTS

POUNDS TO KILOGRAMS (continued)

Based on 1 pound = 0·45359243 kilograms

lb.	0	1	2	3	4	5	6	7	8	9
	kg.	kg.	kg.	kg.	kg.	kg.	kg.	kg.	kg.	kg.
500	226·796	227·250	227·703	228·157	228·611	229·064	229·518	229·971	230·425	230·879
10	231·332	231·786	232·239	232·693	233·147	233·600	234·054	234·507	234·961	235·414
20	235·868	236·322	236·775	237·229	237·682	238·136	238·590	239·043	239·497	239·950
30	240·404	240·858	241·311	241·765	242·218	242·672	243·126	243·579	244·033	244·486
40	244·940	245·394	245·847	246·301	246·754	247·208	247·661	248·115	248·569	249·022
50	249·476	249·929	250·383	250·837	251·290	251·744	252·197	252·651	253·105	253·558
60	254·012	254·465	254·919	255·373	255·826	256·280	256·733	257·187	257·640	258·094
70	258·548	259·001	259·455	259·908	260·362	260·816	261·269	261·723	262·176	262·630
80	263·084	263·537	263·991	264·444	264·898	265·352	265·805	266·259	266·712	267·166
90	267·620	268·073	268·527	268·980	269·434	269·887	270·341	270·795	271·248	271·702
600	272·155	272·609	273·063	273·516	273·970	274·423	274·877	275·331	275·784	276·238
10	276·691	277·145	277·599	278·052	278·506	278·959	279·413	279·867	280·320	280·774
20	281·227	281·681	282·134	282·588	283·042	283·495	283·949	284·402	284·856	285·310
30	285·763	286·217	286·670	287·124	287·578	288·031	288·485	288·938	289·392	289·846
40	290·299	290·753	291·206	291·660	292·114	292·567	293·021	293·474	293·928	294·381
50	294·835	295·289	295·742	296·196	296·649	297·103	297·557	298·010	298·464	298·917
60	299·371	299·825	300·278	300·732	301·185	301·639	302·093	302·546	303·000	303·453
70	303·907	304·361	304·814	305·268	305·721	306·175	306·628	307·082	307·536	307·989
80	308·443	308·896	309·350	309·804	310·257	310·711	311·164	311·618	312·072	312·525
90	312·979	313·432	313·886	314·340	314·793	315·247	315·700	316·154	316·608	317·061
700	317·515	317·968	318·422	318·875	319·329	319·783	320·236	320·690	321·143	321·597
10	322·051	322·504	322·958	323·411	323·865	324·319	324·772	325·226	325·679	326·133
20	326·587	327·040	327·494	327·947	328·401	328·855	329·308	329·762	330·215	330·669
30	331·122	331·576	332·030	322·483	332·937	333·390	333·844	334·298	334·751	335·205
40	335·658	336·112	336·566	337·019	337·473	337·926	338·380	338·834	339·287	339·741
50	340·194	340·648	341·101	341·555	342·009	342·462	342·916	343·369	343·823	344·277
60	344·730	345·184	345·637	346·091	346·545	346·998	347·452	347·905	348·359	348·813
70	349·266	349·720	350·173	350·627	351·081	351·534	351·988	352·441	352·895	353·349
80	353·802	354·256	354·709	355·163	355·616	356·070	356·524	356·977	357·431	357·884
90	358·338	358·792	359·245	359·699	360·152	360·606	361·060	361·513	361·967	362·420
800	362·874	363·328	363·781	364·235	364·688	365·142	365·595	366·049	366·503	366·956
10	367·410	367·863	368·317	368·771	369·224	369·678	370·131	370·585	371·039	371·492
20	371·946	372·399	372·853	373·307	373·760	374·214	374·667	375·121	375·575	376·028
30	276·482	376·935	377·389	377·842	378·296	378·750	379·203	379·657	380·110	380·564
40	381·018	381·471	381·925	382·378	382·832	383·286	383·739	384·193	384·646	385·100
50	385·554	386·007	386·461	386·914	387·368	387·822	388·275	388·729	389·182	389·636
60	390·089	390·543	390·997	391·450	391·904	392·357	392·811	393·265	393·718	394·172
70	394·625	395·079	395·533	395·986	396·440	396·893	397·347	397·801	398·254	398·708
80	399·161	399·615	400·069	400·522	400·976	401·429	401·883	402·336	402·790	403·244
90	403·697	404·151	404·604	405·058	405·512	405·965	406·419	406·872	407·326	407·780
900	408·233	408·687	409·140	409·594	410·048	410·501	410·955	411·408	411·862	412·316
10	412·769	413·223	413·676	414·130	414·583	415·037	415·491	415·944	416·398	416·851
20	417·305	417·759	418·212	418·666	419·119	419·573	420·027	420·481	420·934	421·387
30	421·841	422·295	422·748	423·202	423·655	424·109	424·563	425·016	425·470	425·923
40	426·377	426·830	427·284	427·738	428·191	428·645	429·098	429·552	430·006	430·459
50	430·913	431·366	431·820	432·274	432·727	433·181	433·634	434·088	434·542	434·995
60	435·449	435·902	436·356	436·810	437·263	437·717	438·170	438·624	439·077	439·531
70	439·985	440·438	440·892	441·345	441·799	442·253	442·706	443·160	443·613	444·067
80	444·521	444·974	445·428	445·881	446·335	446·789	447·242	447·696	448·149	448·603
90	449·057	449·510	449·964	450·417	450·871	451·324	451·778	452·232	452·685	453·139
1000	453·592									

Extract from B.S. 350: 1944

FEET AND INCHES TO METRES

Based on 1 inch = 25·4 millimetres

Uncontracted values

Feet	Inches						Differences for sixteenths of an inch	
	0	1	2	3	4	5		
	m.	m.	m.	m.	m.	m.		m.
0	—	0·0254	0·0508	0·0762	0·1016	0·1270		
1	0·3048	0·3302	0·3556	0·3810	0·4064	0·4318		
2	0·6096	0·6350	0·6604	0·6858	0·7112	0·7366	1	0·0016
3	0·9144	0·9398	0·9652	0·9906	1·0160	1·0414		
4	1·2192	1·2446	1·2700	1·2954	1·3208	1·3462		
5	1·5240	1·5494	1·5748	1·6002	1·6256	1·6510	2	0·0032
6	1·8288	1·8542	1·8796	1·9050	1·9304	1·9558		
7	2·1336	2·1590	2·1844	2·2098	2·2352	2·2606		
8	2·4384	2·4638	2·4892	2·5146	2·5400	2·5654	3	0·0048
9	2·7432	2·7686	2·7940	2·8194	2·8448	2·8702		
10	3·0480	3·0734	3·0988	3·1242	3·1496	3·1750		
11	3·3528	3·3782	3·4036	3·4290	3·4544	3·4798	4	0·0064
12	3·6576	3·6830	3·7084	3·7338	3·7592	3·7846		
13	3·9624	3·9878	4·0132	4·0386	4·0640	4·0894		
14	4·2672	4·2926	4·3180	4·3434	4·3688	4·3942	5	0·0079
15	4·5720	4·5974	4·6228	4·6482	4·6736	4·6990		
16	4·8768	4·9022	4·9276	4·9530	4·9784	5·0038		
17	5·1816	5·2070	5·2324	5·2578	5·2832	5·3086	6	0·0095
18	5·4864	5·5118	5·5372	5·5626	5·5880	5·6134		
19	5·7912	5·8166	5·8420	5·8674	5·8928	5·9182		
20	6·0960	6·1214	6·1468	6·1722	6·1976	6·2230	7	0·0111
21	6·4008	6·4262	6·4516	6·4770	6·5024	6·5278		
22	6·7056	6·7310	6·7564	6·7818	6·8072	6·8326		
23	7·0104	7·0358	7·0612	7·0866	7·1120	7·1374	8	0·0127
24	7·3152	7·3406	7·3660	7·3914	7·4168	7·4422		
25	7·6200	7·6454	7·6708	7·6962	7·7216	7·7470		
26	7·9248	7·9502	7·9756	8·0010	8·0264	8·0518	9	0·0143
27	8·2296	8·2550	8·2804	8·3058	8·3312	8·3566		
28	8·5344	8·5598	8·5852	8·6106	8·6360	8·6614		
29	8·8392	8·8646	8·8900	8·9154	8·9408	8·9662	10	0·0159
30	9·1440	9·1694	9·1948	9·2202	9·2456	9·2710		
31	9·4488	9·4742	9·4996	9·5250	9·5504	9·5758		
32	9·7536	9·7790	9·8044	9·8298	9·8552	9·8806	11	0·0175
33	10·0584	10·0838	10·1092	10·1346	10·1600	10·1854		
34	10·3632	10·3886	10·4140	10·4394	10·4648	10·4902		
35	10·6680	10·6934	10·7188	10·7442	10·7696	10·7950	12	0·0190
36	10·9728	10·9982	11·0236	11·0490	11·0744	11·0998		
37	11·2776	11·3030	11·3284	11·3538	11·3792	11·4046		
38	11·5824	11·6078	11·6332	11·6586	11·6840	11·7094	13	0·0206
39	11·8872	11·9126	11·9380	11·9634	11·9888	12·0142		
40	12·1920	12·2174	12·2428	12·2682	12·2936	12·3190		
41	12·4968	12·5222	12·5476	12·5730	12·5984	12·6238	14	0·0222
42	12·8016	12·8270	12·8524	12·8778	12·9032	12·9286		
43	13·1064	13·1318	13·1572	13·1826	13·2080	13·2334		
44	13·4112	13·4366	13·4620	13·4874	13·5128	13·5382	15	0·0238
45	13·7160	13·7414	13·7668	13·7922	13·8176	13·8430		
46	14·0208	14·0462	14·0716	14·0970	14·1224	14·1478		
47	14·3256	14·3510	14·3764	14·4018	14·4272	14·4526		
48	14·6304	14·6558	14·6812	14·7066	14·7320	14·7574		
49	14·9352	14·9606	14·9860	15·0114	15·0368	15·0622		

Extract from B.S. 350 : 1944

FEET AND INCHES TO METRES

Based on 1 inch = 25·4 millimetres

Uncontracted values

Feet	Inches						Differences for sixteenths of an inch	
	6	7	8	9	10	11		
	m.	m.	m.	m.	m.	m.		m.
0	0·1524	0·1778	0·2032	0·2286	0·2540	0·2794		
1	0·4572	0·4826	0·5080	0·5334	0·5588	0·5842		
2	0·7620	0·7874	0·8128	0·8382	0·8636	0·8890	1	0·0016
3	1·0668	1·0922	1·1176	1·1430	1·1684	1·1938		
4	1·3716	1·3970	1·4224	1·4478	1·4732	1·4986		
5	1·6764	1·7018	1·7272	1·7526	1·7780	1·8034	2	0·0032
6	1·9812	2·0066	2·0320	2·0574	2·0828	2·1082		
7	2·2860	2·3114	2·3368	2·3622	2·3876	2·4130		
8	2·5908	2·6162	2·6416	2·6670	2·6924	2·7178	3	0·0048
9	2·8956	2·9210	2·9464	2·9718	2·9972	3·0226		
10	3·2004	3·2258	3·2512	3·2766	3·3020	3·3274		
11	3·5052	3·5306	3·5560	3·5814	3·6068	3·6322	4	0·0064
12	3·8100	3·8354	3·8608	3·8862	3·9116	3·9370		
13	4·1148	4·1402	4·1656	4·1910	4·2164	4·2418		
14	4·4196	4·4450	4·4704	4·4958	4·5212	4·5466	5	0·0079
15	4·7244	4·7498	4·7752	4·8006	4·8260	4·8514		
16	5·0292	5·0546	5·0800	5·1054	5·1308	5·1562		
17	5·3340	5·3594	5·3848	5·4102	5·4356	5·4610	6	0·0095
18	5·6388	5·6642	5·6896	5·7150	5·7404	5·7658		
19	5·9436	5·9690	5·9944	6·0198	6·0452	6·0706		
20	6·2484	6·2738	6·2992	6·3246	6·3500	6·3754	7	0·0111
21	6·5532	6·5786	6·6040	6·6294	6·6548	6·6802		
22	6·8580	6·8834	6·9088	6·9342	6·9596	6·9850		
23	7·1628	7·1882	7·2136	7·2390	7·2644	7·2898	8	0·0127
24	7·4676	7·4930	7·5184	7·5438	7·5692	7·5946		
25	7·7724	7·7978	7·8232	7·8486	7·8740	7·8994		
26	8·0772	8·1026	8·1280	8·1534	8·1788	8·2042	9	0·0143
27	8·3820	8·4074	8·4328	8·4582	8·4836	8·5090		
28	8·6868	8·7122	8·7376	8·7630	8·7884	8·8138		
29	8·9916	9·0170	9·0424	9·0678	9·0932	9·1186	10	0·0159
30	9·2964	9·3218	9·3472	9·3726	9·3980	9·4234		
31	9·6012	9·6266	9·6520	9·6774	9·7028	9·7282		
32	9·9060	9·9314	9·9568	9·9822	10·0076	10·0330	11	0·0175
33	10·2108	10·2362	10·2616	10·2870	10·3124	10·3378		
34	10·5156	10·5410	10·5664	10·5918	10·6172	10·6426		
35	10·8204	10·8458	10·8712	10·8966	10·9220	10·9474	12	0·0190
36	11·1252	11·1506	11·1760	11·2014	11·2268	11·2522		
37	11·4300	11·4554	11·4808	11·5062	11·5316	11·5570		
38	11·7348	11·7602	11·7856	11·8110	11·8364	11·8618	13	0·0206
39	12·0396	12·0650	12·0904	12·1158	12·1412	12·1666		
40	12·3444	12·3698	12·3952	12·4206	12·4460	12·4714		
41	12·6492	12·6746	12·7000	12·7254	12·7508	12·7762	14	0·0222
42	12·9540	12·9794	13·0048	13·0302	13·0556	13·0810		
43	13·2588	13·2842	13·3096	13·3350	13·3604	13·3858		
44	13·5636	13·5890	13·6144	13·6398	13·6652	13·6906	15	0·0238
45	13·8684	13·8938	13·9192	13·9446	13·9700	13·9954		
46	14·1732	14·1986	14·2240	14·2494	14·2748	14·3002		
47	14·4780	14·5034	14·5288	14·5542	14·5796	14·6050		
48	14·7828	14·8082	14·8336	14·8590	14·8844	14·9098		
49	15·0876	15·1130	15·1384	15·1638	15·1892	15·2146		

Extract from B.S. 350: 1944

FEET AND INCHES TO METRES

Based on 1 inch = 25·4 millimetres

Uncontracted values

Feet	Inches						Differences for sixteenths of an inch	
	0	1	2	3	4	5		
	m.	m.	m.	m.	m.	m.		m.
50	15·2400	15·2654	15·2908	15·3162	15·3416	15·3670		
51	15·5448	15·5702	15·5956	15·6210	15·6464	15·6718		
52	15·8496	15·8750	15·9004	15·9258	15·9512	15·9766	1	0·0016
53	16·1544	16·1798	16·2052	16·2306	16·2560	16·2814		
54	16·4592	16·4846	16·5100	16·5354	16·5608	16·5862		
55	16·7640	16·7894	16·8148	16·8402	16·8656	16·8910	2	0·0032
56	17·0688	17·0942	17·1196	17·1450	17·1704	17·1958		
57	17·3736	17·3990	17·4244	17·4498	17·4752	17·5006		
58	17·6784	17·7038	17·7292	17·7546	17·7800	17·8054	3	0·0048
59	17·9832	18·0086	18·0340	18·0594	18·0848	18·1102		
60	18·2880	18·3134	18·3388	18·3642	18·3896	18·4150		
61	18·5928	18·6182	18·6436	18·6690	18·6944	18·7198	4	0·0064
62	18·8976	18·9230	18·9484	18·9738	18·9992	19·0246		
63	19·2024	19·2278	19·2532	19·2786	19·3040	19·3294		
64	19·5072	19·5326	19·5580	19·5834	19·6088	19·6342	5	0·0079
65	19·8120	19·8374	19·8628	19·8882	19·9136	19·9390		
66	20·1168	20·1422	20·1676	20·1930	20·2184	20·2438		
67	20·4216	20·4470	20·4724	20·4978	20·5232	20·5486	6	0·0095
68	20·7264	20·7518	20·7772	20·8026	20·8280	20·8534		
69	21·0312	21·0566	21·0820	21·1074	21·1328	21·1582		
70	21·3360	21·3614	21·3868	21·4122	21·4376	21·4630	7	0·0111
71	21·6408	21·6662	21·6916	21·7170	21·7424	21·7678		
72	21·9456	21·9710	21·9964	22·0218	22·0472	22·0726		
73	22·2504	22·2758	22·3012	22·3266	22·3520	22·3774	8	0·0127
74	22·5552	22·5806	22·6060	22·6314	22·6568	22·6822		
75	22·8600	22·8854	22·9108	22·9362	22·9616	22·9870		
76	23·1648	23·1902	23·2156	23·2410	23·2664	23·2918	9	0·0143
77	23·4696	23·4950	23·5204	23·5458	23·5712	23·5966		
78	23·7744	23·7998	23·8252	23·8506	23·8760	23·9014		
79	24·0792	24·1046	24·1300	24·1554	24·1808	24·2062	10	0·0159
80	24·3840	24·4094	24·4348	24·4602	24·4856	24·5110		
81	24·6888	24·7142	24·7396	24·7650	24·7904	24·8158		
82	24·9936	25·0190	25·0444	25·0698	25·0952	25·1206	11	0·0175
83	25·2984	25·3238	25·3492	25·3746	25·4000	25·4254		
84	25·6032	25·6286	25·6540	25·6794	25·7048	25·7302		
85	25·9080	25·9334	25·9588	25·9842	26·0096	26·0350	12	0·0190
86	26·2128	26·2382	26·2636	26·2890	26·3144	26·3398		
87	26·5176	26·5430	26·5684	26·5938	26·6192	26·6446		
88	26·8224	26·8478	26·8732	26·8986	26·9240	26·9494	13	0·0206
89	27·1272	27·1526	27·1780	27·2034	27·2288	27·2542		
90	27·4320	27·4574	27·4828	27·5082	27·5336	27·5590		
91	27·7368	27·7622	27·7876	27·8130	27·8384	27·8638	14	0·0222
92	28·0416	28·0670	28·0924	28·1178	28·1432	28·1686		
93	28·3464	28·3718	28·3972	28·4226	28·4480	28·4734		
94	28·6512	28·6766	28·7020	28·7274	28·7528	28·7782	15	0·0238
95	28·9560	28·9814	29·0068	29·0322	29·0576	29·0830		
96	29·2608	29·2862	29·3116	29·3370	29·3624	29·3878		
97	29·5656	29·5910	29·6164	29·6418	29·6672	29·6926		
98	29·8704	29·8958	29·9212	29·9466	29·9720	29·9974		
99	30·1752	30·2006	30·2260	30·2514	30·2768	30·3022		
100	30·4800							

Extract from B.S. 350: 1944

FEET AND INCHES TO METRES

Based on 1 inch = 25·4 millimetres

Uncontracted values

Feet	Inches						Differences for sixteenths of an inch	
	6	7	8	9	10	11		
	m.	m.	m.	m.	m.	m.		m.
50	15·3924	15·4178	15·4432	15·4686	15·4940	15·5194		
51	15·6972	15·7226	15·7480	15·7734	15·7988	15·8242		
52	16·0020	16·0274	16·0528	16·0782	16·1036	16·1290	1	0·0016
53	16·3068	16·3322	16·3576	16·3830	16·4084	16·4338		
54	16·6116	16·6370	16·6624	16·6878	16·7132	16·7386		
55	16·9164	16·9418	16·9672	16·9926	17·0180	17·0434	2	0·0032
56	17·2212	17·2466	17·2720	17·2974	17·3228	17·3482		
57	17·5260	17·5514	17·5768	17·6022	17·6276	17·6530		
58	17·8304	17·8562	17·8816	17·9070	17·9324	17·9578	3	0·0048
59	18·1356	18·1610	18·1864	18·2118	18·2372	18·2626		
60	18·4404	18·4658	18·4912	18·5166	18·5420	18·5674		
61	18·7452	18·7706	18·7960	18·8214	18·8468	18·8722	4	0·0064
62	19·0500	19·0754	19·1008	19·1262	19·1516	19·1770		
63	19·3548	19·3802	19·4056	19·4310	19·4564	19·4818		
64	19·6596	19·6850	19·7104	19·7358	19·7612	19·7866	5	0·0079
65	19·9644	19·9898	20·0152	20·0406	20·0660	20·0914		
66	20·2692	20·2946	20·3200	20·3454	20·3708	20·3962		
67	20·5740	20·5994	20·6248	20·6502	20·6756	20·7010	6	0·0095
68	20·8788	20·9042	20·9296	20·9550	20·9804	21·0058		
69	21·1836	21·2090	21·2344	21·2598	21·2852	21·3106		
70	21·4884	21·5138	21·5392	21·5646	21·5900	21·6154	7	0·0111
71	21·7932	21·8186	21·8440	21·8694	21·8948	21·9202		
72	22·0980	22·1234	22·1488	22·1742	22·1996	22·2250		
73	22·4028	22·4282	22·4536	22·4790	22·5044	22·5298	8	0·0127
74	22·7076	22·7330	22·7584	22·7838	22·8092	22·8346		
75	23·0124	23·0378	23·0632	23·0886	23·1140	23·1394		
76	23·3172	23·3426	23·3680	23·3934	23·4188	23·4442	9	0·0143
77	23·6220	23·6474	23·6728	23·6982	23·7236	23·7490		
78	23·9268	23·9522	23·9776	24·0030	24·0284	24·0538		
79	24·2316	24·2570	24·2824	24·3078	24·3332	24·3586	10	0·0159
80	24·5364	24·5618	24·5872	24·6126	24·6380	24·6634		
81	24·8412	24·8666	24·8920	24·9174	24·9428	24·9682		
82	25·1460	25·1714	25·1968	25·2222	25·2476	25·2730	11	0·0175
83	25·4508	25·4762	25·5016	25·5270	25·5524	25·5778		
84	25·7556	25·7810	25·8064	25·8318	25·8572	25·8826		
85	26·0604	26·0858	26·1112	26·1366	26·1620	26·1874	12	0·0190
86	26·3652	26·3906	26·4160	26·4414	26·4668	26·4922		
87	26·6700	26·6954	26·7208	26·7462	26·7716	26·7970		
88	26·9748	27·0002	27·0256	27·0510	27·0764	27·1018	13	0·0206
89	27·2796	27·3050	27·3304	27·3558	27·3812	27·4066		
90	27·5844	27·6098	27·6352	27·6606	27·6860	27·7114		
91	27·8892	27·9146	27·9400	27·9654	27·9908	28·0162	14	0·0222
92	28·1940	28·2194	28·2448	28·2702	28·2956	28·3210		
93	28·4988	28·5242	28·5496	28·5750	28·6004	28·6258		
94	28·8036	28·8290	28·8544	28·8798	28·9052	28·9306	15	0·0238
95	29·1084	29·1338	29·1592	29·1846	29·2100	29·2354		
96	29·4132	29·4386	29·4640	29·4894	29·5148	29·5402		
97	29·7180	29·7434	29·7688	29·7942	29·8196	29·8450		
98	30·0228	30·0482	30·0736	30·0990	30·1244	30·1498		
99	30·3276	30·3530	30·3784	30·4038	30·4292	30·4546		

Conversion tables

Extract from B.S. 350 : 1944

SQUARE FEET TO SQUARE METRES

Based on 1 inch = 25·4 millimetres

Sq. ft.	0	1	2	3	4	5	6	7	8	9
	m.^{2}	m.^{2}	m.^{2}	m.^{2}	m.^{2}	m.^{2}	m.^{2}	m.^{2}	m.^{2}	m.^{2}
—	—	0·09290	0·18581	0·27871	0·37161	0·46452	0·55742	0·65032	0·74322	0·83613
10	0·92903	1·02193	1·11484	1·20774	1·30064	1·39355	1·48645	1·57935	1·67225	1·76516
20	1·85806	1·95096	2·04387	2·13677	2·22967	2·32258	2·41548	2·50838	2·60129	2·69419
30	2·78709	2·87999	2·97290	3·06580	3·15870	3·25161	3·34451	3·43741	3·53032	3·62322
40	3·71612	3·80902	3·90193	3·99483	4·08773	4·18064	4·27354	4·36644	4·45935	4·55225
50	4·64515	4·73806	4·83096	4·92386	5·0168	5·1097	5·2026	5·2955	5·3884	5·4813
60	5·5742	5·6671	5·7600	5·8529	5·9458	6·0387	6·1316	6·2245	6·3174	6·4103
70	6·5032	6·5961	6·6890	6·7819	6·8748	6·9677	7·0606	7·1535	7·2464	7·3393
80	7·4322	7·5251	7·6180	7·7110	7·8039	7·8968	7·9897	8·0826	8·1755	8·2684
90	8·3613	8·4542	8·5471	8·6400	8·7329	8·8258	8·9187	9·0116	9·1045	9·1974
100	9·2903	9·3832	9·4761	9·5690	9·6619	9·7548	9·8477	9·9406	10·0335	10·1264
10	10·2193	10·3122	10·4051	10·4980	10·5909	10·6838	10·7768	10·8697	10·9626	11·0555
20	11·1484	11·2413	11·3342	11·4271	11·5200	11·6129	11·7058	11·7987	11·8916	11·9845
30	12·0774	12·1703	12·2632	12·3561	12·4490	12·5419	12·6348	12·7277	12·8206	12·9135
40	13·0064	13·0993	13·1922	13·2851	13·3780	13·4709	13·5638	13·6567	13·7496	13·8426
50	13·9355	14·0284	14·1213	14·2142	14·3071	14·4000	14·4929	14·5858	14·6787	14·7716
60	14·8645	14·9574	15·0503	15·1432	15·2361	15·3290	15·4219	15·5148	15·6077	15·7006
70	15·7935	15·8864	15·9793	16·0722	16·1651	16·2580	16·3509	16·4438	16·5367	16·6296
80	16·7225	16·8155	16·9084	17·0013	17·0942	17·1871	17·2800	17·3729	17·4658	17·5587
90	17·6516	17·7445	17·8374	17·9303	18·0232	18·1161	18·2090	18·3019	18·3948	18·4877
200	18·5806	18·6735	18·7664	18·8593	18·9522	19·0451	19·1380	19·2309	19·3238	19·4167
10	19·5096	19·6025	19·6954	19·7883	19·8813	19·9742	20·0671	20·1600	20·2529	20·3458
20	20·4387	20·5316	20·6245	20·7174	20·8103	20·9032	20·9961	21·0890	21·1819	21·2748
30	21·3677	21·4606	21·5535	21·6464	21·7393	21·8322	21·9251	22·0180	22·1109	22·2038
40	22·2967	22·3896	22·4825	22·5754	22·6683	22·7612	22·8541	22·9471	23·0400	23·1329
50	23·2258	23·3187	23·4116	23·5045	23·5974	23·6903	23·7832	23·8761	23·9690	24·0619
60	24·1548	24·2477	24·3406	24·4335	24·5264	24·6193	24·7122	24·8051	24·8980	24·9909
70	25·0838	25·1767	25·2696	25·3625	25·4554	25·5483	25·6412	25·7341	25·8270	25·9199
80	26·0129	26·1058	26·1987	26·2916	26·3845	26·4774	26·5703	26·6632	26·7561	26·8490
90	26·9419	27·0348	27·1277	27·2206	27·3135	27·4064	27·4993	27·5922	27·6851	27·7780
300	27·8709	27·9638	28·0567	28·1496	28·2425	28·3354	28·4283	28·5212	28·6141	28·7070
10	28·7999	28·8928	28·9857	29·0787	29·1716	29·2645	29·3574	29·4503	29·5432	29·6361
20	29·7290	29·8219	29·9148	30·0077	30·1006	30·1935	30·2864	30·3793	30·4722	30·5651
30	30·6580	30·7509	30·8438	30·9367	31·0296	31·1225	31·2154	31·3083	31·4012	31·4941
40	31·5870	31·6799	31·7728	31·8657	31·9586	32·0515	32·1445	32·2374	32·3303	32·4232
50	32·5161	32·6090	32·7019	32·7948	32·8877	32·9806	33·0735	33·1664	33·2593	33·3522
60	33·4451	33·5380	33·6309	33·7238	33·8167	33·9096	34·0025	34·0954	34·1883	34·2812
70	34·3741	34·4670	34·5599	34·6528	34·7457	34·8386	34·9315	35·0244	35·1173	35·2103
80	35·3032	35·3961	35·4890	35·5819	35·6748	35·7677	35·8606	35·9535	36·0464	36·1393
90	36·2322	36·3251	36·4180	36·5109	36·6038	36·6967	36·7896	36·8825	36·9754	37·0683
400	37·1612	37·2541	37·3470	37·4399	37·5328	37·6257	37·7186	37·8115	37·9044	37·9973
10	38·0902	38·1831	38·2761	38·3690	38·4619	38·5548	38·6477	38·7406	38·8335	38·9264
20	39·0193	39·1122	39·2051	39·2980	39·3909	39·4838	39·5767	39·6696	39·7625	39·8554
30	39·9483	40·0412	40·1341	40·2270	40·3199	40·4128	40·5057	40·5986	40·6915	40·7844
40	40·8773	40·9702	41·0631	41·1560	41·2489	41·3419	41·4348	41·5277	41·6206	41·7135
50	41·8064	41·8993	41·9922	42·0851	42·1780	42·2709	42·3638	42·4567	42·5496	42·6425
60	42·7354	42·8283	42·9212	43·0141	43·1070	43·1999	43·2928	43·3857	43·4786	43·5715
70	43·6644	43·7573	43·8502	43·9431	44·0360	44·1289	44·2218	44·3148	44·4077	44·5006
80	44·5935	44·6864	44·7793	44·8722	44·9651	45·0580	45·1509	45·2438	45·3367	45·4296
90	45·5225	45·6154	45·7083	45·8012	45·8941	45·9870	46·0799	46·1728	46·2657	46·3586

Extract from B.S. 350: 1944

SQUARE FEET TO SQUARE METRES (continued)

Based on 1 inch = 25·4 millimetres

Sq. ft.	0	1	2	3	4	5	6	7	8	9
	m.²	m.²	m.²	m.²	m.²	m.²	m.²	m.²	m.²	m.²
500	46·4515	46·5444	46·6373	46·7302	46·8231	46·9160	47·0089	47·1018	47·1947	47·2876
10	47·3806	47·4735	47·5664	47·6593	47·7522	47·8451	47·9380	48·0309	48·1238	48·2167
20	48·3096	48·4025	48·4954	48·5883	48·6812	48·7741	48·8670	48·9599	49·0528	49·1457
30	49·2386	49·3315	49·4244	49·5173	49·6102	49·7031	49·7960	49·8889	49·9818	50·075
40	50·168	50·261	50·353	50·446	50·539	50·632	50·725	50·818	50·911	51·004
50	51·097	51·190	51·282	51·375	51·468	51·561	51·654	51·747	51·840	51·933
60	52·026	52·119	52·212	52·304	52·397	52·490	52·583	52·676	52·769	52·862
70	52·955	53·048	53·141	53·233	53·326	53·419	53·512	53·605	53·698	53·791
80	53·884	53·977	54·070	54·162	54·255	54·348	54·441	54·534	54·627	54·720
90	54·813	54·906	54·999	55·092	55·184	55·277	55·370	55·463	55·556	55·649
600	55·742	55·835	55·928	56·021	56·113	56·206	56·299	56·392	56·485	56·578
10	56·671	56·764	56·857	56·950	57·042	57·135	57·228	57·321	57·414	57·507
20	57·600	57·693	57·786	57·879	57·971	58·064	58·157	58·250	58·343	58·436
30	58·529	58·622	58·715	58·808	58·901	58·993	59·086	59·179	59·272	59·365
40	59·458	59·551	59·644	59·737	59·830	59·922	60·015	60·108	60·201	60·294
50	60·387	60·480	60·573	60·666	60·759	60·851	60·944	61·037	61·130	61·223
60	61·316	61·409	61·502	61·595	61·688	61·781	61·873	61·966	62·059	62·152
70	62·245	62·338	62·431	62·524	62·617	62·710	62·802	62·895	62·988	63·081
80	63·174	63·267	63·360	63·453	63·546	63·639	63·731	63·824	63·917	64·010
90	64·103	64·196	64·289	64·382	64·475	64·568	64·661	64·753	64·846	64·939
700	65·032	65·125	65·218	65·311	65·404	65·497	65·590	65·682	65·775	65·868
10	65·961	66·054	66·147	66·240	66·333	66·426	66·519	66·611	66·704	66·797
20	66·890	66·983	67·076	67·169	67·262	67·355	67·448	67·541	67·633	67·726
30	67·819	67·912	68·005	68·098	68·191	68·284	68·377	68·470	68·562	68·655
40	68·748	68·841	68·934	69·027	69·120	69·213	69·306	69·399	69·491	69·584
50	69·677	69·770	69·863	69·956	70·049	70·142	70·235	70·328	70·421	70·513
60	70·606	70·699	70·792	70·885	70·978	71·071	71·164	71·257	71·350	71·442
70	71·535	71·628	71·721	71·814	71·907	72·000	72·093	72·186	72·279	72·371
80	72·464	72·557	72·650	72·743	72·836	72·929	73·022	73·115	73·208	73·300
90	73·393	73·486	73·579	73·672	73·765	73·858	73·951	74·044	74·137	74·230
800	74·322	74·415	74·508	74·601	74·694	74·787	74·880	74·973	75·066	75·159
10	75·251	75·344	75·437	75·530	75·623	75·716	75·809	75·902	75·995	76·088
20	76·180	76·273	76·366	76·459	76·552	76·645	76·738	76·831	76·924	77·017
30	77·110	77·202	77·295	77·388	77·481	77·574	77·667	77·760	77·853	77·946
40	78·039	78·131	78·224	78·317	78·410	78·503	78·596	78·689	78·782	78·875
50	78·968	79·060	79·153	79·246	79·339	79·432	79·525	79·618	79·711	79·804
60	79·897	79·990	80·082	80·175	80·268	80·361	80·454	80·547	80·640	80·733
70	80·826	80·919	81·011	81·104	81·197	81·290	81·383	81·476	81·569	81·662
80	81·755	81·848	81·940	82·033	82·126	82·219	82·312	82·405	82·498	82·591
90	82·684	82·777	82·870	82·962	83·055	83·148	83·241	83·334	83·427	83·520
900	83·613	83·706	83·799	83·891	83·984	84·077	84·170	84·263	84·356	84·449
10	84·542	84·635	84·728	84·820	84·913	85·006	85·099	85·192	85·285	85·378
20	85·471	85·564	85·657	85·750	85·842	85·935	86·028	86·121	86·214	86·307
30	86·400	86·493	86·586	86·679	86·771	86·864	86·957	87·050	87·143	87·236
40	87·329	87·422	87·515	87·608	87·700	87·793	87·886	87·979	88·072	88·165
50	88·258	88·351	88·444	88·537	88·630	88·722	88·815	88·908	89·001	89·094
60	89·187	89·280	89·373	89·466	89·559	89·651	89·744	89·837	89·930	90·023
70	90·116	90·209	90·302	90·395	90·488	90·580	90·673	90·766	90·859	90·952
80	91·045	91·138	91·231	91·324	91·417	91·509	91·602	91·695	91·788	91·881
90	91·974	92·067	92·160	92·253	92·346	92·439	92·531	92·624	92·717	92·810
1000	92·903									

Pressure, Stress: UK tonsforce per square inch to meganewtons per square metre (newtons per square millimetre)
Basis: 1 UKton = 2240 lb; 1 lbf = 0.453 592 37 kgf; 1 kgf = 9.806 65 N; 1 in = 0.0254 m.

UK tonf/in²	0	0·1	0·2	0·3	0·4	0·5	0·6	0·7	0·8	0·9
				meganewtons per square metre						
0	0·00000	1·54443	3·08885	4·63328	6·17770	7·72213	9·26655	10·8110	12·3554	13·8998
1	15·4443	16·9887	18·5331	20·0775	21·6220	23·1664	24·7108	26·2552	27·7997	29·3441
2	30·8885	32·4329	33·9774	35·5218	37·0662	38·6106	40·1551	41·6995	43·2439	44·7883
3	46·3328	47·8772	49·4216	50·9660	52·5105	54·0549	55·5993	57·1437	58·6882	60·2326
4	61·7770	63·3215	64·8659	66·4103	67·9547	69·4992	71·0436	72·5880	74·1324	75·6769
5	77·2213	78·7657	80·3101	81·8546	83·3990	84·9434	86·4878	88·0323	89·5767	91·1211
6	92·6655	94·2100	95·7544	97·2988	98·8432	100·388	101·932	103·477	105·021	106·565
7	108·110	109·654	111·199	112·743	114·287	115·832	117·376	118·921	120·465	122·010
8	123·554	125·098	126·643	128·187	129·732	131·276	132·821	134·365	135·909	137·454
9	138·998	140·543	142·087	143·632	145·176	146·720	148·265	149·809	151·354	152·898
10	154·443	155·987	157·531	159·076	160·620	162·165	163·709	165·254	166·798	168·342
11	169·887	171·431	172·976	174·520	176·065	177·609	179·153	180·698	182·242	183·787
12	185·331	186·876	188·420	189·964	191·509	193·053	194·598	196·142	197·686	199·231
13	200·775	202·320	203·864	205·409	206·953	208·497	210·042	211·586	213·131	214·675
14	216·220	217·764	219·308	220·853	222·397	223·942	225·486	227·031	228·575	230·119
15	231·664	233·208	234·753	236·297	237·842	239·386	240·930	242·475	244·019	245·564
16	247·108	248·635	250·197	251·741	253·286	254·830	256·375	257·919	259·464	261·008
17	262·552	264·097	265·641	267·186	268·730	270·274	271·819	273·363	274·908	276·452
18	277·997	279·541	281·085	282·630	284·174	285·719	287·263	288·808	290·352	291·896
19	293·441	294·985	296·530	298·074	299·619	301·163	302·707	304·252	305·796	307·341
20	308·885	310·430	311·974	313·518	315·063	316·607	318·152	319·696	321·241	322·785
21	324·329	325·874	327·418	328·963	330·507	332·052	333·596	335·140	336·685	338·229
22	339·774	341·318	342·862	344·407	345·951	347·496	349·040	350·585	352·129	353·673
23	355·218	356·762	358·307	359·851	361·396	362·940	364·484	366·029	367·573	369·118
24	370·662	372·207	373·751	375·295	376·840	378·384	379·929	381·473	383·018	384·562

Supplement No. 1 (1967) to
B.S. 350 : Part 2 : 1962

Pressure, Stress: UK tons-force per square inch to meganewtons per square metre
Basis: 1 UKton = 2240 lb; 1 lbf = 0.453 592 37 kgf; 1 kgf = 9.806 65 N; 1 in = 0.0254 m.

UK tonf/in²	0	0·1	0·2	0·3	0·4	0·5	0·6	0·7	0·8	0·9
				meganewtons per square metre						
25	386·106	387·651	389·195	390·740	392·284	393·829	395·373	396·917	398·462	400·006
26	401·551	403·095	404·640	406·184	407·728	409·273	410·817	412·362	413·906	415·450
27	416·995	418·539	420·084	421·628	423·173	424·717	426·261	427·806	429·350	430·895
28	432·439	433·984	435·528	437·072	438·617	440·161	441·706	443·250	444·795	446·339
29	447·883	449·428	450·972	452·517	454·061	455·606	457·150	458·694	460·239	461·783
30	463·328	464·872	466·417	467·961	469·505	471·050	472·594	474·139	475·683	477·228
31	478·772	480·316	481·861	483·405	484·950	486·494	488·039	489·583	491·127	492·672
32	494·216	495·761	497·305	498·849	500·394	501·938	503·483	505·027	506·572	508·116
33	509·660	511·205	512·749	514·294	515·838	517·383	518·927	520·471	522·016	523·560
34	525·105	526·649	528·194	529·738	531·282	532·827	534·371	535·916	537·460	539·005
35	540·549	542·093	543·638	545·182	546·727	548·271	549·816	551·360	552·904	554·449
36	555·993	557·538	559·082	560·627	562·171	563·715	565·260	566·804	568·349	569·893
37	571·437	572·982	574·526	576·071	577·615	579·160	580·704	582·248	583·793	585·337
38	586·882	588·426	589·971	591·515	593·059	594·604	596·148	597·693	599·237	600·782
39	602·326	603·870	605·415	606·959	608·504	610·048	611·593	613·137	614·681	616·226
40	617·770	619·315	620·859	622·404	623·948	625·492	627·037	628·581	630·126	631·670
41	633·215	634·759	636·303	637·848	639·392	640·937	642·481	644·025	645·570	647·114
42	648·659	650·203	651·748	653·292	654·836	656·381	657·925	659·470	661·014	662·559
43	664·103	665·647	667·192	668·736	670·281	671·825	673·370	674·914	676·458	678·003
44	679·547	681·092	682·636	684·181	685·725	687·269	688·814	690·358	691·903	693·447
45	694·992	696·536	698·080	699·625	701·169	702·714	704·258	705·803	707·347	708·891
46	710·436	711·980	713·525	715·069	716·613	718·158	719·702	721·247	722·791	724·336
47	725·880	727·424	728·969	730·513	732·058	733·602	735·147	736·691	738·235	739·780
48	741·324	742·869	744·413	745·958	747·502	749·046	750·591	752·135	753·680	755·224
49	756·769	758·313	759·857	761·402	762·946	764·491	766·035	767·580	769·124	770·668

Pressure, stress:
UK tons-force per square inch to meganewtons per square metre

UK tonf/in²	0	0·1	0·2	0·3	0·4	0·5	0·6	0·7	0·8	0·9
					meganewtons per square metre					
50	772·213	773·757	775·302	776·846	778·391	779·935	781·479	783·024	784·568	786·113
51	787·657	789·201	790·746	792·290	793·835	795·379	796·924	798·468	800·012	801·557
52	803·101	804·646	806·190	807·735	809·279	810·823	812·368	813·912	815·457	817·001
53	818·546	820·090	821·634	823·179	824·723	826·268	827·812	829·357	830·901	832·445
54	833·990	835·534	837·079	838·623	840·168	841·712	843·256	844·801	846·345	847·890
55	849·434	850·979	852·523	854·067	855·612	857·156	858·701	860·245	861·790	863·334
56	864·878	866·423	867·967	869·512	871·056	872·600	874·145	875·689	877·234	878·778
57	880·323	881·867	883·411	884·956	886·500	888·045	889·589	891·134	892·678	894·222
58	895·767	897·311	898·856	900·400	901·945	903·489	905·033	906·578	908·122	909·667
59	911·211	912·756	914·300	915·844	917·389	918·933	920·478	922·022	923·567	925·111
60	926·655	928·200	929·744	931·289	932·833	934·378	935·922	937·466	939·011	940·555
61	942·100	943·644	945·188	946·733	948·277	949·822	951·366	952·911	954·455	955·999
62	957·544	959·088	960·633	962·177	963·722	965·266	966·810	968·355	969·899	971·444
63	972·988	974·533	976·077	977·621	979·166	980·710	982·255	983·799	985·344	986·888
64	988·432	989·977	991·521	993·066	994·610	996·155	997·699	999·243	1000·79	1002·33
65	1003·88	1005·42	1006·97	1008·51	1010·05	1011·60	1013·14	1014·69	1016·23	1017·78
66	1019·32	1020·87	1022·41	1023·95	1025·50	1027·04	1028·59	1030·13	1031·68	1033·22
67	1034·77	1036·31	1037·85	1039·40	1040·94	1042·49	1044·03	1045·58	1047·12	1048·67
68	1050·21	1051·75	1053·30	1054·84	1056·39	1057·93	1059·48	1061·02	1062·56	1064·11
69	1065·65	1067·20	1068·74	1070·29	1071·83	1073·38	1074·92	1076·46	1078·01	1079·55
70	1081·10	1082·64	1084·19	1085·73	1087·28	1088·82	1090·36	1091·91	1093·45	1095·00
71	1096·54	1098·09	1099·63	1101·18	1102·72	1104·26	1105·81	1107·35	1108·90	1110·44
72	1111·99	1113·53	1115·08	1116·62	1118·16	1119·71	1121·25	1122·80	1124·34	1125·89
73	1127·43	1128·98	1130·52	1132·06	1133·61	1135·15	1136·70	1138·24	1139·79	1141·33
74	1142·87	1144·42	1145·96	1147·51	1149·05	1150·60	1152·14	1153·69	1155·23	1156·77

Pressure, stress:
UK tons-force per square inch to meganewtons per square metre

UK tonf/in²	0	0.1	0.2	0.3	0.4	0.5	0.6	0.7	0.8	0.9
					meganewtons per square metre					
75	1158.32	1159.86	1161.41	1162.95	1164.50	1166.04	1167.59	1169.13	1170.67	1172.22
76	1173.76	1175.31	1176.85	1178.40	1179.94	1181.49	1183.03	1184.57	1186.12	1187.66
77	1189.21	1190.75	1192.30	1193.84	1195.39	1196.93	1198.47	1200.02	1201.56	1203.11
78	1204.65	1206.20	1207.74	1209.29	1210.83	1212.37	1213.92	1215.46	1217.01	1218.55
79	1220.10	1221.64	1223.19	1224.73	1226.27	1227.82	1229.36	1230.91	1232.45	1234.00
80	1235.54	1237.08	1238.63	1240.17	1241.72	1243.26	1244.81	1246.35	1247.90	1249.44
81	1250.98	1252.53	1254.07	1255.62	1257.16	1258.71	1260.25	1261.80	1263.34	1264.88
82	1266.43	1267.97	1269.52	1271.06	1272.61	1274.15	1275.70	1277.24	1278.78	1280.33
83	1281.87	1283.42	1284.96	1286.51	1288.05	1289.60	1291.14	1292.68	1294.23	1295.77
84	1297.32	1298.86	1300.41	1301.95	1303.50	1305.04	1306.58	1308.13	1309.67	1311.22
85	1312.76	1314.31	1315.85	1317.40	1318.94	1320.48	1322.03	1323.57	1325.12	1326.66
86	1328.21	1329.75	1331.29	1332.84	1334.38	1335.93	1337.47	1339.02	1340.56	1342.11
87	1343.65	1345.19	1346.74	1348.28	1349.83	1351.37	1352.92	1354.46	1356.01	1357.55
88	1359.09	1360.64	1362.18	1363.73	1365.27	1366.82	1368.36	1369.91	1371.45	1372.99
89	1374.54	1376.08	1377.63	1379.17	1380.72	1382.26	1383.81	1385.35	1386.89	1388.44
90	1389.98	1391.53	1393.07	1394.62	1396.16	1397.71	1399.25	1400.79	1402.34	1403.88
91	1405.43	1406.97	1408.52	1410.06	1411.61	1413.15	1414.69	1416.24	1417.78	1419.33
92	1420.87	1422.42	1423.96	1425.50	1427.05	1428.59	1430.14	1431.68	1433.23	1434.77
93	1436.32	1437.86	1439.40	1440.95	1442.49	1444.04	1445.58	1447.13	1448.67	1450.22
94	1451.76	1453.30	1454.85	1456.39	1457.94	1459.48	1461.03	1462.57	1464.12	1465.66
95	1467.20	1468.75	1470.29	1471.84	1473.38	1474.93	1476.47	1478.02	1479.56	1481.10
96	1482.65	1484.19	1485.74	1487.28	1488.83	1490.37	1491.92	1493.46	1495.00	1496.55
97	1498.09	1499.64	1501.18	1502.73	1504.27	1505.81	1507.36	1508.90	1510.45	1511.99
98	1513.54	1515.08	1516.63	1518.17	1519.71	1521.26	1522.80	1524.35	1525.89	1527.44
99	1528.98	1530.53	1532.07	1533.61	1535.16	1536.70	1538.25	1539.79	1541.34	1542.88
100	1544.43	—	—	—	—	—	—	—	—	—

British Standards covering the design and construction of steelwork

A basic list of British Standards covering the design and construction of steelwork is given.

Note Dates in parentheses indicate that the British Standard has been reviewed at that date.

Bolts

BS 3692　1967 Specification for ISO metric precision hexagon bolts, screws and nuts. Metric units.

BS 4190　1967 Specification for ISO metric black hexagon bolts, screws and nuts. Metric units.

BS 4395　Specification for high strength friction grip bolts and associated nuts and washers for structural engineering. Metric series.
Part 1: 1969 General grade
Part 2: 1969 Higher grade bolts and nuts and general grade washers

BS 4604　Specification for the use of high strength friction grip bolts in structural steelwork. Metric series.
Part 1: 1970 General grade
Part 2: 1970 Higher grade (parallel shank)

Corrosion

BS 1501　Steels for pressure purposes: plates, sheet and strip
Part 3: 1990 Specification for corrosion and heat-resisting steels

BS 5493　1977 Code of practice for protective coating of iron and steel structures against corrosion

BS 7079　Preparation of steel substrates before application of paints and related products
Part 0: 1990 Introduction
Part A1: 1989 Specification for rust grades and preparation grades of uncoated steel substrates after overall removal of previous coatings
Part A1: Supplement 1: 1989 Representative photographic examples of the change of appearance when blast-cleaned with different abrasives

Design

BS 449   Specification for the use of structural steel in building
Part 2: 1969 Metric units

BS 2573   Rules for the design of cranes
Part 1: 1983 Specification for classification, stress calculations and design criteria for structures

BS 2853   1957 Specification for the design and testing of steel overhead runway beams

BS 5400   Steel, concrete and composite bridges
Part 1: 1988 General statement
Part 3: 1982 Code of practice for design of steel bridges
Part 4: 1990 Code of practice for design of concrete bridges
Part 5: 1979 Code of practice for design of composite bridges
Part 6: 1980 Specification for materials and workmanship: steel
Part 9: Section 9.1: 1983 Code of practice for design of bridge bearings
Part 9: Section 9.2: 1983 Specification for materials, manufacture and installation of bridge bearings
Part 10: 1980 Code of practice for fatigue

BS 5502   Code of practice for the design of buildings and structures for agriculture
Part 1: Section 1.1: 1986 Materials
Part 22: 1987 Code of practice for design, construction and loading

BS 5950   Structural use of steelwork in building
Part 1: 1990 Code of practice for design in simple and continuous construction: hot rolled sections
Part 2: 1985 Specification for materials fabrication and erection: hot rolled sections
Part 3: Section 3.1: 1990 Code of practice for design of simple and continuous composite beams
Part 4: 1982 Code of practice for design of floors with profiled steel sheeting
Part 5: 1987 Code of practice for design of cold formed sections
Part 6: Code of practice for design of light gauge sheeting, decking and cladding (in preparation)
Part 7: Specification for materials and workmanship: cold formed sections and sheeting

Erection

BS 5531   1988 Code of practice for safety in erecting structural frames

Fire

BS 476 Fire tests on building materials and structures
Part 20: 1987 Methods for determination of the fire resistance of elements of construction (general principles)
Part 21: 1987 Methods for determination of the fire resistance of load-bearing elements of construction

BS 5950 Structural use of steelwork in building
Part 8: 1990 Code of practice for fire resistant design

Loading

BS 648 1964: Schedule of weights of building materials

BS 5400 Steel, concrete and composite bridges
Part 2: 1978 Specification for loads

BS 6399 Loading for buildings
Part 1: 1984 Code of practice for dead and imposed loads
Part 2: Code of practice for wind loading (to be published and will replace CP3: Chapter V: Part 2)
Part 3: 1988 Code of practice for imposed roof loads

CP3 Code of basic data for the design of buildings
Chapter V: Part 2: 1972 Wind loads

Steel

BS 4 Structural steel sections
Part 1: 1980 Specification for hot rolled sections

BS 970 Part 1: 1991 General inspection and testing procedures and specific requirements for carbon, carbon-manganese, alloy and stainless steels

BS 1449 Steel plate, sheet and strip
Part 1: 1983 Specification for carbon and carbon-manganese plate, sheet and strip
Part 2: 1983 Specification for stainless and heat-resisting steel plate, sheet and strip

BS 2989 1982 Specification for continuously hot-dip zinc coated and iron-zinc alloy coated steel: wide strip, sheet/plate and slit wide strip

BS 2994   1976 (1987) Specification for cold rolled steel sections

BS 4360   1990 Specification for weldable structural steels

BS 4848   Specification for hot-rolled structural steel sections
           Part 2: 1975 Hollow sections
           Part 4: 1972 (1986) Equal and unequal angles
           Part 5: 1980 Bulb flats

BS 6363   1983 Specification for welded cold formed steel structural hollow sections

BS 7191   1989 Specification for weldable structural steels for fixed offshore structures

BS EN 10 025   1990 Specification for hot rolled products of non-alloy structural steels and their technical delivery conditions

Welding

BS 499   Welding terms and symbols
           Part 1:1983 Glossary for welding, brazing and thermal cutting
           Part 2: 1980 (1989) Specification for symbols for welding
           Part 2C: 1980 Welding symbols

BS 639   1986 Specification for covered carbon and carbon manganese steel electrodes for manual metal-arc welding

BS 4165   1984 Specification for electrode wires and fluxes for the submerged arc welding of carbon steel and medium-tensile steel

BS 4870   Specification for approval testing of welding procedures
           Part 1: 1981 Fusion welding of steel

BS 4871   Specification for approval testing of welders working to approved welding procedures
           Part 1: 1982 Fusion welding of steel

BS 4872   Specification for approval testing of welders when welding procedure approval is not required
           Part 1: 1982 Fusion welding of steel

BS 5135   1984 Specification for arc welding of carbon and carbon manganese steels

BS 7084   1989 Specification for carbon and carbon manganese steel tubular cored welding electrodes

Weld testing

BS 2600   Radiographic examination of fusion welded butt joints in steel
  Part 1: 1983 Methods for steel 2 mm up to and including 50 mm thick
  Part 2: 1973 Methods for steel over 50 mm thick up to and including 200 mm thick

BS 2910   1986 Methods for radiographic examination of fusion welded circumferential butt joints in steel pipes

BS 3923   Methods for ultrasonic examination of welds
  Part 1: 1986 Methods for manual examination of fusion welds in ferritic steels

BS 5289   1976 (1983) Code of practice. Visual inspection of fusion welded joints.

Index

accidental loading, 47
accuracy, characteristic, 884−5
   see also tolerances
acid pickling, 1004
additives, steel-making, 201−202
aerodynamic behaviour (of bridges), 115, 118
Agrément Certificate, metal-cladding, 3
air-supported roofs, 177
aluminium, effect of, 202, 203
aluminium cladding, 3
analysis methods
   area moment, 279−80
   finite element, 256−7, 285−92, 337−40
   fracture mechanism, 236−7
   grid, 141−3
   moment-distribution, 281−3
   Serviceability Limit State, 21−2
   slope-deflection, 280−81
   unit load, 283−5
   yield line, 697−703
   see also elastic analysis; modelling process; plastic analysis/ theory
anchorages, sheet pile, 849
angle sections, design of, 395
   see also equal angles; unequal angles
annealing, 209
anti-sag rods, 27
anti-vibration mountings, 105
antimony, effect of, 23
arc plasma cutting, 928
arc welding, 219
arch bridge, 118−20
architectural design aims, 38
area moment method, 279−80
arsenic, effect of, 203
atria, 189−90
   environmental influences, 196−7
   fire engineering, 195−6
   form of members and connections, 194
   paint systems, 194−5
   structural aspects, 190−92, 194
austenite, 204

back marks, 668
bacterial corrosion, 1000
bainite, 205

banksman, use of, 976−8
barrel vaults, 167
base motion response, 329
baseplate design, 787−92
bases, steelwork, 785−92
'bath tubs' (open-top box girders), 125
beam-and-pot type floors, 93
beam and stick building costs, 913
beam bearing plates 790−92
'beam-columns', 10−11, 493
   see also loading, combined
beam to beam connections, 723−4
beams, 95−7
   bridge bracing, 423−5
   buckling of, 413, 415−19
   column connections, 714−19
   composite
      advantages, 594
      applications, 593
      basic design, 600−602, 603−604
      deflections, 615−17
      section analysis, 597−602
      shear connections, 596, 604−609, 611−613
         partial, 609−11
      shear transfer, 613−15
      shrinkage, 618
      span conditions, 596−7
      span to depth ratios, 595
      vibration, 617
   deflection of, 411−12
   design of, 408
      at ultimate limit state, 460−68
      continuous
         design tables, 1051−3
         influence lines, 1054−64
      encastré, 296−301
         design tables, 1036−42
         shear forces, 300−301
         supports at different levels, 299−300
         supports at same level, 297−9
      propped cantilevers, see cantilevers, propped
      simply supported, 294
         design tables, 1028−35
      unpropped cantilevers see cantilevers, unpropped
   influence line diagrams, 1054−64

1253

1254   Index

lateral bracing, 421–2
moment capacity 407, 409
plate girder connections, 722
restricted depth design, 426–8
sections
  classification, 406–408
  cold-formed, 428–31
  rolled sizes, 149
shear effect, 409–10
torsion effect, 413
types, 403–405
  cable-stayed, 175
  castellated, 55, 408, 431
    dimensions and properties tables, 1136–9, 1142–3
    forming, 928
  fully-restrained, 419–21
  notched, 704–705
  primary, 53, 57–8, 88
  profiled, 44
  secondary, 53, 88
  tapered, 6–7, 44, 56–7
web, effects on, 413
web openings, 431, 433
see also girders; joints; universal beams
bearing fixings, 818
bearing pressure, ground, 783–5
bearing stiffeners, 468
bearing strength, bolt, 673
bearings, 808
  assemblies, 821–4
  bedding for, 798, 821
  choice of, 808–13
  for curved structures, 827–8
  installation of, 819–21
  types of, 814–8
  use of, 818–21
bedding, baseplate, 798, 821
bending (as fabrication process), 933
bending analysis, 259–61
  and shear, 268–9
  beyond elastic limit, 263–8
bending resistance
  beams, 460–61
  compact sections, 360
bending strength tables, 1152–3
Bernoulli equation, 259
bicycle wheel roof, 177
bimetallic corrosion, 999–1000
black bolts *see* bolts, structural
blast cleaning, 931, 1004
blast loadings, industrial structure, 107–108
blast-primers *see* primers
board fire protection, 991–2
boiler houses, 82–3
bolt anchorages, 794–5
bolt holes, 665–6, 700, 703
bolting
  shop, 924
  site, 981, 983

bolts
  analysis of groups, 668–71, 1164–85
  bearing strengths, 673
  connection design data, 1164–85
  design strengths, 672–5
  geometry of, 665–8
  holding-down, 792–5
  tolerances on, 901–2
  types of, 664–5, 925–6
    black *see* structural
    close tolerance, 925
    high-strength friction-grip (HSFG), 87, 99, 665, 672, 673–4
    structural, 925
Bosporus bridge, 115
bowstring truss bridge, 121
box columns *see* columns, types
box girders
  fabrication of, 141
  for highway bridges, 125
  safety checks, 967
  trapezoidal, 125, 127
braced-bay frames, 65–6, 100
bracing systems, 63–6, 100, 517
  bridge, 140, 423–5
  lateral, 421–2
  plan, 90–91
  safety checks, 970
  tension/compression, 92
  use of trusses, 542, 544
bracket stiffeners, 693
brickwork cladding, 11
bridge bearings, 820–4
bridges, 111–13
  bracing action, 423–5
  plate girder use, 457–69
  span selection, 113–14
  steel grades for, 133–4
  thermal movements in, 134
  traffic loading, 131–2
  types, 114–15
    arch, 118–20
    cable-stayed, 118, 134–5
    composite deck, 138, 140–43, 149
    curved deck, 141
    girder, 121–30
    highway, 123, 125, 131, 138, 140–43, 149
    pedestrian, 127, 130
    portal-type, 120–21
    railway, 127, 131–2
    suspension, 115–18, 134, 175, 983
    truss, 121, 530–34
British Board of Agrément (BBA), 107
  cladding certificate, 3
British Standards, *see separate listing at end of Index*
brittle fracture, 242–5
  risk of, 252–3
b/t (width to thickness) ratio, 341
buckling

beams, 406
  flexural, 376
  flexural-torsional, 384
  lateral-torsional, 413, 415–19
  local, 341–50
  plate girders, 446, 447
  torsional, 383–4
  web, 413
buckling parameters, 1099
buckling resistance (of stiffeners), 692
building envelope tolerances, 885, 903–4
Building Regulations
  external firewalls, 31
  fire safety, 195
  safety requirements, 885–6
Building Research Establishment (BRE) Digest, 8, 332
building services integration, 43–6
buildings
  range of types, 1
  use factors, 41
built-in beams see beams, encastré
butterfly, truss, 529

cable structures
  applications, 174–5, 177
  construction, 184–6
  detailing, 184–6
  loading, 177–84
  side rail support, 27
  special features, 177–84
cable-stayed beams see beams, cable-stayed
cable-stayed bridges see bridges, cable-stayed
cables
  as tension members, 8
  bridge use, 365–7
  construction of, 184, 186
  protection of, 367
  terminations, 184, 186
  types of, 366
cambering, 933
cantilevers
  propped
    analysis, 294–5
    design tables, 1043–50
    plastic failure, 308–10
  unpropped
    design tables, 1026–7
carbon steels, 201
cash flow, 947
castellated beams see beams, castellated
castellated columns see columns, castellated
casting process, 211–12
ceiling voids, 43, 44
cement manufacturing plant, 85
cementite, 204
CEN TC 135 (European Standard), tolerances, 884, 890
channel sections
  design of, 395

dimensions and properties tables, 1124–7
Charpy V-notch impact tests, 216, 250, 253
chemical composition, steel, 200–201
  added elements, 201–202
  non-metallic inclusions, 203
'chord and joint' space frame systems, 170
chromium, effect of, 201, 202
circular hollow sections, 52
  dimensions and properties tables, 1118–9
CIRIA Technical Notes
  No.98, 694
  No.116, 574
cladding, 2–4
  restraint by, 27
  as structural element, 23–4
  support members, 25
  systems, 28–9
    composite panels, 4, 30
    double shell, 3, 29
    single skin trapezoidal, 3, 29
    standing seam, 4, 29-30
Clapeyron's Theorem see Theorem of Three Moments
classification table, section, 349
clayey soils, foundation design, 784
client brief, 38
close-tolerance bolts see bolts, types of
Codes of Practice
  bridges, 130–31
  circulation space loadings, 102
  No.3, loadings, 161–2
coffer-dams, 849–51
cohesionless soils, foundation design, 784–5
cold-formed light-gauge sections, 5, 350
cold sawing, 936
cold working, 218
collapse, progressive, 47
collapse load (on purlins), 26
column splices, 714
  safety checks, 940
columns/struts
  buckling, 383–4
  composite, 642
    design of, 643–5
      concrete-filled tubes, 651
      encased, 645–50
  compressive resistance, 380–83
  cross-sections, 376
  design of, 375–6
  economics of, 395–6
  effective lengths, 385–92
  erection, 49, 51
  fire performance, 991
  safety check, 969
  stability, 512–15
  stiffening, 720–21
  types, 374
    box, 97
    built-up, 97
    castellated, 1140–41

special, 392–5
see also universal columns
combined loading
  bending and shear, 467
  compression and bending, 491–2
compactness check, 505–509
compartment temperature, 995
composite beams *see* beams, composite
compostie columns *see* columns, composite
composite decking, *see* deck slabs, composite
composite floor trusses, 57–8
compound struts *see* columns, composite
compression members, 14
  effective length, 540
compressive resistance, 380–83
concrete
  composite deck slabs, 570–71, 572
  encasement, 992
  floors, 59, 93–4
  shear walls, 63–5, 101
concrete-filled tubes, 651–3
connections
  design of, 15, 21, 98–9, 707–13, 1164–93
  truss elements, 536–7
  types of
    beam and plate girder, 722
    beam to beam, 723–4
    beam to column, 714–19, 719–21
    column splices, 714
    large grip, 675
    long joint, 674–5
    rigid, 63, 66
    semi-rigid, 66
    simple, 66
    Vierendeel girder, 546
  welded *see* welded connections
construction costs breakdown, 38
construction methods *see* erection methods
contact bearing, full, 888–92
continuous beams *see* beams, continuous
continuous cooling transformation (CCT)
  diagram, 207
conversion tables, 1234–47
conveyor systems, 86–7, 937
copper, effect of, 201
corrosion
  data, 1230
  processes, 998–1000
  rates of, 1000–1001
  steel thickness, 134
corrosion protection
  cables, 366–7
  coatings, 212
  internal, 49
  sheet piles, 853, 855
corrosion resistance design, 1007, 1230–33
corrosion -resistant steels, 201, 202
Corten (steel type), 1013
cost considerations, 47–8
  fabrication, 912–5

minimum cost design, 38
  repetition benefits, 48
crack initiation/propagation, 238, 245, 247–9
crack resistance, 248
crack tip opening displacement (CTOD), 249
  tests, 251, 253
cracking, weld, 222
cranes/craneage:
  erection, 49, 947
    layout, 957, 959
    safe use, 971
  types of
    climbing, 953
    mobile, 948–9
    non-mobile, 950–52
    stockyard, 953–4
  permanent, 9–10
    gantry girders, 457
crevice corrosion, 999
critical damping, 326
cropping, 928–9
cross centres, bolt, 668
cross-section classification checks, 505–509
cut-off (sheet pile), 851
cutting processes, 218, 927–31
cylindrical bearings, 817

damping vibration, 336–7
dead loading, 42
deck slabs, composite, 59–60, 457, 568
  advantages, 569
  bridge uses, 123
  concreting of, 570–71
  design of, 572–81, 582
  diaphragm action, 582–3
  erection, 568, 571, 572, 583–4
  fire resistance, 581–2, 991
  profiles, 569
  selection of, 571
  temporary bracing, 544
defects, manufacturing, 213–15
  corrective action, 940
  feedback, 939–40
deflection
  beam and column design, 10–11
  beams, 95, 411–12
  BS limits, 1150
  composite beams, 615–17
  composite deck slabs, 580–81
  floor plates, 875
  floors, 42, 53
  towers, 163
degrees of freedom, 274
design standards, 917
design strengths (BS 4360 steel), 1150
deviations *see* tolerances
diamond truss, 532, 533
dimensions and properties section tables, 1118–23
disc/pot bearings, 817

dome structures, 167, 170
drawing process, 218
drilling process, 218
ductile fracture, 242–5
ductility, 263
  in connections, 710–11
durability, sheet piles, 853–5
Durbar plate, 870–71
dust enclosures, 108
dynamic equilibrium equation, 324–5, 330–31
dynamic formulae, load testing, 842
dynamic loads, 43, 86, 104, 324–30
dynamic performance, 323–4
  damping, 336
  distributed systems, 330–36
  finite element analysis, 337–40
  railays, 131–3
  turbine generators, 83

earth-retaining structures, 843
effective breadth concept
  beams, 603–604
effective length concept, 383–92
  compressive members, 540–41
  of weld, 683
elastic analysis
  composite beams, 597–9
  portal frames, 316–17, 348–9
elastic critical moment, 419
elastic interaction surface, 358
elastic limit, effects beyond, 263–8
elastic properties, steel, 1023
elastic section modulus, 260, 1101–2
elastomeric bearings, 814–5, 819
elastomers, 812–3
electro-slag welding, 921
element analysis, 256–7
element design tables, 1098–1107
encased columns *see* columns, encased
encastré beams *see* beams, encastré
end panels, plate girders, 453–4
end plates, welded, 718, 719
energy release rate, elastic, 248
English truss *see* Howe truss
environmental effects, 40
  atria design, 196–7
  loading on tower structures, 159, 161–2
  on corrosion, 1001–2
equal angles, dimensions and properties tables, 1130–31, 1133
equilibrium phase diagram, 204–205
equivalent uniform moment, 500–502
erection
  costs, 537, 920
  cranes/craneage, 947–59
  design considerations, 97, 941
  method statement, 49–51, 942–3
  programme, 943–7
  safety, 961–74
  site practice, 974–83

special structures, 983
sub-assemblies, 959, 961
temporary bracing, 970
tension members, 365
Erection Method Statement, 942–3, 970, 976
erection tolerances, 887–8, 893–4, 901–10
Eurocodes
  EC2, 130
  EC3, 130, 253, 378, 690–91, 694
  EC4, 130
    composite columns, 645, 646, 649–50, 651
    effective breadth, 603
    plastic hinge analysis, 601
    shear connectors, 606, 610
  TC135, 884, 890
'European Column Curves', 381
external loading, 8–9
external wall construction, 46

fabric-reinforced bearings, 814
fabricated bearings, 817
fabrication techniques, 48–9, 217–23, 911
  costs, 912–15
  economy in, 911
  tolerances, 885, 887–893, 895–900
fan cable configuration, 118
fatigue, 226–9, 323
  bridge plate girders, 469
  connection design, 98
  improvement techniques, 238–9
  life assessment, 236–7
  loadings, 87, 226–7
  railway bridges, 132
  tension members, 32–3
fatigue damage model, 228
fatigue-joint classification, 229
fatigue-resistant design, 239–40
ferrite, 204
film-formation, paint, 1007–8
finishes
  floor, 43–4, 93, 94
  nuclear fuel plants, 85
  see also paints/painting systems
finite element method, 285–92, 337–40
Fink truss, 528, 532
fire engineering method (FEM), 582, 985, 993–6
fire load, 993
fire protection systems, 51, 985, 992–3
  atria, 195–6
  information sheets
    beam-column connections, 1220–21
    block-filled columns, 1212–3
    board protection, 1208–9
    composite construction, 1216–9
    concrete-filled columns, 1214–5
    intumescent coatings, 1210–11
    sprayed, 1206–7, 1229
  limiting temperatures, 1222–3
fire-reinforcement, 571, 578
fire resistance, 992–4, 995

composite deck slabs, 581–2
encased columns, 652–3
testing of, 992–3
firewalls, external, 31
fixed beams *see* beams, encastré
fixing bolts *see* bolts
flame cleaning, 1003
flame cutting, 218, 927–8
flat roofs, weatherproofing, 10
flatness tolerance, 892
flexibility concept, 274–6
floor deflections, 42, 53, 875
floor depths, 44–6
floor-element assemblies, 51
floor finishes, 43–4, 93, 94
floor loading, 42
floor openings, 94
floor plates, load capacity design tables, 1204–5
floor/roof joints, 825
floor trusses, composite, 57–8
flooring systems, 92–5
  concrete, 59
  metal deck, 59–61
  open-grid, 875
  precast, 61, 63
forging process, 211, 212
forming process, 218
foundations
  analysis, 792
  choice of, 40, 49
  design of, 781–5
  holding-down systems, 795–9
  steelwork connections, 785–92
fracture mechanism analysis, 236–7
  general yielding, 249
  linear elastic, 245, 247–9
fracture-safe design, 252–3
frame weight comparisons, 2
free body diagram, 273
frequencies, structure natural, 104–105
frequency, forcing, 104
Freysinnet Hinge, 813
friction grip bolts *see* bolts, types of
Frodingham piles, 1194–1201
fully-restrained beams *see* beams, fully-restrained
fusion welding, 219

galvanizing, 221
  hot dip, 1005
gantry girders, 457
Geiger, David, 177
girder bridges, *see* bridges, girder
girders
  box *see* box girders
  fabricated, 55, 56
  plate
    bridge design, 457–9
    crane gantries, 457
    cross section choice, 445–6
    end panels, 453–4

    fabrication, 141
    highway bridges, 125
    safety checks, 969
    to BS 5400, 459–69
    to BS 5950, 446–57
    web design, 348
  safety checks, 969
  stub, 58
  vertical, 8
  Vierendeel *see* Vierendeel girders
gouging, 930
grain size changes, 200–209
gravity load paths, 88–9
grid analysis, bridge, 141–3
grid, structural, 52–3
grid structures, 167
  domes, 170
Griffith criterion, 247–8
grinding (well improvement), 238
grit blasting, 1004
ground conditions, 40
grouting, bolt, 799
gusset plates, 15, 537, 703–704
guyed towers, 159, 162

half-through sections, 130
hand signals, 976–8
handling and stacking plants, 86–7
handling methods, steel, 933–8
hanger supports, 96
hardenability, 201, 210
harmonic loads, 327–9
harp cable configuration, 118
haunches, use of, 21–2
HAZ cracking, 923
Health and Safety at Work Act (1974), 943
Health and Safety Executive (HSE) 942, 961–2
heat treatment, 203–207
  practical, 207–10
high-rise buildings *see* multi-storey buildings
highway bridges *see* bridges, highway
holding-down systems, 795, 797
hole forming, 921–2
hollow section properties, 1104–5
  see also circular, rectangular and square hollow sections
horizontal diaphragm design, 90–92
housing
  applications, 186–7
  current systems, 187
  developments, 189
Houston Astrodome, 170
Howe truss, 528, 531, 532, 533
HP200 Plastisol coating, 3
HSFG (high-strength friction-grip) bolts, *see* bolts, types of
hysteresis effect, 813

impact loading, 329
impact tests *see* Charpy V-notch impact tests

impact transition curves, 244
imperfections *see* defects, manufacturing
indeterminancy, degree of, 277
indeterminate frames, 315–16
industrial structures, 81, 87–8
    anatomy of, 88–101
    blast loading, 107–108
    conveyors, 86–7
    environmental effects, 108–109
    handling and stacking plants, 86–7
    plant loadings, 102–107
    power station structures, 82–4
    process plant, 84–6
    thermal effects, 108–109
    wind loading, 107
influence line diagrams
    four-span beam, 1061–4
    three-span beam, 1057–60
    two-span beam, 1054–6
inspection, manufacturing, 939
Instantaneous Centre method, 668, 683
insulating foam, composite panels, 30
insulation board, cladding, 29
internal/external joints, 825
International Rubber Hardness (IRHD), scale, 812
intumescent coatings, 992
ISO 9000, quality management, 939
isothermal transformation diagram, 205
isotropic plates, 271, 272

J intergral method test, 252
joints
    internal/external, 825
    lap, 892
    movement, 825–31, 829–30
    pin, 11, 14
    wall, 825
joists
    dimensions tables, 1115, 1142
    properties tables, 1115, 1116, 1142–3

'K' brace, 66
K-truss, 532, 533–4
Kleinlogel, Prof., analysis formulae, 311
knuckle bearings, 816

lamella domes, 170
lamellar tearing, 222
lamination, 215
lap joints, 892
large grip connections, 674–5
Larssen piles, 1202–3
laser cutting, 218
lateral bracing, 421–2
lateral restraint, 12, 14, 47, 90
lateral-torsional buckling, 413, 415–19
    plate girders, 462–4
lattice girders *see* parallel chord truss
lattice structures

costs, 915
frame weights, 2
towers, 157, 162
lattice trusses, 7
leaf bearings, 816
levelling, on-site, 978–9
Liberty ship failures, 244
lift/core locations, 63
    tolerances, 886–7
lifting devices/equipment, 935–7
    special, 955–7
    see also cranes/craneage
lifting procedures, 975–8
    see also erection
limit state design concept, 130
limited frames, 390–91
limiting temperature, 986–7
line elements, 257
    in axial loading, 258
    in plane bending, 259–61
    in shear, 261–2
line-grids, 97
linear elastic theory, 316
liner tray, cladding, 29
lining, on-site, 978–9
load dispersion, 689–91
load/extension relationship, cable, 177, 179–80
load factor, 268
load paths
    in connections, 707, 709–10
    gravity, 88–9
    sway, 90–92
'load ratio' method (fire resistance), 986
load reversal
    on trusses, 534
    wind loading, 27
load-span tables
    composite deck slabs, 578
    purlins, 27
load-testing, pile, 840–42
load transfer structures, 41
loading, 101–102
    accidental, 47
    axial, 258, 268–9
    combination, 10–11, 493–509
    cranes, 9–10
    dead, 42
    dynamic, 324–30
    external gravity, 8–9
    fatigue, 226–7
    floor, 42
    impact, 329
    internal gravity, 9
    lateral, 17, 90–91, 105–106
    process plant and equipment, 102–105
    rotational motion, 106
    services, 9
    variable-amplitude, 329
    wind, 9, 107, 970–71
loading formulae, 311–13

loading intensity, equivalent, 104
locking systems, bolt, 99
long joint connections, 674–5
longitudinal restraint, 15

machining, 930
MAG welding, 918, 920
manganese, effect of, 201
mansard truss, 528
martensite, 205
masts, 157
  in building structures, 163–6
materials monitoring, 945
mathematical modelling *see* modelling process
mechanical bearings, 815–8
member sizing tolernaces, 885
MERO system node, 170
Mesnager Hinge, 813
metal deck floors *see* steel deck/floors
metallic coatings, 1005–7
  spray, 1006
metallury, steel, 199–200, 224
  chemical composition, 200–203
  engineering properties, 215–17
  fabrication effects, 217–22
  heat treatment, 203–207
    practical, 207–10
  mechanical tests, 215–17
  service performance, 224
method statement, erection *see* Erection Method Statement
microstructure changes, 200–209
MIG welding, 918–20
Miner's rule, 228, 234
minimum cost design, 38
MMA welding, 918
modelling process
  dynamic analysis, 339–46
  static analysis, 255
modified Zed sections, 5
modular ratio, 604
modular space frame systems, 170
modular tower, 159
molybdenum, effect of, 201, 202
moment capacity
  of beams, 407, 409
  of plate griders, 448
'moment capacity' method (fire resistance), 986
moment distribution method, 281–3
moment gradient loading, 499–502
moment of inertia *see* second moment of area
moment of resistance, plastic, 466
moment-rotation behaviour, 343–9
movements, 807–8
  in joints, 825–31
  see also deflection
multi-storey buildings
  choice of form factors
    accidental loading, 47
    building services, 43–6
    building use, 41
    construction, 49–51
    cost considerations, 47–8, 921
    environmental 40
    external wall construction, 46
    fabrication, 48–9
    finishes, 43–4
    floor loadings, 42
    ground conditions, 40
    lateral stiffness, 47
  design aims
    architectural, 38
    financial, 38
    technical, 37–8
  design concept
    client brief, 38
    physical factors, 40
    statutory constraints, 39
  steel-frame advantages, 36

National House Building Council (NHBC), 187
National Structural Steelwork Specification (NSSS), 884
natural frequency, structural, 334–6
needle gunning, 939
nickel, effect of, 201, 202
niobium, effect of, 202
'no-sway' frames, 51
node points, bending moments, 14–15
NODUS system node, 170
normalizing process, 210
notation, xxv–xxx
notch ductility, 216
notched beams *see* beams, notched
nuclear fuel process and treatment plants, 85
nuts *see* bolts

oil rig collapses, 244
open-grid flooring, 93, 94, 875
  fixing, 95
orthotropic decks/plates, 112, 271, 272, 876–7
Outline Planning Permission, 39
overseas location factors, 537, 542

painting, on-site, 51, 1013
paints/painting systems, 194–5, 1008–12
  application, 1012–13
  classification, 1008
  composition, 1007
  costs, 913
parallel chord truss, 532
parallel-flange Universal Sections, 18
Parker truss, 531
part-tied arch bridge, 121
pearlite, 205
pedestrian bridge, *see* bridges, pedestrian
peening, hammer, 239
Petit truss, 532, 533
petrochemical plants, 85–6
phosphor bronze bearings, 812

Index 1261

phosphorus, effect of, 203
physical factors, effect on design, 38
piles, steel
  design, 834–9
  installation, 839
  loading capacity, 840–42
  sheet *see* sheet piles
  types, 833–4
  universal bearing, 1116–17
  uses, 833
pin joints, 11, 14
pipework
  pressure loadings, 106
  prestressing, 106
pitting (corrosion), 998–9
plan bracing, 90–91
plane sections, geometrical properties, 1074–6
plane stress idealization, 272
plane structures, 257
plant items
  fixing, 94
  installation, 96
  load paths, 88–9, 90–92
  loading, 92, 96
    lateral, 105–106
  plinth requirements, 94
  removal of, 96
  stiffness requirements, 95
  support points, 95, 96
  temperature change effects, 105–106
plasterboard (in cladding), 29
plastic analysis/theory, 19–21, 266–8, 316–17
  composite beams, 599–600
  portal frame, 317, 348–9
    worked example, 320–21
  propped cantilevers, 308–10
  single members, 304–308
plastic coatings, 212
plastic hinges, 265, 266–8, 304, 317
plastic interaction surface, 359
plastic modulus 1103–4
  design tables, 1077–1079
  hollow sections, 1105
Plastic Properties of Rolled Sections, The, (WI), 1103
plastic strain, 218, 263
plastic stress blocks, 644, 647
plastic zones, 264
Plastisol coating, 3
plate
  design, 871–5
  grades, 870
plate bending
  analysis, 271–2
  element stiffness, 290–91
plate buckling *see* buckling, local
plate-columns, 52
plate elements, 17
plate girders *see* girders, plate
plated beams, 427–8

plated columns, 52
plated structures, 257
plumbing steelwork, 979
podger (spanner), 983
Polar Inertia method, 668, 685
portal frames, 5–7, 15–22
  analysis, 316–21
  apex joints, 722
  costs, 913
  frame weights, 2
  pinned base, 19–20
  special design methods, 509–17
  tied variation, 23
portal-type bridges *see* bridges, portal-type
Pounder's plate theory, 870–71, 873–4
power station structures, 82–4
Pratt truss, 528, 531, 532, 533
precast floor systems, 61, 63, 93
pressure loadings, pipework, 106
prestressed cable, 182–3
primary beams *see* beams, primary
primary frames, 5–7
primers, paint, 1003, 1006
  blast-, 1009–11
process plant and equipment, 84–6
  loadings, 102–105
production control, 933–4
production engineering, 915–7
production standards, 916
profiled beams *see* beams, profiled
programme, erection, 943–7
property matrix, 289
propped cantilevers *see* cantilevers, propped
protection systems, 86, 1016–8
  see also paints/painting systems
prying action/forces, 673, 693–6, 712
PTFE (polytetrafluoroethylene) bearings, 811
punching, hole, 218
purlin systems, 13, 25–7, 431
  service loadings, 9
push-out test, 604–605
PVF2 coating, 3

quality control/management, 217, 938–40
quenching process, 210

radius of gyration, 1101
rafters
  bending moments, 13
  stability, 515–16
railway bridges *see* bridges, railway
reciprocal moments, principle of, 299
rectangular hollow sections, 52
  dimensions and properties, 1122–3
redundancy, degree of *see* indeterminacy, degree of
repetition costs benefits, 48
residual stresses
  manufacturing, 215
  reduction of, 238, 239

welding, 219, 221
resistance welding, 219
response frequencies, 84
restraint
   by cladding, 27
   lateral, 12, 14
   temperature-induced, 105–106
ribbed domes, 170
rigid connections, 63, 66
rigid frame analysis, 311–16
rigidity matrix, 291
rigidly-jointed frames, 63
rimming steel, 203
robustness (in design), 47, 109
rocker bearings, 99, 815
rods (as tension members), 186
roller bearings, 815
rolling process, 210–13
Ronan Point disaster, 47
roof trusses, 532–3
roofing systems, 29–30
   falls/slopes, 6, 17
   through fasteners, 30
   weathertightness, 2
rotational end restraint, 387
rotational motion loading, 106
routing, delivery, 933–8
rubber bearings, 814

safety, 961–2
   checklists, 966, 967
   structural, 969–71
   workforce, 962–7
saw-tooth truss, 529
sawing process, 929
Schaeffler diagram, 202
Schwedler domes, 170
seating cleats, 716–17
second moment of area, 1101
   tables, 1066–73
secondary beams *see* beams, secondary
secondary elements, 4–5
section classification checks, 505–509
section classification table, 349
section dimensions definitions, 1097–8, 1099–1100
section factors, 987
   tables, 1224–8
section properties, 1101–5
seismic design, 85
semi-rigid connections, 66
service core locations, 63
service loadings, 9
serviceability, deck slab, 573, 578, 580–81
Serviceability Limit State analysis, 21–2
services (in composite beams), 593
setting-out tolerances, 974–5
Severn Bridge, 115
shape factor, 265–6
shear and moment interaction, 615

shear buckling, 464–7
shear capacity, plate girders, 448–54
shear connections, 39, 604–13
   composite beams, 596
   partial, 609–11
shear effect
   in beams, 409–10
   in bolts, 672
shear loading, 218
   analysis, 261–2
   plastic analysis, 269–70
shear plates, 718
shear transfer, longitudinal, 613–15
shear walls, 63–5, 101
shearing process, 928–9
sheet piles
   design 844–51
   driving, 852
   durability, 853–5
   Frodingham type, 1194–1201
   Larssen type, 1202–3
   uses of, 843
sheeting, 3
   as structural element, 23–4
sheeting rails, 431
shims, 904, 910
shock transmission units, 824–5
shop-welding *see* welding, shop
shot blasting, 1004
shrinkage, composite structure, 618
siderails, 25, 27
   see also purlin systems
Sigma sections, 5
silicon, effect of, 203
simple connections, 66
simple harmonic motion, 325
simply-supported beams *see* beams, simply-supported
single members, plastic failure, 304–308
single-storey structures
   cladding, 2–4
   primary frames, 5–7
   secondary elements, 4–5
site-bolting *see* bolting, site
site painting *see* painting, on-site
site welding *see* welding, site
skeletal structures, 256–7
   analysis of, 273–8
   statically indeterminate, 277–9
slenderness
   columns, 375–9, 646, 649
   compression members, 54
slenderness ratios, beam, 343–8
slinging procedures, 975–8
slip resistance, 667, 672
slope-deflection method, 280–81
smoothness, 890
S–N curves, 228–9
sniping (stiffener cutting), 697
snow loading, 8–9

software packages, *see* structural analysis packages
soil engineering, 783–5
Space Deck System, 171
space frames, 257
   analysis, 173–4
   erection, 172
   special features, 170–71
   structural types, 167, 170
span-to-depth ratios, 445, 457
spandrel post arch bridge, 121
spangling, 1006
spans, range of, 2
   bridge, 113–14
spherical bearings, 817
splices, column *see* column splices
spray fire protection, 991
spring stiffness, 709
'springiness', floor, 105
square hollow sections, dimensions and properties, 1120–21
squareness, end, 889–90
squash load, 651
stability, portal frame, 317–18
stainless steels, 201, 202, 1025
standard penetration test, 784–5
standards, production, 916–7
standing seam cladding, 4
static equilibrium, 256
statutory constraints, 39
steel, choice of, 224
   for bridge girders, 133–4
steel, elastic properties of, 1023
steel, rolled
   coated, 212
   plate, 877, 1024
   sheet, 212
   strip, 212
steel bearings, 810, 811
steel deck/floors, 52–3, 55, 59–61, 93, 94, 870–75
   composite trusses, 57–8
   fixing, 95
   formwork, 94
steel frame components, 51
steel houses *see* housing
steel piles, *see* piles, steel
steelmaking, 211–13
   manufacturing defects, 213–15
   tests, 215–17
   see also metallurgy, steel
stepped-back facade, 39, 40–41
stiff bay layout, 100
stiff bearing length, 689
stiffeners
   load-bearing, 95, 691–3
   plate girder, 468
   web *see* web stiffeners
stiffness concept, 276, 338, 421
   in bracing, 421

in connections, 710, 713
stiffness matrix, 290, 338
stock control, 987–8
stocky columns, 643–4, 645
stockyard cranes, 953–4
straightening, 933
strength reduction factor, 989
'stress-block' method, 610
stress concentration, 363
stress corrosion cracking, 1000
stress wave analysis, 842
stressed-skin design, 23–4
stringers
   closed, 876
   open, 877
structural analysis packages, 14
structural bolts *see* bolts, structural
structural grid, 52–3
structural stability, 968–71
   in fire, 978–81
structural tees *see* tee sections
struts
   cased, 645
   compound, 393–5
      dimensions and properties, 1128–9
      compressive strength tables, 1156–63
      section selection, 1155
   see also columns/struts
stub girders, 58
sub-assemblies, use of, 959, 961
sulphur, effect of, 203
superposition principle, 256, 271
supports, sinking, 296
surface preparation, 931–2, 1003–5
surface-stressed structures, 177
suspension bridges *see* bridges, suspension
sway bracing, 11
'sway frames', 51, 63
sway load paths, 90–92
sway resistance, 7–8, 15
Swedish Standard SIS 055900, 1003, 1004

Tacoma Narrows Bridge, 115
Talurit Eye, 184
taper beams *see* beams, tapered
technical design aims, 37–8
Technical Memoranda (DTp), 130
tee-connections, 697–700
tee sections, 395
   dimensions and properties, tables, 1144–49
temperature effects, 618, 970, 983
temperature-induced restraints, 105–106
tempering process, 210
temporary conditions, safety considerations, 973–4
temporary supports, 971–3
tensile tests, 216
tension in bolts, 673
tension/compression bracing, 92
tension field action, 446, 450–51

perforated webs, 454
tension flange restraint, 514
tension members
 axial loading, 353–7
 combined with bending, 357–61
 eccentric loading, 361–2
 erection, 365
 fabrication, 365
 serviceability, 322–3
 sway resistance, 8
 types, 352–3
tension structures *see* cable structures
Terzaghi's method, 784
Test Certificate, manufacturing, 215–17
test specimens, 252
Theorem of Three Moments, 295–6, 301–302
thermal movement provisions, 108–109, 134
thin gauge sections, 5, 25
three-dimensional structures, 257
tied arch bridge, 121
tied portal frame, 23
TIG welding, 929
tin, effect of, 203
tolerances, 881–2
 building envelope, 885, 903–4
 classes, 882
 erection, 887–8, 893–4, 901–10
 fabrication, 885, 895–900
 flatness, 890
 implications, 885–6
 setting-out, 974–5
 standards, 883–5
torsion in beams, 413, 415–19
torsion constant, 1103, 1104
torsion modulus constant, 1104
torsional index, 1102
towers
 analysis, 162
 in building structures, 163–6
 environmental loading, 159, 161–2
 serviceability, 163
 types of
  guyed, 159, 162
  lattice, 157, 162
  modular, 159
traceability, steel, 939
traffic loading
 highway bridges, 131
 railway bridges, 131
tramp elements, effect of, 203
transformers, blast dissipation, 107–108
transition temperatures, 244
transport costs, 912
truss boom members, 90–92
truss bridges *see* bridges, truss
truss and stanchion design, 11–15
trusses
 analysis, 537, 538
 applications, 542, 544
 connections, 536–7

design, 539–41
economy factors, 541–2
elements, 534–6
for bridges, 530–32, 533–4
for buildings, 532–3
for floors 57–8
for roofs, 532–3
load reversal effects, 534
types of, 12, 528–9
 butterfly, 529
 lattice, 7
 mansard, 528
 Warren, 528–9, 531, 533
use of, 532–3
see also Vierendeel girders
tube rolling, 212
tubular members, 52, 157
turbine halls, 83–4
turned-barrel bolts, 926
twisting moments, 261

U-frame design, 423–5
ultimate limit state, beam design, 460–68
unequal angles, dimentsions and properties
 tables, 1132, 1134–5
uniform load model UIC71 (railway bridge), 131
universal beams, 149, 503
 dimensions tables, 1106, 1108
 properties tables, 1107, 1109, 1110–11
universal bearing piles
 dimensions, 1116
 properties, 1117
universal channels, 426–7, 503
universal columns
 dimensions, 1112
 properties 1113, 1114
uplift forces, 92

Vacu blasting, 931
vanadium, effect of, 202
vehicle assembly plants, 85
vertical sway bracing, 11
viaducts *see* bridges
vibration
 amplitude, 326
 damping of, 336–7
 effects of, 42–3, 53
 human-induced, 105
 loadings, 87, 104–105
 modes, 331–4, 617
vibration-resistant fixings, 99
Vierendeel girders, 58
 analysis, 431, 545–6
 connections, 546
 use, 544
von Mises yield criterion, 465

Walings 849, 850
walkway loading, 874–5
wall construction, external, 46

Index 1265

wall joints, 825
walls, 3
  shear, 63–5, 101
warping constant, 1103
Warren truss, 528–9, 531, 533
weatherproofing
  cladding, 29–30
  roofs, 2, 10
weather-resistance of joints, 826–7
weather-resistant (WR) steels, 133, 201
weathering steels, 1013–5
  bolting, 1016
  painting, 1016
  welding, 1015–6
web buckling, 413
web cleats, 717
web side plates, 717
web stiffeners, plate girder, 446, 449–52, 455–6, 458–9
web thickness, plate girder, 447–8
wedging action bearings, 824
weld capacities, 1186–1193
weld cracking, 222
weld toe remelting, 239
welding/welded connections, 48, 219, 221, 917–8
  access for, 922
  defects in, 222, 922–3
  design strength, 687–8
  geometry, 678–83
    improvement in, 238–9
  groups, 683–5
    tables, 1186–1193
  preparation for, 921–2
  processes, 6–7
  shop, 918–22
  site, 48, 719, 980
  tapered frames, 6–7
  types, 677–8
wet-blasting, 1004
wide plate test, 252
width of flange association term, 466
width to thickness ratio, *see b/t* ratio
wind loading, 9, 107, 970–71
  load reversal, 27
  multi-storey buildings, 47
  petrochemical plants, 86
  tower structures, 159
wind tunnel testing, 115
wire manufacture, 218
wire rope *see* cables
workforce safety, 961–7
wrenches, use of, 981, 983

yield line analysis, 697–703
yield stress, 263
yielding process, 263

Zed sections, 5, 431
zoning, services, 43

British Standards:
  listing, 1248–52
BS 4, standard sections, 213
BS 131, Charpy tests, 216
BS 449, cased columns, 645
BS 476, fire resistance tests, 581, 992–993
BS 729, galvanised coatings, 1006
BS 970
  Charpy tests, 216
  stainless steels, 1025
  steel properties, 210, 1025
BS 2569, metal coatings, 1007
BS 3692, precision bolts, 664
BS 4192, black bolts, 664
BS 4232, surface preparation, 1006, 1018
BS 4360
  bolts, 675
  notch ductility, 216
  steel grades, 133
  steel pates, 1024
  weathering steels, 133, 1013
BS 4395, HSFG bolts, 665, 672, 673, 674
BS 4592
  floor finishes, 95
  open grid flooring, 875
BS 4604, bolts, 665, 672
BS 5135, weld preparation, 678
BS 5400
  beam design, 407
  bearing design, 810, 811, 812, 814
  bending and axial loading, 360
  bolts, 696
  bridge trusses, 539–41
  composite columns, 645, 646, 649
  compound members, 394–5
  creep reduction factor, 618
  deck design, 880
  eccentric loading, 361
  effective breadth, 604
  fatigue design, 229, 239
  lateral torsional buckling, 415, 417
  moment resistance formulae, 644
  prying forces, 694
  shear connectors, 606, 608, 609
  short columns, 646
  single angles, 395
  slenderness limits, 376, 377, 378, 383
  steel thicknesses (at high temperatures), 253
  stiffeners, 413
  uniform load model, 131
  welds, 677
BS 5502, snow loading, 8
BS 5531, safety precautions, 961
BS 5606, accuracy in building, 884–5
BS 5750, quality management, 938
BS 5762, CTOD tests, 251
BS 5950
  analysis methods, 316–17
  angles, channels and tees, 395
  axial loading, 353

baseplate design, 789–90, 799
beam design, 406, 407, 411, 615
bending strength, 359–60, 1152–3
bolting, 666–7, 672, 674, 675, 696, 704
bracing restraint, 517
buckling resistance, 692
building trusses, 537, 539–61
castellated beams, 431
column design, 67, 646
compactness check, 505–509
composite beams, 592
composite deck slabs, 572, 577–8
compound members, 393–4
compressive resistance, 380–83, 1156–63
crack limits, 618
deflection limits, 42, 67, 1150
design strengths, 1150
eccentric loading, 361
effective breadth, 603
fire resistance, 253, 582, 985–6, 993
lateral bracing, 421–2
lateral-torsional buckling, 415–19
limited frame concept, 390
load factors, 9, 10
plastic hinge analysis, 410, 601

plate elements, 377, 446–57
purlins, 25, 431
section classification tables, 349
shear connectors, 606, 610
sheeting rails, 431
slenderness limits, 375, 377, 378–9, 430, 540
stiffeners, 413, 692
stiffness charts, 387
strut table, 1155
tolerances, 890
welds, 677
width to thickness ratios, 1151
BS 5957, falsework, 973
BS 6399
   crane loading, 10, 104
   fixed plant loading, 103
   floor loading, 42
   snow loading, 8
   walkway loading, 874
BS 8100, lattice towers and masts, 161–2
BS 8110
   concrete encasement, 652
   crack limits, 618
BS EN 10025, 244